ALBERTUS MAGNUS
ON ANIMALS

Frontispiece. A late fourteenth-century portrait of Albertus Magnus, perhaps a copy of an original drawn from life (see Damon, 1930, 102). From Cod. Lat. Monac. 27029, fol. 88r. The text surrounding the portrait reads "Dominus Magnus Albertus. Rerum naturas proprias qui noscere curas, hunc lege tractatum quem ibi reperis bene gratum." (Lord Albertus Magnus. You who wish to learn the proper natures of things, read this treatise which you will find truly pleasing.)

ALBERTUS MAGNUS ON ANIMALS

A Medieval *Summa Zoologica*

REVISED EDITION

Translated and annotated by

Kenneth F. Kitchell Jr. and Irven Michael Resnick

VOLUME II
BOOKS 11–26

THE OHIO STATE UNIVERSITY PRESS
Columbus

Copyright © 2018 by The Ohio State University.
All rights reserved.

Library of Congress Cataloging-in-Publication Data
Names: Albertus, Magnus, Saint, 1193?–1280, author. | Kitchell, Kenneth F., Jr., 1947– translator. | Resnick, Irven M., translator.
Title: Albertus Magnus, on animals : a medieval summa zoologica / translated and annotated by Kenneth F. Kitchell Jr., and Irven Michael Resnick.
Other titles: De animalibus. English | On animals
Description: Revised edition. | Columbus : The Ohio State University Press, [2018] | Includes bibliographical references and index.
  Contents: Volume I — Introduction: The life and works of Albert the Great — Book one: On animal and human members — The first tract, on the diversity of animal members — The second tract, on the disposition of human members — The third tract, on the internal members — Book two: A comparison of the human and other animals — The first tract, on the visible members — The second tract, a comparison of the internal members of animals and humans — Book three: On the origin of the uniform members in animals — The first tract, on the veins, blood, and nerves — The second tract, on the uniform members which are not contiguous — Book four — The first tract, on bloodless marine animals — The second tract, on sense, voice, and sleep in bloodless animals and the difference between male and female — Book five: On the generation of animals — The first tract, on differences in their copulation — The second tract, on sperm production and the formation of young — Book six: The nature, anatomy, and generation of eggs — The first tract, on diversity among eggs — The second tract, on the eggs of fish and swimmers — The third tract, on the generation of viviparous walkers — Book seven — The first tract, on the dispositions and life of animals — The second tract, on health and illnesses of animals — The third tract, a digression on the things introduced — Book eight: On animals' habits — The first tract, on why they fight one another — The second tract, on animal prudence and stupidity — The third tract, in one, on the cunning and cleverness of sea creatures — The fourth tract, on cleverness in ringed creatures — The fifth tract, on the habits of quadrupeds — The sixth tract, solving doubts arising above — Book nine: On the principles and origin of human generation — The first tract, on human origin from a mature seed — The second tract, treating Galen and Aristotle on human generation — Book ten: On impediments to generation — The first tract — The second tract, on the causes of sterility — Volume II — Book eleven — The first tract, on the order of instruction of animals — The second tract, on the ultimate end of animals — Book twelve: On the cause of uniform and nonuniform members — The first tract, on uniform and principle members — The second tract, on the nature of uniform members — The third tract, on the organic members of the head — Book thirteen: On the inner members — The first tract, on the windpipe, esophagus, and lungs — The second tract, on the intestines — Book fourteen: On the outer members — The first tract, on the outer members of ringed animals, shellfish, and the Malakye — The second tract, on the visible members — Book fifteen: On the causes of the generation of animals — The first tract, on the distinction between the sexes — The second tract, on the nature of sperm — Book sixteen: On the powers of reproduction, with respect both to the soul and to the members — The first tract, on the introduction of the soul — The second tract, on sterility in animals and on the generation of homogeneous members — Book seventeen: On oviparous animals — The first tract, on oviparous birds — The second tract, on the ringed creatures — Book Eighteen: On the generation of perfect animals — The first tract, on the differentiation of the sexes — The second tract, on the number of offspring — Book nineteen: On the senses and their accidents — Book twenty: On the nature of animal bodies — The first tract, on what constitutes the body — The second tract, on the formal powers — Book twenty-one: On perfect and imperfect animals — The first tract, on degrees of perfection — Book twenty-two: On the nature of animals — The first tract, on the human — The second tract, on quadrupeds — Book twenty-three: On the nature of birds — Book twenty-four: On aquatic animals — Book twenty-five: On the nature of serpents — I. The nature of serpents in general — II. Understanding the nature and complexion of serpents' venom — Book twenty-six: On vermin.
Identifiers: LCCN 2017045696 | ISBN 9780814213599 (cloth ; alk. paper) | ISBN 0814213596 (cloth ; alk. paper)
Subjects: LCSH: Zoology—Pre-Linnean works. | Physiology—Early works to 1800. | Veterinary medicine—Early works to 1800. | Science, Medieval—Sources.
Classification: LCC QL41 .A3413 2018 | DDC 590—dc23
LC record available at https://lccn.loc.gov/2017045696

Cover design by Laurence J. Nozik
Type set in Adobe Garamond Pro and ITC Galliard

∞ The paper used in this publication meets the minimum requirements of the American National Standard for Information Sciences—Permanence of Paper for Printed Library Materials. ANSI Z39.48-1992.

Theresae, pro plurimis quam
et Albertus ipse enumerare potuisset
KFK

Elizabeth . . . more than all the commentaries upon Aristotle
IMR

If someone should think that the study of animals is an unworthy pursuit, then he must hold entirely the same view about himself.
—Aristotle, *De partibus animalium* 645a27f

For there are wonders to be seen in all natural things.
—Albertus Magnus, *De animalibus* 11.86

CONTENTS

Volume I

Detailed Table of Contents   ix
Foreword   xv
Preface and Acknowledgments   xix
Preface to Revised Edition   xxiii
Guide to Editorial Conventions and Abbreviations   xxv
Note on the Translation   xxxi
The Writings of Albertus Magnus   xxxix

Introduction: The Life and Works of Albert the Great   1

Book One	45	Book Six	525
Book Two	286	Book Seven	586
Book Three	344	Book Eight	667
Book Four	433	Book Nine	774
Book Five	487	Book Ten	827

Volume II

Detailed Table of Contents   ix

Book Eleven	857	Book Nineteen	1332
Book Twelve	894	Book Twenty	1358
Book Thirteen	985	Book Twenty-One	1409
Book Fourteen	1045	Book Twenty-Two	1440
Book Fifteen	1084	Book Twenty-Three	1544
Book Sixteen	1153	Book Twenty-Four	1655
Book Seventeen	1237	Book Twenty-Five	1708
Book Eighteen	1281	Book Twenty-Six	1739

Appendix: Alternative Version of the Beginning of *De animalibus*   1765
Glossary   1769
References   1777
Index   1803

DETAILED TABLE OF CONTENTS
VOLUME II

BOOK ELEVEN
The First Tract, on the Order of Instruction on Animals 857

I   Demonstration in a Science of Animals 857
II   How One Must Proceed in a Science of Animals 860
III   The Number and Types of Causes 867

The Second Tract, on the Ultimate End of Animals 877

I   Concerning a Science of Animals 877
II   Whether One Has to Begin from Universals or Particulars 885
III   Whether and How Physical Causes Should Be Sought 888
IV   The Division of Formal Causes 892

BOOK TWELVE: ON THE CAUSES OF UNIFORM AND NONUNIFORM MEMBERS
The First Tract, on Uniform and Principal Members 894

I   Differences between Uniform and Nonuniform Members 894
II   Uniform Members' Complexions 900
III   The Hot, Cold, Dry, and Moist in Uniform Members 906
IV   Composition and Complexion of an Animal 912
V   Stages of Life 918
VI   Digestion and the Generation of the Humors 922
VII   The Effects of Heat and Cold in the Brain and Marrow 925

The Second Tract, on the Nature of Uniform Members 930

I   The Nature of Blood, Which Is the Matter of Uniform Members 930
II   The Nature of Fat, *Zirbus,* and Marrow 933
III   The Nature of the Brain, According to Aristotle 936
IV   The Brain, According to the Modern Peripatetics, Experience, and Reason 939
V   The Cause and Nature of Nerves 944
VI   The Cause of Bones and Cartilage 945
VII   Superfluities Generated in Animals' Bodies 949
VIII   Flesh, Bones, Spines, and the *Malakye* Genuses 950

The Third Tract, on the Organic Members of the Head 956

I   The Head, Insofar As It Is the Origin of the Senses 956
II   Ears and Nostrils 960
III   Eye, Eyelids, and the Hairs That Surround the Eye 964
IV   The Nose of Certain Oviparous Animals and the Elephant 969
V   Lips, Teeth, Tongue, and Their Uses 971
VI   Nature, Cause, Types, and Number of Teeth 975
VII   The Cause and Nature of Horns on the Heads of Animals 980

BOOK THIRTEEN: ON THE INNER MEMBERS
The First Tract, on the Windpipe, Esophagus, and Lungs 985

- I   The Windpipe and Esophagus and Their Natural Causes   985
- II   Composition of the Windpipe and Lungs   989
- III   Esophagus and Stomach   992
- IV   The Inner Parts and the Nature and Cause of the Heart   995
- V   Why the Veins Arise from the Heart in Animals   1002
- VI   Lung, Liver, and Spleen and Their Natural Operations   1005
- VII   Liver and Spleen   1009
- VIII   Kidneys and Bladder, and Their Absence in Certain Animals   1012
- IX   Diaphragm   1016

The Second Tract, on the Intestines 1020

- I   Belly   1020
- II   Intestines   1024
- III   Usefulness of the Intestines   1026
- IV   The Inner Members and Gall Bladder in Creeping Animals   1029
- V   Generation of the *Mirach* of the Belly, Fat, and *Zyrbum*   1033
- VI   Animals Which Lack the Inner Members of the Belly   1035
- VII   Inner Members in Those Midway between Plant and Animal   1042

BOOK FOURTEEN: ON THE OUTER MEMBERS
The First Tract, on the Outer Members of Ringed Animals, Shellfish, and the Malakye *1045*

- I   Ringed Creatures   1045
- II   Shellfish   1048
- III   Those Called Soft-Shelled Animals   1049
- IV   The *Malakye*   1051

The Second Tract, on the Visible Members 1055

- I   Purpose of the Belly, Hands, Feet, and Neck   1055
- II   The Hand and the Use of Feet in Quadrupeds   1058
- III   Chest and Breasts in the Human and in Other Animals   1062
- IV   Genital Members in Both Males and Females   1065
- V   Visible Members of Oviparous Quadrupeds   1069
- VI   Visible Members of Birds   1072
- VII   Visible Members of Fish   1077

BOOK FIFTEEN: ON THE CAUSES OF THE GENERATION OF ANIMALS
The First Tract, on the Distinction between the Sexes 1084

- I   Sexual Differentiation Is Not Found in All Animals   1084
- II   Differences between Male and Female   1088
- III   Creation of Testicles in Animals   1090
- IV   Speed of Copulation in Animals Without Testicles, with Internal Testicles, or with External Testicles   1093

V	Differences in the Wombs of Animals 1095	
VI	Why Wombs Are Internal in All Animals 1098	
VII	Genital Members in Hard- and Soft-Shelled Animals, the *Malakye*, and Those with a Ringed Body 1100	
VIII	A Digression on the True Understanding of the Genitals 1102	

The Second Tract, on the Nature of the Sperm   1110

I	Whether the Sperm Is Derived from All the Members, As Plato Said 1110	
II	Whether the Sperm Is Derived from the Entire Body, As Anaxagoras and Empedocles Said 1114	
III	The Implausibility of What Empedocles and Anaxagoras Said 1118	
IV	What Sort of Principle of Generation Is the Sperm? 1121	
V	How the Sperm Is Present in the One Producing It 1125	
VI	The Nature of Male Sperm, Female Sperm, and Menstrual Blood 1130	
VII	On Pleasure in Intercourse and on the Female Genital Area 1136	
VIII	Relationship between Male Sperm and the Female Humor 1140	
IX	Union of Male and Female Sperm in Generation 1143	
X	Male Sperm and Its Powers 1146	
XI	Male Sperm, Female Sperm, and Menstrual Blood 1149	

BOOK SIXTEEN:
ON THE POWERS OF REPRODUCTION, WITH RESPECT BOTH TO THE SOUL AND TO THE MEMBERS

The First Tract, on the Introduction of the Soul   1153

I	How Male and Female Are the Principles of Generation 1153	
II	The Power Moving Embryos, According to the Ancient Philosophers and Physicians 1158	
III	A Digression, Following Alexander the Peripatetic 1163	
IV	A Digression, Following Avicenna, Theodorus, Theophrastus, and the Peripatetics 1167	
V	A Digression, Following Socrates, Plato, and the Stoics 1170	
VI	The Solution, Following Aristotle, Who Saw the Truth 1172	
VII	Summary 1175	
VIII	Whether the Sperm Is Animate and Whether the Powers of the Soul Create the Members Successively or All at Once 1178	
IX	Why Sperm Grows Black and Thins in the Cold 1181	
X	How the Soul Is Generated in the Fetation 1184	
XI	The Vegetative, Sensible, and Rational Soul in Semen 1186	
XII	Why Aristotle Says Soul Is in the Semen but Not Intellect 1190	
XIII	How Sperm Is Made Suitable for Generation 1192	
XIV	On Sex Determination and on Similarity to the Father, Mother, or Another 1194	
XV	Why the Female Provides the Body but the Male the Form and Soul 1197	

XVI	The Later Peripatetics' Opinion on Oviparous Animals, Sperm, and the Animation of Embryos 1200	
XVII	Why Male and Female Are Necessary for Viviparous Reproduction and Why the Rational Soul Can Have No Factive Principle in the Semen 1205	

The Second Tract, on Sterility in Animals and on the Generation of Homogenous Members   1207

I	Accidental Causes of Sterility and the Creation of the Embryo 1207	
II	Why the Female Cannot Generate without the Male and the Generation, Differentiation, and Completion of the Members 1212	
III	Why the Heart Is Said to Be the Principle of the Members 1215	
IV	Generation of the Homogenous Members 1218	
V	Creation of the Head and the Members in the Head 1220	
VI	How the Dominant Members Are Formed After the Heart 1222	
VII	How the Growth of the Embryo Is Accomplished 1224	
VIII	On Which Animals Are Impregnated from Copulation and on Sterility 1226	
IX	Sterility in She-Mules, According to Democritus and Empedocles 1228	
X	The True Cause of Sterility in She-Mules 1231	
XI	Summary 1234	

BOOK SEVENTEEN: ON OVIPAROUS ANIMALS
The First Tract, on Oviparous Birds   1237

I	Those Who Lay Many Eggs and Those Who Lay Few 1237	
II	Eggs and the Humors Contained in an Egg 1240	
III	The Shape of Eggs, Their Creation in the Womb, and Hatching 1244	
IV	The Order of Generation of Animals' Members in Eggs 1248	
V	Against Those Who Claim That There Is No Male or Female in Fish 1252	
VI	Completion of Eggs, Specifically Wind Eggs, by the Sperm 1257	

The Second Tract, on the Ringed Creatures   1260

I	Generation of Larvae 1260	
II	Triple Generation of Bees, According to Various Opinions 1263	
III	Three Opinions on the Generation of Bees 1270	
IV	Generation of the *Halzum* 1272	
V	The Moving Principle in the Generation of Shellfish 1276	

BOOK EIGHTEEN: ON THE GENERATION OF PERFECT ANIMALS
The First Tract, on the Differentiation of the Sexes   1281

I	Sexual Differentiation, According to Empedocles and Democritus 1281	

	II	Sexual Differentiation, According to Aristotle and Other Peripatetics 1288
	III	The Heart As the Principle of Sexual Differentiation. Also, Those Which Generate Mostly Females and Those Which Generate Mostly Males. 1291
	IV	Resemblance to Parents or Forebears 1295
	V	Erroneous Opinions on Resemblance to Parents or Forebears 1301
	VI	Monstrous Births Having a Resemblance to the Proximate Genus 1303

The Second Tract, on the Number of Offspring   1307

	I	Body Size Determines the Number of Offspring 1307
	II	Why Many Offspring May Be Conceived from One Act of Copulation 1309
	III	Hermaphroditic Fetations and Other Monsters 1312
	IV	Successive Impregnations and Excessive Desire 1314
	V	Generation of Complete or Incomplete Young 1317
	VI	Impregnated Females and a Mole of the Womb 1320
	VII	Generation, Cause, and Usefulness of Milk 1322
	VIII	Gestation Period in Animals 1324
	IX	Powers of Generation 1327

BOOK NINETEEN: ON THE SENSES AND THEIR ACCIDENTS   1332

	I	Accidental Traits in General 1332
	II	How the First Accidental Trait Is Similar to Sleep 1334
	III	Colors of the Eyes Caused by the Eye's Humors 1336
	IV	On Eye Colors Caused by the Tunics and on Vision 1339
	V	Good Hearing and Smelling 1342
	VI	Hair, Growing Bald, and the Like 1344
	VII	Cause of Grayness and Wrinkledness in the Human 1349
	VIII	Cause of Colors in Animals Other Than the Human 1351
	IX	Differences of Voice in Animals 1353
	X	Origin, Appearance, and Falling Out of Teeth in Animals 1356

BOOK TWENTY: ON THE NATURE OF ANIMAL BODIES

The First Tract, on What Constitutes the Body   1358

	I	Seminal Moisture of Animate Beings 1358
	II	Dry and Earthy Material for Animate Bodies 1360
	III	Nature of the Air Shared by Animal Bodies 1363
	IV	How Fire Is Shared by Animal Bodies 1366
	V	Whether a Fifth Essence Is Substantially Present in Animal Bodies 1370
	VI	That Heavens Are Not Substantially Present in Animal Bodies 1374
	VII	Resolving Arguments That a Fifth Essence Is in Our Bodies 1378
	VIII	The Mixture of Material Principles in Animal Bodies 1381

IX	How a Triple Moisture Is the Binding Agent in Mixed Bodies	1385
X	How the Mixing of Animal Bodies Is Completed	1388
XI	Complexions of the Humors	1391

The Second Tract, on the Formal Powers 1393

I	Power of the First Cause	1393
II	Formal Heavenly Power and How It Acts	1396
III	Formal Principle of the Body from the Soul	1398
IV	Formal Causes of an Animal's Body from the Four Elements	1402
V	How All the Principles Mentioned Arise Out of One	1404
VI	How Animal Powers Differ and How the Human Differs	1407

BOOK TWENTY-ONE: ON PERFECT AND IMPERFECT ANIMALS
The First Tract, on Degrees of Perfection 1409

I	The Most Perfect Animal, the Human	1409
II	On Types of Perfection of the Soul and Body, and on the Pygmy	1414
III	How Animals, Especially Monkeys, Are Capable of Instruction	1419
IV	Teachability of Quadrupeds	1422
V	Capacity for Instruction among Birds	1425
VI	Cleverness and Shrewdness of Aquatic Animals	1428
VII	Prudence and Shrewdness of Serpents and Creeping Animals	1433
VIII	Shrewdness and Perfection of Ringed Creatures	1435
IX	Distinctions among Imperfect Animals	1438

BOOK TWENTY-TWO: ON THE NATURES OF ANIMALS
The First Tract, on the Human 1440

I	Human Intercourse	1440
II	Quality of the Seeds in the Womb	1441
III	The Seeds, the Young, the Humors, and the Heavenly Signs	1443
IV	Harmful Effects of Intercourse	1444
V	Natural and Divine Properties of the Human	1445

The Second Tract, on Quadrupeds 1447

I	Nature and Traits Quadrupeds Have in Common	1447

ALCHES 1449	BONACHUS 1455
ALFECH 1449	BUBALUS 1455
ALOY 1449	BOS 1456
ANA 1449	CAMELUS 1456
ANABULA 1449	CANIS 1457
ANALOPOS 1450	CAMA 1464
ASINUS 1450	CALOPUS 1465
ASINUS SILVESTRIS 1452	CAMELOPARDULUS 1465
APER SILVESTRIS 1453	CAPER ET CAPRA 1465
ALZABO 1454	CAPREOLUS 1467
HAHANE 1455	CASTOR 1467

CACUS 1469
CATHUS 1469
CEFUSA 1470
CERVUS 1470
CHYMERA 1473
CYROGRILLUS 1473
CUNICULUS 1473
CRICETUS 1474
CYROCHROTHES 1474
CATHAPLEBA 1474
DAMMA 1474
DAMPNIA 1475
DAXUS 1475
DURAU 1476
ELEFAS 1476
EQUI 1477
EQUICERVUS 1504
EALE 1505
ENYCHYROS 1505
EMPTRA 1506
ERICIUS 1506
ERMINIUM 1508
FALENA 1508
FURO 1509
FURIOZ 1509
FELES 1509
FINGA 1510
GLIS 1510
GALI 1510
GENETHA 1510
GUESSELIS 1511
IBEX 1511
IBRIDA 1511
ISTRIX 1512
IENA 1512
LEO 1512
LEOPARDUS 1514
LEPUS 1515
LEUCROCOTHA 1516
LEONCOPHONA 1516
LACTA 1516
LAMIA 1516
LAUZANY 1517
LINX 1517
LINTISIUS 1518
LUPUS 1518

LUTER 1520
MULUS 1521
MONOCEROS 1521
MOLOSUS 1521
MARICON MORION 1521
MANTICORA 1522
MUSQUELIBET 1522
MAMONETUS 1522
MIGALE 1522
MUSIO 1523
MUSTELA 1523
MUS 1524
MARTARUS 1526
NEOMON 1526
ONAGER 1526
ONAGER INDICUS 1527
ONOCENTAURUS 1527
ORIX 1527
ORAFLUS 1527
OVIS 1528
PARDUS 1530
PANTHERA 1530
PIRADER 1531
PEGASUS 1531
PILOSUS 1531
PAPIO 1532
PATHYO 1532
PUTORIUS 1532
PIROLUS 1533
RANGYFER 1533
SIMIA 1533
TIGRIS 1535
TAURUS 1535
TRANEZ 1538
TRAGELAFUS 1538
TROGODYTAE 1538
TALPA 1538
UNICORNIS 1539
URSUS 1540
VESONTES 1541
URNI 1541
VULPES 1541
VARIUS 1542
ZUBRONES 1542
ZILIO 1543

BOOK TWENTY-THREE: ON THE NATURE OF BIRDS   1544

  AQUILA   1547
  ACCIPITER   1553
  ARPYA   1555
  AGOTHYLEZ   1555
  ARDEA   1555
  ANSER   1556
  ANAS   1558
  ACHANTIS   1559
  ASSALON   1559
  ALAUDA   1559
  ALCIONES   1560
  AERIFYLON   1560
  AVES PARADISI   1561
  BUBO   1561
  BUTEUS   1561
  BUTORIUS   1562
  BISTARDA   1562
  BONASA   1562
  BARLIATES   1563
  CALADRIUS   1563
  CYNAMULGUS   1564
  CIGNUS   1564
  CARISTAE   1565
  CICONIA   1565
  CHORETES   1566
  CALANDRIS   1566
  CORVUS   1567
  CORNIX   1567
  CORNICA   1567
  CUGULUS   1568
  COREDULUS   1568
  COLUMBA   1568
  CHARCHOTES   1569
  CHORTURNIX   1569
  CARDUELIS   1570
  CROCHYLOS   1570
  DYOMEDICA   1571
  DARYATHA   1571
  EGITTHUS   1571

  I Shape of Falcons   1572
  II Particular Color of Falcons   1573
  III Characteristic Behavior of Falcons   1574
  IV How Falcons Call and Are Called   1576
  V The Seventeen Falcon Genuses and the First, Called "Sacred"   1577
  VI Gyrfalcon   1578
  VII "Mountain" Falcon   1580
  VIII Peregrine Falcons   1581
  IX Gibbous Genus of Falcon   1583
  X Black Falcons   1585
  XI Nature of the White Falcon   1586
  XII Nature of Red Falcons   1587
  XIII The Falcon Which Has Azure Feet   1589
  XIV The Small Falcon Called the *Mirle*   1590
  XV The Three Genuses of Lanners and Their Rearing   1590
  XVI The Four Genuses of Interbred Falcons   1592
  XVII Domestication, Boldness, and Health of Falcons   1593
  XVIII Cures for Falcon Illnesses, According to William the Falconer   1597
  XIX Cures, According to Emperor Frederick's Falconer   1603
  XX Goshawk Illnesses, According to Emperor Frederick   1607
  XXI Regimen for Goshawks, According to William the Falconer   1611
  XXII Hawking and the Domestication of Hawks   1616
  XXIII Aquila, Symachus, and Theodotion on the Illnesses of Birds of Prey   1618

XXIV Two Other Falcon Genuses and Their Habitats 1622

    FATATOR 1623
    FENIX 1623
    FETYX 1624
    FICEDULA 1624
    FULICA 1624
    GRIFES 1625
    GRACOCENDERON 1625
    GOSTURDUS 1625
    GRUS 1626
    GLUTIS 1627
    GALLUS 1627
    GALLINA 1628
    GALLUS GALLINACIUS 1629
    GALLUS SILVESTRIS 1630
    GARRULUS 1630
    GRACULUS 1631
    IBIS 1631
    IBOZ 1632
    INCENDULA 1632
    IRUNDO 1632
    IPSIDA 1633
    KYRII 1634
    KARKOLOZ 1634
    KOMER 1634
    KYTHES 1634
    LARUS 1634
    LUCIDIAE 1635
    LUCINIA 1635
    LINACHOS 1635
    LAGEPUS 1635
    MILVUS 1635
    MAGNALES 1636
    MELANCORIFUS 1636
    MORFEX 1636
    MEMNONDIDES 1636
    MEAUCAE 1637
    MERILLIONES 1637
    MUSCICAPAE 1637
    MEROPS 1637
    MERULA 1638
    MONEDULA 1638
    MERGUS 1639
    NISUS 1639
    NOCTUA 1639
    NEPA 1640
    ONOCROTALUS 1640
    OTHUS 1640
    OSINA 1641
    ORYOLI 1641
    PELLICANUS 1641
    PORFIRION 1642
    PAVO 1642
    PERDIX 1643
    PLATEA 1644
    PLUVIALES 1645
    PICA 1645
    PICUS 1645
    PASSER 1646
    PASSER SOLITARIUS 1646
    PHYLOMENA 1647
    PSYTACUS 1647
    STRUTIO 1648
    STRIX 1649
    STURNUS 1649
    TURTUR 1649
    TROGOPALES 1650
    TURDUS 1650
    TURDELA 1651
    VESPERTILIO 1651
    VANELLI 1652
    ULULA 1652
    UPUPA 1652
    VULTUR 1653
    ZELEUCIDES 1654

BOOK TWENTY-FOUR: ON AQUATIC ANIMALS

I The Nature of Aquatic Animals Set Forth Alphabetically 1655

    ASLET 1659
    ALLECH 1660
    ANGUILLA 1660
    ALFORAZ 1661
    ASTARAZ 1661
    ALBYROZ 1662
    ARIES MARINUS 1662
    AUREUM VELLUS 1662

ABARENON 1662
ACCIPENDER 1662
AMIUS 1663
HAMGER 1663
AFFORUS 1663
AUSTRALIS 1663
ARANEA 1664
ABYDES 1664
HAHANE 1664
BELUAE 1664
BARCHERA 1665
BOTHAE 1665
BORBOTHAE 1665
BABILONICI 1666
CETUS 1666
CLANCIUS 1671
CONGRUS 1671
CARPEREN 1671
CAPITATUS 1672
CORVI MARIS 1672
COCLEAE 1672
CONCHA 1673
CANCRI 1673
COCODRILLUS 1675
CAHAB 1676
CRICOS 1676
CELETHI 1677
CHYLON 1677
CANES MARINI 1677
CAERULEUM 1677
DRACO MARIS 1677
DELPHINUS 1678
DIES 1679
DENTRIX 1680
EQUUS MARIS 1680
EQUUS NILI 1680
EQUUS FLUMINIS 1681
EXPOSITA 1681
HELCUS 1681
ESCYNUS 1681
EZOX 1682
HUSO 1682
ERICIUS 1683
EXOCHINUS 1683
ERACLEYDES 1683
FOCA 1683

FASTALEON 1684
GONGER 1684
GOBIO 1684
GRANUS 1684
GALALEA 1684
GARCANEZ 1684
GLADIUS 1685
IPODROMUS 1685
IRUNDO MARIS 1686
KALAOZ 1686
KYLOZ 1686
KOKY 1686
KYLION 1687
KARABO 1687
LULIGO 1687
LUDOLACRA 1687
LOLLIGENES 1687
LOCUSTA MARIS 1688
LEPUS MARINUS 1688
LUCIUS 1688
MURENA 1689
MUGILUS 1690
MARGARITAE 1690
MEGARIS 1691
MULLUS 1691
MULTIPES 1691
MURICES 1692
MUS MARINUS 1692
MULUS 1692
MILAGO 1692
MONOCEROS 1693
MONACHUS MARIS 1693
NEREIDES 1693
NAUTILUS 1693
NASUS 1693
ORCHA 1694
OSTREAE 1694
PURPURAE 1694
PYNA 1694
PUNGICIUS 1694
PECTEN 1695
PORCUS MARINUS 1695
PAVO MARIS 1695
PERNA 1695
PISTRIS 1696
PLATANISTAE 1696

POLIPUS 1696
RANA MARINA 1697
RUMBUS 1697
RAYTHAE 1697
SALMO 1697
STURIO 1698
SPONGIA 1698
SCOLOPENDRA 1698
STELLA 1698
SUNUS 1698
SOLARIS 1699
SCUATINA 1699
SALPA 1699
SEPIA 1699
SPARUS 1700
SCORPIO MARIS 1700
SCARUS 1700
SERRA 1700
SERRA MINOR 1701
SYRENAE 1701
SCILLA 1701
SCINCI 1701
STINCUS 1702
TESTUDINES 1702
TYGNUS 1703
TESTEUM 1703
TORTUCA MARIS 1704
TORPEDO 1704
TREBIUS 1704
TRUTHAE 1704
TIMALLUS 1705
VULPES MARINAE 1705
VIPERAE MARINAE 1705
VENTH 1705
VERGILIADES 1705
VACCA 1706
ZEDROSUS 1706
ZYDEACH 1706
ZYTYRON 1706
XYSYUS 1707

BOOK TWENTY-FIVE: ON THE NATURE OF SERPENTS
I   The Nature of Serpents in General   1708
II Understanding the Nature and Complexion of Serpents' Venom   1711

ASPIS 1716
ANFYSIBENA 1717
ARMENE 1717
ASYLUS 1717
ANDRIUS 1718
ASFODIUS 1718
ALTYNANYTY 1719
ARACSIS 1719
ARUNDUTIS 1719
AHEDYSYMON 1719
ALHARTRAF 1719
HAREN 1720
BASILYSCUS 1720
BOA 1721
BERUS 1722
CORNUTA ASPIS 1722
CEREASTES 1722
CAFEZATUS 1723
CAECULA 1723
CERYSTALYS 1723
CELYDRUS 1723
CENCRIS 1724
CAUHARUS 1724
CARNEN 1724
CENTUPEDA 1724
DYPSA 1724
DRACO 1725
DRACO MARINUS 1727
DRACONCOPEDES 1727
EMOROYS 1727
FALITUSUS 1727
JACULUS 1728
IPNAPIS 1728
HYDRA 1728
IRUNDO 1728
ILLICINUS 1728
LACERTAM 1729
MILIARES 1730
MARIS SERPENS 1730
NATRIX 1730
NADEROS 1730
OBTALIUS 1730

PRESTER 1730
PHAREAS 1731
RYMATRIX 1731
SALAMANDRA 1731
SALPIGA 1732
STELLIO 1733
SCAURA 1733
SITULA 1733
SIRENES 1733
SERPS 1733

SPECTAFICUS 1734
STUPEFACIENS 1734
SABRYN 1734
SEYSETULUS 1735
SELFYR 1735
TORTUCA 1735
TYRUS 1736
TYLIACUS 1737
VIPERA 1737

BOOK TWENTY-SIX: ON VERMIN

I The Nature of Vermin 1739

APIS 1741
ARANEA 1742
ADLACTA 1744
BUFO 1745
BORAX 1745
BLACTAE 1746
BOMBEX 1746
BLUCUS 1746
CICENDULA 1746
CENOMIA 1746
CINIFES 1746
CULEX 1747
CANTARIDES 1747
CRABRONES 1747
CIMEX 1747
CICADA 1748
ERUCA 1748
ENGULAS 1748
FORMICA 1749
FORMICALEON 1749
FORMICAE INDIAE 1749
LIMAX 1750
LOCUSTA 1750
LANIFICUS 1750
MULTIPES 1751

MUSCA 1751
OPIMACUS 1752
PAPILIONES 1752
PHALANGYAE 1752
PULICES 1752
PEDICULUS 1753
RANA 1754
RUTELA 1755
STELLAE FIGURA 1758
SPOLIATOR COLUMBRI 1758
SETA 1758
STUPESTRIS 1759
SANGUISUGA 1759
SCORPIO 1760
THAMUR 1761
TAPPULA 1762
TESTUDO 1762
TINEA 1762
TEREDO 1762
TATINUS 1763
URIA 1763
VERMIS 1763
VERMESCELIDONIAE 1763
VESPAE 1763

Appendix: Alternative Version of the Beginning
of *De animalibus* 1765

Glossary 1769

References 1777

Index 1803

HERE BEGINS THE ELEVENTH BOOK ON ANIMALS

The First Tract of Which Is on the Order of Transmitting the Teaching on Animals

CHAPTER I

Why Two Things Are Necessarily Present in Every Opinion, Noble or Common, Concerning Animals

With all the differences in animals having been treated this way, another beginning must now be made with respect to finding the causes of the things we have mentioned. Now above, we set out the substantial differences [*differentiae substantiales*] of animals and we pursued the substantial differences of all their members and parts.¹ Also above, we further laid down the differences which occur in them both in common and individually. Now it is necessary, as far as we are able, to find the natural and true causes of each of them. For in every opinion which we form about something, whether it be noble (that is, on a noble subject) or common (that is, on an ignoble subject), there are at least two types.²

In one, we are instructed as to what the things are about which we are making inquiry by way of a definition. In the second, there arises in us knowledge of the accidents which are in them.³

For accidents are very useful in acquiring knowledge of what a thing is.⁴ Just as it has been proven that these two types are necessary for the perfect study of the soul, so are they necessary for the perfect study of animals. Nor are we here calling knowledge that which is the conclusion of a demonstration, for we cannot have that concerning

---

1. *Differentia* can mean simply "difference" as it is used ordinarily to distinguish any two things, or it can have the more technical sense of a defining difference that, when added to the genus, provides insight into the essential nature of a thing.

2. "Types": *modus,* which also implies mode or method of investigation.

3. "Knowledge": *scientia,* a word of many meanings, several of which may be implied simultaneously when the word is used. These include something as general as well-founded knowledge, or something as limited as "scientific knowledge in the narrow and proper sense of the word, as distinct from *ars, intellectus, prudentia,* and *sapentia,* i.e., the absolutely certain knowledge of a derived truth in the speculative order or certain and evident knowledge acquired by reasoning strictly from indisputable principles" (Deferrari, 1960, 939). It may also imply a process, i.e., the organized study and investigation of a body of material (i.e., *scientia*). For a good discussion of conceptions of science and scientific demonstration in the twelfth and early thirteenth centuries, see Burnett (1992).

4. This section and the one following represent A.'s addition to the text of Ar. (cf. Ar. *Part. An.* 639a1f.).

the particular natures of animals, but we can rather form an opinion based on things that are plausible.[5] While this may in part treat noble and beautiful things, such as the lives of the animals and the activities of their soul, it may elsewhere seem to be about common things, such as excretion, urination, and the like. Nevertheless, it is on the whole useful, for the study of the nature of animals could not occur without them.

It is clear, then, that the two types mentioned are present in every opinion. For one of them is knowledge of the thing investigated, which is the holding of conclusions from which opinion arises in us concerning the accidents that exist in animals. The other is the instruction and teaching which specify by means of prior things "what" each thing is and "for the sake of which" [*quid est et propter quid est*] it exists in a substantial way. For a definition indicates not only "what" a thing is but also "on account of what" [*propter quid*] it exists, since it is given through substantial causes. For this reason too knowledge is, in this area, instructive and theoretical.

That these two types necessarily exist in every opinion is proven from the activity of the wise listener. It is his job to pass judgment concerning the definitive terms of one who is speaking about something by way of definition.[6] It is also his job to pass judgment concerning the statements of one thinking discursively and proving that some one of the shared or individual accidents is present. For he judges whether there is any truth in the words or whether the speaker errs in them. It might indeed seem true while it is in fact false. These two types, moreover, are adequate, since it is agreed from all the things in the books that deal with natural science [*scientia naturali*] that natural things exist, and it is superfluous to prove this, as we have said in the second book of our *Physics*.[7] It is also agreed that many things are in them and only these two types of knowledge are necessary, so that we might know what a thing is via its substance, and that we might know why the things that have been said to be present in these animals we have enumerated *are* present in them.

We should believe with certainty that every learned person who teaches, listens to, and passes judgment about the things he hears ought to have this disposition in the passing of judgment as well as this power of habit which rules and informs the judgment he makes in the knowledge of animals.[8] For these things have to do with the act of teaching [*operatio documenti*] and with the entire mass of judgments he should

---

5. "Opinion": *opinio*. Opinion is to be distinguished from necessary conclusions arrived at by the deductive method. Such knowledge (or *scientia*) is the result of a demonstration. Opinion, however, is a plausible conclusion held with the conviction that it best interprets the available evidence. The outcome of a process of deliberation, weighing the evidence for one conclusion as opposed to some other *opinio*, is specific to human beings and distinguishes them from other animals. But, again, it must be emphasized that the arguments [*rationes*] supporting an opinion for A. "non sunt demonstrativae, sed probabiles" (*De anima* 3.1.7).

6. A definitive term, *sermo diffinitivus*, is not the last word on a subject but rather one which seeks to define, i.e., set the boundaries, or *fines*, of something else.

7. *Phys.* 2.1.6. Cf. Ar. *Phys.* 193a.

8. With a rare stylistic flourish, A. has inserted *docens* right beside the *doctus* of his received text. The resulting combination, *doctus docens*, emphasizes the responsibility of one who has been educated to educate others, a hallmark of the Dominican order.

have on the particular natures of animals.[9] For the teaching [*doctrina*] of a wise person not only concerns a universal nature but also should include some knowledge of the natures of any given thing and the particular knowledge of its particular and individual accidents. Then in this way of looking at things, another wise person (be he the same person or yet another) will be learned and wise in understanding the natures of a second singular and individual thing. There is, however, but a single general science of physics [*scientia physicorum*] which concerns movable body in general. Now this is only sufficient if individual sciences are also included which deal with things which are individual and individualized in nature. For every particular science on particular things has known *differentiae* by which it is separated from another particular science.

Likewise, every part of a particular science has known *differentiae* by which it is separated from the parts of another particular science. For example, one particular science is about the bodies of animals and another is about their souls, and the parts of the science of their bodies are in the particular books of that science. Those things, however, which are the parts of the science of the soul are in the books we have handed down *On the Soul* [*De anima*]. And the *differentiae* they have with respect to one another are known. Similarly, in all the sciences about a particular nature there also exist those parts which further include under themselves some parts which are more particular than they themselves are.

From all the things that have been introduced it is clear that anyone who wants, by teaching, to tell of and convey what he has come to know by reason and what he has seen through firsthand observation [*per experimentum*] regarding the natures of animals should have definitions which are known *per se,* so that the intention of the one who is speaking about the natures of animals may be ordered according to these definitions, because they are the means used to prove every other thing that is investigated in natures. When certain common or individual accidents are said to be present in animals, then one must judge by means of these definitions whether this is true in fact, if it can be demonstrated, if it is nearly or almost true, or if it is based on things that are plausible. For one cannot have a science in all things by means of demonstration, but rather in some cases it is necessary to conjecture, and we believe plausibly that those things which are not incompatible with the nature of animals are present in them. For teaching about natures perfectly does not consist solely in teaching about the common nature, as we have said, but rather treats separately, through an individual definition, each and every natural thing *per se,* and says what each and every substance is. And it then teaches the individual accidents that are present in it, which it will want to discuss by means of reason, just as it teaches case by case the substance of the nature of the human, the lion, the bull, and of other animals differing in species.

If, however, it desires to pass on a common teaching and will desire to speak of the accidents which occur in all animals in a common way, then it will posit a definitive term, one generally common to them insofar as they are animals, and through that definition it will speak of those things which occur in these same ones through-

---

9. *Operatio documenti* could also suggest the establishment of proof. "Teaching" is, for A., more commonly *doctrina,* as in the next sentence.

out their genus. Now many accidents are common to a great number of animals and belong to them throughout their genus, even though the animals themselves may be of different subalternate genuses. Such are the accidents "sleep" and "wakefulness" which belong to the genuses of animals such as the genus of walkers, flyers, swimmers, and creepers, which are subalterns to the common genus (that is, "animal"), with respect to which an animal is an animate, sensible substance. We have already pursued such studies of common animal genuses and accidents in our books on natural science [*libris naturalibus*] wherein we have described growth and corruption, life and death, breathing, and accidents of this sort in animals.[10] Along with these we have discussed other accidents which pertain not only to the body but also to the soul, those which pertain to the knowledge of the senses and sensibles and to other things of this sort, such as the motion of animals, their memory, and all such things.

9   Nevertheless, a discussion held about this sort of science of animals in general (as far as relates to that teaching in which we are now engaged) is latent and neither manifest nor determinate. I say that it is latent according to the principles laid down in those books, principles which are exceedingly common to this science about the natures of particular animals. I call it not manifest, further, according to the science of logic [*scientia conclusionum*] because to know the natures of things in a universal way is to know them only potentially. This is because a discussion coming from this kind of teaching is indeterminate and not individualized to the individual natures of animals and their accidents. Thus it is clear that it is necessary that we bring forth here another science, which should be in accord with the individual traits which belong to each of them. Otherwise we will not have transmitted perfectly a teaching about natures.

CHAPTER II

According to What Order of Teaching One Must Proceed in the Scientia *of Those Things Which Are Attributed to Animals*

10   It is clear, moreover, that if we establish our study for the purpose of indicating those things which belong to each and every animal *per se* and according to its form and species, and thus describe its accidents, then in this way we will discuss the many accidents of animals many times over. The reason for this is that the same accident appears in many animals, since many accidents appear in them which appear in every one indeed, but which do not belong to one alone.[11] And we will thus return to the same discussion many times over, because almost any of the accidents we discussed in

---

10. "Natural science": natural science is for A. a broad term for which this work, *De animalibus,* represents the end or completion (*Phys.* 1.1.4). In the most abstract sense, natural science treats all those bodies that involve change or motion, with respect to either their form or their matter. A listing of those books which A. views as constituting natural science in a special way will be found at *Phys.* 1.1.4.

11. Cf. Ar. *Part. An.* 639a23f.

the previous books of this science on animals is present in horses, people, dogs, and many other animals. For that reason it follows necessarily that anyone who wishes to describe the accidents of any of these genuses of animals and to attribute any accident whatsoever to some animal *per se* would mention the same accident many times over.

Perhaps it happens that in the science of animals certain ones of those which are being studied are, in one way or another, the subject of a single, individual, definitive term because they are distinguished with respect to a difference of specific form from all the others. Examples exist in the genuses of animal which are distinguished one from another in types of subaltern genuses that are not positioned subalternately, as when "animal" is divided into "flyer," "swimmer," "walker," and "creeper." Some of their attributes are different and some are the same. Now since certain attributes are shared among animals and certain ones are individual and singular, let us certainly not be unaware of how we should proceed properly (that is, according to the knowledge and the order of the intellect) in transmitting a teaching concerning them.

Let us inquire, therefore, whether, proceeding from things that are prior by nature, we ought first to consider the accidents of the common genuses of animals, for it is from these that one ought to consider the individual properties [*propria*] of a given genus or, on the contrary (as long as we are proceeding from prior things), we ought to begin by describing the natures and dispositions of the individual accidents, which belong to a given animal according to its individual nature. However, making a secure determination about them by this method is neither determined nor known since it eludes our knowledge because of the multitude and infinite number that exists in such things. If, however, it is a matter of speaking and treating such things in a general and common fashion, then it is necessary to inquire whether the consideration of natural objects is (as is the case for mathematicians and astrologers) a consideration which is linked in another way to physics, as has been shown in the second book of our *Physics*.[12]

These astrologers and mathematicians first pose those matters which form the basis of the inquiry, such as an eclipse of the sun or the moon, or that a triangle has three angles equal to two right angles, and then afterward they adduce their causes which are the means of proving something is necessarily true. Thus, it may be asked whether a natural philosopher [*physicus*], speaking of the common properties of animals, ought first to consider natural things and the visible activities and passions of the common types, that is species, of animals, and should assign causes for them afterward, or whether he ought to proceed in a contrary way concerning the matters we have described.

Let us suppose, then, on the basis of all the things introduced, that those things ought to be described first which pertain to the visible activities and passions of animals, just as we did in all previous ten books. But now we ought to introduce the causes of those things we have enumerated and which we have said are common to the genuses of animals.

In the following books we shall assign many causes, seeing that, according to nature, there are many causes of generation or of those generated, as has been shown

---

12. Ar. *Phys.* 193b22f.; cf. A.'s *Phys.* 2.1.8, where A. inquires whether astrology belongs to physics or is a separate *scientia*.

in the second book of our *Physics*.[13] For example, we will state the final cause, which is the cause of causes, and we will state the efficient or moving cause, which is the principle of motion, for these two coincide with the form in all things movable toward form.[14] Then, wishing to create a perfect science, let us distinguish which among these are first causes (these are causes of other causes, like the final cause) and which are secondary causes, deriving their causality from those others so that they are not causes without them.[15]

15 As we have said, however, it seems that in all things movable toward form, the first cause (which is defined as that on account of which all other things exist) is identical with the form from which there arises the true definition of a thing in all those things whose generation and conservation [*sustentatio*] in true being is brought about through the nature which is truly the form of the thing.[16] And it is similar in all those things whose existence and persistence has been fashioned in the being of something wrought through the form of an art. The definition of such things is received from the sensible form, which has been conceived with matter, and from those that distinguish the thing defined from other things. Now the natural philosopher undertakes the study of sensibles [*sensibilia*], which were conceived with matter, just as we taught in the second book of our *Physics*.[17] Thus does a physician (who is a particular kind of natural philosopher) work when he defines medicine in terms of sensible matter, saying that it is the science of the healthy, the sick, and those who are neither, because it is the science by which the sensible dispositions of the human body are known with respect to those by which it is healed and those by which it is made ill. In like manner does the woodworker work who imitates the natural philosopher when he defines a knowledge of carpentry in terms of a form of art which conceives sensible matter.

16 Now both of these people, in their definitions, are discussing final causes for the sake of which efficient causes produce in the material, sensible subject everything they produce. The best and the ultimate intent in the activities of nature and craft is the cause for the sake of which [*causa propter quam*], that is, the final cause. The factor of necessity, however, which is in all the things of nature, is found in them in the same

---

13. Ar. *Phys.* 194b17f.; A. *Phys.* 2.2.1–4.

14. "Movable" here implies, as it did for Ar., the possibility of not only change in location but change in a larger sense, namely, even that change by which a thing takes on its form or appearance. "Motion" [*motus*] includes: generation and corruption, increase and decrease, alteration, and change of place. Cf. Dähnert (1934), 194.

15. Cf. Ar. *Part. An.* 639b11–20.

16. Reading "et" instead of Stadler's "est," following the emendation in Hannes Möhle's German translation (2011, 245). *Sustentatio* implies maintenance. Thus, once one has being it is also necessary for one to be maintained or "conserved" in being.

17. Much is unclear in this sentence. For example, *naturalis*, while best translated as natural philosopher or natural scientist, will be followed immediately in the next sentence by the more common *physicus*, also a natural philosopher or scientist. The difference may be because the second usage is A.'s interpolation. Or it may reflect the common usage until the end of the nineteenth century in European universities, where the study of physics was found in the faculty of natural philosophy. *Physicus*, then, may be more specific than *naturalis*. In this book, both will appear as natural philosopher. The translation offered also demands an unusual rendering of the verb *accipit*.

way it is found in the arts and in medicine. Moreover, all who provide definitive terms in books on natural science strive for an ultimate end by means of their definitive term, just as in woodworking and medicine.[18] But a satisfactory determination has already been made concerning all these things in the second book of our *Physics,* and it may be sought out there.[19]

It is necessary to repeat here, however, from what was said in books on natural science, the number of ways in which something is said to be necessary. It is said to be necessary in two ways. The first is necessary absolutely [*simpliciter*] in eternal things (which are without motion) and this is the necessity that exists in disciplines subject to demonstration.[20] In the second way, however, something is called necessary in a transumptive mode (that is, by supposition of the end) and this is the necessity that exists in all movable things which are attributed to generation. Just as necessity is said to be in artificial things when artisans want to define their works of art in terms of the artificial forms which they bring to the matter, so too is necessity said to be present when a definition is assigned of a house or of another object like it created by art.[21] For in this case it is not absolutely necessary that there be wood or building stone, but once the form which completes the house has been supposed, it is then necessary that there be stones, wood, and other building materials. And in cases such as these that thing is first which was built first (for example, wood and stones) if there is going to be a wall. Next is that which, in turn, is nearer to the end than this last one, as it is necessary that a wall exist if there is going to be a house. And thus suppositional necessity exists in all things ordained to an end.[22] In this case it is the completion of the house, since the activity of all the things ordained to that end has occurred for the sake of that which the woodworker intends to be brought about last.

In all these things it is necessary that that which is first in the work be supposed last in intention. And that which occurs in nature has to endeavor by necessity in entirely the same way, and the only difference is that our statement about what is made by art is now posited for what is made by nature. In things subject to opinion it is necessary to do this through a conviction based on plausible explanations, and in things capable of being known it is necessary to do this through the understanding of principles. For something is necessary in one way in things subject to opinion and subject to change (such as are natural things) but in another way in intelligibles and things subject to being known (such as belong to the demonstrative sciences). Now we have already determined in the second book of our *Physics* that in some things, such as in those capable of demonstration, that which exists [*id quod est*] first is the prin-

---

18. For definitive term, see 11.4.

19. Cf. *Phys.* 194b33f.; A. *Phys.* 2.2.4ff.

20. Cf. Ar. *Part. An.* 639b21f.

21. "Artificial": the Latin *artificialis* has no negative connotation but merely refers to something that has been created by *ars* rather than by nature.

22. On A.'s conception of suppositional necessity, cf. Wallace (1980a); (1980b). In the latter, Wallace (1980b, 125) notes that "Albert was the proximate source from which Aquinas derived his knowledge of suppositional necessity . . ." although "Aquinas was the authority who became better known in university circles . . ." Wallace (1976) has traced the impact of suppositional necessity on later medieval thought.

ciple of other things, just as premises are the cause and principle of a conclusion. In some things, however, such as those things which are subject to generation and which, according to opinion, are capable of being altered, the principle is not what *is* (since this is matter) but rather what ultimately *will be*—this is first and is the principle of all the others.[23] For this reason, all that is necessary in them is necessary according to the intent which is presupposed, and not otherwise. This is just as if we were to say that health has to be brought to a person by the activity of a physician, then that which will come to pass by the activity of the physician (e.g., such as giving a potion or the regulation of diet) is necessary.

19   Health, however, is not necessary because there is this work of the physician (for example, the potion) or because there was before this another work of the physician (for example, the use of an enema). Or again, health will not be necessary on account of this work, because after health is supposed the preceding things are necessary for the sake of it. And health does not follow necessarily from those already mentioned because it is their cause and they are not, to the contrary, the cause of it. Rather, they are caused by it just as by the first principle of all other things. We cannot attribute the necessity of the exposition of an eternal object to an animal in order to say that because this exists in an animal, then that exists there as well, or necessarily will be there. For the necessity of this exposition, of which we have spoken here and in the second book of the *Physics,* is the necessity of the consequence [*necessitas consequentiae*] that once one has been given, it follows that another exists in accord with what must necessarily happen.[24] And this necessity of the consequence in eternal and neces-

---

23. "Subject to generation": reading *in generabilibus* for Stadler's and Borgnet's *in generalibus,* as does Wallace (1980b), 122. Cf. Ar. *Part. An.* 640a4f.

24. At *Phys.* 2.3.6, A. attempts to distinguish the *necessitas consequentiae* and the *necessitas consequentis,* noting that "although the knowledge of the conclusion may be the goal for the demonstrative disciplines, nevertheless the premises are not necessary on account of the conclusion but have a necessity in themselves, and on account of their necessity the conclusion is necessary. And although these premises have the necessity of the thing, it is not the necessity of the consequence [*necessitas consequentiae*] alone that is applied to them but rather the necessity of the thing which follows, which is called by some the necessity of the consequent [*necessitas consequentis*]. There is, then, in these a movement from the premises to the conclusion, and the conclusion is necessary on account of the premises and not vice versa. And if one proceeds from the conclusion to the principles, there will not be a demonstration *propter quid* but a demonstration *quia,* and that demonstration posits only the necessity of the consequence [*necessitas consequentiae*], since the last things are not true and necessary because of the first, but vice versa, because the effect contributes nothing to the cause but the cause contributes to the effect." (A demonstration *propter quid* is one understood to be true *a priori,* while a demonstration *quia* is true only *a posteriori,* or moving from the effect to the cause.) The *necessitas consequentiae,* then, may be understood as a hypothetical or conditional necessity. This is, indeed, the sense with which Boethius endowed the term as well, when he identified the necessity enjoyed by the conclusion of the hypothetical syllogism as a *necessitas consequentiae* (cf. *De syllogismo hypothetico, PL* 64: 843B). Thus, when a conclusion or the end of a series exists, then the premises or prior terms also exist necessarily. For example, if one should say that a human being exists, then *ex necessitate consequentiae* an animal exists, and something animate, a biped, having an intellect, etc. (cf. Boethius *Interpretatio Topicorum Aristotelis, PL* 64: 628C). *Necessitas consequentiae* may also imply the formal logical consistency of the syllogism (cf. Dähnert, 1934, 195). For *necessitas consequentiae* and *necessitas consequentis* in Aquinas, see Deferrari (1960), 692–93. Also, for this passage, cf. Ar.'s *Phys.* 199b33–200b5.

sary things is a necessity which is capable of demonstration in those which are eternal and necessary.

Now this sort of necessity exists in these things because the first is the cause of the second, the second of the third, and so on as far as the last, just as is shown in geometric theorems, and this is not convertible. However, in a science of animals it does not occur in this way but rather in a contrary fashion, because once the last one is given, it follows by necessity that the prior one is or was, but this is not convertible, just as was shown through those things which were discussed earlier and in those treated fully in the *Physics*.[25] For there we already made a detailed statement of necessity and determined in what things such a consequence will exist and in what things not, and which of these types are convertible and which not, since negative statements [*sermones*] are convertible in such consequences but not affirmative ones. For one does well while affirming something in things capable of demonstration because, if the principles exist, the conclusion will exist as a consequence and by necessity.

But if the conclusion exists, it is not for its sake that the principles exist with respect to the thing. And if it should follow that these do exist once the conclusion has been supposed, then this will be the necessity of the consequence [*necessitas consequentiae*] and not the necessity of the consequent and of the thing [*necessitas consequentis et rei*]. But it is convertible negatively: if principles do not exist, no conclusion will exist, and if there is no conclusion, premises will not exist. In things subject to generation it does not follow that if an efficient cause and matter exist, then on account of this the last thing is, or will be, generated. But to the contrary, given that the last thing has been generated, it follows that there is an efficient cause and matter. However, it is convertible negatively, just as in the [demonstrative] disciplines, since it will not be generated if matter and the efficient cause do not exist. Just so, when there is no conclusion, there will be no principles put forth first. All these things were extensively treated in the second book of the *Physics*, where we discussed which of those types are convertible and which not, and on what basis those which are convertible are convertible and on what basis those which are not convertible are not convertible.[26]

It is necessary, moreover, to know that the ancient philosophers before us used the consequence of necessity, of which we spoke here, determining how and in what manner any of these things "capable of generation" will exist through necessity. Now for them to show, as far as possible, that the necessity of the type belonging to the demonstrative disciplines exists in generated things was more important than to show how natural things exist according to principles and how they exist in reality [*veritate*

---

25. Cf. Ar. *Phys.* 199b-200b and *DG* 337b15f. Here Ar. explains that once a house exists it is necessary that the foundation for the house must already exist. But it does not follow that when the foundation exists, the house must also come into being as a consequent. The order of "causality" is not convertible. However, as he suggests in the next sentence, if there is no foundation one can say with certainty that no house will come to be. Similarly for A., in a science of animals, one can establish with a certain necessity (i.e., suppositional necessity) the existence of the cause based on the presence of the effect, but one cannot claim that in the presence of the cause the effect will be necessary, since other things may block its agency and frustrate its causal power.

26. *Phys.* 2.3.5f.; Ar. *Phys.* 200a19–30.

rei] and according to nature. For they wanted everything to generate from mathematical principles and according to mathematical necessity, and they considered this especially important, not caring much about natural reality: how principles exist and how natural things exist from them. And yet the difference between these two things is not insignificant.

23 It is, moreover, clear from all the things already mentioned that we must begin in the manner of which we have spoken. That is, we should first describe the natural objects evident to sense, throughout each genus, and we should then return and establish the cause of those things which have already been made evident to sense. This should be done especially in the science of the generation of things subject to generation and which is altogether similar in a knowledge of woodworking, just as we said. For in such matters the prior ones exist for the sake of the last one, and not the last for the sake of the prior. Thus do generation and those doing the generating exist for the sake of the form, which is the substance producing the thing generated. It is not the case that, on the contrary, the formal substance exists for the sake of generation and for the sake of those doing the generating. This is why Empedocles erred in his opinion that there is no purpose in natures but that each and every thing comes to be through a cause accidentally. Now, he said that among the ranks of the animals it happens that things are one way, and it often happens that they are otherwise, and for that reason the human is human by accident and chance, and the ass is an ass through chance. Likewise, he said that in their creation and shape the vertebrae do not have a break in the middle on account of a natural cause, but rather such a break occurs in them when the principles arranging their position at the place of the opening have completely changed and have been transferred to another position by chance.[27] For Empedocles did not say that there is a final cause in natural things.

24 However, Empedocles did not know that the sperm of generation has to have the sort of formative power which leads to *this* form and no other, and that that which made this power in the sperm existed before it did, just as it is the human who is generating who gives the formative power to the sperm. As a result, the form and end in natures precede the potential not only by definition but even in time, just as was proven in the ninth book of the *First Philosophy*.[28] Certainly a human generates a human, so that the form of the one generated is produced from the form of the one generating, and thus it is in all natural things, which Empedocles erroneously believed are generated *per se* without a cause. They are, however, generated from the form of the first efficient cause, for the art of making statues (which is the creator and efficient cause of bronze images) acts just as does nature. Moreover, that art acting formally is a sort of definition and is a form without the matter of the form of that which is in the bronze. Therefore the statue is not made *per se,* without a formal cause preceding it, but the form which fashions and the thing fashioned do not participate in matter.

---

27. Cf. Ar. *Part. An.* 640a20f. Empedocles was trying to account for individual vertebrae by saying that the twisting of the fetus broke the backbone into discrete units.

28. Cf. Ar. *Metaph.* 1050a5–2b.

Sometimes, however, just such a form might happen to exist in a similar disposition. Nevertheless, one should not say on account of this that art has no cause, because something is rarely brought about by chance but is frequently brought about by art. Now, to be sure, one who is not a statue maker occasionally produces a statue, and sometimes by chance one who is not a musician does make music. And, just as a thing exists through art, so too a thing exists through nature. For that reason we speak properly when saying that the final cause, which is the cause "for the sake which," is present in the generation of the human. For the human has these things (that is, natural causes) because without them he cannot exist anymore than he can without this or that member. And this is how it almost always is, or very nearly so, even if sometimes [a human] comes to be by chance without them and through a lapse of nature.

Now generally speaking, natural beings cannot exist in a different way.[29] For these things follow one from the other, because when the last thing is necessary, then such a thing occurs necessarily and not by chance. Specifically, that which is prior—and this is matter or potency and efficient cause, and a member will serve as an example—exists for the sake of that which will be. After this there will be another which is composed from it, and then there will be lastly the perfect whole, just as if I should say that because a human will be the last thing generated, it is thus necessary that the uniform members [*membra similia*] exist first, and it is necessary that from these there exist composites, and from these it is necessary that a human exist. And, once the last generated thing is supposed, this whole consequence necessarily exists. Therefore the division of the prior and the posterior in all natural things will occur through this type of necessity.

CHAPTER III

From How Many and from Which Causes Those Things Which Are Being Investigated Concerning Animals Must Be Caused

On the assumption, then, that one must proceed from causes in natural things, let us investigate from which causes it is necessary to proceed. Now the ancients [*antiqui*] and all those who have spoken philosophically concerning natural things considered only matter as a cause among natural principles.[30] Concerning matter they determined the kind of causality such a cause would have, and they considered what the mover is of that which we call the efficient cause, from which a thing exists.[31] They also considered whether it is different from what it ought to be, as happens when [the

---

29. "Way": *modus*. Again, the Latin *modus* may mean way, type, mode, or method of investigation. Cf. 11.1.

30. On the meaning of *antiqui* and the conflict with *moderni,* see especially De Rijk (1962), 1:14–17. Also, cf. the ancient philosophers of 11.22.

31. Cf. Ar. *Part. An.* 640b5f.

cause] fails in its working, or whether the activity of a thing (which is said to exist *per se* when its work does not depart from the correct path) is said to exist *per se* (that is, through nature). Now they said that all natural beings come to be through the nature of the matter and they determined the properties of matter, saying that the nature of fire *per se* is hot and that of earth cold, and that the nature of fire is *per se* light in weight and the nature of earth is heavy. And they relate the generation of the whole world, giving causes in accordance with these sorts of dispositions of matter.

28 In like manner they speak of the generation of animals and trees, saying that in no respect does there exist a form and an end but that all things are generated from a combination of such qualities of matter. This is just as if one were to say that when water flows in a mixture into the body, there will be a wind in the body due to the evaporation of the water, and thus where the water evaporates there will be a hollow [*profunditas*], and from just such a hollow, which is produced from the qualities of the matter, there will exist in the animal a receptacle for food and for the superfluities of the moist and the dry, which are urine and excrement.[32] Again, this is just as if one were to say that the nose is not divided and perforated through the formative power by the power of the form, but rather that it is perforated so that the breath, which has been drawn in with the air that has been breathed, may go out again there. And on account of material qualities of this sort they suppose that air and water, thus hollowing out the earthy bodies, are the matter of the generated bodies; that is, that they sustain the nature of the generated bodies which exists as a whole from those bodies which are called elements. All the ancients say that such are the causes of natural things.

29 But this is not sufficient, for on the assumption that a human and other animals are natural substances (since natural substances exist only through form) it will be fitting and worthwhile for us to describe and declare by means of the formal definition that this member is flesh, this one is blood, this one bone, and similarly of the other homogeneous parts. And, in the same manner, on the basis of the form and end we shall determine the nonuniform parts, which are called organs [*organa*] because they are accommodated to the functions of the soul. An example is the face, which is composed of many organs, and the hand and the foot, which, like the face, are composed of similar and dissimilar parts, being composed from fingers (which are dissimilar parts) and of flesh and bone (which are uniform parts). Now these are all parts *propter quid* (that is, for the sake of an end) and it is necessary that we indicate by which power of the soul such parts may exist and have being, since that power is the form for the sake of which they exist, just as it was said in the study *On the Soul* that if the eye were an animal, sight would be its form.[33] Thus it is not sufficient to assign only the matter, as Empedocles did, in order to state the things from which something exists (such as fire and earth).

---

32. "Wind": *ventus*. While this rendering is secure in both St. and Borgnet, Sc. has *venter,* or stomach, rendering Ar.'s *koilia* (*Part. An.* 640b13). A. has added confusion to an already difficult passage.

33. *De anima* 2.1.3; cf. Ar. *De anima* 412b18f.

Now if we should have to indicate both the definition and the disposition of a bed or of another thing made by handicraft in which art imitates nature, we should use a distinct definition to assign its form rather than its matter, saying that the bed is wood and bronze (if the bed is made from wood and bronze). If, however, we should not define a bed in the way we mentioned (that is, in terms of form) we will define it in this manner when describing the entirety of its being, saying that a bed is such and such, namely, wood and bronze. But this is unsuitable since according to this procedure if a statue is made of wood and bronze, then the statue will be a bed. If, however, we want to describe it, then a definition exists with reference to shape because in things made by craft the shape is the form.[34] We will then describe its form as made distinct by its definition, and we will reveal this form through its defining characteristics [*definientia*], because that nature which is known through form is nobler than that which is known through matter.[35]

If, therefore, every member and every animal ought thus to be known through shape and color, then Democritus spoke well, since it seems from his remarks that his opinion is that form ought to be investigated. But he did not say that form is substantial, but accidental, being the shape and color caused by the position of atoms, just as is revealed in the first of our books *On Generation* [*Pery geneos*], where we explained the view of Democritus.[36] Therefore one who erroneously believes that the shape and color which is seen in a human suffices as the natural form says this only because he thinks that shape is known *per se* and thinks that, without a doubt, it is the principle of the knowledge of things formed. For form is indeed known *per se* and is the principle of the knowledge of things formed, just as was shown in the *First Philosophy*.[37] Nevertheless, in the natural sciences [*in physicis*], which have forms conceived along with matter, form alone ought not to be assigned, for an imagined human has the form of a human without matter, and although it would be human according to the shape and the color imagined, nevertheless it is not truly a human.

Moreover, it is similar with respect to a natural hand and a hand that is shaped or drawn: these are equivocals, agreeing in name but not in definition [*ratio*], just like a drawing of a physician which has only the shape of a physician, but is not a true physician.[38] This is just how a woodworker works. When he wants to depict a hand or some other thing made from wood he provides a form in matter, through the definition of art. A natural philosopher acts similarly when he wishes to define generation (i.e., the

---

34. Reading Borgnet's "voluerimus hoc narrare, est definitio secundum figuram . . ." for Stadler's ungrammatical "voluerimus narratio, hoc est deffinitio sermone figuram . . ." The text remains corrupt. Cf. emendations suggested by Möhle (2011), 258, 260.

35. At *Phys.* 1.2.8, A. describes a thing as adequately defined [*diffinitum*] when it is fully marked out by its defining terms, as man is marked out or defined by "animal" and "biped," i.e., by genus and *differentia*.

36. Cf. Ar. *DG* 315a35ff.

37. That is, the *Metaph*.

38. Ar. *Part. An.* 641a1f. For a good discussion of A.'s understanding of equivocation, univocation, and analogy, see Tremblay (1996), 277–92.

thing generated from this matter), assigning the efficient causes, forms, and powers of the forms which act in this matter. Moreover, the woodworker, when indicating the instrumental causes by whose power there comes to be that which comes to be in matter, indicates the axe and the saw, which have form in this matter and are the instruments by which work is done. The natural philosopher indicates earth and water as the matter in which the first qualities exist, and it is into these, once they have been moved by the efficient cause, that the natural form is introduced.

33 We have often taught in the various books of natural philosophy that the first qualities have the powers of the substantial forms by which they are moved. Thus, the first qualities which are in sperm have the power of the human form, which is called the "formative." Nevertheless, the woodworker, in the definition by which the instrument is posited, defines better and more perfectly than the natural philosopher who only makes mention of the matter. This is so because when the craftsman's instrument is employed in the work, it makes the wood deep in one place and plane and level in another and thus, through the instrument, something is known of the form. It thus says something about the cause for the sake of which the wood is shaped in this shape, and why [*propter quid*] it is made this way, and he fashions his work through the application of an instrument made so that it would be only of such a form or of such and such a shape.

34 It is clear that the natural philosopher assigning matter alone [as a cause] does not speak correctly, since he ought to state in a definitive statement why an animal exists and what it is, so that he might refer to the form and the end, and, while defining it, to relate (that is, state) any member which exists, and why it exists, just as he relates the form of the bed through a definition. And since the soul is the formal and final principle of animals, he ought to state whether this is the soul that falls under the definition of the animal. He ought also to state whether it is the soul or part of the soul that falls under the definition of the members of the animal. And he ought to show that when an animal expires and loses its soul it will not have the essence [*ratio*] of an animal. It will neither remain an animal after this nor will a given member remain a true member. Rather, it will endure only in the shape and exterior form of the member, just like the animals of which poets sing in their proverbial expressions and which were changed into rocks, for these too retain only the shape of animals.[39]

35 If, therefore, matters are as we have described them, the natural philosopher [*naturalis*] ought to describe and teach the disposition of the soul and its parts as far as he is able when fashioning a definition in the study of animals, because the soul is the principle of animals, as we have said in the first book *On the Soul*.[40] And he ought to give his account by assigning the disposition of any soul and the disposition of any sort among the parts of the soul, and to define what an animal is and to show whether the soul is a part of the animal. Then he ought to explain the accidents which occur in the animal and in the substance of the soul, which is of this or that sort.

---

39. The example is from Ar. *Part. An.* 641a21. Neither Peck (1983) nor Balme (1972) cites a source.
40. *De anima* 1.1.1; cf. Ar. *De anima* 402a6.

Animals are spoken of in this way, through a definition, according to two types of causes. One of these is just like matter and the other is just like form or substance, providing the being and the definition. Now nature, just as we said in the second book of the *Physics,* is said to be just like motion (that is, a generation) which occurs on matter, and it is said to be just like a completion.[41] But generation should not be said to be nature, except in the sense that it is related to nature, which is a completion. It is therefore clear that in natural matters we ought to make a consideration following such a division of causes and that we ought to give more consideration in the natures of animals to the disposition of the soul than to a consideration of matter. For matter is not said to be nature because of the disposition of the matter, but rather by an analogy to form. For a bed and a tripod are certainly said to be from wood, for the wood is said to be these things in potency but not in act. In like manner, in the natural sciences matter is said to be the nature of things that have been generated.

When someone reflects upon the natures of animals in accordance with what we have said, perhaps he will be in doubt whether the natural philosopher should speak of every natural soul in general, or of some individual soul which is the principle of animals that are generated and corrupted. This is because if he should speak of every soul, in this respect there will be no difference between him and the first universal philosopher who speaks of the souls of the world. This is because a first philosopher's understanding seeks a science of those things which are truly theoretical.[42] If the natural philosopher seeks the same things, then in this respect there is no difference between them. However, it is clear from those things which were well said in the book *On the Soul* that the natural philosopher seeks to know all natural things through their formal principles, and it pertains to his science to make a consideration in some manner from the intellect, and from the thing understood, although these two are separate and distinct.[43]

Now, there is one science (and one opinion) concerning all things when they are interrelated. So, as the intellectual soul is referred as act to a physical body, to this extent there will be one science (or opinion) if it treats both the body and this type of soul on the basis of plausible arguments. Now the science and opinion of sensation and of things sensed is thus the same, because the one is related to the other by reason and by relation. Not every soul, however, is the principle of local motion or the

---

41. *Phys.* 2.1.1; cf. Ar. *Phys.* 193a28f. Elsewhere in the *Phys.* (3.1.4) A. explains that "with respect to that which is in potency, it has no perfection other than motion because form is the perfection of a thing existing in act, and when it has form it is not in potency toward it but in act. But when it is in motion then it is still in potency to form. Therefore the completion of that which is in potency, insofar as it is in potency, is motion, although the completion of that which was in potency, insofar as it is no longer in potency but in act, is form."

42. "Universal philosopher": Sc. has "natural philosopher." Note that A. returns to "natural philosopher" in the next sentence.

43. Möhle (2011, 264) emends Stadler's *divisiva* to *divina,* so that the final clause reads "even if these things are separate and divine." Cf. *De anima* 1.1.1. There A. explains that although the soul and its activities fall outside the proper subject matter of the natural scientist, viz., movable bodies, nevertheless the soul is the essential principle of all such bodies and therefore must, to some extent, fall under the consideration of natural science.

principle of the motion of animals' members. Rather, a particular one is the principle of growth, like that principle that is in trees. And a particular one is the principle of alteration (with respect to some senses), which occurs in other animals that are immobile with respect to place. The intellect, however, is not the act of a body and nothing of such a kind exists in any way in any of these animals.

39   Now, then, it is clear to us that since we have to speak of the formal causes of animals and their members, we ought not to speak of the whole soul or every soul, but we ought to speak of it in accordance with the way by which it is the principle of animals that are generated and corrupted. For the nature and the principle of animals are not in the whole soul, but a certain part of it is a principle of this type. Now there are many parts in the soul, and an opinion of the natural philosopher [*opinio naturalis*] cannot be about things taken up from every soul or from the soul considered in itself, since from the things said it appears to us especially true that the natural philosopher understands everything he understands concerning the natures of animals only by learning of the cause that is in them *propter quid*. Not every soul is this sort of a cause of animals because superior things live with a more noble soul which is not the principle of animals. Art does the same in things produced by handicraft (for art only considers the form insofar as it is the principle of the thing produced by handicraft). So too does the natural philosopher work when considering the form and members of animals.

40   Along with this we say that in things generated there is another principle and another cause beyond that one which is both in us and in animals, and this is matter, which is the warm, the cold, the moist, and the dry. This matter is a part of the whole that is generated. On that account we also say that if the heaven was generated, as some have affirmed, it is worthy that it be generated from the matter of all generated things, which is the matter we have described, namely, the warm, the cold, and the other elements. If, however, heaven is generated from such a cause (which is matter) it is worthy that it, more so than animals, be attributed to such a cause as we have described (that is, to the final cause) and to such a soul as that which is the principle of animate beings. But then it will not be as Empedocles said, for he said that heaven was generated but that it nevertheless exists by chance and *per se* without any formal or final cause. For everything well-ordered and praiseworthy appears to agree more with the heavens than with us. However, that which exists at one time and at another not, and at one time in one way and at another time in another, and exists just as fortune determines, appears to occur more in corruptible things than in heaven, in which nothing is disordered. Moreover, there is a discussion of these things in the second book of the *Physics*.[44] We do not say, however, that that which we call an end and for the sake of something [*propter quid*] exists in all things and in every place (that is, in every place of the things in which it appears), that there is a completion and an end, and that the motion is toward it if nothing should impede it. For in things subject to motion [*mobilis*] there is an end, but not in those things not subject to motion [*immobilis*].

---

44. *Phys.* 2.2.11; cf. Ar. *Phys.* 196b25.

From all these things it is clear that nature pertains to form somewhat. For the generation of this one will not be from *any* seed, for with respect to form this one is from this seed and not from any other, just as a human always or at least very often comes from the seed of the human, and we know and establish this through clear proof. Therefore, the body from which the seed is divided and goes out is the first principle of that seed which goes out from it, and this one, through the form it gives to the seed, is the craftsman of that which comes from the seed. For generated things arise from a seed and are from the nature of form which is just like a craftsman in the seed. For that from which the sperm is divided (for example, the person generating it) is a cause anterior to the sperm.

Now the sperm is the formal, active generation, and is the active completion and the active substance by means of the formative power that is in it.[45] Indeed, the sperm causes all the things mentioned in the thing generated: namely, generation, completion, and substance. But for all the things which are in the sperm, there is an anterior cause: the one generating, from whom the semen goes forth and is divided. Now semen is so called in two ways, by way of comparison. For it is both the semen of the one from whom it comes by division and it is also the semen of the one to be generated and for whose sake [*propter quid*] it exists.[46] For example, the sperm of the horse is attributed to the one from which it first goes forth. So too, the seed [*semen*] (that is, the seed [*granum*]) is attributed to the plant from which it has gone forth. This is so even though it does not belong to the two of them in the same way but rather to one of them in one way and to the other in another.[47] The seed exists not only as form and artisan [*artifex*] but also as the potentiality of the thing generated. But I only say that that exists in potentiality which tends toward act through the formal principles it has in itself, doing so without any other extrinsic mover. And this is that in which form is already present as the artisan.

It is therefore clear that all such things have two causes, namely, the one for the sake of which [*propter quid*] and the necessity of matter, which is necessary once the end is supposed, just as we have already said quite often. Now many things exist from such necessity just as from a cause. All things capable of generation proceed from such causes, which are very numerous.

Someone will have doubts, however, regarding the types of necessity brought to bear here and in the natural sciences [*in physicis*] seeing that perhaps they are not sufficient and that it is still necessary to find another type beyond the ones mentioned. Perhaps someone may say that one type of necessity is spoken of and found in what has been proposed (where we must examine the causes not only of the generation of animals but even of the shapes of their members) and this type of necessity cannot

---

45. Reading Borgnet's *substantia activa* for Stadler's *substantia activae*.
46. Cf. Ar. *Part. An.* 641b33f.
47. "Grain": Sc. has, correctly, *semen ginni*, i.e., the seed of a mule. Cf. *Part. An.* 641b35f. Ar.'s point is that the semen or seed is both "of" the horse, which is parent to the mule, and "of" the mule, which is the progeny of the horse and is generated from its seed. A. has read *granum* for *ginni*, requiring an insertion on the nature of plants!

be described under one of the two types we have mentioned. With respect to one of these there is a necessity in an absolute sense [*simpliciter*] which exists in the [demonstrative] disciplines. With respect to the other, suppositional necessity is mentioned, and this exists in natural acts of generation. But one should add still a third type from among those types established in the fifth book of first philosophy.[48]

44 Let us briefly touch upon the types established there in order that this might be better understood. There are four types, the first of which is that necessary thing without which it is not possible for something to exist or to come to be. For example, that which is a concomitant cause with being or becoming, such as breathing and food, are necessary to the animal because without these the animal does not exist. In this type is that suppositional necessity which we said exists in the generation of natural things.

The second type of the necessary is said to be in those things without which it is not possible for an intended good to come into existence, and without which it is impossible to expel evil. In this way, one must drink medicine to keep from getting ill, just as it is necessary to sail to the city called Egyna in order to get money.[49] We need the type of necessity just mentioned also if we have to provide the final cause for the shape of the animals' members.

45 The third type of the necessary is said to be that which compels force, just as one who is compelled is said to do what he does from necessity. We require this type of necessity only in moral science, and not in the natural sciences.

The fourth necessary is the one which cannot be otherwise, just as a conclusion is necessary because of the necessity of the premises. Moreover, the other types are reduced to this necessary because they imitate this type of necessity. This is the type of the necessary in the [demonstrative] disciplines, and we require it in the discipline of natural science [*in disciplina physica*] if we should have to argue syllogistically.

But here in the present discussion we need to add to the necessary mentioned in the second type. This is also a suppositional necessity just like the first type, but the first type is for being and causes concomitant with being, while the second is for the useful and the good and those causes concomitant with the useful and the good. So, if the usefulness of the operations of the organic members should be realized in the animal, it is necessary that the members be of a certain shape. As a result, if the animal has to be ambulatory it is necessary that the feet be of just such a shape.

46 Returning then to the case in question [*ad propositum*], let us say that this is called necessary according to the third type, beyond the two brought forth in the second book of the *Physics*.[50] This necessary is also in things which have a principle of generation, as is the one we described above. And we can say that this necessity is so called

---

48. In the following section (11.44), the discussion of the various types of necessity is paraphrased from Ar. *Metaph.*, 1015a20-b15; on the third type of necessity, cf. also *Part. An.* 642a7.

49. Egyna: Aegina is an island near Athens, and often its rival, which originated the standard upon which Athenian coinage was based. The transliteration of the name accurately reflects modern Gr. pronunciation.

50. *Phys.* 2.3.5f.; cf. Ar. *Phys.* 199b34f.

according to one of those types we described, insofar as the first and the second types share in a genus, because each has suppositional necessity. For example, since food is not said to be necessary by one of those types we mentioned in the *Physics* (seeing that it is not for the generation of the animal as earlier it was for the purpose of introducing form, but is rather for the purpose of obtaining the good and the useful) it is not then necessary—in the way in which we stated—that an animal can exist or be generated only if the matter would be such and such. Rather, as we say, supposing that the axe has to split the wood, then of necessity it has to be hard, and if it be sufficiently hard to split the wood, it must be of iron or bronze. For otherwise the uses intended for it would not be realized.

And we speak similarly when we call something an instrumental and organic member in a body, since every member whatsoever exists for the sake of [*propter quid*] some activity and they arise only from such a shape and such a matter. And it is the same for the whole body. Let us then describe this necessity in this way. As a result, something made in this way (specifically, that which has been given shape this way) therefore exists necessarily and the utility of its activity (for the sake of which it exists) is also for the sake of this thing. Through these two types of the modes of the cause of generation there will exist a final cause, on account of which there will be, properly speaking, a generation of natural things, and without which no generation will occur, just as we explained previously.

47

It is clear from the things said that it must be demonstrated that they (of whom we have already made mention) do not speak rightly and truly about nature who establish only matter as a cause, saying that matter is the most worthy thing in nature. According to this view there will be no necessity in nature, because there will be no suppositional necessity when one removes and takes away that on the basis of whose supposition another whole thing is brought about as necessary. Now the truth unavoidably leads Empedocles himself (and he had denied that a form and end exist) and forces him in several places to describe that cause which we call the form and end.[51] On account of this he says somewhere that the substance and form of a thing are the definition and the "whatness" [*quidditas*] of the thing, which gives it being and definition [*ratio*]. In just this way, someone who wishes to define bone will not say that it is matter (which is only one of the elements) or that it is two or three or all of them, but rather he will say that bone is a mixture of elements of such and such a kind, and thus will strive to reach the form. He who wishes to define flesh or wishes to define any member whatsoever will act in a similar fashion. Now the reason that prevented the ancients from applying themselves subtly to a consideration of that cause which we have described is that they did not rightly define the nature of things generated, nor did they rightly explain what generation generates.

48

But among the ancients Democritus is said to have been the first to approach the subtle nature of this cause in some manner.[52] He did not accomplish this by virtue of

49

---
51. Ar. *Part. An.* 642a18.
52. Ar. *Part. An.* 642a26.

having imitated the claims of the natural philosophers who had preceded him, but rather the very truth itself unavoidably directed him to this consideration. Yet Democritus nevertheless thought that substantial form and accidental shape are the same thing, and he said that generation is alteration. This opinion concerning the formal cause prevailed at the time of Socrates, however, and it arrested the ways of inquiring after a knowledge of nature, which could not find peace in the sole knowledge of matter. Natural philosophers were next satisfied with and set themselves to examine the truth of Aristotle, which alone is helpful in seeing the truth and alone is that which rules in the knowledge of the natural body.

50   Therefore, we ought to form our philosophical discourse in accordance with this disposition, by saying that in animals the soul is that cause which is called the cause for the sake of which [*propter quam*] or the final cause. Moreover, that which is a natural body and member exists necessarily because of those things which we have mentioned, which are the form and the end. Now that which we call necessity in natural things shows and indicates that a thing which exists in natural things exists only *propter quid,* because otherwise it would have no necessity whatever.[53] Therefore, the ultimate completion and the end is that for the sake of which there will be that which occurs in natural things.

51   When speaking in this manner, therefore, these natural beings will necessarily exist just as they exist in their shapes and materials, and it is necessary that they exist according to that ultimate, final, and formal nature by which they are given the kinds of natures they have. Now the form forms the nature of everything which is given a nature in natural things [*in physicis*], since the form and the act which are intended ultimately are prior to the potency of matter not only in definition and substance but even in time, just as has been said.[54] And once it is supposed by a suppositional necessity, everything which exists in natural things will be necessary. Therefore, heat will necessarily enter and depart from the being of certain animals, and its entrance will not be opposed to its exit, once we have supposed the ultimate form which requires such an entrance and exit. According to this type of necessity, the airway and its entrance to the interior of the body in the respiration of certain animals are required by the same necessity, because air quiets the heat which is in the interior of the heart. It is thus necessary that the air which enters and has become heated should exit, and that cold air that is outside should enter, because otherwise the animal will not be preserved in the completion of its being.

This, therefore, is the innate quality of the highest and chief conception of natural things, and in accordance with this it will be fitting to ascertain causes for every natural thing when something belonging to nature shall have acquired something through nature which is necessary for completion.

---

53. "In natural things": *in physicis*; "in physics" or "in natural sciences" are also possible.
54. Cf. Ar. *Phys.* 193b7f.

The Second Tract of the Eleventh Book on Animals in Which Is Investigated Which Form Is the End of Animals and Their Members

CHAPTER I

Concerning a Science of Animals[55]

Further, in addition to all the things which we have introduced, it still remains necessary to investigate how the *differentiae* of the things which are ultimate forms and ultimate ends may be learned.[56] Certain persons, philosophers among them, observing anything (that is, a species or genus of animals), divide that genus into two *differentiae*. They want the two *differentiae* mentioned by them to be the last two forms of two specific species. This is not troublesome in one way, that is, when dividing through affirmation and negation; in another, however, it is impossible, since the power of the genus cannot otherwise be limited.

Now it is agreed that there is only one ultimate *differentia* in those things which belong to specific species, and this is convertible, just as it is demonstrated in the seventh book of the first philosophy.[57] Whatever other *differentiae* are assigned are superfluous, for they exist in more than the species constituted by the *differentiae*. Such things are superfluous which are not required for the establishing of species. For example, if the *differentia* of a human is said to be "biped" or the existence of a split foot, then "animal" will be divided between "biped" and "not-biped," or between "having a split foot" and "not having a split foot." Now these are not *differentiae* convertible with the species which must be grasped through division.[58] Rather, they are individual

---

55. This heading is found in Borgnet's edition but not in Stadler's. The translators have supplied the chapter title, "Concerning a Science of Animals."

56. Ar. *Part. An.* 642b5.

57. Ar. *Metaph.* 1038a18f.

58. While it may seem odd here to identify the *differentiae* of the human being as "biped" or "having a split foot," a look at 2.8, 25, reveals that for A. the human *does* have a split foot, clarifying somewhat this passage.

Division, furthermore, represents the process of establishing a definition. As Ar. remarks (*Posterior Analytics* 91b27f.), the "champion of division" will begin with an inquiry about the nature of a genus—e.g., "Is man animal or inanimate?"—and, after eliminating one of the two proposed contraries, proceed, while working his way down, to articulate as many of the attributes of a man as possible. When we reach a point when further division is impossible, then, he assumes, he has arrived at a true definition, and (true) definitions can provide true knowledge. In response to the question, "What is the essential nature of a man?" the "divider" will respond, says Ar., "animal, mortal, footed, biped, wingless," although such a definition is for Ar. inadequate. For A.'s critique of division, see Tkacz (1993), 130–55 and Tkacz (2013), 516–20.

properties, which either are not *differentiae* or are hidden and concealed, because species is understood in the individual, and the *differentia,* which is the last form, is only grasped as it lies concealed in the species, because the *differentia* is only understood in the species in the way that what is prior is understood in what is posterior.

54 If, however, someone should offer a contrary view and say that it is not just as we have said, then necessarily the same thing will fall under a definition many times, and we shall return to the same discussion many times, since the subject (which is "animal") falls both under the definition of a biped and under the definition of that which has a split foot. And if, moreover, biped and having a split foot fall under the definition of their subject (because these are established as the *differentia*) then the discussion will return to the same point in a circular fashion, and the same thing will be said many times. This is just as if I were to say that a human is a biped animal and a biped is a human having two feet, for then a human will be a biped human animal having two feet, just as has been adequately proven in the sixth book of the *Topics,* where it is shown that they who define something in terms of an individual property are in error, because the species falls under the definition of the property.[59] And if the property then falls within the definition of the species, a circle will exist in defining it, and the same thing will be said many times, because the species will fall under the definition [*ratio*] of the property which is posited in the definition of a species. And it will be said again and again, as often as this is posited. This will go on into infinity, with the same thing being said many times.

Further, every individual property flows from the genus of accidents; that, however, which is of an accidental nature cannot be the end and the form of the substance of animals and their members.

55 Further, it is not right to divide any type of the genuses of animals using division-related *differentiae* which constitute a species with the same thing. For some types agree and differ only in shape. Similarly, the definition of the animal is different in each and every case, just as we said in the book *On the Soul.*[60] For this reason, if some forms of *differentiae* emerge from one genus through division, then that genus is conjoined in every respect to one of another definition, rather than being conjoined to another according to being, since potency has a definition by analogy to form. There is not one analogy for one form and the same analogy for another, and, entangled by such equivocation, it is not fitting to divide certain types of birds placing some, by way of division, in one class [*cursus*] (that is, in one similarity of form), and certain ones in another, as if we were to say that one class of birds is aquatic and another land dwelling. For that which a bird is in general (if it is understood according to being) will not belong to one definition in the division just mentioned. Likewise, there can be no valid division if I should divide the genus of painted birds. For it happens that a "bird" would be divided which has the name of a bird only by virtue of the shape and exterior form if we were to say that one of the painted birds is a waterfowl and one is a forest-dwelling bird. For a bird which is artificially fashioned is only called a bird

---

59. A. *Topica* 7.1.3; 7.2.6.
60. Cf. *De anima* 1.1.4.

through a likeness of shape. Such birds have other *differentiae* into which they can be divided, just as if I should say that some lack blood, and some have blood, and there are other, nameless *differentiae* as well.

And, generally speaking, in no *differentiae* whatsoever will there exist a term that is entirely univocal throughout every genus. If, then, it is not right to divide something that is equivocal, unless its equivocation is at least determined earlier, then it is clear that one who strives through division to learn the natural forms which are the ultimate ends errs and is wasting his time.

And, moreover, this same thing is shown in the case of the *differentia* that make divisions, since after one has divided a multiped animal throughout its genus into an aquatic animal and a terrestrial animal, some multipeds are found which are arranged in accordance with both of these, seeing that they are sometimes aquatic and sometimes terrestrial.

Therefore it is necessary (if the *differentia* that make divisions should distinguish well) that these animals be grasped in terms of a privation separating them from each of the prior *differentiae*. Then there will be one privation separating them from aquatic multipeds and another from terrestrial ones. This is because when such animals are intermediate between these two, and they are not intermediate through composition (because a third animal is never composed from two animals), it is necessary that they be intermediate through privation and the negation of each of the extremes. And these animals are thus constituted and formed by means of two privations, but between the two privations there is neither any *differentia* nor a form, since that which does not exist does not have forms or species and *differentiae*.[61] Now if I describe one privation as "footless" [*non habere pedes*] and another as "featherless" [*non habere plumas*], it is necessary that these two privations have no part in the *differentia* of one having feet and one having feathers, since otherwise contradictories would be true in one and the same thing. And thus privation will not be a *differentia,* since every *differentia* which makes division has forms and species, since the parts of a definition are in the greater, and the whole is in the equal, just as is said in the *Topics*.[62] Now if the *differentia* did not have species and forms under it, then there would not be attributed to the whole what is commonly divided, but there would be attributed to some individual what is indivisible and incommunicable. For what issues through division from the power of the genus is communicable to many, just as is the genus itself.

If, however, it should be said that there are some *differentiae* that have other *differentiae* and species under themselves, we thus say that winged birds have forms and *differentiae* because some have undivided membranous wings while others have divided, feathered wings. This is just as we say that some of the biped animals have a foot split in the manner in which the foot of one whose hoof is split in two is divided, and that

---

61. Cf. Ar. *Part. An.* 642b20ff. By definition a privation is not an existing thing, but the absence of a predicate or property where one properly belongs. So, as "that which does not exist," a privation has neither a form nor a species. Nor does it have *differentiae*. Balme, in his note to *Part. An.* 642b20ff. (1992, 108), suggests that for Ar. a privation "cannot stand as a general *differentia,* because it is not further divisible" and "it cannot stand as a specific *differentia,* because it is the same for different species."

62. St. refers to A.'s *Topica* 6.2.3.

others do not have a split foot, but have a solid foot. An example is an animal which has a hoof [*ungula*] that is not split, but rather has a single hoof [*solea*]. In cases like this it is difficult to learn the forms of animals through division. Now if, as some say, they must be learned, thus, through division it is clear that any animal whatsoever in the species ought to fall under some of the *differentiae* learned through division, and none of the same species ought to fall among several *differentiae*. This is because every animal has some form by which it differs from others, and none has two forms, as if I should divide animal into "winged animal" and "animal-lacking-wings." In this case, it is agreed that the division has not been well established, because one animal is found in a species which is under both *differentiae*. An example is the ant, for certain of them fly and others do not, and the one does not differ from the other in species, but only by age. Similarly, one firefly flies and another does not, differing only in age.[63]

59 The same will be true if *differentiae* are established otherwise, and for that reason it will thus be difficult to learn the proper forms of animals, which are their final causes. And the division of those lacking blood is the most difficult of all the divisions, and perhaps here it is said more correctly that it is impossible to learn the forms of the species of animals through division, since whatever *differentia* may be learned through division will appear to fit many cases in various ways. And thus, what is a property in the individual and fits one species will seem to be the *differentia,* as if I should say that lacking blood and having blood is a *differentia*. Lacking blood has so many modes even in a single species that the species itself will not be without *differentiae,* especially since we learn the *differentiae* in the natural sciences with respect to being. In this mode they vary greatly, and take on more and less, and prior and posterior. Further, it will follow that one which is under one *differentia* may even be under a *differentia* contrary to that one. For example, one having blood with respect to one member exists as not having blood with respect to another member. This is clear in many of the animals introduced above, which have blood with respect to one member and with respect to another do not have it. It is clear, then, that it is impossible to learn the forms and *differentiae* of animals in this way. Now the form is that which is predicated absolutely [*simpliciter*] of that thing whose form it is, and whose virtual parts perfect all the parts of the matter which are the members of animals.[64]

60 If, however, someone should say that no substantial form is one and indivisible, then it is necessary that every form be divided into things having further *differentiae,* and none of these will lack *differentiae* through which it is divisible, and division will proceed forever through the substantial *differentiae,* such as is the *differentia* between birds and humans. It then will follow that privative *differentiae* have no *differentiae* at all, because every *differentia* pertaining to division is also a *differentia* constitutive of something. A privation, however, constitutes nothing, for which reason neither *differentia,* nor species, nor the form of a privation exists. Thus, "footless" in an animal has

---

63. Cf. Ar. *Part. An.* 642b43. Ants fly when swarming, after being hatched. Fireflies produce luminescence both when larvae (glowworms) and flying adults.

64. "Virtual parts": *partes virtuales,* which, in the case of the human, are the powers of the soul.

no *differentiae*. This is nevertheless unsuitable, for while a privation posits nothing on its own account, logically, the negation of a given position, established on a contrary, can nevertheless have species and *differentiae*.[65]

But we are not speaking in this manner here, since we are investigating here only whether the final forms of animals can be learned through the division of genuses. This is what the Platonists have said, claiming that every form is common and has *differentiae* because form always remains common unless there is something appropriating it. What appropriates it, however, does not belong to form and, thus, it will always remain common on its own account. Many things, however, which are under that common one are separated only by means of *differentiae*. And it therefore remains that according to these men every form is divided into *differentiae*, and according to this view no form will be ultimate. For if animals are divided into having blood and lacking blood, in an animal having blood there will be further *differentiae*, since to have blood is a certain form and an unappropriated, substantial *differentia*. If, however, they reflect upon these, then a privation (which is not a form) will not have *differentiae* and species (that is, forms). But since the *differentia* is the cause of division, that which is learned through privation will be indivisible and will not have a *differentia*, while that which exists this way is equivalent to an individual. Privative *differentiae* are therefore individuals because individuals are not divided, and none of the privative *differentiae* is common, just as we said.

If, however, someone should say that a certain *differentia* is indeed common and nevertheless is not divided but rather is indivisible, this cannot be. For if it is common it is clear that it has that commonality only through some common nature and principle which is present in those things for which it is itself common. However, those animals in which that common nature is present have number and diversity among themselves. Since diversity stems from form, it is necessary that they be different in form from other animals of the same nature. It is thus clear that there is not some one ultimate form, common to individuals having a *differentia* between them. If, however, as has been said, there will not be one ultimate common nature, it is necessary that there be all sorts of *differentiae* and forms fashioning number and *differentia* under a common nature. These, like parts which produce division, may be attributed to the same common *differentia*. However, the same individual ought not to be attributed as subject to diverse *differentiae* and opposites, which belong to the number of *differentiae* of which we have spoken, since they are substantial forms.

It is clear, therefore, that no form whatever is found to be common which is not divided by means of the same type of division as are those which first divide animals

---

65. The difficulty here seems to be that because a privation (e.g., sightless, bloodless, etc.) signifies only the absence or lack of some property, and does not point to positive existence, it cannot have a form, or species, or difference, as Ar. points out at *Part. An.* 642b20f. But this does not seem to follow for negations (e.g., not-man, not-bird, etc.), which from a logical standpoint appear to operate as species and genus. So, for example, under the "genus" not-man one may place the "species" not-animal. For an extended discussion of this problem for both privations and negations, cf. Odo of Tournai's *De peccato originali* (Resnick, 1994, 43–44).

into two in their genus through *differentiae* placed immediately opposite one another. Thus, they are in error who say that the *differentia* and form are indeed common and nevertheless are not divided. From their claims it follows necessarily that all genuses of animals which are not divided according to form are at the end of the *differentiae*, in which, according to them, division ceases.[66] Now, they say that when some animal genus (from among those genuses which, through the most common nature, are the principles of *differentiae* and division) is divided into the modes of *differentiae*—just as if whiteness were divided and some dividing mode were divided further into other modes and *differentiae*, and that one again into others—then this process continues until it gets to those which are not divided.

Such ultimate differences, however (in which, perhaps, division ceases), are four or more and also have parts which, nevertheless, are duplicated from one first, common genus by means of division, seeing that a good division is twofold and comes about through opposites. In this case ultimate, specific forms will be as numerous as the number of differences. It is necessary, however, that these ultimate forms and differences exist in a proper matter. For no animal member is without matter, nor is there some one of the members that is only matter.[67] Moreover, as has been said many times, neither will the whole collection of members which is called a body be an animal without form. Thus, then, forms are grasped in their matters and can be learned only through division if one is to arrive at the ultimate forms which have been appropriated through the being [*esse*] that is in matter. But this is difficult and perhaps impossible, especially in privative *differentiae*, just as we have said.

Further, however, if substantial form has to be learned through division, it ought not to be a division *per accidens,* like the division of a subject into accidents. For example, if one should wish to divide shapes into the substantial forms of the shapes, he will err if he makes his division by saying that one of the shapes has angles equal to two right angles, while another has angles equal to many right angles. For it belongs to a particular figure, which is the triangle, that it has angles equal to two right angles.

Still, however, since the division of forms occurs through opposite forms, and opposite forms are contraries, it is necessary that if forms have to be understood through division, the division occurs through contraries, just as whiteness and blackness are contraries, and straightness and curvedness. Now these *differentiae* have a type of contrary form in that they center around [*circa*] the same genus and in the sense that generation, insofar as it is a motion, is between contraries, just as we said in the science of the book of the *Physics.*[68] Thus, therefore, it ought not to be divided so that some would be a weight and others a measure, since these are not contraries, although they are found in as many things as there are bodies. For the genuses of animals are

---

66. Stadler's text needs the *sunt* here supplied to it from Borgnet's edition. This sentence and the one following were badly served by the *post illa* style. The received text, while difficult, was grammatical. A.'s additions complicated the difficulty and lessened the grammatical coherence.

67. Ar. *Part. An.* 643a25.

68. Cf. Ar. *Phys.* 188b25f.; *Phys.* 1.3.1ff.

not divided according to the mode mentioned. So, then, substantial forms are learned only through division by a circumlocution, just as a substantial form is sometimes understood from its many conjoined accidents.

For animals are sometimes divided according to activities common to the soul and to the body, just as often we explained above when we said that some animals are walkers and some flyers. In such a division sometimes nothing prevents the same genus from falling under each of these dividing *differentiae*. For we said above that the ant and the firefly are both walkers and flyers according to a different stage of life.[69] Similarly, when "animal" is divided into domesticated and wild this is not a division according to the opposition of forms. This is because every domesticated animal is also a wild animal, examples being the human, the dog, and the cow.[70] Since in the land of India pigs, goats, sheep, and all the types of domesticated animals are wild, they are domesticated and wild by both a common and a univocal name.[71] If, then, these types of animals are the same in substantial form, the *differentiae* cannot belong to opposite forms when a division occurs between domestic and wild, but rather the division occurs through the accidents and the activities of the soul and the body, just as we said.

Such an accidental division as we have described, however, occurs to everyone who wishes to divide a genus correctly according to its *differentiae*. Now an understanding of animals must not be learned according to their species, but according to the genuses we have described, because those genuses conjoined with *differentiae* are the formal quiddity [*quidditas*] of the animals, through which animals are understood best. Many people do not do it this way. Rather, they divide the genuses of birds and the genuses of fish according to diverse *differentiae,* and again they divide each of these *differentiae* into other *differentiae*. And they do not divide correctly, using that mode by which any genus has to be divided into two immediate *differentiae*. There cannot be such a mode of division in any case. For, whatever the *differentiae* that have to be considered, if it should happen that the same one falls into many genuses, just as we said of the ant and the firefly, then it happens that contraries fall into the same part of the division according to their substantial forms.

Therefore it is evident that if one does not find a division of the genus and *differentia* by means of a true and substantial *differentia,* it will necessarily follow that the division will go on continuously, because accidental division does not come to rest but marches on forever. And it will be just like a discourse, bound together, which proceeds from one thing to another without continuity.[72] Moreover, this follows from the claims of those who say that that division, which lacks the art of dividing, has to be this way, just as if we should say that one animal is winged, and another is white,

---

69. Cf. 11.58.

70. The animal equivalent would be, e.g., the wolf and the *bonachus*. For the human, we can cite the monstrous races (1.87), the pygmy (1.44; 2.12), and the so-called hairy men (2.50).

71. Ar. *Part. An.* 643b5ff.

72. "Discourse": *sermo*. The temptation to think not only of a "discourse" but also of a "sermon" was probably as strong for A. as it is for moderns. The word will bear either translation.

and another black, and another domesticated. For even if it is conceded that domesticated and white are *differentiae,* they are not *differentiae* immediately opposite winged. Rather, they will be *differentiae* of something prior which comes before winged. However, by such an artless division, one thing is divided into many *differentiae* which nevertheless do not bear the name of true *differentiae*. And in this way even a privation will be a *differentia,* just as a privation of winged will be established by a *differentia* of some kind. But in a division which occurs through two immediate substantial things, this does not happen.

70 Now it is shown from such acts of division that it is impossible that any one individual thing be found *per se* without the other of those making the division. Now, all those which are in the community of the genus fall under one or the other of the two dividing things. For example, some people have erroneously thought that it is impossible that only one *differentia* be found by itself in individual things (whether it is a simple *differentia* or a composite *differentia*) but rather that each one falls under many *differentiae*. However, they call it a simple *differentia* (whether or not it is divided further into the *differentia* which it has) when it is one simple form. An example is as when bird is divided and compared to the *differentia* of having a split or a divided foot (such as those have which do not have skin between their toes).

71 They call composite, however, one which has a composite form, such as the one that has a split in its foot. Now the *differentia* by which a bird is a bird, when, like a common subject of the *differentiae,* it is conferred on a bird having a divided foot and a bird not having a divided foot, is the one thing which makes a bird be a bird. This is the first *differentia* in the genus of bird, and because it is the first it is necessary that it be simple. However, all those coming after it are said to be composite, because although they may be simple essences, nevertheless the powers of all the higher ones are in them. When the forms are natural it is necessary that its quiddities be defined in terms of all the superior ones, for the reason that definitions begin from the first one, just as demonstrations do. And through this type of immediate *differentiae,* there will be a continuity in the division which begins with the first genus, until it reaches the *differentia* of the last composition. This is because the universal is just like one thing which is in many and of many. But because of the nature of definition, which reveals that the quiddity of a substance is convertible with it, it happens that the ultimate *differentia* alone is the *differentia* which is converted, just as if we should grasp that the *differentiae* of animal is split into many others or "having two feet." For example, if we should say that multiped is split into many others, then an animal is either a multiped or a biped. Now, then, biped and multiped come from a number of *differentiae* which are under other *differentiae* of animal, and they are not first *differentiae*. For it is clear that such *differentiae* continuously grasped under others cannot be many, for if someone should correctly divide into immediate and substantial ones, then he will approach the ultimate division and the form in which division ceases.

72 All divisions, then, will be composed in this way, because if someone will divide the human (which is constituted in terms of the ultimate or other most specific spe-

cies) by saying that the human has two feet, or is a multiped, or has a split or solid foot, he will see that the human falls into only one *differentia* and not several, since he is only a biped. And so it is concerning the nature of every division, that the same thing will not fall into several, contrary *differentiae*. The one into which the ultimate species falls, then, will be a single *differentia*.

Therefore it has been shown that the many *differentiae* immediately opposite one another which may be attributed to one and the same division must not be *differentiae* of one and the same thing, because they are immediately opposite one another. Rather, there must be only one ultimate *differentia*. The many, however, which are attributed to several divisions can be of one and the same thing, just as winged and biped and having a split foot are, since one division is into winged and not-winged, and another is of the winged into bipeds and not-bipeds, and a third is the division of biped into those having a split foot and those not having a split foot.[73]

CHAPTER 11

Whether One Has to Begin from Universals or Particulars

Someone, however, will be in doubt as to why two genuses of animals, namely, aquatic animal (which is fish) and flying animal (which is bird), are not included under one name.[74] Now these types of animals and others like them, which have contrary accidents, are divided correctly by name.[75] But those whose individual accidents are entirely in proportional agreement seem to be unique, also having one name. Now whatever genuses of animals there are which differ only in their individual accidents in terms of more or less are also joined under one name, just as there is one nature in all things commensurable to one another. For whatever things are commensurable are univocal, just as is shown in the seventh book of the *Physics*.[76] But it seems unsuitable that whatever things are in proportional agreement with respect to the same thing be divided and posited by themselves under different names.[77]

---

73. Examples would be bird vs. human, bird vs. flying insect, and bird vs. ?. Albert does not provide an example of a biped that does *not* have a split foot.

74. Ar. *Part. An.* 644a12.

75. "Contrary accidents": Sc. reads, more sensibly, "shared [*communia*] accidents."

76. Cf. *Phys.* 7.2.2. "Univocal" means "having one name." It also implies having some common nature, however. So, A. remarks at *Phys.* 3.1.3, "*comparatio non est nisi univoci, quod unit una natura.*" For a fuller discussion of univocals, equivocals, and analogous terms, cf. 3.140, with note.

77. "Proportional agreement": this has been chosen to translate the phrase *convenientia proportionis*. The idea is that the things compared differ only with respect to degree in a given category. The phrase *ad unum* could mean "entirely" but rather seems to imply, as it will a few sentences later with different wording, the agreement of two things with respect to one and the same thing. Variants of this phraseology (e.g., *similitudo proportionalis*) will be translated equivalently.

74 For example, between one bird and another bird which are of one and the same genus there is a *differentia* with respect to more and less in the same nature, since both are winged, but one is "more winged" having a longer wing, and the other is "less winged," having a shorter wing. For that reason they are in the one genus, one name, and one definition which is in accordance with that name. But the *differentia* which likewise exists between fish and birds seems to exist in proportional agreement with respect to the same thing, insofar as birds have plumage, and insofar as fish have scales in place of feathers. And this comparison, which occurs with respect to the complete proportion of the two, just as feathers and scales are compared to the same thing, is not found readily but only with special difficulty in all the *differentiae* by which the genuses of animals differ, since not all their *differentiae* agree proportionally.

75 Now those accidents whose presence results in a proportionality to the same thing and which are found in many animals in a genus are substances and ultimate, substantial forms, even though, according to the manner of a substantial form, proportionality would not cause such a *differentia* according to more and less, but rather according to a natural potency consequent to this sort of form. This is just as if I should say that a man is more prudent than an ape. Now a comparison does not obtain in the characteristic [*habitus*], but in the nature of the prudent soul, even though more and less do not befall the substantial form insofar as it is a substantial form, but rather in accordance with natural potencies according to which the soul itself is the principle of activities.

76 And this is why the substantial form is considered in two ways: in one way insofar as it confers being, and in the other way insofar as it is the principle of natural activities. It is not compared with respect to more and less in the first way, but in the second. For this reason genuses compared to one another in this way are not compared in respect to the type of the form, but rather in respect to the form insofar as it is the principle of power and activity. It is just as if I should say that Socrates, insofar as he is a human, is stronger than an ant insofar as it is an ant.[78] Therefore even when proportionality exists in other things, they are commensurable with respect to the same thing and then it seems that they belong to a single genus. Yet this is not true, since the genuses of swimmers and flyers are diverse. Now the solution to this is that those which are compared univocally according to more and less have one common nature, which is their genus. Those, however, which are in proportional agreement with respect to the same thing, do not have one common nature whose name would be the name of the genus.

77 On account of this another doubt arises whether it is necessary for us first to speak of universal things or particulars, seeing that the same thing which is an accident and a point of agreement [*convenientia*] of a genus occurs in many particulars. If we have to speak first of the particular, it will follow that we say the same thing many times; for as many times as we speak, just so many times will we speak about one of the

---

78. Ar. *Part. An.* 644a25–26 uses Socrates and Koriskos. It is possible that Koriskos yielded *formica*, "ant," but another Gr. word, *koris,* means "bug."

particulars to which that accident belongs. And the same doubt exists whether we should speak first of genuses or individuals. For the genus contains the species and the individuals, while the species contains only the individuals. Individuals, however, do not contain anything.[79] Let us say, then, that just as when disclosing the nature of a human one speaks of the human, so too one must speak of birds and other animals when their natures are disclosed. For just as, when speaking about the nature of the "human" nothing is said about any type of human in particular, but instead the nature of the human in general is revealed, so too, when speaking about "bird," one must not speak about any type of bird, but about the common nature of birds. For this genus, which is the bird genus, has types of species which are individual with respect to form, although they are divided according to matter. We must speak of such individual species, such as the sparrow, the crane, and other very specific species of birds.

If, however, someone should contradict this, saying that when treating the natures of animals we ought to speak of some individual first, then it will follow that we will be describing the same accident and individual trait of some animal many times over.[80] For the same accident which commonly occurs on account of nature belongs to all of those whose common nature it is, and it will then be necessary that this be repeated in the treatment of each and every animal whatsoever. This is a foolish error, and it not only is an error but also prolongs the discussion if we speak of each and every particular animal in its own right.

It seems worthwhile, then, that we should speak of all the genuses of animals which were correctly defined by people before us. Now there are some more universal genuses of animals which may be common and have one, common nature. Still, in that nature there are forms and species which participate in it through some proportional commonality of one to the other, so that they are not altogether remote from one another. An example is birds and fishes, which are proportionate in having feathers and scales, just as we said above. So it is, moreover, for any type which does not have its own proper name, but which is under that genus. If there are some, however, which do not have a proportional likeness to a common genus, then it is necessary that these be treated separately, in such a way that one should speak of any of them by themselves. For example, there is "human" and something else which is like it. Those which will have an analogous likeness in the shapes of their members also thus have determinate genuses which are not remote from one another but are in a common nature. Examples are the genuses of fish, birds, and *malakye,* and the shellfish [*halzum*].

Their genuses are proximate although they are different. Nevertheless, they bear a proportional likeness to one another and they have "agreement," just as even the human bone "agrees" with the spine of a fish.[81] Just as there is a wing on a bird, so too there is a fin on a fish, and so it is for others. Thus, then, the corporeal accidents of

---

79. "Contain anything": Sc. reads "contain others."

80. Reading Borgnet's *accidente* for Stadler's ungrammatical *accidentes*.

81. "Agreement": *convenientia,* which simultaneously implies agreement, similarity, and belonging, as well as evoking the *necessitas convenientia* above.

animals in largeness and smallness, softness and hardness, smoothness and roughness, and other accidents similar to these ought to be treated together. For in all of these the universals must be considered, insofar as they appear among the animals to a greater or lesser extent.

81 It has already been shown, then, what the division of animals ought to be like, what the mode is of things which divide and are divided, which division ought to be condemned and which not, and that division will sometimes be possible into two when it occurs through forms opposed to one immediately opposite. Sometimes, however, it will be impossible to divide into two, when the division occurs through accidents which are from the common activities of the soul and body. Now because all these things have been demonstrated, let us begin to speak and we will establish the principle of consideration in the way determined beforehand.

CHAPTER III

For What Things Physical Causes Ought to Be Sought and How

82 First, however, it seems that it must be said that there are certain natural substances which are not subject to generation or corruption. According to some, they are animate. And some have even said that they are animals and divine substances, like celestial essences [*essentiae*]. But our opinion and our discussion are brief and curtailed concerning divine substances, because it is clear to us that the accidents *per se* of that substance are especially few. Moreover, we have a greater ability for speaking of animals, which are before us, and trees, since the study of these is easier because of their nearness to us.[82]

83 Moreover, one who wishes to speak of celestial substances will speak with difficulty and effort.[83] Now the reason for the difficulty is due to two things, namely, that they are distant from us, and that they are the most noble. Certainly we comprehend only a little science of the celestial substances, because of the magnitude of their nobility. Now that which we do comprehend is their quantity and the quantity of their motion. However, we do not perfectly comprehend their powers and their natures, but we may conjecture something regarding them from their effects. And this is the little there is to comprehend about them. Now quantity has a single definition in these things and in as many other bodies as there are, and for that reason it indicates nothing at all of their natural being. But a love of the science of celestial things causes one to strive and arouses the struggle to comprehend the little bit that can be perceived about them.

84 Now a lover of anything, when he loves it a great deal, is constrained and zealous to comprehend whatever little bit he can about that which he loves. Now this is the

---

82. "Nearness to us": perhaps "in our country," but the context seems to demand the broader sense.
83. Cf. Ar. *Part. An.* 644b22ff.

proper characteristic of a lover, and the comprehension of that small part is loved by the lover more than the comprehension of other numerous and large parts which are not loved by him alone. The things which are near to our nature (namely, those things which are generated and corrupted) are better and more perfectly known by us because their natures, as we said, are near to our natures.

Since, then, we have discussed in all the foregoing books what the nature of animals is and we have explained in this book, as far as we perceived it, how one must proceed in the assignment of causes, it now remains in the following books to discuss their natures through the assignment of causes according to the most perfect opinion we have of them, whether it be the most noble animal to which we turn our attention, such as the rational and intellectual, or whether it be an ignoble one. Certainly that nature which, like the cause of causes and like a formative cause, creates animals, will, once it is known, be a cause of great delight among these natural philosophers [*physicis*], who can know the true causes near to their nature, even though they might consider them only in a base and ignoble animal. For that reason we ought to consider the forms of animals and to delight in him who is the artisan who made them, since the handicraft of the one acting is manifest in his activity, just as in the activity of making images there is revealed a knowledge of the artisan making the statues or images (that is, statues).[84]

And for this reason as well it is necessary that we not avoid or turn away from the consideration of base and ignoble animals, since their baseness is not due to the artisan but to matter. And on that account it is necessary besides that a consideration of base natures not be burdensome for us, as it is burdensome for perverse men who, influenced by a base nature, conceive perverse affections. For there are wonders to be seen in all natural things. In them is confirmed the expression of the poet Eradytis who sat in front of the temple of Delian Jove and spoke to those strangers desiring to join his lectures.[85] When they saw him entering into the temple and they could not approach him, he commanded them to enter into the temple where the gods were, and to hear his lectures there, asserting that all things are full of Jove, and that he [Jupiter] himself is whatever belongs to nature, both whatever was and whatever will be, and that lectures concerning these things can be understood only by one who has entered into the divine causes as if into their temple.[86]

---

84. "Statues of images (that is, statues)": *statuas ymaginarum sive ydolorum*; cf. 11.24, where *ydolus* seems to be A.'s preferred word and *ymago* one he inherited.

85. "Eradytis": Sc. reads Heraclitus.

86. This is a wondrous example of what could happen to classical texts and allusions as they made their way through the Dark Ages to the light of rebirth. The origin is Ar. *Part. An.* 645a18f., where we are told that Heraclitus, the Ionian philosopher, was once visited by seekers of knowledge who were disappointed to find him warming himself at his stove. "Don't be afraid," he said, "for there are gods here too." Sc. does read *Eraclitus,* but A.'s *Eradytis poetae,* if it recalls anything, brings to mind "Herodotus the poet," as in 3.179. The mention of gods must have given rise to Jupiter and, perhaps, the temple which now takes the place of Heraclitus's stove, which itself, according to Robertson (Balme, 1972, 123), may in fact be a euphemism for the toilet. One still is left to inquire on what basis A. inserted the epithet "Delian" into this text, for Delos was connected to Apollo and Artemis and not to Jupiter.

87 Accordingly, we ought for this reason to inquire into the natures of any animal whatever and to know that in all animals there is a certain natural cause, noble and divine, because none whatever of these natures was endowed with a nature causally [*naturatum causaliter*] in any way either in vain or without purpose. Whatever, howsoever many, and howsoever much proceed from the work of nature will exist only because of that which is the end [*finis*]. And everything which was, is, and will be, was, is, or will be only because of that which completes it, and on account of this it has a place and a wonderful and noble rank in things of nature. If, then, someone should think that knowledge of some of these is ignoble, let him blame himself instead, since his knowledge linked to love [*affectiva cognito*] is ignoble and base because he himself does not conceive things from which the human is composed apart from the baseness of his condition, as when he thinks of flesh, bone, blood, vein, and similar things. It is the accidents of his soul that are base, and not the knowledge itself.

88 Moreover, one should know further that when speaking of some instrument it is not sufficient to relate the matter alone, nor is its defining statement sufficient in terms of matter alone. Instead, it is on account of the form that there is both being and a definition. For in the definition and knowledge of a house it suffices only to speak of rocks, cement, and wood if along with these the form of their composition is stated, and if the form, which is the whole and complete substance of the house, is also stated.

89 Further, however, the accidents which occur in anything ought not to be explained first, and then, after this, the causes of their accidents given. For we have already said in the foregoing books that many accidents are common in very many animals. Indeed, some of these belong to them in a simple manner (that is, absolutely and without proportion) as do feet, body coverings, and feathers.[87] However, some are present in them according to proportional agreement, and not absolutely. So, just as we said, while some animals have lungs, others do not have lungs, but for these there is a member analogous [*conveniens*] to lungs. And, some have blood, and some not, but these have a humor like blood which has the power of blood (that is, to nourish, in the way the blood does in those having blood).

90 Yet we have already said in the preceding books that if we should speak of every animal singly, such a discussion would cause us to return to the same discussion many times, because things commonly occurring in many are the same in many animals.

On the other hand, we have perceived from things brought out above that any instrumental member of the body exists for the purpose of effecting or establishing the disposition of something. Moreover, we have perceived that in the mechanical art the operation of the saw is not due to the disposition of the form of the saw but, on the contrary, the disposition of the saw and the form is due to the operation itself. And we say the same thing concerning the cause, that the creation of an animate body is on account of the soul and its actions, the members of the body were created as organic

---

87. "Body coverings": *cortices*. *Cortex* often means skin or hide, *corium*, but also evokes scales (1.101; 2.80; 13.61f.) or membranes (1.203).

things [*organica*] on account of the powers of the soul and its activities, and the nature and shape of any given member is suitable for the activity for which it is prepared.

Thus, then, when treating the things under consideration, we ought to discuss first the activities common to all the genuses of animals, and then, working our way down, we will speak of the activities attributed to a given form or species of animal. Now we call those activities common which are suited to all the animals in a genus. However, the activities of any *differentia* and species of animal, throughout all the genuses which are divided by the *differentiae,* are those which approach nearer to one another and are mutually restrictive, and these are spoken of in terms of proportional likeness. These appear to be present in animals according to more and less. For example, we say that the bird, according to genus, and the human, according to genus, are the same thing with respect to genus. And thus some have said that a human, with respect to its genus, is a bird, and a bird, with respect to its genus, is a human. Thus, things are the same with respect to the form of the species in any things whatever whose species is the same. Thus, one lacking some constitutive *differentia* is the same as another thing with respect to the universal definition of genus. For example, human, apart from the *differentia* "rational," and ass, apart from "irrational," are animate sensible substances. Now what is common is in some animals with respect to analogy, and in others with respect to the genus, and in others with respect to the form of the species.

However, even the activities differ from one another in the same manner. Some activities, moreover, only exist in some things for the sake of something else. For these activities precede the things acted upon, which are the ultimate things intended by them. For example, the act of construction precedes the house. Other activities, however, are the completions of and the end of others, just as playing the viol is one intended by itself, and others like it. For this, then, and for such an end the disposition and shape of some member will exist, and it will be necessary according to the third type of necessity set forth above, namely, so that there should exist in animals the accidents of their activities. Examples include coitus, growth, wakefulness, sleep, and accidents similar to these. We ought then to determine the members of animals, like the nose, eye, and the entire face, since all these are and are said to be an organic part. And one will have to speak in the same way about the remaining members.

Let these things which have been said suffice for the clever person and for the procedure [*scientia*] we ought to use for coming to know the natures of animals. Next, let us describe the common and individual causes and let us begin to describe their principles, as we have determined and made distinction among them above in what has gone before.

CHAPTER IV

Also a Digression Explaining That Which Was Said Above Concerning the Division of Formal Causes

93 As evidence for those things which have been treated, let us make a digression, explaining that it is the business of the natural philosopher to show the formal, natural causes which tell the "what" [*quid*] and the "for the sake of which" [*propter quid*] of natural things. If, however, these must be placed in the definition with the genus, it is necessary that they be discovered by hunting them down through the division of the genus. But many of the natural things agreeing in shape have diverse forms of their quiddities. If the same forms are attributed to such things because of their agreement in shape, an error will occur just as in the division of an equivocal. And for that reason one must beware of equivocation of this sort. For although the human, the ape, and the pygmy concur in the shape of many of their members, nevertheless it does not follow that they agree in form and in the accidents which commonly are consequent to form.

94 And, further, since there is but one potency of contraries, it is necessary that the division of the potency occur with reference to contrary forms, seeing that privation is not a form nor can it be a final cause to the extent that something may be made on account of it, and neither does it have properties. For that reason one must beware lest some member be treated while hunting down forms of the division of a privative.

Still, however, every form of a thing which agrees with no other is a quiddity, and for that reason one must beware lest such a division be adopted among whose dividing parts would be the same animal. This is as we said of the flying ant and the nonflying ant. However, this is frequently the nature of a division when the subject is divided into accidents or even when the accident is divided into subjects or, again, the accident into accidents. And for this reason such a division will err as it hunts down the forms.

95 Still again, however, since none of the substantial forms is an accident, that division will err which introduces accidents, for the division ought to occur with respect to forms which emanate immediately from the same power of the genus, and each of these should be a form and not a privation; nor may it be an accident *per se* [*accidens per se*] or an accident in general [*accidens communiter*].

And yet one must be careful lest forms be introduced which are not in matter, as Plato did, since the forms of animals and all their members exist in matter and are brought forth from a potency of matter. Once the forms have been thus found, however, it is necessary to know that the ends are double, since one is the end of generation and the other is the end of the thing generated. The end of generation is the form giving being, on account of which matter and the efficient cause have a necessity, which I call suppositional necessity, about which we spoke above.

The end, however, of the already existing, generated thing is the proper activity [96] of any given species of animal and of any given member of the animal. It is for the sake of this activity that it was made, and for the sake of which its shape and form are necessary by a suppositional necessity, which we said above is the third type of necessity. This is the final cause that must be assigned to the shapes of members, since no substance or part of a substance is lacking its proper activity.

We will therefore have to treat these things in the order determined above, because through that which we have introduced here by way of summary, all those things are plainly understood that appear to have been determined a little bit confusedly in the foregoing on account of Aristotle's wording.

HERE BEGINS THE TWELFTH BOOK ON ANIMALS

Which Is on the Causes of Uniform and Nonuniform Members and on Their Complexion

The First Tract of Which Is on Uniform and Principal Members

CHAPTER I

On the Distinction and Differences between Uniform and Nonuniform Members

1. We have already set forth in ten previous books, speaking in a simplistic way, the members out of which occur a given animal's composition and preservation in being.[1] Now, however, in the twelfth book of this study, we wish to set forth the causes for which each member is created.[2] We will give the cause of each of them separately as we did above when enumerating them.

One should know from the start that there are three types of disposition in the composition of animals' bodies. The first type is that which is called the original composition (that is, the mixture) that arises from those things which are called "elements" by some philosophers. Examples are earth, water, air, and fire; and there is a mixture of these among themselves so that the majority of one is with the majority of another and the minimum of one with the minimum of another, as was stated elsewhere, in our books on natural science.

Now even though these elements have very many powers it is nevertheless right that the *ratio* of the mixture follows those powers which are best suited, both actively and passively, to bring about a perfect mixture, rather than arising from all those other powers of the elements equally.

---

1. "Simplistic": *simpliciter*, i.e., not going into individual cases. "Preservation": *sustentatio*, here avoiding "sustenance," which merely implies feeding. Cf. *sustentamentum* at 12.89.

2. It is difficult to explain A.'s reference here at the beginning of the twelfth book to the preceding ten books—a reading confirmed by Borgnet. Given the fact that A. extensively revised and reorganized his *DA* (cf. Geyer, 1935), one is tempted to conclude that Book 11 was inserted between books 10 and 12 somewhat later. Stadler's note at the end of Book 10 indicates that it concludes with the words: "Explicit decimus. Incipit undecimus liber animalium." Indeed, each of the previous books ends with an *explicit*, even though not all include an *incipit* to indicate the beginning of the next book. It seems noteworthy, then, that Book 11 lacks both an *explicit* and an *incipit*. In its content too, Book 11 stands out, following now Ar. *Part. An.* and introducing a theoretical discussion of the foundation of the physical sciences. Still, at 12.4, his reference to a discussion of various nonuniform members in Book 11 accurately directs the reader to 11.29. Similarly, at 11.13, A. refers to "the previous ten books," leaving our puzzle unsolved.

Now, just as we have mentioned mixture in other places, so we say here that it is accomplished with moisture, dryness, hot, and cold, for these powers are the first matter of mixed bodies.³ The other differences in potencies and elemental qualities act as assistants in the mixture of these things. Examples are heaviness, lightness, having a loose texture, density, roughness, smoothness, and other forms (that is, qualities of elements) that are like these. These are elemental bodies and are consequent, as composites, from elements.

The second composition produces complexion. For complexion, as was shown previously, is an accidental quality arising from the composition of opposite qualities in very small particles which both divide and change with respect to one another. As a result of this, one quality called complexion occurs as an accident in them. And this composition is called the composition of the humors, and it produces from them the uniform members of animals which are generated from the humors. Thus, this composition is said to be produced from those members which have parts that are naturally uniform, such as bone, flesh, nerve, and the like. For although this composition might begin from the humors, nevertheless, because the humor is not in actuality part of the animal as is a uniform member which is generated out of the humor, it is better to say that this composition arises from uniform members.⁴

The third composition is from nonuniform members [*membra etherogenia*], which have nonuniform parts [*dissimiles partes*] such as the face, hand, foot, and others of this sort.

Now we already know about the generation of these compositions from the things determined in the eleventh book. These organic members are first in nature, substance, and *ratio* and sometimes even in time, even though they are not prior in time in the same way. That is first in generation which is thus first like a cause of causes for generation, not because of something else which is last with respect to being, unless it is because it is that for which all other things come into being, and by whose supposition the other things have the necessity of generation.⁵

For we are not saying that a home exists for the sake of stones and tiles but rather, to the contrary, that stones and tiles exist for the sake of the house. *Yle,* which is the subject of generation, is related to form in exactly the same way.⁶

That these things are this way is clear not only for the reason which we will bring forward in what follows and which we have already treated, but also from the very definition of any given generated thing, if it is defined properly with respect to physical causes.

---

3. "Have mentioned": following the reading of some other MSS. Stadler offers "made mixture," a reading of decidedly less clarity.

4. It may be worth noting that "uniform member" here is *membrum simile.* A.'s more regular usage is *membrum consimile.*

5. It might also read: "That which is first in generation is that which thus is first without a cause of causes for generation, (and) not because of another second thing which is last . . ."

6. *Yle*: i.e., the Gr. *hylē,* "matter" or "material."

For everything that is generated univocally is generated according to form from some efficient cause and it [form] is in something much as in the end of generation and is thus from a first to a first. To explain, it is from a first mover, which has the nature and power of generating from form, to a first, that is, to the form of the one first generating univocally. Thus, one who is a human generates a human with univocal generation and a tree comes from a tree. In this way, generation will be from material that is posited and determined through the formative power of each thing.

6 In the same way it is therefore necessary that material and generation be prior in time than other things which are the ends and forms of generation. For the form of each thing, which is the end, is anterior in definition and substance. This is clear when someone talks of generation with a defining statement for then he will, in his definition of generation, posit the form to which generation is directed as if to an end. For in defining construction, the *ratio* of a structure is posited, just as the very word construction implies.[7] But, to the contrary, when a house is defined (that is, a structure), the *ratio* for construction is not assumed. It is much the same in other, similar matters as well.

7 It is thus clear by analogy that it is necessary that the elements which are in the first composition be posited as the matter for the uniform members, which belong to the second composition. Then the elements are the material for the uniform members which are later than them in the process of generation. For in these members and in their composition lie the completion and the end of the generation of elemental composition.[8] Then, however, after this composition, we will further grasp true preservation in true being, which belongs to the third type of composition. For example, the completion of the types and species of generation of many animals comes from the nonuniform [*etherogenia*] members, of all those, that is, which are generated through univocal generation. Animals are therefore composed from all the members of the sorts which have been mentioned, but in the ordering of generation, each uniform member exists for the sake of a nonuniform member, which is called organic or official. For the activities of the soul, which are the ends of generation, can be brought about only in the organic members, which are, for example, the eyes, nose, the face as a whole, the finger, the hand that is called "lesser" and which consists of the *rasceta,* the *pecten,* and the fingers, and the hand that is called "greater," which is the hand and the entire arm.[9]

8 One should know that the movements and activities of animals are very different and have many forms or types [*species*]. In a similar way, the operations of the members are also differently formed and different. It was therefore necessary for animals' bodies to be composed out of different powers, qualities, and corporeal forms. For

---

7. The wordplay is between *aedes* and *aedificatio.*
8. Or "of the elemental generation of composition."
9. For these terms, cf. 1.14, 281, 286f., 298f., 495f. An "official" organ is defined elsewhere as one through which the life-giving activity of the soul is performed (e.g., the heart, stomach, liver, etc.). Cf. *De anima* 2.1.3.

every difference in bodies is due to a difference of potencies in the animals' souls. Thus, softness is fitting for certain operations of the soul while hardness suits others. For example, softness is most fitting for the sensory functions while hardness is most suited for the motor functions. Of bodily powers such as these, however, some suit the actions of the palm with respect to its closing, when it bends in over itself. Others suit the palm's function in extending itself, when it opens up. Each uniform member, therefore, has one of these powers and thus some one of them might be soft like the medulla of the brain, while another is hard like the bones. One might be moist and another dry, one viscous and another not viscous (that is, not sticky).

The organic members, however, are composed from many things. They have many powers and thus one power of the hand is suited for grasping things (that is, closing it), which is a contracting movement, and another power suits compression and extension of the hand.[10] This is why these organic members are sustained and are composed of bones, nerves, flesh, and many uniform members of this sort. Thus, the body is composed in this way, made out of organic members. For it is clear then that the types of composition are as we said, and that they are ordered this way, for it is clear that nonuniform members are composed of uniform ones.

It is still possible, though, that the preservation and being of certain organic members is made up of many things in one way and also from one single thing in another way. For example, there are certain internal members such as the brain, lungs, liver, stomach, and the like, which all have a shape arising from different forms. Examples include the different forms of concave, convex, straight, and oblique. There may be, though, one of them which has one uniform body, for the brain seems to be uniform in its parts while the others we mentioned are made up of flesh, veins, arteries, and certain nerves. I am saying that uniform members are simple in the way that homogeneous is said to be simple. As we have said previously, organic members cannot number among them for an organic member does not have a unified composition, but a pluralistic one. Therefore, certain members are called simple which have uniform parts and certain others are called organic which have nonuniform component parts.

Of all the powers and activities of the soul, sensation seems to be mostly performed in a uniform member. For each sense is attributed to some member which has one type of composition. Likewise, each organ of sense which receives sensible forms is attributed to one simple genus. This is done solely on account of the power at work in that organ which, when it is brought about, is brought about by that which is actually sensed in accordance with the potential of its organ. It is therefore necessary that they—both the one which is the first subject, bestowing the power of reception on the organ, and the one which is the thing sensed—be in the genus which gives the power of receiving to the organ, bringing it a power to be acted upon [*passio*]. Thus, transparency of the eye [*perspicuum oculi*] is the same in genus as the bounded trans-

---

10. The actions of the hand (*acceptio* and *deprehensio*) would seem to amount to the same thing, a fact that led A. to insert his own further definitions. For the original sense, cf. Ar. *Part. An.* 646b24f.

parency [*perspicuuo terminato*] which is of the nature of colors, as was carefully proven in the book *On Things Sensed*.[11] Also the same in genus are the quiet air, which is on the *timpanum* of the ear, and air that is struck, which is of the nature of sounds. It is this way in others as well.

12 For this reason none of those who have spoken on natural matters have attributed any organic part to a single element as if he were to say that the face or any other nonuniform [*etherogenium*] member were only made up of earth, water, or fire. For they assign all the instruments of sensation to one of the elements, as we have said in our book *On Sensation and the Things Sensed*, signifying which thing was assigned to air, like the ear, and which to fire, like the eye.[12] Because sensation occurs simply in simple members, it is worth saying that touch too occurs in uniform members simply, speaking, that is, of touch without its assistants. For it is the sense of touch about which everyone holds the opinion that a person senses in many ways with many instruments, as it were, and also that he senses many contrary types of the species of tangibles, such as hot, cold, wet, dry, and many others which we have treated at length in the second book of *On the Soul*.[13] The instrument of sensing these things, according to some, is flesh. But according to others it is nerve or some other uniform thing.[14]

But this much is certain, that the instrument of this sense is mostly composed of nerve-filled flesh. Because it is impossible for there to be an animal without sensation, such simple members were necessary in the composition of animals, existing in those members which are called the organs of the senses.

13 One should know that although each sense is made in accordance with the power and potency of some uniform member, it is only carried out as it should be with the nonuniform members. For example, while vision is in accordance with the potency of the clear, watery fluid, it is nevertheless only carried out as it should be with the membranes of the eye or with the spherical shapes which are in the parts of the eye, holding in and intersecting with each other. Thus too, while hearing may be due to the power of the air on the nerve which is stretched over the *timpanum* of the ear, it nevertheless is only carried out as it should be with the activity of the cartilage of the ear, its twisted shape, and the disposition of its opening. It is the same for the sense of smell, which occurs on the two breast-shaped appendages of the brain [olfactory bulbs] and which are uniform bodies. But it is only carried out as it should be with a nose and the disposition of the nostrils. It is the same for the sense of taste, which occurs on the nerve spread over the tongue. But it is only carried out as it should be with the tongue, the saliva, and the other things joined to the tongue. It is also the same for the sense

---

11. *De sensu et sens.* 1.1.3. Cf. 9.97, with note.
12. *De sensu et sens.* 1.1.4.
13. *De anima* 2.3.30f.
14. "Some other uniform thing": *vel aliud consimile*, perhaps "something very like it." Cf. the phrase *vel sibi simile* at 12.26.

of touch, which occurs not on simple flesh but rather on nerve-filled flesh, which is midway between the excesses of the sensible qualities.¹⁵

The first thing perceiving such qualities is the heart, then the flesh, and then the skin. It is greatest, however, on the flesh and skin which are at the very end of the index finger. Thus, then, sensation is indeed in accordance with the power of a uniform member and of some element, but it is only carried out properly through an organic, nonuniform member. This is why the senses are passive powers on which the sensibles act in accordance with the powers of their forms. It should therefore happen that also in an organ some one passive thing should be dominant which has power proportionate to the forms of active things, just as actives and passives are proportionate [*comproportionata*] in natural things [*physica*].

However, the activities which are from the soul and not from exterior things are very different with respect to the *affectus* and the powers of the soul and are not produced by an agent [*agens*] which has one form in general. They are therefore present in the organic and nonuniform members by the power of the informing sensation and of the accompanying movement in the animals, as we have said elsewhere, and they are brought about by the power of the movement of the appetite in one and the same member. Therefore, those things must be composed differently, as is the instrument of walking and others of this sort.

It is further necessary that in bodies composed this way there be one primary member which first receives these principles and powers of sensing and moving. Since this is the one to which all things sensed are referred, it should be made up of simple members within it. And because it is the mover and the principle of all appetites, it should be numbered among the organic members. This is the heart, in which flesh is dominant to be sure, but which also has other things in its composition, as we have said in preceding books. In animals which lack blood, but which still have sensation and motion, there is that which takes the place of the heart and which is like it in respect to the powers mentioned.

The heart is divided into uniform parts. I am using "uniform" in the same way in which almost every one of the other internal members is divided into uniform parts, for their substance is made up of matter that is almost entirely uniform. For the matter of all the inner, principal members (these being the heart, lungs, and liver) is sanguineous, since their location is on the veins which are the passages for the blood and

---

15. As Ar. notes at *De anima* 424a5f., each of the senses has its peculiar object, and the object has its own qualities. The object of touch (the "tangible") is, for example, hot or cold. But touch itself is neither hot nor cold, although it must be potentially either one. Insofar as the sense of touch itself is neutral, it occupies a kind of "mean" or midway point, and it has the power to sense what lies on either side of it. The "excesses" of these qualities, however, destroy or corrupt the sense itself. If a sound is too sharp or too flat, it destroys the hearing; if a "tangible" is too hot or cold, it is painful and will injure the sense of touch. Thus, for the sense to perform properly, a proper mixture and *ratio* of the sensible qualities is essential. On this point, and for a definition of *excellentiae sensibilium*, cf. A.'s *De anima* 2.4.9. It is in this way, namely, that a sense must occupy the mean between extremes of sensible qualities, that touch is midway between the *excellentiae sensibilium*.

the hollowness of the veins bearing the blood lies within their substances. In just this way are channels produced in places through which vast waters flow and run as they head for a fall. So too, the members of the body are composed in this way from veins in which blood flows. The heart is the point of origin of the veins and the first creative power of the blood. Blood then should be from the food which the body takes in. Therefore the food must first be given form according to the disposition of the heart, since the principle of the nutritive power is in the heart, as has been stated above.

This, then, is the statement concerning the cause by which blood, which is the food of the inner, principal members, is made and also the reason that certain members are made uniform and certain ones organic.

CHAPTER II

On the Differences in Complexions, and on the Comparison of Uniform Members with Each Other with Respect to Their Complexions

17 Before we speak further here on the complexion of the members, we should say something in summary fashion on complexion itself and on its differences so that we can thus have better knowledge of what the complexion attributed specifically to a given member is.[16]

We have already stated what complexion is. However, a certain one of the complexions is balanced throughout the entire species, and we will only speak a bit here on this.[17] Balance of complexion of animals will not be said to be balanced in its nature if it is balanced following an arithmetic mean with respect to the thing itself. Rather, complexion is balanced according to the exigency of its species and according to a geometric mean, as has been stated in other natural books on natural science. In the human alone among all the other animals there is the harmony of balance which is simply and truly said to be near to the aforementioned balance. For this one is most like the balance of heaven. Of all the animals, the human's body is most distant from the excess of contraries, while in the other animals such balance is never found.

18 It seems that there are eight types of balance found most frequently in the human.[18] In other animals there are types which are analogous to those in the human but which are not exactly the same. Balance might be understood as belonging to the species, but not to this or that individual. Thus we say that the human is temperate compared to

---

16. A. here begins to follow Avic. *DA* 12.25v.

17. "Balance": the Latin *aequalitas* and similar compounds are used to represent the Gr. idea of *krasis*, which implies mixture or blending. The overall idea is one of a proper, mixed balance resulting in equilibrium.

18. Though found in Avic. *DA* loc. cit., the following material is a digest of *Can.* 1.1.3.1, with much material virtually verbatim (trans. Gerald of Cremona).

a horse or some other animal. Now it might be understood in an individual through a comparison based on the nature of an individual property to the species, such as when a person is said to be temperate with respect to human nature and according to the exigency of human complexion. Sometimes, when comparison is made of one sex to the other, temperance is understood in reference to shared human nature. We thus say that man is temperate compared to woman. It is sometimes understood in a comparison within the same sex to individuals of the same sex. Thus this woman is said to be temperate with respect to the female sex and this man to be temperate with respect to the male sex.

Comparison is sometimes made in the same individual with respect to his different ages. Thus a youth is said to be temperately complexioned compared to his boyhood and old age. Sometimes comparison is even made of some one member in two ways, namely, with regard to time and the other members. Thus we say that a given member is temperate now compared to itself in former and future times, and it is thus said that one member is temperate compared to the other members. The first of these types has a certain latitude in the stages of the various complexions, but too great a departure from the norm is not possible, for otherwise the thing complexioned will not be a human.

The second type is of a person who is truly balanced with respect to the exigency of human temperament.[19] And while such a one as this is not balanced with respect to being [*secundum esse*], he is nevertheless found infrequently. For such a person is only temperate when every member in him is in the disposition it should be in with respect to the necessity of his balance even though in and of itself it is not considered balanced. Now, this sort of imbalance reduces the whole to balance. For, as will be stated later, it may be that no part in a person save the skin is absolutely temperate. For the heart is very hot but this is needed for the comparison of the reduction of the others to balance. For spirit and heat, which are the two principles of life, are in the heart. Moisture, which is the principle of growth, is also in it.

Moreover, there are three principal members, although one is more principal than the other two. Of these three, one is the brain. It is cold, but with its coldness it can temper the heat of neither the heart nor the liver. Likewise, and to the contrary, the heart's dryness cannot temper the moisture of the brain and liver. The brain, moreover, is not very cold, nor the heart very dry. But the heart is dry compared to the other two and the brain is cold compared to the other two.

This type of complexion is more restrictive than the first. Still, it has latitude according to the latitudes of the *climata*. For the Indians have their own complexion and live according to it, and the Germans and French have their own as well, according to which they live in the seventh *clima*.[20] One would not live well in the *clima* of

---

19. On temperament, cf. 12.22. A. treats temperament, harmony, and proportion as virtually synonymous in *De anima* 1.2.8.

20. It would seem that A.'s received text merely said *Franci* and that A. divided *Franci* into *Germani et Gallici*. Note other hints of nationalistic pride in 1.133, 623.

the other even though we have often seen certain Moors living very well in the seventh *clima* once they grew used to the diet of the northern regions. Nevertheless, as we have said in our book *On the Nature of Places and Locations* [*De natura locorum et locatorum*], each complexion of each *clima* has a latitude and two boundaries in its region.[21] The midpoint between these two boundaries brings on the fourth type of complexion, which is a state of greater temperateness belonging to men living in that clime.

The fifth type is narrower than the first and third. This is the complexion that is determined in a person and according to which he can live. Here too he has a latitude and two boundaries. Still, one must know, to tell the truth, that every person has his own individual complexion, determined for himself, and, while he might live in the degrees of the latitude of his complexion, it is possible for him to recede so far from it that he will no longer live.

The sixth complexion follows the midpoint of the aforementioned latitude and a person having this middle complexion is one best complexioned according to the condition that has been made his own.

The seventh type of complexion is in the members. According to this type, each member has its own quality determined for its species. It is according to this that we say bone ought to be drier than the other members, the brain moister, and so on for the others.

The eighth type, however, is a temperateness of the member in the best type, not considered simply, but rather of that person whose member it is.

When the species of animals are considered, then the temperateness of the human body will be greater than that of the other animals. If locations are considered, those that are directly below the equator will have greater temperateness unless the sea happens to hinder it. Those with the next greatest temperateness are those in the fourth *clima*. This consideration is made with respect to the latitude of inhabited places. Water and mountains, however, produce many changes in such things.

The principal members are not close in their temperateness, as we have said before. Of all the members, flesh is the most temperate and next after this the skin stands out, since the skin does not feel temperate things. This is because its nerve-filled nature is tempered by the blood in the ends of the veins. For it is said that like does not feel like, as we have shown in the study *On the Soul*.[22] However, the skin of the hand is the most temperate, and of the parts of the hand, the skin of the palm is the most temperate. Further, on the skin of the palm, the most temperate is the skin at the end of the index finger, which should also be used to judge tangible things.

One should also know, that when a given species is called temperate, it is not understood to be temperate simply, nor is it understood to be so like the human is because its temperateness would then be that of the human body. It is rather called

---

21. *De nat. loc.* 1.9f. A. gives an alternate title for his own work.
22. *De anima* 2.4.1f.

temperate when, with respect to a given quality, it does not tend away from that which properly belongs to its temperament.

A thing, then, is called temperate in these ways. Something is distemperate when it is opposite in all the ways mentioned. This generally happens in two ways, namely, without matter (that is, through quality alone) or with matter, this being some humor. It is without matter, for example, when something is overly changed by heat or cold with an eye toward changing only the quality in it, as when things grow warm being near a fire or grow cold when on ice. It is called intemperate with the matter of a humor in two ways. Either that matter, which makes the distemperateness, penetrates into the members, or it is contained in the veins and nerves. Of intemperate things, however, the hottest in the body is the heat and the spirit generated from the heart, and next comes the blood.[23] Although it might be said that the blood is generated in the liver, nevertheless, true heat comes to it from the heart. After the blood come the liver and flesh, which are coagulated blood, as it were. Blood is hotter than flesh because thread-like nerves are mixed into the flesh and these cool it a bit. After the flesh, however, the pulsating veins are next in heat. This is not because of their being nerve-filled, but because they take on heat from the blood and spirit that are in them. Next are the nonpulsating veins and after this comes the skin of the palm, which has a temperate heat.[24]

The coldest thing in the body is the phlegm and then the *sepum* [fat]. Next are the fat [*pinguedo*], and then the hair, bones, cartilage, and ligaments.[25] Then come the cords, membranes, nerves, *nuca* [spinal cord], the medulla of the brain, and finally the skin.[26] The moistest thing in the body is the phlegm and then the blood, followed by the fat, the *sepum,* the marrow, the *nuca,* and the flesh of the breasts, testicles, lungs, liver, spleen, and kidneys. Next come the muscles and finally the skin. Galen also orders them this way.[27]

The lungs, however, are not very moist in their complexion, for every member resembles its nourishment in its natural complexion and it resembles the member lying next to it in its accidental complexion. Now the lungs are nourished by very hot blood into which a great deal of bile is mixed. But their moistness is caused by many superfluities that come together in them from the vapor from nourishments and from catarrh from the head. This is why they are naturally drier than the liver. Their softness is accidental. Moisture of a sanguineous member is due to the fact that the blood is buried deep within it, for it has to do with the nourishment of the member. Although

---

23. Avic. *DA* 12.26r, but also found in *Can.* 1.1.3.2.

24. The Sc. translation does not contain the statements found at this point in Avic. *Can.* 1.1.3.2, listing the muscles, spleen, and kidneys as next in order of degree of heat.

25. A. has many words for bodily fat. In general, *pinguedo* will be rendered as "fat" and the Latin will be given for the other terms. See Glossary.

26. *Nuc(h)a*, the "vicar of the brain," is discussed extensively at 1.257f.

27. Galen *De complexionibus* 1.9; Galen being cited at this point in both Avic. *DA* 12.3, fol. 26r, and *Can.* 1.1.3.2.

phlegm might be moister than blood, it is still necessary that much of its moisture be lost before it becomes blood, for natural phlegm is the matter for blood.

26 Hair is the driest thing in the body. For it is generated from a dry vapor whose moisture has evaporated away, and it coagulates dry. Next after hair is bone, which is moister than hair because it is nourished by the blood. Certain animals are nourished on bone, but none are nourished on hair.[28] Next after bone in dryness is cartilage and then ligament. Then come membrane, quiet veins, pulsating veins, motor nerves, the heart, and then sensory nerves.[29] Motor nerves are the coldest and driest. While it is true that the sensory nerves are cold, they are not overly dry, and in fact almost approach temperateness. Last of all is the skin.

Certain members are naturally moist, like blood, *sepum* [fat], *zyrbum*, and sperm. This is true in animals that have blood and flesh or something similar to them. I say "in those having something similar" because not all these members can be present in every animal. Instead, some animals have members that are analogous to these.

27 Further, dry and hard members arise from uniform members. Examples are bone, spine, nerve, veins, and others like them.

Moreover, differentiation and diversity exist in this division of the members. Some of those we have mentioned are those which are known to be different from those which share the name with the whole. Thus, vein as a whole and pieces of veins share the name. Thus, everything which is divided up in the division of something homogeneous shares the name of the whole. In this way a piece [*pars*] of flesh is flesh, and a piece of vein is vein. Those things which are known to be different from these homogeneous ones are heterogeneous members, and a piece into which one of them is divided does not have the name of the whole. When we say that the piece shares the name of the whole, we understand not that it shares the name only in an equivocal way but rather that it shares in both the name and in the *ratio*. Thus, one face shares both the name and *ratio* with another face.

28 The types of causes of dry and moist members are many, for some are the matter for the organic members. For each organic member is sustained and composed from them (namely, from bones, flesh, nerves, and the like), so that some of the uniform members are suited to the substance of the organic members and others to their activity. Certain moistures are food for the uniform members which are in the composition of the organic members. For all members take on growth from fluids. It is also the intention of nature that some be receptacles for the superfluities of food. An example is the superfluity of earthy, dry, heavy food, which proceeds to the bowel [*secessus*],

---

28. At 22.34–35(16), A. repeatedly states that the dog's complexion is drier and hotter than that of most animals, and he says that the dung of a dog that "eats bones" is especially good as an astringent for the stomach. A. probably knew, of course, that certain animals who gnaw bones are, in fact, seeking the marrow, and marrow is listed above as the fifth-moistest substance in the body.

29. A. elsewhere distinguishes between nerves carrying sensory impulses and those that carry impulses leading to motion. Cf. 1.113, 221, 267, 301f.

and the superfluity of moist, heavy, coarse food, which proceeds into the bladder in animals that have one.[30]

Further, in these fluids, which we have enumerated, there are certain superfluities which work toward the betterment or the worsening of the complexion. Examples are the various differences in blood, when one blood is compared to another, in that one is thinner and another thicker, one is cleaner and another more clouded, or one is colder and another hotter. Furthermore, this is sometimes the case not only in different animals but also in different members of the same animals. For the blood that is in the upper members differs from that in the lower members, having the differences we spoke of. The differences of the other fluids also follow a similar disposition.

Generally speaking, some animals have blood, and others have some fluid which is analogous to blood and is in its place. Thick blood feeds the body more and has less sensation, for which reason the things fed off it also have less sensitivity. Thin blood is more suitable for intellect, and the same is so for cold blood.

I say the same for the part that is analogous to blood in those that do not have blood. This is why bees and other types of certain animals which have a thin, cold fluid in place of blood have more prudence and intelligence than many animals which have thick, cold blood or very thick, hot blood. Some blooded animals have very cold and thin blood and these are wiser than those with blood that is the opposite of this. Best of all are those which have thin, clean, hot blood. For such an animal is most suited for wisdom, boldness, and things like these.

The upper members differ from the lower ones with respect to this difference. Likewise, the members of the male vary with respect to these differences when they are compared to the female's members. The opinion that should be held, then, is that the same differences exist in various members, both organic and uniform. Certain dispositions in all of them point out the goodness and betterment in them while others show the contrary. This is seen in the creation of the eye. Certain animals have hard eyes and others have soft. Some do not have eyelids, and others have them so that their vision can be sharper and more refined by means of the eye's softness and an ordered disposition of the eyelids. The most potent thing, however, in causing goodness in the members is the blood and that which takes the place of blood. For this reason all animals either have blood or some other fluid analogous to blood and taking its place. The natures of blood vary, as we have already said above.

---

30. *Secessus*: cf. 1.24.

CHAPTER III

In Which Way Something Is Said to Be Hotter or Colder Than Another in Uniform Members and in Which Way It Is Said to Be Dry and Moist

We ought first to divide hot and cold and to distinguish among the types of blood. Then we will consider the causes of the natures of members and of animals, for the natures of many natural things are attributed to heat, cold, and blood as if to their principles.

Many of the natural scientists [*physiologi*], holding opposing views, seek which animal is cold, which hot, and likewise which member is cold and which hot.[31] Some have felt in general that a water animal is hotter than a terrestrial, walking one.[32] They reason out a proof for this from the fact that an aquatic animal living constantly in cold water does not suffer from the coldness of the water and this can only be because its nature is very hot. Nevertheless, we should know that an animal with blood is hotter than an animal without it in this respect.

Borgyan the doctor holds the false opinion that women are hotter than men and certain others have agreed with him, reasoning out their proof from the fact that women have menstrual flow which, they say, is only due to an abundance of heat in them. Empedocles, however, said the opposite of this, namely, that women are colder than all men even though they are more wanton than men due to the movement of the fluid which is in them and which is easily formed into every cause of wantonness.[33]

Further, some of the ancients have said that blood is cold, as is bile. Some, however, have contradicted them, speaking better and more truthfully. Likewise, some of the ancients had doubts about the natures of hot and cold and they held different opinions concerning their natures, just as they did about the natures of the other things we spoke of.

For hot and cold are the most apparent to us of the other things known through the senses, since they are active more vigorously than the others. It seems, moreover, that the doubts mentioned stem from the fact that "hotter" and "colder" are divided up and spoken about in a number of ways. Thus, each of those speaking about hot and cold thinks that in his own particular division of things he says something different, when actually he may say nothing at all contrary to the truth.[34]

---

31. The identity of these *physiologi* is not indicated here. At *Phys.* 4.2.1, however, A. identifies the *physiologi* with the atomists Democritus, Leucippus, and others like them. Democritus was born ca. 460 B.C. in Thrace and may have been a pupil of Leucippus. What little is known of these two is from the remarks of later commentators.

32. "Water animal": *aquosum*, not the normal *aquaticum*. Used also at 12.87.

33. "Borgyan": Ar. *Part. An.* 648a29f. attributes the first view to Parmenides, whose name became Gorcion in Sc., and the second to Empedocles, whose name suffered a kinder fate.

34. Cf. Ar. *Part. An.* 648a37f.

So that it might be understood better, we must not lose sight of the fact that some 34
natural things are hot and some cold, while some are moist and some dry. This is
especially so since it is already clear from our other books that these are the cause of
life, death, waking, sleep, youth, old age, health, and sickness. However, the causes of
these things are not roughness, smoothness, lightness, heaviness, nor any other thing
other than the ones we have stated. Now this happens entirely properly and reasonably because these qualities (namely, hot, cold, moist, and dry) are the principles and
natural potencies of the first mixables in the body.

We must first consider whether something is said to be hot simply or in many 35
ways, and we must see what the proper operations of "hot" are and whether they are
few or many. Something is called hot in one way when it warms that which is touched
by it or even what only comes close to it. This is so whether it be hot essentially or
accidentally. Now this is one that is hot in actuality and it alters and changes things
near it into the form of heat [*ad formam caliditatis*]. Of the things that are hot essentially, fire acts this way, while a heated stone or piece of iron does this among those
things that are hot accidentally.

A thing is called hot in another way, whether it be hot in and of itself or accidentally, when it does not alter and change the things near it into the species of heat
[*ad speciem caloris*] but nevertheless does clearly produce a change in sensation when
touched by an animal that has the sense of touch. This is because it has more heat than
the flesh that is touched and when it touches the organ of touch, it is mixed in with
the organ of touch at the touch and exceeds it in heat.

Something is called hot in a third way also, when, in accordance with the actual 36
heat it possesses, whether it has it essentially or accidentally, it produces each and
every sensation associated with heat even though the sensation might be weak and
not very clear. Now this type might seem false to some since it is not heat alone which
changes what is touched in this way, but sometimes also vision, seizing on something
visible such as bright light that causes pain in the eye. This pain is the result of heat
since the breaking up of the brightness in the eye brings on heat and dissolves and
melts the lens [*cristallinum humidum*].[35] This objection is resolved because the visible
does not produce heat by means of the form of the visible but by the accidental trait
of breaking up the rays in the lens which is rubbed and polished. But we are here
speaking of heat which is active in the bodies joined to it or in the senses by using the
form of its own proper heat.

Thus, these three types have been comprehended. If someone were to make a 37
comparison between them, then the one called hot in the first way is hotter than the
second and the third, and that called hot by the second way is hotter than the third.
The third, however, has the lowest grade of heat. The heat in something exceptionally
hot is one that melts or burns that which it touches. For this not only makes a change

---

35. On the "crystalline humor," whose other name, the "icy" humor, suits the metaphor of melting
perfectly, cf. 1.197–205.

into the species of heat but also, by the great power of fire it has in itself, it dissolves moist things and burns dry things.

One thing is thus said to be hotter than another in one way when the two things proposed have a form of heat that is one, equal, and equally intense and burning, but one of them is greater in size than the other. Thus a great amount of hot water is said to be hotter or to have more heat than a small amount of water, even though the small amount has the same burning heat as does the great amount of water.

This is so because heat is a material and positional form [*forma materialis et situalis*], which is present in the majority of its larger subject and in lesser amounts in a smaller subject.

Moreover, one thing is called hotter than another in a second way. This is the case if, when it is hot, it grows cool slower than the other thing. Thus, a heated rock is said to be hotter than hot water, even though they might have equal burning heat. This is due to the unequal solidness and thickness of the subject in which the heat is present. Sometimes, to the contrary, a thing is called hotter which becomes hot more quickly since it is more susceptible to heat. In this type, that which has a thin, loose substance is hotter and nearer heat than something dense. In this way wood is said to be hotter than stones. That it grows hot faster from its loose texture is clear from the fact that cold, which is the opposite of heat, will be removed faster from it and that heat will approach it faster. For heat has a natural similarity with it, for heat is in that which is thin, loose, and loosely textured.

Something, then, is called hotter than another in the ways we have stated. All the ways stated cannot exist in the same thing and in the same number or species. For hot water warms more than does cold, since the cold only warms by surrounding and rousing the natural heat at the bottom, as we said in the *Meteora*.[36]

Moreover, hot water burns only if it is boiling, for then it cooks. But fire is hotter than water since it burns those things it touches and is a better melting agent than water. Of the types of water, boiling water is said to be hotter than other things, because it changes more with respect to the form of heat. It might also be called less hot since it grows cool more easily, for it cools more quickly than a small fire. Fire cannot grow cool of its own accord, whereas water grows cool by its very nature.

Further, boiling water is hotter to the sense of touch than a small fire since it changes more fully. Nevertheless, it cools off and solidifies more quickly. The reason for this was given by us in the *Meteora*.[37]

Further, in a similar way, blood is hotter to the sense of touch than is oil. I am speaking of uncorrupted blood, such as is understood to be in the complexion of the animal, for this way it is hot, and it is moist in actuality and in natural property. In this same way hot water is said to be hotter to the touch than unheated oil. And yet, in another way, water is colder than oil. Blood is also colder than oil. This is because

---

36. *Meteora* 2.1.29.
37. *Meteora* 2.1.31.

water solidifies forcefully and cools off faster than oil, and, similarly, blood solidifies and coagulates faster than oil.[38] Also according to the ways given, a stone and iron grow hot slower than does water and yet, after stone and iron are heated, they burn neighboring bodies and things touching them more strongly than water.

Of the things we have stated as warming adjoining bodies, some have accidental heat and others essential heat, and there is a large difference between them. For that which is called accidental heat is close to the true sense [*ratio*] of heat. This is because it has the form of heat in actuality and truly. It is thus said most truly that each of the things mentioned is hot in and of itself in the sense that each of them is, through the form of its subject, truly active and changing with respect to heat. Thus, if I were to say that a musician has a fever and is suffering from his heat, or if I were to say that a musician had more blood or heat, then, while all these things are said by way of accident, nevertheless, the heat is in the musician as form is in a subject, exercising the proper activity of the form.

It is known from these things, then, that one thing might be hot essentially and another accidentally, and that there is nothing preventing something that is hot accidentally from changing the sensation more than that which is hot essentially, and that in this respect it can be called hotter than the other. In another way, however, that which is hot essentially is hotter than that which is hot accidentally in that it is always hot, just as we said that fire is hotter than boiling water.

It is therefore clear that the judgment to be made between two things, as to which is hotter than the other, is not a simple one, to be spoken of in one way alone. For in one way one thing might be hotter in one act of heat and in another way the other thing might be hotter in another act of heat.

There are certain things in which none of these things pertains, for they can be properly called neither hot nor not-hot. This is, namely, when the subject is not hot but rather, by being close to the hot thing, or by having the hot thing put together with it, is soon rendered hot. Thus, while a stone and iron are not called hot, when they are near fire or are put together with it, they are rendered hot. Blood is truly hot both in and of itself and according to its nature, and in addition to this it may happen to burn with accidental heat, such as in anger or fever. In all things such as these, which are hot of themselves, cold is not a sort of nature but rather is a privation and corruption. This is so even though universally, in all things, cold is somehow a privation and heat a *habitus* and form, as we have taught in our book, which is written about the natural science of hearing.[39] But we are not concerned about this sort. Here we are saying that cold is the privation in all bodies to which heat belongs in a way other than that of accident.

---

38. So, like hot water, heated blood cools faster than heated oil.

39. *Phys.* 2.1.7. A *habitus* is, again, the opposite of privation, in that it signifies the presence of the property or characteristic that is made absent by privation.

44 It is proper too that fire, when it is in foreign, earthy matter, should be numbered among those things which are called neither hot nor not-hot. For fire spoken of in this way is flame or a lit brand (that is, a burning torch).[40] Now, to be sure, flame, which is lighted smoke or vapor, is always hot. The brand (that is, the torch) which is the subject of the fire is not hot in and of itself but rather is made hot by being put with the fire. For after all, it is sometimes put out and it then becomes cold.

Oil, however, and a pine brand (or a firebrand of any other wood) are cold.[41] And it is the same for all bodies which are heated by a heat that is added to them. This is the case for ash, which is the remaining superfluity of a fire, and likewise for all the superfluities which remain after digestion in the bodies of animals, and also for the yellow bile and other bodies that burned. None of these things grows hot except through something that remains in it of the power and substance of the fire. The subjects, however, of the fire, which, following the method given above, are not called hot, are called hot in another way. Thus a pine brand and other inflammables are called hot since they are easily changed and altered by fire to the state where they are ignited in actuality and set ablaze.

45 Some have felt that the hot melts and coagulates all bodies, although this is not universally true. Rather, the bodies which have a great deal of water in their composition and have but little that is earthy—metals, for example—are coagulated by cold. Those things, however, which have a great deal of water and earth in their composition are coagulated, concentrated, and gelled by the moisture which evaporates because of the heat. A brick is baked this way because most of the earth which composes such bodies quickly dries out when the moisture evaporates. But we have already spoken about these things most clearly in another part of these natural science books, namely, in the fourth book of the *Meteora*, where we treated things that are able to coagulate and melt.[42] For there we pointed out the bodies which melt and gave the cause for those things which coagulate, gel, and become concentrated.

46 From all that has been introduced, it is clear that the inquiry into the learning about hot and hotter insofar as it deals with the complexion of natural bodies composed out of contraries is vague. For heat is present in such things in a variety of ways and neither is it in all these things in one and the same way.

Thus, one should be careful to make distinctions when saying that a certain thing is hot *per se* and another accidentally. That which is hot *per se* is that which is always hot. The one other than this, namely, that which grows cold when left to itself, is hot accidentally.

Something is also called hot potentially and in act. Each thing is further called hot according to the type of its heat, so that one thing is called hot because it produces a

---

40. "Brand": A. glosses the received word *taeda* with the slightly more common *fax*.
41. Here A. adds the very unusual word *ticio* (CL *titio*) for brand.
42. *Meteora* 4.3.1.

greater change in sensation than another thing does with respect to a change of heat. The one that is said to have the greatest heat is that which is ignited by fire.

Since we know from what has been said that something is called hot in many ways ("hot" being a *habitus*), we know from the same thing that cold too, which is a type of privation, is spoken of in a number of ways. For if one of a pair of contraries is multiple, then the other is as well. Likewise, if one of the contraries is clearly understood, than so is the other.

Consequently, it is necessary to speak about the multiplicity of dry and moist, insofar as they are mixed in physical bodies, for these are spoken of in many ways. Things are called dry potentially and also in act. For ice, which is wet by nature, is dry both in act and accidentally, even though it is wet *per se*. Earth, however, and ash, and others like them are moist accidentally when they have liquid poured on them. But they are dry potentially and *per se*. When such things are not kept away from moisture, they are said to be wet in act, since a great deal of the wetness of water is mixed in them. They are called dry potentially due to the great portion of earth that exists in them. In this way, then, things are called dry and wet both in many ways and cryptically. The same is true for hot and cold and each type is a type of truth in natural, mixed bodies.

With these things determined and established, it is clear that blood is hot *per se*, but that it might be rendered very burned and possessing great heat accidentally as happens in those suffering from *causon*.[43] I, however, am using "hot" *per se*, since hot is contained in its own definition, just as white is contained in the definition of white. For blood is a hot, moist humor, which exists as the last nourishment in those that have it. As we have said, it is sometimes rendered hotter than it naturally is through the accident of fever or putrefaction. In this heat it is called hot not *per se* but accidentally. In this way also we say that the things which are naturally hot and moist, when they rot and are separated from their nature and the body in which they naturally were, are coagulated by the cold and in this way they seem cold.[44] Such is the case for blood that has been removed from the body of an animal. For things which are naturally hot and coarse are changed when they are separated from their nature and from the body in which they naturally existed, becoming altered and taking on contrary dispositions. Bile, dry by nature, is rendered wet when it is made cold, while blood, wet by nature, is greatly dried out when made cold.

It is therefore clear from all that has been introduced that the sharing of contraries and their generation in mixed bodies is a matter of more or less, as is the generation of all the natural qualities. Therefore, we have declared to the extent possible how things are called hot and wet in the blood and in what way and to what extent the blood shares with the contraries which are in nature.[45]

---

43. *Causon* is a disease thought to arise from an excess of bile and blood (Latham[2]). It is from a Gr. stem indicating burning.

44. A. employs *algidum* here for "cold" as opposed to his more common *frigidum*.

45. "Shares with": perhaps "participates in" (*habet . . . communicationem cum hiis contrariis*).

CHAPTER IV

Which Is a Digression Setting Forth Those Things Which Must Be Examined Concerning the Composition and Complexion of an Animal

50 In addition to all the things that have been said, we should call to mind again those things which have been argued at greater length in previous books of this investigation and in the books *On Places and Things Placed* [*De locis et locatis*] and *On the Ages* [*De aetatibus*]. For unless some of these things are examined here, the instruction will not be sufficiently easy.

Let us say first, then, that there are three compositions in the bodies of animals. The first is called mixture [*commixtio*] and consists of the four elements and their qualities altered in respect to one another.[46] In this composition, the first active and passive qualities, with their excesses broken down, are led to a single act of mixture [*mixtum*]. This act takes on different gradations according to the difference in the animals' bodies. This mixture [*mixtio*] of elements is continually extended in alteration until the four humors are made from it, these being blood, phlegm, and both red and black biles.

51 The second is the complexional composition of the humors after a change has been undergone, and from this change of the humors it produces the uniform members. This activity is thus directed toward the species of uniform members and this composition is properly called complexion.

The third is the constitution of an official member out of the uniform members and the constitution of the entire body from the uniform and official members. Although these might seem to be compositions of two things, they are in fact the same composition. For the uniform members are not altered from their forms in the composition of an official member. Further, the official members have the same type of composition in the composition of the entire body.

52 The first of these compositions is indeed prior in time and generation. The second of them is in the middle and the third is last. But according to substance and *ratio*, the last is prior, for it is from this that the animal has that which is animal, both the sensation and motion of an animal. For if the uniform members were enough for fulfilling the animals' activities, the composition of the official members would not be carried out. For nature does not make the official members for beauty or for anything which has to do with the well-being of the animal, but rather for the necessity of the animal's activities and the necessity of its natural potencies and individual passions. The uniform members differ from the official members since the uniform members are dominated by some one element. Thus there are bones in which earth is dominant

---

46. "Altered in respect to one another" (*alteratis ad invicem*): this phrase is repeated at 12.54.

and the same is so for flesh and certain others. The official members, however, cannot exist under the domination of one element.

Someone might object to these statements, saying that sensation and the activities of the soul are completed through one simple thing, for example, touch through the flesh or the tactile nerve, hearing through the *timpanum* of the ear, taste through the nerve that is spread out on the tongue, and smell with the two breast-shaped bits of flesh on the front part of the brain. I respond by saying that while the sensitive power might be primarily in one simple thing, nevertheless sensation is completed in its best way only by means of an organ. Thus, touch is at its most sensitive neither in the flesh nor in the nerve, but at the end of the index finger. Hearing is not good or best without the ear, nor vision without the eye, for which not only the crystalline humor [lens] is needed, but also the other humors, the tunics, and the shape of the eye. Smell requires the nose as well as the muscles of the chest which bring the air to it and it is the same for the other activities of the soul. It is thus clear that an animal needs to be composed of nonuniform members.

If, then, someone should wish to search out the natural powers of the soul carefully, they will be situated in a uniform member and the vital (that is, spiritual) and animal powers will be situated in organic members. However, the principle of all the members is the heart, just as we have said before, and the others have their principle in it.

Moreover, if someone should give attention to the components, there will be humors which are food in an animal and the others will be members. Of the humors, the principle food is the blood and the thicker the blood the more it feeds. Still, such thinner humor as there is in animals will also be "thinner" in the operations of the soul and such thicker humor as there is, which is the final food for the members, will also be "thicker" in the animal operations.[47] For this reason, those animals which do not have blood, but rather some humor "thinner" than blood, are "thinner" in their operations than certain others that have blood. This is as we have said in part above and as will be stated again below. However, as was determined in the book *On the Nature of Places and Locations,* complexion is a quality that occurs in the bodies of animals arising out of the activity of opposite qualities insofar as the qualities themselves are altered with respect to one another and the elements are divided into the smallest possible bits, so that the smallest amount of each is with the smallest amount of the other.[48] Thus, then, in elements which are altered with respect to one another and are divided, there occurs one quality which is called complexion. We should not speak here about the types of balanced and unbalanced complexion or of what is balanced throughout all its species, for this has already been sufficiently determined in our previous books on natural science.

---

47. The word played upon is *subtilis* in its adjectival and adverbial form. As an adjective it implies a physically thin nature. As an adverb it implies subtlety of action.

48. *De nat. loc.* 2.2f.

55 However, concerning balance of animal complexion, one should know that a thing is not balanced in the nature of that which is called the "balance of the arithmetic mean" but rather it is derived from that balance and is taken according to a geometric mean.[49] Nevertheless, among all the various sorts of temperateness of balance in animals, the balance of the temperateness of the human is closest to natural balance. This is not to say that the human consists of balanced amounts of all the elements, for this is impossible, but rather that the human's complexion is furthest from the excesses of the contraries and is closest to the temperateness of heaven in which, of all the natural bodies, there exists the greatest balance of shape and nature. For it is not this way in other animals.

With balance thus understood, it can be spoken of in many ways. Sometimes we will say that the human or some animal is temperate or balanced compared to other species of animals, which, according to the nature and complexion of their species, are less balanced. Thus the human is said to have a balanced complexion compared to that of a horse or an ass and the flesh of a chicken is said to be balanced compared to that of asses or fish.

Something is called balanced in a second way within its own species when it approaches the balance that belongs to its own species. It is called unbalanced when it is removed from it.

56 It is also called balanced in a third way, pertaining to sex, such as when one sex is compared to the other. Thus the male is called balanced compared to the female or, oppositely, the female compared to the male. For if the male is excessively warm and dry for his species, the female will be tempered in a balanced way in respect to these. If, however, he is excessively cold and moist for his species, the male sex will approach the mean of temperament.

According to the fourth method, an animal is called balanced in a given thing when compared to an individual of the same species and sex. Thus a given male is balanced when compared to the males or a given female when compared to the females.

In the fifth way, it is called balanced in one individual concerning the different time of life. Thus, a youth will be temperate compared to his own childhood and old age or he might even be healthy compared to a time when he was sick.[50]

57 Sixth, a given member is called balanced in different ways, namely, when compared to another member or even to itself at different times. It is called balanced in many other ways as well.

The first type of this balance, in accordance with the nature of a species in comparison with other species, has latitude since not all the individuals of one species are in the same balance. Rather, there is a certain state which a given individual attains

---

49. On the arithmetic and geometric means in the composition of the animal, cf. *De anima* 3.5.3. The sense there is that while the arithmetic mean involves a balance between two elements (say, earth or water), the geometric mean establishes a balance among the numerous elements and humors in the body: not only earth, air, fire, and water but also the hot and the dry, the cold and the moist, and so on.

50. The Latin of this sentence leaves much to the imagination. The sense is derived from 12.18.

most simply and most often, and the other individuals are moving by gradations to that state. There can also be such a removal from balance that the species is not saved and is destroyed. Thus the second type of balance is understood according to the simplest type of balance and the simplest type of such latitude. We might thus say that a given person attained the greatest and simplest balance that is possible for the species and nature for the human. Aristotle says in his first philosophy that this person is the measure of all other people, since he participates most simply in the nature of the species of the human and that other persons participate in the same nature of the species insofar as they come near it.[51] It is this way in all the natures of the species of animals.

True balance of this sort is rarely found because such a person or animal must be, both as a whole and in all its members, in the most simple disposition and balance which ought to exist in it naturally. For not every member of such an animal has to be absolutely balanced and temperate. For example, the heart is very hot and the brain very cold. It might be that no member in a person is absolutely temperate except for the skin alone, as will be stated below. Since, however, there are three or four principal members as we stated before, the brain cannot temper the heat of the heart and liver with its coldness nor can the dryness of the heart temper the moistness of the brain and liver. Nevertheless, the brain is not too cold and the heart is not too dry. Rather, the heart is dry when compared to the others and the brain is moist when compared to the others.

This type of complexion, however, is narrower than the first type and yet it has differences of latitude. But this latitude occurs with respect to regions and the latitudes of *climata*, especially in *climata* that have many and contrary qualities, as do the second and seventh *clima*.[52] For an Indian who possesses an entirely middle and balanced complexion that suits his habitation would not live well in England, Scotland, or Dacia.[53] And a Dacian who nears balance of complexion for his habitation would quickly die in India or Ethiopia, and it is the same for other differences in *clima*. As was stated in the book *On the Nature of Places,* each *clima* has latitude and a mean with respect to equidistance between two bounds [*medium per aequidistantiam inter duos terminos*].[54] Now, that is balanced in terms of complexion that is a mean between two bounds, and this is a particular balanced thing specifically mentioned in addition to the types introduced above.

Again in addition to the types mentioned before, there is a balanced complexion which ought to be singled out here. One can live according to this one and if he changes from it the complexion is rendered unbalanced. This one too has a certain lat-

---

51. Stadler's *ei alii* is marginally understandable as "these others," but it is very awkward. The translation offered reads *et* for *ei,* with Borgnet. At *Metaph.* 1062b14, Ar. attributes to Protagoras the claim that "man is the measure of all things" but without the explanation A. provides here.

52. There are two senses of latitude at work in this sentence: (1) with respect to how much leeway a given definition has, and (2) geographical latitude.

53. Dacia was located in the loop of the lower Danube.

54. *De nat. loc.* 1.9.

itude and two boundaries [*termini*] and the midpoint is the balanced one, according to which its own particular life is best. If it is a bit removed from it, it is good but not the best and is unbalanced and bad in proportion to the distance it is from it.

60 Further, there is a special type of balance according to which each member is in the disposition in which it should be. Thus, bone is dry, the brain moist, the heart hot, and so on for the others. It has latitude and a midpoint, namely, when the member is in the best disposition possible for it. When, therefore, we speculate on the complexions of all animals, the complexion of the human will be the most balanced, and when we consider the complexions of regions, the most balanced will be the one of the first *clima*, unless the moisture of these and the Meotydae marshes hinder it excessively. Next after this the one most balanced is that of the fourth *clima*.[55] However, in all of these, accidents occur which are hindrances due to the seas, mountains, and other accidents, as we have determined in other books.

61 Further, as is clear from what has been said before, the principal members are not close to that which has been called, simply, temperateness. Rather, of all the members, the one nearest temperateness is flesh, and next is the skin. For the skin does not feel temperate things due to the similarity it has to them, and the cause of its temperateness is that it has nerves woven into the skin, and it is also tempered on account of the blood which is at the end of the veins. It is a saying of wise natural scientists that like does not feel like.[56] The skin of the hand, however, is more temperate and balanced than the skin of the rest of the body. On the parts of the hand, the skin of the palm is the most temperate and balanced, and on the parts of the palm, it is the skin on the fleshy tip of the index finger which is the most balanced. Therefore, this is the best and most discriminating judge of touch.

62 Someone will possibly have reservations concerning the things just said, saying that there seems to be a single balance for the many species in nature. For we say that food and the species of plants and roots are healthful and balanced for the human and this would happen only if they had the human balance. But those things which have the human balance have a human's complexion and thus many things seem to have balance with the human.

This problem is solved with reference to the fallacy of equivocation, for when food or some other species of thing is said to be healthful and balanced for the human, it is not so called because it has the same balance as the human but because it thus contributes to balance with the human by the effect it has in the human body. Balance, then, is spoken of in these ways.

63 Those things[57] which depart from balance in the bodies of animals, especially in the body of the human, which is the noblest of the animals, are as follows. To be sure,

---

55. Meotydae: This is ancient Lake Maeotis, the modern Azov Sea. A. makes mention of it also in *De nat. loc.* 3.3–4, 7. It was fed by the river Tanais, the modern Don.

56. Cf. 12.23, with source. For the principle that like does not feel like, cf. A. *DG* 323b5f. and *De anima* 417a1f.

57. Stadler's text does not have a section 63. We have supplied the number here.

there is a departure with respect to warmth since the hottest things in the body are spirit and the heart from which the spirit itself arises. The next warmest after this is the blood. For even though it is said by some that the blood is generated in the liver, it must necessarily still derive a great deal of heat from the heart, as was proven in the earlier parts of this investigation.[58] The next warmest after the blood is the flesh, which is a sort of coagulated blood, and which is thicker, since there are, so to speak, thread-like nerves mixed into it. The next hottest after this are pulsating veins, not, to be sure, because they possess nerves, but because they take on heat from the thin blood and spirit which pulses in them. The next are the nonpulsating veins and after these is the skin of the palm, which is the most temperate.

The coldest thing in the body is the phlegm and then come the *sepum,* fat, hair, bone, cartilage, ligament, cord, membrane, *nucha,* medulla of the brain, and skin.[59] According to Aristotle, there is one which surpasses in coldness all those things which are parts of the animal in actuality, and this is the humor of the eye.[60]

Further, of all the things in the body the moistest is the phlegm. The next is the blood, after which the moistest is the fat, the *sepum,* the medulla of the brain, and the *nucha.* There then follow in order of moisture the flesh of the breasts, testicles, lungs, liver, spleen, and kidneys. The next in moisture are the muscles and the skin. Galen set it down this way as well.[61]

64

However, it should be known (and this is just as we remember saying in what has gone before) that the lungs are not very moist in their complexion.[62] Generally, each member resembles its nourishment in its natural complexion, but in its accidental complexion it resembles the member that is near it. The lungs, moreover, are nourished by warm blood into which a great deal of bile is mixed. There are also joined to them, in an accidental manner, many superfluities from the vapor of the nourishment which comes to them and which is converted in their porousness to wateriness. A great deal of catarrh also drips into them from the head and on account of these things they are rendered soft and moist. They are thus naturally drier than the liver since, as we have said, their moisture is there due to accidents. The moisture of the liver, however, is there because the blood which is involved in nourishing the entire body is at first watery in the liver. The moisture of a sanguineous member, however, is caused by the fact that the blood involved with the growth and the nourishment of the member is hidden deep within it. While natural phlegm is moister than blood, nevertheless, because natural phlegm is the matter for blood, it must give up much of its natural moisture before it is converted into blood by natural heat.

65

---

58. Cf. 12.24, much of which is quite close to the text here.

59. Cf. 12.25 for close parallels for this and what follows.

60. Stadler suggests *GA* 744a13, but the passage is marginally connected to this one. Consider, however, that the *glacialis humor* of the eye implies, by its very name, extreme cold. Cf. 1.170, 197.

61. Stadler suggests Galen *De complexionibus* 1.9. The original citation is from Avic. *DA* 12.26r and *Can.* 1.1.3.2. Cf. 12.25.

62. Again referring to 12.25.

66 The driest thing in an animal's body is hair. It comes from vapor, the moisture of which has evaporated and coagulated in the surrounding air. The next driest after the hair is the bone, which is shown to be moister than hair in that some animals are nourished on bones, but none whatever on hair. It is the same for birds' feathers, and thus hair and feathers that are swallowed by animals are cast out undigested either upward or downward through the anus.[63] The next after bone in dryness is cartilage, and after this come ligament, membrane, nonpulsating veins, followed by the pulsating veins, followed in turn by the motor nerves, the heart, and the sensitive nerves. The motor nerves are colder and drier, while the sensitive nerves, which are naturally cold, are not very dry, and in fact come quite close to temperateness. Skin, however, attains the final grade of dryness.

To be sure, all these things have been said above, but they are repeated here so that we might show more easily the reasons for the things we have said and so that the instruction might be the easier.

CHAPTER V

Which Is a Digression Setting Forth the Things Which Should Be Investigated Concerning Ages

67 In the same way, then, we should reiterate some of the things stated in the study *On the Ages*.[64]

There are four ages in general, the first of which is the one called childhood [*aetas puerilis*], although it is called the age of growth [*aetas incrementi*] by some. It lasts for up to thirty years. The second is the age of stability [*status*], which lasts until thirty-five or forty years old. The third is the age of lessening [*aetas diminutionis*], which, however, is said to be accompanied by a retention of vitality since in this age it is more that the causes of vitality begin to lessen than that there is any noticeable reduction of vitality. This is called the age of manhood [*aetas virilis*] and it lasts to, or near to, fifty.[65] The last age is old age [*aetas senilis*], with a noticeable lessening of substance and vitality.[66]

Childhood is divided into infancy [*infantilis*], which is the age of the newborn, the age of movement [*aetas motiva*], in which the child begins to move, and the age which is prior to strength [*aetas ante fortitudinem*]. This is the one which is called the "implanting of the perfect teeth" since the teeth are already beginning to change and

---

63. Owls' pellets and cats' fur balls would serve as a readily observed example of this "upward" casting out.

64. *De iuv. et sen.* 1.1.2. Cf. 2.44–43 for further citations and terms.

65. Avic. *DA* 12.26r gives "sixty."

66. Cf. discussions on this and similar terms at 2.44 and 8.38.

come back.⁶⁷ It is further divided into the age which is prior to the lust of the fully grown and mature seed and the lustful age [*aetas luxuriativa*].

The age of infancy has a sort of tempered heat and is overflowing with moisture because of its nearness to the sperm. For this reason no small number of the physicians are in doubt over the heat of childhood and youth, wondering which of these heats exceeds the other. For some say that the heat of childhood is hotter than that of youth. For children grow more than do youths and have natural functions that are more completed with regard to the appetite for nourishment and of digestion. These would not be in this age unless the natural heat, acquired from the sperm and the spirit of the sperm, were greater and stronger during this age. Others contradict them, however, saying that the heat of youths is sharper and their blood, in which the heat is strengthened, is thicker. An indication of this is that at this time a great deal of blood flows from the nostrils due to this heat. For the complexion of this age is very choleric, while the complexion of childhood tends more toward phlegm.

Another reason is that the movements of youths are more forceful and motion is due to heat.⁶⁸ This is also the reason that the human heart has very many movements. Youths also have a stronger digestion and they digest harder foods. All of these things are caused by nothing other than a stronger heat. These people also say that the appetite which grows in children for food is not due to heat but rather to cold. Thus, those who have a canine appetite have it due to cold.⁶⁹ An indication that youths have a more natural appetite and better decoction of their food is that they do not vomit out their food and that they take great pleasure in what they eat.⁷⁰

Moreover, the fact that this age tends toward bile is indicated from the hot, choleric illnesses, like tertian fever, that very often befall them.⁷¹

In addition, at this age the accident of vomiting is choleric. To the contrary, however, the illnesses of children are generally cold and moist, their fevers are phlegmatic, and, for the most part, their vomit is phlegmatic. These people also say that growth in children is due not to the strength of the heat but rather to the great moisture that abounds in the children and which, with a little heat, is extended into growth. They also say that the prodigious appetite is due to the lessening of the heat, as we said before.

As we have said before, Galen seems to resolve all these doubts by saying that the heat of both ages is balanced in different ways, namely, in essence and radically, but

---

67. On ages and teeth, cf. 1.195f., with notes.

68. The grammar of this passage leaves much to be desired. The sense, however, seems secure. On the heat of youth, cf. *QDA* 3.13. The claim that sense and motion do not occur without heat is attributed to Ar. at *QDA* 1.41.

69. Cf. 1.232 for a "canine mouth"; also note the modern expression "to wolf one's food." A. remarks that while, generally speaking, heat is the principle of digestion, the cold and the dry are the cause of appetite. Cf. *QDA* 14.2.

70. The point of comparison is, apparently, babies who regularly regurgitate what they eat.

71. On this fever, cf. 7.172 and 9.32.

that the heat in children is greater in quantity while that in youths is greater in quality. An example of this is if one and the same heat were found, radically and essentially, in stone and in water. For this heat, penetrating into the dry rock, would be sharper in quality but less in quantity because of the restriction and the solidifying action of the matter. It would be greater in quantity in the water, however, because of the abundance and the fluidity of the matter, whereas it would be duller in quality due to the fluid which softens the acute angles of the fiery heat.[72]

An indication that what Galen says is true is that children are generated from sperm, in which there is great heat, and that there occurs in them no cause for the cooling off of this heat. For the child is always growing and does so in continuous steps and has in itself no reason for the heat to diminish. In youths, however, there is no further reason why the heat should grow in them, and neither do they have a reason to give up the heat. Rather, the natural heat is kept in them by the radical moisture which is in them.[73] This does evaporate, however, and is there in smaller amounts than in children, and the heat that is in it is thus sharpened. When the natural moisture is considered carefully and is compared to the natural heat that is in it as if in its subject, and is compared to the growth that is caused by it, it will seem to be small in comparison to both of these, especially in comparison to the growth that occurs from it. But it will appear abundant compared to the heat, especially in children, for in them it is needed both to conserve the heat and to provide growth. In youths, however, in whom there is less moisture, there is not enough for both of these, but it does do one thing, which is to conserve heat. Not much of it provides growth, however.

Now, one who asserts that there is enough moisture for growth and not for the natural heat will be speaking entirely falsely. For it is impossible that growth be based on a particular moisture upon which the efficient cause of growth cannot be based, this efficient cause being natural heat. It therefore follows that moisture in youths is the cause and the subject of heat, and not the cause of growth.

The statement of some that the cause of growth in children is moisture without heat is simply false. For moisture is nothing more than the material cause of growth and matter does not act on its own without an efficient cause. Even though the nutritive (that is, vegetative) soul is the efficient cause of growth, nevertheless, nothing is effective without natural heat extending the moisture.

The statement that appetite in children is caused by cold is false. For this sort of appetite is not accompanied by good digestion, while the digestion in children is the best. An indication of this is that were not their digestion the best, more would not be returned to their bodies through nourishment than is lost through dissolution. That children sometimes suffer from bad digestion is because they are gluttonous and eat many moist things. And in this way the inordinate movement that is in children

---

72. Cf. 12.86.
73. On radical moisture, see Glossary.

brings on corruption. This, then, is the complexion of children, and also of that in the age near to childhood.

In the age that follows childhood, the natural moisture begins to be consumed by the heat of the surrounding air, assisted by the heats caused by the corporeal and animal movements, as we have said elsewhere. There is especially an exhaustion in the natural moisture which cannot always resist the natural heat. In fact, it has a boundary up to which it can resist just as do all the other natural and material powers, as was shown in *Heaven and Earth* and in the book *On the Ages.* It must happen, then, that after the period in which it is able to resist the heats at work in it, its loss is carried out continuously. Also, that which is restored to it from food is less and baser than that which is lost from itself, and constant loss continues until the whole is lost, whereupon the animal suffers a natural death.

Further, extraneous, undigested moisture is caused by lack of digestion, which increases in the bodies of animals as a result of the cooling of the heat, and it is a phlegmatic moisture. From the first cause, then, life lacks heat due to the lack of its subject, in which it is cared for and preserved. From the second cause, however, it is due to an unnatural, suffocating moisture which swallows up the unnatural, opposite qualities. These two, then, as was said elsewhere, bring on a natural boundary to the life of every individual. But there are other accidental boundaries to life, such as death by beheading and the like, but the causes of these things are not sought by us here using natural philosophy.

It is clear, then, from the things that have been said, that the bodies of children and youths are equally hot, but that the bodies of the children have more moisture. An indication of this is the softness of their flesh, the tenderness of their bones, and the flexibility of their limbs and joints.[74] This view is likewise secured by the fact that whatever things are moister must be nearer the sperm since the sperm is the entire cause of generation. Children, however, are nearer the sperm and the smoky, moist spirit which penetrates the members and softens them at this age.

The age of manhood and especially that of old age is dry. An indication of this is the hardness of the members and the looseness and wrinkling of the skin. This wrinkling is due to both cold and dryness, for the dryness is the cause of the looseness and the coldness is the cause of the wrinkling, since the cold folds over and contracts one fold of skin over another, thus causing wrinkles.

The fieriness and the airiness which are the causes of the fineness of the members are scarce during these age periods. For the heat that is in them is involved in evaporation, and as the heat evaporates it takes the moisture with it, and the bodies of men, especially old ones, stay cold and dry as a result.[75] The bodies of the youths are dry compared to those of children and are moist in comparison with those of men and

---

74. "Limbs": *membra,* more commonly "members," but the current context favors the translation given.

75. Throughout this section, A. specifically uses *vir,* reserved for the male, as opposed to his more generic *homo.*

old men. Likewise, the bodies of men are warm compared to old men and are cold compared to youths. The bodies of old men are simply bereft of natural moisture and warmth, but they have phlegmatic moisture clinging to the outside of the members and softening them. This accidental moisture brings on the chill of death and suffocation of the natural heat.

CHAPTER VI

Which Is a Digression Setting Forth the Manner of the Digestion of Animals' Foods, the Generation of the Humors, and the Manner of Their Generation

75 We must now add to what has gone before something about the types of digestion in animals and on the humors which are generated from digested food. Before we speak, one should understand about the cause of animals' members, for from these the causes of many members are understood.[76]

One should know that food undergoes a certain sort of digestion in the mouth during chewing. The cause of this is that the digestive power is in the inner membrane of the stomach and this is identical with the membrane of the tongue and palate. Thus, as soon as the food is cut up by chewing, it begins to perform its function in the mouth. The saliva assists in this digestion, since there is a certain heat in it which produces a certain amount of decoction. Because of this digestion, chewed wheat brings an abscess to a head quicker than that which is cooked in water or ground.[77]

76 A sign of this digestion is that due to chewing, the food loses a great deal of the taste and smell that was in it formerly. For although the digestion may begin in the stomach when the food comes to the stomach, it is digested there with a complete digestion, not only by the stomach's heat but also with the assistance of the surrounding organs, like the liver on the right and the spleen on the left. For while the spleen does not warm it on its own, it nevertheless does so through the many pulsating veins which come to it from the heat and through the nonpulsating veins which come to it first from the heart and then from the liver. The stomach also warms it using the *zirbus* that surrounds it. The *zyrbus* warms easily due to the fattiness [*sepositas*] that is in it and, with this conceived heat, it warms the stomach from the front, while the heart warms it from above since it comes near the stomach there with the heat of the diaphragm acting as a medium.

77 The digestion of food occurs in the stomach after the food has been mixed with liquid taken in as drink. This needs to be mixed in so that moisture can be generated from all the matter at once to enable it to pass on easily, slipping and pouring itself

---

76. A. here begins paralleling Avic. *DA* 13.27r and Ar. *Part. An.* 650a3f.
77. This belief is repeated at 1.230.

through the veins of the liver. After the digestion in the stomach is done, the stomach absorbs some of the moisture and sends some through the veins to the liver, and the stomach passes on the rest, mixed, through the pylorus [*portanarius*] to the intestines, from which, using the mesenterics, the liver absorbs its moisture.

The mesenteric veins are slender, hard veins connected to many intestines. This moisture passes and flows through them, always getting thinner and trickling until it is separated out into the hair-like veins in each part of the liver. It could never be poured into such small vessels unless it had been rendered passable by the liquid taken in through drink. There is no other use save this for the liquid which is drunk in. The hair-like veins scattered throughout the entire body of the liver come together in the protuberance of the liver. Here they form a large vein and during this decoction, as occurs in everything which is cooked, it happens that moisture is mixed in. A certain thing floats on it, rather like froth, and this is matter for the natural bile. Another settles out in it, earthy and coarse. This is matter for the melancholy. Another is thin and pure and this is matter for the blood. Still another stays incomplete and watery and this is matter for phlegm.

Still another floats, like a heated froth, and this is matter for unnatural bile. Another falls, settling out, and this is heated and turned to ash, and this is matter for unnatural melancholy. Another stays viscous and cannot be separated off by the heat. From this one comes unnatural phlegm. Another stays pure and bounded, and this is the one from which blood is generated.

All these things are done in the protuberance of the liver, in the large vein beneath the heat of the heart [vena cava] or the member which takes the place of the heart in animals.[78] Afterward, the superfluous wateriness is separated off through the *vena emunctoria* [renal vein?] which descends to the kidneys.[79] It is sent to the kidneys with a portion of the blood which is diverted there to nourish the kidneys. This larger vein, which comes from the heart to the protuberance of the liver [vena cava], is divided at the protuberance into two branches.[80] One of these ascends and the other descends and branches out through the members. The blood is borne to the members through these veins and seeps out at the extremities of all the veins. Once it has done this, it is drunk in by the members and nourishes them, as has been determined in previous books and as will be spoken of again in a place to follow.

It is clear from the things that have been said that humor is a moist, flowing body, altered through decoction for the nourishment of the members. Humor is from the substance of food for a sanguineous, phlegmatic, choleric, or melancholic member. Humor is sometimes superfluous, and then the body must evacuate it even though it is not bad or even though it might be convertible into good nourishment.

---

78. The identification is based on 1.597 and the term *vena magna* (Fonahn, 3463).

79. At 1.414f. and 612f. the *venae emulgentes* are the renal veins (Fonahn, 3428). For *emunctoria*, however, Fonahn, 1186, offers "ureters."

80. Cf. 1.581.

Further, some of the bodily fluids are first and some second. The first ones are the four humors of which mention has already been made. The second ones are certain superfluous fluids which are at a distance from their principles and which are indeed powerful enough to pass over into the members but which have not passed over with complete effect. There are three types of these fluids. One is scattered throughout the members like a kind of dew that is drunk in by the members. This fluid is capable of being poured into the members for two functions, one of which is that it can pour into members which undergo some degeneration and the second is that it can moisten the members when for some reason dryness occurs in a strong member.

Moreover, there is another fluid which is nearly coagulated. It is food, altered for the members with respect to the species of complexion of that member, but it has not yet reached completion.

The third fluid is that which enters the members from the principle of growth, the principle and origin of which were from the matter of the semen in the embryo, to the extent that the principle of the embryo was from the humors. Thus, then, both the good and the superfluous fluids have, in general, four types, namely, blood, bile, phlegm, and melancholy.

Blood, which is warm and moist, has two types, namely, natural and unnatural. Natural blood is red and has a good odor and is sweet.[81] Unnatural blood has two types. One becomes unnatural not through the admixture of anything else with it but rather due to the change which occurs in it as a result of heat or cold. The other becomes unnatural through the admixture of some other humor with it and this happens in two ways. One occurs when an internal humor enters the blood and corrupts its substance. An example is when something frothy is mixed into the blood, like bile or the residue of something else, like melancholy or phlegm. It occurs in the other way when one of the things mentioned is generated in the blood from some part of the substance which remained within the blood from its first origin. Both these types vary according to varying mixtures and generations of the various humors so that sometimes a white residue occurs and sometimes a dark or red residue, and to the extent that blood is changed in color by such causes, so too it is changed in odor and taste.

Natural phlegm, the matter for blood and phlegm, is slightly sweet and cold compared to the body, but it is very cold compared to the blood. Galen says that nature did not design a receptacle for phlegm in animals (and especially in humans) as she did for bile and melancholy, because phlegm is necessary for the entire body.[82] Its area of flow is like that of the blood.

We have already said in previous books of this study that there are two reasons for this. One of these is necessity and the other is utility. The necessity is the restoration of the blood and the nourishment of certain members. The utility is the resoftening

---

81. The Latin might first be read, "has a good, sweet odor." But cf. 3.104f. on natural and unnatural blood.

82. Galen *De virt. nat.* 2.9, cited from Avic. *DA* 13.27v.

of the joints lest they become hardened by a lot of movement. For a piece of chalky phlegm is caused by the fact that its thin portion evaporates because of the heat produced from the movement of the joints and the coarse, chalky thing is left behind. The cause of salty phlegm is a balanced mixture of burned, earthy, bitter parts into its watery moisture. If, however, such burned, earthy particles were many and too numerous, they would then make the phlegm bitter and this is just the way that those who work in saltworks make salt.[83] Galen, however, said that phlegm can become salty through putrefaction that occurs alone within it. Nor is this unbelievable, for during putrefaction, as the heat evaporates, the fine moisture is led away and then that which is earthy is burned in the thick fluid that remains, and this becomes the cause of saltiness.

One should know, however, that the passage that is directed from the gall bladder [*kystis fellis*] to the lower intestines is sometimes blocked. Then, because of the lack of bile, the excrement turns white from the phlegm and sometimes the illness called colic is produced.[84] Sometimes, moreover, the bile is burned to ash in itself and its ashiness is joined to its wateriness and what results from this is a very bad thing. Again, unnatural bile is sometimes generated in the liver and sometimes in the stomach. But that which is generated in the liver and is burned to ash is finer. That which is generated in the stomach is green and it is called leek-green [*prassina*] because of this green tinge. Its generation is a result of the burning of the bile which is called egg yolk-yellow [*vitellina*], whereupon the entire burned mass turns dark. Then the natural bile is mixed with it and the whole thing turns green. This leek-green bile is then burned forcefully and turns bright green, like the tarnish on brass. This is the deadly bile, and it is almost poisonous.

CHAPTER VII

Which Is a Digression Setting Forth the Actions of Heat and Cold on the Blood and Fat in the Bodies of Both Blooded and Nonblooded Animals, Especially in the Brain and Marrow

Next, to enhance the understanding of the things that were said on the increase of the heat, cold, wet, and dry, and for the easier understanding of what is to follow, let us present here a chapter on the inner food of animals (that is, on the humor which is the universal food for animals), saying that every animal which is fed either has blood or some other fluid that takes the place of blood.

---

83. Cf. *De min.* 5.2 (Wyckoff, 1967, 240–42).
84. The very name *colica* derives from the name of the bowel it attacks.

Every blooded animal is warmer than one that is not. In the blooded animals, the male especially is warmer than those that are not. Some of the ancient physicians, such as Martynyon and his followers, said that females were warmer than males, and they said that an indication of this was that they undergo menstruation, with the menses returning either at the new moon or at the full moon. But Protagoras, whom they call Abrokaliz corruptly, said what was contrary to these and in this he spoke the truth.[85]

Further, some of the ancients held the opinion that blood and bile are cold humors, but in fact it is agreed through their very effect that these humors are warm. For the blood provides growth, a thing it could do only through the action of heat, and bile stimulates and opens the veins, which it would try to do in vain unless it were biting and stimulating with its heat. Thus did Pitagoras say well that that which stimulates is hot, since it is composed of sharp (acute) angles.[86]

As we said before, a thing is called hot in many ways. That is hot which is warm to the touch, like fire or something set on fire. That is also called hot potentially which, when it is taken into the animal's body, is aroused by the natural heat and then gives heat. It is also called hot in the many ways we spoke of before. Thus, a thing is sometimes called hot which, having been multiplied in its species, warms something when it is ingested even though it might not warm when only a small amount of its species is ingested. But if it is just a bit of its species it does not warm. Thus, parsley is called hot since a great deal of parsley warms when taken in, but a small amount of it does not. A thing is also sometimes called hot when it is slow to lose the heat it has taken on, like lead. Oppositely, when burned, ashy bodies are heated, they produce fire in themselves, but when they are washed they are made cold again.

Something is also called hot both essentially and accidentally. Likewise, a thing is called dry essentially, as in a rock, and is called dry accidentally, as in ice.

Enough has been said about these things above, but they are repeated here so that their effects in animals might be known. For aquatic animals that have thin, watery blood are very timid.[87] Those with thick blood are bold, since that blood holds more heat, just as warm earth holds it longer than does water. This is why pigs, camels, oxen, and lions hold the heat of anger a very long time, and it is the same for a person with thick blood.

---

85. Much of this chapter is based on Avic. *DA* 13.27v. Avic. cites "the first teacher" (*magister primus*) and then cites one *Marinum* and his opponent, *Abocalim*. The original of the text, Ar. *Part. An.* 648a29f., cites Empedocles and Parmenides, the last half of whose name may be the source of *Martynyon*. The present passage is a doublet of 12.32f., in which Parmenides emerged as Borgyan while Empedocles came through the process unscathed. Protagoras, of course, was a sophist during the same period as Socrates.

86. The origin of Pitagoras is unclear, as no parallel exists in the Gr. One must wonder if A. did not supply the name, spurred both by the mention of the like-sounding Protagoras above and by the mention of angles. Another possibility is that he supplied the name because of the doctrine elsewhere attributed to "Pythagoreans," namely, that the natures of all things must be explained in terms of mathematicals like the curved, the straight, line, figure, and even acute angles. Cf. *Phys.* 2.7.8.

87. "Aquatic": *aquosum* should, by normal usage, mean "watery." But the word is an import from Avic., who tends to use it to mean "aquatic." This interpretation is assured by comparison with 12.32.

Moreover, an animal that does not have blood has neither *sepum* [fat] nor *zirbus*. There is earthiness in *sepum* and *zyrbus* and earthiness causes things like these to congeal. *Sepum* and *zirbus* are especially present in an animal that has no teeth in its upper jaw. For the *sepum* of such animals solidifies faster when melted since the animals are very earthy. An indication of this is that their horns and hooves have a great deal of earthiness. The *sepum* of other animals does not solidify this fast. For fat is a product of watery, cold blood, and is easily solidified in the cold. Thus, when it becomes too abundant in the body, it extinguishes and solidifies the natural heat and kills the animal. This is also why bones and *sepum* do not have feeling, for they are a sort of solidified blood, deprived of heat, since their digestion is not like that of flesh. Cold is therefore strengthened in such things and afterward will exert its power over the body, and death results.

Moreover, it happens for the same reason that, after fat increases in the body, generation is curtailed by the chilling of the blood. Further, the marrow in the bones is a result of the dissolution of the thin and watery blood which does not come to the perfect digestion. For that which is perfectly digested from the blood is that which obtains the appearance [*species*] of flesh. Marrow, however, somewhat resembles sperm in its essence. The marrow of a child is bloody, but that of youths is generated from sharper blood than is the marrow of old men. Marrow is the sustenance and the wetting agent for the bones and is generated from the food that is superfluous to the bone and which is strained from the body of the bone, from the inner part of its hollow. And yet it is the nourishment for the bone. However, these are not contrary things, namely, that a single thing may be said to be both the superfluity of the bone's food and the food for it. For since there is a great deal of marrow some of it will be able to be converted into food. Generally speaking, marrow is blood which has been altered to the nature of bone. An animal which has narrow bones and many serrated teeth does not need sustenance and nourishment for the bones. This sort of animal has no marrow in its bones. An example is the lion, and its great heat is of help in this as well.

An animal that has no bone has no marrow, except for the *nucha* [spinal cord], which is surrounded by the backbone.[88] And this marrow of the back, although it is the basis for the nerves that are proceeding out from the backbone, is still strong enough for a few other things. For it is a sort of connector, rather after the fashion of a ligament, for the vertebrae. Likewise, the vertebrae's function is not only to cover the *nucha* but also to be a point of origin for the bones upon which the body is based. Some have held the opinion that the nature of the *nucha* is the same as the brain's, since it is connected with it. This, however, is an error, for the nature and the com-

---

88. As in the anatomical sections, the phrase *spina dorsi* will be translated here as "backbone." This usage is noteworthy for being plural in form, probably the result of A.'s addition to his received text. The sense remains singular, however.

plexion of the brain are cold, but the spinal marrow is hot.[89] An indication of this is its fat. It acquired the heat it has from the heart, which is situated in front of the spine. It is not dried out by this heat because it receives constant moistening from the brain.

91 Some have held the opinion that the substance of the brain is sentient and that it has the power of touch.[90] This is false, however, for its nature is just like that of the marrow in bones. The brain feels no pain from an abscess in its nature [*natura*], but rather from one in its pellicles.[91] Now it must not be denied on this basis that there is something in the brain in which the powers of the senses reside and in which are the vital and the animal spirit. Nevertheless, this is not a thing which is sensitive to the sense of touch. Quite similarly, the brain is the principle of sight and yet does not see. It is the principle of the power of voluntary movement, and yet it does not move. Surely, the principle of these powers is rather the spirit which is in the brain, and the brain itself is a receptacle for the powers which are borne on the spirit and which flow to the members connected with the marrow of the brain. This is just as the skull is itself the receptacle of the medulla of the brain.

92 Nor is it necessary that a receptacle and a point of transit (that is, a pathway) of a power share the same power. Thus, it is not in the two hollow nerves which are called the optic nerves that the visual powers arise, but in the brain, for the brain balances and tempers the heat of the hot spirit, making it visual, and does so for that of the other senses as well, for the brain balances the sensitive and motor spirit. Thus, too, the vital spirit is in the heart as in its receptacle and point of origin, and it is made balanced for the pulse that is in it.

Or perhaps every spirit arises naturally from the heart and still is tempered in the brain and is made balanced for the operation of sensation and motion in it. In the liver, however, this is done for the operation of the natural powers. For the purpose of the brain is not to sense, but to have the sensitive spirit within itself, as in something that balances and tempers it, and thus it need not have sensation itself.

93 Galen testifies to this effect, saying that only that has the sense of touch which has balance with respect to the complexion of tangible qualities. This same one says that the complexion of the brain departs from balance and tends toward coldness and that the result of this coldness is its insensitivity. For although the heart tends away from balance as well, its tendency is, nevertheless, toward heat, and greater heat (as long as it does not corrupt the complexion) is a cause of greater sensitivity. Nor is the heart kept from sensing by the heat, but rather it is hindered when something happens to it with respect to the instruments which serve it, such as the spirit. For when this is chilled, it greatly interferes with the heart's sensitivity.

---

89. This paragraph provides an interesting example of the medieval natural scientist's lack of a consistent vocabulary. Within the space of a few sentences, the spinal cord has been called the *nucha,* the medulla of the back, and the spinal medulla.

90. Lit., "the sensitive power of tangibles."

91. Cf. 3.142a, on the lack of sensation in the brain and its "vicar," the spinal cord.

The best thing that can be said in this matter is the statement that the first thing acting on sensation is spirit and that it in no way follows that the receptacle of the spirit or its protection or that which tempers it and balances it for sensing should be in any way sentient. It should rather receive the complexion and the substance which can balance and temper the sensitive spirits for the operation of the senses. Therefore, the best thing to say is that the sentient member is that which is the most balanced, like the flesh, and that a substance which is excessive in its cold and watery fluid, such as the brain, will not be able to possess sensation. The spirit, however, is suited to bear the powers. It is hot because cold dulls the powers.

If someone should object that inhaling is cold, and that this nevertheless strengthens the spirit, we say that inhaling does not have the purpose of balancing and tempering the spirit, but rather that of keeping the spirit from becoming too hot. For if it were strengthened, it would dissolve the spirit and reduce it. It is therefore cooled by it and kept from superfluous evaporations. Therefore, of the sensory members, the most balanced is the most sensory. The brain is the coldest, however, so that at a balanced distance from the heart, it can hold its heat and boiling in check.[92] It also helps the spirit, which comes to it from the heart and is too warmed, to move toward a certain balance of temperament, a balance which is suited to the movements and operations of the brain's powers. For the motor and sensitive power comes from the heart to the brain, borne on spirit. The spirit, proceeding from the heart into the substance of the brain, rectifies the brain's substance in many ways, such as in being nourished, growth, and many others. For all these are due to the power of the heart.

The spirit, then, insofar as it proceeds from the heart, has many powers and activities. Once, however, it has entered the liver or the brain, it is adapted to the complexion of the liver or brain and it then has only the power suitable for that member. Nor is it hindered by any other power coming from the heart, for it is always adapted to the activity of this member. All of this is caused by the heart. For since there is one soul in one body, the soul is adapted to one prime member which is the principle of all the others and is placed in the middle of them. The heat which is in it is the instrument of all the powers which it draws into all the members. These powers are determined and specified according to the balance and complexion which the spirit receives in all the members. These, then, are the things we wish to say about the activities of hot and cold.

---

92. Boiling in the body entails the idea of bubbling up and producing froth, but heat is implied. Cf. 1.597, 611; 3.110, 115.

The Second Tract of the Twelfth Book on Animals,

In Which the Nature and Cause of Uniform Members Is Treated

CHAPTER I

On the Nature and Cause of the Blood, Which Is the Nature of the Uniform Members

⁹⁷ Let us move on now to the opinion of Aristotle and say that everything that grows is necessarily fed on dry and wet, with respect, that is, to the matter of its nourishment.⁹³ Digestion of this sort of nourishment is accomplished through the strength of the heat. It is therefore necessary that there be some principle of natural heat in all things, plant and animal, which grow. This would be so for the reason we just gave, even if another reason for it should not be found. But this is not the case, for in animate objects heat completes almost all the operations, and this is clear by the things said in the foregoing chapter.

⁹⁸ There are very many instruments (that is, organs) suited for using food. The first to be thought of as suited for using food is the mouth and the things in it such as the tongue, teeth, and the like. This is especially true for those animals which must have their food cut up into small pieces when they take it in. Now this cutting is not, to be sure, the cause of complete digestion, but it is still a reason the food is digested more quickly and that the pure liquid in it can be absorbed on all sides by the mesenterics, since the food has been cut and ground thoroughly in this way. The upper belly, however, which is the stomach, and the lower belly, from which the mesenterics draw, digest the food with complete digestion, using the natural, digestive heat. It is much as we have said very often in the previous books.

The mouth is the pathway through which food enters into the place of digestion and when it enters it is undigested, even though digestion does begin in the mouth. It is the same for the *mery* (that is, esophagus) through which (in animals that have an esophagus) the food passes after the mouth. For food comes to the stomach through this passage.

⁹⁹ There are other members as well which are principles of the entrance of the food in this way. There is the crop in birds, the upper opening of the stomach [cardiac orifice?], and the like. These must be the principles of the taking in of food, since the

---

93. Ar. *Part. An.* 650a3f.

entire body takes in food by means of them. Trees take in cold food through their roots from the earth. It is completed in that digestion which separates the pure from the impure. Plants absorb cold food from the earth and thus food for plants does not have the superfluity of dung or urine to cast off. Because plants only absorb a moisture that has been completed for the plant's nourishment, nature gave them no mouth opening like that in animals. Rather she gave them small roots, which are analogous to the mesenteric veins in animals. The earth and the heat enclosed in the earth from the rays of the sun and stars act as the belly does in animals. For with this heat the pure nourishment is raised up toward the roots of the plants and the impure descends away from them due to the weight of its own earthiness. All perfect animals, and especially those which walk, have a belly in which the pure is separated from the impure. From this the entire body takes its food through the veins or other pathways that take the place of veins. Thus, the roots of the tree take in pure food from the earth through their roots, and, as if through mesenteric veins, the food travels to the entire tree, through the trunk, branches, and twigs, until the digestion of the plant's food comes to its final and complete end from the heat of the sun and the plant.

Just as the operation of the mouth is necessary to project the food into the belly, so must there necessarily be another, internal member, which takes the food from the belly through absorption. This member is the mesenteric veins, which stretch to the intestines midway between the pylorus [*portanarius*] and the rectum [*longaon*]. They begin to absorb below the belly and are drawn separately throughout the entire body, coming from the heart and liver, just as is known through our discussion of anatomy, treated at length in our previous books. In all animals, moreover, there must necessarily be a member that receives the food and another member which receives that superfluity of it which is impure and must be cast off. 100

Veins are certain vessels for the blood through which the blood or some fluid which takes the place of the blood flows to the members. From this it is clear that blood is the last or the purified and digested nourishment which is within the vessels and which flows to the members. This is in animals that have blood, but an animal that lacks blood is fed by an analogous fluid which takes the place of blood in it. An indication of this is that when an animal is not fed enough of this food, it becomes thin and has a bad disposition. But when it is fed enough, it grows and takes on a good disposition. If this food is clear and good the body will be healthy, and if it is bad the body will be sick. Therefore, the blood in animals is for nothing other than food, as is clear from what has been said. It does not, therefore, have sensation, because it really is not a member in act, but only potentially. So, too, the other superfluities which are not converted into body do not have sensation. For the food does not have sensation like that of flesh. For when flesh is touched, it clearly has sensation. The blood is not connected with the flesh, but is stored up, as if in a vessel, in the heart and veins. The way in which the members receive sensation from the heart and how they are fed from it has been well discussed in other, preceding books in this study. There we said that the generation of the members is from the heart. It is enough here 101 to say that the blood exists for no other reason than to be food for the members and we need only speak here of the nature and utility of the blood.

102 Further, in the blood of some animals there are found certain fine threads shaped like hairs and stretched throughout the blood. But they are not found in the blood of other animals, for example, in the blood of deer and certain others. This is why the blood of such animals does not coagulate. If, however, blood of this sort is very earthy, it does coagulate because of the lack of moisture which is in it. Further, the intellect in the human and the estimative power in other animals are purer and better or less pure and worse according to the dispositions of their blood and of that fluid which takes the place of blood.[94] It is not due to the coldness of the blood but to its greater or lesser thinness and clearness. When it is thin and clear it lacks earth, for which reason those possessing a great deal of earthy blood are heavier and duller in their conceptions and activities of the soul.

An animal with a thinner, purer, natural moisture has better sensation. For this reason, too, many animals that lack blood are more refined in their perceptions of sensibles and in certain other activities of the soul than are certain other blooded animals. Examples are ants, bees, and the like, the goodness of whose perceptions, as we said, is not caused by heat and cold but rather is due to a fine and clear or a coarse and opaque natural moisture.

103 Animals whose natural humor has a great deal of moisture and can be quickly cooled are more timid since fear is due to a departure of blood, heat, and spirit to the heart and this chills the body. If the moisture is then solidified, the heat does not quickly return to dissolve it. Thus, such animals stay frightened and unmoving for a long time and an animal whose heart has this sort of complexion, namely, one with cold and watery moisture, often suffers at the slightest provocation this accidental sort of fear. For cold water, especially that which is viscous, solidifies quickly. For this reason too, a bloodless animal is more fearful than one with blood. And the result is that such timid animals are also immobilized, moving only slightly due to the strong passion and solidifying action caused by their fear. Some of these bloodless animals change their color when beset by fear, due to the great loss of heat and spirit in their outer members.

104 Further, the blood of some animals sometimes has small, hair-shaped bodies stretched out through it. These animals have earthy blood which, when heated, retains heat for a long time. Therefore these animals are very irritable, they hold on to their anger for a long time, and they are also bold. Their wrath moves the heat because of the desire for revenge, during which the heart moves as in *dyastole*, blowing out heat, blood, and spirit from itself. For such blood that stays is heated for a long time because earthiness holds the heat for a longer time than does the moisture of water. The thread-like, earthy bodies in the blood of such animals are rendered like charcoals in a fire during a period of anger. This is the reason for the common saying that the

---

94. On the estimative power (*estimatio*), cf. 7.2, with note, and 8.227.

bull and the pig are very wrathful animals, with an indiscriminate and burning anger. For in their blood there are many such thread-like bodies, especially in the bull's blood. An indication of this is that its blood solidifies faster than all other blood.[95] When these parts are strained out of the blood it does not solidify at all. For the earthy part, taken out of the watery moisture, brings it about that the moisture neither solidifies nor dries out after that. Moist mud solidifies easily in the cold, for the heat is pressed and forced out of it in the cold. The moisture will go out along with the heat, since it is the subject of the heat. Thereupon the earthy dryness sets up and solidifies, as we said in foregoing books on natural philosophy, especially in the fourth book of the *Meteora*.[96] This, then, is the cause of its solidification (that is, its coagulation). It does not coagulate from heat but rather from loss of moisture.

There are also other fluids in the bodies of animals, but the humor of blood is greater and is more abundant than all the other bodily fluids. This is as it should be, for it is the body's food, and there should be a lot of it in the animals' bodies. For the food is the matter for the whole, flowing in and out of the body. Blood, as we said, is the ultimate food, and therefore it is right for there to be a lot of it. Within blood there are great differences of heat and cold, caused either by the fact that it is not adequately digested or, perhaps, because it is corrupted by some accidental trait, which we have enumerated in what has gone before.

105

CHAPTER II

On the Nature and Cause of Fat [Pinguedo] *and* Zirbus *and on the Cause of Marrow*

Zyrbus and fat differ according to the superfluity of the blood. Each of these is blood with some decoction, since it is from food that is more quickly and better convertible.[97] That which, because of wateriness and airiness, is not adequately rendered like and assimilated to the last food of the animals and which has easy decoction and conversion becomes fat, *sepum*, or *zyrbus*.[98] An indication of this is that fat and *zyrbus* are only found in animals whose body is fertile—that is, easily nourishable and having a good disposition. For part of fat is shared with air and fire, for it quickly takes on heat

106

---

95. For this reason bull's blood was considered a poison in antiquity (Kitchell and Parker, 1993).
96. *Meteora* 4.2.6.
97. Ar. *Part. An.* 651a20f. begins to discuss the differences between *pimelē* and *stear*, terms that have traditionally been translated "soft fat" and "hard fat." Peck (1961, 140) opts for "lard" and "suet." In any case, A.'s usages are so complicated that it is always better to reserve "fat" for *pinguedo* and to keep his many other words for fat in the Latin, remembering, however, that *zyrbus* is often the omentum.
98. "Rendered like": *affinitatur*, an extremely rare verb, if the text is not in error. One might read instead *affinatur*, "is purified."

and retains it forcefully. This is the reason that these things are not found in bloodless animals.

107 An animal, however, which has thick blood, nearing the earthiness of its body, has *zyrbus*, for *zyrbus,* as has been stated above, is earthy. This *zirbus* solidifies like that blood in which there are the many fine, thread-like parts about which we spoke. Also, a broth in which the flesh of such an animal is cooked will congeal and set. For there are many earthy parts in such animals and few watery parts. For this reason we see that *zyrbus* is especially abundant in the bodies of animals that lack teeth in their upper jaw and are horned. For this animal is clearly filled with an earthy element since its horns and hooves are generated from an element of this sort. An animal, however, which does lack teeth in the upper jaw, but which lacks horns, is also earthy, but not to such an extent. It thus has the fat called *sepum* in place of *zyrbus*. When it is dried out, this fat is not as friable as *zyrbus* because its nature is not as earthy.

108 When there is but little fat of this sort in the members of an animal it will be good, for it does not hinder good sensation and it leads to health and strength. When, however, it increases, it is very harmful and might even corrupt the animal, since if the entire body is covered with fat or *zyrbus* the animal necessarily dies, for then the fat and *zyrbus* cover the member in which the sense of touch is.[99] They are without sensation, as is the blood from which they are generated, because, as we said before, they are nothing more than blood of easy decoction and conversion, which is not assimilated to the members which are members in act. It is therefore clear that if the entire body either becomes this way or is covered with such things, it will lose sensation and die. Thus, a very fat animal ages quickly, for its blood, since it is passing over into fat, is scanty. Old age and corruption are due to nothing save a paucity of blood that is well assimilated to the members, for everything which has little blood has many corrupting flaws since it is quickly and easily changed by cold and heat.

109 Further, an animal that has a lot of fat has little generation and copulation or perhaps none at all for the same reason. For the blood which would have become sperm turns into fat before it is assimilated to the body and descends to the seminal vessels [*vasa seminaria*].[100] This, then, is what has been said about the blood, wateriness, fat, *zyrbus,* and their causes.

The marrow which is in the bones has the nature of blood. It is not true, as many of the ancients have believed, that it comes from the power of the sperm from which the embryo is generated. This is clear in youths whose uniform bodily members are sustained and fed on blood. The marrow, however, as we said before, is the superfluity of the nourishment of the bones and is also their food. But the food of embryos is menstrual blood and the marrow in the bones is blood which is superfluous to the food for their bones. The color of the marrow of embryos and children is bloody, but

---

99. That is, the skin.
100. For this term, cf. 1.61.

it matures and grows white as the children grow. During the growth of animals every hot thing takes on change and the marrow of the youths is changed in the same way.

110 The marrow of a fat animal is very fatty. When the animal's blood is well digested, it does not become fat or *zirbus,* as is clear from what was said before. *Zyrbus,* however, is blood that is digested more than in the case of fat. Thus, marrow, which is from blood that is better digested than is so for fat, is like *zirbus.* For this reason, too, the marrow of a horned animal with teeth only in its lower jaw resembles the *zirbus* of that same animal. But an animal with teeth in both jaws has marrow that is soft and resembles fat. The marrow of the vertebrae is like this because it must be continuous and pass through all the vertebrae. If the marrow of the vertebrae were like *zirbus,* it could not be continuous throughout the entire length of its passage through the vertebrae because the hardness and the intervening membranes would interrupt it. Moreover, it would be moist when it was warm and friable when it was cold.

111 Further, certain animals do not have much marrow. Specifically, these are the ones with strong, hard, thick bones, like the bones of the lion in which there is only a small amount of marrow. For this reason, too, some have felt that there was no marrow in them at all, especially in the neck bones. But it is entirely necessary that in a perfect animal there be either bones or members that are analogous to bones and which take their place (for example, spine in an aquatic animal and in serpents), and there must be marrow in these bones and spines since some of the food from which the bones are generated must be retained (even though it is superfluous in the bones and spines) to assist in their being kept moist and in their nourishment. It has already been determined above that blood is the food for all the members and that from it fat and *zyrbus* are generated. Marrow, moreover, resembles the generation either of fat or *zyrbus.* Thus, marrow must necessarily be either fatty or *zyrbus*-filled in taste and touch, for it is digested from the heat of the bones which is suited to it and, as has been said above, digested blood also, on its own, turns into fat or *zyrbus,* just as on its own a broth in which meat has been cooked sets.

112 It is reasonable, however, that there is no marrow in a thick, strong bone, and that if there is it is minimal.[101] For the food in such bones passes over—either entirely or for the most part—into the bones. In aquatic animals and in others which lack bones, possessing spines in their place, the only marrow found is that which passes through the vertebrae. For such an animal naturally has little blood, and the only place the marrow has in such animals for its generation is as much as extends through the vertebrae. For in this place these animals have need of marrow to strengthen the nerves and ligaments which arise from the vertebrae and to which the ribs are attached. This is the reason that the marrow of the vertebrae has a different color than the other marrows.[102] This is clear from the things determined in the earlier parts of this study.

---

101. Thus, at 1.286, A. specifies that the bones of the wrist lack marrow to enable them to be hard.
102. At 1.257, the marrow of the vertebrae is called "white marrow."

Marrow is viscous and nerve-filled since it is only in the vertebrae in the place of and in the stead of the brain, passing through the inside of the vertebrae.

In this way, then, we have declared the reason why there is marrow in the bones of the animals that have marrow, what marrow is, and that it is the superfluity of the blood from which the bones and the spines of animals are fed. This, too, is the reason that it is kept inside the bones and that it is only generated from the blood of a good, easy decoction.

CHAPTER III

On the Nature of the Brain, According to Aristotle's Teaching

113 A discussion now follows on the cause and nature of the brain. As we have said above, many of the ancients held that the marrow of the vertebrae, called the *nucha*, is connected to the brain and has the same nature as it does. But the truth contradicts these men, for the nature of the brain differs from that of the marrow of the *nucha*. The brain is very cold, more so than all the other inner members. But the nature of the *nucha* is hot, an indication of which is found in the fat in it. The reason the marrow of the *nucha* is connected to the brain is that nature always strives to order the members in the best way possible. Since, then, the brain is cold, she connects to it a member of a contrary disposition so that in this way she might give it a power that assists it and expels the distemperateness which arises from the quality of coldness in the brain and from the quality of heat in the *nucha*. She therefore places one near the other.

114 It is clear for many particular reasons that the marrow of the *nucha* is hot, especially from the reason given based on the coldness of the brain. For otherwise nature would not reduce the order of the members to temperateness unless she placed the hot thing near the cold and connected it to it. An indication of the brain's coldness is that the brain is very cold to the touch and that it is lacking more in blood than in all the other fluids of the body. For almost no blood is visible in it. In addition to this it is very dry. But the brain is not a superfluity, being in fact a principal member. Neither is it in any way one of the connected and joined members. Instead, its nature is a certain nature which is eternal as long as the animal survives. It is reasonable that the nature of the brain is this way, because it is the point of origin for and the receptacle of the activities of the animals. Moreover, the brain has no connection with any of the sensitive members, as is apparent to the eye during dissection [*anatomia*].

115 Again, the brain has no sense of touch whatever, just as blood or some other superfluity of an animal does not. It is in the animals' bodies for no other purpose than that of preserving their nature. Nor is there truth to it when some of them pretend that the soul is fire or some other corporeal power of an element which is like the heat of fire.[103] They say that this is in the brain and that for this reason the brain is hot. The

---

103. Peck (1983, 150) identifies the ultimate source of this belief as Democritus.

argument against these people is similar to that in *On the Soul*.[104] It is better to say that the soul is the principle and cause of the body (that is, the sustenance), containing it, and that it is this way in this sort of body which is hot and possesses the other powers of the elements. Or it is better to say that the soul is the sort of cause as was mentioned belonging to this sort of body and that these qualities of the elements are its instruments. Thus, to say that the soul is fire has the same force as saying that the saw is the woodworker. For these reasons, and those like them, then, the soul must necessarily be an act of the body, having the heat of the fire that is in the body just as a mover and an artisan does in a power of this sort.

As we said previously, all natural, mixed things need a contrary juxtaposed to them to temper them so that there will not be some sort of distemperateness in them. Nature made the brain cold and placed it opposite the heart, which is the hottest member since a member which departs from temperateness is not preserved on its own without the opposite effect of another member tempering it. For this reason, wise nature cleverly placed the brain opposite the heart, as we said. For the nature of earth and water was dominant in the brain, these being very cold elements, and it was untempered. This mutual tempering [*contemperantia*], then, which is the result of the ordering of the members, is the reason that every blooded animal has a brain. No other animal, except for the bloodless ones, has a brain. They might have a part that is analogous to the brain and which performs the brain's functions. Thus, according to Aristotle, the octopus [*multipes*] and the animals like it are very hot animals due to their lack of a brain.[105] For the brain corrects the complexion of the heat and the boiling action of the heat and the heart so that the heart might also have tempered heat in its two veins, namely, the vena cava [*vena maior*] and the aorta [*adorti*], as we said in the section on anatomy. For these veins, throughout their branches, come to the web that contains the brain. This arrangement keeps any harm from befalling it from the heat which would occur if the hot veins actually penetrated its substance. Thus, these veins hold it in with their interweavings. They are thick and slender and are in place instead of a few large veins.[106] Their openings nonetheless pour out the blood to it and nourish it.

This blood is clean, clear, and thin and is not thick and cloudy. For if this blood were colder than a temperate complexion, or even if the brain itself were, catarrhs would immediately befall it. For the vapor of the food, ascending upward and traveling through the veins, comes to the brain. When a vapor of this sort is overly cooled, catarrhs immediately occur because of the strength of the cold which is in that place. This catarrh is phlegmatic from the watery, cold blood, and when those things that occur in a human (who is a microcosm [*minor mundus*]) are compared to those which

---

104. *De anima* 2.2.2.

105. *Multipes* is very often a general term in A. denoting, simply, "multipeds." But it is often given as a synonym of his more normal name for the octopus, *polipus* (4.64, 84; 5.19; 8.122–26). Octopi are known for their intelligence, being capable of working out fairly complex problems.

106. A.'s additions to the text have muddied it. Ar. *Part. An.* 652b31f. clearly states that instead of being few and large, the vessels are many and small.

occur in the macrocosm [*maior mundus*], this accidental trait will occur, like the generation of rainstorms from vapors. For the vapor from the land and water, the lower elements, rises upward and stays suspended in the air above (which contains this air surrounding the earth) and it is converted by the coldness and falls in the form of rain.

Now, then, the cause we are giving for the illness which is called catarrh is, to be sure, insufficient, but it is sufficient for us in terms of the natural philosophy we are now discussing. A physician, however, has to speak about all its types and causes.

118 Moreover, sensation is in the brain. However, the brain is in truth only found in a blooded animal. In bloodless animals there is a member analogous to the brain and which is analogous to it in its functions. For it cools the food which comes to it from the rise of the blood or the humor which stands in place of the blood. This can exist in a member which stands in place of a brain for reasons similar to these, such as the coldness and heaviness of the head.

In every animal, whether or not it has blood, the head is heavy during sleep, a thing that can be due only to cold. For when the food is cooled, the heat leaves with the heavy, cooled blood as it makes its way downward. Because the heat is driven inward and downward, sleep will result. Thus, an animal with an upright body cannot keep its head up during sleep, especially one with a heavy head like the human or pygmy. Likewise, quadrupeds with heavy heads lay their heads down during sleep. But we have spoken on these matters carefully in the book we wrote *On Sleep and Waking*. There we made a complete distinction as to the reasons for sleep.[107]

119 Moreover, the brain is composed of earth and water, as is proven by the fact that it becomes dry and hard when cooked. For then the watery part of it dissolves in the heat and the extensive earthy part that remains grows firm.[108] When the moist part is set free from a part of this sort, it stays very hard. Now the human has a very large brain when compared to the brains of the other animals and a man's brain is larger than that of a woman.

120 The area near the heart and lungs is very warm and it causes the body to be upright by propelling the spirit and blood upward. Thus, the human, since he has these hot places, has an upright body. For the human has a hot complexion and the heat and the spirit, making their way upward, produce greater size in the upper members and an upright stance. Set contrary to this heat of the heart and of the complexion are the moisture and coldness of the brain. Due to the great amount of cold and moistness that is in the brain, the part near the brain, namely, the bone which is at the first suture of the *sinciput*, does not harden quickly in young, for the vapor and the heat that come from the heart and the spiritual members exit through this area.[109] This therefore only happens in animals that have blood and a head with many sutures, such as the human.

---

107. Cf. *QDA* 4.1.
108. The verb is quite graphic, being used also for puddings that "set."
109. Cf. 1.111, 116, remembering that A. had difficulty with the terms *occiput* and *sinciput*.

This accidental trait is also found analogously in the females of humans, for the same reason, namely, so that the heat can "breathe" [*respirare*] better, especially when the brain is very large compared to the size of the body. If it should happen to be consolidated and harden faster than it naturally should, it will not complete its work perfectly. Rather, in such cases, the vapor is raised up and, not being able to breathe out [*exspirare*], becomes cold, thus chilling the head and the body. Illnesses are the result of this, as are madness and loss of intellect.[110] For the heat of the heart and of a first member is great and it is easily harmed by the other members when they are blocked and cooled.[111] Its sensation, which is very refined and forceful, changes quickly when something comes to it of the chilled blood that surrounds the brain. We have, then, spoken this way about the fluids generated in the bodies of the animals so that no thing about their shared dispositions escapes us.

CHAPTER IV

Which Is a Digression Setting Forth the Causes of the Disposition of the Brain According to the Modern Peripatetics and According to Experience and Reason

These, then, are the things said following the ideas of Aristotle on the nature and disposition of the brain. But, with an eye for greater knowledge, one should know that the reason nature made the brain in animals is definitely double, specifically, from necessity and from utility.

The reason from necessity is itself double. One relates to the temperateness of the composition of an organic body, the other to its activities. We have already touched above on the one that is due to the tempering of composition, and this is, namely, that the heat of the heart and the liver be tempered, lest they burn the upper members and dissolve their composition. For the hot and cold powers of an animate being should be mixed together in the way that the powers of Saturn, Jupiter, Mars, and the sun are mixed in the upper members to produce temperateness.[112]

Thus, just as the heart, liver, and lungs are made excessively hot, they correspond to the powers of the sun, to which the heart is proportionate; the powers of Mars, to which the liver is proportionate; and the powers of Jupiter, to which the lungs are proportionate. It was necessary, then, for one cold, moist thing to be positioned above these, proportionate to the moon, and which would bring them to temperateness.

---

110. One of the earliest operations known to us is trepanning, dating from Neanderthal times, in which a hole is bored into the skull to relieve what we would call pressure but that certain other cultures call bad spirits (Favazza, 1996, 88–91).

111. On a first member, see 1.75; 3.145. There it is apparent that the three principal members of the body—the liver, heart, and brain—can also be viewed as first members.

112. This spot marks A.'s third term for temperateness: *temperamentum, temperantia,* and *temperies.*

123 The force [*vis*] of the seminal vessels is proportionate to Venus and the powers [*virtutes*] of the vessels are proportionate to Mercury because of their great and multiple conversion.[113] The spleen is proportionate to Saturn in straining out the melancholic humor, in both moving and drawing it. For necessity, which is for the sake of an activity, exists because animal operations need a clear, shining, and abundant spirit, so that they can have more multiple powers. The spirit of a hot member could be nothing other than smoky and cloudy, and therefore there was need for great cold in which the spirit which represents the animal forms could be clarified, thinned down, and refined, and for this purpose the brain was made cold and moist. It is moist so that it can receive the spirit and cold so that it can thin down and clarify it.

124 The cause based on utility is determined according to well-being. For otherwise the organs of sense would not be well kept since, if the brain were hot, it would constantly cause melting in itself and the cold humors would flow from it into the eyes, ears, and nose area and would destroy the activity of these organs.[114] There is another most powerful utility, namely, that when the vapor is chilled in the brain, it brings on sleep, which is the natural rest for animals. But enough has been said about this in the book *On Sleep and Waking*.[115]

The brain was made cold, then, for these reasons.[116] I say that every animal has a brain or a member that stands in place of a brain. Those that have blood have a true brain and even some that do not have blood have a brain, such as certain marine animals and a segment of the animals called *malakye* [cephalopods] in Greek. But of all the other animals, the human has the largest brain. Since the human has the most powers and animal operations, it is necessary that the brain be this size in the human.

125 The disposition of the brain in the human is that part of it tends toward marrow and the rest toward pellicles (that is, membranes). A third part tends toward certain ventricles filled with spirit. Nerves stem from the brain rather like branches, not as if they are of its substance but rather as if they extend up to it as they leave the heart. This is as has been proven elsewhere. The entire head has three ventricles lengthwise and it has a barrier rather like a line drawn lengthwise through the ventricles and membranes. There is a benefit in this empty spot, as we have determined above, although the sensitive spirit is more apparent in the front ventricles.[117]

126 The substance of the brain is created smooth and oily.[118] Its oiliness is so that it can cause oiliness and viscosity to flow into the nerves that are connected to it. Accord-

---

113. The phrase *propter multam et multiplicem sui conversionem* is difficult to interpret because of A.'s recurring problems with pronoun usage. If *sui* refers to Mercury, the phrase may refer to the planet's quick rotation. If it refers to the male vessels, it may recall their convoluted nature. Neither option is strictly grammatical.

114. "Nose area": *olfactum*, an odd use.

115. See *De somno et vigilia* 1.2.4f.

116. A. now begins to follow Avic. *DA* 13.28r.

117. Cf. 1.518, 527f. These benefits are spelled out at 12.131.

118. Cf. 1.518ff.

ing to Galen, the smoothness (that is, softness) exists so that, as a result of its smooth shape and its soft quality, it might be employed easily for sensible and imaginative forms.[119] This is because that which is moist and soft easily receives images [*imaginationes*] and the impressions of forms. This, however, does not seem to be entirely true. For while that which is moist does readily receive forms, it does not do so in every way. Rather, it receives them with movement and a certain division of the bodies that enter it and divide its parts. Imagination, however, and other sensitive powers are not produced by bodily movement nor by any division which occurs in it.

Let us say, then, that the cause of its softness is so that it might soften the hardness of the nerves and might gradually present good nourishment to the nerves. For a soft thing presents a better substance and nourishment to something viscous than does a hard thing. Thus, it also happens that the nerves coming from it are softer when near it, where they receive their powers, and then, as they proceed away, they gradually harden and become strong. Moreover, a soft substance is more abundant in spirit, which it dispenses to the motor and sensory nerves. Moreover, its substance was spongy and cold so that it could be lighter and filled more with spirit, for a soft, spongy substance is lighter and frothier than one that is solid and hard.

127

Further, the substance of the brain was made diverse and in parts. The front is softer and moister but the rear is harder and drier. There is a hard membrane dividing the brain into sections. The front section is softened by the marrows which are there in the hollows of the bones. This is because many sensory nerves come from the front part, especially those of sight and hearing, and these are naturally moist and soft. It is right that vision is to the front, for the protection of the entire body is given to vision as if to an overseer.[120] The motor nerves are directed more to the rear, especially toward the *nucha,* which is the messenger and vicar of the brain, taking its place in its descent through the back. Now the motor nerves require more hardness than do the sensory nerves. The *nucha,* however, is separated and covered with a membrane so that there is a difference in the feel of the *nucha* and the vertebrae.

128

The brain, as has been said above, is covered with a double membrane which has the greatest utility. Within these membranes are held and entwined the nonpulsating veins which come together there. The ends of these veins pass to the rear toward the *lacuna* and the *planum* of the head where they pour forth their blood.[121] Some are even sunk in the brain, having branched out and, at their end, having been made like the substance of the brain. They pour out blood as its nourishment and afterward they are again united as if into two veins. This is done so that they are strengthened, so that from them might arise the ligaments of the brain with the thick membrane

129

---

119. *UP* 8.6 (May 1.398; Kühn 3.637).

120. This metaphor is repeated from 1.521.

121. Compare this and following passages with the descriptions of the brain at 1.111f., 187, 214, 410, 523, and 535f. The *lacuna* is also called the *trochular* and the *vacuitas*. The *planum* seems unparalleled in this text, but, in both the sense of a wide, open space and in the similarity of name, it might be akin to the *platea* of 1.410 and below at 12.170.

which is equidistant from the serrated crown of the front part of the head.[122] Also in the front section there arise from them breast-like appendages through which smell occurs [olfactory bulbs]. They are slightly removed from the softness of the brain, but these appendages are still not as hard as the nerves.

130   A double membrane, then, covers the brain. One is called the *dura mater* and is on top, toward the bone, keeping the brain from injury from the hardness of the bone. The other is underneath and is called the *pia mater*, because it receives the brain softly. This is made up, as it were, from an interweaving of veins and therefore it enters the substance of the marrow in many places and proceeds to its ventricles, ending to the rear since there is no further need for it. The hard membrane is never attached to the brain or to the lower membrane. Rather, it is constantly kept aloft. Connection between the two membranes is by means of a vein which passes from the hard one to the fine one. The hard one is joined to the bone by means of certain membranous ligaments which arise from that very vein and the membrane and are joined with the bone at the teeth of the serration lest it fall down onto the brain. These ligaments pass through the substance of the skull all the way to the outside where it is divided up and the membrane which covers the skull is woven from it. This is the cause of the good, strong binding of the *dura mater* with the skull.

131   Now the brain has three ventricles lengthwise. Each ventricle has two sections laterally and is thus double-chambered, as it were. The anterior one is divided in two more clearly than the others (that is, the ones to the right and left) and these chambers have the same size. The hollow to the front of each part aids in smelling and for the easier exit of phlegmatic mucus through sneezing and blowing. It also helps in the division of the visual spirit and in the internal activities of the formative (that is, imaginative) power. The posterior ventricle is large because of the size of the empty skull at that point and so that it can be the principle of the large member (that is, the *nucha*) which descends from it. In this too the power of memory [*memorialis vis*] flourishes. Although it is large, it is nonetheless smaller than the anterior section, for the brain grows smaller gradually, beginning at the large portion, so that it is made narrow at the end, in the vicinity of the *nucha*, coming down to the size of that which leaves it. The middle ventricle is a sort of passage and pathway from the anterior to the posterior one. With it as a medium, the anterior spirits, bearing forms, are connected to the posterior powers. This ventricle is covered with a spherical covering that is hollow inside, so that it does not have any misfortunes from outside and so that it might be strong enough to sustain contact with the membrane. In one area the anterior ventricles are joined to it so that it can send to the posterior one through it.

That the animal powers are located in them is signified by the fact that illness of these cells hinders or destroys the activities of the powers that are in them, as if they were in organs.

---

122. "Crown": *serrae coronae anterioris partis capitis*, reminiscent of the *os coronale*, or "prow bone," of 1.111, 187, 214, 537. *Serra*, or "saw," refers to the jagged edge of the suture at that point.

Moreover, in the anterior part of the brain there are jagged [*serrosae*] pathways through which the spirit passes as if through cells. For the ventricles are not always wide open, especially when the spirit is meager. The spirit comes from the heart to the brain and it is digested there for the animal operations, just as it is digested in the liver for the natural operations, and an irregular [*serratilis*] division is better suited for digestion than is a division into chambers.

There are two veins between the anterior and the posterior ventricle. The intersection of these has, on each side, two glandular pieces of flesh which fill in the hollows that result from their branching off. Then these two veins are woven into a sort of a net beneath the substance of the brain. The jagged [*serrosae*] hollows, which lie above the separation into chambers, are filled in with a certain small bit of flesh which is worm-shaped and which contracts and dilates to open and close. A membrane, moreover, floats above the two bits of flesh just mentioned and is bound to them. When the membrane dilates, the bits of flesh contract, and when it is contracted they expand. In this way, the cells open and close, and when they are open the spirit passes through, flowing out to perform the function of the animal powers, the functions of which are dulled when they are closed. This is why a person, when he is astonished and is thinking, moves and shakes his head, specifically so that this movement might help open up the cells. The posterior ventricle is smaller than the anterior one and these two vermiform appendages resemble two grapes. This is why they are called the "grapes" [*uvae*] by some.

Further, there are two openings in the brain for the expulsion of superfluity. One is anterior and it passes through the nose and the other is in the middle of the base of the brain, passing into the mouth opposite the windpipe. There could be no opening in the posterior ventricle, however, since it is small compared to the others and superfluities do not come to it. The posterior and the middle ventricle come together at a passage of the middle ventricle, which divides them. There, two paths, proceeding transversely, come together at one opening. The beginning of this passage is in the thin membrane called the *pia* and passes into the tough [*dura*] membrane. When it has passed through it for a while it meets a certain passage that resembles an acorn and is like a sphere compressed at either end.[123] This lies between the hard members of the bone and the opening at the base which opens onto the windpipe.

Our statement above, following Aristotle, that there is no blood in the substance of the brain, should be understood as saying that there is, in truth, no blood present in the form of blood, for it is already changed into another humor. Likewise, our statement that there are no veins in the substance of the brain is understood to say that they are incorporated into it. Yet the openings of the veins come to the brain and from them it absorbs the necessary nourishment. But in the actual substance of the brain there is no network of veins like that in the flesh, heart, and liver. The skull is therefore thicker on top so that it can defend the brain more easily from accidental mishaps.

---

123. "Acorn": *glandula*; cf. 1.246, 3.40.

Further, the brain is the coldest of the principal members and this can be felt by touch. When the motor nerves are stretched out very far from the brain they grow rigid and cannot move. The *nucha* was therefore made to descend through the vertebrae. It is through propinquity to this that the motor nerves take on the power of movement. The *nucha* also helps to hold the vertebrae in their place and in order. Another reason the *nucha* was made is that if all the nerves were directed to the head, it would be made enormous and would thus weigh down the entire body with its bulk.

CHAPTER V

Which Is a Digression Setting Forth the Cause and Nature of Nerves

135   Lest we cause our instruction to be confused by speaking of the natures of all the members mixed together and at once the way Aristotle did, it seems that we should now interpose discussions about the natures of animals one by one and then we will return to speaking of them mixed together.

Let us speak, then, of the nerves, for although according to many physicians they originate in the brain, in truth they arise in the heart and then divide off from the brain and *nucha,* as we have worked out in previous parts of this study.

Let us say, then, that one utility and cause of the nerves is *per se* and another is *per accidens*.[124] That *per se* is to provide sensation and movement to the entire body, with the brain and the *nucha* serving as the medium. The nerves assist in an accidental fashion [*per accidens*] in strengthening the heart, brain, and every member around which they are twined and into which they are woven. They impart sensation to the liver, lungs, and brain in their pannicles, so that they can perceive when there is a harmful windiness in them or anything which extends the membranes. The point of origin for the division of the nerves is the medulla of the brain, and their end is where they are woven into the substance of the skin.

136   The principle for the division of the nerves from the brain exists in two ways. For some nerves descend from it without a medium and others through a medium.[125] This can be seen through the nerves that descend from the head to the inner members, like the liver and lungs.

Now nerves which lie at a great distance from the brain (from which they divide and branch off) are supported in the muscles. Nature thus provides them with a certain body which is midway between nerve and cartilage but which in substance is like nerve. She gives this to them at a place where they curve, such as the epiglottis, and there is another at the roots (that is the beginnings and ends) of the ribs. There is another one like this at the low place where the sternum [*os pectoris*] ends.

---

124. Avic. *DA* 13.29r.

125. The Latin preserves a nicer sound to the contrast: *quidam descendunt ab ipso immediate et quidam mediate.*

Nerves, however, which bestow sensation and movement, descend in a straight line and have no such things strengthening them since they are nearer to the point of origin from which they divided. There was no need for twists and turns in these to strengthen and harden them. For this reason, too, that which produces sensation arises from the anterior portion of the brain because softness assists its sensation, but the motor nerves arise from the posterior portion of the brain because it is drier, as we have said before.

Now as to how many pairs of sensitive and motor nerves there are, how they are divided from the anterior and the posterior portions of the head and from the vertebrae of the neck and the back—all this was amply covered for the purposes of this instruction in the first book of this study, where we spoke of the anatomy of the nerves.[126]

CHAPTER VI

Which Is a Digression Setting Forth the Cause of Bones and Cartilage

Bones and cartilage are made to be the shield, defense, support, and foundation of the entire body. There are certain animals which seem to have neither joints moving their bones nor shells surrounding them, having instead tough skins surrounding them. Examples are those called the *calapagii* in the Greek language.[127] Some of these have their hard part on the inside and their soft part externally, over the hard part, like the genus of *malakye* [cephalopods]. In ones such as these, nature made the flesh midway between true flesh and nerve. Thus, its flesh cannot be split lengthwise like other flesh but rather sideways, circumferentially, as if it grew in rings. Nature did this so that the nerve-like thing that is its substance might sustain it more firmly.

Ringed creatures, moreover, have their hardest part outside and it is midway between nerve and bone. Rings are placed in this hard part, bound together in such a way that they can move away from each other a bit and then be joined again, with one slightly entering the other, as it were. In animals of this sort different principles are not found for the veins and arteries. Nature did this so that they could be protected better. The bones of lions are very hard and most of them have no hollow for the marrow. When they are struck together, a spark leaps from them as it does from stones. Many animals have cartilage in place of bone and animals like this need a great deal of roundness.[128] They have a great deal of moisture in their substance and are so watery

137

138

---

126. Cf. 1.355–80.

127. The name is from Avic. *DA* 13.31r, in the form *calapagil*. Stadler suggests the ultimate source is Ar. *Part. An.* 653b35f. While this passage of A. is clearly indebted to this passage in Ar., there is little there to help us find the origin of the name.

128. "Roundness": *circumvolutio,* lit., run in "rolling around." The idea seems to be the same as just above where cephalopods, having no external source of protection, rely on the fact that their tissue is formed in circular patterns for strength. Cf. Ar. *Part. An.* 654a13f.

that there is but little earthiness in them. These animals are ill able to endure contact with hard things. Nature helps herself, however, in such animals by supporting the body with spines. Cartilage, which is bone-like although softer, is also found in fish.

139 Spines, hooves, and horns are all bony. Some of these are very hard since they were made by nature for support and for the protection [*munitio*] of the feet. These support the entire body as do the hooves [*ungula*] in certain brute animals. Others among them, however, are made strong so that they can serve as weapons for defense and attack, like horns. But hooves are also weapons, such as the hooves [*calx*] of horses. Others are bound together, as it were, and serve as foundations, like the vertebrae upon which the entire body is put together. They are to the body what the keel is to a ship.

Others, however, serve as a sort of helmet, such as the skull bone.[129] Others act as a surrounding wall and a fortress, as do the spines on a hedgehog. Still others fill in a void, like the small sesamoid bones in the loose joints of the hands and feet. Some are put together above members to close them in, like the bone at the base of the tongue [hyoid bone] which resembles the Greek letter *lauda* (Λ), as we have said in the anatomy of the bones.

All bones keep the body upright and push it out so that it can keep its shape. The bones that are for defense are often not hollow in animals, but those for movement have a hollow so that they can retain their nourishment and will not be dried out through movement.[130] These bones are also made hard so that they do not break easily.

140 A bone made for movement alone has a small hollow, whereas one made for movement and smoothness has a large hollow, for then it is long so that it can produce a straight and balanced position in the part in which it resides.[131]

The *os colatorium* ("strainer bone," cribriform plate) was made full of holes so that the superfluity could be led out through it and so that harmful vapors could be cast off through its holes.[132]

Moreover, all bones are connected one to the other except insofar as there is cartilage between some of them. Some, however, do not need this, such as the two bones of the lower jaw which come together at the chin. A given joint might be loose and another tight, but we have already said enough about these in the first book of this study, where we treated the anatomy of bones at length.

141 We must not neglect to say that some bones are joined with joints that are designed for movement and which are joined loosely, so that one can move without the other.

---

129. "Skull bone": *os cranei*, a term not found in Fonahn. But cf. *caput cranei* with the same meaning at 1.181.

130. That is, by being hollow, they have marrow, which is a wetting agent.

131. "Smoothness": *lenitas*, an odd trait for this context. One might well wish to read this as an error for *levitas*, lightness, which is, in fact, the reading of Borgnet.

132. Fonahn, 934–35, 2336, offers three identifications for the term *colatorium*. Of these, two, the cribriform plate and the sphenoid bone, are possible here. But A. seems fairly consistent in identifying his *os colatorium* as the cribriform plate, associated mainly with olfaction but, at 1.526, also with purgation.

Others, however, are joined with a joint that is difficult to move, so that one cannot be moved without the other. Such are the bones of the *rasceta* in the hand and foot.[133] Others are made solid so that one is fixed in the other, such as the ribs in the sternum [*os pectoris*]. Others are tenoned so that the appendage of one enters the hole in the other, such as the teeth in the upper and lower jaw. Others are joined in saw-like fashion, like the bones of the skull. Some are joined widthwise at their bases, like the vertebrae, and others lengthwise at their surfaces, like the bones of the *asseid* [radius/ulna] on the arm.[134]

One should know that plants neither have the bones nor the other members animals have. Therefore, they do not have a point of exit for their superfluities, such as an anus or a bladder. For the nourishment they draw from the earth is separated, as is the nourishment that is a result of the third digestion in animals. Once drawn up by the plants, the nourishment flows and rises through the plants' passages. For if it stayed in one place it would weigh down and hinder any further drawing action. Then, cooled, the nourishment would descend and block up the roots (for it is due to the weight of the excrement in animals that the member for the exit of the superfluities is placed at the bottom of animals). Plants, therefore, do not have such instruments, and neither do they have a skull [*os capitis*], which is made of strong bone and placed in animals to defend them from harm. The skull is made up of many bones for many reasons. One of these is that if a mishap befalls one, it might not befall the other. The second is that one bone cannot differ in hardness and softness as happens in the skull and as is clear in our anatomy of the head. The third reason is so that there might be a place for the nerves, quiet veins, and pulsating veins to leave through the sutures and also so that the *dura mater* itself, leaving through the sutures, might be forcibly raised up so that it does not touch any part of the brain.

The shape of the head is that of a rounded oblong, coming to a bit of a point to the front and to the rear. The reason for the roundness is that of all the symmetrical shapes, round is the one that is most capacious and least susceptible to mishaps.[135]

The head is somewhat oblong so that the nerves leaving it will be less crowded together, the places for the sensory nerves can be kept apart from those for the motor nerves, its cells can be kept separate, and the spirit, flowing further, might be better digested and thinned out. The head is brought to a point up front accompanied by a widening of the forehead and this happens so that there might be a center of sensation in the organ of the common sense [*sensus communis*] so that the sensibles of all the

---

133. On the *rasceta,* cf. 1.286f.
134. On the *asseid,* cf. 1.284f., 488f.
135. "Symmetrical": *isoperimetricus,* from the Gr., lit., "having the same external measurement."

senses might be confined in one place and thus not flow off.¹³⁶ It is pointed to the rear so that it might narrow down for the generation and departure of the *nucha*.

It has two false, serrated sutures sideways from ear to ear and three true sutures. But enough has been said about these things in the section on anatomy.¹³⁷ The first ones are called false because they do not extend through to the inside. The others are called true because they do penetrate through the bone itself.

It has three unnatural shapes. The first is when it is contracted on the *sinciput* side and almost loses the coronal suture [*serra coronalis*?]. In this condition the apprehensive powers are injured. The second is when the contraction is on the *occiput* side, and the motor and recorditative powers are harmed.¹³⁸ In this condition it almost loses the suture which resembles the Greek letter *lauda* (Λ) [sutural bone?].¹³⁹ The third occurs when both are lost by a contraction of each side and the head becomes short and round like a globe. Apart from these shapes, there is no other that lasts long in life, but these four shapes (namely, the one natural shape and the three unnatural ones) are found among the living. But we have already spoken about these things—the number of bones in the head and their boundaries—in the anatomy section.¹⁴⁰

The head lacks fleshiness for two reasons. The first is that if it had it, it would not retain its proper complexion, namely, cold and moist, for the flesh would give it heat and humor and as a result of this the animal spirits would be confused and mixed up. The second reason is that it would weigh too much and could not be held upright. Further, all senses are in the heart with respect to point of origin, but touch is there also with respect to the first organ of sensation, while the others in the brain and head are there with respect to their own organs.¹⁴¹ But this has been determined elsewhere.

---

136. At *De anima* 2.4.7, A. adds that it is the Peripatetics who locate the common sense in the anterior part of the brain. Indeed, *De anima* 2.4.7–10 provides a lengthy discussion of the common sense, which represents not so much a "sixth" sense as the common source for the powers of the individual senses (sight, hearing, taste, smell, and touch). Through the common sense, one is aware of the operation of the individual senses—namely, *that* we see when we see, and so on. It is also the common sense that "composes" and "divides" the data received from the individual senses in order to establish a complete image of the sense object. In this way, the datum of the olfactory sense (e.g., "sweet") and the datum of the sense of touch (e.g., "hard") can be joined together in one object.

137. E.g., 1.181f.

138. These powers are involved in recollection.

139. Cf. 1.183, with notes.

140. Cf. 1.181–82, where (from Avic. *Can.* 1.1.5.2) this scheme is attributed to "Hippocrates and Apollo." Such a fourfold classification is found in Hippocrates's "On Wounds in the Head," sec. 1 (Withington, 1928, 6–9).

141. That is, the organs of sight, smell, and hearing are located on the head and in the brain. Touch, however, employs the flesh as its proper medium. It requires the heart's influence (that is, heat) more than the other senses, as indicated at *QDA* 12.20. Touch is also principally situated in the nerve-filled flesh of the heart, as indicated below at 12.148. Taste for A. is usually described as a kind of touch, even though the tongue is its instrument. Cf. 1.237; 12.168, 174; and *De anima* 2.3.23; 3.5.2.

CHAPTER VII

On the Type and Number of Superfluities Generated in Animals' Bodies

Having inspected all these matters, we will say, along with Aristotle, that the superfluities generated in animals' bodies come from the superfluity of their food.[142] An example is the superfluity of the food which is called "of the belly." There are dry superfluities, such as excrement, as well as the superfluity of the urine, expelled through the bladder. There are other superfluities as well, like the sperm, decocted in the testicles, and the milk, which is white and is decocted in the breasts in animals that have milk and breasts. All the superfluities of food differ from the sperm and milk, for all the others come from an impurity, rejected by nature.

We cannot make an adequate determination on these matters here. Instead, determination will be made concerning the material of the superfluity along with the decoction of the foods. For the instruction is easier because each superfluity will be determined, along with its decoction, in one and the same statement, namely, that the superfluity of the first digestion which occurs in the belly is a heavy, earthy superfluity which descends from the intestines and is discharged through the anus. Because every animal requires earthy food, every animal therefore has this sort of superfluity.

The second superfluity comes after the second digestion of the food which occurs in the liver. This one derives from the superfluity of the moisture which is drawn in along with the food and without which the food could not flow through the veins or be distributed throughout the small pathways of the liver. It is strained from the food in the protuberance of the liver.

The third superfluity is produced at the ends of the veins as a result of the digestion which occurs there. It is discharged by flowing back to the protuberance of the liver through the veins and from there it is washed away through the *emunctoriae* [renal veins] and is led via the kidneys to the bladder.[143] It is then discharged in *ypostases, eneorrima,* and clouds in urine.[144]

The fourth is produced in the members. Its superfluity is sometimes sperm and the impurity which is there leaves sometimes in the form of sweat and bodily impurities.[145]

145

146

---

142. Ar. *Part. An.* 653b9f.

143. Cf. Fonahn, 1186, where one would expect this form to mean "ureters." The identification here is based on the fact that these objects clearly lie before the kidneys and not between them and the bladder and by comparison with other, similar forms at 1.414, 449, 466, 612f.

144. *Ypostasis* is from the Gr. *hypostasis* and refers to sediment, the heaviest part of something. *Eneorrima* is a fairly good transliteration of the Gr. *enaiōrhēma,* from a verb meaning "to float around on top of," and it generally refers to scum or suspended matter floating in the urine.

145. This production of sperm in the members would be an example of the theory of pangenesis. Cf. Cadden (1993), 91–92.

Now superfluities cannot be known well without also knowing the decoctions of their foods. Thus, along with the decoctions, the generation of the superfluities is known, as is the reason they exist.

All animals have the first type of superfluity, as we have said already. But only those requiring a great deal of moisture for the flowing of food through their veins, as do the quadrupeds, have the second. Those which are either moist in and of themselves like the fish or those which have very loosely textured bodies, and thus have great numbers of venous passages compared to the size of their bodies, drink only to temper their food. They therefore have no urine, for they emit the superfluity of their drink along with their excrement, as do birds.[146] The third superfluity is converted in animals like these into scales and feathers, and they therefore do not need it. Almost all animals, however, have the fourth, which is for the preservation of the species.

CHAPTER VIII

On Flesh, Bones, and Spines and on Those Called Sepion
and on the Malakye *Genuses*

It follows, then, that having determined these things, we must now make determination of the other uniform members.[147] First is the disposition of flesh, which is the most common member in animals, especially perfect animals. For there is flesh or something analogous to it in all animals. It is either the organ or the medium of touch and is *per se* the body of the animal insofar as it is an animal.

This is clear from the definition of animal. For an animal is that which has sensation. But every animal has prior in nature [*prius autem natura*] the first sense, which is touch. The instrument, however, of the first sense, touch, is this member, flesh. For flesh, which is the organ of touch, is nerve-filled. Therefore, this member has the sense of touch in its principle, which is the heart, and the principle of this sense is especially the nerve-filled flesh of the heart, just as we say that the pupil of the eye is the principle of sight. For we are not seeking after a general organ here, but rather the first organ and the principle of all those which have touch. This is nothing other than the nerve-filled flesh of the heart.

Therefore, just as we say that the pupil is the principle of sight (since the first disposition of sight is in it), so do we also say that the nerve-filled flesh of the heart is the principle of touch. For it is first in nature and cause [*primum natura et causa*] and the disposition of touch is in it. For it is necessary that the sense of touch, more so than the other senses, be in a body such as this, or in a corporeal instrument. For all the

---

146. For a fuller discussion of this trait among birds, cf. *QDA* 2.30.
147. Cf. Ar. *Part. An.* 653b19f.

other sensory members are created only for this reason, and it was therefore suitable that this sense be in a member which is the principle of the generation of all the other members. This is the flesh of the heart first, and then, accordingly, the flesh of the entire body is homogenous to the flesh of the heart.

We will now discuss the types of bones, nerves, skin, hair, nails, and the like. The bones then, as we have already said, were created for the well-being of the soft, fleshy body. The nature of bone is very hard. In animals which lack bones, there is a member analogous to bone, such as spine in fish. In certain animals there is cartilage inside the body and in others, lacking blood, it is outside. Examples are those which have a soft shell, the crab and the *karabo* (that is, the sea locust), and also the hard-shelled animals, like the *ostrea* and *concae,* which are called *halzum* in Arabic.[148] For in all these animals the part which contains the sense of touch is created out of an earthy part and is internal, but there is another part, shell-like, which guards this sensitive part, and it is external. This was done in this way because there is only a small amount of heat in the nature of these animals since they do not have blood. Therefore, the shell contains the animal's body inside of itself and thus guards this heat, keeping it from becoming cooled by external mishaps.

Some have held the opinion that the tortoise does not have a moderate amount of heat but that it is quite hot indeed, and they say the same for the animal called the *amydon*.[149] They base their argument on the hardness of its shell, saying that it could only have been decocted by a great heat. This was Empedocles's opinion, but it is not true.[150] For although the shell is decocted in heat, this heat is nevertheless held inside within a viscous earthiness and it is held in check by the coldness of the water. Thus, strengthened in the parts of the viscosity, it decocts the shell. This is then distributed externally by the formative power so that it can protect the moderate heat that is naturally in it.

Although in the animals mentioned the shell is to the outside and the soft part inside, in the *malakye* and in certain of those that have a ringed body, it is the other way around. For in the bodies of these animals there is neither bone nor any hard, separate, earthy part which is external *per se*. For there is no shell on the outside of this sort of animal. I am using the term *malakye* for every animal which has, to the outside, soft flesh, which is, as we said above, almost midway between flesh and nerve, and whose exterior members are created solely from flesh.

150

151

---

148. On the *karabo,* perhaps a lobster, cf. 4.5. On other terms, see Glossary and Kruk (1985). *Ostrea* may be specific here, designating, for example, oysters.

149. Ar. *Part. An.* 654a8f. lists the turtles/tortoises and the *genos tōn (h)emudōn,* "the genus of the pond turtles" (*GF,* 167). This name, in its singular form of *(h)emus,* also gave rise, through Pliny, to the *mus marinus* of 24.44(81). The idea for Ar. is that the shells of these animals act as an oven, trapping the heat.

150. The Empedocles reference seems to be A.'s addition. Cf. 1.119, where heat is especially assigned to bone.

152 To keep this animal from corrupting swiftly, like an animal created from soft flesh, nature established that its flesh be solid, like a substance midway between flesh and nerve. Thus, a member might be soft like flesh, but it might also be able to be extended like nerve. This flesh does not split equally along straight lines lengthwise but rather in circles, as we have explained above.[151] For flesh like this is stronger and more solid. Moreover, beneath the actual flesh in such an animal there is a member that is like fish spines. In Greek this member is called the *sepion,* and for this reason it is proven that the cuttlefish [*sepia*] is a species of *malakye.*[152] Moreover, in an animal of this same genus, but of another species, called the *tobyz,* there is a member resembling this one in substance.[153] This thing is called the sword because of its resemblance to the shape of a sword. The genus of animal called the *multipes* (that is, the *polipus* [octopus]) has no such member whatever. The head of this genus of animal is very small, but its other members are long. Nature measured the animal out this way so that its body could be straightened and so that it is not easily bent, as are the animals that lack bones.[154] Thus, nature has made bones in some animals and in others has made spine.

153 The creation of a ringed animal is contrary to that of both this animal and also of a blooded animal, as we said a bit earlier. For in the uniform members in the entire body of a ringed animal there is no one thing distinct from another in shape the way bone, flesh, and nerve are in the other animals. Rather, the entire body of a ringed animal seems to be a single member, uniform in shape. Its entire body is hard on the outside and, compared to flesh, its nature is that of a hard member like bone. But when compared to bone, it is like flesh. It is earthier than is flesh, however, and it is less earthy than bone. This was done so that its body might not be easily cut.

154 In joining, the nature of bones is like that of the veins in their joining and the same is true for the joining of the nerves. For the joining of all of these is like the nature and the joining of the heavenly bodies, as is clear in the things determined in *On Heaven and Earth.* For all these things are connected by that type of connection by which they are called "continuously connected" [*continua . . . compaginata*], exiting from one principle. Thus, then, bone is neither one in number, entirely *per se,* nor are there bones that are entirely separated. Rather, there are bones connected to one another by the type of connection in which a thing bound is said to be connected exiting from one principle so that nature uses it as one connected bone, or as two or more separate bones, which move by means of an expansion that, however, proceeds from one principle.

155 In an animal with a ringed body, there does not exist an individual vein *per se,* but rather all its veins are parts of one common vein. Therefore, in the body of this ani-

---

151. Cf. 12.137.
152. *Sepion*: Gr. *sēpion*, cuttlebone.
153. Ar. *Part. An.* 654a21 speaks of the *teuthis,* the squid or calamary.
154. This is an odd statement for anyone who has ever seen an octopus move. It arises because something has been dropped from the Ar. text, which ceased speaking of the octopus and moved back to the cuttlefish and calamaries, whose need for straightness is then discussed.

mal there are not uniform members distinct in shape, for their job is sharing through binding [*communicans per colligationem*]. Thus, if the body were not arranged in the way described, namely, being midway between bone and flesh, this animal would not perform the work nature does with bone. For if it were connected in no way whatever, it could not expand, dilate, and contract. Rather, when one part expanded, a gap would appear within the body. Moreover, if there were spine in the body of this animal or that genus of spine called "arrow" (arrow being spine which does not cling to the vertebrae, as is found in the flesh of the tail of certain fish), and if, as I say, something of these were in the body of the ringed creatures, it would only be of harm, for it would hinder the wrinkling up and extension of their rings.[155]

Further, if there were veins in such animals, there would be blood in them, or a humor taking its place. Thus, if a vein distinct *per se* were in them, separate from the other members of the body in shape, the blood that was in it would be corrupted, for the putrefying heat would prevent it from being preserved in being. For blood that is separate and is not connected to the principle doubtless putrefies. Therefore, in such animals, just as in all others, if there were a vein separate *per se,* the blood or the humor taking its place would putrefy. For the principle of the veins is the heart in those that have veins, and in those that do not have veins it is the passage which is the passage for the humor which takes the place of the blood. This passage is also connected to the heart or to that which takes its place.

Just as the principle of the veins is the heart, so the principle of the bones is the spinal column which descends through the body in an animal which has bones. The entire nature and joining of the bones is connected to the vertebrae, for the vertebrae preserve the entire uprightness of the body. Since it is necessary that the body of an animal bend forward and backward, the vertebrae were therefore bound together out of many pieces and are not one solid bone.

Other bones, though, are connected by being bound to the vertebrae in all the animals which have feet and hands. For at the ends of the bones of their hands and feet there are places and shapes suited for this binding together in such a way that one enters the other. Thus, the ends of some bones are hollow like a *pixis* and the ends of those joined to them are projecting and rounded like a *vertebrum*.[156] It is almost as if they had been handcrafted so that one enters the other, for otherwise a member bound together by bones could not extend (that is, move straight out) or contract (that is, curve in) without breaking.

These bones, inserted this way one into the other, are bound together with insensate ligament, nerves, and cords. At the ends of some, moreover, both bones have hollow ends where they are bound together. In between them there is an appendage

---

155. Ar. *Part. An.* 654b7f. says that an isolated bone would be of no help but rather would be like a thorn or an arrow stuck in the flesh.

156. A *pixis* is a box and a *vertebrum,* "turner," is a rounded projection on the end of a bone that fits into a *pixis*. In our terms, A. is describing a ball-and-socket connection. For some other examples, cf. 1.111, 278, 283–84, 292–94.

rather like the *cahab* which is a filler between the two, like the bone called the sesame, so that there might be a moderate extension and contraction of the bone.¹⁵⁷ For without this filler in the middle, the bones would grow too far apart when the member bent. An example is seen in bending the fingers. There would be no such joint arranged in this way if the bones had been prepared in another manner, nor would the motion of the members be well suited and light.

Some bones which are bound together at the ends have ends that are rather similar so that one end does not enter the other. At this place in these bones nature has laid down cartilage with virtually the softness of wool, lest the bones be worn away by striking one another. This is to be seen in the two *asseid* bones [radius/ulna] near the elbow. However, we have treated at length all these types of joints and many others in foregoing books.

159 The flesh is held in place around the bones by slender ligaments. This is especially so for the flesh of the muscles, for in them there are hair-like nerves. The bones were created because of the softness of the flesh, as we have said above, and as is clear in sculptors who make figures of animals using mud or some other wet, soft material. When shaping the figures, they first form something hard and rigid underneath the clay to serve as a basis or support for the wet clay, keeping it from falling due to its own softness. Then they arrange the mud or other wet material over this. Nature acts entirely the same way in her use of bones, placing them under the fleshy parts. In members which are moved often, however, she also set down cartilage for the rubbing together. In members which do not move, she set down the bones for protection. An example is the rib structure which contains and protects the chest. This rib structure was not created for any purpose save the well-being of the members to the front which contain the heart. The bones in the rib structure near the belly are small so as not to hinder the expansion of the belly. This is due to two causes, namely, from food in male and female alike, and from pregnancy in the females alone.

160 Further, the bones of a viviparous animal, whether internally or externally so, follow in all cases one and the same disposition with respect to hardness and strength. The bones of this animal will be larger than those of a nonviviparous animal, provided proportionate body size is considered. In many places there are very large animals, such as the elephants in India and in other hot, dry lands, and, in the North Sea, the great whales. When an animal has a large body it needs a framework which is stronger and harder because of the animal's weight. This is especially so when the animal is very

---

157. *Cahab*: Ar. *Part. An.* 654a22, *astragalon*, "knuckle bone," which generally means the complex of the ankle bone (e.g., 1.300) but occasionally other multibone complexes are meant, as at 1.470. Here the plain meaning is of a single bone that sits between two others, helping to effect their juncture (Fonahn, 654). The sesamoid bones serve somewhat the same function in the wrist. Cf. 1.118–21, which bears no small resemblance to the present discussion.

strong and lives by hunting, like the unicorn [rhinoceros] and the panther.¹⁵⁸ But the bones of the males are stronger than the bones of the females, especially in carnivorous animals. Thus, the bones of the lion are very hard, so much so that when they are struck together they give off a spark, as do two rocks struck together.

Likewise, even though they are aquatic animals, the dolphin and the whale [*balena*] have bones instead of spine, for they are viviparous. In nonviviparous animals which are blooded, nature changed creation a bit. This is clear in the birds, for birds' bones are weaker than other bones. In oviparous fish nature created spine and the bones of serpents also have the nature of fish spine. If they are very large serpents, it is an exception, for large serpents need a strong framework, just like that for a viviparous animal. Nature created the *celeti* out of spines and cartilage. It is necessary that the *celety* be moist and therefore its support must be somewhat soft as well. The earthy part which is in this animal passed over into skin. In a viviparous animal there are many bones which resemble cartilage in composition. In the spots where this occurs it is so that the hard part is not entirely soft and mucus-like because it is only held by flesh.¹⁵⁹ An example is in the earlobe and in the very tip of the nose. However, the nature of bone and cartilage is the same, and they differ only in being more or less hard.

When the cartilage in a walking animal is cut into, it is found to be without marrow. For marrow cannot exist outside of its own vessel, separated by itself. It if were separated off by itself it would grow hard and would, in the cold or the heat of the air, become a cartilaginous substance or it would remain soft and mucus-like. The nature of the vertebrae of the *celeti*, however, is cartilaginous, as is that of the fish called the *sturio* and the *huso* [sturgeons], for there is a bit of marrow in it.

Further, the feel of bone is like that of the uniform members, which are hooves, claws, *soleae*, the skin of certain creased animals, and the beaks of birds.¹⁶⁰ The creation of all these things in the bodies of birds is for the purpose of strengthening and assisting and they are all homogeneous, so that both the whole and the part share a name. For a *solea* and a part of a *solea* share a name and it is the same for the part and the whole of a horn and for all the others. For in these parts nature cleverly gave each animal this for its well-being and for defensive and offensive armament.

The nature of teeth belongs to this genus of member. They are used for the grinding and dividing up of food in all those animals which have teeth. In some, however, they serve the further purpose of strengthening and arming the animal. Such is the case for the strong teeth of wolves, and the tusks of boars and certain other animals. All these things are in the animals' bodies naturally and the nature of all these mem-

---

158. The ferocity of the unicorn was legendary, especially when it did battle with the elephant. Cf. 22.144(106). One can be fairly sure from passages such as 12.224 that A. has the rhinoceros in mind and not the fabulous creature of tapestries.

159. Cf. Ar. *Part. An.* 655a28f., which says that the parts are projecting and thus cannot be too stiff or they would snap off when bumped.

160. On *solea,* cf. 3.77f., 135, and Glossary.

bers is hard and earthy. For they are created for no other reason than strength, as is the case for the armament given the animals. It is thus right that these things exist in viviparous quadrupeds, whose nature is earthier than the others. The sole exception is the human, for he has less earthiness. But we will pursue these matters in the following discussion.

164   These members are next in line of earthiness: skin, the bladder, web, hair, and feathers. We will treat them in what follows, when determination is made on the organic members and their causes in animals. For the natural dispositions of the organs are known only from their activities. The names of the ones we have listed, namely, web, hair, feathers, skin, and the like, are the names of homogeneous members, for their names are shared with the whole and the part. All these homogeneous members are principles from which others are composed, such as bone, flesh, and others like these.

The nature of sperm and milk belongs to the homogeneous ones in name and *ratio*. We will treat these below as well, when we speak of other fluids like them. For it is suitable to speak of these when we speak of the condition and generation of nature. For the principle of generation is due to one of them and the other is the food for the animals generated.

The Third Tract of the Twelfth Book on Animals,

on the Organic Members of the Head

CHAPTER I

On the Head, Insofar As It Is the Origin of the Senses

165   We now wish to speak of the organic members which we recently put off discussing.[161] We say that every animal necessarily has two organic members, one of which is a member for taking in food and the other the member through which superfluity exits. For an animal can neither be generated nor grow without food. Inside the uterus it is fed on menstrual blood or a thing taking its place, such as the yolk of an egg. Outside, however, it takes in food through its mouth. Now trees are alive, for many have held the opinion that plants are alive. Yet, they do not have a place for the casting off of superfluities, since superfluity does not exist in plants. They have no need for a point of exit for the superfluities since they take food from the ground which has been decocted and completed in the first digestion. Both animals and plants alike

---

161. Cf. Ar. *Part. An.* 655b28f.

do, however, emit the superfluity which pertains to the preservation of the species and they have points of exit and members through which they emit things of this sort. An example of this sort of member is the exit of the seed and of fruits.[162]

In addition to the two members mentioned, there is a third member in all animals in which the principle of life resides. This is midway between the two members mentioned and is positioned in their midst. The nature of trees, and of plants in general, has few organic members, and things which have few operations also have few organs. Therefore, the disposition of the trees has been considered by itself in the book *On Plants* [*De vegetabilibus*].[163]

A plant is fixed in the earth and has few natural operations and as a result it has few organs. An animal, however, because it has a sensate life, has, in light of this, different shapes for its organs. This is especially so for an animal that has spines. For this one has a greater diversity of shapes than any other, since its generation is easier. Therefore their activities are very diverse.[164] Animals differ not only in a certain natural cunning but also in goodness, in a certain divine element of intelligence and art, and in the external behavior of their life. Still, of all the animals known to us, only the human genus participates in things of this sort. Therefore, too, the external members on the human are known to have a certain difference from all the others. In these points, however, he bears an analogy with the universe, that is, the upper world.[165] For there is no animal with an upright stature and an erect body save the human.

From this it necessarily follows that he should have a head without much flesh. This is for the reason we gave above, based on the complexion of the brain.[166] The opinion of some, who have stated that if the head of a human had a great deal of flesh he would live a longer life, is nothing.[167] For the human head would not have been created without flesh unless it were so that he could have better, swifter, and more perfect sensation. Some people assign this to all the senses, feeling that all the senses have their organs in the head. But the opinion of both of these is false, for the reason the head lacks flesh is not as they say. For a person's life would not be longer if his head had a great deal of flesh. These statements stem from an ignorance of the reason the human head lacks flesh. The true reason for this is that if there were a lot of flesh over the skull, the functioning of the brain would be contrary to the function for which it was created. We have shown this previously. It would become hot and moist and in this state it would not be able to perform the functions of the animals' powers since, being very hot, it could not then be cooled.

---

162. "Seed": an ambiguous usage for *semen*, which might refer equally to plants or animals.

163. *De veg.* 6.1f.

164. A glance at Stadler's apparatus shows that this passage is badly corrupted from the original Gr. and from the version of Sc.

165. Cf. 1.73; 3.59, with note.

166. Cf. 12.144.

167. E.g., Plato, *Timaeus* 75b.

168 Further, if the brain followed the fashion of certain superfluities in the body, there could be no sensation in it at all. For then it would no more be the organ of any power of the soul than is sperm, excrement, phlegm, or another humor. Those speaking this way, then, are in doubt, and are ignorant of the reason why the senses are in the head.

We have already spoken about and set forth in the book *On Sense and the Sensed* that all the senses are directly connected to the heart.[168] Concerning the sense of touch and taste (insofar as it is a sort of touch), it is clear to see that the heart is the first to feel with the sense of touch. Three senses, as far as their organs are concerned, are in the head to be sure, but they are directed to the heart. Smell is in between the other two. Sight is higher than hearing in all animals, since even certain animals which have ears on the top of their heads still have the opening directed below and to the side, near the *occiput,* where the *timpanum* of the ear is. Beneath both of these, in relation to the disposition of the nose, is the sense of smell. These senses are especially disposed this way in the types of fish, since they do hear and smell even though they do not have any externally visible organs for these senses. In those which have the sense of sight, it is rightly and reasonably near the brain. For the nature of the sense of sight is cold and moist, for its nature is water, and water is one of the very clear fluids.[169] Its clarity is perpetual, fixed, and permanent.

169 Now for clarity of sight it must have a very refined organ and member, whose blood is purer and better. This could only exist in something very cold, for the great motion that exists in hot blood interferes with the operation of sensation. Its boiling disturbs the animal spirits.[170] Thus, the instruments for these senses were in the head because the brain is cold.

Not only is the front of the head deprived of flesh but the rear as well. The reason for this is that the member which is the head (in all animals which have a head) is very upright and erect. If there were a great weight of flesh on it, it could not be upright and erect. For a fleshy head would be heavy and would hang down. For these reasons, then, it is clear that there should not be flesh on the head, especially, as we said, because of the disposition of the brain.

170 There is also no flesh at the rear of the head. Neither is the brain at the end of the *occiput.* Rather, there is the *platea* which we spoke of in the anatomy section.[171] This is because hearing is located in the rear of the head, bordering on this area, especially in the animals with ears on the sides of their heads, like the human and the monkey.

---

168. *De sensu et sens.* 1.1.15.

169. It is clear at 12.181 that sight is not itself water (*aqua*), but it requires a watery fluid (*humiditas aquosa*). Cf. also 19.11.

170. The animal spirit conveys the forms, in the sensory process, to the apprehensive powers of the soul, as indicated at *De anima* 1.2.3. The animal spirits, then, are necessary for sensation and are generated in the brain from the vital spirits (*spiritus vitales*) ascending from the heart. From the brain, the animal spirits proceed to the three senses located on the head (sight, hearing, and smell) whence they receive sense data. Cf. *QDA* 12.17.

171. Cf. 12.129 for discussion and references.

In other animals, especially the quadrupeds, the *occiput* is turned upward at the rear of the brain since their heads are facing the earth. They therefore have ears at the top of their heads. That which is called the empty opening of the ear is full of air.[172] We thus also say that the instrument of the sense of hearing is airy. The pathway of the eyes (which are the places for the eyes) extends to the veins which are woven around the brain. The meninges [*miringae*] of the eyes are composed from these.[173] Likewise, pathways leave the external ears and make their way straight to the rear of the brain and the head.

No animal whatever has any sense without blood or a humor which stands in its place. Blood is not, however, the instrument of sensation, as some have said. Rather, the instrument of sensation is that which is generated from the blood or that which is in its place. An indication of this is that in a blooded animal no member has sensation when deprived of blood, but rather becomes numb. Yet the blood itself does not sense, since it is not a member possessing blood.

Moreover, every animal which has a brain has it in the front part of its head. Since every sensation is from the heart, the heart is in the front of the body. Thus, the brain too, which is a certain principle for some senses, is in the front of the head. For the principle of the senses, as we have already said, exists throughout members infused with blood (in those that have it), for without blood every member grows numb in a blooded animal. The entire rear of the head is bereft of veins and as a consequence is deprived of blood. For this reason it cannot have sensory organs.

Nature arranged the sensory instruments wisely when she placed the organ of hearing in the middle of a round head. For the organ of the sense of hearing does not hear only along straight lines, but on all sides, circularly, all around. She positioned the organ for the sense of sight to the front, since an animal sees to the front along straight lines and it is a very refined and spiritual sense, more so than the others. Thus, it could only look directly forward if its eyes were placed to the front. The instrument of the sense of smell is below, between the eyes, so that it can be assisted by breathing, a portion of which must pass into the windpipe.

---

172. An interesting textual problem. A.'s received text read: *hoc enim quod vocatur vacuum*. It clearly refers to the situation he outlines in 12.129. He did not notice the connection, however, and added to the text, changing its sense completely.

173. Cf. 3.149; 19.17; and Fonahn, 2079. Meninges are, in fact, membranes of the brain.

CHAPTER II

That Each of the Senses Is Double With Respect to the Disposition of Its Organs and on the Disposition of the Ears in Particular and of the Nostrils

172 In addition to all the things that have been said in general on the senses, we should now pay attention to the fact that each of the sensory organs is double. The skin, which is made up of an interweaving of nerves that touch one another, is double in location, since one is to the right and the other to the left.[174]

This duality is clear in the sense of touch since the flesh is not the first instrument of the sense of touch, and neither is the member analogous to it in those which have no flesh. Rather, it is the nerve which is within the flesh, as has been determined adequately elsewhere. For while nerve is quite multiple it nevertheless is bent according to two locations, namely, to the right and to the left.

While there are other differences of positions in the body, such as up, down, front, and rear, nevertheless, the position of the sensory organs is not caused in these. None of the sensory organs is positioned beneath another in such a way that one is above and the other below. Nor is one so disposed that one is to the front and the other to the rear. Rather, it is always such that one is to the right and the other to the left. Because the sensible power needs pure spirit, especially in those senses which take in something through an external medium, none of these sensory organs could be at the bottom since obscure, hot spirit is directed downward. It cannot be to the rear because the rear part of the head is empty.

173 Thus, since two organs were necessary, one had to be on the right and the other on the left. There are two senses, however, which take in sensibles through an internal medium. One of these is not only a sense but also a substantial form through which an animal is an animal and is spread throughout the body. Specifically, this is touch, since the entire body is constituted from things capable of touch and is divided into right and left. It is directed to the heart which is the "first sensitive" [*primum sensitivum*], since the two ventricles of the heart which bestow sensation are to the right and to the left.

174 Taste, which also accepts sensibles through an internal medium, is double as well. For while the tongue is the judge of tastes, there is still more of the sense of touch in it, since taste is a sort of touch. The duality, however, is seen in the fact that the tongue, in all those which have one, seems to be split in two along a line running lengthwise down its middle. In some, such as the lizard and the serpent, it is also actually split in two. In certain others, such as the human, it is as if it were made up of two tongues

---

174. This "right" and "left" skin is a result of misreading Ar. *Part. An.* 656b34f., where Ar. bolsters his argument for sensory duality by saying that the body itself is double, with a right and a left.

joined together.¹⁷⁵ This is done so that it will not present a hindrance in the mouth during eating or forming sounds.

In the other three senses the duality of the organs is clear. Thus, the sense of sight is in the eyes, namely, a right and a left. Likewise, the sense of smell is in the nostrils and the breasts on the front part of the brain [olfactory bulbs], both in two parts. The sense of hearing is likewise in the two ears although they are disposed on the head differently from the eyes, in which vision is located. For if vision were not to the front of the head, it could not perform its job satisfactorily.

In all the animals which have a nose, smelling is only accomplished through the drawing in of air. This member is in the middle of the senses and is to the front part of the head, and therefore nature positioned this member in the middle of the three instruments of the head, that is, the eyes, ears, and mouth or tongue. And this part is disposed like a curtain is disposed, lying in the middle of the nostrils and of the breast-shaped appendages [olfactory bulbs].¹⁷⁶ This was so that they might be moved by the movement of breathing and then the breast-like cones [*coni mamillares*] receive the odors.¹⁷⁷ If, however, they lacked a curtain (that is, a covering), all the air and odors would freely flow to the brain and would corrupt it. These instruments have been carefully and well disposed in other, nonbreathing animals as well, for they have a drier and harder brain, and are thus not corrupted as easily.

The best and suitable position for these three senses which perceive sensibles through external mediums is that which, not only in the human but also in quadrupeds and others, is in accordance with the individual nature of the shape of each of them. For in quadrupeds, the external ears are positioned on the top of their heads, as is clear to see.¹⁷⁸ For such animals are bent over and parallel with the earth and do not have an upright body. The movements of the ears of this sort of animal are manifold, so that the ears move in different directions the better to receive a variety of sounds.¹⁷⁹ They therefore have them on the upper part. Birds, however, do not have external ears, but only openings and passages for hearing. They have tough skin due to the feathers which they have and this is not easily formed into the shapes of ears without presenting a great hindrance to flight. Also, there would be such a noise created in their ears as a result of flight that it would necessarily corrupt their brain. Therefore nature made only passages for them. Similarly, a scaly, oviparous animal has hearing passages that are quite invisible and which keep water from entering and corrupting the brain.

---

175. Cf. 1.237f.

176. "Curtain": *cortina*. Sc. has the reading *una trutina,* a pair of scales, a reading that might be explained from Ar. *Part. An.* 657a10, *epi stathmēn,* meaning something like "according to a plumb line."

177. These appendages still refer to the olfactory bulbs and must not be confused with the similar language often used to describe the actual breasts.

178. *Auricula* most commonly means "earlobe" but in this context refers generically to the difference between the external structure of the ears and its inner hearing mechanism. There is no facile one-word equivalent in our language.

179. On the nature and position of ears in general, cf. *QDA* 1.34–37.

177 The sea animal called the *koky*, or the sea calf [*vitulus marinus*, seal], also has such passages. For the same impediments that ears would create for the flight of birds (if they had them) would occur in swimming animals if they had ears. In general, no oviparous animal has ears, and it has only hearing passages. The mole, which is a quadruped, has neither ears nor sight. It has hearing passages only. It is the same for the mouse which lives in the earth and has a long body, a short tail, the color of a rabbit, and which is called the *cicel* [ground squirrel] by the Germans.[180] It much resembles a rabbit, but it has the teeth of a mouse. This animal has no ears, but only passages, like the mole, because they would interfere with its burrowing in the ground. Neither would they be of any use to them, for these animals spend the majority of their time underground. Ears are given so that the fan shape of the ear might gather in the sounds that are coming in on the air from all sides and might direct them to the inside organs of hearing.[181]

178 One should know, then, by way of summary about the ears and the hearing of animals, that hearing occurs on either side of the head. For sight, rather like a lookout, occupies the front part of the head, as does smell (because of breathing, which must be joined to the mouth).[182] Because of the sounds, hearing cannot be in a very moist, marrowy object and therefore cannot be to the front. Rather, it is positioned to the side since the rear of the human head is empty. It tends to the rear instead, where the nerves are harder. For a hard, dry drum makes a better sound and the ears are also positioned better on the side in a human.[183] In quadrupeds, however, with their heads hanging down, the ears—for those having them—are positioned above for the reason we mentioned, namely, that they can be moved in every direction to grasp sounds. Those placed on the side, in toward the head, however, cannot be moved. Ears are cartilaginous so that they can better cause the sound to reverberate and the sound is thus strengthened near the ear. This is just like an echo which occurs when sound is beaten back, much as when a ball is thrown against a wall. The ears are convoluted and concave for the same reason and so that they can hold the sound. The passage of the ear is twisted so that it can be longer and is curved lest the cold air be able to enter the brain quickly and bring it harm. Rather, it must first stay still and grow warm in the ear and also so that there might be quiet air in the ear, better suited to receive every sound. The nerves of the *typanum* of the ear are hard because, as we said, a hard thing gives the best sound. There is a passage between the ear and the palate but it is not visible. It is because of this that sharp sounds strike the teeth.[184] The reason that certain animals do not have ears has been adequately stated in previous books.

---

180. Cf. discussion of this animal at 2.63.

181. Or "and hearing directs them toward the inside." The "fan-shaped" portion of the external ear is the *ventilabrum*, technically a winnowing fan. Cf. its use at 1.29. It is worth noting that A. now begins to use the more common term *auris* for ear.

182. The "lookout" reappears at 12.182.

183. *Tympanum*, a part of the ear that even for A. (cf. just below) means drum.

184. Cf. 1.174.

In viviparous animals the organ of smell is suitably located between the two breast-like appendages of the brain [olfactory bulbs].

The usefulness of the nostrils is found to be threefold. One is that they might be the pathway for the air that is breathed in, a great part of which is led to the lungs, although much of it comes to the brain. This is because an odor is carried best in the air. It also serves to cool the brain and, by its movement, it serves to expel the superfluities in the brain by sneezing. The second usefulness is so that part of the air, held in the mouth, might be passed out through the nostrils. The passage of air through the mouth is bad, because it is mixed in with fumes from the stomach and is made foul. The third is that the superfluity of the phlegm, descending from the brain, might be hidden from view and then ejected through the nostrils.

We have spoken enough on the composition of the nose in the anatomy section, but it should be known that the middle piece of cartilage which separates the nostrils to the front is harder than the edges on each side of the nose and this is why the outsides can be distended and the opening of the nose can be rendered larger. For if the middle piece were extended it would necessarily always close or contract one or the other nostril. There are two of them because if one happens to become blocked from catarrhs or for some other reason, the other can serve the animal. This is the general reason that every sense is double.

Further, because the elephant does not have a long neck and head, it therefore cannot move them much. It is a very large animal and is aquatic, standing for long periods of time in the water. It thus sometimes holds itself up on three feet, giving the fourth a rest. Nature thus gave it a long nose, which they call the trunk, through which it breathes when it is sunk beneath the water, with the nose projecting up out of the water. Its nose is also midway between hard and soft. It has some softness so that it can be bent better, for it uses it to take up food and place it in its mouth. It also uses it to take in water and does other chores with it. It has retained some hardness so that it is not easily injured by mishaps, for it uses it to fight and to defend itself. Some say, as Avicenna attests, that there are also some cows which have trunks like these and that they use this to take food and place it in their mouths.[185] This is why the animal's neck is shortened and it itself is large and tall. It is necessary that there be something long in front through which a bending to the ground can be accomplished, for the animal's food is there. Now the mouth cannot be elongated this way, for the jaws would be too weighty. The lips cannot be elongated because they would hinder the movement of the jaws, which is especially necessary for chewing.[186] What is left is that the nose had to be elongated and, because the nose was long, the face had to be placed on the bottom of the head, toward the chest, since it had a short neck. For if the mouth had been under the nose, the length of the nose would have covered it over. Birds, however, have narrow nostrils on top of their beaks, since in birds the upper beak takes the place of a nose.

---

185. Avic. *DA* 13.32r.
186. One would, at this point, be interested to see A.'s reaction to his first view of an anteater.

Let these things, then, be those said about the nose, for the other things have been said in the anatomy section.

CHAPTER III

On the Disposition of the Eye, the Eyelids, and the Hairs [Cilia] *That Surround the Eye*

181 Speaking of sight, which as we said in the second book of *On the Soul* has more of the *ratio* of sense than do the other senses, we say that, just as we have pointed out in the anatomy of the eye, sight requires a watery fluid which is able to receive visible forms because of its transparency [*perspicuitas*] and its moistness.[187] Air, while transparent [*perspicuus*] and moist, cannot be made into flesh or a body and remain in its own species. We have shown this in other books. The reason for this is that air is only the pathway by means of which the visible forms arrive and it does not hold them. It is the job of the eye to hold and receive them. It is thus necessary that the substance of the eye be crystalline and somewhat thick. The visual power and spirit come to the eye along the hollow nerves which are called optic [*optici*] in Greek, that is, "visual." They are hollow so that the visual spirit might pass through them quickly. They are connected with the eye in the way we mentioned in the anatomy section.[188]

There are three humors in the eye. The middle one is the crystalline [lens] and is pressed into a flat surface in the front toward the ball of the eye.[189] This is so that it might receive the forms of things as they are, much as a flat mirror does. For if it were convex and rounded, it would receive things smaller than they actually are. The functions, name, and positions of the three fluids and of the tunics of the eye have already been established in the anatomy section.

182 To be sure, there are two eyelids in the human, but only the upper one moves. It is not necessary for the lower one to move, because the entire opening and closing action of the eye is fulfilled by the movement of the upper one. The upper one is best suited for movement, moreover, since it is nearer the nerves and muscles which produce movement. Also the line of vision is not directed downward usefully since only that which is near could be seen.[190] A better lookout is one who sees far off so that precautions can be taken against harmful things, and this is done more suitably by the movement of the upper eyelid. The hairs of the eyelids are fixed over a cartilaginous

---

187. The anatomy and function of the eye is discussed in detail in 1.197–213. The current version parallels Avic. *DA* 13.31vf.

188. Cf. 1.200.

189. The crystalline humor is also called the "icy" humor and is identified with the lens; cf. 1.197–204.

190. "Line of vision": *oculi acies,* a term that in CL also can indicate the pupil of the eye.

edge and have been created as a defense against harmful things which might fall into the eye.¹⁹¹ They are naturally dark to temper the light coming to the eye and thus assist in sight. They stand in a cartilaginous place so that they can stay rigid and upright. For if they fell, drooping, they would hinder vision and harm the pupil.

Further, an animal with hard, scaly, or wrinkled skin does not have eyelids. For skin like this could not move over an eye, opening and closing it, without injuring it with its hardness. Thus, nature gave such animals strong, tough eyes, which are not easily injured by a harmful thing falling into them. Oviparous animals also have a prickliness on their feathered skin, and thus they do not have an upper eyelid, for since the feathers are buried in the skin, the root of the feather would make the inside surface rough, and it would then destroy the eye when moved.

The manner of closing the eye differs in animals. The human closes it by moving the upper eyelid. Quadrupeds, however, do so by moving the lower eyelid since they have long heads and the muscles that move the ears are above, as are those moving the eyes. Thus, so that nature's work might be distinct, she made the muscles that close the eye joined to the lower eyelid. Birds close their eyes in neither way, but instead the eyelids are below with a certain white membrane which runs over the eye and wipes off the moisture flowing on it. When they sleep, however, they close their eyes with the lower eyelid.¹⁹²

Oviparous quadrupeds do not, as do birds, have a skin of this sort through the movement of which they clean the eye. They do close their eyes, however, since they have no need in their eyes, as do birds, for the thin, clear fluid by which remote things are contemplated. For the place for acquiring its nourishment is nearby.¹⁹³ Birds that are not good flyers and go about on the ground, like the ostrich and certain others, are close in nature to the oviparous quadrupeds. Fish, however, have no need of eyelids because of the hardness of their skin and that of their eyes.

Summing up, then, and following Aristotle, we say that oviparous quadrupeds and birds have a covering for the protection of their sight.¹⁹⁴ But a viviparous animal has eyelids, as does a bird with a heavy body. Further, a quadruped closes its eyes using the lower eyelid; the bird closes them with a membrane attached to the corners of the eyes. For the nature of its eyes is created from a very fine moisture, and it therefore requires this protection, for the reason that its sight will be sharper. This, then, is the reason that the eye has been created according to the disposition we have given. In feathered animals with a tough skin, if the covering were any different from what has been stated, many harmful things would fall into the eye from outside and because of these the sharpness of its vision would be hindered. The skin which moves over the

---

191. "Hairs": *spinae,* an odd choice to say the least.

192. Cf. 1.213, 307.

193. That is, whereas an eagle requires long-distance vision to facilitate hunting, a reptile does not, eating what is at hand.

194. Ar. *Part. An.* 657a25f.

top of birds' eyes is there for the sharpness of their vision and is thus very thin and fine. Thus, the eyelids have been created for the well-being of the eye.

186 Everything which closes an eye closes it so that nothing can fall into the eye from outside and this closing is not voluntary, but is natural. The human closes his eyes many times since he has the thinnest skin of all the animals and therefore the movement of the eyelid cleans the eye without harming it. There is skin over the eyelid of the human alone which holds the eye in, but which lacks flesh. For this reason it does not grow back and join together, just as the foreskin does not. This is because these pellicles lack flesh. For that which has no flesh does not grow back since growth comes from the flesh both as from its matter and its vessel. This alone of the uniform members flows in and out during growth and diminution, as we have shown in *On Generation* [*Pery geneos*].[195]

187 Birds which close their eyes do so only with their lower eyelid. An oviparous quadruped likewise closes its eyes using the lower eyelid due to the hardness of the skin surrounding its head. It is to this that the upper eyelid is joined, as we said above. Further, nonflying birds, because of the weight of their body, have very thick skin. This is because the thick plumage on the flyers passes over into thickness of skin in birds like these. This is because the matter of feathers tends toward and is drawn toward skin, that is, skin [*pellem sive corium*].[196]

Thus, every animal like this closes its eyes with the lower eyelid. Doves and many other birds resemble this sort of animal. An oviparous quadruped which has a scaly skin has a skin harder than does any animal with eyelids. That part of its skin which is near its head is very hard and, therefore, it cannot have an upper eyelid. Instead, it has an eyelid possessing flesh on its lower part. Because of this it has in its eyelid both a hardness and extension suitable for closing the eye.

188 A bird with a heavy body closes its eye in the manner we have related, namely, with a membrane suited for this. Now the movement of its eyelid is slow and a closing of the eye which cleans the vision ought to be fast for the sake of sharpness, so that the light of the eye can return quickly to directing its flight. When a bird both closes and cleans its eye, however, it closes it with a membrane which is attached to the corner of the eye at the cone of the nose.[197] It moves from this starting point to close the eye more suitably since it is close to the motor nerves and muscles. This membrane has a very swift movement.

In the human, what follows the eyelids is the principle of the eyebrows. For an oviparous quadruped does not close its eyes with eyelids placed beneath eyebrows,

---

195. *DG* 1.3.6.

196. A tricky passage. A. defines *pellis* so that no one will mistake it for a membrane, and he has been very careful in this passage to use *corium* for "skin" for just this reason. Thus, the "skin" of the eyelid above was *corium*, a term usually reserved for tougher things than human skin. On *corium* as a term for the outer covering of sea creatures, cf. 12.208 and notes to 4.6.

197. On "cone of the nose," cf. 1.138.

because they do not need eyebrows for the sense of sight. For these are earthy and do not have the thin fluid, for they are land dwellers.¹⁹⁸

Birds, however, necessarily need sharp vision, for they must catch sight of their food from a distance and from the loftiest heights. This is the reason that an animal with curved talons has sharper vision, namely, because it has to see its food from remote places. This is why this genus of bird raises itself higher in the air than another bird, namely, to see its food from a distance. Birds, however, which do not rise up from the earth and fly but little, such as the hen and those like it, do not need vision this sharp since they feed on almost every type of food and it is found nearby.

Moreover, fish, animals with tough skin, and certain ringed creatures as well are very different from these with respect to their eyes and to one another. Almost all of them have hard eyes, but some of them have eyes and lack eyelids, for if one of their number has tough skin, it has no eyelids at all. For the work of the eyelid should be fast, and tough skin cannot move fast, as we said above. In place of this protection and of the eyelid their eyes are made hard, and because of this hardness they have weak vision. Because of this, and to make up for this lack, nature made their eyes extremely mobile. Animals with hard skin especially have this sort of eyes, for example, those of certain quadrupeds with hard skin, so that their vision takes on sharpness because their eyes are called up quickly and directed to the light which they receive falling directly on them from the air.

Fish have moist eyes, for they need a great deal of vision since they are animals always on the move. Those animals which live in the water have nothing blocking and covering their vision, as do those that live in the air. Those that live in the air have an almost infinite number of things falling into their vision and blocking it. This is the reason that the eyes of fish have no eyelids, as do the animals that dwell in the air. For nature does nothing to no end.

Moreover, in addition to all the reasons introduced, the eyes of fish are made hard because of the heaviness of the water, for the very coldness of the water hardens and solidifies the eyes. Because they are in water, they also have very large eyes with a great deal of fluid.

All hairy animals have eyelids to cover their eyes. Generally, birds and animals with scaly bodies do not have eyelids to cover their eyes, because they do not have hair. In what follows, however, we will speak of the disposition of the ostrich, and we will give the cause of the disposition of its eyes, since some people say it has eyelashes.¹⁹⁹

Of those animals which have hair, some only have hair on the upper eyelid. The human, however, has hair on either eyelid and this is to protect his eyes better, for nature prepares greater aids and protections for the noblest animal. That which is most useful in nature is the most natural and most possible in nature, and this is the

---

198. On the purpose of the thin fluid, cf. 12.184.

199. A. never fulfills this promise although, as 23.139(102) makes clear, he had personal experience with the birds. A. was unsure whether, all in all, the ostrich belonged to the bird genus or was some sort of cross between a bird and most normal, terrestrial quadrupeds. Cf. 14.75, 87.

reason that none of the quadrupeds has hair on its lower eyelid. Rather, beneath this eyelid there grow spiny hairs like on the eyelid of some animals, much like the hairs around the mouth and nose in the cat, dog, and lion.

192 Moreover, there are no special hairs whatever apart from the general hairiness growing on the groin area [*pecten inguinis*] or under the arms in any other animals, as there are in humans. Rather, instead of these hairs, the entire back of quadrupedal animals is hairy. This is seen in the genus of dogs and of *mussiones* (that is, cats [*catti*]).[200]

Moreover, some of these animals, which have bodily hairiness instead of the hairs on the groin, have hair like the horse. The horse has hair that is smooth hair, although it does have a mane on its neck. Some, however, have hair all over their neck like the male lion.[201]

In some, moreover, nature has arranged for long, hairy tails in place of the hairiness of the groin. It is this way in horses. She made the hair of the tails of those with small tails small and she measured this out according to the nature of each animal's body. Thus, a very hairy animal, such as the wolf and fox, does not have a tail with hairs as long as those of the tail of the horse, ass, and cow.

193 It is proper for the human to have many quite long hairs on his head. No animal has hair on its head as long as the human's. This is because he has a moister brain than the other animals and because of the many sutures which are in his skull [*testa capitis*]. For the fumes escape through these and are converted into hairs. Now, where there is a moisture that is evaporating and a heat that is raising up this vapor, there must necessarily be many hairs to cover the brain and guard it as much from strong heat as from strong cold. It needs the greatest protection, too, for the human brain is very moist and it is quickly chilled and quickly boils up. Animals and members with contrary complexions, however, receive the mishaps of external illness with nothing approaching this ease and thus do not need so much protection.

194 We have spoken here of the hairiness of the head, not intentionally, but incidentally, insofar as those hairs are like the hairs on the eyelids. The eyebrows, like the eyelids, were created to assist in vision. For the eyebrows, by the angle at which they lie, and since they project, hinder the descent of fluid from the head, over the forehead, and into the eyes. This is the same as the eyelids having been made to protect the eyes from things falling into them from outside, like a fence which prevents entrance into a garden.

The eyebrows are positioned over the angular bones of the forehead and of the eye sockets. There are certain passages through the bone itself and the humor especially stands still at the angle as it descends from the head. This is why eyebrows grow long and increase in size in old age, for old age abounds in phlegmatic humor at the front

---

200. Cf. 22.121(78).

201. In a tour de force, A. uses three words in this paragraph for hair. While the Latin *may* be making a slight distinction between the types (e.g., the word for the lion's mane, *coma*, may have overtones of "hairdo"), the terms (*pilus, crines, coma*) seem fairly interchangeable. In the next section he will add *capillus* to the list.

of the head. They are also long in those who are very studious or think a great deal, for the animal spirit and the attention draw humor to the inside. They sometimes grow so much that they need frequent cutting. They grow over the end of the veins of the forehead and where the completion of the skin of the forehead occurs, for the completion of the skin of the forehead and of the veins of the forehead occurs in the same place.²⁰²

This, then, is the necessity of the creation of the hairs.²⁰³ For no suitable operation of nature should be lacking which is useful for the operation and the defense of the noble members. Therefore, generally speaking, there was a necessity for the growth of hairs in the human in the places mentioned above.

CHAPTER IV

On the Disposition and Cause of the Elephant's Nose,

That of Certain Oviparous Animals, and of Others

We have thus treated the organs of the senses that have been introduced. Now, it is necessary that the organ of smell in quadrupeds and in other viviparous animals be put in a spot where it has a location that is most suited to its nature. This instrument is preserved in a member which is called the nose in animals such as these. Such an animal has long jaws [*fauces*] which are the two jawbones [*mandibulae*] led out beneath each of the cheeks.²⁰⁴ In quadrupeds the nose is extended to the full length of the jawbones, but in other animals, such as the human, one nostril each is split toward either cheek and the nose remains small, above the animal's mouth.

Of all the animals, the elephant has the individual trait of having an extremely elongated nose. This is, as we have said, because the instrument of its smell has many powers and is used for many functions. This is the member with which this animal brings food to its mouth and it uses this instrument in almost all respects in place of a hand. It thus has not one but many functions. It sometimes places this member on trees and uproots them and in our time a certain elephant used this member to throw an ass along with its burden over a certain house. As we have said, it uses this trunk [*promuscida nasi*] in place of a hand for all its needs.

195

---

202. Cf. Ar. *Part. An.* 658b20f., which says only that the skin of the eyelashes comes to an end and it is there that the eyelashes begin.

203. *Cilia* forms a puzzle here. Note that it is repeated from the title of this chapter and cf. the discussion of the term at 1.139. In short, our options include "hair" and "eyelid." The former has been chosen because of A.'s preference, in this area of his text, to use *palpebra* for eyelid, and because of the use of *cilia* in the chapter title.

204. Reading *quod* for the typographical error *quad*. *Fauces* is not heavily used by A. and tends to indicate more the ravening jaws of a predator. See Glossary, s.vv. Mandible and *Maxilla*.

196   It is natural for this animal to be in desert places and watery places. For it feeds in the swamps and often sinks into the waters. Since, however, it numbers among the walking, blooded animals, it must have air to breathe. Its breathing increases according to the heaviness of its walking through wet places because it sinks deeply due to the weight of its body, and for this reason too it seeks wet food to cool itself. When it sinks, it helps itself breathe using a device like those who dive under the water and remain there for a long time. These men arrange a hollow instrument around their mouths which covers both their noses and mouths. They fit it tight so that the water cannot enter and in the middle of this thing they attach a very long reed, the opening of which is always out of the water in the air. They draw in air through this and can thus stay under water for a long time.[205] Nature has shown the same cleverness in preparing a long nose for the elephant. For since, due to the weight of its body, it sinks far beneath the water, it holds its nose upright out of the water and breathes air through it. Its composition, as we said above, is midway between soft and hard cartilage, or flesh. It had to have some softness so that it could be flexible enough for work and for taking up food.

197   There are certain men who say the same thing about the horns of certain cows that feed backward, but this seems to be a fabrication.[206] But it is certain that nature made this disposition for the elephant's nose in such a way that it uses it for many activities, using it much as a hand. Some animals use their front feet this way, especially the *varium* [squirrel], the *spiriolus* [squirrel], the mouse, and the monkey.[207] For these animals do not use their front feet for walking alone, but they also pick up food with them and perform other activities with them much as the human does with his hand. This is the reason that such animals have many long toes, so that their foot could be made ready for use like a hand. Elephants, however, because they do not have many toes on their feet (which are not split in order to stay strong to hold up their heavy body), use them for nothing else. Thus, nature placed both the usefulness of a hand and the ability to breathe in its trunk [*promuscida nasi*]. Because the animal is aquatic, the nose had to be long since when it dives it would, due to the weight of its body, be totally destroyed unless it had a long trunk to carry on breathing. She made it bendable so that it could be used in place of a hand, just as certain animals use their front feet.

198   Birds, serpents, and also oviparous, quadruped animals which are blooded have their nasal passages either in their beak or near the mouth, and they are not easy to see and are not separated. For this reason some have said that these animals have no noses. This stems from the fact that these animals do not have two end points at which they have nostrils which separate, *per se*. Rather, the final separation is that of their beak or jaws. This happens most suitably in birds, for they bend to the front of their body and

---

205. According to Stadler's indications, A. has added much to the original description of Ar. and we thus may have a description here of a medieval snorkel.

206. This extremely old tale was first related by Herodotus 4.183, and it appears here via Ar. *Part. An.* 659a17f. The oxen reportedly had horns so long that they had to graze backward in order to eat.

207. The *varius* [sic] is described at 22.149(111) and the *spiriolus* at 22.134(97); cf. 3.96.

thus their head happens to be small and of little weight so as not to droop and press down on them. Others have a long and narrow body, because of which their heads must be narrow and pointed. Therefore, an end for its nose cannot exist *per se,* for it would make the head thick and disordered. Moreover, because birds need food which they must snatch up and seize with vigor, nature therefore made their beak out of a bone-like substance. This beak, moreover, is narrow because of the small size of the head, and it has its olfactory passages above on this beak. It would thus be superfluous for them to have other noses.

Previously, and concerning animals that breathe in another way, we have given the reason that they breathe in a particular way. Just as some, which are true breathers, breathe through a windpipe and lungs, so do others breathe through gills, these being aquatic animals which breathe in a particular way.

An animal with a ringed body breathes contrary to the two ways given, using the web of its diaphragm. This is why they buzz in a wondrous way when they move through the air, because thin spirit comes through their rings to the diaphragm.[208] They also sense odors in this way, and thus, in animals such as these, smelling is accomplished in the heart. For, due to the very great smallness of their head, they cannot have the organs of all the senses in their head. Every animal that breathes has a special cause for the breathing, namely, the cooling of the body by means of cool air entering and hot air leaving. This is as we have said in the book *On Breathing In and Breathing Out* [*De Inspiratione et Respiratione*].[209]

199

CHAPTER V

On the Nature and the Cause of Lips, Teeth, and the Tongue

and on the Particular Uses of These Organs

The lips are created beneath the nose in those which have a nose, and there are lips in those animals which have teeth and blood.[210] Birds, however, as has already been said, have a beak instead of teeth and lips. This is because of their food and for vigor, so that it can take the place of armament. The beak is therefore made hard and solid for two purposes, so that it can take the place of both lips and teeth. The beak of a bird, then, is as if someone should arrange, like a sculptor, an upper row of teeth so that they bend down over a lower row and then arranged that all the teeth in the upper jaw be solid bone and likewise that those in the lower jaw be equally solid. He would

200

---

208. "Diaphragm": *paries,* cf. 1.549.

209. *De spiritu et respirato* 2.1.3.

210. At. *Part. An.* 659b20f.

finally see to it that both rows ended in a point. This would then have the shape of a bird's beak.

In other animals which have them, lips have been made for the well-being of the teeth. Therefore, the lips have been divided so that in this way they can protect the opening of the mouth and each pair of teeth.

Human lips are soft and fleshy so that they can open, move, and close neatly for the protection of the teeth and for the formation and the exit of speech [*sermo*].[211] For nature did not create the tongue of the human only for the use which the rest of the animals make of it. Its use is double, for she suited it both for taste and for speaking [*loquela*]. Just as she did this, she formed the lips and teeth not for a single use as in the other animals, but for feeding and speaking, meting out in her allotment to each of these a double use.

The tongue of the human, the pygmy, and perhaps of certain others which speak is created for a double use, namely, for tasting and for creating speech [*ad sermocinanum*]. The lips are also created for this and for the protection of the teeth. For when an utterance [*locutio*] leaves the mouth it is literate and articulate.[212] If the lips and the tongue were not soft, an uninterrupted formation of letters could not be pronounced in diction and speech [*dictio et oratio*].[213] For the tongue must, for the pronunciation of vocalization [*vox*] of this sort, strike and bend in different ways, and the lips must sometimes compress and sometimes open. For these movements the organs of this sort must move quickly and easily.

But one should make inquiry of the grammarians and the poets rather than the natural philosopher about the types of pronunciation of letters, the sounds of accents, and meter. But it should be known that these organs are soft and capable of every movement in the human, because the human alone of the animals has a refined sense of touch. Therefore, he alone can express the many conceptions he has to another through the formation of speech [*sermo*].

The tongue is placed beneath the palate in the mouth of an animal. This position for the tongue is according to one and the same mode in all blooded, walking animals. However, in the other animals it is according to a mode which differs both from this one and among themselves, for all of these do not share the same sort of position for the tongue.

The human tongue is very broad, soft, quite loose, movable, and flexible. This was done so that it is suited for the two operations we mentioned, namely, that which deals with the sense of taste and that which is involved in the interpretation of speech. For taste is not accomplished without some sort of touch, and it is agreed that a humor, which is the subject of taste, is more easily diffused on something soft and broad. If it is loose and broad as well as soft, it is also suited for the interpretation

---

211. Compare this and what follows with the definitions of speech at 1.234f., 4.90, 7.62.
212. "Literate and articulate": i.e., formed into letters and distinct syllables.
213. Reading Borgnet's *indistans* for Stadler's *in distans*, which he found in a marginal correction attributed to the hand M2.

of speech [*sermo*], for it can be easily extended, contracted, and moved circularly for the formation of the elements of a sound [*vox*]. An indication of what we have said (namely, that a broad, soft, loose tongue is well suited for proper speech) are those who have bound tongues, that is, the *trahuli* and the *blesi,* or those who have some other impediments leading to poor interpretation. We have spoken of this in the anatomy section.[214] When this occurs, it results in the impediment of certain letters, due to a large and thick or large and wide tongue deficient in pronunciation. Among Latin speakers these letters are "r" and "s," for, as Pythagoras said, the small is in the large and not, oppositely, the large in the small.[215] This is also why a bird with a loose and broad tongue can make the sounds of certain combinations of letters more easily than can a bird with a narrow tongue.

A blooded, viviparous quadruped does not have a voice which can be divided up and shaped into the forms of letters. It also has very few variations to its sounds. For, even though some of them might have a tongue that is broad enough, such animals have a hard tongue which is not loose. Small birds, however, have many calls and many differences in them so that some of them recognize their conditions of pleasure or pain by the type of their call. It even seems that all birds perceive this from one another, even though they differ in this regard by understanding it more or less. Thus, it has been experienced, that certain birds learn to imitate the voices of others, either of humans or of other animals. Examples are the starling, magpie, raven, and especially the parrot.

The tongues of blooded walkers (that is, creepers [*reptilia*]) that are oviparous are useless for calling. For they are very hard and bound and are only useful for tasting. The tongues of serpents and lizards, though, are long and split, with two branches to the front. An indication of this is that sometimes these animals extend their tongues quite far out in front of their mouths and are then seen to have forked, sharp parts to their tongues, whose ends are like fine hairs. Taste is sensed at these ends on the two parts due to the division of the tongue. Similarly, a bloodless animal has a member that is like a tongue and through which it senses tastes and fluids.

Moreover, certain fish and certain aquatic animals have some very viscous flesh instead of a tongue which they use to sense tastes and fluids. An example is the *tenchea* [crocodile] and the freshwater river eel. This member is bound below in the *tenchea* since it has a moveable upper jaw.

In some animals the member is bound for the sake of necessity, for if it moved it would be destroyed since the area of the mouth in such animals is very spiny. This is seen in the fish called the pike [*lucius*]. Nor is there any necessity that this member move in such animals, for the sensation of the fluids and of the tastes stays in their mouth but little. For their food moves to the belly very quickly, more by way of gulping than chewing. It does not stay in their mouth long enough that its inner

---

214. These impediments are discussed at 1.236.
215. Cf. Ar. *Part. An.* 660a28f., but note that Pythagoras is not mentioned by name.

fluid, which is the subject of taste, might be extracted by chewing, pouring out over the tongue. For they take their food in water and they could be choked by the water. Therefore, in such animals, the member is so bound that unless the mouth is tipped to one side, nothing of the aforementioned member is visible. This is especially because the entire mouth area is spiny and their mouths seem to be composed of nothing other than the spiny composition of the gills.

However, the *tenchea* has a small and short tongue, curtailed because of the movement of the jaw. For its lower jaw does not move at all and the tongue is joined to it. The *tenchea* moves its upper jaw, something contrary to all other animals in which the upper jaw does not move at all save accidentally. The tongue of the *tenchea* is not connected with part of the upper jaw, since its food enters its body in a way contrary to the entrance of food in the other animals. Likewise, its lifestyle does not agree with that of the others. For although it is one of the walkers, it has a lifestyle like that of a fish. Thus, its tongue was not connected with its upper jaw.

Most of the marine animals have a fleshy palate. Likewise, in most of the river fish there will be a very soft, fleshy palate, like that of the fish called the *kohery* in Greek and that of the scaly river fish called the monk [*monachus*] by the Germans.[216] Thus, many who see and touch this fish think that its tongue has been divided into its palate. For the reason we have given, however, fish have neither a divided nor a loose tongue. Rather, they have a member which resembles a tongue for use in sensing food, but which is not at all like the composition of a tongue with respect to its whole disposition. But it is somewhat like a tongue at its very end, since this is the case only in the fish and not in other animals.[217]

All animals have a sense in the body for food, and, as we have said elsewhere, appetite is only for the sweet and the tasty. Still, the member employed to sense food is not the same in all. In some it is loose while in others it is joined to the lower jaw. In some it is soft, in some hard. An animal with a soft shell, like a *karabo* [lobster] and a sea urchin [*hyricius marinus*], also has a member of this sort and such a member is found in the mouth of the animal called the *malakye* (the cuttlefish, for example) and in the mouth of the octopus [*polipus*]. An animal with a ringed body has such a member inside its mouth, as do the genus of ants and many animals with a shell-like covering [*testeum corium*].[218]

---

216. Ar. *Part. An.* 660b37 lists the *kyprinoi*, a type of carp (*GF*, 135–36). On the problems surrounding the monkfish, cf. notes at 1.228.

217. This sentence is very murky in Latin, and the text as it came to A. was not very clear. A solution may be at hand. The Gr. of Ar. *Part. An.* 661a7 states that this tip alone is unattached in the mouth, *aphōristai monon*. This would have come into Latin as something like *hoc solum absolutum est*. Yet the received text read simply *hoc solum est*. It can be conjectured with profit, therefore, that, due to the similarity of the forms, the word *absolutum* dropped from the text to its detriment.

218. This compound Latin term is discussed in the notes to 4.6.

Certain of these animals have this member so that it is very long and extends out beyond their mouth, much as a lance-like stinger does.[219] Now nature did not do this to no end, for the animals taste with this, sensing a fluid and drawing it in by sucking on the food as if through a hollow reed. For it has the nature of such a hollow thing. This is visible in the flies, bees, and those like them.

Further, this member is also in certain animals that have a shell-like covering. The member has such sucking power in the shellfish [*ostreum*] called the *bokoky* in Greek that it uses this instrument to pierce the shells of certain *conchylia,* which are generally called the *halzumi* in Arabic. Examples are the dog fly [*cinomia*], the *cinifes* [mosquitos], and certain other flies.[220] They use this part to pierce the hides of beasts and of humans as well.

This, then, is the disposition of the tongue in these animals. These animals use this member much as elephants use their trunks. The tongue of these animals is made long so that it can perform the chores which the elephant uses its trunk for. So too has the tongue of these animals been created after the fashion of a stinger.

The disposition of the tongues of all animals, then, is in accordance with the mode given.

CHAPTER VI

On the Nature, Cause, Types, and Number of Teeth

It follows next to speak of the nature of teeth and of the other things contained in the mouth.[221] For the mouth contains teeth which are like the circle of wood which surrounds the millstones in a millhouse and holds them in place.[222] The tongue is like the hand of the miller which collects the grain beneath the stones. Thus, too, quadruped animals which do not have teeth in their upper jaws take food from the ground with their tongues and then bunch it together, so that it can be cut with the scythe of their lower teeth. The universal use of teeth in animals is for food. In some, however, in addition to this, they are used to provide the strength of armament against harmful things.

The nature of teeth in some, such as in wild carnivores, is that they might act on and not be acted upon by contrary things, while in others it is to aid in eating, being

---

219. "Stinger": *aculeus,* on which cf. 1.96, with notes.
220. It is a tribute to A. that when he is as wrong as this, one is surprised. A. is quite aware that *conchilia, ostreum,* and *halzum* are names of shellfish (see Glossary) and uses them correctly time and time again. *Boboky* is corrupted from *porphurai,* murexes, Ar. *Part. An.* 661a22. The two insects are found and discussed at 1.96, to which this passage is indebted.
221. Cf. Ar. *Part. An.* 661a34f.
222. The following conceit is very similar to that of 1.225.

used to cut and grind the food. Some of the teeth are called wide molars [*molares lati*] and are suited for grinding.²²³ Between these and the front teeth are the ones called the canines in the human. These are sharp and their natural disposition is midway between teeth and molars, since they resemble both.²²⁴ They are sharp in one way and in another way they are broad. The disposition of the teeth in all the animals which do not have all sharp teeth is like this.

In the human, in addition to the duties mentioned, the quality of the teeth (which exists in their shape), their quantity (which is in their number), and their size are suitably adapted for speaking [*ad loquendum*]. For the tongue strikes them in different ways and forms different types of speech [*loquelae*] and letters. The front teeth are used for this.

Certain animals, however, have teeth used only for food, as we said above, and others have them, in addition to this, for strength of armament. Thus, some of them, like boars, have two tusks, the sole use of which is for fighting. Some have sharp teeth which fit into each other in a saw-like fashion and their teeth are thus called strong, since their strength is in their teeth due to their sharpness. All the teeth of strong, biting animals fit together like the teeth of a saw so that they are not easily dulled from striking one another.

However, not one of the animals has both sorts of armament at the same time, namely, sharp, saw-like teeth in its mouth and, additionally, tusks that project outward like those of the boar. For nature does nothing superfluously. One of these two types of teeth is for strong tearing and the other is to fight and strike blows. One type of armament is adequate for defense and attack for one and the same genus of animal, and the animal's mouth would be overly burdened with two types of armament. Nature thus did not give armament to the weak and timid sex, for the sow does not have projecting tusks and she bites only using the sharpness of her front teeth, called incisors.

It should be known in general that by analogy with the natural things we have determined, and which we will treat thoroughly below, nature does only the best possible job. Thus, she never gives armaments to animals to strengthen them unless they need them. We are calling armaments such things as stingers, claws, horns, projecting tusks, and the like. Because the males of animals are stronger and more spirited, nature might happen to give these arms only to males or, if she gives them to the females, she gives stronger ones to the males than to the females. Thus, male deer have horns and the females do not, and bulls have stronger horns than do cows. If it should happen otherwise, it is a result of the earthiness of the animal's nature.

However, the members which are necessary for life are equally present in females and males, although the females' members are often smaller than those of the males. But the others, which are not needed by necessity but rather serve as armaments, are either not possessed by the females or are present in them differently from the males.

---

223. Perhaps "wide grindstones." Cf. immediately below where once more one cannot be sure if *molar* is used more as a metaphor or a proper name.

224. Perhaps "midway between teeth and grindstones."

Thus, there is a difference between the horns of bulls and cows throughout the entire genus of bulls, and there is a difference between the horns of ewes and rams. It is the same for the horns of he-goats and she-goats and in all the others.

The teeth of all fish are very sharp, with the exception of a few which have jaws along the sides of their throats, examples being the river fish with large scales, such as the barbel, the monkfish [*monachus*], and certain others.²²⁵ For these have wide teeth slightly more useful for grinding than for cutting and they are thus all equal in size. Nature therefore hid them in the throat so that when their mouth was closed, the water would not suffocate these fish as they chewed. The fish called the *carpo* [carp] especially has teeth of this sort. Other fish, though, especially the marine fish, have saw-like teeth with the sole exception of one genus.²²⁶ The reason for the sharpness of their teeth is that the fish must necessarily take in water with their food. Their food thus comes suddenly to their bellies and cannot stay in their mouths for a long time. For otherwise the water would enter their bellies, fill them up, and kill them. This is the reason that the teeth of all fish are sharp and numerous, so that because of their sharpness and numbers they can cut food into many pieces. They have this arrangement as a substitute for chewing and grinding. The teeth of many fish curve like hooks since they thus have strength for holding.

The nature shared by animals needing the activities mentioned commonly makes teeth in them, just as the nature shared by animals produces organs of breathing in those which walk and are blooded. For, as has been determined previously, nature uses many members in the common activities of animals. This is clear in the mouth.

Every animal eats with its mouth. But the strength of armaments properly belongs just to certain animals, just as breathing is not common in the same way to all animals. Nature, for the sake of brevity, gathers these activities into a single member, which she nevertheless differs by varying type throughout the individual species of animals. Thus, certain animals have a small mouth and others a large one. In every animal which needs a mouth for strength and battle, there is a large mouth opening. The opening of the mouth opens wide, for otherwise it could not seize things forcefully and, once seized, it could not hold them forcefully. All the strength of this animal is in its mouth. For this reason too all animals which have sharp, saw-like teeth and which are carnivorous also open their mouths very wide so that the act of biting can be carried out better and more forcefully. This is clear in the mouth opening of the wolf, dog, and lion, which is known to be very large. Likewise, carnivorous fish with sharp teeth have a mouth opening which is very large compared to their bodies.

Birds have beaks in place of lips and teeth. Their beaks differ according to their differing needs and uses. Thus, the beaks of birds which live by hunting flesh are curved in front like a hook. For this way it holds on to and rips off the portion it eats.

---

225. Cf. 12.207 and 1.228.

226. From Ar. *Part. An.* 662a7f., we find that this fish is the *skaros*, or parrot wrasse, which, in addition to having these wondrous teeth, chewed its cud and had an audible voice (*GF,* 238–41).

The strength of this sort of bird lies in its beak and talons and thus its beak is made very hard.

The beak of every single bird has a shape and disposition that suit its lifestyle. Thus, the beak of the bird which bores into trees, called the *pica* [magpie] in all its genuses, is made very hard and sharp. In the genus of ravens, the beak is strong and hard as well, used for breaking food and for fighting. On the other hand, the beak of the small birds is delicate and sharp, suited for plucking out and seizing seed. The same sort of beak is possessed by small birds which live by hunting *cinifes* [mosquitos], as the swallows do, and the similar small birds which hunt vermin.[227]

Further, birds which have skin between their toes and which live in or near water have beaks which suit their life. Some of these have wide beaks, like the duck, since the width of its beak is suited for digging and for straining, as it were, the mud that is in the water. For in this are the roots of grasses [*herbarum*], the seeds of aquatic plants [*plantarum*], vermin, the eggs of aquatic animals, and other things of this sort which form ducks' food. Some birds have a long, pointed beak for stabbing into the mud and pulling out those things fixed deep in the mud. There are many such birds known to us. Those, however, which have a wide beak include the duck throughout its genuses, the goose throughout all its genus, the swan, and certain others. The width of the beak is adapted for the digging up of soft things. An indication of this is that pigs, whose lifestyle centers around digging to upturn roots, have a mouth that is wide in front. In the middle of this, between the nostrils, there is an appendage which has almost an acute angle. They use this to dig and upturn the earth to get the roots out. The birds mentioned above are like the pigs in that they have a lifestyle centered around digging in the mud.

The summary of all the things we have said is that, as we said in the anatomy section, there are thirty-two or twenty-eight teeth in humans.[228] In the upper jaw and to the front of the mouth there are two wide teeth which different people call the "broad teeth" [*ampli*] because of their size, "duals" because of their number, and "incisors" because of their sharpness. After these, on either side, are two teeth called the "fourths" [*quarti*] and there is an equal number of teeth corresponding to them on the bottom. These are commonly called incisors. After these, above and below, there are two canine teeth, which are used for the breaking and tearing of the food. All the other teeth, both above and below, are used for grinding.

In addition to all the teeth mentioned, there are four teeth which occur in a few people. They come in so that one occurs at the end of each jaw. These are called "wisdom" or "intellect" teeth, since they most often come in at the age of intellect [*aetas intellectus*], which is thirty years old. Still, they have been found to come in at different times, from the time when a person begins to produce sperm up until the time when

---

227. *Cinifes*: cf. 12.205. "Vermin": *vermes*, and thus, possibly, grubs, insects, or worms. See Glossary.
228. A. begins to follow Avic. *DA* 13.33v. With this section, cf. especially 1.195 and 2.46–48.

his growth stops. I myself have seen a man who had these teeth come in when he was in the eightieth year of his old age or more.²²⁹

Teeth have sharp roots with which they are set into the jaw. Over each tooth a certain thickness grows which is directly over the opening in the jaw in which the tooth is set. This helps to keep the tooth from being set further in during chewing, thereby piercing and injuring the jaw. There are strong bonds at the root and the tooth is held by these, keeping it from falling out of the jaw.²³⁰ All teeth, except the molars, have only one root. For they are small and their movement is not repeated over them many times, but only once when they cut into the food. The molars, however, because they move very often and forcefully over the food when they grind it, have many roots. There is a minimum of two for each tooth in the lower jaw and those molars in the upper jaw have three or four roots, especially in those teeth to the rear and in the "intellect teeth." This is so because the teeth hang down and thus require many roots to keep them from falling out. The lower teeth, however, are standing and do not need this much binding. Thus, too, an upper root does not reach out in a straight line but rather curves somewhat away from its natural location. This is so as to anchor it more firmly. Avicenna says that he saw a beaver and that all its teeth were red, curved, and long. The reason for this was that this animal is a hunting animal and unless it hunts it has no food. For this reason nature made its teeth long to attach firmly to the prey and she made them curved to hold on forcefully.²³¹ The reason for the redness was the influx of blood into them. For as Galen says, this type of bone is sensitive, being midway between hot and cold from its nearness to the brain.²³²

Teeth are generally given to animals for use on food, but to the human they are given specifically for the formation of literate and articulate speech [*loquela*]. They are also given to some as weapons, and thus those that live by hunting flesh have sharp, separated teeth, such as the lion, dog, wolf, and the like. Those that pluck plants or roots from the earth have their teeth even, set in one row, for a saw-like arrangement would impede the cutting off of these things. Saw-like teeth would pierce, but they would not cut off evenly. Because the females of many animals are weak and timid, the teeth that are armaments are not given to many of them. An example is the tusks of the boars, for the sows have none. And a thing much like this is said to exist in the camel. This is the reason that the stag has horns and the doe does not and why the he-goat and the ram have horns which are larger or stronger than those the females have. Fish which feed on the flesh of other fish, since they do not have a neck with which to bend their head toward their food, have many rows of saw-like teeth in their mouth, as does the pike. This is done so that they can cut the food quickly with one

---

229. Cf. 2.54.

230. "Bonds": *ligamenta,* elsewhere, ligaments, but cf. *ligamen,* "binding," just below.

231. This is a troublesome passage, for A. clearly knew enough of beavers, as at 22.39f.(22), to know that although they had a fierce bite, they were not carnivorous. The word used, *castor,* is the usual word for beaver.

232. Galen *UP* 16.2 (May 2.685; Kühn 4.271).

closing action of their mouth, keeping the water from entering the inside of their stomach, as we have said before. Others do not need this sort of chewing, for they use a sucking action, or else they have teeth almost in their throats, moving toward one another with the movement of the gills. Thus, too, they can move both jaws. Certain animals have a mouth that is only for vocalizing [*loquendum*] or eating and they have no need for many rows of teeth in their mouths.

220 Just as there are sharp teeth in the walking animals of prey, so in birds of prey are there curved beaks and sharp, curved talons. They thus can seize things on the wings, hold them in their talons, and then use their beaks to tear at what they are holding. Such birds as pick seeds have straight beaks which are thin at the ends. Such birds as browse in the mud have wide beaks, as we have said before. The beak is rather like the two rows of teeth, upper and lower, after they have been made solid, and as if the two jaws had been elongated and brought to a point.

Let, then, the things we said in the section on anatomy, along with these things, suffice on the nature and cause of teeth.

CHAPTER VII

On the Cause and the Nature of Horns on the Heads of Animals

221 Although we have already determined much on the members of the head, we should know that that part of the head which lies between the throat and the front part of the head called the *sinciput* is called the face [*facies*] or countenance [*vultus*].[233] With respect to this part no other animal whatever, except for the pygmy, is like the human. For the genus of monkeys is composed out of the shapes of a human and of other animals, and thus has a marked dissimilarity to the human in its face.

Some animals have horns projecting at the top of the head, on the *occiput* or the *sinciput,* and we should not pass over their nature. Animals which are viviparous have a head which properly displays the shape of a head. Certain other imperfect animals are said to have a head in a transferred sense only. They also participate in having horns in this improper and transferred way. For example, the forest scarab [*scarabeus silvanus*], which eats tree leaves and not dung, has horns which bear a resemblance in shape to stag horns.[234] These are on its head and are movable, and it uses them to seize and hold.

222 However, horns which are true horns are given as armament, as is the case for stags' horns. Some animals, though, have weak horns, and are thus of little or no help to the animal in a fight. Such is the curved horn which is on the head of the moun-

---

233. Cf. Ar. *Part. An.* 662b17f.

234. A. is carefully differentiating between the stag beetle (cf. 2.96) and the dung beetle, *Scarabaeus sacer,* best known from Egyptian scarabs (Beavis, 1988, 137–64; *GI,* 83–85).

tain goat, which the Germans call the *gemeze* [chamois].²³⁵ For this horn is small and weak and is shaped more to hold the animal. Thus, when the animal is scaling a cliff, it holds itself with its points as well as when it falls from the cliff.²³⁶ It is generally the experience in our lands that no animal whatever which has many splits in its foot and whose foot is divided into many toes has horns on its head. The reason for this seems to be that the armament of such animals lies in their teeth and claws, and a horn would therefore be superfluous. Some, as we have said, are strong in their teeth, some in claw, and some in horn. Nature, again, never gives what is superfluous. Others, however, have a horny sole [*solea cornea*] on their hoof [*calx*] and have their strength in this. Examples include the horse and the ass.²³⁷ But animals which cleave their hoof in two have special strength in their horns. For a foot cloven in two is made for planting the foot strongly and for standing, and not as armament.²³⁸

Nature has given many types of defense to animals. Some of them are defended against animals attacking them by the size and strength of their bodies, as are the elephant, the animal called the *seraph* [giraffe], the camel, and the like.²³⁹ Others are defended by their teeth and some of these, such as pigs, have tusks while their other teeth are even. Some have many sharp, saw-like teeth for seizing to the front of their mouths where the mouth opens. Some have hooves [*ungues*] and some have horny-soled hooves [*calces soleatas corneas*].²⁴⁰ Some have speed, like the hare, and others have two of these things. Thus, the horse has a hoof [*calx*] as well as speed, and the deer and the *hinnulus* have speed along with horns.²⁴¹ They fight weaker animals with their horns, but they flee stronger ones using their speed. There is a certain genus of wild cow (that is, ox) which has horns that bend back in on themselves and are not suited for fighting. Nature gave this one the power of casting its dung to the rear, far and forcefully, into the eyes of dogs and hunters following it. It casts it very far off and it is sticky so that it is not easily wiped off. The dogs and hunters following it are thus detained by wiping it off.²⁴²

As we have said previously, nature gives many types of strength and defense to the animals, and sometimes she gathers many types into one and the same animal. Thus, certain animals have both "soled hooves" [*calces soleatas*] and horns, as does the Indian ass [rhinoceros]. And because the nature of the right is very different from the nature of the left, there are sometimes two horns on an animal's head and sometimes only

223

224

---

235. Cf. 2.22, with notes, on this animal.

236. This story is repeated at 22.38(20).

237. Cf. 12.139 for wording and see Glossary for terms.

238. On the terms for feet used here, cf. notes to 2.8.

239. *Seraph,* as Schühlein, 1661, points out, is from the Arabic *zarāfa* and is, in fact, the origin of our word "giraffe."

240. *Ungues* can also indicate claws.

241. Cf. 2.23 for a discussion of the *hinnulus,* probably a form of deer.

242. This description ties this animal to the *bonachos* of 22.24(12), the European bison or, as they are called at 2.23, *bubali* or *wisent.*

one. Thus, the Indian ass has a solitary horn on its nose, and the animal which some of the ancients called the *archos* (that is, a sort of "prince"), but which we call the *unicornis* in Latin and the *rynnoceros* in Greek, has one horn of very great size.[243] It is solid like the horn of a deer and I have measured it as exceeding ten feet in length while its diameter at its base was more than a palm and a half. When an animal has one horn, this is located in the middle of its head since the middle of the head, whether on the forehead, the *sinciput,* or even on the nose, is the middle and shared boundary of the two extremes of the sides. Animals such as this, with one horn, have soles [*soleae*] or hooves [*calces*] with a whole or a cloven hoof [*ungula*]. But it is not a reversible statement, for those which have a horny hoof [*calces corneas*] for the most part lack horns. This is because nature has placed the material for the horns in the four feet, and one type of armament is adequate for it. She makes this placement in large and heavy animals, for they need strong support.

225 She placed the horns on the head because they arm the animal more fittingly and better there. Zeno, however, was wrong in criticizing nature because she did not put the bull's horns on its shoulders.[244] He said that the bull could draw its yoke more forcefully if it had its horns on its shoulders. But nature did not give it horns for pulling. Rather, the function of the horns is to be considered for defense and this is done best by the head since it is easily spun and moved against harm from any direction because of the mobility it takes from the neck. They would be a burden if they were on other members like the feet, shoulders, or the like, and there would be a great hindrance created by the bones on which the horns would have to be anchored. On the head, however, they both strengthen the cranium and assist in the most suitable defense. If not on the head, it could not be put in the mouth, for it would interfere with feeding. Therefore, the most correct position is either in the middle of the head so that they can have equal weight or on each side of the head so that both sides can have an equal weight.

226 Some horns are entirely solid, like those of deer and of other animals which have a nature like the deer's, such as the elk and *hinnulus.* Other horns are hollow near their base but are solid to the front at the tip they use to fight. Solid horns are heavy, however, and such animals shed their horns. This is an easing and a purgation for

---

243. Cf. the etymology offered at 2.27 where the animal is shown to be a sort of gazelle. The Gr. stem *arch-* does, in fact, refer to rule, as, for example, in the nine archons in charge of ancient Athens.

244. The name Zeno belongs both to a philosopher born at Elea ca. 490 B.C., who succeeded Parmenides as head of the Eleatic school, and to a later figure who was the founder of the Stoic school. This latter Zeno, a student of Xenocrates, had emigrated to Athens ca. 320 B.C. A. was certainly familiar with the numerous logical paradoxes of Zeno the Eleatic, which focused Ar.'s attention on the problems of local motion. Cf. *Phys.* 4.1.3 and 4.1.6. Among others, Zeno's famous paradox regarding the flight of an arrow—namely, that although apparently in motion it must in reality be at rest at each stage of its flight—is treated extensively at *Phys.* 6.3.1–4. The source of the claim that Zeno criticized nature's formation of the bull, however, remains unclear. Indeed, Sc. had offered A. the form *Altinoz,* whereas the original of Ar. *Part. An.* 663b2f. tells of one Momus, a character in an Aesop's fable, who criticized a bull for having its horns on its head. A version of the fable can be found in Babrius *Myth. Aesop.* 59.8–10 (Perry, 1965, 74f.).

them for they are very earthy and are purged of the earthiness by the generation of the new horns, especially in the head where the brain is cold and there are many bones. Thus, there is a great deal of earthy, melancholic superfluity there. There is generally a large earthy part in a large-bodied animal which has sebaceous fat [*seposa pinguedo*]. It therefore has many large bones and the superfluity of its earthiness passes over into horns. There is no animal whatever found in our lands which is a runner and is swift and which has horns as large as does the mountain goat, which they call the *ibex* in Latin.[245] For the horns of this one extend from its head to its rump. When it falls from a height it protects its entire body from striking anything with its horns and it absorbs the shock of large rocks with its horns as well.

The work of nature must be determined solely with respect to all the animals of the same species or to most of them. For there might sometimes be found a horned horse, just as a man is found with two heads.[246] Monstrosities are also born in many other ways and these do not represent nature's intent, but rather such things occur due to the flaw [*occasio*] of some error of the natural principles, as we have determined in the second book of our *Physics*.[247]

As we have said, an animal with a large body and many large bones has great earthiness, as is easily seen by anyone using reason to consider natural things.[248] Since earthiness of this sort is abundant in the matter, the food taken in is converted into its likeness. When the overabundant earthiness is thus multiplied, some part of it must pass into some member or other. In some it descends into the *sotulares* of the feet, whereas in others it descends into tusks, into sharp, long, saw-like teeth, or into horns. Because horns are often large, they draw to themselves so much earthy matter that it is taken away from the matter of the teeth of the upper jaw. Thus, animals with horns do not have front teeth in their upper jaw, since the matter is not adequate for both at once. But the lack of chewing their food is compensated for by rumination.

Nature does the best thing in all cases and she thus does not place the horn to the rear, for the animal could not see to direct the blow it was giving with it. If it were placed on the rump or the shoulders, it would impede the movement of the members and it would injure the bones on which it was anchored. She therefore did the best thing in putting them on the head. Now, no animal except for the Indian ass is found which has a solid, horny *sotular* and which also has a horn. This is because the matter for the horns descends to the feet and this too is why the ass has but one. Just as this is so, so too no animal save the unicorn is found with a split foot and a single horn. But the unicorn's horn is so huge and long that there is not enough matter for two of them. The smallest of the horned creatures seen in our land is the mountain goat,

---

245. See 22.105(54) for more on this animal about which A. shows intense interest, presumably from firsthand acquaintance in the mountains of his homeland and in the Alps during his travels.

246. On monsters, see Glossary.

247. Phys. 2.3.3.

248. Or "considering natural things with respect to (their) *ratio*."

which has red eyes, is said to have very sharp vision, and has a curved horn. In our tongue it is called the *gemeze* [chamois] and it travels in herds.[249]

230 One should know that the horns of the deer serve more as purgatives of its nature than as protective armaments. They therefore shed them. Because the females have less heat and earthy dryness, they thus do not have horns, since the horn is more an unnatural purgation than a natural defense. If it were entirely natural, it would be given to the females as well as to the males, for they have the same nature as the males, even though the horns of the females might be smaller than those of the males.[250] This is why in other animals the horn is strongly rooted in the bone and is not shed. It rather grows hard and is given proportionally to the males and the females. For their nature is common, one and the same, and for this reason nature gave large teeth to the males and females in those animals whose armament lies in their teeth. Thus, while the sow does not have tusks due to her great moisture, she nevertheless has long teeth to the front and with these she tears and fights a great deal. Projecting tusks are in the males in the same way the horns are in the horned animals.

Let, then, these be the things said on the cause and the nature of the members of the head.

---

249. Cf. 12.222.

250. In this regard, cf. Cadden's learned discussion (1993) of the medieval conceptions of whether female and male had the same natures.

HERE BEGINS THE THIRTEENTH BOOK ON ANIMALS

In Which Is the Cause and Nature of the Inner Members

The First Tract, Concerning the Windpipe, Esophagus, and Lungs

CHAPTER I

On the Windpipe and Esophagus and Their Natural Causes

We have already set forth the members of the head and distinguished among them according to the forms proper to each of them. Their individual causes have also been given following the method promised. Once the things said here are joined to those said in the section on anatomy, then this will be sufficient for the present investigation.[1]

The member which comes next beneath the head and which connects the head with the body is generally called the neck [*collum*], taking its name from the fact that a *colla* is a sort of connection and this member, as we have said, connects the head with the body.[2] Yet this connection is accomplished more to the rear on the vertebrae and nerves [sinews?] of the neck, which are called "cervical" [*cervices*], and thus the rear part is what is properly called the neck. The front part, which is more a passage for air that is breathed in and of food which enters through the mouth, is more properly called the throat [*guttur*].

As we said before, a neck is not present in all animals but only in certain blooded ones whose head must necessarily have its proper movements apart from the movements shared with the body. Thus, nonoviparous quadrupeds and flying animals are the only ones found to have a neck.

In the front of the neck, in the area which has the name of throat, there are the parts which compose the throat. There is the trachea [*trachea arteria*], which is also called the *canna,* and the member which is called the *mery* in Arabic but the *ysophagus* in Greek. Because the trachea is to the front of the throat, it must be spoken of first.[3]

The trachea is created for breathing, just as the esophagus is created for eating. A blooded animal needs these two, namely, for air which enters from the outside for cooling and for the food which enters the stomach through the mouth and esophagus for nourishment. Thus, passages for these two things should be connected to the mouth. The place of entrance and exit for the air for cooling is through the trachea

---

1. Ar. *Part. An.* 664a12f.

2. Isid. *Orig.* 11.1.60 connects the word to *columna. Colla* can refer to "glue" and is a direct transliteration of a Gr. word with the same meaning. The etymology, however, is not in the Ar. passage.

3. Cf. the following with 1.240f., 419f.

during breathing. Not all animals have lungs, however, and these animals, such as the fish, also do not have a throat.

3   The esophagus is the member through which foods enter the belly. For this reason it is clear that if a given animal does not have a neck, lungs, and a throat, it has no need of an esophagus. For in the ordering of the members in such animals, the location of the stomach (that is, the belly) can directly follow that of the head, but the location of the lungs cannot, in the same way, be connected after the location of the head without some intervening passage, for then the cold air would come into the lungs undigested and would corrupt them. For this reason there must be an intervening member between them, like a reed or a pipe [*canna*], in which the air is first tempered and is then used to divide the air throughout the hollows of the lungs and into the arterial veins which come to it.[4] The lungs, when their composition is complete and perfect, are divided into two parts so that the entrance and exit of air into and from them can be perfect. Because the instrument (that is, the organ) of breathing is long, for the reason given, it thus also necessarily happens that the esophagus is long. For the nutritive members are below the diaphragm and the spiritual ones are above it, and the esophagus is the connection with and the path for the stomach to the mouth, reaching the belly through the throat and chest. It has a fleshy and nerve-filled substance and has the power to expand so that it can widen when food enters.

4   The trachea, however (that is, the windpipe [*canna*]), is a rough arterial vein composed of cartilaginous circles. However, above, where it touches the esophagus, these are just parts of circles so as not to hinder the expansion of the esophagus and so that they do not cause choking during the swallowing of food. It is composed of hard things, so that it is always open for the entrance of air and also because its purpose is not only breathing but also voice [*vox*], which can only be produced from air striking a solid flat surface. For voice is produced from a hard and smooth (that is, flat) thing which is struck.[5]

This arterial vein, called the *canna,* is arranged in the throat in front of the esophagus so that it is kept from taking in food. This is reserved for the esophagus, and the food is directed to the rear, into it. If a bit of food fell into the great arterial vein called the *canna,* the result would be choking, illness, and cough, and perhaps death, since the food might enter the lungs and putrefy or might, during breathing, extinguish the heart.

5   Now it is ridiculous to say that the windpipe is the passage for water and fluids that are drunk.[6] For the lungs below are of a piece and are closed. No passage proceeds from the windpipe to the belly, although we see that fluid that is drunk is taken into the belly and the bladder.

---

4. *Canna* is, of course, a name for the trachea.

5. A. apparently wishes to clarify his received *lenis*, which indicates a smooth surface, from *planus*, which can indicate a flat one. But cf. his statement just below at 13.10f. and cf. the description at 1.419.

6. Peck (1961, 229) traces the origin of this belief to Plato *Timaeus* 70c.

The passage which is the esophagus does proceed from the mouth to the belly and this is patently visible, for the superfluity of food and drink can be seen leaving the belly. Likewise, when vomiting occurs, it is clear enough that the fluid drunk leaves the belly and not the lungs. Further, it is clear to anyone that fluid drunk is not received by the bladder straightway, but rather is first received into the stomach and belly so that it might flow through them and so that the food may be given the power to flow. It then goes from the belly to the bladder and it is clear from this that it comes to the belly not through the windpipe, but through the *mery*. Further, the excremental superfluity of the belly is colored by the dregs of dark wine that is drunk thick and it is clear from this that the wine that was drunk passes to the belly and not to the lungs.[7] Moreover, those who are wounded in the belly lose from their stomach the fluid that was drunk.

It is clear, then, from all these things that drink travels through the stomach's passage (which is the esophagus) and not through the lung's passage, the trachea. In any event, it is unfortunate to speak in rebuttal of this, using reason against the statements of fools who do not use reason to confirm what they say.

So let us resume by saying that the large arterial vein, since it lies at the beginning of the throat and since nourishment passes directly over it from the mouth to the esophagus, is exposed to great harm from food that might fall into it. Wise nature therefore cleverly fitted a covering over its opening. This is a certain member which stands over the base of the tongue on the upper part of the throat and is called the *epyglotis*.[8] This member does not exist in all viviparous animals, but only in an animal with lungs which does not have a scaly skin. Neither is this member present in flying creatures with feathered wings. For such animals, although they have a windpipe, do not have an epiglottis. Rather, in these animals the actual opening of the epiglottis itself opens and closes as needed during breathing. It acts just as in the others whose windpipe is opened and closed with the epiglottis, which lies over the base of the tongue, to keep food from falling into the windpipe. For if an animal carelessly breathed while eating or drinking, and if a bit of food or drink fell into the windpipe, a grave cough would result with choking, and it might even kill the animal.[9]

As we said above, nature was most clever when she created a tongue over which the food passes toward the esophagus. For at its base she created a member which closes the windpipe above until the food has passed by. The movement of the tongue is like that of a miller's hand, for with its movement it works for the chewing and grinding of the food in the mouth, seeing to it that the food to be ground stays within the teeth. When the food descends, it passes straight past the windpipe, never varying from its course, and the windpipe thus needs to have a closure. The animals we have mentioned before, however, lack this member, which is the epiglottis, created on the

---

7. "Thick wine": *spissus potatus,* presumably meaning "uncut" by water.

8. The term *epyglotis* can refer to the larynx (Fonahn, 1199) or it may be used in its more narrow, modern sense. Cf. 1.242f., 317f.

9. Cf. 2.74.

base of the tongue. This is due to the dryness of their flesh and hardness of their skin. For if this member existed in such animals, it would not move easily and would neither open nor close the windpipe quickly since its substance would be of dry flesh and hard skin, and these would greatly hinder the movement of this member. In animals such as these, when there is need of air, the actual opening of the windpipe opens and closes more quickly without this member than with it. If it were on the tongue, it would be made of tough flesh and dry membranes, since the tongue of such animals is very hard, and it is with its own hardness that it presses on their windpipe and closes it. They therefore do not need an epiglottis.

8   This, then, is the disposition of the windpipe in animals that have a neck. We have already given the reason why some animals have a windpipe and an epiglottis and why some do not. For nature invented something that is good and healthful for life in that she saved the windpipe from accidental illnesses.

The reason that the windpipe is placed to the front of the throat is that the heart is placed in the middle part of the front of the chest, although it does tend a bit to the left. For the heart is the principle of life and the principle of all movement and sensation lies within it. Sensation and all movement have their beginning at the front part of the body where the heart is. For it is by means of the heart, which is the principle of life, sensation, and movement, that the middle of the body and the furthest point of the body, to which the power of the heart can be carried, are distinguished.

9   Since the lungs are the fan for the heart, the lungs must also be in the front part where the heart itself is.[10] Breathing occurs through the lungs because of the principle of heat which lies in the heart. For the heart, through its heat, is the principle of life, movement, sensation, and nourishment. To temper this heat, then, and nourish it, the air breathed in first travels to the front part of the chest and of the heart. Because the air passage is the windpipe, the windpipe and the epiglottis necessarily move when air enters and leaves. It was therefore necessary that the windpipe and the epiglottis be arranged in front of the esophagus so that whatever air enters through the windpipe might, as we have said, come directly to the lungs and to the heart.

That which enters the esophagus comes to the belly. Now this indeed is the noblest and best member in an animal in which there is no other nobler member preventing the esophagus from being in the upper part of the animal. In such a case the animal has no windpipe and the esophagus is uppermost.[11] When, however, it has a windpipe, the windpipe is, as it were, nobler than it is in front, on the upper part. For that which is nobler is better and is more worthy of being higher up and to the front, not below and to the rear. It is also more worthy of being on the right than on the left.

---

10. "Fan": *flabellum,* which also can mean "bellows." Elsewhere, the idea of a winnowing fan is used (e.g., 1.29).

11. Ar. *Part. An.* 665a23f. states simply that that which is better and more worthy tends to be above rather than below.

CHAPTER II

*Which Is a Digression Setting Forth the Nature, Cause,
and Type of Composition of the Windpipe and Lungs*

Although we have already spoken a great deal about the nature of the windpipe and the lungs in our discussion of anatomy, which we placed above, we are nevertheless repeating certain things here to facilitate understanding, so that the reasons for their dispositions might be more easily understood.[12]

I am saying then that the esophagus and the windpipe, which lead air to the heart and lungs, are beneath the head. The point of origin of the windpipe is the epiglottis. Let us speak here of these members as they exist in the human, for they are also present proportionately in the other animals which have a windpipe and an esophagus.

Once an animal had blood, it needed two external things, namely, air and nourishment. For this reason nature made an individual passage for each of them. She adapted the trachea for the passage of air and in certain animals there is another member, like this one, to take its place. She made the esophagus, however, as a passage for nourishment. In those which do not have an esophagus, there is another member that is like it. But because air is a very thin thing, if the sides of the trachea were joined in a flat surface, they could not be expanded by the weak and thin air and the passage could not be widened if the opening were narrow. For this reason the windpipe was made circular with two surfaces. But for the esophagus it is enough to be made of very soft flesh and membranes. Thus, when nothing is passing through it, its sides are closed in from their softness and they are thus brought together into a smaller area. However, when it must be open, the hard nourishment opens it up with its weight and the weight and the separation caused by its mass as it passes to the stomach.

Further, since the nutritive members emit a foul and evil-smelling odor from the corruption of the food that occurs in them, nature arranged the diaphragm between the spiritual and nutritive members so that the pure, uncorrupted air entering through the windpipe might not become mixed with this foul vapor. The stomach is below the spiritual members since nourishment is heavier than air and its movement is downward. Furthermore, since the stomach, when it has food, discharges excrement downward to the anus, and the lungs do not, the stomach thus has to lie below and the lungs above the diaphragm. For this same reason the trachea must be above the esophagus, for otherwise the vapor would get into the mouth and would taint the air entering the lungs, for the trachea and epiglottis are instruments of breathing.

The epiglottis is made up of certain hard segments, some of which are circles and some of which are portions of a circle.[13] They are arranged so that one is over the other

---
12. What follows is based upon Avic. *DA* 13.34r.
13. Cf. 1.419; 13.4.

as if all their centers had the same axis. The part nearest the esophagus is that which is called the unnamed [*innominata,* cricoid] and it is almost a semicircle.¹⁴ The cutoff part of the circle is the surface which touches the esophagus and is not cartilage, but membrane. The front part, however, is a portion of a circle and is cartilage. This portion bears the proper name of "unnamed." These pieces of cartilage, having the shape of circles and portions of circles, are covered with a smooth, dry, hard membrane. The windpipe itself is made up of many parts. It is made up of cartilage because if it were made up of soft parts, then its walls would sometimes collapse because of its softness and breathing would be blocked and interrupted. They are hard so that the windpipe can be better protected from harm since it is to the front of the throat, an area upon which external harm falls. Its own hardness aids in the emission of voice in the way described above.¹⁵

Its composition consists of things loosely joined together, composed out of membrane and circles, so that a person does not feel pain during a great and sudden emission of air. For that which is loosely joined stretches and can emit a great deal of air suddenly with no difficulty. It is composed of many things so that if an injury should occur to one of them, another would not be injured, and also so that it can be bent and moved more easily. There are portions of circles above at the narrow part of the throat, where it touches the esophagus, lest it hinder the swallowing of food. For eating is not joined to the air breathed in since the esophagus is closed above when the *bolus* passes by.¹⁶ This is because when the lower part of it expands, the upper part contracts, and the passage is closed by means of the "cymbal-like" one [*cimbalaris*], about which we will speak below.

The inner membrane of the windpipe is made hard so that it is not easily harmed by the sharp catarrh which flows from the head and also so that it is not harmed by the sharpness of the air leaving the inside of the body. It divides below into two branches since the lung has two parts. The branches of the windpipe go along with the quiet veins so that they can take their nourishment from them. The openings of the windpipes which enter the lungs, however, are narrow so that they do not lose the spirit contained within them but give it over to the arteries. It also keeps the blood from entering their own openings, for otherwise the animal would always be spitting up blood.

The epiglottis completes vocalization [*complet vocem*] and sustains respiration. Within it, in the opening of the windpipe, there is a tongue resembling the tongue of a reed and which is even called the "tongue of the reed" [*lingua fistula*].¹⁷ The epi-

---

14. Cf. 1.243, 318, 419.

15. Cf. 1.420

16. The use of *bolus* in the Latin here, in its technical sense of food mass, is noteworthy. Cf. 8.124, where, corrupted from the Gr., the word appears to mean squid's ink, and 23.87(40).

17. Cf. 1.248, 421. The language here seems to indicate that the "reed" is a reed instrument or a pipe. In that case, the "tongue" would be the mouthpiece or what we today would call the "reed." If this is the case, the vibrating reed, as in a clarinet, would be an apt analogy as the vibrating piece past which air flows to produce sound. A. is fond of musical analogies. Cf. 10.14 where the bagpipe and shepherd's pipe are discussed and 19.43 for a reference to contemporary organs.

glottis contracts appropriately along with the trachea and the esophagus. When the *mery* contracts to take in food, the epiglottis is raised up and its cartilages contract while its membranes and muscles extend so that it is covered by the "cymbal-like" one [*cimbalaris*]. And when the *tibia* (that is, the windpipe) is near the esophagus, it stands on the surface of the lower palate over the base of the tongue and the epiglottis is closed. The epiglottis is composed of three parts, the anterior of which resembles a shield [*scutum*] and the inner of which is called the "unnamed" [cricoid].[18] Over this there is placed the connecting part called *cymbalaris* ["cymbal-like"]. The *cimbalaris* is not joined to the *peltalis* (that is, the *scutalis*), but they are loosely bound together. Between the *cimbalaris* and the one which has no name (that is, the "unnamed") there is a double connecting member. On the *cimbalaris* there are two hollow recesses which two appendages of the "unnamed" one enter and the "unnamed" one is sometimes joined to the *peltalis* when the epiglottis is narrowed. But when it expands, then they too expand. When the *cimbalaris* is raised, air is let out and when it is not raised, air is not let out. There is, near the epiglottis, a triangle that resembles the letter *lauda* (Λ). The function of this bone is to allow the fibers of the muscles of the epiglottis to arise upon it. However, enough was said about these muscles in the section on anatomy.

The lung is composed of parts, namely, veins and arteries without tunics.[19] The third part of its composition consists of the small branches of the nonpulsating veins. Flesh fills in between all of these. It is light, soft, and somewhat white from the great deal of air it takes into itself.[20] More complete animals (those with blood) have loosely textured lungs so that they can absorb air in order that it be digested in them and that the superfluity be expelled from them. The lung is created to complete the spirit, just as the liver is created to complete the nourishment. The lung has two parts, a right and a left, and the left part itself has two parts and the right three.

Speaking summarily, the function of the lungs is breathing. The function of breathing is so that that which is necessary to produce spirit might be drawn in and that which is superfluous might be expelled. The function of retained air is to cool the heart and so that there might be enough for the emission of a strong voice, something that can be done only with retained air.

Retained air also helps when the animal is travelling in a foul-smelling place and holds its breath in order not to breathe in the corrupted air. It is also of use when the animal might occasionally be submerged for a while and also because some of it enters into the vital spirit and nourishes it, being rather like something that resembles the vital spirit very much in complexion even though it is not exactly like it. Thus, those who have thought that only simple air can be changed into the vital spirit have erred, for it is not true. Every food, both of spirit and of body, is a composite substance. Because the thick vapor which evaporates from the heart into the lungs cannot rectify the spirit any further, it must be expelled.

---

18. The "shield" is a translation of the Gr. *thyreoeidēs,* "shield-shaped," a Galenic term. Cf. 1.244 where *pelta,* and not *scutum,* was the Latin term for shield and for further discussion of the terms.

19. Cf. Avic. *DA* 13.34v.

20. Reading, of necessity, *et* for Stadler's *e.*

The arteries which branch into the lungs and the branches of the windpipe share in every complete breath. The nonpulsating and the pulsating veins share in the feeding of the lungs with the clear, thin blood which comes from the heart.

17 The function of the flesh of the lungs is to fill in the empty places which lie between the nerves and quiet and pulsating veins. It also serves to unify the branches of the windpipe. This flesh is loose to enable it to take in air which enters not only the branches of the windpipe but also the substance and flesh of the lungs. This is done so that more of the air can be taken in as if into storehouses. Its substance is contracted in the substance of the lung and this helps to press out vaporous air. For such a substance is suitable for two movements, namely, contraction and expansion. The lung is white, as we said, because a great deal of air passes over into its substance. There is always air in it. We have already spoken of the division of the lung. The third branch of the right side performs no great function in breathing, but rather acts as a pillow laid down beneath the vein which comes from the liver [vena cava]. The lung is also covered with a membrane with which it has sensation. The lung itself is like a pillow and a fan for the heart.

CHAPTER III

Which Is a Digression Setting Forth the Nature of the Esophagus and Stomach and Their Natural Causes and Positions

18 The esophagus is composed of flesh and, internally, membranous tunics.[21] There are long fibers in the membranes and by use of these the food is drawn in easily during swallowing. This is done more easily when long fibers are contracted. Over the outer membrane, however, there is another membrane, woven around with many fibers which aid in the expulsion of food, as is known from our anatomy section.[22] Swallowing is accomplished by the drawing power of these two tunics. Vomiting, however, is accomplished by the operation only of the exterior one and its strands (that is, fibers), and it is thus more difficult and painful than swallowing.[23] The location of the *mery* is directly over the vertebrae of the neck for its best protection. When it descends to a point opposite the fourth vertebrae of the back, and not of the neck, it is directly opposite the chest. It passes a bit beyond this and then bends to the right to allow passage to a vein ascending to the head. It then descends over the four remaining vertebrae of the chest and reaches the diaphragm. It then stretches out to encompass the entire opening of the stomach.

---

21. Avic. *DA* 13.36r. Cf. 1.539f.
22. Or "through dissection."
23. Cf. 1.550.

The body of the stomach comes after the esophagus. Above, in its membranes, it has an opening that is wider and harder than its lower opening. Its body, as far as its internal membranes are concerned, is midway in composition between these two in nature. Near the bottom of the stomach the membrane is softer, and in the intestines there is a pannicular-membrane, which is softer still. One of the pannicular-membranes comes from the hollow of the stomach and this one covers the tongue, the inside of the mouth, and the inner part of the esophagus. Its drawing power is aided by this continuity and it especially helps by raising the epiglottis when the *bolus* passes by. It also helps in a smooth passage when the esophagus is retracted voluntarily.[24]

The body of the stomach is pyramidal in shape at its high point near the diaphragm and esophagus, but it is rounded with respect to its exterior surface so that it can be better attached to the back. It has two tunics, in such a way that the inner one has long strands (that is, threads) for the drawing power while the exterior one has latitudinal and transverse ones for expulsion. This arrangement is because the drawing action precedes the expelling one, and this is why the longitudinal fibers are inside and the others are outside. Certain fibers (that is, certain strands) are mixed in to the exterior tunic, and these assist in retention as well.

The bottom of the stomach is fleshy so that it can better conceive heat from it. The opening is more nerve-filled so that it might have sensation and it has large nerves there with which it might sense its emptiness and hunger.

It decocts the food not only with its natural heat but also with that of the exterior members, which are the liver and spleen. They lie beneath it because if they were over it, it would be overcome by their weight. The liver surrounds the stomach with its branches and the spleen with its breadth. The liver is very large compared to the spleen since the spleen is only a receptacle for one humor produced in the liver. Its opening thus must bend to the left so that the liver over it can expand and grow large. Thus its bottom is joined with the liver. The opening of the stomach bends not only to the left but also somewhat downward so that the spleen can achieve its requisite size. Because right and above are better, the liver was put to the right and above. The spleen, however, is below and to the left. *Zyrbus* is added and it properly lies over the intestines of humans since their intestines are weaker than those of the other animals and a human's food needs more decoction.[25] It is thin, however, so that it can be light on the belly and is fatty [*seposus*] so that it can preserve the heat from the front. For fat [*sepum*] takes on heat and retains it due to the viscosity it possesses. Over the *zyrbus* there is a certain membrane, and after it comes the *myrach*. After this are the muscles of the belly, the ligaments, and the *sepum* which lies in the area of the back. After this comes the back and beneath it the large, warm, arterial vein. We have spoken enough about all these in the section on anatomy.

---

24. "Smooth passage": may refer to a simple swallow, when no food is present.
25. Compare this section with 1.470f. The terms are not always used consistently.

21   The membrane which covers all the nutritive members, confining them below toward the area of the back, is connected above with the diaphragm and below with the *anchae* at the very end of the *ypocondria,* which are commonly called the *ylia.*[26] The function of this membrane is to separate the *myrach* and the intestines. This is mostly done so that the muscles of the *myrach* do not interfere with the operations of the belly.

The function of the *myrach* is, using retained spirit, to contract the stomach during the effort involved in the expulsion of excrement. In the same way it presses on the bladder in the effort involved in giving impetus to the ejection of urine. In women it uses contracting movements to help in bringing forth the child.

22   The aforementioned membrane binds the intestines to each other and binds them in turn to the back so that they are almost one with it. Because its heads are connected with the heads of the diaphragm, and it itself is connected to the diaphragm in front, then, directly in front of the back, there arises a certain residual remnant of the membrane. This divides into two parts, one of which proceeds to the part of the opening of the stomach which is visible while the other part passes beneath the opening to its hidden side. There they are joined into one. After this it covers the stomach, wrapping it in a somewhat curved web. This is its function. It also binds the stomach above to the internal parts which are to the inside, in front of the back. With the remainder of the two aforementioned parts, it is joined to one branch of pulsating veins and to another of nonpulsating veins, and a membrane [*pellis*] is formed which stretches over the stomach and which is the origin of the *zyrbum.* This is composed of two thin, *sepum*-filled webs. The *zyrbum,* however, covers the stomach, the intestines, the spleen, and the mesenteric veins. The part which envelops the stomach comes to the part which is joined to the back.

A much broader determination has been made about all these things in the anatomy section we placed in the first book of this study.

---

26. Cf. Fonahn, 3644–51. The area at the ends of the hip bones and the thighs would seem to be meant. Fonahn cites Mundinus as stating that the right *ypocondrium* is where the liver is located while the left contains the spleen, and the right and left *ylia* are beneath their corresponding *ypocondria.* Cf. Avic. *DA* 13.35v, "hypocondria inferiora tenera quae a rusticis dicuntur ylia." The original Gr. *hypochondrion* is defined in L&S as "the soft part or parts of the body below the cartilage and above the navel, abdomen." Yet cf. 1.292, 343, for a sense of how slippery these terms are.

CHAPTER IV

On the Disposition of the Inner Parts in General and on the Nature and Cause of the Heart

Having then made a determination about the neck and the inner parts of the throat, namely, the windpipe and the esophagus, we still must speak of the other inner parts which are proper to the more perfect, blooded animals.[27]

In some animals there are inner members and in others there are none, although in place of some of them there might exist certain other members that are like them with the powers and activities of these members. Examples include the heart, liver, and the like. Democritus was in error when he claimed that all members possessing the same species and shape are to be found in all animals but that they escape notice in some of them because of their smallness. This is clearly false because in many blooded animals these members are visible at the very beginning of creation when they are very small. Thus their hearts and livers are visible when they are scarcely three days beyond conception and their size is almost that of a pinprick. It seems that there is a place in the semen for formation of these members. The beginnings of these members appear very small but there is no impediment afforded by their very great smallness at that time. This is also frequently seen in a miscarriage, which is about the size of an ant, as we have said in previous discussions about human conception.[28]

Having dismissed this as false, we say that just as the use of the external members is varied and that it is thus suitable that there be many exterior members, so too is the use of the inner members multiple and so too is a multitude of inner members required. They also differ in the animals depending on differences in the types of their lifestyles. The inner members, however, are found in a blooded animal according to the perfection of its form. For each of the members mentioned is created, kept in being, and nourished by the blood. This is indicated by the fact that these members are found to be very sanguineous in children that are not far removed from the time of their generation. For they are found in them to be rather large and filled with blood for their age and size.

The heart, which is first and principal of these members, is present in all blooded animals. We have given the reason for this before, namely, that since there is blood in an animal, it is necessary that it be formed in a vessel like this. Nature therefore produced the heart with the following device. For since many veins form the passages for the blood, she could not nourish the body by herself everywhere. It was therefore more suitable to make a single principle for this multitude, for it is more suitable that

---

27. Cf. Ar. *Part. An.* 665a27f.
28. Cf. 9.35.

there be one principle of many things than many principles. That the heart is the principle of the veins is clear from the fact that the veins leave it and do not pass through it. This is just as we have discussed expansively in earlier parts of this study.[29]

26. The nature of the creation of the heart seems, because of the heart's hardness, to be from the substance of the veins. The heart itself seems to be in the classification of veins insofar as it is the vessel of blood and spirit. The location of the heart, which is suited to the fact that it is prior and a principle, seems to indicate the same thing. For it is positioned above and to the front, since a nobler member is naturally consigned to a nobler location. What has just been said is seen most clearly in the human, who is the noblest animal, and in the other, more perfect animals, in which nature positioned the heart in the middle of the body in a place where there is the greatest need for it. Its powers can thus be equal in respect to the extremes. Now I am calling the middle of the body that part of the body which is hollow and in which the inner members are contained, and whose end is at the anus, through which the superfluities leave.

Different members are variously located in various animals according to their various dispositions. Nor are they necessary for life, for there are many animals alive which have no liver or spleen. For many of the animals which have a uniform body live on even after one of their members is cut off.[30]

27. Those who have held the opinion that the point of origin of the veins is in the head have been in error.[31] One reason for this mistaken statement is that they posit many principles, separate from one another, and they posit the point of origin of their sensation as the *sinciput* of the head, which is in the area of the forehead. However, since cold is a deadly quality and is the cause of immobility and since cold befalls the front area of the head most quickly, it is clear that the principle of life cannot be here. The location of the heart, however, has a contrary disposition, as is clear from the things which were often said above.

Further, as we said just a short time ago, veins pass through the other members of the body, but not through the heart. From this it is clear that the heart is the first part and is the origin of the veins. An indication of this is that it is hollow to receive the blood and that it has thick, tough flesh to guard that which is the principle of all vital and animal movement.

28. Further, the only member in which there is blood without veins is the heart. The blood, which is in the other members, is in veins and this indicates that the blood leaves the heart as if from its principle and pours into the veins. Neither does the blood come to the heart from any other place, since the heart is the blood's principle and spring, and is the first member, a sort of vessel to receive the blood. This is clear from the things which have been proven above, as determined in the section on anatomy.

---

29. Cf. 1.575–95.

30. A. means animals whose bodily substance seems to have no differentiation to it. This would include simple animals such as a mollusk.

31. Compare this discussion of the origin of the circulatory system with that at 3.1–65.

This is also clear from the manner of the heart's generation for, of all the other blooded members, its generation appears first.[32] Moreover the heart is the one in whose movements there first appear the things which are pleasurable and unpleasant through its expansion and contraction.[33] Generally, the movements of each sense begin from it and return to it. All these are, moreover, the properties of a principle. For a principle should be the cause of other things and it should be alone (that is, uniquely) the principle of all. It is also placed in the middle, for the middle is the most suitable place for a principle. The middle is unique by being equidistant, and by it the power is spread out proportionally into all the members in one and the same way. We know, however, that not one blooded member has sensation on its own and the blood itself does not have it. The heart, however, has sensation and it is necessary that the principle be a member in which the blood is present, both by way of being in a place and by way of being in a vessel. It is clear that it is entirely necessary that the root and the principle of the blood and life be like each of them.

Now this is clear not only by means of reason but also by using the senses. The heart is the first thing which is seen forming in the body of an embryo at the beginning of its creation and, of all the other members, it is the first seen to move with animal and vital movement. This is because it is the principle of the nature of a blooded animal. From this comes evidence that the heart is present in all blooded animals, in the ways we mentioned.

Nor can anyone reasonably say that the liver is the principle of the entire body or the principle of the blood. For it is not positioned in a place suitable for a principle. Moreover, it has, in all animals that exist with a good and perfect creation, another member positioned opposite it which is contrary to it in complexion, this being the spleen.

Moreover, the liver has no hollow place to receive blood as does the heart. Rather, the blood in the liver is contained with a watery substance in the veins so that it is made suitable to be presented to the heart for completion. It is also contained in this way in the other members to which it flows in veins.

Moreover, the veins pass through the liver, but not through the heart. Now since, according to all the physicians, it is necessary that one of these two members be the principle of the veins and of the blood, and since the liver cannot be the principle, we are left with the fact that the heart is the principle of the blood and veins. This is also subject to sensory proof, as we said a bit earlier.

Again, the first member of an animal (which member, I say, has sensation) has blood, and this is the heart since it is the principle of the blood and it is found to be blooded and to be formed in embryos from the blood earlier than all the other members.

---

32. Viz., the heart appears first and has blood in it once it does appear. A good example would be the chicken egg.

33. Ar. *Part. An.* 666a12f. merely states that all sensations of pleasure and pain must arise from the heart.

31   Moreover, at its upper end, the heart is pointed and harder than the rest of the heart and it is placed toward the front of the chest. Generally speaking, the heart is placed in the front of the chest to keep it from being easily chilled. This is why the chest is created with little flesh at its front while to the rear, in the area of the back, it has more flesh. This is because a warm spot needs the covering action of the back area.

    In all animals save the human, the heart is placed in the middle of the chest, but in the human it tends a bit to the left so that the left side, which is cold, might be tempered, since the liver is on its right and the spleen is on its left. This is done so that the heat of the right might be equal and tempered by the heart's being off-center and so that the coldness of the left might be tempered by the fact that the heart tends to the left. This sort of misplacement of the heart is cold. This is especially so in the human, because a human, having a wide body, has a colder left side than other animals. Thus his liver lies at some distance from the left side in proportion to his bodily size. The heart is placed in the same way in all the fish (namely, in the middle) as it is in the other animals, as we have said above. This, then, is the statement on why the location of the heart in the human appears different from other animals.

32   The hidden side of the heart, which is its base, is opposite the head.[34] There is a continuous systolic and diastolic movement in the heart and there is thus a great multitude of veins woven together in it. This is because all movements come from it. It sends out its powers along the veins and nerves and draws all the members to itself as it wishes. For this work the heart especially needs great strength, as we said above on the anatomy of the heart. Thus the heart is, in the nature of the members, what the soul is in bodies of those who have souls.

    Bone is generally not found in the heart of any animal save that of the horse, according to what Aristotle claims to have proven through experience, and in a particular genus of oxen (that is, cows) called the *wisent* in German.[35] A bone is generally found in the heart of these two genuses of animal when they are full grown. It is there because of the largeness of their bodies. However, I and many others have experienced that a bone is often found in the heart of an old deer and in the human's heart, as well as in that of many animals when they are old. But it is not commonly found in all of them. When, however, a bone is found in the heart, it is agreed that it is there to support and strengthen the heart, as is the case in all bodies that have bones placed in them.

33   The heart of a large animal has three ventricles (that is, three cavities), whereas that of a small animal has only two.[36] There surely must be a ventricle in an animal's heart, however, since it is the receptacle of pure blood. For we have often said that blood, as far as its perfection and completion are concerned, should first be in the heart.

---

    34. "Hidden": it is interesting to note that Sc. reads *acuta,* "pointed end," accurately rendering the *to oxy* of the Gr. This has come to A., however, as *occulta.* Note that A. has it correct at 3.18.

    35. The *wisent* is the European bison (cf. 2.35; 6.102; 22.146). On the "heartbone," cf. 1.118, 300, 583; 2.86; 3.54.

    36. On the use of *ventriculus* and *fovea* for ventricle, cf. 1.578f.

One should also know that the principles of the veins lie in the heart.[37] There are two veins from which all the others take their origins, namely, the great vein [*magna vena,* vena cava] and that called the *orthy* [aorta]. Each of these is the principle of various veins with respect to the differences according to which they are distributed throughout the body.[38] We have spoken about these before and we will speak of them again later when we review the things that have been said. The reason that the principles of the veins are different is that the blood itself is divided at least in two, in accordance with the two ventricles of the heart. It is clear in one and clouded in the other, depending upon what nourishment the various members need.[39] This is also why two places for the blood are visible in animals with a large body, since these have more power and potential for separating locations for the blood than do others. For their hearts have a size that is suited for the formation and the separation of the ventricles.[40]

It is best that there be three ventricles in the hearts of such animals so that one might be a common principle for the formation of the blood and the two to the outside might be for the clear and the less clear blood, respectively, as we have said on the anatomy of the heart.[41] The size of these ventricles should be proportionate. In the right ventricle the blood is very warm and thus the right side is warmer and fuller and more lively than is the left side.[42] The blood in the left ventricle is less and colder. The middle ventricles have blood that is temperate both in quantity and quality and it is very moist. The blood in the member wherein the principle of power and the first power lie should be of this sort.

Moreover, there is a division in hearts which resembles a suture, but this division is not one in which the pieces are connected to one another in a manner such as when something is joined together from many different parts. Rather, as we have said, it resembles a jointed division, although it has but one substance. Among animals, however, those which have the best and finest sensation have uniform hearts with a jointed division, as is clear in the hearts of humans.[43]

The hearts of animals with little blood (for example, the hearts of pigs) show a more obvious jointed division.

Hearts differ, moreover, in being either large and small or soft and hard. For the hearts of animals with fine sensation are soft. Those that have a heart that is proportionately large for their body are timid and those that have a moderate heart are bolder.

---

37. Throughout this section, A. uses many forms of the stem *princip-*, rendered into English variously as "principle," "take origin from," or "point of origin." The Scholastic sense of a *principium,* of course, entails many meanings at once.

38. Perhaps referring to whether they are distributed as arterial or venous vessels.

39. On this blood, cf. 1.578f., 608; 3.117.

40. Cf. Ar. *Part. An.* 666a26f. for the original sense of this now confused passage.

41. Cf. 1.379; 3.55.

42. In this sentence A. switches to *venter* instead of his more normal *ventriculus.*

43. The Latin at this point is confusing. The translation attempts to give the sense of the passage as derived by comparison with Ar. *Part. An.* 667a7f.

These accidental traits stem from no other reason than that an animal which has an enormous heart for the size of its body has little heat. Thus the first, seminal fluid, from which the heart was formed, has overflowed into its large size.[44] This is why the moderate heat in it does not fill it up. This is why slight heat in large hearts is found lacking and cannot warm the blood that flows back to it when it is afraid. It thus remains cold. Hearts, however, which are proportionately large for their bodies are found in hares, deer, asses, goats, mice, and other similar timid animals.[45] For just as a slight fire warms small homes or stoves more than it does large ones, so too a slight heat warms large hearts less than small ones.

36. In large hearts, in which the heat is spread out and scattered far and wide, the types of movements which are present in a warm body are chilled and slow down unless the heat has itself hollowed out the heart more fully with its power. For then, in a heart which is not enormous but is hot and dry, the ventricles will be large due to the great heat which is opening them up and the vital spirit will be greater and stronger in such open places. Thus, every dry and hot heart has large chambers, while a soft, watery heart, loose in substance, has small chambers for the opposite reason.[46] This is why no animal with large veins and large ventricles in its heart is found to be fat. Rather, every animal with small ventricles and slender, hard to see veins has, for the most part, fatty flesh.

37. Of all the inner members, only the heart cannot endure any pain whatever from a wound or a violent injury. Neither can it endure a disease for long. This is reasonable, however, for in no way can the other members be of assistance in the life process once the principle has been corrupted. For all the others take their power from the heart, whereas the heart takes from none. An indication of what we have said is that there never appears any pain or disease in the hearts of slaughtered animals. But injuries often appear on the other inner members just as stones, wounds, or lesions often appear in the kidneys of those that have been killed. It is the same in the case of the liver and lung and also in those near them. Many illnesses especially appear in the lung and they are often present in the great vein, called "hollow" [*cava,* vena cava]. Similarly, they are sometimes seen in the liver at the place where it connects with the large vein on its protuberance, by which it is connected to the heart. Yet none of these prob-

---

44. "Its large size": the antecedent is not specified, but presumably A. is referring to the large size of the heart.

45. Of the animals listed, the hare and deer are the only ones with hearts proportionately larger than man's to any notable degree. Spector (1956, 163–64) lists the following relative heart sizes, given in terms of grams/100 grams of body weight: man, ranging from 0.42 to 0.66; ass, 0.55; white-tailed deer, 0.97; hare, 1.02; meadow mouse, 0.680, but 1.03 for the leaping mouse, *Zapus hudsonicus.* The size of the goat's heart is not listed. Shrews, chipmunks, lions, weasels, horses, and zebras, among others, all have equally large (proportionately) hearts. It should be pointed out that the numbers are based on a small sample base.

46. In the same sentence A. uses *ventriculus* and *camera* for ventricle. Elsewhere he has used *venter, fovea, capsula,* and *thalamus* and, at 13.38, will use *archa.* This serves as an excellent example of the potential for confusion caused by a multiplicity of sources and the lack of an accepted scientific terminology.

lems befalls the heart at all, for it is destroyed before a lesion occurs on it. This is what happens in the case of an abscess of the heart, for it kills the heart before it is felt by it.

This, then, is the statement on the disposition of the heart and the reason that the heart is present in animals that have one. We have also stated why there is no heart in some animals.

But we should not pass over the fact that a heart with a delicate substance and suture is an indication of the finest operations of the soul, whereas one with a coarse substance and suture is dull in the operations of the soul.

Further, animals with a soft heart are more "capacious," but they easily change the things that have been conceived.[47]

Those which have tough hearts are less "capacious," but they retain the things they have conceived longer. We should add to this that the heart has two auricles, which relax when the heart contracts and contract when the heart expands. Their function is to assist in the retention of that which is in the heart, and there are two chests [*archae*], as it were, taking from the vessels of the veins and giving to the heart as much as is proper for it to have. They were made delicate so that they might have better "comprehension" and "occupation" of the heart, and they are tough in substance so that it is difficult to cause them to suffer things contrary to themselves.

The heart feeds itself with its own natural powers and expansion. For blood enters into its bottom part at the same time that air also enters its bottom.

This, then, is what we have said on the nature and the natural causes of the heart. If the things said here are joined to the things said in the anatomy section, the nature and disposition of the heart will be known quite well as far as suits our present purposes.

38

---

47. Cf. Avic. *DA* 13.35v, *Can.* 3.11.1.1. At *QDA* 13.3, A. explains that softness in the heart can be understood in two ways: either as the result of a defect of heat and power or from an abundance of blood. The latter is the sign of a good disposition for, as Ar. indicated (*De anima* 421a25–26), things having a soft flesh are better suited for sensation, since the soft has the appropriate mixture of the moist and the dry (cf. A., *De anima* 2.3.23). For these reasons, A. concludes, a soft heart is better than a "tough" or "hard" heart. Common usage continues to reflect the notion that a "soft-hearted" person is more sensitive or compassionate than a "hard-hearted" person. From the Bible, A. would have been acquainted with the view that the Jews or Israel were "blind" to the truth because of their hardheartedness (*duritia cordis,* as in Mk. 10:5 and Mt. 19:8; *durum cor,* as in Ezek. 3:7). For the heart as the organ of sensation, cf. 12.14, 95, 148, 168. At 12.173 the heart is described as the *primum sensitivum*. For its influence upon the reception of forms, cf. 8.231.

CHAPTER V

On the Reason the Veins Rise from One Principle,

This Being the Heart in Animals or That Which Is in Its Place

It now follows that we should speak of the nature of the veins, which arise from the heart.

39 We will speak of the vein called "great" [vena cava] and of the one which is called *orthy* [aorta] in Greek, or *adorty* in Arabic, a form which puts the Arabic article in front of the Greek word. These two veins are the first to take blood from the heart. All the other veins arise from these at their stems and branch out further from them. We have said before that the veins were created for the blood, but now we intend to speak of the reason that these veins have been made and why they are caused by one principle as well as why they are divided throughout the entire body.[48]

40 It is possible to speak quite reasonably about the cause of this, for the soul of each and every animal, which is the principle and cause of life in it, is one in act. It was thus necessary for there to be one principle for the entire body from which all the others arise. It was necessary for there to be one member in which this power rests, as in a first source and seat. This is one and is prior in every animal that has blood, both potentially and in act.[49] In the others, which lack blood, it is only prior in act. The heat, which is necessarily first and which digests and limits the material, is present in the first member, where the formative power is also preserved. If certain animals are cold in act (as most of those are which do not have blood), there is still present potentially a heat which limits and which is in the member which is taking the place of the first member, which is the heart. For the limiting and formation of material can occur only through heat which is present potentially or in act, or at the least potentially. In those which have blood, this heat is the cause of bringing warm, moist blood into being. For the principle of life and sensation comes from the same member, and they are produced by a life-giving heat. It was therefore necessary that the heat of the blood also come from a single, first member. Thus, there are two sides, right and left, with respect to the two ventricles in those animals having blood and which walk. In all animals of this sort, other differences of position have been determined, namely, to the front, to the rear, above and below. To the extent that the front is nobler than the rear, by just so much is the great vein nobler than the *adorthy*. For the great vein, as far as most of its branches are concerned, is positioned to the front, but the *adorthy*, with regard to most of its branches, is positioned to the rear.

---

48. "Reason that": *propter quid,* perhaps to be taken in its formal sense of "on account of which" or "for the sake of which." Cf. 11.3, 29.

49. "This" is neuter and thus could refer either to the principle or to the member.

Further, in animals with blood, the great vein is visible in many places, while the other is only weakly visible in some that have blood and only when they are quite emaciated. In others, fleshy and fat, it is hidden entirely.

The veins are distributed throughout the entire body because the blood is the material and food for the body in those that have it. It is therefore necessary that it be borne through the veins into all the parts of the body. In those which have no blood, there is a fluid analogous to it, which also must be carried throughout the whole body through those things which take the place of the veins. It is more suitable to speak of some things—how the animal is fed from the belly, and from what proximately and from what first, and how food undergoes digestion—when we have determined how each member is sustained by the blood. This is as is clear in the anatomy section, which we pursued previously. It was necessary for the blood to be borne to the whole body in veins if the body was to be nourished universally by the material of nourishment. For this is just what those do who irrigate their gardens from one spring.[50] They divert two or more large canals from the primary spring and then divert many other small canals from these throughout the entire garden. In just this way, the great vein and the *orthy* are diverted from the heart and then from them come many veins throughout the entire body. This is seen in bodies that are very emaciated, for in these the veins are more visible.[51] In dried out bodies such as these, all that can be seen is a networks of veins, like those in the leaves of the vine, fig, or other broad leaves that have been dried. They look just like the networks of the pathways through which the nourishment flows throughout the entire body of the leaf.

Something quite similar happens in canals and in veins. For when water is not flowing in the canals, canals through which water repeatedly happens to flow still remain open. The small canals are obliterated by the mud which the water carries along with itself, but the large ones stay open and cannot be obscured by the mud. This is exactly what happens to the veins, which are the canals of the blood. The small ones are filled with flesh and fat [*adeps*] so that they seem to be nothing but flesh. But they are still veins potentially. When starvation eats away the flesh, they become veins in act as before. The large ones carry the nourishment, which is the blood or the humor taking its place. Thus, without flesh, there can be no member through which the blood flows. In fleshy members small veins are not visible, just as small canals are visible only during the flow of water if, as we said before, the material blocking those canals is removed.

All the veins proceed in the body from large to small. The further they stretch out in the body, the more they become narrow and are filled in with the thickness of the blood. At their ends they exude a humor which nourishes the members and at these same ends a watery superfluity called sweat exudes. This is why sweat is often produced only when the body is hotter than it should be and the openings of the veins

---

50. The elaborate irrigation metaphor which follows derives from Ar. *Part. An.* 668a14f. Ancient Greece had elaborate water systems, some of which are still partially in existence in Athens to this day.

51. Cf. discussion in notes to 3.5.

are opened by the heat. Then the thick part of the blood is retained, but the watery part flows out. This is sweat.

43 A bloody sweat occasionally occurs in some due to a bad complexion. Their body was prepared and loosened for the flow of the openings of the veins and their blood was very moist and watery because of the weakness of the veins' heat, which should have decocted and thickened the blood. The heat is weakened when it is suffocated by superfluous food. Therefore, when a moderate heat is at work on a great deal of food, it does not decoct it, and it remains watery. If it were to be sufficiently decocted, it would be thickened. The cause, then, of bloody sweat is a small heat acting on a great deal of watery food.[52]

Food is multitudinous compared to the heat in two ways, namely, in quantity and in quality. Sometimes the food is good, but its quantity blocks the heat, so that it is not digested adequately. Then the body sweats since it is overly nourished with incomplete nourishment. It hinders it with quality when, although the food is moderate, hardness or some other flaw stands in the way of digestion so that it is not decocted and is again incorporated into the members undigested, and at this the members sweat.[53]

Moreover, the passages for the blood are very wide and the blood is thin, and thus the flow of the blood through them lacks pain and heaviness. For it flows into the body as blood flows from a vein during a bleeding.[54]

44 As we said above, the great vein and the one called the *adorthy* change positions at a certain spot and wrap around each other in net-like fashion in order to contain the entire body and support it with their extension to the feet and hands. Moreover, they then become wide and branch out. One pair proceeds from the back to the front and the other pair from the front to the back. At a given place they come together in the same area, as happens to bodies which are wrapped up in one another.[55] For it thus happens to be wrapped up by veins, with the one which is to the rear being bent to the front, and the one which is to the front being bent to the rear. The same thing happens in the veins which stretched from the heart upward.

This intertwining of the veins and the way in which they are separate from one another is understood well only through the anatomy which we pursued in the earlier part of this study. There we spoke of the natural disposition of the members of all animals. For now, then, let that which we have said here about the veins and the heart along with what was determined in the study of the anatomy be sufficient.

---

52. Cf. 3.145.

53. This explanation should, most likely, be referenced to the statement of 10.10.

54. This odd statement, only slightly better in Sc., derives from Ar. *Part. An.* 668b16f., which states that hemorrhage is most prone where vessels are most numerous, as in the nostrils, gums, or the anal region. A.'s received text read only "from a vein," to which he added a sensible explanation.

55. This sentence and the confusing ones that follow appear to be all that is left of Ar.'s simple but terse statement that the nexus of veins resembles what happens when various strands are plaited to make something bound more stoutly.

CHAPTER VI

On the Reason for the Lung and Its Natural Accidents.
On the Liver and Spleen and Their Natural Operations.

What follows the things that have been said is to treat the natures of the other inner members using the method determined in previous statements.

Let us say, then, speaking of the lungs, that a walking, blooded animal has a lung. This is necessary because, as we see, it requires the cooling brought about by breathing. A blooded animal needs a lung most since it has greater heat than other animals. A bloodless animal, however, does not have a lung, since the amount of cooling which occurs in it through the agency of the natural spirit beating in it is sufficient. A breathing animal is cooled by the airy spirit that enters it. It is for this reason, we say, that these animals have a lung. We say that all perfect walkers breathe and that they therefore have a lung. Not only the walkers breathe but also certain aquatic animals like the dolphin, *balena* (that is, *cetus*) [whale], and likewise the large animal which belongs to the *malachye* genus, and the sea wolves which draw in air [seal].⁵⁶ All these breathe and have a large body. There are also many animals, both aquatic and land, which live in the water most of the time, just as there are many land animals that are sometimes in the water. All these have a common nature and they generally breathe as do the terrestrial walkers. The spirit itself gives completion to their lives, since it serves to cool the vital power which is in their heart.

Now the lung is an instrument, that is, it is an organ which serves another member organically. The principle of movement for the lung lies in the heart and is prepared in such a way that it can be increased in size when the air is drawn in and enters it. It has the power of taking on air because of the emptiness of its cavities, its softness, and the size of its substance. Therefore, even when it inflates, filled with air, it does not rise up much and thus does not press on another member. In the human, the heartbeat occurs especially because the passions of fear, hope, joy, and sadness move his heart more than it does that of other animals. Thus, while there might be a beat in all animals, variations in it are not as noticeable in the others as they are in humans. The conclusion reached from these things, then, is that the lung serves as a fan [*flabellum*], serving the heartbeat and breathing organically.

---

56. *Balena* is used either as a variant of *cetus* or for a female whale. Ar. *Part. An.* 669a7f. lists *phalaina* (whale), dolphin, and all the *anapneonta kētē*, a term translated as "spouting whales" by Peck (1961) and "spouting Cetacea" by Ogle (1912), but which may mean simply "breathing" or even "breathing on top." In any event, the appearance in A.'s text of the *malachye* is not easily explained. The phrasing of the Latin almost reads as if it were at one time a gloss on a now lost, garbled name. Clearly cephalopods do not have lungs. Perhaps the origin is in stories, like that of 8.128 (cf. Pliny *HN* 9.47.85), of the octopus going out of the water onto dry land. The "sea wolf" is extensively discussed in a note to 1.33. The specificity here in describing the animal as "the sea wolves which draw in air" assures us that the seal is meant.

47 There is, moreover, great diversity in animals' lungs. Some animals have a lung that is sanguineous and large, whereas others have one that is empty and solid (that is, compact). For the most part, a viviparous animal has a large lung for its body and one with a great deal of blood. This is because of the heat of the nature of its complexion. The lung of an oviparous animal, however, is small, dry, and can inflate and swell a great deal. It is the same in an oviparous quadruped, such as the tortoise, the lizard, and those like them. It is much the same too in the bird genus. For the genus of birds has a hollow (that is, porous) lung that resembles froth which is easily produced in great amounts from a little water and which is made small from being great when it dissolves and turns into water. The lung of animals of this sort is small after the fashion of a web. Thus all types of birds have but little thirst and can refrain from drinking for a long time due to the paucity of heat in their bodies. They are cooled internally by the motion of the lungs whose expansion motion draws in a great deal of cooling air from the outside to the inside.

48 An indication of this is that animals of this sort are small in size and have short bodies. Heat is the cause of growth and size and it is reasonable that those which have but little heat should be small. Lack of blood also follows upon lack of heat, for blood is the material of growth and size. It is also heat which straightens and raises up the bodies of animals and, because this heat limits his moisture better than all the other heats of other animals, the human alone has an erect, upright body. Also, a viviparous animal is more upright and has a more erect body than the other quadrupeds. Such animals cannot live anywhere but in the open, cooling air. Thus, for the most part, a viviparous animal does not live in rocks or cracks in the ground if it also has feet and walks upright on them. Mice, however, throughout all their genuses, do not walk upright. Rather, they drag their entire rear foot (from the middle bend in the leg down) along the earth. Generally speaking, then, the lung was created for respiration.

49 One should know that certain of the inner members are divided in two and others are not. I am calling "divided" those which are divided in place and subject like the kidneys. I am calling "whole" and "undivided" those which are the same in subject even though there might be some signs of division in them. Examples are the heart and lung. There is doubt about some of them, like the liver, which has many interconnected divisions and thus it seems to be a single thing.

It should be known generally that the inner members are all, in some way, double. This is because the body is divided in two in all directions: namely, below and above, front and back, right and left. Thus it is held that each member is divided in two (or into more) along one of the directions and that the sensory organs are double. The heart too is said to have many ventricles for this reason. The lung is divided in two in viviparous animals and thus some have felt that such animals have two lungs. The separation of the kidneys is clear for anyone to see but doubt exists concerning the liver and spleen.

50 The reason for this is that many feel that the spleen is a sort of impure liver. According to this opinion they say that the liver is divided in animals that have a

spleen. In an animal that has no spleen at all, however, or which has a very small one, rather like pinpricks and dots, the liver is found to be divided in two.[57] The greater part of the division is on the right but the smaller is on the left. The location of these divided parts is clear to see. However, it is not equally clear in all the oviparous animals. Neither is it equally visible in a given area in one and the same species. Rather, in some the liver is found divided in various places and in others it is not. This is what they say about the animal called the *decheonos* in Greek. Some feel that this one has two livers, just as certain fish and the *celeti* are said to have two livers.[58]

Moreover, because the liver is located on the right, the spleen must be located on the left.[59] For the right and the left sides are the greatest causes for the division of those members which lie below and near the wall [*paries*] (that is, the diaphragm).

Moreover, the inner members are created with the veins at all points, and they are bound to these by the members themselves for the sake of their strengthening, much as ships in port are bound with certain strong ropes to their anchors. These bindings occur on the parts of the veins that are extended furthest from the great vein in the direction of the liver and the spleen. For these two members are like something secretly set aside which retains and keeps the entire body in being. For the great vein proceeds to the liver and the spleen and to one side of the inner members. However, the veins which leave the great vein only travel directly to these two members.

The kidneys lie to the rear of the body and to these come many veins, not only from the great vein, but also from the *adorthy*. Two veins leave the *orthy*, one of which comes to the right kidney and the other to the left. The liver and the spleen are joined to the stomach, for they have the potential of decocting food since they are from blood.[60] For while the spleen is the vessel of the melancholy it still has, from the veins that are in it, heat which assists in the digestion of food.[61] Because these members have heat, they decoct the food. The kidneys, though, serve to strain the moist superfluity which flows to the bladder.

These members, therefore (namely, the heart and the liver) are in the bodies of animals of necessity, for the sake of the principal heat which exists in them. The inner members of an animal must be like a furnace in which the natural heat is produced and this heat must be conserved in these two members. For that heat rules the entire body as a prince rules over the area which constitutes his kingdom. The liver is for

---

57. "Pinpricks": *signa et puncta*. Cf. 13.53 for *signum punctale* in a similar context.

58. Ar. *Part. An.* 669b34f. speaks of hares (*dasypodes*), which, in some geographically different locations, are said to have two livers. He does not mention the selachians, but rather cartilaginous fish whose name (*selachōdeis*) bore a close enough resemblance to "selachians" to create the error.

59. This paragraph number is omitted in St.

60. "From blood": *ex sanguine*. Ar. *Part. An.* 670a19f. states that these organs are *enaima*, that is, they possess blood, and are therefore hot and ready to decoct the food. The next sentence in the translation is A.'s addition, apparently included in an attempt to clarify his murky source text.

61. At *QDA* 1.47, A. explains that the spleen is the vessel of melancholy and is found on the left side of the body. That is why the heart is a little left of center, viz., to counter with its heat the cooling influence of the spleen.

decocting and digesting. Therefore, every blooded animal needs these two members and all animals of this sort thus have a heart and a liver. A breathing animal, however, needs a third member—the lung.

53 The spleen is present in the bodies of animals accidentally and not necessarily, for not every animal has one and it does not serve a principal operation. It is the vessel for the melancholic superfluity and resembles the vessels of the other superfluities of the belly and bladder, even though the superfluity of the spleen has the function which we assigned it above in the anatomy section.[62] This is why in certain animals the spleen is very small, especially in birds with a warm belly, such as the dove, hawk, and kite. It also happens this way in quadrupedal, oviparous animals and likewise in many skinned and scaled animals which lack a bladder and which also have very small, almost nonexistent spleens. This is because the melancholic superfluity in them is diverted through the flesh into feathers and scales. In those with a large spleen, however, the spleen draws the superfluity of the melancholic humor from the belly via the liver and decocts it. For it is created from fetid, melancholic blood. It sometimes also happens to draw a great deal of moisture from the belly via the liver and then it is the reason, accidentally, for drying out of the belly and for hardness and blockage of the spleen, all due to the building up of melancholic moisture in it. A similar hardness of the belly occurs in those who urinate too much, for too much watery fluid is separated from the belly, leaving behind dry excrement which cannot flow through the intestines.

If, however, in some animals the melancholic superfluity of this sort is found to be small in the place where these humors are generated, then their spleen is very small, as happens in fish. For in the genuses of fish a certain animal is found which has a spleen so small that it appears to be but a pinprick.[63] The same thing that happens in oviparous quadrupeds happens in birds and fish, for their spleens are found to be very small.

54 Moreover, the spleen is created from tough, earthy flesh, much as the kidneys are. The lung is created soft and hollow (that is, porous) and with little digestion. But the spleen and kidneys are created from flesh with an opposite disposition. The aforementioned animals have spleens that are small since the melancholic superfluity which is in them turns into the body's skin and scales, just as in birds it turns into feathers, plumes, and wings. But in an animal with a bladder and a blood-filled lung, the spleen is found to be moist for the reason we gave, namely, because the continuous conversion of the melancholy, which is moist in act, comes to it.[64]

---

62. Cf. 1.608f.

63. "Pinprick": *signum punctale.* Cf. 13.50.

64. A difficult passage. The translated terms "turn into" and "convert" are based on a somewhat rare interpretation of words based on the stem *declin-*, sometimes even used in the sense of a religious conversion (cf. *LLNMA,* s.v. *declino* 3b). Another interpretation of this verb which may work here is simply "to go toward," "to migrate toward." The words are translations of Gr. words based on the verb *trepō*, lit., "to turn," which Peck (1983) translates as "applied for the benefit of." The last sentence, where A. has added the oddly worded explanation of the reason, reads "*declinatio melancoliae quae est in actu humida, declinat ad ipsum.*" *Ipsum* can refer in this sentence only to the spleen.

Moreover, the spleen is located on the left since the nature of the left side is generally colder and wetter. For each of the contrary bodily positions is divided and attributed according to some genus of the elements which suits it. Thus, right is the opposite of left as hot is the opposite of cold and these *differentiae* of the position of a body are the first elements of the body in the way we have already set forth elsewhere.[65]

CHAPTER VII

Which Is a Digression Clarifying the Understanding of the Things Which Have Been Said about the Liver and the Spleen

To make the things said here clearer, we will say a few things about the nature of the liver and the spleen, even though we said much about them in the tract on their anatomy.

Let us say, then, that the liver is the member in which blood takes on its red color, although the mesenterics also have some of this power to effect this sort of alteration in the blood.[66] For the power for this alteration flows into them from the liver, even though Galen seems to deny it.[67] The liver has power of this sort because it is flesh, red as coagulated blood, and lacks in nerves. There are veins spread throughout it, some being thicker, and these are the origins and principles (that is, roots) of those which are slenderer. These veins absorb from the stomach and the intestines through the medium of those branches that are on the concave part of the liver, customarily called the *sima* of the liver.[68]

The part of the liver toward the diaphragm turns away from it lest the diaphragm's movement hinder it. It does so in such a way that it contacts it at almost a single point, as when a flat surface touches a rounded sphere. The outside of the liver is rounded so that the ribs can curve over it better. There is a membrane over the liver which provides it with the benefit of sensation and strengthens it by tying it to the other inner members. A pulsating vein comes to it from the heart, providing it with an increase of heat.[69] This comes from the area of the *syma*. For the protuberance, in the area of the diaphragm, is sufficiently warmed by the movement of the diaphragm.

---

65. Stadler's *in prima elementa* is untranslatable and ungrammatical. The translation offered follows Borgnet in dropping the word *in*.
66. Cf. 1.596f. and Avic. *DA* 13.37r.
67. The Galen reference was apparently added by A. Cf. Galen *UP* 4.12 (May, 1.220f.; Kühn 3.296f).
68. The *sima* is the lower part of the liver, resting on the stomach. Cf. 1.397, 554, 597.
69. Stadler has supplied the words, "A pulsating vein comes" from Avic., greatly improving the Borgnet text.

Moreover, there is no wide place in the liver in which the mass of food that has been drawn in is divided up and generated into blood.[70] There are very many such places in the liver divided in such a way that the material of the food is divided and altered more quickly through them. For this reason, moreover, the veins which are nearer to the liver form a thinner web.[71] The altering heat is received more quickly in such places. Moreover, the membrane which envelops the liver binds it to the membrane covering the stomach and intestines. It also binds it to the diaphragm with a strong bond and it binds to the sides at the rear with a slender and thin bond.[72]

The liver is also connected to the heart through the great vein and that vein is held firmly to the liver by the strong binding action of a tough, thick membrane which also sends out the thinnest of its substance into the substance of the liver. This occurs because whatever amount is in the liver is in a safer place. Having originated principally from the heart, the veins pass through the liver and then two veins leave either side of it. One, through which nourishment is drawn in, is at the *sima* and the other is at its protuberance. It is through this one that food is distributed to the entire body and for this reason it is hollow and is called hollow [*concava*, vena cava]. Enough has been said in the anatomy section on the branches of the liver and the veins that divide off it.

Plato, however, and many of his followers, said that the principal seat of a given soul is in the liver and he said that this is the concupiscent soul.[73] Since this has two parts, namely, one that is concupiscent and the other which is irascible [*irascitiva*] (that is, the opposite of that which is sad), he said that the concupiscent (that is, desirous) soul is in the liver and the irascible [*irascibilis*] soul is in the gall bladder.[74] He gave as a reason that the first act of attraction, which is concupiscent and desirous of that which naturally preserves life, exists in the liver, whereas the expulsion of that which hinders and must be rejected is in the gall bladder. For this reason the irascible soul has a small gall bladder, located at the lower intestine, and there, by its biting action, it expels that which is opposite and corrupted. He said that when the spirit is set free from the liver, desire is brought about everywhere in the body and that when the bitter spirit evaporates from the gall bladder, indignation is brought about everywhere in the body. He also said that the desire, through the power of the liver, forms that part of the food that is desirable with the goal of making it desirable for the entire body, and thus it is drawn out and led to the members.

---

70. Once again, the "no" is added to the text by Stadler from Avic., drastically changing the sense. What, though, did A. think of the text without the negative? Cf. 1.598, where the negative is in place.

71. Reading *quae* with Borgnet for Stadler's *qua*.

72. Perhaps "it binds its rear portion to the sides . . ." But cf. 1.598.

73. Plato *Timaeus* 71d.

74. Cf. A.'s *De anima* 1.1.4.

Galen has followed this teaching and attributes the principal and first powers as well as the natural spirit to the liver. But we have shown elsewhere that these are fallacious ideas and we should not discuss this matter here.[75]

Of the spleen, it should be known that it is doubtless the vessel of the melancholy which purges the blood of its earthy residue. This is proven because it lies facing directly opposite the liver, linked to it by straight ducts. Nor does it have any duct save that to the stomach for the purpose of appetite and the one to the liver for the purgation of the blood.

The same thing is also proven by the fact that fetid and impure blood is found in the passages which face the liver in slaughtered animals. This would not happen unless it were drawing the residue of the blood through these passages from the liver.

Further, as has been said above, certain animals do not have a spleen, such as the particular bird which has a head shaped like that of an ass and for this very reason is called the "ass" [*asinus*].[76] This is due to two reasons together or to one or the other of them. One is that the birds of this genus do not eat melancholic food or else these birds are very earthy and have many earthy parts. For this reason, fish and birds have very little in the way of a spleen, for the earthiness passes over into scales and feathers, as we said above.

Every animal which has a small spleen also has a small lung, since it drinks but little watery fluid. Further, an animal with a large spleen has little fattiness [*crassities*], for fattiness is a product of the liver, and when the liver is large and has great power, the drawing power of the spleen is impaired. As a result, the nourishment is not purified of the dregs of the melancholy and thus the members shun it and do not draw it in, and starvation results. This is also why those who are melancholic are thin and dark, although another cause of this is small natural heat. One should know that, as Nicholas the Peripatetic puts it, the spleen and liver are like two posts on either side of which the veins are attached.[77]

Let, then, these things, along with those said in the anatomy section on the liver and spleen, suffice for now.

---

75. The first six chapters of Book 3 are devoted to Galen and his opposition to what was and was not the principal member.

76. Apparently the *onocrotalus* is meant here. Cf. 23.131(86), where it alone of the birds is said to have no spleen. The word is normally translated "pelican" but in the passage just cited A. claims there is both an aquatic and a terrestrial version of the bird. Thompson (*GB*, 212) lists some options for the land identification. It is called the "ass" here because of a long history of associating the bird with that animal, partly through etymology (Gr. *onos* = ass) and partly because of its braying voice. A. himself discusses the aquatic pelican under the name of *osina* at 23.131(88), giving us the MHG name of *volmarus* (Sanders, 424).

77. Perhaps Nicholas of Damascus, a Peripatetic active in the second half of the first century B.C. Nicholas actually served as secretary to Herod the Great. He wrote a massive universal history and, among other works, commentaries on Ar. that are no longer extant. However, many of his works were translated into Arabic (Sarton, 1931–48, 3.432–33; 1952–59, 2.443–44). The work entitled *De plantis,* which A. believed came from Ar.'s pen, belongs in fact to Nicholas. Cf. our introduction, n. 108.

CHAPTER VIII

On the Nature of and Reason for the Kidneys and Bladder in Animals Having Them, and on the Reason That They Are Not Present in Certain Animals

60 While the liver and heart are noble members with noble operations relating to being and life, the kidneys do not operate for being and life, but rather for the removal of interfering superfluity. For through the veins which are called the strainers [*emulgentes,* renal veins], they draw off the salty, watery superfluity from the protuberance of the liver along with a certain amount of blood, which is assigned to nourish the kidneys. This superfluity descends further into the bladder and is cast off. There is a great deal of this superfluity in certain animals and they therefore must have strong kidneys. The kidneys thus were created principally to make the operation of the bladder better and more perfect. Because the activity of the kidneys and the bladder is the same, we should say something about the activity of the bladder, even though for now we will ignore the nature of certain of the upper and internal members, like the diaphragm and certain others.

61 Not every animal has a bladder, nor did nature give a bladder to any other than the blooded animals. For we see that every animal which has a lung is very thirsty and drinks a lot. It does not need dry food, but rather a great deal of moist food, much more than dry. It is not necessary for all this moisture to be incorporated into the members. Rather, it is there to serve as a medium so that the food can flow through the mesenterics and the slender veins of the liver. For if it were all diverted to the members, they would become bloated with water, like the limbs [*membra*] of those suffering *yposarca* [dropsy].

It is thus necessary that after its purpose is accomplished, this moisture be separated from the bodies of the animals. Thus, it must be drawn off from the protuberance of the liver (the place from which blood is directed to nourish the members) and be drawn through the kidneys to the bladder. Thus, as we said above, it is necessary for there to be a bladder in the bodies of all animals which have a sanguineous lung. An animal without a lung also lacks a bladder, since this animal does not take in water in the form of drink to cause its food to flow, but rather for the tempering of its food. Examples are birds, fish, and those with a ringed body. Thus, an animal with feathers or scales or a hard horny scale does not have a bladder because of the little they drink and because their watery superfluity passes over, along with the dry and salty earthiness which is in it, into feathers, scales, and horny scales.[78]

62 The tortoise is an exception to this, for the lung of this animal is very fleshy and "blooded," resembling that of a cow. But since the tortoise is double (that is, both

---

78. "Horny scale": *cortex,* a term that often can indicate skin or hide but which seems here to stand for the *pholidōs* of Ar. *Part. An.* 671a12f., indicating a tougher sort of scale than that of a fish.

land dwelling and aquatic), the land tortoise's lung is larger than it has to be compared to its body. The reason for this is that its body is contained in a very thick shell and moisture cannot pass over into this. Rather, it is generated along with it and grows on its earthy nourishment. Thus superfluous moisture does not pass over into it as it does in birds, fish, and other animals with a scaly skin. This is why there necessarily is a bladder in the tortoise, so that, like some sort of vessel, it can take in salted, watery fluid. Nevertheless, the bladders of the land and sea tortoise are found to be very small as are their kidneys. For only that moisture is collected in these organs which cannot pass over to be material for the bladder which it has above itself, because of the solidness of its shell.[79] No animal whatever which has feathers, horny scales, or scales has a bladder or kidneys, except the tortoise, both land and sea.

In other animals, the flesh from which the kidneys are created is separated into a kind of stretched out flesh, the composition of which is like that of kidneys in that it draws off watery fluid and separates it into feathers, scales, and horny scales.[80] The animal called the *amoz* in Greek, however, has no kidneys but does have a bladder because of the softness of its skin.[81] For this softness is the reason that the watery fluid is drawn to the bladder without kidneys and the *amoz* therefore does not have any of the aforementioned members. But other animals with a sanguineous lung have kidneys, as we have said. For nature only employs the kidneys because of the strainer veins through which they draw and for the leading out of the superfluous moisture through the bladder. There is a certain deep place in all kidneys along with an opening, because of the multitude of the food, and so that the urine might be better drawn to these hollows.[82]

The composition of a human's kidneys is like those of cows. They are very tough, more so than the other members, in order to be hurt less by the sharpness of the urine and so that the power of attraction might by better preserved in them. A human's kidneys resemble those of the cow in that they are, as it were, made up of many small kidneys. They are not made of flesh with a smooth surface, but flesh which is ridged, uneven, and connected like those of the sheep and other quadrupeds.[83] This is why it is difficult to cure a problem in human kidneys, for the actual cause of the illness is

63

64

---

79. The sense is garbled here due to the vagueness of the Latin. The "above itself" would seem far more readily to apply to the animals' shells than to their bladders, and one is therefore tempted to emend the text at this point. Still, the main idea seems to be that the excess moisture would normally pass over to become material for the shell, but the shell is so tough that this is precluded. The origin of the shell is discussed also at 12.151. The fact that the tortoise/turtle had a bladder was of some interest to medieval natural scientists. Cf. 13.95, ThC 8.40, and Barth. 18.106.

80. Compare what follows on kidneys with 1.612f.

81. This too is a sort of tortoise, representing the Gr. *hemys* of Ar. *Part. An.* 671a32f. The name appears as *amydon* at 12.151.

82. This "deep place" (*profunditas*) is probably the hilum, the indentation which gives a kidney its characteristic shape, described at 1.615 as resembling a ventricle and called there a *profundum*. The wording of what comes next is noteworthy for its clumsiness.

83. "Connected": perhaps "solid."

spread around as if on many kidneys separated from one another by the angles of their surfaces. Thus, when progress is made in one area, it does not pass over to another.

The venous passage which proceeds from the great vein [vena cava], which is diverted from the protuberance of the liver to the kidneys, does not come directly to the deep place on the kidneys. Thus, coagulated blood is not found in the hollow of the deep place on the kidneys as it is in hollows of the liver. For if it were there, it would be found coagulated in slaughtered animals.

Further, two passages [ureters] leave the hollow of the deep place on the kidneys. No blood is found in them and they come to the bladder. They are tough and strong since sharp urine flows through them. Certain other passages leave the smaller vein, which they call the *adorthy* [aorta], and these travel to the bladder. They are strong and hard for the same reason. They come one after the other like two veins so that through these passages the fluid superfluity might leave the great vein for the kidneys, and then be carried through others to the bladder.[84]

65 There is also residue found in the kidneys because of the straining of the watery fluid which occurs in the kidneys. This is also why the pit of the kidneys is larger in some animals than in others, for they have great need of this straining. It is also why the kidneys smell worse than the other internal members. The fluid superfluity leaves about from the middle of the kidneys and travels in the urinary passages as far as the bladder.[85] These passages are strong and tough for the reason we gave above. This, then, is the reason for the creation of the kidneys.

Moreover, there are powers in the kidneys of attracting the watery fluid and expelling it to the bladder and these suit the nature and substance of the kidneys.

The right kidney is higher than the left in all animals which have them. There is a twofold reason for this. One is that the movement of attraction exerted by the right side is stronger, and thus the kidney in which the power for this movement resides is higher. For natural movement makes a passage and a place above for all members on the right side. The power is lacking in the left kidney, however, so it sinks downward. An indication of this is that a human has a higher eyelid to the right and above.[86] Thus, then, the right kidney is higher than the left because it is drawn higher by the power of the right as is the eyelid. There is, however, another, truer reason. For the liver is on the right and warms the kidney, thus raising it. In all animals which have kidneys, the liver touches the right kidney.

66 There is much fat [*sepum*] in kidneys, more so than in other members. This is due to the straining of the watery superfluity in which fat [*pinguedo*] is generated from the watery blood.[87] Being thus well digested, the watery blood stays there and is converted

---

84. "Come one after the other": *succcedunt sibi,* an enigmatic phrase that caused A. to add the explanation comparing them to veins. The original sense in Ar. is that they are continuous, without breaks, being solid tubes.

85. "Urinary passages": *viae uritides,* an unusual use.

86. "Eyelid": *palpebra* (cf. 1.139). Ar. *Part. An.* 671b33 has "eyebrow."

87. On the varieties of fat, see Glossary, s.vv. "Fat," "*Zirbus.*" In this passage "fat" always translates *pinguedo* unless the Latin is given.

into fat and *zirbum*. For, just as in burned things ash is the residue of the burning, so, in fluids decocted in the liver, their superfluity is diverted to the kidneys as *pinguedo* and *zirbum*. Some of the heat thus remains in both of them (that is, the ash and the *pinguedo*). This heat has worked long in it as if it had been led in for a purpose, for ash is sharp and fat is light, inflammable, and floats on liquids. There is no fat in the actual substance of the kidneys, however, since they are made of thicker flesh than is any other substance of any other inner member. But the substance of the fat is outside the body of the kidneys and surrounds them, if the animal is the sort that has fat. If, however, it is an animal that has *sepum,* then *sepum* or *sepum*-filled *zirbum* will surround the kidneys. The difference between *zirbum* and fat has been given above.[88] The right kidney has less, since it is warmer and dissolves more of the fat produced around it. The kidneys have a great deal of fat for the reason given.

The kidneys were also created for health and they therefore had to be created hot. For, in those that have kidneys, the kidneys are among the most distant of the inner members. If they were not hot they would not draw moisture and they therefore need great heat. For this same reason, the part of the body which is in the area of the back is very fleshy so that with its warm flesh it might serve as a covering for the members that stand about the heart. At the outer part of the back, however, and below, there is not much flesh, since the fat of the kidneys is a sufficient covering.

Moreover, when there is a great deal of fat on the kidneys, it will preserve a great deal of heat, and the kidney therefore separates and digests the humor with a good digestion, since there is heat left behind in the fat which digests the humor. This, therefore, is the reason that the kidneys have much fat.

Moreover, as said a bit earlier, the right kidney in all animals has less fat than the left. It is also higher since the nature of the members that are on the right side is lighter due to their greater heat and movement. The movement, however, dissolves using the heat and acts contrary to the congealing power of the fat since it dissolves the fat. Nature, then, aiming at the best, amassed a great deal of fat around the kidneys for the reasons introduced above. As a general accidental trait, sheep die when the fat all around their kidneys grows very hard. This is because the sheep is a very moist animal as the multitude and softness of its wool shows. Thus, fat, which is generated from the more watery blood, is very moist in moist animals and it readily putrefies and corrupts. When it multiplies in this way in a sheep, it kills it.

Moreover, pains befall kidneys from windiness and these are not all of one and the same kind. The reason for the pain is windiness, which rises up from a blockage in the kidneys. For this reason too, kidney pain is very harmful and disabling in all cases and when kidney pains worsen, they quickly lead to death. This thin windiness arises from the evaporation of fluid when it is surrounded by warm fat, and very thin windiness extends to the liver and heart, corrupting the liver and extinguishing the heart. Thus the animals die, especially sheep. For the fat which is on the kidneys of the other animals is not very thick and is not like the thickness of the fat of the sheep since

---

88. Cf. 12.106f.

the sheep's moisture is more congealed and concentrated. Moreover, sheep fat is more plentiful than that of other animals and sheep grow fat quickly in the vicinity of their kidneys. The fat around the kidneys grows thick because the surrounding moisture is burned in the kidneys and it evaporates into a dry windiness which is thin and penetrating. As a result of the pain of this sort of windiness, the sheep dies. For the windiness, once aroused, passes through the great vein [vena cava] and the *adorthy* [aorta] coming to the heart. This is because the passages of these veins, following upon the kidney veins, are connected all the way to the heart and the sheep is thus snuffed out.

We have thus spoken of the natural disposition of the heart, liver, kidneys, and spleen and have given the natural reasons for these members, which, although they are inner members, are nevertheless separate one from the other.

CHAPTER IX

On the Nature of the Diaphragm and That of the Other Webs in the Body.
We Will Now Have to Speak about the Diaphragm.

69 One should know first that Aristotle says that some people call the wall of the diaphragm the intellect. This is not the case in our language, which is Latin, but rather in the Greek and Arabic languages, for in these they call the intellect a distinctive power.[89] This name is taken equivocally for the distinction the diaphragm makes between the spiritual and nutritive members and the distinction which the intellect makes among intelligibles.[90] Thus, in this way, the wall of the diaphragm is called an "intellect" [*intellectus*]—an "internal selection" [*intus lectus*], as it were—since it separates the heart and the lung (these being in the region of the spiritual members) from the other members which are beneath them in the region of the nutritive members. Every animal thus has a wall, just as it also has a heart and a liver.

The position of the diaphragm is between the heart and the belly. For nature intended to protect the heart (which is the principle of the sensible soul), keeping it from any misfortunes of disease which might befall it because of the evaporation of foods from the belly or from external mishaps that might befall it from the heat or the coldness of the surrounding air. For this reason nature prepared this member as a sort of defense wall before the entrance to a house and used it to separate the nobler, upper

---

89. Ar. *Part. An.* 672b11, calling it the *phrenes*, from *phrēn*, which began as an anatomical term, apparently, but moved on to indicate the mind since the midriff area, containing the heart, was thought to contain the intellect. The "wall" (*paries*) is commonly used alone to indicate the diaphragm, as at 1.549, 595.

90. "Nutritive": the use of *pronutritiva* for the more common *nutritiva* is noteworthy.

members from the baser, lower ones.⁹¹ For the upper members are those through which all the good operations of the soul are exercised and which pertain to betterment. The lower members, however, are there only for the sake of the upper ones, so that the upper members might take food from them.

The wall [*paries*] was made for these reasons, then, and one should know that the wall is fleshier at its edges where it is joined to the other members, for it requires strengthening there. It is thinner and finer in the middle for in this way it better suits the extension of the spirit and is of more assistance to the heart, which is the principle of all life.

Further, the wall has two passages coming to it from below. This is indicated by the onsets of illness which sometimes occur around it. For sometimes it draws a warm humor from below and when this fluid is changed and becomes corrupt, it moves as a vapor toward the head, changing sensation and intellect, insofar as intellect is common to sensation in the sensible spirit. For although intellect is not present in the wall as if in an organ (since it is the action of no body), the wall nevertheless participates with it there insofar as it is an operation of the intellect. It therefore clearly alters the intellect on occasion.⁹²

This accidental trait and others like it prove that the wall sometimes draws a corrupt fluid into the area of the spiritual members, and that this, moving as a vapor to the seats of the animal powers, disrupts and disturbs their operations. Now the only reason that the wall is fleshy at its extremities near the ribs and is thin in the middle is so that it can be strengthened at the edge and so that in the middle it can be well suited to the spiritual members. Another reason is that little of the humor drawn in for nourishment comes to its middle, for otherwise the spiritual members would be disrupted quickly by thick things. It is bent back toward the edges, however, and gathered there, and thus the wall is nourished more and grows thicker at its edges.

As we have said, this is the most suitable way. For if it were warm and fleshy in the middle, it would attract there a great deal of moisture and would destroy the operation of the spiritual members. For when the wall is heated, it suddenly changes sensation and dissolves the entire body into joy or sadness.

An indication of this is the tickling sensation that results from a light touch on the wall. Laughter follows immediately upon a touch on the web of the wall or on a member that is near the web, for the web is moved by this touch.⁹³ Movement of this sort strikes the wall and the wall, warmed by the movement, pours forth and sets loose the spirit. At this the intellect, which was keeping the entire body contained, is changed and as a result a person laughs even at those things that should not be laughed at and

---

91. "Defense wall": *murus* instead of the normal *paries* is used for the diaphragm proper. A. seems to be doing more here than displaying his normal penchant for varying vocabulary, for a *murus* is generally, but not always, a more substantial thing than a *paries*. The latter more likely indicates an internal wall or partition.

92. At *QDA* 13.15, A. discusses further the power of the diaphragm to affect *per accidens* the intellect.

93. Cf. *QDA* 13.16–17.

does so with a disproportionate laugh in which there is a great expulsion of spirit and great loosening of the nerves because of the heat poured into them. The reason that a human alone is changed by tickling of this sort is that he has a very thin skin, with delicate flesh and a refined sense of touch. Thus no animal save the human laughs.[94] In the human, though, tickling brings on laughter from a movement of this sort and from the movement of the skin that is near the area of the armpits, for the wall moves there. Thus some have held the opinion that if a person is struck and wounded in these places he will feel no pain and that it will instead produce laughter in him because of the tickling heat aroused by the blow.[95]

72 This statement certainly seems more likely to be true than that of the followers of Hesiod who say that the head of a person sometimes speaks after it is cut from its body. They say that Homer related this tale among the wondrous deeds of Jupiter.[96] This story is believed and maintained in Arcadia. For there they are of the persuasion that a certain person, having had his head cut off, spoke to a considerable audience through the divine power of Jupiter.[97] Tales of this sort, however, are false with respect to natural philosophy [*in physica*] for it is entirely impossible for any speech to emanate from a head that has been separated from its windpipe and lungs. Nor is it possible in any way that speech occur without the movement of the lungs.[98] Further, it was the experience of the Greeks and barbarians, among whom many people are beheaded, that in their parts nothing of this sort ever happens.

Other animals do not laugh at the movement of the wall because of the thickness of their skin and also because they are not naturally given to laughter, as is the human. The movement of the body, and its rolling about from place to place after the head is removed from the body, is possible because of the force of the spirit which causes the body to roll. For this reason too a bloodless animal which has uniform members and a viscous fluid lives for a long time after its head is removed. We have assigned the reason for this already in other places in the natural books.[99]

---

94. One must wonder what A. would make of Koko, the famous "talking" gorilla, who frequently requests, through the use of sign language, that she be tickled. Pictures of her reaction to the tickling can only be described as showing laughter. See Patterson (1978), but cf. the cautionary summary of the research by Linden (1986, 115–29) and Hill (1980).

95. Ar. *Part. An.* 673a11f. Both Peck (1983, 280) and Ogle (1912, ad loc.) identify this as the *risus sardonicus* arising from sudden death when the diaphragm is ruptured. Note that *risus,* translated throughout this paragraph as "laugh," can also indicate merely a smile.

96. The followers of Hesiod are also mentioned at 6.99. The idea of a talking head is reminiscent of the story of Orpheus whose head, torn off by Maenads, floated away, singing, until it came to rest and founded an oracle. The reference to Homer is from Ar. *Part. An.* 673a14f. and refers to *Il.* 10.457 and *Od.* 22.329.

97. Ar. *Part. An.* 673a17f. tells the story of a priest of Zeus who was killed and who, despite his head having been cut off, named the criminal.

98. While A. insists that this is impossible with respect to the power of nature, it certainly remains possible by virtue of that power beyond nature. One need not assume that A. would in any way doubt the miracle recounted in Chaucer's *Prioress's Tale,* according to which a young Christian martyr, whose throat had been cut through to the back of his neck, continued to sing *Alma redemptoris.*

99. Cf., e.g., 4.66.

Thus, the reasons why there are individual inner members have been determined.

One should know that it is necessary for all the inner members to be placed at the ends of veins that come to them. For the "food humor" [*humor cibalis*] oozes out of the ends of the veins, solidifies, and is formed into the substance of the member. For this reason too, all the inner members appear sanguineous and uniform and in this way too they have a nature that differs from the other bodily members.

Further, all the inner members are created within a web containing them. They need a soft covering for their preservation and this is the nature of the web. Specifically, the result is that the web is strong and thick so as not to suffer easily from injuries befalling it. And it is not fleshy so that it does not draw superfluous humors to the member, and it is thin and smooth lest the member be weighted down and burdened by that which surrounds it. Of all the inner members, the strongest and most numerous are the webs of the brain and heart. For these two members, because of their noble operations and the softness in the brain, need great protection. Further, the power of life is especially in them.

Further, in the bodies of some animals all the internal members are present and in others they are not all present, for in some, such as in the creased, ringed ones, there is only the member used to take in food and the member which is simultaneously the principle of life, digestion, and sensation.[100] Immediately after this there is the member through which the superfluity of excrement is discharged. We have already spoken in previous books of the members which are in such animals and for what reasons they are there.

Members which agree in number and operation still differ in some animals according to being and definition. For the hearts of all animals are not of a single type and definition and it is the same for the other members. In some animals the liver is split by many creases and in others it has no splitting whatever, especially the liver of fish and of oviparous quadrupeds. Birds' livers resemble those of oviparous animals in that the color of birds' livers is like the color of the liver of fish and of the other oviparous animals we mentioned. The reason for this is that birds have good respiration and light flesh and that which is in them is easily dissolved. Nor is there much bad superfluity in them which is not quickly dissolved. This is why many animals lack a gall bladder. The liver is suited to a good bodily complexion and to health. The perfect completion and perfection of all these things lie especially in good blood. The liver has much blood, as does the heart. Apart from the heart, it has more blood than all the inner members. For the most part, birds, fish, and other oviparous animals have, as we have said, the same colored liver. The liver of some animals is very bad since they have a bad bodily complexion. Examples include the liver of the tortoise, raven, and of others like them in their genuses.

The spleen of horned animals with split hooves, like the ram, is rounded. It is generally this way in the sheep and the goat unless it happens to be elongated because of

---

100. Creased and ringed are two terms normally used separately for insects. See Glossary, s.v. "Ringed creatures."

some accidental growth of the spleen and body, reaching the size of a cow's spleen. The spleen of all animals with many splits in their feet is long. Examples are the pig, man, and the dog. But the spleen of animals with hooves [*soleae*] is midway between these two in size. Examples are the spleen of the horse, ass, and mule, unless this sometimes occurs accidentally, as we have said. Nor do these members differ in the bodies of animals only in size or quantity of dimension: they also differ according to location. Some are external and some are internal, since the nature of some is common to that of veins and these should be internal. That of others is not, and these are external. Some are above and some are below, such as the spiritual, vital, and animal members, which are above, whereas the nutritive and natural members are below. Some are to the right and others are to the left. These do not differ in shape, but have a great difference in motive power, since those on the right are more movable than are those on the left.

The Second Tract of the Thirteenth Book on Animals: The Intestines

CHAPTER I

On the Disposition of the Belly, the Number of Bellies, and the Differences in Animals' Bellies

76 In the bodies of certain animals the belly lies inside the animal's body, after the wall of the diaphragm. In animals possessing an esophagus, the belly is positioned at its end. In those lacking an esophagus, however, the belly is attached to and lies directly after the mouth. However, after the belly (which is what I am calling the place in which food is taken in for digestion) an intestine is placed. It is either single or multiple according to the different natures of the animals. It is clear, moreover, from what was said before, that not one of these members exists for any reason other than food. Food must be digested, for if it were not, the moisture could not be absorbed by the body.[101] Food, however, and the useless and impure superfluity are not of the same type. Thus, certain of these members must be for taking in food, some for altering what has been taken in, and others for taking on the impure superfluity which must be expelled. Each of the things mentioned must have a member in which it can be kept in its own place. Nevertheless, it is more suitable that we speak of these things in the place where

---

101. The moisture of the digested food is meant.

we will treat the foods of animals and the generation of their members. Here we are inquiring only into the reason for the members.

Let us, then, consider now the diversity in the bellies of animals and the diversity of the members which deal with the digestion of food. As far as appearances are concerned, the size and the shape of a belly do not have one and the same disposition in the types of the species of animals. Rather, an animal which has teeth in both jaws is blooded, is viviparous as well, and has a single belly. Examples are the human, the lion, the cat, and others like them which stand on feet with many toes. Having a belly similar to these is an animal with hooves [*solea*], like the horse, mule, and ass, and the same is true for an animal which has a hoof [*ungula*] split in two and also has teeth in both jaws, like the pig. All these have a single belly of almost the same sort.

However, an animal whose food consists of woody and spiny material, that is, one which gnaws at the ends and twigs of trees, like the goat, has many bellies. Now, I am calling a belly every receptacle in which food is stored before that which is impure is separated from that which is pure.

The camel is an animal of this sort as is every animal with horns which lacks teeth in its upper jaw. This is also the reason that, even though it lacks horns, the camel is not like an animal which has teeth in both jaws. And because the camel takes in food this way, it was necessary for its belly to be so disposed as to have many bellies. Thus the camel's belly resembles those of animals lacking teeth in their upper jaw, for the composition of their teeth was like that of the teeth of a horned animal. For since the camel's food was thorny, it had to have a tough palate. For this reason it was proper for it to have a fleshy tongue on which the taste of its food could be sensed.[102] Thus, in the camel, nature employs the palate like the earthy part of the teeth.[103] The camel, therefore, ruminates as do horned animals and its bellies are like those of horned animals. Each of these animals has many bellies, as do the cow, sheep, goat, and others like them.

These bellies are given the animal by nature so that the use and the digestion of the food might be completed perfectly, with the result that the nature of the liver might extract from it whatever purity it has and that it might be possible for it to be converted into nourishment. Since animals of this sort seem to be deficient in teeth for grinding and separating food, nature gave them one belly after the other in which that which seems lacking in the employment of teeth might be completed in the bellies. For these animals eat tough, woody things that they scrape from the earth and in this style of eating, the front, upper teeth cannot be hard, for their material would vanish from rubbing against the hard objects. The lower teeth, however, are not fixed as strongly as are the upper teeth, and they also stand upright on their roots while the upper teeth hang down. Thus the upper teeth would not be useful for long for food such as this.

---

102. The reason given is A.'s, who misses the point that a tough tongue handles tough food better.

103. Ar. *Part. An.* 674b4f. states that nature took the hard, earthy material that would have gone to compose teeth and directed it instead to form the palate.

For this reason nature gave such animals a belly in which the food is taken in in rough form and is softened afterward. It is then brought back to the molars and chewed. This is the form of eating called rumination. In this case the second belly receives the digested food in such a way that the digestion begins in it through softening and separation. This second belly is the stomach. The third belly, however, is the intestine, which is connected to the pylorus [*portanarius*] of the stomach and into which food is taken for better digestion. For it has already been separated and its pure part has been partially drawn off to the liver from the stomach. Nevertheless, it is further decocted in this intestine and the pure part is absorbed through the mesenterics attached to it. The fourth belly is the last intestine, to which the mesenterics are attached, after the food has left the cecum [*orbum intestinum*]. This is the one called the *ieiunum* and in which the food undergoes the final digestion, for the last bit of that which is pure is separated from that which is impure and the impure, now entirely corrupt, is expelled through the rectum [*longaon*] and the anus.

These bellies are diversely named in Greek and we have given their names above. Anyone who wants to understand perfectly the aspect of their location should read above in the first book of this study where we set forth the order of all the inner and external members in the section on anatomy.[104]

80

The members taking in the food in birds also differ for the same reason—for the complexion of the food. For a bird does not have a mouth adapted for grinding and chewing food with teeth and nature therefore gave certain birds a member in which the food is first softened and altered before it descends to the stomach. This is the member which some call the *papa* but which is called the "bladder of the throat" [*vesiculum gutturis*] in Latin.[105] By a good many, however, it is called the bird's crop [*struma*].

This member softens the food and thus makes up for the lack of a mouth where food is softened by being mixed with saliva and by chewing. However, in some birds which eat softer things or which take their food in the water, there is an esophagus but no crop. In others another member is also found in which such a utilization of food is completed. Some have a large esophagus in which the food stays for a long time, descending gradually so that it can be softened and become warm. This is the case in the crane, stork, ostrich, and birds of this sort. Some have other appendages in their belly and some have certain appendages protruding to the front of their bellies in which they deposit the rough food until it is altered and softened. Others, however, have a stomach which is very thick, fleshy, and tough. Many birds have such a stomach, in which they can lay up a great deal of food for a long time for the purpose of digestion. In these, nature makes up for some of the lost operation of the mouth by using the power of the heat of the stomach. Some birds have almost none of the things that have been mentioned. They have instead a very long crop [*papa*] (that is, a "bladder of the throat") as well as a stomach. Examples include the aquatic birds

---

104. Cf. 1.549f.
105. On *papa*, cf. 2.105.

with a long body and neck, such as the one which is the large black diver [*mergus*], the coot, and the like. Their food is digested very swiftly since birds of this sort have very moist stomachs. Their food is thus expelled undigested and for this reason these birds are very gluttonous.

The genus of fish does not need a belly to fulfill the action of the mouth for almost every genus of fish has teeth. The exception is the small genus called the *astaros* in Greek, so called because it alone ruminates and "astaros" in Latin means "ruminating."[106] Small animals which have teeth in both jaws ruminate. The fish just mentioned ruminates, however, because it lacks teeth. All the teeth in fish are sharp and none are like molars with the sole exception of the river fish called the carp [*carpo*] by the Germans and French.[107] This one has teeth in its throat and this one alone of the fish known to us has teeth that are broad as if they were molars. Unlike the many other genuses of fish in the rivers of Germany and France, it does not have a single tooth in its mouth. But other fish have sharp teeth and many of them, so that they can divide up and cut food with such a division as they can. But this is a bad division and an imperfect one since the food cannot stay in their mouths for a long time. For this reason too they do not have molars and if they had them they would be superfluous and useless. The exception is the carp, which has jaws in its throat and grinds its food with its mouth closed using the movement of its gills.

81

Further, certain fish, such as the sturgeon [*sturio*] and sturgeon [*huso*], have no stomach at all, having in its place the intestine called the *ieiunum*. This is why no food is ever found in their bellies. Certain fish have a stomach to be sure, but have a very small and truncated one. It follows that they have this for assistance in the decoction of their food. Some also have fleshy and thick stomachs like those of birds, an example being the fish called the *tarycos* in Greek.[108] Some have fibrous stomachs, with longitudinal and transverse fibers. An example is the fish the Germans call the *welre* [sheatfish] and which has a very large head and mouth and a belly that lies directly next to its mouth.[109] Other fish have parts projecting to the front of their stomachs and they deposit their food in these. These projections vary in both the genuses of fish and birds. These members are found above in fish and below in birds, beneath the throat at the beginning of the intestine which leads to the stomach. Certain viviparous animals have projections of this sort in the lower part, below the throat and they serve the same purpose we mentioned. All the types of the species of fish suffer from having their food undigested because of the smallness of the members that work to soften and digest their food. As a result they expel an unfinished superfluity from their food. This is also why they are very gluttonous. It is much the same for all other animals which

82

---

106. *Astaros*: Ar. *Part. An.* 675a4, *skaros,* a parrot-wrasse. Cf. 6.67 where *astaroz* refers to a dogfish. The etymology is fanciful. On the rumination of the parrot-wrasse, cf. GF, 239.

107. On this and other similar fish, cf. 1.228; 12.213. Cf. also the form *carperen* at 24.20(26) and Sanders, 439.

108. *Tarycos*: Ar. *Part. An.* 673a12, *kestreus,* a mullet.

109. Cf. 1.92; 2.91; and Sanders, 436.

have straight intestines which are neither long nor convoluted. For the food leaves all these quickly, before there has been adequate absorption and thus, directly after feeding, further desire for food is generated in them.

83 It has been determined above that an animal with teeth in both jaws has a small belly, smaller than those animals which lack teeth in their upper jaw. Generally speaking, their bellies differ in two ways. The belly of some of them resembles that of a pig and the bellies of certain others which number among such animals resemble the belly of the dog. The belly of the pig, however, and bellies like it are the larger ones. They have many convolutions in their intestines so that the food can be better decocted and absorbed, as it takes a long time to pass through its convoluted and lengthy pathways. The belly of the dog, however, and those like it are the smaller ones. Nor, after the stomach, is there much left in it from which absorption can occur. For the intestine to which the mesenterics is not connected (it is only the point of exit of the excrement) is positioned directly after the stomach in all such bellies. This is why dogs have large appetites and why people who have inordinate appetites are said to suffer from "dog hunger" [*fames canina*].

CHAPTER II

On the Types of Difference in the Intestines, Following the Differences among the Bellies

84 The intestines also differ according to the differences among the bellies. Sometimes the intestine is wide near the pylorus [*portanarius*] of the stomach and is narrow at its end at the extremity of the belly. This is the arrangement of the dog's intestine and of others like it and is the reason why dog excrement exits accompanied by a bending of its body, an extension of its legs, and a pain in the body. In many other animals, however, it is the opposite. These animals have very convoluted intestines and emit their excrement easily due to the width of their intestine and anus. Such are the horned animals, for these animals have both large bellies and large intestines due to the great size of their bodies. Nearly all the bodies of horned animals are large since they eat moist, green food and they graze especially in marshy places. The exception is the genuses of goats which delight in mountainous, dry pasturages. For this reason, too, cows which graze constantly in marshes grow huge, as is seen in the cows of Frisia, many of which grow almost to the size of large horses. This size is due to their utilization of moist food, as we have said.[110] These animals possess a *colon* and that part of the intestine which is called the *orbum* [cecum]. After these two there comes the narrower intestine called the "convoluted" [*involutum*] to which is connected the straight intestine

---

110. Cf. 3.176 and 7.53, for more on these and similar large cattle.

[*intestinum rectum*] called the *longaon*. This is connected to the anus, through which the superfluities leave. The *longaon,* however, the last intestine, is fat in some while it is thin in others. A wise nature, however, formed all these cleverly according to the operations that are suited to digestion, ingestion, and egestion, which some call the taking in of food, the decoction of the same, and the exit of the superfluity.

Further, however, after the food begins to descend and arrives at the wide paths of the intestines, it does not stay there, but leaves. This is so especially if the animal has a good mouth or lip for taking in food. For then it is not overly dried out in it, but remains moist, lubricating the pathways and making for an easier descent. Food has a good lip (that is, a good and great eating) either because of the size of the body or because of the strength of the heat.[111]

The narrow intestine therefore receives food from the upper belly, which is the stomach, or else from that intestine called the *colon,* thus passing from a wide place to a lower place, and from there it goes to the narrow intestine. Then much is absorbed from the superfluity and it is rendered dry. It thus comes to the convoluted intestine which is convoluted so that the superfluity might be held up there lest it depart suddenly, until it can accumulate and depart all at once. Thus no animal is gluttonous which does not have an intestine that is very wide at the part near the lower belly and which in addition to this has convolution in its intestines and does not have straight intestines. An animal which does have wideness of the intestine is gluttonous, especially if along with this it has straight intestines. For a straight intestine brings back the desire for food soon and often. It will therefore be gluttonous because of the wideness and it will be hungry and empty often due to the straightness of its intestine.

Further, because the food is fresh in the upper belly and the excrement on the other hand is corrupted, foul, and dry in the lower belly, there must be some intervening member in which the food can be altered from a fresh into a corrupt state. For this reason nature gave all animals a member (that is, intestine) which is called the *ieiunum* and in which the food is thoroughly corrupted and all that is pure is separated from it. This is midway between the upper and lower belly. For in the upper belly the food is not yet fully digested while in the lower belly it is corrupted and it changes in the middle from one disposition to the other. This intestine is present in every large-bodied animal. Food that has been taken in is never found in it since it is a thing that links two areas of the belly, namely, the upper and lower. It is always found to be empty in the dead, as if the animal had eaten nothing at all. We have set forth the reason for these things in the anatomy of the belly and intestines.[112]

Moreover, Aristotle says that the time for the alteration of food from one disposition into another is different in males and females.[113] In females the food begins to

---

111. "Food has a good lip (that is, good and great eating)": this odd phrase is A.'s addition.
112. Cf. 1.563f.
113. Ar. *Part. An.* 676a4f. says that the location of the intestine called the *nēstis, ieiunum,* is gender-specific.

corrupt over the intestine commonly called the *ieiunum*. In males, however, the corruption takes place before the opening of the intestine, as it enters, and this is before the lower belly. If this is true, it is because the heat is less in the female and it corrupts quicker than it completes the operation. Completed heat, however, will be the cause of the opposite in males.

Further, rennet [*coagulum*] is sometimes found in the bellies of some animals which are still suckling. All animals having a single belly lack rennet with the exception of the animal called the "hairy foot" [*pilosus pes*], like the hare.[114] While rennet is found in animals with many bellies, it is found only in those which have three bellies and is never found in the first two but only in the third, which is directly before the belly called the *ieiunus*.

All animals of this sort have rennet except for those that have only one belly. For an animal with one belly has thin milk which does not coagulate in the belly of its young when it suckles. The milk, however, of a horned animal coagulates in the belly of its young as it nurses because of its thickness. That of a hornless animal does not coagulate, due to the opposite of what we have said. An exception occurs sometimes when it is a hairy-footed animal. For an animal with a hairy foot for the most part lacks rennet. Only the young of the hare is found to have rennet. Hornless animals do not have rennet since they feed on moist plants and they themselves are also moist. Wetness of this sort blocks the coagulation of milk in the belly of the *conceptus* (that is, the young) as it nurses.[115] For this reason there is a proverb among the Greeks to the effect that there will only be rennet through the belly called the *harmyos*, which translates as "many bellied." However, there is still some question concerning this and which must be determined in the *Problems*. We will resolve this in what follows.[116]

CHAPTER III

Which Is a Digression Setting Forth the Nature, Order, and Usefulness of the Intestines

To serve the understanding of the things that have been said we will speak a bit on the names, order, and use of the intestines, and thus the instruction will be easier.

I say, then, that there are many intestines in the human and that they can also be found in other animals analogously to him.[117] Some of these are very convoluted so

---

114. "Hairy foot": a literal translation of the Gr. *dasypous*, "hare." Cf. 14.47. It would seem that A. thinks that "hairy foot" is a class name.

115. See Glossary, s.v. "Embryo."

116. The foregoing is badly confused from Ar. *Part. An.* 676a12f.

117. Cf. Avic. *DA* 13.36v.

that the nourishment might undergo a slowing up in them until the liver, through the mesenterics, has adequately absorbed from it. For otherwise if it had one, straight intestine, the animal would constantly desire to eat, for the nourishment would always leave it and it would always be defecating. In this way it would be hindered in the performance of the many operations it has which are designed to help life.

Further, since the mesenteric veins absorb from the stomach but do not absorb enough for the animal there, nature, who gives all that is necessary, cleverly designed convoluted and slender intestines so that the nourishment would be divided throughout them into small bits and would thus afford the mesenterics the opportunity of absorption from all sides. Thus, that which the first ones miss, those that follow can absorb and so on, in order, up until the last mesenterics, so that in this way all the nourishment may be absorbed.

The intestines, which are so called by the true name of intestine, are six in number. The first of these is called the *duodenum* and this is a straight intestine which is connected to the pylorus [*portanarius*] of the stomach.[118] After this is the one called the *ieiunum,* after which slender, convoluted intestines follow. After these comes the one called the sack [*saccus*] or the *orbum,* and after the sack is the one called the *colon.* The last is the *longaon.* All these are bound to the back by ligaments that are appropriately strung as befits the location of each of them. Of all these, those which are above have the thinnest substance since they contain very moist matter which cannot rend the intestine. The reason for the thinness is so that the power of the liver can work better in it as it digests and absorbs that which is pure and has been decocted.

The lower intestines have a thick pannicular-membrane and are tougher. These tough intestines begin from the *orbum* since there is dry excrement in them from which the water has dried out. This is also why these intestines are *sepum*-filled on the inside, so that with their softness and viscosity they may counteract the dryness of the excrement and provide easier passage for it. For this reason too it is clear that the excrement does not putrefy due to the action of the intestine, but rather from its own putrefaction, within itself. The upper intestines are not *sepum*-filled, but they are not without a light windiness and a viscous fluid which make departure easier, much as the presence of the *sepum* does in the others.

Further, the *duodenum* intestine is connected to the bottom of the stomach so that its opening is the opening of the stomach called the *portanarius* [pylorus].[119] It is directly opposite the esophagus, but on the other side of the stomach. Just as the esophagus serves to draw something into the stomach, so, oppositely, does this serve to draw out of the stomach that which has been digested in it and has not been drawn off by the liver. This intestine is narrower than the esophagus, however, since only soft nourishment which has already been partially digested passes through it. There is no danger of such food rending the intestine, but hard things pass through the esoph-

---

118. Cf. what follows with 1.557f.
119. Cf. Avic. *Can.* 3.16.1.1, for the beginning of this section.

agus and these could tear it unless its large size provided passage for hard things as they make their way through it. There is still another reason for this, namely, that the esophagus only functions using the single power which draws things to the stomach. The *duodenum,* however, functions using two, one of which expels things in the stomach away from it and the other which draws along that which is in the intestine. It is thus pushed along quite expeditiously. It therefore was necessary that the single power in the esophagus be aided by the wideness of the intestine.

91 Moreover, this intestine is straight and descends from the stomach so that the descent of the food might be easier, and the actual mass and weight of the food assists in this with a natural movement.[120] Twistedness would afford a hindrance to descent and the excrement would then hinder the members around the stomach—that is, the liver and spleen.

92 After the *duodenum* the one called the *ieiunum* follows. This name was allotted to it because in every dead animal it is found to be empty. There are two reasons for this. One is that it is near the liver and many mesenterics are attached to it which, assisted by the heat of the liver, quickly absorb that which is pure. The rest is expelled by the same heat into the intestine that follows. The second reason is that the bile duct [*porus fellis*] leads to this intestine and this bile is pure, unmixed with other fluid. It arouses the expulsive power by its pungency. The bile duct is attached most usefully to the place where convolution first begins. For since expulsion begins there, the pressing action is accomplished in all the other subsequent ones. But if it were connected to the subsequent ones and the expulsion were accomplished in them, it would therefore not occur in this one. The *duodenum,* however, because of its straight descent, does not require this sort of stimulation by the bile.

93 A long, convoluted, slender intestine is joined to this one. For this reason some people call the "slender intestines" [*gracilia intestina*] the ones in which the food is divided into bits so that it can be absorbed from all sides. To this one is connected the one called the *orbum* [bereft], *caecum* [blind], or *monoculum* [one-eyed], which angles away from the belly a bit at an angle to the rear on the right side.[121] It has only one opening and in this the food is collected until it can all exit together. For it is expelled more easily and suitably together than it would be in small bits. It is mixed again there, so that whatever in it might be pure can be separated. In this one the food begins to corrupt, turning to the disposition of excrement. Worms and *ascarides* are generated in this and are held there because, if the excrement did not stay above, there would be a danger for the narrow, lower intestines since it could tear them with excessive windiness.[122] This intestine, because of its great size, is also sometimes called the sack [*saccus*] or the purse [*mantica*]. This is the one which, in those with ruptures,

---

120. The placement of the paragraph number is by deduction, as it is omitted in St.
121. Cf. 1.558.
122. *Ascarides*: a pure Gr. plural of *askaris,* a generic term for intestinal parasites or for a stage in the lifecycle of the *empis.* See Beavis (1988), 231–32, and Keller, (1909), 451–52.

passes into the scrotum for, of all of them, this one is the least supported by the mesenterics, as we have stated in the anatomy section.[123]

The *colon* is connected to this one. The ailment of colic takes its name [*colica*] from it, since it occurs most often in this member. When it is a bit of a distance from the "purse," it turns to the right, to the liver so that that which is pure in it and which is superfluous to the feeding of this intestine might be absorbed by the liver's mesenterics. After this it turns to the left and then turns around again as it descends on the right. There it is connected to the *longaon* so that the excrement might descend gradually and with no force. This intestine is narrow, thick, and *sepum*-like. The reason for this has been outlined above.

The *longaon,* the last intestine, is connected to this one. This expels the excrement through the anus. Various muscles assist it by raising the buttocks, extending the anus, and contracting the intestine, as was stated in the anatomy of the muscles of the anus and sphincter.[124]

Let, then, the things that have been said about the intestines suffice. For from this their causes can be adequately known when this is added to that which was said about them in the anatomy section.

94

CHAPTER IV

On the Inner Members of Creeping Animals and of Those Like Them Throughout the Length of Their Body. In Which Also Is Treated the Nature of the Gall Bladder in This Sort of Animal.

The disposition of the belly, intestines, and of the inner members is of this sort in all viviparous animals.[125] There are certain animals, however, which lack feet and have long bodies, such as serpents. With respect to bodily length these resemble certain ones which have feet. Thus, if a lizard lacked feet it would be a serpent.[126]

95

Certain animals of this sort have lungs, since they breathe and dwell in the air, creeping or perhaps walking. An example is the lizard which breathes, the salamander [*stellio*], and the like. Others, however, do not have lungs, but have gills in their place, as do the fish. Not one of these types has a bladder except the tortoise. Thus fish do not have a bladder since salted fluid in them passes over into scales or a scaliness of their skin. The moisture drunk in by these animals is very small. Thus, it is not because their lungs lack blood like those of birds but rather because they neither urinate nor

---

123. Cf. 1.447, 461f., 479.

124. Cf. 1.341.

125. A. returns to Ar. *Part. An.* 676a22f.

126. Thus the fragile lizards of the genus *Ophisaurus* ("snake-lizard") are commonly called glass snakes instead of glass lizards because they lack legs.

have a bladder that the salty superfluity of urine turns into feathers and scales. The true reason that animals of this sort have no bladder is that the fluid separated from the food in them is small, as it is in birds, and the fine part of this moisture passes into flesh while the coarse part passes into scales and feathers. These two substances are present in urine, as is proven when urine is allowed to stand in a container.[127] For then a certain amount of earthy, salted matter settles out onto the bottom.

96 The same sort of diversity of genus exists between serpents and *tiri* as between fish and the *celeti*. The sea animal called the land *celeti* bears a live young and, to a certain degree, the *tyrus* does the same thing.[128] For this one produces a live young to be sure, but first, as has been often stated, it conceives an egg inside itself. All the animals of this sort have bellies that are like those of animals with teeth in both jaws and they have small and narrow front members like animals that lack a bladder.

The inner members of serpents are long and very narrow as suits the shape of their body. In this respect they do not resemble the members of other animals which are pleasing to see and have a full shape on account of the fullness of their body and of the places where the members are formed. All these animals have the intestine called the *ieiunum* and certain other intestines as well. They also have the wall (that is, diaphragm) beneath the heart. Likewise, all blooded animals have these members, but not all have lungs and a windpipe. Fish are an exception for they surely have no lungs or windpipe and there are many as well which lack these same things. The location of the windpipe and esophagus has the same disposition in all animals that have these two members. We gave the reasons for this above.

97 Further, most animals which have a gall bladder have it over their liver. In some, though, it hangs down from the *ieiunum* intestine since the nature of the gall bladder deals mostly with the lower belly, serving to stimulate it and assist it in expulsion as we said above. The gall bladder is especially visible in fish for none of the fish known to us is found to lack one. This is due to their coldness. From this comes the bile they need for the process of expulsion. In some, however, it lies over the intestine and it is sometimes found woven, as it were, over all the intestines as it is in the sea animal some call the *araia*.[129] This is the source of error of those who think that the gall bladder was created for sensation. For these people have said that the gall bladder stimulates the liver into giving it good sensation. But the liver, like the intestines, is insensate.

---

127. The observation is A.'s.

128. The *tirus* represents a word commonly translated "vipers" in the Gr. and the live births mentioned would seem to further this identification. It seems clear that the local snake A. calls a *tyrus* is not the same as the eastern, poisonous variety known better to antiquity. Cf. 25.42(59).

The "land" selachian is another matter entirely. Selachians are sharks and rays. *Agreste* could also be translated "wild" or "of the field," the former being superfluous (there are few tame selachians) and the latter absurd. The overall translation offered has been simplified for the sake of sense. A.'s interpolations at this point seem to have missed the point of his received text, which simply said that a parallel diversity exists among these two sorts of animals based on how they give birth. If A. actually had a specific animal in mind for his "land selachian," he probably envisioned some sort of viviparous serpent.

129. Ar. *Part. An.* 676b22, *amia*, "a sort of small tunny or large mackerel" (*GF,* 13).

Certain animals have no gall bladder whatever. Examples are the horse, mule, ass, and elephant. The camel has no defined gall bladder, but rather has one scattered on small veins. The *koky,* however, which is the sea calf [*vitulus marinus,* seal], and the dolphin have no gall bladder.[130]

Sometimes, however, a gall bladder is found in an individual of the genus of one animal but is not found in another, such as occurs in mice and in humans. For some people have a gall bladder visible above their liver and others do not. This is why the confusion arose concerning the sensation of the gall bladder, as we said, and it thus happened that, along with goats and sheep, all humans are felt to have a gall bladder. It sometimes happens that it seems to be so large in size that it is miraculous and monstrous. For example, it happened in ancient times that in the seventh clime there was a person whose gall bladder took over almost all his liver. This was in the city which Aristotle called Halzyz.[131]

Further, as we have said, it is sometimes found that a distance separates the gall bladder of fish from the liver. All these variances in the gall bladder provide material for those that hold different opinions on the gall bladder.

I, however, think that those who share Anaxagoras's opinion are in error when they say that the gall bladder is the cause of severe fevers.[132] These people say that if the gall bladder grows large, it flows to the lungs and ribs and to the bottom of the diaphragm and it thus spreads a severe, feverish heat throughout the entire body. This is proven false, however, since it has often been experienced that these infirmities befall certain people who have no gall bladder whatever. For if the gall bladder had been spread out as these people say, it could be seen through an incision and it would be clearly visible through dissection. Instead, that which is clear through dissection [*divisio*] and experience does not agree with what they say. The gall bladder [*kystis fellis*] therefore seems to be a sort of residue stored in a vessel, just as the excrement which is collected in the belly and intestines has its own vessels.[133]

It is fortuitous that in some animals nature sets aside certain purged superfluities in bladders [*in kystis*].[134] For example, she stores up melancholy in the spleen and bile

---

130. Ogle (1912, ad loc.) points out that this list is remarkably accurate with the exception of the seal. Above, the elephant represents an addition by Avic.

131. Ar. *Part. An.* 677a3 lists the place as Chalcis, on the island of Euboea.

132. Note that A. is using *fel,* which can mean either gall bladder or bile. Since his most recent usage has favored the former (using *colera* for bile), *fel* is so translated here. Yet the use of such verbs as *curro* and *diffundo* brings to mind more the bile itself. The picture is muddied further by A.'s use just below of *kystis fellis.*

133. The use of *kystis fellis* to mean gall bladder is certain from 1.605, 611, and 12.84, and the translation reflects this. Yet the easiest reading of the Latin would be that "the bile seems to be a certain residue . . ." and it might be best to take "gall bladder" as an error for "bile." Cf. Ar. *Part. An.* 677a13f.

134. As noted above, *kystis,* meaning pouch, bladder, or sack (cf. 1.605), is common in this context. QDA 2.30 is explicit that the *cistis fellis* is the receptacle for bitter, red bile (*cholera*). In some animals, A. says, *fel* is present in such a receptacle; in others, those lacking the sack, it is dispersed throughout the body.

[*colera*] in the gall bladder [*in felle*]. She uses these for various purposes about which enough has been said above. But this necessity of supposition in nature should not escape our notice.[135] For all things of this sort occur for some reason. Thus in those animals in which the nature of the liver is very healthful and the nature of whose blood is very sweet, and is not bubbling but temperate, then no gall bladder is found there, or if it is found it is found scattered throughout very slender veins. It might also be found in certain individuals of a given species but not in others. Thus the liver of those which lack a gall bladder has good coloration and is sweeter than other livers. Sometimes, however, when an animal has a gall bladder, the part of the liver found underneath it is very sweet. For the bitterness of the bile [*colera*] is drawn away from the gall bladder.[136] If, then, the substance of the gall bladder is as stated, and it is generated from but little, bubbling, superfluous bile, it will necessarily be opposite to its food and for this reason is separated from it. For every superfluity that is purged from food has a complexion contrary to that of the food. Thus the bitter which is cast off is contrary to the sweet which is drawn into the food.

100   It is clear, then, from all these things, that the gall bladder [*fel*] is generated from the superfluity of the bile [*colera*] and that this is done by nature for some other purpose.[137] For this reason what the ancients said is amazing and humorous for they said that lack of a gall bladder was the reason for longevity. The reason for this error is that they considered that certain animals, for example the stag and the like, which have hooves [*soleae*] and hooves [*ungulae*] and which lack *fel*, live a long time.[138] They likewise considered those which do not have *soleae* of this sort—like the dolphin and the camel, which has a hoof [*ungula*]—and which also live a long time. They did not, however, consider other animals which have *fel* and which live a long time. If they had considered these they would not have said that lack of *fel* is a cause of longevity. For perhaps the reason for longevity is that the *fel* is not purged from the food or that it does not stimulate the intestines, as we said previously discussing colic, and how it is brought on in many cases from a lack of or shrinking in the *fel*, or because its duct, leading to the intestines, is blocked.

101   Further, it would be more correct to say that the nature of the liver is the cause of a long or a short life. For the liver is a principal member in all blooded animals and it is the member which either has such a superfluity as this or lacks it. The other members which are principal do not have a superfluity of this sort. Neither the heart, the lungs, nor the brain has *fel*. Indeed, it is impossible for a bilious [*fellea*] fluid to come near the

---

135. On suppositional necessity, cf. Wallace (1976; 1980a; 1980b).
136. Or "drawn away by the gall bladder."
137. Ar.'s point throughout is that bile does not occur for a purpose, but represents only a superfluity. Note that in what follows A. gradually seems to move from thinking of *fel* as the receptacle and begins to treat it as the fluid. For this reason, the Latin is left in the next few sentences. It should be pointed out that despite modern translations to the contrary, the Gr. of Ar. is equally unclear.
138. A. has added the reference to *ungula*. See Glossary, s.v. "Hoof." This argument about the gall bladder's relationship to longevity is treated also at *QDA* 14.3.

heart since it can endure nothing in the way of these strong, debilitating agents. The liver, though, is one of the members which requires the presence of such a superfluity in it. Thus the *fel* belongs to the liver only accidentally. It is irrational to think that the crude phlegm and the residue of excrement are not superfluities of the belly and then to say that the *fel* is superfluous. In these things, there is no diversity according to the *ratio* of superfluity except for the places in which they are present.

The things said by us, then, represent the reason there is *fel* in the bodies of certain animals and not in those of others.

CHAPTER V

On the Generation of the Mirach *of the Belly, and on the Generation of Fat* [Pingeudo] *and* Zyrbum *within the* Myrach

In Arabic every web is called *myrach*, but we are here speaking of the *myrach* that covers the belly.[139] Let us say, then, that the *myrach* is the membrane in which there is *zyrbum* in those that have it or fat in those that have fat. *Myrach*, however, has the same sort of position both in those that have one belly and in those that have more than one. It begins from the middle of the belly below the diaphragm and covers the entire rest of the belly below this. Likewise, the multitude of intestines is the same in all blooded creatures whether they are land dwellers or are aquatic.[140]

Myrach is generated in the bellies of animals out of necessity, for in all cases that which has its generation last of all and which spreads out over the edges, arising as a result of heat decocting a mixture of dry and moist, is the hide [*corium*]—that is, the skin [*pellis*]. This is generated from viscosity, at the outer, surrounding edge of a mass. The area of the belly is filled with food, hot and dry, mixed together, and the actual generation of the members is produced from this sort of mass. Thus, the generation of such skins surrounds the things in the belly. Because of the thickness of these same skins nothing can be strained from the sanguineous food except for the fat which surrounds these skins. This is generated from the very thin fluid which is watery, airy, and which causes fat to grow.[141] When this fat is well decocted by the heat of digestion which lies close to these webs, it remains more earthy and dry. This will be *zyrbum*

---

139. What follows is based on Ar. *Part. An.* 677b15f. For a parallel passage, cf. 1.477f. *Myrach* is variously used for the abdomen, abdominal wall, muscles of the abdominal wall, navel, and peritoneum (Fonahn, 2077). Although definitions change throughout the breadth of A.'s work, in this section it would seem that *myrach* is the omentum, *pinguedo* is regular fat (and is so translated), and *zyrbum* is suet or lard. See Glossary, s.v. "Fat."

140. Badly garbled from Ar. *Part. An.* 677b20–21, which says that the omentum covers most of the intestines as well.

141. Or "which grows fat."

instead of the fleshy, sanguineous substance which, in other members, is generated from food. Sometimes, however, when the food is very moist it will be fat and not *zyrbum.*

The generation of the fat and *zyrbum,* then, follows this pattern. Nature uses this *zyrbum* and fat for the betterment of the digestive heat—for it is the heat which digests. The fat is warm and is a guardian of heat and thus, since *zyrbum* is fatty, it aids digestion. This is also why it begins from the middle of the lower belly. For the digestion of the food is due to the heat of the liver in that area. Thus, to the extent that it can be stated, its cause is a function of the *myrach.* The intestine is wrapped in the middle of this membrane and the membrane extends continuously along the entire length of the intestine, until the membrane itself reaches the large vein [vena cava] and the vein called the *adorthy* [aorta].

The reason this member is present in blooded animals is clear. For it is obvious to anyone carefully investigating the matter that an animal which takes in its food from outside must have receptacles of this sort for the food. It is also clear that the blood, or that which takes its place in bloodless animals, is produced from the food in the last digestion. The blood flows in the members via the veins which come from and branch off from the belly as it travels to all the members. This is like the sap of trees as it comes from the roots to all the branches. Trees have their roots in the ground and from them they take in food for themselves, their branches, and all their parts. But all this has been set forth by us already in the books *On Plants* [*De vegetabilibus*] and in a better fashion than can be thought of for expression here.[142]

The position, then, of the middle intestine was devised by nature for this purpose. For the veins which pass over into it and which are gathered together in the hollow of the liver and then branch off from the liver throughout the entire body are like the roots. There, as we have said, they take on the food, and from this point the food enters the veins and then the body.

We have thus made a determination about these members in blooded animals. The disposition of almost all their inner members has been clarified and we have come in our treatment to the individual members which have a fixed shape and location. As far as we have been able, we have given their causes as well.

---

142. *De veg.* 1.144; 2.13.

CHAPTER VI

On Those Animals Which Lack the Inner Members of the Belly That Have Been Mentioned and Discussing What the Members Are Which They Have in Place of the Aforementioned Members

Before we speak of the outer members it seems we must speak about the members having to do with generation and how they differ in the male and the female.

Before we speak of these things generally, however, we should say that the genuses of *malakye* and those of the soft-shelled creatures do not have the diversity in their inner members we mentioned. Neither do they have the diversity of the other members blooded animals have.

Further, besides these, there are two genuses of bloodless animals which have a hard shell and a ringed body. These also do not have the inner members we described before, for the substance of the nature of these animals, throughout all their genuses, is different from the substance of the nature of blooded animals. For it has been established and proven previously that some animals have blood and some do not. This was set forth based on the defining discussion which marks the distinction between the types of the species of animals and their substances.[143]

Moreover, these types of animals do not have in them the reasons for the inner members which are the intestines. They do not have veins or bladders, and they are not breathers. For this reason they are found to have only that member which takes the place of a heart. For the member which is the principle of sensation and the cause of life and movement has to be the principle of all the members and the principle of the entire body in all animals, no matter what their genus. These types have members suited for the taking in of food, but they differ according to location insofar as the places through which they take in the food differ. In the mouths of the *malakye* there are two teeth and a fleshy member that takes the place of a tongue. It uses this to sense the taste of its food. Likewise, a soft-shelled animal has teeth and a member analogous to a tongue. Generally, all those that have a hard shell have this member for the reason we gave for the blooded animals, namely, for the taking in of food.

Moreover, in the same way, a ringed, bloodless animal takes in its food through an appendage sticking out of its mouth. All do not do this, however, just certain species of this genus of animal, such as the genuses of bees, flies, and the flying insects called *erucae* or *bombices* [butterflies].[144] Those which do not have an appendage of this sort in their mouth, looking like a stinger, have another member in their mouth

---

143. "Defining discussion": *sermo diffinitivus.* Cf. 11.4, with note.

144. *Eruca* is often a caterpillar, as at 5.32 or 26.15(17), whereas *bombex* is most properly the silkworm. However, the same names are occasionally used for the later, winged stages of the insects' lives.

which they use to take in food. Examples are the ants and genuses like them. Certain types of these genuses have teeth which do not, however, resemble the teeth of other animals. Examples are the ant genus and certain bees. Others, however, have none of this at all for they feed on fluid food and merely suck. Most animals with ringed bodies, however, have teeth, to be sure, not because of their food but for their weaponry, to strengthen their defenses. Some marine animals with a hard shell have a strong tongue, as we have said in previous discussions. The animal called the *Rogale* has teeth along with a strong tongue, much like those with a soft shell.[145] The *malakye* genus, however, has a long stomach behind its mouth and after the stomach it has a bladder resembling the crop in a bird. Then comes the belly and next it has an intestine that stretches out, arriving at the point of the exit of its superfluity.

107 The members of the animal called the *sepyon* [cuttlefish] and those of the animal called the *multipes* (that is, the *polipus*) [octopus] are very alike in shape and touch. They have two members for the taking in of food and these are very like bellies. But one of them resembles a bladder or crop, and the other bears less of this resemblance. These two differ from one another in shape, for the entire body of this animal has soft flesh for its foundation. Therefore, the members of these animals have almost the same disposition so that the members of one are like those of the other. There is the same reason for this in all of them. For neither of them can grind or chew its food or soften it in its mouth. For this reason it has a crop (that is, a bladder) before its belly, just as in the case of the birds.

108 The animals of this genus, that is the *malakye,* have as their own trait that they hold their sperm amassed in membranous receptacles for the preservation both of themselves and of their species.[146] For the sperm of these animals is sometimes a defensive weapon and in each of them the sperm is attracted to the place where the belly's superfluity leaves. It is like the member called the *canale,* which is toward the rear of the back in long animals.[147] This member is present in all the species of *malakye,* and especially in the *sepia,* since this one has a great deal of sperm, which in this one is a black humor. When it is afraid it ejects this black fluid and uses it to build a sort of wall around itself, using it to thicken and darken the water. The animal called the *multipes* or *polipus* and that called the *taccydos* (which is *calamare* in Latin) [squid] have sperm in their upper part, above the member called the *bosnyz* in Greek.[148] The *sepiae,* however, have this fluid lower down, near the belly. The reason there is more of this fluid in the *sepia* is that it uses it for the purposes of both defense and copulation.

---

145. *Rogale*: A.'s version of Sc.'s *kogile,* readily seen as the *kochloi* (sea snails) of Ar. *Part. An.* 678b24. The capitalization is also A.'s.

146. Throughout much of what follows the ink of these animals is confused with their sperm. How this occurred is not readily apparent when one compares the original Gr. with the Latin.

147. *Canale* would usually mean merely channel or intestine. Cf. 1.392, *canales in circuitu colon.* Here it stands for the Gr. *aulos,* a pipe or reed, in referring to the anal vent.

148. Cf. Ar. *Part. An.* 678b28f. *Taccydos* is a version of the Gr. *teuthides,* calamaries.

It occurs this way in it because its dwelling place and lifestyle are for the most part near land. Nor does it have any other defense, the way the *multipes* has its scaliness.

When the *multipes* is frightened, it envelops all that approach it, and another result of fear in it is that it changes color. This is the same as the release of the sperm in the *sepia* due to fear. The one called the *ynankys* is also one of these types of sea-going *malakye*.[149] But in the *sepia* there is a great deal of this fluid, more so than in the others, and this is why it is located below. It leaves it when it has grown abundant, so much so that it reaches far out into the water, staining it around the *sepia*. This superfluity is like the white, earthy, lime-filled excrement produced by certain birds. The sperm, then, in this animal is amassed in this way because it has no bladder and it only leaves mixed in with much earthiness. There is much of this fluid in the *sepia*. The fact that there is a lot of earthiness in the *sepia* is indicated by the earthiness of its skin. The *multipes*, however, does not have this sort of skin. The *inankys*, however, has a cartilaginous and thin skin. We have designated above the reason this part is present in some animals and not in others.

Now it happens that bloodless animals, and likewise other animals, undergo flux when they are afraid, and emit the superfluity of their belly. And it happens that some, which have bladders, urinate when fearful. Nature thus uses this superfluity for protection and help. Thus crabs, the animals called soft-shelled, and those that resemble the lobster [*karabo*] have two teeth from among the first front teeth.[150] Between these is a certain member resembling a tongue. Then is the mouth and after it is a stomach which is small compared to the size of its body. After the stomach is the intestine which is its lower belly. In the crab, however, there are also other teeth below the first ones. For the first, exterior teeth are not adequate for the cutting up of its food. After the belly, however, there is an intestine stretching through the tail and coming to the point of the exit of the superfluity. These members are present in every hard-shelled animal.

In some of them, however, the members are more distinct and in others less so. In an animal which has a larger body, this distinction is quite clear. The animal called the *rogalio* has hard, sharp teeth, as we described above.[151] The member which is between its teeth is fleshy like that in the *malakye* and the soft-shelled creatures. It also has an appendage, as we said above. The composition of the appendage is midway between that of a tongue and a stinger. After its mouth it has a member rather like the crop in birds and then it has a stomach and after this the belly. In this belly is the superfluity

---

149. The octopus is not, of course, scaly. Somehow the *plektanas* (tentacles) of Ar. *Part. An.* 679a13 became *squamositatem pedum* ("scaliness of feet") in Sc., and mere scaliness in A. The *ynankys* is also somewhat puzzling because by position it would seem to derive from *teuthis*, a calamary.

150. "Thus": the Ar. text has the simple *de*, which merely means that a new topic is being introduced. In the process of transmission this has become *propter quod etiam*, indicating a direct causal link, however improbable, between the casting off of superfluities in the name of self-defense and the dental arrangement of lobsters, on which cf. 4.28.

151. Cf. 13.106. Although A. probably did not know it, he was speaking of a sea snail.

which is called the *micon* and after the belly is an extended intestine which begins in the area of the *mycon* and proceeds to the anus.[152] A superfluity of this sort is present in all hard-shelled animals. Some feel that this is something eaten by the animals mentioned. The disposition of certain other animals (these being like the animals called the *astonyoz*) is almost like the disposition of the types of lobster (that is, sea locusts). These are like the *barcora* and the one some call the *kykyz*.[153]

There are very many types among the genuses of hard-shelled animals. Some of these types resemble the *astornyos* as we said. Others, however, have two rooves that are like two shells [*conca*].[154] Others have one roof (that is, shell). The animal which resembles the one called the *astarynoz* resembles one type that has two rooves in that it has coverings over whatever of its flesh is exposed. This is found in all the animals of this type from the moment of their generation. Examples are found in the *barcora,* the *kyrikez,* the *byrinon,* and those like them.[155] If this shell [*testa*] did not cover its flesh everywhere, injury would befall it from all the external things that touched it.

An animal, however, which has one roof is better protected, since its shell is hard and strong at the rear of its body. This shell emerges over its flesh like two rooves which are not long but are close to each other to form a contiguous covering. The animal called the *babarez* is of this sort.[156] An animal that has two rooves is protected in that it brings together its two rooves (that is, shells [*conca*]) and closes them as does the shellfish called the *amnyuz* and the one called the *moez*.[157] An animal which resembles the *asturonoz* is protected by the member which resembles a covering. This is one which has, as it were, two rooves, although at first it has but one.[158]

We are using the word "roof" here for a shell in which shellfish are enclosed, even though the use of "roof" is improper. But Aristotle used this word before us for "shell" and we therefore use it too.[159]

---

152. Cf. 13.108, with notes.

153. *Astonyoz,* as well as the forms following (e.g., *astornyos, astarynoz*) derive from *strōmbodē,* spiral-shelled whelks. *Barcora,* as often, is derived from *porphyrai,* murexes, and *kykyz* is from *kērykes,* whelks.

154. "Rooves": the Latin is *culmus,* which for A. most commonly means tusk. Occasionally it means some long, projecting appendage such as a lobster's antennae (4.32, 77), the points on a stag's antlers (8.29, 38), or the pincers on a crab's claw (14.58). Its use here clearly troubles A. The translation offered, "rooves," is defensible since the word took on this meaning by analogy with *culmen* (Latham²), but one must ask if A. thought the Latin indicated roof or "tusk-like object." The latter option, in fact, is more probable, leading to his constant glossing of the term with *conca* and his explanation at the end of section 112 (see below) of why he kept it in his text. Cf. 4.39 for a similarly complex problem brought on by trying to translate the Gr. terms *mono-* and *dithyron,* "one- and two-doored," for mono- and bivalves.

155. Ar. *Part. An.* 679b20f.: *porphyrai* (murexes), *kērykes* (whelks), *nēreitai* (a "spiral, univalve shellfish of undefined species"; *GF,* 176).

156. Sc. offers *superez,* ostensibly from the Gr. *lepades,* limpets.

157. *Amnyuz:* Gr. *hai ktens,* scallops. *Moez: myes,* mussels.

158. Cf. Ar. *Part. An.* 679b23f., where Ar. says that an animal like a limpet clings to something that in turn covers its exposed parts, rendering itself in this way a sort of bivalve.

159. This statement says much about the trust A. had in Ar. In fact the words Ar. uses in the parallel passage are *monothyron* and *dithyron,* "one-doored" and "two-doored."

The sea urchin is warmer than all these animals, for in addition to a shell surrounding it, it is girdled with spines. As was said above, this is a trait peculiar to it of all the hard-shelled animals. For the nature which is in it is sort of midway between that of a hard- and soft-shelled creature. It is the contrary of the *malakye* genuses for, as we stipulated above, there is a hard, earthy part outside and a soft substance inside some of these genuses, although in others the soft part is outside and the hard is inside, as in the genuses of *malakye*. In the sea urchin, however, there is no fleshy part, for all its members are as we described far above.[160] All the other hard-shelled animals have a mouth, a thing like a tongue, and a belly, although they differ somewhat in respect to them.[161] One who wishes to seek to know this perfectly should read the earlier parts of this study where the members and the anatomy of these animals are enumerated.

The sea urchin and the one called the *rybo* in Greek have a trait particular to them among the hard-shelled creatures we have mentioned.[162] For the sea urchin has five teeth and among them there is a somewhat fleshy member. Then come the stomach and the belly. Their bellies are divided into many parts as in an animal that has many bellies. These bellies are stretched out, filled with the superfluity of food. These bellies, however, leave their stomach and hang down from it. Their location is near the point of exit for the superfluity. There is no fleshy part whatever in these bellies and in them there are sometimes many separate eggs, each of which is placed on its own within a single web. These eggs have a black color. The genuses of urchins do not have individual names, for there are many genuses of them. The eggs are present in all their genuses, but none of them are eaten save those which float.[163] Urchin eggs are small and this is generally the case for all the eggs of hard-shelled animals. For their flesh is not eaten any more than are the eggs. The flesh of some is eaten, however, and of others it is not. It is the same for the superfluity of those animals called the *mycon* and which the ancients called the *murycon*. That which is present in some of them is eaten although that in others is not.[164]

Further, in those animals called the *astarinoz*, there exists a convolution of the intestine. In those, however, which have a single roof (that is, shell) the parts of the intestines resemble the intestine of the one called the *bonyz*, but in one which has two rooves (that is, shells) the intestine is at its end, so that its eggs are on the right side and on the left is the point of exit for the superfluity.[165]

Now these cannot be called "eggs" in the true sense of the word, for the things called eggs in this animal rather resemble the fat in a fattened, blooded animal. Thus,

---

160. Cf. 4.53f. for a more detailed exposition.

161. Omitted by A. from Sc. is "and a member from which the superfluity goes forth."

162. Cf. Ar. *Part. An.* 680a4f., where it is clear that somehow *rybo* is a derivative of *tēthyōn* (gen. pl.), sea squirt. Somewhat analogous corruptions are found at 4.40, 65.

163. "Float": this odd statement has its origin in Ar. *Part. An.* 680a18, *epipolozontōn*, which itself has troubled translators of Ar., as witness the notes of Peck (1961, 327) and Ogle (1912, ad loc.).

164. *Mycon*: the *mēkōn*, rather like a liver.

165. *Astarinoz*: Gr. *strombōdē*, whelks. *Bonyz*: *lepasi*, limpets (dat. pl.).

these eggs are especially found during those periods when the animal is growing fat, that is, in the spring and autumn. During the cold of winter, however, and the heat of summer, the good disposition in hard-shelled animals is harmed since these animals can endure neither excessive heat nor cold. An indication of this is that this fatty egg-like part is generated suddenly, especially during a full moon. This is not because it eats more, as some have erroneously held, but because the nights during the full moon are warmer due to the great light of the moon. For since this animal does not have blood it is greatly harmed by cold and it needs the heat, which moves its moisture and does not dry it out. Thus this animal grows fat everywhere except in the Boryoz Horynoz Sea, since that sea always has winter-like traits.[166] This fattening occurs this way because all these types find a great deal of food in spring and autumn, especially because during this time the fish abandon these parts and then these animals are the only ones feeding there.

116 Moreover, there are a fixed but uneven number of eggs in all the inner parts of the urchins, five to be precise. In the same way it also has five teeth and five bellies. The reason for this seems to be that which is called an egg is not truly one, as has been said before. Instead, it is a particular object resembling the fat of a fertile, fat animal which has a good disposition.[167] This is only on the right side. The urchin has a rounded body all around and does not have a shape like that of the *halzum* (that is, shellfish [*ostrea conchilia*]). It has neither different angles in its body nor is it rounded on one side and not on the other, as are many shellfish [*conchylia*]. Rather, it is rounded on all sides, and for this reason too it is suitable for its egg to have the same disposition.

117 The head of all the types of animals we have mentioned is in the middle of the body, with the exception of the urchin, whose head is at its bottom.[168] Now the eggs cannot be connected any more than they are in the other types of shellfish [*ostrea*].[169] For the only trait the urchin has for itself, apart from the others, is its rounded shape and its fixed, uneven number of eggs. If the number of its eggs were even, each individual egg would have to be opposed to another. But such an arrangement of its eggs is not suitable in any genus of *halzum* (that is, shellfish [*ostrea*]) nor in that of the animal called the *pecten* in its genuses. All these animals have their eggs in one area of their roundness and for this reason it is suitable for them to be unpaired, either five or three. But if there were three of them, they would be too far apart from one another due to the large size of the body. However, if there were more than five, they would press on each other too much and would be touching and would be destroyed. Now, because it would not be the best thing in nature for there to be only three, and because it would be impossible for there to be more than five, it is therefore necessary that

---

166. Boryoz Horynoz Sea: Ar. *Part. An.* 680a36 helps us to identify this as the Strait of Pyrrhos in Lesbos.

167. According to Ogle (1912, ad loc.), these objects are the animal's ovaries or testes.

168. A statement directly opposite that of Ar. and Sc., whose text reads *in superiori parte* and not A.'s *in inferiori parte*.

169. Cf. 13.114, where the eggs are said to be *distincta*.

there be five. The disposition of the belly into the same number is for the same reason and the same holds true for the number of their teeth.

Further, each of the eggs is like a body that is suited to its type of life. The egg is present of necessity, since it must first take on growth and nourishment from the egg. If, indeed, the egg were single it would necessarily be either very far removed if it were small or would fill up its whole body if it were large. Then, if the urchin were pregnant with the egg it would have a sluggish movement and would also be filled up with it, would not take in food, and would die. Since, then, the number of teeth is five and there is a bit of a space between them, the eggs are thus suitably five as well, for nature is accustomed to give pairs of members in this way. This is clear in the leaves of the flowers, almost all of which are produced following some kind of equability or other. This can be seen in the flowers of the borage, spinach, columbine, and in others which are made up of many, both even and uneven.

It has already been stated that the urchin has five eggs. But because certain urchins are better and larger than others, their eggs are also larger, whereas those of others are smaller.[170] Heat can decoct and digest certain of these animals. Thus, those which are edible are filled with superfluity but the heat of nature gives these animals greater movement and they are better at finding food and are thus better fed. An indication of this is the filth that is found on the spines of this sort of animal since they move about a great deal in the mud and urchins use their spines much as animals use their feet and hands.

Let, then, these be the things said about the inner members of this sort of animal. We have also spoken above at length concerning such animals and their anatomy.

118

---

170. This sentence offers us a nice portrait of A. as text critic. A.'s received text read *Quia vero quidam yricii meliores aliis, ideo et ova eorum sunt maiora et quorumdam minora,* translating as "But because certain urchins are better than others, their eggs are also larger, whereas those of others are smaller." A. was troubled by *meliores,* since it is immediately followed by *maiora* and not *meliora.* He therefore suspected corruption and inserted into his text *et maiores sunt* after *meliores,* being careful, however, not to delete the original, received text. Curiously, his insight was closer to the original text, as Ar. *Part. An.* 681a4f. stresses size.

CHAPTER VII

On The Inner Members of Those Which Are Midway between Animal and Tree and How They Differ from One Another. Also on the Inner Members of the Ringed Creatures.

119 The sea animal which is called the *chyboz* in Greek differs but little from the trees.[171] But with respect to nearness to animals they are nearer to life than is the sponge or the cloud, which is an animal with almost a hemispherical shape and is very imperfect.[172] Therefore, this animal has a power like that of the trees. For when nature is passing over from inanimate objects to animate ones, and from plants to animals, she takes intermediate steps. And then she changes toward only those things which, although animals to be sure, are not perfectly and truly so. These imperfect animals are midway between plant and animal and differ only slightly from one another. This is because they are very near to one another. When the clouds are attached to their spots they live and they move with a contraction and expansion motion. When they are removed from their spots they die. For this reason some believe this animal's disposition is like that of a tree.

120 The animal called the *colobya* in the Greek tongue and the one called the *raham*, as well as certain other types of sea creature, are unattached in the sea.[173] They do not belong to a single genus and they live like trees except that they are unattached. Some of these types also live in earthbound trees and are generated from them much as a tree is generated from a tree. Still others of them are unattached like the animals which are generated from trees and leave them after they have been generated. These are called the *barnotheos* and by some Greeks are called *astarnyoz*.[174] It is possible, moreover, that there is a type of imperfect animal called the *rycho* which is somewhat similar to these.[175] For all these are attached to the places where they live and in this regard are like trees. But because they have a fleshy part and a contraction and expansion movement, it is said that they are clearly alive. In this regard they are like animals. In this

---

171. *Chyboz:*. Gr. *tēthya*, ascidians, at Ar. *Part. An.* 681a10.

172. The "cloud" is probably some sort of sponge. Cf. the *nubans* at 1.32, which, while close in description to this animal, is referred to as a shellfish.

173. These two animals are somehow derived from the *holothuria* and *pneumones* of Ar. *Part. An.* 681a17. The former are oyster-like, but they live a detached existence, whereas the latter, whose name means, lit., "lung," is a jellyfish (GF, 181, 203).

174. Stadler suggests that these names have come from the parallel Gr., which speaks of Mt. Parnassus. One must wonder, however, if this is not a reference to barnacle geese. At 23.31(19), these are called *barliates* and LLNMA lists a reference to them, s.v. *barnax*. This would account for *barnotheos*. *Astarnyoz* is just another version of the usual corruption of *strombōdes*, spiral-shelled whelks. Cf. notes at end of 13.111.

175. By position, *rycho* represents the Gr. *tēthya,* ascidians or sea squirts.

way, then, the species mentioned come nearer to other animals than to the genus of trees. These animals do not have the superfluities of excrement and urine and neither do trees. Neither do they have anything in the middle of the body, in the heart area, except a single web and this has the power of life.

The sea animal which some people call the *alfidez,* but others the *halkylidez,* does not have a hard shell.¹⁷⁶ It is outside the nature of all the genuses of animals we have mentioned, and is almost common to the nature of both trees and animals. It is not attached to a place and it feeds on and senses things suitable for it and uses the roughness of its body for self-protection and defense. In these traits it is like an animal. But because it does not sleep and is often attached to rocks, it is like trees. This genus of animal does not have a visible superfluity which leaves its body and yet it has a mouth opening, and when it happens up on one of the species of *halzum* or shellfish [*ostrea*] it sucks its moisture out, and in this is like an animal. This resembles the genus of marine animal about which we have spoken above, in one of the books of this study. And this is like the animal which the Greeks call the *malakye* and like a soft-shelled animal. The same statement holds true about the hard-shelled animals for they must have a member that is analogous to a sentient member in which the sense of touch resides. They must also have something like a principal member, namely, the heart, which is in blooded animals.

The *malakye* is found to be placed in a moist web. Thus their stomach is found stretched all the way to the belly and is attached to the back area. This member is called the *bastyz* by some people.¹⁷⁷ The hard-shelled animals also have a member like this. It is called the *bastyz* in Greek and is a moist body which extends to the area of the belly. The stomach is between them. If the back were between these two members it could not take in food because of the back's roughness. An intestine is found over and toward the outside of the member called the *bastyz* and its sperm is near its intestine, but at a distance from the opening through which food enters. These animals are at a distance from the brain in which is the location of the first and loftier power, which is the animal power. For the better a thing is, the more worthy it is as well.

This animal, moreover, has a member which is in place of a heart, as is clear from the fact that this member occupies the heart's position in the middle of the body. This is also indicated by the sweetness of the fluid in that place, for it seems to be a bloody sweetness since it is decocted from a member which is analogous to the heart. In the same way there is found a principal member in which the sensitive power is first present, even though it is not especially obvious. This member, in which lies the sensitive power, should always be sought in the middle of the body, midway between the member that takes the place of a mouth and takes in food and that through which

---

176. Cf. Ar. *Part. An.* 681a37, where we get two Gr. acc. pl. forms: *knidas* and *akalēphas,* both types of sea anemones. Ar. goes on to discuss the starfish next and their traits follow below.

177. *Bastyz* derives from Ar. *Part. An.* 681b20, *mytis,* an organ more like the liver than the heart (Ogle, 1912, ad loc.).

123 This is to be observed in an animal that is affixed to its spot and does not have progressive motion. In walking animals, however, the heart is always in the middle between the right and left sides. In those with a ringed hide, however, the midpoint, in which lies this first power, is found between the head and the rest of the body which contains the head, as we have related in a previous discussion. This member is found singly in most genuses of animals we have mentioned but in some there are many of them, as in the long-bodied animals called the *haroleos*.[178] For this reason this animal lives for a long time after it has been cut in two. Nature especially intended to place this member in all animals. If she cannot place this member actually, divided and perfect, she places it with great potential. When it is divided, it becomes many in actuality such as in the *haroleos*. The *haroleos* is a black animal which travels on pathways and seems to have two heads. When it is cut through, one part goes in one direction and the other in another, as if there were two perfect animals. The German speakers call this the *werra*.[179] An indication that nature always intends to form this member is that this member is found formed in all animals, although in some it is quite visible and in others it is more difficult to see.

124 Further, the members that are designed for food are also present in all these animals. But they differ since some of them have a stinger whose powers are composite, possessing the power of a tongue and the power of lips. In some, also, this stinger is among the teeth and after this member there immediately follows a straight intestine which reaches to the place where the superfluity leaves. This intestine is found in certain of these animals. Some types of this genus of animal have a belly after the mouth which is a convoluted intestine.

The genus of serpent, however, which is a certain animal called the *alchehar* in Arabic, has its own proper nature.[180] For it has the aforementioned member, a mouth opening, and a tongue, and using these it takes in food much as trees take it in through their roots. Thus its food is only fluid. All bloodless animals eat but little because of the smallness of their bodies and their weak heat.

---

178. Ar. *Part. An.* 682a5, *ioulōdesi* (dat. pl.), a centipedal creature.

179. Cf. Sanders, 446, who identifies it as *die Maulwurfgrille*, or mole cricket, *Gryllotalpa vulgaris* L., a large burrowing insect. Cf. *Werre*, in modern German.

180. Schühlein, 1655, points out that this cannot be the name of a snake. The content of Ar. *Part. An.* 682a18 deals with cicadas. Stadler reports that Sc.'s text read *genus vero [serpentis] animalis strepentis*. The confusion between *serpentis* and *strepentis* ("buzzing") is an easy error to make.

HERE BEGINS THE FOURTEENTH BOOK ON ANIMALS

In Which the Outer Members of Animals Are Treated

The First Tract of Which Is on the Outer Members of the Ringed Animals, Shellfish, and the Malakye

CHAPTER I

On the Outer Members of the Ringed Creatures

Now that we have made a determination on the inner members of all the varieties of animals, the discussion must return to the external members of these same varieties of animals. So that we will not be forced to repeat things already said, let us speak first of the external members of the ringed creatures, whose internal members we discussed at the end of the preceding book.

These animals have few senses and are weaker than the other perfect blooded animals. Consequently, there are not many external members on such animals even though some of them differ greatly from the others. Some of the ringed creatures have many feet because of a natural slowness, which is caused by a natural coldness and serves the purpose of lightening their movement through the use of many feet. This is the sole reason that cold animals are for the most part multipeds. It is also due to the length of a cold animal's body, as in the genus of animal called the *boloz,* which is made ringed because of the large amount of its flesh.[1] For this reason too it has many feet.

Likewise, a given member of a genus of such animals which has little flesh and a rather small body also consequently has fewer feet. This is especially so for the bird called the *solitaria,* this being a certain fly which is winged and very large. Ringed birds which fly in a group and dwell together in civilized bands have four wings and a light body.[2] Examples are the bees and those like them. They have two feet on the

---

1. Ar. *Part. An.* 682b4 makes it clear that the *boloz* is the "genus of centipedes," *ioulōn genos.* Note that this same genus gave rise to the *haroleos* of 13.123.

2. Stadler's apparatus to this text holds the clue to its decipherment. Ar. *Part. An.* 682b7f is speaking solely of insects. Ar. says that some have fewer feet, but they have wings to make up for it. They live nomadic lives (*nomadika*) as they search for food. *Nomadika* was apparently misread as *monadika,* "solitary." This was rendered as *solitaria* in the Latin, which then became not an adjective but the name of an animal, the *passer solitarius* (solitary sparrow) of 23.137(99), and Stadler notes that the phrase "which by some is called the solitary sparrow" was deleted directly after the name *solitaria,* leaving A.'s level-headed identification as a fly intact. Yet the next sentence still began with the word "Birds." A., ever loathe to change his received text, left it intact, but he made the best of the situation by adding the word "ringed," leading to the unlikely identification of winged insects as ringed birds.

right and two on the left. They do not have more than six feet lest the multitude and weight of their feet keep them from feeding. Thus, whichever of these animals is small has only two wings, like the genus of flies. Whichever is small but has a stationary and fixed lifestyle has many wings, like the wings of the bee. The result is that wings are membranous, as is the case for the genuses of wasps and those like them. This is for the protection of their wings.

Because this animal is immobile and lives a fixed existence with little movement, it more quickly suffers problems with its wings than do other animals which have good movement. The wing of this animal is split in two and it does not have that member which is analogous to a shaft.³ Neither are its wings feathered, being membranous instead. That membrane necessarily separates from the body when the fleshy part in them grows cold.

That this animal has a ringed body is due only to the reasons given above. For by being ringed, an animal which was long is rendered short, something that would be impossible if it were not for rings on its body.⁴ A member of this sort of animal which is not convoluted is made very hard when its body contracts and solidifies. This is to be seen in the *iohalh* wasp, for when this one is afraid it becomes unmoving and very hard.⁵ This animal was necessarily created with a ringed body, for it is in its nature to have many hardenings and it somewhat resembles trees in this. This animal also lives a while after being cut, although for just a brief time. But when trees are cut they heal perfectly. For this reason many trees arise from one tree even though it has been cut.⁶

Moreover, certain ringed animals have a stinger which nature gives them as a weapon, for strength, and for defensive purposes. Some of them have the stinger near the area of the tongue and some on the tail. Just as in the elephant where the trunk suits two functions, namely, as a weapon and for the acquisition of food, so these ringed creatures have a stinger-like tongue for a double purpose. They are armed with it and they also suck up their food with it from the inside of the skins of animals, piercing them with it. They also use it to sense their food. Those which number among the ringed creatures but which do not have a stinger-like tongue of this sort have some teeth, one of which moves to the right and the other to the left. With these they sense and take in their food as do ants, *ataci, opimaci*, bees, and wasps.⁷ Others

---

3. Cf. 4.69, for similar thoughts and language.

4. That is, it can curl up. But the language is quite tortured and it is uncertain whether the medieval reader got this sense from it.

5. Apparently this wasp derives from the dung beetle (*kantharos*) of Ar. *Part. An.* 682b26.

6. The original reference was to splitting a single plant to obtain two.

7. The list is largely A.'s invention. *Ataci*: the bumblebee at 8.143f., whereas *attacus* is a type of locust in the Vulgate, cf. Lev. 11.22. *Opimaci*: at 26.20(27), *opimacus* is said to be an insect which can kill a snake. Scanlan (436–37) notes that Lev. 11.22 also mentions the *ophiomachus*, usually translated as locust. The Gr. means literally "snake fighter" and Beavis (1988, 68–69) traces the history of how it came to be identified as a locust. Both *attacus* and *ophiomachus* are glossed in the *Expositiones in Leviticum* of Rabanus Maurus (*PL* 108: 358C), and in the later *Glossa ordinaria* (*PL* 113: 330A) included in Migne among the works of Walafrid Strabo. In the *Epitome commentariorum Rabani in Leviticum* (*PL* 114: 815C), both the *attacus* are the *ophiomachus* are described as *contrarius serpentibus*. For the likely date (perhaps 12th century) and contributors to the *Glossa ordinaria*, cf. De Blic (1949).

have a stinger on the rear of their tail, some having it inside their bodies while others have it projecting externally. Those with it inside are the genuses of bees and the large yellow wasps. Those which have it projecting outside are the genuses of scorpions. If the stinger of the bees and wasps were outside the body, it would be easily ruined, for it is slender and fragile and it would also cause heaviness. But the genuses of scorpions, which have their stinger to the rear, have it outside because of strength and because they do not fly and thus are not laden down by its weight.[8] All these animals with their stinger to the rear on their tail have it solely as a weapon. There is no animal which has only two wings and which has its stinger to the rear. This is because of its weakness and because it lacks blood. The type of animal with a small stinger to the front is very numerous both in numbers and in species. Examples are mosquitos [*cinifes*], bedbugs [*cimex*], fleas, and the like.[9] Whichever sort of stinger it has, an animal bears the smallest it can and still be sting-bearing, and it therefore bears it in the front of the body where it is the most strong. Yet it still stings with a stinger in the front only with difficulty and exertion on account of its weakness.

The type of ringed animals with many wings have a multitude of wings because of the size and strength of their bodies. This animal has strength in its rear members and thus has its stinger to the rear. For it would be best if there were one member suited for one function and another for another, unless necessity forced a situation in which it had two functions. Thus, such a sharp stinger fulfills only the task of weaponry. But a stinger which is tongue-like is used for drawing in food. Nature indeed, when she can conveniently do so, uses two members for two functions where one does not interfere with the other. For she does nothing trivially. Sometimes, however, she uses one member for two functions much as a craftsman sometimes uses one tool for two functions. Thus he produces a lantern which serves both as a reservoir and as a candle-holder.[10] Thus too when nature can conveniently do so she uses the same member for two functions and then she makes one member and does away with the other.

Further, the front feet of these types of animals are larger, for the animals have hard eyes and do not see well. They must therefore have a strong grasping power in their front feet in order to hold on to things they grasp. For they grasp only that which they desire through the power of their front feet. It is clear to see that these animals do what has been said. It is seen this way in the genuses of flies and the types of all animals which resemble bees. The rear feet are larger than the middle ones. The rear feet have two purposes, namely, for walking, during which the rear feet especially support the body, and also to give firm footing when they rise up from the earth either in a leap or in flight. Strength is also required for this. This is clearest in those leaping animals such as the locust and the flea. For these first bend their rear feet and then

---

8. "Strength": the Latin is such that the phrase *propter fortitudinem* could mean either that they have the strength to hold it up or (more likely) that they get strength from having it there.

9. On the identifications, cf. "mosquitos," 1.96 and 26.14(15), with notes.

10. The exact nature of A.'s *lanterna quae est et vas reservatorium et continens candelam* would seem to be one which holds oil and has a recess to accept a candle. It is his version of Ar.'s "spit and lampholder in one," Ar. *Part. An.* 683a25, on which cf. Ogle (1912), ad loc.

suddenly extend them and thus force their entire body in executing a leap up off the earth. During such a leap their feet must be bent to the inside of their body and the knee of their leg must be bent to the outside. Such a bending can never occur in the front feet. All these types have six feet, with two long rear feet to perform their leaping.

This, then, is the disposition of the outer members of the bodies of ringed animals.

CHAPTER II

On the Outer Members of Shellfish

7 A hard-shelled animal does not have great variation in its outer members for it is one of the animals which is fixed, virtually dwelling in a single spot. If it does move, its movement is slow and slight. Those animals, however, which have great variation in their outer members are those which have many functions and many movements. For this reason they need multiple organs. Certain of the types of shellfish [*ostrea conchilia*], however, do not move at all. Others move but little and for this reason nature provides for them, keeping them from being crushed or eaten. She placed hard shells around them within which they might protect themselves.

Further, some of them have a single shell and others two. Some are like the *saurinos*, as we said above. Some have a twisted spiral in their body and in the shell, like the one called the *kyrikoz*, a particular type of *halzon* (that is, shellfish [*conchilium*]). Others, though, have a round body like the sea urchin.[11]

8 Further, some of those which have two shells open and close them as do the shellfish [*ostreum*] called the *pecten* [scallop] and that called the *moez*.[12] Some, though, have two shells that fold toward one another, attached in the middle where the shell is largest and strongest. An example is the one called the *solinez* in Greek and which means "the youth" [*iuvenis*] in Latin since it is always fresh and does not live long.[13]

Generally speaking, however, all these animals which are called hard-shelled (they are called *testudines* by some) are in truth the ones which in our lands bear the name of *ostrea conchilia*.[14] All of these, then, are like plants in that they take their nourishment as do the plants. They do not have heads with a distinct shape, and they absorb their nourishment through a certain web as if through a root. Thus too it is said by

---

11. Cf. Ar. *Part. An.* 683b13f. *Saurinos: strombōdē*, spiral-shelled whelks. *Kyrikoz: kērykes*, whelks. *Halzon* is a variant of *halzum*, see Glossary.

12. *Moez* is fairly close to *myes*, mussels, in Ar. *Part. An.* 683b16.

13. *Solinez*: again, fairly close to *sōlēnes*, razor clams. According to Stadler's markings, the etymology came to A., who expanded upon it. It is specious, however, for the Gr. *sōlēn* indicates a pipe or tube.

14. *Testudo* (pl. *-ines*) has been used variously for "snail" and "turtle." Cf. notes at 4.38. Here, however, it would appear to be a synonym for shelled creature. On the multiplicity of terms, see Glossary, s.v. "Shellfish."

some that their mouth and head are lower down on their bodies. For this reason it also happens that those members lower in location and situation are upper insofar as they perform the functions of such upper members as the mouth and head. On the other hand, using the same reasoning, the upper members are lower.

Every animal of this sort is in a web that it uses to strain out the fresh water which it sucks in for nourishment. These types all have no head whatever except in the way just described. The other members of its body are not named since the only member in it that is named is the one that takes in the food. Those members through which the superfluity leaves are located above in them.

Whichever of these animals moves, they have no organs for movement. Rather, their movement is accomplished by stretching themselves toward the place to which they wish to move and by dragging the rest of their body along behind. This is clear in the snail's [*testudo*] movement. As has been said, they seem to resemble plants in how they get food. For some of them are fixed below the sand and they only feed on the moisture of the sea as it passes over them. At this they open up, take in strained seawater, and from this they grow fat, especially when the moon is on the wax. These, then, are most imperfect animals.

CHAPTER III

On the Outer Members of Those Called Soft-Shelled Animals

Soft-shelled animals are not fixed but rather walk about in the water. All these have large feet compared to their bodies. We are using the word "foot" here to refer to the entire organ of progressive movement and upon which walking is accomplished. That is, the leg and the foot together. There are four genuses of this type of animal, namely, the lobster [*karabo*], the *cesy*, the *hakokul*, and the crab.[15] Each of these genuses has types which differ not only according to shape but also in quantity and size of body.

Of these animals the types of lobster have the greatest similarity as far as the appearance of the shape of their outer members is concerned. Each of these types has two things in front which are scissor-like and which seem to have been formed as if for the activities of hands and lips, namely, for grasping. They are heavy, for in animals of this sort they are not used for walking but rather for grasping and holding, taking the place of a hand. For this reason too they seem to be almost added on, as if located outside the area the other feet. The roots by which they are affixed to their body reach deep down into the body of the animal below. The other legs, however, are rooted in the round part of their flank on either side. These appendages bend according to the requirements of their functions. With them they take in certain things that have to

---

15. Ar. *Part. An.* 683b25f. lists *karaboi*, *astakoi* (lobsters or crayfish), *karides* (prawns and shrimps), and *karkinoi* (crabs). A *karabo* will be translated here as "lobster," but cf. 4.5, with notes.

do with feeding and bring them to their mouths. The crab and lobster, however, are alike in their other legs, but they do differ, for although the river crab [*cancer fluvialis,* crayfish] has a tail, as does the lobster, the sea crab [*cancer marinus*] has no tail whatever.[16] This is because the lobster does not actually have a tail for swimming, for it uses it only to travel backward, and this is not in accordance with natural motion. Rather, the lobster props itself up with its tail and swims using its large feet like the *kelatoz.*[17] The crab, then, since it is not a deep sea animal and lives in the cracks and fissures in rocks lying near land, does not need a tail to prop it up.

12   Such members of the crab genuses as are in the deep sea have very weak feet and walk about but little. An example is the crabs called *hahaz* in Greek and those called the *hakekyokyz* which only walk a small amount and which have, for their defensive protection, bodies like the bodies of the *halzum* (that is, *conchylia*).[18] This is the reason that the types of the *hahaz* are made with slender feet. The type, though, which is called the *karkylyekyz,* has short feet. Two types of other small crabs, caught along with small fish, have many wide feet adapted for their movements. For by being wide they take the place of fins for swimming. The *hakokiz* and other types which resemble crabs differ. For the *hakokyz* have tails.[19] There is also deviation among those which resemble the lobster, for they do not have the hand-like appendages the lobster has. They lack these appendages because of the multitude of their feet. The *hakozyn,* however, has many feet, since it uses them like oars to swim and it does not walk on its feet. The members which are in the area of the head and back are those which are adapted for the taking in of water and then spewing it out. These are gill-like and have many folds. These members are more present on the females than on the males, especially in the male lobster. The members of the female crabs, however, are thicker and stronger in the "parts of the gates" it uses to open and close them.[20] Thus, because these spots are wide, they lay their eggs and hold them there and do not lay them far away, the way fish and other oviparous types do when they lay.

13   Further, because the body is biggest on the right, the right appendage is thus also larger in all crabs and other crab-like types. In the lobster, however, the right appendage is also the larger and stronger, for the right side in all animals is larger and stronger. For nature always places the greatest strength on the side where it is used most

---

16. The distinctions are A.'s. Ar. *Part. An.* 684a2f. merely differentiates between lobster-like creatures and crab-like creatures. On the "river crab," cf. 4.9.

17. Ar. *Part. An.* 684a3–4 states that the tail is used like an oar, but the source of *kalatoz,* which A. must think is an animal, is not readily apparent.

18. Ar. *Part. An.* 684a7f. *Hahaz: maiai,* the large, spiny-legged crab which is usually taken as *Maia squinado,* a spider crab. *Hakekyokyz: hērakleōtikoi karkinoi,* Heracleotic crabs; identification is not certain, but taken by Thompson (*GF,* 105) as the common edible crab, *Cancer pagurus* L. *Karkylyekyz* is apparently another version of the Heracleotic crabs.

19. *Hakokyz:* from the Gr. *karides,* prawns/shrimp, as is *hakozyn* below.

20. Ogle (1912, ad 684a19f.) discusses these appendages, which have more to do with reproduction than with respiration. The cumbersome "parts of the gates" (*in partibus portarum*) represents the Gr. *epiptygmata,* "flaps."

to function. This is clear in the right serrated teeth, horns, right claws, and in other things which are given for help and defensive strength.[21] The animal which the Greeks call the *aceh* has two hand-like appendages to the front, the larger of which ends in a hooked claw both in the males and in the females.[22] This is because this is one of the animals of this sort which possesses appendages, although they are sometimes generated by chance. The nature of this animal is easily injured and it is thus given these members for defense. The nature of each and every thing uses its appendages for the functions for which they were created.

From the things we have said above on the visible members of the animals and on their anatomy, running from the first to the fifth book of this study, the location of these members is easily seen, as is the diversity that exists among them when they are compared to each other. From these things too can come a recognition of males and females.

CHAPTER IV

On the Outer Members of the Malakye, *throughout Their Species*

We have in previous books discussed the disposition of the inner members of the *malakye*. This animal's outer members, however, are neither differentiated nor set. Rather, its feet are for the most part on the inside and to the front of its body, that is, around its head. Near its eyes and mouth there are certain teeth for biting those things which it entwines with its feet or "tails" (since its feet are more like tails than legs or feet).[23] Certain animals of this genus have two feet (that is, double feet) and they have feet to the rear as well as to the front. Others also have feet on the side, bound to their *femor*, as do all blooded animals.

The *multipes*, however, which is called the *polipes* [octopus] by its Greek name, has as a particular trait its feet on the front of its body. This is because in this animal the front and the back parts are joined together as if in a single spot. Its feet emerge from its head like a set of lines coming from one central point. This also happens in the animal that is like the *astarinoz*, a particular trait this animal has that it does not share with any hard-shelled animal.[24]

In general, a hard-shelled animal resembles the *malakye* in one way, and in another way it resembles a hard-shelled animal in that its shell is very hard. Yet its

14

15

---

21. "Claws": perhaps hooves, but cf. what follows immediately.

22. *Aceh*: from Gr. *astakoi,* lobster/crayfish, Ar. *Part. An.* 684b32f.

23. The translation of the two previous sentences reflects Stadler's wise choice to take *hos* as equal to *os,* "mouth." Otherwise, the translation runs as follows: "the front of its body, that is, around its head, and near its eyes and around these there are certain teeth . . ."

24. *Astarinoz* reflects the Gr. *strombōdesi* (dat. pl.), spiral-shelled whelks. For similar forms, cf. 4.38, 42.

fleshy part, inside the shell, resembles the flesh of a soft-shelled creature.[25] The shape of its body is like that of the *malakye* and especially that type which resembles the *astarynoz* and which also exhibits convolution.[26] The nature of these two types is as we have described, and they thus walk with an even gait, as do many quadrupeds and humans.[27] The human, however, has his mouth in his head, in the upper part of the head. Then he has his stomach and then his belly. After the belly is the intestine which reaches to the point of exit for the superfluity. In blooded creatures these matters follow the disposition given. After the head in such animals is the "oven" [*clibanus*] (that is, the chest and that which is near it).[28] It is for the sake of this type that the front and rear members of hand and foot exist.[29]

16 Further, following the method we have set forth before, there is a similarity between the soft-shelled animals and those with ringed bodies with respect to their front members. But they differ from a blooded animal with respect to their outer, movable members. The *malakye*, however, and that which seems in many ways like the one called the *sanyrinoz* differ when they are compared to the hard-shelled animals.[30] For in the creation and production of their body these animals are very much alike. But the rear portions of these animals' bodies are bent, curving toward the front portions of their bodies as if the line acb were bent back to point C, thusly:

$$\overline{C} : \underline{a\,c\,b}$$

Thus, the location of the members to the front of the body of these animals follows this disposition and the body of the *malakye* is especially created this way.

17 In the sea animal, however, which is commonly called the *polypes* [octopus] using the Greek term and which translates into *multipes* in Latin, nothing seems to be present of the front members save the head alone. For the shape of this animal is as though eight snakes had one head from which their bodies emerge in a circle as if from one center point.

---

25. Cf. Ar. *Part. An.* 684b17f., where the crustaceans are said to resemble both the cephalopods and the soft-shelled creatures.

26. Badly misconstrued from the intent of Ar. *Part. An.* 684b20f. The animals in question are the *strombōdē*, whelks, and the convolution is that of their shells.

27. It is worth noting that Sc.'s Latin is used at this juncture to help impart sense to the corresponding Ar. *Part. An.* 684b22f., which is itself quite corrupt.

28. On *clibanus*, cf. 1.110, 419, and below at 14.39, where it is defined as the entire hollow part of the body.

29. The Latin is probably in error here, and we can gain understanding by reading *motum* (motion) for *modum* (type) and thus rendering the members instruments of progressive motion.

30. Perhaps "and that which seems like it in many ways, called the *sanyrinoz* . . ." The name is yet one more corruption of *strombōdē*, whelks.

In a hard-shelled animal one of the forward members is called the *astarynoz*.[31] They do not seem to differ in any way save that nature placed hardness in the *malachye*, putting it inside a bit, under the flesh that surrounds its body. But in the animal with the hard shell she placed the hardness entirely outside, encasing all its body within it to protect it, for it cannot escape harmful things because of the slowness of its movement. For this reason the *malakye* discharges its superfluity from its mouth, just as it also comes out of the mouth of the animal that resembles the *astarynoz*. Thus, too, the feet of the *malakye* are made in contrary fashion to those of other types of animals.

Moreover, the *sepion* [cuttlefish] and the one called the *tonydez* [squid] differ when they are compared to the *polypus* [octopus] (that is, the *multipes*).[32] For the two types named first swim, while the *polypus* occasionally walks. Moreover, the two types mentioned have six feet which have small teeth to the front. The outermost of these six feet are the largest and one pair of these six feet is to the bottom, and this pair is very strong. For just as the rear feet are the largest and strongest for propelling the body, so too is this pair of feet, in this type of animal, the strongest for holding up the body's weight. The outermost pair of feet is stronger than the middle pair since it uses this pair more. The *polipus* (which is called the *multipes* in Latin) has four middle feet which are very large. All together there are eight feet in this animal. The feet of the *malakye* are generally fleshy and the feet of the *malakye* are short and attenuated in the *sepyon* [cuttlefish] and *tonydez* (that is, the *calamare* [squid]). In the *polipus* they are very large and long, so much so that it sometimes holds back a person with its feet, sometimes even a small ship. The bodies of those two types which have short feet are large and the body of the *polypus* has very little substance. For what nature takes away in bodily size in the *polypus*, she compensates for in foot size and what she takes away in foot size in the *sepyon* and *tonydez*, she compensates for with bodily size.

In some cases nature makes feet not only for anchoring purposes in those which stay unmoving (as the *sepyon* and *tonydez* do using their feet) but also for walking, as in the feet of the *polypi*. The feet of the other two types are of no assistance in walking because they are small and the bodies are large. They thus receive no help from their feet for either grasping or walking, which occurs when they let go of their rocks. The *polipus* receives help in all of these things from its feet. A common trait shared by these three genuses of animal is that nature gives each of these animals two appendages which the aforementioned animals use to anchor themselves when the sea's waves and storms grow stronger and winter and bad weather increase. They move on them just as ships do at anchor and they use them to catch on to something that is forcefully rooted in the sea and attach themselves to it. These appendages are found only on the *sepyon* and the *tonydez* (that is, the *calamare*), and the *polipus*, for only these animals' feet are suited for this sort of grasping and anchoring action.

---

31. The sentence is poorly written, but it does seem that A. here has forgotten the *astarynoz* is supposed to be a creature in its own right.

32. Ar. *Part. An.* 685a12f.: *sēpiai* and *teuthides*.

20 Further, the types of *malakye* which have openings on their feet for sucking also seem to have veins the length of their feet, as does the *polipus*.[33] They also have a scaliness and a sort of roughness on them. All of these things are present to serve as weapons to strengthen the animal. For this reason such animals defend themselves by wrapping their feet around those things they seize. The wrapping action resembles a wrapping action the ancients used to do and, by so doing, they strengthened their arms and legs with rough things, using them to hold on to the things they touched with their arms and legs.[34] These genuses of animal draw in whatever they seize and touch, using the wrapping action they sometimes carry out on their slender, nerve-filled parts. Thus, the wrapping action is accomplished using that which is soft, suited for seizing and grasping. After it has been grasped, they strengthen that by which it wrapped up the thing grasped. Thereupon the wrapping action becomes strong and most often the part of the animal that was used to do the wrapping becomes thick around the thing that has been grasped. This can be seen in the *polipus,* which, when captured by a man, immediately wraps itself around his arms and clings to him so forcefully that it even scrapes the skin off his arms sometimes before it is pulled off. This sort of wrapping action on things seized in this way occurs in no other animal save in certain genuses of *malakye*.

21 Certain of these types take in their food using an appendage of this sort. For they use the appendages of this sort in place of hands. Some of these animals have one member and some two. The reason for this is the flimsiness of their body and its natural length and slenderness. Because of the length and slenderness of these members, nature composed its actual body out of a single bone. Nature did this not because it is best but because otherwise its body could certainly not be held at so great a length.

Moreover, all these types have fins around their body. These fins are connected to the body in other species, but in the animal which is called the *tany,* they are also connected this way with the body.[35]

22 If one of these types has a small body, as does the *tonydez* (that is, the *calamare*), it has a wide fin which is not narrow like that of the *sepyon* and *polipus*. This is because the principle for the generation, power, and growth of these animals comes from the middle of their body and does not surround their entire body. This animal has these fins solely for the purpose of supporting itself on them, much as the breast in birds is that upon which almost the entire body is borne. This is also the case for the tail in animals that have one, according to Aristotle. For in birds the breast supports the entire bird in that the wings are connected to the chest.[36] The tail, however, supports

---

33. The origin of these veins, an error for the suckers themselves, is explained in the notes to 5.19.

34. Ar. *Part. An.* 685b5f. compares the intertwining tentacles to *seirai,* woven tubes not unlike the children's toys today called "Chinese finger puzzles." Ar. says they were used by "doctors of old" to treat dislocated fingers. Cf. the lucid note of Ogle (1912), ad loc. A. has added considerably to his obscure, received text, trying to make sense of it, but with little luck due to the odd grammar of his addition.

35. *Tany* stems from Ar. *Part. An.* 685b18, *teuthōn* (gen. pl.), calamaries.

36. The analogy is clearer in Latin since both "wing" and "fin" are *ala* in this passage.

with a sort of binding action, and thus, too, these animals are supported by their fins with a binding action.³⁷ The fin on the *polipus,* however, is so small that it is not noticeable at all or can be seen only with difficulty. There is no need for it to be large, for its body is elevated adequately by its feet.

The Second Tract of the Fourteenth Book on Animals,

in Which Is Treated the Reason for the Visible Members

CHAPTER I

On the Reason for the Disposition of the Belly and of the Extremities of the Hands and Feet Below the Head and Neck in All Genuses of Animals

Having thus determined the disposition of the animals with ringed bodies, those with soft and hard shells, and the *malakye,* and having determined the reason for the disposition of their inner and outer members, it follows that we now pass over to a consideration of the disposition of the members of blooded, viviparous animals.

Proceeding, then, with this consideration, we will begin by considering the remaining exterior members, namely, those we did not touch upon above when we dealt with the inner members of these animals. After we have treated these individually, we will determine the outer members of blooded, oviparous animals, running through, in exactly the same way, the outer members of individual animals, much as we have made determinations about the inner members of these same animals.

We have already determined the disposition of the members which are near the head and neck of animals. We will thus begin there and say that every blooded animal has a head. But the head in certain animals having one is distinct and separate, while in others, like the crab, it is not.

Further, every viviparous animal has a neck. Some of the oviparous animals have a neck and some do not. Further, every animal with lungs has a neck, but an animal which does not breathe in air from outside does not have a neck. The head is created only for the sake of the brain, for this is the member which is necessarily present in all blooded animals. The brain should only be above, opposite the heart, for the reason we gave above.

---

37. "Binding action": *colligatio,* implying also a formal relationship such as that of a treaty or an alliance.

Further, nature placed certain senses in the head since the nature of the brain is temperate and its complexion is tempered by certain senses. We have already determined the reason for this. This was done because these senses desire quiet and delicate spirits and these things are to be found only in a cold complexion that is temperately moist, like that of the brain.

25 Nature placed a third member, separate from the head and neck, below the neck. This serves for the entrance of food and is the belly. It was reasonable that the members be arranged this way, for this way the belly is in a place that is suitably arranged and tempered for it. For it would not have been suitable for it to be over the heart, in which resides the first power of life, sensation, and nourishment. Neither would it be suitable to place the belly over the divine member which has the most excellent operation. I am saying, then, that the operation of the divine member is nothing other than that of intellect and sensation. For if this member must protect such operations, it could not support another member above it since it could not bear the weight above it. For a weight of this sort would weigh down the movement of the intellect and would change the common sense, especially if a very heavy body like the stomach and belly were placed over the brain.[38] Moreover, it was necessary for this third member to slope toward the ground below the heart and the chest in order to protect it and the upper members from being crushed.

26 Nature, further, placed the front feet in place of the hands in quadrupeds. The rear feet, however, were necessary for walking. Because of this, there were four feet in all animals of this sort so that they could carry the weight of their bodies.

In all those animals which have four extremities, the front of the body is larger than the rear, with the human being the exception. For the upper part of a human's body is moderately sized compared to the lower and it is much smaller than the lower. But in children which are still bent over from the weakness of their age, it is the opposite of the way it is in those who are of advanced age. For in children, the upper part is heavier than the lower. This is why children go forward bent over and, during the first age, cannot raise up their bodies. They instead creep on their hands and feet due to a great flow of nourishment to their upper members, especially to the shoulders and head. For in them the upper part is larger than the lower. When a person has grown old, however, and the moisture of youth has dried out and the members through which the passage of nourishment is effected have shrunk, not as much nourishment is brought to the upper members. Thereupon the lower members take on greater growth and the body is raised more upright. The disposition of the quadrupeds, however, is the opposite. For at the very first age of their lives, their lower part is heavier than the upper since it is closer to the *umbilicus*.

27 During the second age, when the child matures and the moisture has dried out, a shrinking occurs in the members through which the passage of the nourishment is effected. At this point the upper part takes on greater strength and grows more than does the lower. This is because such animals carry their entire body on their front

---

38. For the position of the common sense in the head or brain, cf. 12.143.

legs and are terrestrial. Thus they are bent to the front and they experience a greater movement of the nourishment to the front than they do to the rear because of the heat of their heart and liver. This is especially so since, due to their earthiness, they can never, at any age, have an upright nature and an erect body. Still, in many horses, the movement of the nourishment to the front makes the front part higher than the rear part. Moreover, in the animal the Arabs call the *seraph* [giraffe], the front part is much elevated above the rear, so much so that they seem almost to have an upright bearing. Further, in those monkeys which bear the clearest likeness to humans, this movement of the nourishment is so effective that they sometimes become upright like a human. The pygmy, which is the most like the human of all the irrational ones, has an entirely upright bearing as does the human. This, then, is the natural disposition of the quadrupeds and of those with hooves split in two.

In an animal such as the lion and the dog which has a foot split into many sections, the front is greater than the rear due to the heat of the first age. But when they enter the drier age, the rear part grows proportionately more than the front. The genus of birds and fish have at every age more flesh to the front than to the rear. For the entire force of their heat is forward and in the rear there is nothing but the intestinal passages that carry excrement to the anus. This disposition suits their movements of swimming and flying, for in movements such as these with a thinner part going first, a heavier rear part would be drawn along with great labor. 28

A flow of nourishment of this sort brings it about that all animals have less intellect than the human. Children, too, in whom the movement of nourishment and spirits is not settled, have less intellect than do their elders, as we have said in the seventh book of our *Physics*.[39] For whenever the weight of the nourishment is multiplied above, it diminishes the intellect and the operations of the animal powers. This is because then a great deal that is extraneous becomes mixed into the brain, which, like an organ, is the principle of animal operations. But when this is mixed in, the organ is rendered heavy and corporeal, not possessing refined spirits. In this case there will be but little and scanty movement in the operations and movements of the sensible and intellectual soul. An indication of this is that people become insane in whom the melancholic, earthy part takes over the seat of the intellect. People also become epileptic, forgetful, and slow to discover reasons for things when a phlegmatic humor has overflowed in the brain. Blood and bile are often converted by the coldness of the brain to the qualities of melancholy and phlegm and the same would happen to all nourishment if there were a great, superfluous flow of it to the brain. The accidental elevation of bile or of too much blood leads to various problems in the brain in other, more imperfect animals, like the ringed animals and the shelled animals, be they hard- or soft-shelled, or of the genus of *malakye*. 29

The body must be smaller in direct proportion to the extent that the natural heat is less and the material, earthy part is greater. And because the weight of the body is so much heavier and because this moderate heat cannot make it lighter, it has to 30

---

39. *Phys.* 7.1.9.

have many feet. Further, it is possible for the natural heat to decrease so much and the earthiness to increase so much that it will lack feet entirely and will be one that crawls on the earth. Also, there is a member in which the most worthy power of the soul resides, which is the cognitive and ruling power. This is the head, which will be tipped toward the earth, scarcely ever raised up from it. Some are so deficient that they also have no sensation whatever save for a slight and dull sense of taste with a bit of touch mixed into that taste. The power of these animals is below in the earth, as it is in trees, and in them this lower part is much larger than the upper, as in trees. They are midway between animal and plant, possessing larger, lower parts while their upper members are diminished like the tips of branches. They sometimes take on the shapes of perfect animals but without their powers, as is seen in the mandrake, which has the shape of a man and which also seems to exhibit differences of sex and other accidental traits. But we have spoken of these things in the last book of *On Plants* [*De plantis*].[40]

Thus, then, the reason has been determined why some animals have many feet, some four, some two, and others lack them entirely. The reason has also been set forth why the human had no need of front feet, and why nature gave him hands in place of front feet.

CHAPTER II

On the Reason for the Shape of the Hand, the Reason for Its Use, and How It Is an Indication of Intellect. Also on the Use of the Quadrupeds' Feet.

31 The fact that a human has hands is an indication that he has more intellect than the other animals and has a greater natural capacity [*ingenium*] (that is, reason [*ratio*]). However, possessing hands is not the reason for his intellect, but rather, to the contrary, the possession of intellect is the reason he has hands.[41] For the hand is not causally related to the intellect but rather is like an instrument. For this reason, since this instrument is operative (that is, practical), it is properly an indication of the practi-

---

40. *De veg.* 6.379–81. The mandrake generated a great deal of interest, in part because of its reputed powers as an aphrodisiac, based on a reading of Gen. 30:14–16 and Song of Songs 7:13. Its presence in the Old Testament, moreover, assured that it would appear in medieval biblical exegesis, which expands upon its description and notes its resemblance to the human body as well as additional medicinal uses (e.g., as an anesthetic). Cf. Haimo of Halberstadt, *Commentarium in Cantica canticorum* (*PL* 117: 349B); Rupert of Deutz, *Commentaria in Cantica canticorum* (*PL* 168: 949A); and Honorius of Autun, *Expositio in Cantica canticorum* (*PL* 172: 472C). Hildegard of Bingen, *Phys.* 1.36 (*PL* 197: 1151–52), is quite interested in the mandrake root, claiming it comes from the land where Adam was created and adding several detailed recipes for its use.

41. This line of argument curiously parallels modern discussions. Did we evolve more rapidly because of our opposable thumbs and stereoscopic vision, giving rise to increased brain activity, or did increased brain size (or bipedalism) come first and make the development of such human "tools" more profitable? For a brief discussion, cf. Leakey (1977), 38–45 and Farb (1978), 76f.

cal intellect. But because the practical does not exist without the speculative, it consequently indicates that the speculative intellect is present. For since the hand is an organ with not one operation but operations that are multiple in species and genus (for in general, operations of every genus are performed by the hand), this will be an indication that it is the universal principle of every genus of operation.[42] This, however, is neither the nature nor the power of the soul operating after the fashion of nature. For these do not have every genus of operation but perform an operation of one form and species, such as the fact that fire burns or that every swallow builds its nest like every other swallow. The hand, however, is used to perform weaving, sewing, house building, and all sorts of arts, not in one way, but in every way. The hand thus will be an indication of the principle which is generally the principle for the conception of all forms of every genus of operation and contemplation.

As we said, nature and form operating after the fashion of nature always remain in the same disposition while functioning. This, in all in which this alone is the principle of operations or even in those in which the principle is the power of the soul operating after the fashion of nature, the general instrument of every operation, which is the hand, is missing, but in its place an instrument is given it which is deficient in the operation of a hand, just as the hand has a principle that is deficient in the operations in nature or the soul.

Because the human alone has the most intellect of all the animals, he suitably takes from nature an organ suited for many movements and all its other functions. For pipes are given to a piper rightly and reasonably since he has in him the principle which is the art of piping and which is suited to using the pipes. Just as it is in artists who imitate nature, so it is in nature. For nature, since she does what is best in all cases, supplies that which has less to the greater.

She does not, however, give the nobler and the greater to the lesser, for in this way it would abound in superfluities and in those cases she would not always be doing the best thing possible from among those things nature herself has in her power. Thus, since it is less to have an organ than it is to have the principle which is the power operating in the organ, nature gives power to each one and adds the organ to that unless an accidental flaw interferes. But the one to which she does not give the principle and the power is therefore placed in a lesser position. If she gives it the power she gives it an instrument [*operatrix*] having only one function. She does not add more or better to this in order to give it a universal organ possessing all functions.

It follows from all that has been said that the human is not intelligent because he has hands but rather that he has hands because he is intelligent. Because he is very intelligent, possessing multiple forms of intellect, he has need of an instrument which is multiple in use even though it is single in substance.[43] Thus, with it he might per-

32

33

---

42. Because of its many functions, A. describes the hand as the "organ of organs" (*organum organorum*). Cf. 1.287, 299, and *De anima* 3.3.12.

43. The relationship between the hand—an organ especially appropriate in humans—and the intellect is so intimate, claims A., that a person who really understands and wants to communicate a subject to another person will naturally gesticulate animatedly with his hands. Cf. *QDA* 14.11.

form many functions properly for the conceived form. The hand, however, while it might be one in substance, is still not a single instrument with regard to power, but is multiple. It is like an organ that receives into itself the forms of many uses and functions that are multiple in genus and species.

34. There are those, then, who hold that the human body is not suitably disposed by nature but has the worst disposition of all the animals since it is bare, lacking hair, and is born weak in body and unable to go forth and raise up its body in its first age. It has neither hooves to protect its feet nor horns or other natural weapons suitable for a strong defense. But these people are making a grievous error, for as we have already said, other animals have only one type of strength and can have no other. Neither can they change their weapons to suit the purpose. Instead, whatever the need, there is one principle in them of doing it. Rather, it is as if we say that a sleeping person and a shoemaker who makes sandals or boots are doing those things proper to sleep or to the art of shoemaking.[44] For the sleeper, even if he had soul, or a principle of soul, other than that which is operating in his sleep through the digestive heat, would still not diminish that heat which operates in his sleep through evaporation so that another soul or another principle of soul would be operating on his behalf. Rather, it remains bound and the heat operates only around the place of digestion.[45]

35. Likewise, even though the shoemaker might have many principles of soul, nevertheless, only one, that related to the manufacture of shoes, operates through the art of shoemaking. This is how it is in the rest of the animals which, while they might have many principles and potentialities of soul, are, according to Plato, individually different souls, such as sensation, imagination, memory, and the like. Nevertheless, because all these things operate after the fashion of nature, they are operating only in one mode which follows the form of the operating nature. Thus, the organs of such animals are designated for a single operation.

The human, however, because of the universal principle of soul which he has, has many types both of strength and assistance and can interchange these types one among the other according to the forms of his arts and conceptions. He thus produces external weapons and tools for himself of whatever size and sort he wants and in producing and manipulating all these things he employs his hand. A human's hand, then, takes the place in him of a hoof and of a sharp, curved claw. The same is true for a lance, sword, and, generally, all his other instruments. For he can make and manipulate all of these with his hand.

36. Further, the hand is made to be suited for all operations in that it is extended for grasping and is divided into the many parts of fingers, articulations, and joints. A human can use one part alone, two or more, or all at once and can change these modes of use into different ones. The mobility of the fingers is suited for grasping and holding. The thumb is short and thick because of its strong actions in closing and

---

44. It is instructive to compare this example with Ar. *Part. An.* 687a23f. to see how much the text could change as it passed through the ages.

45. On the powers "bound" to the soul of one asleep, cf. *Phys.* 7.1.9.

strengthening the hand. For this reason it is called the greater finger [*digitus maior*], even though it is smallest in length. For a person gets no help from the other fingers without the thumb. As we have said, if there were not a hand there would be no grasping and in just the same way if there were no thumb there would be no holding. The thumb holds by having its own bottom part being placed on top of the other fingers.[46] But the other fingers only hold by contracting from their bottom. From these two a very strong holding action results. The thumb, therefore, is strong and its strength is seen to be equal to that of all the other fingers in that it is placed over them to strengthen the grip. For this reason it was made short and thick, for it would be entirely useless for the aforementioned purpose if it were long.

The last finger of the hand is reasonably small and short. For the grip is terminated in it when the hand is closed, a closure which could not be accomplished through the use of a long finger or a short and large one.

The middle finger was properly made long for breadth and capacity in grasping. Thus, too, are the middle oars on a ship longer than the other oars so that they can catch a lot of water for pushing the ship along. On either side, the index finger and the leech finger (that is, the ring finger) are proportionately shorter so that the hand can be tightly closed into a fist and in this way the hand's holding power is strengthened along with its dragging action, especially by means of which all grasping is done.[47] For this one, however much it holds, it necessarily holds very well when the other fingers are shortened proportionally around the middle finger, as has been said.

Further, it was a clever and wise nature who strove to act in the best and proper way in the creation of the nails. For the nails of certain other animals are only curved, but the human has nails designed to cover the tips of his fingers. Because the hand is an indication of intellect, as we said, the disposition of the nail will be, according to Palemon, a sign of the disposition of the intellect. However, we have spoken of this once already in the physiognomy section in the first book of this study.[48]

Further, the *asseyth* (that is, the reed [*harundo*]) in a human's arm is connected to the flexion of the hand. A person's *asseyt* are somewhat curved to the inside.[49] Its flexion and curving are for the purpose of making its movement more suitable to bring food to the mouth and to suit the other functions of the hand. The *asseith* of other animals are created in a different way and they bend differently in walking, quadrupedal animals.[50] But of these animals, a quadruped which has many toes uses its front

37

38

---

46. "Bottom part": the confusion is due to A., who did not understand his received *ab inferiori parte* as meaning "from below" and added *sua*. For the original intent, cf. Ar. *Part. An.* 687b14f.

47. The *medicus digitus,* known in English as the doctor's finger or leech finger, is a direct translation of an earlier Gr. term. The name derives from the fact that a vein was believed to travel from this finger directly to the heart. This is also why, in the West, wedding rings occupy this finger, giving rise to A.'s *digitus anularis.*

48. Cf. 1.504f. and, for Palemon, notes to 1.128, cf. *ADP* 60.

49. *Asseyt(h)* is a variant of the more common *asseyd,* the radius or ulna. Cf. this passage with 1.284f.

50. Poorly stated, but the sense is that a human's arm bends toward the mouth to bring it food while the front foot of a quadruped bends in the opposite direction for locomotion.

feet in place of hands and does not use them solely for walking. This is clear to see, for it fights with them and scratches itself and keeps harmful things away with them. An animal which has hooves [*solea*], however, fights with its rear feet since its front feet are not suited for use as hands but rather for bearing up their body. For this reason, too, an animal with many toes has five toes on its front feet and four on its rear. Examples are the lion, lynx, dog, and wolf. This is so that out of this multitude of toes more suitable "works" might arise from the multitude of toes on the front feet which take the place of hands. Animals, however, which have many toes on their feet and five toes on their rear feet make their way creeping over the earth. For they place their entire hind leg on the ground from the knee down. This is because their rear part needs to be raised up on many claws so that its movement might be swifter and lighter to raise its sides from the earth and thus lift up its head.[51]

This, then, is what has been said on the reason for the hands and for those things which take the place of hands in animals.

CHAPTER III

On the Reason for the Shape of the Chest and Breasts in the Human and in Other Animals

39 Between the clavicles [*humeralia*] and the front feet in quadrupeds lies a visible part of the body called the chest.[52] This part in the human, however, lies between the clavicles [*humerus*] (to which the arms are attached) and the belly. For if all that is between the clavicles and the feet in a person is considered, then that which contains the entire hollow part of the body is called the "oven" [*clibanus*] and it is also called this name in the other animals.[53] In this way, the area of the spiritual members, which contains the heart and the lungs and which lies over the diaphragm, is properly called the chest or the thorax. And the lower area, beneath the diaphragm and as far as the beginning of the thighs, is called the "oven" [*clibanus*] using the common name. This is because those things which have to do with nourishment are cooked in it.

Considering the chest this way, the chest of the human alone of the animals is broad. Now this is reasonable, for the clavicles are attached to the sides above, namely, the right and the left, and these expand the chest a good deal.

---

51. Cf. Ar. *Part. An.* 688a7f. Ogle (1912 ad loc.) identifies these smaller quadrupeds with such creatures as rats, weasels, moles, martens, etc., but he points out that elephants and bears also fit Ar.'s description at this point.

52. *Humeralia*: an adjectival form of *humerus*, which can mean variously shoulder, clavicle, or humerus bone (Fonahn, 1567). The choice of "clavicle" here for this unusual form of the word *humerus* is therefore somewhat arbitrary, but cf. 1.278f. and 3.21. *Humerale* clearly stands for the *ankōnes* (elbow, arm) of Ar. *Part. An.* 688a12.

53. Cf. 1.110, 432, 437; 14.15.

Further, since the spirits in a human are nobler and more refined, the area for the spirits, in which they disperse, should be roomy and broad.⁵⁴ Such breadth of chest, however, cannot exist in quadrupeds, for these animals, for the reasons given before, face the earth in the front and they need their front feet more to hold up their body, whereas they need their rear feet mostly for walking. Front feet which are spread apart do not carry a body well, since the weight of a large body would press down on it and hang down in between them.⁵⁵ The front props thus ought to be fixed close together. And since these are joined together the chest must be narrowed and pointed in the middle, almost at an angle. Thus, the human chest has greater breadth and less depth, whereas the chest of the quadrupeds has greater depth and less breadth. However, in both, the length of the chest from the fork of the neck [*furcula colli,* clavicle] to the "oven" is proportionately equal.

A further result of this is that animals which are quadrupeds cannot have breasts on their chest. For these would keep their front legs from being close together and would hinder their walking since spread apart front legs totally hinder walking and running in quadrupeds. But rear legs which are moderately and temperately separated and spread apart in them both effect and assist in walking and running. In the human, however, it is the other way around, for legs which are spread apart hinder walking and those which are close together aid it. Thus the breasts of a woman are placed on her chest and are moved away from the groin, but in quadrupeds they are placed on the groin and are removed from the chest. Because the chest of a woman is fleshy, with loose flesh, she thus has fleshy breasts which hang down and consist of loose flesh. For a similar reason, quadrupeds also have this sort of breast on their groin.

Although the shape of the breasts finds a reflection in the males, they do not have perfect breasts. The women do, for nature uses women's breasts for a grand function, without which the species would not be preserved, namely, the feeding of the young, and nature does not use the breasts in males for this purpose. The reason that there is a certain reflection of the breasts in males has been treated in the tract on the anatomy of the breasts.

There are two breasts in women, separated from each other because there are two sides to which they are rooted and attached. But with the exception of the monkey, which is like a human, it was not possible for quadrupedal animals to have breasts on their chests.⁵⁶ For if they had them on their chest, they would weigh them down, hindering their walking and movement as we have said above.⁵⁷

Further, animals which have few offspring and horny, unsplit hooves [*solea*] on their feet or those also which have a hoof [*ungula*] split in two have breasts on their

---

54. Compare what follows to 2.11.
55. As in the feet and legs of heavier reptiles, such as crocodiles.
56. At *QDA* 2.14, the elephant is mentioned alongside the monkey as an animal having breasts on the chest. But the more usual placement of elephant breasts is discussed at 2.12f. and below at 14.43.
57. Cf. 2.11–12, 36–37.

groin beneath their hips.⁵⁸ They have most often two breasts joined into one and two breast-cones [*coni mamillarum*].⁵⁹ A few others, though, seem to have four udders [*ubera*] joined into one and four breast-nipples [*papulae mamillarum*]. An example is the domestic cow and certain other cows.

An animal, though, producing many offspring which has a foot divided into many parts has many breasts. Examples are the dog and the pig, which, although it has a hoof [*ungula*] divided in two, nevertheless has many offspring. Some also have feet divided into many parts and few offspring, and have only two breasts. Such is the lioness which has only two breasts in the middle of her belly. Although the lioness has few offspring, this is not why she has few breasts. Rather, it is because she has little milk due to the heat and dryness of her complexion. For her food passes over into bodily tissue due to the power of the heat. For it is a hot and dry animal, which eats infrequently and then, when it does eat, it eats flesh, which is quickly converted into bodily tissue because of the affinity which it has to it.

43 The female elephant has only two breasts at her armpits. The reason for the small number of breasts is that she produces only one young. There are nonetheless two, since there are two sides to which the milk flows. They are under her armpits because of the great distance between her front legs created by the size of her body. The reason that they are not near the hips is that this animal has a split foot, even though the divisions are closed up to strengthen the foot and especially because it is suited for her to have very thick milk possessing a great deal of earthy nourishment. Such milk must be thickened using a great deal of digesting heat and for this reason its vessel must lie near the heat of the heart.

Certain animals, however, have many breasts and much milk, like the sow. She has as a particular trait also that she gives her first breast to her first young, the second to the second, and so on for the others. I am calling the first breast the one that is closer to the right armpit since in this breast the milk is better decocted.

This is the reason the elephant has but two breasts, in the place we stated.

44 Further, the breasts of an animal which produces many young are on the belly, for an animal that has many young needs many breasts. Since they cannot be arranged breadthwise in more than twos (for there are only two sides, a right and a left), the location of the breasts had to be lengthwise in two rows on the two sides of the belly in the area between the front and rear feet (that is, hips). An animal which lacks many divisions on its hands and feet and which produces few young at once has more separated breasts lying beneath the rear hips. Examples are the she-camel, the she-ass, and the mare. For these animals rarely produce more than one young. The disposition of an animal that does not have a *sotular*, but rather has a hoof split in two, is much the same. This is seen in the doe, cow, she-goat, and those like them. The reason for this is that the greatest part of growth goes on in the front parts of the body and these thus need the most food. The lower part of the body, however, has a contrary disposition.

---

58. Perhaps "beneath their thighs." Cf. 1.18.
59. On the terminology used in this section, see Glossary and 2.36f.

Nature therefore placed the breasts there since where less food is converted into bodily tissue, more is converted into milk. Moreover, the natural movement of the food, since it is heavy, is to this lower part. There its conversion into milk is easier and its reception into the breasts is fuller.

Further, both males and females in the genus of humans have breasts to some degree. The males in many other animals, however, do not have any sign of breasts. Stallions, when they resemble their mothers in color and other dispositions, sometimes have them. But when they resemble their fathers, they have no signs of breasts.[60]

This, then, is the determination on the dispositions of the breasts.

CHAPTER IV

On the Reason for the Disposition of the Genital Members in Both Males and Females. Also on the Tail in Animals and on the Disposition of Their Feet.

The middle of the chest is, for the reason we gave above, connected on either side to the ends of the sides. This is done so that the passage of food through the middle of the chest into the stomach will not be blocked. At the furthest edge of the chest lie the members which serve as an exit point for the dry and wet superfluities. These are generally at the end of the member which is commonly called the belly. Nature employs the same member for the exit of the moist superfluity and the sperm in both the male and the female. An exception exists in very few species (that is, types) of blooded animals. The reason for this seems to be that nature deals with few things whenever she can and she thus expels the superfluities, which are of one, common sort, through one and the same member. Now the moist superfluities like sperm, urine, and menstrual blood are of one, common sort in females, and nature thus expels all these through one and the same opening, which is the females' vulva. We are, however, disregarding the sperm and the menstrual blood here and will treat them at length below. For below it will be necessary for us to understand what the disposition of the sperm is, what accidental traits belong to it, what differences exist between male and female sperm, and all about impregnation. All of these things will be understood in the discussion on generation.

It is clear that the shapes of these members, called genitals in the males and females, suit their functions. The males' member is designed for intercourse and has many differences. The fact that every male penis is nerve-filled is a trait commonly found in all male penises.

Further, it is commonly found that this member is the only one of all the other bodily members which grows larger and shorter without any damage to its substance.

---

60. Cf. 2.12.

Its growth occurs when it swells, whereupon it is ready for intercourse. It shortens when it is not extended but rather subsides. It is then ready for other things. If it were not shortened occasionally, this would hinder its many functions. For its substance is such that these two things can suit it, namely, growing and shortening. It is composed of nerve and cartilage and thus grows long as the sperm and windiness pass through its veins. This is so that its passages are opened by the windiness for the free departure of the sperm.

47 Certain females of the quadrupedal animals urinate to the rear, since such an arrangement and shape for their vulva suits their intercourse. Numbered among the male quadrupeds which urinate to the rear are the elephant, lion, and the cat, which is leonine in shape (that is, a sort of reflection of the lion much as the monkey is a vague reflection of a human). The camel also urinates to the rear as does that animal whose proper name is *pilosum* ["hairy," the hare].[61] None of the animals with hooves [*soleae*] urinate to the rear if they are male.

That member which is at the rear of animals, near the hips and legs, is called the tail. In the human, as far as his construction is concerned, this has a composition opposite that of the other quadrupeds. For all quadrupedal animals have a tail, large or small. Not only those which are viviparous but also those which are oviparous quadrupeds have a tail, large or small. Also, some have hair on their tail and some do not. The human, however, has no tail whatever, just as no quadrupedal animal has *anchae*.[62]

48 A human does not have a tail because the tail is essentially for covering the anus and keeping it from becoming cold. A human, however, has very fleshy thighs which cover the anus adequately and this fleshiness draws the substance of the tail to itself. Moreover, since the human has an upright body, a tail would hinder his movement, his walking, and his sitting. Thus, it is best that a human lacks a tail. The other animals, however, lack flesh on their buttocks in that they do not have as much flesh on them, proportionately, as a human does. Viviparous animals might have thighs and legs composed of bones and nerves, and they might have a great deal of flesh on them as do fat horses. Yet, because they do not have an upright body, their anus is not covered by the thighs and buttocks. This is why in almost all cases a single reason obtains, namely, that, as we said, of the animals only the human has an upright body. Therefore, in order that there might be a regular attitude to his body and a regular passage to his lower part, nature, for the sake of lightness and smoothness, took flesh from

---

61. The identification of the hare is based on 13.87, q.v. with notes, and by comparison with Ar. *Part. An.* 689a30f. The cat is an addition by A. to this list and is correctly retromingent. It replaces the lynx in Ar.'s listing of male retromingent animals.

62. *Anchae* is a difficult term, clearly referring to the pelvic area, but with varying overtones. A comparison with Ar. *Part. An.* 689b7 shows it stands here for *ischia*, in the sense of buttocks, which A. will mention just below using the more usual *nates*. Cf. 1.292, where *ancha* is said to refer to the back of the pelvic area and cf. below at 14.73f., where the term may refer more precisely to the bony structure of the hips. On all the meanings this word can have, cf. Fonahn, 353–54.

the parts of the body that are at the descent and added it to the buttocks, making the muscles there larger than in any other part of the body. For this reason the *anchae* in a human are very fleshy as is the area near the hips, and the inside of the leg, both above and below the knee.

Nature strove to do this so that there might not be too much bodily weight over the slender legs and so that the human's upright body might taper proportionately. In this way it is borne about more easily and is more suited to its functions. The thighs, however, and the legs of the quadrupedal animals are made of strong nerve and bone to carry the weight of the body. The four feet act as four props, to the front and rear. For this reason a quadruped stands with no effort when it is born but the human cannot stand erect for a long period of time. Instead he requires rest, both because of the tenderness of his component members and also because his entire body rests on only two props. The thigh, then, as is clear from what has been said, and the inside of the legs in the human are very fleshy for the reason given. He thus does not have a tail because all his food passes into the nourishment of the thighs and legs. Because he has fleshy *anchae,* he also lacks a tail, for he would necessarily need it, if what has just been said were not so.

A quadrupedal animal and other genuses are contrary to this. For the majority of their flesh and weight lies in their foreparts. For this reason their *anchae* and legs are not very fleshy compared to the rest of their body. Neither are their legs very thick and hard compared to the muscles of the buttocks and calves below the knee. So that the member of the anus from which the superfluity exits might otherwise be thus protected—and also the genitals, which have a great need of protection from the cold and harmful things—nature made a tail for them. The monkey, however, has no *anchae* since its form is composite, being made out of that of a human and that of a quadruped. Its rear mostly resembles a quadruped.

Moreover, tails differ greatly, for nature does not use them for one job alone, covering the anus, but for many ancillary tasks as well.[63]

This, then, is the disposition of the tail and the genitals in animals.

Moreover, there is a great diversity among the animals' feet, which, as we have said, have much in common with the tail.[64] Some have only *soleae* on their feet while others have *sotulares*. Still others have a foot split into two or more toes. The *soleae* are present in animals that have a large body and in which a great bit of the earthy part, directed to the nourishment of the large bones, passes over into the *soleae* of their feet instead of into horns. Now, I am using *soleae* for that horny part of the foot which they commonly call the *calx*. Those animals, since they do not have tusk-like teeth or horns, pass on all their earthiness to the *calces* of their feet. For they need a well-protected foot due to the weight of their body and so that in it they might find a defensive

---

63. Cf. *QDA* 2.27, where the question why humans have no tail is treated as well, with the additional information that a tail helps many animals in their motion, including an unattributable story about a dog which, having lost its tail, was afraid to cross a bridge lest it fall off.

64. See Glossary, s.v. "Hoof" and "*Solea.*" For possible equivalencies, cf. Ar. *Part. An.* 690a5f.

weapon. *Soleae* are found in them instead of many nails, all joined together, so that it is rather like a single nail.[65] For this reason too animals of this sort do not naturally have a *chahab*. I am using *cahab* for the rear nails which face downward on the rear of the legs in animals which have hooves split in two.[66] This is also the reason that these animals bend their foot to the rear with difficulty, for they do not have a *cahab*.[67]

52 A foot with a *cahab* is like one with two angles.[68] Then it has two bending points and easily opens and closes, that is, extends and contracts. One that has but a single angle does not have a *cahab* and contracts and extends with difficulty. The *cahab*, however, is more for the planting of the foot than for moving it. Thus, animals which do not have a *cahab* bend their rear feet with difficulty. The *cahab* is like an external member placed away from the foot, between two members, namely, the knee and foot. It would bring on weight unless it had some special function, but it does effect the planting of the foot, as we said. This is why no *cahab* is given to the front feet, for they have a lighter and stronger planting.[69] Feet should not be so disposed that the rear feet are stronger and have a firmer planting. This is because the animals use them more to turn and to whirl about their steps than to plant them. Animals, however, which have a hoof divided in two (that is, which have two *sotulares*), have a *cahab* on their feet.[70] This is because their feet have a lighter movement. Those, however, which have neither *soleae* (that is, *calces*) nor a hoof split in two and have instead many feet have no *cahab*.[71] If they had one they could not have many toes, for the toes dominate the area of the *cahab*. For this reason there are *sotulares* on the feet of many animals which have a *cahab*.

53 Moreover, it is reasonable that the human has large feet for his body size, for by being this way they can carry the weight of his body. This is also why they have length, breadth, and a number of toes. The length of the toes is contrary to the length of the fingers. For the function of the hands lies in grasping and holding, and they are thus longer. If this length were not in the fingers and the joints they would be loosely joined and then the grasping and holding would not be the best. The function of the foot, however, is to plant the foot and to carry the body. The toes are therefore created shorter and strong, for in this way they plant the foot better and carry better. It is

---

65. "Nail": *unguis,* most commonly "claw" or "nail," may be virtually the equivalent here of *ungula,* "hoof." Cf. Ar. *Part. An.* 690a9–10.

66. *Cahab* (the spelling used here is unusual) is variously used throughout the *DA* (see index). Here it corresponds to the Gr. *astragalos,* often translated as the huckle-bone.

67. Note that just below it is clear that this phrase should read "bend their rear feet. . . ."

68. Interestingly, A. uses *angulus* here instead of the *ungula* Sc. offered him and, in so doing, is closer to Ar. *Part. An.* 690a13.

69. This illogical statement is due to a textual problem, for A.'s *fixionis,* "planting," should be the *flexionis,* "bending," of Sc. Thus, the front feet would not have a *cahab* in order to facilitate their movement. A.'s text apparently offered him *levioris fixionis,* for he inserted *fortioris* to try to bring it some semblance of sense.

70. This sentence seems to equate *sotulares* with the sections of a cloven hoof.

71. "Many feet": has A. slipped this comment into the text? Might he not mean "many toes"?

proper that the beginning (that is, the end or the front tip of the foot) be divided into toes. Thus should some mishap of a disease befall one toe, it does not befall another. Thus too the nails were made at the end of the toes for their protection. This separation of the foot produces a stronger, firmer planting of the foot as well.

CHAPTER V

On the Visible Members of the Oviparous Quadrupeds and Their Natural Causes

Having thus made a determination about the members of blooded animals and of certain oviparous animals, we will say again that some animals have four feet, and some lack them entirely, as do the serpents. We have touched on why serpents lack feet in what has gone before, where we spoke of the movement of serpents. We will speak of it again, however, in what follows when we specifically pursue the nature of creeping animals.[72]

Certain other animals have a form like that of an oviparous quadruped. This animal also has a head and the members in it are there for the same reason given for the others. Blooded animals, however, have a tongue in their mouth, with the exception of the *tenchea*. The opinion is that this animal does not have a tongue but rather something in its place. Now, many aquatic animals lack a tongue, as has been stated before. A tongue is visible in fish only by careful investigation. For they are very ravenous and their tongue is neither unattached nor projecting. They have only slight need of a tongue for tasting and none for chewing purposes. Thus, the member which takes the place of a tongue in them senses fluids (that is, tastes) and the taste of the food will occur in the fish during the descent of the food inside. For fish, and those animals like them, sense warm, fat things but sense few other tastes while swallowing.

Viviparous animals have this sense and almost all sense the tastes while swallowing, for they perceive the flavors of the food during the opening of the esophagus. Because fish have a weak sense of this sort, they are very ravenous for food and have huge appetites. So much so that some either cannot abstain from food whose taste they have sensed or can just barely do so. Fish, then, and those like them, sense in the way we have mentioned.

Of the oviparous quadrupedal animals, the lizard has a tongue split in front into two branches like that of the serpent. The ends of these branches are as slender as hairs. The *koky* [seal], however, also has a tongue split into two branches.[73] This is why

54

55

---

72. "Creeping animals": *reptilia,* which can also be translated "reptiles."

73. The statement is from Ar. *Part. An.* 691a8f. and, although it sounds implausible at first, it is basically true in that both eared and noneared seals (the superfamilies *Otaridae* and *Phocidae*) have a tongue which is notched at the end. This is not true for the walrus. Cf. King (1983), 15–17.

all these types of animals—the oviparous quadrupeds—are made with sharp teeth. For these animals' teeth are like those of fish.

But all the instruments of sensation are found in these types of animals as they are in other animals. They have a nose, which is the organ of smell; eyes, which are the organs of sight; and ears, which are the organs of hearing; but they do not have protruding ears.[74] Neither do ears protrude in birds. Thus in these animals there is only an auditory passage. We have said previously that the reason for this in birds is the toughness of their skin and it is the same for these.[75]

For birds have feathers and feathered wings and their skin is toughened by these. These animals, however, have hides [*cortex*] or scales which are naturally hard, as is clear in the tortoise, the large serpents, the *tenchea,* and other river crocodiles. These scales are even harder and stronger than bones and they partake in the nature of bone. For this reason we said above that all these types lack an upper eyelid. Thus too the birds do not close their eyes with an upper eyelid but with a lower one. But the animal we spoke of never closes its eyes since its skin is harder than that of the birds and rubbing this over the eyes would harm the eye's substance. Birds, however, close their eyes with a membrane as we said above since they require sharp vision.[76] Otherwise, they could not catch sight of their food from far off and their lifestyle would then be a bad one. The type of animal mentioned has tough eyes, with dull vision because it has no need of sharp vision. This is because it lives in rocks, holes, and cracks in the earth and in buildings.

Furthermore, the head of this animal is split in two, namely, into an upper part of the head, which is totally of a piece and is a single, solid bone, and into a lower jaw, which is attached to the bone of the head.[77]

A human, however, and every quadrupedal animal, moves its single, lower jaw in a triple movement, namely, from bottom to top, from top to bottom, and sideways, from right to left or vice versa. This is the action and function of the jaw, for the movement from bottom to top and vice versa is for the purpose of separating the food. The sideways movement is to crush and to grind the same food. Fish and birds, however, move their lower jaw only up and down. For the sideways movement, as we said, suits chewing and grinding, that is, softening, of food during the pressing out of the food's own moisture and the mixing in of the salivary moisture. This, however, is not suited to fish, for the reason we have given often above.[78] Nature has thus deprived them of this movement and of the molars that are attendant upon it since, for the most part, nature does nothing superfluously. Generally, every animal known to us moves its lower jaw except for the *thenchea,* which moves the upper jaw. This is because its feet

---

74. A. uses *auricula* in this sentence in the sense of "ear" although it more generally refers only to the external portion of the ear, the earlobe.

75. Cf. 12.176.

76. Cf. 1.213; 2.72.

77. Hence the mistaken belief throughout antiquity that crocodiles could move only their upper jaw.

78. Probably that fish must close their mouths quickly and swallow in order to avoid being drowned.

are very small in proportion to its body and are not suited for grasping and holding food. To compensate for this lack, nature adapted its upper jaw for movement. Thus with one and the same instrument it can seize, hold, and cut up its food.

It is clear that a blow which comes from above is far stronger than one which comes from below and thus, so that one and the same instrument might be suited for grasping, holding, and cutting (since it has neither hands nor feet with which to seize or hold), nature made the crocodile's mouth strong enough for three functions. These could not be accomplished by a movement of the lower jaw and she thus saw to it that the upper jaw moved for them. This is also why the crab moves that part of its claw (which it has instead of a hand) which closes the claws from the inside (working from the inside of the claw) so that it can grasp and hold.[79] The claws on crabs are organs suited for grasping and not for biting. The first and principal function of biting is the cutting up and division of food. Thus, in crabs and in all those animals like them, called soft-shelled in general, the faculty of seizing things is found to be accomplished by using the two pointed parts [*culmi*] of their claws.[80] A similar faculty exists in all animals whose dwelling is not generally in the water, however. All these have the capacity of grasping by using hands or feet and they have that of dividing their food in their mouths. However, it was necessary that the *tenchea*'s mouth be suited for three functions at once and it was therefore suitable for it to move its upper jaw.[81]

Further, all these types of animals have a neck because of their need for lungs and because they take in breath through the *canna,* which is the *trachea arteria* ("rough artery," trachea). Now I am calling a neck that member which lies between the head and the shoulders [*spatulae*]. Many hold the opinion that the serpent does not have a neck, having instead another particular member in place of a neck which is marked off at the edges of the head and the shoulders, things which have already been noted by means of what has been said before.[82] The serpents, however, throughout their genuses, seem to have as a particular trait that in moving their head they can turn it to the side and to the rear of their bodies while the rest of their body is at rest. The reason for this is the same as for a ringed body. This is, namely, that the ring-bands of the vertebrae (like the rings of a ringed body) are made of cartilage so that they can bend well.[83] This cause is necessary for this accidental trait. The accidental trait itself is useful for the well-being of the serpents, namely, because they can thus look back to the rear over the entire length of their body. The body is exposed to many mishaps as

---

79. Sc.'s text says, more clearly, that the crab "closes its appendages and does not move the lower part (of the claw)."

80. For *culmus* in this context, cf. 8.29.

81. Cf. 1.227, where A. relates his observation of crocodiles' mouths. Here he clearly indicates that the issue is not so much that they cannot move their lower jaws but rather that the upper jaw, so often part of the skull in other animals, is hinged in this animal.

82. Heavy interjections by A. make this sentence something of a muddle. Cf. 13.1.

83. The "ring-bands" are *armillae* as at 3.19f., 71f., here translated a bit differently to differentiate them from the "ringed body," *anulosus*. The former is an analogy with an armband, the latter with a finger ring.

it is drawn over the ground since no species of serpent is found to have either feet or hands, or any other organ for seizing food and holding it and for protecting the rear of its body from harm. Neither would they be helped by raising their head unless they could look behind, for their entire body is behind the one doing the looking.

60 Moreover, in animals of the sort we are discussing here, there are members which are analogous to the chest, yet they do not have breasts either on their chest or in any other remaining part of their body. For example, these members are not found in birds or in any species of fish, for no species of oviparous animals has milk. This is because they produce eggs and the soft fluid which deals with nourishing the young lies hidden away in their eggs. A viviparous animal, however, has milk and therefore lacks eggs. We will speak of the reason for these things in what follows, when we treat the generation of animals.

Further, all the species and types of the animals mentioned before have a tail, but they differ in that they have larger or smaller tails. We have established the general reason for a tail in what has gone before.

61 Moreover, there is a certain special animal called the *hameleon*, which in Latin means "earth lion."[84] According to Avicenna, it has the shape of a lizard, always lies on the ground, is cold and very timid, and is slower, smaller, and thinner than all other oviparous quadrupeds because it has very little blood. It is also not very fleshy and when it is afraid it draws in its spirit so far inside that its color changes into many colors and it is rendered motionless. Its fear is increased at the slightest reason because of its small amount of heat and blood.

Let this, then, be the determination made about the disposition of blooded animals, oviparous quadrupeds, and those that have no feet, and of their members, both inner and outer. For what has been said about the reasons for these things is enough for the present investigation.

CHAPTER VI

On the Visible Members of Birds and on Their Natural Causes

62 Birds show a great diversity in their members. What is immediately apparent is that some have very long legs, some short ones, and some have legs that are midway between the two. Likewise, some have a broad claw on their toes while some have a narrow one.[85] There are differences like these in the other members as well. Now differentiation is found not only between dissimilar birds but also in those that seem to be alike. For differences are found in their members, although their degree of differentiation may be less.

---

84. Avic. *DA* 14.46r. The etymology of the Gr. name, nor the Latin, is a detail added by A.
85. "Claw": *ungula*, which replaces the tongue in the description of Ar. *Part. An.* 692b6, and which is properly translated in Sc. The error seems to be an orthographical one, writing *ungulam* for *linguam*.

Birds differ from other animals according to the figures and shapes of their members. That which is common to all birds and which is peculiar to them is the possession of plumes and feathers. As we have often said, just as certain animals have scales and others have hides, birds have wings and plumes.

Furthermore, the wings of some birds have feathers at a distance from one another so that the wings do not give the appearance of being solid.[86] This can be seen at the ends of the wings of the large eagle. Some, however, have wings made of feathers rather like a wall of shafts [*cannalia*] so that one shaft [*canna*] is directly near another and a third touches both of these. Some, though, do not have wings of this sort.

Further, birds have a member on their head which is found only in a few of the other animals. This is the beak, which birds and a few fish possess. Just as the elephant has a nose in place of a hand and certain ringed animals have a tongue in place of a hand, so the birds have a beak, created from bone, in place of jaws, teeth, and hands. They also have a long neck for the reason we gave and which is shared with the other animals as well. For there is one common reason that a neck is given animals.

Some birds have a long neck and some a short one. But for the most part the size of the neck follows that of the legs, so that a bird with long legs has a long neck. If it has short legs, it will have a short neck, except in aquatic birds which have membranes between their toes, for some of these have a short neck and long legs, but they also have a very long beak. Examples include birds which walk about in the water in search of food. If this were the case for birds living on land, it would not be possible, but in birds that have skin between their toes, it is possible, for they dwell in the water and thus do not have a long neck. You should know, however, that it happens very rarely that a bird with long legs has skin between its toes and a short neck. But some birds with skin between their toes do have a very short neck and short legs, like the swan and the very large, white bird which has a red sack in front of its neck and about which we spoke previously.[87] But when it happens that birds with long legs have a short neck, their shortness of neck must be compensated with length of beak. Few, however, of the birds known to us are so dispositioned.

Moreover, birds with hooked talons, called birds of prey, cannot have a long neck since they do not have long legs. Certain aquatic birds, with skin between their toes, do have a long neck like the swan, for this suits its activities. For such a disposition of its neck is more suited for seizing food from the bottom of the water. The legs of these birds are short, suited for swimming.

Moreover, the beaks of these birds are divided according to their types and their lifestyles. Some birds have a straight beak and some a curved one. A straight beak will be on birds that need it only to grasp food. A curved beak, however, will be on carnivorous birds, used to tear away raw flesh, which is tough and sticky. Its beak has to be made this way since it acquires its food by hunting animals.

---

86. From Ar. *Part. An.* 692b12f. it is clear that the distinction is actually being made between the wings of birds and those of insects.

87. A. recalls his discussion at 7.39 of the *volmarus,* a pelican.

66 Birds that are not birds of prey, however, and which are peaceful, living on plants and seeds in the mud, need to have only a wide beak since this is suited for digging out and cutting plants. For this reason too the end of a beak like this is denticulated, as if little teeth have been cut into it, as when little teeth are cut into a sickle. This can be seen in the goose's beak. Certain birds of this genus have a long beak because of their long neck, and because they seek their food on the bottom of the water. Some of those which have simple skin between their toes sometimes also have a member by means of which they seize things in the water. This bird's neck is like a fishing pole and is thus long. Its beak is like the hook, taking up the food. The bird called the *mewa* [seagull] in German is like this.[88]

67 Further, in birds there is the front of the body, that which is called its rear, and the member called the chest. This chest in all quadrupeds fills in solidly between the front legs, hanging over the *asseyd* of the front legs [tibia, fibula]. Birds, however, have a member between their neck and chest particular to them, namely, the wings. These lie between the shoulders, near the upper limit of the back.

Moreover, birds have two feet as does the human. This is necessary because they are numbered among the blooded animals and also have wings. Now, an oviparous quadruped only has four organs for movement, for it has members like those of the walking animals. The birds, however, commonly have wings in place of front feet and the *asseyd*. In these lie almost all their strength for the movement of flight. Thus, they need to have only two feet. For in this way they, like nearly all walking, blooded animals, have four organs of progressive movement, namely, two wings and two feet.

68 Again, all birds have a chest that is rounded, is sort of fleshy, and comes to a point. For the more pointed it is, the better the bird's flying will be, since, if the chest were wide, it would displace a great deal of air and its movement through the air would be sluggish. But it is fleshy since that which is pointed would be weak unless it were covered with a lot of flesh.

Moreover, beneath the chest of birds there is a belly reaching to the point of exit of the superfluity, just as there is in the quadrupeds and in the human. These are the members of the birds which are between the wings and legs.

Moreover, every viviparous animal, as well as one that first lays an egg and then gives birth to a live young, has an offspring that possesses an *umbilicus* at the time of its birth. When it has grown, however, the *umbilicus* is not noticeable, so much so that after a while it is not visible at all. For it is connected to the intestine by means of a section of a certain vein we spoke of in the anatomy section. There has to be something like this in the birds, not outside to be sure, but inside, and through which they draw the nourishment from the fluid of the yolk.

69 Some birds are good flyers and have good, large, strong wings. Others, however, are not good flyers. The good flyers are especially the birds of prey with hooked talons, and the falcons most of all. Those birds are also good flyers to which nature has given

---

88. Cf. 7.38; 8.16, 71; and Sanders, 435.

good flight for safety's sake, since they go back and forth from area to area. There are bad flyers which do not fly well because of the weight of their body. Those which live on the ground and whose food is there, for they eat seeds, are also bad flyers. Others also are bad flyers which live on the water, for they are heavy and cold.

Further, the bodies of the birds of prey are small and slender except that they have large wings since their food passes over into wings. Their strength lies, for the most part, in their wings. The bodies, however, of birds which are poor flyers have a contrary disposition. They are large and heavy and their wings are small in proportion to their bodies. Some of these, though, have another defensive item to make up for wings, since they sometimes have long claws on their legs as the rooster does.

No bird whatever has claws on its legs and feet at the same time since nature does nothing superfluously. Birds have no use for the claws on their legs if they have curved claws on their toes. For the claws on the legs are more suited for those birds which live a fixed existence and go about on the ground.[89] Nature gave them these claws for weapons and these birds have a weighty body and do not have curved claws, for they would not be of use in walking and would instead be a hindrance, since they would become tangled in bushes and in other things they walk over. This is the reason a bird with curved claws walks so rarely and does not sit on the small branches of trees. For the natural disposition of its claws would be a hindrance to these two activities, namely, walking and sitting of this sort.

Heaviness is present in birds with a heavy body because of their earthiness. Since this earthiness gravitates toward the legs, the long claws on their legs are produced from it. This is also the reason for the size and strength of the claws. If this earthy part and superfluity were not present in the birds they would have a very weak nature.

The membrane which is between the toes of certain swimming birds is generated from an earthy part of this sort, mixed with water, with the result that they have wide, connected toes. Birds have these members for the reasons given previously. The plumage of these birds, covering their members and reaching down to the knee-joint on the feet, is suited for swimming because the bird is borne up on the lightness and airiness, floating on its feathers. The skin of the feet acts as oars, as do the wings (that is, fins) fish have near their heads. Thus too there are no feathers between the feet of birds, since the feet would be borne upward by the lightness of the feathers.

Moreover, some birds have long legs because of the weakness of their life. For all the instruments nature gives, she gives according to the suitability they bear to the function. She does not prepare the operation according to the instrument but rather acts to the contrary, giving an organ that fits the operation. Because birds of this sort walk about in moist places and do not swim, nature rarely has placed a membrane between the toes of the birds of this sort. Sometimes they are bad flyers because they are heavy and dwell on damp ground near their food. Thus the earthy food which

---

89. Cf. Ar. *Part. An.* 692a12f., where separate words are used for "talon" and "spur." Here, A. uses *unguis* throughout, giving an odd quality to his sentences. Elsewhere, "talon" will be used to translate *unguis* as appropriate to the species of bird.

should have passed over into their tails and chests passes over into long legs and they thus acquire long legs. Thus too, while flying, they use their feet in place of a tail for a rudder during flight, extending them to the rear. In this way the creation of long legs suits them.

73 Some birds have small feet and short legs. When these fly, they gather their legs up to their belly lest they keep them from flying. Again, the feet possessed by the birds of prey have curved talons and are aptly suited for grasping and holding prey. All birds with long feet and which have a thick, heavy neck fly with their neck stretched out. But if they have a long, slender neck, they bend it and tuck it in while flying as does the crane.

All birds have *anchae*, even though many are of the opinion that they do not have them.[90] The *anchae* extend as far as the middle of the belly. For whatever animals are bipedal must necessarily have two *anchae* and thus birds do not lack *anchae* the way a quadrupedal animal does. The two *anchae* in birds are toward the area of the anus and the thighs and are attached to them above so that they can lift their entire body at the same time. This is the reason that the human, a biped, can lift up his body, while a quadrupedal animal faces the earth because of the weight of its head. Birds, however, cannot raise themselves up at any given time. But this is because of their need to seek food.[91]

74 Moreover, birds have only two feet since they have taken on wings from nature in place of other feet.[92] For this reason nature made the *anchae* in birds very long, almost in the middle of their body. This serves to support their body so that their weight might be equal to the rear and to the front, and so that they can fly and walk, and also so that the legs themselves might provide good footing in carrying the entire body, which is almost divided in two at the *anchae*. This, then, is the reason given that birds have only two legs.

Moreover, there is no flesh whatever on birds' legs and the reason for this is the same as why quadrupeds do not have flesh on their feet. This has been established above.[93]

75 Further, there are four toes on the foot of every bird except the ostrich. But we will speak of the ostrich, found in the province called Nubia, below. Some have the opinion that this animal is in no way a member of the bird genus because of the difference in its construction, just as we said once in the earlier parts of this study.[94] For it has

---

90. Cf. 14.47. The term here seems to mean a formal, upright hip structure, perhaps even the elevated iliac crests.
91. A rather confused version of Ar. *Part. An.* 695a7f.
92. "Other feet": *aliorum pedum,* another misreading of Sc., who has, correctly, *anteriorum pedum.*
93. Cf. 14.49f.
94. Perhaps at 12.184. Cf. 23.139 (102).

only three toes on each of its feet because of its wings and because of its great weight.⁹⁵ It has more flesh on its legs [*tibia*] than is normal for the nature of birds as well. That which it has most in common with the birds are the accidental traits that are found in long-legged birds. It is especially like the birds called the *gothoz* in Greek.⁹⁶ This is a genus of bird which also does not have many toes.

This, then, is the disposition of the toes of birds. Now the animal which is called the *gotez* in Greek has only two feet in front.⁹⁷ This is because its body does not bend to the front as much as it does to the rear.

Further, all birds have two testicles and these are within their body. For this reason their intercourse is swift. We will determine the reason for this below when we speak of the generation of animals.⁹⁸

Let these, then, be the things said about the visible member of birds.

CHAPTER VII

On the Visible Members of Fish and on Their Natural Causes

The fish genus is quite imperfect in its outer members. For it does not have wings, feet, or hands, and its body runs continuously from head to tail without any break. However, their heads are not at all alike. Some of them seem very close to being alike while others are entirely different.⁹⁹ Some wide fish have a long, spiny tail. From this the breadth of their body constantly increases. The body of the fish called the *acro* in Greek is like this.¹⁰⁰

There are some fish which have much flesh and are short for the same reason their tail is very short and fleshy.¹⁰¹ In the frog, however, the opposite of what we said

---

95. The ostrich actually has two toes, as correctly noted at 14.88 (cf., however, 2.71). The error apparently arises from Ar. *Part. An.* 695a16f. Ar. dismisses the ostrich until later in the book and then says the four-toed birds generally have three toes in front and one pointing backward. The sense is apparently that the ostrich's great weight and weak wings put more pressure on the toes, forcing them to be fewer but stouter.

96. Ar. *Part. An.* 695a21 makes it clear that this is the *krex*, itself unknown (*GB*, 177). It is said to have the requisite number of toes, but some of them are stunted.

97. Ar. *Part. An.* 695a23, the *iunx* (Sc. *ocoz*), the wryneck, much celebrated in Gr. mythological/magical tradition (*GB*, 124–28; Pollard, 1977, 130–31). It has of course two *toes* to the front, not two *feet*.

98. Cf. 15.19f.

99. The first few sentences of this section are quite distorted from Ar. *Part. An.* 695b2f., for the translator has incorrectly assumed that an adjective in Ar. refers to "head" whereas it refers to the tails of the fish.

100. A very interesting corrupted form of the Gr. *narkē*, the torpedo or electric ray. The Gr. of Ar. *Part. An.* 695b9 reads *esti narkais*. The first word can also take the form *estin* and so the translator read *estin arkais*. The fish's name was then rendered (incorrectly) into the nominative and its consonants were transposed, leaving the form *acro*.

101. Lit., "tail was very short . . ."

occurs.¹⁰² For the breadth of its front part has more flesh than the rear, for nature added to its front whatever she took from its rear and tail.

Fish do not have perfect members. Their nature is for swimming and nature does nothing superfluously.¹⁰³

Moreover, the substance of a fish is phlegmatic with watery blood, if it has blood. It swims and thus has wings, but it does not walk because it does not have feet.¹⁰⁴ For the possession of feet is suited only to a particular movement called walking. Likewise, it does fish no good to have four wings and two feet, or two wings and two feet, or to have any other member whatever involved in walking. For if it had one, it would, for this reason, only have cold and minimal blood, and then it would move slowly and with difficulty.¹⁰⁵

There is, however, a certain fish which has two membranous wings and which flies for a short distance and then falls into the sea. The Italians call this the sea sparrow [*irundo maris*].¹⁰⁶ Its wings are membranous and it flies for a short distance. There is also a fish in the sea of Flanders and Germany which they call, in their language, the *aslec*.¹⁰⁷ I have studied it diligently and have found that it has two feet and four wings which move bent to the inside of its belly and chest like those of a bird.¹⁰⁸ There are two wings in front near its gills and two to the rear in front of its tail. The wings are membranous and are thick very close to the body in the part [of the wing] which takes the place of the helper of the arm [*adiutorium,* humerus], but they are thinner to the front part of the wing which takes the place of the *asseit* [radius and ulna].¹⁰⁹ The rear wings also have this disposition. The gills of this fish do not have a split reaching to the belly or chest. Rather, it has gills bounded over the shoulders of the four front wings by squared openings, in such a way that two are nearer the head and two nearer

---

102. Ar. *Part. An.* 695b14 makes it clear that this is the fishing frog (*batrachos, GF,* 28–29) which A. usually calls *rana marina* (e.g., 1.31, 92; 2.78) but occasionally shortens it, as here, to *rana*. There is no need, therefore, to think this is an error and that A. is mistakenly discussing the amphibian. Cf. 24.50(101).

103. Cf. 12.190.

104. As elsewhere (cf. 1.89), "wing" or *ala* indicates a fin. But in this context and given that A. will soon (14.79) differentiate *penna* and *ala,* it is best to continue to translate it as "wing" despite the incongruity to modern ears.

105. This sentence is heavily interpolated by A. In its original form it merely said that if they had feet and wings they would have no blood. That is to say, they would be insects.

106. This entire section appears to be A.'s addition. The fish is probably a flying gurnard. Cf. 4.92; 14.87; 24.36(62).

107. Cf. 24.7(1), where the form is *aslet* and A. professes personal knowledge of the fish. Stadler (1907, 249) had identified this fish as the *Chimaera monstrosa* but this was rejected in favor of *Rhina squatina* by Balss (1928, 22). Sanders, 440–41, discusses the difficulties surrounding a more precise identification.

108. Compare the following description with that of 24.7(1). *GF,* 221, has a drawing of a monkfish (*Rhina squatina*) from above which does show four "wings" and, on 26, a ventral view of the *batis* (skate) showing the sexual "claspers" which exist in fish of this sort and which may be A.'s "feet."

109. *Asseit* is more commonly found as *asseyd*.

the right wing.¹¹⁰ The gills on the left side are arranged in a similar fashion. The head of this fish, however, the color of its skin, the shape of its body, and the taste of its flesh are quite close to those of the fish called the *raia* [ray].¹¹¹ Its legs, however, are cartilages lacking joints. Beneath, at the feet, it has grooves [*foveae*] so that it can be planted better.¹¹² Its tail, however, is not that of a *raia* but resembles that of other fish except that it is somewhat longer. In front of the tail, in the area of the back where the solid part of the body that comes after the hollow of the belly narrows out, it has a fin [*pinna*] which is like that on other fish, but it is very large for its body.

Aristotle also says that the types of fish called the *feidolez* have gills and feet but do not have wings, having instead a broad, thick tail.¹¹³ Moreover, the wide-bodied fish, like those called *botoz* in Greek, have four wings, two on the belly and two on the back.¹¹⁴ Here, these wings are called fins and this exact arrangement is also found in the fish which we call the sea hares [*lepos maris*], but which the French call the *gornays*. For these have two fins on their belly and two directly above them on the side toward the back.¹¹⁵ Some have a long fin or wing along the lower length of their back and it is above the solid part behind the hollow of the body. Examples are the *encheliz*, which is the eel, the *murena* [lamprey], and the one we call the "nine eyes."¹¹⁶ The fish called the *henchetoz* in Greek has the same thing and a particular genus of *fastoroz* is said to be like them in this as well.¹¹⁷ These are found in the sea of eastern Greece. This type has a long body and is composed like the serpents called the *ascaiorymy* in Greek.¹¹⁸ This type does not have wings and only swims by bending its body like a serpent. For it moves in the water the way serpents do when they creep on the surface of the earth. Moreover, serpents swim in the water in the same way they creep on the ground.

---

110. The sense may be rather that the openings are arranged in a square pattern.

111. Cf. Sanders, 440.

112. *Fovea* is an odd word in this context, implying some sort of a recess. At 24.7(1) the soles of the "feet" are said to have confused *divisiones*.

113. A. here rejoins Ar. *Part. An.* 695b25f. The *feidolez* is the *kordylos*, probably a newt (*GF,* 127).

114. Ar. *Part. An.* 695b29 cites *batos* and *trygōn* (skate and stingray). Only the former made the transition to A.'s text.

115. Cf. 24.39(72), where A. discusses several interpretations of the "sea hare," saying that one common name for it is the *gernellus*, or gurnard. The modern Fr. name is *grondin*, bur Tobler and Lommatzsch (1925, 4:454) cite the older form *gornal*. *GF,* 286, has an excellent drawing of a flying gurnard showing two upper alar fins and two ventral fins. Here, at least, it is clear that A. prefers to think of the upper fins as the "wings."

116. This river eel is also mentioned at 24.41(74) as a type of lamprey. Its nickname exists today in the German *Neunauge*, used for the river lamprey. Scanlan (360) discusses the origin of the name; cf. Sanders, 413, and Diefenbach (1857, 372) for earlier forms of the German name.

117. Ar. *Part. An.* 696a5f. lists the *enchelus* (eel), the *gongros* (conger eel), and the *kestreus*, which normally comes to A. in some form resembling *fastoros*.

118. Ar. *Part. An.* 696a7 states rather that these creatures are snake-like and gives as an example the *smyraina*, or moray eel.

80   There is but a single reason that this type of fish does not have wings and why serpents do not have feet, and we have already determined this reason above when we spoke of walking and the movement of animals.[119] For if it had feet, its movement would have to be poor movement, since it should have more than four feet. Since it has a narrow body, its wings, if it had them, would have to be very near one another and it would thus have very sluggish motion. If, however, it were granted that its wings were set at a distance from one another, it would still have poor movement since then there would be a long expanse of body between one wing and the next and this would weigh it down. If, however, they had many spots upon which organs of movement could be placed, then they would not have blood.[120] This is the reason that these fish and those like them have neither wing nor foot. For their construction, which is one of bodily flexibility, takes the place in them of two wings and for this reason they creep on the ground and live a long time.

81   Further, it is sometimes found that some fish have only two wings on their back, as in those fish which are not kept from swimming by the width of their body. Some also have wings directly near their head since they do not have a long body. The type, however, called the *batoz* in Greek, and those like it, have width instead of rear wings and using this width they swim by contracting and straightening themselves out. This is done too by the type called the *gyddoz* and the sea frog [*rana marina*].[121] For all these swim with the lower width of their bodies.[122] The wings which are near their head do not keep them from movement because of the fish's width. And in a case where the wings are so arranged, these wings will be smaller than the rear wings.

    The type called the *gyddoz* has two wings near the tail and uses its width in place of other wings.

82   Further, fish have a member peculiar to themselves which is not present in the other animals, especially the blooded ones. This member is in the nature of gills. We have already given the reason for and the usefulness of this member.[123] Of the fish with gills, some have a covering entirely over the gills since their gills are spiny and the covering is likewise spiny. All the types of the *celeti*, however, have cartilaginous gills. The spiny composition of the gills, which is as if they were made of red soft spines that are fleshy, is more suited for movement. For the movement of the covering of the gills, which is the bone over the gills, should be swift so that they can use the gills suitably.[124]

---

    119. The cross reference is from Ar. *Part. An.* 696a12, and it refers either to *De incessu* 709b7 or *Part. An.* 690b16f.

    120. "Spots": *signa*, a literal translation of the Gr. *sēmeion* used in the sense of a mathematical point.

    121. *Batoz,* as above, is the Gr. *batos,* a skate or ray. *Gyddoz* is less clearly seen in the Gr. *narkē* (electric ray), but the identification is secure by its position. Cf. Ar. *Part. An.* 696a26f.

    122. The original Gr. states correctly that they swim by moving the outer fringe of the bottom of their bodies.

    123. Cf. 2.76f.

    124. All badly confused from Ar. *Part. An.* 696b13f.

Moreover, certain fish have few gills and others many. They are sometimes present in certain very small fish. Sometimes the gills are very delicate, but they protrude. This is known from dissection [*anatomia*] of the fish and is to some degree known from the discussions we entered upon in previous books of this investigation, treating the anatomy of fish.[125] The reason for a great or small number of gills is the heat of the heart. For greater heat of the heart requires greater and swifter movement of the heart. Those that have these kinds of gills have this kind of nature. For they have greater strength in the heat of their heart than do those with smaller and protruding gills. For this reason, too, those with many gills live for a long time outside the water on the ground. An example is the *encheliz* (that is, the eel) and all the types of fish whose makeup is serpent-like. All these types do not need much cooling.

Further, the types of fish differ widely in their bodily makeup. Some have a mouth to the front, directly to the front of the body, where the body is exposed to external things and others have it on the bottom of their head, as do the sturgeon [*sturio*] and especially the fish called the *caliga*.[126] This one has its mouth below, almost in the middle of its body. Others, however, have it almost on the back, like the dolphin and all the types of animal called the *celeti* which take food only when turned over on their back. It is clear that nature made this sort of arrangement in them not only for their well-being but also to slow down their taking in of food. For this type of fish has a round, thin appendage on account of which it cannot cut off or cut into its food well. If, then, it had an easy time of eating, it would quickly die from being too full.

Further, some types, having openings in their upper area, have mouths [*orificia*] that open wide while those of others open a small amount. Those which are carnivorous have openings that open well. Those, however, which are not, have a small opening that opens but little.

Further, some of these animals have a skin that is filled with horny scales [*cortex*] (like a sort of scale) and others, differently, have a rough skin like the one called the *cohabytoz*.[127] There are very few of these animals which have a smooth skin. The *celeti*, however, have horny scales and their skin is rough, since it is made up of cartilage and spine and since its earthy part naturally passes over into skin. None of these types has testicles either inside or outside. Likewise, no footless animal is found to have testicles. But serpents have containers and ducts in which the superfluity of their sperm flows and from which it leaves, just as do other animals which are viviparous and are quadrupeds. The serpents, though, do not have a passage for voiding urine, since they do not have a bladder.

---

125. As often, the reference to *anatomia* is from Ar. and might mean merely the study of anatomy.

126. The *caliga* poses many questions in that, in one sense, it means a leather shoe sole and thus evokes a flatfish like the sole. Stadler's apparatus at this point also notes that a later hand (M3) changed the original *caligo* ("darkness") to *caliga* in a marginal gloss. While Stadler (1912) identified this hand as belonging to the 15th century, Sturlese (1985, 258) identifies it with Berthold of Moosburg.

127. Cf. 13.61, with notes. Here too *cortex* apparently contrasts the superficial, easily dislodged scales of fish [*squama*] with the thicker, more deeply set scales of reptiles. The *cohabytoz* seems to be the *rhinē kai batos*, the "monkfish and skate" of Ar. *Part. An.* 697a6.

This, then, is the difference between the genus of fish and the genuses of the other animals.

The dolphin, however, the genuses of whales [*balaena*], and those like them which are large-bodied animals, do not have gills. They have, instead, since they have lungs, a windpipe [*canna*], which is called the trachea [*trachea arteria*].[128] They do not hold their food in their mouth for a long time since if they held it a long time the water would enter (for they take their food in the water) and it would not be spewed forth through the gills.

86 It is clear from the things that have been said that gills suit the types of aquatic animals which lack a windpipe, lungs, and breathing. We have already determined the reason this is so, when we spoke about the members of fish.[129] Not one of the types of fish can have gills and breathe at the same time. But only a breathing type has a certain passage for the casting off of water. This is reed-like [*canalis*] and this passage lies at the front part of its brain.[130] The reason why a fish of this sort has lungs and breathes is that it numbers among the large animals that require a great deal of heat since otherwise it could not move its body. It thus has lungs filled with blood and natural heat. Even though it is aquatic, this animal is like a terrestrial animal in one way, namely, because it breathes air like a land animal. But, like an aquatic animal, it lacks feet and takes in food that is actually moist.

87 The type which in Greek is called the *kokaz* and the *rapaz* (that is, the *hyrundo* [swallow]) is also, so some say, common to both land and aquatic animals. The *koky*, which is the sea calf [*vitulus marinus,* seal] is surely like both land and sea animals.[131] This is why it has a spine. Now the *koky* has feet to be sure, like both aquatic and land animals, and it also has fins [*ala*]. Its front feet are very close to the feet of the fish.[132] Moreover, all the teeth of this animal are sharp like those of the fish and they stick out outside. The *hyrundines,* however, have bird-like feet, but they lack a tail and in this they resemble land creatures.[133] Thus it is that what we said earlier happens, namely, that the wings of the *hyrundines* are membranous and that they do not have a tail. If they had a tail it would interfere with the movement of their wings since their wings are not divided as are feathered wings.

---

128. It is possible here that *canna* is being used not in its technical sense but merely to mean "reed." Cf. the adjective *canalis* immediately below. Moreover, while *canna* is a good translation of the Gr. *aulos*, its sense at Ar. *Part. An.* 697a18, as common with these animals, means "blowhole."

129. Cf. 2.78f.

130. "Reed-like" may be rendered "like a windpipe." The translation offered also entails changing Stadler's ungrammatical *qui* to *quae*, with Borgnet. Note that if *canna* is understood as the Gr. *aulos*, "blowhole," the placement of it at the brain makes sense.

131. Ar. *Part. An.* 697b1f. is speaking of seals and bats. *Koky*, as always, is the seal. The bats [*nyktirides*] are not easily detected in the corrupted names. *Yrundo*, at 14.78, was a flying fish.

132. A. has become confused. He is trying to say that the front flippers of a seal, while technically feet, more closely resemble fish fins.

133. Here it is clear that we have the bats of Ar. and that *hyrundo* (swallow) has evoked an affinity with the birds. It is also clear from the next sentence, referring as it does back to 14.78, that A. thought he was speaking of the flying fish and thus we once more have fish with feet.

Further, the ostrich is also an animal that has similarities with various genuses of animals. It resembles the birds in the composition of its body and in another way it also resembles the quadrupeds. For just like a quadrupedal animal it neither rises up nor flies in the air. This is because its wings are not suited for flight since its feathers are delicate, as if they were strands of coarse wool or of some other hair.

Moreover, just as a quadrupedal animal has upper eyelids, so part of its head is divided, as is the area near its neck, and for this reason the eyelashes are slender and are not like plumes or feathers, but like hairs.[134] But because it is also like a bird, it has true plumage lower down on its body.

Moreover, it has two feet like a bird and has two toes to the front like a quadrupedal animal that has a hoof split in two. This is because the size of its body does not resemble that of the birds, being more like the size of quadrupeds. The size of birds' bodies is necessarily and generally smaller throughout their genus since they cannot at the same time have both a large body and swift movement through the air.

We have used this method, then, to determine the disposition of the members, and the reason each member is present in the bodies of the animals throughout each genus and each species of animals. Next we will investigate the generation of animals, assigning their natural causes.

---

134. Ar. *Part. An.* 697a14f. says that it has upper eyelids because it is bald on its head and on the adjoining neck and as a result the material for hairiness went over to its eyelids. How "bald" (*psilos*) became *divisa* is almost as difficult to understand as the sense of the resulting sentence. The general sense would seem to be that the body is clearly marked off from the head and neck area, a true enough statement.

HERE BEGINS THE FIFTEENTH BOOK ON ANIMALS

Which Is on the Causes of the Generation of Animals

The First Tract of Which Is on the Distinction between the Sexes in Animals

CHAPTER I

That the Causes Involved in the Generation of Animals Must Be Given and That a Distinction among the Sexes into Male and Female Is Not Found in All Animals

1. Now that we have determined the digestion and the type of the members of animals in what has gone before, both universally, by each genus of animal, and specifically, by what belongs to each, we have determined for each and have assigned to it the reason that it exists.[1] This is the cause which is called "that for the sake of which" [*propter quid*] and the end [*finis*]. For there are four natural causes of things. Namely, the reason for the sake of which [*causa propter quid*], which is the end [*finis*], and the definitive reason [*ratio diffinitiva*], which is what its being was [*quae quid erat esse*]. These two causes coincide in one thing in natural objects, although the mode of causality is not the same.

   The third and the fourth causes are, on the one hand, the material and, on the other, that from which the principle of movement comes.[2] We have already spoken of these causes above. The *ratio*, however—both that *propter quid* and the *finis*—are forms of the same thing, namely, in animals and other natural things. The members, however, are the material of all whole entities [*tota*] (that is, of composite things). All whole entities (that is, composites) are like each other in this regard.

   We mentioned the genital members above, but we have not treated them thoroughly.[3] Here it remains for us to give their motive cause and to say what the principle of generation is in each animal.

2. Let us therefore begin to speak here about the generation of animals and we will then state what the motive (that is, the efficient) cause is in generation and then, in following books, we will speak of the first principle of generation. There we will show that all these things take their principle of life and generation from the revolutions of the orb and stars.

---

1. This book begins the commentary on Ar.'s *GA* 715a1f.
2. That is, the motive or efficient cause.
3. Cf. 1.341, 467f.; 2.39f.; 5.9–11; 14.46f.

We will begin, then, by saying that all animals in which there is a male and a female are generated by the copulation of the male and the female.[4] The male, however, is that which generates from his own seed in another of his species, producing an animal like him in appearance.[5] Now this might be produced from a uterus or from an egg. The female, however, is that which generates within herself using semen taken in from another of her species. One which generates totally on its own has neither male nor female, and this is how the plants do it. But we have spoken of this in *On Plants* [*De vegetabilibus*].[6]

But male and female do not exist in each animal in the ways just mentioned. For in those animals in which these sexes are present, they are perfectly differentiated in some and imperfectly in others. These sexes are perfect only in blooded animals. In those that are bloodless, male and female are found sometimes while in others they are found imperfectly. Every blooded animal generates something like it in genus.[7] An animal, however, which is produced without copulation from its own putrefaction is a ringed animal and this most often does not generate one like itself. Generally, however, every animal which moves by walking from place to place, by flying on a feathered wing, or by leaping on feet, and which is blooded, generates one like itself. Thus, walkers and the bird genus, be they those that generate things like themselves or even oviparous, have a male and a female. This differentiation between the sexes is present not only in the animals mentioned but also in certain bloodless ones. This might be in an entire given genus, such as the differentiation of the sexes found in the *malakye* genus and that of the soft-shelled creatures. And those which generate young like themselves likewise have sexes.

Animals which have no perfect sexual distinction and do not have copulation are generated from the putrefaction of their own bodies. Now to be sure these also generate like the others, but they first generate a thing of another sort [*genus*]. In these there is no male or female as far as a perfect distinction between male and female. Some of the ringed creatures generate in this way, whereas others have a distinction of the sexes and copulate. This is a reasonable occurrence. For if they generated ones like themselves throughout their genuses, then their generation would necessarily be continually successive and the propagation would be of one from the other. This, however, occurs only in perfect animals generating ones like themselves. The animals just mentioned, however, are not of this sort for they are generated from their own putrefaction or from other putrefying superfluities. In such animals, then, what is generated is not

---

4. Avic. *DA* 15.47v.

5. Or perhaps "in species." Both senses are undoubtedly present in the Latin *in specie*.

6. *De veg.* 1.12–13.

7. Throughout the *DA*, phrases that indicate that an animal produces a young like itself are generally translated as "viviparous." See Glossary, s.v. "Viviparous." In this section, such a translation would be misleading, as the text is more literal at times, specifically saying that an animal such as a dog produces a young that resembles it (*in specie* or *in genere*), whereas an insect, for example, might first produce a caterpillar that cannot make this same claim. At times, then, the more cumbersome, literal translation will be maintained.

like that which generates, but rather they are *gusanes* [larvae], which are, so it seems, unlike the generating one in shape and species.[8] But the process of generation stops in these since these dissimilar ones do not generate others that are unlike themselves and their parents. For if generation always progressed through dissimilar ones, this dissimilarity would go on infinitely. Nature rejects this since what is infinite is imperfect and nature always strives for perfection and the perfected. The *gusanes,* then, is generated with a dissimilar shape. The *gusanes* is an imperfect animal, just as an egg is. After a time it is perfected and it then returns to a likeness of its parent.

5   The most plausible statement in this view is that such ringed creatures as bees, lice, and fleas first generate eggs from which the little worms [*vermiculi*] called *gusanes* come. Thus the louse's egg is the nit, and that of the flea is a nit of almost the same color and shape. After a while, if they do not die, lice and fleas are generated from these. Among such animals, there are many which are not seen, and the male and the female differ little in them.[9] For in some of them eggs are present in them all and these have the power of plant seeds. In others, however, a male and a female exist, as is the case in bees and flies.

Animals, however, which do not move their location, such as the genus of hard-shelled creatures, and generally those which live attached to rocks, do not have a distinction of male and female. Their lives resemble that of trees and they are called animals only because of a certain similarity. They differ from trees only slightly, specifically in having a contraction and expansion movement. This sort seems to be like the trees with respect to their generation, for some trees produce a fruit from their seeds which is like themselves with respect to being and species.[10] Others, however, do not produce fruit which is like them in species but rather produce a fruit which has a certain likeness to them and which returns through particular stages to the species of those trees from which the fruit first came. Thus, wild figs produce fruit which can be helped along by cultivation in order to return to the generation of good figs.[11]

6   Much the same thing happens in certain wasps and certain other ringed creatures. For just as certain trees are generated regularly from a seed and some are generated by chance and not by a cause that is generating univocally, so too are some of these generated from the putrescence of the earth or the putrefaction of certain other parts.[12] For certain wasps and certain genuses of flies are not generated from the earth but from the putrescences of certain trees or fruits, especially from that of the glutinous trees. We have given consideration to the natures of such trees especially in [our] book *On*

---

8. Or "in shape and appearance."

9. Perhaps A. means that there are many that are not seen to copulate.

10. Again, *species* may involve both the sense of "appearance" and "species."

11. Cf. Ar. *GA* 715b22f., where the fig is extensively discussed as having both sorts of plant contribute to the production of the fruit. A. discusses the matter further at *De veg.* 6.103–4.

12. On the nature of a univocal cause and univocity in general, cf. 3.140, with note. Also, cf. 5.3, where A. expands on the nature of univocal generation.

*Plants.*¹³ In this investigation, however, we have proposed to speak of the cause of the generation of animals in a discussion that is in harmony with those things which have been said in the earlier parts of this study, *On Animals*.

Thus, as we have said, the generation of animals does not generally have a male and female principle. But in those in which there is male and female, the male is the principle of movement in generation and the female is the material, as it were. This seems to be reasonable since we will show in what follows that the semen of generation is that whence (that is, from which) generation occurs.¹⁴ This is because all natural things take their principle from a motive cause of this sort. Thus it happens that the principle of generation in perfect animals comes from the male as if from a motive cause and from the female as from the one providing the material. The sperm of generation comes from both and is mixed in the womb, but the power of the sperm produces a creature and a creation [*creatura et creatio*]—that is, the formation and a production [*formatio et factura*] of an embryo. Thus we say that "male and female" is the principle of generation. The male is the one, as we have said before, who generates in another of his own species or a closely allied species, such as the horse who generates in a she-ass. "Female" is the name for the one which generates in herself from another. Thus, it was even the belief of the ancients that in the creation [*factura*] of the whole world the material of the earth and of the other things was like its mother, and Jupiter, whom some used to call the "god of gods," was like its father. It was from this "god of gods" that the name of the star now called Jupiter was appropriated.

7

---

13. "Glutinous trees": an *arbor glutinosa* would, presumably, be one full of sap or other sticky substances such as pitch. Ar. *GA* 715b30 makes it clear, however, that this is somehow derived from Ar.'s mention of *ixos*, mistletoe. It would appear that the Gr. verb describing the mistletoe's action upon its host (*sunistatai*) was read by its translator not in the technical sense of "coming together" (Peck, 1942, lxi–ii) but rather more simply as "binding together." Yet A. says, at *De veg.* 4.84, that fowlers use mistletoe to produce birdlime. Cf. *De veg.* 5.12, 24, 31.

14. "From which": *unde sive a quo fit,* troublesome in its terseness, for it can be translated "whence or by which it is done."

CHAPTER II

On the Difference between Male and Female with Respect to the Definition and Shape of Their Members

Male and female are not only differentiated by definition but also exhibit external differences of shape that are perceptible to the senses, for they have genitals and breasts which differ perceptibly.[15]

The difference by definition is that the male, as we have said, is the one who can generate in another. The female, however, is she who generates in herself. The shape of the members of some of them differ perceptibly, for we have often proven that all natural functions require a suitable instrument. That which is suitable in generation occurs when one part can be enclosed within the other in such a way that the organ of the one that generates in the other is enclosed in the organ of the one that generates in itself. This is necessary for generation. This is also why copulation is necessary in the generation of those such as these and nature has furnished various members for copulation which are suited to each other. This is also why the members are different, with the result that the male's member is differently shaped from the female's member.

Even though the generating parties consist generally of a male and a female, they are nevertheless not alike in everything, but they differ in power and in certain members. It is the same for a seeing animal, insofar as it sees, and for a walking animal, insofar as it walks, for, as is clearly perceptible, they differ in their power and organs.[16]

The organ of generation in the female is called the womb, along with the neck and opening attached to it. In the male, however, there are the testicles and those things that are associated with them, these being a sort of channel, ensuring that the sperm flows properly, and the penis. This difference in the members is seen in all blooded animals, for although some of them might have testicles, some seem not to have them. They have instead a part that is like the testicles in effect and function. Males and females, then, differ with respect to these members.

---

15. "Definition": although *diffinitio* is clearly the *logos* of Ar. *GA* 716a19–20, Ar.'s Gr. was a bit unclear at this point as well. Platt (1910, ad loc.), glossing his translation that the animals differ "anatomically," adds: "Lit., 'according to sensory perception.' The *logos* is not visible or tangible, but the organ, which corresponds to it, is." Thus, following Stadler's markings, A.'s received text read, "Male and female differ not only by definition but also exhibit diversity in their members according to sensory perception." A. had problems with this and interpolated heavily. The resultant sentence, however, is open to several translations. The one presented here is the most likely in the light of the original Ar.

16. The corresponding section, Ar. *GA* 716a26f., is itself obscure and somewhat corrupt. The sense is that an animal is not male or female with respect to its entire self but rather with respect to the organs that make it so, and the same is to be seen in an animal called "seeing" or "walking." A. may suggest here that although generation requires generally a male and a female, differences appear with respect to the members employed and the power of generation.

Further, the members of the male and the female differ in shape. It is known from our other books on natural philosophy that a small change made in the first principle produces a great change in those things consequent to this principle. This can be seen clearly in those who have been castrated. For when some males are castrated, having their penis and testicles cut off, the whole appearance of their bodily complexion seems to change and this change creeps constantly forward, increasing until he takes on the characteristics of a female. At this point there is but little difference between the castrated one and a female.

From all these things, then, it is clear that the male and the female are principles of generation in the animals in which they are present. When a member suited for intercourse is ruined, the rest of the body will change from the characteristics of the male sex, just as happens when things whose existence depends on something as if from a principle are changed when that principle is destroyed in them.

The members suited for copulation are the testicles and the womb. Their arrangement differs in blooded animals. Of animals which generate, some males lack testicles entirely. Examples are the fish and serpent genuses, for these and the genuses of those like them have only two passages in which the semen for generation is present. Some have testicles under their body near the kidneys and from these come two passages travelling to the place of the exit of the sperm. At this place they come together again and this place is above, at the anus. This is the arrangement in all the air-breathing oviparous animals which have lungs, and it is found this way in the tortoise, lizard, and all quadrupeds with a scaly skin, that is a horny-scaled hide.[17]

A viviparous animal has testicles below at the front of its body since the penis is not to the back, which we call the rear. Some of these, however, have their testicles internally at the end of the belly. Such are the dolphins. But they do not have passages, having instead their own particular instrument which emerges after the fashion of a penis, just as does the instrument of copulation in bulls. Some, however, have their testicles near their anus, like the boar and the pig. Others have them hanging, like a man, the ram, the goat, and the like.

In all animals the womb is divided in two just as the testicles are in males. This is as we have said in the anatomy section. They are placed near the genitals near the hips and in all of them the necks of the wombs extend to the mouth of the vulva.[18]

Further, the wombs both of women and of other viviparous animals are internal. This is true not only for their wombs but also for those of oviparous animals and of all others, both those of fish and of other female animals. But certain of these ani-

---

17. "Scaly skin, that is a horny-scaled hide": *squamosa pellis sive corticosae cutis*. A. is making the same distinction as at 14.85, q.v., with notes.

18. This passage is to be compared to others (e.g., 2.76; 8.106) where the confusion caused by translating the Gr. *arthra*, which can mean both "joints" and "genitals," is explained. Cf. Ar. *GA* 716b34 and, below, 15.24, 27. The unanswered question is exactly what A. thought when he saw the word *iunctura* in such a context. Did he understand that it was a synonym for genitals, think it was a shortened expression indicating the hip joints, or have no real idea what the word meant in this context?

mals have a womb so far inside that it is near the wall (that is, the diaphragm). It is this way in birds, oviparous fish, and in those that generate live young from eggs. The wombs of these animals and of others are split in two. The arrangement of the wombs in soft-shelled animals which generate and of the *malakye,* throughout all its genus, is much the same. For that which we call fish eggs are in membranes which resemble the membranes of wombs. This member, however, is not easily made out and it does not have its own location in such a way that it seems separated from the other members in those animals which are multipedal (that is, those which have more than four feet). The reason for this is not bodily size but rather that such creatures have a body that is homogeneous in almost every part. The wombs of creatures with ringed bodies are divided in two. This is very visible in those with a large body, but this is hard to see in the small ones because of its smallness.

The arrangement of the generative members in all the animals is the way we have stated.

CHAPTER III

On the Natural Cause of the Creation of Testicles

in Animals That Have Testicles

13 One who wishes to study the extent to which the organs of sperm follow nature must necessarily first consider the causes for which the testicles have this sort of substance and nature. Since, then, nature does all things either out of necessity or because it is best done in this or that way, then the substance of the testicles will have been perfected by nature for one of these two reasons. For it is clear that testicles are <not> necessary for the being and substance of a single individual.[19] Neither are they necessary for generation in and of itself in every genus of animal. For if this were so there would have to be testicles in all animals that generate. Yet it is known both from sight and from the things we have said that serpents and many genuses of fish have no testicles whatever. For it is clear to see that the pathways of fish and serpents are filled with sperm during the mating season, at which time the testicles also appear if they happen to exist in these animals. It is therefore clear that the creation of the testicles is "for the best."

14 Generally speaking the emission of sperm is an activity found in very many animals. And just as we have said that the natural activity of trees and plants is the emission of fruit and seed, and as we have said above that an animal with a straight intestine is greedy for its food, so too do we say that every animal that has straight passages for the emission of sperm and which lacks testicles is swift in its copulation. An animal of a contrary disposition is slower in its copulation. For just as an animal

---

19. Note that the negative in angle brackets has been supplied by Stadler from Sc. It is also missing in Borgnet's text.

with a convoluted intestine is less greedy than another animal, so too does the convolution of the spermatic passages bring about a slowness of movement toward copulation. From these things, then, it is clear that the nature of the testicles is like the nature of the bones—that is, it is for the best. That is, it is so they can be the vessels of the superfluity of that which we call the sperm. This, then, is the reason sperm stays in these vessels and does not flow constantly to the point of exit in the penis, and it is also the reason that animals having testicles do not move quickly toward copulation.

However, an indication that the testicles are the vessels for the sperm in humans is that during the pleasure of intercourse, when the sperm is poured forth, the testicles contract and move inward, so that as they press and pour forth in this way, the sperm might be discharged into the channels of the penis. That the testicles of such animals form no part of the parts of the spermatic passages but rather are positioned beyond these passages is known from the movements the animals perform during intercourse, producing spurts so that the sperm might be forced upward for exit. This is rather like something done by ladies who weave. They place long pieces of wood transversely on the warp of the cloth. The threads are bent back over these pieces of wood in a straight line so that the upper ones are drawn downward on them and the lower ones upward.[20] This is done so that by weaving between them they might catch the woof. Now, in those animals which are more perfect and are most dissimilar in body, the sperm requires a greater concoction [*coctura*]. Therefore, its departure is held back from a straight path so that its power might be given greater potential differentiation in the seminal vessels, because the body must be made very diverse in its members. When, however, the testicles are cut off, the passages of the nerves and veins, in which the sperm flowed to the testicles, contract upward, just as when a cord that is stretched tight snaps back when it is broken. This is why a castrated animal cannot produce a young, since the spermatic passages are not attached and do not end at the proper passages for the exit of the sperm. However, as we said before, there was a certain bull in times gone by which, after it had been castrated, mated with a cow and reproduced. We have given the reason for this above, namely, that the sperm was already in the passages of the penis and these passages had not yet shriveled up.[21]

All the genuses of animals which have testicles and which are viviparous have a place to receive and to ripen the sperm. This is why the sperm leaves them more slowly, and it is clear to see this in the birds. During their mating season their testicles are very large and in the birds which copulate at one set time the testicles become very small after the mating season, so much so that they are not even visible in some of them. Yet, during the mating season, they grow to great size. Further, all birds which have their testicles inside their body have shorter, quicker copulation. Animals which have their testicles on the outside do not eject their sperm before the testicles contract upward. This is just as we have said about the other animals whose testicles are external.

---

20. This represents a fascinating "updating" of Ar. *GA* 717a36f., which describes stretching the warp threads with the stone weights that are found everywhere on Gr. sites.

21. Cf. 2.115.

17 Moreover, there is an instrument suited for copulation in quadrupeds, for it is possible for this to be in them due to the composition of their bodies. In birds, however, and in animals that are not bipedal, this is not possible, since the legs of the birds are under the middle of their bellies. In animals that are not bipeds, however, and are not quadrupeds, there can be neither testicles nor penises present since the penis is created so as to hang between the rear legs in all those which have a penis and testicles. Thus the genitals are always attached beneath the belly, and during the time for mating the nerve of the penis, which has its foundation on the nerve of the leg, has to be extended. Because the penis is made of nerve and because it is impossible for this to be in such animals which drag almost their entire body on the earth, it necessarily happens that these types of animal lack a penis and testicles. For if these animals had these members they could not be in this location. For movement of the testicles and penis would have to take place there and this cannot happen since at this spot they would interfere with the movement of the body.

18 Moreover, in animals that have external testicles, the penis happens to become warm through the movement of copulation. Thus, too, the animals move themselves by rubbing the penis when they copulate, and when the penis is warmed it draws the sperm to itself. This amasses there and then it exits with a spurt, virtually boiling, at the time of copulation. For it is amassed there and is forced along by the heat.

It can be gathered from all the things that have been said that, as we have said, all these animals have their testicles below and to the front of their bodies. The sole exception is the hedgehog. For its testicles are near its kidneys for the same reason that the testicles of the birds are located near their kidneys. They must, therefore, necessarily copulate swiftly. Neither do they copulate like other quadrupeds, with the male mounting the female. Instead, the female raises herself up, and the male raises himself up opposite her, and in this way copulates with her. This is because their spines allow copulation in no other way.[22]

The reason, then, that animals have testicles and the reason that some have them internally and some externally has thus been determined by us.

---

22. This belief, at least as old as Ar. (cf. *GA* 717b27f.), was not refuted until 1948 by the careful and patient observations of Stieve. The female lies flat, facedown on the ground, flattening her spines and raising her pelvis. The male mounts from the rear and intromission is very quick (Grzimek, 10.207–8).

CHAPTER IV

On Why Those That Have No Testicles Are Swift in Their Copulation and Why Those Having Internal Testicles Are Swifter in Their Copulation Than Those That Have Them Externally

Animals which do not have testicles do not lack them for the best but rather because, due to the necessity imposed by the composition of their bodies, they are, as we have said, swift in their intercourse.[23] Examples are found in the nature of fish and serpents. Fish, however, copulate by touching one another quickly at the bottom of the belly. At this touch they eject the sperm quickly. And just as it happens that the human and all other animals hold their breath during the moment the sperm drips out during intercourse, so too does it happen that fish do not spew out water through their gills when they eject their sperm. For when pleasure takes over their entire body the breath is held to assist in the ejection of the sperm. However, if they did not hold themselves this way, neither taking in nor emitting water at this time, they would quickly be destroyed by the entrance of the water which their power, undone by the pleasure of intercourse, could not resist.

Further, during the moment of copulation, the sperm in animals of this sort is digested just as it is in walkers and oviparous animals.[24] There is therefore need for retained breath. For during this time the sperm flows in their passages and is completely digested. Its digestion does not occur fully at the moment of copulation, but at this time it has been totally prepared to leave.[25] Thus, because the sperm is adequately digested in such animals, these types do not have testicles and have instead small passages that reach directly to the exit point in the way a small member extends from the testicles of small quadrupedal animals into the channel of their penises.

Further, some parts of the testicles are passages for blood and some are not. These are those which receive the moisture of nourishment. When, however, the sperm leaves through these passages during copulation, then the male and female separate.

Further, fish are arranged like the human and other animals in that their sperm is at the furthest part of their bodies, subjoined to their bellies. This is especially so for that of the female, for the place for its exit in such creatures is, in many animals, lower than it is in the male.

---

23. Cf. Ar. *GA* 717b33f. A.'s received text faithfully reflects the Ar., but its terseness prompted additions by A. The given translation is an attempt to bring sense to the fairly inferior Latin; other options exist. Cf. *QDA* 15.18, where A. clearly states that testicles are "propter bene esse et non propter esse."

24. "Moment of copulation": *in tempore coitus,* a phrase that could equally be translated, as elsewhere, "mating season."

25. The point of Ar. *GA* 718a3f. is that viviparous land animals "mature" (i.e., decoct) their sperm during intercourse, but fish cannot and must have it matured in advance. Cf. Platt (1910), ad loc.

The serpents, however, lack testicles and wrap around each other during copulation. For they have neither testicles nor a penis. They have no penises since they do not have legs, but they lack testicles because of the great length of their bodies. They have passages like those of fish.

21 Further, if they had testicles, the sperm in them would become too cold. For the sperm is created near the heart and diaphragm in such animals and if it were received below where they are very cold, it would grow cold, with the result of slowing down its exit. This is because of their cold bodies, which create a distance from the source of the heat. This is proven because even in the hot animals like the human, some of them have abundant sperm and have extraordinarily long penises. Yet these animals end up reproducing but little or they do not do so at all. Those which have average-sized penises do generate for the most part. For in those that have long penises, the sperm evaporates as it flows and becomes cold by being a great distance from its principle. Those animals, however, with a penis that is too short do not propel the sperm to a place from which the womb can absorb it. Thus, it most often flows away and lacks the effect of generation.

22 In what has gone before we have set forth the reason some animals have testicles and some do not.[26] We are therefore passing over it here. Serpents belong to those that do not have them and these wrap around each other while copulating. This is because their construction is not suited to another type of intercourse. Because of the length of their bodies, they are not suited to intercourse without difficulty. For the body of one cannot stay next to that of another except by wrapping around it. Even though they emit their sperm quickly, nevertheless there arose from this the belief that they separate more slowly than the fish, not only because of the greater length of their passages but also because of the construction of their bodies which, apart from these passages, does not allow them to touch the way the fish do. This, then, is the reason they are held back from separating from one another even though, as we said, they copulate quickly.

23 Those animals, however, which have testicles, but have them internally like birds, copulate swiftly, but not as swiftly as those which have no testicles whatever. The swiftness of the copulation in such as these is because the testicles of such birds have the sperm conceived in them before they are moved to copulate. When they copulate, they pour it forth easily since they have spermatic passages. The sperm, moreover, that has been poured forth flows easily toward its exit since its vessels are not suspended but rather there are passages which descend directly and lead to the point of exit. Its flow is thus a natural motion. This is helped along by the force of the heat and spirit and, thus moved, it is hurried along, for the sperm is moved by four movers, namely, nature, soul, heat, and spirit.

In all the other animals, copulation is slow since the vessels are outside the passages and are suspended. Because they are outside the passages, the movement of the sperm

---

26. Cf. 2.108f.

is turned awry and is drawn out. Because they are suspended, it must rise unnaturally, using the power of the soul, heat, and spirit to do so.

These, then, are the reasons for swift and slow copulation.

CHAPTER V

On the Reason for and Natural Arrangement of Wombs and on Their Differences in Animals

Someone might reasonably have doubts in considering the makeup and form of the wombs and the types of their shape. For the wombs of animals differ greatly with regard to these things.

The wombs in viviparous animals differ. For the same type of womb is not found in all animals of this sort, that is, below at the genitals [*iunctura*].[27] In the *celeti*, the womb is found near and below the wall [*paries,* diaphragm], much as it is arranged in oviparous animals. The wombs of the fish, however, are at the bottom of their bellies, as are those of women. In viviparous quadrupeds, there is a diversity in the womb according to what is suitable for each.

Further, the oviparous animals also differ since some lay perfect eggs and some, like the fish, lay imperfect ones. The eggs of fish both take in nourishment and grow once outside. This is because the fish have many eggs and in many natural things their actions are quite like those of trees. For nature, because of the necessity of a uterus, cannot complete these eggs inside the fish because they are so numerous. Their great number is the reason that they are packed in so tightly that the entire womb seems to be a single egg. This is especially so in small fish since they have more eggs for their body size than do other fish. In this, however, the eggs of fish are like the seeds of plants. For in all these it happens that growth in size passes over from them into seed and this seed afterward is completed when it falls to the earth.[28]

However, birds and oviparous quadrupeds lay complete eggs and these eggs, by nature, need to be hard-shelled. For while they grow they are soft so that they can take on nourishment. But when they are complete they are given a hard shell by the victory of the heat that dries out the superfluous moisture in the egg and hardens the earthy part that is in them. The place for the generation of the egg is hot for the reason given, and it is therefore suitable for it to be near the wall of the diaphragm. Now this place is hot since it is the one that digests the food. Since it is necessary for the eggs to be in the womb, the womb must be beneath the diaphragm in such animals, since otherwise they could not be completed at the heat. Conversely, it is suitable in

24

25

---

27. The problems surrounding *iunctura,* here in the singular, are outlined at 15.11.
28. That is, what would have gone to growth goes instead to the production of seed.

other nonoviparous animals to have their wombs arranged at their lower part, so that it might be more gently cushioned over the lower intestines and so that the young might come forth more easily from it. Thus, in such animals the womb will naturally be at a lower, not a higher spot. For none of the natural powers of the upper members does the two things mentioned, whereas the lower members do them most fittingly.

Further, the end [*finis*] of the womb is in the lower part, where the point of exit for the young exists. Now, the end of the womb is its function, for the sake of which it exists. For the womb does not exist for its own sake but rather in order to perform the function of the birth of the *conceptus*.

26 The wombs of viviparous animals also differ. Some animals do not form the live young outside their wombs but inside, as does a woman, the mare, the she-ass, and, in general, every hairy animal. Of the aquatic animals, the dolphin does the same, as do the whale [*balena*] and certain others that have a large body. The *celeti*, however, and certain genuses of serpents generate by giving birth to live young externally, having first conceived these same young in eggs internally. At the time of birth, however, a live young emerges from them. When the time for birth has arrived in such animals, the eggs burst and the live young come forth from the shells into the womb. For they are oviparous in that they conceive perfect eggs, and they thus generate not from eggs laid externally but rather from eggs kept warm in their wombs and not completed externally. Such animals do not lay externally because of the coldness of their natural complexion, and it is not due to their heat, as some have falsely believed.[29] Certain animals lay soft-shelled eggs since their moderate, inner heat cannot dry out the shell. If a soft-shelled egg remains outside it is quickly ruined. Many genuses of animals of the bird genus produce soft eggs in this way.

27 Further, as we have said before, certain animals are internally oviparous and yet a live young emerges from them. They actually give birth because when the time for parturition has arrived the eggs descend to their lower parts, near the genitals.[30] Here the live young leaves them, as happens in viviparous animals which conceive live young from their first formation without eggs. This is the reason that the wombs of viviparous animals and those of oviparous animals are found to differ. These animals, which both lay eggs and give birth, have similarities with two types, namely, animals that lay and those that give birth. The womb, therefore, in all the types of animals that are like the *celeti* are found at the lower part and also by the diaphragm.

Now perfect knowledge of the wombs of these or of whatever other animals is not gained without also knowing the anatomy of their members. It is also known somewhat through the discussions we have held in the earlier books of this investigation.[31]

---

29. Cf. Ar. *GA* 718b35f. Peck (1942, ad loc.) believes the reference is to Empedocles, who felt that aquatic animals lived in the water to temper their natural heat.
30. "Genitals": lit., "joints"; cf. 15.11, 24.
31. Cf. 6.119–25, and perhaps 108–18.

It is therefore clear from what has been said before that animals which lay complete eggs have wombs in their upper parts while those which do not lay complete eggs have wombs in their lower parts, as we have said above. However, animals which first conceive eggs and then produce live young have wombs that bear similarities to each type. But their wombs are mostly at the lower part, since wombs that are arranged this way interfere with no function of nature whatever.[32]

Further, in addition to what we have said one should know that the creation of the animal does not occur near the diaphragm. For every embryo necessarily possesses weight, and movement of the womb or its accidental raising toward the diaphragm quickly brings on death. This is because the diaphragm can in no way endure the weight of an embryo.

Further, if the womb of such animals were near the diaphragm, great pressure would necessarily be attendant upon birth because of the long delay in the departure of the young formed from the sperm. Thus, too, it sometimes happens in women that due to an upward push of the womb, there is a sudden opening of the mouth.[33] If it often made such movements, so that the embryo were drawn upward, the pressure during birth would be intolerable. For a womb that is lifted up and which stays aloft frequently leads to suffocation.[34]

Further, wombs that bear and form live young are necessarily tough and strong. Thus, too, all wombs of this sort are made fleshy. However, wombs that are under the diaphragm are rather like certain membranes. This, however, occurs in the wombs of animals which are impregnated in two ways. For the live young leaving the egg internally are generated in the lower part of the womb. An animal which conversely does not generate but rather lays eggs conceives its egg in the upper part of the womb, as we have often said already.

Let, then, these be the reasons determined by us that the wombs of animals vary and that some are in their lower parts, whereas others are in their upper parts near the diaphragm.

---

32. I.e., if placed higher up near the wall or diaphragm, the fetus(es) would interfere with the vital functions of the heart and lungs.

33. A confusion. Ar. *GA* 719a21f. says that yawning can draw the uterus up, yielding a feeling of suffocation.

34. For more on suffocation of the womb, cf. Ps. A. *DSM,* 131–35.

CHAPTER VI

Why Wombs Are Inside the Body in All Animals Even Though in Some Animals the Testicles Are Internal and in Others They Are External

30 The reason the wombs of all animals are inside the body even though some testicles are internal and others external is that the wombs have to be inside the body because the embryo is formed in them and it needs great care and a heat which is both forming and warming, both life-giving and natural. An external location, however, is cold and is open to sudden harm leading to mutations.[35]

In some the testicles are internal but in others they are external. This is because this member requires a special covering for its protection and for the digestion of the sperm. They can have such a special covering outside the body more suitably when it is made of skin like that which covers them in the walking animals. However, in those animals which have a tough skin not suitable as a covering because of its hardness, like the skin of the fish and similar scaly animals, they must necessarily be inside the body. This is also why the testicles of the dolphin and of all large-bodied marine animals are inside the body. The skin of the birds is also tough and is not suited for enveloping such a delicate member as the testicles. These reasons, then, along with all the ones we gave before, are clear, just as the accidental traits which occur during the time of intercourse are clear. Because of these reasons the testicles of the elephant, tortoise, and hedgehog are inside their bodies. For the skins of these types of animal are not suited to cover a soft, delicate member.

31 As we have said before, the location of the womb varies in animals that produce live young externally and in the oviparous animals which have their wombs beneath the web of the diaphragm. This is after the fashion of the way the womb differs in the fish, birds, oviparous quadrupeds, and animals which have similarities with either type, namely, those which conceive eggs internally and generate live young externally. For animals which generate and conceive live young have their wombs above their bellies, like the cow, she-goat, bitch, and those like them. For there must not be any weight whatever above the womb, for the protection and growth of the embryo.

The passage, however, which travels through the neck of the penis, away from the testicles, is present in all animals for the sake of the better. The superfluity of the sperm leaves through this passage and it is much the same in the female animals. For testicles are found in the females just as in the males. In the testicles there is a passage through which the male's spermatic fluid leaves and this passage is at the upper part of the penis, over the urinary channel [*canalis urinae*] and to the front in the body,

---

35. "Mutations": *est et velocis occasionis ad immutationes*; although clumsy, this might also mean "sudden harm due to changes (in the weather)."

lower down and in front of the anus, which is the place through which the superfluity of the food leaves.³⁶

Just as in fish, in the oviparous animals which produce imperfect eggs the womb is beneath the belly and below the area of the kidneys. This is because in this place the growth of the eggs will not interfere with any natural function. For even though the eggs begin internally, nevertheless all eggs of this sort are completed externally. For after a small amount of growth at conception they immediately descend to the lower part. In animals of this sort, the same passage is consigned to generation and to the exit of the superfluity of the food. This is clear to see in all the oviparous animals even if they have a bladder, as in the case of the tortoise, for it has two passages for generation just as do certain other animals.³⁷ But it does not have those passages for the voiding of moist and dry superfluities, for the nature of the spermatic superfluity is moist and it shares the same passage with the food's moist superfluity. This is shown in that all animals have semen, but they do not all have the moist superfluity of urine.

In such animals as these the sperm leaves, for the reason given, through a passage very near that of the exit of the food. The spermatic passages in the males, however, should be fixed and unmoving, just as wombs should necessarily be fixed.

Wombs, moreover, are necessarily arranged at the front of the body and not in the area of the back. In viviparous animals they are in the front part of the body for the sake of the embryo. In oviparous animals, however, they are in the area of the back and the kidneys. In animals that are internally oviparous and externally viviparous, the wombs are found according to both these arrangements. This is due to the similarities these animals share with both the animals mentioned. Specifically, they are oviparous in conceiving and viviparous in giving birth, and their wombs are both in the upper part where the eggs are (above the diaphragm and near the location of the kidneys) and in the area near the back. When birth is prolonged for a while, they descend to the lower part over the belly.³⁸ There they give birth to the live young.

The passage for copulation and that for the voiding of the moist spermatic superfluity is the same.³⁹ For we have said in what has gone before that none of these types of animals has testicles. The path we mentioned is hanging and is bound both in the males that have testicles and in those that lack them. Therefore this passage hangs with the wombs and is attached to them in the area that is near the back, next to the vertebrae. For the womb of animals such as these should not be shifted from place to place but should be complete. The arrangement of the rear of their body is the same.

---

36. This confusing description is based on Ar. *GA* 719b29f., itself a passage in a state of confusion. The intent of the original Ar. is simply that males have a single passage for urine and generative materials, whereas females have two.

37. An obvious misstatement of Ar., who says that its passages are double.

38. The text of Ar. *GA* 720a18f. makes it clear that the eggs shift downward as time passes, not when a particular birth is prolonged. Although the Latin can be made to say this, it is not a reader's first choice.

39. Cf. what follows with Ar. *GA* 720a22f.

When these animals have testicles which hang down, they are found with the location we gave before.

Moreover, these passages come together outside in the area of the penis. In the dolphins the passages follow this mode and the testicles lie hidden inside, below the body of their belly.

The members, therefore, that are suited for generation, and the reason each of them exists, have thus been determined by us.

CHAPTER VII

On the Disposition and the Cause of the Genital Members in Four Genuses of Animals: Soft-Shelled Animals, the Malakye, *Those with a Ringed Body, and the Hard-Shelled Animals*

35 In bloodless animals the members pertaining to generation are disposed differently than in the blooded animals.[40] There are four genuses about which a determination must be made: namely, the soft-shelled creatures, the *malakye*, those with a ringed body, and the hard-shelled creatures. The disposition of the generative members of these animals, however, is not sufficiently clear, but what is known is that most of them do not copulate.

We must speak of the manner and method in which the soft-shelled animals act during copulation, for their copulation is better known than that of the others. For this animal copulates in a way somewhat like an animal that urinates to the rear. For one lies passively on its back and the other is on its belly. They change these positions and thus copulate. If they did not copulate in this way their tails would prevent them from doing so.[41] When they are thus attached, they bend their tails toward one another and copulate. At this time in the males there are found thin passages filled with semen while in the females there are found web-like wombs filled with eggs on either side near the intestine. Their wombs are divided in two and are filled with eggs, as is clear to see in the river crayfish [*gamarus*].

36 The *malakye* genus, however, copulates when the members holding the semen are intertwined one to the other with the other members which are analogous to them. For these animals gather together the parts of their spermatic superfluity near the

---

40. Note that the paragraph number 35 is added here; it is omitted in Stadler's text.

41. Compare this troubled passage with Ar. *GA* 720b9f. (trans. Peck, 1942): "one lies prone and the other supine and they fit their tail-parts one to the other. The males are prevented from mounting the female's belly to back by their tail-parts which have long flaps attached to them." Perhaps the positions that are described in A.'s version as changed are the normal "belly to back" positions taken up by most animals.

opening.⁴² When one bends its members to the members of its mate, it then pours forth its semen just as we have said before where we treated these animals.⁴³ In the females of this genus, however, the member that is the womb is visible. At the first conception, however, she conceives an undivided mass of eggs which then divides into many eggs. Yet none of these eggs is completed internally, being instead completed externally after they have been cast forth. In the passage for the superfluity, however, there is a member in these animals which is like the womb in the animal called soft-shelled.⁴⁴ But in this genus it seems to be at the very place from which the superfluity leaves at the front of the body, namely, at the place where there is a separation in the covering and where the seawater makes its entrance. It is also in the place that copulation is accomplished, and it is the place toward which the sperm of the male is poured forth. It is on a member which has a power analogous to that member through which the other passages we spoke of approach and are directed to the womb. The passage, however, which lies in the member with which the male intertwines with the female passes through as far as the channel.

This is seen clearly in the animals called, since they have eight feet, the *polipedes*. This is also why fishermen think that these animals copulate by intertwining with this same member. For this member shows by its makeup that the animal should be intertwined in it and in this way something that is outside the passage and outside the body seems to be suitable for generation. It is probable, however, that this genus copulates as follows, with the male and the female lying on their backs. For it seems that the *malakye* genus could not copulate otherwise. Nevertheless, the method of copulation of these animals has still not been perfectly observed. Therefore, this member is disposed this way either for generation or for some other reason that escapes us.⁴⁵

Further, certain animals with ringed bodies both copulate and are generated from animals that are like themselves. This sort of generation is found in most bloodless animals, namely, locusts, wasps, and ants. However, some of the ringed creatures copulate and generate one unlike itself in genus. For these generate larvae [*vermes*], which are not like themselves and which some have called *gusanes*. Some of this type of animal are not generated from animals but rather from putrescent fluids, and others are generated from dry, putrefied bodies, like the bee, fly, and *cantaris*.⁴⁶ Others are almost

---

42. "Opening": *orificium,* regularly meaning either mouth or opening in *DA.* Cf. Ar. *GA* 720b16f.

43. Cf. 14.14f.

44. The translation given follows the text of Borgnet, which inserts a much needed *est* into the sentence, supplying it with a main verb.

45. Ar. *GA* 720b33f. rejects the fishermen's stories that this arm is a generative organ, which in fact it is. In some species of octopus it even detaches and remains in the female. See Platt (1910, ad loc.) for an extensive discussion and references.

46. Ar. *GA* 721a8–10 lists, as results of spontaneous generation, fleas, flies, and *kantharides,* the latter often identified as a blister beetle (Beavis, 1988, 168–73; *GI,* 92–93). The dung beetle (*kantharos*) was said to arise from the bodies of asses (*GI,* 84–85). The belief in the spontaneous generation of flies is easy to put down to faulty observation. That of bees, which were supposed to arise from bull carcasses, might, in fact, be due to ancient ritualistic memories (Kitchell, 1989).

never generated from animals and neither do they copulate. Examples are the bedbug [*cimex*], the *musciliones,* and certain other genuses of animals like them.⁴⁷

39 Further, certain genuses of ringed animals which copulate have females that are larger in size than the males. Also, in males such as these the passages for the sperm are not visible. Neither does the male seem to have any breadth whatever to the member which should enter the female. Rather, a certain lower and broad female part enters the male from below.⁴⁸ This is known through visual confirmation [*experimento visus*] to be present in many types of these animals, but it occurs also in a few types of other animals. That the females have a larger body size than the males occurs also in many other animals from among both oviparous fish and oviparous quadrupeds. It is also found in birds of prey, as we recall mentioning in what has gone before. For a larger size in the female is suited for impregnation and there is also another reason which we have set forth before.⁴⁹

40 Further, the wombs of the females in these animals suit the shape of this genus of animal. They also lie near the intestine as they do in other animals. The fetus is conceived in the wombs, as is seen in the large locusts and in all the animals of this genus which have a perceptible size. In the others which have a ringed body, however, this is not seen because of the smallness of their bodies.

This, then, is the disposition and reason for the members of animals which pertain to generation which had been omitted above, where we treated the members of animals. It remains, however, to speak of the sperm and the nature and quality of milk, for these, too, have to do with generation.

CHAPTER VIII

Which Is a Digression Setting Forth the True Understanding of All the Things Which Have Been Introduced in Previous Chapters on the Causes of the Genitals

41 Beginning again from the beginning, each of the causes for the previously introduced members have been discussed and we say, to make understanding easier, that an animal which generates in another is male and is the active force in generation. For the active potency is the one which acts in another.⁵⁰ However, that which generates in

---

47. Ar. *GA* 721a10 lists the *empis* and the *kōnōps,* which may be gnats and mosquitos. On the bedbug, cf. 26.14(15), but it is not cited elsewhere as an animal that neither copulates nor comes into existence spontaneously. Might not the pl. form *cimices* be an error for *cinifes,* which elsewhere seems regularly to replace a Gr. word indicating a mosquito (1.96; 5.33; 7.151; 12.209, 215; 14.4)? On the *muscilio,* a kind of biting fly, cf. notes at 1.96; 4.68, 77.

48. "Breadth" and "broad part" both represent the Latin *latitudo,* which seems to make no sense. It may well be a mistranslation of the Gr. *epi to pleiston,* "for the most part." Cf. Ar. *GA* 721a14.

49. Cf. 4.106.

50. Cf. Avic. *DA* 15.47v.

itself by means of another is the female, for to be thus changed by another belongs to the passive potency. If, however, there were some animal which generated in itself and from itself, as some depict the phoenix as doing, then there would be neither male nor female, rather like a plant.[51] However, in my opinion, this situation cannot possibly exist in animals unless they are exceptionally imperfect and tiny. For the principle of life, which is the sensible soul, wants to have something in the generating principles which the vegetative [principle] in plants does not have. And if the intellect were a power in an organ, it would have to have something in the generating principles which the sensible [principle] does not have. Now, since the power is separate, it is given by a separate substance operating in nature which is the first cause or, according to some, the intelligence. But we will make inquiry into these matters in the following book.

The masculine and the feminine, then, are the principles of generation. However, all blooded animals, whether they are walkers, flyers, or even leapers, are differentiated into the male and female sex. Some of the ringed creatures are differentiated by sex and others are regenerated through the putrefaction of others. Those, however, which number among the ringed animals and which have sexes have some sort of intermingling of the sexes during copulation. But some of them do not give birth to a creature like themselves. The ancient Arabs called this sort of offspring a *gusanes*.[52] Those, however, which do bear one like themselves number among those which are viviparous. When it generates one that is unlike itself, this generation is not infinite, running without end through an endless number of unlike ones. For nature sets an end and desires a boundary in all things, abhorring the infinite. We thus say that some animals generate with a complete generation and others with an incomplete one. For some of the oviparous animals lay complete eggs, as do the birds. Others, however, lay incomplete ones like the fish.

The eggs of fish grow after they are discharged from the uterus and it is thus reasonable that a ringed creature be generated in the earth via putrefaction. Examples are the red scarabs [*scarabeus rubeus*] which gnaw on leaves in the month of May at the beginning of summer.[53] These do not generate any young but rather produce eggs. These, however, change later into an unformed young which is called a larva [*vermis*] or a *gusanis*. And it is probable that these *gusanes* also revert to a resemblance of their first species and take on the species of their first parent. In this way the generation of this one does not stop in the *gusanes*, for generation returns, according to this manner of generating, to a univocal type. For otherwise there would be no power preserving its genus and species. And this indeed is certainly true, for I have already investigated a nest of bees firsthand [*experimento*].[54] There I found many which were partly larvae

---

51. See 23.110(42).

52. "Offspring": *partus,* and thus perhaps "birth." See Glossary, s.v. "Embryo."

53. Stadler, in his index, identifies this as *Melolontha melolontha,* and ad loc. cites a fourteenth-century gloss giving the Germanic name *maycheuer.*

54. In this instance the terse *experimento* might also be rendered "for the sake of gaining experiential knowledge."

and partly bees that were beginning to form from the *gusanes*.⁵⁵ In the case of certain marine creatures which are very imperfect, however, further investigation is required.

44 An indication of what has been said about the ringed creatures is an event which, according to Avicenna, occurred in a Saracen city in the province of Coratheny, in a town called Scealikan in their tongue.⁵⁶ After a heavy rain which occurred there, silkworms [*bombex*] appeared in the many thousands, covering the earth.⁵⁷ Each of these wove silk over itself and then later, tearing through this silk, they left it from the side, flying. They made semen. However, the silk they made did not consist of continuous threads and it was very fragile. Neither could it be turned into connected threads through any sort of curling iron, that is, the instrument or the art by which silk is joined together.⁵⁸ For this reason the women native to that land did not bother to gather up the semen. However, it is probable that if mulberry leaves had been given to the young of the silkworms, they would have returned to the complete likeness [*species*] of silkworms and would then have generated good silk.⁵⁹

45 *Gusanes*, however, are generated by bees, those which fly around candles, certain locusts, and silkworms. These, after they have been completed, return to their own species [*ad suam speciem*], doing so specifically by the eggs of the perfect ones. Now, one that generates "toward a species" [*ad speciem*] is called perfect.⁶⁰ The generation occurs this way because these animals first produce eggs and from these come the *gusanes*. From the *gusanes* is produced the first generating species. Some, however, who have not seen all these intermediate steps have held the opinion that only the *gusanes* are generated from animals such as these. Avicenna, moreover, says that he had a friend who produced scorpions from the putrefaction of certain woods and these scorpions generated others.⁶¹ Now it is in no way believable that certain animals should be generated by means of some generative power working in putrefied things and then not be generated by another, more natural, method. For although generation exists for the preservation of the species, still it only generates an individual which begins to generate following its own completion.

46 If, however, we were to posit that humans ceased from generation (voluntarily, as the result of some plague in the air, for whatever reason, or even that they all died) and then we posited that humans were produced by whatever means and that man had his woman, then he would generate humans by means of copulation, despite the fact that

---

55. This is a fine description first of the larval stage of the honeybee and then of its pupal stage. Cf. pictures offered by Morse and Hooper (1985, 196, 310).

56. Avic. *DA* 15.47v.: "in una civitate saracenorum quae dicitur scealikam."

57. *Bombex* can refer to the moth as well as to the larval stage.

58. A *calimistrum* is normally a curling iron, used on human hair. Its odd use in Avic. leads him to define it in his own terms. A distaff or the like is probably meant.

59. The supposition is Avic.'s.

60. As noted elsewhere, *species* simultaneously connotes the idea of appearance and species.

61. Cf. 26.33(39) for another version of scorpion reproduction. For the spontaneous generation of scorpions, cf. Beavis (1988, 27).

he himself had not been generated by copulation. For this human action assists generation in that it produces a generating agent which has the potential for generation.

Thus, in entirely the same way, we will say that certain animals are not produced via copulation, which nevertheless afterward generate through a true mixing together of their semens, thus preserving their species.

In a certain stream called the Iacton in Arabic, the beaver sometimes appeared. This stream lasted but a small time, however, and it was a very great distance from all the places where there were beavers. Therefore, it is most probable that the beaver was procreated in that same place by the power of the stars.[62] This is often seen when fish are born in new waters in which they had never been before and in which they had not been placed by humans. These fish afterward generate other fish by means of semen.

Passing on from these things let us return to the matter at hand and say that the male differs from the female in the location of the testicles and the womb. When the testicles are cut from the body of males they change the complexion of their body. This is because the male complexion drips from a single member into all the members save the heart. It drips into the members from the testicles, especially during the proper and complete operation of the testicles, which is intercourse with a woman. Thus, when this member is severed, the complexion of all the members must necessarily change. If the amputation occurs before the rise of the body hair, the hair will not arise in the places where it does arise in males and where it does not in females, such as the beard. The voice also changes from any resemblance to a male's voice. If, however, the castration takes place after the appearance of hair that is rooted in place, then the hair is not totally lost and the voice does not change totally.

47

There are some males in the genuses of animals which have no testicles whatever. These are very swift in their copulation since the semen in them descends suddenly and directly.[63] Examples include the fish which have no testicles since in these there must be two straight passages through which the sperm descends.

Moreover, in some animals which have testicles, they are not a part of the spermatic passages but are separated from these passages. They have, moreover, the beginning of a penis from the ligaments which arise from the bone of the *pecten* and from their flesh. It is a penis of very loose substance and thus easily extended by the spirit that enters into it when there is work for it, and when the work is over it falls back down. It becomes erect to accomplish two functions: namely, that it might enter the vulva more easily and that it might eject the semen more directly toward the internal opening of the womb. For an even emission of sperm is straight into the middle, into a place from which it can be absorbed. It is not sent to the side, from which it could not be absorbed by the womb.

48

---

62. Cf. Avic. *DA* 15.47v., where the river is called the Iactam. The attribution of the power to the stars would seem to be A.'s.

63. Cf. 15.19f.

Moreover, the two extremities of the head of the penis are bent to one and the same side since the cord by which it is attached is short.⁶⁴ This cord was created for appearances and to support the penis lest it have a continuous erection if it were made of hard material like bone. The knob on the penis and the incision which is behind its head is "for the better" so that it can be better surrounded by the vulva and all the semen can be drawn out of it.

49 Moreover, one whose penis is very long has little generation since, because of its length, the semen is changed (for it is very easily changed) when it travels away from the place of its generation. The womb, moreover, is placed to the rear of the bladder and in front of the intestines, so that it might be cushioned between these two and safeguarded for the sake of the creation being formed within it. The muscles which move the penis have been adequately described in the anatomy section.⁶⁵ However, there are two principal pairs which move it. The muscles of one pair are on either side of the penis and when they extend, the passages widen and straighten out so that the sperm can exit swiftly. The other pair arises from the bone of the groin [*os pectinis*] and is connected transversely with the base of the penis. When its extension is even, then the raising of the penis will be straight. Thus, the raising of the penis always follows the movements of the muscles.

50 Further, the testicles are made suspended so that the arrangement of the spermatic vessels might be more useful. These vessels were made twisted so that the material of the sperm might be held up in them and be better digested so that the generation resulting from this sperm might be more perfect. In addition to this, this is also useful in that the sperm might thus be ejected in an orderly fashion and might not hasten to leave too rapidly at each and every lusting of a person, a thing that occurs in some brute animals. Sometimes, however, it brings on heaviness since its ejection is held up. For this reason too this is not absolutely an aid to nature but rather is done principally for the reason given, namely, that it might be better digested, being more useful for generation in this way. For it is not unsuitable for the same thing naturally to have many functions. When we previously said that the testicles are away from the spermatic passages, it was not our understanding that they do not contribute to the generation of sperm. Rather, what was said was that the substance of the sperm is different from that of the vessels and of the passages connected to the testicles. For the testicle is not a passage but rather is a support for the vessels since the passages are directed to the testicles, as the physician truly says.⁶⁶

51 Further, when animals are castrated, the nerves and veins contract upward and become separated from the penis. For this reason the material of the sperm does not flow to the penis. As we said above, however, a bull that had been castrated copulated

---

64. A. seems to read *ad unum* for Avic.'s *ad imum* (downward) and *curta* for his *certa*, with confusion the result. Cf. Avic. *DA* 15.47v.

65. Cf. 1.341f.

66. The reference to the *medicus* is from Avic. *DA* 15.48r, probably referring to Galen, e.g., *UP* 14.10–14 (May 2.641f.; Kühn 4.183f.).

immediately thereafter and still generated since the material had descended previously toward the testicles [*didimi*].⁶⁷

Further, every one which lacks legs also lacks a penis and testicles. Thus, fish do not have a penis and their eggs are completed externally by the semen of the male, for lacking a penis he cannot cast his semen into the female. It is because they lack members of this sort that trees complete their seeds externally.

Further, the instrument the female animals have is the womb. It, too, is a certain principle involved in the generation of the creation, and the penis on the male corresponds to it and is analogous to it. But in one of these the material for generation is completed and comes forth to the outside, whereas in the other it is diminished and held inside. As we have said in the anatomy section, the scrotum [*oceum*] in the male is like the womb in the female, the penis is like the neck of the womb [*collum matricis,* vagina], and the vulva and the two testicles of the woman are like the male's two testicles.⁶⁸ In the males these are greater in size, longer with some roundness, and held externally. But in the women they are smaller, more flat than rounded, short, and internal.

Further, just as the males have spermatic vessels between the place for the exit of the urine at the base of the penis, so too the women have certain testicular vessels [*vasa dindimalia*] between their testicles and the opening of their wombs. But in the males the sperm begins in the testicle, is raised up, and firmly lodges itself in a certain hollow from which the testicles hang. Afterward it descends in a twisted fashion and is completed in accordance with the spermatic digestion (thus having reversed in a curve) until it exits through the passage that is at the base of the penis on either side of it.⁶⁹ This place is near the neck of the bladder. This is long in the males, but it is short in women, since the neck of the bladder in women is short.⁷⁰ In women, however, the sperm proceeds from the two testicles like two horns and proceeds in a spread-out fashion.⁷¹ The heads of these two passages are connected with the testicles and contract during the pleasure of intercourse and during the desire for intercourse which precedes the pleasure. They reach out equally to the opening of the womb. When they contract on two sides, the opening of the womb opens, widens, and draws in the sperm. These are shorter than the ones like them in the males.

52

---

67. *Didimi* represents a fairly literal transcription of the Gr. *didymoi*, "testicles," lit., "the twins." Cf. Fonahn, 1117, and Latham², s.v. *didymus*, who indicate that it might also be more narrowly interpreted as the epidymis. A. relates the story of the bull and that of a cleric at 2.115, where he says the sperm had already left the testicles and had made its way into the passages connecting the testicles and the penis itself.

68. For the scrotum, cf. 1.446f. On *collum matricis,* cf. 1.414, and Fonahn, 950.

69. Compare this admittedly confused passage with Avic. *DA* 15.48r.

70. "Neck of the bladder": *collum vesicae*, thought of generally as part of the urinary tract, but a case can be made for taking it to mean the clitoris. Cf. 9.132.

71. The Aristotelian tradition of horns, promulgated by Galen, endured throughout the medieval period, extending even to Andreas Vesalius in his *Tabulae sex* of 1538 (Singer, 1957, fig. 62, 83). Cf. also Ar. *HA* 510b7f. (Thompson, 1910, with ill.).

53 They differ further in that the spermatic vessels in the women are attached to the testicles and pass on to the horn-like appendages, each arising from one testicle and spewing forth the sperm into the vessel. These two, arising this way from the testicles, are called the "sperm spewers" [*vomitantia sperma*]. The spermatic vessels in women are connected with the testicles since these vessels in women take on softness and smoothness. For there was no need for them to be tough or even that their membranes be tough, for they are well covered and have no need of length, as they do in the males. But in the males harm would result if these vessels were connected to the testicles since they would cause injury by their contraction. For this reason a medium was placed between them.

54 Further, according to Aristotle, the spermatic vessels in women spew forth the sperm in front of the inner opening [*orificium*] of the womb, near the opening [*foramen*] of the penis, which is the neck of their bladder.[72] This, however, can be checked thoroughly only through dissection [*anatomia*] or by the examination of women. But an indication of this, however, is that when a woman moves the end of her penis by drawing her hand across it, she senses the pleasure of intercourse and has an emission. When, however, the sperm is cast forth there, the opening of the womb hastens to draw it up as soon as it senses its own sperm. Along with this it also draws up the sperm of the male since they both fall into the same place. If, however, the womb attracted the male's sperm only by itself, without its own sperm, and thus naturally only attracted the male's sperm, the result would be that the womb would always intend to draw up the sperm of the male in the same manner. It then would not be necessary for the woman to emit any sperm during intercourse. Now this is false according to Galen, although it seems that according to Aristotle it is true.[73] We will look into this below, just as above some things have been said about it. It is thus necessary that the male's sperm be that which arouses the womb's own sperm and the womb thus accomplishes two functions at the same time, namely, the ejection of its own sperm and the taking in of that of another. In this matter, however, great trust must be put in women who are discreet in matters relating to the nature and manner of copulation.

55 Further, some of the wise ancients related that the passage for the exit of the sperm, especially in women but also in men, is very narrow and that glandular flesh surrounds and covers it. This glandular flesh and the neck of the bladder [clitoris?] emit from themselves certain warm fluids which come to the opening of the womb and stir up lust. These fluids are white and thin.

---

72. As mentioned above, the "neck of the bladder" may mean, in such contexts as these, the clitoris. Likewise, the female penis (*virga muliebris*) would normally be taken as the clitoris (Fonahn, 3598) and is also mentioned at 1.446 and 1.466. The interpretation of the term as the clitoris is supported by the next sentences, which obviously describe masturbation. Yet the overall sense is marred by A.'s clumsy insertions into his received text. It would appear that the original Avic. merely said that the spermatic vessels in women spew out the sperm at the opening (of the womb?) near the opening of the neck of the bladder (urethra?).

73. Galen *UP* 14.7 (May 2.632f.; Kühn 4.165f.); *De sperm.* 2.1 (Kühn 4.594).

Moreover, the sperm in males is more digested and it comes to the testicles through many veins that are twisted rather like branches. These veins carry blood to them which is digested and altered until it resembles the testicles and the fluid that is in them. It especially, however, resembles the airy spirit that is generated there and through this it grows white like a delicate froth.

Further, the womb is created from many veins branching into its substance and from which drips out sustenance for the embryos, for the menstrual blood descends through them. The womb is bound to the back with strong bonds to keep it from slipping.

Further, it is made of a nerve-filled substance since it often needs great extension when the fetus grows, and it needs to shrink when it leaves. Because of this it can press on the fetus at the moment of birth, pushing it to leave. Moreover, its hollowness would not be filled if it were not nerve-filled and web-like. This is just as the hollowness of the two breasts is filled in by their own size. If these were not the operations of the fetus and birth, there would be no need for the womb and the breasts.

In humans, the womb has two tunics and two divisions, as we said in the anatomy section.[74] In other animals, however, it is divided according to the number of breasts.

Further, oviparous animals also have wombs.[75] But some of the oviparous animals complete their eggs internally while others lay them inside but give forth live young outside. Still others complete them entirely outside, as do the fish. In those, however, which first produce eggs and then afterward bear a live young, the live young leaves the egg at the bottom of the womb. For if it left it at its top, it would press on the diaphragm and, because of the length of the passage, the birth would be burdensome.

Further, those which have a soft skin produce eggs and lay them outside. One with a tough skin, however, does not have eggs laid outside, but such animals have certain pouches around their shell-like skin into which they put and cover their eggs.[76]

Further, nearly all oviparous animals are so arranged that they do not have a separate passage for the urine and one for the excrement. The exception is the tortoise, which, as we said before, has a bladder.

Further, some oviparous animals clearly copulate, for example, the birds. All that some others do, however, is propel themselves at each other and touch at the point of exit for the eggs and sperm. This contact of their extremities serves them for copulation. Some also wrap themselves around one another, whereas some of the ringed creatures have no form of copulation. Rather, they generate themselves out of themselves, like plants.

---

74. Cf. 1.446f.

75. Cf. Avic. *DA* 15.48v.

76. This odd sentence is A.'s own and the phrase *in circuitu testeae pellis suae* is troublesome with its reference to "shell-like" skin.

Further, certain of the oviparous, copulating animals have a smaller size for the female sex than for the male, as is the case in many birds of prey and in certain others.[77] But generally, when birds copulate, the female places her back beneath the male who then twists his tail beneath that of the female. The female raises her tail up, making ready the opening through which she conceives, and then draws up the sperm of the male from the outer opening.

From the things said, there comes a clear understanding of all the things which have been said from the beginning in this book about the genitals of both males and females.

The Second Tract of the Fifteenth Book on Animals,

Which Is on the Nature of the Sperm

CHAPTER I

On the Questions That Exist about Sperm and Whether the Sperm Is Derived from All the Members in the Way Plato Described

It now remains to speak of the sperm and its nature.

It is clear to the senses that sperm exits some animals, but whether it exits others escapes notice.[78] All blooded animals produce sperm, but those which have a ringed body and those of the *malakye* genus are for the most part not clearly found to have sperm. It therefore must be asked at the start whether all male animals produce sperm. If it is conceded that some do not, we will then have to ask why some male animals produce sperm and others do not. Further, it must be asked whether the females produce sperm and if they do, why this is and what contribution their sperm makes to generation.[79]

---

77. "Smaller": an apparent error, as "larger" is what is needed here, as found by Stadler in some codices and as printed by Borgnet. A. knew that female birds are larger (4.106; 8.5), so the text should probably be emended.

78. A. begins to follow Ar. *GA* 721a31f.

79. A difficult passage. A. is asking whether the female has any *convenientiam ad principium generationis*. Normally, this would seem to ask whether the female's sperm has any "likeness," "suitability," or "similarity" to the principle of generation. But comparisons with the relevant Ar. and with 15.101f. below seem to make the present rendering preferable.

Again, the contribution of the sperm to generation must be considered along with its nature in a general way and what the nature of the menstrual blood is in those animals from which a fluid of this sort comes.

One particular opinion on these matters belongs to those who have said that *genitura* is the principle of generation, using the term *genitura* for sperm.[80] These say that everything which generates is born from sperm and that every sperm trickles down from the begetters and that generation can never occur without a trickling down of this sort. For this reason too they claim that Venus was born from the foam of testicles and that this goddess is the principle of all generation.[81] They say further that the sons, grandsons, and fathers of the gods were born from a mixture of this sort.

Let us ask at the start, then, whether both the male and the female produce sperm or whether it is only one of them and also whether the sperm generally leaves from all the members of the body as a whole.[82] For some say that either it ought not to be derived from the entire body or that it ought not to have a double nature, that is, be derived from the male and the female. Thus, in the midst of these doubts, we should consider and ask whether it is derived from the entire body and how it does so. For, as we said before, many have the belief that it is derived from the body as a whole and they bring forward a fourfold proof for this.

The first of these is the strength and vehemence of the pleasure. For these posit that the pleasure which comes from all the members is stronger and more vehement than that which comes from a few of them. Since, as they say, pleasure is the basis for and the generation of something that pertains to the sensible soul (so Plato says), then the greatest pleasure will be in accordance with the generation of something that pertains to all the organs of the sensible soul.[83] These, however, are all the members of the body and the sperm therefore is derived from all the members of the body.

The second proof stems from something which might occur as an accidental trait in those generated as a result of generation. For many times something imperfect is generated by an imperfect begetter. Thus a dog which has no tail many times generates an offspring without a tail, and it is much the same for birds. For sperm does not depart a member the begetter lacks and thus, they say, this same member is missing in the one generated.

The third proof is the similarity between the one generated and the one generating. This frequently occurs to the entire body and not just to a part of it. They say that there can be no other cause for this similarity of all the members than that the sperm is derived from all the members.

---

80. The insertion of the term *genitura* seems to be A.'s addition.

81. This well-known story is first told in Hesiod's *Theogony,* 188–206. He derives her name from the foam (*aphros*) that arose when the testicles, severed from Uranus by Cronus, fell into the sea. Botticelli's famous painting depicts Aphrodite on a shell, washing to shore after this incident.

82. The theory that sperm came from the entire body, pangenesis, derives from the Hippocratic school and was frequently debated. Cf. literature cited by Cadden (1993), 91–92, and Zirkle (1946).

83. Plato *Philebus* 32b.

61 They take the fourth proof from a syllogism. They say that it is right that the sperm is derived from all the members. For the whole is the same thing as all its parts. However, it is agreed that the sperm is derived from the entire one that generates first and proximately, since that one has all the sperm and it has sufficient abundance of it to be adequate for the generation of all the members, both uniform and nonuniform, in the one generated.

Those who support this position say that these proofs are enough, especially the one which was given about the similarity between the one generating and the one generated. For these are often alike not only in matters of substance but also in matters of accidents which are consequent to substance. For many that are born have on their bodies the flaws of their begetters and even have them in the same places. Thus sometimes the scars from the wounds of the father are seen to be etched in the body of the young and are in the same places they occupy on the father's body. According to Aristotle, a certain child once had a mark on its arm exactly like that that had been on its father's arm, an event which, he says, occurred in the province called Alkydoz.[84] The mark seemed divided and not unified, scattered over the arm of the child just as it had been on the arm of the father. From these reasonings, then, these people believe in what was just said, namely, that the sperm is derived from all the members.

62 These people seem to posit the statement that determinate parts of the sperm have the powers of the members from which they are derived and that through the specific and partial powers of the parts of the sperm, each one acts on one member, acting from the part of the sperm in which it is.[85] Thus, there is not one, universal, formative power in the sperm which acts on the whole, acting from one thing which is the principle of the whole the way the heart is said to be. For, otherwise, unless they understood that which they introduced in this way, the proofs introduced before would have no weight.

Now this was the position of Plato and his followers who posited that there were many souls in each body according to the division of its principal parts.[86]

63 If anyone considers these things carefully, he will find beyond doubt that they are the opposite of what these people say. Neither is it difficult to contradict all the things they brought forth. The argument upon which they rely the most, that based on the similarity of the offspring to the begetter, does not support this position. For often the child is like its begetter in hair, nails, voice, and parts of the kind from which it is impossible for the sperm to be derived. Moreover, many begetters have no hair and no beard and if the sperm is derived from every member, uniform and nonuniform alike, how might we say that the children have hair and a beard? Yet we often see this happening in the children.[87] Moreover, it happens very often that the young resembles

---

84. Ar. *GA* 721b33, Chalcedon, opposite Byzantium.

85. All of this section is a digression by A.

86. Demaitre and Travill (1980, 420) point out that A. shared the common (erroneous) belief of his time that Plato was the originator of the basic doctrine of pangenesis. Note, too, that the tripartite soul (*Timaeus* 69f.) is rejected by A. above at 9.115.

87. I.e., bald-headed men have children with hair; beardless fathers produce children who grow beards when adults.

neither its father nor its mother, but rather the grandfather or the great-grandfather and yet no part of their seed is in it. Further, this similarity is sometimes traced to one who existed many generations ago. This is said by Aristotle to have happened in the region called Ylas, where a certain woman copulated with an Ethiopian and bore a white daughter. Then, this white daughter conceived from her husband, a white man, and bore a black daughter.[88]

A further objection against those who defend this position comes from the similarity offered by trees. For if the seed [*semen*] of trees were derived from all the parts of the trees, it would seem unsuitable, for some parts of the tree are abbreviated and others have not yet arisen on them, while still others are not attached to them on the branches from which the fruit and seed emerge. The seed would also have to be derived from the coverings which contain the fruit and the seed, since such coverings function in a different way if it is posited that the seed is derived from every single part. All these things are absurd.

Further, we should ask whether it is derived only from uniform members—to put it in a word, whether it is derived from flesh, bone, veins, nerves, cords, and ligaments—or if it is also derived from the organic members such as from the face, hands, and feet. For if the sperm leaves from this second sort of member, then, according to this position, the young ought to resemble the begetter only with regard to these. Thus, if it leaves the official members, then the young will resemble its begetter only in its official members, namely, the hands and the feet. Yet this is absurd.

It is therefore clear that similarity is not because the sperm derives from the entire body, but for another reason which we gave earlier in this study.[89]

For if it is said that the sperm is derived from all the uniform members, since the uniform members come before the official ones on the path of composition (this being because their composition is in the face and hands), then the young one should always resemble his parent in these members and will thus be like him in flesh, nails, and other like parts.[90] If, however, it be said that the sperm comes from both members, then it must be asked of these people what the manner of generation is. For we see that official members are composed of uniform members. If, therefore, the sperm comes from both, it ought first to derive from the uniform members and then later from the official members. Indeed, it is more fitting that it come from simple members than from composite ones, for composite members are nothing more than an assemblage of the simple ones. Thus, if we were to say that something comes from a noun or a sentence, it should come from syllables, which are the parts of a noun and a sentence. Likewise, that which comes from syllables should come from the elements of letters. Likewise, if it is said that the sperm comes from the entire body, it should come from the uniform members and from the nonuniform members. If it comes from these, it should come

---

88. Ar. *GA* 722a10, placing the story at Elis. The story is also alluded to at 9.63.
89. Cf., e.g., 3.131f.; 9.130.
90. On the priority of the uniform members over the official members in the process of composition, cf. 12.51–52.

from the first parts of these parts and it will thus come from its elements. Because it leaves these, there need not be any similarity of the one generated to the parent.

The claim that the sperm leaves the entire body is therefore not proven by the similarity of the young to the parents, and the belief of Plato contains no truth.

CHAPTER II

Whether the Sperm Is Derived from the Entire Body in the Way Described by Anaxagoras and Empedocles

66 There are still two other beliefs which profess that the sperm is derived from the body of the generators, namely, those of Empedocles and Anaxagoras, and we should make inquiry into them.

Empedocles says that the sperm derives from the entire body because all the uniform and nonuniform members are in it in actuality, but they are neither connected together nor visible.[91] Thus, in the sperm for twins there are present the heads for each of the twins, but they are not connected to their necks. For this man is one of those who posits that all things are in all things, whether they be uniform or nonuniform. However, Strife and Friendship, by breaking up and gathering together, join each member to that which suits it and they separate it from that which does not suit it.[92]

Anaxagoras, however, said that the sperm is derived from all the uniform members and that all uniform members are in it in actuality. In much the same way he says there is invisible flesh [present in the sperm] from the visible flesh of the generating one and blood from its blood, bone from bone, and so on for other things. Because these things lie hidden in the sperm he therefore says that the intellect separates and joins suitable things from unsuitable things.[93]

67 If, then, we follow what these men say, then those members they mention should be in the seed, whether they are divided or connected and joined together.[94] If, how-

---

91. I.e., the "seeds" are too tiny. Empedocles, of the fifth century B.C., was a teacher of Gorgias and espoused a sort of early atomism. He was often quoted and studied by Aristotle.

92. "Strife": *lis,* translating the Gr. *neikos,* a centrifugal force for Empedocles that caused the individual units to remain separate or to fly apart. Likewise, Friendship, *amicitia,* is Empedocles's *philia,* which caused things to come together to form wholes.

93. Anaxagoras was the first philosopher to reside in Athens, coming there in 480 B.C., perhaps with the invading Persians. He became both teacher and friend of Pericles. He seems to have believed that at its inception the world consisted of a mixture of the seeds of every natural substance, be it organic or inorganic. In A.'s terms they would be the ultimate uniform elements. Anaxagoras had asked how hair could arise from "nonhair" and blood from "nonblood," and this led him to believe that a bit of every elemental substance must be in the sperm. Being very tiny there, it is "invisible."

94. The text now begins to follow more closely Ar. *GA* 722b3f.

ever, they are divided, they cannot be alive, for life is found only in an organic whole.[95] And if it is not alive, then it is not from a living whole [*totum vivum*] and, according to them, a living whole is not generated from it. If, however, the members are joined to one another, then the sperm will be a living, organic body. An animal is just such a thing. Thus, the sperm would be a small animal, which is absurd.

Further, how is it that the sperm of two is joined together, as when the sperm of the female is joined to that of the male? For if each of them is derived from the entire body of the female and of the male, then from the sperm the young should bear a definite resemblance to both the female and the male in all its members. Thus, as we said before, there is not one animal, but two, and from these two, two animals will be generated.[96] For, according to these men, one does not act in the other. Rather, those things which are in the sperm have the formative power, that is, the power to make distinctions [*virtus formativa sive distinctiva*] from the members and not from the testicles.

If, moreover, someone should speak in this way, then the statement of Empedocles will doubtlessly be the suitable one. For he feels that there is a certain nature in the male and the female which bears an essential resemblance to nature, where male and female exist, and that the entire sperm does not depart from just one of those generating but from the entire body of each. In fact, however, the nature of the members is distinct and what flows from distinct natures seems to have a distinct shape, power, and function.

68

Further, what is the reason that the female does not produce from herself alone if the sperm comes from the members and has its power from them? Indeed, the female should conceive and give birth alone, since the sperm has the formative power from the members and since in the female herself the receptacle of the womb exists, in which the embryo can be formed and completed. There is nothing else needed for generation.

It seems, however, that one of two things must necessarily follow from these [statements?], namely, that the sperm does not come from the entire body or, as we said, that it comes from each parent, namely, the male and the female. An indication of this is that generation does not occur without intercourse, so that both sperm might be together. Then the inconsistencies [*inconvenientia*] we spoke of above will follow.

Further, this seems to be unsuitable, for since there is one sperm for many twins (that is, offspring), it should happen that the distinct members are preserved in that sperm, even though they are invisible. And because distinct members have powers and potencies for vital functions, they should be animate. For just as the entire soul is the act [*enthelechia*] of the entire body, so the parts are the acts of the parts.[97] If

69

---

95. "Found in": *convenit*, perhaps "is suited to." "Organic whole": *organicum coniunctum*, lit., "an organic (thing) that is joined together."

96. "Animals": perhaps "live young."

97. At *De anima* 1.1.4, A. provides a definition of "entelechy," noting that what the Greeks call *endelechiae*, "Latine sonat perfectiones sive actus . . ."

the eye were an animal then vision would be its soul, as we have said in the book *On the Soul*.[98] Empedocles, however, said this, claiming that the heads of twins were present in the semen without being connected to the neck and that afterward they were connected through the agency of Friendship. It is clear, however, that soulless, unconnected members cannot be preserved in the semen. For then they would not, according to Empedocles, be functioning according to the potency of life. It is likewise agreed that it is not an animal. For, according to Empedocles, in reality, an animal exists only when made up of many connected and composite organs.

70   Nor can it be said again that these members are many animals, so that, as we said, each one is an animal in and of itself. For if the eye were an animal, then vision would be its soul. Thus it cannot be said that the members which, according to Empedocles, are in the semen are numerous animals connected and arranged one to the other to make up a single thing composed out of many animals. For such a statement is only posited by one who believes that the sperm leaves the body in the substance of the members as we said.[99] Because something which has been closed up in the earth and then is dug out and brought to the light arises from it not in material so much as "in place" [*in loco*]. This is the sort of arising of the sperm which Empedocles claims occurs from all the members gathered together at once, saying that it is the result of Friendship gathering the members together into the single substance of the sperm. According to this position the sperm will be an unseen member which has left the members of the body. Against this, however, the members cannot be connected in such a way that that which leaves them is a sort of connected unity [*continuum adunatum*], as Empedocles says. For although they might be connected in the animal's body, their forms are still different. For one is flesh, the second nerve, the third bone, and so on for the others. That which leaves those such as these retains the form of that which it leaves, and it is not connected and joined in one place and "act" [*actu*] of the spermatic fluid, since, as we have said, the members are not joined together. For it seems ridiculous to say that the upper, lower, right, left, external, and internal members are distinct and yet that they are in the one act and place of the sperm. Therefore, all the statements of those who hold this belief are entirely false.

71   Further, certain members have distinct and principal powers, as has been determined in the preceding portions of this study. Some, however, do not exist for themselves but rather accidentally, that is, for something else, since they serve principal members. The manner in which this distinction of the members is preserved in the sperm cannot be assigned, even if one uses one's imagination.[100]

Further, if we wish to speak about this position in greater depth, we find that the nonuniform members are distinguished from the uniform ones in that these, because they are official members, can perform their own actions. Thus, the tongue is for speak-

---

98. *De anima* 2.1.3.
99. The translation offered depends on reading Borgnet's *membrorum* for Stadler's ungrammatical *membro*.
100. "Uses one's imagination": perhaps, more coldly, "even if one were to make it up."

ing and the eye for seeing, whereas the hand and foot perform their own activities. The uniform members, however, are distinguished from one another by hardness, softness, and the other proper, accidental traits which especially belong to them. It is impossible to find these differentiated differences in the sperm.

It is therefore clear that it is impossible that the sperm is derived from the body of a person and has a share in all the uniform and nonuniform members, specifically, that the blood be from the blood and the flesh from the flesh.[101] The only ones who make this statement are those who say that it comes from all of them. For it can be from a member that has a single nature [*ratio*] even if we do not concede that blood is from blood, for according to Anaxagoras, all the others are in each one of them. For if all are in all, why is it that the sperm cannot be from some other single uniform member? For this position is like that of Anaxagoras who holds that the sperm comes only from the uniform parts, as we have said above. But Anaxagoras holds to this statement only on the generation of animals which are found to have uniform parts like this from which others are composed, which cannot exist from nothing. Therefore, these must have precedence in the semen, even if they are not visible.

Empedocles, however, supports that which he says from all the beings and movements in general which are composed solely through Friendship, out of those things which were prior in the act of the form even though they were not prior in the act of composition. For he says that in growth, things which existed previously are added through the agency of Friendship and that they are taken away during diminution, and that it is the same for other movements as well.

But let us ask from these men how the members of the embryos grow to their requisite size if there is nothing in them except that which is first derived from all the members of the one generating. For they cannot say that the members are in the sperm in the size they have in the young when it is first born, for in that case they would be visible and distinct as far as location and being seen is concerned, a thing that is clearly false.

It is also for this reason that the statement of Anaxagoras is more bearable than that of Empedocles. For he says that flesh is added to flesh during the growth of the members and that it is this way for the other uniform members, although hand is not added to hand or foot to foot.

Empedocles, however, and his followers contradict him in saying that flesh is surely made greater and grows as a result of the addition of food, but this food is not changed into flesh since this food is not suddenly rendered sperm, which is like flesh and the other members in form. Rather, so they say, all the members are generated from the sperm, since it has all the members in itself and it leaves from all of them as well. When the food passes through the members it gathers together invisible members and, fed on these, the body grows and in this way produces flesh, blood, and the other members. This position is based on the statement that the substance of the members emits another substance from itself without any diminution to itself and that this substance

---

101. I.e., blood is not blood in the sperm, and the like.

contains unseen all the other members in it, unconnected and indistinct, as if the very substance were multiplied in itself with an eye to the emission of this sort of humor or substance.

Even today many people of our time, filled with error, defend this error. But this statement cannot stand since we do not see that that which leaves the bodies of the animals does so in a mixture. Thus wine, previously unmixed, is mixed with water, and after the water was put in the wine it caused it to have a thinner and more penetrating substance. When these things are mixed, however, they are indistinct, but are united in the act of that which was mixed. Thus, whichever things are in the sperm, are indistinct and united in the act of that which was mixed.

CHAPTER III

On the Improbability of the Statements of Empedocles and Anaxagoras, Based on the Reason Maleness and Femaleness and on That of Those Who Generate Those That Are Not Like Them in Genus

75 His statement which claims that the sperm is actually flesh, bone, and any given other member (nerve, for example) and that it is therefore composed from nerve, flesh, and other members is entirely contradictory to our statements on natural matters and to our position on the cause of male and female sex, especially as it relates to what Empedocles says. For Empedocles says that the cause of the generation of a female or a male lies with the time of the impregnation.[102] For if at that time the sperm which exits from the whole body of the male conceives, the *conceptus* will be male and the sperm of the female flows away. If, however, the sperm which leaves the entire body of the female conceives, the embryo will be female and then the male's sperm flows away. If, however, both sperms conceive, she will conceive twins, of which one is male and the other female.

76 But we have proven the opposite of this experientially [*per experimentum*]. For we have seen women who do not conceive at first and then, following some change in their bodies due to age, medicine, air, or locale, do conceive. And we have also seen those who for a given time always bore males and then, afterward, females. Some are disposed the other way around. And it is irrational to say that one of the seeds always flows away and is never conceived. Therefore femaleness and maleness are not caused by the fact that a sperm which comes from the entire body of the male or the female is conceived, but rather for the other reason we assigned earlier in this study. This is the temperateness or intemperateness that is present in the sperm of the man and woman as they are mixed together. From the things, then, that have been said here and in the earlier parts

---

102. Ar. *GA* 723a24f. cites Empedocles as saying that the woman is formed when the semen encounters cold (Diels, 1951, frag. 65).

of this study, it is clear that that which is amassed from the two sperms is a thing which properly belongs to the generation of the member which shows maleness or femaleness. For from one sperm that arises out of two that have been mixed together with respect to their substance, there can sometimes be a male and sometimes a female. It is not because there comes from it some member which previously lay unseen in it. Rather, it is because there is a formative power in the sperm suited to the generation of the male or female member, the result of various degrees of heat or cold which are present in the complexion of the sperm. We say entirely the same thing for the other members as well, for if it is proven that the sperm does not necessarily have to emerge from the womb of the one generating for the generation of a womb then this same type of conclusive argument [*rationis probandi*] will exist for all the other members, and thus it is not necessary that the sperm come from any member at all.

Further, as we have said above, some animals are born during the first generation and in appearance [*in specie*] are not like those generating them but do not, however, remain terminally unlike their own genus. Examples are certain flies and bees. Those which are generated from them are not immediately like them since they generate larvae [*vermes*]. It is agreed, however, that those which are differentiated and unlike in their generation this way are not from sperm which, coming from all the parts of the one generating, has in itself members that are like those of the one doing the generating. For if they were born from sperm like this, they would certainly immediately resemble the ones first generating them. This would especially follow from those who say that similarity is a sign that the sperm comes from all the members and from the entire body.

Moreover, some animals generate many young from a single coupling. This happens, however, most often in plants which generate many fruits and seeds from one simultaneous rising of the sap which occurs one or more times each year. If, then, the seed is that which produces all the members (for it is derived from all of them and has, invisibly, both the members and the powers of the members in itself), it should happen that many invisible members exit along with their powers at the exit of a single sperm which exits with a single movement. Yet this is most absurd. How can it possibly be? For if the sperm is derived from the entire body as they say, it follows that one and the same sperm leaves with but one movement and a single opening of the members. For one and the same sperm flows forth from one coupling and one division of the members.

Nor can it be said that it is divided up in the womb, into the members and powers of many, for they say that the sperm is caused to divide into members because it comes from all the members of the animals. This, however, has already been proven to be impossible.

Moreover, if branches are cut from a tree and are set in the earth, there are sometimes found those which bear fruit and produce seed from which a complete appearance [*species*] of a tree germinates.[103] Nevertheless, it cannot be said that this seed comes

---

103. The Latin, *germinat tota arboris species,* is a bit stilted and could merely indicate that a new tree, entirely alike in species with the original tree, sprouts from the branch. Cf. A.'s further discussion of this process at *De veg.* 2.48 and 5.40–41, with judgments as to which trees best respond to this sort of rooting process.

from all the parts of the tree, for the sperm (that is, the seed) comes from these branches on its own, without any connection to the tree as a whole. It is clear that the power of bearing fruit is in them before they send forth roots from themselves, for if they did not have the power before, they could not send forth roots from themselves. It therefore follows that the sperm for this fructification does not come from the tree as a whole.

An indication of what we have said is something which is quite commonly seen in certain ringed animals. Now it does not occur in each and every genus of the ringed creatures, but it does in most if one were to consider the matter closely. The female has a genital member that she uses like a penis to conceive, placing it within the male's body and drinking the sperm from him.[104] This is proven because in every animal, that which is placed below during copulation is the female. Yet in these ringed animals, while she is placed below, she raises up a penis that is shaped like a tail and that is, namely, the lower part of her body. She puts this into the body of the [male] ringed creature lying on top of her and drinks the spermatic fluid from him. As we said, this does not happen in all the ringed animals but in most, and it is clear that she drinks in only the fluid found in the lower portion of the body.[105]

It is therefore clear that the reason generation lies in the sperm is not that it comes from the entire body. Rather, there is another reason for this and we will investigate it later.

A more suitable and better statement is that the reason for the generation is not that the sperm comes from all the members and from the entire body, but that it is from the operation of the formative power which serves as the artisan and craftsman of the entire body. We thus also say that a footstool, with respect to its form, exists because of the carpenter and not from the material which has within it the parts of a footstool in a general way. These men bring forth a remarkable fantasy in their search for the reason the one generated resembles the one generating. It is as if someone were to say that the son is like the father because he has members that are like those of his father.[106]

Moreover, the reason for the great pleasure in intercourse is not the one they bring forward, namely, that it is because the sperm comes from the entire body and all its parts. Rather, the reason for this pleasure is the deep and strong movement of the sperm. This touches the inner, nerve-filled genital areas, especially the area of the foreskin of the penis in the male and the nerves of the womb in the female.[107] The rubbing also contributes to this by an opening action and by leading on the heat. An indication of this is that those who copulate a great deal take less pleasure in it since

---

104. The female's penis is presumably the ovipositor; cf. 15.123.

105. The activity described does in fact occur in many insects. In dragonflies, for example, the female's head is held by special claspers at the end of the male's body, allowing her to dangle below, curling her ovipositor beneath the male to take on the sperm. A. returns to this in some detail at 15.123f.

106. Ar. *GA* 723b31f. is more direct: "As it is, they talk as though even the shoes which the parent wears were included among the sources from which the semen is drawn, for on the whole a son who resembles his father wears shoes that resemble his" (trans. Peck, 1942).

107. "Foreskin": *praeputium* may also refer to the entire tip of the penis. Cf. 1.467.

they are cooled by their practice of intercourse and the passages through which the semen flows have become loose. Thus, the semen does not touch them acutely.

Moreover, pleasure occurs only at the end of intercourse and in the genital members. Yet, according to what these men have said, it should be in the entire body, equally throughout all its members. For there should not be pleasure in some members first and in others afterward, for if it were this way, the pleasure from intercourse would last a long time before it passed through all the members successively. But this is not true, for the pleasure lasts only a short while. And yet this is wondrous, for if the pleasure were in all the members equally and occurred all at the same time, the sperm would have to descend from a member that was a great distance away with the same swiftness that it does from a member lying near its point of exit. Yet this is impossible, for all the members are moved to pleasure with one movement and one mover.

Moreover, if what these men say is true, the offspring of those who are maimed and have imperfect and truncated members will always be maimed, imperfect, and truncated in their members. But we see with our own eyes that this is false. In what follows we will look into the reason for each of these things, namely, the reason for the resemblance of the young to their parents and also for the truncation of the members which is sometimes present in the parents but not in the offspring. For this is a question common to each of the accidental traits mentioned.

The reason, then, for the things that have been said is not that the sperm comes from the entire body and from all its members. Nor does any charge of falseness follow the things we have said, even if we admit that sperm comes from the female. For we admit that the female is in some way a cause of generation through her sperm and the conception of her womb, as we will explain later.

CHAPTER IV

What Sort of Principle of Generation Is the Sperm When It Is Said That Generation Occurs from Sperm, and What Is It According to Nature?[108]

Having determined these things, and having proven that the sperm does not come from all the members of the body, we should first inquire into the nature of the sperm and come to know it definitively.[109] For if we know this perfectly, we will easily come to know all the operations of sperm and the reasons for the accidental traits in it both in and of itself and naturally [*per se et secundum naturam*].

Let us say, then, that sperm, from which generation occurs, should be that in which the sustenance for the entire offspring [*genitura*] is naturally present. This hap-

---

108. This cryptic chapter heading is clearer in Borgnet's version: "What Sort Is the Principle of Generation When It Is Said That Generation Occurs from Sperm?"

109. There are two typographical errors in the Latin of this sentence in St. They are easily remedied, however.

pens only if it is the first thing laid down as a subject in generation and is that from which comes that which is the first foundation and sustenance of the creation [*creatura*] that is to be formed. This can be seen in the sperm that comes from the human, for in power this is like a human in act, and it is like a human from which the operation of a human first comes. The same thing is also true in other animals.[110]

One should know that something is produced from another thing in many ways. One way in which one thing is said to be produced from another is found in the case of a sculpted image produced from bronze or stone and when a footstool is produced from wood. This is generally so whenever something is said to be produced from something else in the sense of being from it with respect to its material.

The second way in which one thing is said to be produced from another is found in the example of a person being rendered nonmusical from having been musical or being made ill from having been healthy. Thus it is, generally, when a contrary comes from its contrary.

The third way in which something is said to be produced from another differs from the two mentioned so far. Thus, Antyfon said that *olimpia* are produced from incense [*ex incenso*] or also that battles are produced from sulfuration [*sulfuratio*] or from something else from which a battle is the consequence.[111] For when sulfur or burning [*incendium*] is smelled, the soldiers, knowing that the enemy is nearby burning the fatherland, move out to fight. When a battle is produced this way, as a result of burning [*ex incendio*], one thing is not being produced from another either in the

---

110. The Gr. is corrupt at this point also. Cf. Ar. *GA* 724a14f. However, A. apparently means that the sperm has in power (*virtus*) or potential what the human is in act.

111. Sc. gives the name Anformez, and A. has changed it, possibly having in mind Antiphon the sophist. At this stage Ar. *GA* 724a28f. actually cites Epicharmus (a Sicilian comic poet) as saying that from slander comes abuse (*loidoria*) and from this abuse in turn comes a fight. Sc.'s version gives this as "pugna fit ex susuratione," "a fight comes from whispering." The whispering has changed to *sulferatione,* leading to a fairly lengthy discursus on military matters. A. has apparently added the information about the *olimpia,* but it is not immediately clear what they are.

At *Phys.* 3.2.10, A. says clearly that the *Olympia* are *agones,* that is, games (*ludi*), and equates them to *torneamenta,* claiming that the Olympics were held every five years and had as their purpose the training of young men in warfare. He is following Ar. *Phys.* 206a9f., which states that in one sense the Olympics provide an example of infinite succession. At *Phys.* 2.2.1, A. parallels the wording of the current passage by saying that the Olympics are produced from the *palaestrae,* that is, wrestling schools. In neither passage is there any mention of burning.

If the games are not meant here, an interesting possibility exists. The normal meaning of "celestial events" or "heavenly phenomena" for *olimpia* does not impart sense to this phrase. Blatt (1959, s.v. *olympius, -a, -um*) cites two uses that are intriguing. One is to refer to a black stone, spotted with white, which, when struck by the sun, sends forth a fire the way sulfur does. This is called the *olympius lapis.* Perhaps A. has in mind *olimpium saxum,* which would give the requisite ntr. form? Unfortunately, the stone is not mentioned in his *De min.* A second use is the term *antidotum olympium,* whose other name is given as *hydrotopion,* which is used as an antipyretic. Note that elsewhere the compound term *incensum sulphureum* occasionally means merely "incense" (*LLNMA,* s.v. *incensum*). We must also ask if *incensum* here is "incense" or merely "something burned." Nowhere, however, does A. seem to know that the ancient Olympics began with the lighting of a torch. See the following note.

sense of being from its material or as coming from its contrary. Rather, it is from it as if from a symbol of it.[112]

In whichever of these ways one thing is said to be produced from another, it is necessary that it have some principle of motion. Some of the types mentioned have a principle of motion in themselves, as do the two types which we mentioned last. For burning, that is, the throwing of sulfurated fire, is the principle that moves to battle.

Some things, however, have an external cause for one thing being produced from another. For example, we said that art is the principle of the *apothelesma*—that is, the thing produced by the art [*artificiatum*].[113] This is the case when a statue is produced from bronze, and so we say that a candle, which is no part of the house, is still the principle of illumination in the house.

It is clear, however, that the sperm falls into one of the types mentioned when we say that the sperm is the principle of generation in the sense that it is that from which generation is produced. That which is from sperm is from it only as from a material or as from a mover and it cannot be that it signifies one thing is after another as when we say that midday comes from morning and as Homer said that the ship was sunk by the navigation of Atynez.[114] Nor can a creation [*creatura*] be from sperm in the sense of one contrary coming from another. Neither do I say that a contrary comes to be from a contrary and that something comes to be from an element or subject in the same way. For a contrary is never an element or a subject since an element can exist in and of itself without either of the contraries. But none of the contraries can exist without a subject or an element. For in contraries there must always be something that is the subject from which will come that which is produced besides the contraries.

85

From the fact, then, that sperm is either like a material or like a mover or, perhaps, is like both, one should consider the nature and disposition of the sperm as to whether it is like a material. For if this is so then it is like that which is acted upon and is moved. We must also consider, if it is not like a material, whether it is like a form which produces [*forma faciens*], just as there is an art and an artisan.

However, it is clear from the very nature and operation of sperm that it is equally the principle of generation in each of these ways. Thus, sperm will be both these things equally, namely, that some of it is like material and some other of it is like the form and artisan. Generation, however, that comes from contraries is natural to be sure, for it is a product of first, simple generables as we have said in the book *On Genera-*

86

---

112. A. has frustratingly moved from *incensum*, "incense," to *incendium*, generally "torch" but in light of his military context perhaps he means a special siege weapon consisting of incendiary devices laced with sulfur. This may even hearken back to his understanding of the Olympics as military training games. "Soldiers": *tirones*, perhaps "squires."

113. *Apothelesma*: Gr. *apotelesma*, a finished product.

114. Cf. Ar. *GA* 724b1f., which says that it is not a case where "from" means "after" as when "from the Panathenaian festival there comes a sea voyage," which refers to the sending of a sacred ship to the island of Delos following the Panathenaia in Athens. Thus, Atynez seems to be a remnant of the name of the city and, oddly, the insertion of Homer's name is due to A., much as when, just above, he inserted the name Antyfon.

tion [*Peri Geneos*].¹¹⁵ These contraries, which are the qualities of elements, are in the sperms. From them, as if they are undergoing transmutation, comes the generation which is from the sperms to the extent that it is a natural [*physica*] transmutation. However, the substance of the sperm lies in the nature [*ratio*] of these principles and it has in itself the transmutating contraries, sometimes from male and female, but at other times it is emitted by only one of them, as is seen in the seed [*semen*] of trees and, generally, of all plants. It is also this way in certain animals about which we made mention previously. For in them there is no differentiation of the male and female sex. In such a case it is produced from one of them but has the powers of either sex, as if it had come from both male and female. This is as we have stated in the book *On Plants*, that the powers of the sexes are united in the egg.¹¹⁶

87 The sperm is the substance which, throughout all the species and types of animals which generate—whether they copulate or not—comes from the one generating. It is naturally the first subject in which the powers of both generating ones are united. Whether they copulate or not, they still have the powers of the sexes mixed together, as do the trees and the animals in which there is no differentiation of the sexes, as we have said. In these the sperm is as if it came from the male and the female and has the powers united as they are in the egg, as we have stated carefully in the book *On Plants*.

In such seeds as this each power is present, namely, the male and the female, and in some it comes from two of them, divided. But in the plants there is sometimes a difference between the seed and the fruit and yet sometimes, with respect to subject, they are the same. For fruit is that which comes first from the plant, like apples, pears [*pira*], quinces [*citonia*], *melangi,* and the like.¹¹⁷ Therefore the fruit is a certain principle of the seed. The seed, however, is that which has been completed last and is the end [*finis*] which the plant produces to be something that will germinate into something like itself. If, however, this differentiation is not found in some plants, then in that plant the fruit and seed are the same thing. It is this way in wheat and in the other seed grains which do not grow in a fleshy or in any other sort of substance.

From what has been said, then, one can gather that because the powers of both sexes are united in it, the sperm is the principle of generation and its relation to generation is as material and artisan.

---

115. *DG* 1.1.24f.
116. *De veg.* 1.39–50.
117. *LLNMA* (s.v. *cydonia*) cites literature to equate *cydonia* as a sort of pear and a type of fig. This is also Jessen's identification (1867, 710). Yet A. has just used *pira*. It is perhaps better to take it here in its older sense, however, for in antiquity the Kydonian apple, named after a town on Crete, was undeniably the quince. Jessen (1867, xviii) comments on this passage in a footnote and wonders if the *malangi* might not be the *melangula* or *citrangulus,* the lemon tree (cf. *De veg.* 3.97).

CHAPTER V

In What Way Sperm Is Present in the Body of the One Producing It and in What Way It Leaves It

Moreover, the sperm is that in which there is first present the entire nature of that of which it is the seed, according to the power and according to the potency of the material. But I am calling what is called the sperm in Greek the nature of that of which it is the seed. For everything which is in the body from which the semen is derived is either one of those things which are like-natured to the body or is one of those which are not like-natured. I am using like-natured [*connaturalis*] for those things which compose the body, like the uniform members or those members which are made up of uniform members. I am using unnatural [*non naturalis*] for the superfluities which are in the body and are not like-natured to it. For they have not attained to any likeness of a member which is actually a part of the body. An example is some superfluity or a collection of humors in the body or even the superfluity of food. If, then, the sperm is in the body and leaves it, it must be in it either as like-natured or not like-natured. But it is clear that it is not like-natured to the body as is some other part of the body. For the parts are either heterogeneous or homogeneous. The sperm is not a heterogeneous part since all the parts of the sperm resemble each other while the heterogeneous parts are dissimilar. It likewise follows that it is not a uniform part, for nonuniform members are composed of uniform members as various organs are made up of flesh and nerves. However, not one of the members is made up of sperm.

Further, every member, be it uniform or nonuniform, is distinct in the power, form, and operation within the body of which it is a part. Yet the sperm does not have this sort of distinction. All other things that are in the body are there, as we said, possessing a distinction of the sort we mentioned. Likewise, sperm seems to be from none of those things which are not like-natured to the body. Neither does it seem to be a thing brought on by disease or the dominance of a *discrasia* in the body, for it is in all bodies which are in a natural power and condition and it is from the sperm that the nature is sustained and endures.[118]

It is also clear that it is not in the body in the way food is. For food is put into the body from without and is distinct from flesh even though it is through it that it happens that that which is potentially flesh and potentially nerve and bone become flesh, nerve, and bone in actuality.

Perhaps someone will wish to say, in the way Empedocles and Anaxagoras did, that it is a certain superfluity that comes from all the members from the dissolution of the

---

118. For a discussion and definition of *discrasia*, cf. 7.93, with note.

members rather like sweat.[119] For these men say that it leaves from the entire body due to the heat aroused by the motion of intercourse and that it has the power of those things from which it comes by means of this dissolution. If this is the case, then the sperm must be a kind of superfluity that comes from the members of the body and the sort of superfluity which is the result of dissolution. For every superfluity of this sort comes from food and is the sort of thing which is not needed for any of the natural functions. Rather, it is an impediment to nature and is thus expelled since it can be useful to no natural function. Instead, nature is needed in order to consume the majority of this sort of superfluity which is not needed for natural functions. An indication of this is that the members are relieved by consuming a superfluity of this kind. Yet it is clear that the sperm is not a superfluity of this sort. For there are many such superfluities in certain ill persons who have a weak complexion and whose members have weak powers which cannot cast off the superfluities that are generated in the body. There is only a bit of sperm or none at all in these individuals and if there is some it is not suited for perfect generation, for if it does generate, the offspring will be sickly and will die since part of the bad superfluity is mixed in the sperm coming from bodies like these.

91 Moreover, the superfluity of the first food, which is drawn from the stomach to the liver and which stems from indigestion, is phlegmatic or something like phlegm since phlegm is the viscous superfluity of good, unburned food. This is needed as we said above, either for the nourishment of the phlegmatic members or also because it is convertible into food through the agency of the digestive heat. An indication of this is that it is mixed in with the blood, which is food, and as time passes, and when it has been dissolved by the heat, it will be like food and will be consumed, even turning itself into blood. This happens when the body is at work and is aroused by the heat caused by the work. Now if a member of the body is fed on this, this will be the last member to take on food. These are the members which are fed from that which is digested last and from the nourishment. The first member which takes nourishment from the first food is not fed by this phlegmatic humor.

92 The member that is last in position is also fed by the phlegmatic humor, and this is the brain. The first, which is the heart, is not fed by it but rather by the warm, choleric blood. That superfluity which is left last, after all the digestions, is very slight even if the first food from which it is derived is abundant. This is often converted into growth of the members and this occurs either in length, thickness, or width. For we know from the things which have been determined in our previous books of natural philosophy that all animals and trees grow little by little until they arrive at their natural limit, which is the boundary of their size. It therefore stands to reason that the sperm

119. *Dissolutio* and its derivatives are difficult to translate in what follows. The Latin stands for the Gr. *syntēgma* or *syntexis*, a breakdown of tissue into noxious fluids (Balme, 1972, 146; Peck, 1942, lxvi). While the Gr. noun is traditionally translated "colliquesence," the Latin stems simultaneously imply not merely liquefaction, but also deterioration, and the English "dissolution" seems to capture both senses. Both meanings should be kept in mind, as the words occur in what follows. Note the wordplay also at 15.105.

is not a superfluity of the sort which only resembles sperm, for sperm is viscous and watery. The choleric and melancholic superfluities have no resemblance to sperm and thus we should not speak of them. Blood, though, is not a superfluity, but it is food, as is clear from the things determined above in this study.

Thus, from all the things that have been said, it is clear that sperm is not a superfluity of this sort and that it does not come from all the members like something expelled, set free from the body by heat.[120]

Having proven these things, then, we should speak of the nature of sperm in a way quite contrary to those who have spoken on sperm before us and whom we mentioned previously. For they all agreed that the sperm came from the entire body. We, however, say that the sperm is that which has a nature which, from the active form and material, is suited to the whole, but which does not have any similarity to all the members of the body.

Moreover, they said that sperm is a certain dissolution of the body or a humor which comes forth as a result of a dissolution of the body like sweat does. We, however, say that it is indeed a superfluity of the fourth digestion which is needed for the preservation of the species as concerns the existence it has in a succession of individuals.[121] Thus it is better and more suitable to say that it is the last superfluity, that is, it is left over from the last digestion and is added to that which is assimilated to the last body. The superfluity is added this way to the individual and has the potential of being made like the entire body in actuality.

This, therefore, is that which serves generation, for it already has the powers of all the members of the body and has potentiality [*posse*] from that similarity which it has undergone. This is like the potentiality art and the artisan have over the material of the thing produced. Thus, Polyclytus has the power [*virtus*] of making statues shaped like men.[122]

In speaking this way we will not say that sperm is something which leaves the body through dissolution. For everything which leaves an animate body in this way is corrupted and changes nature. Sperm, however, is not corrupted but instead, more than all natural things, accomplishes marvelous functions to the extent that it is thought of as a divine thing by some because of the creative and formative power which is in it.

That it is emitted from the body according to the nature [*ratio*] of this sort of superfluity and not according to the way or method of dissolution is signified in that large-bodied animals have little generation and semen in proportion to their species, whereas those with a small body have great generation and much semen. Yet there must necessarily be a great deal of corrupted fluid set free [*dissoluti*] by the members in large-bodied animals. There is but little superfluity from the fourth digestion since

---

120. I.e., in Aristotelian terms, it is a residue of digestion, not a product.

121. "Indeed": *quidem*, but perhaps better read, with Borgnet, as *quoddam*, "a certain superfluity." On sperm as a product of the fourth digestion, cf. 3.151–52, 161; 9.109; 16.32; and 17.3.

122. Polyclitus was the leading sculptor of the last half of the fifth century B.C., famed for his sense of proportion and fine finish.

almost all which is completed by that digestion is drawn to the large members which need a great deal of nourishment. Thus, little of the superfluity remains. In animals with a small body, however, it is exactly the other way around.

95 Further, the superfluous humor that has been set loose [*dissolutus*] (that is, that which is set loose by dissolution) from the members of the body does not have a natural vessel and location (that is, a receptacle) in the body. It is expelled instead, but before this it wanders randomly among the members and weighs them down. But all the superfluities of all the digestions have natural locations and vessels to which they are directed. These locations and vessels are known in the bodies of animals. The superfluity of the digestion which occurs in the stomach is directed below to the intestine and the *secessus*.[123] The superfluity of the digestion occurring in the liver is directed to the bladder in those possessing one, and, in those that do not, to the intestines. A certain choleric superfluity of the digestion which occurs beneath the heart is directed to the gall bladder [*kystys fellis*] while that which is melancholic goes to the spleen. Phlegm, however, is convertible into blood, as we said above. In this way the superfluity of the fourth digestion, which occurs at the ends of the veins, is, generally speaking, directed to the womb in women, to the seminal vessels [*vasa seminaria*] which are the members which serve copulation. In the same way, milk, which is a superfluity of this sort, is directed to udders.

96 Further, it is proven from the accidental traits that are consequent to the discharge of sperm that it is not set loose [*dissolutum*] from the body, but that it itself is a humor in which there is the greatest natural assistance. For if more sperm comes from the body than is fitting, it brings on a great wasting and dissolution of the body and a clear weakening.[124] This is because it then does not have in itself the superfluity which is left over from the food completed at the last completion. It thus does not have the wherewithal to restore that which has been lost by its members. However, in a few people, quietude and a sense of easing occur after the discharge of sperm. But these are the young ones, and they have a lot of sperm but little or no intercourse and the sperm multiplies in them beyond measure. When the sperm has multiplied, it brings on a heaviness like that which occurs in a body that has been filled with the best, suitable nourishment. This is no flaw with regard to any complexional quality but rather the condition is solely the result of superfluous quantity, leading to heaviness in the members and the body as a whole. When the reason for this heaviness is removed through voiding, the body is relieved and restored, given a good disposition. The sole reason for the relief which is sometimes felt after the discharge of sperm is not the discharge of the sperm. For not only pure, good sperm leaves a man. Often the sperm leaves mixed with another superfluity of one or another corrupt humor and this superfluity is bad and leads to ill health. Thus, many men who have sperm mixed with bad humors in this way do not generate. Therefore, the relief to the body after

---

123. Cf. 1.24. Here *secessus* would seem to indicate the lower extremity of the tract, if it does not mean, lit., "privy."

124. At 3.151, A. compares this condition to a sort of anemia.

the discharge of sperm of this sort is sometimes due to the bad humor which has been mixed in and from which the body is freed and not to the discharge of the sperm.

Further, there is no sperm during the first age, which is the age of infancy and childhood. Neither is there any in the last age of old men nor in those who are very ill, especially in those with high fevers. It is consumed in the sick by their weakness and heat. In the old men, however, it is not digested because of their coldness and turns into a viscous humor clinging to the outside of their members. In the young, however, it passes over into growth. An indication of this is that we see that in the first five years of life a person generally takes on half the total height he will attain his entire life.

Animate creatures like trees and animals have, after the fashion mentioned, many differences among themselves in their possession of sperm. Not only do the genuses share these differences when compared to one another, but so too do individuals which are alike in species and in the shape that follows upon their species, such as when a human is compared to a human or a horse to a horse and the nobility of the complexion of one is compared to the nobility of complexion of the other. For some have much sperm and some but little while others do not have it at all, not because of weakness or illness but rather for the opposite reason. For some are healthy, possessing a great deal of flesh and fat, and these produce only a small amount of sperm and desire intercourse but rarely.

This also happens in many fertile plants whose fruit rots and falls, or perchance does not form because a great deal of its food remains uncompleted and undigested. This even happens in animals, especially in goats, for they do not copulate much when they are very fat. For this reason clever shepherds keep the goats in their herds moderately thin so that they can effect conception in as many she-goats as they can through intercourse. By a similar trick tillers of fields sometimes keep their plants moderately thin so that they might be rendered more fertile. This is the same reason that men who are excessively fat reproduce infrequently, less often than those with bodies midway between thinness and fatness.[125] A similar thing also occurs in the bodies of women. The reason is that something superfluous in fat bodies is digested quickly, passing over into fatness since it is very convertible, watery, and thin. Further, for this same reason, certain trees do not bear fruit, like the willows, or bear it but scarcely, like the Roman nut tree [*nux romana*].[126] For plants of this sort have a large body and draw in a great deal of nourishment.

As we have said, there can also be other reasons for accidental traits of this sort. A lack of sperm is sometimes due to weakness and an abundance of sperm is sometimes due to strength. Further, it sometimes happens that many bad superfluities amass due to overretention of sperm in the bodies of the sick. It does them no good but rather hinders their bodies. Thus, when bodies cannot be purged because of the amassed sperm, too many infirmities and even sometimes death occur from sperm that is held

---

125. "Men": *homines,* more commonly rendered as a generic "persons," but specifically contrasted with *mulier* in the next phrase. See Glossary, "Human."

126. Ar. *GA* 726a8f. lists the willow and the *aigeiros,* the poplar.

in immeasurably long. Thus, some are made healthy by intercourse and others die, as often occurs in different people or races who either overly abstain from intercourse or overly indulge in it.

Moreover, in many animate bodies, the passages for the discharge of the superfluities are not the same. This is the case in the human, in whom there is a dry superfluity from food and a wet superfluity from drink. These do not leave by the same passage, but in this case the wet superfluity leaves through the passage for the discharge of the sperm, and the dry superfluity has another passage of its own, separate from this one.

It is clear from this that sperm is not a dissolution, for a dissolution is not expelled through an established passage. It is expelled through all the members, just as it wanders about in all the members.

100 Further, as we said before, when any superfluity is excessive, its discharge relieves the body. However, the discharge of a dissolution leaves behind a certain weakness even though it does relieve the body. This is clear in those who have sweated excessively, no matter what the reason for the sweating was. Sperm, then, is not a dissolution.

Further, during the discharge of the sperm, two things sometimes leave at once, namely, the sperm and the dissolution, due to a mixture of the good food with bad humor. If, however, as some have held, it is sperm which is dissolved through a dissolution of heat from the members, then it would be entirely a superfluity composed of bad humor. Yet we see through experience [*per experimentum*] that sperm never acts at all like a superfluity set free [*dissoluta*] from the members. This is clear from the things said before. Therefore, the statement of those who hold that the sperm is separated from the members like a superfluity set free from them by heat is false. For this reason, all that remains is that the sperm is the superfluity of food completed during the fourth digestion. This is clear from all the things that have been introduced previously.

CHAPTER VI

On the Nature of the Sperm and the Menstrual Blood, Possessed by Both Male and Female—a Nature It Has Insofar as It Is the Principle of Generation[127]

101 What remains, then, is that we must set forth what this superfluity of the food is and that we should inquire into the disposition of the menstrual blood. For once this has been well established, it remains and is known in corollary fashion what the disposition is of female animals, whether they produce sperm like the males and whether that which has the creative, formative, and generative power is a mixture of the two sperm, specifically, that of the man and that of the woman, and, generally speaking, of the male and female. Or is it not a mixture and the sperm comes only from the male? If it is

---

127. This poorly written summary does not imply that A. thought males possessed menstrual blood. It is the sperm that is common to each, to the extent that he specifies.

conceded that it does not come from the female, then we will have to inquire whether the female contributes to the principle of generation or makes no contribution to it.[128] If it is said that she makes a contribution to generation in the sense that she takes the sperm into its designated vessel and, acting as an assistant for it by way of preparation, warms it, holds it, and provides nourishment to the creation, if, then, this be said, we will ask whether she has any other contribution to the causality of generation in any other way.

By way of beginning our inquiry into these things, we say that we have already stated above that the blood is the last food in blooded animals. In nonblooded animals, however, the last food is a humor which is analogous to the last blood and takes its place. We have also proven in the earlier parts of this study, *On Animals,* that the sperm is the superfluity of the last food, that is, of the food completed in the last digestion. This food, however, is only blood or a fluid that is decocted and completed which is analogous to the last blood in substance.

For no member is nourished and perfected except through the digestion of the blood and when sperm is digested and decocted fully to the point of its requisite exit and is altered, it is not like blood in color, as we have said above, where we determined the generation of sperm. It is, however, from the substance of blood and an indication of this is that if it should, for whatever reason, occasionally leave incompletely digested, it comes forth the color of blood. This occurs when a given male is overly engaged in intercourse and is moved to intercourse for the slightest reason.[129] An example is the *stincus* [starfish] or another of this sort.[130]

From this it is clear that sperm, with respect to substance, is nothing other than the superfluity of the last food, which is the blood scattered through the members during the last digestion. For this reason it endows the body with very great energy and weakens it greatly when removed, and if it departs immoderately it makes the body weak. Thus, too, those who copulate a great deal grow old quickly and die more quickly. Therefore, because it is scattered this way throughout the members, it is reasonable that the young very often turn out like their parents.

For the sperm, which is a part of nourishment which leaves from the members, resembles in power that which has stayed behind in them. Thus, just like that which has stayed behind, it resembles the members through the power it has already taken from them. Thus, that which exits has the power of doing things like these members and thus the sperm which comes from the hand, face, and the animal as a whole has the power of doing things like them. For although it may not be said that the sperm has a separate hand, face, or an entirely separate animal, nevertheless, in power, potential, and in the form of the artisan, as it were, it does have a similarity to each member

---

128. "Contributes to": cf. 15.58, with notes.

129. "Male": actually *vir,* implying human behavior, and Platt (1910, on Ar. *GA* 726b9) notes that this could arise from bleeding of the prostate in such a circumstance. But since A. gives an animal as his example, "male" seems preferable.

130. See 24.55(122) for details and references concerning the aphrodisiac qualities of this animal and an elaboration on its blood-producing effects.

and to the animal as a whole. Thus, sperm is, in potential, like that which every member is in act.

104   Yet, it is still not sufficiently clear to us from this determination whether it is like each member as far as substance is concerned, or only as far as power is concerned. Neither is it sufficiently clear from what we have said here whether the man alone, without the woman, is the efficient cause in generation through his sperm, much as the artisan is the efficient cause of the full completion of something [*apothelesma*] through the form of his art, which he has within himself.[131] However, it is still not clear from what has been said whether the cause of generation and the cause of the size and shape of the body are the same or whether it must have a certain disposition and motive principle from the soul. For we know that it is in no way possible that the hand or any other member of the body be in the form of a hand without the power of the soul. For any given member shares commonly in a given animal power and this is the power of the soul. For this reason, too, both the member and the entire body share a common name since they are called "animate." If the member were divided, the power of the soul which is in it would be its soul.

105   From the things said before, it is clear that the dissolution and weakness which occur in men from a great discharge of sperm is like the dissolution due to the setting free [*resolutio*] of a superfluity.[132] This sometimes happens when the sperm is dissolved right away by some heat, much as a fumigation agent is dissolved when it is placed on the fire. Thus, in the members and in the body the last food is sometimes dissolved this way by some hot disease which befalls the body. For in the bodies of the sick many superfluities are necessarily produced since the members of sick people are weak when it comes to expulsion. The digestion of this superfluity will be lesser and weaker since those who are ill have but little natural and digestive heat, since it is set free by the illness.

106   While speaking in like manner of the bodies of females, we say that just as was declared above, the nature of females is weak and has less heat than that of males. It thus necessarily follows that the last food in females is less well digested and decocted. Thus, too, the superfluity of the fourth digestion has, in them, more of the nature [*ratio*] of blood than that of sperm. For sperm by its name and its nature implies the formative power. This is not present in the superfluity of the woman but rather in that of the man, as we have stated previously where we spoke about the sperm. Thus, although that which comes from the woman is like the sperm of the male in color, nevertheless it is not alike in power.[133] For it is watery and thin, more fit to be formed into a young than to form one. It therefore has the nature [*ratio*] more of menstrual blood

---

131. *Apothelesma*: Gr. *apotelesma*. Cf. 15.85, with note.

132. On the terms, cf. 15.90. *Resolution* is rendered "setting free" due to the overtones of the English "resolution."

133. Here A. departs from Ar. *GA* 726b31f., who is discussing the female contribution as menses. A. seems to be referring rather to the lubricating fluid secreted during intercourse. Cf. 15.109f., where he calls it "white" menstrual blood and expands on its role in conception.

than of semen, that is, sperm. The fact that the discharge of the menstrual blood occurs in the woman in much the way that the sperm does in man seems to be indicated in that almost the same accidental traits are shared by each. For sperm begins to be present in men and menstruation in women at about one and the same age and at that time their voices also change and the fleshy portion of their breasts rises up.[134] Again, it is at the same point of old age that menstrual flow ceases in women and that the generative power ceases to exist in the sperm of men. From these indications, and those like them, we have confirmation that the menstrual blood is a superfluity present in the bodies of females analogously to the sperm in the bodies of the males.

Because the menstrual flow is great in females, they are thus often freed from other flows of blood. For example, females do not often suffer from hemorrhoids of the anal veins.[135] Neither do they often lose blood through their nostrils after they have their menstrual flow and they are free of the sciatic pains when they are cleansed suitably and proportionately through the menstrual flow.

Again, it is due to the same menstrual flow that the bodies of females are loose, soft, smooth, and have little hair and that the bodies of the males are the opposite. For the superfluity which would fill up the flesh and generate hair leaves along with the menstrual blood. Now according to the true way of looking at things, this is the same reason females' bodies are small in size, since in every genus of viviparous animals the females' bodies are generally smaller than those of the males. For the menstrual flow is found only in these genuses of females and is not present in the oviparous ones in which the females are sometimes larger than the males. This is most noticeable in women, for more menstrual blood leaves them than from all the other female animals. Thus their veins are less prominent and they have a smaller body, as we have said.

It is therefore clear that the menstruum in females is analogous to the sperm in the males in the viviparous animals. Now it is not possible that there be a discharge of sperm from the female in the true and perfect sense [*vera et perfecta ratio*] of sperm. But it is clear from these things that the sperm of the female does not possess the same causality for generation that the male's sperm possesses. Neither does it contribute to generation in that way.[136] That which is sperm does not have the nature [*ratio*] of menstrual blood and, conversely, that which is menstrual blood falls short of the nature [*ratio*] of sperm.

The manner in which sperm is called a superfluity is clear from the things that have been said. This is attested to by the accidental traits that are shared by those producing sperm. Fat animals, as we said above, have little sperm since the fat is generated from the superfluity which would pass over into the substance of the sperm if the fat were not there. For sperm is a superfluity from blood that is well digested and decocted. Although menstrual blood is analogous to sperm, its generation and that of

---

134. On these "breast apples," cf. 1.441.
135. But cf. 15.114.
136. "Contribute": cf. 15.58, with notes.

the sperm do not occur in one and the same way. For this reason, too, impregnation is easy in all the genuses of the *malakye* and of those animals called hard-shelled.[137] This is because they have neither blood nor fat, but they have something else analogous to fat and the superfluity of this is their sperm.

109 As, then, we have said, these two superfluities come from the bodies of the males and females, one from one and the other from the other. When the menstrual blood turns white, some say that this is a sign that the female is producing sperm. But this is a false sign, as is clear from what was said before. For although the white fluid that comes from the female's body during intercourse might contribute more generation than the red menstrual blood, it nevertheless has no analogy to sperm but rather to material, as is the case for menstrual blood. Neither is the young procreated from a balance between the two sperms but rather there is one which acts and the other is that which takes on and sustains its actions, as is seen in what occurs in milk reacting to rennet.[138]

An indication of this is that women who are experienced in conception say that quite often a woman becomes pregnant without having pleasure during the act of intercourse. She is also impregnated when she has pleasure during intercourse. When, however, there is no pleasure, it is a sign that the white fluid which some have called, poorly, sperm, did not descend. But it is truly said that when the man and the woman produce sperm at the same time, then generation occurs more readily from them, not from two, but from one, due to the joining together of an active and a passive agent which have analogy both to each other and to generation.

110 This process, however, is impeded by many accidental traits. For if the temperate moisture (which is derived, as it were, from the residue of the white menstrual blood which descends during intercourse) remains in the womb, then it easily conceives a good offspring. If, however, a great deal of this sort of fluid has left and that which is impure has remained, it does not conceive. This is why a woman who menstruates a great deal does not give birth easily.[139] For either the food does not remain in the womb or else not enough remains of that which could sustain a *conceptus* in women such as these, even when, from the power of the sperm, they conceive. Sometimes, too, the moisture is overly increased and it will then ruin the *secundina*.[140] Then the *conceptus* will not be held in and will die. But when the fluid is temperate the superfluity will come forth from it. What is left behind is adequate and contributes to generation and to the fetus. From this the live young [*animal*] is procreated and sustained.

111 Moreover, when there is no menstrual blood, there is no generation. Again, if there is a great deal at the beginning and it immediately slacks off, the *conceptus* will

---

137. I.e., oysters and clams as contrasted with soft-shelled creatures such as crabs and lobsters. Cf. 14.7–13.

138. "Balance": *examinatio*, implying the equality of two things on a balance scale.

139. *Parit*: lit., "to give birth," but it is probably better understood here as "conceive" or "breed."

140. *Secundina* is normally thought of as the afterbirth, but usage elsewhere shows that it is also readily seen as any sort of enclosing membrane, appearing in the eye (1.201), brain (1.530), and even eggs (6.15f.).

die. It is the same if there is a great deal at the beginning and in the middle but it is cut off before the end.¹⁴¹ For this reason women perform an abortion by being bled often, especially beneath the ankle on the inside part of the foot.¹⁴² The reason for all these things is because sometimes such an overabundant superfluity is left in the womb that it leaves the opening of the womb, softening its opening. For this reason too the wombs of very many women hang down after the purgation of their menstrual blood. When this overabundance is consumed, the womb does not remain at all dry and it then receives very easily sperm put into it during intercourse, and impregnation is the result of this.

Further, a woman sometimes finds some menstrual blood after impregnation. This is not generally bad, for it has been experienced that menstruation occurred in some pregnant women after the time in which they became pregnant, but the flow of blood was moderate. Since it occurred but rarely and was due to a bodily infirmity, it thus did not interfere with the *conceptus*.

Further, when the menstrual blood is limited proportionately in quantity and quality, it will be natural and in such a case the impregnation will be good.

From all the things said, then, it is clear that the female contributes to her impregnation according to the analogy of material, which is the substance of the menstrual blood, and that the menstrual blood is nothing other than the superfluity of which we spoke.

The opinion of the unskilled crowd of physicians which says that the woman has the causality of the moving principle in generation, because the fluid which she ejects during the pleasure of intercourse resembles sperm, is false. What is true is that at this time a certain white fluid comes forth which, however, does not have the power and nature [*ratio*] of sperm. Rather, it is proper to the womb and is discharged in some women but not in others. It comes forth most often from white, pale women and from those who most often bear females. It comes forth rarely from dark or swarthy women who bear males. The emission of this fluid, called "sperm" by these people, is a sign of a great deal of moisture in them. For this reason too a difference in food provides a basis for differentiation in this fluid. For fluid of this type is increased or decreased as a result of moist or dry food, as it is by those foods which generate sperm. Examples are the seed of the garden rocket plant [*eruca*] and certain others which we have mentioned in the book *On Plants*.¹⁴³ So, too, does the opposite happen in horses on the run, from whom a great deal of fluid comes. Their bodies are dried out because of the strong, frequent movement and because of the rush of the fluid downward to the places through which superfluities exit.

112

---

141. A careful look at A.'s problem with grammatical gender in this section reveals the perils of the *post illa* style.

142. "Perform an abortion": lit., "snuff out the fetus," *extinguunt conceptum*. The comment is A.'s and may be taken to reflect contemporary practice.

143. "Garden rocket": *Eruca sativa*, cf. *De veg.* 6.329, where it is said to increase the flow of milk in women and to cause erections in men.

CHAPTER VII

On the Reason for the Pleasure in Intercourse and on the Type of Fluid Which Flows from the Genital Area in the Female

113   The reason for the pleasure associated with intercourse is not only the exit of the sperm. It is sometimes caused by the discharge of the spirit-like windiness and of the spirit. An indication of this is that a great deal of spirit is then present in the sperm and is almost a substantive foundation for it since it is the instrument and vehicle for the formative power. The fact that sometimes all that comes forth is spirit is indicated in youths who cannot produce sperm yet, but still take pleasure from rubbing due to the touch and movement of the spirit in the genitals. The same thing happens in many men, especially in those whose heat is so great that it dries out the spermatic fluid in them. For pleasure does occur in similar cases as a result of the rubbing.

A woman's complexion, however, resembles a boy in whom the heat is dulled by excessive moisture and has not come to the time in which it can thicken his sperm for the sake of generation. For a woman emits a fluid of this sort, but the difference lies in that the woman does discharge and the boy does not because his fluid passes over into growth. Thus, in truth, a woman is to be compared to a man who cannot generate because of the great amount of moisture which he does emit during intercourse but which is nonetheless imperfect since it has not been completed by the bounding heat [*calor terminans*]. This happens to him because of the weakness of the heat which then cannot digest the sperm which is left over from the last food, this being the blood or the fluid which, as has often been said above, takes the place of the blood.

114   Further, just as the flux of the belly called lienteric flux occurs because of a weak digestion, so does flux occur in the undigested blood in the veins.[144] In such a case there will be a discharge of the menstrual blood which flows sometimes from the openings of the anal veins and which are called the hemorrhoids. It flows almost the same way as the menstrual blood except that it is a flow of natural blood and the hemorrhoids are an unnatural flow.[145] Because of the like-naturedness, however, of the blood to the sperm it so happens that those whose hemorrhoids flow produce little sperm and rarely have nocturnal emissions.

It is clear from the things that have been said that generation reasonably occurs in some way from menstrual blood. For menstrual blood is impure blood which needs further digestion and thus it does not have an active role in generation. It is the same way with menstrual blood as it is with the fluid from which the generation of trees is accomplished. For this is sometimes incomplete and undigested and then the trees

---

144. The term *lientericus* refers to the spleen.
145. There is an untranslatable set of wordplays at work here between "flux" (*fluxus*) and "flow" (*fluit* as verb, *fluxus* as noun).

need something else operating in them, digesting and purging the humor, such as the heat of the sun. Then, with the sun digesting, it is converted to the principle of generation and passes over into seed or branch and, generally, into plant. In entirely the same way, when the sperm of a male is completed and is mixed with the menstrual blood, as if with a bad, incomplete food, it digests, purges, and cleanses it. Then there will be effective generation for the sperm, but the food and the material will come from the menstrual blood.

An indication of what we have said, namely, that the female does not always feel pleasure during intercourse from the fact that she is producing sperm, is that the male also often takes pleasure and delight from contact alone without any emission of sperm. Further, emission of this sort of fluid does not occur in all women but only in those who are rich in this sort of blood. Nor does this occur in all female blooded animals, but in those which properly do not have a womb raised up beneath the diaphragm [*paries*], which is the place for the generation of the eggs, as we have said above. Further, the emission of a fluid of this sort during intercourse does not occur in the females of bloodless animals. For as we have said very often, just as there is blood in some animals, so, too, in some is there another fluid taking the place of blood in those animals that have no blood.

Further, the dryness present in the bodies of a number of humans and other animals is the reason that the bodies of their females are deprived of this sort of purgation of blood and of the emission of the fluid, and this is not present in nonblooded females or in those in which the womb is raised up beneath the diaphragm, as is the case in oviparous animals. For dryness of this sort diminishes the superfluity of the humor, for the humor is present in ones such as these only in quantities sufficient to provide the substance of the generation. Neither can anything exit from ones such as these during intercourse. It is therefore clear that there is not always a descent of this sort of moisture due to the pleasure felt during intercourse.

Viviparous, nonoviparous animals, however, such as the human and those quadrupeds which have their rear knees bent to the rear, eject a fluid resembling sperm in color.[146] All the types of these sorts of animals are viviparous and not oviparous. There is an exit of the fluid in all those of this sort unless they are flawed in their generation due to a generation of unlike species, such as occurs when the genus of mules is generated. Thus mules only generate in certain lands.[147]

But this menstrual flow is less visible in the females of other animals compared to what is seen in women. For the menstrual flow in women is very copious, just as the sperm of the man is copious compared to his body. For sperm increases in a body whose complexion is warm and moist, since in it the heat digests well and it provides the moist material. For this reason a great deal of superfluity, naturally well digested,

---

146. The added detail on the bend in the leg is from Ar. *GA* 728b9f. and has generated comment and even emendation. Peck (1961, 433; 1942, 104–5) has explained it well, however, and the point here would seem to be to exclude other nonmammalian quadrupeds.

147. Mules are, in fact, barren. The exception, offered by A., may stem from Ar. *HA* 577b19f. or Pliny *HN* 8.69.171f.

should be present in a complexion such as this. Woman, however, above all the other female animals, has much moisture, since she does not have those members into which the moisture in other animals is converted. For the other animals have either hair or, along with this, horns, and the moisture is converted into these. Since woman, however, has none of these things and is thus very moist, she must have a great deal of menstrual blood through which her body is cleansed.

117 As we have said before, however, menstruation occurs in a woman only at the time during which the sperm occurs in the men. This is because the genital areas, where the vessels that receive the two aforementioned superfluities are located, open up and pour them forth at the same time, as if men and women had the same nature of a common species. At this time, because of the diversion of the fluid to the places mentioned, hair begins to appear on the groin of both the men and the women and at this time the women's breasts begin to rise up. For the most part, menstrual blood begins to flow when the breasts rise up to a height of two fingers. Also at this time, the testicles of the women begin to swell as if they were abscesses.

118 In animals in which the males and females are not differentiated according to the shape of their genitals, many conceptions result from one and the same sperm.[148] We have given the reason for this in what has gone before. The nature of sperm differs in trees and animals.[149] For while each is a seed [*semen*], they are alike in that just as in grain plants many plants sprout from a single seed, so too many live young arise from one seed in animals. An indication of this is that one seed is usually the result of one act of copulation. However, from one copulation, many live young are usually generated in females of those animals whose nature it is to produce many at once, such as the dog, cat, and the like. It is also proven from this that the sperm does not come from the entire body. For it could not suddenly leave from one act of coition, already divided so as to generate multiple conceptions [*conceptus*], and neither could that which emerges into the womb all at once be divided into so many conceptions according to a separation of forms which it has in it from the members from which it flows. For that which flows all together only comes forth one time and that which comes forth at one time can take on only one form of the members. Thus, it happens most correctly in the way that has been stated, namely, that the sperm of the male provides the form and the principle of motion while that which comes from the female provides the body and the material. What happens in the coagulation of milk as a result of the rennet is like the relationship that exists between the sperm which descends from the male and the fluid which flows into the womb from the female. We have determined in what has gone before the reason for multiple conceptions and why the sperm sometimes divides into many parts. For it sometimes divides into many conceptions, sometimes into a few, and at other times only a single conception is conceived from it.

---

148. "Conceptions": *conceptus*; see Glossary, "Embryo." Here the sense of "conception" seems preferable to that of "embryo."

149. As often, the word "trees" for A. seems generic for "plants."

We will say to those who desire a more precise knowledge of this matter that in truth there is no differentiation in form with respect to that which forms all conceptions of this sort. For that which is divided is one, with equal form in all regards.[150] Division does occur, however, with respect to material, and since that which is doing the forming is in it, it too is divided. The division is accomplished by the areas of the womb, by its movement, by various introductions of sperm, or by a great deal of sperm and its powers that come together in various places when it expands throughout the womb, as we said above.[151] However, in itself sperm is neither many nor few absolutely, nor does it require another, additional digestion from the womb. It is also not prevented from sustaining and holding the menstrual blood it touches. With this the situation of the sperm, since it may be one in substance and multiple in power, there will be many conceptions in the womb for the reasons given and in each there will be generation from the principle of the sperm that is sustaining the material.

It is clear enough from all the arguments [*rationes*] we have presented that the female does not contribute to the principle of effective generation from any sperm which comes from her and which would have the true nature [*ratio*] of sperm. Rather, the contribution she makes is from her menstrual blood, which all female blooded animals possess. In other animals it is the fluid which takes the place of the blood. Whoever applies this statement generally to all animals and not to the human alone will find that what we say is the sole truth. Some have erred because they have taken into consideration only the type of generation that occurs in the human and because they thought that the woman herself has a sperm because she sometimes emits a white fluid during intercourse. In reality, she has no sperm save that which would take the name of sperm equivocally.

The argument [*ratio*] through which our statement is proven syllogistically is that in every act of generation there must necessarily be that which generates formally and effectively and that which is generated passively, upon which that which is doing the generating acts. These need not be entirely different but rather must be of one genus. For otherwise, that which is passive would not be born to undertake the actions of that which is active. They do, however, differ with respect to the form which provides being and gives definition to the humors present during generation. For in all things which possess distinct powers and bodies, that is, distinct subjects and differing natures, there must necessarily be different potentials for operating—that is, being active and being passive. For if it is the male who is the one moving and acting in the generation and the woman who is the passive one, it is clear that the sperm of the female does not correspond to the sperm of the male with respect to the same type of causality in generation. Rather, her relation to it is one of material. From the things we have said, it is clear that the first and proximate material of generation is that which has the nature and the true sense [*ratio*] of menstruum.

Let, then, these things which have been said about the power of the sperm be enough for now.

---

150. Perhaps "in all cases," or "in all animals."

151. "Come together": *constant,* implying a congealing or coagulating action.

CHAPTER VIII

On the Way in Which the Sperm Which Comes from the Male and the Humor Which Flows from the Female Are Related to Each Other

122 It follows that we must next investigate and consider the disposition of the sperm, namely, how it is the cause which sustains the material of the one generated as well as its form and how it establishes it in being and shape. We must ask specifically whether it is like a particular part of the body and of the material which is rooted in the material which comes from the female, or whether it has no share whatever with the material and the body, providing nothing whatever of the material but rather providing only the power and moving principle for generation. Rather, that from which it is sustained and receives food is entirely material and this is the menstrual blood. It is true that the substance of the sperm is that toward and into which the menstrual blood is drawn, and it is formed in it as if in an organ and is converted to the form imparted by the sperm. It is clear in all cases that this way of operating is present in all which operate organically and officially. One who considers the matter in a general way will see that that which is effected is effected only by the operation of that which operates naturally, since it is operating in a passive one. For the entire power is not from a mover which exists entirely alone, on its own, for it would not have that upon which it could operate. The subject, therefore, must have the power of operating and an organ by which it might operate. It will then operate upon that which is the subject of the power, subject, and organ.

In this way it is clear that the male operates through his sperm and the female is passive. In the same way the form of a footstool or tripod is in the carpenter effectively and is in the wood by way of material. It is thus clear that it is not necessary for the generation of all the members that the sperm exit from the entire body, as those people we introduced above have held.

For it does not exit this way so that it might be a part, or the material parts, of the thing generated, but rather it will be like the operative form in which lies the principle of motion, which effects the entire thing generated. This is the relationship that "health," taken as something in the complexion of one who was sick, has to the "health" that is the knowledge and *ratio* of the physician.

123 This is known because in certain males and females copulation is so arranged that the female places the member for conception within the body of the male, as we have said is the case in certain ringed animals.¹⁵² For we know that in all genuses of animals the female is placed under the male during copulation since this sort of arrangement suits both the filler and the filled. It is suitable for the one lying on top to fill and for the one lying below to be filled, as Plato says quite well.¹⁵³ However, in ringed crea-

---

152. Cf. 15.79.
153. The source of this supposedly Platonic belief is unknown.

tures such as these, we know that the one that is placed below raises up the end of its body and places it into that which is placed over it during copulation. From this we know that the female places her genital member in the male, a thing that would be suitable in no way if the female ejected any fluid whatever from herself during intercourse. For if we grant that she emits, then this falls into the male and the male would then generate in himself from the female, which is contrary to the true nature [*ratio*] of male and female, as given in the beginning of this book. If, however, you were to say that she ejects first and then absorbs it up again, this would not be suitable, for in this way nature would not always be doing what is suitable and it would be entirely superfluous to first cast something out and then absorb it back again. Therefore the female places her member in the male when she takes on the sperm from him. These animals wrap around each other when they copulate, and they are attached to one another this way for a long time, remaining unmoving, during intercourse. But a short while after conception [*conceptus*] they produce eggs and an incomplete larva. For all the types mentioned generate larvae, as we have said previously.

An outstanding indication of the things we have said is what occurs in the oviparous genuses of birds and fish. For it is known with great probability that in these animals the sperm does not come from all the members. For in these the male does not produce sperm which, in the one generated, is rendered with respect to parts similar to those in the one doing the generating. Rather, the entire *conceptum* takes on the species, being, and life of the animal by the power of the male's sperm. For it is as we said before concerning the nature of the ringed animals in which the females place their genital member into the males. So too does something similar occur in these. For when the females have already conceived wind eggs from their menstrual fluid, if a change occurs in them as a result of contact with the sperm of a male, before the eggs have been altered from yellowness to a white that surrounds the yolk, then, without any sperm of the female coming forth again, the eggs become suitable for generation after they have been changed from wind eggs by the sole juxtaposition of the male sperm.

Further, if such females, after the sperm from one male has been conceived, copulate with another male of their own genus and his sperm touches some of the yellow of the egg they are holding in their wombs, the egg will be completed solely by the contact of this sperm and the result will be a chick that resembles the male whose sperm last touched the egg. Birds are produced with different colors and natures by means of this device.[154] First, males are put to the females and their sperm falls into the womb and begins to draw in the material of the eggs, as we have said previously. But before the eggs are completed, other males are put to the hens and then variety is produced in the ones born as a result of the powers of the different sperms. That which comes from the first act of coition is mixed into the properties of that which came forth last. This occurs from a mixing of power and could happen only because the sperm operates on power alone and not because it has in itself members that are

---

154. What follows would appear to be A.'s observations of contemporary practices.

like the whole because it comes from the entire body, as those say who claim that the sperm comes from the entire body of each, namely, male and female. For the sperm of the man (that is, the male) does this on its own power and does not act along with the sperm of the female. Rather, it acts in the female's fluid as if in the material which is the subject of its operation. It suits it to itself like food, rather after the fashion of the food on which embryos are fed.[155] Thus, that which is taken in last over the material is that which makes it like itself with respect to power and not with respect to act. For it has the act of no member even though it has the power of them all. The sperm, therefore, of the male is that which digests and warms the material humor of the female since it takes in food for its own growth, taking it from the egg as long as it finds the opportunity [*facultas*] in the egg's fluid.

126 The same thing happens in the genuses of the oviparous fish in which the female suddenly scatters her eggs with the male following and ejecting his sperm over them. Whichever egg is touched by the sperm produces fish as a result and whichever is not touched is not rendered fertile for generation. But these things have been fully treated in the previous books of this study.[156]

Therefore, from the things that have been said it is clear that the sperm of the man (that is, the male) generally contributes little or nothing to the material quantity of that which is being generated, contributing rather to the quality (that is, the form) using the formative power which it possesses. This is because it has already been proven that the sperm does not come from the entire body. But it is true that it comes from the body of a male who is producing sperm and not from the body of a female, for the female does not have the same relationship to the cause of generation that the male's sperm has.[157] From the male comes the principle of movement in generation, whereas the material comes from the female. This is also why the female does not generate on her own but desires the male as a principle of movement in generation. In many animals the female forms nothing at all, and instead, at a given time, emits the material. This is the case in the human and in hairy quadrupeds, these being the more perfect animals. In other animals, though, she does form something, albeit imperfectly. An example is found in birds, which produce wind eggs. The male casts only sperm into the female or onto the material of the female, the way fish do. The material from which the fetus is created is in the female. However, this material is quickly coalesced and is taken over to be the substance of the embryo. It is then coagulated and formed by the sperm of the male, whereupon the other material of this sort is served with the first [material?] as food for the embryos that have been conceived.

127 Conception must necessarily be in the female since the sperm operates on the material of the female much as a carpenter operates on wood or a potter on clay. Based on this argument [*ratio*] it is proven that the male contributes to generation much as

---

155. The pl. *embria* is notable.
156. Cf. 6.58f. for one example.
157. "Have the same relationship": *habet convenientiam*, a most difficult phrase that implies simultaneously analogy, suitability, and even contribution toward something.

art makes a formative contribution to the material. This has already been repeatedly stated.

Now, not every male animal produces sperm, but the sperm of those which do is not a material part of that which is conceived any more than a carpenter is a part of a wooden house on which he works. Rather, a carpenter is said to be its species and form (these being in his mind), since by his movement shape is brought to the pieces of wood. The form of the house and the art are in the soul of the artisan and through this art his hands move with suitable movements and through these movements the form is introduced. The nature of the sperm which comes from the male acts in exactly the same way. For nature uses this sperm formatively, like an instrument, as it were, and with it acts on the conceived material. The movement of this sort of power of the sperm acts as does art upon the material.

In this way, then, we have spoken about the way in which the sperm, of such animals as produce it, contributes to their generations.

CHAPTER IX

On the Way the Sperm of the Male and the Sperm of the Female Join Together in the Generation of the Conceptus

But the types of the animal that was mentioned above, namely, one in which the male does not place his genital member into the female, but, to the contrary, the female does so to the male, also bear a slight similarity to the operation of artisan and material. For they have movement that is mixed in with the material as it is in art that is impeded. For an indication of the weakness of the sperm of these animals is that nature and the formative power cannot bring about a complete thing.[158] Rather, it produces a *gusanes* from the egg, as was said above.[159] Movements of this sort are produced with difficulty when nature is oppressed, as if she were a litigant on the other side in a suit, and yet nature accomplishes something in the suit.[160] This same sort of thing is seen in magical activities and in curses in which one nature acts for the hindrance of another.

This activity has no resemblance to one which nature brings about by means of an unimpeded instrument. Rather, she operates in such matters with almost partial

---

158. Reading *natura et,* with Borgnet, for Stadler's *naturae et.*

159. E.g., 15.38, 42f.

160. A very garbled sentence. Sc. (Oppenraaij, 1992, 54) says that in these cases, due to the weak males, nature cannot bring about a completed thing and the males can move a thing only with binding (*ligatione*) and with the constant supervision of nature. This is fairly close to Ar. *GA* 730b28f. Sc.'s binding (presumably of the male to the female) is the source of the problem with A.'s text, for *ligatione* has been shortened, probably through poor abbreviations, to the verb *litigat* and to *lite,* implying litigation or, perhaps, merely "quarrel."

operations, scattered throughout many mediums so that it is first an egg, then a larva, and then a live young, as we have said above.

129 But in all animals which walk about with their bodies raised off the earth, the male is found distinguished from the female. But they agree in the form of the species and in the nature of genus and species. We thus say that "human" and "horse" are both male and female, since one nature is common to both the male and the female. But the sex which is distinguished in the animals just mentioned is entirely mixed together in trees. For in these, and in plants in general, the powers of the sexes are mixed together and they are not divided into male and female according to the act of generation. Thus, too, they bear fruit on their own and one does not cast sperm into the other but rather conceives seed [*semen*] in itself and produces it, casting it to the earth as if onto the mother of all plants. Protagoras speaks this way, and well, and thus, too, the statement of Empedocles is praised. He said that tall trees which are full-grown produce eggs on their own (for he calls the fruits of the trees "eggs").[161] For the tree bears fruit just as the animal generates. Its fruits, however, come from two powers, one of which is rather like a formative one and the other is that which provides the material and the food for the fruit. It is thus clear that the operative power of the plant is in a certain part of the sperm and the rest is consigned to material and food.

This, however, occurs mostly in the genuses of animals in which the males are distinguished sexually from the females, not by means of the vessels for the testicles but by means of another, weaker differentiation. For when these need generation, they wrap around each other and do not separate so that they might contain the powers of either sex in a single place. This is how these powers are joined together in the plants.

In order that this might occur, nature provides allurement by placing pleasure in copulation so that it might thus be sought with frequency for the preservation of the species in order that one animal of like species might be engendered from the mixing together of the powers of male and female.

130 For the same reason, certain types of animals are bound to one another for a long time, until the *conceptum* is rendered complete by the sperm of the male. For this sperm sustains and bounds [*terminat*] the entire *conceptum*.[162] Animals with ringed bodies copulate in this way, with some of them inserting certain members for intercourse into the females while others, to the contrary, take the members of the females into their bodies. They copulate for a long time due to the coldness of their complexion and the viscousness of their humor. This is the reason that the sperm which sustains a *conceptum* only matures after many days. However, after the sperm has left, the male and the female separate from one another. An animal of this sort, however, is very much like a tree or a plant which is separated with respect to the powers of the sexes. For, as we have said, the seed-producing powers [*seminativae virtutes*] of the sexes are

---

161. Cf. *De veg.* 1.45 and 48 (with Jessen's notes) for A., Protagoras, and trees. Empedocles and his "tree eggs" (olives) are ultimately from Ar. *GA* 731a4f.

162. See "Glossary," s.v. "Embryo." Both the act of conception and the actual thing conceived are referred to at once.

known to exist in a plant just as in the nature of an animal. For nature creates and forms each one of these animals reasonably. Nor do they have any other substantial operation save the generation of sperm, in the way we have described. This emission is completed through the copulation of the male and the female as they mingle and conjoin their powers together, one to the other during conception. We have investigated the reason the male and female sexes are not distinguished from one another in the study *On Plants*.[163] But it is enough to know here that the operation toward species and form among animals is nothing other than generation pure and simple. All animals have this operation in common.

They agree to a greater or lesser degree, however, with respect to knowledge and judgment. They all have sensation, but they have it to a greater or lesser degree. Sensation, moreover, is a sort of knowledge of sensibles. They also differ greatly among themselves with respect to the diligent operation of these functions.[164] Many seem to have surpassing diligence while there are also many almost lacking all diligence in these operations, so much so that memory does not seem to operate in them. Rather, they seem to share only in touch and in taste insofar as it is a sort of touch. Thus, a remarkable differentiation exists among the animals in the parts and powers of the soul. For every animal, because it is an animal, desires to have a natural participation in sensation and, to the extent possible, a knowledge of sensibles. It desires this because it is the noblest thing in an animal's life and no animal wants to have the position of a dead thing in the hierarchy of being, because it would have no perception of sensibles.

Animals differ even further in their sense perception since certain animals have strong senses and others have weak ones. For example, hard and soft eyes differ in their vision, yet they all are alike in the possession of this sense. All animals also have in common an operation that aims at the preservation of their species. Thus, when they need the operation of life which aims for the permanence of nature, they copulate and mix together, therein mixing the powers of the sexes in one *conceptus*. As we have said, this resembles the way a tree has these powers unified in itself. Hard-shelled animals are midway, as it were, between animals and trees in their manner of generation, for they have something in common with both genuses. In that they lack a differentiation of the male and female sex, they resemble the trees and are thus not viviparous. In that they are like other animals, they do not bear fruit-like plants but rather generate from a moist, fatty material, as we have said in what has gone before. We will also come back to these types of animals in the books of this study yet to come.[165]

---

163. Cf. *De veg.* 1.39–50, 84–93.
164. Cf. 8.67f.
165. "To these": the repetition "ad ad hos" is found in both the text of Stadler and Borgnet, senselessly.

CHAPTER X

Which Is a Digression Clarifying the Things Which Have Been Said about the Sperm of the Male and Its Powers

133   Resuming again, we say, for easier understanding, that just about the best thing through which the doctrine of the generation of animals can be most understood is the sperm. According to the ancient Peripatetics, namely, Aristotle and his followers, especially Porphyry and Theophrastus, generation is completed from the sperm of the man (that is, the male) even if no sperm is emitted from the female.[166] Thus, some of them have passed along that women and female animals have, properly speaking, no sperm. They say that this is proven mostly by the fact that a *conceptus* is sometimes formed when the female emits nothing during intercourse. They also say that the womb conceives only in the menstrual blood, swallowing down the male's sperm and that no *conceptus* at all results from the fact that they both emit sperm from their entire body. For, according to them, the semen is that which is formative and the menstrual blood, apart from all the other things the woman emits, is adequate material and food for generation. They also say that one should pay careful attention to whether the sperm comes from the entire body in such a way that it comes partially from the flesh and partially from the bones, and so on for the other members of the body. For some of the ancients claimed this and, as we have said, found proof for it in the pleasure of intercourse that occurs at the exit of the sperm, since it pervades and encompasses the entire body.

134   They also give as proof the resemblance of the young to the mother or father who generated them. This likeness is sometimes total and at other times only partial. At other times there are flaws or defects in the members of the ones generating and which are reflected in the ones generated. This has been determined in things said previously. These men say that the only reason for this resemblance is that the sperm has the powers of the members and that it could have their powers only if it were something that has flowed from them.

135   The Peripatetics used to give an objection to this argument [*ratio*], saying that the resemblance of the young and its parents sometimes occurs in the claws, hair, and teeth, and yet it cannot be said that the sperm flows from these.

They have also lodged an objection based on the fact that this resemblance is sometimes to the grandfather. They tell how a black Moor generated a white daughter, and

---

166. The text begins to follow, loosely, Avic. *DA* 15.48v. Theophrastus was both the student and successor of Aristotle in the Lyceum. His specialty was botany. Porphyry (A.D. 232/3–ca. 305), the biographer of Plotinus and editor of the latter's *Enneads*, not only published a very influential commentary on Aristotle's *Categories*, the *Isagoge*, but also a treatise in four books on vegetarianism that borrowed heavily from Theophrastus.

she then bore a black son who resembled the grandfather of the one being born. Yet there was no sperm from the grandfather in the child since, so they say, he was dead.[167]

They further bring the argument to an impossible point, for if the sperm were parts of all the members resembling the powers of the members, then the sperm would be a little animal.[168] Moreover, this animal either has members that are suitably arranged in order or it does not. If it has them suitably arranged, then without doubt the sperm of the human is a little human and the sperm of each animal is a little animal, a thing that appears to be ridiculous. If, however, it does not have members suitably arranged, then it could not live but would be dead and would not generate, for that which is dead does not generate but rots.

Further, if the sperm of the woman shares in the name and nature [*ratio*] with that of the male (for the young often resemble her) then it will follow that when these two sperm descend at the same time during a single act of intercourse, two animals will be created at the same time.[169] From this it follows that twins will always be born, with different sex, a thing which we see is false.

They also add that nothing then prevents a woman from bearing a child on her own, without a man. For she has sperm in herself, in a suitable place, and it is a little animal and impregnation is due to it since the womb only provides it, according to them, with food until it reaches the requisite size.

Further, we see that some males first generate females and afterward males, whereas others do the opposite. Nor can it be said that the sperm does not always come from him when he generates. What then is the reason that he sometimes produces one that is like him with respect to his genitals and at other times one that is dissimilar? Or again, what will the reason be that some produce those unlike them in genus, as do the bees and certain flies? Again, some produce many offspring as the result of a single mixture during copulation and some of these are male and some female. Therefore, we must ask what the reason is for the young's dissimilarity when it is the result of a single act of copulation, and it will not be possible to find it.

These men object, saying that a transplanted branch puts forth fruit even though the branch does not come from fruit and does not have it in itself. If it is said that the fruit is scattered throughout the branch, we will say that this fruit is not "of" the entire tree, since we have posited that the branch be cut from the tree. If, however, it be said that the fruit is scattered in the tree but is only in the branch in a material sense and is in the entire tree in the sense of potential and power [*potentialiter et virtualiter*] and that it therefore germinates from itself a thing like the entire tree and not only like a part of it, then the proposition stands. For in this case nothing precludes that the seed which descends from the animal's body flow materially from a single member and yet have the power of the entire body. It thus generates toward a likeness of the whole. Generally speaking, the material is not the cause of the one generated in any act of

---

167. Cf. 9.63.
168. Cf. 15.59f.
169. "Animals": perhaps "live young."

generation but rather the formative power in it, as the reason for a footstool is the carpenter and the reason for its entire shape is the form of the craftsmanship in him.

137 What they say about the pleasure is not as they say it is. For the pleasure is present only at the end of copulation when the sperm has already flowed into the spermatic vessels. Vessels of this sort are nerve-filled, sensitive, and swollen by the strong spirit which extends them. Thus the pleasure is due to the sperm which is touching it inside and to the spirit which at that time flows through them most forcefully. A light rubbing is good for this, such as the tickling sensation that is produced in abscesses which are swollen and when something light is led across them, especially after the putrescence in them has come to a head.[170] Nor is the pleasure present equally in all the members. But its greatest intensity is primarily in the members which are the spermatic vessels and this intensity flows over into the other members because they are bound to these.

The sperm then is a substance, moist in fact, which comes from the uniform parts and is derived from the body. But it is not dissolved as we have said previously, for dissolving is unnatural and sperm is the most natural of things. There is also a natural assistance to be found from sperm, whereas in something that is dissolved from the body there is no assistance, since it is by way of a corruption, unless it be that the body is sometimes eased by having this thing set free from itself.

138 Further, many acts of setting free [*resolutio*] are produced by the bodies of those who have no sperm at all. For the sperm is a particular superfluity of food which is not the result of the corruption or the liquefaction of the members and is not a superfluity of the first or second digestion. For phlegm and bile [*colera*] are superfluities of this sort and thus, after the second digestion, vomiting of bile and blood is found. Many examples can be brought forward for this. But sperm is the superfluity of the last digestion and it is abundant when there is a lot of nourishment in the body. Because this superfluity is like all the members in power, every member is thus made from it. Nevertheless it is not a thing drawn from every member. Since this is the way sperm is, it is not set free from the body in anything like a liquefaction. For if it were, then an animal with a large body which has many such acts of setting free [*resolutio*] would have a great deal of sperm. But this is not the case. An indication of this is that a large-bodied animal has few conceptions [*conceptus*]. The reason for this is that the food of an animal of this sort is cleaned of the superfluity of the second digestion, which is the urine, and is then disbursed to the entire body. Because the body is large, little remains to provide the substance of the sperm after each part has been adequately nourished. Thus large-bodied people and lofty, large trees have little generation. Of similar light generation is an animal with a great deal of fat since its good nourishment passes over into *sepum* or *pinguedo*.

139 Further, the process of bodily liquefaction has no vessels in which it might be better digested and formed. But sperm does have vessels in which it is received to better

---

170. The same analogy is used at 9.104.

it. Thus the milk in women has vessels in which it is completed and cleansed, for milk is like sperm to some extent.[171]

Moreover, sperm is lessened by the setting free of superfluities. It is therefore not after the fashion of a setting free but rather of a digestion. Thus, too, it is diminished in the sick and is not found in boys because of their growth. For in boys the third digestion is very strong and because of their growth they necessarily have a great deal of food.

Moreover, every liquefaction of the members produces illness, whereas a great emission of sperm does not lead to any illness even though it might lead to weakness. However, if the sperm should happen to be altered toward corruption, then its emission will be somewhat in the manner of a setting free [*resolutio*].

We approve, however, of these arguments of the Peripatetics that are brought against the Stoic sect.

The statement of the Stoics and the physicians that the sperm comes from the entire body is not entirely false, something even we have said previously where we treated the nature of the sperm. But we are not saying that it comes from the members in the way the part comes from a whole, but rather that it is derived from the fourth digestion, which has already taken on the power of being assimilated from the members before it was drunk and made one with them. For there are four things in the humor of the fourth digestion. One of these is digestion; the second is assimilation. The third is the apprehension of the humor by the species of the member when the humor, already divided, is drunk by the members. The fourth, however, is the uniting, when it is united.[172] The first two happen in the humor, external to the members, and the second two occur in the member when the humor is divided and distributed throughout the members. The semen is derived from them after the second. Because this assimilation is universal, especially in the uniform members from which the composite, instrumental members are entirely and everywhere composed, it is thus generally said in this way that the sperm is derived from the entire body.

140

CHAPTER XI

Which Is a Digression Making Clear the Difference between Two Sperms—
That of the Male and That of the Female—and of the Menstrual Blood
Which Comes Forth from the Bodies of Certain Female Animals

It is therefore certain that the sperm is the superfluity of the last digestion which forms the *conceptus,* which brings spirit into it and also generates in it a likeness to the spermatic vessels and which turns white because of the strength of the digestion. That it comes from blood is shown in that when copulation takes place with too much effort,

141

---

171. Cf. 3.151.
172. "Quartum autem est unitio quando unitur."

the sperm comes forth bloody. Thus, in much the same way the menstrual blood in a woman is also the superfluity of the last digestion, although it is not so strongly digested that it takes on a white color and thus does not come close to the sperm's digestion.

There is also another humor which is called the woman's sperm by some and, according to Galen, it is drawn from her testicles and is emitted within the womb during intercourse through the horns of the womb, or so it seems to Galen.[173] Or it might be emitted, according to Aristotle, inside the womb before the opening at the neck of the vulva [*collum vulvae,* vagina], at the place into which the sperm is cast. We have pursued each of these opinions above. This humor, according to Aristotle, is present neither in all women nor in all the females of other animals. Also, in those in which it is present, it is not always necessary for conception. But when it is present, then it is agreed that it is the best material there is for conception, for it is better digested than is the menstrual blood and a more beautiful and nobler conception is formed from it.[174]

142 Galen, however, believes that this fluid is always present in a woman for conception and provides the informative power for it, just as has been established in the earlier parts of this book. He states that without this there could be no conception and that this should be emitted at the same time the sperm of the male is emitted in order to have impregnation occur. Almost the entire mass of physicians follows this line of thinking. But it is indeed wondrous if this humor is not found in the females of other animals and yet is necessary for conception in a woman. For, of those things which have one common genus, it seems that there is one type in the genus of material principles, unless it be said that this humor is present in the females of all animals even though it is not visible in some of them. And this seems the most likely thing to me.

However, as we proved above, this humor does not have the perfect nature [*ratio*] of sperm. And this is because it does not achieve the digestion of sperm and cannot do so, for the woman is generally weaker than the man in natural powers as are females to males. Thus, too, since their coldness contracts in them, their veins are narrower, their flesh moister, and their bodies are often smaller. This is also why a great deal of undigested superfluity occurs in them.

143 As we have said before, however, the time for the movement of the superfluity of females is the same as the time for the sperm's movement in the males. Sometimes, however, it hurries a bit, since the female power is weak to digest all the food and to convert it to growth during the time in which all the food in the males is converted to growth by the strong power of the males. Further, it happens as a result of this that

---

173. *De semine* 1.3 (Kühn 4.516f.). The "horns of the womb" (*cornua matricis*) are identified by Fonahn, 997, as the broad ligaments of the uterus, but one might ask if the fallopian tubes are meant. Cf. 2.116.

174. This passage and what immediately follows form an excellent study piece for the comprehensive nature of the term *conceptus*. In one phrase it seems to refer to the fetus, in another to the act of conception. Its contrast with *impregnatio,* just below, is noteworthy.

the increase in the superfluity in them is intensified when it is held in the body due to illness in them.

Further, a decrease in the menstrual cycles in females is sometimes caused by the loss of a great deal of blood through some other member. If we were to say that the sperm in a woman is of the same sort as it is in the male and is intensified in the area of the genitals, it would happen that the woman would be reproducing on her own and it would then seem that she did not require the menstrual blood to be the principle of conception [*conceptio*], since the digestion of the menstrual blood and the digestion of the sperm have a certain contrary quality. For the digestion of sperm is complete and that of menstrual blood is incomplete and imperfect. For this reason, too, males whose complexion resembles that of females become fat and their sperm is but rarely generative.

It is thus clear from the things that have been said that the sperm does not descend from the woman in the same sense that it does in the male. However, I do not think that it does not descend in any way, but in truth it is probable that the white humor descends which sometimes does so prior to intercourse and is guarded in the womb until the sperm comes upon it. At this time it takes on impregnation in the way that material is impregnated by form. Thus, some women say that they have become pregnant from intercourse during which they had no pleasure resulting from the emission of sperm. This is like a wind egg which the hen first conceives and which is then completed by the sperm that comes over it through intercourse with the cock. Now it is not necessary that, if the woman takes pleasure in the intercourse, the pleasure always be due to the descent of the sperm. For sometimes the spirit produces this pleasure by its stimulating action. Then the woman's sperm, when it is emitted during intercourse, is the particular fluid which descends from the glands we have named. This indeed is more complete for generation than is the menstrual blood, but it is still nothing more than material. For the menstrual blood necessarily has further digestion before it will feed the creation.

But that fluid about which we are speaking in a woman is immediately converted into the material substance of the *conceptus*. Thus, the woman's sperm is like that of a boy who emits sperm before it is digested and like the sperm of a tree which, for the digestion of its seed, requires completion by another mover through further digestion to come to where it is able to produce fruit. According to this, generation is a result of the sperm of the woman and that of the male, but the sperm of the male is not material, but it is totally mixed into the operation (and is the subject of the operative power) and is mixed in the spirits which are the instruments of the operation. The sperm of the woman is like the clean material, whereas the menstrual blood is like the unclean material. I am calling "clean material" that which takes on form immediately, without alteration. By "without alteration" I mean that which has greater digestion. But the menstrual blood is like a not fully assimilated material. This is the same as when bread is called food and blood is called food. For bread is not an assimilated food, but blood is food which has been assimilated.[175] And this is the reason too that wind eggs, which are like the sperm of birds and fish, are completed when they are touched by the male's sperm.

Since, then, we have said above, following the belief of Aristotle, that generation occurs from the sperm and that the feeding is from the menstrual blood, we must accept that generation is, materially, from that which is called the sperm of the woman and the feeding is from the menstrual blood, as we have already said.

Let, then, these things be those said by us about the two sperm and their comparison to one another.

---

175. "Assimilated": used here to translate both *similis* and *assimilatus*.

HERE BEGINS THE SIXTEENTH BOOK ON ANIMALS

Which Is on the Powers That Produce the Thing Generated, with Respect Both to the Soul and to the Members

The First Tract, on the Arising and Introduction of the Soul

CHAPTER I

In What Way the Male and Female Are the Principles of Generation

Having set forth in the previous book the powers of the sperm and what relation they have to generation and menstrual blood, it remains to investigate in this volume the reason that there are male and female in the genuses of many animals and what powers they have in generation. For we know from what has gone before that this separation of the sexes is due to natural necessity. In those things which generated, these things are procreated from a first and a proximate mover which is in the sperm, and in the way the material is disposed by necessity.

As we have said previously, this occurs in the universal nature (which is the cause of order for the better) and on account of the cause which is called *propter quid* and which is the end.[1]

This is known because we see that in natural things [*in physicis*] there are certain divine things which last eternally (that is, unchangeably). Even if their substance moves, that substance is still reiterated the same in number and species in that movement, always remaining incorruptible and ungenerated, much as the substance of the celestial spheres and stars.

There are also certain natural things which have contingency and a movable substance. This is like the substance of those things whose substance is generated and corrupted, and which is not reiterated the same in number but is reiterated as the same in form and species. Examples are the human, the ass, and whatever things that can be generated and corrupted. They have the divine trait of remaining the same in form and species. They seek this naturally, for all beings desire permanent existence. For this reason, too, that very divine thing which they seek is, in them, the reason for the betterment of their existence. For everything which exists has equally the potential for being or nonbeing and is only made better in its being through a divine permanence in which it participates with respect to form and species.

---

1. "Universal nature": *natura universalis,* a general substance or essence.

It is known well enough from all the things which we have determined in books on natural philosophy that the soul is nobler than the body. For this reason, too, that which is animate is nobler than that which is inanimate because of the soul in which it participates. Thus, being [*ens*] is nobler than nonbeing and life is nobler than death (that is, than the privation of life).

Therefore, the generation of animals was invented by nature for the sake of being, soul, and life. For the nature of this genus does not endure eternally in number and substance. But the way in which it can participate in the divine and the eternal occurs through generation. It cannot be eternal through number since the substance is received in each one according to number, is corrupted, and does not remain the same. For if it were to remain the same according to number, then it would participate in eternity in the individual as do the heavenly bodies. It therefore seeks eternity in form, for it is possible to attain this. For this reason the generation of animals was invented by the ordering of the universal nature. This is the cause of generation in both trees and animals.

3 Since, then, a certain principle of generation in most animals exists as male and female, both male and female exist for the sake of generation. Between these sexes, that which is the first moving cause is better since that cause moves only via form, and form is worthier and better than material. Since, then, the male moves with respect to form during generation, it is nobler than the female.

In all natural things [*physicis*], when they are perfect, that which is nobler is always separated from the more ignoble. Thus, since life is visible and complete in animals, it was necessary for the male to be separated from the female according to subject among the animals. But in plants, in which life is difficult to see, these differences are united and joined together in a single subject. The reason for the separation is that the male is nobler in generation because its power is close to the factive and operative form while the female is close to the material. This is because that which is derived from her comes to be and takes on the impression of the male's sperm. Now I am saying that these powers are close to the form and material because each is in a subject which undertakes some part of the operation. The material subject in the sperm of the male is formed into instruments such as spirit and the like. The material subject in the humor of the female has a great propensity to and preparation for form. Thus, the sperm of the male is neither entirely lacking in material nor is the humor set free from the female entirely without form. Rather, as we said, the one is closer to the form and the other is closer to the material. In generation, however, both are joined together in one and then that which is joined together from these two separate things simultaneously shares in the possession of the powers which emanate from the male and the female. For this reason, neither the male nor the female has the potential and power for life and soul on its own, but one which has been joined together from both of them does. Thus, too, we say that plants share in life because of the joining of their powers. For in the generation of those things that are soulless there is no necessity for a composition of this sort.

Now sensation in general, and especially sight (which, as has been proven in the book *On the Soul* [*De anima*], especially possesses the true nature [*ratio*] of sensation), is not found as a common trait among living things.[2] On the contrary, a requirement for its generation is that the power of male and female be separate. For, as was said previously, male and female are found in all animals which share in this sense. For if the sexes were not separate but were united as they are in the plants, then none of the seed would be freed from opaque earthiness into such a state of purity that it would be susceptible to the sensible forms apart from matter. This especially is the cause of the separation of the sexes so that that which is close to form in the male might purify and digest some parts of the material so that they might be susceptible to this state of purity. For it can only digest in this way if it separates them through some sort of separation from the less pure ones. Neither could it cause separation if the masculine power were everywhere mixed in with the female power throughout the subject.

This, then, is the main cause of the separation of the sexes in animals.

However, as we have said in what has gone before, some animals produce sperm during intercourse and others do not.[3] Those which do produce sperm during intercourse are numerous compared to the others and are, generally speaking, more perfect and larger than those which do not produce it. This is because a nature which is sufficient for the perfection of the animal is warm and moist and possesses vital heat. For those things which are larger and more perfect are moved by a greater power. And thus, also generally speaking, blooded animals are larger and more perfect than those that are bloodless and the same is true for those that move and those that do not. This is because they are completed by a greater heat. These animals, then, must have an adequate principle for their generation because they must produce sperm while generating. The others, however, since they are imperfect, and especially those which are the most imperfect, also have a more imperfect principle of their generation. Thus, ones like these which do generate either practice no form of copulation or else do not produce sperm during copulation, as will be made clear from what is to follow. For an animal produces sperm because of the moving heat [*calor movens*] and because of a great deal of suitable humor, providing the material of the sperm. And this very sperm, in such animals, is rendered the cause and the principle in those which are generated.

The reason for male and female has been clarified briefly, then, and by way of summary. Some animals, as has been said often, generate in a way in which they give birth to a live young externally, like those which are called viviparous. Others, however, give birth to live young which are not formed and differentiated with respect to form and shape. If one of those which generates in this way has blood, it produces eggs. But one which lacks blood entirely produces larvae (that is, *gusanes*). Fish, however, following this division, have some blood, albeit a small amount. This is seen especially in those

---

2. "Common trait": *communiter*, implying that not every living being participates in this sense, viz., sight. Some insight on this passage may be found in *De anima* 2.3.7.

3. Cf. 5.44f.

which have attained full growth, for around their heart there is some blood. However, generation which is accomplished from eggs or larvae differs in that the egg forms the live young and its members from one part of itself while another part of the egg is relegated as food for the creation. A larva, however, passes over entirely in and of itself into the substance of that which is being generated.

Further, some animals suddenly give birth to their young [*fetus*] externally as do the human, the cow, and the like. Of the sea animals, the dolphin, whale [*balena*], and the like do the same. Others, however, do not give birth straightaway to a young which has been put into shape and form, and they lay an egg instead. Others bear larvae which they hold internally first and then the live young, completed in form and shape, come forth from the eggs and larvae. This can be seen in many animals.

7 Moreover, some oviparous animals produce eggs that are complete as far as size is concerned. Examples are the birds and certain quadrupeds like the lizard and the tortoise. But others produce incomplete eggs like the fish, animals with a soft shell, and the *malakye*. For all the eggs of these genuses grow in size after they have been laid.

Further, animals which generate equivocally (that is, producing those that are not immediately like themselves) are sometimes blooded. For all blooded animals either generate live young univocally or they are oviparous. Bloodless, ringed animals have a generation appropriate to them and peculiar in itself, and in this generation they differ greatly throughout their species. For bipeds are not all solely oviparous and neither, again, are all quadrupeds solely oviparous. For the horse and cow are quadrupeds and they generate live young. Neither, as we said, do all quadrupeds lay eggs, as do the lizard, the *tenchea*, and the tortoise.

8 Moreover, there is further diversity among these. For some which lack feet generate live young as do the *tyrus* and the *zeleti*.[4] Others, however, lay eggs first, as do many genuses of fish and serpents.

Again, some footed animals are oviparous internally first and afterward bear live young externally, as we have said many times. Some footed animals generate live young both internally and externally, whereas others that are footless generate live young, as do the dolphin and the whale [*balena*]. These generate a live young complete in shape, which resembles the ones generating it in species and shape and which takes air into its lungs. In all the differences mentioned, as we have said, the most perfect of the animals is that which has the greatest heat, the purest and greatest moisture, and is not overcome by a great deal of coarse earthiness. Now the boundary for natural heat is none other than the lung, for the lung exists to temper and to set bounds for it. Further, the lung is present only in those which abound in natural heat. For the lung is sanguineous as we have said, and if it is soft, this is accidental. For this reason, every animal with a lung is warmer than another animal that lacks it and thus an animal possessing lungs has blood and a great deal of heat. Now this is a completed and perfected animal, whereas eggs and larvae are not completed immediately.

---

4. *Tyrus*: a viper; cf. 1.102. *Zeleti: celeti*, selachians.

We will say, then, speaking philosophically, that that which is completed by the perfection we are speaking of is naturally generated from that which is complete. But a winged animal which is a flyer has a dry complexion, as its many feathers indicate. Thus, it also has a dry lung. These animals, since they are dry due to this imperfection, are oviparous. Oviparous animals, however, which are colder than these, are internally oviparous and externally viviparous. Birds and animals with a scaly skin do indeed complete their creation, but they lay eggs due to their dryness even if they do produce them complete. But the *celeti,* since it has less heat and more moisture than those just mentioned, has something in common with each of the genuses in that it produces eggs internally and afterward bears a live young outside. For, because it is cold and moist, it produces a live young. Moisture is the cause of life, but earthiness, which is dry and coarse, is removed from animation. Wings and scaliness are signs of an earthy nature. Because of the humor, therefore, which is dominant in this genus of animal, it produces soft eggs. Thus, too, wise nature guards and cherishes them internally, for if she emitted them externally, they would be corrupted by whatever chance befell them since they do not have a covering shell or a tough, thick web. If one of the animals is cold and dry with great coldness and dryness, it produces incomplete eggs, but they are harder, since the very nature of an oviparous animal is earthier. For this reason, too, their eggs sometimes have a hard shell or a hard skin which defends them from corrupting influences. The species of *malakye,* however, have a viscous body and for this reason they produce incomplete eggs with a viscous shell. Still, because of the great amount of viscosity around the egg, it is protected from corrupting influence. Almost all animals with a ringed body produce larvae. For all animals with a ringed body are bloodless and for this reason it is generally true that every animal which generates larvae externally is bloodless.

Further, not all animals with ringed bodies generate in one simple way, any more than those that are bloodless generate in one simple way. For there is diversity among the ringed animals. The types of oviparous animals and those which do not complete their eggs within a uterus include the scaly fish, the *malakye,* and the soft-shelled ones. The eggs of these types resemble somewhat larvae with respect to their creation, specifically in that they grow outside the animals. Other types of animals produce larvae which, after a few days, produce a live young and in this they resemble eggs. Whichever of the animals has completed heat throughout its genus produces an animal which is like itself in species and shape. Thus, such a one is complete with respect to quality, but it falls short of completion with respect to quantity until it gradually takes on size through nourishment. Generally, all animals produce young which take on quantitative growth after generation. Therefore, some animals generate by conceiving live young internally, while others conceive eggs internally and then, later, a live young comes forth from them. Still others do not generate a complete animal, but a complete egg. But those which have a colder nature produce incomplete eggs, which will nevertheless be completed externally. Examples are the genus of fish and of soft-shelled creatures, as well as the types of *malakye.*

11 The types of hard-shelled creatures, however, are very cold and thus do not lay eggs. But this accidental trait occurs in them apart from other animals, as has been determined previously. Those which have a ringed body produce larvae first and then, after a few days, they bring about the same result as eggs. They are treated like the eggs of the bird which is called the *horokyz* in Greek, but the *hornezich* in German [hornet]. This is a large yellow bee which is long and has a very strong stinger in its hind parts.[5] The eggs of this bird are like eggs in color and they have power like that of an egg in that a live young is born from it after it has taken on the completion of generation, that is, creation. However, as we have said before, certain other, more complete animals are born from sperm in the way that an animal is said to exist effectively as a result of sperm. Generally, however, all animals which are born as a result of the intercourse of male and female are completed in form, shape, and sustenance [*sustentatio*] from the sperm and from this they take on form and species. Some of them, however, receive these perfections in the interior of the animal during conception and impregnation. Others, however, produce eggs from which a complete animal will come.

CHAPTER II

On the Power Moving Embryos, As to Whether It Is Intrinsic to the Sperm or Extrinsic According to the Opinions of the Ancient Philosophers and Physicians

12 A serious question arises concerning these types of generation: Is like generated *per se* from something like itself, and not that an animal is generated from the sperm of a tree, or is the opposite true?[6] For everything which is generated is generated both from some efficient principle and in some material principle of its generation. But that which is generated or produced, again, is either an animal or the proper material for an animal, much as an egg is the proper material for the animal. This is produced and generated in the female of some animals and is found impregnated and conceived within her. Some too, after birth from their mothers, take on, for a long time, the material of their growth, as do those which suck milk and as do all animals which are viviparous and those which give birth outside but conceive internally.

Let us then ask not after the material from which the animals are generated but after that which is the effective and formative principle of the members. For it is important to find this out, but it is also most useful for an understanding of the natures of animals. Let us therefore ask whether this thing is intrinsic, acting as the fashioner

---

5. Cf. Ar. *GA* 733b14. A.'s source has first misread the Gr. *chrysallis*, our chrysalis, making it a bird (*avis*). If Stadler's markings are correct, A. identified the animal as a hornet, apparently thinking *avis* is an error for *apis*, "bee." But the matter is complicated in that A. immediately reverts to *avis*. Cf. Sanders, 446.

6. Cf. Ar. *GA* 733b24f. for the original intent of this passage.

[*operatrix*] of the members, or whether it is a thing intrinsic to the sperm like some part of the soul.

As we have said in the second book of *On the Soul,* that which has a soul in act, like an act operating for the potencies of life, will be called animate and is an animal.[7] In this case the sperm would be an animal and would have the name and *ratio* of an animal. But this is entirely false and impossible since no animate thing is composed entirely of uniform parts. For we have already shown that sperm is composed of uniform parts and is not a small animal. But to say contrariwise, as some of the physicians have said to the unskilled populace, that sperm is an extrinsic fashioner of some or all of the members is entirely false and impossible. Aristotle's argument on this is that that which is external to the sperm would begin to move the sperm only if it touched it. For alteration of what is moved follows upon some moving force only if it touches it. But so that this argument might be understood, we should set forth this disposition at greater length.

Some of the physicians say that conceived sperm first move in the manner of plants, toward nourishment, conversion, growth, and formation, and they say that plants move by the celestial power and by the power of the earth toward the changes mentioned. They thus say that embryos (that is, conceived seeds) move by virtue of the mother's soul toward conversion, configuration, and nourishment at first, that the soul is infused afterward, and that it moves then by the movements of the soul proper. But this is entirely absurd according to the knowledge of all the Peripatetics. The most persuasive argument set forth is found in the sayings of Aristotle. For since there is no medium between that which moves and that which is moved, that which naturally moves must naturally be joined in the way nature is joined to that of which it is the power and form. For natural movement comes from an intrinsic and essential principle. Thus, if the mother's soul will move the embryos which are converted and moved internally, that which moves must be the internal and essential principle. But in no way is this the mother's soul, for when the celestial power moves it changes something only if it is rendered the one informing the internal moving powers of the material.

Further, every moving thing which generates, converts, and forms only moves material toward itself. But to say that the mother's soul might move, form, and convert embryos toward itself is totally absurd.

Further, according to this, if the mother's soul is that which configures and converts, then the sperm of the male would be superfluous and the female could conceive and generate on her own. But this is absurd, for as is clear from what has gone before, the sperm of the male is needed precisely for conversion and configuration of the members.[8] Therefore, embryos are not moved toward the conversion and to the form of members by a mover that is extrinsic to themselves.

---

7. Cf. *De anima* 2.1.2.
8. Cf. Ar. *GA* 734a1f.

It is therefore clear that this power is in the embryo, moving by converting and forming the sperm. For if it were not in it, then it would necessarily follow that it was separate and divided from the conceived sperm by subject and place. But to say that a moving power or soul is divided in this way is false, as is clear from arguments introduced earlier. In addition to all the things introduced above, let us ask if there is a power existing externally and effecting the conversion and formation of the sperm, so that this power is not the mother's soul (as those already discussed have said) but is the power of some part that is separate from the sperm and acting in it.

Now if it is separate this way, we will next ask whether this is corrupted after its work is completed or remains instead in the one born in the disposition it previously had. The first thing against both of these, namely, whether it remains or is corrupted, is that once we have considered all the things which are a result of intercourse in those that produce sperm and once we have considered those things which are the result of all the sperm of both trees and animals, it appears to the senses that nothing is totally differentiated. For we see that all things emitted by trees into plant seed and by animals during intercourse are united under one form of sperm and we see that nothing is divided from it which acts on it by way of converting and forming it.

15 Returning, moreover, to the division introduced, we say that the statement of one who asserts that that which is divided from the sperm is corrupted after its action in the members is false. For a power which acts on one particular member is corrupted after it has acted on that member in the same sense as a power which acts on all things is corrupted after it has acted on all the members. Therefore, if the power which acts on the heart is corrupted after it has acted on the heart, let us ask from them what it is which acts on the rest of the body after the power which has acted on the heart has been corrupted. For it is agreed that the powers of the other members have reference to the power of the heart, just as the other members have reference to the heart. Thus, once the power which acts on the heart has been corrupted, all the other powers must necessarily be corrupted and thus none of the other members whatever will be made by the heart. But this is a thing we see to be false. Therefore, the operative power [*virtus operativa*] of the members is not corrupted after it has acted on the members. The idea that it might remain and subsist outside the sperm is entirely absurd, for either it must be that a single member of a single species has two powers or else there must be two animals, one making and one made. For these powers of the members separated externally would, when gathered together, possess all the powers of the animal and such powers do not exist without members and they would thus be a small animal producing another small animal, as we have said in the preceding book.[9] But all these things are absurd.

16 Therefore, the statement of one who says that that which makes and forms the members of the embryos is outside the body of the embryos is entirely false. For it is clear that that which is the effective principle of generation, whatever it is, makes a given

---

9. "Powers": *potestas* instead of the more usual *virtus*.

part of the embryo. This comes forth suddenly into form in the substance of the sperm. However, a given part, when formed, cannot exist without soul. For every shaped member of an organic body has a part of the soul which would be its soul if it were a separate [*divisum*] animal and perfect in itself. Nevertheless, this part, formed suddenly by that external power, is inanimate. For an external power is not its soul since the soul is the internal principle of life.

If it is posited, then, that a power which shapes a single member is external, we would ask by what power the other members come to be. Therefore, according to those who say that all things come to be from external powers, it is necessary that all the members come to be at once. For one power is as efficacious in acting as is another. For, according to the argument introduced, the heart, lungs, liver, eyes, and other members are all made at the same time. Thus, whatever is made by an external power will also be inanimate. This is as those have said who claimed the external powers flow out of the members of those generating and act on the sperm. They said that it acts on the sperm and that at length, after the formation of the members has been completed by the giver of forms, various souls are led into the members according to subject and form.

This was the opinion and belief of Plato and Pythagoras, who believed that there were many of the immortal gods present in a single body. The gods, however, were the seed of the celestial and immortal gods, sown in the formed members of the bodies.

Homer the philosopher-poet, however, weaves together philosophy from wondrous tales and says that the generation and formation of the members come from powers extrinsic to the members.[10] But their generation is not parallel because the generation of some members, like that of the principal members, is anterior to the generation of the nonprincipal members. He holds that the generation of the members of animals is after the fashion of the weaving of certain insects like the *bombices* [silkworms] or the weaving spiders. This occurs just as links are successively made which are in the substance of a net in which one link is bound after the other and is then cut off when the net is done.[11]

17

Homer bases his argument for this statement on what is clear to the senses. For we see with the senses that the creation of all the members is not simultaneous in an embryo. During the first creation of an embryo it is seen that certain members are clearly formed at the start while others are not. Nor can it be said that this is because of their small size. That is to say, it is not that they have all been formed but that some cannot be seen. For some of those not seen at the beginning are larger than those which are visible. For the lung is larger than the heart and yet it remains unseen even after the heart has been formed. Thus, if smallness lay at the cause, the heart would be less visible than the lungs. It must be then that, as Homer says, one is formed after the other and thus some members appear formed earlier and others later. He does not say that the reason for this is that one member forms another. For he agrees with the

---

10. Homer: Orpheus, the semi-mythical poet, is cited by Ar. *GA* 734a19.
11. "Link": *macula,* in CL a spot or flaw, can be used at this time to refer to the link in a chain.

others that all are formed by separate powers, as we have said. He therefore does not say that one is the cause of the formation of another, but only that one is prior in time to the other and that one comes after the other in generation.

Thus the heart does not make the liver nor does the liver make any other member. It is only that the liver will come after the heart in time, much as a youth comes to be after a boy and after the youth comes an old man. It is not that one comes from the other as from the cause of its generation. Thus he [Homer?] does not want that that which exists in potential should come from that which exists first as it were in actuality but only that it should exist after it, saying that this is so in all things, both natural and manufactured.

18 We, however, see it the opposite way. For in all natural and manufactured things, the posterior will always exist in the prior potentially. For the form through which it is known and through which it is the thing it is exists in the prior potentially. For this reason, once the posterior has been posited necessarily, it is that which is prior. Thus, the form of the liver exists in the heart potentially and in its formative power. Thus, too, the statement of Homer resembles a poetic tale constructed following Plato's belief. For nearly all tales follow his beliefs.

Likewise, false is the statement of one who says that there is one part in the sperm, separate from the rest, created first by the power of the one generating, and which afterward creates all the other members. This, though, is impossible and belies the senses, for powers come from the sperm. It is clear, however, that whatever might be said to be prior in these parts was emitted only from the one generating who produced the sperm. However, that which is emitted first from one generating is nothing other than the sperm in which, with regard to form, no part is from another substance separate from the sperm. For the entire substance of sperm is a certain operation of the one generating and emitting the sperm, which is homogeneous. We thus cannot say that some other member is created suddenly in the sperm of the first generating one and that this member, using its own power, creates and forms the other members. Neither can we say that the principle which forms and creates is external. However, it seemed necessary to many of the ancients to make one of these two statements.

CHAPTER III

Which Is a Digression Setting Forth a Summation of the Difficulty of Solving the Previously Introduced Questions, and of Their Solution as Found in Alexander the Peripatetic Who Was a Follower of Empedocles[12]

We should, however, work toward a true solution of these questions. Also, so that this solution might be more readily intelligible, let us stress again in summary fashion the difficulty of the question.

The question is, in summary, what is that which performs the first operations in the sperm, namely, those of conversion, separation [*distinctio*], and formation of the members? For since both the soul in and of itself, and every part of the soul is the substantial form [*entelechya*] of an organic body, neither the soul nor any part of the soul can operate where the organs have not already been formed. However, those organs which have been formed are not in the sperm since it is homogeneous. The natural corporeal powers—these being heat, cold, and the like—do not lead to the soul, since it has already been proven in the book *On the Soul* that the soul is not the result of such powers as these or a mixture of them, and it is not a consequence of a mixture.[13]

Moreover, the formation of sperm into the shapes of the members is not one of the active and passive bodily qualities. For since action follows form, a formation of this sort does not suit the form of any one of the first qualities. Neither does it suit all of them mixed together, as is clear to all who have even a small understanding of natural matters. For all such bodily forms act only on one thing, differing in this from the soul, which acts on many things. The operation of the sperm, however, ends at many forms and shapes. Since, then, there is nothing and there can be nothing internal to the sperm except for the power of the soul or bodily power—be it active, passive, or both—and since neither of these powers can act on the formation of sperm, it does not seem likely that sperm undergoes conversion into blood and into uniform and official members as a result of any of the internal powers.

And if, for this reason, it is conceded that sperm undergoes these things from some external power, then this power either will or will not be joined to the sperm, and in whichever of the two ways it is stated this cannot be a bodily quality for the reasons stated. Therefore, it must be either the mother's soul or some other power which to

---

12. Empedocles (ca. 490 B.C.–ca. 435 B.C.) identified the four elements—earth, air, fire, and water—as the four uncreated, simple, root elements of all things, from whose mixture all living things emerge. Alexander the Peripatetic—presumably of Aphrodisias—commented on both Plato and Aristotle toward the end of the second century and the beginning of the third century A.D. His treatise *De anima* was translated into Hebrew and Arabic and greatly influenced Maimonides. His *Treatise on the Intellect* was translated into Latin by Gerard of Cremona (d. 1187).

13. *De anima* 1.2.8.

some degree leaves the substance of the father along with the sperm but which is nonetheless separate from the substance of the sperm. Each of these ideas has been destroyed, however, in the preceding chapter through incontrovertible syllogisms. This, then, is the true sense of the difficulty in the question before us.

Alexander the Peripatetic solves this, however, by saying that the soul is the harmony of a mixed and complexioned body. Thus, he says, the powers of the elements have wondrous efficacy in mixtures in which they are divided into their smallest parts, so that the smallest bit of each of them is with the smallest bit of the others. Then, he says, wondrous actions [*operatio*] come about from the natures and proportions of these smallest bits, which are thus intermingled and bound to each other through the act of being mixed.

This happens when one or more of them is dominant (not entirely, but according to one of the arithmetic or geometric proportions) or shares in an arithmetic, geometric, or musical mean. He says that the numbers of the elements are bounded at one of the means mentioned, according to the proportions of mixables [*miscibilia*], and he claims that whereas each quality of an element, while it is simple, performs one thing only, nevertheless, when taken with another in a numerical proportion, it is bestowed with many powers. It then performs many things, both on its own and insofar as it is not simple, but it is mixed in with the numbers and proportions of others. This man so far trusts the power of proportions of this sort that, as we said in *On the Soul*, he says that even the possible intellect may be such a preparation of these bodily qualities and powers.[14]

21   This, however, is the worst sort of error and interferes with all true understanding of natural things. For it follows from this that souls are only bodily forms and harmonies of bodily qualities. But we have already shown that this is false in our book *On the Soul*. Still, we should introduce a few things here as well. However much the bodily qualities might be bound in a mixture with regard to number and proportion, they are nonetheless always bodily. And if they are rendered as having multiple powers by being mixed proportionally, all this multiplicity nonetheless comes under the genus of bodily powers. For a bodily power, be it simple or multiple, has two convertible and proper traits. One of these is that a larger thing is in a larger individual subject and a smaller in a smaller. The other is that it acts only in a bodily fashion, whether it acts with a simple or a composite power. Therefore, every soul has these proper traits and every soul will thus be an action of the body that is larger in the larger and smaller in the smaller. Yet this is an absurdity and cannot be understood in any of the souls.

Further, to say that every soul acts in a bodily fashion is also entirely false, since no soul acts after the fashion of a bodily power.

Also, to say that there is no action of the soul save as a result of bodily harmony, and not only in accordance with bodily harmony, is all absurd and hostile to the truth. For this reason these things must be renounced.

---

14. *De anima* 1.2.8.

He further does not agree that the sperm is moved by powers external to it. For he saw that if it were moved by some external principle, it would not bring about a natural form. For every natural form is brought about by internal principles since it is inchoate [*incoata*] in the material and is inside it. That which acts externally, like an art, brings about only accidental form since the material is changed from the species in which it exists to another one only by means of things which alter and enter the material.

This, then, is a summary of the statements of Alexander, excerpted from the things he said in the book he wrote on material and form. He does admit, though, that the elemental powers, which through various levels of mixture cause the species of things which have been generated and corrupted, take on finer and simpler powers from the heavenly light. This falls on them in many ways from heaven and these many ways are caused by the many stars and constellations in the heavens which differ in species and nature. A diverse mixture is produced in these lights just as it is in elemental qualities. The cause of this diverse mixture of the lights is their location and movement with respect to conjunction, precedence, distance, nearness, and their multiple relationships, one to the other.[15] Again, it is also produced according to the different way their rays fall on the material of things which are generated or corrupted. Their rays fall on this either perpendicularly, in a tangent line, at an obtuse angle, or at an acute angle.[16] There is a very great difference indeed between their acute and obtuse angles.

Moreover, some touch with a straight ray beneath a single location, whereas others do so with a curved ray at a double location of the ray.[17]

The elemental qualities are variously given form [*informantur*] by these and many others by the heavenly powers. In this way powers are produced in them which are sharper or duller, simpler or more multiple, finer (that is, more formal) or more material, more wondrous or baser.

In whichever of these ways it might be said, this fellow wants the bodily forms in generated things to be only mixtures of qualities and the consequences of heavenly lights. Thus, he says that all substantial forms, whether of inanimate objects or of the soul itself, are harmonies or are the harmonious consequences of bodies. It also seems that he wants only certain proper, accidental traits of the material to be mixed into them in this or that way. Now, while this is absurd to every Peripatetic, this is not the place to disprove it. This must be determined in first philosophy because that is where principles of substance, insofar as it is substance, are considered.[18] For here, however, that which we said in the book *On the Soul* is sufficient, for there it was proven that no soul is a harmony of the body or is consequent upon a harmony.[19]

---

15. "Precedence": See 6.103.

16. On *recta diametro* as perpendicularly, cf. *LLNMA* D.393 (*diameter*).

17. The straight ray (*recto radio*) might also be translated as perpendicular. Likewise, curved (*reflexo*) might be either bent back or reflected.

18. "First philosophy": i.e., metaphysics.

19. Cf. *De anima* 1.2.8.

Now it might be objected to Alexander that, according to this, there should be no male or female in animals, because a mixture of this sort is completed in the female alone in conjunction with the heavenly powers, and nature is contained in as few as possible.[20] Nature therefore produces only one that generates in itself and this is the female, with the result that the male is entirely superfluous in nature.

Alexander will then respond to this by saying that this is not true because generation is accomplished only by active and passive powers which variously surpass each other or are surpassed by one another. That which is formed in generation must have powers which receive and hold the form. The power which receives the form is a humor. Moreover, that which holds it is the same humor which causes it to become firm by means of the cold which somehow compresses and bounds it by means of dryness. However, these powers must be digested this way because digesting heat does not sharpen the dry, which bounds the moist (which must take on form) and does not expel the cold at all, which compresses the moist so that the form might be sealed in it.[21] However, just as there is need of that which takes on form in generation, so too is there need for that which impresses and seals the form, and this should be moving, causing an impression, and sharp. These powers are brought about only through sharp heat and penetrating, dry heat. These things should be digested so that the cold might disappear and not interfere with the movement of the heat and so that the moist might be overcome and not interfere with the sharpness of the dry.

These digestions, then, have opposite means [*modus*] and ends [*terminus*] and thus cannot be produced in one and the same complexion. Therefore, as he says, male and female were created. In the male is completed that which seals these qualities. In the female is that which is sealed.

All of these things, however, are mistreatments by Empedocles, who denied that substantial forms and souls exist. He said that there is one true sense [*ratio*] of mixture in bone and another in the flesh. That what these people say is of no account is clear because according to their statements, male and female should exist in each one that generates. For every type of quality involved in the act of sealing and receiving, both that which is surpassing and that which is surpassed, is present in every act of generation and not only in the generation of animals.

This, then, is the summary of the statements of Alexander, given in short, following Empedocles, not Aristotle.

---

20. "Is contained in as few as possible": or, another possibility, "is content with as few things as possible."

21. "Sealed": *sigilletur*. The metaphor is that of a signet ring.

CHAPTER IV

Which Is a Digression Setting Forth the Solution to the Question Introduced, Based on Avicenna, Theodorus, Theophrastus, and the Peripatetics

Avicenna, however, speaks fairly well on the solution to the questions before us. He claims that the generative powers are given to the sperm of the male from two sources. To be specific, they are given in a general way by the members in that the sperm is that which is separated off for generation from the last digestion. For that which is digested last is like all the members with respect to the proximate power [*propinqua virtus*]. Thus, the nearest potency has the action of all the members, but none in respect to act. For such a nearest potency does not exist without some power of the members. For, as was determined in the book *On Nourishment* [*De nutrimento*], assimilated food has motion toward the members because it generates in the member the form of the members and that form is already putting the powers to work in itself when it moves to the vicinity of a member. As a result of this, then, it has the formative power at the beginning of each member, although it is not yet complete in its substance with respect to act.

The second source is the seminal vessels in which are the formative powers which complete the powers begun [*virtutes incoatae*] in the fourth digestion so that they might act on and form the members. They thus complete the sperm by digesting it further and introducing into it the spirit which has the power of the soul and which forms and separates the whole into parts and into the shapes of members. These powers are thus internal to the seed with respect to subject and are external to it according to origin. On this point Avicenna agrees with Alexander, who also said the formative powers are internal to the seed since they are in it as in a subject, while their origin is in the male and the female as they are digesting.

But Avicenna differs from Alexander in that Alexander posits that all these powers exist of the body with respect to subject and origin. Avicenna, however, does not say this, but rather that they are in the body according to origin. This is because they arise from the soul, which is not a bodily power, whether the bodily power be simple or mixed. If the bodily power be mixed, then again the soul is neither a mixture of it nor something consequent to mixture, as has been proven in the first book of *On the Soul*.[22] Rather, the warm, moist, and other powers in the semen are the instruments of that power which is poured into the spirit of the semen by the soul. We call this the formative power. This directs those instrumental qualities to the end it intends, that is, for the formation of the members.

---

22. Cf. *De anima* 1.2.8.

Therefore, the power of the soul, which is in the soul, is not the soul, as has already been proven in the second book of *On the Soul*.²³ Rather, it is something of the soul and its subject is the spirit held inside the semen by its thickness and viscosity. This, then, is why thin, watery sperm does not generate, because it easily gives off the spirit in which lies the formative power within it. The instrument of this power, therefore, is the heat which digests and converts the material of the sperm and of the whole offspring [*genitura*]. This heat has the power of the soul much as a tool has the power of its mover. Thus, it moves toward life and has, due to the material in which it lies, the celestial and elemental powers within itself.

This material, which is the subject of the heat, comes from elements and thus has elemental power, not simple, but mixed, and this is multiple, as was made clear in the previous chapter.

From the same material it also has the celestial powers in addition to the things mentioned before. Since these powers have vivifying light, it has from them the ability to establish and to give substance to that upon which it acts. Therefore, from the powers of the element it has the ability to alter, refine, and amass homogeneous things, and to separate heterogeneous things. From the celestial powers it has the power of establishing in permanent being and in congruent shape and to make substantial the substance of the things which have been formed. Thus, too, the celestial things themselves, having been established in permanent being and in the shapes of constellations and of rays, fall upon and permeate the material of generated and corrupted things. And these are substances subsisting in themselves which are perpetually formed. But this same heat, insofar as it is of a soul whose nature is above the nature of the world (because the world was created in the shadow of a lower soul, as Isaac says), has the ability to introduce life into a formed sperm.²⁴ This, then, is how that which is conceived in sperm is rendered animate, be it in an animal or in a plant genus.

However, Alexander, since he says the soul is only a bodily power, says that the same powers which form the sperm into members are present in the formed members and that they move them, being like souls or parts of souls in them. Avicenna, however, because he says that none of these powers is a soul (and this is the truth of the matter), says that after formation is completed, the formative powers pass on and do not remain. Rather, he says, the soul which has been introduced then moves the members in all their movements, namely, natural, vital, and animal. In this he differs from Alexander.

---

23. Cf. *De anima* 2.1.7.

24. It is possible that A. may have in mind here Isaac ibn Sa'id, a Jewish philosopher in Toledo in the thirteenth century who headed a group of Jewish scholars that produced the "Alphonsine Tables" (1252–56) for King Alphonse X of Spain. However, it is more likely that this is a reference to Isaac ben Solomon Israeli (ca. 855–ca. 955), a Jewish philosopher influenced by Neo-Platonism. This latter Isaac was an important physician as well whose medical treatises were translated or adapted in Latin versions by Constantine the African in the eleventh century. His *Book on the Elements* was also translated by Gerard of Cremona in the late twelfth century. The "lower soul" A. mentions may be the World Soul, which guides the lower world and is itself an emanation from the Universal Intellect.

It is thus clear that these powers are internal [*intrinsecus*] to the sperm and that their operation is bounded in the direction of the animate and the living. Then he says that they pass away and do not remain. But the soul, which has already been brought forth to act, afterward serves as the principle of all movements of nourishment, growth, sensation, and generally all other movements which exist in trees and plants. If, however, the question is when this soul begins to operate and if it is one or many, then this will be determined later.

This, then, is the summary of the statements of Avicenna. However, Avicenna has taken these words from the philosopher whom the Arabs and Greeks call Theodorus.[25] This one said that the semen has the act of soul, an act which some call soul. They, however, call this act the formative power of an animate thing, impressed on the seed [*semen*] by the soul of the one generating.

Theodorus, however, said this act is formal, giving form to the celestial powers which are in the semen. These, given form in this way, give form to the elemental powers which are in the same seed. It is because of the formative actions of this sort that the entire action [*operatio*] of the heat and of the spirit in the seed operate on the form of the animate one, configuring the organs to the operations of the soul and leading forth the soul (which is the form of the animate one) from potency into act. When this act of the soul has led generation to its end and completion, it ceases to operate. Just so too does every individual generating cease to move when it has achieved the end of generation. When, then, powers of this sort are not operating, they do not exist. For that which is useless does not exist in nature. And thus too, according to this man, these powers are exterior to the sperm in origin but are internal in subject and being, though, as has been said, they are not permanent, being transient instead.

This entire philosophy is taken from Theophrastus, who wrote a book *On Animals* in which he also passed along the theory [*sententia*] just mentioned on the powers of the sperm.[26] He adds only that in the sperm there exists the power of all the parts of the soul, impressed by the parts of the soul of the one generating. And because the power of the parts of the soul produce the animate one and the instruments of the animate one, the powers of the sperm operate like members in all respects.

All these men agree that these powers of the sperm are acts and impressions of the soul, but they are neither the soul nor a part of the soul, rather being powers in the heat, spirit, and moisture of the sperm, and, because they are acts of the soul, they have powers of the sorts of actions which have been mentioned. For something which has been caused formally by something has the operations of that by which it has formally been caused, since every operation is the result of a form. They admit, however, that male and female are the principles of generation and that the male is the active

---

25. The identity of this Theodorus is uncertain.

26. Theophrastus of Lesbos was a pupil, collaborator, and successor of Aristotle. His works on plants are extant. The "book *On Animals*," *De animalibus*, may refer to a section of the now lost *Doctrines of the Natural Philosophers*. For a discussion of A.'s sources and knowledge of the views of Theophrastus, see Sharples (1984).

principle and the female is the passive principle, as has been said. They also say the seminal vessels are necessary for generation, these being the place where the heat and spirit in the semen are matured, digested, and formed into something which can act.

30 Those men also claim that the womb must be the place for generation because the vital heat is fomented only by this sort of heat, and vital spirit is fomented only by the spirit of life.[27] They call nourishment and coherence "fomentation." For this reason they say that embryos (that is, the conceived sperm) cling to the womb above the openings of the veins and arteries. For just as the vital spirit is nourished and fomented by breathed-in spirit, so too is the formative spirit in the semen nourished by the spirit which pulses through the openings of the arteries of the womb. The heat is nourished by the blood and heat which ooze into it through the openings of the veins. Nevertheless, the spirit and blood of the womb neither form nor perform any action in embryos save that of fomentation and nourishment (as has been said). So too does the food entering into a nourished body neither form anything nor do anything. Rather, it is formed and acted upon by the powers of the body. Being ignorant of this distinction, certain physicians have said that the spirit and heat of the womb move and form the semen before it has a soul.

This, then, is the theory of Theophrastus on the operation of the sperm and on the solution of the questions we have set forth. This is extremely close to the opinion of Aristotle since Theophrastus was himself taught by Aristotle and was the most excellent of his students.

CHAPTER V

Which Is a Digression Setting Forth the Solution to the Questions Introduced Above, According to Socrates, Plato, and the Opinion of All the Stoics

31 Even though certain things have been said above concerning the opinion of Plato on the power of the semen, nevertheless we must add certain things here. For what we will speak of had its origin in Socrates, the one whom Plato, along with the entire Stoic school, followed.

These all commonly teach that the powers of all the members are present in the semen in act but are latent, and that the instrument of these powers is the heat of the semen. They say that in order for suitable heat to be rendered the instrument of the aforementioned powers, it must be digested in the testicles.

They do not say that this digestion is necessary in order for it to take on in that place the power we call the formative power but rather, as has been stated, that the

---

27. "This sort of heat": following Borgnet's reading (*in tali calore*). Stadler's reading of *vitali calore* yields "vital heat is fomented only by vital heat."

spirit and the heat might be digested and refined in the testicles so that they can be fit for action. This is just as a builder sharpens his adze so that it can better cut and chip away at the building with a sharper edge.

Those who posit that all the powers of the soul are present in the semen, but are latent, are not speaking of a latency of the sort of which Anaxagoras speaks.[28] For he said that all things are in all things, by reason of the homogeneous from which the heterogeneous are composed. Thus he said that flesh, blood, and all things of this sort are present in sperm but that they are latent. These men, however, say that no power of parts of this sort is present in the sperm, and that only all the powers of the parts of the soul of the one doing the generating are present. I say all, not just of the homogeneous but also of the heterogeneous, and that the powers are in the sperm in such a way that they stay outside the members they are shaping, but they produce in them other powers like themselves.

They say that the origin of these powers does not come from the material or from the fourth digestion of the one generating but rather from the separate forms which, like some sort of seal, press forms like themselves into the material of the ones generated. They have said that these sealing forms are present in the immortal, celestial gods. Plato meant this when he addressed the celestial gods, who are the gods of the earthly gods since they press their own forms into them, saying "Gods of the gods of whom I am the father and fabricator."[29] Afterward he adds something on the terrestrial gods, saying, "I will make a sowing of these, and will hand them over to you. It will be right for you to complete it."[30]

Thus, the sperm is well purified by the fourth digestion and likewise by the heat of the testicles (these being the spermatic vessels both in the male and the female) and it is also prepared by the digestions to enable it to be sealed upon, just as wax is mixed and softened to ready it to receive the shape of a seal. But it does not receive the form of the powers of the soul except from that sowing which has been given by the gods of the gods to be fulfilled. They then fulfill it when they make an imprint by sealing in forms similar to the material of the things subject to generation. In this way, Plato says, souls come into bodies from heaven.

He provides a proof [*ratio*] for this in that heat does not lead out any living thing insofar as it is the heat of fire, but rather insofar as it has been formed by the light of heaven. This is as was set forth in the preceding chapter.

Further, since every power of the soul is in accordance with a potency of life, every power of life must necessarily arise from that which is the font of life. It then follows that the fountain of life is present only in the mover of heaven and in the celestial powers.

---

28. Ar.'s text does not mention Anaxagoras here, although his name does appear earlier at *GA* 723a11f. For Anaxagoras's claim that "all is in all" in act, albeit in latency, cf. A.'s *Phys.* 1.2.11–12.

29. Plato's text (*Timaeus* 41a–c) has occasioned several different translations due to the difficulty of the original. Cf. Cornford (1975), 367f. A. would have known the *Timaeus* in the translation and commentary of Chalcidius (Waszink, 1962).

30. *Timaeus* 41c.

All the Socratics solve the questions introduced above by saying that the figurative powers of the sperm are external and celestial in origin. Those which shape the members directly are to this extent also external since none of them is brought forth from material but instead enter from without, from heaven, into the material of the sperm. Even though these cling to the substance of the sperm, they are still not internal to it, as has already been shown.

34 Aristotle attacks this opinion as erroneous, for in truth that which adheres to something externally does not transmute material with the sort of transmutation by which it moves to form. Since movement to form is a flowing of that very inchoate form toward completed act (as has been stated and determined in the *Physics*) and since movement is a completed act only if there is something of form in it, it follows that a being which exists totally in act and which adheres only to the outside, produces nothing toward the movement of generation.[31] This is better than all the reasons by which Aristotle does away with this opinion of Plato.

Further, if these forms, clinging externally, are to change the material, they should move it with respect to something. It cannot be found that it moves even by one wanting to pretend, since heat is not its instrument. For it is the instrument of the material, and the quality of the moved, moving, and transmuting material is not the same. This is clear to anyone on his own since there is no movement whatever toward form among similar things.

These, then, are the things which, in the aforementioned way, do not allow the aforementioned questions to be solved. However, these things must be carefully investigated in first philosophy and so let us put them aside for the present. It is enough that here we have set forth two necessary arguments [*ratio*] by which this position is demolished. Now let us follow Aristotle and the truth in pursuing the solution of the questions that have been set forth.

CHAPTER VI

In Which the Solution to the Questions Introduced Above

Is Introduced, Following Aristotle, Who Saw the Truth

35 We will say, then, along with Aristotle (who, it seems to me, was alone in giving a true solution to these questions), that it is reasonable that the sperm has a relation to that which is formed according to one of the ways introduced in the question.[32] For it is necessary that the mover be either internal or external to it and we will say that the

---

31. On the *inchoatio formae* as the potency of form in matter that is led to act in the process of generation, cf. *Phys.* 1.2.12.

32. Ar. *GA* 734a29f.

mover is not simply external, outside the sperm, but that that which moves the sperm is, to a degree, external to it. For it is impossible that the power moving within the sperm be external to the sperm at a given time (this being at the time of generation). For the formative power in the sperm moves toward the forms of the members of the one generated and at that time it must be internal to the sperm. Yet, in another way, it is possible for this power to be external to the sperm, at another time, namely, when it is in the last digestion. For then it is sealed into a likeness of the power of the members by the members' powers which are external to it.

Thus, then, these powers are external with regard to primary origin, but with respect to the subject they are internal, as was stated above in the opinions of Alexander and Avicenna. Likewise, the statement of one who says that the sperm forms and creates the members is possible in one way and impossible in another way. For it is impossible to say that it forms and creates using the powers latent in it which belong to all the members. But it can be possible that it forms and creates the members insofar as there is in it the principle of the movement of the formation and the creation of the members. For when we say that one thing can move another and this one still another, it must be that the movement of that which is moved by a prior one comes as if from a prime mover, from wondrous things which exist in and of themselves. For every order of movers must necessarily be grounded in some one, prime mover, from which the motive power exists in all the others. And this mover is loftier and more wondrous than all the other moved movers which are of the same order. Such a prime mover is the power in the members. For since this power first moves a thing external to itself in the last digestion (by external I mean neither joined nor united with respect to act), then there is necessarily created, suddenly and with respect to a certain act, a certain similarity to that power in that which is moved by it.

This is done as happens in things which move themselves and in moved movers. For in these a second thing moves a third because it was moved by a first. Nor is it necessary that when a second moves a third, the body of the second be in contact with the body of the first, but it is necessary that it have touched already in the past. In exactly the same way, a power which is sometimes quiet and in no way moves in the members will at another time move and impress upon the sperm when it is the superfluity of the fourth digestion. This, when it is moved further, moves toward the form and figure of the members, once it is projected into the womb. Nor is it necessary that it then come in contact with the power of the members.

Something similar to this is to be found in the knowledge of Polyclitus the carpenter.[33] For in this example it is clear that that which moves to form and trim the parts of the building is not a thing complete in the shape and form of a house. For according to this line of thinking a thing complete in the shape and form of the house would exist before there was any complete house, something impossible. There is a house, to

---

33. Cf. Ar. *GA* 734b18, which only mentions the building of a house. The name Polyclitus is A.'s addition. Polyclitus was no mere carpenter but one of the more famous sculptors of the end of the fifth century B.C. He is explicitly identified by A. as such at *Phys.* 2.2.8.

be sure, in power, but not in act. That which first moves toward the building of the house is the art of house building, which exists in the soul of Polyclitus, but this art is not a house with respect to completeness and act. The art in the soul, then, moves the hands and tools, presses them toward that which is the species of art in them, in the sense of an operative power. It is in this way, then, that we must speculate as to how each member is produced in the embryo. And, so that this might be done most suitably and intelligently, let us resume the discussion from its beginning.

38 We think, then, that each thing produced by art is most worthy of existence since the work lives on in the art, receiving and retaining a similarity to the power of the art. This is just as if we were to say that the art in Policlitus's soul is present in his hands and tools, for in them it retains a similarity to the power of the art, since it is operative in them as it is in the artisan. But the art which is drawn into the stones and beams has no similarity with art save in shape. For in these it is operative toward the art of no given thing.

Therefore, following the analogy of the hand and the tools of art, sperm is produced by the powers of the members and for this reason has the movement of forming and shaping the members and has the principle in itself of this sort of movement. When this sort of formative movement ceases, a given member will be made, with respect to form and shape, for movement of this sort stops only at the form which is the terminus and end of its movement. This happens when the given member is animate, so that a certain member becomes the face, possessing the shape, form, and power of the face. And those which are like the face are called nonuniform. As a result of a formation of this sort a given member becomes flesh or another uniform member, possessing the form and act of a uniform member in which the formative movement of the sperm has ceased. For both uniform and official members are made by the movement of the sperm. For we cannot say that the principle of this sort of formation is some member like the hand or foot, but rather that it is that which has been moved and been impressed with a seal by the powers of the members, retaining in itself the form of the first mover, thus producing hand and foot, and likewise flesh, nerve, and other members, both uniform and nonuniform.

39 Thus none of these members will exist without one forming and operating this way, nor will it exist without the operation of such a power. For even though all things are made from the operation of natural heat, nevertheless, because the heat is the instrument of this sort of power in the sperm, the heat thus necessarily acts and operates in the power of this same formative power. But hardness, softness, and viscosity are effectively caused in the members by the elemental qualities, which are heat, cold, and likewise other natural accidental traits. But the mixture from which a given part becomes flesh and another bone and the like occurs in no way from the extent to which heat and cold are of this sort but rather from the movement which arises from the formative power of the generating sperm, as we have said. For that which exists potentially is produced from that which exists, to some degree, in act. This is just as can be seen happening in manufactured things. For the hardness and softness of iron

occurs due to the heat and cold which are the elemental qualities. However, a sword, saw, or the like are only produced by the principle of the art which is the principle of movement toward these forms. And they are made by the instrument which is the hammer, tongs, anvil, and the like insofar as the skill of the artisan moves them. For art is the factive principle and the form of that which is produced through art.

This, then, is the shortened summary of the aforementioned solution, namely, that the sperm, because it is rendered like the members during the fourth digestion, resembles the powers of the members in virtual act. And because it exists in the virtual act of the members, it can lead forth the material of the semen and of the woman's menstrual blood which exists potentially for the act of forming the members. And heat is the instrument of this in a threefold power. For heat has elemental power and thus digests, both gathering together the homogeneous things and separating the heterogeneous ones. It is also in the heat of heaven and thus establishes and stabilizes in substantial and firm being that upon which it acts.

It is likewise present in the power of the soul or in the power of the soul which moves it and it thus can vivify and produce animation in accordance with every part of the soul which, for it, is the entire principle of life in the body. And just as art is successively in three things, so too is the formative power. For the art in Policlitus's soul is like the power of the members in the members of the one generating. However, the art in the axe, hand, and adze is like the formative power in the sperm. Yet the art led into the stones and beams is like the formative power in the formed embryo.

This, then, is the summary of the belief of Aristotle.

CHAPTER VII

Which Is a Digression Setting Forth a Summary of All the Things Which Have Been Said, Bringing Them Together in Summary Fashion from All the Things Which Have Been Mentioned

If anyone should choose to make a determination on the questions that have been mentioned, doing so by gathering together those things which are true from the statements of all the philosophers we have introduced, then that person alone would have the most perfect knowledge of the power of the sperm. For while Alexander in many instances speaks at variance with the truth, he nevertheless speaks quite truly when he holds forth on the mixture of elements, which is especially present in sperm which passes through many digestions. For food is not a simple element, but it is a sort of mixture and thus food, before it is an animal's food, has had many digestions and "ripenings" [*maturationes*] of its component parts. Moreover, after the animal eats, the food is digested after the separation of the dry and earthy superfluity in the stomach.

And that which is taken in is assimilated. It is also digested by the separation of the watery superfluity from itself in the liver, and that which is taken in, digested, washed, and cleansed also takes on, beneath the heart, the heart's vital and motive power for the members. As it flows through the veins it is struck by the spirit and heat and is also digested in the veins. That which is most foul settles out from it and that which is most viscous and cold stops, not progressing to the absolute end of the veins. That which is lightest floats off and evaporates. That which is the cleanest and purest flows to the members. Fourth, it is digested and thinned by the power of the members during the actual oozing out of the veins because, as we said above in previous books, the ends of the veins are narrow and only that which is very thin passes through them. Then at last its superfluity is allotted to semen. Yet this takes on still further digestion in the seminal vessels.

42 In all these there are three powers imparting form to the digestive heat. The power of fire is present, for example, in the way in which the power of fire is thinned in the manner that was mentioned, having been mixed thinly.

The heavenly power imparts form to this power, giving foundation and substance to those things digested, since it is the power of a perpetual body which retains its substance in a single way. This power has multiple powers, based on the locations and movements of the conjunctions, oppositions, and of other things spoken of above. Last of all, the power of the soul imparts form to this heavenly power as well, just as the movement of the instrument of the builder is given form by the art of building. Through this giving of form to [*informatio*] the soul, it begins to work toward life, or, as is more correct, toward the living and animate. These powers are gathered together in the spirit and heat of the sperm. And because they work toward an animate substance (insofar as they are from a heavenly body and the soul), these powers are therefore said to be like an art, and they are present in semen just as art is present in the tools of the art. And because they belong to the soul and work toward an animate thing, therefore, just as a sculptor (with respect to the entire power of the art of sculpture) is present in his hands and in the tools with which he works the statue, so then are the soul and the heavenly influx (with respect to the entire power of animating and forming) present in the spirit and heat of the sperm.

This is what the cleverer Peripatetics said, namely, that the soul is in the semen not like the substantial form [*enthelechya*] of the body, possessing the potential for life, but rather as an artisan and art are in the instruments by which a manufactured item is produced.

43 To these statements we should add that every work of nature is a work of intelligence (as we have already determined in the other natural philosophy books). For that is especially a work of intelligence in which the power of intelligence terminates the work of nature at the form nearest and most like itself. For although the work of nature and intelligence are one and the same (since nature acts through intelligence and intelligence acts in nature), nevertheless the end of the work is sometimes closer to nature and at other times closer to intelligence. For when the forms which are

related to being [*esse*] and power [*posse*] are material and the limits of this sort of work are bodily, then certainly the boundary of the work is closer to nature and those qualities prevail in the work which are the prime elemental qualities, and those which are consequent are caused by a mixture of them. When, however, form is raised over the conditions of bodily material, so that it is raised more and more, thus it also more and more approaches a likeness to the intellectual nature, that is, intelligence. Then the powers of intelligence prevail in a work of nature.

The first level at which it is elevated above the nature of the body is in the vegetable, for it acts on many things and not on one, like a bodily form. The second, higher level is in those participating in sensation, which take in the forms of the sensibles without any material. This is just as wax takes on the shape of the seal, as we have said in our book *On the Soul,* and it is something none of the powers of the bodily forms could do.[34] The highest level, though, occurs when the very form itself is such that it is neither an act of any part of the body nor does it act toward harmony of the body or the power of harmony. Rather, it is like one that participates in reason, something which is separate, almost the likeness of an operative intelligence.

There is no natural cause for this, except that in generation the powers of the intelligence overcome and prevail over the heavenly and elemental powers, bringing the form which is the end of generation as near as possible to their likeness and nearness. But we will treat these matters more fully in the book *On the Immortality of the Rational Soul* [*De immortalitate animae rationalis*], which, with God's help, will be our next work.[35] What we have said here, however, was said because the last and the most simple of the powers in the sperm is that which the prior intelligence, which is the simplest and most formal power, is first in moving, and forms the power, both heavenly and elemental, of the soul and the members.

In summarizing, then, the powers of the sperm, we will say that the one moving first is that which pertains to intelligence and next the power of the heavenly motion is moved along with all the diversities and proportions of the locations, constellations, and movements, both of the heavenly bodies and of their rays, as have been determined above. And these are the determinations that those who aim at understanding the causes of generation via the heavenly powers make. After this is the power of the soul or of an animate body insofar as it is animate, and after this power is the power of the members which digest the sperm in the course of four or five digestions (for it undergoes a fifth in the spermatic vessels). After this is the power of the mixed food, insofar as it is a mixture of mixables and a unity of changed things.[36] Now the powers of simple things are bound in numbers and proportions so that they do not act with

---

34. *De anima* 2.1.3.

35. "*On the Immortality of the Rational Soul*": an apparent reference to *De natura et origine animae*, which, as Geyer concludes based on this passage, A. intended to bring out, at this point, as a separate work following the last (i.e., nineteenth) book of *DA*. Cf. Geyer (1935), 579f.

36. "Insofar as it is a mixture of mixables": perhaps "insofar as a mixture (is) a union of altered mixables" or "insofar as a mixture of mixables (is) a union of altered (things)."

the action of simples but with a harmonic action, as was explained above in the statements of Alexander. Thus, each quality takes on harmonic levels as the more skilled of the physicians determine the levels of their simple medicines. The last accessory of all is the power of the simple qualities in the mixture, preserved according to some first being [*primum esse*], as we showed in the book *Peri geneos*.[37] This, then, is the belief we had to summarize on the powers of the sperm.

The questions of how the sperm of the male has the formative power, how the sperm of the female does not have the full sense [*ratio*] of sperm but has the informative and passive power, and also how the sperm of the female is most often found in the womb even though the female does not always produce sperm during intercourse have already been stated by us satisfactorily in previous books of this investigation. Let, then, these things about the power and activities of the sperm suffice for now.

CHAPTER VIII

In Which Two Questions Are Resolved Which Arose from the Things Said about the Sperm. One of These Is Whether the Sperm Is Animate and the Second Is Whether the Powers of the Soul Create the Members Successively (That Is, One after the Other) or All at Once.

46   From all the things which have been introduced, a question arises which is quite serious to solve. For since nature is spoken of in many ways, as we determined in *Physics*, it is agreed that the nature which both is and is called a nature with respect to being prior [*natura per prius*] is the nature which is form.[38] This is act and it produces in act the being which it perfects and in which it serves as an agent and an active force. From the things that have been said, however, it seems to be a given that this nature is, in animals and plants alike, the soul. It acts in the semen, as is clear from what has been introduced above.

We will therefore investigate whether the sperm of plants and animals is animate or inanimate.[39] For if it is called animate, then the sperm of an animal should be a small animal and that of a plant a small plant. But this has been disproven in what has gone before. If, however, it is called inanimate, then it seems that it ought not to have the operation of the soul. In that case the operation of the members which occurs in conception is not from the semen but from some external bestower of forms. And this too has been disproven in what has gone before.

---

37. Cf. *DG* 2.2.15.
38. Cf. *Phys.* 2.1.3f.
39. Cf. Ar. *GA* 735a5f.

The question is much the same concerning the members formed in the embryo and in conception. From the things carefully stated in the book *On the Soul,* we posit that there is a single soul in an entire animate body, even though it has a multiple diversity of organs.[40] However, each member of an organic and animate body has an essential sharing with the soul, since one of the powers of the soul is its substantial form. For we are not speaking here of members which are called members equivocally because of a similarity of shape, as when the eye of a dead person is called an eye. Thus, with these as our suppositions, it is clear that in each member there is some potency of the soul which is a potency for life. Now in one it is nearer and in another more removed, as in bone it is more removed and in flesh it is closer. Again, it is clear in the heart but is more hidden in the liver and brain. For these are insensible and for this reason both uniform and nonuniform members are related unequally to the potency for life and for this reason they are also related unequally to the soul.

We can thus say that some members are related to the potency of life and to the soul in the same way that a writer's habitual potency for writing is related to him when he is sleeping. For, overcome by the heaviness of sleep, it neither acts nor functions and yet it is still in him. So too is the potency for life, sensation, and enlivening [*vegetatio*] present in the brain and in the bones and in the other insensible things, and yet they have some share [*communicatio*] with the soul, just as the faculty of writing has some share with the one sleeping. Other members which possess the operation of life share with the soul just as does the faculty of writing in one who is awake and who works when he wishes. Thus, then, we cannot say for all the members which have a share with the soul that a given member which exists in act in the sperm is the cause of the generation of all the members which share with the soul of the embryo, for this has been disproven above in an earlier book of this study.

Moreover, since any given member which is said to exist in the conception is generated as are the other members, it will be necessary that that member generate itself. Thus, the same thing would be both the one generating and the one generated and thus at the same time would both exist and not exist, a thing that is entirely impossible. By the same necessity, we are convinced that all the members, existing latently and actually in the semen, cannot generate themselves. Thus, whether it be said to have only one member lying latent in the semen or all the members, it cannot be said that a small animal is the one generating which is said to transform itself into semen and thus sets generation in motion. But if it be said that this is fed and increased by the semen and takes on the appearance of the members, then this can also be proven to be false, for everything generating is the principle of growth in that which generates in every animate body, whether it be plant or animal. However, the principle of growth in plants and animals comes about through the nutritive power that is in them. Now by the one generating I mean one producing one like itself. For this activity [*operatio*] generally follows the path I have outlined both in plants and in all animals. Genera-

---

40. *De anima* 1.2.15.

tion, however, is spoken of in two ways. For the human generates a human from the material of food univocally while the sun generates equivocally, as we have said in the *Physics*.[41]

50 With the supposition, then, that human generates human, we will say that the sperm is not animate, but rather that in the sperm there is the act and effect of the soul just as art is in the instruments of art. Thus, the sperm is in actuality an effect of this sort, on account of which it functions and acts by forming, animating, and vivifying the members. It is not, however, constituted in actuality by the soul as if by the substantial form [*entelechia*] of the body according to the potencies of life of the one acting, as is clear from the things said both here and in the book *On the Soul*.[42] And we will say that just as the members, constituted differently, refer to the soul according to the potencies and operation of life, so too the parts of the sperm refer differently to the generation of the parts. Thus, the heart is formed first by the formative power, and it is the seat of life and of the soul. From it are formed the other members, both above and below, just as has been determined in the place where we spoke of the formation of the conception.[43]

Thus, then, the formative power makes the heart first in those particular animals which have a heart, and in those which do not have a heart it first makes the member which takes the place of a heart. Then the principle of creation of all the members will be from the heart or from the member that takes the place of the heart, as we have said so often in previous books.

Thus, we have set forth what is the cause of the generation of a given animate object and what is the prime mover, that is, the principle of generation in the sperm. Also, all the difficult questions introduced above have been fully resolved. For it is clear that the mover most proximate to generation is internal to the sperm and is in act with respect to power. However, its first origin is from the one generating, from whom the sperm is derived, and the power of that one remains in the sperm. Neither does it take the power from the soul of the mother, although it does take on material growth from her menstrual blood and spiritual nourishment from the spirit which pulses in the arteries of the mother. For this reason, too, it is attached in the womb to the openings of the veins and of the arteries, sucking on them, as it were. And this is clear from the things set forth above. For this reason too, certain physicians have accepted this opportunity for error, thinking that the entire vegetative power is the sperm in the soul of the mother. But this has been disproven previously.

---

41. *Phys.* 2.2.22.
42. *De anima* 2.1.3.
43. Cf. 9.115f.

CHAPTER IX

On the Resolution of the Question Why Sperm Grows Black and Thins in the Cold, Whereas It Is White and Thick in the Animal in Which It Is Generated

Someone might ask what the natural disposition of sperm is, namely, whether, in regard to its first and proper subject, it is of the nature of water or of earth.[44] For we see that sperm is thick and white when it leaves the animal. But when it has been deposited outside the animal's body and cools, it turns moist and dusky like water. For this reason it seems falsely said that the prime subject of sperm is water. For those things whose first material [*prima materia*] is water gel and thicken in the cold and are dissolved by heat. This is as we have shown in the fourth book of the *Meteora*.[45] However, sperm is thickened in the heat while it is inside the body of the animal and is dissolved by the cold when it stands in the open outside the body of the animal. For this reason it seems that it is more of the species of earth than of water.

Moreover, in these animals the sperm, the majority of which is earthy, gels when boiled and coagulates like milk. This is in no way in accordance with something that is of the nature of water. For if it were of the nature of water, it would gel and thicken when chilled, and it would liquefy when heated. Yet the exact contrary of what has been said is seen in sperm. Moreover, the entire difficulty of this question is that it in no way seems true that a species of water ought to be thickened by heat. Rather, it ought to liquefy and dissolve. Yet the sperm thickens from a hot boiling in the body of the animal and dissolves and liquefies when it is outside its body.

If, however, someone were to say that the first material of the sperm is not water but rather is that which has a goodly amount of earthiness in its material and that it thus gels and thickens both in heat and cold, this would be contrary to the commonly held position of all the ancients who posited water as the principle of generation of animals. They based their argument on the birth process, calling the birth process the sperm.[46]

The resolution of the question before us is found in those accidental traits which have been clearly determined about sperm above. Sperm is a watery, buttery, fatty substance, in many ways like the substance of oil. There is in it the fine earthiness that is in the fats, yet in those that are like this, it is the spiritual, moist air that is dominant.

---

44. Cf. Ar. *GA* 735a30f.

45. *Meteora* 4.2.3f.

46. This translation is an attempt to make sense of an overly terse sentence. A. implies that the ancients' observation of the birth process, which is accompanied by a flow of fluids or water, provided confirmation that water is itself the ultimate principle of generation and of the sperm.

In those which have a substance of this sort, the thickening is not caused only by the earthiness getting up in them but also by the very fine, spiritual, windy air which lies shut up within them by the heat and motion. This renders the entire material substance bubbly and this is then made like a foam. To the degree that this bubbling is smaller and less obvious, then to that extent the substance is made the whiter. It is also an indication that to this extent the spirit enclosed in it will be finer and nobler. The material of the air and of the spiritual moisture is easily converted into spirit of this sort. It is for this very same reason that we see oil thicken, for it thickens when the air in it is multiplied by heat and cold.

53 In much the same way, when, in alchemical activities, litharge, that is, silver residue [*scoria argenti*], is mixed with water and oil and is shaken and struck forcefully, it is thickened into bubbles and grows quite white, even though it was black or dusky before this.[47] Now the reason for this is the enclosed air which is mixed in with it. For the same reason it grows white, since it is translucent itself and causes the substance of that into which it is mixed to grow clear. For air is the cause of both growth and whiteness in that material. Thus, too, foam and snow take on a greater size and whiteness by being rarified. For snow is a particular type of congealed foam.

Much the same thing is visually proven to happen in common situations when water is mixed with oil. For from such a mixture the water is thickened and grows white due to the bubbles which the airy moisture makes in it. The air enters into a mixture of this sort for a double reason, namely, due to motion and to the oil. For in the fattiness of all oily things there is little or nothing of water and earth, but there is a great deal of air in it. This is why everything that is fatty floats on water. The air enclosed in the fat bubbles causes them to float and carries them off as if in an inflated hide. This, then, is the cause of its lightness and for the same reason oil grows thick in the cold but does not coagulate, just as we have said in the fourth book of the *Meteora*.[48] For the enclosed air is not coagulated and is only properly thickened through a type of suspension and compression into the substance of the watery fluid to which it is pressed by the cold compressing it.

54 It is from these causes, then, that we obtain the reason the sperm exits thick and white from the body of the one generating. For it is generated by the internal heat and is aided by a very warm spirit which thickens and whitens its substance. When, however, it exits and becomes cold, the heat in it dissolves and the spirit flies off with the heat. Then there remains the watery, thin, dusky substance, and thus what is left is black (or rather dusky in color) since all that remains in it is water and that earthy part in it, being the other (that is, middle) part of its material. This is just like what happens to phlegm when it is chilled, and especially to phlegm which is watery. However, what has been mentioned happens to sperm especially when, having been chilled, it begins

---

47. Litharge, *lytargium*, is lead oxide, PbO. In *De min.* 2.3.2 (Wyckoff, 1967, 132f.), A. describes its use in producing a material called virgin's milk by alchemists. A. also discusses it in *De sensu et sens.* 2.2, giving a fuller recipe.

48. *Meteora* 4.2.11.

to dry out. For it is composed of wind and water. The wind is warm air which, having been cooled, flies off from the sperm. This wind has as its material the thickened humor of the sperm, for the causes which are in it and about which we have spoken.

Antiphon, however, has spoken most falsely concerning the sperm of the elephant when he said that it is black because of the earthiness of the elephant, claiming that it is very dry for the same reason and that as a result the bones generated in the elephant's body are many and large.[49] He says that it turns black like black *karobe* and is dry.[50] But what he says is impossible. Rather, it is necessary that especially the sperm, in accordance with the nature of the animal, be earthier than anything else in accordance with the diversity of the complexity of the animals. The sperm of an animal whose body is earthy is rendered earthy, white, and thick because of the mixing in of air, as is also true of other sperm.

Likewise, the philosopher who is called Brocotoz states falsely that the sperm of the Ethiopians is black in color.[51] For since he surely saw that the Ethiopians' teeth are exceptionally white it is remarkable that he claims their sperm is black. For the cause of the whiteness of the sperm is the frothiness in it. We already know from what has gone before that the froth has a white color, especially that whose substance is made up of small, windy parts and of tiny, hard to see bubbles. The same thing occurs in a mixture of oil and water when they are agitated forcefully together, as we have said above.

The ancients who related tales were not ignorant of the fact that the sperm is buttery and frothy. This is why they said that Venus, the goddess of copulation and the one who nursed Cupid, was born from the froth of the testicles.[52] From this same power of the sperm, it is clear why sperm does not gel and coagulate in the cold but rather liquefies and dissolves, as we have stated previously.

55

---

49. Antiphon was the name of an Attic orator (ca. 480–411 B.C.) as well as of an opponent of Socrates who was known as a sophist. Neither individual is meant here, for the name somehow is corrupted from Ar. *GA* 736a3f., where Ctesias, the physician at the court of the Persian Artaxerxes, is cited. The reference must be to Ctesias's *Indika* (Henry, 1947; Lenfant, 2004).

50. Stadler suggests that *karobe* is a corruption of Ar.'s *ēlektron* (amber), to which the dried sperm is compared.

51. "Brocotoz": *Corrupted* from Herodotus, cited by Ar. *GA* 736a10. The passage of Herodotos in question is 3.101 and its influence was long lived (Friedman, 1981, 54–55, 64–65).

52. The tale is as old as Hesiod, explaining the name Aphrodite from the froth (*aphros*) given off by the genitals of Uranus, served by Cronus.

CHAPTER X

On the Disposition of the Female and Male Sperm, How They Are Related to the Conception, and How the Soul Is Generated in the Fetation

56   We must now, after the things just discussed, make inquiry into many weighty questions. A determination is made for some of these already in things considered earlier. One of these questions concerns the disposition of the *gutta* (that is, humor) which has the name sperm from the humor and which is produced in the females of all generating animals.[53] For if that sperm which is the sperm of the male, and which has the true sense [*ratio*] of sperm, is not made part of the conception after it enters into the womb, then some ask what use there is for the glandular and corporeal substance which is in the sperm.[54] This question especially arises if, in accordance with what has been said before, the entire sperm of the male works only with the power of the heat and spirit in it. If it works this way, then it seems that it is totally set free into the spirit which is spread throughout the viscosity of the female's sperm (that is, the feminine one) unless some of its humor remains to serve as the subject of the heat which operates instrumentally and from which it takes the nourishment of the formative spirit. In this way the material for the entire body would be from the female but the forming spirit and heat would be from the male. Avicenna concedes this and it seems very probable that one can say in this question that this is undoubtedly a belief of the Peripatetics.[55] For, as we have said in earlier parts of this study, small bubbles rise up in conceived sperm after conception.[56] From these a spirit arises which puffs up the entire substance throughout the body so that it can carry the power for the separate formation of each member. Since this is so, then the entire substance of the sperm must be set free, better and thinner, into the substance of this spirit and that which is residue becomes the radical moisture of the heat which impels the spirit and continuously regenerates from the same fluid, lest the first principles which form and distinguish the members of the body throughout the entire body should be lacking.

57   We should therefore first distinguish whether that which enters into the womb of the females and is maintained there receives something else joined to it. For it stands to reason from the things said above that it first receives the sperm of the female and then afterward the menstrual blood or that which is in place of the menstrual blood and on which, with a sealing action, it impresses the form of the creature and of its

---

53. Cf. Ar. *GA* 736a24f.
54. "Glandular": having small globules. Cf. 16.132.
55. Avic. *DA* 16.50r.
56. "Bubbles": *vesiculae*, as opposed to A.'s previous *ampullae*. The word normally indicates a bladder or blister.

members. Because this is how it works on the fetation.⁵⁷ If, however, it is this way, then we must investigate the disposition of the soul, through which there is life and sensation in the conception, asking whether it itself is present in the conceived sperm. For no one can say that the fetation has no soul at all and that it is entirely bereft of the principle of life. For the seed of a tree and the seed of every plant universally germinates, just as the sperm of animals produce animals. This can be only because they have some share in the principle of life. For a conceived sperm has the potential for an animal to be produced from it and a plant seed has the perfected potential to germinate into a plant after the time established for its conception, during which time there is added to it a spirit which distends the seed so that it can be formed. Now this distension and formation cannot occur without food, drawn in from outside. The seed of the plant draws in this food from the ground, since, when the spirit and heat are made from the radical moisture, the seed certainly germinates and sprouts, forming upward a shoot of the plant, or a branch, a leaf, or something else of this sort. Then, the substance of the seed, having been emptied out, is left behind, except that when moisture is drawn in from the earth, it replenishes that part which has been raised up. But it would not have the means by which that which has been raised up could grow unless it were fed continuously by the moisture that is drawn in. However, drawing in food, digesting, and converting it into members are functions of the nutritive soul. We cannot say that the nutritive soul does not exist where its essential operations are present.

The fact, then, that the nutritive soul is present in the sperm of animals is proven in exactly the same manner. For in it too moisture is set free by the heat and consolidates the spirit which, lest it slip away, is held in the thickness of the sperm, strengthened, and, by inflating, raises up bubbles in the sperm. Blowing forth from these bubbles it perforates and distends the sperm. In this way the sperm is deprived of moisture in its own midst, and, unless this is restored, the operative power [*virtus operans*] is also destroyed and there will then be nothing from which those things which must be formed can be formed. Therefore it absorbs the menstrual blood from the womb as nourishment, digests it, converts it, and joins it to the material which is being formed. These are the essential activities of the nutritive soul which exists in plants and animals. Thus, from the things which we have carefully disputed in our book *On the Soul*, it is clear that the vegetable soul is present both in plants and animals.⁵⁸

Further, for the parts and activities which have been proven in the same book to belong to the vegetative soul, it is perfectly understood that the vegetative soul is present in some manner or other in conceived semens. For as we have said, the essential operations of this same soul are present in them. Now since these operations precede all sensation in conceived semens, it seems that the vegetative soul is present via the nutritive and growth part, before the soul, which is the principle of sensation and

58

---

57. *Sic operatur conceptum*: this phrase allows several translations, including "Because this is how a fetation works." Moreover, *conceptum* may be a contraction of *conceptum sperma,* which is used several times in this passage.

58. *De anima* 2.2.9.

movement. For the soul, which is the principle of sensation and movement, will be present when formed members have taken on a certain amount of growth and form. In this way, then, that which has been conceived will be called an animal, and none of the operations of reason are present in it yet. For this reason the conception will not be animal and human at the same time nor will it be animal and horse at the same time. This is as Aristotle seems to say, namely, that completion will always be at the very end since in every animal completion has a particular time for its own generation.

But if Aristotle meant by this that the rational soul or the rational substance is other than the substance of the sensible soul in a human, then it should also be a substance different from the substance of the sensible soul which, in a horse, makes it a horse. For just as he says that it will not be human and animal at the same time, so he says that it will not be animal and horse at the same time. We, however, will clarify these statements in clear precision in what follows.

For now, however, we say that from these considerations we must move on to inquire after the disposition of these operations and completions or acts. Thus, we might know how and when the formed creation of the fetation becomes an animal. We must inquire also after the disposition of how each thing takes on the manner of its completion through the soul. All these questions are admittedly ponderous, but they are necessary for knowledge. When, however, we are investigating these questions, we should seek to understand their truth in accordance with the power of nature.[59] For they so exist in nature that no sort of true knowledge clarifies reality unless they depict it according to the truth and power which lie in its nature. For the most suitable and intelligible thing to do will be to call to mind for a while the things which have been previously established. Let us digress a bit, then, so that that which was said might be understood.

CHAPTER XI

Which Is a Digression Setting Forth How the Soul Is in the Semen and in What Order of Priority the Vegetative and Sensible Souls Are Present in It. Also in What Way Only the Rational Comes to It Externally.

We resume, saying that the work of nature is the work of intelligence. It can only be the work of intelligence if it belongs to that which spins the earth. For the earth is spun by one first intellect to which all other movers refer, much as the powers of the members refer to the power in the heart. Thus, all of nature is the work of the intel-

---

59. Sc. reads that we should seek to understand these questions according to our power (*secundum potestatem nostrum*) and not according to nature's power. The shift from *potestatem nostrum* to *potestatem naturae* has forced an awkward explanation.

lect of that one which is the prime mover. For otherwise there could not exist one principle of the entire universe. It is much the same in a microcosm [*minor mundus*] which is a perfect animal like the human or the lion. Here the principle of all operations is that one thing which is the heart, just as we have set forth in earlier parts of this investigation. Moreover, the heavenly powers in the seeds of plants and animals are wondrous. For they are very many and multiple, coming from the multitude of the heavenly bodies, their locations, and movements, and from the multiplicity of the rays, and the angle of the rays which they acquire in every way, whether as they intersect one another or whether they are falling on the material of the thing generated, or whether they reflect back toward some given place of generation.

The power which is more powerful among all these, however, is that which is in them because they are the instruments of the first intellect and of the lower, moving intellects, just as the instruments of the soul move in accordance with the powers of the soul. For every power which belongs to a composite entity (this being one which is composed of mover and moved) is more in accord with the power of the mover than with the power of that which is moved. An indication of this is that we see that each and every thing in the bodies of animals follows the powers of the soul more than it does those of the body. Therefore, these heavenly powers are wondrous and they are all in the spirit and heat of the semen. They are in the spirit as in the subject upon which they bestow form, and they are in the heat as in the instrument by means of which they fulfill their activities in the material of the conception. From what has gone before it is clear enough how there are present in these same things the powers of the soul and of the members which are moved by the soul. It is also clear how the powers of the simple elements are present in them as well as the powers of the mixed, to the extent that it is mixed.

I therefore say that it is the power of the fiery heat to gather together a thin, homogeneous substance to itself and to separate off the thick earthiness. When this is done, the fine moisture that has been drawn is terminated in the manner and form of a mixed thing. There results from moisture of this sort a spirit which beats in it much the way a mover does in a movable. The celestial powers, then (which are in it to the extent possible), raise it up and assimilate it to heaven's balance, removing its complexion from the excess of active and passive contraries and basing it in a medium as similar to heaven as possible. This, then, is near so that it can undertake life and the activities of life. These things could not occur without the power of the soul and the powers of the members. Therefore, the power of the soul and of the members then informs the spirit and the heat so that they will form the shapes of the organs through which the activities of the soul and of life are accomplished. When these things are perfected in shape, the soul there is created in it by the power of the first intellect, and this performs the activities of life. These activities are established for the multiplicity of organic powers by the intellects which move the lower spheres, that is, by the powers of those things in the semen.

All the powers mentioned, then, are in the seeds of plants and animals, much as art, functioning in act, is present in the instruments of art, just as we have often said.

In this way the soul is said to be present in the seed just as we have said previously. When these activities become manifest, then there is a successive manifestation of these powers since first to be made manifest are the fiery powers of heat, of the mixed, and then of the celestial ones. Afterward, in the activities of the soul and of the members, those activities begin, which belong to the powers which are the principles of life. And when they make manifest the activities of life, the completion of the soul will be through the intellect.[60] Thus Aristotle says that each of these things does not exist at the same time as its completion.[61] For just as the seminal fluid is not attracted and digested at the same time as the configuration of the members, so too the activity of life is not in it at the same time as the species and the perfection of the plant through the vegetative soul. In animals the activity of life is not present at the same time as perfection with respect to sensation for the genus and species of the animal. But in plants two gradations succeed one another, whereas in animals there are three and in humans four. In others, there are gradations of mixtures and these are prior to life and the principles of life. In plants the activity of life is from that which is like art in the semen when it forms the organs of the plant. The perfection of the plant, however, occurs when the vegetative soul is in these formed organs. In animals, however, the activity of life is present in the way stated. Because the same organs most often belong to both life and sensation at the same time (since in animals to sense is to be), then when the sensory organs are made, the power of animalism [*animalitas*] is present in them.[62]

After this comes the attainment of the specific, perfect form according to a determined form which gives the horse, ass, or any other animal which is generated its being and *ratio* within the species. Aristotle thus understands that the living and sensible are not simultaneous and that the animal and the horse are not simultaneous.

63   Since, then, it has been proven that there is one soul in one body and that there is one quiddity of the one which is in the species, and one form, as we have shown in our book *On the Soul,* then if the semen first had the vegetative soul as substance and afterward acquired the sensible soul, it would be changed from one substantial form to another substantial form and from the complexion of the vegetative soul to that of the sensible soul.[63] But all these things are absurd according to any right-thinking philosopher. Moreover, if someone should say that the vegetative substance is different from the sensible one, but that they are united in the one act of animating the body, this is the fantasy we have disproven in the study *On the Soul,* and things that have been proven there ought not to be repeated here.[64]

The rational (that is, intellectual) principle of life is identical in a human subject with the vegetative and sensitive, but with respect to being it is different. For one sub-

---

60. Reading Borgnet's *manifestant* for Stadler's *manifestat.*
61. Ar. *GA* 736b1f.
62. "Animalism": the state of being endowed with the soul.
63. *De anima* 2.1.9.
64. *De anima* 2.1.7.

stance is the rational soul from which flows the vegetative potencies, as well as the sensible and intellectual ones of which some are implanted in the body and some are not.

Thus, those which are not attached to the body have none of the power which could lead them out of the bodily material. Rather, they are a certain likeness of the light of the intellect acting in the nature and principles of the sperm. For this reason it is entirely from outside the material of the sperm and of its powers, and it is by the light of the intellect (which, according to Anaxagoras and Aristotle, is the one acting first on all the powers mentioned) that the rational soul is led into the fetation as is the intellectual soul which is completed afterward through the forms of speculation. This is as we have demonstrated in our book *On the Intellect and the Intelligible*.[65] This is also the order of events because it is not alive and sentient at the same time and it is not sentient and human at the same time. This is the understanding of Aristotle's statements and of Avicenna, Averroes, Theophrastus, and all the more expert Peripatetics who agree with this exposition of them.

If, then, the question arises whether the principle of life, sensation, and reason is that the soul is in the seed, I say that the principle of life and sensation is spoken of in two ways.[66] There is the principle of life which is the substantial form [*enthelechia*] of the organic body with respect to the potency for life. In this way the principle of life is not in the seed and neither is the principle of sensation. For if the principle of life were in the seed of the plant or animal in this way, then the seed would be a plant when it was the seed of the plant and it would be the animal when it was the seed of the animal.

64

But if the principle of life is said to be that which operates on the organs of life, then the principle of life is in the seed in the way in which art is in the instrument of the art, as we have stated often enough. In this way too the soul is in the seed, like art and not as the substantial form [*enthelechia*] of the organic body. For art is the factive principle which acts according to reason. But it is more accurate to say that that which is in the seed is something of the soul and is not a soul, for it is the act of a soul, as is clear from what has gone before.

It stands to reason, then, from all that has been said, that since reason is not joined to any part of the body as if it were the act of any given member, then what must be is that it is led into the fetation neither out of material nor by the bodily instruments but rather by that which is not mixed into any material of the body nor into any powers which are acting in the material of the seed. Therefore, the principle of this one is nothing other than the light of the first acting intellect. For this intellect is pure, unmixed, and impassible, as we have shown in the third book of our *On the Soul*.[67]

These things, then, when joined with what has gone before, point out the correct understanding of the things that have been introduced above.

---

65. *De intellectu et intelligibili* 2.1.

66. In the next few sentences *semen* is translated "seed" despite the context to emphasize the flow of thought and the interconnectedness of its points. Elsewhere, of course, animal *semen* is rendered as "semen."

67. *De anima* 3.2.12.

CHAPTER XII

Concerning the Fact That Aristotle Says That the Soul Is in the Semen and That Intellect Is in It from Without

65 Let us return then to the matter proposed, saying that the nutritive soul is in the conceived semen of an animal, in the way that has been stated, because the semen is from the animals. But it is clear that this soul is in the two sperm, joined together, in potency in such a way that potency is called a power which is an act of the soul of that one which generates the things in the semen. But insofar as it is the act or the perfection of an organic body, it is not in the sperm unless food has first come to it which, once it has been drawn in, converts in the convening member, much as an animal, when fed, converts its food. In this case it is performing first the operation of a true soul. For it seems in the beginning that every fetation lives as do plants, in the way that life is called an operation of that which is the act and likeness of the powers of that which has the vegetative soul as its act and its completion. One can say the same thing about the disposition of the sensible soul, for just as the vegetative soul is a power in the body, so too is the sensible soul, throughout all its parts, a power in the body. It is thus an operation of the power which is in the semen for, as we have already said, there is no form of generation which does not follow the powers of the material which are in the sperm. The sole exception is the rational soul, for neither the spirit of the sperm nor any heat seems to act on this one. It seems rather to be caused entirely by the light of the first intellect which acts in the work of nature, since it is the act of no body or of any part of the body.

66 Let us say, then, that the vegetative and sensible soul are in the semen in the way given. We will say then that both are present anteriorly and potentially. We call this potency formative, along with the powers of the spirit, the heat, and the sperm, about which we have spoken. Afterward, then, they are rendered active at the end of generation, when the fetation is completed. For since these souls (or forms) are not present first in the semen and next in the fetation, it is necessary that they be made in such a way that they are led out of the potency of the material toward act, since they are powers in the body. But this is not true of the rational soul. For it is not a power in the body, as we have said so often. For when these were souls universally they came to be such, or one must say that some vegetative and sensible souls were thus made and others were not. For if it were said that the vegetative soul is in the semen, and that this is the one which makes the vegetative soul of the fetus, then there will be a certain vegetative soul which makes and was not made. The same will hold true of the making and made sensitive soul. Plato spoke in this fashion, as was clear above when we spoke of his opinion.

However, these souls are made in the material in the aforementioned way, even though we do not grant that they, inasmuch as they are souls, enter into the semen, having been transmitted in the semen of the male, as Plato said. For the act of the soul is transmitted and it is of the male and is a likeness of the powers of his members and is not a soul properly speaking.

One might ask whether, in the vegetative or sensible soul in plants or animals, it is necessary that the souls enter having been transmitted along with the semen from the souls of the ones generating. And someone might doubt whether all the principles of life of this sort are at first in the male outside the sperm and are afterward transmitted along with the semen, or whether some are transmitted from without and some are not. For Plato said that they are all from without. Some, though, have said that some are from without and some from within. They said that the vegetative soul is present from within since they did not see there that one fills out another by ascending just as the male fills the female. They said that the sensible soul is transmitted in act [*actualiter*] from without, for in such cases one male fills many females. They thought that they transmitted souls which existed in the semen in act, albeit latently.

It is clear, then, that it is impossible to say that all the principles of life are present in act in the semen beforehand, having been transmitted. This is clear for the very reasons which we brought forward in the book of this study before this one. For certain principles act on bodily operations and these principles are the perfections of certain organic bodies and cannot exist without them. I am calling bodily operations those whose instruments are bodily organs. An example is walking, which can never exist without feet. Thus, the progressive power [*vis gressibilis*] can exist only in feet and legs. All powers of this sort, then, in no way come into the seed from something external. Rather, they are generated as follows: they are led out of the semen from potency to act, as has been stated previously. It occurs in this way for the body of the sperm is the superfluity of the food altered into the likeness of the power of the members. In it is the formative power, working toward the act of the members of the vegetative and sensible soul. Because the intellect is neither a bodily power which is greater in a larger body and less in a lesser body nor a power which acts in the body like those powers whose working is toward the harmony of a bodily organ in the way that vision [*visio visus*] is for the harmony of the eye, it follows of necessity that it alone be given externally to the material by the principle of generation because it is not mixed into the material. As Anaxagoras has said, this is the intellect whose work is that of nature, rather like the one first moving and causing. We have explained his statements above. Thus the intellect alone is divine, that is, perpetual and incorruptible, for, as we have shown in the book *On the Intellect and the Intelligible,* it alone has the sort of operation which has nothing in common with the operation of a bodily organ in any way whatever.[68] And this is the point to which we wished to arrive in all the chapters we have written before on the powers of the soul and sperm.

---

68. *De intellectu et intelligibili* 1.1.6.

CHAPTER XIII

On the Principles Which Make the Sperm Fitting and Suitable for the Generation of an Animate Thing

68 There is a certain other power of the soul, apart from the one which is intellectual and rational, and these powers are the same in subject.[69] It has an operation which is shared with and not separate from the body and its organ shares something in common with certain of the elements the way vision shares in water and hearing shares in air and so on for the others, as was determined in the study *On Sense and the Sensed*.[70] For this reason they differ greatly in nobility of soul. For the sensible is nobler than the vegetative, and the rational is nobler than either of them. And because for every diversity of the movers there is a proportionally corresponding diversity of the movables from these movers, there is therefore also a difference of nobility in the nature of the complexion of the organic bodies which are perfected by these souls. For this reason it is entirely necessary to posit in the sperm certain powers which make it suitable for generation. In this way it might be perfected by an ignoble soul, a noble one, or the noblest one.

69 These powers exist especially in accordance with that in the sperm which is called the natural heat and is generative of that which is conceived. For this power does not exist in accordance with cold or anything like cold, for cold retards movement and causes a state of nondigestion, as was shown in the fourth book of the *Meteora*.[71] This heat is especially held in the spirit which is in the frothy sperm, just as is perfectly clear from the many things which have been said before this. For this nature, which is the complexional heat of the sperm, is not suited to the heat of the element which marks its limit, this element being fire. For the heat of fire dissolves, dries out, and does not generate animals. Nor can the heat of fire be endured outside its own place, unless it is fostered with the moist air on which it feeds or in some dry, flammable place such as in dry wood. Apart from the aforementioned materials it has no substance within which it can be held outside its own place. But the heat of the sperm is not of this sort. Rather, over and above heat is the power of the sun and of the entire heaven and also the power of the soul which is in the body of the animal which emits the sperm. Because the power of this soul is made in it, it is clear from this that the heat and power of the soul are also found in superfluities, such as in dung and urine, which are not made like the body and the members. This is why much greater heat of the soul is

---

69. This sentence is marred by problems of number and gender agreement which make clarity elusive. For the original intent, cf. Ar. *GA* 736b30f.

70. Cf. *De sensu et sens.* 1.1.14.

71. *Meteora* 4.1.8.

found in sperm, which is a superfluity made to some degree like the members. With regard to those things which have been called celestial and animal, the principle of life is in that semen which is thus emitted from the body of the animal.

From these reasons, then, it is clear that the principle of life which is in animals is not there in accordance with the sole property and nature of fire. Neither is the fire which exists in the sperm the principle of the vegetative and sensible soul, as many have erroneously said, stating that fire is the soul because of the natural heat which they see as the principle of life and the operations of life. Now the body of the sperm is that with which and in which the spirit exits from the animal producing the sperm. This spirit is the most powerful power and is the principle of the soul which is led into the fetation, led forth from potency to act. This spirit is separate from the body of the one generating and has the powers of the members. This spirit is a divine thing since it has the divine power of forming and creating. Thus, this spirit is called the practical, formative intellect, just as the instruments of the artisan are called art, being art since through them the form of the art is led in. The principle of all of these is the practical intellect. Just as this is so, then even more will the spirit of the sperm, separated from the body of the one generating, be called the intellect since it is the instrument of the operative intellect, whose entire work is that of nature. For this moves to introduce form and is rather as if the entire power of the art came into the adze from the artisan and that thus the adze were to enter on its own into wood and stones and, without any contact with or movement by the artisan, were to make the house. An adze of this sort would not be distinct from the power of the practical intellect, since it would move toward parts of the entire work and would work as a whole by distinguishing, determining, composing, dividing, and forming in every power of the practical intellect. Thus, as we have often said in previous books, the spirit moves in the sperm entirely in this way, in the power of the practical intellect which is the operator and perfecter of all nature. For it is in this way that the work of nature is called the work of intelligence or intellect. For all natural things exist in the agent or practical intellect, just as things made by craft exist in the operative intellect of the artisans. This spirit, then, for these reasons, is called a divine thing and an intellect.

This spirit is enclosed and held within the sperm by the viscosity of the sperm and is not separate from it. Rather, it distends in it, widening out, as if it were undoing and loosening its parts and distending them. With a piercing action it fills the entire sperm with a fine and spiritual airiness, for it has a moist and watery substance and this "breathes" easily. For this reason too it is absurd to ask how the formative power enters into the sperm from without. This is what Plato asked when he said that this power was given the sperm by the giver of forms, from without. For this spirit is bodily and is not part of the species and form through which a thing is what it is, that which the Arabs call the aspect and form, since the observation and the aspect of a thing occurs through it.

On the other hand, this spirit is not properly a material part of that which is conceived and formed. For it has the same relation to it that coagulant has to coagulated milk. For the coagulant is not a part of coagulated milk. When it is poured in, it is used up and mixed in with the milk. And when it is heated in the milk and evaporates away as spirit it diffuses throughout all the milk and coagulates it. In this way the coagulant is like a coagulating spirit, whereas the milk is the material coagulated. The coagulant enters all the milk much as the spirit enters the body. In entirely the same way the spirit of the sperm enters into the material which is conceived and which comes from the feminine *gutta* and the menstrual blood.[72]

The sperm of the male, breathing, enters into the spirit, distends it, coagulates, forms, makes it distinct, and sustains it by keeping it in check. This is what we said above, namely, that the spirit is given by the male, that is, by the father, whereas the body, with regard to the mass of material, is given by the female, that is, the mother.

Let, then, this be the declaration and distinction made concerning the disposition of the soul and how the soul is present in the fetation during generation. For we have stated how the conceived semen has a soul and how it does not have it. Likewise, we have said how the sperm has a soul and how it does not. For it has a soul potentially but not in act.

CHAPTER XIV

On the Cause of the Sex of the One Born and of Its Similarity to the Father, Mother, or to Someone Else

As is clear from what has gone before, the sperm is the superfluity separated and divided from the last food, when the body is moved to growth and to take on nourishment by the power of the members which are digesting the last food. When this sperm falls into the womb of a female suited to it by nature it immediately grasps it and holds it, sustaining the *gutta*, which is like it and which is the proportional superfluity of the female. It moves it as it makes it distinct, sets it in order, and forms it, much as the practical intellect moves the material of a thing made by art. This superfluity is like it and is proportioned to it as the passive is made proportionate to the active and as material is made proportionate to art. This is as we have stated many times. All the members are present in this superfluity potentially, whereas no member whatever is present in it in act. With respect to the suitability or difference of these two sperm, there occurs in the fetation a differentiation between male and female, and there occurs a reason for its similarity to the father, mother, or both of them. This is as we set forth a good time ago when we spoke of the sperm.[73] There we said that when the warmth of the sperm of the male prevails (this being that which bounds, forms, and confers the proper powers [*proprias virtutes*] to the members), then the masculine

---

72. Cf. 16.56.
73. E.g., 9.63f., 87f.

sex will be brought into the one fetation. But if the moisture which is being formed suffers any undigestion, it will be cold (this being the reason for undigestion). This complexional coldness and moistness will result in the female sex.

A great many people have questioned whether the coldness of the entire body does this or whether it is only the coldness of that part in the groin area, occurring when the generative member is formed in that place. There are great men who hold reasons and opinions for either side. But since the other members of a given woman are frequently warmer than those of a given man, it seems that the moisture which is the material of the groins and of the genital members must first suffer *molynsis* or it must also have heat which terminates completely.[74] If the moisture has suffered from *molinsis*, then the feminine sex is formed. But if it is completely digested, it is formed into the male sex. Then, the characteristics of the female sex are spread from these places throughout the entire body and all the other members. And the same is true of the masculine sex. An indication of this is that when males are castrated, the characteristics of the female sex spread throughout the members which formerly had masculine characteristics.

The cause, then, of the male sex is not that the power of the father's genitals is present in the semen and that the power of the mother's genitals forms the female sex. Rather, as often happens in bodies flawed by some infirmity, it is that flawed sperm produces greater corruption in the newborn. Thus, sometimes one who is cancerous produces one that is leprous, but sometimes this is the case and at other times it is not. For the sperm of one who is infirm in a certain area will not always have this kind of corruption. In entirely the same way, it happens that sperm undergoes some degree of undigestion in the manner mentioned previously and it then generates a female. If it does not suffer it, it produces a male. An indication of this is that those who abound in watery and cold blood, like those who are fat and the elderly, most often produce females. For this reason a female is a male that has suffered a flaw. For, as has been well shown in previous parts of this investigation, male's sperm, which is formative and factive for a fetation, always aims to make one like itself, always to produce a male unless it is hindered by some flaw which corrupts the instrument with which it works, and this flaw is heat. Or it may be hindered by the intractableness of the material which it forms and makes, this being the humor. When either or both of these suffers *molinsis* in the area of the genitals, it forms a female lest the work of nature be rendered totally useless. Although she does not generate as such, she is still a helper to the male and necessary for the purpose of generation just as the passive is necessary for the active and just as material underlies works of art. This is why too we have said in the *Physics* that the material desires form as does the female the male and the ugly the beautiful.[75]

73

---

74. *Molynsis*: from the context, apparently a version of the Gr. *mōlysis*, which indicates an imperfect boiling or heating. *LSJ* lists the form *molynsis*, which is to be found in certain MSS of Ar. *GA* (e.g., 776a8). The word also appears in the LXX (cf. Jer. 23.15), rendered *pollutio* in the Vulgate, and indicating some sort of corruption, as below at 18.93.

75. *Phys.* 1.3.16–17.

74   Further, with regard to material the menstrual blood is the same as semen which is not perfectly cleansed and purified. It is missing that which is truly called the principle of nature in generation, this being the effective and formative principle. This is why there are wind eggs in certain birds, having been deprived of this principle of generation. This is also why nothing is ever generated from eggs except by being formed from the semen of the male in that genus of birds. When, however, the sperm of the male is mixed into and joined to the female humor, which is the seminal superfluity, and then the two are as one, then a fetation and impregnation are completed from these two.

This conception occurs in the following manner. For generally, when moist, viscous things warm up, the exterior surface of that which is decocted dries out and draws around it like a skin. Its viscosity keeps all the bodies in the semen, whatever they may be, from evaporating and causes the spirit to circulate within it. As time passes the viscosity of the surrounding skin grows drier and tougher, to the point where it becomes a nerve-like substance, forcefully surrounding its surface. It thus forcefully retains and contains the entire body of the animal with respect to all the members produced from the material of the semen. Now when I said that it becomes a nerve-like substance, I mean that it resembles nerve, but not that it truly and in all ways belongs to the nerves. For in some animals they are truly nerves, whereas in others they are not nerves, but are a member that is analogous to the nature of nerves. For nerves, veins, and membranes [*tela*] do not differ with respect to their primary material, save to the degree of greater or lesser viscosity. For all of these are created from a viscous material.

75   Moreover, just as we have said that the female sex is created through a defect, active or passive, in the principle of generation, so too do imperfect animals come to be through a defect of the principles of those generating. This is as we have stated above. Thus one animal conceives an egg internally and yet later gives birth to a live young with the shell of an egg or a membrane because of an imperfection in the principle of the ones doing the generating. But the fetation is never completed in the best manner of its conception save when a male is generated from it in those genuses of animals in which there is a differentiation between male and female. For we have said above that some animals exist in which there is a difference between male and female. We have also said above that some animals exist which generate neither male nor female and there are others which do not come from copulation. These are very imperfect animals indeed and we will mention them later. In the complete animals, however, which are viviparous, a complete fetation is formed, hanging from the womb, until it is born externally.

There are two types of animals which give birth externally and which conceive internally. Some first conceive eggs from which live young suddenly emerge within the uterus. These are emitted outside the womb. Some, however, are conceived in the womb and remain there for a long time since they are fed in it on the food conceived in the egg at the very beginning.[76] When this food is used up, then a completed live

---

76. A.'s vague use of the terms *conceptus* and *conceptum* (see Glossary) makes this passage unclear. It may say, "Certain 'conceptions' remain in the uterus for a long time. . . ."

young emerges from the egg, much as one born from a womb emerges. In all of these, as we have said before, the cause of maleness and femaleness is none other than the cause we have stated. Moreover, the cause of similarity to the father or mother is the dominance of the qualities of the mother or father in the semen. Sometimes, however, the power lies hidden in the members of one of the grandparents.[77] This is assisted by certain accidental qualities of the superfluous digestion and is brought back into action. Then the result is a similarity to one of the grandparents. Thus, the cause is not always the power of the members, which is latent and existing in act [*in actu*] in the sperm.

CHAPTER XV

That There Are Particular Receptacles for the Superfluities in the Bodies of Animals and That It Is from These That Generation Occurs. Also That the Female Provides the Body but the Male the Form and Soul. That Not Every Male Produces Sperm, Whereas Every Female Has Menstrual Blood or Something in Its Place.

Sea fish differ with respect to the principles of generation. We have mentioned this and it is especially so in the *celety* genus. But there will be time to speak of the disposition of their generation after.

We will begin here, however, with those which are prior and more perfect in nature. Those animals which are first in nature participate most nobly in the being of an animal and are the most noble. These are those which are viviparous. Of all these, the noblest is the human, since he is the principle of all animals. The fact which one should know first concerning the principles of the generation of perfect animals is that in all animals each of the natural superfluities goes to its place without any need of anything to expel it to that place. Thus, it has no need of any spirit to expel it nor of any other cause, which is what some have held. While the spirit and the other causes produce the expulsion of the quicker and better ones, thus too the sperm, which is one of the superfluities, descends on its own, without any propulsion, to the seminal vessels.[78] It is not true, as some men say, that the testicles draw it in the way a cupping glass draws in blood. For it is possible for the aforementioned superfluity to leave without any necessity for violent expulsion or attraction. Both the moist superfluity

---

77. "Grandparents": perhaps merely "ancestors."

78. "Produce the expulsion of the quicker and better ones": for this somewhat awkward phrase one might prefer to read, with Borgnet, *expeditiorem* and *meliorem* for Stadler's *expeditiorum* and *meliorum*, to mean "produce a quicker and better expulsion."

and the dry superfluity move this way, for when they have been gathered together in their receptacles as if in storage places, they pass through their pathways. An enclosed windiness goes along with them, making their exit faster. But this is not so necessary that this sort of exit cannot occur without it. For this sort of exit is common to all superfluities. They can exit well without the force of any such windiness. This is clear from when they exit from the bodies of those who are asleep, whose bodies are very full and from which they exit because of overabundance and not because anything expels them and brings them to the point where they must exit. This is especially clear in trees in which the sap of the seeds moves due to the heat of the sun and to the natural moisture held within it along with the spirit. The fruits thus emerge from them at the right spots. The reason for this is the one we have given, namely, that every superfluity has a place to which it moves as long as a need for it is discovered. Thus, the blood has veins and flows in them for its proper vessel.

77 The superfluity of the menstrual blood of females comes from two large veins, namely, from the one called the great [vena cava] and from the one called the *adorthy* [aorta]. These are divided and branch out into many thin and invisible pathways whose openings arrive at the womb. When these pathways of the veins are filled with more food than they should be and when, due to cold, this food cannot be digested and the blood cannot exit because of the thinness of the veins, it flows back to the womb. When the womb cannot take in this blood because of its amount and because of the narrowness of the place, then the flow of blood occurs in women. This is why the menstrual flow does not have a set cycle or a fixed time in all women. But for the most part it flows most often at the waning of the moon when it is hidden from view and unseen because of its conjunction with the sun. This time is most suitable because animals' bodies are colder then since the surrounding air is colder during this time. Then the blood is weighed down and the superfluous menstrual blood is generated, exiting according to the cycle of the month, as we have said. If, however, the superfluity were not digested and decocted at a given time, then the menstrual blood would flow constantly, bit by bit. This is the reason that the menstrual blood of young women is often white in color.

When, then, these seminal superfluities are tempered in the male and the female, then their bodies will be made healthy and healed when these superfluities are voided. For if they remained in their bodies, they would cause disease and if they were kept in the bodies longer than they should be, they would harm the bodies. For this reason, too, when the color of the menstrual blood is frequently white and the blood is present in large amounts, this serves to inhibit the bodies of the young women from growth. The superfluity of the blood in the veins then needs to exit, and when its exit is increased it exits from the slender and thin veins. For nature is burdened down by this superfluity and only directs it to the womb to order it for the better. This is the fact that generation occurs as a result of the semen and from menstrual blood of another one which is like itself in species but is a separate individual. This occurs from the semen and the menstrual blood since that is potentially like the body from which it was emitted.

This superfluity is necessarily in all females of animals which generate, since it is the material of generation. It is more present, however, in blooded animals than in others and is most abundant in women. As we have stated, this superfluity must exist in all females, for some males do not produce sperm, but they create offspring through the movement and spiritual effect of the sperm, as generation occurs from the sperm of those producing sperm. Females, however, do not create that which is sustained and brought about from the material, for the creation of this sort is due to the sperm or the spirit of the male. The location of the conceptions of all of these is in the area below the diaphragm [*paries*] and is lower down in all the animals which have a diaphragm. The heart, in those that have one, and the member which corresponds to a heart in those which do not, is the principle of all nature. What is added to the substance in the entire body is due to the heart, which gives life and power to all.

The reason that there will not be a superfluity generating materially in all males, even though there is a material superfluity in all females, is that an animal is an animate body and in this composite entity the female must provide material while the male must provide that which creates and forms. For we know that the powers of each of these sexes differ according to these two things. We have already shown above that this difference exists between male and female. The female necessarily therefore provides the material moisture for the body while the male does not do this necessarily. For it is not necessary that the father be in the ones generated materially but rather as the artisan. Moreover, the body, as far as it is the mass of material, is solely from the mother, while the soul, which is the form and the mover, is from the male since the soul is the formal substance and the *ratio* of the body. The material often draws in the factive principle and conquers it, whereas at other times it is conquered by the factive principle, which is to be seen clearly when animals copulate which belong to species closely related in nature but not at all of one and the same species. For they then are suitable in bodily size or are of approximately the same bodily size. In these the young are not generated with a likeness shared variously with the mother or the father. Rather, they are mixed, as in the case of the product of a mare and a he-ass, or in the product of a fox and a dog. In birds it is to be seen in the one that is generated from a partridge [*cubeg*] and a rooster.[79] If these then generate in their own right, the similarity tends more to the mother because of the amount of her material principle. This is such as occurs when certain seeds are first sown in unaccustomed ground. They are noble at first and afterward are rendered less noble. Finally they are rendered like the other plants which usually grow in that earth. An example is Roman cabbages [*Romani caules*], which are sown in France or Germany. They eventually degenerate into the sort of cabbages grown in France or Germany. We have spoken of such

---

79. Cf. 16.130.

things in our book *On Plants*.⁸⁰ The reason is the same as the one given there. For in the generation of plants the earth is like a mother and provides the material. The seed [*semen*], however, provides the formative power in those bodies. For this reason the genital member of females is not prepared just to be a passageway through which the sperm of the male might be received. On the contrary, it is a receptacle of that sort of material which will provide a body for that which is conceived. As has been said, then, every superfluity in the body has its own place in the body in which it is kept.

These, then, are the true reasons why those superfluities which are the principle of generation come forth from the bodies of animals. When the sperm has thus left the male and has been received within the womb of the female, it seizes and gives substance to the cleaner superfluity which it finds there. It cleanses it since the greater part of the female superfluity is foul and this is not suited for generation. The fluid in the menstrual blood, however, most suited for generation is that which is very moist and watery in the sperm of the male. The first menstrual blood is unsuitable for generation in many animals as is the first sperm in human males and in other animals since the heat of the soul is still minimal in them and is not adequately digested. For semen which is adequately digested thickens.

In women from whom, for whatever reason, the superfluity of the menstrual blood does not depart and likewise in other females of animals which do not emit menstrual blood, the menstrual blood is retained within the womb and is in all ways like that emitted during menstruation. It is from this that the material for the bodies of the embryo is furnished. That which seizes, sustains, and forms it is from the sperm which comes from the males, as has been stated.

CHAPTER XVI

Which Is a Digression Setting Forth the Opinion of the Later Peripatetics on the Diversity of Oviparous Animals, the Power of the Sperm, and the Animation of Embryos

Since, then, the entire usefulness (or, rather, the better part of the usefulness) of this study on animals concerning the power of the sperm and the cause of the soul in animals rests on this point, let us pick this point up again, speaking about the power of the sperm and the cause of the soul in animals, doing so in accordance with what the later Peripatetics had to say about these matters.⁸¹

They are in agreement in stating that a complete animal is that which is completed in the natural heat and in the humor which is the subject of physical heat. Such an

---

80. *De veg.* 1.185f. discusses the concept of degeneration.
81. Cf. Avic. *DA* 16.1 fol. 49v.

animal, moreover, is complete if its body is complete according to the qualities which are the principle of generation, even if it is not yet completed according to the quantitative size. This is because it does not have sufficient nourishment in the seed of the mother and father for quantity. This is an animal which has blood in its entire body and which has been generated from completed blood. This has but two types, one of which is a creature whose young [*fetus*] is not completed in the womb, but rather from which there exits material which has been prepared but which is both completed and formed externally. Oviparous flyers which lay completed eggs are of this kind, since that material, before the young [*partus*] is formed, weighs down the narrow belly and forces departure. Therefore, a strong membrane is prepared for it along with a shell. Within these it is kept safe from mishaps lest it perish once outside the belly and before it is formed. This sort of egg laying is the norm for blooded animals.

Those which do not possess blood generate incomplete eggs, however. Of those which generate incomplete eggs there are some from which white *gusanes* emerge when they are completed externally. Others are internally oviparous and produce young from the eggs only internally since their eggs are soft and, if they were to emerge incomplete, they would be corrupted by various mishaps.

Further, there is much earthiness mixed into animals of this sort. For these animals are earthy and have but little heat. They therefore must lay and hatch internally. For an animal which has great earthiness, little moisture, and moderate heat has dry and cold sperm and this is slow to complete that which has been conceived unless they have many things warming it externally. Therefore the animal does not complete the eggs externally, and they must instead remain within it. Neither can the material remain in such an animal until it is seized and formed by the power of the sperm, for this is slow in acting due to its coldness and dryness. Therefore the material must be gathered together inside an egg. This is the true reason why the viper and those like it both lay and complete their eggs internally. Since it is a cold animal and since their shells could only harden through great heat, these eggs are necessarily soft and viscous.

Having determined these things, then, concerning the differences in oviparous animals, let us speak of the sperm, asking whether the animal power [*virtus animalis*] is in it. Because the sperm moves toward generating the animal and does so not through an external power but through itself, it is necessary that it have in itself the formative and nutritive power by means of which it can have the material for its nourishment from which it might form and complete its members. For the generation of the members by the sperm is not simultaneous. For it is clear enough from the inspection of miscarried fetuses that the heart is formed first in all animals. In animals in which there is no heart that member is formed first which takes the place of a heart. Just as the heart is first to participate in life, so it is the last to die. For there is life in it longer that in any of the other members. Nor is it true, as some say, that the liver or the lungs are formed first, before the heart, and that they escape notice in a miscarried fetus just because of their smallness. For if smallness were the cause of their escaping notice, the heart would escape notice more than the liver or lungs since it is smaller than they. Yet in a miscarried fetus it shows up quite clearly when the other internal members like the liver, lungs, and brain do not. For there is a dominance [*excessus*] of power in the seed of the father over that of the female in natural generation, and these two are joined together as mover and moved.

84. That which acts on and generates the blood, from which the semen is derived, is, according to some, the liver. This is what Galen says, but according to Aristotle the operator of the blood is the heart, and the sperm derives from it.[82] This is clear from things said before. But the sperm which is separated from the blood is not completed by the heart or the liver but rather by the seminal vessels. After it has been completed, it is ejected from the body of the male into the womb. It then moves the *gutta* found there in such an order that it first creates in it, before creating the other members, the seat of life, and this seat is the heart. Then the power which is the principle of creation and formation of all the other members spreads out in order from the heart as if from a first member. Then the soul is introduced since the masculine power is buried in the *gutta* of the female just as the spirit is buried in that of which it is the spirit. This is just as we have said, that the masculine sperm is totally, or almost totally, converted into spirit and that the body of the embryo is taken from the *gutta* of the mother. After the entire being has come together it will be animated and will be moved by the soul toward completing the members. This is the soul of which we said that it is the act and the likeness of soul. An especially good example is the nutritive soul since the effect of another soul is not visible in it at that time. There is whiteness in the sperm due to the multitude of spirit generated in it. Therefore, when it emerges from the body and cools, it becomes dark, thinner, and watery, not because its thickness is due to air but rather due to the bubbles of spirit, as was said above.[83]

85. When, moreover, the *gutta* of the female has rested and has been overcome by the operation and terminating action of the sperm (for these are principally determined toward the sensible soul), then the sensible soul is given it. And since the sensitive power is like a potency of the rational soul, then, without doubt, the rational and the sensitive powers are the same in subject and one is completed by infusion with the other. Thus, too, all the members pertaining to sensation and to reason are completed together since both these powers of the soul are occasionally based in the same member with regard to their operations. But it is not this way for the nutritive members and the reason for this is that all the members of animals do not have a share in sensation, but they all share in the taking in of food. This is why the *gutta* which is formed must participate in nourishment. For as we will show below, the nutritive power which first draws in nourishment, before the nutritive members take on shape, is derived from the mother's substance. But that which comes afterward and which is present in the members which have already been formed and which are called the

---

82. Stadler cites Galen *De iuvamentis membrorum* 5.3, the name given to a twelfth-century Latin translation of an Arabic abridgement of the first ten books of Galen's *UP.* May (1968, 1:6) calls the end product "a wretched substitute" for the actual text of Galen.

83. Cf. 16.52f.

nutritive members is, beyond doubt, from the father. This is because it is the same in subject with the sensitive power in animals. The nutritive power first comes from the mother and then passes on. That which remains is from the father. Moreover, when the heart and brain have been formed, then the rational soul is infused, even though it is then superseded by the sensible soul. For the rational soul still remains material since it has not yet been differentiated. Rather, in such cases it will be present as it is in a drunkard or in an epileptic. It is then completed by an extrinsic power continuously bestowing the intellect. This has been shown in the book *On the Intellect and the Intelligible*.[84] Other sensible powers are completed by the completion of the organs and other bodily things. For if a boy be said at first to have sensible substance, and afterward to gain possession of a soul or rational substance, he will be changed from species to species and from one substantial form to another, an absurd thing to say.

Moreover, the principle which has been prepared for the introduction of the soul into the fetation cannot belong to the nature of the heat of fire, but rather it must be from the heat which the heavenly bodies produce. This is the heat which is very uplifting and ennobling to the complexion, so much so that it makes it analogous to the equality of heaven through the power by which heat is formed from heaven. A test for this [*experimentum*] is that the heat of fire dries out eyes and other moist bodies, but this heat forms them. For this reason it also has an effect in conferring life, an effect which the heat of fire cannot have. This heat, therefore, is the one which bestows life in that it adapts lower bodies to heavenly ones. And when they have been adapted, they then take on life. This life is spread about commonly by this heat in the dry bodies of plants and in the moist bodies of animals.

Moreover, there is a certain first substance in the sperm which first takes on this life, and this substance is the spirit beating within the sperm. This spirit bears away the heat we have spoken of. Thus, it is rendered the formative cause of all the parts of the sperm by decocting it and adapting it using a power separate from the body, this being the power of the soul. While this may be separate to the extent that it is not a corporeal power, it is still not so separate that it does not operate in the body in any way whatsoever. For if it were to depart from a physical body in every way, it would be infinitely drawn off, a thing that cannot be in physical powers. This spirit, then, is a divine body and its relation [*proportio*] to the other powers and parts of the sperm is like that of the practical intellect to the other powers of the soul and to the parts of the material of a fabricated object. Just as in all things active and operative the practical intellect is the best and most efficacious among the incorporeal powers, so too this spirit is the best and the most efficacious among the corporeal powers and substances. This substance of the spirit never leaves the sperm as long as it stays healthy in the womb. For it changes the spermatic moisture within itself making it thinner, and it fills the *gutta* of the female's sperm with a spiritual windiness and not with an inflative windiness, as some of the unskilled physicians have taught. This spiritual wind is the spirit which,

---

84. *De intellectu et intelligibili* 1.1.8.

by piercing the *guttae* of the female, bears the formative powers into every part of the sperm. This is much as a coagulating agent acts in milk, as we have stated above.

87 For this reason too this sperm is not truly a material part of the members but rather is a principle of the spirit passing through the members and acting upon them. For it should not be thought that this sperm is worn away to nothingness. Quite to the contrary, it should be known that it is converted into that which is noblest in the fetation, namely, the spirit. Thus, the sperm of the male touches the humor of the female, is mixed in with it, and directs and moves it. It is also moved by it in turn as everything moved physically is moved by that which it moves. For this reason too it is eventually overcome by the material on account of the great amount of material and it then tends toward the properties of the material which comes from the female. Thus too seeds which are sown many times eventually tend toward the nature of the earth in which they are sown, as we have said above when we spoke of Roman cabbages.[85]

Further, the entire sperm does not always come to the womb. Neither is all which comes to the sperm functional. Rather, it is some part of it which operates more by its power and quality than by its quantity. It is sometimes so small an amount that it brings about a very small fetation. Thus, I have seen a girl in Cologne who was nine years in age but only had the size of a one-year-old child.

88 Further, there are certain animals, such as the oviparous birds, which do not have vulvas and which instead take in the sperm beneath their diaphragm where they are impregnated by it. This sperm is ejected by the male and is drawn in by the female who, although she may have strong attractive powers, is still aided greatly by heat in its ascent.[86] For every semen requires a double assistance, namely, that of suitable external nourishment and that of suitable air which surrounds without any other thing surrounding it. An example of this is the belly or the earth for plant seeds. Further, every fetation requires a double nourishment, namely, a first and a second. The first one is conceived in the very seed itself, both of plants and of animals, while the second is drawn in.

Again, the first member produced in the fetation is the heart. This is made from the nourishment inside the sperm. In some, however, it draws in nourishment from outside, drawing it first from a nearby place and then from one far off. It is possible, however, that in certain genuses of fish the powers of the sexes are joined together as they are in plants. In such fish the male is not found to be distinct from the female, but rather each power is found together in them. Therefore the power of the male prepares, enables, and forms the material, as is clear from what we have said before.

From all that has been said here we are not disproving the statement of Galen that the sperm which comes from the woman has the informative power. All we are saying is that this is the power which prepares and enables the material so that it may undertake work from the operator, which is the sperm of the male.

---

85. Cf. 16.79.

86. Borgnet's *habeat* is more grammatical than Stadler's *habeant*.

CHAPTER XVII

Which Is a Digression Setting Forth Why Male and Female Are Necessary for Generation of a Live Young and Also That the Rational Soul Can Have No Factive Principle Whatever in the Semen Because of Its Separation

Having dealt with all those things which seemed necessary to say about the power of the sperm and the animation of embryos according to the earlier and later Peripatetics, it is still necessary to understand better in what way male and female are necessary for generation.

In order to understand this well, one should know that the forms present in generation are diverse. Some of them are material, accomplishing nothing whatever apart from those things resulting from the principles of the first qualities or from those things which are caused by a mixture of them. These are closed in and absorbed into the material and they thus do not have the true sense [*ratio*] of soul. An example is the shape of a stone. These are perfected by coagulation or liquefaction or some other action upon [*passio*] the active or passive first qualities which we have set forth in the fourth book of the *Meteora*.[87] Some, however, are not absorbed by the material but are raised above the powers of the principles of the material of the ones mixing. They thus do not act on one thing through a natural potential, but on many. An example is the vegetative soul which has no operation of a simple form whatever beyond the qualities of its material. For nourishing is accomplished only by a corporeal, material body, and the same is true for growth and generation. Therefore, there are forms of this sort in the material which do not require a separate thing acting as agent and generating them outside the material, but rather they require a mixed agent in the material. In just this way the forms too are mixed with respect to their substance, being, and operation.

This is why, in a generation of this sort of form, there is a power which generates and forms things other than that which is formed. But that mixed power is still present in the same material. This is as we have said in the book *On Plants*—that these powers are joined together into one.[88] But there are certain forms which have operations and passivities [*passiones*] which have no share at all with the material. An example is the sensible soul, which takes on the form of the sensible entirely without material. It does this much as wax takes on the figure on the ring, apart from the silver or gold material of the ring. And it moves with the message delivered to it about the movement of the appetite through a form of this sort, which it takes on without material. These forms cannot be absorbed and joined to the material at all, and also for this reason that which is formally generating and acting as agent on forms of this sort

---

87. *Meteora* 4.1.1f.
88. Stadler cites *De veg.* 2.1–19.

should be distinct from the material. And since the material is in the female, it must be that the active principle [*activum*] of generation of this sort of animal should be outside her. The male, therefore, had to be created. This is what Aristotle says, namely, that the male is due to the sensible soul in animals and not to the vegetative soul. The enabling and effective principle could be in the female for the vegetative soul but is in no way in it for the perfect sensible soul. For the habitual ability [*habitualis habilitas*] in the woman's sperm is called the informative power by Galen, and this is what Aristotle speaks of when he says that a wind egg is alive potentially but that a bird does not come from it and that the *gutta,* conceived in the woman's womb, lives potentially the life of a plant, but no live young is created from it without the sperm of the man or male. For this reason too, in those animals which are imperfect and have only dulled and confused taste and touch, there is no male and female. For their sensible soul has no passivity or operation outside their material. Thus, the entire form comes from a principle which is intrinsic to the material.

91 From these things, then, we can tell that the intellectual soul, which not only has *passiones* and operations in things which lack material (such as sensation and imagination) but which has them without any bodily organ, cannot in any way come from any material principle. Rather, it is influenced by the light of a separate intellect, which is the first principle and the most powerful agent [*operator*] of nature's entire work.

This, then, is the true reason for male and female. For if an effective principle were in the very substance of the material, it could in no way have passions and operations in those things which lack material, such as the forms of sensation and imagination. But since the effective principle of these is external, it is further refined and digested and the whole is rendered spiritual and animate. This is clear from the things which we have introduced in many chapters already, and for this reason too we have said that the entire male sperm, or almost all, is turned into spirit and enters the female sperm much as the vital spirit enters into a body which it vivifies. For when the heat decocts the two sperm at the same time, the exterior viscous portion begins to congeal on the surface. Within this the spirit pulses and it flows over the material of generation and brings to it life for the organs which have formed within that membrane which contains the spirit and keeps it from evaporating. This is sufficiently clear from things we have already said many times over.

Let the things which have been said, then, be sufficient concerning the cause of generation.

The Second Tract of the Sixteenth Book on Animals Which Is on the Cause of Sterility in Animals and on the Generation of Homogeneous Members

CHAPTER I

On the Accidental Causes of Sterility Attributable Both to the Male's Member and to the Female's Womb. Also on the Manner of the Creation of the Embryo in the Womb, from What It Is Generated, and in What Order the Members Are Differentiated within It.

We must next speak of the causes of sterility and of the generation of homogeneous members from the sperm. For the cause of sterility which is a cause *per se* is a departure of the sperm from the generative power which is naturally in it. There are many accidental causes of sterility, however. It is sometimes due to the length of the penis in that it cannot extend to the area in front of the opening of the womb whence the womb can draw up the sperm. Or perhaps the animal has no penis whatever, having instead another member which has a power like that of the penis. This occurs, as we have said, in certain animals with ringed bodies.

We have also said above that the fluid which is discharged by females during the pleasure of intercourse does not have the power of the sperm for impregnating. But it is created by the same cause during intercourse and in nocturnal emission (that is, emission during dreams) in both men and women. This, however, is not a sufficiently sure sign that there is a humor of the same nature which discharges during both dreams and intercourse. For the discharge of a certain humor occurs in young boys from rubbing of the genitals when they approach one another in intercourse before they produce sperm or can do so. This humor, however, does not have the same nature as the male or female sperm.

Nor can there be impregnation due to the superfluity of the sperm of just one of them. Thus, impregnation does not occur solely from the man's sperm without the woman's menstrual blood. Impregnation is especially impeded by the penis when it holds back discharge too long since then the spirit in the sperm evaporates on the way. Likewise, when it remains inside the body for too short a time, the sperm does not fall into the place whence the womb can draw it in. Likewise, the female is only impregnated when the place for impregnation is suited in quality and location for the movement of impregnation and when the womb descends below with a suitable motion when it accepts the sperm. Impregnation occurs in the womb properly when the opening to the womb is not reversed and the womb itself is not paralyzed.

Moreover, whenever the pleasure of the male and the female occurs at the same time during intercourse impregnation is easier. When, however, both do not have pleasure at the same time, it will be more difficult. For the male does not emit his semen inside the womb as many believe, because the opening of the womb is very narrow. Rather, he emits it outside the opening much as certain females emit the superfluity of food which leaves from them, emitting it in front of the inner opening of the womb. Sperm which has been cast in front of the inner opening of the womb remains there and is tempered. When the complexion of the womb is good, then it draws in either a small part of it or, if the sperm is good, all of it. If, however, a part of it is bad, it rejects that part. Animals which have their womb near the diaphragm, as do birds and fish which generate from eggs, cannot project their semen into their females from without because they have no penises. Rather, the semen is projected in front of the area through which it is taken in and then of its own accord it makes its way to the womb. The womb then draws it in because of the heat aroused in it. For there is great heat in such a womb, which is near the diaphragm. In this way it draws in the semen via its own heat. This is just what happens when a jar which is not full is moistened with warm water and is turned on its opening, for it will then draw up water beneath itself. This is how the drawing power of the womb takes place.

94 There are, however, some who say that this attraction occurs only by means of an instrument suited to intercourse. They say that it does not take place in the same way, but in many ways, even though the opposite of what they say is what occurs. They are speaking in much the same way as those who assert that women produce sperm, saying that women eject sperm outside the opening of their wombs. They say that this is eventually drawn within to be mixed with the semen of the male to produce impregnation. They say that the way in which the semen is drawn into the womb is the same and they assert this on the basis that nature does nothing uselessly, and for this reason they claim that the fluid released by the female is drawn in. For otherwise it would be useless and its emission would be useless unless the sperm of the male were tempered by it when it was emitted so that, thus tempered, it could be drawn in by the womb. The humor (that is, the *gutta*), which exits from the female into the womb, is held, controlled, and congealed by the power of the male's sperm. What happens to it is what happens to milk which is coagulated using a coagulating agent. For when milk is coagulated there is a spiritual heat present in the milk which solidifies and draws together the milk. The male's sperm does much the same thing in the menstrual blood, for the nature of milk and of menstrual blood is the same, as we have set forth in what has gone before. When, however, it solidifies and congeals that which is most corporeal in the menstrual blood, the watery fluid then exits to the surface. When the more earthy parts at the surface dry out, webs and *secundinae* are produced from them and these surround the fetation. They are necessarily dried out by the heat decocting

them. Afterward they are cooled like the skins of animals. It is reasonable that a conceived animal not be immersed in a watery fluid which is distinct from the material of the animal conceived, but rather that it be kept apart from them by *secundinae*. This is the reason then for *secundinae* and for other webs generated from the surrounding earthy, viscous parts, which differ according to more or less. For they are not of the same type in animals which produce live young and in those which lay eggs.

Further, when the *gutta* of the female is touched and held by the sperm, it is then that what happens to plant seeds happens to it. For they have within themselves the power which moves toward generation. When that moving power, which is first present potentially, emerges actively, then branches, leaves, and roots arise from the seed. It then draws in food through the roots. Trees and plants share in nourishment, growth, and generation, as we have said in our book *On Plants*.[89] Growth is accomplished in entirely the same way for all the members of an embryo. All these are produced by the first formative power which is in the semen. This is the reason also that the heart is like the seat of this power. The heart is made before the other members in act, although all the members are in it potentially. Now this is noticeable not only to the senses but also to reason. For the fetation is solidified, coagulated, and formed by the semen of the male out of the *gutta* of the female. When it is rendered one thing distinct in itself, it has to rule itself, much as a son who leaves his father's house. For a son leaving his father's house takes along a part of his substance with him. He augments this first substance and thus purchases a home. Thus, then, that which is first in the substance of the parents is converted into the substance of the heart. But another thing from which nourishment occurs is placed in the furthest members, namely, the lower ones, which are at the furthest part of the body.

This is how it happens in all animals which are generated. For if the first substance leaves and is placed at the furthest part of the body, someone might ask a difficult and insoluble question, namely, when does this happen? For nothing is placed in the furthest part of the body except through the agency of that which is in the first part of the body. Therefore, from what is the first part made? For this reason all those who, like Democritus, say that the lower members, which are at the furthest part of the body, are made first are speaking falsely and are speaking of an animal that is like one made of wood in which the artisan begins from those parts which lie furthest away from the body.[90] This is what occurs in animals of stone and wood, whose principle of generation is not internal but external. For all animals have a principle which is moving internally in their generation and thus, in all which have blood, the heart appears first among all the members, formed and separate. For the heart is a type of principle for all the other homogeneous members, that is, if it is proper to name something as the principle which is not the prime mover in generation. But it is especially a prin-

---

89. Stadler cites *De veg.* 2.1–19.
90. Ar. *GA* 740a13f; cf. Diels (1951), 2:124, frag. A144.

ciple in those which require food.⁹¹ For the ultimate food in those with blood is the blood and in those which do not have blood, it is that which takes the place of blood. The blood, however, originates from the heart, just as has been set forth long ago in previous parts of this study. The vessels for the blood are the veins and for this reason too the heart is the principle of the veins, as is clear from our discussion of anatomy which we introduced above.

As long, then, as the animal exists potentially and is imperfect, existing in the material of the first generation, it necessarily takes its food from the womb the way a plant which is first sprouting forth from its seed takes it from the earth. Thus, when the heart has been formed, nature draws out two large veins from the heart from which all others split off. From these come the thin veins which are directed toward the womb. From their intertwining comes the member called the *umbilicus*, for the *umbilicus* is a vein, and in some animals it is a single vein while in others there are many. There is a covering of skin around the veins in the *umbilicus* since these veins require protective devices due to their weakness. The veins of the *umbilicus* are attached to the womb like roots and through these the fetation absorbs food. This is the reason that the animal remains in the womb, and the sole reason it remains in the womb is not, as Democritus said, for the sake of impregnation and creation (that is, the production of members).⁹² For this can be done outside the womb, as is clear in the egg layers in whose eggs the members are created and formed outside the womb.

Perhaps someone might ask a difficult question which arises from what has gone before, saying that if blood is the food and the heart is the first thing which is generated, and if all blood is drawn in from without (specifically, from the womb), then where does the first blood come from out of which the heart is formed? For since the heart is flesh and all flesh comes from blood, the heart will then necessarily be from blood. This blood cannot come from without, since there is as yet nothing which might draw it in. Therefore, it is false that all blood, and especially the first blood of an embryo, comes from without. Rather, the first blood is generated of a sudden in the sperm by the conversion of the seminal *gutta* into blood, just as happens in a plant seed in which the intrinsic material of the seed is converted of a sudden into soft food from which the shoot is formed before it draws up anything by way of food from the ground. However, once the semen has been distended and passages for the veins have been pierced through it, then it grows from the food drawn in through the *umbilicus*. We have spoken about this above, but we will speak of these things again in places and times suitable to this investigation.

The differentiation of the members within the embryo does not occur as some people think when they say that it would be natural for that which is similar in shape and form in the fetation to come from that which is similar in shape and form. For many difficulties are attendant upon this statement, as we have partially determined

---

91. Cf. Ar. *GA* 740a19f.
92. Democritus is quoted in Ar. *GA* 740a37; cf. Diels (1951), 68 A 144.

while refuting Plato above. For if this were true, then it would be necessary that each of the homogeneous members would at the least be separate unto themselves in the semen. Thus, the flesh would be there in and of itself, the bones in and of themselves, and so on for the others. And since this is shown false by the senses, it is more suitably said that the *gutta* of the female is potentially of such a nature as the nature of animals is in act and that all the members are present in the *gutta* potentially but none is actually. For the artisan and the art are the male (that is, the masculine) sperm. When the artisan and the material of the manufactured item are joined together by touching, then the operation comes forth gradually and not at all suddenly. This is how generation emerges when the masculine sperm touches the female *gutta*.

It is thus clear that the material comes from the female and the principle of movement is from the male and, just as a work of art is accomplished only by using the instrument which is suitable, so the same thing occurs in generation. For it is reasonable that the manufactured thing is made only through the movements of its instruments. For art is the operative form of those things which are the works of art, and much like this is the power of the soul which is present in the seeds of trees and animals and is called the formative power. From this power comes formation, nourishment, and growth, which comes after the taking in of food. This power uses heat and cold like instruments since movement of the material occurs through these qualities. To speak in a general way, each member of the body will come to be in this way and in this way occur the sustenance and bounding of that which is made according to nature. For the material from which the body gives subsistence to the fetation and from which it is nourished is the same. Therefore, the power which does this and that, which are one and the same thing, is greater than if it did but one thing, namely, if it did only the first. If, then, this power is the vegetative, nutritive soul, it is also the same as the generative. And in this, naturally, lies the nature of trees and of all animals, for they are other parts of the soul. It is clear that in some bodies which are alive, in trees for example, the male and female sexes are not differentiated. In animals, however, they are differentiated, although not in all of them. But they are differentiated in some and in these the female requires the male for generation, as we have said.

CHAPTER II

Why the Female in Animals Cannot Generate on Her Own without the Male and on the Generation, Differentiation, and Completion of the Members in Generation

101   Perhaps someone will raise a question based on what has been said before. For if the vegetative and sensible soul is in the female and the female *gutta* is a superfluity having the impression of her soul, how is it that the female requires a male and does not generate on her own without him? The reason for that which is asked is that there is a difference of sensation between animals and trees, one which the animals have in addition to what the trees have.[93] For a hand, flesh, and the like are only what they are if they have been perfected with respect to sensation, either potentially or actually. For without sensation they are only members equivocally like a member of a corpse. If, then, the male, as has been held above, is the producer [*factor*] of the sensible soul, there can be no generation of this animal when the male and female are separated. For the power of the male is formative, as we have related previously while resolving the serious doubts introduced above. An indication of this lies in wind eggs, for these signify that that which the female generates has the potential for turning into something incomplete and indeterminate.

102   One may again raise a doubt on this matter, saying that wind eggs have some sort of life even though they cannot be like those from which chicks are generated (for if they were of this sort, then doubtless some sort of winged thing would be created out of them). On the other hand, they are not at all like wood or stone for to some degree they share a certain participation in life. It thus seems that somehow there is a soul in them potentially. Now in all things in which a soul is present potentially the nutritive power (which seems to be present in eggs such as these) will somehow necessarily be present, much as it is in trees and animals. If, then, it is a question of why neither the members nor the entire animal is formed in an egg such as this, we will say, following what was said before, that this sort of formation of members requires the sensible soul, since the members of animals or organs have sorts of operations different from the organs of plants. This is why the female requires intercourse with a male and male and female are therefore differentiated in animals of this sort. Thus it happens that chicks are generated even from wind eggs if the sperm of a male is joined to the material of the egg by intercourse before eggs of this sort are removed from the female's belly. But these things will have to be matters for discussion later.

---

93. As often, "trees" should be taken generically as "plants."

If, however, there were some female genus of animal and the male were not differentiated from the female but rather both sexes existed in the same individual, then perhaps such a female could generate on her own. But no person who speaks the truth and whose statements merit much attention has ever seen such a species of animal. Thus, doubt arises about the disposition of certain fish in which the male is never seen. They are called the *aryonobo* in Greek and all the fish of this genus appear to be female because they are found filled with eggs.[94] An indication of this is that in certain lands neither the female nor the male of this genus is found. The *enkeliz,* which we call the eel and which is a river fish, seems to be similar to this genus. In the genuses, however, in which the male and female are differentiated with respect to their subjects [*per subjecta*] the female cannot generate on her own. For if the female could generate on her own in such fish, then the creation of the male would be useless and it is generally agreed that nature does nothing uselessly. Thus, in the generation of ones such as these the male, who completes the generation, must be present, especially for the sake of the sensible soul, whose generation is not sufficiently perfected without the sperm of the male, as we have shown above. Those things which are generated without a male are imperfect and slimy land creatures. This is why they live for a long time when they are cut in two, for they have their spirits and souls deeply imbedded in their material, and they have bodies that are uniform virtually all the way through. Neither are their activities [*opera*] very clever.

In those which have sexual differentiation with respect to their subjects, then, it is from the male that the sensitive soul is produced in the fetation, either from the spirit pouring itself in or from its sperm. For as we said above, not every male casts sperm into a female. The sensory members are in the material potentially and when the principle which generates and moves formally is joined to them, then that wondrous and noble thing occurs which occurs in those which do not have differentiation of sex on their own, namely, completion with respect to the sensible soul. There were certain of the ancient natural philosophers who did not believe that the members were changed by the power of the sperm from their own dispositions into dispositions related to hardness, softness, and the other accidental differences which those members possess that are called homogeneous or uniform. All these things are done in act after they were present in potency. Those men, however, held that they proceeded from latent act to visible act.

However, in the transmutation which they have from potency into act, the heart is the first member which is generated, in those which have a heart. In those which do not have a heart the first is that thing which is analogously proportioned to the heart. This has already been said many times and it is also clear to the senses in miscarried fetuses. It is also clear because this is the member which retains life after all the other members. For the vital spirit remains in the heart longer and for this reason all the other members die before the heart. It is as if nature placed in the act of life a final cir-

---

94. Ar. *GA* 741a36–37 lists the *erythinoi,* perhaps a species of sea perch, which is often hermaphroditic, secreting both eggs and milt (Peck, 1942, 204; *GF,* 65–66).

cle of life, enclosing the principle of life within that same member, namely, the heart. For generation is not circular, since it lies between the two terms of a contradiction. For generation is a process from nonbeing to being.

Once the heart has been generated, however, then the internal members are generated before the external ones, for they are the vital members.[95] Thus, in miscarried fetuses the bulk and size of the internal members are apparent before those of the external ones. In both the former and the latter, however, those which are above the diaphragm come into being before those below it and the upper ones are larger and the lower smaller in comparison with the size of the whole body. For this reason too the lower members sometimes appear undifferentiated when the upper members have already taken on clear distinction.

This is clear if one were to inspect the miscarried fetuses of all animals with blood. For in all animals the upper part appears distinct from the lower, save in an animal which produces larvae. In this distinction, the upper members take on more growth while the lower ones stay small by comparison with the upper ones. The upper and lower parts have the clearest distinction in walking animals with blood. They are present indeed in the others, but not in the true sense [*ratio*] of "upperness" [*superioritas*], since all the parts of the bodies of animals of this sort are drawn along or are impelled over ground, and one does not rise up over another. It happens in the same way in trees where the seeds send forth roots first and then branches. The roots are similar to the upper parts in the walkers. This preeminence of the members is especially due to the power of the sperm which holds sway in the upper members, just as certain of the natural philosophers have held well and suitably. An indication of this is to be seen in birds, fish, and animals with ringed bodies. For these types are generated by sperm which comes forth from the parents who are generating them and their eggs are completed outside the womb of the ones generating. They then take on the differentiation of their members and their division into a bodily location of front or rear, above or below, to one side or the other. It is the sperm, then, which performs this differentiation and not the power of the one generating.

Further, certain animals do not breathe and produce eggs or larvae, and some of these generate live young like themselves.[96] But all breathers take on the differentiation of the members in shape and bodily position while they are still in the womb of the mother. While they are in the womb they do not breathe until their lungs are completed. But there are various teachings on this matter, and we have treated them above when we spoke of human generation. The lungs, however, are differentiated among the members which are created before the *umbilicus* and before nourishment is drawn in. They are therefore made of the substance of the *gutta* of the female.

---

95. "Vital members": *vitalia,* i.e., pertaining to life. Cf. 7.173; 8.202.
96. Ar. *GA* 742a3f. states that some animals do not breathe but are produced as larvae or eggs (e.g., fish and insects), while others do breathe and become differentiated within a womb but breathe only once their lungs are completed (most vivipara). The passage confused A. and he interpolated.

Further, each quadruped animal which has feet with many divisions, like the lion, dog, wolf, and fox, generate young which are blind from the first moment of their birth. Their eyes are completed externally, whereupon they open and they see. Whenever there is a mixture of hot and moist, then there necessarily will be one that is acting and operating while the other is passive and being operated upon.

CHAPTER III

On Distinction in the Cause on Account of Which a Thing Exists. Which Cause Is Sought in Eternals and Which Not, and in What Way and for What Reason the Heart Is Said to Be the Principle of the Members.

The ancient natural philosophers, since they did not have true experience of the generation of the members, believed that members existed in the sperm in act and that from these others like them were generated. These men did not say in how many ways something is said to be prior to another and one thing anterior to another. Yet these things are spoken of in a number of ways.

The first way is that on account of which there exists that which is.[97] That on account of which there exists that which is, however, is itself spoken of in two ways, one of which is first with respect to generation while the other is first with respect to substance. The efficient cause is that on account of which there exists that which is and is the principle of movement in generation. Form, however, is also an end on account of which there exists that which is, and this is first with respect to substance and *ratio*.

Moreover, that which is for the sake of something with respect to the *ratio* of the efficient principle is spoken of in two ways. One of these signifies that which is the principle of movement, as we have said. The other, however, signifies that which it uses in moving, like an instrument. That one, however, on account of which it is generated and made (as if on account of an instrument belonging to things which generate and operate) is both one and the other, although it is not entirely the one and the other in *ratio*. Because an instrument sometimes exists before and sometimes after. With respect to being made and generated it exists before the act of being made and generated, but with respect to the use of that which has already been generated, the instrument exists after.

In much the same way the teacher exists before the student, whereas the pipes come after the student has been made into a pipe player. For the end of the one teaching is that the student be made a musician and then that he begin to act on the pipes

108

---

97. "On account of which there exists that which is": *propter quod est id quod est*. On the basis of 16.109, this phrase might be rendered "on account of which a thing is that which it is" (*propter quod est res id quod est*).

according to music. With respect to another cause, however, the pipes also exist before the student, but these are the teacher's instruments.

109 In general, therefore, there are universally three things on account of which a thing is that which it is and one of these is the one that is called completion [*complementum*]. And principally among these three is the one on account of which there exists that which is. There are two others, however, on account of which a thing is that which it is. One exists just as a thing exists on account of the one first generating and moving which is borne along by its own movement into the passive one and into that which is generated. The third, however, which is led to the efficient cause from the two just mentioned, is that which is needed, like an instrument, in the movement of the efficient cause.

According to this distinction, then, there will necessarily be members which first bestow power on that in which the principle of movement resides. On this they bestow being, the principle of motion, as if by way of an instrument in generation. And because this is the sort of relationship the heart has to the other members, it will necessarily be prior to the other members since it holds great control over the creation of these other members. After this the member will be entire and complete with respect to the progress of generation. For a member such as this, which is a principle in the way stated, will necessarily be present in the bodies of the generated animals, in whom the principle of vital movement resides. It is therefore necessary that this member be created first, before all other members. The reason it must be first is that insofar as it is moving it must be first, and it must be first also insofar as it is a "member on account of which" [*membrum propter quod*] just as all the other members exist on account of one moving. Likewise, however, and for the same reason, each organic member which naturally generates certain others (much as the liver, lungs, and brain come before many other members), must be prior to them since it is rather like the principle and cause of their generation. However, a member which does not have this sort of disposition must be generated and completed last.

110 For this reason, one who does not carefully consider these matters cannot easily differentiate the members in order that he might know which of them exists on account of another and which of them exists on account of itself or *per se*, even though it may be necessary for the understanding of the generation of the members so that in this way we may know through inquiry what is prior and what is posterior to something else. For we have also said that the completion of certain members is accomplished afterward while that of others is done before. It is necessary first for the member to be made in which lies the principle from which another is made and after that member the upper part of the body is made. An indication of this is that the head appears first in the creation of the embryo along with very large eyes. But that which is below the *umbilicus,* like the thighs and legs, shows up as very small in the beginning. This is because the lower parts are only due to the upper parts.

Therefore, the ancients did not speak well concerning the necessity of that cause on account of which there is that which is. Democritus announced without any distinction that that on account of which a thing is what it is does not exist in eternal things, which are forever and without movement. And he said as if it were the truth that these things do not have a principle and end of movement. But that on account of which there is that which is must be the principle of generation and of movement, as Democritus says.

He says further that that which always exists has no end. For this reason he believes that asking after that on account of which any given eternal exists is like asking why that which has no end has a principle. For that which has no end has no efficient principle on account of which it is, for every movement of the efficient principle is toward a determined end which it sets for itself in its movement. But it is true that in necessary, eternal things, as has been proven elsewhere in our *Physics,* nothing is demonstrated through the principle and end of movement since all such things lack movement.[98] But, as we have said before, the cause, which is said to be on account of which a thing is what it is, is spoken of in many ways, and it is found in many cases with different shades of meaning. Some of these are physical, considered with movement, whereas some are eternal, abstracted from motion and matter. For it is always the case that the three angles of a rectilinear triangle are equal to two right angles. It is therefore not fitting to seek a principle of motion in all things in which there is, nevertheless, a principle *propter quid* in accordance with which the form (which exists *propter quid*) is the medium of the demonstration.[99] Thus a principle of movement should not be sought in those things which are not generated and which are eternal, existing without movement. For we call such things eternal.

These things do have, however, a principle in another sense of principle. The understanding of principle in one of this sort is the intended end. For through a principle we intend to arrive at the true knowledge of a thing. Thus, proof [*demonstratio*] is not the end of the principle in mathematicals but rather knowledge, which is the effect of proof. In such things, which are unmoving and which are removed from motion, matter, and time, there is no principle save that which is [*quod quid est*].[100]

In things, however, which are generated, there are many principles which are thought of in many different ways. For there will not be only one type of movement in them and thus, according to one type of principle, the heart will be prior and will be the principle in those that have a heart. That which is in place of a heart will be the principle in those that do not have one, as we have stated above. Those that lack blood have something analogous to a heart and not a true heart.

---

98. *Phys.* 4.4.1f.

99. "Medium of the demonstration": *medium demonstrationis,* perhaps "middle term of the demonstration."

100. That is, a thing's essence.

CHAPTER IV

On the Manner of the Generation of the Homogenous Members from the Heat of the Heart and the Complexional Cold

113 Veins leave the heart first, stretched out from it in a way resembling hairs. They are both pulsating and quiet. Around these occurs the generation of the material since they carry away from the heart the spirit and blood needed for the generation of the members. The generation of the homogeneous members is accomplished by complexional heat and cold, for some of them are coagulated by heat and others by cold. We have already established the difference in them in the fourth book of our *Meteora* where we also spoke of which homogeneous bodies are moist, which liquefy, and which do not.[101] From the veins then, and the passages for the veins which are in every area of the body, food is directed to each member. Because the food is moist and flowing and because members are sometimes generated which are dry and not fluid, what happens to potters' products must happen to the material of the generation of the members. These have moisture mixed into them, being softened by it, and then afterward, by means of a baking heat, the superfluous moisture is led off and the clay vessels are thus completed. In the same way the superfluous moisture in the material of the generation of the members is also led off and then the members are made perfect in their shapes.

114 The first of the homogeneous members which is generated from the food around the veins is the flesh, or that which takes the place of flesh in those animals which have no blood. The bounding of the flesh is due not just to cold but to complexional cold. For as we have stated elsewhere, cold is to some extent a deadly quality. An indication of this is that when a dry heat is acting on flesh it seems to undergo a particular form of dissolution since it exudes complexional moisture from itself. The earthy members, however, in which there is very little moisture and little heat, have as a particular trait that when they grow hot they evaporate and exude the moisture which holds them together. They thereupon dissolve. If a baking sort of dry heat should prevail they will be hard and very earthy. This is how claws, horns, *sotulares,* and *soleae* are created and generated. There is a small amount of oily moisture in these, however, and on account of this they soften when heated in the fire, but they do not dissolve since this moisture is held in forcefully by the earthy parts which are in the material of these members. Some members of this sort, eggshell and teeth for example, are softened by sharp moisture such as urine and vinegar.

115 Veins and bones, however, are dried by an internal heat by means of which the superfluous moisture in them is dried out. This is also why bones do not soften or

---

101. *Meteora* 4.1.2 and 4.1.3f.

liquefy in fire having been first readily dissolved.¹⁰² This is because they are solidified with a dry heat like those things which coagulate and solidify due to a roasting heat in a furnace. For the reasons for coagulation and dissolving are not the same as we have shown in the fourth book of the *Meteora*.¹⁰³ Bones solidify as a result of the roasting heat of generation which people call *optesis*.¹⁰⁴ And that heat does not generate flesh any more than it generates bones in a given material. Rather, it performs this operation in a suitable material which is proper to bones, doing this when and where it should, much as does every natural potency. For the generation of these kinds of members is the result of a mover and from one in which they are not present in act but in potential. And it is not possible for them to be generated from material in which they are present in act, because if they were generated from it they would exist in act and not exist in act, all at one and the same time—something that is impossible. This is similar to when we said that a carpenter cannot make a chest from boards unless they lack the form of a chest made by carpentry, for otherwise he would produce and generate something that already exists in act, which is impossible.

The heat which brings these things about is present as if in a subject, in that superfluity which is the sperm.¹⁰⁵ Movement and activity are produced along with this heat as if with an instrument, and it acts according to a tempered quality and acts when and where it is suitable, as we have said. For in this way it is suited to the creation of each and every member. For if there were excess or lack in any of these, it would make a bad and incomplete member, like flawed members. Something like this is seen in things made by external heat, such as in works of art which are always incomplete and flawed from a lack or excess of any of the principles on the part of the efficient cause or the material. Tempered heat, then, is suited to the food from which the generation of the members occurs. Thus, nature uses the tempering action of the heat to make it suited to the movement of that which is moved. This is much as we do in the arts where one thing is decocted longer and with greater heat and another is decocted in less time and with less heat. We observe this especially in the works of alchemists, which imitate nature more than all other creative activities [*artificialia*]. For, as we have often said, the principle of movement in natural things comes from nature and follows the nature of the one generating. The cause which changes material on its own is cold and heat, but this is not excessive cold, but rather that which is present in a watery fluid mixed in with a boundable substance. For this is the kindling [*fomentum*] and food of natural heat. Nature uses these two, namely, a heat and a cold of the sort that chills the heat.

It is therefore clear that, according to the nature of these things, one of the above-mentioned members must necessarily be made in one way and another made in

---

102. "Readily": A. has misread Sc. to the detriment of the text. Sc. had written that bones and veins do not dissolve much as a vase (*fictile*) does not. A. reads *faciliter* for *fictile*.

103. *Meteora* 4.2.4.

104. On *optesis*, cf. 6.12.

105. Perhaps "This heat is operative, as if in a subject, in that superfluity . . ."

another way. But in those things which exist on account of something just as on account of an efficient principle, it happens that some are warm by nature and some cold. Thus, each member will exist according to the method that suits it. Thus, flesh is made soft, but bone hard, and while nerve is dry and elastic, bone is hard and breakable. Skin, however, is made from the dryness of the flesh in which are woven together the nerves, just as occurs in the decoction of a viscous substance over and around which arises a viscous, dried-up skin. This is because the dry part is raised over the water of the decoction and is seized by the moist viscosity and cannot evaporate. It thus thickens and creates a skin. This viscous substance turns into shell in very dry and earthy creatures in which there are hard or soft shells. For these types are earthier and drier than are those which have blood. In some which lack blood, however, this surrounding viscosity turns into fat and thus their skin will be viscous and soft. Much of this sort of viscous fattiness is gathered beneath the skin of such animals when they are earthy. All these members are necessarily generated in the way we have stated, each according to the method that befits it and each will exist on account of something effectively proportionate to it. Thus, the upper part of the body is formed and differentiated this way earlier from a principle such as this at the time of generation. The lower part, however, is formed later in all those with blood.

118 Generally, however, all members take on shapes and natural bounds to their size and they then receive two natural colors and natural hardness or softness. For nature acts like art, which imitates nature. But we see that a painter first sketches out the shape and afterward superimposes the colors. Nature acts in exactly the same way. According, then, to what has been said, there should first be present in the heart the principle of sensation and movement and of all the other natural operations so that all the animal powers will arise from it. It therefore should be created prior to all the other members.

CHAPTER V

On the Creation of the Head and the Members Which Are in the Head

119 Because heat comes out of the heart and the veins arise from it, nature established a cold member in opposition to the heart, and this is the brain. It is also for this reason that after the heart is formed the head is formed. The size of the head is larger indeed than that of the other members because the brain is large and moist from its earliest formation.

But a question arises that is not to be ignored concerning the eyes. For the eyes, at their earliest formation, appear very large in walkers, swimmers, and flyers alike and they thereafter constantly grow smaller. They are formed first in the head but are completed last. And during the time which lies between the first creation and last formation, they grow smaller to their allotted size and are also sunken into the head. This is because the organ of sight is elevated on high above the organs of the other senses of taste, smell, and hearing. Taste and touch are bodily parts of the animal because they do not exist through some external medium.[106] But the organs of hearing and smell are passages connected to the outside air, filled with a spiritual element which reaches to the fine veins extended from the heart to the brain. It is through these that the report of the forms of sensibles is made. The eyes, however, alone of the other organs, are made of a cold and moist body, for they are made of the humor which contains the brain. For a watery humor [*humor aqueus*] comes from the brain—from the webs which contain it and from which the *myringae* of the eyes are also made—and extends to the material of the eyes.[107] An indication of this is that no member in the body is as cold as the brain and the eyes. Because the eye is moist, it necessarily is large during the first creation and then gradually grows smaller afterward in the decocting heat. The same thing occurs to the brain and for the same reason. At first it is very moist and large and it afterward solidifies unto itself.

The reason given for the size of the eyes at the beginning of their formation is the reason for the size of the brain. For it is due to the size of the brain that the head too is large compared to the body as a whole. And it is due to great moisture that the eyes at first appear large and, solidified at the end, are completed to their requisite size. A long period of time is required for this since the brain is bounded by the natural heat with difficulty and, due to its great moisture, resists the natural heat. This necessarily happens in all animals which have a large, moist brain, and especially in humans. It is also for this reason that the area of the first suture in the *sinciput* is soft for a long time in children. It is due to the humor of the brain and it becomes completely hardened only after the period of generation and after the period set aside for the drying of the superfluous fluid which is in the anterior part of the head.

As we have said, this occurs in human infants because of all the animals the human has the largest and softest brain in proportion to bodily size. In humans the heat of the heart is tempered and is very clean, not mixing materials and functions. An indication of this is that the human has good intellect and discretion and thus does not undergo swift, violent drying. For this same reason infants cannot bear their head upright in their earliest age since their brain is heavy from its size and its humor.

In other cases, it is the same for movable members that are weighed down by a humor. For they droop straightway from the weight of the humor and are rendered immobile, as are the members at the bottom of the body where there is a humor, and the eyelids also have a similar disposition due to the humor pressing down on them. This is why a person cannot raise his eyelids when oppressed by a sleep-inducing moisture or by drunkenness. For we have known for some time that nature does nothing

---

106. "Bodily parts": *partes corporeae,* that is, areas where sensation occurs rather than defined and separate organs, though A.'s explanatory comment seems to miss the point. Cf. Ar. *GA* 744a2f.

107. On the "watery humor," cf. 1.200f. *Myringae*: a misreading of Ar. *GA* 744a11, where the *mēninga* (acc.) or membrane of the brain is mentioned. Fonahn, 2174, lists an instance of *myringa* as the tympanic membrane. A. may thus have in mind the tunics of the eye in the present passage. Cf. 19.14, 17f.

uselessly, neither in the earliest nor the last age. The eyelids therefore must split apart and open and can move for the reason established, namely, when the motion of the drying heat overcomes the superfluous moisture which is in them. And we know this from a similar thing which occurs in those whose heads are heavy due to epilepsy or drunkenness. These people cannot close and open their eyelids except by the descent and digestion of the superfluous moisture.[108]

Enough, then, has been said on the determination of the eyes and the brain and why the eyes are completed last of the members.

CHAPTER VI

How the Dominant Members Are Formed After the Heart and How,

After These, the Other Members Are Formed

122 All members are created out of food, and the members which are nobler than the others—the dominant ones which are formed and bounded by the first mover of the sperm—take nobler food, which is first, principal, clean, and has good decoction. The members, however, which they need to serve them and which are created for their sake, have worse food. For nature is a sort of controller and arranger of works and she gives clean, good food to the nobler members. That which is left over from the nobler food and which is bad, or less good, she allocates to the secondary members which are subservient to these nobler members. For just as a rector and steward act in arranging and dispensing external matters, so does nature act in dispensing those things which pertain to the distribution of natural things. Thus, the sensory organs are created from the cleaner material, while hair, horns, claws, *soleae*, and similar things are procreated from and nourished on the superfluities and, as it were, remnants and purgings of this material.[109] This is why members of this sort are completed last, for this is when this sort of superfluity begins to be abundant in the body.

123 As was said, then, the creation of the substance of animals occurs in the first members from the spermatic superfluity and much of this passes over into the dominant members. What is left behind, the baser part of the superfluity of the food, passes over into the other members. Food, therefore, is that from which the generation of all members occurs. But we will speak again more clearly of these matters in what is to follow.

The material generation of nerves thus follows this pattern, for they are generated from these same things we have discussed, namely, from the spermatic superfluity and the nutritive power. This will also be the method for the generation of claws, hair,

---

108. Cf. 14.29.
109. "Claws": *ungues*, which here and just below could also indicate "hooves."

soleae, birds' claws and beaks, and of all sorts of similar things. All of these are generated from food through the nutritive power which is in the embryo from the power of the sperm (that is, the *gutta*) of the female.

Bones, however, are from other material. An indication of this is that they do not grow continuously, but rather they grow to the size allotted them by nature even though they have the principle of the movement of growth. For they do have an allotted size in all animals. For if bones grew continuously, then the animal in whose body the bones are would also grow constantly. As the bones grew, the other members which are supported by them and in which animate life shares would also grow. The size of an animal is due mostly to the length of its bones. We have already determined elsewhere the reason that an animal does not constantly grow in length. As an animal gets older it does not grow because of the cold superfluities which are retained in its body and which the natural heat cannot consume. At this point the food left over from the impurity which is excreted, and which should be nourishing the principal, dominant members, is very meager, due to old age and to infirmities in the members which cannot then convert food to themselves. As the remaining superfluity of this sort of food diminishes in the body, the hair is first diminished, bereft of the fluid of the food. And for the same reason horns grow smaller. The bones, however, do not shrink easily because of their hardness, but when they do, this occurs from a dwindling of melancholic food which should be separated out for them from the common nourishment of the members. Hair has a special trait in that it grows because of the earthy smoke of the nourishment. But when the root area falls out, it is in no way regenerated.

Someone may be dubious about the nature of teeth, asking if their nature is like that of bone. For we see that hair and horns grow from a superfluity of food rooted in the skin. For this reason too they vary their colors according to the colors of the skin, and skin colors are white or black. Teeth, however, do not change color at all, neither according to the skin nor to anything else. They are white in all cases. This is an indication that they are generated out of the material of bones. This is seen in all animals which possess bones and teeth. If they are from the nature of bones, then it seems right that they should not always grow. But we know as a singular fact that of all the bones an animal possesses, only the teeth grow until the end of its life. This is proven by the test [*experimentum*] of sight, in the teeth of those creatures where one tooth slants away from another so that it will not directly block the other tooth. In these animals the teeth grow in one after the other. The reason nature orders their growth is that they continually function in grinding food. For if they did not grow continually, they would continually be used up and thus there must be a steady stream of material from which the teeth grow and are nourished. This is the reason the teeth of animals which eat much and often are greatly worn down, especially if they do not have very large ones. Thus, a wise nature cleverly adds in more and more teeth with age and if its life were to last to its thousandth year, nature would have accomplished many renewals of its teeth as its age wore along—more in fact than she did in its first young age.

The reason, then, that teeth grow continuously has thus been set forth. But the nature of the teeth and of the other bones is not the same. All bones are created out of the first radical moisture of the body which is called the seminal moisture. But the creation of the teeth is ultimately from the nutrimental fluid and not from the radical. Teeth also have as a special trait that they grow back after they have fallen out. Therefore, new teeth occur and are generated in some animals. Their generation has in common with others the fact that teeth come from only melancholic food as do bones.

126 Further, certain animals have teeth or that which takes the place of teeth only at the first moment of their generation out of a uterus through an accidental, contrary happening. In what follows we speak of why some animals have stronger teeth than others and why teeth fall out in some but not in others. It is clear from what has been said that all animals of this sort are generated from superfluous melancholy in the food which is left over from the food for the nobler members. For this reason too the human body is soft and smooth, since it uses the nobler food. It is smoother than the bodies of all other animals and has small and thin nails compared to its body since the earthy superfluity in the human is extremely small. This superfluity comes from undigested and unconcocted food.

We have, then, in this way, determined the reason that each member is generated.

CHAPTER VII

How the Growth of the Embryo Is Accomplished through Cotyledons

127 The first growth of each type of member and of animals is created by the *umbilicus*, as we have stated above. For the nutritive power is generated suddenly in animals that have been conceived and the *umbilicus* sends out cotyledons to the womb which are rather like tree roots.[110] The *umbilicus* is created out of the bark (that is, the skin) which contains its veins and these veins are numerous in large-bodied animals. In animals with an average-sized body there are only two veins and in those with a small body there is a single vein from which the embryo receives its sanguineous food. Wombs are, in fact, nothing more than the intertwined ends of many veins. In animals which do not have one large vein in their womb, stretched along its inner surface, but which have many small ones in place of this one, these numerous little veins will follow close upon one another in the womb and are called by their proper name, "uterine" [*matricales*]. They are connected with the openings of the umbilical veins. For the veins are connected to the *umbilicus* at one place and in another resolve themselves into the surface of the womb. And at their ends there are openings of slender veins and of the webs in which they are wrapped.

---

110. On cotyledons, cf. 9.56.

The round part, then, of these veins will be around the womb while the part that is concave will be toward the *umbilicus* toward the embryo. Between the womb and the embryo there are webs and *secundinae* for the purpose established in previous books on human generation. But the more the embryo grows within the womb, the more the openings of the veins are constricted. When the embryo is completed, they begin to disappear entirely. For these are the passages through which food flows to the embryo as does the sanguineous food. And nature indeed gathers that food together in the veins we mentioned, just as, afterward, she gathers milk together in the breasts. Because it is gathered and drawn to the openings of the veins gradually, it happens that the openings of the veins swell within like small pustules. While the embryo is small and unable to take on much food, the veins remain full and swollen. When the embryo has grown and takes much food, they are narrowed and shrink, settling into themselves.

Further, wombs do not have visible veins in many animals which have teeth in both jaws. Rather, the *umbilicus* extends into one vein and that vein divides and extends, encompassing the entire womb. These things are known especially through the anatomy of pregnant animals, and this anatomy has already been set forth in the first book of this study on animals.[111]

It is therefore clear from what has been said that the embryo hangs from the *umbilicus*. The *umbilicus* in turn is connected to a vein and the vein extends throughout the womb. Blood flows through the vein to the womb much as water flows through a canal. Around each embryo there are, as we said, webs and *secundinae*. Those who hold that the embryo is fed on small bits of flesh that leave from it are in error and misstate the truth. For if this were the case in the human, it would happen also in other animals. But this has been proven through observation [*per inspectionem*] never to happen in the other animals, as can easily be determined through dissection. In this way we know that every embryo—those of walkers, swimmers, and flyers—is wrapped in delicate webs and *secundinae*. These webs separate the fetation from the womb and the fluids, as has already been stated.

From this discussion, then, it is clear that all who agree with Democritus's opinion, as introduced earlier, are misstating the truth.

---

111. "Anatomy": *anatomia*, which may possibly evoke the idea of dissection as well. Cf. 1.444f.

CHAPTER VIII

Which Animals Are Impregnated as a Result of Copulation and Which Are Not and on the Causes of Sterility

130   Animals which are similar in species and genus naturally intermingle through copulation. In addition to these there are those whose nature is close to, though not exactly in, the same species, and this is so especially for those which do not have a great difference in shape and size and which are alike in the length of time their pregnancy lasts. All the things of this sort which produce compatibilities are present in some animals which are otherwise different in species but which have natures that are close to one another.[112] For example, foxes, dogs, and wolves have comparable natures. This is seen sometimes also in monkeys which copulate with a dog and also in birds when the partridge [*cubeg*] copulates with a hen.[113] It is also the opinion of some that hawks of different shapes and species copulate with each other and generate something one from the other. It is even said to have happened in the animal called the *raniez* in Greek.[114] Some also believe that the tall hunting dogs of India, which resemble greyhounds, are generated from bitches and tigers, as we have already said.[115]

131   And we see that in all the animals mixed this way both the females and the males can generate and preserve the species. Yet he-mules and she-mules generate but rarely, even if they copulate with their own or a close species. From this arises a general question about the cause of sterility in animals and especially what might be the cause of sterility in women and men.

For we see that some women are sterile and that some men are nongenerators. A very similar thing happens in all genuses of animals. But the mule genus is universally sterile with rare exceptions.

The reason generation is withheld in other animals is complex. One reason is that sometimes the members suited for copulation are weakened, flawed from the moment of their birth. This is the case for certain women who have no menstruation and cer-

---

112. The translation offered takes some liberties with the Latin, which itself came to A. as a garbled version of Ar. *GA* 746a33f. A. attempted to improve the text he received, with mixed results. We read *facientia* for Stadler's *facienta*.

113. The monkey and dog combination almost surely is an attempt to explain baboons, whose Gr. name, *kunokephalos,* means "dog-headed" (Keller, 1.7–11). A.'s description of the mythical race with this name at 2.57 is illuminating as well, and Friedman (1981, 24–25) relates this race to baboons. A more remote possibility is the dog-headed ape, *Simia cynocephalus*; cf. 22.137(99) and Scanlan ad loc. On the partridge and chicken, cf. 16.79.

114. Ar. *GA* 746b7 cites the *rhinobatos,* apparently a cross between the *rhinē* and the *batos,* both probably types of rays. Cf. *GF,* 222–23, and Peck (1942), 244–45.

115. Cf. 7.129. Might this actually be the cheetah, which, while admittedly a cat, is long-legged and is used for hunting?

tain men who have no beard but are beardless as eunuchs their entire life through. In those of this sort, the defect is inborn while in others it occurs in youth from some accidental trait of their food or something else. For women who are well nourished on suitable food after their youth are rendered very fat and fully fleshed. Then the nourishment passes over into flesh and nothing worth consideration is put in their semen.

Also, the menses of some women is thin and watery and neither gels in the womb nor undertakes an ordered coagulation. These women do not become pregnant.

In the same way the watery sperm of certain men cannot coagulate. The cause for sterility of this sort is sometimes due to an illness on account of which the man's sperm is rendered cold and watery—when, that is, the illness is in the man. When the illness is in the woman her menses become watery and unclean, filled with bad superfluities.

It also occurs in many men and women because of a natural flaw occurring in the areas and members related to copulation. Some of these infirmities are curable and others are not. Sometimes, the sterility of both men and women lasts, as we have said, throughout an entire life. They have this problem from their earliest birth, due to a defect in one or another of the principles of generation. Some women are tomboys, having little of the female sex save in their groins, and men are sometimes effeminate. In this case this sort of woman has a warm and dry menses, but the type of man we mentioned has very thin and watery sperm.

This is known from trial [*experimentum*]. For if a man's sperm be cast over moderately cold water it immediately dissipates and spreads out over the surface of the water if it is thin and watery. But sperm suited for generation sinks under the water when cast upon it and holds itself together since it is nodular and thready and these are the vehicles which do not allow themselves to dissipate and to spread out over the water.

Women who are knowledgeable concerning copulation know from experience [*experimentum*] whether the lower passages are open to receive sperm. They make suppositories for the opening of the vulva, which are sharp and made out of garlic and certain other things. If the taste should travel from below to the saliva in the mouth, they know that the womb is open to receive the sperm.[116] They test [*experimentantur*] the upper passages (through which the majority of the sperm descends) both in themselves and in men by using things which have a very fine substance and which have a strong coloring effect, such as saffron. They place such things above the eyes and if the color penetrates and colors the saliva, they know that the sperm has a free descent from the brain.[117] For the brain has a strong link to the sperm and to the eyes, a fact proven because the eyes of those who copulate a great deal are sunken and shrunk in size, and vision is weakened as well from frequent copulation. This is because the sperm is very much like the brain and, as some say along with Hippocrates the physician, descends mostly from the brain. Whatever may be the truth of this belief it is

---

116. Professor Helen King kindly supplied much information on this use of strong odors in a fertility test. Cf. especially the Hippocratic *Diseases of Women* 2.146 and 3.214 (Littré, 1853, 8:322, 414–16) and King (1985), 140.

117. Hippocrates *De semine* 2.

certainly true that the heat of copulation enters into the brain and dries it out and the eyes as well. The purgation of the menses, however, occurs from below the diaphragm since that place is nearest the womb and the places natural to it. For this reason the movement of the menses occurs from the genitals ail the way to the chest.

This, then, is the flaw of sterility in men, women, and in many other types of animals, as we have said before.

CHAPTER IX

On the Cause of Sterility in She-Mules According to Democritus and Empedocles, and on Disproving Their Opinions

134 The genus of she-mules is, for the most part, entirely sterile. Democritus and Empedocles have offered some quite faulty causes for this fact, yet Empedocles has spoken less poorly than Democritus.[118]

Empedocles believes that the cause of sterility is that the sperm cannot mix well by means of digestion during copulation because they do not possess a similarity of genus. He said that from this there arose a "stammering" mixture of the sperm like the mixture that occurs when copper and lead are mixed.[119] Democritus says that the passages of the wombs in these animals are partially corrupted. He says that the reason for the corruption is that the first generation was not the result of animals which share the same nature in genus and species.

Now neither of these men speaks the truth, as is proven from the fact that oviparous animals of differing genuses mix together in copulation. What is born from these animals is as fertile as if it had been born from animals which shared the same species. For if what they claim to be the cause of sterility in mules is true, then this same cause should induce sterility in all which are born from the copulation of animals which differ in species.

135 As we said, however, Empedocles claims that the cause of sterility in she-mules is a stammering mixture of the first sperm of the ones generating. He says that the sperm of the mare and the he-ass are soft and thus one does not sink deeply into the other as do those which mix well together. For the sperm of the he-ass is strong, he says, as that of the mare is weak. Now weak things do not mix well with strong things since one overcomes the other too strongly. If such things are mixed together, then the whole

---

118. Ar. *GA* 747a23f.
119. "Stammering": *balbutiens,* often rendered "lisping." There is a triple metaphor at work here. Just as words are rendered slurred by the speech impediment, so the result of the mixture of sperm is slurred, yielding no well-mixed finished generative product, just as an improper mixing of copper and lead yields a soft, ineffective bronze. The softness is discussed following this passage. Cf. the original intent of Empedocles and Ar.'s rebuttal at Ar. *GA* 747a34f.

becomes hard. Sterility therefore is caused by a corruption of creation during the first mixing, according to Empedocles. It is much like the stammering mixture of copper and lead, each of them being soft, or like the stammering mixture of the alchemists made from the stone which the ancient philosophers call *mez* and the Arabs *falcardam*. We call it red arsenic [*arsenicum rubeum*] in Latin.[120] This stone is first calcined and then sublimed and it is then calcined and sublimed a second time.[121] It penetrates all metals and imparts a color to bronze [*aes*] when it is sprinkled over glowing sheets of it. The red arsenic is mixed in with bronze as well but only forms a stammering mixture with it for it dries out the bronze too much and makes it subject to heat.[122] Empedocles says that a confused mixture of this sort is what is produced out of the sperm of the he-ass and the mare and therefore what is born is rendered sterile.

They say that Aristotle spoke of mixtures of this sort in his book *On Problems* in which he gives the reasons for many such things, arguing by means of conjectures about unknown matters based on things that were known, and by showing which hard, dry things sink deeply into other soft things and which do not, and which things when mixed hold together.[123] For one single thing is made out of things which hold together such as when water and wine are mixed. But to speak of matters such as this is a more difficult inquiry and requires a higher intellect than the matter we are now addressing. There is no need now to make an in-depth study into mixables and into how wine and water are mixed.

Thus, the statement of Empedocles is shown to be false in its basic proposition. For if the imperfect and soft are always generated from the copulation of animals of different species, then for such animals a fertile male should never be generated. Yet this is false, for just as it happens that both fillies and colts are generated from a stallion and a mare, so too a he-mule and a she-mule are generated from a male ass and a mare. This is just the same as for others where both male and female offspring come from the male and the female. Thus, if this mixture were stammering this way and were hindered from entering into one because of the thickness of the passages of the other, then every young born from a he-ass and a mare should be sterile. Yet this is false since a male mule, born from a mare and a he-ass, generates and can produce offspring.

136

---

120. "Copper" here is *aes* although A. used *cuprum* just above. Wyckoff (1967, 221–23) explains the usage and comments upon A.'s lengthy section on copper and its mixtures in his *De min.* 4.6. Arsenic is mentioned prominently in A.'s discussion of copper there and cf. also 2.2.6 and 5.5. The origins of *mez* and *falcardam* are unclear but cf. Schühlein's (1658) efforts.

121. To "sublime" or "sublimate" a substance is to raise its temperature, even to the point of vaporization. "And *arsenicum,* when calcined, changes from red to black; but afterward, if sublimed in an *aludel*—which is a covered vessel with a long neck, as we have often said—it again becomes white as snow. And if such calcination and sublimation are repeated a number of times it becomes extremely white and sharp. And because of its sharpness, [*arsenicum*] added during the fusion of copper penetrates into it and changes it to a shining white" (Wyckoff, 1961, 224 = *De min.* 4.6).

122. "Subject to heat": lit., "roastable," *assilis*.

123. A. takes his reference to the *Problems* from Ar.'s own self-reference at *GA* 747b5, but a matching passage cannot be found.

137 Further, from what has gone before we know that all sperm are soft and able to flow, and that a he-ass can copulate with a mare and that a he-mule or a she-mule is generated. Likewise, a stallion can copulate with a she-ass to generate a hinny, male and female.[124] These sperm are always mixed to generate young and the mixture is not stammering and does not then hinder generation right away. Further, generation occurs from a stallion and a mare and generates a fertile animal, and yet we know that the sperm of the stallion and the mare have a different complexion. If the mixture of the sperm were the result of the coupling of just the male or the female horse, then it could be said that the he-mule or the she-mule born, although composite, did not resemble the entire sperm.[125] In this case it would have a resemblance to both the mare and the he-ass and, according to Empedocles, it should thus be perfect for generating. But we see the exact opposite of what he has said. For we see that the mixture is not stammering in the she-mule, but rather that what has been generated has a secure mixture from both. Her actions and shape show this, for they resemble both the he-ass and the mare.

138 Shepherds hold the opinion that a he-mule is able to generate when it reaches seven years of age. The she-mule, however, will in no way generate (except very rarely and in very few warm locales).[126] They say that the reason for the sterility in the she-mule is that she cannot create and nourish the fetus up to the point where it is completed for birth. They claim that if the she-mule could care for and nourish the fetation in her womb, then the sperm would conceive well for her, saying that there is no hindrance to conception.

This is a logical determination and is based on probables, and it is more satisfying than that which Empedocles and Democritus said. For these men say that the further something is removed from its natural principles, the more it is weakened. And this is universally true, to be sure. For we know that a male and a female which are alike in species will generate an offspring like themselves in the same species. A male dog and a female dog will generate either a male or a female dog.

139 It is, then, a plausible inference that from a male and a female which are different in form and species, a young will be generated which is different in form and species from each of the ones generating it. Thus, if a male dog were to generate out of a lioness or a lion from a bitch, the young would differ in form and species both from lion

---

124. Hinny is the English term for the result of such a union. The Latin is *burdo*, a term A. uses at 19.43 for the organ on a pipe that is still called the bourdon. Since the term also was once used for the drone of a bagpipe, the comparison must be to the bray of this mule.

125. As a result, it would be infertile. The text has been amplified at this point for the sake of clarity. A.'s Latin, literal, and overly terse, says: "if the mixture of the sperm was only from one." Cf. Ar. *GA* 747b21f.: "If only one sex of the horse united with the ass, it would be open to Empedocles to say that the cause of the mule's infertility was the dissimilarity of that one sex to the semen of the ass" (trans. Peck).

126. It is worth noting that the parenthetical material is added by A., as if he knew personally of exceptions to the general rule.

or dog. Thus, the *lunza* is said to be generated from a male leopard and a lioness.[127] And if similarity is a cause of fertility then a male or a female mule should be generated from a male and a female mule and it should be fertile and like its parents in form and species. But we do not see this happening. Instead, out of a mare and a he-ass which differ in form and in species, there is generated a male mule which resembles neither of the animals mentioned in form nor in species. Out of a she-mule, however, which is somewhat like the he-ass, and out of a he-ass, nothing else is generated. Yet, as we have often said, something is generated from each of the two when the ones generating differ in species and form.

This statement, then, which explains sterility and fertility on the basis of likeness or dissimilarity of forms, is base and false and that which follows upon base principles of this sort which are falsely put forth is entirely false itself. In geometry and in all other fields of learning, a statement is called false which is said to exist in a way that is other than the way it is in fact. Insofar as a thing is or is not, the statement is true or false. And we see the opposite of what they have said, for many animals which are dissimilar in form generate an offspring which is unlike both the parents and which is yet fertile for further generation. We have given examples above in the cases of many animals.

CHAPTER X

On the True Cause of Sterility in She-Mules, Based on a

Summary Consideration of the Collective Nature of the Mare and the He-Ass

We should then find the true reason why a she-mule is sterile, based on a consideration of the nature of the mare and he-ass.[128] In this way we will understand why what has been described happens. For in all cases we see that things generate one for the other only if they either share the same genus or possess close genuses or natures, and the females are then impregnated by the males' sperm. For this reason too the males are kept away from copulation for a long time since the females, once impregnated, cannot endure copulation very well.

The mare has only a small amount of menses for her body size. For this reason only a small amount of menstrual superfluity emerges from the mare, less in fact than from other female quadrupeds. Moreover, mares do not easily retain semen and neither do they easily conceive. For this reason, when they have copulated with the stallions, they

140

---

127. At 22.15(2), A. tells us the *alfech* is the Arabic name for what the French, Italians, and Germans call the *lunza*. Scanlan (ad loc., 70) suggests it is a "panther or some other species of *Felis pardus*." The modern Italian for leopard is *lonza* and Lexer lists *lunze* as a MHG term for lioness (1.1984) and *linse* as a term for an animal born of a wolf and a dog (1.1928).

128. Cf. *QDA* 16.19–20.

are struck with whips to keep them from urinating. For they are accustomed after copulation to urinate away the sperm they have received.

141 Now the he-ass is an especially cold animal. An indication of this is that it is not generated in cold areas or it does not do well if it is generated. Therefore there are no asses in Sweden or in the neighboring regions to the north of Sweden, such as in the Holsatic region and in Dacia.[129] For the cold of these regions is quite severe. For this reason too asses do not copulate before the summer equinox and they also give birth in warm weather. Likewise, they copulate in May when the sun is near the star which is called Aldeboran and which is the eye of Taurus and they are pregnant for a year. Since the ass is a cold animal then, its sperm must necessarily be cold as well and an indication of this is that when a stallion copulates with a she-ass which has previously been impregnated by a he-ass, the earlier fetation is not corrupted by the subsequent coupling. This is due to the coldness of the sperm of the he-ass. Therefore, when a he-ass and a mare copulate with each other, or when a stallion and a she-ass do so, and when the semen are mixed, the mixture will be preserved since it is tempered because two opposites have been mixed. For the mare's semen is warm and that of the he-ass is cold and because that which is made from both of them is tempered, a fetation arises from them. Because of the equality in the complexion of one species which is not comparable to that which exists in the other species, the fetation is preserved by two warm sperm, such as when a stallion copulates with a mare, or by two cold sperm such as when a he-ass generates out of a she-ass.[130] Equality of complexion is spoken of in many ways as we have said in previous books, where we spoke of multiplicity of complexion.

142 When a she-mule has been conceived and born from a he-ass and a mare it is not suited for generation since the powers of opposite sperm are in it and they hinder one another. The accidental characteristics of coldness which we mentioned occur in the he-ass and the he-ass rarely generates. For if he has not begun to generate after the loss of the first juvenile teeth he will never generate for the rest of his life.[131] In this way it is said that he-asses are not very well suited for generation. In the same way too the mare is not very well suited because of the small amount of her menses and the stallion is not well suited for generation because of the excessive heat of his sperm. He would be more suitable for generation if his sperm were slightly cooler and this does happen to it when it is mixed with the sperm of the he-ass, whereupon it generates.[132] Likewise, if the sperm of the he-ass were slightly warmer it would be more suited for

---

129. In this and the following sentences A. has done an admirable job of adapting the specifics of the Gr. text to ones with which his audience would be more familiar.

130. Reading Borgnet's *comparatae* for Stadler's *comparati*.

131. On juvenile teeth in horses and asses, cf. 16.149.

132. A. must have in mind a situation in which both mount the same female as above and the sperm thus mix. This passage is badly confused because A. is continuing to specify the animals by gender (e.g., *asinus* as he-ass and *equa* as mare). But a comparison with Ar. *GA* 748b8f. shows that Ar. was speaking in general about the relative complexional temperatures of horse and ass. As often, A. was confused by the text he received and he interpolated in an attempt to make sense of it.

generation and this happens to it by being mixed with the sperm of the stallion. Thus, although each of them is ill suited to generation, if some unnatural flaw befalls them, then their fertility is easily impaired. However, this unnatural flaw occurs in one that is born from them and especially in one of the female sex. Thus, it is right that one thus born should naturally be sterile. It should thus not be said that anything renders her sterile save that flaw which is the thing conceived contrary to nature. Therefore the generation from a he-ass and a mare is completed with difficulty.

The bodies of she-mules grow large because, since they are sterile as we have already said, their entire menses passes over into their bodies and into hair. Since animals of this sort, which have horny hooves [*soleae corneae*], remain pregnant for one year and during this time feed their fetation on menstrual blood and afterward on milk, none of these things is removed from the she-mules since they are sterile. They therefore grow large because all their menses is converted into food for them and into their hair. If there is a little bit of their menses left over, it is cast off in their urine. This too is a further reason why he-mules do not sniff the urine of the she-mules in the same way that other hooved beasts smell the urine of their females. When they do sniff it, they raise up their head, split their lips, and show their teeth as if they were laughing at their urine.[133] But it is possible that the she-mule might conceive in certain lands. This is rare however and happened during time gone by. But it has also happened recently during our own times. When she conceives she rarely creates and nourishes it all the way to completion, producing a miscarriage instead.

But a male mule can generate in a given time before his old age since the male is naturally warmer than the female. That which is generated out of a male of this sort, however, will be flawed since the one generating it is a flawed animal (as we said before) having itself been generated out of a mare and a he-ass. In the same way pigs are sometimes flawed in the womb and take on a corrupted shape. In this case names other than "pig" are given them.[134] In the same way short dwarves are born flawed with bodies only one cubit high. This birth is the result of a flaw which befalls their members in the womb. This is just what happens to the pigs and to the animal which is called the *dorreg* in Greek.[135]

This, then, is the cause of sterility in she-mules and it is clear from what has been said before that it is not the same cause as in those generated out of a fox and a dog and the others which we mentioned above and these are therefore found to be fertile and to preserve their species. It is just this way in graftings, for a graft of a like thing onto a like thing is most likely to prosper and it is the same way for the same sort of mixture in animals.

143

144

---

133. Again, the observation is A.'s as is his comment below about mules that conceived.

134. Ar. *GA* 749a3 calls this deformed pig a *metachoiron,* "after-pig," implying the one born last and thus the runt.

135. *Dorreg* must be from *ginnoi* at Ar. *GA* 749a1. Ar. *HA* 577b19f. defined the *ginnos* as the offspring of a mule and a mare as well as a diseased mare's offspring. As Platt (1910, ad loc.) points out, there is no such result of a mule and a mare's mating, but Ar.'s diction seems to indicate it was a freak occurrence.

CHAPTER XI

Which Is a Digression Summarizing the Things That Have Been Said Above in This Tract and Assigning the Causes So That What Has Gone before Might the More Easily Be Understood

145 The things which have been said before, throughout the entire tract, are three in kind: namely, on the generation of the homogeneous members; on the causes of sterility in general; and specifically on the sterility of she-mules. So that what has been said might be better understood, we are summarizing these three things in one chapter, saying that the first member generated in an embryo is the flesh and the fat.[136] These are coagulated by cold and dissolved to some extent by heat. Claws, however, and veins come from earthy material, but the moisture in them is somewhat oily and this is the reason that they do not liquefy in fire but soften instead. After softening, they are turned to ash when the moisture holding them together is set free from them. If any part of them liquefies it is very small. The skin is made from the viscous material which surrounds the generation of the flesh. There are some members which are generated quite large at the beginning and which afterward grow gradually smaller when their moisture consolidates. Such are the eyes, which at first are very large in the embryos. This is because this member is moist and very watery and thus resists the heat working within it for a long time before it is reduced to its proper size and shape.

146 Because the front part of the brain is marrowy, moist, and resists even the bounding heat, it is necessary that the front part of the bone be quite smooth and soft. There is another reason for this, though, namely, that during the early stages of birth the brain has many superfluities. For this reason too its top part stays loose and has many sutures, both lengthwise and transversely, so that the vapors raised to the top can evaporate. Enough has been said about the teeth in our anatomy section in previous books and here we must therefore consider only that of all the bones only the teeth grow continuously if at a stealthy pace. Otherwise, through constant wearing away, they would be used up. Moreover, in certain animals the *umbilicus* contains two veins while in others only one. This difference follows the differing size of the bodies of the embryos. Further, those animals copulate with each other and then conceive which are of the same species or of species which are close to each other. This is especially so when the period of their generation is the same or is almost the same as the other's period. Examples are dogs and foxes, horses and asses, and the like.

147 The essential cause of sterility is due to the two sperm. The accidental causes, though, are many. Sometimes it is on account of the womb in women, especially when the woman has masculine traits and her womb has been rendered unsuitable for the

---

136. Avic. *DA* 16.50v.

reception of sperm. Sometimes it is due to the male penis, for either it does not emit sperm correctly or else it might be that the instrument is short and cannot accomplish ejaculation to the required spot. Or it may be that it is too large and convoluted, this sort in a membrane that impedes the flow of the sperm. Certain men also emit watery sperm. When good sperm is placed on water it sinks beneath it, whereas watery sperm immediately dissolves and floats. Further, certain women do not care for intercourse.

To speak generally, the cause of sterility is either from the two complexions of the sperm or from the organs of copulation as has been said elsewhere in an earlier part of this study. The reason which is due to the dissolution of the sperm occurs when the sperm is watery. This occurs mostly in the sperm of the female when it meets the sperm of the male at the same time and on the same path but does not undergo the action of being seized by it. Rather, it corrupts it and causes it to recede from the equality of complexion. The male sperm sometimes does this as well, for a multiple cause, such as diminished size or receding from equality of complexion.

Further, the complexion of the womb will sometimes be bad and will corrupt the sperm that enters it. Its internal opening is sometimes blocked. Sometimes, the vessels and canals for the sperm are injured and doubly so, namely, either from a bad complexion or from an illness which overcomes it, an illness which is instrumental and not complexional. Further, the spermatic passages are sometimes blocked, for most of the sperm descends from the anterior part of the marrow of the brain. This is what Hippocrates says.[137] The effect of copulation does not prove this directly, but it proves well that either the sperm descends or that some of it does or its spirit does. For those who copulate a great deal sense an emptiness of the brain and their eyes become sunken. A clearer indication is that if the two veins behind the ears and which are called spermatic have been severed in men wounded in battle, then it happens that they do not produce sperm.[138] There is also an indication that some of the sperm comes from the heart, because at the moment of pleasure, when the sperm comes forth during intercourse, breathing quickens and occurs faster, as if the heart is at that moment busy and intent on expelling something from itself. Now this is not necessarily an indication, for according to this the sperm departs the heart and the body at the same time and this is not plausible since it is a long distance between the heart and the seminal vessels from which the sperm departs during the pleasure that occurs during intercourse.

148

Democritus speaks of the cause of sterility in the she-mule and says that the reason is that the sperm is changed from its complexion, since it is not generated out of two that are alike in species. If he were speaking the truth, then that which is born from a dog and a wolf and a dog and a fox should not be able to generate. Empedocles, however, said that the cause of the sterility of the she-mule is an overly great softening and smoothing out of the two sperm, for each undoes the other. Thus, that which is born out of them is ill and infirm and easily takes on flaws due to the stammering mixture we spoke of above, a mixture like that of copper and lead. Empedocles takes the cause

149

---

137. Cf. *De semine* 2.
138. Cf. 3.161; 10.44.

for what he said from an oppositeness of the sperm of the stallion and the she-ass. Yet it seems to me that even though Empedocles may speak plausibly, Aristotle intends to say otherwise when he says that the stallion's sperm is very warm. An indication of this is that the mare's womb does not seek after it on account of its heat. This is why their keepers strive to keep them from urinating it out, beating them with whips so that they cannot urinate. The mare also has little superfluity of the menstrual blood while he-asses have a great deal of sperm which is very cold, and complexional cold leads little to the members. An indication of this is that asses do not do well in cold regions and also that they are moved to copulation only during periods of the year which have heat and moisture, when nature's moisture is strongly moved by the sun's heat. Also, they are moved to copulation only during the time of their youth which comes between the loss of their first teeth (they are called juvenile teeth in horses and asses) and their seventh year.

150 Therefore, the sperm of the he-ass is suited for generation only if it is helped along. When a she-ass conceives from a he-ass, it is helped by a likeness of species, the heat of the weather, and age. When, however, a mare conceives by a he-ass, it is helped by the heat of the *gutta* of the mare and is reduced to a tempered level. Likewise, when a stallion generates out of a she-ass, the sperm of each tempers the other's sperm. Mules increase in size beyond that of asses because of the large amount of material on the part of the ass and because of the formative and well-moving heat on the part of the stallion. But what is born is, for the most part, rendered sterile or if it does conceive the young will not do well since what is generated is an unnatural thing. Even though at the first moment of conception the sperm might somehow be reduced to equality, nevertheless, decline sets in afterward toward the nature of the mother, and she provides the material for the fetation, as we have said above. Thus, in this case the two powers will not be proportional and instead one will prevail over the other, and this will be the mother's power. The other will be like an impediment to her in her activities. And whatever is like this cannot be long preserved in nature. This is what Empedocles intended to say.

151 It is not this way in other ones which are born as a result of mixing the species, for these species are near to one another in the complexional type. The mean is then but little distant from the extremes and is almost like the extreme and it therefore does well. But this is not the case in the generation of the mule. For while the he-ass and the stallion have the furthest difference according to complexion, the mule is too far distant from each of the extremes. Therefore, it cannot do well due to the powers of the extremes which are in it. The coldness of the he-ass impedes progress the most and an indication of this is that when a he-ass copulates with a mare after an impregnation caused by a stallion, the fetation is destroyed because of the coldness of the ass's sperm on it. But if she should copulate with a stallion, then the fetation is not destroyed because the warm sperm falling over it does not harm the warm fetation. But it would harm it greatly if cold sperm were to fall on it. We are, however, saying all these things conjecturally. Perhaps what we have said is the true cause and perhaps there is another one which escapes our notice and which, should God reveal it, we will discuss in what is to follow.

HERE BEGINS THE SEVENTEENTH BOOK ON ANIMALS

Which Is on the Cause of Oviparous Animals and of Eggs

The First Tract of Which Is on Oviparous Birds

CHAPTER I

On Those Who Lay Many Eggs and Those Who Lay Few.
Also on the Natural Cause for Them.

The cause for sterility in animals has thus been determined in the preceding book. And we have posited all the diversities among the animals which generate internally and externally in a univocal generation, which is because it generates one like itself in form and species.[1] What remains to be determined in this book concerns the oviparous ones. Blooded oviparous animals possess almost the same principles of generation as do the walking animals and they generate those that are like themselves. For both of them are generated from intercourse and from a male and a female. Therefore, a similar philosophical investigation is to be employed in determining their causes. For while there is great diversity among animals of this sort, still all are alike in that they are generated through the copulation of a male and a female in which sperm passes from the male into the female.

The most perfect animal in the genus of oviparous animals is a bird which lays perfect, hard-shelled eggs unless some flaw befalls the bird due to an illness, with the result that the egg is laid in a membrane, lacking a shell. Bird eggs, however, are alike in that they all have two colors to their humors, white and yellow. The *celety*, however, and those like them, lay internally, as we have often said, and the live young exits from the egg within the uterus when the egg has descended from the upper part of the womb to the lower. All the eggs of this genus are soft and have but a single color. Nor is there a single species in this genus which does not lay many eggs. The sole exception is the type called the frog [*rana*], which lays but one.[2] We will designate the reason for this in what is to follow. There is a single color also in the eggs of all oviparous fish, namely yellow. For they lay incomplete eggs which, after being laid outside, grow prior to producing live young.

---

1. For univocal generation, cf. 16.7 and *QDA* 15.3.
2. "Frog" is a direct translation of *batrachos* at Ar. *GA* 749a24, which is *Lophius piscatorius*, a skate-like animal with a dangling appendage with which it lures its food. A. more frequently calls it the *rana marina* (e.g., 1.31). Cf. 24.50(101) and *GF,* 28–29, with ill.

In previous portions of this study we have made distinctions among the wombs of oviparous animals and we have shown the differences in their shapes as well as the reason this member is in animals. There we said that of those animals which become impregnated and have live young in their wombs, some have their womb located below, near the joints of the hip [*anchae*] or thighs [*ilia*].³ Others, however, have them above, near the diaphragm [*paries*]. The *celety* has its womb near the diaphragm, whereas an animal which both conceives and generates a live young has its womb below, like the human, the horse, and those like them. Likewise, certain egg layers have their wombs below like the fish while others have them above like the birds.

3 It happens that certain birds conceive eggs on their own without the benefit of copulation. These eggs are called wind eggs. Aristotle says that this is not found in birds which fly well and often and that neither is it found in birds of prey with curved talons. Rather, it occurs in birds with many eggs which do not fly a great deal and do not have many large, strong feathers. For in birds such as these the superfluous spermatic fluid passes over into the wings and feathers and is consumed by the intense labor involved in flying a great deal while hunting. Thus the eggs in these are only conceived by the sperm of the male attracting the material. Yet I myself have seen a domesticated bird of prey which they call a *sperverius* or a *nisus* [sparrow hawk] produce many wind eggs. But this one was domesticated and was not hunting at the time, standing instead in a basket where it was renewing its plumage.⁴ It therefore had an abundance of this sort of superfluity. For we have said previously that the menstrual blood and the sperm come from the superfluities of the fourth digestion and nature cannot mature these two and cause them to be abundant in some cases, for the reasons we have already mentioned. For this reason birds of prey neither have many eggs nor copulate a great deal. Instead, their bodies are dry, small, and sharp.

Heavy birds, however, which are not good flyers, such as pigeons and those like them, generate a great deal. Chickens, the partridge, and others like them are both especially bad flyers and have few feathers, and have a great deal of egg superfluity [*ovalis superfluitas*]. For this reason too the males of these genuses copulate a great deal and the females have a great deal of egg material [*ovalis materia*].

4 Of those types of birds which produce many eggs, some produce many all at once before incubation while others produce them successively by interspersing the incubations. The first type includes the chickens and the partridge, ducks, and geese; the second, the pigeon. For the pigeon type is just about midway between the birds of prey with curved talons and the heavy birds. For they are good flyers as are the birds of prey and they have a body containing many humors as do the heavy birds. Because they are good, swift, frequent flyers, their food must pass over into feathers and wings and is partially used up by the exertion of flight. They therefore do not lay many eggs all at the same time. But because they have a lot of moisture and a warm belly and have

---

3. A. is careless with terms such as these. Cf. 1.292, 2.69.

4. Although *sporta* generally means basket (cf. 6.54), at 7.103 it is used to mean a fish trap. It may indicate a sort of wicker cage in this context.

a lot of food which can be easily found (this because they eat seeds), these birds must necessarily lay many times. This is one reason for the paucity of eggs in birds of prey, for they come by their food only with great difficulty.

Further, unless something else hinders, small-bodied birds engage in a great deal of copulation and have many eggs. For this reason there is a genus of chicken—the chickens of the Deiamos region, according to Aristotle—which have very many eggs.[5] For their food passes over into material for eggs. And the closer chickens come to this species, the more eggs they will have because their food passes over into material for eggs. The same thing occurs in trees and in other animals, namely that they change food for growth into seminal material. This genus of chicken has a great deal of moisture and a small body. The bodies of other types of birds which are dry and warm produce more spirited and wrathful birds. For strong wrath is present, for the most part, in very dry bodies.

Moreover, thinness and weakness of legs result in greater copulation and far more material for eggs. Thus, in people thinness of the legs is often a sign of dryness and warmness and, when it is not a result of lack of eating or disease, it then signifies complexional heat which moves one to intercourse. For the food which in others passes over into thighs and large legs passes instead into spermatic material. Birds of prey, with their hooked talons, have strong legs and feet which are firmly fixed for holding strongly that which they seize. For they have a hunting lifestyle. They therefore have but little sperm, copulate but little, and have only a little material for eggs. The one exception is the genus of birds of prey which the Greeks call the *fieriz*.[6] This one species of bird of prey alone has many eggs, and this is due to its great moisture.

Moreover, birds of prey with curved talons drink but rarely and then little. Moreover, every humor, both natural and accidental, provides material for the eggs. And this is also a reason they have few eggs, for they are deprived of such moisture to drink. The number of eggs in these birds is four or five at most and only the smaller ones like the sparrow hawks and the kites come near this number. The species of bird called the *kokokoz* in Greek but the *guguli* in Latin [cuckoos] is not a bird of prey and these birds do not have curved talons. Yet they have but few eggs, so much so that these birds are said to lay but a single egg, laying this in the nest of another bird.[7] This is because the bird is very cold, not suited to incubation and egg laying. An indication of this is that this bird is exceptionally timid and flees all birds, even small ones. For all the other birds pluck its feathers because it breaks their eggs.

The genuses of birds which resemble pigeons, such as the ringdoves, the turtledoves, the *fethe*, and the like, lay but two eggs.[8] Nor is there a bird which lays only one

---

5. "Deiamos": Ar. *GA* 749b28 offers *Adrianikai*, Adriatic fowl, perhaps a bantam hen (*GB*, 39).

6. Ar. *GA* 750a8, of the *kenchris*, the kestrel.

7. Cf. 6.52–53; 8.91–94.

8. *Fethe* is probably another of the many variants for Gr. *phatta*, a dove, and this sort of bird fits the current context well. Cf. 1.44, with notes; 23.111(43), which specifies that the *fetyx* produces few eggs; and *fehyte* at 7.37. This would suit the context better than this being related to *fethis*, a kite at 8.15.

egg except for the genuses of cuckoo, which we have discussed. And on occasion even this bird may happen to lay two eggs even though the second is not found, for the eggs of this genus often perish since it seeks to lay them in the nests of others and can enter these nests but rarely. Those, however, which are like the pigeons lay two or, on rare occasions, as many as three.

7   In all those which produce many eggs, it is clear that this is because their food passes over into material for sperm and eggs. This is clear from the accidental traits of plants. Trees which produce a great deal of fruit are dried out more quickly. This occurs in trees which produce fruit every year as do the hedges. For their food passes over into seed and since they have many seeds it is necessary for them to lose substance constantly and to dry out. In this way too, certain chickens lay many eggs (so many that they may even lay two a day) and these birds die very quickly. This happens commonly and for the same reason in both trees and birds.

This is also why the lion generates but little. For the lioness which is producing her first brood has five or six young and afterward has this reduced by one a year so that in the second year she generates four, three in the third, two in the fourth, and afterward stops at one. For in her youth the food passes over into sperm and once youth has passed by, the movement of the sperm ceases or is reduced. Then she either does not give birth or generation in lions is diminished.

Let, then, this be enough said by us about why some birds have many eggs and others but few.

CHAPTER II

On the Nature, Generation, Differentiation,

and Disposition of Eggs and of the Humors Contained in an Egg

8   Eggs are produced due to the spermatic material which is in the females. The menstrual flow does not occur in the bird genus as it does in other animals which have blood. There are some animals which have a substantial flow of menstrual blood, so much so that the flow is even visible. But this flow occurs in neither fish nor birds and for this reason eggs are conceived in the bellies of many fish without copulation. Thus too it happens that eggs are conceived in some birds without copulation but this is not as apparent in fish as it is in birds since the nature of fish is colder than that of birds. The exit of sperm and menstrual blood which occurs in animals occurs only during times within which the seminal superfluity is moving. For the place for the conception borders on the diaphragm, especially in birds. It is therefore warm. But the egg that has been conceived will not result in generation unless it has male sperm joined to itself. The reason for this has already been mentioned previously. Eggs conceived without the male's sperm are wind eggs and they are not frequently generated in birds which fly well and often. The reason for this is the same as the reason that

birds of this sort do not produce many eggs. For the seminal superfluity in birds of this sort, as well as in birds of prey with hooked talons, is scanty and it does not easily amass in the womb except as a result of the heat of the male's sperm. Thus, in birds of this sort, wind eggs are not produced. However, in heavy birds which fly infrequently and which have food prepared for them, there are more of these wind eggs than there are completed eggs since they are conceived more easily and are smaller in size than completed eggs. This is because there is no sperm in them to forcefully attract the material. They are not as delectable and sweet to the taste for those eating them as are completed eggs since there is undigested humor in them. In every case that which is digested is both wetter and more delectable than that which is undigested.

Further, it is clear from what has gone before that fish eggs are conceived without males but are not completed without them. For as we have already said, conception occurs in many fish without the male's semen, even though this is not as visible as it is in certain birds. According to Aristotle this has been experienced in the river fish which are called the *arrubabo* in Greek.[9] For the eggs in this genus of fish appear suddenly, before they begin to copulate or to rub together.

Generally, however, in all bird eggs, the female takes in a great deal of sperm by frequent, successive intercourse. As Avicenna says, if frequent and ample copulation were not present, then the eggs would revert and be made into wind eggs in certain birds which have a great deal of moisture.[10] For when the womb is at rest after intercourse, it draws in the egg moisture [*ovalis humiditas*]. The passages are opened through this drawing action and this accidental trait occurs in birds more than in other animals since the menstrual blood in birds gradually moves to the womb with none of it exiting externally because of its paucity. It is also assisted in this by the fact that the wombs of birds are above, near the diaphragm. Therefore this superfluity remains still, held in the womb, and it grows in it almost in the same way in which the embryo grows in the womb of an animal which conceives and generates live young. For just as an embryo of the one sort grows through taking in food which comes to the *umbilicus* from the veins of the womb, so too does an egg cling to the womb through a sort of *umbilicus* and thus draws in the food by which it grows. Because many eggs cling to the womb at the same time, the spot for the *umbilicus* of the eggs is commonly called the root of the eggs [*radix ovorum*].

Further, after birds have copulated once, there are always eggs in them afterward. The eggs are very small at the beginning. This fact has led certain people to say that wind eggs arise from the residue of the sperm which is left over from the completed eggs. However, experiences [*experta*] prove this to be false for we have often seen hens and small geese possess wind eggs in their body prior to all copulation. This is especially visible in the females of the partridge which, whether they copulate or not, are filled with eggs when they see, hear, or especially if they smell a male of their genus.

---

9. Somehow this form may have arisen from Ar. *GA* 750b31, where some MSS offer *erythrinos,* a red, deep sea fish (*GF,* 65), although it is clearly out of place in the context.

10. Avic. *DA* 17.51v.

Some of them even suddenly lay eggs near the males due to an odor of this sort. This is much as happens in men when they have an emission of semen merely at the sight, touch, or sound of a pretty woman.[11] It also happens that other animals are moved by desire at the sight, sound, or smell of their own females. For even though, as we have stated, eggs are conceived in a more orderly fashion by means of copulation, they are, nevertheless, conceived without copulation by the birds we have mentioned. These types of birds naturally have a great deal of sperm and the superfluity in them is thus moved to conception by even a small movement of desire to bring about conception and the descent of the fetation.

Further, flyers which lay eggs externally produce eggs complete as to quantity. Fish, however, lay eggs externally which are incomplete as to quantity. This is because they grow when outside. For the fish genus is one that has many eggs and therefore the eggs cannot be completed internally as to size. This also is why fish have their wombs connected to their genitals [*iuncturae*] and they therefore pour out their eggs more easily.

Further, the eggs of all birds have two colors, whereas the eggs of fish have but one. There are two colors in bird eggs because of the strength of the bird's members. They therefore have the white color in the outer humor on account of a similarity with the members and to the inside they have the yellow humor for nourishment.[12] This second humor comes from the blood and from that which takes the place of menstrual blood in the conception of an egg since a bird does not conceive a live young. It is therefore necessary that the material of its food be provided externally to the animal which is generated within the egg. We have already said above that the menstrual blood is the material for the bodies. It is therefore necessary that there be something in the eggs that corresponds proportionally in overall appearance and likeness to the things which are produced during the generation of live young. Therefore in ones such as these the sperm of the male is like that which forms and from which the spirit is made, while the white is that which is formed into members. The earthy part which is in the middle is the yellow and is the sustenance for the body. It is like nourishment, even though it is quite separated from the other humor from which the substance is made. This is the reason that eggs have two colors to the humors held within the shell. Therefore the principle of generation in birds comes from the white and the principle of the sensible soul is in the circle which surrounds the yellow. The food is produced from the yellow.

Further, in animals with a warmer nature, these colors are distinct. One humor is white and the other is yellow, but the yellow is more earthy than the watery white.

---

11. It is odd that A. uses *homo* here in a male-specific context. He is normally quite careful to reserve this term to a more generic use.

12. "White color": St. reads *calorem album,* lit., "the white heat." Since A. has just mentioned colors (*colores*), one is tempted to emend *calor* to *color* here, following the reading of Borgnet. Yet cf. Ar. *GA* 751b3f., where the text upon which this is based specifically refers to the outer fluid as *to thermon,* "the hot." One suspects then, that St. may accurately reflect the Gr. with its *calor,* but that *album* was added somewhere as a gloss, explaining just what this "hot" was.

In those which are moister and are colder, it is the other way around, for the white in these animals is larger and is more watery and thin. This accidental trait occurs in aquatic birds where there are trees around the water. For these birds are colder and moister in general than other, land birds.

Further, because the yellow is earthy, the yellow is multiplied in the eggs of these birds. In others, however, the yellow color is the lesser one for the opposite reason. Birds of this sort have a colder and moister nature than all the oviparous animals which have blood.

The white is not separated from the yellow in fish eggs due both to its small amount and to the coldness of the fish. Thus a fish egg is a sort of mixture made of white and yellow. Wind eggs, though, have two colors and are imperfect since they still lack the male's sperm.

Nor is it true as some say that the reason for this separation of the colors is due to male and female, so that the white is from the male and the yellow from the female. Rather, both are superfluities from the female and one part is warm and the other cold. When there is a great deal of heat in the eggs, then these two colors are separated and divided. When, however, there is but little heat in them it cannot divide the colors and they will be mixed. The sperm of the male only holds and sustains the humors. Thus, in the first stage of generation the eggs appear white and afterward they gradually appear yellow after a few days because a great deal of blood is by then mixed into them. During the completion of the egg, however, the separating heat prevails and the white is then generated around like a fluid, bubbling forth thinly around the yellow. The white is naturally moist and the heat of the soul is in it just as it is in an individual subject. This is why it is divided and is placed around the yellow.

The yellow, however, since it is earthy, is placed within, like a center. Thus, just as on earth the land is in the middle of the water, air, and fire, so the yellow is in the middle of the water in an egg.[13] An indication of this is that if someone takes many eggs, breaks them, mixes the yellow and white together, and then places them in something membranous like a bladder, and if he then cooks this over a slow fire, he will find the yellow in the middle and the white around this unless the eggs were old and rotten. Yet, when a fresh egg is broken, the yolk floats on the white, even though it is naturally heavier than the white.[14] This is because the principle of the movement of the nourishment is already in the yolk. Thus, because of this principle, it has motion upward toward the members which must be nourished, since it is already prepared for attraction. This is also why eggs which are taken in as food are very convertible. The yolk is more nourishing than the white and this is also the reason the chick is generated in a circle which connects the convexity of the yolk with the concavity of the white.[15]

---

13. Cf. Avic. *DA* 17.52r.

14. Note that A. uses *citrinum* for "the yellow" and *vitellum* for "the yolk." Just below he will use the compound *vitellum citrinum*. He also has two terms for "the white," viz., *album* and the more modern sounding *albumen*, the former being the more frequently used.

15. Perhaps "the chick is generated in a continuous circle between the convexity of the yolk and the concavity of the white."

The reason has thus been determined that some eggs have two colors and others one. The principle which is from the sperm of the male is separated out in eggs and stands on the side where the eggs absorb a humor from the womb. It is positioned in the egg extending out from the white (where the entire sperm of the male is located) up to the yellow yolk from which the nourishment is absorbed when the members are being formed and completed.

CHAPTER III

On the Shape and the Reason for the Shape of Eggs, Both Nonrounded and Round Ones. And on the Creation of the Eggs in the Womb and Their Leaving the Womb. Also on the Departure of the Chicks from the Eggs.

16 Eggs which have two colors are not perfectly round but rather are pointed at one end, having an acute spherical angle, as if they were composed of two hemispheres: at one end extended to an acute angle, and at the other end made of spheres not extended at the place where the pole of the egg is.[16] This differentiation is due to the differentiation of the principles which is in them. Because of the white, in which lies the principle of the generation of the animal, it is necessary that it be pointed. And it is necessary that this angle be harder than the lower angle so that the formative principle might be better guarded within the egg. For here is the sperm of the male and here are formed the upper members of the chick. Also, during its exit, the acute angle of the egg exits last since it is stretched out toward the inner parts of the womb, toward the diaphragm, where the egg is connected to the womb during its generation. For this is just as in trees where the point of the seeds [*semina*] is turned toward the cotyledon to which the seed is attached as if to a womb. Still, certain seeds [*grana*] are connected to the branches of plants, others to the barks of cuttings or of little branches.[17] Other seeds are made up of two halves as are beans, the acorn, and the like. Where both halves meet at the cotyledon, then there is the principle of nourishment which completes the seed [*semen*] and the principle of generation of the seed. But we have fully treated all such matters in our book *On Plants* [*De plantis*].[18]

17 Perhaps someone will ask a question that is difficult to resolve, namely, from what does the egg grow and increase in the womb as if from its instrumental principle. For

---

16. On the position of the poles in a sphere, cf. *Phys.* 6.3.3.

17. "Cuttings": *quisquilliae,* a rare word that normally indicates trash. Its use here is to be compared to *De veg.* 3.18, where A. mentions seeds which mature in straw or in cuttings of straw. Cf. also DuC s.v., where a single gloss is cited to define *Quisquilias frumentorum purgamenta vel paleas creentatas* (*creentata,* itself rare, seems to refer to chaff). In sum, then, we seem to have a reference to a process for germinating seeds in a matrix of loose vegetable matter. The choppy style of the sentence is due to A.'s interpolations.

18. *De veg.* 4.85f.

it cannot be said that it grows on its own like a larva, for animals which have been conceived in a womb undergo growth and food from the *umbilicus*. If the egg were said likewise to undertake growth from some similar thing, then it would be asked what becomes of the *umbilicus* of the egg when the egg is completed. For the egg exits at that time and we see that it is entirely contained within a shell and nothing of an *umbilicus* of the egg is to be seen. This question reasonably arouses doubt, for the drawing in of nourishment by which the egg grows is hidden from view. Nothing of the egg save the shell is visible and this is at first soft and becomes hard afterward. In the beginning it is not very hard in the bird because otherwise it would harm the womb of the bird, but after the egg has come forth, the shell hardens further in the coldness and dryness of the air.

In response to the question, then, we must say that there is a particular part of the web which envelops the egg and which is at first soft inside the bird. It is bent toward the womb after the fashion of the *umbilicus* in animals and this "egg umbilicus" is at the acute end of the egg. From there it accomplishes its absorption and when the egg is completed, it remains within the shell and dries out, extending out to become the web which envelops the egg. The fact that eggs do absorb blood from the womb of birds is indicated in miscarried eggs. For either due to cold or to some other flaw, birds occasionally miscarry. At such a time the miscarried eggs are found to have a bloody color and at the pointed end small passages are found in them like the passages of an *umbilicus*. As the eggs grow larger and larger, the passages become more and more stretched out and the eggs grow more and more yellow since this was from the bloody color. This change of the *umbilicus* and of the colors occurs especially near the time of the completion of the egg. When they are completed, the closing up takes place in the same spot, whereupon the passages between the yellow and the white are closed up. When the creation of the egg is completed, the entire egg turns white outside and is totally closed off and divided from the womb. Then, reasonably, those passages which resemble the *umbilicus* in other animals are not to be seen.

Moreover, emerging from the eggs is opposite to the departure of an animal which is conceived and formed in a uterus. For the natural departure from the vulva of a conceived animal is onto its head and onto the beginning of its body. But that of an egg is onto the feet and onto the rear part of the egg (that is, that part in which the feet and rearmost members are formed). We have spoken of the reason for this above, namely that it is connected to the womb at its front end and turning it around would be difficult and useless.

Moreover, the creation of birds from eggs is not finished in the uterus but rather requires that the bird sit on and incubate the eggs. With this heat the eggs are formed into birds. Then, because there are two parts in the egg, the animal emerges formed from one part and fed from the other, as we have said. Since the bird is not completed within the uterus, it is necessary that food be prepared for it outside the womb along with the very substance out of which the animal is created. Otherwise the generating animal provides nourishment to the embryo in the womb. But when the fetus is newly born it has food that is separate from itself, namely the milk in the breasts. But as we have said, birds carry with themselves their own food which they take on from the mother's body, shut up in the egg with their substance. This method of generation of the chick is the opposite of the opinion of the philosopher whom some call Alkumon, from the city Betonya.[19] He was wont to say that the white was in the egg like milk, whereas the yellow gave substance to the chick's members. But according to the things we have said, the white is not like milk. Rather, the yellow has an analogy to the milk in that it is food prepared for the chick.

The chick is thus generated from the hen incubating the egg and this should be done in a suitable place and in suitable weather conditions. For when these two are present then the chicks will emerge, and if the heat and the place are suitable then the eggs will hatch without the hen's incubation. For this is how all the eggs of egg-laying quadrupeds hatch. They lay all of them on the earth and the eggs hatch from the heat from the rays of the sun on the earth. If any of the animals of this sort is found sitting on her eggs, she is doing so to guard them and not for the purpose of incubation.

20 Further, the eggs of quadruped animals have a hard shell as do those of birds and they have two colors inside. They are also generated near the diaphragm as are birds' eggs. Virtually all of the accidental traits of birds' eggs, both inside and out, occur as well in these and thus there is one opinion and understanding for all the things which have been said about both kinds of eggs. But the eggs of quadruped animals are warm and they therefore are completed by the heat of the weather on account of the strength of the power which is in these eggs. The eggs of birds on the other hand need warming because of their weakness.

Moreover, a most intelligent trait is found in these oviparous animals for there is a great deal of concern about their young. After completion of the young a friendship remains between parents and children as it does among humans. This is also found in certain quadrupeds. Birds, however, show this care all the way up till the time their chicks are full grown. Thus, if the female should on occasion be away and not incubate the eggs, the eggs move toward a bad state as if they have been deprived of some natural principle by the female's absence, a principle which has to be generated with them or which must be present at their generation.

The chicks are quickly completed within the eggs in warm weather since warm weather aids in the digestion of the creature to be formed. And the females have, by a sort of natural cunning, the ability to pick in advance a period that is advantageous for the incubation of her eggs.

21 In some hot weather the eggs of certain birds are ruined and this occurs especially when the food-humor [*humor cibalis*] which is in the egg is disrupted.[20] It is for this reason that many eggs are ruined. For in this weather the yellow is disrupted since it

---

19. Ar. *GA* 752b25 cites Alcmaeon of Croton, whose work on natural science was known to Ar. and Theophrastus.

20. "Disrupted": *turbatur*, implying also a muddying or darkening of the fluid, as in the text just below in reference to wine.

is from the earthy part which is disrupted and smokes during this weather. This is also why wine is disrupted in this weather for the dregs move around and become mixed with the wine. Wine and eggs are corrupted only because of the dregs and the yellow which is in them. This reasonably occurs most often in those which lay many eggs because, as was stated above, these are moister. They corrupt more readily because they easily have a sufficiency of heat.[21] Thus if the heat is increased and if it be southerly and disruptive when the sun is ascending (for its power is greatest then), then it will doubtless disrupt the eggs. And because the yellow in eggs has a natural affinity with the dregs of wine and the white has one with the wine, then it happens that when eggs are immersed in wine along with sand and chalk this clarifies the wine. For the sand and the chalk perforate the substance of the wine and the yellow of the egg attracts the dregs. This is why when water is decocted into salt, the salt, which has an earthy nature, gels up only by using eggs or blood.[22] For blood and the yellow of eggs have the same nature, as is clear from the things said before.

Further, birds of prey with hooked talons lay few eggs. This occurs in them for the same reason as the others. Some of them, such as the golden eagle [*herodius*], lay only one egg and if they do lay two, one is most often found to be corrupted. Fowlers of our land, which is upper Germany, have experienced this for themselves. Over a continuous, uninterrupted period of eight years, they never found more than a single chick in the nest of a *herodius*. Now when I say *herodius* I mean the large, black eagle which is called *herodius* because it is the "hero" of birds.[23] This type of bird has a very hot nature and it therefore incubates its eggs, warming the egg until it boils, as it were.

The nature of the white has almost a sort of opposition to that of the yellow and for this reason the yellow congeals in cold weather. Afterward, when it is warmed again, it is released from the cold and turns fluid. We have said above that the yellow is the food for the animal by which the animal (which is being formed to the front at the rounded part of the egg) is fed and sustained. Because the yellow has this sort of a nature, it therefore does not grow hard when cooked (unless it is burned up) but rather is softened as is wax, though it does not liquefy.

This is therefore the reason it softens when it is heated and is easily corrupted if it lies for a while in the superfluity of fluids of weather or place.[24] The white of an egg is not easily congealed by the cold but rather is made more moist. When it is cooked it

---

21. *Calorum sufficientem* is literally a "sufficiency of heats" and one suspects that *calorem* should be read as in Borgnet's text, yielding "sufficient heat."

22. Cf. *De min.* 5.2 for various kinds of salt and some observations on how they are made. Blood and eggs are not mentioned. The verb translated "to gel up" is *constat*, which most regularly refers to the solidification of a liquid.

23. *Herodius* is a transliteration from the Gr. where it most often means a heron, but sometimes refers to a sea bird (*GB*, 102–4, with a discussion on etymology). Neither fits the current context, however. It is generally accepted that for A. it should be seen as the golden eagle. Cf. A.'s lengthy description at 23.7(1).

24. The text received by A. simply states that they are corrupted by an excess of superfluities. A. has muddied the issue by his addition.

becomes harder. During the generation of the chick it is thickened into the substance of the members since the entire generation of the chick's radical members comes from it. The yellow, however, is the food for the chick and for this reason the white and the yellow are separated in webs, due to the differences in their natures.

One who seeks to understand this matter with careful and perfect reason, and to understand how one humor begins from another and how they are related to each other, and how they are related to the making of the animal's substance, and what the disposition of the webs and *umbilicus* is, should read those things we said here along with those things which were said above about the anatomy of birds and eggs. For we have established these things in the earlier books of this investigation where we wrote about the manner of the generation of animals.[25]

CHAPTER IV

On the Order of Generation of Animals' Members in Eggs,

Be They the Eggs of Birds or of Other Animals

It is sufficient here with respect to the generation of a bird in an egg to know that the first member which is generated in the egg is the heart, from which is produced the great vein [vena cava]. From this comes the branching off of the other veins into two *umbilici,* one of which moves to the web containing the animal and which is under the shell's web, while the other proceeds inward to the yellow. The chick takes in food from the yellow through one of these *umbilici.* During the incubation period the yellow is very prominent since it grows moist due to the heat, whereupon it spreads out like softened wax. This has to be a fluid then since it is the food for the tender members which could be nourished only by a fluid. In the same way the members of other animals are fed only by a fluid in the first stages of generation. For the principle of that which is vivified in eggs and trees is in some way similar. For both take in food and growth from material which is connected to them.

One *umbilicus,* as we said, is extended and connected to the web which contains the chick and this resembles the *secundina.* But the disposition of the yellow is like that of the womb by which an embryo in the womb is fed. For after the eggs have been laid, they cannot take in food from the belly of the bird and nature thus must give them a part that is analogous to food. The outer web on the chick, and which contains it, is bloody since its disposition is like that of the womb. That which is around the yellow, however, is like the *secundina.* The shell of the egg resembles the web which surrounds the womb and which is a sort of covering encasing both embryo and womb. For the embryo must be in the womb to be guarded. In animals which

---

25. Cf. 6.1f.

generate and conceive live young, the womb is positioned in the anterior portion of the belly. It is the other way around in oviparous animals, toward the inside. In the case of eggs, it is as if someone were to say that the womb is the mother and that from the yellow comes the food for the embryo which is from the mother. It is for this reason that the food of such ones cannot stay in the mother. When the chick has grown, the *umbilicus* shrinks since the animal must soon leave the egg. The remnant of the *umbilicus* which comes to the yellow leaves last of all. For a newborn animal must have food right away and, since it cannot have it from its mother as other animals have their milk, and since it cannot straightway seek food out, it must retain something of its food in the yellow near the *umbilicus*. For this reason, with nature's help, there will still be some yellow within the animal along with the *umbilicus*. Animals are generated in this way which are generated from complete eggs, laid by birds or by quadruped animals. In large animals these things are noticeable and apparent to sight. In small animals, however, they are invisible because of the smallness of their bodies.

The genuses of fish produce many incomplete eggs for the reason we gave above. The *celeti,* throughout all its species, lays a complete egg inside its uterus and then bears a live young outside the uterus. The exception is the one they call the frog [*rana*] and which belongs to this genus.[26] This species lays only one complete egg outside its uterus. The reason for this is the nature and disposition of its body. For the head of this animal is many times larger than the rest of its body and its head is spiny. For this reason it cannot hold young internally and neither could they exit if they were inside. The eggs of all animals which resemble the *celeti* in nature are soft and cannot be dried out or hardened externally since they are colder than bird eggs and thus remain soft in this way. The eggs of the frogs are hard and strong because of the safety they need when outside. The generation of the young which are produced from the eggs of the frog and from eggs which are completed externally, as well as those which are completed internally, is generally the same.

But there is a difference between the eggs of these types of animals and the eggs of birds in that the eggs of these animals have no *umbilicus* which extends to and comes to the *secundina* which is below the web containing the egg. The reason for this is that the eggs of these animals do not have a hard shell since they have no need for one. This is because a hard shell is only for protection of the eggs and is thus present only in those eggs which come forth from a uterus. These eggs have a hard shell because of the many types of accidents which can befall them. The eggs of these animals differ because the formation of the young is, in the egg layers, inward from the lower extremity of the egg, where it has a spherical pole. But in the eggs of birds the generation of the young is at the sharp end of the egg at the part where the egg is connected to the womb. The reason for this divergence of generation is that the eggs of birds are separated from the womb before the formation of the young, whereas the eggs of the animals we have mentioned are not separated from the womb at all. For while these

---

26. A fish, not the amphibian. Cf. 17.2.

animals do lay complete eggs, they are nevertheless connected to the womb and the *umbilicus* of the egg remains attached to the womb all the way to the end as it does in completed eggs. For incompleted eggs are in a membrane as we said above and none of them can have an *umbilicus*. In this regard these eggs resemble that which is emitted from the womb in the form of an animal. For these also cling to the womb through an *umbilicus* all the way to the end of a pregnancy, this being the time of giving birth. In certain of the types of *celeti*, as in the sea frog species which we mentioned above, an egg is emitted when it has been completed internally.

28 Someone may ask what the difference is between the generation of birds and fish with respect to the formation of the eggs or live young from eggs and the type of generation which we said is in the *celeti*. Because each of these animals is generated from an egg, one must investigate the difference in their generation. As to this, however, we say that the eggs themselves are different according to the humors contained within them. For the white and the yellow are separate in bird eggs but are not separate in fish eggs, being rather mixed together. For this reason there is nothing preventing the first creation of animals from occurring in a way opposite to that of these eggs. For the egg of the *celeti* is not fixed but rather is connected to a spot on the womb which is inside and is opposite the lower spot where the generation of a live young would occur. These eggs lightly draw in their food from the womb as is clear in eggs which have not separated from the womb. For the eggs of certain types of *celety* are not found separated from the womb but rather are connected to that part of it which is to the rear of the womb. They move to the lower part so that the live young can be generated from them there. And it is a trait of the eggs that when the generation of the live young which is in them is completed, they suddenly are set free from the womb. For the moisture which was in the egg and had been drawn in from the womb has now been used up and consumed, as it were. At this point the egg and animal are set free at the same time.

29 It is clear from what we have said above that the passages in the egg, through which it draws in nourishment, are like the passages in an embryo in a womb. This especially occurs in the species of animal which the Greeks call the *hahabe* and in that which is called the *humloch*.[27]

The generation of fish and that of birds, then, differ for the reasons we have stated. Another trait which we have mentioned, however, occurs in almost the same way in egg layers.[28] For the *umbilicus* which comes to the yellow is in the eggs of fish just as it is in the eggs of birds. This is how it is in all fish eggs, but in their eggs one part is not white and the other yellow. Rather, the entire humor has the same color and it is from this that the animal is fed. When the humor begins to be consumed into spirit, then flesh arises around it on all sides.

---

27. The parallel passage, Ar. *GA* 754b33, talks of smooth dogfish, but the origin of A.'s names is not clear.

28. St. (*Alia quae . . . accidit*) is ungrammatical as it would require a ntr. pl. noun to take a sing. verb. It may just be an overly literal translation of Ar. *GA* 755a1 (for Gr. has this grammatical quirk), but it seems best to follow Borgnet and read *quam*.

This is how generation is accomplished out of the eggs of those which lay completed eggs internally and generate live young externally. Some fish, however, lay incomplete eggs externally, with the sole exception of the sea frog, as we have already said. But the animal which has fourteen feet has one solitary duplication which is toward the inside and has another, small duplication at its end, which is toward the outside. We have already spoken of the reason for this above where we discussed the members of sea animals.[29]

The generation of young from these fish eggs will follow one common method, just as there is a common method for the generation that occurs from the eggs of those which lay eggs internally and bear a live young externally. All these eggs, however, grow quickly after they are set free. Their growth and increase bear a similarity to the increase of larvae. For an animal which generates larvae generates them quite small, whereupon, once born, they grow on their own. The reason for this increase resembles the increase of yeast. For a small bit of yeast grows large because the dry element which is in it is infused with moisture and this moisture then fills it up. Now that which most capably creates and forms larvae is the heat of the soul which lies in the moisture. The yeast increases because of the heat which is in the moisture of the yeast. Thus the reason that eggs grow is their moisture is like the moisture of yeast. The young of the fish cannot be completed within a womb because of the great number of the eggs and the narrowness of their wombs. The eggs are small, therefore, because they are so many, but they do possess the trait of growing or corrupting quickly. Although many of them are corrupted, this poses no hindrance because it is necessary that out of so many eggs many should reach completion. Thus a wise nature made many eggs in fish so that she might recover by sheer number what is corrupted in them because they have no external shells. Still, certain genuses of fish grow before they lay eggs, as does the genus of fish which they call the *aculeus* [pipefish].[30] This one naturally increases its eggs in size but in the same degree it decreases them in number. Thus, this genus has but few eggs. The reason that eggs grow quickly outside has thus been stated.

30

---

29. Might this creature be the same as the "forty-four" (a centipede-like creature) discussed above at 4.7? At 2.83, A. tells us there is a marine version of the animal as well. The odd *duplicatio* mentioned here by A. might refer to the double body rings at 4.7.

30. The identification of the fish is based on allied forms in 8.78–79, where, as here, they stand for the Gr. *belonē*, a pipefish or, with less likelihood, a gar. Cf. Ar. *HA* 567b22f. and *GA* 755a30f., where the fish was actually supposed to burst due to the burden of its eggs, referring to the fact that the eggs are carried in a pouch.

CHAPTER V

On Disproving the Error of Those Who Claim That There Is No Male or Female in Fish

31 The assertion that fish produce eggs and that there are both male and female in them is based on an irrefutable indication, namely that fish which generate live young externally, as do those in the *celeti* genus, first conceive eggs internally. They would not do this unless some material of generation were in the egg. It is therefore clear that in each genus of fish there are eggs made for the sake of generation. There exist differences in this in that the eggs of some fish are not completed internally but rather do so after they have come forth. These eggs belong to those genuses of fish in which there is male and female. When the eggs leave the females the males follow along and cast their sperm over them to enable them to grow.

Some have thought, however, that all the species of fish, except of the types of *celeti,* are female. But what they say is false. These people have also said that there is no real difference between the fish we call males and those we call females, much as is the case for the differences between fruit-bearing and sterile trees which have the same form and shape. One example is the wild oleaster and the fruit-bearing olive and another is the garden, fruit-bearing fig and the wild, "empty" fig.³¹ They say that all fish differ from one another in this way except for the types of *celeti,* concerning which they state without doubt that they truly possess females and males.

32 Based on all the things said above, we however know that this belief is false. For by means of the things which have been said we know without a doubt that many of the other fish are divided into male and female. For the passages of those fish in which there is sperm resemble the passages of the males of the *celeti* and others. During mating season the sperm appears in the males of oviparous fish and when these passages become enlarged, the "sperm of the wombs" exits from them.³² When the females are filled with eggs, their wombs begin to be visible. For wombs do not exist only in the genuses of fish which produce few eggs, but are also present in other genuses which produce many eggs. The wombs of oviparous animals differ in much the same way that the wombs of women differ and of the other animals which bear live young like themselves and which have a hairy tail.³³ Some fish do not lay eggs, as we have said

---

31. At *De veg.* 6.167, A. discusses the true infertility of the *oleaster,* which Jessen identifies ad loc. as *Oliva europaea.* At 6.155, he talks about conditions under which olive trees do not produce fruit. The compound term used here, *ficus fatua agrestis,* is echoed in *De veg.* where *ficus fatua* and *ficus silvestris* are both used. Jessen identifies it as the "goat fig," *Ficus carica* L., var. *caprifica.*

32. *Sperma matricum,* "sperm of the wombs," is a new term for A. In Ar. *GA* 755b15f., it states that semen can be seen oozing from the males during the mating season and then it adds that the females have uteruses. It is almost as if there is a lacuna in the text A. received.

33. Modern editors admit that the relevant passage at Ar. *GA* 755b18f. is corrupt.

above. Rather, without female and male, they generate on their own. In all the genuses of fish in which the male and female sex are differentiated, semen is found in the male and in the female there is a womb filled with eggs. This is clear from observation [*per experta*] in all the genuses of fish save for two, namely in the *corioz* genus and in the *hashy* genus.[34] In the *enkeloz* [eel], however, neither male nor female is visible according to the above-mentioned indications of male and female.[35] In all cases the males have vessels in which there is sperm and the females have wombs.

For this reason too, some think that it is an easy matter to solve an issue which nonetheless contains a weighty question on the differentiation of male and female in various animals. For these folk acquit themselves easily through negation, saying that male and female exist in animals only in a single way, in that this is present in animals which produce live young through univocal generation, in which a penis is visible in the male and a vulva in the female, and which employ copulation in which the male emits sperm into the female. But these people take something false as their supposition, believing that no animal which has copulation of male and female can complete as great a number of young as the number which comes from fishes' eggs. And they take as a sign of proving this the fact that all other animals, be they those which generate live young or which lay eggs, never arrive at this great a number of young or eggs. In this matter alone are they speaking the truth. For none of the other ones which lay complete eggs or which produce live young in any way approaches such a number of young or eggs as does one which produces small, incomplete eggs like in the fish genuses. These people have not pondered that fish eggs are entirely different from all the acts of generation of other animals and from the types of birds and of others.[36] For all birds and oviparous quadrupeds as well as those of the *celeti* genus lay complete eggs which undergo no growth after they emerge into the air outside the body of the one laying the eggs. But fish lay incomplete eggs which, once deposited outside the body, grow quickly. Soft-shelled animals and the *malakye* also have this method of egg laying and all these types copulate together openly. This may not be known to everyone, for they copulate only after a long period of time, but the male and female exist in these animals with a clear sexual differentiation.

Now it is not reasonable to seek masculine and feminine in all genuses of animals using one and the same differentiation and power through which there is participation in the sexes in those which generate live young like themselves by means of a univocal generation. For male and female exist in all the types of aquatic animals which we have mentioned. The reason for the statements of those who contradict the things which have been said is simply a lack of knowledge of the types of copulation which these animals have and which differ from other animals. For the types of copulation are quite varied and numerous among various animals. These people have, however, considered but few of many and hold the opinion that the method of copulation in all

---

34. Ar. *GA* 755b22 lists the *erythrinos* (cf. 17.9) and the *channē* perhaps a sea perch.
35. A. has added the information on the eel, using a name form he has used frequently before.
36. A.'s additions to the text are cryptic and repetitive at this point.

animals is similar. This is also the reason for the error of those who hold the opinion that the female fish swallow the sperm which the males eject. For the sperm is found in the males at the same time during which eggs are found in the females. The closer the female comes to the time for laying, then the more the eggs are multiplied in her womb and the sperm in the males. Therefore this increase of the eggs in the female and of the sperm in the male occur at the same time. But if the eggs were growing in the female as a result of the sperm, then the sperm in the male and the eggs in the female would perforce be present at different times. The female, however, does not scatter her eggs successively. What was said therefore occurs reasonably, for eggs can be best sprinkled by the sperm of the male successively.

35   Further, just as it happens in bird eggs that some are wind eggs and that these are more infrequent than eggs completed through intercourse, so too does this very often happen in fish eggs. For fish eggs are no more suited for generation on their own than are wind eggs. But those eggs over which the male casts his sperm come to completion. In this way too wind eggs in birds cannot be completed externally because when they emerge they are completed, enclosed within a shell through which sperm cannot enter. Since, however, fish eggs are incomplete when they are laid, they can be completed externally by the sperm of the male just as they also take on growth on the outside. At this point they receive the formative power and become suitable for generation. Now the sperm of the male fish is cast out and is consumed in this method of scattering the sperm over the eggs. Further the sperm is missing from the bodies of the males at the same time as there is a loss of eggs from the females' bodies. It is therefore clear that male and female are present in all the fish genuses with the exception of those we have previously mentioned. In these any differentiation of sex is totally unknown because the principle of generation is that they are generated on their own, as we have said.

36   Moreover, one of the things which nurtures them in the aforementioned error is the speed with which fish copulate. Now the copulation of many genuses of fish escapes the notice of many fishermen, for these men are not viewing fish in order to acquire knowledge of the nature of fish. Therefore many things are seen in fish about which the fishermen know nothing out of a lack of interest. The dolphin genus copulates by pressing one to the other just as other fish copulate which some cause or other keeps from perfect and well-attached copulation. But dolphins separate more slowly while oviparous fish separate very quickly. The copulation of many fish cannot be observed for this reason and many uneducated fishermen thus agree with those unlettered people who say that the females swallow the sperm or that the males swallow the eggs. Therefore this error has become so widespread that the poet Hermodicytes has written in his proverb-like writings that fish are impregnated by swallowing sperm.[37] Nor have they given much thought to the impossibility of what they are saying, for they could easily find out, if they paid careful attention, that whatever comes through the mouth travels to the stomach. For in those which have no lungs, there is no passage

---

37. Herodotus 2.93, cited by Ar. *GA* 756b6, where he is called a *mythologos*, "storyteller."

from the mouth save to the stomach, and it does not pass on into the womb. Thus, everything which passes to the stomach necessarily turns into food and is digested. Yet the wombs appear full of eggs. Neither can they ascribe a means by which the eggs might pass from the mouth and stomach to the womb.

There is a similarly erroneous statement made about the generation of certain birds. Many people commonly say that the ravens and the birds called the *anyz* (which many say are partridges) copulate with their mouths together.[38] They also say that the one called the *hus* takes on its fetation by being impregnated through the mouth. This was a story passed on by Anaxagoras.[39] But this is only said because they have taken too superficial and light consideration of these matters, and because they have not based their observation on experience, for the copulation of these animals is not observed by one who gives it only superficial consideration. Avicenna, however, considered the matter diligently and says that he saw two ravens copulate just as other birds do and that they did so frequently as he watched.[40] Now ravens copulate rarely while people are watching and thus when they see the ravens kiss, they think they are copulating. But I say that perhaps animals of this sort have a time for copulating in the morning, before daybreak, or in the twilight when they are not seen by people. Therefore, their copulation is not observed. For every genus of raven kisses often, as is clear from the tame ravens brought up in a home. Similarly, every genus of pigeon kisses one another frequently, but because the copulation of pigeons is open to view, people do not hold the belief about them which they hold for the others. Moreover, the raven genus generally copulates and generates but little. Yet, as we said before, there are many who have seen ravens copulate openly. As we have stated, the great cause of their error is that they have a very lax sense of care when they investigate natural matters. For how could it be possible for the sperm to come to the womb through the stomach, which digests everything which passes through it?

Further, those types of bird which they assert do not copulate have wombs, and eggs are to be seen beneath the diaphragm of the females in the first stages of generation as they are in other birds. A quite similar error exists about certain animals. There are those who say of the hare (which some call the *hyzum*) that it is sometimes male and at other times female and that it sometimes conceives and at other times impregnates.[41] And yet the females of this animal have wombs as do other quadruped animals and its young emerge from this womb during birth. But because the female of this animal gives birth to very small and incomplete young (as do almost all the females

---

38. *Anyz* is the ibis at Ar. *GA* 756b15. The partridge was notoriously easy to impregnate and had a lascivious nature (Rowland, 1978, 123–27).

39. Although *hus* looks like the Gr. for "pig" it represents the *galēn* (weasel, acc. s.) of Ar. *GA* 736b16. The weasel was thought to conceive through the ear and give birth through the mouth.

40. Avic. *DA* 17.52r.

41. Despite the similarity of *hyzum* to the MHG *hase*, "hare" (Lexer 1.1192), it would appear that *hyzum* reflects an Arabic pl. of *hyz* and is thus really still speaking of the weasel, as is Ar. at this point. The information concerning the hare is from A. On the hare's hermaphroditism, cf. Neckam, 134, and Rowland (1973), 91.

of animals which divide their foot into many parts), the opinion has arisen that she carries her young from place to place many times and that it is sometimes male and sometimes female and that the young are completed outside of her. This error was aided in being commonly stated as well about the bird which is called the *acrasylus* in Greek and also about the one called *astryboth*. And yet it is also false when they assert changeability of sex in their cases.[42]

Further, the philosopher Arybaros, born in Arcadia, says that the lobster [*karabo*] has two members, both male and female, and says that the *karabo* copulates with the *acricyloz* as well as with the one he calls the *azaro*.[43] He also states that each year they change the copulation so that the incubus becomes the succubus and the one impregnating becomes the one impregnated. They say that it is visually confirmed that the *azaro* sometimes has male members and sometimes female members. But what they believe about the hare, which the Arabs call the *adhab*—namely, that it has each set of members on alternate months—is not true.[44] Rather, it has beneath its tail certain lines which resemble the clefts of vulvas and these lines are always found in both males and females. Avicenna even says that these clefts increase in number according to the number of the years of its life and that thus one was captured which the hunters believed had eight vulvas.[45] But this is nonetheless false, for those lines are found both in males and females alike, but since males are caught more often, it is more often thought that they are hermaphrodites than any others. So these things which they say are false and the things which we have mentioned are the sources of the error.

---

42. Ar. *GA* 757a3 discusses the *trochos* and the *hyainē* (hyena). The identity of the *trochos* is unknown, but someone before A. has identified it as a bird, undoubtedly by the closeness of its name to that of the *trochilos,* a Gr. name for a wren which comes to A. as *trochilus* (cf. 8.22, 25, 70).

43. Ar. *GA* 757a4 cites one Herodorus of Heraclea, but is still discussing the *trochos* and hyena. The origin of the names in A. is obscure. Cf. Barth., 18.59 for the hermaphroditism of the hyena, reflecting the bestiary (White, 1954, 31) and *Physiologus* (Curley, 1979, 53) in its language. Cf. McCulloch (1960), 131, and McMillen (1996), who explains the physiology leading to the belief in hermaphroditism.

44. Schühlein, 1655, suggests that *adhab* hides the Arabic word for hyena, yet, according to Stadler, the name has been added by A. and does not appear in the parallel passage of Avic. *DA* 17.52r.

45. Avic. *DA* 17.52r.

CHAPTER VI

On the Completion of All Eggs, Specifically Wind Eggs and of Others, by Means of Joining the Sperm to Them

Perhaps someone might ask why female fish do not lay complete eggs and males do not emit their sperm into their females. On this we should say that the genus of birds does not have much sperm at all in proportion to the size of its body and that the wombs of the female birds are beneath the diaphragm.[46] Male birds differ from other males and their females from other females, even though for the most part the colors of the females are like those of the males. Now, to be sure, eggs are found in the females and sperm in the males and this sperm is more than is necessary for copulation. Perhaps nature does this so that it can be cast upon the wind eggs conceived in the wombs of the females.

As we have often said, bird eggs are completed inside the females, whereas fish eggs are completed outside. Thus something happens to fish eggs which is like that which happens to those which generate larvae. For those which are generated out of larvae are incomplete and afterward grow and are completed.

But the completion of bird and fish eggs, as far as the generative power is concerned, comes about only from the sperm of the male, so that the eggs are completed by the sperm, those of birds internally and those of fish externally since fish eggs are laid incomplete with respect to size. What happens to these two types of eggs due to the sperm is entirely similar. This is why wind eggs, conceived in the womb of the bird before completion, are sometimes rendered suitable for generation by the frequent copulation of the bird with the female while these eggs are still in the womb. Avicenna says that unless the male copulates a great deal and frequently, the eggs revert and lose the generative power and will then be wind eggs.[47] If, however, in addition to the copulation of one male who copulated over the eggs a number of times from the beginning, another male should copulate with the female before the eggs are completed and are enclosed in a shell, then the eggs will change from the nature of the one who copulated first to the nature of the male who copulated second. When eggs have been conceived in the womb and the male does not copulate with the female, then the eggs do not grow quickly to the requisite size. This cannot happen all the time but it can happen when the yellow, which is within the white, has already been conceived or else is undergoing the change during which it generates the white from itself.

40

41

---

46. As Stadler points out, the present section is "*plane confusa.*" One reason is that A., despite posing a question about fish, goes on to insert the word "birds" into his received text whenever he can. For original intent, cf. Ar. *GA* 757a14f.

47. Avic. *DA* 17.51v.

42   Nothing of this sort seems established in fish eggs. Rather, the female fish lay the eggs and the males follow, scattering their sperm over those eggs so that they may quickly be saved and completed by the sperm. The reason for this is that there are not two separate colors in fish eggs but rather one, incomplete humor. They thus do not have an established time such as that for the birds' laying of eggs. For when there is white and yellow in the egg, then there can also be contained in the egg the moving principle in generation, which comes from the male. For this reason wind eggs are made suitable for generation by the male's sperm, provided that it falls on them after the first conception from the male's sperm and before the final completion and enclosure within a hard shell. For wind eggs first grow in the womb from the egg fluid [*humor ovalis*] and it is impossible for them to come to completion in such a way that an animal comes from them without the sperm of the male. For, as was proven in the preceding book, an animal needs the principle of sensation and this can come only from the male's sperm. But the power of the nutritive soul is in the females as it is present in all who possess life, because, according to Dionysius the philosopher, it is present in all things due to the soul's heat, this being the first principle of life.[48] Thus these eggs are completed in the womb with respect to nourishment and growth much as fructification, generation, growth, and the nourishment of trees are completed through the vegetative soul. They are not completed with respect to the principle of sensation as is the conception of animals. If there were no males in these birds, then it would be possible for something to be produced from these eggs by means of that type of production by which generation is accomplished by certain fish eggs if, that is, there exists in the fish a certain genus which can generate something without a male. For we have said above that there are such genuses of fish. But we said above as much as is sufficient for our present purposes when we declared that nothing has been experienced with certainty about this matter until now. But it is known for certain that there are females and males in all genuses of birds.

43   Because wind eggs are like trees, they are therefore laid completed and for this reason, after they have been perfected, they are not changed or altered by copulation. For they are truly not like an animal and they are thus not completed through formation and neither does a live young come out of them. Thus they are not simply like trees nor yet are they entirely like animals, which are both generated by and generate through the means of copulation. Now a chick does not come out of a bird's egg unless it has been completed through copulation. That is to say, this is an egg whose interior is white joined to yellow, and this happens when the male who copulates places his sperm near the two humors of the egg as we have stated above.

44   The generation of the *malakye* from its eggs is like this, as is that of those like it, namely the cuttlefish, the octopus, the lobster, and those like it which have soft shells. For all these are generated only from eggs which have been completed by copulation with a male. The males of these genuses are often seen copulating with the females

---

48. For a discussion of the influence of pseudo-Dionysius the Areopagite on A.'s psychology, see Park (1980), 528f.

and for this reason, those who say that all fish are females and are not generated as a result of copulation are in error and are saying the opposite of what is quite clear to the senses. And when they say that certain types are the result of copulation, like the *celeti* types, and that others are not the result of copulation, it is a wondrous thing and signifies nothing more than the ignorance of the one making the statement. The copulation of all the types of *malakye* and soft-shelled ones is slower than the copulation of other fish. The copulation of ringed creatures is slow in the same way. This is reasonable since they lack blood and as a result their natures are cold. For this reason some of these animals lay but two eggs in the summer. The cuttlefish type does this as does the one called the *ramyz* (that is, the squid [*calamare*]), since its womb is divided in two.[49] In the octopus, however, no more than one egg appears, just as only one appears in the perch [*pertica*].[50] The reason for this is the narrow, long, and rounded shape of its body and because of this shape the divisions of the womb in animals of this sort escape notice when the womb is filled with eggs.

The womb of the lobster and of all those like it is divided in two. All these animals also lay incomplete eggs and for the reason we have stated, all types of female lobsters lay their eggs internally. For this reason too those involuted members on which the eggs are entwined are more often found on the females than the males.[51] The males only guard eggs that have already been laid. The *malakye*, however, lay externally and the male spreads his sperm over the eggs as do male fish. This sperm, when emitted, is viscous and connected, clinging strongly to all that is touched by it. But nothing of this sort is seen in the types of lobster. For this one's eggs are within the females, with a very tough membrane [*pellis*], and here they grow. The *malakye*, moreover, lay externally as do other types of fish.

Further, the young of the cuttlefish is contained in the front portion of its egg. It could have no other disposition since the front part is suited to the same place of the egg.

Let what has been said by us concerning the generation of animals, flyers, and swimmers, and even of certain of those who get about by creeping, be all there is. For what has been said is sufficient for our present purposes.

---

49. "Squid": *ramyz* is corrupted from *teuthis* (calamary) at Ar. *GA* 758a7. A marginal note in the MS identified it as *calamare*.

50. The intrusion of the perch (more commonly *perca*, but *pertica* is attested) came to A. with his received text. It is not readily accounted for by corruption of some word at Ar. *GA* 758a7f.

51. On these appendages, cf. 4.24 and 5.72.

The Second Tract of the Seventeenth Book of Animals on the Generation of the Ringed Creatures

CHAPTER I

On the Generation of Larvae or of Those Which Have Their First Generation from Larvae

46 It now remains to speak of the generation of animals with ringed bodies and of those which are called hard-shelled animals. We will discuss these in the same way we dealt with those above. But of these two genuses of animal, we will speak first about those with ringed bodies.

We have already said above that certain members of this genus of animals are generated as a result of copulation, but others are generated on their own, and that some produce larvae, but others produce eggs. But perhaps all the animals of this genus will at first be somehow incomplete so that their offspring is at first incomplete much as the offspring in animals which generate live young is at first incomplete at least as to size, even though it is complete as to form. In just this way then, eggs are at first complete in certain oviparous animals and these do not take on any growth. Still others produce incomplete eggs.

47 Let us say, then, that the nature of animals is also divided this way so that some generate complete animals, whereas others generate incomplete ones like the insects.[52] Some animals therefore produce incomplete eggs which afterward become completed externally, as we have said above with regard to many types of fish. But those which generate live young which are conceived internally are, after the first conception and retention of the sperm, like eggs to some degree. For much as an egg is the material of the chick, contained beneath a web with the principle of movement which is in it, so too is the embryo of animals conceived in a womb. Thus, a miscarried fetus which has gone bad and which has been ejected in its covering is commonly called an "egg." But animals with ringed bodies, whether they generate larvae as a result of copulation or whether they are not generated as a result of copulation but rather do so on their own, are generated at first from some sort of wrapping in which it is sustained lest the material of generation flow away, having with it the principle of movement for its own

---

52. Or "others (produce) incomplete ones, like they were larvae."

generation. Therefore we have said above that the generation of all animals is from eggs at first.

It is thus reasonably said that caterpillars, which are larvae generated in cabbages and other vegetables, and those which are generated in the sand, have a principle of generation that is like the ones mentioned.[53]

It is likewise reasonable that this type of generation and many other types of generation mentioned above, contrary to the type of creation of larvae, be perfect in the ordering of the members generated from the material in which the principle of generation is first present. This is especially so due to the length and roundness of the larvae which is not as well suited to the order of creation of the members as is that in the animals mentioned above. For we should not make our consideration based on the shape of these animals' bodies nor on their hardness or softness. For some of them generate hard young and some soft. We should rather first consider them in general and not in particular on this or that point. For that which generates larvae and has the shape of a larva, when it casts this off and it will be made large, is again rendered immobile and remains in a long membrane rather like an egg. The skin encasing it hardens as can be seen in the caterpillars which cling to the walls. For their entire body is turned into something like an egg and at this time the larva is made immobile. This is also to be seen in the larvae which are generated from the putrescence of fish, in the wasp, and in many others which are generated from animals and from vegetables.

The reason for this is that nature prepares the creation of eggs before the time for laying them. And she does this in the larva which has not had final completion, for it is finally complete only when it flies. But the flying animal is only made from the larva which, after eating, lies immobile and is changed into the nature of an egg with a hardened skin.[54] It is from this egg that the flying creature emerges. This one later, in the fall, makes a nest out of webs and produces eggs. The larvae at first grow using leaves and vegetables as food. After this growth has been accomplished, they generate from themselves an egg in a shell in which at first only a soft humor is contained. From this, little by little, the animal is formed. This occurs as well in all these types of vermin and fish which do not generate as a result of copulation. Examples are the vermin which are generated from wool and the corn weevil [*gurgulliones*] and termites which are generated from seeds and wood.[55] Once all these types have taken on the nature and growth of a larva, they return, become immobilized, their skin hardens, and their humor is changed into egg material [*ovalis materia*]. Thereupon a flying animal is formed in such material, having more than two wings and a minimum of

---

53. "In the sand" is an interesting corruption whereby *harena* has emerged from Ar.'s *arachneōn*, "spiders."

54. The "flying animal" is *animal volans* and, in this context, also evokes overtones of "flying young." *Animal* is consistently used by A. for young animals, most often those of viviparous parents. Just below, A. will make the comparison very clear.

55. On the *gurgullio* cf. 1.96, with notes. Note also that A. uses *vermis* throughout this passage to mean either vermin (cf. Glossary) or larva. At times either or both senses might be at work.

four if its body is long. The hard skin then splits and then the entire animal emerges, completed in shape and size, much as a live young emerges from the womb of any animal that bears live young. The flying creatures of this sort often are larger than they were at first when they walked. But this is a reasonable occurrence, even though it may be wondrous to many people that a given animal should take on food and size after its first generation, then become unmoving and be changed to the nature of an egg. But we see that this happens not only in the vermin mentioned but also in the gold-colored bees which have the shape of large flies.[56] Likewise, it is from the larvae of wasps and bees that those which are called *harynos* in Greek are generated, these being certain long bees.[57]

50   In generations of this sort, nothing whatever is present which does not contribute to generation. For the nature of the eggs of these animals is such that they do not grow the way incomplete eggs do. Rather, in the first generation the larvae grow and take in food and size. They then stay still this way until the inner substance begins to be separated from the skin in which the larvae are. Then the eggs have a completed size.

Further, some larvae have food within themselves to feed on and they cast off the superfluity from the eggs within which they are. The young of the wasps do this. What they cast off is like a kind of black dust and it is found in little sacs of the web in which they are wrapped during the generation period. Some, however, take on food from outside, like the vegetable larvae which we have mentioned and also the larvae of other types which are like them. We have already stated above the reason these types triple, as it were. For at first they are extremely small eggs, and from these larvae are generated which are again changed into the material of eggs. Then from these will come the flying creatures. They thus have a triple change from the egg: namely, into a larva, from larva into the nature of an egg, and from the nature of an egg into a given flying creature. We have said above why they are rendered unmoving when they are emptied of the humor that is in them. We have also declared above that some of these are generated as a result of copulation, much as birds are generated from copulation and many genuses of fish. Some of these, however, are generated on their own, much as plants are generated on their own without copulation.

---

56. The origin of these golden bees is fascinating. Ar. *GA* 758b28f. discusses bees and wasps while speaking about the *chrysallis*, or chrysalis. This could easily be misread as a form of *chrysos*, "golden." The golden bees came to A., who then added the identification of them as resembling large flies.

57. *Harynos* is less easy to trace as it apparently corresponds to *nymphai*, or pupae, in Ar.

CHAPTER II

On the Triple Generation of Bees According to the Various Opinions of Various People

The matter of the generation of bees is a weighty one and requires great consideration. For both Aristotle and Avicenna confess that they do not have certain knowledge of the generation of bees.[58]

Certain fish seem to generate fish from themselves much as plants are generated without copulation since in such fish, in one and the same individual, both the masculine and feminine power are present. Perhaps bees are generated in this way. For we must not give our agreement to certain people who say that bees bring the semen from which they are born from elsewhere and do not produce this semen from themselves. For these people say that the semen of bees is not born *per se,* but is generated by some other animal and that the bees are generated out of it afterward. Thus the one generating this semen and the one delivering it are different in nature and species.

There are also those who say that all bee semen is sown by the king bee and is brought by the bees into their individual homes and is turned into bees. In this way one will generate the semen and another will care for it. They say that an indication of this is the size of the king's body in which there is a great deal of humor for generation.

Certain people say that all bees are generated by the males which exist among the bees [drones] and they therefore remain idle in their masculine power, forming little bees. The others that do not carry on the work of generation bring in honey and wax.

Some people say that they are generated by means of copulation and that bees conceive from the males' semen when they copulate together much as other animals of another genus copulate. Thus every genus of these animals is either generated on its own by the power of the male and female both existing in the same place, or else an animal of its own genus, like the king, generates all the other animals of one swarm. Or they may be generated one from the other via copulation since bees copulate one to the other, or it may be from the copulation of some other genus of animal.

It seems most probable to us that bees either are generated from bees (for it seems to some that they copulate together) or that they are generated by their males which are called *kyrykez* and which are like them in species, but different in sex.[59] Or one should say that they are generated from bees of another type or that all are generated

---

58. Since bees mate on the wing, some thirty to one hundred feet off the ground, and only with the queen, the difficulty in observing the process is understandable (Fraser, 1951, 16–22). Morse and Hooper (1985, 246–48) describe the process and provide some remarkable pictures. Cf. Ar. *GA* 759a1f. and Avic. *DA* 17.52r. Most of what follows is garbled to greater or lesser degree from the original Ar.

59. *Kyrykez* (a marginal correction for a deleted *kirikes*) would appear to derive from *kēphēnes,* drones.

from the kings or from those which are called the *kyrikez,* which are the male bees, and from the female bees by means of copulation. Some people hold the opinion that there are different sexes in bees, namely male and female, and that the males are called *kyrykes.* This means "masters" in Latin and is due to the fact that they are not assigned menial tasks in the swarms for they carry neither honey nor wax.[60] They believe that the other bees, which have a sting, are females, and that the little bees are generated from the copulation of these sexes.

53 Therefore, there are four main theories about bees: namely, that they carry their semen from elsewhere and that they produce bees by caring for this; that all the bee semen comes from the king; that they come from the *kyrikys* who have in themselves the power of both sexes; or that they are due to the copulation of the *kyrices* with the stinger-bearing bees since the *kyrykes* do not have a stinger.

None of these statements seems able to stand if we want to follow the things we can learn from the testimony of those who keep bees and busy themselves with tending their swarms. For from these people we learn many accidental traits which are proper to bees and occur in them alone as well as many common traits from which we will also learn the nature of bees. For as we said in the book *On the Soul,* accidental traits contribute greatly to the understanding of the substance and nature of each and every thing.[61]

Let us demonstrate first, then, that the semen of bees is not brought from other places. For if we say that bees bring the semen from some other place, then this either will be the semen of bees or it will not. And if it is the semen of bees, then there must be some other bees which do not bring semen into the places in which there is semen.

Therefore, it must be that when they transfer semen from place to place bees are generated from some of it but not from another bit of it, for they are generated from that which has been moved and are not generated from what has not been moved. But this is not observed, for we see that the semen of all ringed animals stays in one place, whether what is generated from it be an animal of the same species and form or whether it be different in species and form, as we said above when speaking of the semen of caterpillars, silkworms, and of many others in which the one seminating is winged, but what is born is a creeping larva which, having turned into an egg-like form, again produces a winged creature.

54 Further, it follows that if this semen is brought in from elsewhere, then there must be other bees which are emitting this semen from themselves during copulation. Otherwise, the bee which is generated out of the collected semen would not be generated out of it. It has been shown above that there must be two powers joined to all semen, namely the male and the female, and it seems that these can be joined into the same semen in animals only during intercourse.

---

60. "Masters": A. probably has the Gr. term *kyrios,* "lord," in mind, which had worked its way into medieval Latin. It is regularly used as an equivalent of *dominus.*

61. *De anima* 1.1.5.

There will be, further and necessarily, an error belonging to nature in this idea of bringing in the semen. Let us ask why the bee collects and brings in semen even though it is not his own. For as much as nature may teach every animal to be as concerned as it can about birth and young, it still teaches that animal to care only for its very own young.[62] It will therefore be an error on the part of the universal nature to collect another's semen and to take pains over the young from this semen as if they were their own.[63]

Further, according to this position, there should be no sexual differentiation in the bees even though we say that some bees are *kyrikes* and others are females. Now it is a given that sexual differentiation is in accordance with the differentiation of genital members, which nature creates not out of concern for the offspring but for the sake of generation. Yet differentiation of members is found in the bees, for the female has a stinger but the male does not, a differentiation that is entirely useless and which would thus have nature doing something pointless. According to this they should all be males, for the males show the most concern for the young, whereas in many animals it is the females who are the most concerned with generation. But in bees it seems that perhaps it has to be the opposite of this, since, according to the things that have been said, all bees would have to be females. It is not generally customary for males to have concern for the young, especially in the bees. For experience [*experimentum*] shows that the female bees do this. Often, in hives in which bee semen is nourished to generation, there are only females present and the *kyrikes* are not present at all. For as we said in the earlier parts of this study, the males are occasionally killed or expelled by the females due to insufficiency of food or for some other reason.[64] In such a case it should happen that the bees would not multiply if the *kyrykes* had so great a concern for the generation of the young.

If it were to be said that the semen which is collected and brought in comes from some other animal, then it would not contribute to the generation of bees. For although animals of this sort are sometimes born from putrefaction, generation never happens from semen unless the one generating and the one generated are from the same species, either immediately or through a medium. This is as we said above, that the generation of vermin ultimately comes down to univocal generation. For otherwise it would go on into infinity through dissimilars, and nature abhors infinity.[65]

Because bee semen is never found in their hives without the presence of king bees, some have said that the bees bring in from elsewhere only the semen of the males which they call *kyrykes,* using the Greek name. For all the *kyrykes* are of approximately the same size as the king and because of this they say that the semen is found only where there is a king. But the opposite of this seems to be true. For it is already clear from what has been said before that the bees do not generate the semen which

---

62. Yet A. knew of the habits of birds such as the cuckoo. Cf. 23.38(30).
63. "Universal nature": *universalis natura,* i.e., a general ordering principle. Cf. 9.134; 16.1.
64. Cf. 8.159f.
65. Cf. 15.4f.

is brought in by using any sort of copulation in which the two genuses of the *kyrikes* and kings could be said to copulate with the bees and to generate the semen which is brought in. For the same conclusion would follow about this semen which is brought in this way as was concluded previously for the semen which is brought in from elsewhere. For the same reasons we mentioned before, it is thus not possible that they bring in only this semen. For we have proven above that there is no necessity for this bringing in if one similar thing is said to happen in every generation of bees with regard to the bringing in of the semen for all of them from elsewhere. For this semen either will not contribute to generation like that of other animals or else will not be universally brought in. It is, rather, cast off by some bees and collected by others. And this is disordered and irrational, with nothing of a necessary ordering or natural care.

57   Moreover, it cannot be said reasonably that bees are generated by the particular semen generated through copulation and that thus certain bees are males and others females. For even though nature makes distinction between male and female in the generative members of almost all animals, this does not seem to be the case in bees. For the general opinion is that this semen is found only in the hives where there are kings. Now, much earlier in this study it was stated that the bees sometimes kill the kings when they become evil.[66] Therefore, if the semen of generation of bees were generated though copulation, there would be no reason it is never found in a single hive without a king.

The statement which seems most well suited to the generation of bees is that of the one who said that the generation of bees is commonly from bees and from the ones they call the *kyrykes*, as well as the kings. But this is in no way done through copulation, but by the male and female power being joined together in one of these three which are found in almost every hive unless it be flawed somehow, through the loss of the king or the *kyrykes*. In this case the bees ally themselves to another king and they create or lead in other *kyrykes*. This statement is more universal than the others according to the testimony of beekeepers who say that copulation in bees is never observed. Now if male and female did exist in the bee genus then copulation could not help but be observed many times in the great multitude of bees that exists in swarms.

58   There remains the statement of one who may say that bees are generated from the copulation of the kings and *kyrykes* and that these produce the semen while the bees carry it in collected from elsewhere and that the bees could not collect and bring in this semen unless there were a leader present in their hive. But this is absurd.

It has been proven until now, then, that they are not generated from copulation. What remains is to state that it seems most probable to us that bees are generated in the manner mentioned previously, just as it happens that certain fish generate without copulation.

According to what was said before, bees are generated without copulation and the *kyrikes* are likewise generated without copulation. It seems necessary to say that all these are generated out of the bees without copulation and that the *kyrikes* and kings

---

66. Cf. 8.156.

generate nothing whatever. It is for this reason that they are also commonly called "males." The bees are therefore called the females because they produce bees, *kyrikes*, and kings from themselves. The so-called female bees are quite numerous in a given hive and they possess a generative power compounded of two powers, as in plants and trees. The powers which are joined together are masculinity and femininity. For this reason too there is present in the bees a generative member in which resides this power which generates semen or larvae. We call the bees females through a misusage since it is not reasonable to speak of a female in any genus in which a sexually differentiated male is not found.

If the sense of our last statement, namely that the generation is accomplished by the bees without copulation, should appear the most plausible, then this must necessarily be the last statement of all those that have been produced, namely that both bees, *kyrikes* and kings, are generated by the bees without copulation.

Nothing from the questions debated above contradicts this statement save that experts say that semen is not found in hives which lack kings. But we think this is false. If the semen appears in beehives without the presence of kings and *kyrikes*, it is, according to what we have said, necessarily so that bees are generated from bees without copulation. But all these things are unsure, for we have never seen one who is an expert in swarm observation claim anything about this with assurance. Thus, of the things said, what is left is that the stinger-bearing bees generate all the offspring or young. Now since the kings are said by some to possess a stinger, some say that the kings generate young bees as well. For we know from what was said before that this properly takes place in the bees because of the generative member which is present in them but not in the *kyrikys*. Thus the generation of these properly is to be found in those which are commonly called the bees. Nor does what has been said seem unsuitable since it is found that many other animals generate and are generated without copulation.

But one thing seems to hinder this statement. For we see that there is no genus of animal which generates anything save one that is like itself in shape and species, whether it does so immediately or through a medium. This is as we have said above. If, however, the bees generate, they will seem to be generating things unlike themselves. For the *kyrikes* and the kings are unlike the bees in shape and in certain members. We have shown above that an unlike generation goes on into infinity since unlikeness has no boundary and deviation from the natural boundary is a thing for infinite types.

Nor can there be any objection to what we have said (namely, that every animal generates one like itself) based on those particular animals which seem to have dissimilar first products of their birth. These are called the *arothatylo* in Greek, which means "little wheel," or ringed, and the *nycer*, which means certain insects.[67]

We, however, have shown above that the generation of certain ringed animals reverts to a like thing via a triple participation. Thus the *arothatylo* generates an *aroth-*

---

67. Cf. Ar. *GA* 760a8f. These names derive from the same fish, *erythrinos* and *channē*, mentioned in the notes to 17.32. The strange etymologies must be marginal glosses of some sort which, according to Stadler's markings, came with the text to A.

atylo like itself and the *nicer* generates a *nicher* like itself. Thus too the generation of bees should come back to something like itself but it does not come back at all, for never is a bee produced from a king and *kyrikys* never change into bees.

The solution to this, which shows the reason for this sort of generation, is that bees and certain flies are not generated by univocal generation one from the other as are those which are generated from copulation. They are, rather, generated from other semens which nevertheless have some natural suitability. For we concede, along with Plato, that everything which is generated is generated from one suited to it and in which lies the principle and material of its generation. But it need not be that it is generated from one that is entirely like it in shape, size, and appearance.[68] For this reason too in the generation of bees from semen, there is some suitability between the ones generating and those generated, even though it is not as much as is present in a truly univocal generation brought about by copulation of a male and female. For the kings resemble the *kyrikys* in size, though they are somewhat larger, and in having a sting these same kinds resemble the bees. For there should be some differences among those generated in a generation of this sort, for otherwise the same thing would always come from the same thing—alike in shape and size. But this is contrary to the evidence of the senses. If this were possible in bees, then the entire genus of bees would have to be kings since nature always aims at the most perfect member of a genus. The bees, then, resemble the kings in the energy of their works and in generation. The *kyrikes* resemble the kings in size, since if the *kyrikes* had a stinger, they would be kings.

61 On this matter, and before the full determination is made on the question at hand, we must ask another serious question that arises from what has been said. For it has been said that the kings resemble two genuses, namely the *kyrikys* in size and the bees in the stinger. Yet it is necessary that the kings be generated from some first principle of their generation. They seem to be neither from the bees nor from the *kyrikys* due to the dissimilarity they have to each. It would seem, then, that the kings are generated on their own and it also thus seems necessary that the bees are generated from the kings and the *kyrikys* as well. For in this case each of the ones generated agrees partially if not totally with the one first generating it. This is just as we see that sometimes an individual is born flawed and does not totally resemble the one first generating it. Thus, when the king generates a king, it is a generation of total similarity, but when it generates a bee or *a kyrikes,* then it is partially similar. This is the most persuasive opinion of those who say that the entire swarm is created from the king's semen while the kings are generated entirely on their own.

62 Following what was said above, we say that the kings are not numerous in the hives in which they dwell. It is thus not probable that the entire swarm is produced by their generation. If then we were to say that the kings are generated on their own, there should be no reason we could not say also that the genus of bees produces something as well. And just as we could say that the kings generate something that is unlike

---

68. "Appearance": perhaps "species."

themselves, so would there be no reason why the bees also could not generate other things unlike themselves. Thus, through those which were generating different things, there would be an infinite generation of unlikes. This is why nature deprived the bees of an infinite generation of this sort. For in everything nature does, nothing is done save in orderly fashion and thus it is necessary that the *kyriky*, born as they are different from the bees, be deprived from a further generation of something different. For otherwise the dissimilarity would become infinite. Clearly, nature brings about this privation. For the *kyrikes* are generated from others and generate nothing in a further generation, having instead an end to this generation in the individual *kyrike*. In this way generation can be bounded in the requisite and proper natural way so that the first generation stays bounded and they do not depart from an extraneous multitude into which they might come through a generation of unlike individuals.[69]

According, then, to what seems to us most probable, all of what has been said comes to the fact that the kings, *kyrikes,* and bees are generated out of the bees without copulation.[70] The semen with the greatest energy produces the kings, and that which lacks it produces the bees and the *kyrikes*. This lack is in either material or power. If it is in power and if the material is abundant and is not well bounded by the power, then *kyrikes* are generated. They possess neither the potential for generating nor the power for doing work, for a watery fluid is abundant in them and this is not well bounded by natural heat and natural power. These animals are therefore sluggish and do not bear the name of *androrum,* that is, of men, since there is no virility in them.[71] They rather possess the name *kyrikes* because they are not busy with work due to their sluggishness and because the other bees serve them as if they were impaired. They only have the name of males among the unskilled mass of country folk who have given them this name because they saw that they did not generate out of themselves and this is a male attribute among those which generate by means of copulation.

When, however, the material is diminished and the power is abundant, then bees are generated. This provides the reason that there must necessarily be few kings. For since bees lack blood it can happen to very few of them that both power and material are abundant in one generated. More often the material abounds and most often it is the power, for this is the most natural of the three in those which lack blood. If, however, both power and material should be lacking, then there will be no generation.

63

---

69. This odd sentence may be traced to a misreading of *excedunt* for *exeunt* in the Sc. text, which itself renders no very clear sense. The general sense seems to be that nature precludes the possibility of an infinite series of dissimilar procreations.

70. What follows represents A.'s own opinion on the matter.

71. *Androrum* is an interesting hybrid. The *andr-* stem is the proper stem for the Gr. word for man while the *-orum* ending is an appropriate Latin gen. pl. ending. The term would seem to be A.'s.

CHAPTER III

On the Multiplication of Bees and Why There Are More Bees Than Kings and Kyrikes. *On the Comparison of Three Opinions on the Generation of Bees.*

64 From this too we know the reason for what the beekeepers tell us, namely that in dry, warm years, honey and the *kyrikys* multiply in the hives, whereas in rainy years the young bees multiply.⁷² In dry years, however, their sweet moisture and humor are multiplied and this is the reason for the honey and the *kyrikes* and there is no strong heat in it. But in rainy years the humors in the bees' bodies are multiplied and much semen is produced. If there is one in which power and material abound, then a king is generated from it, while bees are generated form the others. Because a great deal of watery humor is not bounded by a small amount of heat, then the generation of *kyrikes* is either totally or mostly lost. In years with a temperate complexion, however, the number of young bees slacks off. Aristotle says that the reason for this is that bees are smaller in size than the other two types and that for their generation they therefore require more by way of a good complexion. This is so that the seminal humor might multiply in them, as we said a little while before.

65 Because, as was said, the kings were created out of a suitable generation, in accordance with both principles of life, they therefore hold a sort of monarchy and stay within the hives. They do not work and they have a large body because of its suitability for their generating principles. The *kyrikes,* however, which are the male bees, are watery and have no member suitable for fighting due to the heaviness of their bodies. The bees are of an average size that is moderate in comparison to the two types already mentioned. This type resembles in their front part those who are their children, generated from them. In the rear parts, however, there is not so great a similarity. Thus a generation is accomplished that approaches similarity to the one doing the generating.

There is a common occurrence which indicates that the generation of the kings and *kyrikes* is from the bees. This is that they are allowed to remain in the hives, not working. Unless they were the young of the bees, this would not be allowed, for every animal fights for its food and home and only allows one that is born from it to eat its food and to live with it in its home.

We see further that the bees occasionally strike the males and cast them out of the homes. Now it is more suitable to the nature of all animals that the parents strike the young than, to the contrary, the young strike and chastise the parents. Fathers chastise worthless sons and, as we said, this is natural. Therefore the bees are the ones generating, so it seems, and the kings and *kyrikes* are the ones generated.

---

72. Ar. *GA* 760b2f.

Someone might ask why the kings are few in number and the bees are many, especially in light of the opinion of those who say that they are generated from the kings. For this is a reasonable opinion even though we would defend another. We will respond to this opinion by saying that it is similar to what happens to a lioness who, as we said above, bears five young in her first conception and then bears fewer each successive year until, at last, she stops at one. We must say the same about the kings. The first time, they produce many perfect offspring who are the kings and then the size continually slacks off and diminishes until it stops at the size of the bees. This happens when the seminal superfluity in their bodies diminishes. Therefore, the generation of the bees occurs in the way we have stated and it can be understood from the accidental traits which occur in the bees as we said above. And when a given reason is bolstered by those accidental traits which are apparent to the senses, then it is all the stronger and more believable. The two opinions given, however, find strong agreement with those things which are seen to happen in the bees.

Those who say that bees are generated from copulation take as an argument for what they say that they see that when the bees are first born from the semen, they are exceptionally small. For very small things are found in the openings of their hives. They also bring up on their own behalf that a thing which belongs to the ringed animal genus and which is generated as a result of copulation is generated as a result of a very prolonged coupling. For the ringed animals stay copulating for a long time before the sperm descends. But they are generated quickly because of the uniformity of their bodies. Moreover, an animal of this sort which is generated by means of copulation resembles a larva having a columnar, oblong size.

All three of these accidental traits occur in the generation of bees, for the first thing generated is generated quickly, is small, oblong, and resembles a larva.

Further, something occurs in the generation of bees which is like that which occurs in the generation of animals which are in the genus of ringed animals and are like the bees in that they have two wings and gather honey.[73] Examples are the many genuses of yellow wasps. It is common knowledge that in the types of wasp mentioned, there is no seminal superfluity of this sort out of which wasps could be generated without copulation. For according to the opinions introduced above, there is nothing present in them of the sorts of things that are present in the bees and out of which they might be generated without copulation. For they are generated only out of the female wasps, which are called mother wasps. Those are the ones which, like nest-building animals, first prepare their homes and put them in order. After, when they have copulated together, the young wasps are generated out of them. Nor can this be denied, for the copulation of wasps has been observed many times. The difference which lies between the genuses of wasps and bees is very small, however. It therefore seems that bees too are generated as a result of copulation.

---

73. Cf. Ar. *GA* 761a2f.

These, then, are the opinions about the generation of bees and wasps and of all those that are like them. When these statements are joined to the ones made in the earlier parts of this study about the generation of bees, then a more perfect knowledge will result of the generation of bees and wasps. Let these things be sufficient on these matters then.

CHAPTER IV

On the Manner and the Cause of the Generation of the Halzum,

Which Are Shellfish [ostrea conchylia]

69 The types of animal which are called hard-shelled have almost the same generation as trees and plants for they are not generated through copulation. All the types of this genus of animal are generated on the surface of the earth just as vapors coming to the surface of the earth are the cause of the generation of plants, as was previously determined in our other books on natural philosophy.[74] They have no difference save insofar as it is truly said that earthy moisture is the material nearest to the dry life such as that of earthworms and animals of that kind. These are dry in effect although they are moist in actuality. But the moisture of water is a material nearer to the life of the shellfish [*conchylia*] and the shelled animals than to trees.[75] For trees do not have as much of life as do these animals since trees have their life hidden. The animals mentioned, however, have their life open to view. Thus the relationship of a tree to earth is like the relationship of those which have a hard shell to watery moisture, for trees are related to earth as are the shellfish which are called the wild *halzum* in Arabic (as if something could be a wild *halzum*). Also, vice versa, the *halzum* (that is, a shellfish) are a sort of aquatic tree insofar as their participation in the material principle is concerned. For this reason these animals live and are created on the surface of the watery earth much as plants are rooted in dry earth.[76] For a moist nature mixed into earth is more suited to the formation of these animals than is earth alone.

70 For the most part sea animals live this way on the land, for as we said much earlier, the water which they drink is fresh and therefore it must be strained through the surface of the earth. Therefore, shellfish of this sort stay fixed on the ground beneath the water. For this reason too no warm animal that is blooded is found in still lakes. Neither can any animal in which the saltiness is overpowering be generated in moisture. There are but few animals of this sort in fresh water which are like the soft- and hard-shelled animals and the *malakye*. For the nature of all these animals is cold and they

---

74. Cf. *De veg.* 4.43f.

75. Cf. Glossary for the profusion of mutually confusing terms that A. uses for shellfish. In this section, "shellfish" will be used for *ostrea* or *conchylia*, whereas meaningful variants will be given in the Latin. Note too that "shellfish" may also include such creatures as snails.

76. "Watery earth": i.e., the bottom of a lake or ocean.

do not possess blood. Therefore some of them are occasionally found in lakes which resemble small seas and in places where rivers issue forth. Animals of this sort come to these places when they seek heat and food. But in the ocean, animals of this sort are generated to be moist and corpulent more often than occurs in fresh water. This is because the nature of the sea is warm and has a partial share with all the elements which are mixed into the bodies of animals. That is, it has a share of earth, watery moisture, windy moisture, and airy spirit. It has been said reasonably in the book *On Plants* that plants have a share of earth, whereas the aquatic animals we have spoken of have a share of air.[77] There is a diversity among these types in the manner in which they participate in the powers of the elements mentioned. It is a wondrous and multifaceted diversity, with respect to more and less, nearer and further.

The fourth type of element is fire and is the opposite of the elements mentioned. This is because it is the most refined of the bodies in appearance, possessing a violently active and destructive heat. Now we should not make inquiry into places of this sort concerning the generation of animals even though it might be in the assigned place for its species, which is the place for fire.[78] For fire is the fourth element, but its form and appearance are not apparent in and of themselves at all in a place for the generation of plants and animals. For it has its own form through which it is separated from the other elements. It therefore only appears in the middle place (which is the place for generation) in the form of some other, mixed body. For every body that is ignited this way rarifies and takes on a form like that of air or smoke before it ignites. Or if it does not vaporize or dissolve, it looks like earthy material that has been ignited and heated, the way fire appears in a lit stone.

But the question of the generation of these animals does not require an inquiry into the nature of fire. Let us rather inquire after the disposition of the moon in its effects, for the nature of the moon is quite analogous to the substance of the earth insofar as the earth is the place of generation and thus too a shade appears on the moon because it is nearer the nature of earth than the other stars. But a consideration of the nature and disposition of the moon must be pursued in another investigation which is the other part of astronomy. In this the effects of the celestial bodies on earthly bodies will be investigated. What is sufficient for now is that we know that the bounding of the material of certain hard-shelled shellfish [*ostrea durae conchae*] is accomplished on their own without the copulation of male and female, using only the effect of the moon and the heavenly bodies.[79] As will be clear in what is to follow, the bounding of certain shellfish is accomplished by some power which comes forth from the shellfish themselves, not through copulation but through a certain seminal power mixed into the material which clings to their shells. Very often, however, it is sustained and bounded with respect to being *per se* by means of the first method.

---

77. *De veg.* 4.2f.

78. I.e., the natural position of fire is above since, being light by nature, it moves upward. Cf. *Phys.* 6.2.1.

79. Compare this discussion with 5.60f.

73 These two types of generation are also found in the trees, and the shellfish mentioned have a generation close to that of the trees. Certain trees are generated on their own; others, however, from a seminal power. Some of these are generated from a cutting from the tree, while others are from branches which have been split off and planted, as is the case for garlic. The shellfish which is called the *hoz* in the Greek language is generated in this way, namely from different parts of another.[80] It is therefore first generated small and grows larger afterward. The shellfish which is called the *kyrokez* and the *barcora* (called the *purpura* and the *murex* in our lands) and certain other types of shellfish are generated from a superfluity.[81] I am calling a superfluity that mucus-like material which clings to the outside of the shellfish mentioned. But these types in no way possess true semen, but rather by virtue of the formative power which enters this mucus-like superfluity along with the power of the moon, many shellfish appear of a sudden. They are not generated as small as if they had been born out of semen, but are of an average size. After, when the moon wanes along with some other drying agent, these shellfish also lose this sort of superfluity. As a result they are almost completely gone and are not to be found. It happens therefore that these types are not generated on their own, but rather from the superfluity of certain shellfish, as we have said. After some are generated in this way then suddenly, from the superfluity of the ones that were born, a multitude of shellfish of the same species is produced. They are generated not through copulation, but from a mucus-like superfluity.

74 Further, everything generated in this way only seems to be generated in the earth beneath the sea or near the sea, for all these are generated from a mucus-like fluid mixed in with watery mud. That which first emerges from this sort of material is thin and sweet and it is at first contained and sustained by the bounding heat and the surrounding earthy, glutinous material. What remains on the inside of the superfluity takes on the form of a shellfish and nothing is generated save what is in accordance with the worth of the putrefied material and which has been made analogous to that material through digestion. For everything which is generated is generated out of a principle that is suited to itself, as we have said above. For in such animals, there can be no generation of a whole from a whole as occurs in univocal generation. This is what happens when a human who is a whole as to matter and form is generated from a human who is the one generating and is a whole as to matter and form. In these animals, though, it is done rather as in things produced by art in which a part of the one generated is from a part of the one generating. For the form of the thing produced by art is from the form of the artisan, but the material is not from his material. In the arts it is such that the thing produced is made from the superfluity, so I may call it, of the art and which is the factive form of the thing produced. The artisan brings this to it.

---

80. *Hoz*: Ar. *GA* 761b30, *myes*, or mussels.

81. *Kyrokez* is a reasonable version of the Gr. *kērukes*, whelks (cf. forms at 4.45; 13.112), and *barcora* commonly represents the Gr. *porphyra*, the murex. Both *purpura* and *murex* should be taken as synonyms of *barcora*, although the grammar could allow them to be parallel synonyms of the first two names.

In similar fashion nature places its superfluity in semen and in this are present the power and material of its own power. By way of imitating these principles, these shellfish [*ostrica conchylia*] animals are generated on their own in like material and in an imperfectly similar power, in earth and in watery moisture mixed with earth. For in earth of this sort there is a bit of water mixed in and in water of this sort there is a bit of air mixed in. And present in these is the heat of fire and of heaven just as in an airy, vaporizing nature. There is also present the heat of the soul insofar as this superfluity clings to the animal and it is from these principles that shellfish are generated, as we have said. Trees too are generated by means of similar principles, both on land and in the water.

Furthermore, speaking in another way about the same nature, we say that all material superfluities of this sort are filled with powers and by these powers they are quickly contained and bounded to form through that power, namely through the spirit which is extended through the body of its matter. Thus too we have said that the power of the sperm is extended throughout the *gutta* of the female during the formation of the embryo.[82] This power is held within the material by its viscosity and it begins to be dispersed when the humor of the material heats up and is filled with the spiritual wind through which it is made froth-like. Thereupon the power begins immediately to convert, bound, and form the material.

Further, there is diversity among the types of shellfish caused by the differences in the material, power, and place in which they are generated. Some types of shellfish are nobler and others are less noble or baser. This, as we have said, is present in them due to the power of the soul which is held in them from the beginning and this is their formative power, and is also due to place and the body, this being their material. For the bodies in which this power first arises and is retained are the reason for this diversity. There is a large, burned, earthy part to the ocean. This contains these animals and is hardened around them, congealed by the dryness of the ocean and by heat. It becomes a hard shell much as horns and bones in animals are congealed. Because they are hardened first by this hardening and drying heat and afterward by compressing cold, they therefore do not liquefy or grow soft in fire.

Moreover, none of these animals whatever is seen to copulate save for the genus which is called the *kahylion* in Greek. This is a certain genus of snail [*limax*] (that is, snail [*testudo*]) which has the power of progressive motion.[83] Still, there is no surety about this one as to whether the generation of this same genus is a result of copulation or happens without it.

---

82. On *gutta*, cf. 16.56.

83. *Kahylion*: a reasonable version of *kochliōn*, "of snails." Curiously, Stadler's markings indicate that A. added the identification. This identification itself causes problems since *testudo* can also mean tortoise or turtle. Yet, based on such passages as 4.89, it seems the best choice here.

CHAPTER V

What Is the Moving Principle in the Generation of Shellfish and Which Part Is Formed First?

77 It is now proper to ask which part is that which is generated first in these sorts of animals, generated externally and without copulation. We should ask this about these animals since we already have learned previously full well that the material for the *conceptus* is in the superfluity of the females and that this material is moved by a moving principle which is in the sperm of the males in all which are generated by means of copulation. It is necessary then to ask also in regard to these animals what the moved material is and what the moving principle in the generation of those which are generated without copulation.

Responding to the second question, we say that the superfluity out of which generation occurs in the animals doing the generating comes from the food which the animals take into their bodies and which they concoct through digestion. After the heat of the soul of the one generating has bounded, digested, and differentiated it through the powers of the members, it is then suited to have a *conceptus* made from it. The seed in trees is made in entirely the same way. The only difference lies in the fact that in plants and in certain animals the first mover of the material need not be from the male since in these the powers of the male and female sex are mixed together in one and the same material. These powers are assimilated to the plant or animal generating that seed. But in the material of many animals there is a need to take on the principle of movement from the male.

78 There is a further difference in that the food digested in the first digestion of plants and of certain animals that resemble plants occurs in the earth and in the water. But the food of the other plants, from which the material of generation is taken, enters into their body and is there decocted and assimilated so that it can become the material for generation. The heat of the weather and the surrounding air assists in the digestion in trees and those like them much as the animal's heat, which is called "natural," acts in the digesting and assimilating of food.

It is therefore clear from what has been said that the heat of earth and sea, along with the celestial power which it has, digests, alters, differentiates, and bounds the material of the body in the generation of the types of shellfish mentioned above. When this heat is held inside the material along with the spirit then the conception of the shellfish occurs in bodily material and it will be the principle sustaining and bounding the first animate thing which has to be generated from this material. It is also this which brings about the principle of the movement of generation in this bodily material. The sustenance and the bounding of trees which are generated on their own and not from seed has entirely the same form as the ones mentioned. For in all the ones of this sort the generation occurs in such a way that from one material part are made the parts which are the partial principles of generation. From this material part comes the first food and from this the parts which are extended out into forms are fed. The first food is that from which the first radical members are formed and is in the very material of things which are being born. We have said this often above.

Further, some animals generate larvae and belong to the number of those which lack blood and which are not generated by means of copulation. An example is the one called the *fastaryos* in Greek (a species of caterpillar) and some certain genuses of river fish which resemble eels.[84] One of these is the one they call the *feliz* and which naturally has very little blood and which has a heart in which and around which this little bit of blood is generated.[85] We have shown above, however, that the heart is the first member formed in creatures having blood and which are generated in a uterus or in an egg. But the one called the "earth's intestine" (that is to say, an earthworm) has a method of generation like that of other vermin.[86]

In every generation of this sort the upper body is generated first. Since in this sort of generation food must be drawn in before the soul, then it is necessary that the members through which food can enter from without be formed first and these are in the head. Therefore, in all that are generated in this way, the head is formed first along with the upper body and after this are formed the inner members, which form the pathway for the nourishment, and lastly come the lower members. In this way too it formed the body of the *enkeliz* (that is, the eel), generated not from copulation or eggs but from a type of larva, in the mud.[87]

To speak universally, this is necessarily the manner of the formation of all those which are not generated in the uterus of those copulating or which do not have their generation in an egg or in a type of egg as we discussed above concerning the flying caterpillar.[88]

Further, if what they say is true and people and quadrupeds are created by the power of the stars, outside of any uterus of those copulating (and certain astronomers tell us that new worlds are filled with animals this way), then one must believe that in these the heart was indeed not first. Instead, the head and the upper members of the body were first. The inner members were formed next and lastly the lower mem-

79

80

---

84. *Fastaryos*: the mullet (*kestreus*) of the Gr. This form and its kin appear regularly when "translating" *kestreus*. Note the following: *fastaleon,* 6.81; 24.34(54); *fastaniz,* 5.36; *fastarce,* 8.124; *fastaroz,* 7.76; 8.113, 120; 14.79; *fastonoz,* 6.81; *fastoreon,* 6.85, 90; *fastoreoz,* 7.98; 8.32; *fastoroz,* 4.82, 100; 5.37, 40; 6.91; 7.24; 14.79. Despite the regular appearance of this *fastor-* stem, A. did not recognize it here as a fish.

85. *Feliz*: by comparison with Ar. *GA* 762b24, this should be derived from eels, but they have just been mentioned and A. (and his sources) regularly use the form *enkeliz* as the equivalent of the Gr. word for eel.

86. "Earth's intestine": *intestinum terrae* is a literal translation of the Gr. at Ar. *GA* 762b28 explaining how eels were generated out of worms which themselves spontaneously generated in mud.

87. This translation follows Borgnet's *sed* instead of Stadler's difficult *si*.

88. Cf. 17.49.

bers, all for the reason we gave above. Things formed in this way do not have food assimilated within them as do those which are in the uteruses of animals and which are formed within eggs. For this reason the material for forming the members must flow into them from without. It must first form passages for this influx so that that in which form and life are first sustained will occur in one of two ways. It will either be in the way by which those things are formed which come from something else which contains them, as in the case of an egg, or it will be in the manner we described. In such cases it is necessary that there be food which is the material for forming the members, for they do not have this food within themselves as do those which are formed either in a uterus or in an egg. For there is nothing which naturally takes on growth without food. But whatever the opinion of the people mentioned before, anything generated this way, in the usual course of nature as we know it, has either a larval generation or one that is like it. One generated in this fashion necessarily takes food for the growth of the members to be formed from a place other than the heart because in those generated in this fashion the heart is the principle of the formation. For this first member either generates the first member, which is the heart, or will be like that through which food comes, this being the mouth and head of the *conceptus*.

81   The principle of generation must be present in all animals in one of these ways if there is to be life in them. The first act of life is to use nourishment. The first sustenance and retention of the material for the body to be formed may be closed in like the matter which is contained in an egg, and in this enclosed material food will necessarily be present for the growth of the members which will be formed in that which has been conceived in this way. The second way is the generation of a larva, in which case it will take in its food from another place outside of itself. It is then necessary to form a place of entry for this food first. Therefore, the power and the formative principle in ones of this sort are clear to us from what has been said.

82   No other possible way by which food for the *conceptus* may be taken in other than in eggs or outside of eggs is known to us in any sort of animal generation. The first way, which is through the involution of the material as if in an egg, only appears in certain bloodless animals such as some of the types of ringed creatures. It is also to be seen in those we have just been discussing, the ones called the hard-shelled shellfish [*ostrea conchylia*]. During their first formation their food is contained within as is the food of one formed from eggs and their growth is like that of larvae so that the upper members are produced first, the inner members next, and the lower members last of all. Larvae begin at the upper part and the food comes through this to the lower members. This type of food is somehow like the food which is in eggs, for neither of them is consumed entirely in the formation of one part of the body. In larvae, once the upper part has been formed by the food for the involuted material, then that part is separated off and differentiated from the lower part. The food then remains below due to its own weight and the lower part of the body is formed out of it. For it is just as necessary for food to be present for the end of the body as it is for the first part. The end of the body is that which is below, under the diaphragm. It is clear from what

has been said that some things are like larvae in their first creation and are fed and grow in the manner mentioned. This is clear in the first creation of bees and those like them. For in their first creation their upper part appears large and their lower part very small and minimal in comparison to the upper part. Exactly this type occurs in the hard-shelled shellfish and those like them and this is most clear in the sea animal which resembles the *achyherynos*.[89] For this one grows first in its upper part and the front part of its upper part is generated first. The member which is called the head on this animal is larger than all the rest of its body. The method, then, according to which these shellfish are generated has thus been determined.

It is clear that the first of these types are generated on their own, from the power of the stars and from suitable material. This is especially so for the *ostrea* called hard-shelled. An indication of this is that hard-shelled shellfish [*ostrea*] are often generated from the frothy filth that is scraped off ships. Thus, in many places they are sometimes generated all at once, especially those in which there is slimy earth, infused with a great deal of frothy moisture. But in places such as this the type of shellfish generated most frequently is the one they call the *labastynes*.[90] An indication of this is the story told about a certain leader who sailed the sea with an army and then cast anchor in port. When the anchor was raised, a clay vessel was cast onto the shore. After this had lain on the shore for some time it was filled with mud and within this mud was generated the shellfish they call the *asturie*.[91] It is a type of *halzum* (that is, a hard-shelled creature).

In these types there is no spermatic or egg superfluity which emerges from them so that these animals can be generated and which is suited to generation. This is proven by experience. For when a recent expedition of warships came to Libya, certain folk took many hard-shelled shellfish of differing species and deposited them on the shore near the city they call Arym. It was said to be near the equinox. These lived there a long time and were found not to be greater in number than when they had been placed there. But these same ones had grown in size and had become fatter. But this sort of fat is not to be found in animals which possess blood, for they were rendered tasty to eat as a result of this fat.[92]

An indication that what has been said is likely is that there are always found in these shellfish certain things which resemble eggs, like those found in the ones they call *karuka*.[93] They will differ, though, because sometimes their eggs are larger and at other times they are smaller. Moonlight always causes growth in all such animals and the moon's waning diminishes them greatly. Nevertheless it does not appear that

---

89. Ar. *GA* 763a22–23 is the parallel passage, referring to spiral-shelled creatures. But no facile explanation of A.'s form presents itself.

90. Ar. *GA* 763a30, *limnostrea*, "lagoon oysters," which Peck, ad loc., identifies as barnacles.

91. Ar. *GA* 763a34, *ostrea*, "oysters."

92. Ar. *GA* 763b1f. tells of people of Chios who sailed from Pyrrha in Lesbos with oysters and "seeded" some straits there with them.

93. Ar. *GA* 763b9, *kērukes*, whelks.

generation takes place through these eggs and thus, what appear to be eggs, rather are certain nodules in their flesh. These eggs appear in these types more during springtime when moisture abounds and this is especially so for the type called the *reteo,* the *moez,* and the *liberatelen.*[94] For every type of aquatic creature has a time suited for its fertility and for the goodness of its disposition.

This happens not only in those which have been mentioned but also in certain types of fish. In fact it occurs in almost all of them except in the type of fish which some call the *tybo.*[95]

Should someone wish to know in what places in the seas these types exist, let him read the statements which we made long before concerning the shape of the members of animals and their anatomy and generation. Then let him join to these the things which have been said here on the causes of their generation. He will then find that this suffices for understanding their natures.

We have thus determined the manner of the generation of animals, both in general and specifically. We have spoken of the disposition of each and every generation in animals in and of themselves and have done so in sufficient degree for the present investigation.

---

94. Ar. *GA* 763b13f. offers *ktenes* (scallops), *myes* (mussels), and *limnostrea,* "lagoon oysters."
95. Ar. *GA* 763b15, *tois tēthyois* (dat. pl.), sea squirts.

HERE BEGINS THE EIGHTEENTH BOOK ON ANIMALS

Which Is about the Manner of the Generation of Perfect Animals

The First Tract of Which Is on the Cause of the Differentiation of the Sexes and of Likeness among Animals

CHAPTER I

On the Cause of the Differentiation of the Masculine and Feminine Sex According to the Opinions of Empedocles and Democritus. Also on the Disproof of Their Opinions.

Having determined then all the things which had to be said pertaining to the generation of animals both generally and specifically, and since it has been shown that in all complete animals male and female are differentiated through different individuals, it now remains to speak of the powers of male and female.[1]

We will therefore say that these powers are first in every generation and that they are the principles of every generation both of animals and of plants. These principles, though, are separate in some, while in others they are not.

Let us speak first then about the cause of the generation and differentiation of these sexes. For without doubt the male sex is differentiated in shape and nature from the female sex before the embryo is completed in the uterus. There is some doubt whether this differentiation of the sexes arises from the male or the female, or whether it is due to the areas of the womb. Some have said that a male is conceived on the right-hand side of the womb and that a female is conceived on the left-hand side. Others ascribe the reason for the differentiation of the sexes to the womb, saying, along with Empedocles, that it is not the place but the quality of what is in the womb that is the reason for the differentiation of the sexes. For they say that the sperm falling into a hot womb is formed into a male, whereas the sperm falling into a cold womb is formed into a female. They say that the reason for this heat or cold is the menses, which, by its heat or coldness, changes the sperm into the sex-linked differences which have been mentioned. Democritus, however, believes that the cause of sexual differentiation is not in the womb. Rather, he believes that the sperm which is derived from

---

1. Cf. Ar. *GA* 763b20f.

the male is the cause of this sort of difference, and he says that this creates the differentiation between male and female.

2   The opinion of Empedocles is weak and has but little reason in it.[2] When he said that the cause of the difference between male and female is only the coldness of the womb and nothing else, he gave little thought to the great difference which the other members which are suited to generation also produce in the one born. For there is a great difference between the womb, the other members suited to generation, and all other members. The sperm holds their powers in itself because it is the principle of generation. For all the other members, whose powers are in the sperm, seem more to produce a difference of this sort than does the womb or the members suited for generation. For if only the womb and the members suited for generation did this then the following would necessarily be so. If we should posit two animals conceived complete in power but not in act, one of them being male in power and the other female in power according to the formative principles which are in them (so that one would, through the formative power, have the male member and the other, through the formative power, have the female member), and if we then should say that the womb in which the male lay is chilled and the womb in which the female lay is warmed, then generally the male will have to be changed into a female and the female will have to be changed into a male. For, according to Empedocles, the embryo is in the womb much as we might say that a clay pot is in a furnace, because it takes its entire completion from the furnace and has none of it from itself. This is how he says that the embryo has nothing from itself or from the sperm, but takes its entire formation and completion toward masculinity or femininity from the womb. Since this is entirely unreasonable, it is clear from this that the statement of Democritus is better and more correct. For Democritus seeks the differentiation of generation in terms of the substance and he finds differentiation in the internal principles of the one generated. He therefore speaks more reasonably. In what is to follow we will show what Democritus's statement has in it of truth or falsehood.

3   Further, if the heat or coldness of the womb are the causes for the differentiation of male and female, then clear indications should follow as to the cause of male and female. And this would be easy to see because, as certain physicians say, the indications of a warm or cold womb are clear. For these physicians say that they know the cause of male and female based on such things and from the signs of such things, saying that the differences of the sexes are clear.[3] Yet it is no easy thing to discover the principle of the generation of male and female truly and according to a definite cause based in the principle of a cause which one adduces concerning the coldness or heat of the womb. For it is quite unreasonable to say that the sperm, which is the principle of generation, necessarily follows the disposition of the womb, so that when it grows warm in the womb, then not a given member but its opposite will result. In this case, then, if the

---

2. Diels (1951), 2:336, frag. B65.

3. This translation relies on Borgnet's *ex talibus et talium signis,* an odd phrase, but one which is at least more grammatical than Stadler's *ex talibus et talium signum.*

sperm is the principle of the female and grows warm in the womb, it would not be this member but would be its contrary. The same is so if the sperm is the principle of the male and grows warm in the womb. In this case also it will be its contrary, that is, female, and in this way the members suited to copulation, in which male and female differ, would be formed according to the disposition of the womb. This is, as we have said above, very unreasonable.

Further, we have often seen twin fetuses in one womb or in one chamber or separate area of a womb. One of them was male and the other female even though the womb was not cold and warm at the same time. We see this especially in animals which produce many fetuses at once. But it would seem, according to the statements of Empedocles and of certain physicians, that only one member suited to copulation should be present in these fetuses many times over. For we have seen such twins in one chamber of the womb frequently by means of dissection both in walking animals and in certain swimming fish which generate live young, like the genus of the *celeti,* the dolphin, and the large whales.[4]

The author of this statement therefore erred seriously and was deceived if he did not see these difficulties as being attendant upon it and still adduced this position. But if he did foresee these difficulties, he erred all the more intolerably in stating the opinion, saying that the cause of male and female was what he said it was. For he said that the cause of this was the heat and coldness of the womb. Now according to this, all twins, especially those in the same chamber of the womb, should be either both male or both female. For it cannot be said in the case of twins of this sort that those members in which male and female differ are generated by a womb that is separated into parts which contain the twins, that the warm part generates the genital members of the male and the cold part those of the female, and that there are thus certain genital members in the male and others in the female so that the generation of these members is referred to the neighboring parts of the womb. For this sort of quantity of chambers in a womb containing differing twins must necessarily be separated in the generation of these twins and must then again be united into the same power when it conceives twin fetuses of the same sex.[5] But this is entirely unreasonable and cannot be assigned as the cause for the repeated separation of this type, nor for the repeated reunion. For this separation and reuniting are not due to the cold and heat even though these qualities occur in the womb as a whole. And the menstrual blood which flows into the womb is either hot or cold and cannot have each quality to excess at the same time.

In previous books we have said that the true sort of cause of masculine and feminine lies in the disposition of the sperm and, according to the truth as it was presented in earlier books, the statement which Empedocles puts forth about the cause of this matter is false. For if the disposition of the sperm is of the nature we have determined

---

4. "Dissection": as elsewhere, *per anathomiam.* Yet in this case "study of anatomy" seems an unlikely translation. In any case the dissections are reported by Ar. *GA* 764a35.

5. Cf. Ar. *GA* 764b4f. A. has not understood the difficult text which came to him and has changed it to his liking.

above, then, from the things said above, we have at hand reasons with which we can contradict the arguments of Empedocles and Democritus. For there it was also shown that the sperm does not come from any member possessing the power of every member with respect to act, and it was shown there nonetheless that the sperm of the male is not the material of the body of the embryo but rather is converted into spirit. If anyone else should agree with them and say something similar, we will contradict him using the same arguments introduced above concerning the sperm. For it cannot be, as Empedocles said, that there is a given spermatic body in the female, possessing the power of the female members, and from which the female is to be made, and that there is also another one in the male, possessing the power of the male's members and from which the male is to be made. This has been well proven above in the fifteenth book of this study.[6]

Now, the natures of the members are not separate in the sperm and neither do they exist in it with respect to act.[7] It is rather that one part overcomes another in digesting and bounding, as we will show below, and thus a given sperm will be male and another will be female. Generally speaking, it is better and more natural to attribute the differentiation of sex to a cause such as this, as the result of the victory of a member which is the principle of the formation of all the other members. Thus, following the manner of this member, the animal is rendered either male or female, for this is in accordance with the formation and generation of the members and of nature. For it is necessary that every multitude which is in a single thing be reduced to one as to its cause and we should thus reduce the entire multitude of members which is in the body to one alone which is the cause of all the others. Thus, attributing the cause of male and female to a victory of one member in digesting and bounding the material is better than attributing it only to the heat or cold of the womb and the menses.

Those who introduce these opinions raise the question of what the means are through which it happens that the copulatory member in one generated individual has a disposition contrary to the one it has in another, so that in one it is like a receiver (like the womb and those allied to it) and in the other it is like a deliverer (like the penis and those allied to it). They ask this question as if it were difficult and as if it were one that could have no response—other than the one we offer. For we say that the power of the members of the father and mother are present in the sperm and that the powers of the womb work the copulatory member toward its likeness and the power of the father works the copulatory member to his likeness in appearance and shape.[8] They thus say that the members always follow one upon the other.

---

6. Cf. 15.1–2, 66f. A. rarely cites a book number when cross-referencing. But cf. 12.1, with notes.

7. The translation offered ignores the doubling of the phrase *in spermate,* an apparent error on A.'s part.

8. "Powers of the womb": *virtutes matricis* may be an error for "powers of the mother," even though the reading is secure both in St. and Borgnet. A. has just spoken of the powers of the mother and the father (*virtutes membrorum patris et matris*) and it would seem an easy emendation here to change *matricis* to *matris.* Indeed, just below, A. will use the phrase *virtutem membrorum matris.*

In contradiction to them, we will say that if this occurs because of a nearness of certain members to one another (for example, if the sperm comes near to the father or the mother through a similarity of power), then this similarity should be found not only in certain members but in all the members, both external and internal, throughout the entire body. This should happen to the degree that they are near one another in the father or the mother as we have said, by means of the sperm which either has the power of the father's members or the power of the mother's members. This is an invention of those who hold that this is the reason that a female embryo resembles its mother and that a male embryo resembles its father.

We, however, see that this is false, for we have seen often that the sole similarity to the mother is in the child's groin and that in almost all her other members she resembles the father, or vice versa. Therefore, such a nearness of members cannot be the cause of sexual differentiation, as we have said. The cause will rather be the alteration of the entire body. This alteration will doubtless have the heart as its principle and will flow through the veins, both quiet and pulsating, which are positioned beneath the heart and from which they arise. It will then flow through veins which branch off from these and which, using the food which seeps out of their ends, will feed and nourish the flesh with food which has in itself the power of the heart.[9] Now these veins are not created for the sake of the womb as if for the principle of their origin and neither do they have the quality of the womb in nourishing. Rather, and to the contrary, the womb in its natural disposition will exist for the sake of the veins which are carrying this sort of power of the heart (or some other power) in the blood. This is much more reasonable and is nearer the truth since we see that the womb naturally is nothing more than a receptacle for the blood and is not one of the principal members which have the powers of creating and forming others, like the heart, liver, and brain, as was determined previously. The creation of the veins which carry the power of the heart is naturally prior to the creation of the womb, and the sperm is prior to the cause of generation which thus necessarily takes place out of a womb, because that principle of generation which is the sperm has some quality that is substantial to itself from the one first generating. Because of this there occurs from it this sort of differentiation of the sexes in the ones generated and this is a lack of similarity in the members which pertain to copulation, when they are compared to each other.

Now it should in no way be posited that the cause and principle of this differentiation is what these philosophers introduced above said it is. It is in fact some other cause. For once it is posited that the sperm is the principle of generation it will have posited in itself, even though impossibly, that it has not exited from the male and is sustained not in the womb but somewhere else, wherever that may be.[10] In one in which there is a cause of generation we must seek the reason for the differentiation of

---

9. "Feed": *lactant*, evoking images of nursing an infant.

10. This difficult sentence begins *Posito enim quod sperma sit sperma principium generationis*. The second *sperma* seems clearly redundant, muddies the sense, is not in Borgnet's edition, and is omitted in this translation.

those generated, and the principal cause will not exist in another even though there may be concurrent causes which are causes which arrange the true cause.

This statement then is suited for the contradiction of those who hold the opinion that a male is one whose sperm falls on the right-hand side of the womb and that a female is one whose sperm falls on the left. For these people say that the areas of the womb are the sole causes of differentiation, even though there is no fitness for this or that sex based on the fitness of the material. Yet even if we were to posit that the sperm is suited to the differentiation of sex, then those who hold this position would still not be speaking correctly. For that fitness which is in the material would then be some sort of disposition and the completion of the formation of the members which produce sexual differentiation would come from the areas of the womb. And this cannot be. Using this same statement we ought to contradict Empedocles who, as has been stated, places the cause for the separation of male and female in the heat of the womb.

11 Those who posit the cause of the distinction between male and female as based on the nature [*ratio*] of the powers of all the members (which powers, they say, are in the sperm of the woman and the man) are acting much the same way. For these people say that the two sperms fall into the womb, but that only one or the other is formed, namely the sperm of either the male or the female. They say this is caused by its falling on the right or the left. For when both fall at the same time on the right side, the sperm of the male is formed into a male because of the fitness of the place, while the sperm of the female perishes because of the opposite nature of the place. If, however, they both fall on the left of the womb, then, because of the suitability of the place, the female's sperm is formed, while, due to the contrary nature of this same place, the sperm of the male perishes. If, however, the sperm of the male falls onto the right and the sperm of the female to the left, there will be twins of different sexes. But if what they say is true, namely that the generation of a male or a female is due to these two parts of the womb, right and left, and is not due to the body of the womb as a whole as Empedocles says, and if we then say that the semen is in the womb and that the embryo which is created in the womb will be generated without the action of the womb (that is, without the action of the power of the womb which is in the semen), then the power of all the members need not be in the semen of the females. By this same line of reasoning, that which is formed into a male by the power of the right side will be formed without the power of the male member which is in the semen. And according to this, the sperm is, through one acting externally, converted into the shapes of the principal members, since the genital members number among the principal members. But this is an exceptionally absurd idea to any person who knows the nature of natural generation.

12 Further, as we have also said before, it has already been found through dissection [*anatomia*] that a male is formed on the left side of the womb and a female on the right. Twins were also found on the same side of the womb and they were of different sexes and this is not a thing found only once, but one that has happened many times, especially in animals which generate many young at once. Therefore, what they say

about the right and left side as the causes of sexual differentiation amounts to nothing at all.

Further, some say that the cause of sexual differentiation is that the sperm exits from the right or the left testicle. They say that sperm leaving the right testicle becomes a male and that from the left a female. They prove this because if one binds and ties off the passage which lies between the right testicle and the channel of the penis, and if copulation then occurs, a female will be generated from this sperm. But if one binds and ties off the passage which lies between the left testicle and the channel, then a male will be generated from sperm poured forth this way. They say moreover that Homer also believed this when he said that those who have been operated on and do not have their right testicle generate females and those who do not have their left, males.[11] But this is not true, for I myself saw one with his right testicle amputated and after the castration he first sired a son and then a daughter. Therefore what they claim does happen, but the cause is not the one they gave. In the same way diviners say that they have seen with their own eyes what has happened and that they foretell this. Yet that which they foretell is not the cause, but it occurs rather from something else as if from a cause and it happens according to the divination. These things, then, and those like them are stated out of ignorance of the true cause.

For the members which they say are the cause of sexual differentiation are without doubt not suited to the animals in the generation of the differentiation of the male and female sex. A clear sign of this is that many animals participate in the differentiation of the male and female sex and generate males and females, and yet they have no testicles. Examples are ones which have no feet, such as the many genuses of fish and serpents.

Yet the opinions stated have something in common. For those who say that the cause of the generation of male and female is heat and cold (like Empedocles and Democritus) and those saying that the cause is the exit of the sperm from the right or left side agree in the remote cause. For they only said that the exit of the sperm from the right testicle is the cause in that this side is warm. It thus produces a male and the other side, because of its coldness, produces a female. For heat is the principle of digestion and therefore even these men, however remotely, have still grasped something of the truth of the cause of the differentiation of the sexes. But none of them has set forth the perfect and absolute cause.

13

---

11. Ar. *GA* 765a25 cites Leophanes (ca. 440 B.C.), an authority also cited by Theophrastus (*De caus. plant.* 2.4.11). According to Ar., however, the procedure entailed merely tying off and not excising the testicle.

CHAPTER II

On the True Cause of the Differentiation between Male and Female According to the Teaching of Aristotle and the Other Peripatetics

14 As we approach the truth of a matter, we should understand the cause based on the principles nearest to it. Above, we have already established the disposition of each member along with a declaration of the cause which makes the member what it is. In accord, then, with the things that have been established above, the male will be differentiated by the strength of a given power and the female will be caused by the weakness of the same power.

In this light, then, we say that sperm is powerful, able to digest and to bound well the material it comes near, using the formation which is in it and which is the principle of generation. This will without doubt be the cause of the male gender in the fetation. Now I am not calling a "principle" the material from which a body will come to a likeness of the one generating it, but rather the principle which moves and forms the material, if it can accomplish movement of this sort on its own or with the help of certain others in such a way that it is still the one acting upon and bounding it. For the material of the female, which descends from her, is passive, as it were, and not active. So then, if this is the case, and we already know from the things handed down in the fourth book of the *Meteora* that that which is well digested and bounded is only necessarily digested by heat, and that the material in males is better digested and bounded, then it will necessarily follow that the males of animals, insofar as they are males, are warmer than the females insofar as they are females.[12]

15 Further, the females of certain animals may sometimes be weakened by a great deal of menstrual flow, due to the cold and the weakness which befall them, and not to heat. Thus, some are deceived when they hold that females are warmer than males because the menstrual flow occurs only in females and not males.[13] For these people say that the menstrual flow only occurs because of an abundance of blood and heat in the females beyond that in the males. They are wrong, however, when they say this, for not all blood is said to be warm in one and the same way. Yet these people hold that every fluid the color of blood has a warm nature, making no carefully honed distinction between clear, clean blood and other blood which is neither clear nor clean. For clear, clean blood is generated from the best, well-digested food. Through this type of digestion a small amount of good food emerges during the last digestion, coming from a large amount of food that is neither completed nor assimilated.

---

12. *Meteora* 4.1.14.
13. At *QDA* 15.6, A. attacks a view that he attributes to unnamed physicians, namely, that females are warmer than males in complexion.

This is clearly seen in the workings of the ripening fruits. In these a very small amount of juice is drawn off from a large amount of food for the formation and digestion of the fruit. For in all those of this sort, when the last food, completed in the last digestion, is compared with the first food, that which has undergone the last digestion will be the purest and quite modest in amount, just as that which is strained out of a lot of food will be the purest. It happens much this way in the body. For the members of the body receive food from one another according to its proportion of decoction. The food in the first member is the least decocted, while that in the one to follow is more decocted. In the third it is still more decocted and in the last members it is the best decocted. Thus, at the end of decoction, the food will be quite small in amount and clean. This is the blood in those that have blood, while in those which lack it, it will be a humor analogous to blood and taking its place. Now this last food cannot always be equally digested or equally strong. But, certainly, attendant upon the outcome of the digestion, it will sometimes have strong power and sometimes weak. The clean superfluity which leaves this humor to become the substance of sperm has entirely the same disposition.

For every power which performs and enacts some act, or which ought to do so, takes on an instrument by which it can accomplish that activity which it naturally has, for those which are smaller and more compact than it is. Therefore, the sperm with the stronger power which forms the material into a male necessarily bestows a member suited for male copulation and the weaker sperm, which forms the material into a female, gives a womb to the female for the reception of the material which can be formed into the fetation. In this way one sperm is passed over into another much as the male casts his semen into the female, and for this purpose the copulatory members have suitability in either sex. By this necessity, then, the male has an instrument suited to him and the female has another which corresponds to it analogously but in an opposite way. The one which is in the female is the womb and in the male there is another which has the ability of placing sperm into a womb. An exception is found in a very few animals in which the female has a member which is placed into the body of the male. Since this animal is very cold and its sperm cannot flow, it is therefore necessary that the female draw it out as if from a sort of cistern.

We know already from the many things said earlier in this study that nature gives a power to every official member along with its form so that with this it might accomplish the task for which it naturally exists.[14] For in all things nature does the best and the best that can be is that the official members have the sort of disposition that has been mentioned. Therefore, all the locations and shapes of the members are distributed according to the powers suited to them for the accomplishment of their tasks. Likewise, the superfluities which come out of them are also perfected by the powers. Now I am calling superfluities those things which are abundant in a thing into which the members have impressed their powers and I am not using the term for those things

---

14. "Task": *officium*, thus producing a play on words with the "official members," *officialia membra*, these being members which are designed to perform a special task. Cf. 1.13.

which are cast off as corrupted. The fact that members are completed by powers and are necessary for activities is proven by an inductive consideration of all the members. Vision, for example, is not completed without eyes and the eye is nothing in the true sense of eye without vision. It is the same for the other members. We know from antecedent considerations that the completion of the creation and formation of the belly, bladder, and all the members like them occurs only through the appropriate exit of the superfluity. This is why they have powers.

19 Moreover, we have long ago proven in other books on natural philosophy that generation, nourishment, and growth come about as the result of the same things.[15] Every member, then, must have food in itself by which it is nourished and it must be of the same sort as that food from which it is generated or was generated. In a member this food is a superfluity of the member like that superfluity by which the member was generated. And because it resembles that one, this superfluity can also bring about another member of this sort during generation.

There is no obstacle to this in the opinion of some that nourishment comes from a contrary and not from a similar thing, for in another way this too is true. "Contrary" is taken in the two ways which we have determined elsewhere—specifically, at the beginning of decoction or at its end. For if everything which is corrupted is changed into its contrary and if every alteration is a kind of corruption, then alteration will have to be movement from one contrary to another. Decoction, however, is the alteration of food and is thus moved from one contrary to another. Thus, in the beginning food is dissimilar and contrary, while in the end it is similar and suitable.

20 Once these arguments are thus taken as having been determined elsewhere, it is right to declare in greater and clearer fashion the cause on account of which one of those things which is generated in the womb turns into a male and another into a female. The reason is certainly this: when that thing is weak which is first and is the principle of the movement of that whose movement is to digest, bound, and form the material of the fetation, it thus cannot overcome the material perfectly. This is because it is weakened for digesting due to the weak heat it possesses and it will in no way be able to form the embryo in accordance with the form of the formative power which is in it. In this case, then, the material of the fetation is necessarily moved over into something contrary to that which first formed it, although it does remain in the same species. This is nothing other than the complexion of the female. For as we have shown above, the female has a complexion contrary to that of the male. For that which is in the womb will necessarily be either male or female on account of the unity of the form. But these two sexes have different powers and therefore it necessarily follows that the instruments which correspond to the opposite powers be different themselves.

21 Further, when one of the principal members is altered in complexion, then the entire body must necessarily be altered in power since in some way the entire body is sustained by a principal member. Therefore, it is necessary that the animal's body be totally altered in accordance with the alteration of its genital members, and that the

---

15. Stadler cites *De nutr.* 2.1.

bodies of males and females have contrary complexions. An indication of this is that this leads one to understand that when a given animal is castrated, its complexion is changed throughout its entire body in such a way that there is but little difference in complexion between castrated ones and females. The reason is the one we have stated, namely that certain members are principal in the body and others are secondary. When the principal members have been changed, then many of the consequent members must necessarily also be changed. In every genus of animal the male is the principle of nature and power according to Pythagoras. He is called the male, who, having perfected the potential of nature, has power. But to the female is attributed weakness and the role of assisting nature's actions by supplying the material. Now strength and weakness are caused only by digestion and nondigestion of the natural heat and by the power of a soul which bounds and completes the last food. Thus, sexual differentiation is caused in those which are generated by the heat of this sort which is in the superfluity of the last food.

CHAPTER III

In Which the Heart Is Proven to Be, among the Members, the Principle of Sexual Differentiation. Also on Those Which Generate Mostly Females and Those Which Generate Mostly Males.

Having thus determined the cause of sexual differentiation with respect to the power which is in the sperm and which is the moving principle in generation, we must ask further what its principle is in the members. For we have shown above that there is one member from which the rest originate. According to the things which have been well established above, the cause of sexual differentiation is the first power which is present in the member in which, among the members, the principle of the natural heat is present first. For this is the reason that the bounding and the formation of the heart is anterior to the bounding and formation of the other members, both in the male and the female. In animals which have no heart, that member is formed first which is analogous to a heart and which takes its place. For there must first be a cause of male and female before this sexual differentiation takes place, and the differentiation of male and female is due only to differentiation in those members which play a role in copulation.

Thus, as we have said before, these members have their own proper activity, just as all the others do. Just as the eye has sight and the ear hearing, so do these members have copulation because they belong to those members formed second, after the heart. They take the principle of their creation from the heart and, as we said above, they pour out throughout the entire body the power of sex which they receive from the heart.

Returning now to those things we spoke of initially, we say that the sperm is the superfluity of the last food. By last food we mean that which has already passed through the powers of all the members in the fourth digestion. This passing to the members creates a likeness in the one born to the father who is generating, a likeness in the form and appearance [*species*] of the members. In what we have said, the difference between our statement, that the sperm passes over to each member, and the statement of Plato and certain others, who say that the sperm comes from each member and possesses its substantial power, becomes clear.

24 The difference which exists between the semen of the male and that of the female is because in the semen of the male there is the power which first moves in the generation of animals and which can digest the last food which is the material of the fetation, causing it to pass over to the womb of the female. But the material out of which the embryo is created is in the female. Thus, when the semen of the male is perfectly victorious and the material from the mother is entirely subservient, then the fetation will resemble the father since it leads the fetation to a form which is in it due to the power of the father. When, however, the father's semen is overcome by a recalcitrance in the material or by its innate weakness, then it will move during generation to a contrary form, namely to a likeness of the mother. For the female naturally has a complexion which is contrary of the male's. The male, insofar as it is a male, is warm, since heat is dominant in the form of the male, whereas cold is dominant in the female. The sanguineous food, which is naturally warm, is chilled only because of a loss of heat.

25 Again, as we said before, nature gives each superfluity a suitable instrument to take it in. Now sperm is a superfluity and it thus has as its proper receptacle the womb, in which it accomplishes generation. But the superfluity which is sperm is greater in the males of animals which have greater heat. These are the blooded animals which have a good body. The members which receive it in the bodies of the males are nothing other than the passages and veins for this superfluity and thus the sperm in males is shunted to the seminal vessels. In females, however, due to poor and imperfect digestion, the superfluity is quite sanguineous and is not digested. For this reason it is necessary that in the female there also be a member receiving this great superfluity. This member should be unlike the male's and, due to the large amount of the superfluity, it should have some size. For this reason the womb is formed in the shape it is. The difference between male and female is due only to the difference in the nature of those members which play a role in copulation and generation.

We have thus set forth the reason on account of which male and female are differentiated in the womb.

26 The accidental traits which befall generation and those generating are also witnesses to the truth of what we have said. For we see youths (in whom the humor of the first age abounds due to the dullness of the heat in them) generate daughters more often than those who have progressed further along in years, even to the point

of mostly drying out the first fluid and reaching the peak of their heat. But we have spoken of these people in the book *On Youth and Old Age*.[16]

Further, we often see that old men in whom the heat has lessened and a humor abounds (a humor which moistens externally and does not cause growth) generate females out of young women. For the heat which is in the younger ones is not yet complete, while that in the older ones is diminished.

Further, those with moist, phlegmatic bodies and whose complexion is feminine generate females more often than those whose semen is globular and granular, congealed, as it were. For such a bodily disposition is due to a reduction of natural heat.

Moreover, if the north wind should blow, then males are most often conceived. For this wind closes the pores with its coldness so that the natural heat, turned inward in this way, digests the sperm better. When, however, the southerly winds blow, then females are most often conceived.[17] For their bodies grow moist and the heat which is in them is turned loose in the warm, southerly wind which, with its own heat, opens the pores and loosens the bodies. For this reason too in southerly winds there is more sperm in the bodies of animals than there is when north winds blow. And when sperm is multiplied it is not digested well.

It follows from a similar reason that the menses in the bodies of women flows most during the waning of the moon because those days are colder and moister than the other days of the month due to the final lessening of the light. For a lessening of heat follows upon the lessening of light. Now the sun causes summer and winter during the course of the year but the moon causes four different weather patterns in every month. For from its first glimmerings of light to its shape as a quarter it is warm and moist.[18] From the quarter to its "precedence" [*praeventio*] when it is full the month is warm and dry.[19] From the *praeventio* to its second shape as a quarter it is cold and dry and from the second quarter to its conjunction it is cold and moist.

Further, shepherds claim that males are generated not only from northerly winds and females from the southerly winds but also according to what they look at. For if copulating sheep look to the north, shepherds say that they frequently generate males, and if they look to the south during copulation, shepherds claim that they very often conceive females. It is clear from this that a small alteration in the first principles which cause changes in bodies will frequently be the cause of heat or cold in the sperm

---

16. *De iuv. et sen.* 1.3–5.

17. It is interesting to note that A. glossed Sc.'s *austrini* with *meridionales*. As both words mean "southern" the redundancy is omitted in the translation. Elsewhere (e.g., 4.97; 5.31; 10.54) he appears comfortable with either word.

18. According to Macrobius (*Somnium Scipionis* 6.60), moisture regularly accompanies the new moon. According to Stahl (1990, 11 n.61), the many references in classical literature to the moisture-producing powers of the moon can be traced to the fact that the greatest deposit of dew occurs on cloudless nights. As a result, any visible phase of the moon was associated in folk belief with the production of moisture. The second phase of the moon, according to Macrobius, is hot, the third is dry, and the fourth and last phase is cold.

19. For *praeventio*, see 6.103.

and that heat and cold occurring afterward cause differences in generation. Thus, the differentiation between the generation of females and males is due to these causes.

Further, both males and females require a tempering of their complexion. For all official members are produced only by a certain bounding of the temperament and by an established harmony. Heat which conquers and overwhelms this harmony will dry out the natural moisture and the heat must therefore be reduced to a tempered state. For just as a stronger and larger fire dries and chars, so also to the contrary is a slight, waning fire unable to digest, and when it has died down a great deal it cannot bound and complete the humor. For this reason it needs to be helped in the tempering process. However, if it is not suited for tempering, what happens to a fire happens to it.[20] For, as we said, a large fire dries and chars and a small one can neither digest nor complete moisture. This is why both the male and the female as the ones generating require tempering, so that the sperm of the male might be made proportionate to the female's humor, much as an active, completing, digesting, and forming agent is made proportionate to one that is passive, while completing, digesting, and forming. On the other hand, the woman's humor is to be made proportionate in the same way to the sperm of the male.

29 From this arises the fact that many women and men who copulate do not effect impregnation since their sperms are not made proportionate with respect to harmony, but when they are separated and the man copulates with another woman and the woman with another man, they both will be fertile and will generate. From this it also occasionally happens that youths do not generate but that when they have progressed in years they will generate when the heat which exceeds the temperate level abates somewhat. The remoteness of regions also brings this about sometimes due to the differing complexions of climes, as we have stated in the book *On the Nature of Places and of Things Placed*.[21] The reason for this diversity is multiple, and a watery spermatic humor is therefore rendered more or less suitable in a man and woman for many reasons. Food sometimes causes this and sometimes it is a disposition of the body which accidentally becomes excessive or at other times it is the complexion of the surrounding air. But that which was stated is most properly caused by types of food and bodily dispositions. Watery, moist foods especially cause this for they cause sperm and the nutrimental humor to multiply. For the watery humor of food has the greatest resemblance to the accidental traits we have mentioned. This is also why cold, icy, hard waters are very harmful to the body, for they are not soft and they thus sometimes bring about sterility and at other times cause the generation of females.[22]

The things which we have mentioned, then, are the causes of sexual differentiation.

---

20. The implied subject here is the heat mentioned earlier.
21. *De nat. loc.* 2.2.
22. Cf. Hippocrates *Airs, Waters, Places* 4.35 (Jones, 1923, 79).

CHAPTER IV

On the Reason for the Resemblance of the One Generated to Its Parents or to Previous Ancestors. And on the Reason for a Lack of Resemblance of the Same One to Its Forebears.

The cause of resemblance of the young to its father or mother or to one of its ancestors, as well as the cause of any lack of resemblance to them, is also derived from these causes.[23] For some young resemble their fathers in their entire body and some are so like their parents that they are not like their ancestors at all. But in others the contrary of these things occurs and ones which resemble the ancestors and not the parents are generated. Sometimes there is even a casual resemblance to each of them. Again, it also sometimes happens that the males are like their mothers in some members and the females are like their fathers in all save their groin. Sometimes too they resemble neither of their parents but still preserve the shape of the species in that they are humans. But at other times they do not even retain a human shape or that of those that generated them, but take on instead a monstrous and wondrous form. An offspring which is in no way like its parent, either in the nature of the species or individual shape, is monstrous and is called a wonder of nature. For this offspring, according to the nature it possesses, was never in the power of the parents. For, by its own genus and nature it is not analogous to its parents save in that there is generation of male or female according to an analogy to its nature and species. For nature necessarily requires a generation of this sort if that genus of animal in which male and female are distinct and separate as to place and subject is to be preserved.

The reason for all these things is taken from the harmonic proportion of the complexion of the sperm to the nature of the one conceived and vice versa. At one time the power of the male which is in the sperm prevails and bounds perfectly, due either to the power of the sperm considered in and of itself, or because it is reduced through age to a tempered state, or for some other reason. When tempering of this sort occurs, then it happens that the animal necessarily generates on account of this sort of complete proportion. This cause (on whose account generation occurs in this way) will be a result only of those things which are either essentially or accidentally suited to those which are generating, or from either of the ways we have mentioned.

The first cause of resemblance or lack of it should then be found in this cause. For when the superfluity which is the sperm is well concocted, it can act well and suitably on the superfluity which is the menses, and the form of the embryo will be completed into the likeness and form of the male if the material does not resist. This

---

23. "Ancestors": *avi* is more properly "grandparents," but usage in this chapter sometimes demands a broader translation.

is accomplished by the movement of the male. For we do not make any differentiation in saying (a) that only the first mover acts on this complexion in generation, or (b) that the sperm does this, or (c) that it is the movement of each member (which is in the sperm in power) and that it is the movement of the creating process and of growth which accomplishes this function. For in all the statements the intention is one and the same. For the formative power is the prime mover in the sperm and the sperm is its subject. Also, in it lies the power of the members. It is therefore clear that when this movement prevails, the offspring will be male and will resemble the father that generated it. In this generation there will be no female which resembles the mother. When, however, it can in no way prevail, it will not resemble her according to a different sort of nonresemblance.

32 Sometimes, it will not only be male like the father, but it might also have the nonresemblance which the Greeks call *korceokes,* which possesses a subcontrary resemblance, namely, this is when it possesses the form of a person with respect to the nature of those who generated it, but it has a resemblance with respect to genealogy.[24] In this way, some resemble their near parents and some their remote. This generation of resemblance occurs when generation is accomplished in an essential and not in an accidental way. In this case the resemblance to the parents or the ancestors is produced. For as we have said, if the sperm is entirely victorious, then it will resemble the father, as if another father came into existence, as if the father had begotten another self in his son. If, however, the material should draw the operation of the sperm to itself, then the resemblance will be to the mother. This can occur in three ways. First, it may occur in the principal and secondary members at the same time or it may be that the material draws sperm to itself in a principal member, but that sperm is not found in the secondary members. In this case the one born will be like the mother in the complexion of the heart and in sex, but will resemble its father in the other members. If it should happen to the contrary that the sperm leads the material in the creation of the heart and not in the other members but rather is drawn off by the material, then the one born will resemble the father in the heart and in sex, but will resemble the mother in its other members.

33 It sometimes happens, almost always in fact, that there are many powers lying unseen, potentially, in the sperm. For the power of the ancestors is in the members of the great-grandchildren up to the fourth generation and occasionally further. For the power transmitting the semen is present in the transmitted power until it is extracted from it through digestion. The power of the ancestors is thus present potentially in the bodies of those generating and when it is helped by resemblance either of food or of place, it functions in actuality. In this case it forms the sperm into the form of one of the ancestors, either a near or a distant one. But we will give a fuller reason for this further below.

---

24. Ar. *GA* 767b24f.: "'Faculty,' as applied to each instance, is used in the following sense. The generative parent is not merely male, but in addition a male with certain characteristics, e.g., Coriscus or Socrates" (Peck, 1942, 403). Clearly, A. has travelled far from Ar.'s text.

But it is clear from all that has been said that the strength of the entire generation lies in the sperm and this accomplishes that which is present in everything which has been generated. One which they call *korceoken* (since it has a subcontrary resemblance of sperm) is still a person and retains a resemblance to the nature of the one generating it and is quite like it in species. Nor does it come to the point that it only has a resemblance to it in genus. Thus, in a case where a person is the one generating, the result will be a subcontrary man and not merely an animal. For even though in such ones the nature of both the genus and of the species is generated, still, the nature of the species is the one that is most victorious. Generation thus achieves its end at a resemblance to the species. In this case it does not stay in a resemblance of the genus, but proceeds to a resemblance of the species of the one which generates it. For it is true that the one generating is the substance and the nature and thus is that which this thing is, having a determined being, size, quality, and substance. It will generate one like itself in all these respects if the sperm is totally victorious. For we have said previously that the principle of movement of all such things is only in the sperm of animals in power and potential. And in this way too the power is in them from the grandparents and great-grandparents, great-great-grandparents, and *subavi,* but each of these powers is not equally at hand.[25] If, however, it does not act as a result of the essential powers of the parents, grandparents, or great-grandparents, but is instead from the accidental powers, then the form introduced will not have a definite resemblance to any one given human individual. It is as if we were saying that it will be a likeness of Socrates if it should happen that an accidental trait were disposed to form his likeness. This is what we have called *korceoken* above, that is, one that has a subcontrary likeness.

As we have already said often above, everything which moves is changed if it moves physically. When it is changed, it is not moved to a given form but rather to a contrary disposition within the same, common nature. Such a movement necessarily happens to sperm when it is moved not by an essential power but by those things which happen to it. In this case the father and the power of the male do not prevail, but rather the material and the accidents do. Therefore, if only the power of the male is weakened, then that which is generated will be rendered female. If it is an accidental trait and is moved by a subcontrary resemblance, then it will be rendered a single, solitary *koceoke.* It is as if Socrates were not like his father or any of his ancestors, quite outside any likeness of genealogy. The power of the male, moved according to a weakness, thus sometimes moves to the female sex (for the power of the mother is entirely

---

25. The relatives listed in the Latin which came to A. are *avus* and *proavus,* grandfather and great-grandfather. A. has added *abavus* and *subavus.* The CL progression is *pater, avus, proavus, abavus, atavus, tritavus.* One is tempted by the position of *subavus* to make it "great-great-great," but *subavus* does not appear in the lexicons and no parallels can be found in Lexer (cf. Diefenbach, 1857, s.v. *Tritavus*). The prefix *sub-* may also indicate "approximately" and is so used frequently in color terminology (e.g., *subalbus,* "off-white"). *Subavus* could thus be taken as a very distant relative, as in the "remote parents" of 18.44, but cf. 18.37 where the order of the terms is not the same and *subavia* precedes *proavia.* Likewise, at 18.45, A. changed his received *proavis* by adding *remotis* to it. In sum, it is clear that A. had no small amount of confusion concerning the terms.

contrary to the father with respect to things which produce complexion). When the same sex is retained, much the same thing brings about a corrupted likeness of males. For the activities of the sperm exist only along with the principles of movements and of the powers which are the powers of species and genus, such as the powers of humans and animals. Therefore the power of the offspring and parents and ancestors produce such activities.

36 Further, when the young moves in the womb, it will be changed and it will doubtless move toward the contrary. The powers created in the members of the young will move toward powers which are nearer the sperm. To put it in a word, if the female power of the one generating is weakened by the power of the male and draws the material toward itself, then it will move toward the form of a mother nursing and nourishing the fetation out of the nourishment which the one born takes on and the form itself will change to a remote form. This is as we have stated concerning the oppositeness and distance of male and female. Thus, because the father and the male power are one and the same power, it necessarily happens that this power is sometimes victorious in generation and at other times is overcome. It is thus not difficult to assign a cause of both forms occurring in the fetation, namely, that of the father and that of the mother.

37 A double power can also be victorious, namely, that of the male and of the members of this particular one which is generating insofar as it is generating. Thus it may be both Socrates and the man who is generating. In this case the one born will be both male and like its father. Because this happens most often, those which are born male most often resemble their fathers and those born female are like their mothers. For in the way just mentioned, a change occurs in both their powers. For the female is the contrary of the male and the mother the contrary of the father. And, as we have often said, movement is always toward a contrary. If, then, the movement of the male prevails as to sex, and yet the movement of the one generating (such as Socrates) does not prevail as to the powers of the members, and yet there will still be a change with respect to the movement which is from a given power of the father, then the resemblance in the fetation will be to one of the ancestors from whom the father is linearly descended. This will be the grandfather, the great-grandfather [*proavus*], or the *subavus,* even unto the fourth generation as we said above.[26] But if this movement is destroyed due to the victory of the material and is made not from the power of this particular mother but from her sex according to the essential power of the material and not insofar as it is from this woman (who is, for example, Cleopatra), then this movement will be in accordance with the power of one of the female ancestors from whom the mother is descended. Then the resemblance will be to the grandmother, great-grandmother, *subavia,* or *abavia* since it rarely goes beyond the fourth generation.[27] This is the way

---

26. Cf. 18.34, with notes. As in that passage, A. received *avus* and *proavus*. To this he has added the odd *subavus,* omitting *abavus.*

27. Ordinarily, *abavia* is great-great-grandmother. Yet by position it would seem to be great-great-great-grandmother. The order of the puzzling *subavia* is different from the order shown in 18.34.

in which likeness occurs in the members. For certain members will resemble the members of the father and others those of the mother. Perhaps some will resemble members of the grandfathers or grandmothers who are the "old mothers," since, as we have said many times, certain members are present in actuality while others are present potentially.[28]

We should accept these types of resemblance according to nature and according to genus (that is, according to the nature of the individual, of species, and of genus) for the powers of these three are in the sperm. The first reason which we mentioned is that there is a certain movement of the sperm, in accordance with the principle moving in it and which is there potentially, and there is another there in actuality. If it should happen that this principle is destroyed and hindered, then the sperm will move in accordance with the nearest essential or accidental power. And if that one is hindered, it will move after this to the next nearest one. If the destruction is greater still, so that most of the ones nearest in order are hindered, then it will move off to a further one and finally, when the movement of all the essential powers that exist in the sperm from the parents is destroyed, then there will be no likeness in the young to anyone in its genealogical tree. And yet a person will be produced, retaining the nature and likeness of its species. The reason for this is that the nature of the species follows all its individuals and is present in all individuals. For everything is generally a human which in some way is related to his generation. For the father of him who is Socrates and the mother of her who is Fronesys, are humans.[29]

The reason for the weakening and destruction of the movements of these powers which act in the sperm is that something which acts physically in material is affected by its contrary which is itself generating in the same material and which is held, warming the material. What it thus does in the material is also chilled by its contrary which is in the material.

Yet it is true that not everything which moves generally is moved so that it is removed from the power moving and overcoming it. For sometimes something weak in its action is acted on more than a strong thing. Thus an active thing is not always acting toward its power, for it is either overcome or perhaps it acts with a lesser action than it would have if the contrary power were not acting in a contrary direction. Now all these things have already been determined from those things which were said in the first book of *On Generation* [*Pery Geneos*], where we made distinctions among the physical operations of the dispositions, forms, and contacts and where we also set forth in which things there truly exist a physical operation and contact and in which things they do not exist.[30]

---

28. "Old mothers": *antiquae matres*. While the term is in the Sc. text that came to A., it reads very much like an explanatory (and incorrect) gloss on the *progonōn*, "ancestors," of Ar. *GA* 768b3.

29. *Fronesys* is intriguing because it is a single-worded addition to the text by A. In Ar. *GA* 768b15 it says, "of the mother, whoever she was," and Gr. is not close in form to this name. In fact, *Fronesys* is close to the Gr. *frōnesis,* "prudence" or "thought," a suitable name for Socrates's mother, who, in fact, was named Phaenarete.

30. *DG* 1.1.f.

Likewise, many differentiations exist in the area of the passive material. For while the passive is always moved by the sperm, it is not always overcome but draws to itself its operator. This can be due to many causes. It will either be due to a weakness of the operative power or to the coldness and indigestibility of the material which is being decocted and digested. Thus, there occurs a defect either on the part of the operating power which is weak or on the part of the passive one because it is not suited to be acted upon by the agent, either fully or in this way. These are the causes on account of which that which is formed and bounded within the womb is rendered multiple in form, shape, and species with regard to its external appearance.

40 The accidental trait which we said occurs due to diversity in the material to be bounded finds a similar example in what happens in the nourishment of those who have, from their youth, been employed in business activities where gain is difficult and in strenuous activities which are accomplished through heavy bodily motions.[31] These people need to take in a great deal of food and because of their movements the digestive heat evaporates from them and their moisture is consumed. Therefore, their growth cannot be accomplished in suitable fashion, for their bodies do not grow to requisite size and their members do not grow proportionally. This is because more nourishment is drawn by heat to those members which strenuous movements exercise. In such a case it is necessary to take it away from other members. Thus, those who have been fullers from their youth show increase in size in their legs and those who have been smiths from their youth have enlarged arms and hands.[32] In this way too the material in the one conceived does not take on equally the action of the sperm and bends the form which has been introduced toward some other accidental likeness. For then the common form does not remain in its natural disposition and neither do the forms of the members. They rather take on diversity according to the dispositions present in the one being acted upon. The reason for this is that the actions of the active ones are in the one being acted on and being disposed and are there not with respect to a faculty of those acting but rather to a faculty of those being acted upon. Perhaps that which has been stated is the only cause of diversity, namely, that that which is generated by something else does not take on a likeness of one of its ancestors due to being related to it.

41 Further, the disease to which the Greeks give the name of *satyriasis,* and which lies in an inordinate movement toward and an overabundant desire for copulation, is sometimes the cause for a lack of resemblance in those which have been generated.[33] This is due to a superfluous descent of the sperm to the seminal vessels of the male. This is necessarily undigested and incomplete. And it is right that this occurs, for

---

31. Reading Borgnet's *eorum* for Stadler's impossible *eo.*

32. The medieval trade of fulling cloth involved long hours treading upon cloths in tubs and vats filled with a variety of solutions. The process is described at length in *DMA* 11.705f.

33. It is worth noting that A. has added "Greeks" to his text as well as a definition of the condition, named after the perpetually erect satyrs which adorn so many Greek vases. Cf. the discussion of the *satiriasis* after death in the *Prose Salernitan Questions* (Lawn, 1979, 175–76).

sperm such as this is drawn by the violence of the heat and windiness which results from undigested food.

The causes, then, of differentiation of the male and female sex have thus been established along with why sons most often resemble their fathers and daughters their mothers. We have also established why they sometimes resemble their grand- and great-grandparents of old and why some bear no resemblance to their relations at all. We will return to all these matters in the later parts of this book, when we gather again in summary fashion the reasons for all the things we have introduced.

CHAPTER V

On Disproving the Error of Those Who Have Ascribed Other Reasons for the Resemblance and Nonresemblance of Those Which Are Generated

Certain of the natural philosophers [*physiologi*] have not assigned these reasons for the things we have mentioned, but have introduced other reasons for the resemblance of the young to their parents. [42]

They seem generally to attribute the reason for this sort of resemblance to two things. Some of them attribute this to a resemblance of the sperm, saying that the resemblance of the young will be to the parent from whom the most sperm has left during the act of copulation during which the young was conceived. According to these men this will be the cause of the resemblance whether the resemblance is throughout the entire body or just with respect to certain members. For these are the ones who say that the sperm comes from the members and that the member from which the most sperm comes will take on the greatest resemblance in the one generated.

These people also say that when the sperm leaves the male and the female in equal quantities, then one overcomes the other. In this case the one generated will have no resemblance to either parent.

But this statement is entirely false since even if we posit that the sperm comes from the entire body it would still not be, as they say, the cause of resemblance or nonresemblance. For according to the statement they make, a female can never be like her father or a male like his mother. And yet we often see this happen. These people say that the one generated can be made female only if more of its sperm has come from the female and it can be male only if most of its sperm has come from the male. Thus, this statement cannot, according to the cause they give, be distinguished by the bounds which are clearly present in the generation of like and unlike things. [43]

Those who speak as did Empedocles and Democritus attribute the generation of male and female to heat of the soul. This is clear from things said above. Their statement is impossible, a fact proven in another way which we mentioned above where the statements of these men were disproven.

But now we are speaking against those who have said that the cause of resemblance to one of the parents is that more sperm comes from the parent to whom the young has a resemblance. What we have said before stands in objection to that, for we often see that a female is made like her father in some members and a male like its mother in some members. Yet it is impossible that more sperm came from both at once.

44 Further, according to these people, a reason cannot be given why the young is sometimes like grandparents and great-grandparents and will resemble only its remote parents. For no sperm whatever has come from these.

There are others who speak differently and attribute the generation of male and female to another cause. Their statement is more tolerable. These say that the sperm is a unity but that this one sperm is rather like an aggregate of many sperms in a single place, this being the vessel which receives the sperm in the one generating. When the entire mass has been gathered, they say that during generation the formative power takes on the material of generation out of the entire, assembled mass. It can thus happen that it in no way takes up an equal mixture from that mixed complexion. For the fluids of the sperms are mixed together naturally and equally but the formative power which they say is the soul and which prepares organs for itself takes from this mixture sometimes more and sometimes less. They claim that this is possible because the entire completed mixture is made up of many fluids. They say that the fluids of the grand- and great-grandfathers and grand- and great-grandmothers remain in both parents and descend into the seminal vessels of the ones generating and are mixed together in the womb. When the soul takes on a lot of moisture from each of those whose moisture has been poured together during copulation, and has taken on little or no moisture from others, then the resemblance in the one generated will be to the one from whom the soul took a great deal of moisture during its formation. They thus determine all the causes of resemblance and nonresemblance based on the quantity of the moisture taken up.

45 This statement, however, is neither clear on its own nor has it been proven by them. It rather resembles an unreasonable statement. All those who maintain this opinion are certainly speaking falsely since, as is clear from earlier books of this study, the causes and types of resemblance and nonresemblance are not present in the sperm in actuality but only in potential.

Further, all types of resemblance and nonresemblance cannot be caused by what they call the gathering together of the sperms. For from what they say it is not perfectly clear why sometimes that which is generated is female and at other times male. Neither is the reason clear why one born female sometimes resembles the father and why at other times one born male resembles the mother. Nor is the reason very clear why it sometimes resembles the ancestors and, further, why it sometimes resembles its remote ancestors.[34]

---

34. On *proavi remoti*, see notes to 18.34.

Further, it is not clear from their statements what the reason is that the young sometimes bears a resemblance to a human only in species and bears no resemblance whatever to any of its parents.[35] Sometimes, moreover, it does not even retain the resemblance of the species and will be a monster and the reason for this is not clear from their statements. It thus is clear that the positions held by all these natural philosophers [*physiologi*] are false.

CHAPTER VI

On Monstrous Births, in Which There Remains Only a Resemblance to the Proximate Genus

Having determined these things, then, it follows now that we speak of the reason for monstrous births.[36] In the sperm there is an assimilative power of the one that is generated and it sometimes causes resemblance to the individual, the species, and the genus. At other times, however, there is a resemblance only to the species and genus and not to the individual generating. At still other times it is not to the species, but is only to the genus of animal. It keeps this resemblance at a minimum, for no animal is found which has ever given birth to a plant or a stone, but at a minimum the genus is preserved in all things which are generated. We should, then, speak of such births here for enough has already been said about the other matters with respect to our present objective.

Concerning monsters and their causes, then, we should first note that when the moving principles and the movements of those present in the sperm due to the nature of the species and the material of the individual are destroyed due to some cause acting strongly in a contrary direction, then only the universal power of the genus will remain and this cannot be removed as long as the material is preserved, for otherwise nothing is generated from material such as this.

When, however, this universal power remains in the material which has a balanced complexion, then the generation will be monstrous. For the universal power of the genus moves toward life and a special cause acting toward the contrary of the species will bring on another form. This is mostly caused by two things, namely, the heavenly power and the material which has been conceived.[37]

It is due to the heavenly power because there are certain parts of the heavens in which the light-bearing bodies of heaven gather which especially move one toward

46

47

---

35. Reading *dictis* for Stadler's impossible *dixtis*.
36. Ar. *GA* 769b10f.
37. On monstrous births resulting from celestial causes, cf. Price (1980, 175f.); Demaitre and Travill (1980, 434–40).

generation. When they do so there can be no human generation. But it will be more proper to speak of these things in another study.

It is due to the material for example, when the material of many things is poured together into one, or also when a defect of certain members is present in diminished material.

Due to a power which is acting contrarily, a human birth might have the head of a ram or a bull, as is said of the Minotaur in the tales of the poets.[38] And the judgment is the same for analogous cases. But such flaws occur more often in other animals than in humans. Sometimes a calf is generated which has the head of a human or a lamb which has the tail of a bull. Occasionally, this sort of generation comes about out of certain bodily parts which are not flawed so that the entire part is a perfect member, fulfilling the function of the member and not corrupted or flawed in any part of itself.

Certain physicians attribute all these monstrous resemblances to shapes which are seen in two or three species of animals which seem to be mixed into the monsters. They relate that these things are due to the mixing together during copulation of the sperms of two or three animals which are of differing species.

48

But we hold the view that the sort of monster they speak of cannot arise from the copulation of greatly dissimilar animals because one of the sperm would corrupt the other. Rather it is due to the copulation of similar ones that the generation of which we have already spoken above occurs. An indication that generation of this sort does not occur is the difference in the gestation period of such animals, an area in which they differ widely. For the human, the dog, and the bull have vastly different gestation periods and yet we see that certain monstrous births are due to a jumbled mix of the members of the animals mentioned.[39] There is one which is a monster because of the multiplication of the number of members due it according to its species and these are different in form and shape. An example is one generated with many feet or many heads. The causes of these monsters are near to each other and are quite similar, for those which are monstrous and which possess flawed members have closely related natural causes. For a monster has suffered some flaw.

Democritus held the opinion that monsters are generated because two sperm do not fall into the womb at the same time but one after the other, whether they be the sperm of the same or different species.[40] The first has remained there and has acted imperfectly, whereupon the second one, falling into the same place, is mixed in with it and also acts. Thus the formation of the members is changed. This, he says, occurs especially in egg layers (a true-enough statement) and he says that sperm which fall successively into the womb this way move the eggs from the places in which they had begun to be formed by the first sperm. This happens especially in birds so that many eggs are thus generated by the one sperm attracting the egg material. Sometimes however it is collected together from one act of copulation, as Democritus says. In our

---

38. The mythological allusion is A.'s.
39. Strange as it sounds, A. actually used *taurus* when discussing pregnancies here.
40. Cf. Diels (1931), 2:124, frag. A146.

opinion, this belief—namely, that in the case of the birds, eggs are generated as a result of much copulation and that many sperm are gathered together into one egg—is much more likely than a belief that many are generated from a single sperm. For that which is said in the natural-philosophy way [*physice*] and with some likelihood should not be rejected any more than that which is set forth sophistically and falsely should be accepted. This necessity especially occurs in the bird genuses, whose sperm fall into the womb not all together, but successively.

It is generally better to believe of all monsters that which is most natural and likely, namely, that the cause lies in the material and in the types of impregnation and creation of that which is conceived and generated. An indication of this is that monstrous births occur rarely in those animals which generate only a single fetation. In those animals, however, which conceive and generate many young at once, more monstrous births are found, especially in the birds which lay many eggs like hens and pigeons. For this reason many hens produce eggs in which two yolks are found, due to a joining of two sperm which are the result of two acts of copulation all in one and the same egg. The two sperm are held connected in the same egg because one falls next to the other in a neighboring area of the womb. This is as we often see happen in fruits when two or more fruits are formed under a single skin.[41] When two sperm are joined this way in a single egg, a monster will be generated unless they are separated by a strong membrane. If the membrane separates well, then two chicks will be born and neither of them will have an excess number of members. If, however, the sperm are connected and there is no separating wall in the middle nor one which generally separates them, then a monstrous chick will be generated from them, possessing perhaps one head, a single body, and four feet.

We ourselves once saw a two-bodied goose.[42] Its bodies were joined only at the back and it had two heads, four wings, and four feet, and it went in whatever direction it turned itself. But it did not live long. For the upper part is formed earliest out of the white and is fed off the yellow and the lower part is then created. The food in ones such as these is unified and is not separated into two yolks. Aristotle thus says that for the same reason there once appeared a serpent with two heads. For the serpent is a member of the oviparous animals which produce many eggs. Yet monstrousness is rarely found in serpents because the shape of their womb is long and narrow and thus their eggs are separated in them and are not pressed together into the same area. For this reason too bees and wasps are rarely monstrous, since each of their young is in its own separate hole. But the opposite happens in hens. From all these indications, then, we form the opinion that the cause of monstrosities lies solely in the material.

For this reason monstrosities are found most often in animals which generate many young and are found but rarely in the births of women. Yet in certain regions

---

41. The term "fruits" in Latin extends to the products of many kinds of plants (cf. *De veg.* 1.142, 199f.). Thus peas in a pod might fit A.'s description here.

42. According to Killermann (1944, 266), A.'s testimony here may be the first report of Siamese twins in anthropological literature. For A.'s understanding of the birth of twins in general, cf. Thijssen (1987).

the women generate multiple young more often, especially in parts of Egypt and in regions that have complexions like that of Egypt. Monstrosities are found more often among the quadrupeds—in goats, sheep, and pigs—than in other animals. This is because these animals conceive many young, and goats and sheep very often conceive twins. It likewise occurs in animals whose feet are split into many toes, since these animals conceive many young at once. This is also why many of them bear incomplete young, as does the dog whose young are produced unseeing. We will speak below of the cause of large or small numbers of young. That the dog bears incomplete young, a lessening of completion, indicates to us a path moving toward monstrosity because the material of adjoined fetations easily flows together and the members of the newborn which nature cannot complete within the womb are easily flawed. For monstrousness in the form of excess, lack, position, or shape of the members is due to some error in the operation of nature. Since it happens that animals of this sort conceive many young at once, nature changes the manner of creation into a monstrous one.

52 A wondrous creation of this sort is most common in pigs. Sometimes there is in them an astonishing and unnatural diminution of members or sometimes an excess. Sometimes there is a perversion of their location or shape—not always, to be sure, but more often than in other animals. In some this will never occur because that which is unnatural cannot happen always or often. But it does occur when the natural formative power does not perfectly overcome the form and the material.[43] For this reason many people do not call such a birth monstrous, for its generation occurs many times. The same thing occurs in trees and especially in the genuses of vines. The vine genus which is called *summosum* (which some also call the *soreth*) quite often bears black bunches of grapes.[44] It is thus not held to be monstrous even though this genus of vine naturally belongs to those which bear white grapes. And yet when it changes to bearing black grapes, this is not a natural change. Such a monstrosity occurs in animals which generate many young since the multitude of the young hinders the completion of the creation, as has been said.

53 Someone perhaps will ask what the cause of monstrosity is in those which produce many young. For in these young the members are sometimes less in number and at other times are too numerous. These animals sometimes produce a young which has many digits. I have thus seen a person who has six digits each on his hands and feet. His brother, born after him, likewise has six digits on his hands and feet. Others seem deprived in number, having but one digit. Similarly, all the other members sometimes are both too numerous or too few, sometimes being superfluous and at others diminished. Sometimes the fetation is born possessing both sexes, with a man's penis and a woman's vulva. Such a fetation is called a hermaphrodite and this monstrosity occurs

---

43. A. offers only *naturalis formativa*. The text of Borgnet supplies the noun *virtus*.

44. *Summosum* is A.'s version of *fumosum*, "smoky," in Sc., a perfect translation of Ar. *GA* 770b20, *kapneon*. According to Stadler's notes, *soreth* is A.'s addition but its origin is obscure. It does not appear in *De veg.*

sometimes too in the goats which the Greeks call the *virreagaryez*, since they possess the members both of a male and a female.[45]

Once, before the time of Aristotle, a goat was seen who had horns on its legs. And I have seen a ram which bore four large horns on its head and two long ones on its legs, rather like goats' horns. I have also seen a goat which had only front feet and which walked about on them, carrying the rear portion of its body on them in the air and not dragging it along the ground.

But this alteration and diminution is found not only in the external members but also the internal ones. For the creation of certain animals is diminished with respect to these members while in others it is too great. Sometimes the members are also changed with respect to position, out of their proper places. Sometimes some lack certain members entirely. But an animal has never been found which lacks a heart. But an animal has been found which lacks a spleen and another which had but one kidney. But no animal has ever been seen which lacked a liver. Sometimes, though, a gall bladder is not found in some of those which have one by nature and in others many gallbladders are found. Sometimes the spleen is found on the right and the liver on the left. All these things have been found in complete animals.

These, then, are the things which should be said about astonishing, monstrous animals.

The Second Tract of the Eighteenth Book on Animals on the Cause

of a Large or Small Number of Offspring

CHAPTER I

That Largeness or Smallness of Body Are the Causes of a Large

or Small Number of Offspring

As we have said above, there is great diversity among offspring and on occasion a great mixture of differing sperm. That which emerges first of such sperms into the womb and is generated before the time allotted it lives for a brief time. For that which leaves its place of generation before the time for its completion cannot live, especially if there is no alteration of its nature in the principle members which are the principles of life.

---

45. Ar. *GA* 770b35 offers *tragainai*, hermaphroditic goats. Sc.'s version, *traganez*, is closer than A.'s. For A.'s other remarks on hermaphrodites, cf. 4.102; 9.102; 17.39; and *QDA* 15.5; 18.2.

What I call alteration of its nature occurs when those members have been altered through an alteration of the formative power and heat to a state of completion. If they have not received alteration before birth, the animals live for but a short time. For this reason we desire to find out in this chapter whether one may reasonably believe that there is one and the same cause for an animal which generates a single offspring and for the alteration which presides over the diminution of a given member and if there is one and the same cause for a large number of offspring and for too many members. Or should we not, in the way indicated, make inquiry into the cause of too few or too many members and why some have only one offspring, such as the elephant, horse, and camel, while others have many offspring?[46] But we should make inquiry into the specific cause of this.

56   Every animal which has horny hooves [*corneae soleae*] most often generates only one offspring. Dogs, wolves, and all animals with a foot split into many clefts bear many young, and small animals, such as the genuses of mice, likewise bear many young. On the contrary, though, animals which divide their foot into two hooves [*ungula*] have few young. An exception is the pigs, which belong to the genus of those which produce many young because of the great amount of sperm and humor which is in them. It is, moreover, reasonable that large-bodied animals generate only a few young at the same time. For that from which they grow causes deprivation in the material for generation. And a reduction in generation is doubtless created by large body size. For food passes over into nourishment and bodily growth and therefore little is left of it to be material for generation. To the contrary, however, in the small-bodied animals, little is put into the nourishment and growth of the body and therefore most of the food is relegated to the material for generation.

Further, the sperm from which large animals are generated should be ample if they are going to produce newborn which are complete in members and size. Thus, if these animals generated many young at once, there would have to be an overabundance of sperm which could overflow into the nourishment of the body. But small-bodied animals can generate many young from a little sperm which is easily adequate to nourish the body.

57   In addition, many small animals can be in the same part of the womb together as it covers them with blood for their nourishment. But many large-bodied animals cannot exist together in a single womb containing and nourishing them. For even if all the mother's blood were to pass over into the womb, it would still not be enough for many young of this sort and size.

Further, animals which have horny *sotulares* [*cornei sotulares*] for the most part generate only a single young or have few young. Those animals which divide their feet into many clefts often have many young. The reason for this difference is, as we have said, the largeness and smallness of these animals' bodies, and these differences bring about a large or small number of young. For *soleae* and *sotulares* and splitting of feet into

---

46. We follow Borgnet in making this sentence a question.

many toes are not the causes of these things, even though they are generally concomitant traits. An indication of this is that the elephant is very large, in fact, larger than all walking animals and its foot is split into many clefts. The camel, however, is also large and it divides its foot into two parts. Yet each produces a single offspring.

Nor does this occur only in the walkers, but also in those aquatic swimmers and in flyers. Whichever of them has a small body often has many young and whichever is large has few. The same thing also occurs in trees and plants.

We have thus stated the reason that some animals generate many young and others few, with still others in-between.

CHAPTER II

On Why on Occasion Many Offspring Are Conceived from One Act of Copulation

What occurs in females who conceive many offspring at the same time seems remarkable. For it happens that from a single act of copulation they conceive the entire multitude of their young. For whether the sperm of the male is suited to mixture and to the formation of the material which comes from the seminal *gutta* of the female at the time of conception, or whether it be said that it is not suited for this and rather only gathers it up much as a coagulating agent does to milk, it seems that it should produce only one out of all the material, for it is one and the material is one. In this case someone will wonder why it produces such a plurality of fetations.

There are those who say on this question that the reason for the plurality is that the sperm is divided into many areas of the womb and at the many vein openings which end at the womb. But this is false. They say that the vein openings draw the sperm toward themselves and since they are diverse, the sperm is thus divided off to diverse places. This view seems to agree with the statements of Galen.[47] But this is proven to be false through what we have said above, namely, that through dissection [*anatomia*] many young are found in a single chamber of the womb. This has been seen many times over in animals which generate many young when the womb is filled with numerous offspring.

The true cause of this is that every genus of animal has a boundary to its size, between smallest and largest, which it cannot overstep, as we have said elsewhere where we showed that for all things which exist there is naturally a boundary and a plan for their size and growth. Now, nature never goes beyond those boundaries unless it is done as a result of the error by which monsters are created. I once saw a girl nine years of age who did not even have the requisite size of a one-year-old infant. Therefore

---

47. Cf. *UP* 14.4 (Kühn 4.150; May 2.625).

it suits the nature of a human to observe a boundary just as it suits the nature of all animals. Thus the material from which an animal is created does not have unbounded size, but has a boundary and end discernible in nature. Therefore all animals which conceive and generate many young at the same time have males and females which produce more sperm than is needed for the creation of one young. Thus, from this great superfluity the material for generation is divided by the formative power for the generation of many young.

60 This power of the soul has the power for dividing the sperm and for making it proportionate both to a suitable number of members and also to that which suits the number of fetations in accordance with the established number. This is the governing power, governing the distribution of the sperm much as the governing power of food distributes food in the body according to the number and suitability of the members. Thus the sperm of the male and the formative and governing power which are in the sperm do not take up more or less from the material of the female's moisture than is proper in accordance with the boundary of its natural size. And even though the male might produce much sperm, still this power does not take up more or less for one fetation than has been naturally determined according to the plan for the size of the members. Indeed, to the contrary, if it should happen not to find in the womb the humor that it would form into another fetation, the contrary occurs. For that which exceeds a level of sufficiency for the fetation is dried out both by the heat of the womb and by its own heat and is corrupted. Thus, we see that fire dries out the material in which it is present when it passes through it as it loosens the boundary of rarification that is requisite to that material. So too, if water is heated excessively, then its material is extended more than suits water's species, whereupon it evaporates into the form of a more rarified element and dries out. For when the material of water is extended it will stop at the boundary of rarification that is requisite for its nature and species. Nor does it extend beyond whatever might be added to the fire. But if it is extended it is corrupted and evaporates and thus at length it is consumed and dries out. Much the same thing occurs to the sperm in the womb if it is superfluous and does not find the seminal fluid of the female to grasp and be contained in.

61 It is, moreover, clear from the things said above that the sperm of the male and the superfluity of the female, which is the material of generation, must be tempered in all animals which generate many young. And this tempering occurs according to both quantity and quality. When such a tempering occurs, the sperm of the male is able to grasp and to bound the superfluity of the female which is divided into many parts. This is divided by the governing and the distributive power according to a proportion by means of which many young can be formed from it. But the example which we gave above about the coagulating agent and the milk does not apply to this sort of sperm, for the sperm does not treat the entire quantity as the coagulant does in the milk. Rather, it establishes a proportion of quality and power in the material, according to which it is made proportional to a sufficient quantity in the embryo, and this distributes the quality to the members and to the fetations according to a geometric, divided

mean which is coagulated and bounded. But a coagulating agent treats only the entire quantity of the milk, not changing its quality in a distributing and forming action.

We have thus stated the reason many young are caused by one act of copulation in the wombs of those animals which naturally conceive many young, as well as the reason one large, mixed young is not created out of the entire conceived material.

When, however, the sperm is so scanty that it cannot grasp the material of the female's humor, nothing whatever is caused by it. And if there is a great deal of it compared to the passive one, whose power is set (as in the case of the power of active heat), then there first must occur proportioning and tempering of one to the other through division. Tempering is also an accidental trait found in the sperms of large-bodied animals which conceive a single young, since in such animals many young cannot be created out of a great superfluity—only one can. The powers of those acting and being acted upon in all animals have bounds that are known naturally and thus such animals do not produce sperm beyond what is sufficient for the fetation. Neither does the superfluity of the female grow beyond the boundary of this sort of natural sufficiency and this is due to the cause we have mentioned before. In the large-bodied animals only one offspring is created from this superfluity. If, however, due to some accidental cause it should sometimes happen in such animals that the male produces excessive sperm and the female with whom he copulates should overabound as well with a great deal of superfluity, then from these, two or more twin fetuses will be created.[48]

The human is an animal lying midway, as it were, between two genuses, one of which naturally generates many young and the other which naturally generates one young. For the human naturally generates a single offspring, but due to an abundance of heat and humor, sometimes generates two or more young. For when the sperm is warm and moist, it is naturally multiplied by moist and warm things. These things occur in the human more than in other animals. This is also the reason that the gestation period in the human, more than in other animals, is not hard and fast. In other animals gestation periods are well established. There are many differing gestation periods in the human, for a woman gives birth in the ninth month, the seventh, and in times in-between. But a child born in the eighth month often dies and the reason for this is known from the things which we have mentioned earlier in this study. For we have spoken fully of this in what has gone before and that is adequate for our present purposes.[49]

According to the things said before, the reason for the generation of twins is, generally speaking, the same as the reason for the generation of superfluous members. The accidental traits which occur in each of these occur only during the first conception, for then, in one of the members, material which is abundant and superfluous is encompassed by the power and then that member grows larger than its requisite size.

---

48. That is, twins, triplets, quadruplets, and the like. The awkwardness of diction is because of A.'s insertion of words into his received text which merely said "two fetuses will be created" and because the Latin *gemellus* means any member of a multiple birth.

49. Cf. 9.46f.

We often see this in the digits on the hands and feet of many people as well as in other members, especially those on the extremities. If it should happen that that material is divided, then two or more members will be formed. We see this happen also in the roots of trees. At first there is but one root and afterward the flow of moisture is divided by the spiritual windiness and then there will be two roots, even though the first bounding and completion of the tree will come from one root and one formative power. It happens entirely the same way also in the creation of multiple young from one primordial power which is afterward divided when the material divides in the manner we described above. There is still another cause of this, namely the movement of the body of the womb and the split absorption of the sperm by the womb, as we determined in earlier parts of this study. But the principal cause is the superfluity of the material when it was distributed in such a way that more than one young could be adequately created from it.

CHAPTER III

On the Cause of the Generation of Hermaphroditic Fetations As Well As That of Other Monsters, Resulting from a Multiplication of Members

65 There are accidental monstrous traits which occur in certain generated ones which people call hermaphrodites because during the first generation they take on both the male and the female members. These are likewise caused by the superfluity which is the material for generation, specifically when it tends to one side more than the other. For this reason too one of the members is principal and the other has little more than shape. It will therefore not be principal and is like that which is generated from food which is taken in but is not naturally suitable. This putrefies and is converted into an abscess if it bursts forth into the outer skin. So it is for superfluous members of this sort. They are in the body after the fashion of growths but they differ in that such superfluous members are generated from an overabundant radical moisture, while growths are generated from superfluous nutrimental moisture.[50] Thus the growths start out as putrescent because of the unnatural heat which is in them. But there is no power in them or it is very weak.

66 The hermaphrodites about which we are now speaking have the cause of their generation we have mentioned. For if the impregnating sperm should find abundant moisture and should overcome it perfectly, it will divide it totally and make two twins, resembling the male in sex. If, however, it is equally overcome both in its entirety and in its parts it will produce two twin sisters. If, though, in one part which has been divided off it is overcome and in the other it overcomes, then the twins will be different

---

50. Following Borgnet's *tamen in hoc different* rather than Stadler's *non in hoc differunt*.

in sex, one male and the other female. This cause is common to the whole and to the partial members because there is no difference in this with respect to the overcoming of the power or its weakness. If the material is abundant only in a single member which is near the groin, then that which is overabundant in that same place will be divided up. If the power in one is overcome and overcomes in the other, then a hermaphrodite will be generated. Sometimes the shape of each member is so complete to sight and touch that a person cannot tell which sex is dominant. And there is nothing preventing such a young also from having two bladders and that it emit urine through each of them or that during intercourse it play both the active and the passive role, lying both on top and below. But I do not think that it can both impregnate and become impregnated. Certainly, however, that sex will be the more principal which is aided by the complexion of the heart. Yet sometimes the complexion of the heart is so much in the middle that one can scarcely discern which sex is dominant.

The causes of male and female have already been determined in what has gone before, as have the reasons that certain extremities have exterior members in a way other than what is natural.[51] These creatures with a different sex in the same individual have one and the same cause as that which is the cause of a certain sort of miscarriage. In this the woman or female does not miscarry, but rather the fetation becomes monstrous in shape. Great diversity exists among ones such as these since sometimes the alteration occurs in the small members, like a finger, but at other times fetations are found altered naturally in their large, principal members. Certain animals sometimes have two spleens, while others have many kidneys. Changes of this sort are sometimes caused by a movement of the womb and the material and then, with no wall to separate, the animal will be monstrous such as when two are occasionally connected and have either one or two hearts. Such is the case of the goose we mentioned seeing above.[52] Further, many worthy of trust have related to us that they have seen such a man who was in fact two men joined at the back. One was rash and wrathful while the other was gentle. They lived more than twenty years and after one had died the other lived on until he himself died from the putrefaction and stench of his dead brother. Sometimes though they have one heart and they have other multiple members as a result of abundant material.

Further, it occurs in some that from generation on, certain passages of the visible openings (like the mouth, anus, urinary passages, ears, or nose) are blocked and it is found that they are closed off at their end or perhaps that the passage has moved from its own location into another. This occurs also in women when the exterior openings of their wombs, called vulvas, are closed off and blocked from the time of their generation up to the time of puberty. When the time comes for their menstrual flow, the passages of the womb swell with blood and a great heat is aroused which sometimes in and of itself splits the lips of the vulva and casts out the menstrual blood. Some females die for this reason but others are aided with the help of surgery.

---

51. Omitting *fiunt*, which Stadler brackets as an intrusion into the text.
52. Cf. 18.50.

69   Sometimes a male is born whose urinary tract departs from its proper path in the penis, this being at the lower part of the penis, and such a man must therefore urinate while seated. In a certain man born in our time the testicles were contained higher up within the skin in such a way that their outward bulge gave the suggestion of the two lips of a woman's vulva. There also seemed to be a split in the middle which was closed over by skin. Since the parents thought he was a girl, and that the split should be opened to ready her for intercourse, an incision was made and out leapt his testicles and his penis. He afterward took a wife and bore many children from her.

    Sometimes there is also generated one who has a blockage in his anus, through which the dry superfluity leaves. We have knowledge of this happening twice, once in the time of Aristotle and a second time in our day. Aristotle says that this occurred also in birds and other animals. For in Greece a cow was born and the superfluity of her food departed her through the place through which the superfluity of the bladder should depart. Noting this, they opened her anus, the passage for the departure of the superfluity of food. But the opening immediately joined together and was covered again. Though it was opened many times, it did no good since it immediately closed up again.

    Let it be, then, that the cause of multiple or few fetations and the cause of too many or too few members has thus been determined by us in this way.

CHAPTER IV

On Those Which Are Impregnated with Successive Impregnations and on Those Which Desire Copulation More Than Do Others

70   One should know that it is not a trait of large animals to become impregnated after impregnation.[53] For certain animals do become impregnated after impregnation and some of these form and bring to completion the one undertaken second along with the first. Others, though, do not bring the second to completion and it is corrupted.

    An animal, however, which is not impregnated by a second impregnation on top of the first is one which conceives and bears only one young at a time. Such are the animals which have horny hooves.[54] These do not become impregnated with a second

---

53. Superfetation, while rare, does occur in the animal world. In some cases ova produced at the same time may be fertilized by successive copulations and in others ova produced at different times and stored are impregnated. More succinctly: "some twins can undoubtedly arise from different fathers who have had intercourse with the same woman within a short period of time. It is interesting that in many archaic societies it was believed that all twins were conceived by this mechanism, called superfecundation. Superfetation, which implies differential fertilization of two oocytes released at a several week interval, has not yet been conclusively demonstrated in humans" (Leroy, 1991, 306).

54. *Soleae corneae*: cf. Glossary, s.v. "Hoof," and note that A. added the stipulation "horny."

impregnation on top of the first, with the exception of the mare, as we will mention below.⁵⁵ The larger animals as well are not impregnated by an impregnation of this sort. The reason for this is that the fetation of such animals is large in size, as are the animals themselves, and the superfluity which can be taken away from the mother's large body is scarcely adequate to nourish the first fetation, which has already done well in the womb. It draws the menses to itself to such a degree that nothing can be delegated to the second. An indication of this is that the young of an elephant grows almost to the size of a bull in the womb.

But animals which conceive many young at once do become impregnated by a second impregnation on top of the first because of the small size of the fetation, which thus draws in little food. The human, though, who is, as we said above, somewhat midway between these two, occasionally is impregnated by a second act of intercourse on top of the first impregnation, and the fetation does well. Just as has very often appeared in times gone by and just as we have said long ago in this study, the cause of this is the same as we stated in the case of a woman who conceives many young from a single act of intercourse where a great deal of sperm and an abundant *gutta* from the female are divided into many parts. For when the humor is abundant, after the first fetation a woman sometimes conceives as a result of a second act of intercourse even though this happens but rarely. Just as it happens that large-bodied animals generate only one offspring because all their nourishment is barely adequate for one (especially when the embryo has already done well in the womb and is drawing nourishment to itself with a strong attraction), so it is for this reason that such animals are not impregnated with a second impregnation, as often occurs in women. For while they may resemble those who conceive many offspring at once in the abundance of that humor which is the material for generation, the first fetation still draws that humor to itself most often and presents an impediment to a second impregnation.

Among all animals none is seen which takes on a second fetation after the first save for woman and the mare. The reason for this is the capaciousness of the wombs of mares and women and the toughness of the flesh of these females compared to other females. For of all the female animals woman has the most capacious vulva in proportion to her body size, even though man has the smallest penis of the animals in proportion to his body size. For an adult girl sometimes has an opening to her vulva which is longer than that found in any cow or large mare prior to giving birth. The reason for this is the size and the roundness of the head of the fetus which would otherwise only find exit through the death of the mother. Therefore when a capacious womb exists they conceive a second young over the first. But this one rarely does well since the first and stronger embryo draws the nourishment to itself.

Likewise, these two types of females desire copulation more than do others. It is the toughness of their flesh which creates (or rather gives indication of) this appetite. All animals which have tough flesh have full flesh because blood does not flow out

---

55. A. has discussed this earlier at 9.52f.

from them and the last food is poured into the members. They thus desire a great deal of copulation, especially woman and the mare. For after the first act of intercourse in which they conceive they do not menstruate and the embryo, as yet small, does not draw blood to itself. Thus the blood which is drawn in excites the nerves and the groin area and moves toward a greater lust than it produced before the conception. This lust is not quieted through intercourse, but may even be more strongly desired from the rubbing. For it is through rubbing that the seminal humor is further drawn in and causes further excitement. Likewise, very little or no superfluity at all leaves mares during the time for menstruation and that which excites therefore remains behind. All animals with tough, full flesh desire copulation for the reason given, since the nature of such females is almost masculine and because the animal only retains the humor through heat which digests well. But, more so than others, those females desire it from whom adequate blood does not leave during the time of menstruation.

73. This is the same reason that virgins think more about intercourse and desire it more than do those who have been spoiled and that women who have not been impregnated desire it more than those who have been impregnated often. For these women have flesh which is more filled with spermatic humor than that of the other women. For when we say that men and women with tough flesh desire intercourse more, we have in mind those whose toughness is caused by fullness and not those whose toughness is caused by dryness. A sign of this fullness in the body is when flesh pressed between the fingers bunches up and springs back from the fingers and does not fold up when pressed by them. The reason in all these matters is that the exit of the menses from the females resembles that of the sperm for the men and males. For, as we have said above, menstrual blood is nothing other than undigested sperm. For this reason too women who at first desire intercourse greatly desire it little or not at all after they have given birth to many children. This is because they have been emptied of their great store of seminal humor just as males who have been emptied of sperm do not desire intercourse. Women's wombs, moreover, are above, toward the diaphragm in certain women. But the testicles of the males are drawn up below since it is the male's business to eject the sperm. It is the business of the female, however, to draw the ejected sperm upward to the place for generation. Those women who have a great deal of sperm desire a great deal of intercourse.

74. The reason has thus been set forth why certain animals are impregnated with a second impregnation on top of the first and why some bring the second fetation forth and others do not. Also determined is why they are not all impregnated by a second act of copulation. The reason some desire intercourse greatly and others do not has also been shown.

Further, certain animals are impregnated with a second act of copulation even when a long time has intervened between the copulation of the first impregnation and that of the second. Yet they still will complete the second fetation. But this cannot occur in animals which lack a large body but it will be found in large, warm, and moist ones which naturally have a great deal of semen and many offspring. Because they have the capacity to generate many children, the wombs of the females have to be very capacious. And because the nature of males like this is suited to a great deal of sperm, they cast off much sperm during copulation. The humor which is the material for generation in such females is also great, for the hearts of such animals are small and do not consume moisture. For this reason menstrual blood in such animals is multiplied over and above the amount of food needed for the one conceived in the first act of copulation. It can nourish the second fetation on this and can complete it and bring it to birth.

In addition, the wombs of certain female animals are sometimes blocked at their openings since a great deal of menstrual blood is choking the opening of the veins there. This occurs in certain moist women and they also happen to have menstrual flow after their impregnation, lasting from the beginning of conception to the end of impregnation. This accidental trait does not occur naturally and that is why the fetations of such women are often flawed. When this same accidental trait occurs in animals not through a flaw, it is there by nature, as is the case in an animal with a hairy foot.[56] This one happens to become impregnated again after its first impregnation even though it is not one of the large-bodied animals but is one of those which conceive many young together and has many clefts in its feet. For every animal which divides it foot into many parts has many offspring and much sperm. An indication of this is that it has a great deal of hair, especially on its foot, as do the hare and the *cirogrillus*, which some call the *cunicus* or *cuniculus* [rabbit].[57] This animal has a lot of hair as well in the area of its thighs. An animal which is very hairy like this one is not only impregnated on top of another impregnation, but even sometimes gives birth afterward to the fetation but nevertheless completes the second fetation since, in accordance with its disposition, the impregnation remains after the birth of the first fetation.

75

CHAPTER V

On the Reason Some Animals Generate Complete Young

but Others Generate Incomplete Young

Further, of the animals which generate live young, some generate complete young but others incomplete. Thus all animals which have horny *soleae* [hooves] generate complete young.[58] So too does every animal which divides its hoof in two. But most of those animals which divide their foot into many clefts generate incomplete young.

76

---

56. "Hairy foot" is a good translation of the Gr. *dasypous*, hare, Ar. *GA* 774a32.

57. On the *cyrogrillus* and the *cuniculus*, cf. 22.46(28–29) and the notes to 3.143.

58. On the hooves, cf. note at 18.70 (where, as here, A. has added the adjective "horny"). As indicated in the Glossary, s.v. *Solea*, A. generally replaces the Gr. of *monycha* ("solid-hoofed") with *solea*. Cf. Ar. *GA* 774b5f.

This is because animals which have *soleae* generate a single fetus, while those which divide their hoof in two sometimes generate one and sometimes two (though they hardly ever surpass two). Therefore, because these animals and those like them generate few young it is an easy matter to complete them. But animals which generate many young at the same time can, to be sure, form them in the beginning when they need but little moisture. But afterward, when they have grown and require a lot of moisture, they are unable to complete them. Thus, they generate imperfect young, as do those which generate larvae. Therefore, some of them occasionally bear young which are not defined and are flawed. Examples include the fox, the lion, the wolf, and the dog. Some will be blind like the offspring of the dog and the animal which the Greeks call the *aycoron*.[59]

The sow, however, even though it divides its hoof [*ungula*] in two, generates many offspring in the manner of those which divide their foot into many parts. But certain pigs have solid, not divided, horny *soleae* such as is found on the pigs of certain regions. The sows thus sometimes generate only one young, while at other times they generate many. Frequently, however, they generate many young since the pig is a moist and cold animal and the food which should be passing over into bodily nourishment passes over into spermatic fluid. For this animal is not large like the animals which have solid *soleae*. Now the sow forms and increases that which she conceives in her uterus until the offspring is completed and she does not bear an imperfect one. This occurs due to the goodness and superfluity of the moisture which is in her body. Thus, too, a tree planted in land that is rich and has a lot of food completes its fruit.

Similarly, certain birds hatch out their young in a blind and incomplete state. These are the ones which produce many young and lack bodily size, such as the swallow, titmouse, sparrow, and small birds like them.[60] Similarly, the birds which hatch out few young but do not generate adequate food in the eggs for their complexion do much the same thing. Examples are the *kartagla*, the *fetta* [ringdove?], the pigeon [*columba*], and many other birds in our lands like the crow, the jackdaw, and many other genuses. Some therefore say that if the eye of a swallow chick is extracted while it is still imperfect, then it will grow back again, since the material of the eye is still incomplete internally. Nevertheless, this is not probable unless it were possible for the eye to be extracted without harming the nerves and the veins from which the material for the eye comes.[61]

From these things it is thus clear that certain animals generate incomplete young due to a lessening of food. This is also clear in the young which women generate in the

---

59. *Aycoron*: Ar. *GA* 774b17 lists the dog (*kyōn*), wolf (*lykos*), and jackal (*thōs*). Scotus's *noz* is fairly close, but it is hard to see a source for A.'s form, which has somehow become substituted for both wolf and jackal.

60. Compare this list and the one that follows with Ar. *GA* 774b28f.: *korōnē* (crow), *kitta* (jay or magpie), *strouthoi* (sparrows), *chelidones* (swallows), *phatta* (ringdove), *trygōn* (turtledove), and *peristera* (pigeon).

61. The disclaimer is A.'s.

seventh month. Because these young are incomplete, it often happens that they have defined and divided passages for the senses and visible openings for the nose, ears, eyes, and others.[62] Sometimes they do not have a clear definition to these but when they have grown they become defined. For this reason imperfect young of this sort are wrapped in soft layers of unsalted fat and are covered with skins. Milk is gradually dripped into them until they grow strong and are completed.[63] Many of those born this way grow strong and live, but they also frequently die.

In human generation males are stricken with this sort of flaw more often than are females. But this is not so in other animals. The reason for this is that there is a difference in natural heat between male and female. Males are hotter and thus as they leave they move both more and more swiftly than do females. Because of these males' many movements, their members, while still tender, are stricken with more flaws of breakage and injury, as well as those of other sorts than occur in females.

Heat, then, is the cause of the difference in the completion of producing the differentiation of male and female in women's wombs. This is because their wombs are not the cause of the differentiation of male and female in the manner mentioned. In other animals, however, no difference in the creation and completion of male and female is apparent, for the male in these animals is not completed more swiftly nor the female more slowly.

Further, males take on defined and completed creation in less time than do females in the wombs of the women. But after birth, human females grow and are completed more swiftly than the males and they also grow old more swiftly. This is because females are weaker than males due to their colder nature. For as was proven in earlier parts of this study, one must hold the opinion that the female sex is a sort of natural flaw with which a fetation is stricken. Therefore, when it is within the womb it must take on complete definition of its members slowly because of its natural coldness. For complete definition of the members is a particular type of digestion and heat digests and defines more quickly something more readily digestible, this being that which is hotter.[64] But the female of the human species grows and is increased more swiftly outside the uterus because of the weakness of its nature. All things having a lesser and weaker power achieve more quickly just that sort of completion they can sustain with nature permitting. This is because they are moister and thus they have less power to oppose a contrary force. This happens in all of nature's activities which are bounded by natural heat and for the reason we have often stated previously.

79

---

62. "Defined": *distincti*. Scotus's *indistincti* seems to make more sense.

63. This comment on medieval care of premature babies was added by A.

64. If one follows Stadler's indications of the *post illa* style, it appears that A.'s received text said "For heat digests, and that which has better digestion has greater heat." This tallies well with Ar. *GA* 775a18f., but A. saw fit to "clarify" his received text with interpolations of dubious value.

CHAPTER VI

On the Disposition of Impregnated Females As a Result of This Impregnation and on a Mole of the Womb

80 It seems a trait peculiar to woman that when she is pregnant with twins and one is male and the other female, they grow strong with difficulty. If, however, they are the same sex, they will often be saved. In other animals this does not happen, for those animals which produce many offspring produce males and females mixed together, whereas producing them all the same sex is not natural. But in women there will naturally be no difference of male and female in twins and if there is they will hardly ever do well because of the contrary nature of the qualities existing in the same semen.

Further, in women as well as in other animals a disruption and ruining of appetite sometimes occurs after giving birth. Even though we have designated the cause of this already far above in this study, we should still talk here about its diversity. For a ruining of appetite of this sort occurs more in woman than in the others, for females of many other animals show much more growth during a pregnancy and have more health than when they are not pregnant.

Likewise, in women who toil and perform manual labor the disruption of appetite due to pregnancy is not as visible as it is in others. Perhaps women of this sort give birth or suffer a miscarriage more quickly since their work consumes their bodily superfluities. But the bodies of women who are sedentary generally have many superfluities. Toil causes many superfluities to evaporate and these women thus give birth all the more easily when the time comes. For they can hold their breath more easily and retention of breath makes for ease of giving birth. But a great deal of superfluity that has massed together works toward the contrary effect.

81 These, then, are the things which pertain to female animals as a result of their loss of appetite and complexion due to pregnancy. This is especially true in the case of women since the females of the other animals do not have as noticeable a menstrual flow as do women. For this reason too, when the menses first stops after impregnation and is not yet being drawn into the embryo, it flows back to the stomach and liver and will thus be the cause of the ruining of the appetite and of bodily change, especially in women.[65]

Further, when women are not pregnant and still lose their menses, this will still be a cause of bodily change and corruption and of serious illnesses in certain women. During the early stages of pregnancy, however, some women are greatly fatigued from the retention of their menses, as we have said, for at this time it is not consumed as food for the embryo. But as the embryo grows at the end of the pregnancy, the

---

65. A. has interpolated heavily into Ar.'s text to explain morning sickness.

aforementioned superfluity is consumed and the women will do better. For once the woman's disposition has been made better by the consumption of the menses, then the body will be relieved of its burden. But in other animals there is but little of this sort of superfluity and it is proportionally tempered to an amount adequate to feed the fetation. When, however, a superfluity of the sort which ruins the appetite is consumed and passes over to be food for the fetation, then the body of the pregnant woman is also better nourished for that period which is the end of pregnancy. This is why they require a great deal of food during that time, both for themselves and for the embryo, which at that point is fairly well grown.

The infirmity which strikes many women and which is called a mole of the womb occurs most often following impregnation.[66] A certain woman in Aristotle's time had intercourse with a man and believed that she was pregnant since at first her belly swelled and then the signs of pregnancy appeared in her. When the time to give birth came, she did not give birth and the swelling in her belly did not subside. Rather, she remained in that swollen state for three or four years. After this she was stricken with a dysenteric flux of the belly so that she was brought to the point of death. Thereupon she gave birth to a lump of flesh which was a mole of the womb and was nothing other than blood which had coagulated in the womb. This infirmity sometimes lasts in the womb into a woman's old age and occasionally lasts until her death. I saw a case of this in a woman who had two round masses in her womb until she died. When it has lasted a long time it turns very hard, so much so that it might not be split even with an iron tool. Midwives think that they are the heads of rams or goats, since it comes to a point toward the womb's opening but is rounded toward the concave interior of the womb.[67] When a fetation finds a mole of the womb in the womb then a woman rarely avoids death during birth in the north lands.

We have determined above in this study the cause of this superfluity. For the same thing happens to what is conceived in a womb as a result of bad digestion or decoction as happens to flesh which has been cooked. For this does not happen on account of heat, as some believe, but rather is due to weakness of the heat. For when a nature is weakened, then it cannot fulfill the complexion of that which must be digested and it therefore remains in the womb for a long time without putrefying until, perchance, the woman grows old. Now it is in no way like a completed and natural thing but neither is it entirely outside its nature and it therefore endures. There is no cause for its hardness other than bad digestion and corruption of digestion. Still, there is something of digestion in it there and thus it has a certain special sort of nondigestion which converts, to be sure, but which also does not form into a body that which has been converted. Neither is the nature of the member strong enough to be able to expel it.

We should make inquiry into the reason that this infirmity does not befall other animals lest this totally elude us. As far as we can conjecture, the cause of this is that

---

66. Cf. the discussion of moles at 10.31f.
67. Cf. the discussion of a "mooncalf" in the notes to 10.31.

the womb of a woman is contrary to that of other animals and there is a lot of blood in it which cannot be readily digested. When the fetation is nourished and the nourishment exceeds that which is adequate for it, and when it has a bad digestion and cannot exit because of the closure of the womb, then that lump will be generated in the wombs of women which is called a mole. This then is the reason this sort of infirmity occurs.

CHAPTER VII

On the Generation, Cause, Usefulness, and Diversity of Milk in Animals Which Have It

84 Milk is generated only in the females of animals which generate live young like themselves. It is good and useful during the period of birth since nature has created milk to feed the young outside the uterus as long as it is unable to take in external food on its own. For this reason too milk should naturally not be present in such women before birth nor for long after the period for nursing the young has passed. If it is found otherwise, it will be contrary to nature. From this we take it that since the gestation period of the animals other than the human does not vary, but only does so in woman, then the digestion and completion of milk must also be at a set time in the other animals since in women the time varies. Since the earliest time for birth is in the seventh month, the milk should also be good in the seventh month in a woman. And it is during this time that the milk begins to turn good. It is reasonable that this happens since in this way it is following a natural cause.

85 Because milk has good decoction, it eventually dwindles at the end of the period designated by nature for lactation and is led back to nourish the body while its impurities are purged through the menses. For the food of all animals is naturally nothing save that which is sweet and well digested, as was proven elsewhere where we spoke of tastes.[68] That which is cast off from this sort of digestion is salted and has a bad complexion. When the creation of the young is complete there will be more superfluity than what is consumed and retained as food for the mother but there will be less than the menses was before. Thus, milk that is well decocted will be sweetest and the further it is removed from birth the less of it there will be, but it will be sweeter and more suited for food. For then, that which is thinnest and best is not taken away from the milk since the mother does not need it to create the young. For this reason too the first milk is the worst, since that which was thinnest and best was held back for the creation and feeding of the young in the uterus.

The period of lactation comes after birth, then, because that which previously was separated out for the nourishment of the embryo is not separated out at this time. It

---

68. *De sensu et sens.* 2.6–7.

thus amasses in the breasts since they are colder places, are capacious, and have white flesh. It also takes on its color in them, for the reasons we have given both here and far above.[69] For these places in a woman are on the outside of the body and near the diaphragm so that they can be near the principle where life resides. For an upper location, especially in a woman, is better adapted to receive superfluous food of this sort because, if they were udders below, they would hinder a woman's walking when they grew swollen, filled with the food of milk. Because of its nearness to a member which is the principle of life and heat, the superfluity of sperm mostly exits from elevated places for the reason we stated previously.

The superfluity of the sperm of males and of the menses of females comes only from a sanguineous nature, the principle of which lies in the veins and in the heart. For the veins are the vessels of the blood and for this reason it is necessarily in the veins that the blood is changed into sperm and milk in the vicinity of the upper areas where the heart is located and where the veins arise. An indication of this is that the voices of animals change during the first dripping down of the sperm and of menstrual blood. For the voice is generated from the upper areas from which the sperm and menses flow. For the voice is changed due to nothing other than a change in that which moves and strikes the voice.

Further, the fleshy parts of the breasts [*poma mamillarum*] are more visible and prominent in females than in males since more superfluity leaves females than males. The same thing happens in animals whose udders are on the lower part of their bodies, even though there could be another location for the breasts. This is known from the spermatizations of animals. The time and the spermatization itself differ in humans greatly and the cause of this is due to the females' great superfluity, which is necessary for the material for generation and is more than that which is in the males. Now it is suitable that there exist naturally more of the material superfluity of the formally efficient cause which is present in the males. For that superfluity which is the material one in the females needs to be greater in order to see to the growth of the embryo in the womb so that there will be adequate food for the young when it is outside the uterus as well. Therefore, when the embryo is at the end of the pregnancy and does not take on this superfluity and it cannot leave through the menses, it then is gathered into the empty vessels that are designated for the young's feeding. It is led there along the passage stretching from the womb to the breasts. This is called the *ryvertis vena* [inferior epigastric vein], as we have determined above in the anatomy section.[70]

Breasts are present in the animals we have mentioned for two reasons. One of these is that of usefulness, namely, that milk be contained in them for use in feeding the young. For during lactation a digested food comes from them which is suited to the nursing (that is, suckling) animal. The reason for the good digestion of this food is the one we gave above, namely, that after the embryo has grown and been formed in the uterus, it must have food which it can take in outside, and that which is superflu-

---

69. Cf. 1.439.
70. Cf. 1.416.

ous to it is designated to become milk. And the less of this there is, the better digested it is, especially in one who does not copulate and is not impregnated. This is as we have stated far above in this study. For we said above that milk is a natural food and is prepared by nature for the young. For generation and nourishment are from the same things and the material for each of them is none other than the sanguineous humor in blooded animals. Milk is nothing more than decocted, digested, uncorrupted blood, and thus the statement of Democritus about milk is proven false.[71] For he believed that milk is generated in the eighth month and that in the tenth it was *virus* and not pure milk. For *virus* is generated only from corrupted digestion, whereas milk comes from completed digestion.[72]

We have, moreover, determined in earlier parts of this study why the first milk is virulent.[73]

Further, nursing and pregnant females have no menstrual flow save just a bit and that only in certain individuals. For the milk and the menstrual blood have the same material and nature is not able to provide the material for both of them at the same time from the superfluous food. When the superfluity of the food passes over into one of them, it is taken away from the other, except that in some women it happens otherwise. This is due to a certain accidental trait, not one that is natural to a woman insofar as she is a woman. For that which is for the most part true and occurs in most cases is said to be natural and to be done according to nature's course and ability.

These things, then, which have been said here about milk are, along with those which were determined in previous places, adequate for our current observations.

CHAPTER VIII

On the Length of the Gestation Period in Animals and the Cause for Its Diversity

88   The periods allotted for gestation in animals have been reasonably fixed and are set at a fixed time. For immediately after an animal can no longer accept food in the uterus through the *umbilicus* the food itself is led to the breasts through the *rivertis* vein and is prepared to become milk. Since food no longer passes through the *umbilicus*, the veins which are wrapped in the web called the *umbilicus* are narrowed. Thus, at this point the offspring drops and leaves the uterus. Now the generation of all things is naturally headfirst, so that the head points down prior to its departure. This is like what happens in the balance arms of a scale on which that which weighs the heaviest sinks down. So too in the young the upper part is heavier and this naturally descends and leaves first.

---

71. Ar. *GA* 777a9f. quotes a passage of Empedocles, not Democritus.

72. For other uses of *virus*, cf. 6.34. For this usage, cf. 6.105. Note that "virulent" below is a pun on *virus*.

73. Cf. 3.169, where this first milk is said to be no good at all.

The periods allotted for gestation in animals are fixed proportionally to the length of their lives. Thus an animal with the longest life span should naturally have the longest gestation period. Even though this is not the cause of this it still happens commonly in most cases. Likewise, a large-bodied, blooded animal lives a long time for the most part but it still does not always follow that every large-bodied blooded animal has a very long life. For the human is an animal without a large body and he lives longer than every animal except the elephant. This is known from experience [*experimento*] and yet the human is smaller in size than many hairy animals with tails and is also smaller than many other species of animals. The reason for the length of a human's life is found nowhere other than in his warm and moist complexion which resembles the surrounding air. It is also for this reason that each thing lives longer in air that is naturally suited to it. There are also other causes of long life which we have designated in the book *On the Causes of Long Life* [*De causis longioris vitae*] and which we will speak of again in books to follow.[74]

The cause of the length of a gestation period follows the size of the animal conceived. For creating a large body cannot be completed in a brief time any more than, in the case of things made by skill, a great thing can be completed in a small time. Because of this the horse and those like it which live for a shorter time than do other animals have different lengths of pregnancy even though they live the same length of time. For some of their pregnancies last a year while that of others is for ten months. The female elephant is pregnant for two years because of the enormous size of her fetus.

However, the lengths of pregnancies of all animals are reasonably determined according to their life spans and they are marked by the periods and revolutions of the heavenly bodies. I am calling "revolutions" the turnings [*conversiones*] of days and nights, months and years, as well as of intervening periods. For the moon, which is the queen of the bodies below, has a revolution which marks off those things below according to her conjunctions to the sun when she first begins to have light and also to the completions which she takes on in opposition when she is filled with light in that part of her orb which is turned toward the things that have been generated. She also has this in reference to the "seventh" when there is a half-moon [*dycothomis*], that is, a half-distance in losing or taking on light from the sun, taken through a right angle in its center.[75] For this, a fourth part of the circle is subtended, as is proven in the third book of Euclid's *Theorems*.[76]

---

74. *De morte et vita* 2.2.

75. Cf. Ar. *GA* 777b20f. According to Martianus Capella's *Marriage of Philology and Mercury* 7.738 and 8.864 (Stahl and Burge 1977, 282, 336) the *dichotomon* is the half-moon which is halfway illuminated by the sun at an eastward elongation of 90 degrees. The "seventh" likely refers to the fact that the number seven marks the lunar quarters, which require approximately twenty-eight days for completion. Each quarter, then, requires a "seventh" or seven days, one-fourth of twenty-eight. As Macrobius notes in his commentary to the *Somnium Scipionis* 1.6.45 (Stahl 1990, 109), the number seven is especially suited to the moon, which is the seventh of the planetary spheres and has seven phases: new moon, half-moon, gibbous moon, full moon, gibbous moon, half-moon, and conjunction. At the moon's conjunction, its light is entirely invisible.

76. Cf. Euclid's *Elements* 3.27.

The moon moves to this place in seven natural days after its conjunction with the sun or from its position in opposition to the sun. This is thus called the "seventh of the moon" or the "seventh day of the moon." For all these accidental traits of the various phases of the moon occur according to its varying proximity to or distance from the sun. A common month is the revolution of both, namely the sun and the moon, and especially a month which is called "solar." For the moon only has its light, which is a principle of movement in generation, through the interrelationship it has with the sun in taking its light from the sun. It is therefore rendered like a second sun with respect to moving things generated and the moon thus has a strong share in completing in many ways the things that have been generated. For heats and coldnesses which have been tempered are firstly suited to types of generation and secondly as well to types of corruption. And the principle and the completion of these natural heats lie in the movements of these stars.

91 An indication of what we are saying is that all watery things are quiet or move according to the sort of movement the winds have. But winds only move and are quiet according to the revolutions of the sun, as we have already discussed in many instances in the *Meteora*.[77] For in all cases the movements of lower and baser things necessarily follow the movements of those which are higher and nobler. Thus, to an extent, the winds have the movements of the stars, especially of the sun and moon. And just as it is outside in the world, so too is it in the bodies of animals. Thus, every revolution of the generations and of the completions of size and life in animals will be ordered according to the revolutions of the stars. But because the material of generated and corrupted things can in no way be set down to a single thing, the confusion of material does not allow that the changes in things below be entirely as regular as are the revolutions of the things above. For confused modes of generation and corruption of the material preclude this. For this reason the time for the creation, growth, and corruption of the lower things is sometimes lengthened or shortened. But this works out contrary to the ordering of nature who universally orders things. We have discussed this at the end of the second book of the *Peri Geneos* in a sufficient amount to satisfy the purpose of a natural philosopher.[78]

---

77. *Meteora* 3.1.5.
78. *DG* 2.39f.

CHAPTER IX

Which Is a Digression Setting Forth in Easy Summary Fashion All the Things Which Have Been Introduced Above in This Book. Also on the Levels of the Powers of Generation and to What Extent Consanguinity Is Lacking in the Levels.

In order to facilitate the understanding of all the things which have been discussed in this book in accordance with the wisdom of the Peripatetics, we should once more take up from the beginning and state certain of the things which have already been treated thoroughly.[79]

We thus state here about the cause of masculinity and femininity that although the heat of the right side and the heat of the womb contribute to masculinity, this is still not the universal cause of this sex in a fetation. For according to this a male and a female could not be generated at the same time in the same womb with the same complexion.

Further, when we say that a cold sperm causes and produces a female, it does not follow because of this, conversely, that a warm sperm always produces a male and not a female, as is clear from what has been said before.

The reasoning of those who said that this occurs when the sperm falls onto the right side of the womb or comes from the right testicle is not good reasoning, but it does come somewhat close to the true cause since the right side is warmer. But their error was in positing something remote as the proximate cause of the thing and as interchangeable with the thing itself. Now the proximate and essential cause of this should be sought in the passive, material principle of the embryo and in the operative principle which is in it. When the material generated is digested and is warm, and the effective principle is digesting the warm thing proportionally, then most often a completed male will be generated. If, however, the material is not submissive and if too the active power is weak, it will, to be sure, generate one in the same species as those generating but not a completed one. And it will furnish for it the instruments suited to an uncompleted generation, these being the womb and the vulva, acting passively and materially in generation.

When the sperm moves and goes through the material, whether it be entirely submissive or not, first to be formed is the principal member whose complexion all others follow naturally. And from the complexion of this one the testicles and the seminal vessels for the male or the female sex are formed afterward. This is because the seminal vessels are second in principality and are beneath the heart, which is more

---

79. Based on Avic. *DA* 18.52v.

principal than all others. For the heart is generated necessarily in a given complexion and if that is weak, raw, and moist, or if the heart is generated warm and dry at first and then afterward there comes over it something which changes it toward crudeness before the other members are formed, then the disposition of the female sex will proceed from the heart into the seminal vessels and the entire body. But it is otherwise for masculinity, for even if the substance of the heart first should undergo *molinsis* [corruption] and afterward is strengthened and comes to the digestion of completed heat, it will lead to the masculine sex.[80] Also, through means midway between these two there will be generated women with male traits and effeminate males. For if, because of a given complexion in it, the material is not subservient to that which is acting but still takes on its basic outlines and does so weakly, following the nature of moisture, what is generated will be a male but it will resemble its mother in many points. But if the sperm should overcome perfectly the material and not be changed by that power which it partially possesses, then the fetation will resemble the father perfectly. But if the material is changed in its complexion from that which it has from its mother's body, then it takes on a casual, accidental complexion and falls away from a likeness to its genus. The same holds for a change in the sperm from the complexion which it has from the father.

94   This is confirmed by indications, for a youth and an old man most often generate females due to their humor and crudeness. Those, however, who are between the two ages generate males unless some accident of complexion hinders them. Likewise, those with loose and soft flesh or those who have the south winds loosen their bodies generate females. But males are often generated by those with tough flesh and by the north wind. Heat, however, should be tempered in generation and it thus happens that one with untempered, warm sperm does not generate out of a woman whose *gutta* is also warm but rather from another whose *gutta* is such that his sperm is tempered by it away from its untempered heat. It is similar for a female compared to the male.

On the reason for resemblance, once should know that resemblance is to nature and to a person, as we said above. The first is called natural and the second personal. This second is twofold, namely, common [*communis*] and specific [*propria*]. The common is a resemblance to a race of a given clime such as to the Germans, French, or Romans. The specific is that which resembles a single person of its genus in a determinate way, from whom the generation takes its formative power and movement. For this reason, if the sperm overcomes perfectly and is acting in the power of the person of the father, then there will be a perfect resemblance to the father. And if the material overcomes perfectly and is passive in the power of the person of the mother, then the act of the active powers will be in her only in accordance with the power and faculty of the mother and the offspring will thus resemble the mother perfectly. If however, each of these both overcomes and is overcome, then the resemblance will be a mixed one to both the mother and the father. And because the sperm has powers from the father's ancestors and has material from the mother's ancestors, if the powers of any

---

80. On *molynsis,* cf. 16.72–73.

of these are led back to action through some accident, then the resemblance will be to the ancestors on the father's or mother's side or will perhaps be a mixed one to both sets of ancestors. But this does not extend back beyond four generations because all power is measured in four levels, as is proven by the operation of medicinal simples.[81]

There are four items in the material of generation, these being substance, power, and operation. But substance is really two substances, namely, the material and the efficient. Thus, these four are very often overcome by four mixtures. Thus, in the first, the substance is mixed with something external so that it lacks operation but retains a hindered power. In the second mixture it loses its power but retains substance, although a confused one. In the third the efficient substance recedes from a likeness to the first from which it descended. In the fourth the material is also confused so that it retains nothing of the first which can act on another. But there is nothing external whatever in a line descending directly through the grandparents and the fathers and thus this one is never lost. For this reason Aristotle said that sometimes a likeness occurs to an ancestor the recollection of whom is not in living memory.

But it seems, because of the things that have been said, that, due to these four, the consanguinity of brothers and sisters is lost when they marry people external [to this lineage]. For there is a gradation of power for the four reasons mentioned above, and it is set in four gradations. For because of the material, the power alters and impresses. But because of the efficient cause, it alters and, by seizing it, holds other powers not like itself. When the strength of the efficient cause is increased, it alters, holds, and impedes the operation of the others. And if strength of this sort is increased still more it will alter, hold, and impede their operation and will diminish the substance which is the subject of the opposite powers. And we see all these things in the power which forms generation. Nor can we adapt impediments to it except according to the number of its mixture with people external to the line whose substance, power, and operation do not have a resemblance to them according to generation arising from the generation and the principles of the material of generation in the sperm and menses. For although these things are also impeded by the accidental causes of air and food, nevertheless they are not in the purview of art and neither can an impediment generated from them be known by reason.

And this is the reason, according to the things that have been said, that consanguinity with the grandparents and great-grandparents is never destroyed, while that of brothers, sisters, and other lateral relatives is reasonably thought to spread out over four generations. For each of the grandparents, great-grandparents, and parents has something of himself (which in potential and in power is all that he is) present in his next successor. Thus this bond can never be broken. But brothers and sisters, great-grandsons, and great-granddaughters are not bound to one genus because one of

---

81. In A.'s youth as well, the Fourth Lateran Council (A.D. 1215) reduced the bounds of consanguinity from the seventh to the fourth degree. As a result, a marriage contracted between two individuals was invalid if it fell within these four degrees, that is, if the two parties shared a common grandparent. Cf. Le Bras (1968).

them has something in the other but rather because they have taken that which they are from one particular individual through their original generation. Because they are bound through this thing that comes from someone else and it is not their own, their relationship can be dissolved.

If, however, the power of the sperm and the material falls away from all resemblance to the nature of the species and of the individual, then all that remains is the nature of the genus and they will be monsters, tending to the shapes to which the stars move them. In some members they may retain delineations of their fathers or ancestors, but of these things enough has been said in the earlier parts of this book.

97 Now Democritus has said that the cause of such monsters is one fetation on top of another and one act of copulation over another. While he has not stated the principle cause, he has still stated the cause which frequently applies. For when there are two sperms in the womb, one draws more forcefully at the material of the other and thus there is created either an overabundance in the one doing the drawing in or a loss of material in the weaker one.

Likewise, whoever has stated that the movement of the womb causes this has also touched on a secondary cause even though he has not stated the principal cause. For it is a monstrous thing when many fetuses are conceived in one and the same enclosure and there must necessarily be some cause for the division of the sperm. This is sometimes divided into such small things that the ones conceived do not have adequate size to live. They are fetations about finger size, as we have stated above concerning the woman who conceived seventy or more of them.[82]

98 One should also know that it is generally a convertible statement that those which split their foot into many parts and are not large-sized conceive many young at once. Those which are large-bodied conceive but one.

It is likewise a convertible statement that in most cases one which conceives a single young is also one that does not have one impregnation on top of another. But the mare and woman are exceptions to this because of the strength of their complexion and the abundance of their material for generation, just as we have said before.

We should not fail to mention that the bodies of women who have cleaner and fewer corrupted superfluities bear better and more perfect offspring. Those, however, who have numerous corrupted superfluities in themselves bear imperfect and unhealthy young. For the corrupted superfluities are held in their wombs and they corrupt the bodies of their young. They therefore bear those which are lame, blind, suffering from leprosy, epilepsy, and laboring under other diseases as well.

99 One should also know that of all the animals which have blood, the human naturally has the longest life span after the elephant. But the gestation period and life span in all animals follow the revolution of one or more stars. And because over the circles that are distant from the equator and over the poles of the earth there is one revolution to the stars, then some animals have a life span of one day. The second period is

---

82. Cf. 9.50.

a revolution of the moon over a quarter circle and thus some animals have a life span of seven days. The third period follows the return of the moon to the sun and the first images and shapes of a circle. According to this one some live for a month although others live for a year due to the turning of the sun, while still others take for their life span the year of the circle which is twelve years.[83] Still others take on the years of other planets and stars and constellations and just as they take their life span from these so do they take on the length of their pregnancy and conception.

These, then, are the things which are sufficient for our present purpose concerning the causes of impregnation, the types of young conceived and the times of conception and gestation. Other things should be asked of an astronomer.

---

83. E.g., the orbit of Jupiter is 144 times that of the moon, and therefore its circuit requires twelve lunar years. Cf. Martianus Capella, *Marriage of Philology and Mercury* 8.861 (Stahl and Burge 1977, 335).

HERE BEGINS THE NINETEENTH BOOK ON ANIMALS
Which Is Whole in a Single Tract and Which Is on Those Things Which Naturally Occur in the Sense and Parts of Animals

CHAPTER I

On Accidental Traits in General So That One Might Know Which Things Must Be Investigated in This Book and in What Way It Is to Be Done

1. Previously, in the book which has gone before, we have determined the disposition and the quality of those born within and those born outside of a womb and we have done this with a determination that is universal (suited for all animals) and with one that is particular and individual (insofar as it suits each animal according to its own proper nature). And we have specified the differences according to which the members of animals differ and are differentiated one from the other, with special attention to that which properly belongs to humans, the most perfect animals. We have also made a determination about all the members, both internal and external, of animals. It therefore remains now to consider the accidents which occur generally and individually, according to which the members of animals have certain differences as well.

   I am calling accidental traits things such as the blackness, greenness, and multicoloredness which occur in eyes, things such as sharpness of vision and changes in color and hair, and the differences among feathers and plumes. Some of these accidental traits are present in all, or most, genuses of animals while others have a cause that is peculiar to one genus, such as to the human or to another genus of animal.

2. Further, certain changes in animals occur in accordance with accidental traits. These types of changes are sometimes general, such as change in accordance with ages and first qualities.¹ But other types of changes are peculiar to certain ages, such as a change in the voice during the time of the distillation of the semen and a change in coloration of the hair during old age. For animals other than the human do not grow white visibly during their old age; this is an accidental trait peculiar to the human out of all the other animals. Now to be sure, certain of these accidental traits, such as eye color, follow upon the nature of the animals from the first moment of their birth.

---

1. On first qualities, cf. 11.32–33; 16.19; and 20.75.

Other traits, however, overtake them during youth, like the change in voice, while still others come upon them in old age, like whiteness of hair.

Now it should not be thought that all accidental traits of the sort we have mentioned, and those like them, occur universally in every genus of animal or even that they occur universally within one and the same genus of animal. For those things which occur in some individuals and not in accordance with the common nature of the species or genus do not occur universally and for the sake of something [*propter aliquid*], this being the end in nature. Rather, such things are brought about by a variation of the material. An example of this is that the eye which is, to be sure, connatural to those animals which see according to the nature of their genus, exists for the sake of something in nature. But greenness of the eye or blackness or grayness or certainly multicoloredness of the eye is not for the sake of anything, but rather occurs due to a variation in the eye's material, unless perchance one of these colors is found in all the eyes of a certain genus of animal. In this case it is doubtlessly intended and perfected by nature for the sake of something.

Such accidental traits are not to be ranked with those which are present substantially, according to the common nature of the genus or species. Rather, they are due to a variation of the material or have as a cause the mover disposing the material in this way or otherwise through changes introduced by it.

For as we have said in other places, namely, where we had a discussion on logic, it does not follow should everything exist according to a certain quality of such a sort that generally everything in that genus will exist in a determinate way according to this or that quality.[2] Because in this or that individual there exists variation, although some quality should be present in the genus. This is especially so in the ordered and set actions of nature, for these take on a great number of variations from the material and from changes. For in those which naturally take on generation, it is necessary that in one way or another there be something in the generating principles prior to the substance. Therefore, substance is not for the sake of generation but rather generation and the principles of generation are for the sake of substance as if for the sake of an end. But the diversity of the accidental traits has nothing to do with an end but rather it happens that material exists in this way because of changes in the moving qualities.

Although substance might be the end for the sake of which generation and the principles of generation exist, the most ancient of the natural philosophers have nevertheless held an opinion contrary to what we have said. For they have not looked into all the types of causes. They were content with a consideration of the material and of the efficient or the moving cause, and they handed down nothing certain, clear, or set, even about these. They entirely dismissed the formal and final causes. We, however, should principally investigate these two causes in all matters which are related to generation since generation is related only to the end and the one generated exists only through form and substance.

---

2. *De praedicamentis* 5.1f.

According to the things that have been said, then, the creation of the eye is necessarily for the sake of something. But it is not necessary that the configuration of the eye have a given disposition—of color, of being deeply set or not, and of other accidental traits—but rather this happens because nature performs or undergoes a given act which brings about such an accidental trait out of material and changes.

CHAPTER II

How the First Accidental Trait Which Befalls the Young Born Either in the Womb or Outside It Is Similar to Sleep

5 With these things specified and established in this way, then, we must begin to determine the causes of the accidental traits which are consequent to animals from the first moment of their generation up to the last moment of their old age.[3] These accidental traits are those which belong to the senses or their organs, since sensation perfects an animal insofar as it is an animal.

Let us consider, then, the first accidental trait which bears a similarity to sleep and wakefulness. For sleep and wakefulness belong to all animals and all senses. We should know first, therefore, that all the young of animals and especially those which are born incomplete are, when they are born, in a state similar to sleep or dormancy. This is present in them due to an accidental trait they suffered in the womb, for in the womb they were accustomed to exist as if sleeping.

But a not very easy question arises concerning this, namely, whether wakefulness or sleep first occurs in the young at the first moment of its making, when it takes on life. We, however, say on this question that if sleep is taken in its broad sense and is said to be an immobility of the senses due to an upward flow of nourishment for whatever reason, then sleep is the first thing which occurs to a live born fetation at both its first moment of creation and its making. Wakefulness, however, does not occur in it at first. The proof of this syllogistically is this: every change which has to be made in a thing moving from one contrary to another has first to be made through a middle before it comes to the extreme. Moreover, the change in one generated is made from not-living to living, and it operates according to the powers of life, as we have said in the study *On the Soul*.[4] It will therefore be accomplished through a medium and sleep seems to be rather like a thing between the sort of nonliving and living which we have already mentioned.

6 For it is not said that one sleeping is in no way alive in act, but only potentially. For it is not alive as is one who is awake, especially with respect to that sense accord-

---

3. Ar. *GA* 778b19f.
4. *De anima* 2.1.6.

ing to which sleeping and waking occur. For life properly lies in wakefulness because of using the senses. For wakefulness is nothing other than the expansion outward of spirit, heat, and sense. For this reason too it should not be held that the disposition which befalls the young according to the principle of generation is truly sleep. It is rather something like sleep and is a certain immobility arising from a great flow of nourishment upward reaching the point where the sensory organs are. This immobility is like that of those animals which are midway between plants and animals and which move with progressive movement from place to place. For there are certain principles of the senses in them but they are restrained, preventing them from waking to act and operation. Now, an embryo at its first moments of creation has this sort of a life and in this respect has a life that is like that of plants because there is no movement or operation of the senses in it, only a strong movement of nourishment and growth, as there is in plants. In plants, though, and in those of this sort which are midway between plants and animals which move from place to place, there is no true sleep. For sleep and wakefulness are present in the same way, and because wakefulness cannot be present in such animals, then sleep also cannot be in them in any true sense of the word. But neither can this disposition, which we say is like sleep, be present in trees for there are no principles of the senses in plants.

An animal in a womb for a long time, and especially the human fetation, necessarily has a resemblance to sleep of this type. This is because it does not have free breath (although it has some, as we have shown previously) and especially because of the weight of its upper parts which are weighted down by the great flow of nourishment to them.[5] But we have established the causes of sleep in the other books which we have written, *On Sleep and Wakefulness*.[6]

Further, an embryo sometimes appears to be awake and moves in its mother's womb. This is so not only in the human but also in other animals, as is proven moreover through dissection in oviparous animals in whose eggs the live young move. And after waking they are again overcome with a thing like sleep and again, suddenly, for whatever reason, they are awakened. But movements of this sort are disordered and are like those of an epileptic.

Further, for the same reason, a human infant sleeps for a long time even after it has left the womb. When waking it does not first smile, but rather when waking it emits its first vocalization in crying. It gives its first smile when asleep and when asleep it does not cry. In some it happens that it will cry in the womb with a shrill shout and the soothsayers say that this is a sign that such a fetation will be evil in days to come.[7] However, the reason is that the one sleeping does not feel loss and harm, especially because it sleeps deeply due to the great flow of nourishment, but a slight flow

---

5. St. lacks paragraph numbers on 1248–49. We have supplied numbers 7–9 at natural divisions in the text.

6. *De somno et vigilia* 1.2.3.

7. This is A.'s comment, not Ar.'s.

of nourishment when it touches the diaphragm moves it to laughter since it lightly touches the nerves inside.

8   Dreams also occur in the genuses of animals during sleep and in addition to dreams it also happens that some perform acts of wakefulness in their sleep. They perform these acts without knowing it. Some people walk in their sleep, make distinctions among things they see, and perceive things in their way. Enough has been said about these matters in our books *On Sleep and Wakefulness*.[8] But what we have said here was brought up because it is not customary for children to perform the activities of wakefulness in their sleep, because they do not perform these activities, being overcome with sleep instead. Still, they habitually have sense and life in their sleep (as a disposition [*in habitu*]) until the period of early age has passed when the upward flow of nourishment is tempered and that of the nourishment which is heavier begins to flow downward. For then the spirits and the senses are clarified in the upper parts and the child becomes purged and begins to awaken to the use of sensible life and the senses. In certain other animals this seminal disposition of the earliest age or generation prevails more, because they do not generate young which are completed in respect to their sensory organs. They therefore must take on growth in the upper portion of their bodies (for their sensory organs are here) for a longer time after their birth. For this reason all the whelps of such animals are born without sight, such that both the material of their eyes is not yet solidified to the requisite shape and also such that some of them are found to be as if dead, as are the lion's whelps.[9] Others are born almost unformed and without shape, as are the bear's whelps, while still others are almost blind, such as those of the dog, wolf, fox, and the *musio* which some call the cat [*catum*].[10]

CHAPTER III

On the Colors of the Eyes and Their Change Insofar As They Are Caused by the Eye's Humors

9   The eyes of all children, at the first moment of their generation, are green, tending toward grayness.[11] But then, when they are in use, seeing and awake, when the material and the sight of the eye is solidified, they change to the color which is natural and which prevails in them. In other animals, however, this is not observed since the eyes of other animals have only one color, as is clear in the eyes of cows, which are black,

---

8. *De somno et vigilia* 1.2.3.

9. In the bestiary tradition, the lion was regularly believed to revive its "dead" newborns with its breath, often after a symbolic three-day period (George and Yapp, 1991, 46–49; White, 1954, 8–9).

10. Cf. *DA* 22.123(80). On the unformed shape of the bear's cubs, cf. Ambrose *Hex.* 6.4.18. The story is that the she-bear licks the unformed cubs into shape. Vergil was said, in reference to his slow pace of composition, to have licked his verses into shape as a she-bear does her cubs.

11. Ar. *GA* 779a26f.

and of sheep, whose eyes are more watery than those of other animals. The eyes of some are gray or green and there are those who have one of the other colors which are called the colors of "goat eyes." Such are the falcons, which have yellow eyes. The eyes of humans have many colors. The eyes of some humans are gray, while others have green and still others have multicolored eyes like those of goats.

Likewise, the eyes of horses have the most diverse coloration of all animals and sometimes a horse has one gray eye and the other a color different from it. This is peculiar to horses and does not happen to other animals except in the case of humans, for in the human too sometimes one eye is found gray and the other an opposite or different color from it. This accidental trait, that of varying the colors of the eye, is not seen in other animals in their youth and old age, except in the human. For in children it is seen according to the change in first age and youth. The reason for grayness in the eyes of children is that their eyes are weaker with respect to the humors of the eye than are those of youths. For grayness of the eyes is nothing more than a particular type of weakness of the eyes.

Let us then investigate here generally the cause of eye colors and not only insofar as they are present in children. We will make inquiry into the reason the color gray is in the eyes of some while other eyes are green, and why others are black like the eyes of goats and cows. The belief of Empedocles is that grayness of the eye comes from a great deal of fire, which clarifies the eye, and that blackness is a result of a small amount of fire and a great amount of water, which obscures and darkens due to a lack of fire clarifying it. He therefore says that the vision of a black eye is not sharp at night when it is not aided by external light, since the inner light has been diminished by the water and is extinguished by the external darkness. He says the vision of a gray eye, however, is sharp at night because of the great deal of fire which penetrates the darkness of the night. This is also the opinion of Plato, as Calcidius sets forth in his exposition of Plato.[12]

But these opinions are unsuitable and false. For we know that the organ of sight lies not in fire but in water in all animals which have eyes of whatever color.[13] And we have at hand an explanation by which we might render the causes of the colors and accidental traits of vision in another and a true way. For it is reasonable that the humor composing the eye be the cause of its colors, as was proven in the book *On Sensation and the Thing Sensed*.[14] There we have spoken of the soul, not as to itself but as to the works it does and undergoes. For there we have assigned the reason that it is proper that the eye be made of water—that it is not suitable with respect to its nature that it be from air—and we have also shown there that it is not made of fire. It is thus more suitable to believe that the cause of the colors comes from the watery humors composing the eye.

---

12. I.e., *Timaeus* 45b–c. For A.'s criticism of the view, which he attributes to both Empedocles and Plato as well as to some of his contemporaries, that vision was accomplished by a fiery light emitted by the eyes, cf. *De sensu et sens.* 1.1.4–8.

13. Cf. *De sensu et sens.* 1.1.3–4.

14. Cf. *De sensu et sens.* 1.1.13.

In some eyes there is more humor and in others less, and in some the humor is placed deeper while in others it comes more to the top, that is, to the front of the eye. In some therefore there is a great deal of humor in the eyes and the color black will be deeply sunken in the eye, for blackness of the eye is caused by a great deal of water. But when the humor is moderate and approaches the front of the eye, then the color will be gray. When there is a great deal of humor but it is not overabundant, then the color of the pupil of the eye will be like that of the azure stone.[15] Eyes differ in these colors according to more and less, in accordance with the differences in their humor and their size.

12 But the reason that gray eyes do not have sharp vision during the day and that black eyes do not have sharp vision at night is that grayness, because of the small amount of its humor, is moved by the light of day more than it can bear and the vision is therefore disrupted, since a small amount of humor is overly clarified by a great deal of light. The movement of the eye is the result of clarity and the reception of forms in the eye is due to the power of the moisture. And when it is disrupted by a great deal of light just like a mirror put up to the rays, it does not have a movement determined for visible things. On the other hand, black eyes are less overcome during the day by clarity due to their having a great deal of humor and they thus have a more solidified vision during the day. But, to the contrary, the small amount of humor in a gray eye is better moved to see at night because the light of night is weaker and such light is naturally proportioned to a small amount of humor. A black eye is arranged in a contrary fashion because a small amount of light cannot move a great deal of humor and its movement is thus weak and heavy at night due to the lack of light. When, however, the movement of the light is greater than the faculty of the eye, it dissolves the motion which is less with respect to the power of the eye.

For this reason people who are moved from stronger, well-lighted colors and from the strong clarity that exists beneath the sun into dark, less bright places see nothing at first. For whatever weak movement exists inward as the vision sets out from the sensibles is prohibited from outward movement. For the stronger movement always prevents the weaker and, generally, weak vision does not see very clear things well but rather is acted upon and diluted by them since the visual humor of the eye is moved disproportionately by such things.

13 What has been said is also to be seen in diseases of the eye. For weak vision during the day occurs in gray eyes and weak vision at night occurs in black ones. For grayness is a certain type of dryness of the eyes. This is why defects of vision usually occur for the most part in old people, in whose bodies and eyes there is an abundance of natural dryness, although it may be overabundant with an accidental phlegmatic humor. A defect in night vision is caused by a superfluous humor, and for this reason it occurs very often in women, because their brains are naturally moist. Good, sharp vision is that which comes from a humor of temperate quantity for this will neither hinder the

---

15. "Pupil of the eye": *acies oculi*. Cf. *DA* 4.76.

eyes' movement by the light because of a small amount of humor, nor will it generate too much movement and require too much movement of the external light because of a superfluous amount of humor.

So that the many things which have been said might be better understood, we should repeat some of those things which were established above concerning the anatomy of the eye. We therefore say that when grayness occurs (not on account of the quality of the tunics of the eye) when the humors of the eye are clear and pure and are positioned toward the outside of the eye so that the lens is large and the white in which the light of the eye is diffused is moderate, then the eye will certainly be gray unless something from the meninges, or tunic, interferes.[16] If, however, the other fluids are obscured and the lens is small and the white great so that with its thickness it creates the sort of murkiness that water does when it is very deep, touching and obscuring those things which are sunk in it, then this same thing is caused by the lens's humor being deep. For these two reasons, then, or for one of them, the eye will be black.

CHAPTER IV

On the Causes of the Colors As Caused by the Tunics and on the Differences in Sharpness of Vision

The reason for these colors that lie in the tunics is mainly in the *uvea,* for if this is black then the eye will be black and if it is multicolored or gray then the eye will be multicolored or gray.[17] For the *uvea* grows gray or becomes multicolored because there is not good digestion in it. This is as we often see in plants where at their earliest moments of birth they are white because of nondigestion and they grow green later. This is also the reason that children's eyes are gray, since the digestion in them is incomplete. If, however, it is due to moisture, then it is because the moisture is dissolved. In that case, "shadowing" of the eye [*tenebra oculi*] follows upon such grayness and the pupil of the eye is converted to whiteness.[18] So too do plants grow white when the watery moisture in them is dissolved. This, then, is the grayness due to dryness prevailing in the eyes.[19]

It is for this same reason that the eyes of sick people and of old people grow gray. For old people are bereft of natural moisture and abound in extraneous moisture. And both grayness and multicoloredness are twofold, namely natural and unnatural (that

---

16. Cf. Avic. *DA* 19.53v. On the meninges of the eye (*miringa*), cf. *DA* 12.170.
17. Avic. *DA* 19.53v.
18. A. is, apparently, referring to cataracts. Cf. the disease *albugo* at 22.111(60).
19. Cf. 1.206.

is, accidental), as is clear from the things we have said on the anatomy of the eye. Multicoloredness of the eye stems from the causes of grayness and blackness being mixed.

Further, if the eye were composed of fire as Empedocles holds, then a gray eye necessarily would be hindered and infirm due to the lack of humor in which the fire should be diffused.

Also, there is some black in eyes which either diminishes or hinders vision because it hinders the passage of sight. For if the blackness is due to cloudiness of the white humor, then it will prohibit the passage of the sight by hindering translucence and preventing impressions from being made on the eye as far back as the lens-humor. The same thing happens in an eye with abundant and cloudy humor for these do not readily allow passage of forms as far as the lens and they do not allow vision to exit to the front for purposes of sensation.

16 Further, even though a black eye sees less well at night because of its great humor, still, blackness which is due to the tunic solidifies vision very strongly.

One should know that sharpness of vision is twofold. For vision is said to be sharp when it sees from a distance and vision is called sharp when it grasps well small differences in the thing seen. The cause of the first sort of sharpness is that vision does not have much moisture and thus when the vision is narrowed and contracted it does not see well, but when it is surrounded with proportionate quantity and is assisted by the light of the surroundings, then it sees well. For then the parts of the eye are not acted upon excessively and are not disrupted by the excessive clarity of a nearby visible object. The reason, however, for grasping the differences in a nearby thing is the very clear and fine humor which takes on strong movement from the strong, nearby, visible object and is neither disrupted nor obscured by it. One should know also that when the eye is fixed on something and struggles forcefully to grasp it, it moves outwardly and labors. Thus, after the act of seeing, it returns and moves inward, whereupon it sees less well if the same things are put up to it, since its power has not yet turned outward again.

17 Thus, not only those things which we have mentioned about the humors are the natural cause of sharp or weak vision.[20] There is also the disposition of the skin, or tunic, which is over the eyes and especially over the pupil. For fine vision this must be very clear, smooth, delicate, and light in order to allow vision to move lightly and easily through it. It must be smooth lest there be any wrinkle in it, for a wrinkle produces a shadow over the pupil. This is one of the reasons old people see whatever they see in shadow, for the meninges [*miringae*] of their eyes are wrinkled like the skin of their bodies.

Further, it is right that the meninges are white for that which is black cannot be entirely clear since blackness is nothing other than lack of clearness. This is also why the eyesight of the sick and the old is not clear, for the black, earthy fumes disrupt the

---

20. Ar. *GA* 780a25f.

clarity of their meninges. But the eyes of children at first appear gray either because of a lack of humor or because of the disposition of the meninges, as we said above.

Now it sometimes happens that one eye is gray and the other is not. This occurs in people and horses and the reason for each of them is the same as why of all the other animals, only the human and the horse take on signs of old age. For man grows white in his hair and the horse in its teeth. This gray-headedness and the whiteness of the teeth are a particular type of weakness in the brain, a result of weakness of digestion, led forward to the whiteness of phlegm. Grayness of the eye is caused by the same thing.

Moreover, the conditions of great whiteness or blackness in the eyes have the same cause in material, but not in manner. For one of them is caused by a scarcity of humor but the other from an abundance of this same humor. But in each it is a result of nondigestion since in neither one can the humor be adequately decocted. For when on the one side the heat is weak for decocting and on the other it is strong and decocts adequately, then it will happen that one eye will be gray and the other will not be gray.

Further, as we said above, some animals have sharp vision and others do not. This has two causes, for one is said to have sharp vision in two ways, just as is the case for hearing, smelling, and the other senses. One is said to have sharp vision when one sees well from a distance. In another way one is said to have sharp vision when one grasps well small differences among the things seen. These two are not alike nor do they have the same faculty, because one seeing at a distance does not distinguish well the differences of things, even if one should do so from up close. When, then, the humor in the pupil beneath the meninges of the eye is clean and clear and is not excessive, and the meninges is thin and there is ready movement of vision and of visibles through it, then it will see better from far off than from near. Such vision will be disrupted if it looks from nearby and does not make good distinctions among the differences of the things seen because its spirit is scattered. This vision, though, is better than that of eyes whose humor is clean, to be sure, but which do not have an entirely clean covering. The reason that it does not grasp differences well is that such sharpness of vision, which carefully grasps differences, will be due to other causes that exist in the eye, as we have said above. For a small spot is noticeable on a clean cloth and it is the same in an eye that is fully clean, namely, that certain small things occur on it as long as the power of its spirits is well unified. For just as a small spot is quickly discovered on a clean cloth, so is the small movement of the differences of the visibles quickly and easily discerned in an eye that is fully clean and which possesses unified vision. This is the reason for vision which is refined for grasping carefully the differences of the things that are seen, and these eyes are called "lynx" eyes. They possess a very clear and unified humor as well as a similar visual spirit [*spiritus visivus*] and a clear and fine meninges.

Moreover, there are other accidental causes of these two types of sharp vision which are found in the locations of the eyes. The reason for seeing well from far off is sometimes due to the location of the humor in the eye or to the location of the entire eye in its seat. Bulging eyes do not see well from far off since the vision going out is not unified by the shading of the eyebrows. It is disturbed instead and thus spread about and is weakened as it comes forth from the eye. On the other hand, deeply set eyes whose vision comes forth narrowed by the lids are proven to see better from far off since its motion is neither divided nor consumed but rather the visible power [*virtus visibilis*] leaves unified by it and passes on a straight line to the things seen. For we make no distinction here whether the eye is said to see through the movement of the sensibles to the eye or if the power is said to move to the thing seen. For we have established this elsewhere and it makes no difference for us as to the cause of sharp vision which of these is said.[21]

Further, when the covering which is outside the eye, [made up of?] the *fenica* and the eyelids, is not extended a little beyond and outside the vision, so that it will hold in check the vision going forth or the visible entering in, vision is necessarily weakened and can see from a distance only weakly.[22]

These, then, are the various causes of the compositions of the eyes and of the differences in their sharpness and weakness, which we stated above, and when what has been stated here is joined to the things we said on the anatomy of the eye, then one may know quite perfectly the things which a natural scientist should know about the vision.

CHAPTER V

On the Causes of Good Hearing and Smelling Which Stem from the Composition of the Organ

In the same way, now we must make a determination about hearing and smelling, for there is one common type of sound hearing and smelling, in keenly hearing that which touches the hearing, in keenly grasping the differences of sounds and voices, and in smelling and keenly distinguishing among the differences of the odors of that which is presented to the sense of smell.

There is yet another type of goodness for each of these senses, namely, when the one hearing hears from far off and the one smelling smells from far off. For when the composition of the organs of these senses is good, they will make perfect judgments concerning the differences of their sensibles, just as we have already said concerning vision. This occurs when the position of the ears is clearly well proportioned and there is a web containing the hollow of the ear stretched out over the auditory nerve, called the *timpanum* of the ear. This ear will have sharp hearing and it is the same for smell.

---

21. Cf. *De sensu et sens.* 1.1.5ff.

22. Stadler observes, with no small understatement, that this passage is "plane confusa" when compared to the original at Ar. *GA* 781a8f. The reading *fenica* is not secure and may be *femea*. Borgnet reads *tunica*.

For we have already said in our book *On Sensation and the Thing Sensed* that the passages of all the sense organs lead to the heart in animals with a heart, or to the member which is analogous to a heart in those which have no heart.[23]

Further, the instrument of hearing is filled with a spiritual body, which is air, which is naturally suited to the ear. This spirit, like a messenger, bears the pulse and beat to the ear. Just as this spirit produces breath in the organ of breathing, so it produces the power of hearing in the ear. This is the reason that hearing is a sense subject to training. For just as speech comes formed in the spirit which produces breathing, so too it comes from one speaking and expressing his intention as to the things about which he speaks, so too it comes unbroken and in the same contiguous configuration and intention, and in the same contiguous spirit it enters through hearing. For this reason it is clear that a human comes to know what he hears in entirely the same way he speaks what he intends.

Further, in those who are groaning due to illness, the voice is weakened and made higher in pitch and hearing is weakened also. The reason for this is that the organic principle of this sense is positioned in the head, over that member through which breathing is performed. Thus, when breath is moved, it also necessarily moves internally and the hearing is disturbed in the ears. This is clear in one who, in yawning, suffers from stretching [*extensiones*] and shaking [*alices*] due to a smoky, coarse spirit.[24] For in the opening of the ear and upon stretching the body, one feels a great fullness in the ears arising from the drumming spirit. This accidental trait occurs especially in those which are similar in temperament to the human complexion. In the human the ears especially fill up with spirit, as far as can be determined naturally, because they have a shared boundary [*collimitatio*] with the spirit which is coming from within during breathing.

As we have said, then, the cause of good hearing lies in the grasping of differences of speech, voice, and sounds. Likewise, the cause of good smelling lies in sensing differences in odors and this will be a result of the cleanness of the composition of these organs and of the internal membranes which contain the nerves of these senses. For there are three movements of the senses which come from afar over the sensibles. This is seen in vision, whose goodness lies in the composition of the eye and in the clarity of those things which compose the eye; in the sense of hearing, in the comprehension of remote sounds, voices, and speech; and in the sense of smell, which smells those things which are far off and have an odor.

To speak generally, all these senses have the cause we stated in the case of vision. For the organs of hearing and smell, if they have the instrument of their sense well exposed and projecting outward, will sense from further off when there is no impediment present in some baseness of their composition. The reason is that these senses go forth from their organic instruments much as water goes forth from channels. For this reason, noble, small dogs, which track down hares and other wild creatures using

---

23. *De sensu et sens.* 1.1.15.
24. For these terms, cf. 6.22, with notes.

an odor trail, have a nose that is well exposed and short, and they have a good sense of smell. The claim that dogs with long noses have a good sense of smell does not hold up at all as it is found false by experience [*per experimentum*]. For a good sense of smell is somewhat like that which occurs in vision, as is clear from what has been said before. Good hearing will also be from a similar cause in those which have ears of the best composition with a proportional hollow and involution. This will be sharp hearing.

Further, the human sense for hearing voices and sensing odors from a distance is weaker than that sense in other animals and is also weaker than the other senses in the human. But his sensing of the differences in voices, sounds, and odors is better than in all the other animals. The reason for this is that the organ of these senses, which is the brain, is more clean and refined in the human than in the others.

Further, wise nature always looks for what is best in positioning organs of this sort. This is why she did not make external ears [*auriculas*] in the *koky*, which is the sea calf [*vitulus marinus,* seal].[25] For the *koky* has four feet and generates a completed live young like itself, and it lacks external ears. The reason for this is that its lifestyle consists of dwelling in water. Since external ears (in those which have them) are positioned spread open to the passages of the sensibles precisely so that the movement of the sense to the sensibles coming from a distance might be readily guarded, this animal did not need external ears. For if they were spreading open to the passages of the sensibles, then they would take in much of the water's moisture and they would also, as a consequence, impede the movement of the *koky* swimming in the water.[26]

The things said here about hearing and smelling, along with the things we have frequently said previously in the anatomy section, are enough for our present intention.[27]

CHAPTER VI

On Those Things Which Befall Hair at Various Ages and Other Natural Causes,

Such As Growing Bald and the Like

The hair of men undergoes changes according to their various ages.[28] But it should be noted that hair is, for the most part, present only in viviparous animals even though some of these have spines (like those belonging to the hedgehog genus), for such spines are certain types of hair as are pig bristles.

---

25. "Auriculas": usually earlobes, but A. frequently uses this term to refer to the entire external structure of the ear, as opposed to its internal mechanisms. Cf. 12.176, with note and 14.55, with note. At 12.177, he makes it clear that the sea calf, or *koky,* does have the internal structures necessary for hearing.

26. The sea lion, with its external ears, would have presented a difficulty for this argument.

27. Cf. 1.168f.

28. Ar. *GA* 781b30f.

There is much diversity in hair as to roughness and smoothness, length and shortness, hardness and softness, and abundance and scarcity. It also varies in color according to age, for in youth and old age it changes colors, especially in humans. For the hair on humans multiplies during youth and they begin to bald toward the front of the head as hair falls out. But baldness does not beset youths at all or, for the most part, women. During old age the hair grows white in almost all people, something that occurs in no other animal known to our experience [*per experimentum*]. Baldness occurs to the front of the head and graying most often begins on the temples, while baldness rarely prevails all the way to the rear of the head. I have seen, though, one man who was so bald that on his entire head there was not the trace of a single hair. This happened to him as the result of an acute fever, whereas before he had had a great deal of beautiful hair. But in his beard, and in all the other places where people usually have hair, he had a good deal of it.[29] I have also seen myself two quite bald women.

In animals which do not have hair but something which takes its place, such as the feathers on birds or the scales on fish, this sort of variation in color sometimes occurs, but not in all cases. Above, where we spoke of the causes of the members, we gave the reason nature made hair in animals.[30] But now we are asking why types of differences exist in hair.

We say, then, that the skin in some animals is fine and in others coarse, in some thin and in others thick. The reason for this diversity is a diversity of the humor, for sometimes it is watery and sometimes coarse (that is, thick). I am saying this about the moisture which is outside around the flesh on the visible surface of the body. For when its looser part evaporates into vapor, what remains is a hard, membranous residue.

But hair is generated not from the body but from the skin and from a humor of this sort which is expelled outward. Its generation occurs when this moisture, dissolved, is rendered vapor. That is why coarse hair is generated from coarse skin, due to the great amount of earthiness which is in it. Fine hair is generated from fine skin. Thus, in pore-filled and thick skin there will be coarse hair because of its earthiness and the size of the pores. Contrariwise, when skin is very solid, the hair will be fine and thin because of the narrowness of the openings.

Further, when there is watery moisture beneath the skin which is also easily dried out, the hair will be neither dense nor long. But when there is a coarse, fatty moisture there, the hair of this sort will have an opposite disposition. For that which is fatty is not quickly dried out. This, then, is why an animal with coarse skin has coarse hair like the hair of pigs, cows, and elephants. For in all these the coarseness of the hair is due to the coarseness of the skin. Likewise too the length of the hair on the head of a person is due to the same reason. For the skin on a person's head is very fatty and there is a great deal of moisture in it. The pores are also infrequent because of the sutures and thus the hairs are made long by the moisture that is there which is not

---

29. Also described at 3.89.
30. Cf. *DA* 2.21ff.

easily dried out. Now this mostly happens for two reasons, namely, because of quality and of quantity. For a great deal of moisture is not easily dried out and likewise a fatty moisture is not easily dried out. In that case the hairs are lengthened greatly on a person's head since a human's brain is very cold and moist and a hair rooted in it is anchored in a great deal of moisture.

28 Curliness or smoothness (that is, straightness) of the hair is likewise due to the nature of the vapor which is the material of hair. For when the vapor is smoky, warm, and dry, then the hairs will be curly and twisted since they travel through two paths contrary to one another. For that which is earthy twists by pulling downward and that which is warm twists by forcing it upward. They twist this way continuously. But sometimes it also happens that there is curliness because of a small amount of moisture and a great deal of earthy parts. For thus the earthy part is dried out by the surrounding air which encloses it, becoming crinkled and being rendered curly. For we see this happen in smooth, straight hair when the moisture which is in it is dissolved by heat. And we see this happen to hair because of a nearby fire, for it is crinkled and twisted up by it. Sometimes, however, hairs are neither twisted nor curled by a small amount of natural heat but when they reach the outer air, which then works with the inner heat, they begin to crinkle and curl. An indication of this is that curly hairs are sometimes moister than straight ones, having a smooth position in parts. For the complexion of such hairs is sometimes moist like that of straight hairs and the air surrounding them is also moist. An example is the hair of those who inhabit the sixth and seventh climes and the places which are of greater latitude. The Ethiopians, however, who dwell in hot and dry places, have curly hair, rather like grains of pepper.[31] They have this due both to the dryness of the vapor and to the dryness of the exterior, surrounding air.

Further, certain animals with coarse skin have thin hair. This happens in them for the reason we have stated. For in ones such as these, thin hair arises from narrow pores. This is why the hair of sheep has the disposition it has, for wool is nothing other than a particular abundance of hair.

29 Further, certain animals have soft hair which is not very fine. An example is the hair of hares. The difference between these and the wool of sheep is that the hairs of the hare arise from the visible external surface of the skin. But the wool of sheep comes from deep down in the skin. Thus the hair of hares does not attain the length of wool.

Moreover, the opposite happens to sheep which live in cold areas from what happens to people. For in cold and moist areas human hair is soft, much as the bristles of pigs in those places are softer. But sheep which live in the area called Ceolinata in Greek have tough hairs (that is, tough wool), even though the land there is cold.[32] The reason for this is the accidental trait we have said occasionally befalls humans and certain wild animals. For coldness, when it is excessive, is accordingly sometimes the cause of congealing and hardening and then, while compressing, it dries. And when

---

31. Cf. 3.82.

32. Ceolinata: for Ar. *GA* 783a15, Sauromatika, Saramatian (sheep). Saramatia is in the area between the modern Vistula and Don rivers.

the heat which has been pressed out is released, the moisture leaves with it and what is then left are dry, earthy skin and hair.

An indication of this is what occurs in the sea urchins in the cold land of Kaycos.[33] For these are the sea urchins whose skins weavers use on the uprights of looms.[34] They are the marine type and the reason for all this is that they dwell in the coldest sea, in which, due to the depth, the water's coldness is increased. This sea has a depth of more than sixty feet and the rays of the sun do not reach there. For this reason these sea urchins have a small body with exceptionally hard and large spines. The size of these spines is due to the fact that the nourishment for their body passes over into them because of a scarcity of digestive heat. For since the food is not digested, a large superfluity is generated in their bodies. But hairs and spines and things like them are generated from nothing other than superfluity. But the rock-like hardness of its spines is due to the great coldness which, by pressing out, expels both the heat and moisture from them. For spines that grow this way become very hard, rock-like, especially in the northern regions, more so than in the southern region and in windy areas more than in quiet ones. For in places of this sort the wind cools more and thus expels the vaporous moisture. Thus, spine of this sort is made hard since heat recedes from its nature and cold remains, and along with the heat dissolves the humor, since there can be no natural softening humor without heat. But cold not only hardens, it also thickens the body and condenses it. Heat, however, purifies. And thus the older that animals with hair become, the harder their hair becomes. Thus the wool of lambs is softer than that of sheep. And in vaporous, moist lands the wool will also be softer as it is in England, for there the vapors temper the cold and continually moisten and, by moistening, both soften and narrow the pores. Because of this the wool is rendered fine and soft.

This is also the reason that most animals have in their old age a sort of scabrous husk for a skin once their skin is thickened and hardened during old age due to the loosening of the heat, along with which moisture is also set free.[35]

Moreover, it is from a cause of this sort that of the animals, men have to themselves the trait of baldness. For an accidental trait of this sort seems to be natural to men much as in plants it is natural for them to lose their leaves. Certain trees lose their leaves and others do not and so it is in birds which build nests, for certain of these lose their plumage and change it. Baldness too seems in some way to be an accidental trait similar to losing leaves and changing plumage. For the loss of hair in men happens similarly. There is, however, a difference in that the feathers which are cast off grow back again in birds, but baldness is the name given the loss of hair after which none other grows back. The cause of baldness is a lessening of the humor. Now I am talking of the warm, fatty humor we have already spoken of. For this reason too trees which

---

33. For the identification of the marine *hyricius* as the sea urchin, cf. 2.109 and 4.53.

34. Uprights of looms: *statorii liciorum*. This enigmatic comment is A.'s addition. Cf. Ar. *GA* 783a20f., which says that urchins are good against the urinary disease strangury. Were urchin spines kept on the loom to cut threads upon or as guides to keep the threads separate?

35. Reading *scabrosam* for Stadler's *scabrosam*.

abound in a fatty, warm humor do not lose their leaves. We will speak of this in later parts of this book, however, since baldness also has other causes in addition to the one we have mentioned here.

32 It happens that trees lose their leaves in winter since there is a stronger change in the sap and pores of trees during this season, and it produces a greater change than in all the changes of the other seasons. Something similar occurs in nest-building animals. For the nature of these animals has less humor than does the nature of humans and especially than that of men. The change in humans which is in accordance with age is somewhat like the change which is in accordance with the seasons, these being the changes of winter, summer, spring, and autumn. This is the reason that boys do not grow bald, for they are moist and have fine skin. But after they begin to copulate, baldness besets them, especially those whose nature is colder than that of the others. And because the brain is colder than the entire body and is further chilled by intercourse—for intercourse draws away a great deal of the natural, finest humor—it is thus reasonable that a loss of nature is felt in the brain which is greatly weakened by intercourse. Thus it is necessary that the skin which surrounds a given member be similar in humor, or dryness, to the member it surrounds. Consequently, the hairs which grow from this sort of skin will have a like nature and complexion to it.

33 It is clear from all these things that baldness reasonably occurs during the period when the sperm is exiting and occurs on the front portion of the head from which the greatest moisture for the sperm is set free, for the front part is very moist and large in size, especially in the human. Only a very few women grow bald, for their nature is like that of children, since women and children do not have the cause of creating sperm save in a modest amount. Likewise, those who have been castrated do not grow bald after their castration since they are changed over to a woman's complexion. For when a man is castrated, then either hair does not grow at all in the places on which it does not grow on women or, if he has hair before castration, then he gradually loses it after castration. But he does retain it on the groin and under his armpits just like a woman. It stands to reason that this change does not occur in those who have been castrated unless they are changed to a womanly nature through castration.[36] The reason that other leaves grow back on trees after they have lost them and likewise that some feathers and hair grow back on animals which build nests and renew themselves, and that the hair does not grow back on the head of a bald man after the loss of his hair, is as follows. In trees and in animals other than the human the falling and loss are caused by the season, drying externally and closing up their pores. Then, due to moist and warm seasons, they restore the moisture and reopen the pores and these things grow back again. But in the human this is caused by an internally deprived complexion. For a change in the age of a person is somewhat like the changes in the four seasons of the year. Just as the ages are fixed in the human, so too the accidental

---

36. On the change in castrated animals to a female nature, cf. 3.90; 4.108; 15.10; 16.72; 18.21; and QDA 1.13.

traits of this sort of age do not undergo variation. This is so even though we have said in general that there is a single cause of them all.

The things we have said here, then, along with those which we said far above in this study, can suffice for our present intention about the cause of baldness.

CHAPTER VII

On the Cause of Grayness and Wrinkledness in the Human

In animals other than the human the cause of uniform or varied coloration is the nature of the skin.[37] But human hair does not change in color for this reason, for it varies from the color it has only when it changes to the whiteness or gray-headedness of age.[38] This type of change occurs occasionally due to illness and especially so in beards. For the hairs in a beard grow white from the watery phlegm which turns the hide or skin white, but when the hairs grow white it is due to old age, whereas old age is not a reason that the skin grows white.

The reason for the change due to illness is that the reason for an illness of this sort lies in the skin and without an illness of this sort, when the skin grows white for some other reason, the hairs do not grow white as a result. Gray-headedness, however, is due to a weakness of the skin and the vapor which leaves the body, for old age is cold and dry. For we know from the things established by us in our natural science books that food, insofar as it nourishes, comes to each member.[39] This food is digested to a likeness of the member by the natural heat which is in each member. And the food is corrupted when deprived of heat of this sort and then either impairment or illness will befall the member. The reason for this has been carefully stated in the book *On Nourishment* [*De nutrimento*] and is known from the things we have already said, from the determination we made on the powers of the members in the preceding portions of this study.[40] When, then, the humor in the member has but little heat which cannot bound it completely, and if the humor which then enters the food overcomes the internally digesting heat, it will putrefy from the heat of the surrounding air due to the weakness of its decoction and digestion. For all putrefaction is due to unnatural heat, as we showed in the fourth book of *Meteora*.[41] Unnatural heat, however, is the accidental heat of the surrounding air. Putrefaction, moreover, mostly occurs in water and in earth and for this reason a vapor from the earth or water putrefies easily; then, bodies become wrinkled and corrupted from the watery vapor of this sort. Therefore

37. Cf. Ar. *GA* 784a23f.
38. Compare this explanation with 3.86f.
39. *De anima* 2.1.3.
40. *De nutr.* 1.3.
41. *Meteora* 4.1.8.

earthy food which passes into the hair putrefies when it is not well digested and from these, gray hairs are generated.

But the reason for the whiteness of the gray hairs resembles the reason for the wrinkling of bodies. The reason derives from the fact that in every earthy vapor there is much of the nature of coarse air, and wrinkling of bodies is in some way the opposite of icy congealing. For a congealed vapor will be rather like smooth ice, and if it is not congealed it putrefies and creates wrinkledness when the fine elements have been led off and the coarse ones have been contracted, not through a congealing but by a compressing cold. I am saying this about the vapor beneath the skin which is on the surface of the body. This is why Homer did well when he called old age wrinkledness and ice, for these are alike in general as to cause, but they do not have one and the same form with respect to the same cause, as is clear from what has been said before.[42] For both ice and wrinkledness come from a vapor which exists in dissimilar form, even though both are, in some way, a certain putrefaction of vapor. An indication of this is that experience has shown [*expertum est*] that many people have grown gray due to a disease, and that after the disease their hair returned to its original color. The reason for the grayness was a lessening in the members' heat which occurs in bodies at a time of illness throughout all the members equally, large and small, and grayness occurs from an imperfect digestion of the humor. When health returned the heat was strengthened and it completed the digestion, whereupon the hairs once again took on their original color. And this is an accident much like that which took place in the story about the man who was, in the way mentioned, turned from an old man into a young boy.[43]

For this reason too it is reasonable that this sort of disease is called a sort of untimely and accidental old age and that old age is called a natural illness, since in illness of this sort the same thing happens as does during old age.

Further, the hair on the temples is among the first to go gray since the back of the head is more empty than the front and is also drier and harder. For, as is clear from the things we have said on the anatomy of the head, there is no brain in the posterior *platea*.[44] But in the areas of the *synciput* there is a great deal of humor and a great deal of humor does not putrefy quickly. On the temples, however, a midway situation prevails. For in them there is neither so little of the humor of the *medulla* that it dries them out nor is there so much that it can stave off putrefaction for a long time. Therefore, the hair on the temples turns white before other hair.

This then, in general, is the reason for gray-headedness and for baldness, save that one should know here that when hair falls out due to coldness compressing the pores, the skin thereafter remains solid as ice and through it nothing can penetrate. This is why these people gleam with baldness like so many mirrors. It is also known for this same reason why the other animals neither grow bald nor turn gray. For the brains in

---

42. Ar. *GA* 784b19f. speaks of unnamed poets of comedies. Since Sc. mentions only *versifactores*, the source of the mention of Homer remains unknown.

43. Ar. *GA* 784b30f. does not mention a specific story. A. may have in mind the tale of Medea who used her magic to render Jason's father, Aeson, young again.

44. On the *platea*, cf. 1.410 and 12.170.

their heads are proportionally modest for the size of their bodies and do not have very much humor. Therefore, the digestive heat prevails well in them.

Nevertheless, of all the other brute animals, horses sometimes have white hairs, for the bones in a horse's head are more fragile and are less dense [*rarior*] than the bones in the heads of other animals. An indication of this is that when even a modest blow falls on the head of a horse, especially on the temples, death follows directly, especially if a wound is brought on by the blow. For the humor of the *medulla* quickly leaves the head via the temples, due to the fragility of the bones there.

Further, red hair grows gray and turns white more quickly than does black hair. This is because reddish-brown and red are sorts of infirmities which befall the hair. Everything which is weakened in its natural power turns gray quickly.

37

Further, certain people believe that cranes turn dark as they grow old, doing the opposite of what happens to people. The reason for this is that the feather of a crane naturally tends to be white and when cranes age the increased humor in their feathers quickly putrefies and slips away. Earthiness remains and darkens the feathers.

The fact that grayness is a certain sort of putrefaction is shown by the fact that when hair is covered it quickly grows gray. Thus, grayness is not dryness as some believe, for covering prevents ventilation and ventilation wards off putrefaction. Moreover, covering also prevents exhalation which, if beaten back, quickly putrefies. Thus, water in a jar covered at the top produces dew on the covering. Now an indication of what we said about disease is that smearing baldness with water and oil produces something of a cure, if baldness is not deeply rooted. For the water dries, especially if some drying agent has been cooked in it, but the oil adds fat to the humor which forms the material for hair.

Moreover, in some people the hairs come forth white at their front end but not white near the head. This is because a slight digestive heat loses power more quickly far off than it does nearby.

CHAPTER VIII

On the Cause of the Colors in Animals Other Than the Human

White hairs which are present on the bodies of other animals are there naturally and not as the result of illness.[45] The reason for the color of these hairs is the color of their skin. In these animals white hairs arise out of white skin and black from black, and so on for the others. If the skin is multicolored, then the hairs are multicolored. But in people the color of the skin is not the cause of hair color. For there are some who have extremely white skin but very black hair. The reason for this is that human skin is very delicate, more so than the skin of all the other animals when taken in proportion to

38

---

45. Cf. Ar. *GA* 785b1f.

his body size. For this reason his skin changes in color for the slightest reason but his hair cannot be changed by these reasons. A person's skin is changed by the sun and the wind and takes on darkening. The hair cannot be changed in this way, since it is rooted more deeply than merely in the skin. But in other animals the skin is related to the hair the way the earth is to a plant, which takes its power from the earth. For this reason their hair changes according to changes in the skin.

39   Further, certain animals have one color throughout their whole species as do the lions. For all lions have one and the same color and the same thing occurs in many genuses of birds, fish, and other animals. But others have different colors, so that one has one color and another member of the same species has another. Some of these have the same color throughout their entire body, while others are multicolored, as is the case in cows, dogs, and pigeons. Those which are multicolored, however, are of two sorts. Some are multicolored throughout their entire genus, such as the magpie, *fehit,* peacock (although white peacocks are to be found in lower Germany), and, among the fishes, the *kakata* among the Greeks and, in our sea, the *macarellus.*[46] There are also the panther, the leopard, and many others among the walkers. Others, though, are not multicolored throughout their entire species, such as the cow, goat, sheep, chicken, goose, and the like. For some of these are multicolored and others are not. Further, multicolored animals change their colors more easily than do singlecolored animals. For the multicolored ones sometimes change from white to black and vice versa, for such animals have mixed reasons for many colors and naturally have the ability to take on many colors. The nature in such animals readily transfers one color into another. But the nature of animals having one color has the opposite disposition. It does not change color for any reason save illness and this very rarely. We once saw a white wolf; however, the bears which are above the northern ocean, in Dacia and Norway, are almost all white.[47] Many white sparrows have also been seen. And we once saw an entirely white crow, as white as snow. But this rarely occurs in these animals after the moment of their birth and that which is rare is, in every species of animal, due to a flaw and illness.

40   The change which belongs to multicolored animals is very often due to the nature of the water. For warm water produces white hair, while cold water produces black.[48] This is because warm water has a great deal of spirit and cold does not. A great deal of spirit in water produces whiteness, as is seen in foam. Moreover, this often occurs in plants and this is why plants blanch in warm water. Thus, variation in skin, whether it is due to nature or illness is, as we have said, the reason for variation in hair. Whiteness, to be sure, occurs in hair for all the reasons mentioned. It will sometimes be from

---

46. On the *fehit,* cf. 1.44, with note; 5.41; and 7.93. For the *kakata,* cf. 7.102. And for the *macarellus,* cf. 7.76.

47. The reference here seems to be to the polar bear. As Balss notes (1928, 87), A. is the first medieval author after Adam of Bremen (d. ca. 1075) to mention the polar bear.

48. Note that A. has just above included the polar bear among the white animals. Certainly, he knew that polar bears live in cold water.

the natural heat of the skin, sometimes from an accidental heat. But always, or at least frequently, there will only be whiteness from a vaporous heat which is retained in all the members. This is the reason that the bellies of multicolored animals are more often white than another color, for the vaporous heat is most abundant in that location because of the digestion. This is also why the flesh of the belly is tastier than all other flesh, for the digestion provides the flavor. For this reason red or black animals, in every species of animal, are more flavorful than white, for these colors indicate warmth, and warmth, when naturally abundant in the members, produces better digestion.

Further, age produces a variation in animals' coloration, whether they be simple (that is, of one color) or multicolored. And we have often stated previously the reason for this, namely, that if the skin has differing colors, then the hairs will be of different colors. If the skin is white in color, then the hairs will be white in color.

One should also hold the opinion that the tongue follows the nature of the exterior skin and hair in this accidental trait, even though this may seem wondrous since it is one of the interior members, hidden in the mouth, and is one of the organic ones like the hand, foot, or one of the others. For when the skin of a given animal is multicolored, then the skin of the tongue is found to be multicolored, and when the skin is of one color, then there is a single color to the tongue.

Moreover, certain genuses of birds and wild quadrupedal animals change their color according to the changes in the seasons. The reason for this is that just as change befalls people according to their ages, so too does it likewise occur in these types due to a change in the seasons. For the differences in the seasons are stronger than those of the ages, since the differences in the seasons are the cause of the changes in ages.

Again, animals which use many kinds and types of food are, for the most part, multicolored. Thus there is one color for bees but the coloration of red and yellow wasps is varied. For different types of food cause changes in coloration. This happens reasonably since different superfluities are caused by different foods and it is from them that skin, hair, hide, and plume and feather in birds are generated.

CHAPTER IX

On the Differences of Voice in Animals

There are differences in the voices of animals in that some have a deep [*gravis*] voice, others a high-pitched one, and others midway between.[49] Some have a small voice and others a large one. Sometimes the voice will be pleasing and beautifully modulated, while at other times it will be the opposite. We should now ascertain and assign the causes of all these differences.

---

49. Compare this passage with 4.90f., which uses identical language; with *De anima* 2.3.22; and with Ar. *GA* 786b7f.

One should know that the cause of high- and low-pitched voices very often follows changes in age. For the voice of young animals is higher for the most part than the voice of older animals. For all, or almost all, the genuses of animals have a high-pitched voice in their youth. Of familiar animals, none are found which have an opposite arrangement save for the calves of cattle. For these have a deeper voice than the cows. Uncastrated bulls have an especially high-pitched voice and likewise the voice of the cows is deeper than that of the uncastrated bull. But in all other known animals it is the other way around. It happens that the voice is made deeper from one of two causes, namely, that the mover of the voice is the lungs, which, blowing air to the windpipe, move a great deal of air.[50] This is called the "great mover." In just the same way we see that there is a deep sound in bourdons, which are the great pipes on musical organs.[51]

Another cause is a weakness in the mover, for while it does push along a great deal of air it does not do so forcefully. Thus calves have a deep voice and cows have one deeper than that of the bull. On the other hand, if the mover should move and force it along strongly, dividing the air, as it were, to the parts of the windpipe, and if the windpipe is not fully wide enough, then the voice will be quite high-pitched. Again, if it should move it weakly and produce little movement, then the voice will be high-pitched, as is the voice of the sick.

44 These differences of voice are more apparent in humans.[52] For to the human alone did nature give the power of using the articulate and lettered voice which is called speech [*sermo*]. For the modes of speech are nowhere save in the voice. We have already determined in the study *On the Soul* in what way it belongs to animals to vocalize and what the voice is and what sound is.[53] Here, however, we are saying that a voice is deep due to a slow movement which does not penetrate that which it moves, but rather pushes it along by pressing on it from the outside, as it were. On the other hand, however, a high-pitched voice is the result of strong and sharp movement, that is, one which penetrates and divides that which it moves. But some doubt arises whether movement is the cause of a deep and a high-pitched voice given that its movement is heavy (that is, slow) and swift.

For some believe that everything which is "much" is moved slowly and they say that everything which is "little" is moved swiftly. They say that this is the cause of deep and high-pitched voices in animals. Now what they say is true in one way and false in another, as is clear from what has been said before. For the opinion of those who say that deepness comes from a great amount of movement is reasonable. An indication of this is that vocalization [*vociferatio*] is difficult through a small, deep voice. That is why in modulated chants the lower voice is held with difficulty as it descends considerably. But vocalization with a full, high-pitched voice is entirely impossible, for the

---

50. Avic. *DA* 19.54v.; cf. *QDA* 19.10–11; and *Summa de creaturis* q. 25, art. 4.
51. Cf. 16.137, with note.
52. Ar. *GA* 786b7f.
53. *De anima* 2.3.22.

instrument is not adequate to move toward a full, high-pitched voice while vocalizing in a higher chant, and the sound must be diminished. For this reason too some have felt that deepness of voice is an attribute of the nature of a stronger mover. They take the rationale for what they say from the neumes in music.[54] In these they say that a deep neume is better and more perfect since it has the perfection of growth. They say that deepness is a particular type of growth.

But we know that deepness and highness are opposites and are found both in a large and a small voice in such a way that each is found in each of these voices. For a large voice can be high-pitched and a small voice can be deep. We therefore must distinguish between the types of these voices using another distinction than the one mentioned, so that we may say that largeness and smallness of a voice does not follow the greatness or smallness of those things which are moved by it. For if we say that highness and lowness of voice is caused for the reasons those men say, it will sometimes follow that a voice that is one and the same in number will be both high-pitched and deep, which is impossible. A thing is said to be "large" in many ways, and these are said in a simple way [*simpliciter*] and by comparison of one thing to another. A voice is called large in a simple way when that which is moved is large in a simple way, and a voice is called small in a simple way when that which is moved is small in a simple way. But high and low pitch of voice are spoken of comparatively. Thus in them there are differences of large and small and whenever the power of the thing moved overcomes the power of the mover, then the movement will necessarily be heavy even if that which is moved is small. If, however, on the other hand, the power of the mover overcomes the power of that which is moved, then the movement will necessarily be swift, even if that which is moved is large and multitudinous. A strong movement will sometimes be slow because of its strength and sometimes will be swift because of its winning out. In just the same way the movement of that which is moved will sometimes be weak. For when a large rock is moved by a weak thing then the movement will be slow, and when a smaller thing is moved the movement will be swift.

45

These, then, are the true reasons the voices of young animals are generally high-pitched, even though not all are. Neither are all voices of old ones deep, whether they be of males or females. Sometimes the voices of the sick are high-pitched just as they are sometimes high-pitched in healthy people. Sometimes old ones have higher-pitched voices, even though the age of the old ones is the opposite of that of the young.

46

Further, many females and many which number among the young animals have high-pitched voices. This is because, due to the weakness of that which moves, they move but a little air and overcome it in so moving it. Thus, because it is but little, it is moved swiftly.

This then is the reason that the voice of the bulls is high-pitched, while that of the females and the calves is deep. For the calves, due to their age, and the cows, because of their sex, have a weak member moving the air and with that member they move a

---

54. "Neume": *neuma,* from the Gr. *pneuma,* can indicate an individual note as well as a musical phrase.

great deal of air. The mover does not prevail in the movement, so that the movement will be heavy and slow. Bulls, however, create strong movement with the strength of their instrument and for this reason the movement will be swift. Now when they have progressed in age, the moving member will grow stronger and then their voices will change. The reason for this is that the moving power of all animals lies most powerfully in the nerves. For this reason youth is stronger than old age in all animals, for then their dried out nerves are not yet undone by an extraneous humor and have not yet been made rigid by natural dryness.

47 This is the reason that the young of brute animals are stronger than the children of humans. For they have stronger nerves due to the earthiness of their complexion. But in human children the joints are not strong and neither are the nerves. The same is true of the nerves of the old, but for another reason. Bulls, however, have the strongest nerves of many animals, especially in their heart, which is the place from which, according to the skillful teaching of the Peripatetics, the nerves arise. And because air enters due to the heat of the heart, it is necessary that there be a member in the bulls which moves the air most forcefully, because it shares the nerves with the heart, like a string stretched between both of them. An indication that the nature of their hearts has this sort of disposition is that bones are found in the hearts of some of the stronger and tougher animals and such animals move the air very forcefully.

Moreover, all animals which have been castrated have their voices changed to the nature of females. Thus, the voices of castrated animals have their voices rendered like those of the females due to the weakness of their nerves.

CHAPTER X

On the Origin and Cause of the Appearance and Falling Out of Teeth in All Animals[55]

48 The rise of teeth in animals is for the cutting of food. The statement of some that nursing is the cause of the rise of teeth is false. For nursing does not cause the teeth to arise in and of itself, but rather it is the heat of the milk. The sharp teeth arise before the broad ones. For the sharp ones are needed to cut the food and cutting of food comes before grinding it. Further, that which is smaller arises more quickly from heat than that which is larger and thus it follows that the sharp ones are smaller than the broad ones. Besides this, the jawbone is broader where the broad teeth are rooted than it is where the sharp ones are. Thus nursing does not have to do with the appearance of the teeth, but rather the heat of the milk causes the teeth to arise quickly. An indica-

---

55. Ar. *GA* 788b2f.

tion of this is that those children who nurse on warmer milk have their teeth arise the quickest. For heat is, as we have often said, creative and formative.

Now teeth fall out after their first appearance, but do so for the better. For that which is sharp is easily blunted and worn down and then it must necessarily be changed. But the broad teeth are not worn down this way; rather, they are worn away only by the long passage of time. The incisors, however, fall out from necessity since they do not cling forcibly to the jaw as do the broad teeth, which have many roots and which arise on the strong and broad bone of the jaw. The incisors, though, arise on the thin and weak bone and they thus fall out readily and also grow in again at the proper time.

Those teeth which are latest in time and after which no other teeth arise most often arise before the age of twenty years. But they do arise in some old people on occasion, due to an abundance of that nourishment which the body uses internally. This is especially so in certain women at the end of their life, for in these are born the teeth which we have named "teeth of the intellect" [*dentes intellectus*].[56] The slender ones to the front are nourished and fall out most quickly. For there is no superfluity of the material of generation and of food in that which is slender and it is thus consumed more quickly.

Democritus, however, as we have said above, dismissed the true final cause for the sake of which teeth exist and attributed their generation solely to the necessity of matter and of the efficient cause. Still, though, we know well from what we have determined elsewhere that the appearance of the teeth is for the sake of something, which is their end. For all the things we have said are for the sake of completion and yet nothing prevents there being a first appearance of the teeth followed by their falling out and then that there be a second appearance. For just as instruments are completed by nature for some end, so too, in the process of modifying an object, artisans employ many instruments, such as an axe or an anvil. The vital and natural spirit acts in the same way in those things which exist naturally. But the statement of one who says that natural causes come from absolute necessity is like one saying that the water comes forth from the belly of one suffering dropsy [*ydropicus*] only because of the wound caused by the instrument that does the cutting and not because of the health which the one doing the cutting means to bring about.[57] The cause of the appearance of the teeth has thus been determined, as well as why some fall out and grow in again afterward and others do not, and generally we have said why teeth arise. We have likewise adequately determined the causes of the other accidental traits which occur in each member, whether due to a necessity of the matter and the efficient cause or for the sake of something, that is, for the sake of a final cause which first moves the efficient cause.

49

---

56. A. is referring to the wisdom teeth. Cf. 2.54 and 12.217–18. His received text said "rear teeth" (*posteriores*). He has changed it to read "posterior in time."

57. Elsewhere, *dropsy* is *ydropisis* (1.479) or *yposarca* (13.61).

HERE BEGINS THE TWENTIETH BOOK ON ANIMALS

Which Is on the Nature of the Bodies of Animals

The First Tract Being on Those Things Which Constitute the Body

CHAPTER I

On the Seminal Moisture of Animate Beings in General and in Which Is Also Treated the Nature of the Moisture of Other Bodies Which Are Generated Mixed

1 The following discussion will treat the nature of an animate body. First of all, however, we should investigate the differences between an animate body and an inanimate one, and we will then study the differences animate bodies have one to the other. Although the body of everything generated consists mostly of earth and water (as we have shown in the second book of the *Peri Geneos*), these elements nevertheless differ in one which is an animate body and in one which is inanimate.[1] But that which is the seminal moisture in the bodies of the living should be more mixed in with the airy, spiritual moisture and a heat should be in it, making the spirit in it frothy. For otherwise, this moisture would not be rendered the principle of life, for the vehicle of life is the spirit, whose origin can lie only in the spirit, as has been shown in many places in our natural science books.[2] This is also why Heraclitus said that vapor is the principle of all things.[3]

2 But vaporous moisture differs from breathing moisture and from dripping moisture.[4] The dripping moisture is embodied, mixable, and connective [*continuatum*]. For everything which drips (that is, flows) and overflows must be embodied, because if it were not embodied it would follow the movement and location of the element whose form it took on. Thus we see that vapors and spirits rise up and do not drip down or descend to the ground unless they first return to an appearance of the embodied element in the presence of a compressing cold. This moisture is mixable by virtue of two powers. One of these is fineness [*subtilitas*], by means of which it is rendered penetrating.[5] The other is moistness and by means of this it flows to the limit of the body

---

1. *DG* 2.2.17.

2. Cf. *De spiritu et respiratione* 1.1.1f.

3. For a discussion of Heraclitus's opinion, cf. *De spiritu et respiratione* 1.1.2. Cf. also Diels (1951, 1:159, frag. 36) and Cadden (1980, 325f.).

4. The terms are *humidum vaporativum, spirans humidum,* and *humidum manans.*

5. "Fineness": the words *subtilis* and *subtilitas* suggest several things at once—smallness of size, thinness, and a certain delicacy and refined state.

into which it is mixed.⁶ It is connecting [*continuativus*] through the power by which it is bounded by the limits and spaces of another's body. For, occupying the spaces among its parts, it moves each part to the other because, not remaining within it, it constantly drips and flows from one to another. This moisture, then, is that which has both coarse and fine parts at the same time and has more fine parts which have turned to coarse. Thus, when it is rendered material, it mostly constitutes itself by converting fine to coarse, rather than the other way around. Those coarse parts which are in it are there in act, while the fine parts are there potentially. Thus, when it constitutes an earthy, mixed body, the coarse parts draw the fine parts toward act rather than the other way around.

A clear indication of this exists in all mineral bodies, which, even if they sometimes are mixed vaporously into material, still, when they are drawn to the species and form, the fine parts are seized by the coarse and are coagulated, drawn to the embodiment of earthy solidity in such a way that nothing vaporous or "breathing" remains in them with respect to act.⁷ Because of the connecting moisture which is in them, the space and passages in them for the entrance of another moisture into them are eliminated. This is also why such bodies are not susceptible to nourishment. One cause for this (among others) is that the passages through which the nutrimental humor has to flow to the individual parts are eliminated as a result of the first mixing together of its natural moisture with that which is earthy. For an embodied moisture does not cause the spaces of the body in which it is present to separate. Rather, it fills in the gaps and joins them together. This, then, is the nature of this sort of dripping (that is, flowing) moisture.

A vaporous moisture is composed of fine parts which are able to be changed into the form of air by mild heat. And because the whole is vaporous, then one part hastens to leave before another. For this reason it is rendered with an indeterminate shape and it continuously changes from one shape to another. Thus, continuously rising, it moves until it finds a coldness which thickens and compresses it. It then is turned into the moisture of rain, dew, or some other water which falls from on high.

Now the moisture we have spoken of did not have this nature either in material, in power, or in movement. This is clear because such a one has only fine parts, both with respect to act and with respect to the movement of the fine parts. Although another might have both fine and coarse parts, it has the fine ones with respect to act and with respect to the movement of the coarse parts. Thus, those things which fall from on high are not continuous and they also have a great effect on the fine things. For waters of this sort which fall from on high are inflative and astringent [*stipticae*] and they do not fall with a continuous dripping, but, as we have said, with an intermittent one.⁸ This is the reason as well that either few, no, or small mineral bodies are generated

---

6. Ar. *De anima* 405a25.
7. On "breathing" vapor, cf. *Meteora* 4.3.18. There it is described as appearing especially above bodies of water, created from the reflection of the heat of the sun.
8. Note that a later hand corrects *divisa* (intermittent) with *densa*.

from such waters. But we have fully treated these moistures in the books *On Minerals* and *Meteora*.⁹

5 Frothy moisture is composed of a triple substance. One substance in it flows not according to its own form but to that of another. Another substance is in it which is viscous and which surrounds and holds in. A third substance is "breathing" and pulsating.

The first of these substances is that which flows, not according to the form and nature of a moisture, to be sure, but rather according to the nature and form of the member to which it flows in order to form it. The viscous, surrounding substance, however, is that which contains the spirit, keeping it from evaporating. Examples are the skin and pannicular-membranes surrounding animals and the members of animals when they are formed.¹⁰

The "breathing" substance in it is the spirit, which is the instrument of the formative power and its vehicle into all the things which are formed. This is as was stated in previous books. For this humor alone is the material in the generation of animate bodies. This humor is differentiated, however. One part of it is homogeneous, as it were, tending toward earthiness and from this are generated the bodies of plants. Another part of it is more heterogeneous and more moist and frothy, with an airy frothiness. Animals are generated from this part of the humor. Also in this humor are the earthy, which is generated into bones; the viscous, from which nerves and skin are generated; and the soft, from which flesh is generated. It is thus possible to find many parts in this humor.

This, then, is the differentiation of these humors and if there are others, they can readily be ascertained from these.

CHAPTER II

On the Dry and Earthy, Which Is the Material for Animate Bodies

6 We must now reach a determination in the same way about the earthiness of the bodies of animals. Generally, every earthy thing which belongs to the constitution of the mixed, natural bodies is fine. Yet in one it is finer than in another. For this reason it is to be understood that the earthy element would be mixable only if it were fine, for since cold and dry do not produce mixture, earth has no property by which it can be mixed. Similarly, however, the coarse and the rough resist the binding by means of which a mixed thing has to be bound by the powers of the other elements.¹¹ Therefore, every earthy mixable should have undergone two "passions" by the moist and the warm.¹² One of these is refining and the other is digesting, separating from the

---

9. *De min.* 1.1.5; *Meteora* 4.2.17.
10. Perhaps "surrounding live-born young and their members when they are formed."
11. Following Borgnet's *grossum et scabrosum repugnant* over Stadler's omission of *et*.
12. For "passion," cf. Glossary.

roughness. That which has thus been made fine and has been digested can stop at a triple limit. For if the limit of the refining and digestion should move in the direction of the power of the dry in the first matter [*prima materia*] itself, then it will produce a moisture by itself which cannot cling to something touching. This moisture, thus bound, will exist then between the pores and the spaces of the dry, conferring mobility and mixture to the dry. This earthiness is the material for minerals. An indication of this is that metallic things are coagulated by the cold. For we have proven that the cold coagulates only by compressing moisture. But moisture is not compressed between the parts of dry earth. This sort of earthiness overcomes the matter in minerals.

Another limit for the dry in a moisture moves in the direction of the power of the oily, airy, vaporous moisture. And the earthiness, which is finely mixed into it, is the material for plants. An indication of this is that plants dry out greatly due to the evaporation of their moisture and they are flammable because of the oily and airy moisture. They also float on water because of the airiness prevailing in them.

The third limit, however, moves in the direction of the viscous, frothy moisture, as we have said, with the result that the dry earthiness is bound by it. This earthiness is very digested and quite formable due to the powers of the moisture. It can never be entirely hardened, but is always soft for the most part, due to the moisture that prevails in it. This is the earthiness which is the material for the bodies of animals.

Moreover, this dry has undergone three passions due to moisture. One of these is the binding of its parts to bring about the plasticity [*formabilitas*] it has in various uniform and nonuniform members.

7

The second is the separation from the rough coarseness. For otherwise, the moisture would not follow everywhere. The third is the refining which occurs through digestion and through washing the foulness out of it. Through this it is quite able to be made flat and able to be drawn out into every shape, line, and surface.[13]

The viscous, however, is this way because its parts vary, having been seized inseparably with the moist, doing so to effect longevity. For a moisture that is not readily dryable is a cause of a longer life. And a viscosity, which holds in the earthy parts so that the moisture does not evaporate, and a multitude of moisture that cannot be overcome by earthy dryness tend to prevent drying out.

Earthiness of this sort, then, is the basis for all forms which are produced in animals' bodies. It has differences since one part of it is coarser and drier and another is softer and moist, while another is more able to be drawn out and is more viscous.[14] The bones are generated from the first of these parts and flesh and fat from the second, while veins, arteries, and nerves are from the third, according to the differences which they possess among their own material parts.

---

13. The terms used are reminiscent of metalwork, especially of the jeweler's craft.
14. Following Borgnet's *grossa et sicca,* again supplying the *et.*

8   Further, the earthy, which, as was stated, is the material for animals, at first follows the action of the moist in all respects. For otherwise it would not be receptive of shapes. Nevertheless, in the end, it is limited in the direction of the dry by the evaporating moisture.

The fact that dryness of this sort exists at first, with respect to substance, in the moisture out of which the generation of bodies occurs is clear because although the blood and the seed of procreation [*genitura*] out of which generation occurs may be viscous, they are still easily dried. This could be due only to a dry earthiness which draws to itself the parts of the moist lest they be scattered. An indication of what we have said is that all bodies of animals, as they putrefy, at first turn black and then at length turn to dust as the moisture evaporates from them. This can only be so if dry earthiness is their material principle.

As we have already said before, the dry earthiness cannot be the principle and matter of living things through its own simple properties and potencies, but rather through that which has been mixed in with the moist. For in every living thing, both for the sake of nourishment and for growth and, prior to both of these, for the sake of the formation and the definition of the shape of the members and, prior to all of these, for the sake of bodily extension, one part can never be produced from another in such a way that it remains whole and not divided, unless one part flows from the other through the nature of the moist.

For this reason too it is necessary that in such cases there exist a moisture that prevails and draws the dry along with it. For the moist is a quality and potency of life, while there is no potency for life at all in the earth. Still, as we have shown in the books *On Plants* [*De vegetabilibus*], the bodies of live animals share less in the powers of the earth than do plants.[15]

9   These bodies have four things from the earth. For the members could not retain their shapes if it were not for the earthy dryness and for the cold, which slightly compresses the moist, even though it does not entirely press it out. Again, the bodies would become feeble and vapid unless there were dryness and coldness to curb the moisture that is drunk in. Further, animals' bodies would not be held erect unless the solidness of earth were present to prop up the bodies of animate beings. Rather, they would collapse and flow together like water. This is why the Pythagorean poets tell the tale that Atlas held up the sky, for they used to say that the world was a large animal.[16] They say in their poetry that Atlas is a giant since he has a great deal of earthy solidity. The fourth is the mobility of the members, for there would be no rotating motion in the ball-and-socket joints in the members unless a very strong solidness existed, well concocted out of the earthy element. For these members would be undone by the movement which excites heat. This is why, during paralysis, when the

---

15. *De veg.* 4.2.
16. For Atlas and the Pythagorean poets, see *De principiis motus processivi* 1.4; 2.7; and *DCM* 2.1.3.4. Cf. Ar. *De caelo* 284a18f.

members and joints have dissolved due to watery humor, animals have none of the motion which people call progressive.

For these reasons, then, and those like them, the earthy is present as an element in the bodies of animals. And it has four different types in bodies of this sort. It is, in one part of them, smoky and dry and in another part earthy, coarse, and dried. In another part of it the earthiness is affected a great deal by a coarse, oily moisture pressing upon its parts, grafted to them as if bound inseparably, as in the links of a chain. In another part of them it is bounded in a fine moisture, well digested, and it sets up and is coagulated by that digestion which is called *epsesys* in Greek, or by one like it.[17]

The smoky and dry earthiness is present in the hairs on the bodies of animals. The coarse and dryable earthiness is found in the bones and cartilages and is roasted in that digestion which is called *optesys*. That which is interwoven into the oily, viscous moisture is in the nerves, veins, and arteries, according to their individual differences and passions. But that which is bounded by the moisture digested in *epsesys* is in the flesh and fat, with respect to the differences which we have already determined in earlier parts of this study.

These, then, are the two elements which are materially abundant and the foundations, as it were, which are in animals' bodies. Because of these their bodies have the movements of these two elements, and the repose of them all is in the earth.

CHAPTER III

On the Nature of Air Insofar As It Is Shared by the Bodies of Animals

The substance and qualities of air are shared variously by the bodies of animals. For all things share the form and nature of air in the spirit which flows through their bodies and seeds. But the actual unmixed [*simplex*] substance of air cannot arise from the constitution of a given animal. The reason for this is that air is not retentive of corporeal shapes and if unmixed air is heated by the natural heat of the animal it becomes further rarified and will be less receptive and retentive of shapes. Therefore, the spirit which is in the bodies of the animals is a vapor set free by the seminal moisture during generation. And since this moisture is made up of four humors and elements, this spirit cannot have the unmixed nature of the element of air.[18] For since that spirit which has the form of air is the vehicle for the power and is the soul's instrument by which it forms the body and by which, once formed, it gives it life and sensation, then it is necessary that this spirit be so composed that the power which it bears clings to

---

17. "One like it": see Borgnet (*illi*). Stadler's reading of *illis* is more difficult, "or one like these." On *epsesys* and *optesys*, lit., "boiling" and "roasting," see 6.12, with notes.

18. Cf. *De spiritu et respiratione* 1.1.3.

it. This can be done only through the power of a thick fluid, for since this alone is the principle and subject of life, it can retain the powers which it conveys to the members.

12 Further, since spirit is the instrument of the operations of the soul, it is necessary that its substance be of the sort which can bring into the members those forms which the soul intends to bring into the members through it. This is clear by analogy. For a woodworker, whom people call an "architect," does not make his saw or adze out of wool or lead. He makes it instead of iron and he arranges its teeth in a suitable row as befits the needs of its function. And since the work of nature is more sure and well ordered than that of art, it is all the more necessary that an instrument of nature be suited to its functions. Since, however, the soul leads many powers and very many forms into the members of the body (the first of which is the form and act of life, while next is the form of sensation) and since fineness and mobility are surely required to convey life throughout the body (for this is accomplished through the *diastole* and *systole* of the pulse), it is thus necessary that the spirit possess a great deal of airy fineness and mobility. Moreover, lucidity is necessary for sensation so that the sensible forms borne by it are not obscured. So too should the spirit be clear, very transparent, and not intemperately warm.[19] For that which is intemperately warm has its warmer parts rising and the less warm parts descending and thus the forms do not stay in order and are borne about in it. In this regard, again, it must be airy. For air is mobile in an orderly manner, transparent, and is not intemperately warm. This is how the nature of air is shared by animals.

Further, since the body of every animal is porous, the nature of the unmixed surrounding air is shared in that it enters the empty spots and the spaces of the pores. For moisture could not be everywhere in the body unless the pores had spaces in which it could be received. Therefore the air is also received in these spaces. For this reason, if it becomes poisonous and destructive it harms the bodies, for in such bodies it penetrates to the interior members via the pores, unaltered and undigested into another quality. In these two ways, then, according to the proper form and quality of the air, it is shared in the material of the bodies of animals.

13 There is, moreover, another way in which air is shared, not in its proper form but rather in the way unmixed things are.[20] In this case the moisture shares in the nature of the moist out of which the substances of the members are made. For there are two principles of life, namely, tempered warmth and tempered moisture. Both of these are qualities of air. For when the moisture out of which the substances of the members are made is one which flows continuously outside its own form toward some other member's form, then it should itself be warm, for this causes moisture to move and flow outside itself. But if it were excessively warm, then it would drain and dry out the moisture. It thus should be temperately warm. Likewise, if it were excessively moist, then its parts would flow one into the other and would not flow toward the

---

19. Cf. *De spiritu et respiratione* 1.1.4.
20. "Unmixed things": *res simplicia*. Note too that M3 inserts *secundum quod* before *res simplicia*. Borgnet reads *refracte simplicia*, which suggests that the unmixed things are "scattered" in a mixture.

form of the members which must be taken on from outside of its own substance. It thus should be temperately moist, and this is something that is airy and not watery. For every mixed thing which acts according to the qualities of an unmixed thing in it has this unmixed one and in accordance with its qualities it acts more as a force which is dominant and prevalent in its mixture. Thus, the nature of the air is said to be more dominant in the spermatic moisture and in the nutrimental moisture and it could not be any other way than this. An indication of this is that everything which is formed for life, whether it is only nourished, as are plants, or whether it participates in sensation, is raised up in parts that are formed upward from the area of the earth. This movement occurs only from the nature of the element of air which exists in the ones formed this way. Thus, in this way the nature of air is shared by all living things.

There are certain of the nobler animals which share in air by means of respiration, and there are things which are produced via breathed-in air, such as voice and sound. For many of the aquatic animals which seem to be the larger and nobler ones breathe. Examples are the whales [*cetus*], the baleens, the dolphins, and the ones called the *koky* (that is, the ones called "sea calves" [*vituli marini*, seals]), as well as many other genuses of aquatic animals which we have treated previously. But not a single one of these is found in our lands in rivers or freshwater lakes. For these waters are too cold and the animals living in them are colder than the marine animals are. And since respiration is for the purpose of lessening the heat of the heart and lungs, those which are in the cooler waters have been made nonbreathers. But certain of the larger and nobler ones which are in places with warmer waters have need of respiration. Respiration consists of air which replenishes the spirit and chills it, as we have said in the book *On Respiration and Spirit*.[21] For this reason the passage through which respiration occurs has been made wide and tough, so that it might let in and out a great deal of air, and so that it might endure the blows of the air on its tough sides. It is made without a corner lest the air stand still in its corners. In all such animals the lungs are made having their openings above and not below, for if they had the openings below then the cold air would reach the diaphragm, which in turn touches the stomach and liver. The result would be harmful to digestion due to chilling and, moreover, the heated, smoky, and corrupted air and spirit in the heart and lungs would not be totally driven out and would instead reach to the inner members. The animal would then be corrupted while its inner members putrefied as a result of the corrupted air which reached them. For these members are soft and putrefy quickly. For this reason, then, it is driven out through the tough upper members, which do not readily undertake change and corruption. This is the way, then, in which the element of air is shared by them.

There are certain ringed creatures which do not share in the airy spirit for respiration. They rather have their spirit closed up internally, pouring forth to encircle the diaphragm itself.[22] For in all such creatures the circle of the diaphragm is very small

---

21. *De spiritu et respiratione* 2.2.1.
22. Reading Borgnet's *ipsum* for Stadler's *ipsa*.

and the two parts of the body, namely the upper and lower, are connected to it. And the spirit within their bodies pours forth through the segments to the point of cincture and makes a noise in them when they move. This is why they even make a noise when their heads are cut off.²³ That spirit refines more than it cools the moisture which is in the members of these animals. For this spirit is exceptionally fine and spiritual since a great deal of air is mixed into it. An indication of this is that these animals are more refined than many others in the operations and perceptions of some of their senses. But the bodies of plants share in air in none of these ways except in the mixture of it with the oily moisture and, to some extent, in that it enters into their pores, as we said above. Thus spirit in plants is not as noticeable as it is in the bodies of animals.

CHAPTER IV

How Fire Is Shared by the Bodies of Animals

16 We must now see, following the same method of investigation, how the bodies of animals share in fire. We know that there are many effects of fire in the bodies of animals such as decoction, digestion, the separation of the homogeneous from the heterogeneous, the bringing together of the homogeneous, piercing, widening of passages, consuming moisture, and expelling certain superfluities with force. Anyone will find these and other known operations of this sort in the bodies of animals if he will attentively and diligently give the matter his scrutiny.

All the operations mentioned are caused only by the heat which is the heat of fire. For the decoction of the seminal and the nutrimental chyle can be accomplished only by means of fiery heat. And although we have demonstrated this already in the fourth book of the *Metheora*, nevertheless we will say here as well that every decoction is accomplished by heat either from one's nature or from without.²⁴ For it is agreed that the decoction which occurs for constituting or nourishing the bodies of animals will be brought about through heat. And when decocted things coagulate and thicken, then decoction is accomplished only through the removal of moisture. And the removal of moisture is accomplished only through the heat of fire. It is thus clear that decoction proper is an operation of fire.

17 It is much the same for digestion. For digestion is a sort of decoction, wherein the seminal or nutrimental moisture is completed and bounded by its proper and natural heat, is led to an end which is suited to its nature, and is removed from qualities which are contrary to its nature.

---

23. Clearly a reference to an experiment on the cricket which A. conducted himself. Cf. 4.66f., 91; and *De anima* 2.3.22.

24. *Meteora* 4.1.1f.

Now the seminal moisture will never take on the forms of the animals and of their members unless it should first exist at a limit which is suited to their forms. But it is led to the limit (that is, completion) suited to the forms of the members only by the heat which separates it from opposite passions. Now I am calling the limit the moisture's matter in which the forms of the members of the body as a whole exist according to proximate potency. And they are called forth from it directly by the simple and single movement of the formative power, just as we say that nourishment is already digested when it comes to its last assimilation to the members, after which there follows only conversion to the form of the members. However, this splitting off from the contrary forms and this assimilation to the powers of the members can be accomplished only by the heat of fire. For no power whatever is active in simple fashion [*simpliciter*] save the heat of fire. Coldness, which seems to be an active quality, acts more by curbing and constricting than by splitting some things off and introducing others. That which curbs does not split off, and thus cold things remain raw and undigested. But heat, which causes moisture to boil from its center, changes by splitting off and by introducing certain passions. It is thus digestive and limiting.

Separation is a further, proper effect of fire. Because it acts by splitting, it draws off those things which it splits off from those whose contrary passions are not separated from them. Therefore, separation necessarily follows upon splitting off. For it separates the heterogeneous from the things homogeneous to the nature which it digests. There is an example of this both in crafts [*ars*] and in nature. Now in the crafts, when gold is refined [*digero*] and decocted those things which are not part of the nature of gold, such as stones and mixed-in metals, are separated from the gold. The same holds true for rust and the like. At the same time, however, the gold is drawn off and rises upward. In natural matters we see the same thing, how every fetid thing is separated from the nutrimental chyme and how that which is coarse is separated out through the bowel, or though the vessel which is appropriate to that superfluous humor, such as the spleen and the vessel for melancholy [*vas melancoliae*], the gall bladder [*kystis fellis*], and the gall vessel [*vas colerae*], as was shown in earlier parts of this study.[25] That which is pure, however, is taken up, separated by things of this sort, and assimilated to the body. The separation of the homogeneous, therefore, from the heterogeneous established at the same time is the proper operation of fire and digestive heat.

It is agreed that perforation is an operation of fire because nothing at all among all the elements is an enduring, active, and sharp quality except warmth joined to a fine dryness. But only the heat of fire is like this. This is why the first spirit which is in the substance of the seeds of plants and animals does not perforate by making the passages both porous and separate, unless it does so through the power of warmth. For spirit is, rather, more capable of impelling and pressing upon the whole substance than penetrating it. Thus, we see that perforation occurs both in the formation of embryos and in their nourishment. Perforation during formation occurs in many

---

25. For *vas colerae*, cf. 1.611; on *kystis fellis*, cf. 1.597; 2.87; 3.32, 112–13; 12.84; and 13.98.

ways. Some are perforated with a sort of hollowing action. In these there is a great deal of the warm spirit, as in the heart and lungs. Others are perforated with a sort of boring action so that there are passages for the superfluities which run through them or into them. Examples are the intestine, stomach, bladder, and the like. And some such as the "quiet" and pulsating veins are perforated by a fine perforation. An abundant, sharp, warm spirit works in those which are perforated with a hollowing action. A warm spirit, not sharp but windy, operates in those perforated by boring out, making the extremes distant from one another just as bubbles are separated by an enclosed wind.[26] In those, however, perforated by a fine perforation, a very fine, sharp, small, and warm spirit operates. Fire operates this way in the formation of the seed. In nourishment, however, it is necessary that it pierce and penetrate the member which is being nourished, for otherwise the nourishment would not enter into its substance on all sides nor would it be nourished proportionally in all its parts. This is also why a sharp, pungent heat causes the body to be filled with pores as it penetrates it. This, then, is the manner of the operations of fire in the act of perforating.

20 Further, the heat of fire also widens the passages once they are made and makes them free for passage. For as has been said earlier in this study, a continuous flow of a mixed humor plugs up and fills in passages, just as cracks are filed in by the flow of muddy water.[27] This is why there is need of a burning heat which will widen the passages by removing materials of this sort which cling to the passages, enabling them to pass on. An indication of this is that many people suffer from colic and *hylyaca* when the bile duct, which leads to the lower intestines, is blocked.[28] For then the walls of the intestines near one another are constricted. The superfluity, choked off, stops there and blocks further passage. But when the sharp bile arrives, its walls move apart, the passage is widened, and the choked-off superfluity descends. Heat works in the same way in other passages as well, by widening them.

21 The action of fiery heat is also present during the consumption of moisture. None of the other qualities does this. The heat of air does not consume moisture, but only moves it around. But the heat of fire drinks it up and brings about dryness. It is for this reason as well that all bodies are dry in which the fiery heat prevails. For as a result of its constant attraction of the fine moisture, at length a hard, dry earthiness remains. Heat employs this same action in forming bones and in consuming superfluous moistures present in the members. Otherwise the members would become too softened, could not move in an orderly fashion, and the shapes of the members would become monstrous and disorderly. Excessive heat also acts this way when it severely dries out

---

26. The Latin is unclear. Perhaps A. has in mind the process whereby the surfaces of a bubble move apart from one another as it grows larger.

27. Cf. 13.42.

28. *Hylyaca*: presumably similar to the *passio yliaca* besetting horses and noted at 7.111. For a discussion of *passio yliaca,* see Isid. *Orig.* 4.6.14 and Sharpe (1964, 66), who identifies it as "a clinical syndrome marked by pain, vomiting, and abdominal distention . . . which (has) a potential for rupture and death."

the members in those with hectic fever.²⁹ At length, becoming entirely prevalent, it acts this way in putrefaction and in turning to dust. This is why the edges of rotting things are moist at first and ultimately are dry and then turn to dust. These are all nothing save operations of fire.

Forceful expulsion is likewise an operation of fire. For the heat in a moist, coarse, earthy thing creates a windiness possessed of strong movement. By continually generating it, it increases it and thus causes it to become dominant. For this reason it causes expulsions of such force that it sometimes shatters and bursts its vessels. An indication of this can be found in vessels which have been sealed on all sides and have been placed into the fire. These burst with great force much as a roasted egg or chestnuts sometimes do, along with other things in which a warm vapor is enclosed. These, then, and those like them, are the proper operations of fiery heat in the bodies of animals.

Fire, as are the other elements, is present in mixed things of this sort substantially. For the qualities of the mixables are present in a mixed object only on account of the substances of those things which are in it. For if one changed into another through generation then it would no longer be mixed, but rather it would be unmixed [*simplex*], generated from one or several others. Therefore, fire must be in such things substantially. An indication of this is found in the upward growth of all animate bodies. They have this movement solely due to the lightness and warmth of fire. An indication of this is that animals with warmer, less earthy bodies are more upright and erect. Examples include the human and certain genuses of apes which walk about upright on two feet. Likewise, the genuses of birds are erect because of the great airiness and lightness of their bodies. This is so even though birds have far larger upper members than lower ones.

This, then, is our statement on how fire is present in the animate bodies. Now someone may ask, however, how fire descends to the place of generation in animals and how it is incorporated into mixed bodies. But we have already shown the reason for this in other books. For the rays of the sun, as they strike the areas of generation, cause fiery heat, as is demonstrated in the study of mirrors that burn.³⁰ This is the fire which joins with the elements in mixed bodies. For the sphere of the sun must move fire and transmit its power to mixed bodies.³¹ The fire which is often generated in mixed bodies is also caused by movement.

Fire is present in the bodies of animals in this way and for reasons of this sort.

---

29. On hectic fever, see 1.436, 504.
30. Cf. *Phys.* 2.3.1, 3.
31. At *Phys.* 2.3.3, A. makes it clear that the sun must 'move' fire as the fixed stars 'move' the earth, as the moon 'moves' water, and the other five planets 'move' the air. Cf. too Avic. *De caelo et mundo*, c. 16.

CHAPTER V

Whether the Fifth Body, the Essence, Is Substantially Mixed into the Bodies of Animals. Many Arguments Are Also Set Forth of Those Who Posit That It Is Mixed into an Animate Body.

23 There are those who say that in addition to the things mentioned, which are called the four bodily elements, there is a fifth body mixed into bodies. They say that light is the substance of the heavens and that this is in mixed bodies and is called the fifth body. They seek to buttress this argument with false proofs. They assert that contraries do not stay mixed unless there is something holding them in a mixture. For light things naturally moving up and heavy things naturally moving down will move apart unless there is something to hold them together, keeping them from dissolution caused by their proper and natural motions. That which holds together things which have opposite motions—either toward or away from the middle—cannot be any of the elements mentioned. For as we have shown in the fourth book on *Heaven and Earth,* all the elements mentioned move either away from or toward the middle.[32] They therefore say that that which holds them together is light, which exists because of the nature of the fifth essence or the fifth body, namely, the heavens.

There is a similar argument concerning hot and cold, and whether each is active due to the necessity of its nature. Things which are active due to the necessity of their nature necessarily act on one another when they touch each other. Continuous action is necessarily the cause of dissolution since every passion grown great departs from its substance. When, therefore, a mixture of hot and cold things is not undone, there must be something holding it together apart from heat and cold. This thing, however, which is apart from heat and cold is none of the elements. There must be, then, a fifth element, existing outside the elements, which holds them together. Or so they say.

24 The same sort of argument exists concerning the moist and dry. These are passive subjects [*passiva*] and dispositions of matter. Contacting one another through a natural necessity, they are affected by each other. For this reason, so they say, they require something holding them together. Since that which contains is not identical with that which is contained, they say that it must exist outside of all material principles (that is, elements). Using this argument they say the fifth body is that which contains.

Further, based on movement, quality, and effect, they seem to show that spirits are bodies of this sort, existing from the nature of the fifth essence. For they say that the movement of light moves from a heavenly body in the shape of a pyramid.[33] Such too is the departure of the spirit from the point (that is, peak) of the heart into the

---

32. *DCM* 4.2.8.
33. "Heavenly body": lit., light giver, *luminaris*.

entire breadth of the body. They therefore say that spirit is of the nature of luminosity [*lumen*] and light [*lux*].³⁴ They say, as we have said elsewhere, that light is a body.

Further, they say that there is no element which moves from the principle of its place, going forth circularly along every diameter. Earth, for example, only descends along a diameter [*dyametraliter*], while fire continually rises. Air, on the other hand, descends on one, but never rises on two, while water, to the contrary, rises on one but descends on two along straight diameters.³⁵ But the bubbling forth of spirit from the place of spirit is circular along every diameter and thus does not seem to be of the nature of any element.³⁶ They say that an indication of this is the movement of the heart during *diastole* and *systole*. While continuously blowing out spirit, it moves by expanding itself out toward its circumference and by contracting itself in toward its center. This could only be, so they say, due to the pyramidal exit of the luminous spirit.³⁷ They add to this the shape of the heart which, they say, is pyramidal because of the pyramidal generation of the spirit in it. They say, then, these and similar things, basing their conclusions on motion as they defend their previously introduced position.

To prove this supposition, they also adduce certain things based on the quality of the movement and the quality of the spirit. The speed of the motion (which is a quality of motion) is so great in the motion of the spirit that, as Isaac testifies, spirits are suddenly moved in the body just as illuminations are suddenly produced in the world.³⁸ They say that they could not move suddenly if they were of the nature of the four elements. They therefore say that spirits are of the nature of the fifth body. They also add the conformity of their quality, which they say are clear spirits and certain lights. They thus say that dulled spirits indicate disruption and *discrasia* of the body.³⁹ Therefore, since, so they say, light and luminosity are not of the nature of any lower body (for even fire in its proper sphere and nature does not shine), this light, they say then, is of the nature of the fifth essence, or fifth body.

In addition to all types of proofs mentioned, they also argue the point many ways from effect. They do so first on the nature of the limit [*terminus*] and end [*finis*]; second, on the type of its action; third, on its composition for its movement; fourth, by its analogy to the powers it bears; fifth, by the conformity of the complexion it contains; sixth, by its proportions to that in which there is first life (for example, in subject

---

34. In this book, A. regularly uses these two words for light. To assist the reader, "light" will be reserved for *lux*, whereas *lumen* will be rendered by "luminous light" or some similar term or phrase.

35. For A., a diameter is an aspect of mathematical body, and each body can be divided along three diameters, which measure length, width, and depth (cf. *Phys.* 1.1.3). These are themselves straight lines, however, along an axis.

36. On circular motion and the fifth body, see *DCM* 1.1.4.

37. On the circular motion of the pyramid, see *DCM* 3.2.9.

38. "Isaac": presumably, Isaac ben Solomon Israeli (d. ca. middle of tenth century), one of the foremost Jewish physicians and natural scientists.

39. For a definition of *discrasia* as a state of imbalance in the body, see 7.93, with note.

and cause). They look seventh to the argument of its mediation between body and soul; eighth, to the argument of its contraction and expansion; ninth, to the argument of its growth and diminution; and tenth, to the argument of its diversity.

26 They also argue in this way from the nature of the limit and the end. For they say that the primary qualities have the power of altering material only in order to take on their simple or mixed forms. But the light of the spirits which are in the body leads toward the appearance of life in flesh and other members. For this reason it follows that the light which is spirit in the body of an animal is not of the nature of any of the elements.

Further, the primary qualities which are in the elements have one, simple effect, but light causes many. For example, light causes very many effects in the world and leads almost every form to an effect, just as has been demonstrated in the natural science books in the study *On the Soul*.[40] So too, in the body of an animal, it acts on not one but on very many nonuniform ones. For this reason, so they say, it follows that light is not any of the four elements but rather is a fifth, which is apart from and outside of them.

They show the same things using the type of action. For the light which is of the spirit makes everything without having any opposite. None of the elemental qualities functions in this way. It therefore follows that the light of the spirits is not an elemental form, and it follows that it is not a form without a subject. It is thus necessary that it be either a fifth body or that it exist as the form of a fifth body.

27 They argue again from the composition of this luminosity to its movement, saying that there is no doubt that the light of the spirits is an instrument of the soul and that the body which is created next beneath the soul (in its shadow, as has been shown elsewhere) is a fifth body. This light therefore seems to be of the nature of a fifth body.

They say that this same thing can be grasped by an analogy of this luminous substance to the powers it bears. For all these powers are powers of life. And since that which bears something or bears it in cannot be the same as that to which the power is borne, and since that to which it is borne is an element or is compounded of elements, then that which bears these powers in will be some substance existing apart from all the elements or things compounded of elements. This can mean only that all spirits of this sort are of the nature of a fifth essence. Therefore, this fifth essence will be the one acting in the bodies of animals.

Again, they prove the same thing by the nature of the complexion which these lights contain. For this withdraws from the excesses of contraries and approaches a body having no contrary. This is nothing but the heavens, as has been proven in previous books. Since, then, it leads those things which are from contraries to a likeness of the heavens and contains them in a likeness of an orb, it seems that it can be part of nothing other than the substance and nature of the heavens.

---

40. *De anima* 2.3.11.

Further, according to these people, a nobler life is found first in the heavens, just as in the cause and subject of life. Life, however, is not in the elements insofar as they are elements. Therefore if, as these people say, something composed of elements lives, it must exist first by participation in a living body, and this is nothing other than the heavens. Therefore, the nature and the substance of heaven are in every living body.

Again, since the vehicle for the soul into the members is the substance of light in the body (much as the light of the heavens is the vehicle for the mover of heaven into all matter), light, then, must be midway between the nature of the soul and matter.

Matter, however, is a thing made up of elements. Therefore, the light which is in a body is midway between the soul and the matter of the elements. Now the middle is neither of the extremes, but is below one and above the other. Therefore, the light of the bodily spirits in the body of an animal will be above a body composed of elements but below the soul. It thus can be nothing other than a fifth body.

Again, that substance is contracted to a single point which is at the peak of the heart during its *sistole* and expands during *dyastole* to the expanse of all the members. Such a contraction and expansion cannot occur often and suddenly in any body composed of elements. The light of this sort, therefore, must be apart from the four elements and must be something apart from all things composed of elements. Thus, again, it will be said to be nothing other than the fifth body.

Again, when the elemental body is diminished, this substance sometimes grows and, in contrast, it diminishes when an elemental body grows. This could never be the case if it were one of the elements or one of the things in the body composed of elements. For restriction increases it and at death the spirit seems to leave the body and to withdraw from the members as if it were incorruptible. This is called "agony" by the physicians, or the drawing away of the spirit. The common folk call it the leap of the soul [*saltus animae*], however.

Again, none of the elements or things made of the elements is other than one with respect to form. But the clear spirit which is in animate beings is quite multiple in form, with one being natural, another vital, and another animal with respect to form. Now each of these, and especially the animal, is quite different with respect to all the variations of senses, imagination, memory, and other things. The spirit is also varied with respect to varieties of stimuli [*motivum*]. Therefore, since this diversity can be reduced to something as to a cause, and since this diversity cannot be caused by the elements, it can only be caused by the heavens. This is due to the diversity of forms which are in the heavenly lights, from which that light which is called spirit descends.

It is by relying on these things which are more fantasies than proofs and on things like this that certain people say that light, existing as a celestial body, is present in the bodies of animals. They say that the light of heaven penetrates into all bodies and that it is in the same place as whatever body it is with. They also say that as it exists in bodies, it contains and preserves them. It is the middle substance between body and soul and is the soul itself which moves the body. But we have discussed this assertion elsewhere. Here, however, we have mentioned as much as pertains to the nature of an animal, supporting it with proofs which, although they are not brought forward by those who formulated this assertion, we nevertheless have been at pains to bring forward for them so that when it is shown that those arguments which seem so strong do not reach a suitable conclusion, then this assertion will be rejected as erroneous and as the worst of errors by all right-thinking philosophers.

CHAPTER VI

On Disproving the Statement That Heavens and the Sphere Are, According to Some Part of Their Substance, Present in the Bodies of Animals

30 Now if light is said to be of the nature of a fifth essence in the bodies of animals and is part of their constitution, then this must occur in one of three ways. It will either be generated from the elements, as certain people have especially said already; or the light which is the substance of the sphere or the stars which come from it penetrates the bodies of animals; or it is something from the substance of a fifth body which, descending during the generation of animals, arranges their bodies with the other elements and things made of elements which constitute the animals' bodies during their first generation.[41] Now the first creator of this fallacy, as we have already said, set forth the first of these ideas. He said in a certain book where he spoke about the reason for the transmutation of bodies that it is possible for the bodies to be converted to the nature of the heavens only through heating and refining.[42]

But everyone who is well grounded in the Peripatetics' teaching knows that this is the worst of errors. For whenever there is transmutation of any bodies, one to the other, then they possess a single, common matter. This is as we have often proven in our natural science books. But in no way is there a single matter for the heavens and the elements. It is therefore impossible that the elements are transmuted to the nature of the heavens and vice versa. It was shown in *On the Heavens and Earth* that a single matter for the heavens and the elements does not exist.[43]

31 Further, if the nature of the heavens and the sphere is subject to generation this way, and if everything subject to generation is subject to corruption as was proven in the first book in our *Heaven and Earth,* then it will follow that the sphere and the substance of the sphere are subject to corruption.[44] And yet we have disproven this with many arguments in the first book of *Heaven and Earth.*

---

41. On the difference between the sphere and the stars, see *DCM* 2.3.4. On the number of spheres, cf. *DCM* 2.3.11.

42. The identity of this philosopher is uncertain. Stadler conjectures that A. may have had in mind Averroes.

43. *DCM* 1.1.1.

44. *DCM* 1.1.8.

Again, if the nature of the sphere is subject to generation and corruption, and since everything which is subject to generation and corruption requires a body as a cause for its motion which exists prior to it and which is not subject to generation or corruption but which itself has perpetual motion, then there will exist something prior to the substance of the heavens and the sphere which is incorruptible, has perpetual motion, and which is the cause of the motions toward place and toward form which exist in the nature of the sphere. It has already been shown in the eighth book of our *Physics* that such a body, which is the cause of alteration and generation, exists anterior to other bodies.[45] Now if perchance someone were to say that the heavenly nature is partially and not entirely subject to generation and corruption, then it would follow that it would not be homogeneous, but it would be composed of the perpetual and that which is subject to generation. But this again is the greatest absurdity, for according to this a sixth body will be necessary anterior to the fifth, and this one will be simple and incorruptible.

It will further be the height of absurdity to say that a thing which, through its motion, is the cause of the generation of both simple and composite things is composed of heterogeneous things and that that which is caused is something simple, like an element.

Moreover, if the nature of the heavens is subject to generation, what is the reason that the moon's heaven, touched over so many ages by the convex surface of the fire, has not been altered and changed? Now there have appeared those who are adherents of that sect which claims that there is no fire above, only air. How foolish this is can be seen from the things which we have proven in the book *On Places and Things Placed*.[46] For if it is taken as a given that it is self-evident and verifiable through the senses that the heat of fire is present in the matter of all things which are subject to generation, and that the heat arises from movement if the movement is strong and continuous, then we know for certain that fire cannot be absent in a place where there is so huge, swift, and strong a motion as is that of the sphere as it dissolves and ignites the material of the elements over so many ages. It thus cannot be said that the heavens are in any way partially subject to generation or corruption.

It is entirely impossible for the statement to stand that the substance of the heavens extends itself throughout all things. This is the substance which some investigators call the nature which is in all bodies, both those which are transparent and those which limit sight. This is as we have shown elsewhere, for that which is something continuous is one in number, but that which is one in number cannot exist in bodies which differ in number, species, and proximate genus. For it is agreed that elements differ in number and species and the heavens differ from them as well in matter and proximate genus. Neither is it comprehensible how a thing which is the same in number, substance, and essence can be present in all these things.

---

45. *Phys.* 8.3.1.
46. *De nat. loc.* 1.1.3.

33 Moreover, in the circular generation of elements, one from the other, the clear [*perspicuus*] is brought about from the unclear. Therefore, the clear which is brought about in it is led from potency to act and is thus caused by generation. How then will this be identical with the clear, which is not subject to generation or corruption and which exists in the heavens.

Further, even though we may grant that this enters into the constitution of the bodies of animals, and that the body is thus rendered colored, how will it be proven from this that this is the substance of the heavens or the sphere? For the things said by those who hold this position are completely absurd and unsuitable. Yet certain authors of this sect have chosen this path and have then left it in their writings. For this sect was once united and later split from one position into three, with one holding one thing; the second, another; and the third, a third.

And since we have already refuted two parts of this sect let us gird ourselves against the third, which is supported by certain people who pride themselves on being able to invent incredible and impossible things. And in so doing they fancy themselves philosophers. These say that the light which descends from the heavens and its rays are bodies, are part of the nature of the heavens, and enter into the constitution of generated bodies. Therefore, let us investigate whether corporeal light is split off from the substance of the sphere when it enters the constitution of a generated body. If it is said to split off from it, then there must be something causing the split and there is no agent other than an elemental quality. Therefore, the matter and the quality of the element divides the substance of the sphere, and the element is thus stronger than the heavens—an absurdity.

34 In addition, since the heavens are homogeneous with respect to incorruptibility and corruptibility, then just as it is divided in one part it can be divided in any part. How is it, then, that the fire has not long since divided it up? It is not possible to assign a suitable cause for this. If, however, it is not divided, then we must posit that each thing generated by something else is connected to the heavens, and this is contrary to sense. Further, since that light which is in things which have been generated is not invisible, it is necessary to posit that whatever is generated by some of it is visibly connected to the heavens. And yet this is absurd.

Now, how can it be said that this light is substantially a fifth body and that it descends to the midpoint [*medium*]? One of these is a nature and one is a simple motion. For it has been proven in the *On Heavens and Earth* that everything which is of the nature of the fifth body moves around, and not toward, a midpoint.[47]

Moreover, what moves it to descend? For when the movement is in accordance with nature, this should be moved either by itself or according to nature. Now, if it is

---

47. *DCM* 1.1.7. A.'s point seems to be that descent or ascent belong to rectilinear motion, not circular motion. How can light be a fifth body, whose natural motion is circular, if it descends or ascends? Therefore, it either is not a fifth body or it does not descend. Note too that at *DCM* 1.2.3, a *medium* is the center of a circle, but it is also any point between two extremes, as for example the mean between qualities.

moved by itself it must be animate, just as we have shown in the eighth book of our *Physics*.[48] If, however, it moves according to nature, then three unsuitable things occur. One is that a simple body will have two movements which differ in kind, of which one is circular, around a middle, and the other is straight, toward a midpoint. The second is that every generated thing follows the movement of that which generates it. Thus, that which moves light and a ray downward will naturally move downward, and that is earth. Thus the earth will generate light and will move it downward. The third thing is that since motion toward form and toward place are the same thing, it will follow that just as light seeks a place downward, so too it possesses the form and nature of light perfectly only when it is below. And it is clear to everyone that this is a most absurd thing.[49]

Light, therefore, which exists as part of the nature of a fifth body, cannot be incorporated into things which have been generated.

Now a new fiction of this sect has arisen which states that light is formally generated by the heavenly lights [*lumen*] and that it is form. And they say that the matter of this form is constantly being created out of nothing by a maker or first cause. Nothing resembling this fairy tale has ever been said by philosophers. For Plato's philosophy taught that the forms were bestowed upon existing matter by a maker. Aristotle's stance affirmed that forms were led forth from existing matter, whereas that of Anaxagoras said that all forms lie hidden in existing matter.[50] Never has anyone imagined that matter was led forth from nonbeing to the being of a form passed along by the one generating it. This sort of empty-headedness goes so far that not only does it speak contrary to philosophy, but it also states things the likes of which none of the philosophers has ever said, the likes of which nature has never produced, and which art imitates in none of its types.

Let us further ask them whether what they say can be proven by the senses, by reason, or by both in some trial [*experimentum*]. If it cannot be established by any of these, then it is established that it is contrary to all the principles of nature, for all of these are proven by way either of the senses or of reason. Therefore, this position they hold is erroneous and contrary to nature. But we have argued elsewhere against these fictions and therefore let those things we have set forth here suffice to provide certainty concerning the material principles which make up the body of an animal. That there are but four prime elements making up the body has therefore been proven by us.[51]

---

48. *Phys.* 8.2.3.
49. "Clear to everyone": *tonsoribus patet,* lit., "clear to barbers," but recalling an ancient proverb. See L&S, s.v. 1.
50. Plato, *Timaeus* 53bf.; Ar. *Phys.* 210a21f.?
51. Stadler's *probantum est* is senseless, perhaps an error for *probatum est*. Borgnet reads *probantur*.

CHAPTER VII

On Resolving the Arguments Introduced above Concerning the Fact That the Substance of a Fifth Body Is in Our Bodies

36   It is no heavy task to dispense with the things introduced above. For it is not true that contraries of that which is mixed in act remain in act only by something containing them, and that this is fifth among the elements. For mixture is the union of altered mixables into a single form. This is the act of the mixed and is caused by mixture. We have established the causes and types of this at the end of the first book of our *Pery Geneos*.[52] Because mixables are altered, each suffers something from the other. And because they are united into the single form of the thing mixed, each is seized and held by the other. They therefore stay in mixture. An indication of this is that when imbalance [*discrasia*] befalls animals, the wise physician does not strive to reform the light which is in the body holding the contraries together.[53] Rather, he strives to lead the contraries back to a state of balance in accordance with the notion of the arithmetic or geometric mean. This is just as has been stated in other places where we discussed complexions.[54] And this is the general case for all contraries, be they light and heavy, hot and cold, or wet and dry.

Now, their statement concerning the pyramidal departure of the spirit from the heart is laughable. For, as we have proven elsewhere, this spirit is generated from natural moisture. For this reason the heat causes an exit of this sort rather than the nature of the light. For the movement of heat is from the center to the circumference and thus, when the center rises above the surface which heat delineates by the widening motion of the spiritual moisture, then the shape of a pyramid is produced. Now it is due to the shape of the heart that the center is raised. But we have said enough about the shape of the heart where we spoke about its anatomy.[55]

37   Moreover, that which they set forth about the quality of movement and of that which moves is irrational. For a fifth body does not move quickly and there is no body whatever which can possibly and naturally move quickly. Thus the cause they set forth is nothing. Neither do the spirits move quickly in the body and the accidents of spirit provide testimony for this. These are the following sorts of spirits: undulating, swift, slow, spasmodic, falling, and all the accidents of pulse or spirit.[56]

---

52. *DG* 1.6.
53. *Discrasia* is, again, an imbalance of the humors. Cf. 7.93, with note.
54. Cf. 12.50ff.
55. Cf. 1.589.
56. "Spasmodic": *caprizans*, cf. Latham², s.v. 2.

Further, another witness is the composition of the spirit, which is composed of two movements and two rests, just as we have said in the book *On Spirit*.[57] And the statement that they move like the lights in the world is not made with respect to the time of the motion but with respect to a similarity of utility and end. For just as the heavenly powers are borne by the lights in the world into the matter of things subject to generation and corruption, so too are the powers of the soul and heart borne into the matter of the members. Thus, illumination occurs quickly in the world, just as we established in our study in the book *On Sense and the Sensible*.[58] And we should not repeat that here.

This statement that the natural (that is, vital or animal) spirits are lights [*lux*] or luminous [*lumen*] is an open fallacy. For spirit has more natural affinity with smoke and vapor than with *lumen* and *lux*. Now the spirits are luminous, especially those called vital, due to the very clear nature present in the substance of spirits, just as other bodies, outside the animal, are made clear and luminous. This luminosity is weakened and obscured by unclear, earthy vapors. And in this regard unclarity of the spirits attests to the disruption and *discrasia* of the complexion.

What they set forth concerning the effect of light carries no force. For the limit (which is the form) is brought about by the natural heat which does not act only in the form of fire but rather is formed by the form of the heavens and the soul. This is just as we have adequately set it forth earlier and we will speak more fully on it in what is to come. It is therefore stupid to say that this is an operation of light. That natural heat brings about many things in this way is thus because celestial forms and souls possess multiple powers and not just one. For the celestial form is a form of periodic generation and the form of a circle which is made up of the twelve circles of the twelve houses, and of all these things contained therein and of the relationships of each to the other. Likewise, the power of the soul in the seed is made up of the powers of all the members and the relationship of each to the other. It is thus no wonder if heat, insofar as it is the instrument of these powers, performs many things in many ways. For each instrument of a given mover or art performs as many things as there are forms and powers present in the art of the artisan who moves the tool to the task. Examples can be found in the mattock, hammer, and in all the artisan's tools. Pythagoras says the same thing about musical instruments and an example can be seen in the monochord, in which all the notes sound on one and the same chord as a result of touching it at different intervals.

Their statement that light acts on all things without having a contrary does not hold up at all. For while it is granted that spirit has no contrary in substance, it does have one in the qualities it possesses. For spirit is warm and clear and it possesses a contrary for each of these. For no substance has a contrary, but the qualities are contraries one to the other.

---

57. *De spiritu* 1.1.10.
58. *De sensu et sens.* 2.1.

39 Likewise, it is not true that there is a medium which the soul uses in vivifying and perfecting the body. For soul does not use spirit except when it is joined to a body. It is therefore clear that the soul is, in accordance with nature, joined to a body before it uses spirit.

Again, if there were a medium, it could not be from the nature of a fifth body, because a medium is multiple. It is a medium by being equidistant from the extremes; it is a medium by the composition of the extremes; and it is a medium, not of the substances or the extremes, but rather because something which acts on something else in the way of an instrument is a medium. It is in this last way alone that spirit is a medium of the soul and the body. Everything which is a medium instrumentally has the nature of that concerning which it acts. A clear example is in the tools of woodworkers, smiths, glassworkers, and each of the other craftsmen. The conclusion, then, is that spirit must be of the nature of a body on which the soul acts. This, as we have stated, will be from the elements.

Their statement that that which brings in the power is not the same as that to which it is brought should be phrased to say that it both is and is not the same. For it is not the same in form and being but is the same in origin and common material. Anyone can see this clearly in the case of all the smith's hammers. These share materially with those to which the artisan's form is brought. Yet they do differ in form and being in that one is the form of the tool and the other is that on which it works.

40 Nor is their statement true that light leads complexion to the nature of the heavens. For, as was established in the study of complexions, this complexion borders on the equality of heaven through its own equality and not through any light of a fifth body present in it. Likewise, according to the belief of the Peripatetics, the life which is in celestial things [*superiores*] is not univocal to the life of animals. Still, there is the fact that bodies composed of elements share in life only by means of something resembling a sphere. But, as we have stated before, this is only equality of complexion and not any sort of light, just as we said.

Their statement that the medium must be below one extreme and above the other has already been established previously. For spirit is not the light of the heavens but is a medium, rather like an instrument of the soul. Thus, by sharing in the form of the formative and first mover, it is suited to the mover and, since it is moved by it, is below it. But it is above matter since it moves it and imprints upon it. According to this line of thinking it and not something else is the medium. Therefore, what they say is as nothing, for they suppose that it is the medium with regard to nature. But this is false, because no such medium can exist between incorporeal and corporeal substances.

41 Their statement that the spirit is gathered together into one point in the heart and then expands is not well said, for none of the natural bodies can retract into a single point. Rather, the spirit collects toward a center when bodies are chilled and expands outward when the extremities are heated again. This is due more to the nature of heat than of the heavens. Only elemental spirit grows larger under stress. This is because heat and stress in some way generate spirits through the release of internal moisture.

But spirit of this sort does not remain when an animal dies but rather is breathed out of the dead animal.

Their statement that spirit has a multiple form is in fact true, but it does not have the cause for its diversity that they set forth. Rather, a spirit which is single as to origin (since it arises in the heart) is led off to the principal members, which also possess causative powers, and it is rendered multiple. This is why one is called natural; a second, vital; and a third, animal. And it is not rendered multiple in this way alone, but it also takes on a multiplicity of form in each member since there is no member which does not have its own power and operation. This is as we stated it in the book where we discussed the anatomy and powers of the members.[59]

It is clear, then, from all that has been said, that it is an error to state that a fifth incorruptible body enters into the constitution of animal bodies. Rather, through its movements it causes their generation, but it is not mixed into them.

CHAPTER VIII

On the Plan [Ratio] *of the Mixture of Material Principles in the Bodies of Animals and on the Cause and Type of Their Mixture*

With the material principles of animals' bodies having been thus dealt with, we should, before we treat of the formal ones, establish the mixtures of these things one to the other and the causes and types of their mixture. For the body of an animal is not made up of material things compounded in a given way. Rather, as Empedocles says, there must be some sort of binding agent and a "plan" [*ratio*] for the mixing and composition.[60]

It is the same in the crafted things which are brought together in the construction of a house. These do not stay together without a binding agent and, without a plan, the things that have been joined do not achieve the shape of a house. Thus does Empedocles say that the proper heads do not adhere to the members without a binding agent holding them together and that they do not constitute the suitable shape of the animal unless the composition follows a set plan and proportion. According, then, to Empedocles, a minimum of three things must be present in the material in such cases. These are, namely, the mixture, the binding agent for the things mixed, and the plan of the mixture. Again, three or four things are necessary for the mixture, or so it pleases Democritus to say. The first is the dividing of the mixables into as small pieces as possible. The second is the penetration of one into another to the extent that everywhere the smallest of one is with the smallest of the other. The third is the alter-

---

59. The entire second tract of Book 1 may be meant (1.107f.).
60. Diels (1951, 1:289–90, frag. 34), from Galen's *In Hipp. nat. hom.* 15.32.

ation, breaking up, and imperfect transmutation of each of the things mixed by the other. The fourth, which Democritus adds, is the ordering or locating of the mixed parts. One who wished to investigate perfectly the nature of animals must look into the manner and cause of all these things.

43 One should therefore know that the division of elements into their smallest pieces occurs only when they are divided into vapor. This is why those skills which are more perfect and which change the body more subtly strive after the vaporizations which occur through evaporation. For evaporation both divides things into their smallest parts and raises up that which is noblest and refined in the very thing which is evaporating, leaving behind that which is foul and not readily miscible. The evaporation must be completed by means of two powers which are present in the matter, and a third which is in each and is the cause of the moving and changing. Now one power is in the lower, coarser, and heavier material. This is the heat in the earth and water which raises up in vaporous form, one of these into the other, and then both, together, to the element of air. The second power is the tempered cold of the air which presses together and presses down the air into that which is being raised up from the earth and water. In this way each of the three elements penetrates into the depths of the other. For when they vaporize, they open up and can be penetrated everywhere, and they are then moved by their opposite powers, the heavy ones affected by the rising warmth, the light ones by the downward pressing cold. Entering one another everywhere, they penetrate, touch, and divide each other.

44 The power which causes each of these is the luminous light of the period which falls from the multiple powers of the stars onto the place of generation. This is as we have said in our book, *Metheora*.[61] It is made of the power of hot, cold, moist, and dry stars by means of the many differences in the moving powers. In this way it causes each of the elements to move toward each other along the *radii* of the hemisphere. For the radius of one ascending, striking on a line tangent, touches it and gives it quality. And while the ascending, striking radius constantly rises, it is buried in that which it has struck until it stands erect, fixed into it at the meridian [*in meridie*]. Then, having bent back onto itself, impressing all its power upon itself, it will then position itself from the zenith of the thing to be generated to the very bottom (with respect to its longitudinal diameter) of the thing to be generated. Since, however, that which first ascended later descends, it leaves the tangent in its fall. Thus it penetrates it along the radius of one ascending at its rise and fall, following the latitudinal diameter of the thing generated. This is the reason that members on the left and right side are similar in shape but dissimilar in power and potentiality. For every power in the heavens is stronger in ascent than in descent.

45 The period completes these two powers of longitude and latitude in accordance with the daily movement which runs from east to west over the circles which are par-

---

61. *Meteora* 1.1.2. Elsewhere, at *DG* 2.3.4–5, A. defines a *periodus* as the measure of a circle or circular motion. In the case of celestial bodies, the period of the sun, for example, is its movement from east to west over the poles. The movement of heavenly bodies has an effect on the generative process below.

allel to the equator and over the spinning poles of the world. It is through this movement—a movement of the period from west to east, above the poles of the sphere of the signs of the zodiac and above the circles parallel to the zodiac, and which is also the movement from south to north and is called the radius of the ascending from front to rear—it is through this movement, then, that light also influences the power of the thing which is generated, doing so along the diameter describing its depth from front to rear. In this way the body will have, along three diameters, a plenitude of heavenly power by which it moves to being and generation. The radius [*radius*] of a given heavenly constellation does not penetrate as well the length, breadth, and depth of a thing as does the radius of one ascending, and this is why the qualities of a generated thing are best recognized through the one ascending.

In this way, then, the division of elements and the penetration of one into the other are accomplished. And thus too are found a suitable explanation [*ratio*], cause, and type of the two things mentioned above which are necessary for mixture.

But these things will not alter one another and will break up on account of the powers mentioned. For material is not bound back onto itself by this type of movement of a radius over the material of that which is generated. Rather, it goes forth outside itself. This is why the radius which forms the thing generated must necessarily fall and descend beneath the hemisphere toward the angle of the west, to that of the middle of the earth, and so too it must return to the angle of the east so that that which is at the edge of its matter might thicken in the cold of the shadow and of night and may thus contain within itself, lest it slip away, the power of the warm and the moist, as well as the other powers present in matter. The prime qualities, held in check this way, roll in on themselves and break each other up, distance themselves from the excesses, and are led to a mean which possesses equality in either the arithmetic or the geometric sense. This is as we have set it forth many times. These, then, are the three things needed for mixture. For mixed things are not totally transmuted with regard to substantial forms and species, but they are broken up with regard to the excesses of their qualities in the manner we have given.

To these three, Democritus has added the order of mixed things. For, seeing that all mixed things are divided into the smallest parts possible, Democritus said that bodies called atoms [*athoma*] are the material principle of every natural thing and that they produce differentiation of form by their shape and ordering. And although he did not know how to express himself with properly skillful words, he nevertheless spoke the truth in saying that in every mixture the coarser and heavier things are lower and are more suited for dilation, while the finer things rise more and are more suited for longitudinal extension. Some, however, float in between as if moved transversely. This is just as we see, for every vapor spread out widely below is raised up on high and comes to a point. From the side of a shape like this, some things move to the side through the middle, moving as if along the bottom of the thickness of this sort of shape. Thus, three diameters are established lengthwise from these parts of the body's material, by means of things which are light when compared to the heavy and held by them, though they are not entirely held back. Now, when I say they are held, it is keeping them from flying away but they are not held back from lengthwise extension. The latitudinal diameter is caused by the heavy having been moved by the light and not having been raised up. For if it were not moved, then it would fall in around itself into a round shape. Whereas if it were raised up, then it would extend lengthwise. But when the light is held and held back by the heavy, and the heavy moved by the light, then dilation must occur to the side. In this way, then, the second diameter which is in mixed bodies and which is to the side is brought about. As will be made clear below, this especially occurs in the bodies of animals. There are those which are both light and heavy and when the heat of an ascending ray moves and is held both in and back by that which is both light and heavy, then a third diameter is created which arises from their downward movement. Their movement comes from this heat and their particular property. They do not have simple descent or ascent but rather move to fill the middle.

48 These, then, are the things which are needed for the first mixture of the elements in a natural body of animals, should a person care to investigate the matter clearly. Thus, a double cause has been established for the mixture of the elements according to the three diameters of the body.

Our previous statement was that the radius of one ascending possesses the power of the entire period and that it gives shape to right, left, up, and down by its daily movement, and to the front and rear by its oblique movement over the poles of the constellations. From this statement the reason is also learned that the front and rear do not have the same shape as the right and left and also why the upper and lower have neither the same shape nor potentiality. For the movement of the declination from south to north, just as from front to rear, is not perfect. Therefore, its power does not bound [*terminans*] the work at the same shape. Thus, the bottoms of bodies are only to the rear, and to the front are the operations of the senses toward the strength of the luminous light which is sent down from the heavenly bodies. Now even though the lower ones are moved by a complete *radius* drawn about them in the manner of a circle, nevertheless, because the lower part of the material is earthier and thicker it is more suited for bodily support. Examples are legs and the like. The upper part, which is purer and clearer, is more suited to light and the operation of the senses. The mixture of the elements, then, is of this sort and has this cause.

CHAPTER IX

*On the Binding Agent in Mixed Bodies, Which Is a Triple Moisture.
Also on the Similarities and Differences of These Moistures.*

The binding agent for the mixture of mixed bodies is moisture. There is nothing else except moisture which could hold together things mixed together, keeping them in one mixed form. That moisture which is the binding agent of mixed things is especially present in the bodies of animate beings.

It is a triple moisture. One is drunk in with the mixed elements and this one provides connection. The second flows through them and provides softening. The third is constantly assimilated to them and this one provides nourishment. The first of these is in the substance of the bodies. The second is spread out into the sponginess and the pores of the parts. The third travels through the passages running through the bodies, flows into the parts, and becomes one with them.

When this last one is removed, the second and first ones still remain. When the second is removed, the third still remains for a while. But when that one is removed, all that remains is dust. For this reason these moistures participate in the name of the binding agent in terms of both earlier and later [*per prius et posterius*]. For without a doubt the binding agent is, in the first instance, truly the connective moisture and the other two are binding agents later on. Second, this is true because by flowing into the bottom of the mixture it uses softening to make one part flow toward another. Third, it is because as it restores moisture to the parts, it hinders the dissolving which is caused by the evaporation of the other two moistures.

It should be noted here that moisture is a connective in the body insofar as it takes on size and shape. Moisture is a softening agent in the body to the extent that it is receptive of the sensible forms which are only received with a softening moisture. An indication of this is that the sensory nerves descend from the front portion of the brain where that part of the brain is softer and more marrow-like. The third moisture is present in an animate being insofar as it is animate, for it is there for the nourishment, growth, and restoration of what has been lost—these being the primary activities of an animate body. In this way, then, the binding agent in mixed bodies of animals is, as we said, a triple moisture.

Now, someone might ask how the connecting moisture is present universally in the mixed bodies of animals. For things which have different species and limits are not connected. Now, the parts of animate bodies are flesh, bone, nerve, vein, and the like and these have diverse forms and limits. For this reason, they do not appear connected and there is not a single connecting moisture in them. But this is easily resolved if attention is paid to those things which were established about the multiplicity of connection in the fifth book of the *Physics*.[62] For one thing is completely and simply

---

62. *Phys.* 5.2.3.

connective, while another is so after the fashion of a binding agent. The one that is connective simply has one and the same limit, and since the substantial form of a thing is its proper and essential limit, it is thus necessary that that which is connective in this sense have a single, simple form. It therefore cannot be mixed because it preserves a lot of things in itself which are mixed with regard to first being. Therefore, the connective is, according to reason, anterior to the mixed. According to the example of the binding agent, the connective is bound like stones held into a wall by cement, for this is what Empedocles called the binding agent.

Therefore, the humor which flows into all things which have different forms in the bodies of animals is one with respect to the first thing forming it, which is the heart; it is one in the matter of a mixed body; and it is one in relationship to the one form which is the soul, active in all the parts. Thus, a connective of this sort is said to be one in the same way as a binding agent. The moisture is poured forth from one and the same seed or heart, or from both, and is then borne away and diffused throughout all the parts. There is a connective from all of these to a single thing, which is the origin. This is the binding agent, because it spreads out from a single thing, which is the origin, to all things. A body is therefore truly and properly said to be bound by this.

51 The nutrimental moisture is poured into that one so that the first moisture might be restored by it, retained, and kept from being consumed by the heat. Now, this moisture is generally nutritive with respect to power and is formed by the heart as if by a first thing in which the principle of life and nourishment is present. Thus too is this brought back to the same thing as far as its origin is concerned. For the strength that every nourishment possesses enabling it to nourish comes, as has been proven previously, from the heart.

The softening moisture is that which, before it is converted into the members, is poured into the members by the nutrimental moisture and then, by moistening, dissolves to softness the earthiness and hardness of the members. Because this is natural, it must be rendered natural by the power of the heart. For if it did not have this, it would undo the animals' members inordinately, making them clumsy and unsuited for ordered movements.

All these three moistures are connected to a single thing which exists as their cause, and this is the heart. For although these moistures flow out from the heart they are divided into many members. An indication of what we have said is that all the members of animate beings are generated from the heart, and they could only be generated from the heart by means of moisture which, having been bounded and digested, is poured out of the heart in order both to form and connect the members. The connective does not suffice by itself, as we said, since this does not create the material for the animal's body. For a connected body alone does not necessarily have a penetrating moisture in it.

But every animate body does have this, both for nourishment and growth. But even if it does have this it is not necessarily an animal's body, for the body of a plant has penetrating moisture in it. But since it does not soften the body, the body is therefore not susceptible to sensibles. This is why an animal's body requires the softening moisture. All these moistures are present in the first matter of the seed and are in mixture with it. For in all things, both those which are generated internally within a womb and those incubated externally in eggs, the radical moisture is present. It has been poured into the members and parts and is the nutritive moisture, either flowing continually in the womb, as, for example, the menstrual humor in animals, or piling up as if in a storehouse from which the operative nature can take what it needs to fulfill the formation and nourishment of the members.[63] The third moisture is what is left over from these two and it pours into members which have been formed, to soften them.

Further, these three moistures are ordered one to the other with respect to cause and effect. The first, which is the radical, is primordial and seminal, as it were, in the mixing of the bodies of animals. This has the power of assimilating to itself the other moisture which, as it were, nourishes the parts. Neither is that one converted into parts of the mixture unless it is first assimilated by the power of the radical moisture. That which remains from the one assimilating and the one assimilated, which is neither converted to nor made one with the parts of the mixture, is poured forth throughout the body and softens it. This, then, is the cause of softness in the bodies of animals. For even though certain parts of the bodies of certain animals may be hard, such as bone and cartilage, the inner areas of these parts are not bereft of the softening moisture. This shows clearly in the marrow in these parts. Such a binding agent, then, is proper to the bodies of animals. For the mixture of the bodies of animals is not directed toward an act of something hard, but toward an act of the soft moisture.

This depends on two reasons. One is the movement of the animal, which can in no way be perfected in something hard. The second is the stamping and impressing of the senses which again could in no way be perfected in something hard. An indication of this is that those animals which have hard eyes see poorly.

To be sure, minerals and plants have mixed bodies made of elements. But the act and the completion at which the mixing stops, as if at a limit, is hard and not soft. All minerals are coagulated or congeal due to cold, heat, or both. The operation is always bounded at something hard, whether moisture is compressed by cold or removed by heat, even if it is removed by heat at first and its residue is compressed later by cold, as happens in the case of stones.

Now, in plants coagulation is not of this sort. Rather, the earthy part and the powers of the earth prevail in them. Thus, even though plants are not entirely hard, they are nevertheless rigid and tend toward hardness to the extent that they do not fall flat when touched. Rather, for the reasons given above, the act by which the mixture of elements in the bodies of animals is accomplished is such that they are soft, and hard parts are placed among the soft solely to hold them up. For all operations of the soul are carried out by means of and in soft things.

Moisture, therefore, is, as we have said, the binding agent of the mixtures in animal bodies.

---

63. "Operative nature": *natura operans*, i.e., an individual nature. Cf. *De causis et processu* 2.2.18.

CHAPTER X

On the Plan According to Which the Mixing of Animal Bodies Is Completed

54 As Empedocles says so well, mixture must follow some plan and proportion.[64] It is for us now to note in what way the proportion of this plan is followed, both in the whole mixture and in parts of it. Pythagoras called the plan and proportion "number," by which the elements are bound to the mixture.[65] According to Pythagoras, this plan is followed in the number of the parts of the powers of those things which are mixed with respect to being dominant or being dominated, grasping or being grasped, and holding or being held. Thus, if we say that, since ten *pugilli* of air are created from one of water in mixture, the power of ten *pugilli* of air is equal to that of one *pugillus* of water.[66] Thus the water grasps the air and holds it and is itself grasped and held by it due to the equality of their powers. Otherwise, if the water were superior in number and power, then it would prevail in grasping and holding, would not be grasped and held, and there would be no mixture. He spoke in the same way about the other elements as well. And just as we have spoken about the potency of the loose and the dense when we compared one *pugillus* of water to ten of air, so he spoke of the other potencies of the elements, claiming they were very many. For example, the heat of fire in many parts is greater than that of air. Thus twenty *pugilli* of air are equal in heat to one of fire. Whichever one exceeds the other, then, will grasp, hold, and transmute the other. Then, due to the inequality of their potencies a mixture is not accomplished. Instead, the one element dominated by the other will be transmuted into it and the whole will be one simple element.

55 He said the same thing about the proportion of the potencies of cold, wet, dry, light, heavy, clear, unclear, and of all others found in the elements. For Pythagoras relied heavily on proportions and held the opinion that all things were made according to harmonious proportions.[67] Empedocles agreed with this, saying that bone exists according to one plan of proportion, flesh by another, nerve by still another, and so on for the other parts of mixed, homogeneous bodies.[68] And, as far as can be grasped from their words, they thought that there was no form of things other than the harmony of the mixture which is a result of number and the proportion of those simple ones which came together and are arranged in a mixed thing.

---

64. Cf. Ar. *De anima* 408a19f.
65. Cf. Ar. *Metaph.* 985b30f.
66. *Pugillus*: normally, a dry measure equal to a handful, but clearly here it is used as a liquid measure as well.
67. Cf. Ar. *De mundo* 396b25.
68. Ar. *De anima* 408a19f.

We have held forth this view in our other books.[69] Although we do not agree with their view when they state that there is no form for the matter save that which is the plan [*ratio*] of the mixture, we do agree with their statement that every mixture occurs according to a plan of some proportion or other. But we do not say that proportion is understood absolutely in the number of those things which come together in the mixture. For we have already shown in other books that it is impossible for anything to be so mixed that each of the elements in it is equal to the other. For in this case, nothing whatever of that which is completed by the mixing will have a natural place.

For this reason, we say that the proportion of mixed, animate bodies is understood in reference to the proportion of the organ to the act of the soul which works through it. Therefore the mixture of eye, flesh, bone, and ear is each different. For in the mixture of the eye, clearness should predominate, holding the forms which are moved by the action of the light. The mixture of flesh should be in accordance with the mean of tangible qualities, and so forth for the mixtures of the other members.

If, however, anyone should doubt that this plan for mixture exists, we have said elsewhere that it exists in accordance with the geometric and not the arithmetic mean. And for this reason it will not overcome the dominant element in such a mixture so that it can transmute the others to itself and thus be rendered a simple one out of them all. For, as we have said, the median act [*actus medius*] to which mixture comes restrains the elements to keep the proportion from being ruined. This act possesses the following powers forming it: the heavenly, the animal, and the natural ones by which it can bring about this holding action. It can even bring *discrasia* in the mixture back to equality, especially when the mixture is different in the composition of different members. For then the quality which is excessive in the mixture of one member reduces the cold *discrasia* in another member. For example, a warm liver brings a stomach that is chilled back to a state of *crasis*.[70]

And it is this way too in other mixtures of composite members in an animal's body where contrary mixtures of members are composed. This is as we will demonstrate in later parts of this book.

This, then, is the proportion that is followed with respect to animate bodies. And because they become varied in the sounds of harmony due to a proportion which is varied in relation to act, Empedocles believed that the soul of animals is nothing other than the harmony of this sort of proportion.[71] For just as he saw that from a varying proportion of chords and cymbals there sounded the *dyapason*, the *dyapente*, the *dyathesseron*, and so on for the others, so did he think that vision was produced by a proportion created in the eye and that from another proportion in the ear, sound was produced and came forth.[72]

56

---

69. *De anima* 1.2.8.

70. *Crasis*: i.e., balance or equality.

71. Cf. Plato, *Laws* 10, 889b. A. discusses this opinion at greater length at *De anima* 1.2.8.

72. *Dyapason*: Gr. *dia pasōn*; more properly written as two words, commonly as one. Lit., "through all (the chords), an octave." *Dyapente*: lit., the interval of "fifth." *Dyathesseron*: lit., the interval of the "fourth." For medieval definitions of these musical terms, see the (twelfth-century?) *Anonymi II tractatus de musica PL* 151: 681–86.

57 We will show later that the cause of this sort of act or passion of an organ is not the power of the harmonic form but rather that the power of another form produces the plan of the mixture and the act and the passion of the organ. This is not something to be neglected, for the plan of the proportion which we have mentioned gives a name and being to the organ and generally to the mixed thing which is the animal's body. For this is the body in which the operations of life are carried on. Then, too, it is clear to anyone how the pupil takes its name from the proportion of things which have been mixed together for the act of sight and the ear takes its name and being from the proportion of things which have been mixed together for the hearing of sound.

A similar basis [*ratio*] for naming exists for all the other mixed members. For since every naming comes from the end (for naming is from a proper like-natured and essential act), then the proportion of the mixture should exist and be named with respect to the act and not, as Pythagoras said, with respect to the number of those things which come together in the mixture.[73]

58 Anaxagoras has spoken less than well on this matter, saying that each mixed thing takes its name not from the essential act but from that which is most abundant in the mixture.[74] Now we must not ignore that there are certain acts or operations which we have just named and which show that they do not possess one simple power. Rather, it is necessary that different things flow together during the completion of their acts. Examples are seeing, hearing, and others of this sort. For sight is completed only by a power undergoing the act of something bright [*lucidus*], by taking up the visible, holding it, and leading or directing it inwardly, and by a power making a judgment upon it. And we most certainly know that there is no passive power from the act of the bright unless there is the power of the clear [*perspicuus*]. Taking up the form is the power only of the moist, while holding it is the power of the thick or dry and the cold. Directing it to the inner part of the soul is the power of the bright spirit and the one judging is none other than a power of the soul. How, then, could it ever enter into a person's mind to say that the eye is made up of a simple element when its act is made up of so many powers? Moreover, the same is true for the other operations of the animal and vital powers. This is also why the body of an animal is not some simple element but needs to be mixed and organic. From this it is also clear that none of the parts of an animal's body exists univocally in the animal's body and separated from it. But this matter was settled satisfactorily at the end of the fourth book of the *Metheora*.[75]

These, then, are the things which have been said concerning the mixture, binding agent, and plan of the mixing in the bodies of animals.

---

73. Ar. *De mundo* 396b25.
74. Cf. Diels (1951, 1:83).
75. *Meteora* 4.4.7.

CHAPTER XI

On the Complexions of the Humors and on the Properties of Those Things Which Result from an Induced Mixing in the Bodies of Animals

Complexion and differentiation of complexions result from this mixing in the bodies of animals. From the moisture flowing through the body in the vessels (within which the decoction of this humor occurs), four humors are necessarily generated: namely, blood, bile, phlegm, and melancholy, all accomplished in the way we established in earlier parts of this study. In those animals which do not have blood, it is something in the place of blood. And in those, while there may be something hotter and drier then the other parts, it is still not truly bile, but a thick, viscous phlegm. For this reason too the creatures are, for the most part, slow and remain motionless much of the time. If some of them, such as the bee, happen to have fine, warm phlegm, which is close to the nature of blood, then they move around and are swift and refined in their activities. They are still, however, overcome and slowed down by a moderate chill. Whichever ones are earthy and melancholy are heavier and afford little or no refinement of their activities. But of those which have blood, as many as have light, bilious, frothy blood are agile, exceptionally capricious, and unstable.[76] The same thing is to be seen in people who have this complexion.

Those with coarse, warm blood have bile mixed in with the blood that is either burned or beginning to be burned. This is a particular type of melancholy. These sorts are stable and have constant boldness, and many spirits midway between coarse and fine. For this reason too, people with this complexion have a stable, strong mind and are not rash. As Aristotle says in his book *On Problems,* this sort of black bile has somewhat the complexion of red wine.[77] For clear, red wine is full of fumes. It makes a person be of good hope and, with the weight of its thickness, it stabilizes his spirit and heat. Because this sort of bile is filled with fumes, it multiplies the stable spirit which holds forms well. And because its heat is in a moderately thick matter, its heat does not muddle and mix together its operations. The results are stable concepts and ordered activities of the mind. It is through ascending heat of this sort that the spirits are properly carried into the openings and then those who are in difficulties do not despair and always cheer themselves up. This is what Aeneas did when he spoke to his comrades in danger saying, "Oh you who have suffered worse than this—God will grant an end to these things as well."[78]

---

76. Reading Borgnet's *spumosi* for Stadler's ungrammatical *spumosae*.
77. Ar. *Problemata* 953b23f.
78. Vergil *Aen.* 1.199.

This is why Aristotle says that all men who are outstanding in philosophy and the heroic powers are possessed of this sort of melancholy.[79] Examples are Hector, Aeneas, Priam, and others.[80] For this reason, the lion and other animals of this complexion are more generous and sharing than are others.

61 Certain ones with cold, foul-smelling, murky blood have ponderous, almost unmoving thoughts. The forms of their apprehension are obscure and the forms result in a thick, vaporous, earthy, and cold spirit. They are thus timid and some are so by nature. Once taken on, this fear does not leave them and they admit of no consolation. Rather, such animals and people suffer from madness and insanity and people sometimes find no pleasure in one another. They also become badly suspicious of others and commit suicide. They neither take pleasure nor can give it. They love solitude, which is an evil in the life of humans. They find it pleasurable to be in squalor and many other bad things happen to them. They are often thieves, even when they have no need of the thing they steal. They suffer greatly from insomnia, due to the dryness and coldness of their complexion.

The blood of other ones has a good deal of phlegm. These are slow, sleepy, and soft with a woman's softness. They are very sleepy and have virtually no memory. They are white and, lacking any red mixed in, grow very pale. They are timid because of the cold and are unstable, possessing undigested excreta. Those who especially abound in phlegm have such a complexion and thick skins. They very often have many undigested humors and, as was established in earlier parts of this study, they sleep for a large portion of the year. Their natural disposition is bad, as is their memory. The natural disposition is due to a cold which neither moves nor brings in the spirits but which instead splits them up. The memory is due to their watery humor which has no power of retention. Animals of this sort are timid, have small veins, and are incapable of warming their blood. This is why boldness does not come back to them after their fear.

62 Those which possess clear blood which has good sediment have a most pleasant complexion and present a friendly appearance.[81] They are always of good hope, are not depressed during times of misfortune, possess a good natural disposition, good bearing, and good habits.

Those who are sanguineous are otherwise, for they have good flesh and good bearing to their bodies. Choleric ones are tall and slender; phlegmatic ones are short and stout. Those who are melancholic are slight, short, and dark. Those, however, with warm, burned melancholy are very tall, slender, and dark, and have tough flesh. There are many other accidental traits of these complexions and these have been partially established by the physicians and partially by the natural philosophers. But what has been said is sufficient for our present purpose.

Such other things which were necessary to say about complexion and differences of complexion we have established in earlier places.

---

79. Reading *praecipui* with Borgnet for Stadler's *praecipue,* which, while possible, is awkward.
80. Cf. Ar. *Problemata* 953a10f.
81. "Sediment": *ypostasis.* Cf. 3.117, with note, and 3.119, 121.

The Second Tract of the Twentieth Book of Animals Which Is on the Formal Powers in the Bodies of Animals

CHAPTER I

On the Power of the First Cause.
What It Is, How It Is Multiplied in Effect, and How It Acts.

We have now made a determination, in the preceding tract, about the material principles of an animal's body. In this, the succeeding tract, we must first make a determination on the formal principles before we can treat the composition of the body as brought about from dissimilar parts.

There are certain formal powers in the bodies of animals which are simple and only move. There are others which are made up of both mover and moved. Moreover, some are animal and others are corporeal. All these powers are made up of the powers of a moved mover, except for the first power which is nothing other than the power of the first mover. We should, then, make a determination concerning these powers and the manner in which they act on the matter of animals' bodies if we wish to achieve a perfect knowledge about the bodies of animals.

Let us first, then, distinguish among these powers and make a determination about them. Then afterward let us speak of the manner of the action of each of them.

The first power, then, is that of the first mover, which is the principle of the entire universe. This is the power of that which only moves. It has nothing prior to it which moves it. This is that which is formal and which informs all others, much as heat informs all things hot. This is the one which, as Plato says, possesses universally and beforehand all the forms, as many as are made in all matter. And as has been shown previously, there are forms which are made in material, images, as it were, of that one. This power is communicated to things in which generation occurs according to the proportion of each one which takes it up. It is in them according to the power [*posse*] of generative things, and not according to that which is in the first cause.

We can have no other suitable example of this in bodies except in those which act through themselves and according to their own essences. Examples are the illuminating action of the sun, heat, and the like. In spiritual bodies we have the active or practical intellect. A luminous body so boils forth light that it constantly seems to move with a boiling motion if it is a source of light like the sun.⁸² This is why some have said

---

82. "Boils forth": *ebullit*, but suggesting perhaps a spontaneous act arising from an agent's very essence. For a discussion of this term, see Pagnoni-Sturlese (1980).

that the sun moves with a staggering motion, for they believe the sun staggers when it pours forth light. Now these lights are present in the sun with a single form, nature, and essence. When they fall upon other bodies which are at a distance from the sun they take on a different being according to the differing powers of those receiving them. Thus they become luminous light [*lumen*] in a clear body and in a body which bounds vision they become the white of a clear surface. On a less clear surface another color is produced according to the power of that surface, just as we have shown in our book *On Sense and the Sensed*.[83] And so it is in the fontal cause of the universe. From a boiling forth of its goodness, the lights and the forms go forth which, when received by distant objects, take on a different being according to the differing powers of the ones receiving them. Thus, some come close to a resemblance to the first cause, while others are far removed from it, and still others are vague effects of it on the matter of mixed and animal bodies.

65 Another more suitable example lies in the practical intellect of craftsmen. This intellect takes the form of the craft from itself and pours it forth into the tools and the matter in which it takes on a baser being according to whether it is at a greater or lesser distance from the intellect. It is thus the universal cause of all things, and it boils forth from itself forms which go forth from itself and impart form both to those things which move moved things and those which take on a proceeding form, such as the act and *ratio* the form is. This is as occurs when a hot object emits heat from itself and on its own.

To speak generally, all things which act on their own and which are the first cause of an action in general multiply themselves using the same method of sending forth forms. They do this is in the matter provided [*in materia supposita*] and, by the power of such a form as boils forth from the first cause, they do whatever all other forms do which are of those which move things which have been moved. Therefore, this power is a sort of foundation for all the others and it infuses the power of moving toward form in all the others. For it informs all the other powers, both celestial and material, and is not informed in them. It is separated [*distinguitur*] and determined [*determinatur*] in them—not, to be sure, as genus is determined by species or as material is determined by species, but rather as light is determined by colors. For if we were to consider what it is that separates light into the colors white, green, black, and other colors, we would find that it is nothing which is truly part of the being of color other than the clear [*perspicuum*], shared with light by a bounded and impervious body. Everything which is added by way of dark, smoky, watery, sooty, or the like is more by way of a privation of the true being of color than it is of the nature of color. An indication of this is that color acts on vision in the action of none of these but instead in the action of the clear. But we have determined this in *On Sense and the Sensed*.[84] This is why we have also said that white and black are opposed in this manner more by way of privation than as contraries.

---

83. *De sensu et sens.* 2.1.
84. *De sensu et sens.* 2.2.

We should understand in entirely the same way that the form of the first, simple mover possesses beforehand and absolutely in itself the entire multitude of secondary movers. And when it is determined, it is not determined by means of those things which bring on a new being which does not arise from itself but rather by those which bring on materiality, contingency, and multiplicity. These are things which more move away from true being than toward it. This form, then, boiling forth this way from the first cause and existing as something having its own light, imparts form to all the principles of generation. From it they have the power of producing form in the one generated and because of it the entire work of generation is sure and finite, not a result of chance as Empedocles said.[85] What is said by many and understood by a few is now clear—that the first form of nature is one. This is just as certain of the most ancient authors related. For it is one in the first mover, but with respect to progression from the first mover it is driven more and more into multiplicity the further it progresses from it, moving through the gradations of those things which move and are moved. In relation to beings, then, this form is many and when taken in relation to the first cause it is single. It is, therefore, first one thing and then another according to its relationship to things formed, but in the first cause it is entirely the same and is simple. It therefore causes no multiplicity whatever in it.[86]

Further, since the intellect of the first cause does not act through a particular habit which is different from its light but rather does so through itself (as, for example, shining forth from itself, it brings out the lights which go forth from it), it follows that while the form itself may be rendered different from the first cause by its progression away from it, it still has no diversity whatever in it and thus causes no composition within it.[87] Now in every progression in which it progresses away from it [*ipso*], nothing is added to it through which a new being might be created in it [*ipsa*]. Rather, things which have been added tend to draw away to a sort of nonbeing, such as simple nonbeing, single nonbeing, perpetual nonbeing, abiding nonbeing, and the like. Since this is so, then, it is clear that that which is truly the principle of things is single and is rendered multiple in the way we have stated, by participating in determination and materiality. For this reason those forms which are in matter by participation are more images and results of the first form than they are true forms. This is as we have stated in the preceding book.

Now while these matters might seem to be situated in first philosophy rather than natural philosophy, we have nonetheless introduced it here to facilitate the understanding of the formal principle of the generation of animals. Still, these matters must be carefully established in first philosophy.

Let these, then, be the things said by us about the first, formal principle of the generation of animals.

---

85. Cf. Diels (1951, 1:333–4, frag. 58–59).

86. "In it": Borgnet's *in ipsa* would refer to the first cause, Stadler's *in ipso* to the first mover.

87. "Within it": Once again, Stadler reads *ipso*. In this case, it would refer, grammatically, to *intellectus*. One must remember, however, that A.'s use of pronouns is somewhat cavalier. To aid the reader, the pronoun will be kept in brackets for the next sentence.

CHAPTER II

On the Formal Heavenly Power Which Is Called the Power of the Period. On What It Is, How It Goes Forth from Itself into Matter, and How It Acts.

68 That which moves beneath this one is that which moves the thing formed. It is double, namely, heavenly and elemental or material. Although the heavenly one has multiplicity with respect to every movement of the existing heavens, from the first heaven all the way to the sphere of the moon, nevertheless the powers of all those above flow into one sphere—not, to be sure, the sphere of any planet or of a single heaven. Rather, they flow into that revolution of the sphere which is called the period. This sphere is given its composition by the interrelationship of the twelve circles which are called the twelve houses and toward which all stars are referred. This is just as we have stated previously.[88] Now an upper sphere always infuses form into a lower one, a form which was infused into it by the first cause and which, as it descends by means of participation, is constantly determined and multiplied. This is why it is said that every intelligence which moves a sphere is filled with forms, but the upper ones have more universal forms and the lower ones have those which are more determinate and less universal. Now I am calling them universals with respect to the universal cause which possesses beforehand, formally and absolutely (and not according to a mere turn of phrase as it is most often used), those things which come after itself.

Since, then, power of this sort belongs to the period and since a parabolic circle [*circulus obliquus*] has reference to the period, and since even a true circle [*circulus rectus*] (in respect to which there exists a diurnal revolution) has reference to the period, then every power of the mover has reference to the period as does every power of the circles which are moved and every power of the stars with respect to the fixed points they maintain in the period both with regard to themselves and to one another.

69 The power, then, which descends and is caused by this sort of period is both caused by and descends from the greatest number of powers, belonging both to those which move and those which are moved. Although this power is simple, there is nevertheless a multiplicity of powers in it insofar as it descends and is caused by many things producing causes at once during a single period of generation. Despite this fact, some are stronger within it and some are weaker. And because there are powers in the period of the intelligences which move the spheres (called the divine, separate intellects by the ancient Greeks), certain Stoics therefore said that all lower things are produced by the power of the intelligences, just as all the things which are made in the bodies of animals are made by the powers of the soul. And they also said that intelligence acts on its own only by a boiling forth of its active light, much as the craftsman

---

88. Cf. 20.44.

acts by the boiling forth of his intellectual forms. They therefore state that the lower forms are all made by the luminous light [*lumen*] of the intelligences. For it is not their claim that the luminous light of the intelligences acts on them on its own, but rather that the first cause acts on its own and the intelligences act using the movement of the celestial powers as if they were their tools.[89] But a determination on this matter has to be made in first philosophy.

What should be known here is that, as we have stated, the form of the period is simple in essence but multiple in both act and power. It is this which informs the qualities of the matter of things subject to generation, doing so with respect to the beginning, growth, stasis, decline, and the end of being and life. Thus, the more skilled of the astronomers prognosticate concerning the beginning, the progress, the fortune, and the end of one's life, using the zeal they exercise in understanding the period. As we have stated, this power embraces the cause of life which comes from the movers and from those which are moved. This is that which possesses beforehand all the things which pertain to variation in generation, being, and the end of life of one that has been generated.

Now the oldest of the Epicureans and the Stoics used to say that the power of the period was a numinous power which, below the god of gods (which is what they were wont to call the first cause), directs and rules life. Apuleius, speaking *On the God of Socrates,* gives evidence that they used to first separate it into two types of god.[90] They said the first was incorporeal and the second corporeal. They divided the incorporeal into twelve because of the powers of the twelve houses in the period. They divided the corporeal gods into heavenly and earthly, calling the stars, constellations, and the seven stars, which they said were more powerful (which is what they called the planets), heavenly. The Chaldeans posited forty-eight constellations and then posited many more since among these constellations the planets take on admirable powers and they bear the same relation to them as souls do to their bodies. Just as the soul receives many passions from the body, so do the planets receive them from the constellations.

They used to call the earthly divine powers Tethys, Esculapius, Ceres, and the like.[91] These were nothing other than the heavenly power giving form to the quality of those subject to generation. Thus, according to them, every nature of the gods was either above the period, like the god of gods which is the first cause in the period either spiritually or corporeally, or beneath the period, when the heavenly powers give form to the qualities of the earthly matter so that they may act on them. They therefore said that there were many Jupiters, many gods named Mars, and other things of this sort.

---

89. "Celestial powers": *caelestium,* perhaps heavenly objects.

90. Apul. *De deo Socr.* 1.116, 3.123f.

91. Tethys is a sea goddess and the wife of Oceanus who bore him the sea nymphs. A.'s spelling (*Tethym*) also allows the interpretation of Thetis, Achilles's mother. Asclepius, born to Apollo and Coronis, is technically a demigod. Ceres, of course, is the Greek Demeter, goddess of grain and crops.

71 As I have stated, they made up the period out of *planes* motion and *aplanes* motion.⁹² They called *planes* motion the motion of the declining sphere, over whose poles all the planets, constellations, and stars revolve from west to east. From the twelve houses it possesses twelve divine powers apart from the divine powers [*numina*] in it from the planets and the stars. They called *aplanes* motion that which is rotated from east to west over the poles of the world without any wandering variation. In what has gone before we have already spoken about how the *radius* of the one ascending, rotated in a circle, describes in a causative fashion the diameters of a generated body. But that *radius,* because it is mixed in with many lights of other stars, is rendered as having multiple power. Thus, when this multiple power descends from the period, it possesses a very multiple operation. Because this is corporeal in the period, it is formed by the intellectual mover. For, as we have often said, every action and operation which proceeds from an operator composed of both an incorporeal and a corporeal mover is, with respect to form, from the incorporeal. This is clear in all the operations the soul accomplishes using the body.

Let, then, what we have said about the heavenly power which is active in generation be sufficient. These matters have to be more fully determined in our *Astronomical Matters.*⁹³

CHAPTER III

On the Formal Principle of the Body,

Which Comes from the Soul and Is in the Seed

72 Among the stated causes called the heavenly ones are those which are present in matter. In the matter of animals these are called seminal and are present in the seed, coming from the soul of the one generating. Because they are of the soul, they possess unity [*unitas*]. Because this is but one power among the powers of the members unto which the seed itself is assimilated, they possess a multiplicity of power. But all this has been well enough established previously. This is that power which is in the seed of animate things rather as a craftsman is in the thing crafted. Some philosophers call this power the formative, others call it the separative [*distinctiva*], the intellect, the image of the world, or the soul. Now there are reasons for these names.

It is already known from things presented earlier that this power is elevated above all natural forms and imprints upon them insofar as they are its instruments. For

---

92. For the two types of celestial motion, viz., *motus planes* and *motus aplanes,* or *ordinatus,* see *DG* 2.3.5. The former motion is from west to east and is described as erratic (cf. *DCM* 2.3.10); the latter is from east to west and is a diurnal motion (cf. *DCM* 1.3.5).

93. Evidently a book which A. intended to write but never composed. His interest in astronomy, however, is evident in the *Speculum astronomiae,* which may properly be attributed to him.

example, the digestive heat is an instrument of the soul in an animate body. Since, then, this power gives form to the heat and the spirit which are in the seed and since it uses these as tools, as it were, to form the heart and the members, it is called formative. And once it has formed the members it causes power to flow from itself to each of them. Much has been set forth about these matters in many earlier portions of this study.

This power is called separative [*distinctiva*] because it is the first to make separations in them. One of these is that of the powers, a second is of the operations, and a third is of the organs. Since the natural power is only from a single power (as is clear in the case of heat, cold, and all others), then this power is the first which has many powers. This is clear in plant seeds, which are nourished, grow, and generate, doing so in various ways into roots, shoots, twigs, leaves, flowers, and fruit. It is more multiple in animals in which it is distributed into the sensitive, imaginative, and appetitive powers. It has the greatest multiplicity in humans, in whom, since one has all these things, the power is rendered deliberative, of the intellect, the will, and the like.

It is separative of operations, however, because, although the natural power acts on its own only on a single thing, and does this out of necessity (for example, light gives light and heat heats, and so on for the others), this power acts on many things on its own. Further it acts on none of them out of necessity. For it shows appetite, it senses, has intelligence (when it is intellectual), and so on for the others. It sometimes abstains from these and at other times, when it wishes, it turns toward these acts.

This power forms the separations of the organs which are present first in embryos and, after they have been formed, when the animal has been generated with a diversity of shape and matter, it constitutes the organs. Now the natural power does not do this, for it is always homogeneous insofar as it has one shape unless that shape is accidentally ruined. For this reason, then, some have called this the separative power.

Some call this power the intellect, because its activities are like those of the practical intellect. For, as Anaxagoras says, in all woodworkers we see that the craftsman's intellect separates and forms the activities without being mixed into the material. We likewise see that this power is not mixed into matter so that matter is overcome by it but rather it acts on the powers of the matter. For just as craft causes heavy things to rise up in the building, and just as it is necessary that, contrary to the power of nature, each thing move to the shape of the thing crafted, acting in accordance with the intellect of the craftsman compelling it, so does this power move earthy and watery things upward and to the side, just as suits the design of an animate body.[94] It could only do this if it were naturally raised above the qualities of the matter, just as the intellect of the craftsman is raised above the matter of the thing crafted.

---

94. "Design": *machina*. A. often uses *machina* to designate the whole of the animal body (cf. *QDA* 1.18 and 1.55) as well as the entire design or structure of the universe realized in matter, viz., the *machina universitatis* (cf. *De principiis motus processivi* 1.4). In addition, A. uses the term *machina* in the more modern sense to designate various devices, for example siege engines [*blidae*] and other war machines (cf. *Phys.* 8.2.4).

This power is called the greater and lesser image of the world to the extent that it is more or less perfect. For it is imperfect in plants, is more perfect in animals, and is fully perfect in humans. In the human, then, it has the true sense of image. This sense of image has four parts. Of these, the first is suited to every soul. For just as the heavenly design [*machina*] consists of a certain multitude referred to a single thing, so does this power seek being in a similar multitude. For we see that all the celestial bodies [*superiora*] refer to a single mover and that the lower movers wish to be made like it both in their own particular motion and in the shared, diurnal motion. It is thus clear that all movers are referred to the first mover by means of the form which first flowed out of and proceeded from the first mover, just as we said above. And it is present in just this same way in the power forming an animal where all the powers of the members and the matter are referred to the first, divine power, which principally forms and informs all the others.

75 The second part is present in motion. For we see that all the celestial bodies possess a motion called diurnal. It is roughly circular although with respect to their own motion they have others. Likewise, anyone can see that every motion of change and of any material power is always bounded at the species which exists as suitable to the first power. It is at this species that every motion of matter is bounded, whether it is caused by a change in the first qualities, or whether the heavenly bodies or some other powers are moving it.

The third part is, as Aristotle and all the astronomers say, because the heavenly, luminous lights are all referred to the sun's luminous light. It is much the same in the soul, which is an image of the world insofar as all the strength of a given power of the soul is referred back to the luminous light of intelligence. As we have already determined both elsewhere previously and in our book *On the Intellect,* this is present in them in one of two ways—clear or shadowed.[95]

The fourth and last part is that just as every heavenly motion is circular and spherical, so too do the motions of the soul imitate this circular and spherical motion. We have shown this already in the book *On the Movements of Animals* and will do so again below.[96]

In these four, then, the soul is the image of the world and this proves that the soul (especially that of the human) greatly imitates the first cause and the celestial intellect. The entire celestial order has therefore made its impression on it in dreams, and through dreams forecasts are made about all things which are caused by the celestial motions. Since, then, it is the image of the celestial powers and motions in this way, it is no wonder that it has many powers and activities.

76 In order that these things might be even clearer, recall the previous statements about the power of the heart. For the power of the heart, which is the first to be present in animals, gives all the others both the vegetative and sensory powers and it is to

---

95. *De intellectu et intelligibili* 1.1.4.
96. *De mot. an.* 2.1.2; *De principiis motus processivi* 1.2, 2.13.

the heart that every other power, both natural and animal, must be referred. Likewise, everything which is in the body shares in the motion of the heart in *diastole* and *systole.* For without that nothing would have life or vital spirit. The heart further imparts the first appearance [*species*] to nourishment by which it moves to the form of whatever member is accepting the nourishment. For, as we have shown in our book *On Spirit,* and also in earlier portions of this book, every spirit is from the heart.[97] It is thus clear how, in the four cases mentioned above about the heavens, the first animal power is analogous to those which are in heaven. This especially suits the rational soul, for that one, through the divine power which it has and through its intellectual and animal power, is an image of the first cause, a likeness of intelligence, and an example of celestial life. And through the natural powers of its body, which are moved by the soul as if they were certain organs, it is the seal of natural things [*sigillum naturalium*].[98] For this reason it especially is the very image of the world and whatever things belong to the order of all the causes of the world are impressed upon and worked into it. Thus, as we said just a bit before, there is a certain amount of conjectural knowledge of the future to be found in its images. The sensible soul, however, is a particular outcome of this sort of image, due to the light of the senses. But a most obscure echo of this is present in plants. This, then, is the reason that the form in the seed is called the image of the world.

Some have called the power in the seed the soul, not because the seed is an animal or an animated body, but rather because it acts on life and the members of life by an act of the soul. It is as if we were to say that the soul is in something as an act or a perfection and that this is an animal in some way (for example, as the craftsman is in the thing crafted) and that this one is not an animal. Or it is as if we were to say that the art of the statue is in the bronze and that the art which is in the bronze molds the statue without any external movement on the part of the artist. This, then, is the power of the soul which operates formally in the seeds of animals.

But one should know that this power is sometimes from the animal and sometimes from the stars. When from the animal, it is in the matter of the seed or egg because it descends from the animate being, possessing a power like the powers of the animal's members. But, as we have said before, all plant seeds are like wind eggs. The power of the soul is present from the stars in those which are generated without copulation and conception. These things are all clear from the things we have said before, both in this study and in the other books of natural science which we have published. And as we have stated above, the member formed first differs in the generation of each of these two.

---

97. *De spiritu* 2.1.

98. On the soul as the "seal" of the separate intelligences on animated bodies, see *De natura et origine animae* 2.6.

CHAPTER IV

On the Natural, Formal Causes of an Animal's Body, These Being Present in the Matter Mixed from the Four Elements

78 The proximate causes present in the matter of animals' bodies are the hot, cold, wet, and dry. These are complexional, not simple, because the simple do not act on the animal's nature. Things which are complexional have actions and passions which are different from simple actions and passions. This difference consists in five types, the first of which lies in the manner of their act or passion. The second is in the form of the act, the third in the sharpness or dullness of the act, the fourth in its variation, and the fifth lies in its end.

The manner of the action of simple heat, with respect to fire, is to burn, dissolve, and to reduce that upon which it acts as far as it can be reduced to its own species. Now complexional heat does not do this. Rather, it digests and completes that which is moist and earthy in the body. It does not separate one from the other but rather purifies each by separating the pure from the impure. It also has a further form of action. In the simple one it leads only to the form of fire. In a complexioned heat, however, it is moved by the form of the complexion and by the soul. It therefore leads toward the form of a complexioned and animate being, much as a hammer leads toward the form of the art that is in the mind of the artist and not toward the form of the hammer or of the iron, these being the forms of a hammer.

Further, in simple heat, the heat acts by receding, as it were, since it is lightly moved upward. But in complexioned heat it sharply penetrates the inmost things since it is split up into the smallest parts possible throughout all the inmost parts of the thing. This is clear from the things said before.

79 It also acts on many things by dissolving, decocting, digesting, assimilating, limiting, converting, and uniting. But in simple heat it acts on its own to heat only one thing and if it does anything else it does so by accident.

Further, simple heat ends its action in ashes when it is acting on a moist, earthy thing. But in a complexioned heat it ends its action in a kind [*species*] of flesh, bone, nerve, and of other homogeneous members.

There is a similar sort of differentiation in cold. Simple cold is deadly, causes constriction, and curtails motion. Complexional cold, however, holds forms and shapes, keeps moisture from evaporating, and surrounds the outermost things lest they be dissolved by some warm agent.

Again, cold is quite dull in simple cold and enters a thing deeply only if it is truly excessive. But in complexioned cold, it enters the inner parts and retains them more than it surrounds then externally. This is due to the moisture which is in it and to its being split, for it is split up into the smallest possible bits.

It possesses, moreover, the form of complexion and the moving soul. For this reason it does not curtail motion but it does contain moving things, keeping them from being dissolved. It also has the quite different operations of containing, coagulating, closing up openings, converting fine evaporating things, and keeping the things that are in the members inside them. It does this in certain especially cold members, such as the brain and the eyes.

The simple and the complexional cold also have different ends to their act. The simple causes parts of matter to gel to a resemblance [*species*] of earth. In the complexioned one, however, it holds the principles in the form of the complexion lest the complexioned one be undone.

80

The same situation exists in the simple and complexioned moisture as in all the five differences listed above. Simple moisture flows outside itself toward a limit in another. But it is complexional moisture, instead of the other, which penetrates the dry. Further, in the simple it leads to the form of air or water, whereas in the complexioned it leads to the form of life and of the soul that is moving it, both during the first forming and in the nutrimental moisture. Again, the simple moisture moistens, whereas in complexion the moisture mixes by using its fineness to penetrate the dry. It also accomplishes many things in a complexioned one such as pouring in, softening, mixing, connecting, carrying nourishment, taking on and returning forms, and the like. But in the simple it does one thing alone, this being to flow toward a limit in another. And in the end it is limited at the form of life, because the cause of life is a moisture that is hard to dry up. This is just as we have shown in our book *On the Cause of a Longer Life*.[99] But it is not limited at this and does not end in a simple thing, but rather at the species of air and water.

This same sort [*ratio*] of difference exists between the actions and passions of the dry. Simple dryness lies in the evaporation or expulsion of the moist. In the complexioned it is present only potentially in the moist and dry. Further, it is caused in the simple by the form of fire or the species of earth. But in the complexioned it is moved by the form of the complexion and the form of the soul.

81

Again, in the complexioned it is divided up finely into the moist, but in the simple it has sometimes been pressed out of the moist, as in the case of earth, and sometimes the moisture has been removed from it, as is the case in fire.

In the complexioned it produces thickening of the moist, fixes the power which would be flowing in the moist, limits the movement of the moist, keeps the limits of shapes in place, and keeps the moist from evaporating due to heat. It does many such things in the complexioned, but in the simple it only causes constriction by limiting itself at its own particular limit. In the simple it is limited at earth or fire but in the complexioned it is limited at the species of a live animal. These, then, are the ways in which differences between the complexioned and the simple appear.

---

99. *De morte et vita* 2.8.

82   But as we have often said, special attention must be paid to the fact that all these elemental qualities bear the relation of tools to the powers of celestial luminous lights and they operate only to the extent that they have been informed by the celestial luminous lights. But the powers of the luminous lights bear the relationship of tools to the powers of the soul and the powers of the soul are as tools of the powers of the celestial intellects. The powers of the celestial intellects are as tools of the first cause or of the first power which proceeds from the first cause, operating upon generated things and bringing about being in the manner we have described above. Since, then, all these forms are referred to a single thing, it is clear that every multitude of materials is collected with regard to a unity of the celestial powers and that this is further collected with regard to the power of the first cause. In acting causally, then, the single power of the first cause acts in this way on a multitude.

Now, while there may be many things in matter which lead toward the being of one generated in the way mentioned, nevertheless there are two principal ones which inform all the powers mentioned, and these are heat and spirit. Heat is from the material side and moisture is its subject. Spirit is from the powers' side as if it were the vehicle by which the vital power is borne into every part of the matter to work therein. The way each of these works is clear from the things which have so often been established previously.

Let, then, this be what has been determined in this manner about the efficient causes of the generation of animals in matter.

CHAPTER V

How All the Principles Mentioned Arise Out of One and Are Collected Both Again into a Single One. How the Multiplicity of Organs Is Caused by a Multiplicity of the Powers of the Soul in Animals.

83   All these powers, then, are collected together, both in the one generated and the one generating, into a single part of and a single place in matter. This happens in the seed of animals or even in the matter which has the force of seed even though it is not seed. This is just how all the powers proceed, by means of a distribution from one first power. In part of the seminal matter they are gathered together into the heart's location or into that which takes the place of a heart. It forms that part first and it forms all the other members later.

Likewise, in one that has been generated, the seat of all these powers is in the heart, and they are then borne away from this by the spirits. They all then complete their powers and operations in all the other members. Now, although Plato and Galen are against this belief, this is still the belief of all the Peripatetics.[100] We have shown the reason for this belief earlier where we spoke on the powers of the sperm and the heart.[101]

The reason they must be collected into one place and into one member is that this power is scattered and it is weakened and corrupted by this scattering and separation of itself. But when it has been collected and unified, it is strengthened and preserved. Nor does nature intend only to make the animal. To the contrary, it also intends to preserve it in being and strength for its operations, for otherwise generating it would be pointless.

Again, a multitude which does not proceed from one thing, which does not have reference back to something and is not collecting to one thing, does not one thing alone but many. Since, then, only a single animal is generated, every multitude of its formal principles ought both to proceed from one first thing and be borne back and collected into one thing. In those which have one, this is the heart and in those which do not, it is that member which takes the place of a heart. For this reason too all things which are directed throughout the entire body, like the pathways of the powers, must necessarily arise from it as well. Examples of this are the veins, arteries, and nerves, and this has already been established before.

Moreover, the entire material multitude of the members arises from the heart through generation. For just as in generation no part of the matter takes on the form of a member except by leaving from the heart (within which it takes on and dons the power which forms the member), so too in a generated animal no nourishment is convertible into the form of the member except by using the heat of the heart to don the power through which it moves to the member. Thus the heart is in the members much as the first heaven is in celestial things and the power of the heart is like the power of the first cause in celestial things. Yet just as a given heaven among celestial things and as a given star in its locations each possesses its own powers, so too do the given members have their own powers which have been established for them in their locations in the members. This is as we said when we spoke about the powers of the members. This entire multitude is related to a single thing. In celestial things this is to the first heaven, composed of a mover and a movable, and in the animals it is to a single first member which is composed of a mover and a movable, and this is the heart.

One should know that all differentiation among movables and matter (when movables differ formally and with respect to shape) is caused by differentiation of the movers and the forms. Thus, unless the soul were a mover composed of many natural virtues [*virtutes*] and powers [*potestates*] there would not be any differentiation of the parts of the body in form and shape. Further, its own organ corresponds to each power just as an appropriate instrument is provided for each faculty of the mechanical art. We see, though, that these powers generally differ in four ways. Certain members

---

100. Apul. *De Plat.* 1.13; Galen *De plac. Hipp.* 6.4 (Kühn, 5.533).
101. Cf. 16.84f.

correspond to natural powers, and the instruments of locomotion, like wings and feet, correspond to the powers which provide motion from place to place. Those organs which are affected by objects of the senses, such as eyes, ears, and the like, correspond to the animal powers, such as internal and external senses. Thus, in the crafts, a weaver does nothing with a mattock or an axe, because these are not the tools of his craft. Neither does the pipe player do anything with boat and oar. So too does the power of a single faculty of nature do nothing through the organ of another faculty. Therefore those who are blind of eye see nothing through their ears and the lame do not walk using their ears and so forth.

It is thus clear that every difference in the organs of the body corresponds to a difference in powers of the soul and that both in the soul and in the body every difference must be reduced to a single thing from which all others are informed for motion and operation. For otherwise, the animal could not be one in either soul or body.

86 The statement of some that most moving substances in the body are reduced to a single act of animating, just as many luminous lights are reduced to a single act of illuminating, is entirely unsuitable. For when they say that they are reduced to a single act of animating, then it is either the same act in number, species, or genus. If they say that it is the same in number, then the animating things should likewise be the same in number as the things which are animated. And this is not true, for they themselves say that there are many animating substances. If, however, animation is one in species, then animation will be the same in species as the form of animation.

Now how can this be true, when they themselves say that the animating substances are different in form, because one is vegetative, another is sensitive, and a third is intelligent? Further, if it is granted that it is the same soul in species, it will only constitute an animal that is one in species and not in number. The same thing follows if they are one in genus as well.

87 If, however, they say that they produce one animation through composition and union, then we will ask what unites them, since they would have different forms. For they are not united to themselves as are the luminous lights which have a single form. Such of them as have different forms are separated from themselves and are only joined in that they have the same relationship between themselves as potency has to act. But they do not say this, since according to this they would be one in substance, for act and potency are joined in a single substance. And for this reason their statement is clearly absurd. Rather, the relation of the multitude of powers and members must be made to a single thing in the way which we have established. For it is in this way and in no other that they constitute a single animal.

We have, then, determined in this way the separation of the single first thing in animals into many and have established the reduction of the entire multitude which exists in the soul and the body to a single first thing.

CHAPTER VI

How Animals Differ One to the Other in the Powers Mentioned Above and How the Human Differs from Them All

We must now understand that these powers, divided from each other according to being and subject, constitute the *differentia* of genus and species among those beings which are animate.

If the vegetative is taken with respect to itself, it differs from that which is sensitive. The reason for this is clear from what has been said before. For since every difference among bodies stems from a difference of forms, and since we see that the bodies of plants—both in the simple and the complex ones—belong to an entirely different genus than that of animals, then it also necessarily follows that souls differ in genus.

In the same way it is also clear that animals differ from one another in species. For there is one shape of the body and the organs for the form and species of each one. Since, then, the shapes of the bodies and the organs of animals differ, their souls will also have to be different one to the other in species. This is the same reason that both trees and plants [*herbae*] differ from one another in species.[102] The human, however, seems to differ from the other animals in more than species. For we see that those animals which display no difference in either sense or motion still differ from one another in species. A human differs in species from the others in participation in the senses. This is proven by the shapes of the sensory organs and by the shape of the entire body, with the addition of the difference of reason. Since this is so, he seems to differ from the other animals in more than species. An indication of this is that a person's irrational powers, such as the concupiscent and the irascible, are susceptible to the persuasion of reason. This is not the case in other animals. He therefore differs in more than species from the brutes and he seems to have a certain difference in kind [*genus*] over them since he participates in animality itself in respect to a power which is different from that in other animals.

If, however, someone should object that a genus encompasses many species and that thus a human ought to have many species, it will carry no weight. For form, communicable in and of itself, is not necessarily communicated to many things which are different in species or number. Rather, it may be present in one only with respect to being, as is the case in the sun, the moon, and in other things which have but one individual member under the species because they exhaust its matter.[103] And if we

---

102. Omitting the double *ab invicem,* as does Borgnet.

103. The sun, the moon, and the phoenix were examples evoked since late antiquity to illustrate the paradoxical possibility that there should exist a species with but a single member. Cf. Boethius, *In Categorias Aristotelis PL* 64: 177D, and his *In Isagogen Porphyrii Commentorum Editionis Secundae* 3.6 (CSEL 48, 217–18).

were to posit that there is but one ass and that there is but one tree, there would still be the different genuses of animal and animated body. Thus, the human differs from those which are called brute animals in more than species. The lion and the horse differ in species, but a human differs from each of them in more than species.

Further, the animal genus which is in animals is different, with regard to being, in each of the animals. And since the true definition of a natural genus is given with regard to being, it is clear that there is not one true sense of genus in animals. Rather, the nature of many genuses exists in animals, with regard to the being of the difference, which is the form and act. It therefore is driven toward differences with regard to being, just as the acts themselves are different. The differentiation of shapes in the bodies of animals corresponds to this difference. Thus, no animals are found which differ in species and which have a single shape to their body.

90 There are, moreover, certain powers which are suitable for all animals and which differ greatly one from the other. The power of the sense of sight in all of them has to do with the form of the clear, and thus the organs of sight differ little in shape and composition one from the other. But in hearing they differ a great deal. Thus, dull-witted animals have differently shaped ears from those animals which participate more in the signs of things, like the human and certain genuses of monkey. Others, however, are entirely incapable of being taught and seem to have no instrument of hearing at all. From this it is clear that, as we said, all the diversification which exists in a body is caused by the diversification of the powers in the soul. Of those powers which the soul exercises in the body, some, such as nourishing, growth, using food, and generation, are common to every animate thing. We have already made a determination on these in the study *On Plants*.[104] Others, however, are common to all animals and constitute it as an animal, such as motion and sensation. For touch constitutes an animal for the sake of touch and a certain motion is caused by sensation in every animal, such as local motion for the sake of appetite. At the least, every animal moves with a dilating and constricting motion.

We should, therefore, determine the differences in animals with respect to their senses and locomotion, and we will introduce for this purpose the beginning of another book.[105]

---

104. *De veg.* 1.67f.
105. As Stadler notes, in one MS (C), A.'s *De principiis motus processivi* is placed after Book 20.

HERE BEGINS THE TWENTY-FIRST BOOK ON ANIMALS

Which Is on Perfect and Imperfect Animals and the Reason for Their Perfection and Imperfection

The First Tract of Which Is on the Degrees of the Perfect and the Imperfect

CHAPTER I

On the Highest Perfection of Animal Which Is the Human

It now seems that we must make a determination concerning the perfection and imperfection of animals with respect to the faculties and powers of the soul and with respect to the reason for their imperfection and perfection.[1] For this seems to be a consideration prior to that of animals with respect to the species of perfect or imperfect animals.

Since, however, every perfection and imperfection in animals exists in accordance to their perfect or imperfect participation in the animal virtues, and since the imperfect can be known only through the true nature [*ratio*] of the perfect, we should first determine what the true nature of the most perfect animal is. For we find every degree of imperfection in others by comparison with this one. Thus, one must seek the true nature of the more perfect animal according to the powers of the soul for, as we have already said above, that which is perfect in the powers of the soul is perfect in its organs. This is because nature produces an organ in animals only by means of the power of the soul which moves the organ itself. In this way, then, there can be no more perfect sort of powers in any soul than that the organic powers exist according to the being of the nonorganic ones.

Thus, as we said previously, in all those in which the sensible soul is the ultimate act (that is, the completing substantial form [*enthelechya*]), the vegetative soul exists only as a partial power [*pars potestatis*] and it exists in them according to the being of the sensible soul.

An indication of this is that that on which it acts terminates at sensible life. And since in this same way in anything for which the rational soul is the completing substantial form [*enthelechya*], the naturally existing powers of the vegetative or sensible soul exist according to the being of the rational soul, because the being [*esse*] which is

---

1. In this book A. employs a bewildering variety of terms to express the notion of a power. As a result, we shall render *virtus* as "virtue," while *potestas*, *potentia*, and *posse* will be translated as "power."

the act and effect of the essence and its substantial form [*forma substantialis*] is unique for each individual animal. For that reason, just as "to live" is the mode of being [*esse*] in an absolute sense for plants, so too "to sense" is the mode of being for sentient beings and "to understand" is the mode of being for intelligent beings. I say then that "to live," "to sense," and "to understand" have life, sense, and intellect, because the soul is first an act just like sleep, as we said in the book *De anima,* and later it is an act just like waking.[2] Likewise there are two acts of a sword, to cut off or into something with the shape of the sword. When, however, the sensible and vegetative powers are present according to the mode of being [*esse*] of the most perfect sort that can exist in an organic body, it is necessary that these be the most perfect [powers] and participate in reason in some manner. An indication of this is that in the human, in both the vegetative and sensible powers, there is a rational order [*ordo*] to life. For temperance and chastity are present according to the functions of the vegetative soul; humility and gentleness, however, and likewise fortitude and many other virtues, are present according to the desires and angers of the sensible soul. This could only be the case if the vegetative and sensible souls were in the human according to the being [*esse*] of the intellectual soul, for otherwise the soul would not be receptive to a good rational order.

3   If someone should say, however, that temperance and chastity and modesty and other virtues of this sort exist not as a result of the functions of the vegetative soul but rather as a result of the desire which belongs to the sensible soul, this cannot stand. Since nourishment and generation belong to the vegetative soul, for that reason too the virtues which pertain to these functions can be present only as a result of the vegetative soul. Both the desire, which pertains to these functions, as well as sense and motion pertain to these functions, just as the vegetative soul is in the sensible, much as a triangle is in a quadrangle.[3] And this is why we said earlier that it is present as a result of the being of the sensible soul and further that it is present as a result of the being of the rational soul. For that reason, function takes its order from reason and takes desire from the being of the sensible soul. Nevertheless, function always belongs to the vegetative power.

From these things, then, it is clear that there can be no more perfect animal than a human, seeing that the soul which is the ultimate act cannot be more perfect than that the being and the *ratio* of the others should be separate, and that its natural power be conjoined to an organ.[4] Prior to such a substance, there is nothing at all except what is

---

2. *De anima* 2.1.2. For the claim that the first act of the animal soul is similar to sleep and the second similar to waking, cf. also *Metaph.* 1.1.6. The point of this distinction, especially at *De anima* 2.1.2, is to make clear that even when the animal is asleep and not engaged in sensing external objects, it remains ensouled. Even though the act of the animal is, then, the sensible soul, it is not necessary that animals always be engaged in the act of sensation. The power of the soul "shrinks," as it were, during sleep, and it expands, to take in external objects, when waking. Therefore, the first act of the animal soul is similar to sleep and the second is similar to waking.

3. A. uses this same image at *De anima* 1.1.4 to explain how the *ratio,* or definition, of one soul (e.g., the vegetative) can be in another (e.g., the sensitive)—just as the figure of a triangle is "in" a quadrangle.

4. Following Borgnet's reading for this troubled passage.

separate in being and power. And this does not have the true sense [*ratio*] of soul and for that reason it does not constitute an animal. Now power [*potentia*] can be related to only two things, namely, to the object and to the subject (that is, the substance) whose natural power it is. On the part of the object, however, it has only the power that is permitted to such an object according to nature, or that functions for the sake of such an object according to nature, and these do not bestow and elevate its power [*posse*] or the mode of its power. But from the nature of the subject from which power itself arises (just as every natural power arises from the victory of the form of the subject), it will have without a doubt a power that is much greater in proportion as the form of the subject shall have been higher and nobler and nearer to the first cause, according to whose being it exists and the natural power is present. And for that reason the vegetative soul has a power bestowed in sensible souls, and the vegetative and sensible are the most bountiful powers that can exist naturally in the rational soul. And this is indeed what many say, but which perhaps many do not understand: that whatever an inferior power can do, a superior, more excellent, and more eminent power can do. It is clear, then, from all the things said already that the human is the most perfect animal not only by virtue of the addition of reason but also in terms of all the powers and the manner of the carrying out of the powers, both sensible and rational.

Nevertheless, a question arises from the things already said, seeing that sight and smell and hearing and other things of this sort are certain powers of the sensible soul. And we know that lynxes see better than a human, vultures have a better sense of smell (and likewise dogs), and certain other animals perhaps hear better. What has just been introduced, however, does not present a problem. For to see, smell, or hear better can be understood in two ways. That is seen better which is received more acutely by sight, and such a power of sight occurs from the complexion of the bodily organ in which sight exists. And in this way, nothing prevents certain animals from having more acute senses than a human has. For that reason, one is said to see better by the use of a nobler sight (that is, one of a more copious power) upon any object whatever. In this manner, that sight that conveys the differences of the thing seen in a disciplined way [*disciplinabiliter*] is more copious than that which conveys the same differences only in a sensible way [*sensibiliter*]. And thus does a human have sight and likewise hearing and the other senses, which in him are the source of experience and memory and of universal art and science. In this sense, no animals use senses that surpass those of the human.

Therefore, what we said is clear, namely, that the most perfect of animals is the one whose entire power of soul exists according to the being of a separate nature. And for that reason we said in the previous books of this study that a human differs from other animals not only in terms of a specific difference but also in terms of the being of his proximate genus and in terms of the being of his remote genus. For the proximate genus is the sensible [*sensibile*] and the remote genus is the living [*vivum*]. And we have shown that both the sentient [*sentiens*] and the living participate in reason in some way, although these powers may be, by their own nature, irrational. In the same way the vegetative is different with respect to being in a sensible one and is different than it would be in itself from its own nature. And this is why the *ratio* of animal and of each genus (to the extent that it is a genus of animals) takes its existence in various ways and has varying potential, just as we determined in the first book of *De anima*.[5]

5 There is still another perfection of the human that is beyond all the other animals. Now, the first of the senses on account of which an animal is an animal, and which no animal lacks at all, is touch.[6] And this, if we observe the nerves and sensible organs, is the basis of all the other senses, and it widens and loosens all the nerves to sense its object, because complexional heat loosens a sense, and cold restricts every use of the senses, just as one shows with great subtlety by means of those things which were determined in the book *On Sleep and Wakefulness* and by means of the cause that states *propter quid*.[7] If, then, every sense has its basis in touch, it will have to be, necessarily, that every sense will be present in animals according to the power and virtue of touch. Now, whatever forms exist in something as in a fundamental subject are in it according to the power of the subject and not according to the particular power of the form or the efficient cause, just as we have proven often. For we see the light of the sun in clear air and cloudy air and it is altered in the cloudy only on account of the different power and different virtue of the one receiving it. Every sense, then, is present in all things according to the power and virtue of touch. However, a human has the most subtle and the surest touch, as we proved in the book *De anima*.[8] Therefore, the other senses in the human will with some certainty also be nobler and surer and more refined than they are in any of the other animals. Since an animal is and is said to be more perfect according to the perfection of its senses, the human will be the most perfect of all animals because he participates in sensation most perfectly and surely.

6 Further, in terms of the organs of the senses and the powers of the soul, the human alone among all the animals participates in the hand, which is alone the organ of organs and the organ of the operative intellect, as was determined in preceding books of this study.[9] None of the animals, however, except the human has the organ of organs in his body. Rather, every organ they have exists for one act alone, as all the quadrupeds have their front feet only for walking, and biped flyers have their anterior members only for flight. The human, however, does not use his upper members of this sort for motion but rather employs them generally in the application of art and the tasks of the other members. It is clear, then, that the human participates in certain organs beyond the way in which all the other animals participate, and thus even in the organic composition of the body he is more perfect than all the bodies of animals and more perfect than the animals themselves.

---

5. *De anima* 1.1.4.
6. In addition to numerous references in *DA*, cf. *De anima* 2.3.31 and *Metaph.* 1.1.6.
7. *De somno et vigilia* 1.2.4?
8. *De anima* 2.3.23.
9. Cf. 14.31, with note.

Moreover, the shape of the body reveals this itself. For since three [different] diameters compose every body, that body will be more perfect and more natural that participates in the measure of the natural diameters.[10] The longitudinal diameter measures up and down, and only in the human is it the same above (which is the upper part of the world) and the same below (which is the lower part of the world). It is similar for the latitudinal diameter. For the human alone among all the animals has a body latitudinally proportional to its size, because although vermin are broad they do not have a width that is proportioned to length. For length always ought to exceed width in a natural body, if there is not a flaw of nature. Quadrupeds, moreover, have bodies that are thicker than they are wide. The human alone has a diameter measuring depth that is less than the other diameters. Since therefore the sense organs are positioned according to length as one descends, and the organs of motion are positioned according to breadth, the organs of the body in the human have a greater perfection of distinction than in any of the other animals.

Further, the human alone, as was proven above, is the image and likeness of the world, both in terms of the soul and the body.[11] Every other animal, however, lacks something according to more or less, and a defect arises from the lack of something pertaining to perfection. It is clear then from this that the human alone is the most perfect among all the others. In a case in which some animal does not have a likeness to those things which, generally, cause the world, then it is not imprinted by it in that instance. For this reason, only the human is imprinted by intelligence; nor is there another of the animals that is imprinted by heaven as the human is. Moreover, among all the animals the balance of his complexion belongs more to the nature and balance of heaven.

Besides, the second mover in all things in which there is a mover of the second order is imperfect unless it should be joined to the first mover of the same order. Now the first mover alone is that which moves by itself, and it is the cause of motion in all things, just as was revealed by the things said earlier.[12] According to those things which were correctly determined in *On the Movements of Animals,* it is a commonplace for us to say that the moving phantasy is related to the moving intellect just as a mover of the second order is related to the first mover.[13] However, none of the animals has a perfect and well-ordered motion unless a motion proceeds in it according to the true nature of a first mover, which is a straight motion. Therefore, since the human alone is moved through the intellect, he has the perfection of the motive principles.

According to these modes of perfection, then, the human alone is the most perfect of all animals.

---

10. Cf. *DCM* 1.1.2.
11. Cf. 20.74–75.
12. Cf. 20.64.
13. Cf. *De mot. an.* 1.1.3ff.

CHAPTER II

How Many Types of Perfection There Are in Animals in General, and Which One Would Exist According to the Soul and Which According to the Body, from Which the Nature of the Pygmy Is Known

Theophrastus, however, says that one must speak of the perfection of other animals in two ways.[14] Some of these are more perfect than others in terms of the body, and some are more perfect than others in terms of the powers of the soul. And each of these perfections is double, seeing that that perfection which exists according to body exists either according to the quantity of the body or according to the quality of the balance (that is, complexion) of the body. And that perfection which exists according to the quantity of the body is further divided in two: seeing that the perfection of quantity exists either according to the magnitude of a continuous quantity or exists according to the number of organs, which is the number of a discrete quantity, just as an elephant has a trunk, which other animals do not have, which it uses for many purposes in place of a hand.

And, further, that perfection which exists according to the soul is divided and rendered double. Either it exists by participating in the many interior powers of the soul or by participating in the many exterior powers according to the number of senses. And in terms of the participation of the interior apprehending powers, there are still two types, one of which is according to the number of powers and the other according to their type and quality. Some animals are perceived to have few or none of the interior powers, and some others are so much richer in these that they seem to have something akin to reason. Further still, of those which have virtues and powers of this sort, some have a certain subtlety in them (just like the bees) which others greater in size do not have. And it is necessary that we determine every type of perfection and imperfection of animals, because from the types of their perfection and imperfection the greater part of their nature and their works is known.

We will speak, then, first of the types of perfection with respect to the soul and, treating the perfection of the soul first, we will investigate that one which has to do with the number of the powers and virtues of the soul. Let us use the skill which Aristotle taught in the beginning of the first philosophy, when we say that we see that all animals have some sense and are receptive to sensibles.[15]

Memory occurs in some, however, from sensation, and in others, not. And we know that memory does not occur in some from sensation because memory is that which causes the sensible, in its absence, to return out of what was received earlier by sense, just as we see that vultures once they are sated depart from the place of a cadaver

---
14. For A.'s knowledge of the work of Theophrastus, cf. Sharples (1984), 186–90.
15. By first philosophy, A. means metaphysics. Cf. *Metaph.* 1.1.6.

and later return to it from the memory of the place and the cadaver.[16] And in this way too sheep return to their sheepfold and birds to their nests, and others of this sort. However, we know that any animals which follow only a present sensible and do not return to the sensible when it is absent on the basis of the one received earlier have no memory of the things received earlier. Likewise, there are flies which, when they are driven away, fly back unmindful of a blow received earlier.[17] We also see that they do not keep a fixed dwelling, and we see that they follow only a present sensible.

We see, further, that some animals have a sort of prudence with respect to gathering up things for themselves but they are nevertheless not subject to instruction. This is clear among bees and, likewise, ants, which have great prudence in collecting things and yet are not instructed. It happens, on account of this prudence, that they provide storehouses for themselves. But the fact that they do not respond to human voices and neither fear their threats nor seem to flee from terrible noises is a sign that they are not subject to instruction by means of human teaching. It is for this reason that some even say that they do not hear sounds; however, this has been disproved in what has gone before, since they seem to hear sounds. But whatever the nature of their hearing may be, this is without a doubt true that they do not hear sounds for instruction so that they can be called by name and be instructed as many other animals are instructed, like the dog, the monkey, and certain others.

Now, hearing is possessed by animals in two ways: it is possessed by some only insofar as it is a sense; it is possessed by others insofar as it is a sense capable of instruction.[18] And animals participate in this second manner in two ways as well. This sense is capable of instruction to the extent it uses sounds and voices to grasp hints [*notitio*] of the intentions of the one producing the sound or the voice, for this is how sounds and voices instruct. Moreover, this occurs in two ways. For whenever there are sounds and words, these sometimes produce a sign of the intention which is confused, and sometimes one that is well defined. They produce a sign that is confused in brute beasts, and they produce one that is well defined in the human. For that reason, those animals that have hearing as a sense capable of instruction, and, with this, retain a memory from which the signs of instruction are perceived in either a confused or well-defined way, are capable of instruction and perceive the teaching in either a confused or well-defined way. And thus, many animals do many things in response to human voices, as the elephant bends its knees before the king in response to the command of the one speaking it, and dogs do many such things. But bees and other small animals never perceive the sounds and voices in a manner capable of instruc-

---

16. Indeed, memory may be defined as a treasure house of the forms received previously in sensation, as well as of the intentions of these objects received from the estimative power. Cf. *Metaph.* 1.1.6.

17. At *Metaph.* 1.1.6, it becomes clear that flies lack memory because of the nature of their complexion, which immobilizes the animal spirit bearing sensible forms.

18. A. has quite a lot to say about the differences between those animals that merely hear sounds and respond instinctively, one might say, and those that hear sounds carrying meaning (*voces*), seemingly interpreting them. Cf. especially *De anima* 2.3.22 and *Metaph.* 1.1.6.

tion, although they thrive with a great faculty of memory. This is, then, the cause of the capacity for instruction in some animals. There is especially a lack of this type of capacity for instruction in the very small animals, as in bees, wasps, *achathys*, fleas, and other vermin of this sort.[19]

11   Some of these animals seem to participate in experience in a small measure.[20] For experience arises from many memories, because many memories of the same thing produce a power and faculty of experience. And we see that many animals, in addition to the human, have some measure of experimental knowledge in individual cases,[21] just as we see that a weasel wounded while fighting with a serpent takes as an antidote for the poison an endive leaf, which is called by some pig's snout [*rostrum porcinum*].[22] We have introduced in the preceding books many other such things that animals do. However, they do not participate in experience sufficiently, since they do not approach, through experience, both universal art and reason, but nevertheless they participate in experience according to some measure, as we have already said.

Certain animals, however, are raised up so much in these powers that they have a kind of imitation of art, although they do not attain art. And we see that this happens in two ways in animals. For some seem to be capable of instruction both by sight and by hearing, since they reproduce what they see and retain what they hear, like the monkey. Some, moreover, flourish so much in the instruction of hearing that they even seem to signify their intentions to one another, as does the pygmy, which speaks [*loquor*], although it is an irrational animal nevertheless. For this reason the pygmy seems to be the most perfect animal, in terms of the animal virtues, after the human. And it seems that among all the animals, it preserves its memories so much and perceives so much from audible signs that it seems to have something imitating reason, but it lacks reason.[23] For reason is a power of the soul for running through the experiences received from memories, drawing out the universal from the specific or syl-

---

19. The source for the form *achathys* is unknown. Stadler suggests the Gr. *achetai*, cicadas, or *arachnai*, spiders, but these are guesses at best. No name similar to this one appears earlier in this work and A. is not following a Gr. text.

20. "Experience": *experimentum*, a term which for A. conveys much more than merely passive sense perception and may imply the process of learning by trial and error. Thus, when A. wishes to say that he has learned something through the senses he often uses the phrase *expertus sum* and frequently appeals to *experti* such as fishermen, fowlers, and hunters, whose knowledge comes directly from experience and observation.

21. Stadler gives a reference to Ar. *Metaph.* 612a28. A similar example is given by A. in *DA* 8.47 and *Metaph.* 1.1.6. In both places it is a bird that when it is wounded in a battle with a serpent, takes a leaf of wild lettuce as an antidote to the poison. In the latter example, however, this is intended to show that animals participate in experience through memory only in a material, albeit not formal, sense.

22. For the identification of the weasel (*mustela*), cf. especially 3.96 and 22.122(79). For *rostrum porcinum*, cf. *De veg.* 3.9; 4.145; and 6.331.

23. Although A. had never seen a pygmy, he received a good bit of information about them that suggested their similarity to humans. Nevertheless, A. found that their deficiency in the power of reason—at least, that power of reason that seeks the universal—rendered them more like monkeys than people. For a good discussion and bibliography, cf. Köhler (1991), Koch (1931), and Friedman (1981).

logistic figure [*habitudo*], and bearing from it principles of arts and sciences through similar figures. The pygmy does not do this. It does not separate from the intentions of the sensibles what it receives by hearing, and it commends them to memory as if they are the intentions of sensibles, and in this way it conveys and signifies this amassed material to another pygmy through speech.

And thus the pygmy, although it speaks, nevertheless does not argue or speak of the universals of things, but rather their voices [*voces*] are directed at the particular things of which they speak. Its speech is caused by a shadow echoing in a defect of reason. For reason has two principles, one of which results from its turning back upon sense and memory, where the perception of experience is. The second principle, however, is that which it has insofar as it is raised up to a simple intellect, and this is the one capable of drawing out the universal that is the principle of art and science. The pygmy, however, has only the first of these two. For that reason, it has the shadow of reason because the full light of reason is in the second. Moreover, I also call a shadow that which is an obscure echo of reason that is not separated from the matter of sensibles and from things joined to matter. And for this reason the pygmy perceives nothing at all of the quiddities of things, nor does it ever perceive the figures [*habitudines*] of arguments. Its speech is just like the speech of morons who are stupid by nature because they are incapable of perceiving reasons. But in this respect there is a difference, since the pygmy is deprived of reason by nature, while the moron is not deprived of reason but rather is deprived of the use of reason accidentally on account of melancholy or another accident.

However, perfection of this sort is next in rank beneath the human, so that the pygmy does not watch over a perfect political system or laws but rather follows the impulse of nature in such things, just as do other brute animals. But it walks erect. And it does not use a hand accordingly as an instrument of the intellect and a certain species of instruments (that is, organs), but rather, just as certain animals use their front feet for many uses—as several genuses of mice pick up food and direct it to their mouth with their front feet—so too the pygmy uses a hand for several uses, but not for works of art.[24] And for that reason, then, its hand does not have the complete true sense [*plena ratio*] of a hand. The fact that the pygmy is always erect, however, produces in it a greater clarity of the spiritual members [*spiritualia*].[25] And for this reason it has better apprehension than the other brute beasts, but it does not pay attention to the shame that results from the unseemly or the glory that results from that which is noble.[26] And a sign that it has nothing of the judgment of reason is that it uses neither rhetorical nor poetical devices when speaking, which, nevertheless, are the more imperfect of all arguments [*ratio*]. For this reason it always dwells in the forests, presiding over, actually, no political system. That power of the soul which, above, we

---

24. Cf. 12.197 for the use of the front feet. At 8.229, A. relates seeing a mouse which had been trained to hold a candle while its master dined and he repeats the tale at 21.21.

25. Elsewhere the term signifies respiratory members. Cf. 2.11; 6.122; and 9.23, with note.

26. Cf. 1.51, "No animal except for the human is ashamed over the doing of some foul deed."

called a certain shadow of reason indeed goes without a name among the philosophers, but, while using a sort of circumlocution, we know that this power contributes a measure of power beyond the estimative power. For although the estimative power of brutes may judge the intentions which are received with the sensibles, this power does more since it conveys the object [*res*] of this sort of intention to memory and draws on the experience and uses it later for that for which it is useful.[27]

14 This too ought not to be passed over, namely, that the object of the estimative power and that of experience are referred to the universal in two ways. In one way, in a contemplative fashion, just as the quiddity of things (which is the reality of things) is received *per se* or by a common sign as if from the objects received in sense that are experienced and remembered. The object experienced and remembered in this manner is not present in the pygmy. In another way, the object experienced and remembered is present to the extent that it is this in him which brings him to desire or flight. In this manner the experienced and remembered are received by the pygmy and that is why we said above that the pygmy participates some little bit in experience. Now, the pygmy receives these things only by way of referring them to what is useful or noxious in the object [*res*] that is outside it, just as the other brute animals refer all the intentions they apprehend to them.[28] For that reason, they never separate the intention from the particulars but instead mix them all together. For this reason too, mute animals are called brutes because their conceptions say nothing about the reality of things. The pygmy, then, according to the things already introduced, is midway between the human, who has a divine intellect, and the other mute animals, in whom nothing of the divine light is perceived to exist, insofar as it uses experiential cognition through a shadow of reason, which it alone of the other animals receives. Nevertheless, with respect to nature it is nearer the brute than the human, as was shown in the things already said, seeing that the object known by experience belongs more to the universal and contemplation than to the particular and motion.

---

27. "Estimative power": *aestimativa* [*potestas*], a power of appraisal based on images rooted in sense experience.

28. "To them": in a passage composed entirely of fem. and ntr. nouns, A. has managed to insert the masc. *ad eos*. Its antecedent is uncertain.

CHAPTER III

How Animals Are Capable of Instruction by Some Participation in the Virtues of the Soul, and Especially How This Occurs in the Genuses of Monkeys

The remaining genuses of monkeys, which are especially numerous, seem to participate in the soul in the third degree of the powers and virtues of the soul. Instruction [*disciplina*] occurs in the human, who alone is an animal naturally amenable to instruction, namely, through teaching [*doctrina*] and discovery. Moreover, the instruction which is by means of teaching is double: namely, subject to intellect and to sense. Still, monkeys more than other animals seem to have this sagacity: they are capable of instruction from sensibles. For which reason monkeys are known by a Greek name derived from "play" [*ludus*] because they imitate the games presented to them in sensibles more than the other animals.[29] Moreover, for this reason some wise men have handed down that these animals were made for play, and that whatever they see they imitate with this keenness, and through hearing they gather the ordered intentions of those calling them to play.[30] And this is the first degree of "teachability." Now, first it is necessary that one who is receptive to instruction diligently attend to and convey those things which he sees and hears, and it is necessary to take from those gathered the objects of experience from those retained in memory and from these receive a purified universal, according as it is the principle of science or art or prudence or some other intellectual virtue.

The monkey, however, perfectly obtains only the first of these three degrees, and it obtains this only by its capacity for imagination and for memory.[31] But it does not participate in experience further whether using experience much or very little; in this it falls short of the pygmy. Moreover, imitation is easier for it to the extent that it does not refer that which is sensed or seen to experiential cognition. A reference to experiential cognition is known in the collection of memories (that is, in the memory of things received); not bearing these, it is immediately referred to the sensible. And it is on account of this that the monkey, which has the first faculty for instruction, immediately imitates that which it sees. But the pygmy and the human do not immediately imitate what they see. For the monkey, just as we have said, has only the first faculty of "teachability." An indication of this is that it does not discern in what or whom it imitates.

---

29. A.'s etymology is wrong but is also unusual. As one would expect, the normal derivations are from Isid. *Orig.* 12.2.30, where *simia* is correctly linked to the Gr. for "flat-nosed." Others, Isidore says, relate *simia* to the Latin *similis* in reference to the monkey's *similitudo rationis humanae*, an argument A. would have liked. These etymologies are routinely repeated (White, 1954, 34–35).

30. Reading *ordinatas* with Borgnet for Stadler's *ordinata*.

31. Borgnet reads: "and the second it obtains by imagination . . ."

16 It must be observed beyond all these things which have been set forth that no animal is capable of instruction in any way at all unless with the instinct of nature. Now the human makes progress from instruction toward three things: namely, to the contemplative sciences, to the mechanical arts, and to the moral virtues, and he does not progress to any of these without reason and intellect. And this is clear *per se* concerning speculative science. Moreover, this even occurs with art, seeing that it is together with reason that art is productive of those things it makes. In virtue, however, the mean is obtained that is the mean by a well-defined reason, just as wisdom has determined. For which reason none of the animals is instructed in any of these three. But since the human has two senses capable of instruction, namely, sight and hearing (as was determined in the book *On Sense and the Sensed*), sight is subject to the discovery of instruction.[32] And since the discovery of this occurs only when reason confers memories and things sensed, there is no animal capable of instruction through sight [alone] because it does not have the faculty of discovery from the gathering of sensibles. Hearing, however, is a sense capable of instruction from the signs of things and by another, and for that reason some animals can be instructed through hearing but none through sight alone. And on account of this the monkey is known to have a better faculty than another animal, seeing that it receives instruction through hearing and is capable of representing an imitation of acts received through sight, which no other animal does except the human, the pygmy, and the monkey.[33] And for this reason it is clear that these three animals descend by continuous degrees. Indeed, a human participates in sense capable of instruction in every respect, and in memory, experience, reason, science, and art. The pygmy, however, participates in sense capable of instruction, memory, a little bit of experience, and nothing of reason, science, and art. The monkey, however, is a participant in a sense capable of instruction and hearing, and it perceives by sight something ordered to instruction, namely imitation, and has a memory of sensibles, but it draws out from it almost nothing from experience. Moreover, other brute animals perceive some measure of instruction by hearing alone, receiving nothing teachable by sight.

17 Further, just as it is established by the things considered above, moving forms and speculative forms are in the human. The monkey, and indeed every other animal, receives the instruction of only moving forms, for which reason animals especially are more capable of instruction with regard to things in motion. This is why animals are instructed to walk or stand or seek out something, and not instructed in art or science. And for that reason none of the animals is instructed unless it is moved by appetite to possible tasks, because these [tasks] are particular ones, moreover, received with the particulars capable of being perceived by animals of this sort.

It seems too that the genus of monkeys—more than all the other brute animals—appraises [*estimare*] that which is beneficial and that which is capable of causing injury,

---

32. Cf. *De sensu et sens.* 1.2.

33. At *Metaph.* 1.1.6, A. indicates that the monkey does achieve something in the mechanical arts by virtue of its likeness to the human.

doing so from sense, imagination, and memory. While others appraise things beneficial and capable of causing injury to themselves, either following after or avoiding these things, the monkey appraises those beneficial to it and to other animals, and for that reason when it sees a young it shows the young a breast—not its own, but the breast of the mother who gave birth to it, if it is permitted. Now, it even displays to a child the breasts of women, if it is permitted, and other animals do not do this. By this one knows that the monkey has better powers of appraisal than any other animal. But in all these things it is moved only by a phantasm, and for that reason it errs frequently, as do other animals because, as we said above, when the phantasm is not joined to the intellect error frequently occurs. And in such animals there is no practical syllogism [*syllogismus operis*], but only an imperfect form of argument.[34]

And just as in those capable of contemplation an enthymeme and an argument by example are imperfect forms of argumentation—whose imperfection is rendered perfect, nevertheless, by being reduced to a syllogism—so too are there imperfect practical syllogisms in these animals.[35] These have only phantasm-based appraisal [*fantastica aestimatio*] of that which is capable of being done or desired and they have an appetite that brings about the impetus for the work. But there is a difference in this respect: the enthymeme proceeds from that which is beneath the universal according to whether it is in many or in all things. The phantasm-based appraisal, however, arises only from that which is merely seen. And for that reason it is deceived frequently, just as from a false conclusion [*sophisma*] based on an accidental trait, because the animal only proceeds in accordance with things which are seen in the here and now, thinking that things in which the same accident occurs with reference to the here and now ought equally to be pursued or avoided. The pygmy seems to participate in some measure of inference, however, in that it participates slightly in experience according to the "operative" syllogism, which the Greeks call "practical," but, when it does not progress as far as to receive the universal, it does not perfectly arrive at an inference. Moreover, these animals do not use an argument based on example in any way because an argument based on example cannot be made without some comparative judgment of reason.

These two genuses of animals, therefore, throughout all those contained under them, achieve a human likeness beyond all animals, not in the capacity for contemplation but rather for motion. And for that reason these animals have a more lucid estimative power, certainly, than all other animals. That is why they appraise phantasms better than do other animals, and they draw out better the intentions received with the phantasms better than any of the other animals, so that they seem to have something like reason. They especially seem to have something like that act of reason

---

34. *Syllogismus operis* seems an odd expression, but one may assume that it is related to the more common *syllogismus operativus,* which is a practical argument or syllogism employing a universal major premise and a particular minor premise that should result in some concrete action. Cf. A.'s *Super ethica* 7.3.625.

35. An enthymeme is an incomplete argument, having one or more premises suppressed. An argument by example is one in which the major term belongs to the middle by means of an analogous third term. For a definition of an argument by example, cf. Ar. *An. Pr.* 68b35f.

which gathers together something from the intentions in the sensibles themselves and the phantasms that are received and are not separated from them. And this is the reason that these genuses of animals are called "human likenesses." An indication of this is that they have round heads, as if there were a distinct sphere compressed throughout three chambers, and because they have semicircular and unmovable ears just like the human, a hand that has long fingers, and a bend in the arm at the elbow toward the inside of the chest. Also, they use their hands for many uses as does the human, although they do not use them in the uses of art as does the human. They have a wide but not deep chest like a human and they have a small neck and broad clavicles like a human. The females have their breasts hanging down onto the chest, just as the human female does, and they have large vulvas in respect to their size and these are positioned at the bottom of the belly so that the cleft of the vulva ascends toward the navel, just as the vulva of a woman is positioned. These animals copulate with the female positioned on her back and with the male lying on top of her, just as humans copulate.[36] And this harmony with respect to exterior things signifies as well an interior harmony since, just as we said, their estimative power is more like reason than is the estimative power of any other animals.

Therefore, the powers of the soul are present in these and other similar ways in these animals, which achieve the greatest likeness to a human among brute animals.

CHAPTER IV

Concerning Things That Are Observed with Respect to the Teachability of Quadruped Animals

20 Moreover, all the other quadruped animals seem to have less discretion in their actions than was stated. And for all of them three things ought to be held out for consideration: viz., hearing, estimation, and memory.[37] For, just as we have said in what went before, hearing in the human exists according to three differences of virtue or according to three modes of virtue capable of sensation through hearing. There is one according to the judgment of sounds, and this is the same for all those that have hearing. In the human this is even a sense for instruction, as voices are received by hearing, according to which there are the signs of concepts and passions which are in the soul of one bearing them, and according to which they are related to objects [*res*] signified by names. And this occurs in two ways: in one way as the quiddities of things subject to contemplation are signified by names, and in another as the operable form, for the

---

36. Not true for simian genuses, with the exception of the Bonobo chimpanzee, whose sexuality is surprisingly "human" in some respect.

37. Estimation is an internal sense or power that apprehends the intentions accompanying sense experience. For a brief discussion, see 8.233 and Glossary, s.v. "Phantasy."

sake of which there exists either an effect of the one bearing it or a thing which the animal desires (just as dogs are moved by the voices and by the sounds made by deer), becomes known in the names. And the first of these applies to the sense of hearing insofar as the particular sense has arisen from the common sense, which is the font of the senses. The second applies, moreover, to the sense of hearing insofar as reason is brought to bear upon it. And for this reason this applies to no irrational animal. The third, however, applies to sense insofar as the power saying "flee" or "pursue" or "act" is united to that which is the phantasy and estimation. And for this reason some animals—both quadrupeds and bipeds—perceive sounds and voices in each way. Many animals, now, perceive voices in respect to which they are signs of the one calling, and they are moved either to pursue or avoid that one, just as a lamb hearing a wolf flees but hearing its shepherd, follows.[38]

Moreover, it happens that they perceive voices in two ways, according to the manner in which the voices are signs [*notae*] of the disposition and desire of the one who produces the voices. They do this in one way distinctly, and in another confusedly. Those are susceptible to instruction which appraise distinctly the wish intended by the one who produces the voices. There are many of these and they possess this capacity only through instruction and not from nature. Those which do not perceive the wish itself but perceive some measure of anger or kindness attendant upon it certainly flee or approach the one calling, but they are not susceptible to instruction, except in some very small way. And for that reason we see that dogs and some other animals learn many things. Moreover, the horse, the mule, and the ass are discovered to be more dull-witted in this, although we have seen before horses, asses, and geese which learned some games and mimicry. Small animals, however, are less perceptive and teachable in these things than are large ones. Nevertheless, not long ago there was seen a mouse (which is called a rat and is large), which, sitting erect, held a candle on the table.[39]

It is especially the case in animals that the estimative power is composed from the apprehension of the phantasm and the appetite of desire, and it is present in a sensible soul in the same way that deliberate choice [*prohaeresis*] is present in rational creatures. And when this is especially pure and belongs to a good organ, then it is made to have good judgment of the intentions of things. I say, moreover, that it is pure when its spirit is pure, clear, and its heat is temperate and it is neither sluggish from cold nor confusing its operations from excessive heat. A good organ is not melancholy which receives spirits with the same clarity with which they are carried to it in the spirit. Now, these animals are generally better at discerning sounds which occur through

---

38. "Voices": *voces*. While elsewhere A. discusses the *voces* of animals, in this particular discussion he reserves *sonus* for the sounds or calls of animals and *voces* for the voices of humans. Humans alone are capable of true speech, or *sermo*, according to which the meaning of the speaker is not only conveyed to but understood by the listener. However, while the meaning of human speech may elude the animal, nevertheless, it often will perceive the *intentio* behind the speech. Thus when sheep follow their shepherd when he calls, they perceive his *intentio* but not the meaning of his words.

39. Cf. 8.299; 21.13.

hearing. It is for this reason that those with an earthy melancholy, as if lacking interior light, do not make good judgments about the intentions which are received in sensibles. The others of which we spoke are related to these in the opposite way, and for that reason they make better judgments. And since sounds and voices are both the signs of the dispositions in the animals making the sounds and resemble the signs of actual things [*res*], these animals which have the more lucid estimations make judgments from hearing this concerning both the things [*res*] and the desires of the ones making sounds. The ones which estimate less well do not judge the things but only the ones making the sounds, and they do so only confusedly.

23 Similarly, much consideration ought to be given to the memories of animals. Now, indeed, in the human we see that a power of this sort exists according to four virtues, the first of which exists with respect to holding well the forms received in sense. And another exists with respect to holding well the intentions drawn out from sensibles through estimation. And the third exists with respect to returning well to the sensibles that are received through recollection. And the fourth exists with respect to returning well to the intentions which have been drawn out from the sensibles. Any animal participates in memory well in accordance with all of these and is well capable of instruction if along with this it participates in hearing and has good estimation. Any animal that participates in all of these badly is not teachable. And any one that participates in some of these well but some of them badly, or if it participates in all of them indeed but does not participate in these in accordance with all the powers of them all, will be teachable only in a mediocre fashion. However, the four virtues which we say exist in memory are generally reducible to two powers, since the first two arise from a dry, tempering cold, while the other two arise from heat and moisture among those that are temperate. And for that reason they are said to belong to recollection rather than memory. We have spoken of these, moreover, in the book *On Memory and Recollection*.[40] Now, it often happens that animals with a good memory have poor recollection, and vice versa.

24 A capacity for instruction in animals, however, is more attendant upon recollection than upon memory, and especially with respect to that part that is capable of recollecting intentions. An indication of this is that animals instructed to do tricks, in response to the nods and commands [*sermones*] of their master, perceive his will and do those things which they are commanded to do according to the intention of their master. The fact, moreover, that they use hearing and recollection to apprehend their master's will and disposition rather than their master's thought is indicated by this: that sometimes they do not do the things they have been taught to do unless there are threats or blandishments or there is some sign of fear or blandishment. This is why they act to bring about the outcome desired by the one making something known in his voice, and not in response to his thought. And for that reason they do not apprehend the articulation of the voice, yet still correctly complete the commands given to

---

40. *De memoria et recordatione* 2.6–7.

them, and they are moved from instruction, in the manner in which nature moves, rather than in the manner in which the apprehended and conceived intention of some things moves. This is because to be moved by a concept of this sort is a characteristic of reason; however, to be moved by a habit [*ex habitu*] that moves in a natural fashion is also a characteristic of a brute's sensible power.

Moreover, we see something similar to this in the mechanical arts, whose functions are best performed by those who have a habit from custom and who are ignorant of the true nature [*ratio*] of their work. And those who play the lute or work in glass or do something else from customary habit actually know nothing of the true nature of their work. There is a similar difference between a human and brute beasts moved by the habit of instruction. For brute beasts are moved as if by nature, not attending at all to a voice except as a sign of the desire of the one who calls, perceiving nothing at all from the articulation of the speech by which the master, who commands the works, addresses them. A human, however, does the things he is commanded to do according to the intention of the articulation which is in the speech of the one commanding.

Moreover, it was established above that animals sometimes are moved from the sensation itself to desirable things and, similarly, some of these are immediately obedient from the sensation itself regarding those things for which they have been instructed, on account of the compliant faculty which they have from their habit of instruction. Nevertheless, although every discipline consists of the three things mentioned, the most important is the recollection of intentions because with that returning well, all the things commanded may come to pass, so much so that some animals seem to have something of reason in such things, although in truth they have none because they do not separate the intentions to which they are moved from the sensibles received in sensation.

According to these, then, and others of this sort, and according to the degrees which are present in them, the capacity for instruction in animals is observed to be greater, lesser, or more modest.

CHAPTER V

On the Capacity for Instruction, and On the Utterances and Chattering of Birds

Birds seem to surpass other animals in all these things, because many birds are instructed even as far as the pronunciation of utterances and phrases.[41] And this does not occur in only one genus, but in many genuses of birds, and the parrot is superior to all the others in this capacity for instruction. After the parrot, the starling, magpie,

---

41. "Utterances and phrases": *loquelae et sermones*. A. employs here terms which normally would apply to human speech in order to emphasize the capacity certain animals have to imitate perfectly human discourse.

jackdaw, raven, and certain other birds are found to be subject to this sort of instruction. And all the powers of the sensible soul seem to be present in them, both those that are exterior and those that are interior. Besides, almost all birds have more vocalizations than other animals; but with respect to calling, chattering, and capacity for instruction, the aquatic birds seem to have less capacity than the others, and a lesser capacity for instruction is found in them.

Further, large birds chatter less and have a lesser capacity for instruction than small ones. We know, from the things that were established concerning birds in the preceding books of this study, that birds have a lighter flesh and nature than other animals, and the nature of air is abundant in them.[42] This is why it is fitting that birds have lighter and more subtle spirits than other animals throughout their genus. Now, forms are represented more clearly by the lighter and subtler spirits, and since spirits of this type are easily imprinted and changed by any form whatsoever, it follows that birds, throughout their genus, are subject to many images and have a multiple power of estimation.

27 However, we have already shown in what went before that all brutes receive the forms of sense and imagination only practically and not contemplatively. It follows, then, that birds have many desires due to their multiple forms of estimation. And the excessive chattering of birds is caused by the many desires. However, the production of speech occurs only from voice that has been produced broken down to a determinate configuration. That one, however, which by nature can make many noises can imitate many sounds, and this is the reason that birds of this sort imitate almost any sound they hear. And when they have appropriate instruments—for example, a tongue that is wide, long, soft, or flexible—they break up the noise of their call for the imitation of speech. Yet when speaking they do not conceive the articulation of the voice, but they only chatter, bringing forth the configuration of a word by imitating it.

Aquatic birds, however, almost universally have a sluggish, heavy, melancholic flesh, and especially those among them that are large. It follows, then, that the spirits of these birds should be proportionally the same, and for that reason they are less impressionable with respect to the forms of estimation. And, on account of the humor of the water which is their dwelling and which abounds in their food, they are rheumatic and for that reason too have an impediment to vocalization, because there abounds in them a catarrh-like superfluity, which, when it occurs in birds, is called *pytuita*. The sounds which they make when swimming in the water reveal this. Now, often they make a sound through their nostrils while catarrh is flowing down through them. The illness *pytuita* is so-called from an imitation of this sound, because they seem to make a *py* sound.[43] For this reason they have a not very clear call and have

---

42. 8.66f. deals especially with the cleverness of birds.
43. In fact, pituita is a Latin word of long standing which refers both to a slimy nasal discharge and to the disease listed here. The pituitary gland, in fact, was so named because it was believed to create nasal mucus.

few calls. The large aquatic birds also swim less than the small ones because they are heavier and more earthy than the small ones.

Because of the chilliness in the brain of such birds it follows that they are not very teachable. For there is there a cold and moist humor, a thick spirit, and a fluid just like vitreous phlegm, and this neither receives nor retains well. And with respect to those things which it has received and retained, it is sluggish and not very mobile. And it follows thus that they have powers of the soul in a proportionate fashion. This is why birds of this type are almost universally poor for instruction. Sometimes, even, as occurs in many, they have few noises, do not chatter, especially when they are large, except in the case of great accidents—as the swan emits some cries [*voces*] of the sibyl at the pain of death. A sign of this, moreover, is that birds of this sort are harsh sounding, and when they are angered they seem to puff up rather than to call, just as the swan, the goose, and other birds of this sort do.

Moreover, wild, nonaquatic birds have, as we have said, lighter and clearer spirits and have a long beak like two elongated jaws. And the upper beak closes on the lower one just like one jaw upon the other. Its hardness serves them in place of teeth, and the length of the tongue is proportioned to the beak and its width is proportioned to the fullness of the beak. For this reason, the tongue moves against the hardness of the beak just as the human tongue is moved against the teeth, for fashioning vocalizations. And this is why when they retain a type of voice in the sound of speech, they fashion similar voices and imitate the one speaking. This is why not all birds form voices but only certain ones.

It must yet be noted among these that the hardness of the palate contributes a great deal to the formation of utterances, and among the animals, birds have the hardest palate. And for this reason they speak better than quadrupeds.

Moreover, we see that in the genus of birds some are musical and some are not musical, and among the musical ones we see that some make very diverse sounds and word fragments, just like the *galandra,* the lark [*alauda*], and the nightingale.[44] We see, besides, that some of these produce only one song, as it were, and we have already established this in preceding books of this study. Now, this does not arise from the teachability of these animals, but rather from a multiformity of desires and from the lightness of their blood and spirits. And for this reason birds sing especially in spring and at mating time, when a moisture and heat of this sort abounds in them so that, indeed, when it is drawn off through intercourse and the weakness which occurs in them on account of the care they take to incubate and feed the chicks, then they cease to sing, unless they have abundant food, as does the lark and certain others. And because domestic birds placed in cages are not weakened from copulation and a concern of this sort, but rather food is abundant for them, they sing almost the entire year.

---

44. "Word fragments": *vocum fractiones*. In 21.27 we read that voice is "broken down" into recognizable fragments. Here the implication may be that birds can produce fragmentary calls or songs. The birds named are famous for their varied and beautiful songs. *Galandra*: cf. 23.37(26), where the *calandris* is said to be a larger lark than is the *alauda*.

But nesting is a sign of a greater skillfulness among birds than there is among other animals. For we see that no other animals construct houses as artfully as birds do.

These, then, are the things by which can be known the teachability and skillfulness of birds.

CHAPTER VI

On the Cleverness and Shrewdness of Aquatic, That Is, Swimming, Animals

30 We see, moreover, among the swimmers artifices by which they ensnare one another, as the octopus ensnares another octopus or an octopus ensnares a shellfish and a toothed fish ensnares the tentacles of the octopus, and some others ensnare one another by a marvelous shrewdness. Certainly, these artifices seem to have been constructed for treachery, and treachery has the power of great estimation and memory, and this same skillful treachery even seems to have some anticipation of the future. For unless the octopus foresaw that the shellfish would open the shell, he would not seize it, for he hides himself not far from the closed shellfish and wraps a rock in one of his tentacles placing it inside the shell after the shellfish has opened the shell. And because then the shell cannot be closed on account of the rock, the octopus sucks out the shellfish and devours it.[45] Similarly, moreover, the braize lies in ambush where the octopus swims, and bursting forth suddenly it cuts off the tentacle.[46] It seems that these things cannot be done without foresight, memory, and the shrewdness of great estimation. And we see similar things in other swimmers.

We see too that fish are domesticated to the extent that they gather at the sound of a bell and accept a bit of grain, which would seem to be impossible without instruction.[47] It follows, then, that these animals have that perfection which stems from participation in the motive powers, which are sense, imagination, estimation, foresight, and shrewdness of a certain inference.

31 Swimmers differ a great deal, nevertheless, with respect to these powers, both according to genus and according to species. They differ a great deal according to genus because generally they are more sluggish with regard to inference and estimation than are birds or quadrupeds. And swimmers are more dull-witted in the operations of the soul and the senses. And this is clear because neither sight in a hard eye, nor hearing in a closed ear, nor smell in a thick element, nor touch in a hard, cold scale can be subtle senses with respect to acting for sensation. Since phantasy and other interior powers of the soul are a certain motion produced from sense, these animals cannot have great perspicacity in the acts of these powers. Thus, it is necessary

---

45. Cf. Pliny, *HN* 9.48.90.
46. "Braize": *dentatus*, which is also described at 24.29(42), where it is called the *dentrix*.
47. Cf. 4.79; 24.59(133).

that they be generally dull-witted, and for this reason they are instructed very little and in few things. However, land animals, and especially birds, have all the senses disposed in a contrary manner, and so they are generally more subtle than the swimmers and, as a consequence, they are more perfect in the motive powers of the soul and they are capable of instruction in many things in which aquatic animals are not capable of instruction.

Among themselves, however, they display differences according to natural activities, just as was said earlier. Now, some of the aquatic animals have no cleverness, while others seem to have a great deal of cleverness, as we indicated above by examples.[48] And for this reason, generally, aquatic animals are more easily deceived by fishermen than the flyers are by fowlers, and there is one specific genus of fish that is more easily deceived than another. With respect to this difference, the aquatic animals that bear live young and breathe, such as the dolphin, whale, and the one called the *koky,* which the Latins say is a sea calf [*vitulus marinus,* seal], and others similar to these have more cleverness than the others. And those which are strictly aquatic, on account of the viscous phlegm abundant in them, are more dull-witted.

Moreover, there are differences of natural cleverness and activity in their genuses. Now although there seems to be in them some equality in the participation of any senses (because they all seem to have harder organs and, for this reason, seem to have poor and weak sense perceptions), nevertheless certainly we see that some of them throughout all their genuses move, and others are immobile with respect to place, just like some of the genuses of shellfish. However, we know from what has gone before that local motion is present only if the moving principles and the organs of motion are present through which motion is itself exercised. And, again, because nature does not abound with unnecessary things, it remains that there are no organs of motion and moving principles in those which do not move from their places. For this reason, those aquatic animals which move are more perfect, with respect to activity, than those which do not move.

Further, a well-ordered distinction among the sense organs exists for the sake of the ordered apprehension of sensibles. And this is how we know that those animals that have members in which sense is present, arranged in an orderly manner, apprehend sensibles in a more ordered way than those which have these members arranged in a disorderly fashion.

And thus we say that fish, throughout their genus, apprehend sensibles in a more orderly manner than those which belong to the *malachia* genus, such as the cuttlefish and the octopus. Moreover, it has been proved in what has gone before that sense moves and that motion begins from sense. And likewise it has been proved that disorder in the senses produces a disorder of phantasy and estimation. Now from these things it follows that fish generally throughout their genus are more perfect in the virtues of the soul than are those which belong to the *malachia* genus. Indeed, in the

---

48. Cf. 8.113f.

hard and, as it were, cartilaginous bodies, the soul does not have the balanced spirit and humor with which the soul introduces their powers as it has in softer bodies bearing a likeness to flesh. And since the bodies of all those belonging to the *malakye* genus are hard and approach the nature of cartilage, the vehicle for life, which is the spirit conveying in a nobler way the forms and virtues of the soul, will not exist in them. And neither will there be in them the fluid which is the basis for life, which serves the noble virtues and the receptions of the soul's forms. The different composition of the fish and the *malakye* is an indication of this. For the body of fish, both in terms of uniform parts and in the composition of nonuniform parts, is constituted in a way contrary to the body of those belonging to the *malakye* genus. With respect to a uniform part (which the flesh is, or that which is in the place of flesh), the fish grows and increases along the longitudinal diameter, along which growth proceeds from the heart to the whole length of the body. In the *malakye* species, however, there is growth in a circular manner from the center out, as if its body were not extended from a single point along its whole length, but rather as if it were formed by the addition of surfaces placed one next to the other, as if one surface is placed above another.[49] And this is a sign of an imperfect heat which cannot extend the subject in which it exists outward from a single point.

34 Moreover, in a similar way the fluid that is in the genus of fishes is more fluid and softens more those things which it nourishes. But in the genuses of the *malakye* it seems to be earthy, tending toward the nature of bone. Therefore, imperfect heat does not move well and a thick fluid is not very mobile. This is why it is necessary that these animals be constituted in a proportional manner with respect to the cleverness of the soul. For this reason too fish are shrewder in that genus than those which are under the genus of molluscs. In terms of composition, moreover, fish are composed from spines (which take the place of bones), and from flesh and nerves, and they have a head, a heart, a stomach, and other parts of this sort that are proportioned for those animals having shrewdness and perfection of the sensory virtues. The entire *malakye* body, however, is hard inside just like cartilage, although it might be soft outside. It does not have the right shape for a distinct head, intestine, or heart. And this is an indication that it is also defective in the way it participates in the interior sensory virtues, and thus it cannot be very well ordered for participation in the moving virtues.

35 Therefore, although some of these may be cunning and have some foresight toward gathering things to eat, like the octopus, or even toward the evasion of danger (like the cuttlefish, which, in a storm, ties itself by its tails to rocks so that it will not be cast away by the force of the waves and, when afraid, emits a black fluid by which it darkens the water so that it will not be seen), nevertheless these examples of foresight are found in all animals and are caused by the sensation of what is useful or harmful rather than from the precognition [*praecognitio*] of estimation.[50] Now it is necessary that these animals be imperfect and dull-witted in such powers for the reasons already

---

49. Reading *circulariter* with Borgnet for Stadler's suspect *circularitur*.
50. Cf. Ambrose *Hex.* 5.8.21. The cuttlefish's tails are its tentacles (cf. 8.124).

introduced. That foresight, however, which arises from the sensation of something useful or harmful nearby is not denied to any animal. Now, all avoid harmful areas by their sight or hearing, and certainly they use smell, taste, or touch to pursue favorable areas. Nevertheless, dull-witted and imprudent animals have a presentiment of such things more quickly than do animals that are shrewd and prudent. This is because the less any animals are occupied with their own impressions, the more easily impressions of other things appear in them, as we have said often. And for this reason the cuttlefish senses at its depth the vapors which are the cause of the storm, and for this reason too it provides foreknowledge of the storm when it binds itself to the rocks. In a similar manner, moreover, the octopus knows that a shellfish is nearby from some change the shellfish produces in the water, and then it seizes it.

Moreover, those which are called soft-shelled are, in every respect, more imperfect in their participation in the powers of the soul, and still more imperfect are those called hard-shelled, seeing that these are especially dull-witted and have little enough of cognition and prudence. Now although some have organs for the senses like eyes, nevertheless they have very hard eyes, harder than all other animals with hard eyes. The other organs for the senses are not seen in them, and they have a very weak sense of touch, so much so that they do not sense a moderate or light touch. The noble internal members are not found in them, like the heart, the liver, and the brain, and this is an indication of their confused souls, because just as the formative power which is in their semen produces a confused function in the noble members, so too the sensible soul in them infuses its powers into the members confusedly and produces a confused function in them through its powers. Still, it is necessary that these animals have sense and motion, insofar as motion begins from sense, and have very confused phantasies, estimations, and memories. This is because without any phantasy, estimation, and desire, motion would not occur, just as we have shown many times.

Moreover, the confusion of phantasy and desire in such animals is detected only when it is isolated with respect to sensation. Now if the phantasm had the power of determination in itself, and its desire were complete, they would be moved toward something that was absent. Now, however, they are hardly moved at all except by some nearby sensible that is somehow perceived by the senses. And if they should move without perception of this sort, they are moved in a very disordered manner. And this is an indication, as we said, of the confused estimation and imperfect desire of these animals. This is why they have very sluggish forward motion, which they perform with their feet. And if they have a motion for flight performed by swimming, like the lobster and crayfish, they have that speed to withdraw from something harmful. And when they move to seek out useful things, they move as if testing their feet, as is clearly seen in the lobster, crayfish, crabs, and all aquatic creatures of this sort. This "test" they perform with their feet is a sign they are not confident in themselves on account of a poor estimation and cognition of those things toward which they move. These, then, and those like them, have just such a participation in the powers of the soul.

38   The sea shells [*ostrea conchilia*] have, nevertheless, an even more inferior degree than these throughout all their genuses (which we have identified as four in what has gone before).[51] These are very sluggish in their motion and little of the sense organs is seen in them, except perhaps taste and touch. And for this reason they move just a little around the sea floor for sucking out fluids, just like plants. And what the division and stretching out of the roots provide in plants (so that they may absorb nourishment from a wider area), the little motion which occurs in animals of this sort provides for them. They move themselves from place to place just like a snail, not through the instruments of motion (which are feet, wings, or fins) but rather through expansion and contraction movements. Thus, these animals are more imperfect than all the ones introduced above.

    Yet still more imperfect than these are those which exist in the manner of a sponge and are more like plants than those of which we have spoken. And these belong to three genuses, as it were, seeing that some of these, like the starfish, are just as if midway between the *malakye* genus and a plant.[52] Others, however, like the mobile sponge, are much nearer to a plant and seem nevertheless to approach in some respect the cuttlefish. And still others are right next to the plant and have little of "animality," like the immobile sponge, which does not move from place to place but palpitates in place while expanding and contracting.[53]

39   Now, the starfish seems to occupy a certain midpoint between the octopus and a plant, even though the octopus surpasses the starfish in the number of tentacles, because the octopus has eight and the starfish only five. And the octopus has long tentacles and the starfish has short ones. Yet they are alike in this respect: that each one—both the octopus and the starfish—sucks through the whole body, because along the whole length of all its tentacles the octopus [*polypy*] (that is, the *polypes*, which is the same thing) has round pores which are certain "mouths," as it were, through which it sucks and has the strongest suction. An indication of this is that when it clings to some animal, and especially when it clings to a human, as soon as it wraps the tentacles around him, it draws out blood everywhere. The starfish, however, does not have open, round "mouths" of this sort, but it has a cleft along the entire length of all its tentacles in which there are some small, hard little veins [*venulae*] in the manner of small roots, and it sucks through these.

40   Besides this, in the center of the octopus there is an opening coming from below, and above it there is a hollow sack in which there is a black humor, just as there is in the cuttlefish. And at its extremity is the head of the octopus, and in this its eyes are fashioned. In its interior there is a member which has a horn-like substance, but is very thin and soft, just like the soft nails of small boys. This one has a black or dark color and has a shape just like an eagle's beak, and it is composed of two beaks, lower

---

51. Cf. 4.35 for the four kinds of *ostrea conchilia*.
52. On the starfish [*stincus*], cf. 4.4, 73; and 24.55(121).
53. "Animality": *animalitas,* cf. 4.40 and 16.62, with note.

and upper, in just the way an eagle's beak is composed, having an upper beak curved down over a lower one. And it is not for biting or tearing, because it is very flexible, unless perhaps it should seize something very soft with it. The starfish does not have such a formation in its middle but rather is shaped in the manner of a root that has five divisions. Each one, nevertheless, has the movement of expansion and contraction, but the starfish has less "animality" and more from the nature of the plant. Conversely, the octopus actually has more "animality" and less of the nature of the plant.

Nevertheless, the sponge which moves from place to place is closer to the nature of a plant than the starfish is. And the sponge that is immobile with respect to place is yet closer, so much so that it seems that it is a certain plant participating in something of "animality." However, there is not found any midpoint between shellfish and plants, on account of their extreme nearness to plants. And this is why they suck nourishment just like plants and do not have a place for the exit of superfluities, just as plants do not.

This is why some philosophers said that such animals have no phantasy whatsoever, because for those senses in which they participate, they participate more in a natural way than an animal way. Now Isaac the Israelite philosopher said that taste and touch are present in plants because these are natural senses more than senses of the soul, because they do not occur through an extrinsic medium, nor do they apprehend a species separated from matter, but rather one always conjoined to matter.[54]

These animals, then, and those like them, are imperfect, because they participate imperfectly in the powers of the soul, both those that are interior and those that are exterior.

Moreover, one can know easily the type and degree of imperfection in other swimmers and aquatic animals from the things which we have said.

CHAPTER VII

On the Prudence and Shrewdness of Serpents and Creeping Animals and on Their Perfection and Imperfection

Let us speak now about the creeping animals, both the aquatic ones and the land ones, and see what kind of perfections or imperfections they have. We will establish the creepers and serpents together, because it seems that their nature is close.

We call creepers those animals which, although they have feet, nevertheless creep more with these than they walk, because their feet as well as their legs and tails lie on the ground when they walk, as is clear in the lizard, the newt, the tortoise, the sala-

---

54. "Isaac the Israelite philosopher": i.e., Isaac Israeli (ca. 855–ca. 955), the author of the important *Liber febrium*. Cf. 16.27, with note.

mander, the genuses of crocodile, and other quadruped animals of this sort. However, we call those animals serpents which, having no feet at all, glide with a motion of the body by means of contortions or by means of the movement of the ribs, as do all the genuses of serpents across all their differences, which are very numerous.

All of these are found to have some prudence and shrewdness. Among them, however, the serpents seem to excel in these respects in cleverness. An indication of this is the ambushes that they make. They are not very receptive to teaching, although Pliny tells of a certain head of a household in Egypt who tamed a wild asp which each day came forth from a cavern and accepted its accustomed grain from before a table.

43 It follows, moreover, from all the things introduced above that all the cunning of this sort is caused by a thin fluid of these animals. Those living in the cold do not have heat mixing in with their functions, but rather a complexional heat dissolving them. And for this reason their animal spirits are clear, which is why they become shrewd in cunning. This is also why it is necessary that the perfected ones of this sort participate, in this respect, in both the interior and exterior powers of the sensible soul. Now, all have both exterior and interior senses, and a very clear estimation, that is, one that belongs to a clear spirit. And this is why, in the opinion of some, such animals are more clever than other, larger animals.

Any of these which are aquatic are less clever, because of the cold that congeals the spirit and fluid. Those, however, which creep, even though they are quadrupeds and even though they have all the senses, flourish less than the serpents in estimation and memories, on account of their greater earthiness and less subtle fluid. Nevertheless, all these animals of this sort seem to have this imperfection: they do not have perfected organs for progressive motion. Now, the mobility of the ribs arises because the ribs in serpents are immediately connected to the spinal cord. Moreover, the power of progressive motion arises from the heart and is directed from the spinal cord to the nerves arising from it. Now, if this power were perfect in these animals, just as it is in walkers and flyers, not only would it strengthen those things with which it is immediately connected but it would even form organs for the feet, and movement would flow into these just as it does in others.

44 However, an additional reason for this is that it happens for the best [*propter melius*]. For animals having very long bodies of this type cannot be assisted in carrying the body with only a few feet. Which is why any of these that are shorter, like the lizard, crocodile, and the salamander, creep on four feet and, so that the larger part of the body is raised up, they drag the entire length of the leg upon the ground. This is because if these animals were to fix only the extremity of the foot on the ground, the long part of the body extended between the anterior and posterior feet would hang down as an impediment to motion. Nature, however, always proceeds according to the better way.

The serpents, though, differ from creepers of this sort in that the serpents raise up the anterior portion of their body when they wish to, which the creepers do not. But they agree in this respect: after the head has been turned around and rotated, both one and the other look backward along the entire length of their body. Nevertheless, this does not appear in short ones of this type, like the frogs, the toad, and the genuses of creepers of this sort.

Further, however, all of these that are long live for a long time after they are cut up, on account of the uniformity of their body throughout almost all its composition and complexion. This is why, moreover, tails that have been cut off grow back a little bit.

These animals, however, and those of this type that are scaly, change their skins frequently on account of the hardness of the scales when they grow older, and they cast them off, just as the crayfish changes its shell and the birds change their feathers. And then they become more sensitive. And this happens to them on account of the similarity between the nutrimental fluid and their radical or seminal humor and, just as occurs in trees and plants, their youth returns to such animals according to the manner mentioned.

These animals, then, and ones of this sort, have such motions and perfections.

CHAPTER VIII

On the Shrewdness and Perfection of Ringed Creatures

The ringed creatures, moreover, besides being hard outside and very earthy (especially the genuses of bees), indeed participate in all the senses and in estimation and memory. The genuses of flies are an exception. And the spirit created from the subtle humor circulates in them, and for this reason they are very shrewd. They are not subject to instruction, however, by means of teaching, because they cannot hear the sounds which are the names of things in the manner of a concept, although they do perceive the intentions of things.

Now, sound is heard in two ways: namely, in a simple way, insofar as it produces a sign of the one making the sound; and then insofar as it is imposed upon a thing for the purpose of designating a name. Now, animals of this sort perceive sounds in the first way, but not the second. The cause of this, moreover, is that the name of the thing is the result of someone's intellect imposing a name upon it, and the very form of the thing is simpler, and it is on account of this that these sounds are not perceived by cold animals of this type with that conception, although they are perceived by other animals. And for this reason animals of this sort are prudent in estimation and in the memory of what is useful to life and of benefit to life, but they are not susceptible to instruction through teaching, as are dogs, monkeys, and other such animals.

With respect to progressive motion, almost all such animals are in accord with the shelled walking animals in that they have many feet, and in that all these are attached below, on their chest, under the area of the member which takes the place of a heart in them.

46 The reason for this is that motion arises from the heart, just as from a first mover, and is diffused into the entire body through heat and spirit, as we have shown in the book *On the Principles of the Motion of Animals*.[55] Since, then, all such creatures, such as the shelled creatures, are cold, if the organs for moving forward were attached far from the heart, they would be rendered immobile and useless for motion. This is because every power of a mover is stronger when nearby than when afar, and this is the reason that frequently in such animals the base of the legs is joined with a strong compression so that they may approach the source of the heat. And all these are frequently girded, and the posterior part of their body has no fixed spot for the leg, as is clear in large and small bees, wasps, spiders, locusts, flies, and all those of this sort. And in shelled creatures (soft-shelled as well as hard-shelled) it is similar, as is clear in crayfish throughout their genus and lobsters and in all the walking shellfish.

But there is something remarkable in the caterpillar and the ringed vermin like the caterpillar. For these have a double generation, as was established in what has gone before, and they have a double motion in the first generation: they reach distant places by a viscous thread which they draw out from a moisture in their mouths, drawn along or suspended on that thread. In this they differ from the spider, which for this purpose draws the thread from the anus.

47 Further, however, these animals are seen to use webs woven from these threads only as a nest to guard their eggs. The spider, however, uses a web woven from its threads for hunting. However, these animals have a second motion in their first generation on very short feet fixed along the entire length of their body, by which they move to nearby places in order to feed. In the second generation, when they are again generated from an egg-like nature to become flyers, they do not seem to use the web and the thread at all, or little, seeing that from the beginning of their leavening, when the natural heat first begins to move and mix the fluid of their body toward an egg-like nature, some of them, declining, just as some caterpillars and all the silkworms, weave something woven thickly with thread wrapped around themselves from which silk is produced.[56] The heat in it leavens the moisture into the nature of a flyer which, when it has been born, flies out through an opening which it makes with its mouth. And then it receives long feet, all attached under the chest, because then at last this animal species has returned to the perfection of its species.

48 Moreover, all animals of this sort seem to have all the senses, although among many animals they have a weak sense of touch, and weaker than those having blood. For this reason Origen is deceived when he says that the spider has an excellent sense of touch.[57] All such animals, however, have taste by means of a long instrument which exits from their mouth just like a certain proboscis. They especially excel in smell,

---

55. *De principiis motus animalium* 2.6.

56. "Leavening": *fermentatio*. A. seems to liken the development of the moth or butterfly in the cocoon to the process of baking in which a moist material takes shape under the influence of yeast, expands, and hardens. "Declining": *declinando*. From the Latin it is unclear who or what is actually declining.

57. The source in Origen is unknown.

however, on account of the dryness of the head and the member in place of the brain, for odors occur for such animals on the air, and they move about flying for a long time to the odors of flowers. And this is especially suitable for them when they are changed to the perfection of their species in the second generation, for it is then that they collect the dew infused in flowers. Before this, in the first generation, many of them are nourished by the leaves of trees and of herbs. An instrument of hearing, however, is not viewed in them by dissection. Nevertheless, they perceive sounds through the pores of the anterior part of the body, as was demonstrated, because they flee certain sounds and many of them are drawn to the tinkling of metal objects being struck.[58] The instruments of sight, however, are found in them by dissection, and their eyes are hard and uncovered, since soft eyes would be injured when they are borne flying through the air by a rapid motion. And this is why these animals are said to be perfected as far as the participation of the senses is concerned.

Moreover, the power of the animate is said to differ from the power of the inanimate with respect to two things which clearly belong to life: namely, sensation and motion. They participate in various motions, seeing that at their last generation they fly by means of more than two membranous wings and walk with more than four feet. And, many of them, living as multipeds after their first generation, both walk and are drawn along on a thread, which signifies in them an imperfection of their motive power which is unable to complete movement by means of a member of one shape (as feet on walkers and fins on swimmers) and these creatures are unable to complete a motion by means of a member of only two shapes (as there are both wings and feet on birds). But the power makes one leg longer than another and one thicker than another, and it acts in a similar fashion with regard to the wings of some of these, which do not have wings that fold together upon the back but rather extend from above, with one lying upon another, just as the flying caterpillar does and every genus of silkworm.[59] And all of these have some legs longer to the rear for pushing off when they are raised up for flight. And some of these have longer ones at the front as well, as if for touching and feeling their way to where they might rest and find food. They have legs of medium length, moreover, in the middle, which are simply there to hold up the body. And the genuses of the flies and *musciliones,* which, nevertheless, have two wings attached above the back, agree with these in such a disposition across all the legs.[60]

These, then, have such a perfection of this type.

---

58. From antiquity on, this process, called tanging, was widely believed to attract swarming bees. It does not work as bees cannot, for the most part, hear airborne sounds. For a review of the evidence and a possible source for this belief, cf. Kitchell (1989).

59. "Flying caterpillar": apparently the adult, "perfect" stage, which we call a butterfly. The terms appear also at 17.79. Elsewhere, A. uses *verviscella* (cf. 5.32).

60. *Musciliones*: cf. 1.96 and 4.68.

CHAPTER IX

On Imperfect Animals, and the Reason for and Difference of Their Imperfection

50   Those animals, however, which seem to be imperfect throughout their genus (in members as much as in participation in sense and motion) are certain genuses of vermin, like those which are called the earthworms, which seem to be born more from the vapors enclosed under the earth or from the putrescence of roots than from the seed of male and female. Now, in these none of the distinct organs of sense is seen except touch. Moreover, taste is perfected in them only by way of sucking, just as the roots of plants suck humors. And for this reason only one path for food is found in ones such as these, extended along the middle of its length, just as there is a path for sap along roots. One does not find in them the various digestions, members for digestion, anything in place of these, or any vessels (that is, vestigial intestines), but the entire body is composed, throughout its length, of rings. And there is in the middle a ring having a greater thickness than others, to which the anterior and posterior rings are gathered up in motion. They do not move with feet or wings, but rather by contracting one part of the end along the length of the front half of the body up to the middle ring, and extending the other part from the same ring. They do not always extend one and the same part, but sometimes one and sometimes another, because they move toward each end of their body. They neither see nor hear, unless perhaps the sounds are made by striking, but then they perceive the striking by touch and not by hearing. Moreover, these seem to smell nothing at all nor to have taste, in the sense in which it is the judgment of flavors, but only in the sense in which there is a sense of nourishment, and this is indeed touch. This is the reason, moreover, that they absorb nourishment by sucking. Neither do they seem to emit excretions, just as plants do not. But sometimes one of these is found extended toward another on the ground along the front half of the body and, drawn straight out, it is fitted to it as if one sought heat from the other. Some think that this is the copulation of such animals, which nevertheless is not true in any way [*nulla ratio*] because no seminal paths are actually found in them, nor do they grow moist during such a union as is found to occur in serpents when they fold together, copulating. Avicenna opines that eels are generated from these, however, when they are near muddy waters.[61] And if this is true, then it is necessary that these creatures are the material seeds and eggs, as it were, for the generation of eels.[62]

51   These, then, and those like them, have just such a type of imperfection.

---

61. Avic. *DA* 5.9r (?) and 6.12r.

62. The generation of eels rightly fascinated authors and no male or female was thought to exist in the genus. Cf. 4.103f.

Truly, the genuses of sponges which, out of all the senses, receive only a very confused sense of touch, are more imperfect than these. And they have only the movement of contraction and expansion and they are borne along in water by this movement only accidentally from place to place. An animal moves by itself (that is, through itself) which has an organ by which it conveys itself along the grades of progressive motion. However, progressive motion occurs only by means of such contraction and expansion in water when the water is pushed by the expansion. Then the animal is conveyed to another place by the power of the water, like wood that is floating and just like a person pushing a boat, and conveying themselves accidentally [*per accidens*] to various places in it. For this reason those of this sort remaining on the ground are fixed to the ground, immobile with respect to place, although under the ground they expand for gathering sap and they withdraw from unsuitable things. On account of this these are not animals in the true sense, but they live midway between animals and plants.

They have something in complexion as a means of touch, although very little. And this is why they have touch like one asleep and not by means of evident life, the way touch exists in wakefulness.

In composition, moreover, they have almost nothing at all in common with animals, since they do not display the shape or operation of animals in either the interior or exterior members. They therefore have great imperfection and they participate least in the nature of animals.

In these things one is led to an end, as far as a science of the bodies of animals, and through these animals, although they are imperfect, the whole science of natural things is perfected, with the help of God.[63]

---

63. A.'s nod to divine assistance in this enterprise is quite unusual, appearing previously only at 16.151.

HERE BEGINS THE TWENTY-SECOND BOOK ON ANIMALS

Which Is Especially on the Natures of Animals

The First Tract of Which Is on the Human

CHAPTER I

On Human Intercourse, Ordered for Generation

1   At the beginning of another book we shall set down certain other natures, putting each under their appropriate names specifically and following the order of our alphabet. For although we have said above that this method is not appropriate for philosophy (since it is necessary when using it always to repeat the same thing), we will append just such a tract at the end of our book since we feel we are under obligation to both the learned and the unlearned alike, and since we feel that when things are related individually and with attention to detail, they better instruct the rustic masses. We will begin with those animals which are designated by [each?] particular letter.

We will begin with the class [*ordo*] of the more perfect ones. Since the human is the most perfect of all, we will deal first with him, assigning both the shared traits and the specific traits peculiar to him. Although we may have already indicated the causes of these traits in previous books, we will gather them here again with no need for study in depth.

Human reproduction is universally by means of intercourse in which the powers of the sexes are mixed together both out of the sperm of the man, the creator and maker, and out of the woman's sperm, or *gutta,* and her menstrual blood, which are the materials, as it were. As Constantine of Monte Cassino says in his book *On Intercourse* [*De coitu*], indeed the Creator, wishing the race of animals to remain in a stable and fixed manner and not to perish, saw to it that the race would be renewed both through intercourse and through generation, so that it would not take its renewal from total destruction.[1]

2   Then, as Constantine says in the subsequent parts of the same book, the Creator completed those natural members which would be suited for this task and he also added to them a wondrous power for bestowing pleasure. For if the animals were to shun intercourse, the race of animals would surely die off, since they are weighed

---

1. The reference is to Constantine the African, who was born in Carthage but settled in Monte Cassino, where he died in 1087 (Sarton, 1927–48, 1:769). See his *De coitu* (1536–39, 1.299).

down by pregnancy, have pain in giving birth, and are busied with the care of nursing the ones born.

Since, moreover, all these things are first derived from the soul, it therefore likewise follows, he says, that three things are necessary in intercourse—namely, the imaginative appetite, spirit, and humor. To these Avicenna adds windiness [*ventositas*] as a fourth.[2] Of these, as Constantine says, the appetite arises in the liver, the spirit in the heart, and the humor specifically in the brain but generally from the body as a whole. Windiness, however, arises from heat dissolving moisture into windiness and directing it to the reproductive members.

Although these things might be said generally of the reproduction of animals, they are especially true of the human, whose nature is worthier and whose complexion is nobler. Therefore, intercourse in humans is accompanied by more pleasure than in other animals and in human intercourse the apprehended form of a sex occurs such that the lovely image of a woman comes to a man as presented to the eyes of his heart and the apprehended image of a man who is desired drives a woman to a desire for intercourse, and it is this, among those things which work toward accomplishing intercourse, which provides the first stimulus, as it were.

The semen as it descends is obedient to this, much as the power of the material is obedient to a power serving as a principle. And when it descends to the seminal vessels, heat evaporates part of the humor that is with it and releases it into a windiness by means of which it extends the reproductive members. Since these members are of different sexes, the male's members swell and grow rigid, whereas the female's open and swell, although less than the male's. When this windiness is abundant and dry, the male's penis becomes bone-like, sometimes so much so that the channels through which the sperm should be ejected are blocked up and it does not exit until the windiness has abated a bit. And it is remarkable that suddenly accidents of the intellect are at work in this business, for once he withdraws the focus of his thoughts from the desire for intercourse, the windiness abates and the semen exits. Satisfactory determination has been made concerning these and similar matters in preceding books, however.

CHAPTER II

On the Quality of the Seed in the Womb

According to the testimony of Galen, human semen is the pure, warm, moist substance out of which a human is made. Thus, it is that this same Galen, in his book *On the Care of the Members* [*De cura membrorum*], says that semen is a spirit and a frothy humor. This same Galen says, moreover, in his book *On Sperm* [*De spermate*],

---

2. Avic. *Can.* 3.20.1.1–2. The windiness acts, as is made clear below, as the force expelling the male sperm.

that semen is derived from the humor of the body as a whole and comes to be from the most refined nature of the four humors of the human body.[3] There are nerves and particular veins which draw it from the body as a whole to the testicles.[4] Both the veins and the nerves, as he says in the same place, bring it forth complete and animated. Then, having been brought in contact with the *gutta* of the woman, the sperm is received into the womb and it grows with the heat of the womb and with the spirit that is drawn to it through the thin arteries. This exterior spirit does not allow the interior one to escape and binds it to the task of nourishing and warming the child. Afterward, as Galen says in the same place, another spirit comes into the womb and turns the sperm and *gutta* of the woman to a fleshy nature with the help of the heat of the womb.

5 Galen says in the same place that the spirit which is in the sperm has three particular powers, the first of which is a necessity, the second a virtue, and the third rather like an organ.[5] Necessity has two functions, for it watches over the heat and is also the nourishment for the animal spirit, both of which are necessary. Power is present in that which brings about the completion of the formation. It shows itself to be an organ in that it flows to everything as a result of the operation of the soul, which is in the sperm the way an artisan is in his creation.

The sperm, or *gutta,* of the woman is, moreover, colder than that of the male and acts, as it were, as a nourishment for the male's sperm. When an embryo grows out of these, each and every thing comes into its own particular nature and property. Thus, like is joined to like—point to point, finger to finger, artery to artery.[6] This is what Empedocles called the necks of twin heads, which, he said, are joined in reproduction through the power of love.[7] This all occurs in the womb. For as Constantine says in the *Pantegni,* the womb embraces that which it takes in and closes up to such an extent that, as Hippocrates says, the point of a needle cannot enter.[8] Thus, the interior opening often remains closed until the form of the infant is completed.

---

3. *De cura membrorum* is cited in Constantine the African's *De coitu* (1.299–300). Cf. also the pseudo-Galenic *De spermate* 1.

4. St. is corrupt at this point, and we read, with Borgnet, *a toto corpore ad testiculos* for Stadler's *a toto corpore ad toto corpore ad testiculos.*

5. *De spermate* 12.

6. "Point to point": *acutum cum acutu.* Scanlan's "nail to nail" is tempting, but unparalleled. For *in acutum* as "coming to a point" in later Latin, cf. *TLL* s.v. *acuo,* 1.468.

7. Stadler suggests as a source frag. 61 of Empedocles (Diels, 1951, 1:334; Freeman, 1963, 58–59). Although it is not close to this passage in wording, it certainly shares its spirit in speaking of what happens when disparate parts of the human body are not joined together through the force of love over strife. Scanlan suggests frag. 17, citing it from Wright (1981, 166). The Latin here (*cervices capitum germanorum*) is less than lucid and one is left with the impression that the text has suffered somewhat in the transition from the original.

8. The citation from the *Pantegni* cannot be located. On the needle, see 10.56. The closure is that of the mucous plug, which is created in the womb following conception.

CHAPTER III

On the Disposition of the Seeds and of the Young That Arise from Their Parents, the Humors, and the Heavenly Signs

For this reason Constantine, in his book *On Intercourse,* cites Hippocrates in giving the reason for sexual variation at birth as the causes present in this sort of seed. For he claims that Hippocrates said that if a male sees during his puberty that his right testicle is more prominent than the left, he will produce male children.[9] If the left one is larger, he will produce females. Galen, however, in his book *On Sperm,* says that if the sperm should fall on the right-hand side of the womb during intercourse, the fetus will be male owing to the warm, moist liver and the warm, dry gall bladder that are in that area. If it should fall on the left-hand side, however, the place will, on account of the cold, dry spleen, contribute to the production of females. It is even possible, as Galen says in the same place, for the semen to reach such weakness and confusion that the child will be of both sexes.

For the womb contributes greatly to the disposition of the children. According to Galen there are seven receptacles within it and it will conceive just as many children as the number of channels [*vasa*] into which the semen falls. And if the womb is long and narrow, the fetation will be long and slender; if it is long and thick, the fetation will be long and thick; if it is short and narrow, the fetation will be short and narrow. This is true frequently but not always. For these are no more than contributing causes and, with regard to these same causes, Galen says in the same book that if the sperm of the male is more abundant in the fetation than that of the female, the fetation will be like the father, both in its face and in its other members. If, however, the sperm of the mother is more abundant in both power and quantity than that of the male, the fetation will be like the mother both in its face and in its other members. But all these matters have been carefully disputed in previous books.

Galen also says in the same place that it sometimes happens that the powers of one humor are more abundant than the others in the sperm. The fetation then follows that humor and will be sanguine, choleric, phlegmatic, or melancholy, and in no way like its parents. Sometimes one or another power is lacking and squinters, lame ones, or those otherwise afflicted are produced. At other times the mother or father is beset or is occupied with various passions and the child acquires similar passions from the sperm. It also occasionally acquires similar passions from its nourishment of milk.

Galen and Avicenna say that the child sometimes follows its parents to such a degree that it even follows the accidents that befall their souls [*accidentia animae*].[10] Thus a certain king once imagined a black monstrosity during intercourse and made

---

9. *De coitu* 1.301.
10. *De spermate* 5. Cf. Avic. *DA* 9.20v.

mention of this to the queen while he was having intercourse with her. The fetation, when born, was monstrous and black.

Moreover, Galen even makes the statement that the sperm is so readily changed that, just as bodies change according to the four Hours of the day, the various times of months and seasons of the year, and also according to the paths and relationships of the planets, so, too, are seeds conceived in the womb changed.[11]

If it were otherwise, all men would have had one and the same complexion from the beginning. Hippocrates says, as does Galen in the same place, that all substance is bound together and joined together in the planets, the constellations, and the linkages of the four elements. Thus it is that Nectanabus, the natural father of Alexander, coupled with Alexander's mother, Olympias, when he observed that it was the time of the sun entering Leo and when Saturn was entering Taurus, for it was from these planets that he wished his son to receive shape and power.[12]

CHAPTER IV

On the Harmful Effects of Intercourse

8 Since the seed is derived from the most suitable bodily moisture, Constantine says that those who engage in a great deal of intercourse die the quicker for it, and Avicenna agrees with this.[13] A person can copulate so much that nature does not have the seed to eject and emits blood instead. This destroys the body, causing the person to grow old quickly. This is why eunuchs sometimes live longer than those who have not been castrated. When excessive sperm has been emitted, it especially weakens the brain and those members near it. Thus, Hippocrates suspected that the semen descends from the brain and he put forth still another proof of this.[14] For those who have the veins running behind the ears cut into and cut through can no longer produce sperm. If these people emit anything during intercourse, it is a watery humor and is not effective with respect to reproduction.

These, then, are the facts from which the properties of human reproduction can be drawn. Many things have been said in preceding books and they should be given more weight than these.

---

11. "The four Hours of the day": this does not appear to be a reference to periods of time but rather to the four Hours, or *Horae*. These somewhat vague deities are most commonly depicted as daughters of Jupiter and Themis who presided over the changes of the seasons and kept watch at the gates of heaven. Their names, number, and parentage varied by location and epoch throughout antiquity.

12. Thorndike (1923, 1:551–65) treats this story at some length. Nectanebus (there are many forms of his name) is most commonly depicted as an Egyptian astrologer who was the confidant of Olympias, Alexander's mother. Descriptions of the extent of his role in the future king's conception vary widely.

13. See *De coitu* 1.304 and Avic. *DA* 9.19r.

14. *De semine* 2.

CHAPTER V

On the Natural and Divine Properties of the Human

The most outstanding human characteristic is that which Hermes relates in writing to Asclepius, namely, that only the human is a point of union between God and the world.[15] For the human has in himself the divine intellect and through this he sometimes is elevated above the world to the extent that even the material of the world follows upon his thoughts. We see this in those best-born men who use their souls to bring about a transmutation of worldly bodies and are thus said to perform miracles. Moreover, even in that very part by which a human is bound to the world, he is not subordinated to it but rather is set over it like a helmsman. This also is the source of spells by which the soul of one person works toward the harm or betterment of another through his vision or some other sense.

It is, therefore, just as Hermes attests, that if an individual human should, of his own free will, make himself subordinate to the world, discarding, as it were, the honor of his humanity, he takes on the characteristics of a beast. Through concupiscence he is said to become like a pig; through anger, a dog; through plunder, a lion; and likewise for the other vices. This is what Plato called the second incorporation of souls.[16] If, however, a person perseveres, using the height of his mental power, he draws to himself both his body and the world, for the soul is born to instruct the body and the world. If, however, of his own free will, he subordinates himself to the body, then when the accidents of the body transform the soul, he takes on the characteristics of corruption and the soul does not restrain the body and, broken and bent to the body, it passes quickly through mental images [*imaginationes*] and passions and hastens the corruption of the body.

Avicenna therefore says that imagining red colors increases the motion and flow of one's blood and that one who is saddened and fearful of leprosy will sometimes contract it. Many other things of this sort befall people.

Sometimes the shape of one member signifies the dispositions and afflictions of another, such as when a humped shape of the nails foretells an abscess of the lungs.[17] Also, a particular movement of the brain brings on forgetfulness, as Galen relates in his book *On Accident and Disease* [*De accidente et morbo*]. He relates that Popchydyus

---

15. The work in question is the "holy book of Hermes Trismegistus, addressed to Asclepius" (Scott, 1924, 1:286–311). The exact passage in question is harder to pinpoint. Stadler cites, and Scanlan follows, para. 6a (Scott, 1924, 1:294). There are two other letters addressed to Asclepius as well. For A.'s knowledge of Hermes Trismegistus, see especially Sturlese (1980).

16. Cf. *Timaeus* 42b and Chalcidius's commentary (37, 24). A. takes up this doctrine again at *De anima* 1.2.7.

17. Scanlan (66) points out that this is often a sign of suppurative diseases of the lungs. For other signs one can gather from nails, cf. 1.504f.

the philosopher says that once, after all who were in a particular city had been freed from a pestilence caused by air and corruption, they fell into such a state of forgetfulness that they did not know their parents, their friends, or even their own names.[18] For the soul and the body change each with the other and it is the natural order that the soul should restrain the body to keep it from being destroyed. And it is by reason of this continence that the accidents of the soul change the body. The pleasurable, imagined form moves the body and, conversely, the sufferings of the body serve to disturb the soul, as is apparent in sleep. Determination of these and other, similar matters has already been made elsewhere, however.[19]

11 One of the characteristics by which the human is human is the capacity for being ashamed of an ill-done deed. This belongs to no other animal save the human.[20] This is the reason that incorrigible people are said to be called shameless, because to some degree they have been changed, moving from the honor of reason to assume an irrational, or cattle-like, nature.

Moreover, among the animals, the power of discerning the difference between what is honorable and what is base belongs to the human alone. From this it follows that it is a strictly human quality to pursue the honorable, whereas all brutes merely pursue what is useful and pleasurable.

It is also a strictly human trait to set his passions in order in accordance with virtue and to hold to the limit of reason those things which move in sensation. Likewise, although animals may be friendly among themselves, they have only the appearance [*species*] of friendship which is similar to the one which the Greeks call *ethayrica*, this being friendship that stems from the pleasure held by those who have been fed or nursed together.[21] The human, however, possesses every conceivable type of friendship.

12 Further, only the human has the power of speculation on intellectual theorems and of being delighted by a pleasurable thing that has no contrary. Therefore, only the human is a perfectly conjugal animal since he makes honorable marriages ordained by

---

18. The "philosopher" is Thucydides, describing in his history (2.49) the great plague that swept Athens, killing Pericles, in 430–29 B.C. Stadler suggests *De accidente et morbo* 5.7, but the identity of Galen's work is unclear. The editor of *QDA* (103, n.47) suggests that the title may in fact refer to Galen's *De morborum differentiis* (Kühn, 6:836–80), while Scanlan suggests *Cognoscendis curandisque animi morbis* (Kühn, 5:1–57), but there would seem to be no section 5.7 there and Kühn's index shows no reference to Thucydides in this work. In Galen's *De symptomatum causis* 1.6 (Kühn, 7:290), the plague is described as having arisen from corruption of the air and of putrescence, an explanation that is reminiscent of the present passage. Likewise, his *Quod animi mores corporis temperamenta sequantur* (Kühn, 4:788) bears a reference to Thucydides and mentions not knowing one's relatives.

19. It is not at all clear where A. has treated this earlier, although he may have had in mind passages such as at 9.77, in which he discusses the influence of mental images or phantasms of the soul on our dreams.

20. A. made this point at the very start of *DA* 1.50.

21. Gr. *hetairikē*, indicating comradeship.

laws. Simply put, he is the only civil animal since he has sharing, separation, dwelling together, treaties, and battles, all perfected and ordained by laws of urbanity.[22]

It is also proper to the human to be able to learn by use of his reason, to be gentle by nature due to his civilized character, and to be an animal that can laugh and exult due to the perfect causes of rejoicing that are found only in the human.[23]

We have already made a determination in previous books about the members of a human. But among those things which pertain to the human body is that only the human has a tall, broad, erect body. There is also the fact that the spittle of a fasting human heals abscesses when smeared on them and removes spots or scars. What seems even more wondrous is that if an arrow or sword is touched to the mouth of a fasting person and if another person is wounded with the arrow that has been thus wetted, the other person is infected.[24] Now, this is reported by the experts. Also, if the saliva of a person who has fasted for a long while is well thinned by the viscosity of food and falls into the mouth of or into a wound on a scorpion, serpent, or other poisonous animal in such a way that it penetrates to its inmost parts, it kills that animal.

Let what has been said about the human be sufficient, then.

The Second Tract on Quadrupeds

CHAPTER I

On the Nature and Traits That Quadrupeds Have in Common

By way of determining the nature of quadrupeds, we say, following Galen in his book *On Sperm,* that the genuses of all quadrupeds have in common that in the reproduction of brute animals the sperm does not change according to the progress of the seasons and the workings of the planets and constellations as happens in the case of the human. Rather, they all have the same complexion as befits the first property implanted in the sperm.

13

---

22. A.'s language has overtones. "Civil" is *civilis*; it implies citizenry, and thus an organized, civilized group of people working toward a common goal. *Urbanitatibus* has within it the Latin word for city, and for A., as for Ar. before him, truly civilized people come together in cities.

23. Some of these "causes" of laughter are intellectual, while others are rooted in human anatomy. See especially 13.71 and 9.135, with note.

24. An explanation for this is provided at *QDA* 7.39.

Galen says a bit later, when assigning the reason for this, that their sperm is so fluid and unreceptive to change that it never changes entirely and thoroughly, but only sometimes and partially. This is the direct opposite of what happens in humans. The reason for this is certainly one that Galen does not touch upon, namely, that the bodies of the brute animals are of a material and terrestrial nature, whereas the human body approaches more the celestial likeness. It therefore follows celestial influences more easily, especially since the mover of the human is the image of the human's mover and is a likeness of the world. As a result, all things which come from universal movers and from movable things of the world shine forth more brightly in the human than in others. Those things, however, which are the common influences of the breezes and of the air and come from heaven are felt more easily by others than by humans, as they are less concerned and busy with their own particular affairs.

14. Although there is a very large number of quadrupeds, Constantine still says in his book *On Intercourse* that it is a common characteristic that every wild animal becomes raging before it mates but that it is tamer after it has done so.[25]

Moreover, that animal emits many calls when it is in heat and is kept apart from the female. He uses these to call the female who, due to her coldness, has little lust and calls but little.

One must also observe whether the quadrupeds are dry or wet and whether they are cold. A dry and cold one is melancholic, as are cows. These conceive infrequently due to their coldness and rarely bear twins due to their dryness. Watch, therefore, to see if on occasion the cows conceive and give birth to twins more frequently in the summer, as it will be a sign of a lot of rain in the winter. The same is true for the times when cows are dying at a great rate, for this is a sign of a corrupt humor which, once the moisture has been corrupted, corrupts the bodies of animate beings in their causes. This too is a sign that there will be exceptionally numerous storms.

The same means can be used to prognosticate using sheep that are moist and cold, and the old ones are colder yet. Thus, if the young ones are moved to coitus at the same time as the old ones and the old ones burn for coitus like the young ones, it is a sign of an overbearing heat which will corrupt the moisture in the young ones due to its thinness.[26] They will die within the year. In the older ones the heat will not corrupt the moisture but rather thins it out, and in that same year the old sheep will therefore grow stronger and will prosper.

In addition to causes of this sort there are many other noteworthy things. All animals of this sort are bent toward the ground due both to the weight of their heads and to the earthiness of a body which their complexional heat cannot raise up. Since, though, many things have been said and must be said about quadrupeds in general, let us move on to them one by one, repeating nothing that has been said in preceding books. We will follow the order of the Latin alphabet as we have done in our treatment of rocks and plants.

---

25. *De coitu* 1.300.
26. Perhaps "if the young ones are sometimes moved to intercourse with the old ones."

(1) ALCHES: The *alches,* according to Solinus, is an animal that bears a likeness in shape, color, and size to a mule's.[27] But it has an upper lip that is so extended that, unless it proceeds backward, it cannot pluck plants on the ground with its teeth.

(2) ALFECH: *Alfech* is the Arabic word for an animal which many Italians, Germans, and French call the *lunza.*[28] It is born from a lion and a leopard, and it is very fierce and harmful. It is sometimes tamed, but unless the hunter is very soothing to it when it is led to the hunt, it doubles back and kills both dogs and men. It also kills wolves for pleasure.

(3) ALOY: The *aloy* is said by Pliny to be quite like a mule. It lacks flexion of the knee joints in its front legs, a trait it shares with the elephant.[29] It therefore does not lie down when sleeping, but rather it leans against a tree, and if the tree has been cut almost totally through in one place by a hunter, the *aloy* falls over as the tree does. Then, scarcely able to get up again due to the stiffness of its legs, it is captured. Apart from this it is rarely possible for it to be taken due to the great speed of its flight.

(4) ANA: They say that the *ana* is a very fierce animal of the East, with sharp, long, very strong teeth and sharp hooves.[30] It delights in those of its own kind [*genus*] and moves about in herds. It does not, however, allow those of another kind to join with it. Rather, if some stranger should approach, they gather together and rush at it, either putting it to flight or killing it. If, on the other hand, one of these animals is found alone and is attacked by an animal stronger than itself, it frequently defends itself and escapes using its teeth and hooves as weapons.

(5) ANABULA: The *anabula,* as Pliny writes, is a beast of Ethiopia which some of the Arabs and Italians call the *seraph.*[31] In its head it bears a resemblance to a camel; in its neck, a horse; and in its legs and feet, an ox. Throughout its entire body it has a shiny red color that is mixed in with white spots. Its long neck is marvelously adorned and its forelegs are long while its back legs are shorter. Its skin, because of the col-

---

27. *Alces alces,* the European elk. Cf. Solinus 20.6 (from Pliny *HN* 8.16.39–40, who offers a confused version of Caesar's description at *BG* 6.27). Solinus deletes Pliny's comment that the animal walks backward to avoid tripping on its lower lip. Cf. ThC 4.7 and Vinc. 19.2. Cf. also *aloy* below. *AAZ,* 67.

28. Arabic *al-fahd,* a cheetah (Schühlein, 1656). *Lunza* is of unknown origin; the animal is mentioned as the product of a leopard and a lioness at 16.139, with notes on the form of the name. At 22.107(58), the lioness and the leopard produce a leopard and at 22.113(65) we have the *lauzany.* On the name form see Sanders (420) and cf. Jacob.'s *lanzani* 88 (176).

29. The *aloy* is an error based on Pliny *HN* 8.15.39, which describes the *achlis,* itself an error for the *alces* originally described by Caesar's *BG* 6.27, as an inhabitant of the Black Forest. Cf. *alches* just above and Vinc. 19.2 (*alces*). The reference to a mule [*mulis*] apparently arises from a misreading of Pliny's *multis* and appears both here and in ThC 4.5 (Aiken, 1947, 217). On the elephant's legs, see 22.51 (37).

30. The *ana* may be the same as the *anabula,* just below, and thus a giraffe, though A. surely did not know this. Cf. ThC 4.9.

31. The description is based on Pliny *HN* 8.27.69, where he notes that the Ethiopians call a giraffe a *nabun* and also gives the common name of "camel-leopard." *Seraph* stems from the Arabic for giraffe, *zar(r) âfa* (Schühlein, 1661). Cf. ThC 4.6 and Vinc. 19.3. Cf. 22.37(19), where the giraffe is the *camelopardulus,* and 127(87), where it is the *oraflus.*

oration of the hair, sells at a dear price. During our own times and in our area the Emperor Frederick possessed one of these.[32]

(6) ANALOPOS: The *analopos*, as Jorach says in his own book *On Animals,* is an animal with such sharp horns that it cuts down trees with them.[33] When it tries, however, to cut down shrubs with them, they become entangled in the undergrowth.[34] Bound thus by the horns, they cry out and by their cry they are betrayed and are then killed by hunters.

(7) ASINUS: The ass is a common animal. It is ugly, slow, hardworking, and cold, for which reason it does not live in very cold areas and, if led into one, it neither conceives nor reproduces and even has difficulty with intercourse.[35] It is very hardworking and is a good bearer of burdens, but it carries a burden better over its kidneys than on its back or over its shoulders, for it is a melancholy animal and has stronger and drier bones lower down where the source of melancholy is.

A sign of this is that asses have more melodius bones than other animals.[36] Its skin is tough and thick and it therefore hardly ever moves away from what it desires by way of food or sex, even when greatly beset by blows. The earthiness, coldness, and dryness of its complexion work together to produce this dullness in its senses to blows. For this reason it is only slightly teachable and is always starving and very hard to fatten up, even a little. For this reason also it suffers from heaviness of the head and it more frequently dies with discharge coming from one side of its head and from the heaviness that occurs when a very thick and viscous phlegm is flowing over its lungs from its head. At such a time it experiences difficulty in breathing and develops shortness of breath [*asma*].[37]

On account of its dryness and coldness this animal is not moved to intercourse at the same time as other animals, namely, before or during the vernal equinox. It rather

---

32. Frederick II possessed a large menagerie, which he often took with him in his travels. On this particular giraffe, see notes at 22.126(87).

33. This antelope-like animal was conceived of as having serrated horns. It appears in ThC 4.16 as the *calopus*, which also appears in A. at 22.36(18) and in Vinc. 19.3 as the *aptalon.* Cf. also Arnoldus Saxo 2.4.22. The apparent origin is the *Physiologus*'s *autolops,* where its name underwent many changes (Curley, 1979, 70). Jorach, a Jewish author about whom little is known, is treated by Sarton (1931, 171). Stange (1885, 42–56) collects and studies the Jorach citations of Arnoldus Saxo. See also Draelants (2000).

34. The interpretation of sawing limbs comes from the *Physiologus,* perhaps arising from observations of deer rubbing "velvet" off their horns. The shrubs are the *herecine* of the *Physiologus,* identified by Curley (1979, 70) as plants such as arbutus and azaleas. George and Yapp (1991, 72–73) identify the animal as the Asian blackbuck, *Antilope cervicapra,* and offer a nice picture of it entangled in bush, which, they claim, represents the blackbuck's habit of marking twigs with scent glands near its eyes.

35. References to the climatic preferences of the ass actually go back as far as Herodotus 4.30, which was quoted by Ar. 605a16–22. Cf. ThC 4.2, Ambrose *Hex.* 6.3.11, Vinc. 18.10–12, Neckam 2.160, and PsJF 354–55. See George and Yapp (1991), 103–4; McCulloch (1960), 92; and Rowland (1973), 20–28. On the ass in antiquity, see *AAZ,* 57–59, Gregory (2007) and Griffith (2006).

36. Pliny *HN* 11.87.215 states that flutes are made from these bones.

37. Scanlan (71) identifies the condition as the strangles. Cf. ThC, citing the *Experimentator.* On the identity of the *Experimentator,* see 7.50, with note. Cf. also 22.70 (horses) and n.117 (dogs).

does so in May, when the sun has already ascended to the middle of a right angle from the equinox. It is then aroused most fervently since its thick humor has been loosened and it acts as if crazy, especially if the ass is a virgin and has passed beyond childhood.

The ass's thick skin is a sign of the thickness of its humor. For if the soles of some shoes are made from the skin that comes from the place where an ass carried a burden for a long time, the shoes do not wear out even if the wearer continually in his wanderings traverses rocky terrain. They eventually become so tough that he can no longer endure it on his foot. And this I have seen for myself firsthand.[38]

From its dryness comes the fact that the milk of the female is so thin that it has little cheese to it and it is therefore given to those suffering from hectic fever.[39] Its whiteness, moreover, is said to be good for cleaning and whitening the skin. Pliny says therefore that Poppaea, the mistress of Nero, bathed in warm ass's milk.[40]

It is the custom of this animal almost never to leave its young. If it should be separated by force, the young one does not leave the spot where it was left by its mother by more than three or four paces.[41] A she-ass that is about to give birth is said to flee the light, perhaps because of the weakness of the newborn's eyes.[42] This animal does well in warm lands of low latitude. In those lands, however, which spread beyond the sixth *clima,* the animal does consistently poorly, to the extent that it does not grow strong in lands at the 50 degree mark of latitude.

From those who know by experience comes the fact that the liver of an ass, if roasted and eaten by a sick person the whole day long, and if done many times and for a long time, is good for epileptics.[43] Its hooves function similarly if they are burned and if every day an amount of it is drunk which weighs three *aurei* and one ounce.[44] If a person should anoint inflamed spots with the marrow and fat of an ass, it will be a clear help.[45] If a poultice be made from burned hooves, it eases *scrofula* and cures cracks of the skin resulting from the cold.[46] The hooves are of use if they are ground

---

38. A.'s *vidi expertum* is especially emphatic, and one wonders if he might not have availed himself of a pair of these shoes as he walked across Germany repeatedly on his many travels. At 2.56, he extols the virtues of shoes made from hippopotamus hide.

39. For A.'s earlier discussion, cf. 20.21 and 1.436, 504. See Hall (1971) for a discussion of hectic (wasting) fever and its significance in medical theory.

40. The story is from Pliny *HN* 11.96.238, which says asses' milk is the thickest.

41. Cf. Barth. 18.7, who says that the female has such great love for the foal that she will pass through fire to reach it.

42. Cf. Ar. *HA* 577b1f., Pliny *HN* 8.68.168.

43. Cf. Avic. *Can.* 2.2.55, Barth. 18.7, and Vinc. 18.13–14 for remedies ascribed variously to Dioscorides, Asclepius, and other ancients.

44. Pliny *HN* 28.63.225, who specifies two spoonfuls. The *aureus* was a standard Roman coin weighing ca. 120 grains, or 7.78 grams. The ounce (*uncia*) was one-twelfth of a Roman pound (ca. 327 grams), or about 27 grams. What A. thought when he envisioned these amounts may well have been different.

45. Pliny *HN* 28.75.244.

46. Pliny *HN* 28.51.191; cf. Vinc. 18.13. *Scrofula* was the ancient term for "swelling of the cervical lymphnodes accompanied by respiratory catarrh" (Scanlan, 72).

and sprinkled over open abscesses of the feet or body.[47] If the urine of asses is drunk, it is good against illnesses of the kidneys.

The ass has the habit of urinating in a place where another ass has previously urinated. If it cannot find such a place, it chooses to urinate on feces. The mule, its son, does this as well. As we said before, its urine is especially good for the kidneys when the disease is caused by thick moisture. If a poultice is made from either the burned or unburned manure of an ass, it restricts the flow of blood.[48] If a house is smoked out using the lungs of an ass, vermin flee from it.[49] If the warm dung of an ass is put into acidic wine and if bandages are put into it for a long time until they are well soaked, and if the bandages are made of cotton and are thoroughly wetted, and if they are then put into the nose, they stop the flow of blood.[50] If a poultice made of this is placed on the forehead, it cures a person from flux.[51] If the burned liver is ground into a little bit of *collirium* and is liquefied with oil and the fat of a bear until it achieves the thickness of honey, and if it is then put over eyebrows that have lost their hair, the hairs grow back.[52]

(8) ASINUS SILVESTRIS: There is also a wild ass which the Greeks call the *onager* and which translates into Latin as "wild ass."[53] It is like the tame one but its powers [*virtues*] are said to be stronger in it and in its members.[54] The flesh of this one is good against a pain in the side and in the two hips.[55] If one is anointed with the marrow of a wild ass, it cures gout and removes pain. If women anoint their hair with the dung of a wild ass which is ground up with the bile of oxen it curls the hair.[56] Also, if the dung of a wild ass is dried and drunk with wine, it is good against the bite and sting of the scorpion.

---

47. Pliny *HN* 28.74.242.
48. Pliny *HN* 28.73.239.
49. Cf. Pliny *HN* 28.42.155, "all venomous creatures."
50. Avic. cited by Vinc. 18.13. Cf. Pliny *HN* 28.77.251.
51. Perhaps a flow of blood (Scanlan, 72), but more likely diarrhea, "flow of the belly," as at 8.221. One explanation that favors this interpretation is that many of the cures are homeopathic and the dung would work its sympathetic magic, much as urine is good for the kidneys.
52. *Collirium* is often used to designate a salve, particularly an eye salve, but sometimes (as in Pliny *HN* 28.37.139) a suppository. Cf. Pliny *HN* 28.46.163–66 for other versions of this cure, involving bear grease, burned ass's genitals, and ass's urine.
53. Several wild asses are native to Africa and Asia (Grzimek, 12.556–61), and the proper attribution of the ancient and medieval *onager* is debated (George and Yapp, 1991, 84). See 22.126(83) for other references and cf. Isid. *Orig.* 12.1.39. The Gr. form is *onagros*, "wild ass" (*AAZ*, 136–38).
54. Pliny *HN* 28.45.158 states that potions made from the wild ass are stronger than those made from its domestic parallel.
55. "Hips": *anchae*. For a discussion of this difficult term, see 1.292.
56. Throughout this and subsequent books, gall bladder and bile will be ingredients in all cures. Note that terminology was not yet sufficiently precise to be sure which is meant at a given time since *fel* can often indicate either.

(9) APER SILVESTRIS: The wild boar is a savage beast which some are accustomed to call the "solitary wild one."[57] It is armed for offense with teeth for weapons which are called tusks and armed for defense with the thickness of its hide. Moreover, they toughen their hide with mud and by rubbing it against trees.[58] If it is awakened in the morning before it has urinated, it is captured more easily. When tired, it sits on its hind legs and hides its tiredness while it awaits combat with the hunter. In India they have tusks one foot long but in other places they scarcely reach half a foot long.[59] Ones of this length are sometimes found in our land. The animal runs right over the blow of a spear and, unless mortally wounded, knocks down the hunter, unless he gets out of the way, falling down and lying prone on the ground, hiding behind a tree, or taking refuge in a tree.[60] The boar, since its tusks are curved upward, cannot cut what is beneath it, but it can trample what is lying on the ground.

The female, having no tusks, tears with her teeth. She goes about in herds and does not allow young from any but her own litter to eat with her. The boar's tusks strike like iron while they are in the living animal and they therefore inflict large wounds. The male is very fierce and quarrelsome during intercourse, as is the female in giving birth. This is a common trait of almost all animals.[61] Its flesh is viscous and cold, especially the closer it is to the milk.[62]

Even if they are quarreling among themselves, when they see wolves they help each other as if they were united by treaty. Also, all who hear the cry of a boar in trouble run to its aid together. This, too, is a trait of many animals. Some are even found horned, having small horns.

The domestic one shares these characteristics to a lesser degree.[63] But both have in common that they eat things they have rooted out of the earth with their snouts. The domestic one, however, stays longer in filth than the wild one. When many boars fight together, the one that finally prevails lords over all the others.

A characteristic of this genus of boars which is also common to that of pigs is that, it is said, when one squeals they are roused to a rage.[64] At such a time it is dangerous to enter into their herd, especially if dressed in white and even more so if the females are in heat, as Pliny says.[65] If they lose an eye, they do not live long having only the

---

57. *Sus scrofa,* the European boar. Cf. ThC 4.3, Neckam 2.139, Barth. 18.6, Vinc. 18.5, and PsJF 355. The Latin *singularis ferus* may also be read as "the singularly wild one." In either case it is the origin of the Fr. *sanglier.* See Rowland (1973), 37–43; George and Yapp (1991), 74–75; and McCulloch (1960), 97–98.

58. Cf. Pliny *HN* 8.78.212.

59. Cf. Pliny *HN* 8.78.212. The Indian wild boar, *Sus cristatus,* is slightly taller than the European boar.

60. The attack is vividly illustrated in MS Bodley 764 fol. 38v (George and Yapp, 1991, 74).

61. Cf. Pliny *HN* 10.83.181 and 8.78.212.

62. That is, to when it was a piglet.

63. This *aper domesticus* is a separate entry for ThC 4.4, who cites the *Liber rerum* as his source. Cf. Vinc. 18.78f.

64. A. clearly differentiates between the wild *aper* and the domestic *porcus*.

65. Pliny *HN* 10.83.181.

other one.⁶⁶ Because of the abundance of their humor, they have intercourse often for a long time, since they are not warm in temperament.

A disease befalls pigs which people call *lanchos* and which most often begins in an ear, tail, foot, or, more rarely, in some other part of the body. It then creeps into the adjacent living flesh until it arrives at the lungs through the swelling of some vein or other, whereupon the animal dies. Therefore, the cure is to cut away at the very start that part in which the disease begins.⁶⁷

This animal sleeps a good deal, especially in the summer when the heat loosens its moisture. At this time, unless it is frequently aroused, it suffers lethargy and dies. Also, since sleep moistens the body, its flesh is less firm in the summer. However, in warm lands it grows more and is fattened more than in cold lands due to the motion of its viscous moisture.

This animal has many monstrosities associated with it which have been determined in preceding books.⁶⁸ Therefore, let the things said here be sufficient.

23 (10) ALZABO: The *alzabo*, as is stated in Book 60 of *On Animals,* is an animal of great medicinal worth which dwells in the desert of Arabia.⁶⁹ Its flesh, which is warm and moist, is good against gout when cooked with vinegar. It is also effective for joint pain caused by the cold. The same flesh, when cooked in water and placed in a bathtub, cures, with great relief, a gouty person who enters the bath.

These animals flee one bearing the root of the *coloquintidae*.⁷⁰ They say that if hairs from the neck of this animal are taken and mixed, ground to a powder, and burned along with pitch, and if the anus of a sodomite is anointed with the result, it cures him of his vice. If about one ounce of its bile is drunk with spikenard water, it is effective against dropsy caused by flatulence. Its blood does well against leprosy if the affected areas are smeared with it while it is still warm. If its tooth is suspended over the right arm from the shoulder to the elbow, it is effective against forgetfulness. If its left foot and claws are placed in a linen rag and bound to a person's right arm, he will not forget whatever he may learn or come to know. If the bile of a male is bound over a man's left femur, he will not cease to have intercourse with a woman as long as he shall have it over his femur. If its right paw, along with the skin, is cut off by a man and is hung

---

66. Pliny *HN* 8.77.206.

67. *Lancho*s is a corruption of the Gr. *branchos* ("hoarseness") from Ar. *HA* 603a31. Scanlan (74–75) identifies it as swine fever or hog cholera and lists the symptoms. Cf. Vinc. 18.81. The "cure" is found in ThC 4.4. Compare the diseases at 7.105f., many of which bear similar symptoms.

68. These "monstrosities" are the diseases described at 7.105f.

69. Arabic *al-ḍabu*, a hyena (Schühlein, 1656). Stadler identifies this entry as from Rhazes, i.e., the Persian physician Abū Bakr Muḥammad ibn Zakariyyā ar-Rāzī (850–ca. 932). He is credited with 184 different compositions, bur his *Liber pestilentiae, Liber continens,* and his compendium *Almansor* (*al-Kitāb al-Manṣūrī*) exercised a strong influence on European medicine. It is questionable whether A. recognized this creature as a hyena, but the specificity of the cures argues for this knowledge.

70. *Cucumis colocynthis*, an herbaceous vine often called bitter apple, bitter cucumber, or colocynth apple. Cf. *De veg.* 6.81; Pliny *HN* 20.8.14–17. Also see Pliny *HN* 28.27.92–110 for an entertaining list of some seventy-nine magical remedies derived from the hyena.

on a person who comes to a king or someone else to accomplish some business or other, he will achieve his purpose. Further, if its right paw is cut off with the blow of a man's left hand and is hung on someone, it renders him beloved by all who see him, except by an *alzabo*. They say also that the ground marrow from its left foot is good for a woman who does not love her husband, and that if thrown into his nostrils, she will love him more than anyone else.[71]

(11) HAHANE: The *hahane* is an animal about the size of a deer whose gall bladder, contrary to the custom of other animals, is in its earlobes and it is through these earlobes that its bile is purged.[72] It is similar to the gall bladder of man and is very bitter, arousing violent anger and fierceness.

(12) BONACHUS: The *bonachus* [bison] is reckoned in the bovine genus and we have held forth concerning it in previous books.[73] As Solinus says, it has the head of a bull, but the body and mane of a horse. It has horns that are so curved in on themselves with many bendings that it cannot harm anyone with them, even by running into them. When this animal is in flight from hunters, it turns and, with a flowing motion of its belly, hurls its dung the length of an acre.[74] The odor or the actual substance burns what it touches and with this weapon it defends itself.

(13) BUBALUS: The *bubalus* is of the ox genus.[75] Its body is larger than that of the common ox. It has very strong legs, but they are short with respect to its body. It has long, black, ridged horns like a goat, and although its head is small compared to its body, it has a very thick body. Its entire defense consists of piercing with its hooves, and it trramples those it harms. It does not, however, readily pursue them except in

---

71. Or "her nostrils." As often, A.'s use of pronouns, here *eius,* is vague.

72. The deer called *achainēs* by Ar., who at *HA* 506a24–26 is the original source of this description. Ar., however, says that the deer appear to have their gall bladder in their tail but that it is a resemblance only in color and that the organ resembles a spleen internally. ThC 4.8 places the organ in the animal's ears. Ar. is, of course, describing a scent gland. He is accurate in his statement that deer (as he knew them) lack a gall bladder. Scanlan (76) identifies this animal as a "brocket, or two-year-old stag whose antlers have a single tine," which may be true for the original Ar. We will never know what ThC or A. had in mind. Cf. Vinc. 19.2.

73. Cf. 8.205f. and 22.146(108). This is the European bison or wisent. Stadler identifies it as the *Bos bonasus,* whereas Scanlan (76) chooses *Bison europaeus,* an identification he shares with Gessner (George and Yapp, 1991, 75). Cf. McCulloch (1960, 98) and other descriptions in Solinus 40.10, from Pliny *HN* 8.15.38, ThC 4.11. Note that "bovine genus" here is *genus vaccinum,* whereas in the next entry we find *genus bovinum.* On the problem of multiple terms for wild oxen, cf. *AAZ,* 141–45).

74. Many animals defecate or urinate at such times, as can occur in humans as well. The "acre," *juger,* was a land measure in antiquity of 120 feet by 240 feet. An excellent picture of the *bonachus,* there called the *bonnacon,* with curly horns and caught in the act of hurling dung at two hunters is found in an Oxford MS (George and Yapp, 1991, 76).

75. Perhaps *Bos bubalus,* the Indian buffalo, introduced into Italy in the seventh century. Yet Pliny discusses the *bubalus* at *HN* 8.15.38, saying that this is the common name for some wild oxen of Germany. Likewise, A. says at 1.49 that there is a large wild ox in his land that some incorrectly call the *bubalus,* and at 2.23 and 35 he tells us that the *bubalus* is called *wisent* in German. See the discussion in the notes to 2.30 and see ThC 4.10, Vinc. 18.21, Barth. 18.14, and PsJF 356. The confusion between wisent and aurochs was also widespread (Lengerken; *AAZ,* 140–45).

a straight line. It is black, and it has short hair and a short tail. With a ring passed through its nose, it is led around and draws weights equal to those of two horses. It pursues something multicolored and especially something red. We have said much about this in preceding books.

(14) BOS: The ox is commonly known and all the things which appear worthy of mention have been determined in the preceding books. If it is not castrated, this animal is fierce, especially when it has been provoked.[76]

(15) CAMELUS: The camel is a deformed animal that has two swellings on its back.[77] They have tall legs, a long neck, and when it walks it almost never wears out its feet, for it has fleshy soles that have sets of fleshy masses alternating with lung-like tissue.[78] For this reason they can barely endure rocks when walking a great deal.

It has a double use, both as a baggage and as a transport animal. But it does not readily carry burdens that exceed what it is used to and if they are bearing someone, they do not readily go beyond their accustomed distance.[79] When it is being loaded, it bends its knees to accept the burden after some light taps. They hate the entire equine genus. They can endure thirst for a three-day period but when they find water they drink enough to satisfy both their past and future thirst. They drink water that is either turbid or has been disturbed, but they do not willingly drink from wells.[80] During the mating season, they take pleasure in solitude and do not bear a human to be near them except perhaps for their keeper. Rabies and gout afflict camels and they die easily from them. They do not shed their hooves, but they are troubled by long and stony roads.[81] For this reason tough skins are made into shoes for them for such roads so that their feet might be less worn down. They love barley, a food which they gulp down all at once and then ruminate on all night long. When one stays apart in the stable, the others, as if commiserating with it, sometimes keep themselves apart as well.

The common camels have one hump. Those, however, which are called dromedaries are very speedy and have two humps.[82]

---

76. Compare the lengthy treatment at Vinc. 18.15–20 and cf. PsJF, 356.

77. Jacob. 88 (181), cited by ThC 4.12. Cf. Neckam 2.141–42, Barth. 18.18, Isid. *Orig.* 12.1.35, Vinc. 18.22–26, and PsJF 360–61. See also *AAZ*, 21–23; George and Yapp (1991), 105–7; Rowland (1973), 48–50; and McCulloch (1960), 101–2.

78. Solinus 49.9.

79. Cf. Jacob. 88 (181).

80. ThC 4.12 says it is water that they themselves have disturbed. Cf. Pliny *HN* 8.26.68.

81. "Shed their hooves": a difficult phrase. ThC 4.12, citing the *Experimentator*, reads *non eiciunt ungulas* to A.'s *ungulas non abiciunt*. Elsewhere A. uses the verb *eicio* consistently with teeth and horns to mean shed. Scanlan prefers "toenail" for *ungula*. Camels were thought to have delicate feet, explaining the boots put on their feet for use in war, mentioned just below from Ar. *HA* 499a23–30.

82. Solinus 49.9, from Pliny *HN* 8.26.67. On the dromedary, cf. PsJF 355.

The brain of a camel is said, when taken in a drink with vinegar, to be good for epileptics.[83] In complexion, the camel is warm and dry. Galen says that the marrow of a camel, when taken dried and prepared with wine, is good for epileptics who have the cause of their problem arising from black melancholy and that this is especially true if taken once a month each month.[84] The saliva of a camel, when taken mixed with water, returns a demoniac from drunkenness. If anyone takes dried and ground camel's lung in an amount equal to about the weight of a gold coin [*aureus*], he will incur blindness of the eyes. Camel's milk is the thinnest of all milks and most praiseworthy for its thinning action.[85] For this reason, it thins out superfluity, loosens the belly, and opens the pressures on it, due to the strong natural heat which is in it. It also strengthens the liver and opens its blockages, thins out thickness of the spleen, and is good for dropsy victims if taken in a warm potion, and especially so when placed with sugar from the *alchabiar*.[86]

Many things have been said about camels in the preceding books, and these things here are therefore sufficient since one ought not to repeat oneself.[87]

(16) CANIS: The dog is a familiar animal concerning whose diversity we have had much to say in previous books.[88]

It is an animal that is faithful to its master to the extent that when the master is dead, it is separated from him with difficulty and it sometimes undergoes death for its master. It is said that in Albania there are large dogs that bring down lions and elephants and which scorn bears and wolves, not harming them.[89] Jorach says that they flourish on odors alone and this is true of some of them.[90] They are produced from diverse animals and are therefore diverse in shape and they also differ greatly in size due to the variety of their food. The most ignoble genus of dogs before one's table consists of those that are thought to be on guard but which frequently so place themselves that they keep one eye on the door and one on the generous hand of the master. Since all dogs are born blind, the best among the number of those born is said to be the one that begins to see last or the one which the mother first carries into the room.[91]

---

83. Pliny *HN* 28.26.91. Stadler notes that from this point until the end of the paragraph (*alchabiar*), the passage was written in the margin by the hand M1.

84. Galen *De theriaca* 9 (Kühn, 14:240).

85. Pliny *HN* 11.96.237.

86. *Alchabiar*: Schühlein, 1655, discusses the various interpretations of the source of this sugar, indicating it is a spiny, gum-bearing tree having a tuberous "fruit." Cf. *De veg.* 6.262.

87. Cf. 2.19; 8.209.

88. Cf. 7.109, 8.4; Ambrose *Hex.* 6.4.17, 23, Neckam 2.157, Barth. 18.24–27, Isid. *Orig.* 12.2.25f., ThC 4.13, Vinc. 19.10–27, and PsJF 361–63. See *AAZ*, 47–52; Rowland (1973), 58–66; George and Yapp (1991), 107–8; and McCulloch (1960), 110–11.

89. Cf. Solinus 15.6, from Pliny *HN* 8.61.149.

90. Cf. Arnoldus Saxo 2.4.23d.

91. Cf. Pliny *HN* 8.62.151.

Dog rabies is said to be cured by an admixture of hen's feces in the dog's food.[92] Moreover, the cure for the bite of a rabid dog is said to be the root of the wild rose [*rosa silvestris*].[93]

Every dog seems to have the habit of sniffing urine and posteriors as if they receive some benefit from it.[94] Perhaps this is because they flourish on odors, and they recognize the qualities of their relatives by the odor of the body's excretions.

When sick they eat abominable herbs in order to be forced to vomit the bad humors. They are said to cure both their own wounds and those of others with their tongue and that if they are not able to touch it with the tongue, they use a foot that has been coated with saliva to touch the injured place and cure it.[95]

We have said many things about dogs in preceding books which we will not repeat here. But since this animal is suited for many human uses we therefore wish to pass on the skill by which noble dogs can be had, nourished, and cured if they are sick.[96]

28

It seems to us that it should be pointed out that among hunting dogs that are not greyhounds, those are noble which have large ears that are broad and hang directly down to the jaws. They should also have a fully wide-open mouth, open nostrils, an upper lip that hangs down a good amount, a sonorous voice, a tail that is not too long and which is curved more to the right and is often carried erect, and an anus that is free for the process of evaporation.[97] Therefore, when such dogs are found, take females and males that are, as far as possible, the same size, color, age, and strength. Color matters the least among these. When the female has been chosen, let her be secluded and starved until she is evacuated and likewise let the male be tormented with hunger until he is totally empty. Then give them food into which a great part of butter and fresh cheese have been mixed until they are perfectly full. Then shut them in a pen suited for this purpose for nine days or until you see that the female has conceived a litter. You should be careful that they are not in any way allowed to wander about before this time. But when the female has conceived, you will set the male free and will hold the female in confinement. When you do allow her to walk about, call her

---

92. Cf. Pliny *HN* 8.63.152.

93. Identified by Jessen at *De veg.* 6.43 as *Rosa canina*, the "dog rose." Scanlan identifies it as the French rose or apothecary's rose, and he cites *De veg.* 6.212. At *QDA* 4.9 (appendix), *thyriaca* is recommended as a cure for rabies. *Rosa canina* contains tannins (astringents), which precipitate soluble proteins as well as a wide spectrum of vitamins. Rose hips were collected by children in Britain after World War II to supplement supplies of vitamin C (Mabberley, 1987, 506; Talalaj and Czechowicz, 1989, 337). Pliny *HN* 8.63.153 also says that the only cure for rabies following a bite was to be found in *cynorroda*, "dog rose."

94. Cf. ThC 4.13.

95. Cf. ThC 4.13. Vinc. 19.10 and the bestiary of White (66–67) say the touch of a dog's tongue cures ulcerations.

96. With rare exceptions, the source of the remaining text on dogs cannot be identified. It is not found in any of the classical or medieval sources used and seems, from some of A.'s phrases, to be at least partly the result of his conversations with trainers and breeders of hunting hounds. The remainder may perhaps be based on some unknown or no longer extant work.

97. A tracking dog such as a bloodhound is being described.

back straightway to a place made comfortable with straw. The closer she gets to giving birth, the more you should tend to her food so that the fetus will find what it needs to absorb from the womb. You will serve moist foods like milk, whey with a little bit of butter, and other things of this sort. Mix in also a little bit of bread and cooked meat. When she has given birth, she is to be fed according to the number of whelps, being careful lest she should suffer either thinness or fatness, for thinness brings on defects in milk production and fatness produces a bad disposition in the milk. Dog's milk is very thick due to the heat of its complexion and this heat is dulled and relaxed in the presence of too much fat. Thinness, therefore, by drawing nourishment to the body of the mother, results in diminution of the milk, but excessive fatness results in coldness and thinness of the milk.

At an appropriate time the whelps should be taken from the mother's teats and they should at this time be fed neither to fatness nor to the defect of thinness. They should rather be fed proportionately as is sufficient for sustenance, growth, and strengthening. They should at first be given whey with a little milk and the milk should be buttery. As they grow more and more, less of this sort of humor should be put into their food. Then, when they have completed eight months, they should be fed on bread softened with whey that has but a little liquid in it. This is due to the dry complexion dogs have. Then, in the space of a year, they will be found to be healthy, quick, brave, and cunning.

After the space of a year and a half, they should be exercised for hunting. At first begin with moderate labors so that they exercise themselves more and more from day to day, for if they are exercised with too much running and work at first, the work will dry out their still tender limbs, and once the moisture in them has dried out they will neither be able to nor desire to work, but will rather shrink from hunting. If, however, they are exposed to their labors gradually, the heat of the motion in their legs will draw nourishment to their legs and will not use it up. Then, when they are at length strengthened in their legs, they will be strong and will desire hunting on their own.

In every respect, the same thing has to be done for the reproduction and nourishing of greyhounds except that their nobility is recognized differently. The best of these have long, flat heads that are not enormous, their ears are small and pointed backward into a point, and the upper lip does not hang down over the lower one unless it is very little. It has a long neck that swells a bit to a size a little larger than the head at the spot where it is joined to the head. It has a massive chest that is well pointed below, long and strong ribs, narrow flanks, a tail that is neither thick nor very long, and tall legs that are thin rather than fat. They rarely or never bark, for it is a characteristic of this dog to be scornful of the barking done by small watchdogs. For this reason they do not attack just anybody, thinking this to be unworthy of themselves. When it is being weaned, this dog should be fed more with milk than with whey.

The dog which is called the mastiff [*mastinus*] and which is similar to a wolf has much the same situation when it comes to breeding, but when it is being weaned, it should be fed more with solid and dry food. Of these dogs, those which are born from a greyhound and a fairly large and noble mastiff have speed along with bravery and are therefore the best.

30 The dog is by far the most easily taught animal. They therefore learn the mimetic works of actors. If anyone should wish to find this out for himself, let him take a dog born out of a vixen or from a fox, if that can be done. If it cannot, let him take a red dog from among those that are his watchdogs and let him accustom it while it is young to keeping the company of a monkey. For with her he becomes accustomed to do many human things. And if he should have intercourse with the monkey and the monkey should give birth to a dog, that dog will be the most praiseworthy of all for games.[98]

Not every dog is useful for detecting robbers. Rather, pay attention to that ferocious dog which does not deign to make distinctions among various people. When this dog is being trained, first arm some man with a thick hide that the dog cannot tear, and then urge the dog on against him. When he flees, he should allow himself to be finally taken by the dog, and by falling down in front of the dog he should allow the dog to bite him quite thoroughly. On the next day the dog should again be urged on against another one in the same manner and this should be done frequently since at length, by virtue of his sense of smell, it will follow the steps of anyone upon whose trail it is set. Now the reason a dog has to be chosen that does not make distinctions between people is this. If it learns to be petted by people, it follows the track of those whom it loves and their shape and odor stay with it and confuse its brain to the extent that it does not pick up the trail of the one that it is supposed to be following.[99] Many such dogs are found, however, that deign to recognize almost no one. A proof of this is that even hunting dogs sometimes are confused when they are placed over the track of different animals and they begin to wander. We have determined elsewhere in previous books which dogs have the better sense of smell.[100]

31 Dogs which are good for birds seem to have this talent more from training than from their sense of smell. But they still have a bit of it from each way and are taught as follows. They are first led around some captured partridges a number of times and, by means of threats, finally learn to go in a circle. But they learn to find the partridges by their sense of smell, since they are at first set on the tracks of captured ones.

This, then, is how dogs are nourished, taught, and bred.

The regimen of health for dogs, however, is that they are not allowed to sleep a great deal, since it is an animal of great heat and, when it sleeps a great deal, the intense heat which is around the place of its nourishment attracts bad humors to the stomach and the dog is infected and weakened. Let it, therefore, sleep a little after it is fed and only until the food is digested. Because a dog is dry, let it be fed with moist

---

98. Scanlan (81) believes that A. would not think a fox could breed with a dog and therefore translates the terms as "foxhound." Since the next sentence suggests a monkey mating with the dog/fox, it may be best to take the literal translation.

99. This would seem to be a clear description of the use of bloodhounds to track down criminals.

100. Perhaps a reference to 19.23.

foods, since its complexion is then more tempered. The sign of good purgation is that there is only a little and not much long hair, since then the purgation is done through the hairiness. If, however, it should have long, thick hair, then its skin is corrupted beneath the hair, turns foul, and contracts mange [*scabies*].

Dogs are most frequently beset by nine diseases: (1) scabies or impetigo or perhaps, to call it all by one disease, leprosy; (2) worms in ulcers; (3) swellings; (4) a thorn in some limb or other; (5) rabies; (6) untoward emaciation; (7) sluggishness; (8) fleas; (9) constipation of the stomach.

If, then, the dog is touched by impetigo, scabies, or leprosy, let it first be bled in all its legs from the larger vein which descends on the exterior of the leg. Afterward make an ointment, taking equal amounts of mercury, sulfur, and ground acalips seed.[101] Mix this with a double amount of old lard or butter and smear it on the injured spots and it will make them better.[102] Water in which lupines have been cooked or salted water is also good for this.

32

If, however, worms spring up in wounds or blows, they should be drenched with wild tansy juice [*yppia agrestis*] until they are killed.[103] Such hunters as are experts say that all worms in such brute animals die if yellow lavender is hung from its neck as soon as it is dried.[104] One who is especially experienced told me that this plant, when ground and given in the animal's drink, is very effective against pests afflicting oxen, horses, and all animals. When the worms are killed in the aforementioned manner, anoint the animals with butter made in the month of May and in this way the swellings themselves are reduced and the places themselves that have been anointed in the aforementioned way are gratefully licked. This is especially useful for dogs.

If, however, a swelling springs up in some limb of the dog and you wish to shrink it, mix ground marsh mallow with water until it can be handled like wax and lay it

---

101. "Mercury, sulfur, and ground acalips seed": *argento vivo sulfure semine acalipis trito*. Scanlan (83) has a lengthy note on the possible chemicals involved here. The present translation merely repunctuates Stadler's text, putting a comma between *argento vivo* and *sulfure*. The plant is identified by Stadler as identical with the Gr. *akalēphē*, a nettle. All in all this must have been a most unpleasant remedy. Some stinging nettles inject histamine-like substances when brushed against and that might lessen swelling. If the plant in question is of the *Acalypha* species, A. may have been drawing on the knowledge of the Arabs who were privy to Indian herbal remedies. Bellamy and Pfister (1992, 329) cite the *British Pharmaceutical Codex* of 1902 as listing acalypha as a cure, and Talalaj and Czechowicz (1989, 178) point out that *Acalypha indica* was used externally to treat scabies and ringworm.

102. Majno (1975, 115–20) discusses the use of grease, honey, and lard as topical treatments in ancient Egypt, adducing fine arguments for their use.

103. Jessen (1867, 750) identifies this as *Tanacetum vulgare*. Tansy is an aromatic, bitter herb, the leaves of which are brewed into a tea often used in simples. Historically, it has been used for treating internal parasites in animals. The "worms" may be fly larvae, and tansy has been used to deter worms (Mabberley, 1987, 369) and to keep flies from landing on meat (Grigson, 1955, 382). Talalaj and Czechowicz (1989, 296) note its use against roundworm and threadworm infestations.

104. Lavender: *Lavandula stoechas* (*stycados citrinum*). Identified by Jessen as *Lavandula stoechas*; cf. *De veg.* 6.433. Scanlan (83) points out that it has a foul odor, which would help drive worms (i.e., larvae) away. Brown (1995, 301–2) reports that the plant was used as an antiseptic in ancient times and was used externally as an insect repellant.

over the swelling.¹⁰⁵ A waxed rag that is laid over the area also does good, especially if cumin is mixed in. Likewise, sour milk that is well mixed and placed over a swelling quells the pain entirely and quickly expels the swelling. Also, flax seed that is cooked with fresh blood and is then applied is of use.¹⁰⁶ Also, a hempen rag that is moistened with water and is placed on a swelling from which the hair has been shaved away is of great use. Also, ground white birch [*betula*] does well when applied.¹⁰⁷ Also, groundsel [*seneciones*, pl.] that is ground with fresh lard and applied does the same, draws the foulness from down deep, and closes both the sore and the wound.¹⁰⁸

33 If the dog should have a spine in its foot or a thorn in any member of its body, bran that has been ground with fresh lard and applied draws out the spine and thorn and lays open the foulness by denuding the area. A powder made of swallow chicks that are burned in a new pot with all their inner parts does the same thing.¹⁰⁹ This powder should be kept in a box, for it is effective against such things.

If a dog is mad or rabid, immediately separate him from the others lest he make the others rabid by biting them. The teaching of Armeria, the king of Valentia, is that a rabid dog should be immersed for nine days in warm water throughout the length of its body in such a way that its rear legs barely touch the bottom and with its front legs raised up. Then, taken from the water, it should have its head shaved and thoroughly depilated even to the point of wounding the skin. Then it should be anointed with beet juice and this should be poured on it very often. If the dog is eating anything, its food should be treated with the same juice and it should be given the pith of an elder because this is of use.¹¹⁰ If all this does no good within the space of seven days, the dog should be killed because it will not get better.

If a dog is growing thin, not from the holding back of food but from some natural failure, fill it up three or four times with butter. If it is not noticeably fattened by this,

---

105. "Marsh mallow": *altea, Althaea officinalis*. The part used is its root, as in 22.64. Cf. *De veg.* 6.285. Brown (1995, 236) and Talalaj and Czechowicz (1989, 213) list several external uses designed to reduce swelling, be it from boils, infections, or minor trauma.

106. Brown (1995, 270) points out that the oil of cumin (*Cuminum cyminum*) is antibacterial. If the swelling described here is from an infection, the oil could help. Brown (304) also notes that the flax seed, *Linum usitatissimum,* is used externally for boils, abscesses, ulcers, and the like (cf. Talalaj and Czechowicz, 1989, 140).

107. Perhaps *Betula alba*, a white birch. Cf. *De veg.* 6.107, *fibex*. Some birches contain salicylate in the bark, a relative of aspirin.

108. Jessen in *De veg.* was unwilling to identify *senecio* definitively, perhaps because of the confusion of names in Pliny *HN* 25.106.167. Its modern use refers to a genus of herbs sometimes used to promote menstruation, and Scanlan (84) points out that its Old English name means "pus absorber." Most species of groundsel contain pyrrolizidine alkaloids that can cause severe liver disease, though external use would probably be harmless (Talalaj and Czechowicz, 1989, 161). Vickery (1995, 163) reports that it is used as a poultice for skin wounds.

109. Pliny *HN* 30.12.33 prescribes swallow chicks burned to ashes and mixed with honey for tonsil infections, as a cure for quinsy, and to keep eyelashes from growing (29.37.116).

110. "Elder": *sambucus*; cf. *De veg.* 6.115, 220–21. The common cider, *Sambucus nigra*, has been called the "medicine chest of the people," having all of its parts used for an extremely wide range of ailments (Brown, 1995, 347; Talalaj and Czechowicz, 1989, 131).

it has worms under its tongue.¹¹¹ Draw them out with a needle and it will fatten up. But if it does not grow fat, it is tending toward death from some natural defect.

If, however, you want to change sluggishness into swiftness, fill the dog faithfully with very well cooked oat bread that has been well leavened and it will be swift in the run.

Against fleas, anoint the dog with olive oil and the fleas will retire and not soon return.

There are some discomforts which are obscure and which can befall dogs besides those just mentioned. These are cured by the heat of the sun and the ingestion of grass. You should know, moreover, that as Galen says in the first book of his *De complexione,* a live dog is warm but a dead dog is not.¹¹² It is warm compared to man, but cold compared to the lion. Likewise, it is dry compared to man but moist compared to the ant and bee. Therefore, when dogs are given naturally dry and stringent foods, they become constipated and frequently die.

The sign of constipation is that the dogs groan frequently, run about from place to place, and tremble as if with a fever due to their excessive effort at relieving themselves. This is seen most often in the ladies' small dogs which almost always die of constipation. Let them then be given oatmeal that has been steeped in warm water to the consistency of thick porridge. Or else let them be fed with leavened, soft oat bread and let them be given a little milk whey and they will become loosened and become swift and whole.¹¹³ There is nothing else beyond the things that have been said which deserves much attention with respect to the care of dogs.

If, however, you mean to produce superior guard dogs, select the strong ones and keep them locked up and tied all day long. Let them loose by night and they will be praiseworthy in guarding, acquainted with no one within and admitting no one into your house.¹¹⁴

Let, then, these things, along with the other things that have been determined concerning dogs, be sufficient.

It is said in the sixtieth book of *On Animals* that the flesh of a dog is warm and dry.¹¹⁵ If dogs' teeth are hung on a jaundiced person, they do some good. And those that have the teeth hung on themselves are not barked at by dogs, or so they say. Costa ben Luca says that if a person keeps the heart of a dog on his person, he will not be bitten and dogs which come up to him and sniff him will immediately run away.¹¹⁶

---

111. Cf. Pliny *HN* 29.32.100.

112. Galen 1.5 (Kühn 1:537).

113. Oats would provide high dietary fiber and bulk in the dog's system.

114. This advice is also found in Barth. 18.27.

115. Cf. 22.23(10).

116. Also known as Qusṭā ibn Lūqā, a Christian philosopher who translated Gr. works into Arabic and died ca. 923 (Thorndike, 1929, 1:652f.). This is probably a reference to the work *On Physical Ligatures,* that is, on things hung and suspended upon oneself. Thorndike also notes that this work was attributed to Rhazes in a MS at Montpellier. See *De min.* 2.3.6 (Wyckoff, 1967, 146–47) for a discussion of A.'s familiarity with this work.

Some say also that if the tooth of a black dog is held in the palm, dogs will not bark at the one holding it. Thus it is that night thieves keep such a tooth with them. A woman in labor with a stillborn child will be relieved of the child if she drinks dog's milk with a bit of honey. Dog feces constricts the stomach when consumed, especially if it is the feces of a dog that eats bones and if it has been dried for twenty days in the month of July and is taken before sunrise, in the amount of an *aureus,* together with broth made from an old cock. The same treatment is good against quinsy [*squinantia*] and abscesses of the tonsils because it quickly closes them up and, when ground with coriander seed and smeared over abscesses that are inflamed, it cures them as well.[117]

Some also say that if the filth from a dog's ears is collected and is smeared on bandages of new silk, and if these are then stretched out on the oil in a green lamp, the heads of all present will appear bald.[118] Galen says that the head of a dog which has been burned, ground, and mixed with oil and dog fat will dry up sores on the head and cure mange.

They say as well that if the skin of a dog's penis is wound up on the ground on a spot where someone has urinated, and then it is bound over that person, it prevents him from urinating as long as it remains on him. If the coagulated milk is removed from the stomach of a small puppy and is taken mixed with wine, it temporarily alleviates the distress of colic.[119] Also, filth from dog ears makes one immediately drunk when mixed with wine. They say also that a stone bitten by a dog and taken with wine forces the drinker to shout.

36 (17) CAMA: The *cama* [lynx] is a beast also called by another name, namely, *rufinus* [red one].[120] Its place of birth is in Ethiopia, beneath the summer tropic. In shape it resembles a wolf but has white spots scattered all over it like a pard [*pardus*]. It is suitable for tricks since it is very trainable and quick to understand, a trait it has in common with the dog's nature.[121]

---

117. This use of canine feces may originally come from Galen *De simp. med.* 10.2.19 (Kühn 12:291–92). Earlier (5.17), Galen comments (Kühn 11:760) that dog feces is best as a medicinal cleanser if the dog is fed on bones. Quinsy is the Gr. *kyanchē,* "dog strangles," an inflammation of the throat. See 1.470, with notes. Brown (1995, 267) claims that the oil of coriander seed is fungicidal and bactericidal.

118. Compare this odd party trick with those in *The Book of Secrets* ascribed to A. (Best and Brightman, 1973, 97).

119. Perhaps in the manner of the *oxygala* of Pliny *HN* 28.35.133f.

120. The source of this passage is easily identified as Pliny *HN* 8.28.70, where Pliny states that the *chama* was first exhibited by Pompey the Great (see below) and was the animal that the Gauls used to call the *rufium.* His description makes it clear the animal is our *Lynx lynx,* but the name *cama* is less common in the bestiary tradition than is *lynx* (*AAZ,* 114–15). Cf. ThC 4.15 (*chama,* and similar in content to this entry), and for the *lynx,* Neckam 2.138, Barth. 18.67, Vinc. 19.79–80, and Isid. *Orig.* 12.2.20. Most illustrations are decidedly unrealistic (George and Yapp, 1991, 49; McCulloch, 1960, 141).

121. An interesting example of how the tradition from antiquity can become corrupted. Pliny *HN,* loc. cit., comments that the *chama* was first displayed at the games (*ludi*) of Pompey the Great in 55 B.C., a fact also noted by ThC. A. has misunderstood the nature of *ludi* and has taken it in terms of tricks and games, as it is used in 21.15, 24. Its trainability, therefore, followed as A.'s own conclusion.

(18) CALOPUS: The *calopus* [antelope] is an animal of Syria that dwells near the Euphrates and, because it drinks from the cold water of the Euphrates, it is cunning and swift, so much so that a hunter cannot come near it.[122] It has long horns that bear a resemblance to a saw and with these it is said to overthrow trees. But it cannot break the thickets which fall before the onrush of its horns, however much it tries to do so, and it therefore is often held captive with its horns entangled in them and it cries out. When this cry is heard, it is captured by the hunter.

(19) CAMELOPARDULUS: The *camelopardulus* [giraffe] is a beast from Ethiopia with a ruddy color, the neck of a horse, a camel's head, and the feet of a deer or an ox.[123]

(20) CAPER ET CAPRA: The *caper* (he-goat) and *capra* (she-goat) are familiar animals but have many genuses.[124] The domestic variety has long, sharp horns, a good deal of milk, and is nourished on grasses and twigs from trees. It leaps up mountains and does very well in mountain pastures. It has a very strong forehead and horns, a trait it shares with the ram. Constantine in his *Viaticum* says of this animal that if it is fed for forty days on the leaves of the tamarisk, it will not have a spleen.[125] In his book *On Simple Medicine,* Serapion cites Galen as saying that he had seen she-goats fed on tamarisk leaves and found their spleens to be very small.[126] Galen again says that he saw these she-goats lick skinned serpents and that afterward they aged less and grew less white.

The blood of this type of animal, boiling hot and freshly extracted, softens diamonds.[127] The bile of this sort of animal, if placed over the eyelashes or eyebrows, causes them to disappear by taking all the hair off.[128] This is because it is most bitter and pow-

---

122. A clear variant of 22.16(6), *analopos* (Curley, 1979, 69–70). Cf. ThC 4.16. Scanlan (87–88) thinks the description is sufficiently different from the *analopos* to warrant a separate identification and suggests the Arabian oryx.

123. The name form is more commonly *camelopardalis* and reflects the odd shape of a giraffe (as if from a camel) and its spots (as if from a leopard). A delightful picture of such a composite creature, in no way resembling a giraffe, is reproduced in George and Yapp (1991, 77). Cf. ThC 4.17, Isid. *Orig.* 12.2.19, PsJF 364, and Vinc. 19.9. Cf. 22.16(5), where the giraffe is the *anabula*, and 127(87), where it is the *oraflus*.

124. Compare the long discussion at Vinc. 18.27–33, PsJf 74. For the ancient tradition cf. *AAZ*, 76–77.

125. The citation is found in Arnoldus Saxo 2.4.23a. A.'s plant is the *thamariscus,* the tamarisk (Gr. *myrikē*), a plant of the tamarisk genus. It was a traditional medicine for problems of the spleen (Pliny *HN* 24.41.67). Jessen, however, identifies it as the European yew, *Taxus baccara* (*De veg.* 6.228).

126. Serapion, known as Ibn Sarabi, was an Arabic physician of the twelfth century. His book, *De simplici medicina,* had been translated into Latin at the end of the thirteenth century (Sarton, 1927–48, 2:229; Stange, 1885, 42).

127. Pliny *HN* 37.15.59. "Diamonds": *adamas,* originally meaning a very hard metal, often rendered as "steel." For Pliny *HN* it may have meant also the diamond, as it did for A. (*De min.* 2.2.1). Cf. Barth. 16.9 and 18.58, Neckam 2.92, ThC 4.18, Jacob. 91 (197), PsJF 364, and Vinc. 8.39, perhaps all via Isid. *Orig.* 12.1.14. See the discussion by Wyckoff (1967, 70–71). For a discussion of the sources for the medieval understanding of the diamond, see Herrera (1994).

128. The source of this and the next use of bile is given as the *Experimentator* in ThC 4.18. Cf. Pliny *HN* 28.47.171, who counsels the use of the bile for afflictions of the eyelids, with the hair first being pulled out and the applied preparation then being left to dry. Vinc. 18.30 cites Pliny *HN* 28 as his source of this use to ameliorate alopecia.

erful and does the same thing to other hair. If this bile is laid up in a vessel in the earth, it is said to attract frogs to it as if they were going to find something of use in it.

This animal learns to strike so hard a blow with its horns that whenever it is shown a board or a shield, it immediately butts it, sometimes breaking the shield and knocking down the man.

The blood of an animal of this sort, provided it is fed on diuretics like rock parsley and is given wine to drink, breaks up kidney stones and bladder stones, if it is taken in powdered form.[129] The liver, when eaten, is said to be good against clouded vision.[130]

This animal goes about in herds and if one of the herd is taken away, the others look on from afar as if they had been put into a trance. When they graze by day, they are said to look out to the things that are opposite them in the distance. Later on, however, they look at those things which are positioned nearby.[131] This animal makes use of many grasses and poisonous serpents and prospers from them, but if honey is drunk it becomes very weak and perchance it dies.[132] It is said to extract a thorn, an arrow, or anything else that is stuck in any of its limbs by using pennyroyal.[133]

A salve made from the bile of wild or domestic goats is effective against "webeye."[134] If a poultice is made of burned goats' hooves and strong vinegar, it cures hair loss [*allopicia*].[135] Burned goats' horns strengthen both teeth and gums if rubbed on them.[136] They also say that one who eats two goat testicles and then has intercourse will produce a son from the superfluity of the digestion of this food. This is true unless

---

129. Pliny *HN* 28.41.148 says the blood is so powerful (corrosive) that it sharpens steel. Rock parsley is *Petroselinum sativum* (*De veg.* 6.413), described by A. as having heating and drying properties, a description matching that of Galen *De simp. med.* 8.16.16 (Kühn 12:99), also prescribing its use for stimulating menses and urination.

130. Cf. ThC 4.18, Pliny *HN* 28.47.170–71 and 8.76.203 (the apparent original source), and Galen *De simp. med.* 11.11.11 (Kühn 12:336).

131. Cf. Pliny *HN* 8.76.204, accurately presented in ThC 4.18.

132. Curiously, Pliny *HN* 16.33.79 and 24.53.90 comments that rhododendron is poisonous to goats.

133. Pennyroyal: *polegium silvestre*, modern *Mentha pulegium* (Jessen at *De veg.* 6.422). *Pulegium silvestre* is discussed in Pliny *HN* 20.55.156–57 as being called *dictamnos* (dittany) by some, named in legend for Mt. Dikte in Crete where it grows abundantly. Pliny *HN* 25.52.92–93 comments on the goats' fondness for the plant and says that deer also feed on it when wounded to expel arrows. The Cretan goats losing their arrows by eating dittany are found originally in Ar. *HA* 612a3f. and clearly are to be seen in the indigenous goats found on Minoan art and called *kri-kri* or *agrimi* today. They once roamed wild but now live on small preserves. Note the religious interpretation given this phenomenon in bestiaries (White, 1954, 43).

134. *Tela oculi*: A. usually uses *tela* as fine membrane and it often is used in conjunction with the eyes. But here the meaning may be more literally a covering over the eyes. ThC 4.18.21f. prescribes goat liver for those suffering from night blindness, whom physicians call "nictilopas" after the night birds of the same name. Cf. Galen *De simp. med.* 10.2.13 (Kühn 12:280) for a similar cure.

135. A recipe identical to that of Galen's in *De simp. med.* 11.1.17 (Kühn 12:341). Stadler notes that both this prescription and the preceding one were written in the margin. Cf. Vinc. 18.30. Alopecia is not to be taken in its modern, technical sense, but rather as indicating spotty, random hair loss. The name derives from an old belief that wherever a fox urinated the grass died.

136. Cf. Avic. *Can.* 2.2.182.

there is some impediment in the woman with whom he has intercourse. If he eats only one, the boy child will have only one testicle. Goat fat mixed with goat feces and rubbed on one with gout relieves the pain.[137] Also, the burned dung of a goat, mixed with either vinegar or vinegar-honey [*oximel*], cures hair loss when it is rubbed on.[138]

We have said in preceding books that there is another genus of this animal that is not large and has horns curved in the shape of a hook that it uses to catch itself when it falls from the slopes of mountains.[139] There is also a third genus which is called the mountain goat [*caper montanus*] or by another name, the *ybex*. About these enough has been said in other books.

(21) CAPREOLUS: The roebuck has more of the appearance of a deer in its horns and is said to have keen vision and a weak voice which a hunter can imitate by blowing on a leaf.[140] When he does so he captures or kills the roebuck, which bounds forward as if to some animal of its own genus. Mention has been made of the mountain goat elsewhere and should not be repeated here.

(22) CASTOR: The beaver is an animal which has feet like those of a goose for swimming and front feet like a dog, since it frequently walks on land.[141] It is called the castor from "castration," but not because it castrates itself as Isidore says, but because it is especially sought for castration purposes.[142] As has been ascertained frequently in our regions, it is false that when it is bothered by a hunter, it castrates itself with its teeth and hurls its musk [*castoreum*] away and that if one has been castrated on another occasion by a hunter, it raises itself up and shows that it lacks its musk.[143]

This animal fells trees that are of appropriate size with its teeth and builds its houses on the banks of bodies of waters in front of the dens in which they frequently

---

137. Pliny *HN* 28.62.219 and Barth. 18.23 offer similar advice.

138. Cf. Pliny *HN* 29.34.106.

139. This is shown from 2.22 to be the chamois. Cf. 12.222, 229. The behavior of catching themselves on their horns is often ascribed to the *ibex* in the bestiary tradition, yet here the *ibex* is a separate animal. George and Yapp (1991, 81–82) say that the ibex is often identified as the chamois but that its horns are too small for this story, whereas the animal we call the ibex, *Capra ibex*, has horns of sufficient size.

140. *Capreolus capreolus*, the tiny roebuck of Europe and Asia (George and Yapp, 1991, 80). Cf. 8.41, ThC 4.19, Vinc. 18.32, and Barth. 18.22.

141. *Castor fiber* L., the European beaver. Cf. ThC 4.14.

142. Cf. Solinus 13.2, from Pliny *HN* 8.47.109. The etymological conclusion was that of Isid. *Orig.* 12.2.21. The belief was very widespread and is as old as Herodotus 4.109, who mentions the use of *castoreum*, but not self-castration, which is perhaps first found in Aesop (*AAZ*, 14–15). Cf. Neckam 2.140, Jacob. 88 (180), Vinc. 19.28–32, ThC 4.14, Arnoldus Saxo 2.4.22c, and PsJF 359. See Rowland (1973, 35–37) for its symbolic uses.

143. On *castoreum* as a word for testicles see 7.50. The confusion goes back to Pliny. Yet just below it is clear that the term refers here to the musk secreted by glands near the animal's tail. Compare A.'s reliance, in both passages, on firsthand sources (trappers) with the unfamiliarity of British artists who attempted to draw the beaver, generally with unfortunate results (George and Yapp, 1991, 59–60). Interestingly, ThC 4.14.6 specifically says it is the Poles who deny the stories and, at 7.50, A. states that there are many beavers in Poland. See also McCulloch (1960), 95.

live.¹⁴⁴ It makes these dens two- or three-roomed with terraces that it can go up on or come down from when the water is rising or falling. It carries the wood to structures of this sort, it is said, by reducing to servitude wandering beavers. They load these beavers, lying on their backs, by putting wood between their legs and carefully placing it on their bellies, and then they drag them by the tails to the house site and in this way they build their houses.¹⁴⁵

The beaver's skin is ash-gray tending toward black. Once it brought a good price but now it goes for little. It has thick, short hair. Its favorite food is fish.¹⁴⁶ It also eats the bark of trees.

It has its musk inside its body.¹⁴⁷ The musk is dry, warm, and a source of strength for the nerves, for which reason it works well against trembling of the limbs and paralysis, since it is a good separating and drying agent of phlegmatic humor.

40 The beaver has a broad tail that is very fat, with a sort of scaly skin which it tries to keep in the water almost all the time. The tail is suited for swimming like the rudder on a ship, but it is not true that the beaver never takes its tail out of the water since it does so when the water is too cold with ice. It is therefore false that this animal forces the otter to keep the water around its tail moving in the wintertime so that it does not freeze. Rather, it overwhelms the otter and expels or kills it.¹⁴⁸ It has a very sharp bite. Its entire flesh is abominable, however, except for its tail.¹⁴⁹

The musk, which is extracted in equal amounts with the humor that is around it, is dried and, when dried, it ought to be of a delicate substance and should tend to the color not of black but rather of cloudy blood.¹⁵⁰ If, however, it is multicolored, tending toward blackness, it is poisonous and it sometimes kills on the same day it is taken and sometimes turns the current ailment to another bad illness.¹⁵¹ When it is good,

---

144. Jacob. 88 (180).

145. Cf. 7.50 and ThC 4.14.22f., citing the *Experimentator*. The same passage is also found in Barth. 18.28, without citation. Compare the lesser-known story in some bestiaries whereby lowly badgers are used as wheelbarrows by other badgers (George and Yapp, 1991, 69–70, with excellent illustration).

146. "Its favorite food is fish": this translation is at best a guess. The Latin in St. and Borgnet reads *cibus eius melica piscis est*. *Melica* can mean "buckwheat," but the resultant "Its food is fish-buckwheat" is far from illuminating. *Mel(l)icus*, as an adj., can mean "honeyed," but *melica* is fem. here and nothing else in the sentence can agree with it. In medieval animal lore, the beaver was well known to eat bark and leaves, not fish. For now, the passage remains unclear.

147. One can see throughout that A. is never entirely sure whether *castoreum* refers to the testicles themselves or the musk he thought was their product.

148. Even a large otter is only one-third the size of a full-grown male beaver. Beavers are generally mild in disposition, while otters are well known for their unique interest in play activities as adults. They not uncommonly nest in beaver dams. Their playful activities could have been construed as forced by the ever-busy beavers during the colder months.

149. Gerald of Wales *Topographica hibernica* (Dimock, 1867, 59; O'Meara, 1982, 49) reports that certain monks in Germany ate the tail during times of fasting from meat, claiming that it was of the order of fish, not flesh, a fact echoed by ThC 4.14.13–16, quoting the *Experimentator*.

150. A troubled passage in which Stadler notes an erasure of five letters.

151. Perhaps a reference to the frauds alluded to by Pliny *HN* 32.13.27.

however, it is warm and dry with a delicate drying power and it therefore strengthens nerves and paralytics when it is drunk warm with saliva. When drunk by a woman in labor along with two ounces of calamint and honey and following a bleeding of the *saphena,* which is the vein of the liver that is in the lower part of the arm, it brings forth the fetus healthily, as well as the afterbirth.[152] In this way it also provokes menstruation healthily since if the vein is not bled, it perhaps calls forth less. In the same way, if it is drunk by men, it makes the testicles warm.[153]

(23) CACUS: They say that the *cacus* is an animal in Arcadia which lives in caves on the bank of the river Tiber and which is so hostile to cattle that it sometimes drags three at once by the tails to its cave.[154] It fears man greatly, however, but nonetheless kills and preys upon him. The exhalations of its lungs are so hot and have such a delicate poison that they consume whatever they touch as if with flames.

(24) CATHUS: The cat is an animal named for its ability to capture [*a capiendo*].[155] It is an enemy of mice and is said to be shy in its ways and a lover of beauty. It has a gray color like that of frozen ice. This is its natural color and it has others as a result of the accidental traits in its food, especially the domestic cat. It is very much given to biting, and in many ways is similar in shape to the lion, armed with claws and teeth like the lion, and it withdraws and extends its claws as does the lion.[156]

It has soft, warm, loose flesh, and the flesh of a wild cat is said to be good for the gout when it is spread over it. The bile of the wild cat is a great help against facial pains.[157] If half an ounce of the bile of a black cat is taken and mixed with Arabian jasmine [*zambach*], it produces a sneezing powder. The bile of a cat, especially a wild one, draws a dead fetus from the womb when a fumigation of it is made beneath the woman at the mouth of the womb. It should be mixed with black feces of a cat, either of any cat or of that cat whose bile it is. The bile of a cat acts the same way when it is mixed with bitter apple water, is soaked up into a sponge, and is held by the woman near herself, especially below, near the external opening of the womb.[158]

---

152. The identification of the *saphena* as in the arm is a clear error, whether compared to modern usage or to A.'s earlier understanding, for he correctly describes the pathway of the saphenous vein in the leg at 1.418. Scanlan's reference (92) to the *saphena* at 1.114 is incorrect.

153. Its "heating" properties are described by Galen *De simp. med.* 11.15 (Kühn 12:337–41). See Pliny *HN* 32.13.27 for many other uses for *castoreum.* Its modern use is only in perfume manufacture. "Castor oil," in modern usage, refers to a vegetable oil.

154. In mythology, Cacus was the son of Vulcan, a giant who lived on Mt. Aventinus and a troublesome bandit who robbed even Hercules. The most famous classical source is Vergil *Aeneid,* 8.190f. ThC 4.20 and Vinc. 19.5 have extensive passages on this "beast."

155. One of several etymologies given at Isid. *Orig.* 12.2.38. See Rowland (1973), 50–53; George and Yapp (1991), 115–16; and McCulloch (1960), 102. Cf. Vinc. 19.33.

156. The relationship between the cat and the humans with which it dwelled was never entirely positive during the Middle Ages, but starting in the fourteenth century highly valued cats were imported from Syria (Rassart-Eekhout, 1997; Walker-Meikle, 2012, 10–13, 28–30).

157. Perhaps "facial tics."

158. See 22.23 for bitter apple.

(25) CEFUSA: The *cefusa*, as Solinus says, is a monstrous beast whose front legs bear a resemblance to the arms and hands of a human and whose rear ones resemble his legs and feet. He says that this animal was brought to Rome during the time of Caesar.[159] We have seen this one even in our own day, captured in the forests of Sclavia, a male and a female, as we have said in preceding books.[160] This animal expressed various calls as time passed and belongs to the monkey genus.

(26) CERVUS: The deer is a familiar animal which adds points to its horns in proportion to its age and does so for up to six years, as Solinus says.[161] Then, so he says, it no longer produces points but rather the horns grow larger, in thickness. How many they produce is not certain, for I have seen a horn with eleven points on one and eleven on the other. Pliny, however, relates that when a deer feels that it is weighed down by old age, it draws serpents out of their dens with the breath from its nostrils, and when the poison has spread throughout its entire body and it feels itself becoming warm, it drinks from a clear pool and thus, as if stripped little by little of old age, its pelt grows young.[162] But I do not believe that this is true. The most reliable indication of its age is the great number or scarcity of teeth or that the oldest ones have no teeth at all.

Deer cross a river by swimming and, if the width of the river which they want to cross for pasturing is great, they each put their heads on the rumps of another due to the weight of their horns, arranging it so that the strongest is in the lead and the weakest last, and in this way they support one another.

When they come together for the purposes of intercourse they go mad and fight for the females, even to the point of mortal wounds. They do this at the end of August, when the star Arcturus is rising with the sun. At this time the females prepare their wombs by eating the herb called *silesys* so that they can conceive more easily.[163]

---

159. Clearly a simian, but of what sort? The main source is Solinus 30.20, from Pliny *HN* 8.28.70, who says the *cephus* was exhibited along with the *chama* in the games of Pompey the Great in 55 B.C. for the first and only time. He says they are from Ethiopia. Stadler conjectures a gorilla, but what was a gorilla doing in Sclavia? Cf. Pliny *HN* 8.80.216, ThC 4.21, and Vinc. 19.34. McDermott (1938, 67–68) offers no firm identification. On the confusion of names cf. *AAZ*, 4–5, 118–22).

160. This seems a clear reference to 2.50, where a couple belonging to the "wild men" tradition was found in Dacia. A. seems to speak from firsthand knowledge.

161. Solinus 19.9f. Cf. ThC 4.22f., Barth. 18.29, Vinc. 18.34–43, Neckam 2.135–36, and PsJF 356–59. Pliny *HN* 8.50.112–19 seems to be the ultimate source for much of this section, itself drawn largely from Ar. *HA* 578b6–17 and 611a15–b32. The most likely candidate for *cervus*, and undoubtedly the one A. saw most, is the red European deer, *Cervus elaphus*, but one must remember that many species (e.g., the fallow deer, *Dama dama dama*) across many centuries and lands may have added to the description (*AAZ*, 44–46).

162. Cf. Pliny *HN* 8.50.118, Isid. *Orig.* 12.1.18. For an illustration of a stag sniffing a snake, see George and Yapp (1991), 80. The same statement is found in several bestiaries (White, 1954, 38). Note the great age often attributed to stags by the ancients, but denied by Ar. *HA* 578b23f.

163. *Silesys*: Jessen, on *De veg.* 6.448, discusses the possible identification of *silesys*, which he conjectures to be *Lasperpitium siler*. The name appears as *seselis* in Pliny *HN* 8.50.112 and this is commonly identified as hartwort, *Iordylium officinale*. See Pliny *HN* 20.87.238 and Thompson (1910, on *HA* 611a18), who also provides a number of citations on its medical properties.

For, due to the hardness of the male's penis, they accept the semen only while fleeing. A pursued deer flees to the public thoroughfares used by people rather than to its lair, lest its hiding places be betrayed.[164]

The doe sometimes bears twins and, using the chastisement of her hooves, hides them secretly beneath a thicket lest they go forth. When they are ready for the run, she teaches them endurance for running and how to lead over thickets, lest they get stuck by their horns and captured while running through dense undergrowth.

It is said of this animal that it stands in wonderment of all things rarely seen and heard and that it therefore incurs the danger of capture or death by wishing to see something more closely or from its pleasure at pipes or a song.

The experts say also that both the deer and other beasts expel weapons and thorns stuck in their limbs by eating dittany.[165] When stung by the *phalangius,* which belongs to the genus of spiders, they eat crabs if they can get ahold of them and thus overcome the power of the poison.[166]

They say that these animals have a life immune from fevers and that its flesh is therefore very healthy if taken in the morning, two hours before all other food is eaten, in whatever quantity the body can take in.[167] They also say that Alexander proved that deer live for more than one hundred years by putting gold rings in their noses. These deer were captured more than one hundred years after the death of Alexander and showed no trace of old age.[168]

44

But they do flee poisoned things to the extent that they back away from poisoned weapons that are thrown into their vicinity. When pursued they run into the wind, so that the wind will carry the voices of the dogs away from them.[169] An ointment made of their marrow will, so they say, keep one immune from fevers.[170]

They shed their antlers into water lest their right antler be found, as if they feel that there is some sort of good in it. This, however, I have proven not to be generally true, for I have found a left antler in the forest hidden under some leaves.[171]

Stags that fight in the pastures obey the victor as if he were their lord. For bearing young, they mark out a place to which there is but one area through which to enter, so that by fighting there, they can defend it against a foe.

---

164. Cf. Ar. *HA* 611a15f., who says that hinds intelligently bear their young near roads, so that predators will stay away. Pliny *HN* 8.50.112 says that they fly to men for refuge when pursued by hounds.

165. Cf. 22.38.

166. Cf. Pliny *HN* 8.41.97 and Ar. *HA* 611b21f. The *phalangius* is treated at 26.20(29).

167. Or "for two hours."

168. Cf. Pliny *HN* 8.50.119 (but with gold necklaces, covered by folds of fat after one hundred years), which ThC follows more exactly. The actual life span would be a maximum of twenty-five years.

169. A reversal of Pliny *HN* 8.50.114, who says they run downwind.

170. Cf. Galen *De theriaca* 9 (Kühn 14:241); Solinus 19.16.

171. This is a somewhat confusing statement, based on Pliny *HN* 8.50.115. One must believe that A. feels he has disproven the belief by finding a left antler since the deer sheds both antlers into water only to hide the right one. But cf. Ar. *HA* 611a29f. Scanlan's translation, "and in these cases both right and left antlers were found together," is not supported by the Latin.

They say that if one eats the flesh of the aborted fetus of a deer which was killed in the womb of its mother, it is very effective against poison.[172]

They say also that there are twenty worms in the vertebra of the neck, where the head is joined to the neck and which is called the base of the head [*basis capitis*].[173]

They say as well that in its old age, wasps and ants are produced in its head, within the eye socket, and that they sometimes come forth.[174] They have four teeth in each jaw and below there are four others which are larger in the male than in the female.

45 Platearius says that its heart bone, when dried and crushed, is good against *cardiaca*.[175] Isaac says that its flesh is melancholic and hard to digest.[176] Its brain, however, and its fat are said to be good against cardiac palpitations and pain of the hip joints [*anchae*] and the flanks, and they soothe and give strength for the repair of fractures. If an ointment is made of the fat, it puts worms to flight. The blood of a deer, fried with oil and made into an enema, does well against sores of the intestines and removes chronic diarrhea. If it is drunk with wine, it is good against poisonous abscesses. The brain relieves hard abscesses in the muscles and on the cords of the joints. Serpents flee the odor of a deer's horn if the far end [*caput*] of the house is fumigated with it and its hoof.[177]

They say that if the horn of a deer, cooked with vinegar, is moved about in the mouth, it quiets pain in the teeth and firms the gums.[178] If the horn is taken as a powder and a dentifrice is made of it, it cleans the teeth. If burned and drunk in the weight of two *aurei* and one ounce, it restricts the flow of blood without putrescence and is good against chronic diarrhea and against bladder pain. It also cuts off dis-

---

172. Solinus 19.16 and Pliny *HN* 8.50.118 prescribe instead the rennet of a fawn killed in utero. ThC 4.22.27 says it is the meat.

173. Scanlan (96) suggests that the nerves uncovered by decapitating the animal might be taken as worms. But real worms or maggots could also be a parasitical infestation from the many flies that plague deer. Cf. Pliny *HN* 11.49.135, the source for A.'s comment. On *basis capitis,* see 1.184, 186.

174. ThC 4.22.37f. adds more details. The bone beneath the eye acquires a hole and the wasp is formed naturally within from the superfluous humor, although others say it arises from the marrow. In either case it presages the death of the animal unless it finds a snake with which to revivify itself. Note that "eye socket" here is *os oculi*, rather than A.'s more normal *orbita* (1.189–90, 197, 359–60).

175. Stadler cites Platearius *De med simp.* 30.7f. Matthaeus Platearius was a physician of Salerno (d. 1161), and this work filled with simples and cures was quite popular. See also Dorveaux (1913). The *os cordis* is a tough structure mentioned by A. at 1.118, 300, 579, and 583. *Cardiaca* can indicate many heart-related problems, ranging from mere heartburn to more serious disorders involving either the heart or stomach (L&S).

176. Stadler cites Isaac Israeli, *Dietarum particul.*, fol. 135v. Isaac ben Solomon Israeli (ca. 855–ca. 955), also known as Isḥâq ben Sulaiman (Sarton, 1931–48, 1:639–40), was a neo-Platonic philosopher and physician in the court of 'Ubayd Allāh al-Mahdī, whose very popular medical works were adapted or translated into Latin by Constantine the African. See also 21.41, with note, and 16.26, with note.

177. The phrase *caput domus* is difficult. Scanlan (96) takes it as indicating the snakes are living in the eaves of the house—a fairly unlikely proposition. The translation offered here follows the sense of *caput* given in Latham² no. 13.

178. Galen *De simp. med.* 11.1.8 contains the elements of many of these cures. See also *De theriaca* 9 (Kühn 14:240).

charges coming from the womb and is good against jaundice.¹⁷⁹ If the burned horn of a deer is drunk with a bit of honey, it expels worms. If dried stag's penis is drunk, it is good against the bite of the serpents called the *tyri,* which are the ones *tyriaca* is made from.¹⁸⁰ If dried stag's penis is given to a man suffering difficulty in urinating or one with colic, it removes the difficulty in urinating and calms the colic, provided he is bathed and drinks the bath water.¹⁸¹

The entrails of a deer are very foul.¹⁸² Pliny says the cause of this is that their bile is scattered throughout them.¹⁸³ Some say that it is diffused in the deer's ears and others the tail. Many other things have been said about deer in the preceding books.

(27) CHYMERA: The chimera, they say, is a beast whose front part is monstrously elevated and whose hind part is low and which is found in parts of Babylon in Chaldea. It copies many of the motions of actors and is therefore clothed and is said to glory in clothing when it is covered with expensive things.¹⁸⁴

(28) CYROGRILLUS: The *cyrogrillus* is a small, weak animal that dwells in holes in the ground and is as much the enemy of such "earthy" (that is, land-dwelling) animals as it can be.¹⁸⁵

(29) CUNICULUS: The rabbit is an animal smaller than a hare, but stronger.¹⁸⁶ It dwells beneath the earth in burrows, is the enemy of vineyards when it is able, and in color and shape is almost exactly like a hare. It seeks out caves in hilly land and it uses dust to level its caves lest it be caught.¹⁸⁷ In the evening and in the morning it sits in front of its burrows.

The rabbit copulates with its backsides together since the passage for the sperm and the male's penis are pointed to the rear.¹⁸⁸ It is very productive of its own kind.

---

179. Cf. Pliny *HN* 28.77.246, who prescribes it to purge the uterus.
180. On the *tyrus* see 25.42(59). On the famous antidote see 1.102.
181. Cf. Pliny *HN* 28.42.150, where the dried penis is drunk in wine for good general health.
182. Cf. Ar. *HA* 506a32–5: "their gut is so bitter that even hounds refuse to eat it."
183. Pliny *HN* 11.74.192.
184. This Greek mythological monster dwelled in Lycia (SW Turkey) and vomited fire. It had, in legend, the head of a lion, midsection of a goat, and the rear of a snake. It was killed by Bellerophon. A.'s confused story about the clothing is found in greater detail in ThC 4.23, who says that the Saracens capture this beast and then offer it, covered in priceless garments, to their ruler as a token of their goodwill. ThC then launches into a diatribe against pride in one's clothes. Cf. Isid. *Orig.* 11.3.36, Jacob. 88 (183), PsJF 364–65, and Rowland (1973), 54–55.
185. See the discussion at 3.143 for arguments over whether this is the Syrian hyrax (*Procavia syriaca,* a small, marmot-like mammal) or a squirrel. At 3.143, it seems to be a sort of rabbit. Cf. ThC 4.24 and Vinc. 19.35, with very different content, who cites a gloss over Lev. and makes it the equivalent of the hedgehog, which would, in this instance, explain its hostility to earth-dwelling creatures. To say the identity of the animal is unclear in A. is an understatement, despite the confident comments of Scanlan (97) and George and Yapp (1991), 60–61. Cf. 22.97(43), where A. says it is identical to the hedgehog.
186. Cf. ThC 4.25 and Vinc. 18.44.
187. Cf. ThC 4.25, who makes it clear that camouflage is the rabbit's aim.
188. Cf. Pliny *HN* 10.83.173, cited by ThC. On the classical tradition, *AAZ,* 159–160).

It has white flesh and is a timid animal. Therefore, when injured, it leaves its habitation. Seeing this, the entire colony leaves the place, as if angry over the injuries to its comrades.

(30) CRICETUS: The hamster, so they say, is a small animal that lives in caves in the ground.[189] It has a multicolored head, red back, shining belly, and hair that is so firmly fixed into the skin that part of the skin comes away from the flesh before a hair can be pulled out. It is not easily removed from its hole except with hot water or something else poured into the hole. It has the following in common with the rabbit and the ground squirrel [*cycel*], about which we have spoken in earlier books of this investigation, saying that it has hair like a rabbit but does not have ears, having instead a pathway for hearing like a bird. The *cricetus*, however, is an animal we call the *hamester* in German, and it is an animal that is very prone to bite and is irritable.[190]

(31) CYROCHROTHES: The *cyrochrothes*, as Solinus says, is a beast which imitates the voices of humans, as the hyena does. It never closes its eyes, has no gums in its mouth, rather having one continuous tooth, which, lest it be dulled, is enclosed within boxes, as it were.[191] The *cyrochrothes* is so powerful that it crushes almost everything. It is said to be produced from a wolf and a dog.[192]

(32) CATHAPLEBA: The gnu is a moderate-sized, idle animal which carries its head with difficulty due to its weight. It is said to inhabit the banks of the Nile near the so-called Black Spring. There is such a delicate poison in its eyes that whoever should blunder into the line of their gaze dies immediately.[193]

(33) DAMMA: The *damma* [gazelle?] is a beast the size of a goat with a shape and coat like those of a deer. It has horns that do not branch but are smooth, long, and sharp. It is fast on the run, lives a careful life, and uses its horns against beasts hostile to it.[194]

This beast is called the *algazel* in Arabic. Its flesh is cold and dry and generates hemorrhoids unless it is seasoned with pepper, cinnamon, and mustard. The dung of this animal, prepared with oil, causes the hair to grow and improves it.

---

189. *Cricetus cricetus*, the European hamster of central Europe (*AAZ*, 81). Cf. 1.582, where we get its MHG name, and ThC 4.26, citing the *Liber rerum*.

190. On the *cycel*, cf. 1.582 (German name), 2.63 (ears), 8.104, 12.177.

191. That is, they fit into a recess in the opposite jaw.

192. The more common form of the name is *corocotta*. Originating in Pliny *HN* 8.45.107 as the offspring of the Ethiopian lioness and the hyena, it was passed on by Solinus 27.26, Jacob. 88 (181), PsJF 364, (*centrocata*), and ThC 4.27. Cuvier (*B&R*, 2:296–97) concludes that it is fabulous, but Stadler and Scanlan conjecture it to be the spotted hyena, *Hyaena crocuta*, but even if this is the origin of the beast it is doubtful that Pliny or A. knew what it was. It may be identical with the *leucrocotha* of 22.112(61).

193. Cf. Pliny *HN* 8.32.77, Solinus 30.22, ThC 4.28, and Vinc. 19.33. The name comes from the Gr. and means "down-looker." Cf. *AAZ*, 75–76.

194. Perhaps *Gazella dorcas*, the dorcas gazelle of northern Africa. The general description, Arabic name, and location seem to fit this. George and Yapp (1991, 80) identify it as the fallow deer in the bestiaries, but see the entry that follows on the *dampnia*. The *damma* is mentioned in Pliny *HN* 8.79.214 but with little detail, and at 11.45.124 we learn that the horns curve toward the front. Cf. ThC 4.29 (*demma*).

They say also that if a man's penis is anointed with this and if he has intercourse with his wife this way, she will always adore him. If a throat on which there is a leech is fumigated from below with the dried tongue of this beast, the leech falls off. If its burned hoof is used to fill the hollow of a fistulous ulcer, it soothes it.[195]

They say also that if equal amounts of the bile of this animal, the semen from the testicles of a fox, pepper, and the seed of garden rocket are taken to equal the weight of an *aureus,* and two ounces of honey are added, and an electuary is made from this, and a woman holds it in wool, she will conceive when she lies with her husband.[196] And if the bile was from a female she will conceive a female, whereas if it was from a male she will conceive a male.

(34) DAMPNIA: The *dampnia* or *damula* [deer] is a small, weak beast which, as Isidore says, is named from the fact that it flees from the hand [*de manu*].[197] When this one gives birth, it suddenly devours the afterbirth before it falls to earth. This animal is prey for other animals.[198]

(35) DAXUS: The badger is a very fat animal and its lard is good for kidney pains.[199] It is an animal that has a wide back and short legs. It is very much given to biting and has hair that is more white than black. On its back it has more black hair and on its sides more white, while its head is black in the middle and white on the sides. When it is tamed it is very playful.

They say that the legs on its left side are shorter than those on its right.[200] But I have not found this even though I have examined this animal very often. It does have tough hair and a thick coat and is the size of a fox.

It is found in two genuses. The one that is called the dog badger [*daxus caninus*] has a foot split into many parts like that of a dog.[201] The one called the hog badger [*daxus porcinus*] has a hoof split into two parts like a pig.[202]

It sometimes has a poisonous bite, but not always. It is said that it lives on hornets and other vermin, since it is not a swift hunter.

---

195. "Hoof": lit., "heel"; see Glossary, s.v. *Calcaneus.*

196. Because an electuary melts in the mouth, the presumption is that the syrupy paste was put on the wool, which was then "held" in the mouth by the woman.

197. This is most likely the little fallow deer, *Dama dama,* mentioned just above. Cf. Isid. *Orig.* 12.1.22, Vinc. 18.45, and ThC 4.30 (*damma vel dammula*). Barth. 18.34 cites Papias, an eleventh-century lexicographer who says that *damula sive dama* is a wild goat.

198. Its passivity even inspired a poem by Martial (13.94): "The boar is feared for its tooth, their horns defend the deer. But what are we, the pacific *damae,* besides prey?"

199. Cf. ThC 4.32 (*daxus*). Given as *taxus* in Neckam 2.127, Barth. 18.101, PsJF 394, and Vinc. 19.110. Cf. MG *Dachs.* In the bestiaries it is sometimes called the *melo* and *melota* (George and Yapp, 1991, 69–70). Cf. *AAZ,* 9.

200. From the *Liber rerum,* cited by ThC 4.32.

201. *Meles meles,* the common European badger?

202. *Arctonyx collaris,* the hog-nosed or sand badger of Asia?

(36) DURAU: They say that the *durau* is a cruel, swift, strong beast that hurls its dung at hunters and dogs. This, because of its foul smell and stickiness, prevents the hunters and dogs from following it.[203]

(37) ELEFAS: The elephant is the largest animal among the quadrupeds and sufficient mention has been made in preceding books concerning its shape, intercourse, and reproduction.[204] It has a trunk ten cubits in length, which it uses in place of a hand for battle, feeding, and other tasks. When calling it sometimes does so through its mouth and then the sound is terrifying. Sometimes, however, it sounds through its trunk and then the sound is rendered sweet, like the sound in the hollow of a large reed.

This animal is said to be afraid of a mouse and to fear the grunting of pigs to the extent that it flees them. It is said to be the enemy of the wild bulls. Some say that large-bodied snakes [*dracones*] fight with elephants and, having conquered them, drink their blood and thus, so they say, they become cooled from the heat. But I think this is just a story.

Because of the weight of its body, it has its testicles inside so that intercourse might be quick. Still, at the time of intercourse, it extends them. It has a dark, almost pumice-colored skin that is mangy and into which flies fall. It kills these by drawing its skin into wrinkles.[205] They are said to be chaste animals, ignorant of adultery, and go in pairs during mating time. They are so strong that they carry twelve or more men on themselves in wooden towers, and some are so large that they are said to be able to carry forty.[206] They are said to be of two genuses, the more noble of them exhibiting largeness of body.[207]

Elephants' legs are large, having about the same size below as above, like columns. And while their foot is divided into many parts, nature nevertheless joined the toes together in order to strengthen the foot in this manner. This is why they are said not to have joints in their legs beyond the knees.[208] Perhaps they have rigid, not loose joints

---

203. Stadler suggests this is the tur, *Capra caucasia*, a wild goat found in the Caucasus. George and Yapp (1991, 81) choose this identity for the *tragelephus*, however, which is also said to dwell in the Caucasus. The dung hurling, however, reminds one of the *bonasus* of Ar. *HA* 630b8f., and ThC 4.31 cites Ar. as his source for this animal, which he calls the *duran,* and at the same time he refutes those who would equate it with the unicorn. Cf. Vinc. 19.36.

204. The elephant held a high place in the bestiary tradition (White, 1954, 24–28). Cf. Jacob. 88 (177–78), ThC 4.33, Barth. 18.61–63, Neckam 135–43, Vinc. 19.38–52, Rabanus Maurus *De universo* 8.1 (*PL* 111:220–21), Peter Damian *De bono religiosi status* 26 (*PL* 145:785–86), and PsJF 365–70. Scullard (1974) treats the ancient evidence fully and cf. *AAZ*, 65–67. On medieval treatments, cf. George and Yapp (1991), 89; McCulloch (1960), 115–19; Flores (1993), 11–38; and Beer (1989).

205. A charming belief found in Pliny *HN* 8.10.30.

206. See Ambrose *Hex.* 6.5.33–35 for a vivid account of their use in war, and for these and other details see Druce (1919) and Scullard (1974). A cheerful elephant with its tower is depicted in a MS from the British Library (George and Yapp, 1991, 90).

207. *Elephas maximus,* the smaller Indian elephant, and *Loxodonta africana*, the African bush elephant. Pliny *HN* 8.9.27 reverses the relative sizes.

208. The column simile appears in Ambrose *Hex.* 6.5.31.

and are therefore thought not to have any by the inexperienced, for if they had not joints, they would not be able to have an orderly walk.

Their flesh is cold, dry, and abominable. One who eats it with water, salt, and fennel seed is cured from a chronic cough. If a pregnant woman should drink a potion made of this flesh cooked and liquefied with vinegar and fennel seed, she casts out whatever she has in her womb. If its bile is introduced through the nose in the weight of an *aureus,* it is said to be useful against epilepsy.[209] The outer edge of its liver, eaten with water and citron leaves, is good for liver pains.[210] If elephant dung is smeared on skin on which lice are evident and it is left there until it dries, the lice will not remain but will immediately depart. If elephant fat is smeared on one suffering headache, it is said to cure the pain.[211] It is also said that if an ounce of elephant bone is drunk with ten ounces of mountain mint by one whom the first stages of leprosy have touched, it does him a lot of good.[212] If a house or some other place where there are mosquitos is fumigated with elephant dung, they are put to flight and die.

(38) EQUI: Horses are so called from equalness [*equalitas*] or pairedness, since they were yoked in pairs to chariots in antiquity.[213] It is a familiar animal and admits of reproduction almost everywhere on earth, but the ones produced in Syria and Cappadocia are said to be excellent. Insofar as we can see in our time, the ones largest in body are produced from the third *clima* to the end of the sixth, and especially in Spain. We see stronger ones, and these quite large, produced in the seventh *clima* as well. These are more durable for work than those coming from the third or fourth *clima.*

Now, in horses, four things are considered: namely, form, beauty, worth, and color.[214] Form occurs when the body is strong and hale and has a height corresponding to its strength. A horse with form has long flanks, rounded rump, and its entire body is almost knotted with the density and abundance of its muscles. It has strong, dry legs that extend evenly from knee to foot, with no knots, swellings, or softness, and that are free of sores. The foot is even and with a smooth surface, that is, not rough yet rounded, touching the ground on all sides with a well-rounded hoof.

Beauty occurs when the head is small with respect to the body and is very dry, so much so that the skin adheres closely over the bones.[215] The eyes are large, lying

---

209. Cf. Pliny *HN* 28.24.88.

210. Citron: *Citrus medica.* Cf. *De veg.* 6.189, the citron tree, which produces a lemon-like fruit whose dried rind is often used in baking.

211. Pliny *HN* 28.24.88 says the touch of the elephant's trunk is a headache cure.

212. Mountain mint: *Mentastrum montanum,* today's wild mint, *Mentha silvestris.* Cf. *De veg.* 6.387.

213. Isid. *Orig.* 12.1.41. Other coverage of horses is found in Solinus 45.5, ThC 4.34, Neckam 2.158, Barth. 18.38–39, PsJF 370–72, and Vinc. 18.47f. It is of some interest to note that although the term "yoked" is used, the modern-style and far more efficient equine harness had been developed in the tenth century (Klemm, 1964, 80–84). For the ancient tradition, see Hyland (1990) and *AAZ,* 88–91. For the medieval horse, see Centre universitaire (1992).

214. Compare much of what follows with Isid. *Orig.* 12.1.45f. and Palladius *Agriculturae* 4.13.2f.

215. In older pictures of horses, it is striking how small their heads are in comparison with today's horses.

almost in front of the head. The ears are short and lively, pointed toward the front. The nostrils are wide open, and when it drinks, it sinks them deeply into the water. It has an erect neck, dense coat, and a large, long tail, and a roundness accompanying a firm solidness to its whole body.

53 As for worth, it is thought that the horse should be very bold, pawing and stamping the ground with its feet, whinnying and trembling in its limbs, for this is an indication of strength. It is also thought that it should be easily aroused from the greatest quiet and should easily grow calm and at rest from the greatest state of excitement.

The natural color of a horse caught in the wild is ash-gray, with a dusky line extending throughout its back from its head to its tail.[216] Among domestic ones, however, black, red, and sometimes white ones are found to be good as gray ones which have black coats intermixed with scattered small white circles, as it were. It is said, moreover, that when the young of a horse is born, there is born with it a bit of flesh on its forehead which is so effective a poison that it kills immediately. But the mother horse negates this by licking it and taking it off her young.[217]

In our lands there are, among the broken horses, four types of horse: war-horses [*bellici*], which are called chargers [*dextrarii*]; palfreys [*palefridi*]; racehorses [*curriles*]; and the ones called workhorses [*runcini*].[218]

War-horses are not castrated because they are rendered timid by castration. It is a trait of these horses to delight in musical sounds, to be excited by the sounds of arms, and to gather together with other chargers. They also leap and burst into battle lines by biting and striking with their hooves. They sometimes care so much about their masters and grooms that, if they are killed, they grow sad and pine away, even to the point of death. In sadness they sometimes cry and from this there are those who forecast concerning a future victory or defeat.[219]

54 The palfreys are used in that type of conveyance called equestrianship [*equitatio*].[220] These, too, should not be castrated lest they become effeminate.

The racehorses are used for escape and pursuit. These, lest their tendons [*nervi*] grow too hardened from the heat of running, are castrated to enable the dryness induced by the heat aroused from movement and running to be met by humor and coldness.[221]

---

216. An addition by A. On the wild horses that inhabited Europe in olden times, cf. Scanlan (104).

217. The *hippomanes* of antiquity in the sense of Pliny *HN* 8.66.165, used as a love potion. Compare the *hippomanes* of Vergil *Georgics* 3.280f. See also 6.99, 100, 102, with notes; and 7.115.

218. A *runcinus* could also indicate a "nag" (Latham), much as an inferior racer will derogatorily be called a "plow horse."

219. This belief, and perhaps the origin of this particular passage, goes all the way back to Homer, for whom the horses of Patroklos were said to weep at his death.

220. *Equitatio* could, in fact, have several meanings. Latham suggests a sense of cavalry service. Scanlan (105) chooses a sense of sport riding.

221. Cf. 19.33.

Workhorses are those which are kept for work with burdens or for pulling wagons or carts. Nonetheless, others are sometimes relegated to these chores as well.

Likewise there are four types of motion in horses: galloping [*cursus*], which in the horse consists of leaps; trotting [*trotatio*]; the pace [*ambulatio*]; and walking [*peditatio*].[222] Galloping is accomplished when both front legs are raised and the horse propels itself forward with both of its rear legs.[223] The trot occurs when, more swiftly than in its ordinary gait, it raises one front and one rear leg on opposite sides at the same time.[224]

The walk occurs in this way, but it occurs without any exciting of the horse.[225] A pace occurs when it lifts one front and one rear leg on the same side at the same time.[226] It occurs more smoothly, however, when it leads its feet along not by lifting them far off the ground but by sort of dragging them along, and it is better when it plants its front foot a bit sooner than its rear one. The more the motion deviates from one like this, the more difficult it is. Thus it is inevitable for the best pacing horses to stumble frequently, especially on a rough road.

The most appropriate food for horses is hard and not inducing flatulence. This is oats, wheat, and sometimes spelt. Barley is less appropriate and rye is the least appropriate of all because it causes flatulence.[227] Those who wish to fatten a horse quickly cook cakes from soft fodder and from this the horses take on a great, if false, weight gain.

55

---

222. We should be careful not to impose modern precision on A.'s descriptions. Modern terms have arisen after a centuries-long tradition of equestrian and cavalry training. Still, the descriptions that follow are not terribly far from their modern sense. Our profound thanks to Jacques Charton, former member of the famed Cadre Noir, trained at Saumur, and currently owner and trainer, Classical Riding Academy (Chattanooga, Tennessee), for his hours of help in this regard, including demonstrations of the various movements. Note as well that before the advent of photography, precise analysis of the movements of a horse was rather difficult. A.'s descriptions are thus notable and one must imagine that yet again he has had access to the *experti*.

223. There is a large number of formal movements in which the horse propels itself off of its rear legs with its front legs raised off the ground. A few terms include *levade, croupade, galopade,* and the extreme *courbette* and *cabriolle*. Mr. Charton, however, points out that what is described here is the preliminary movement just prior to the full charge, a most useful and common movement for medieval knights. In this movement the horse seems to gather itself, push off, and leap ahead. Indeed, *cursus,* in its etymology, indicates a running charge.

224. This is an excellent description of the trot, even in its modern sense.

225. While one is tempted by the etymology of the name *peditatio* to call this a walk, it is probably simply a slower trot.

226. *Ambulatio* has survived in the French name for this movement, the *amble*. Note that it is an unnatural movement for the horse, learned only by having its two legs tied together. The result, however, is an extremely smooth journey. It is used today in this country in harness racing. The drawing of a horse from Bodleian Library MS Ashmole 1511 fol. 32v clearly shows a horse exhibiting this gait (George and Yapp 1991, 114, ill. 73).

227. "Rye" is rendered for *siligo*, which can also, depending on the author, indicate fine white wheat flour. But in light of such passages as 22.69, it seems best to pick "rye" for *siligo* and be consistent with its translation throughout this section on horses.

This animal is one beset with many infirmities, which the groom in charge of caring for them should recognize.

In a horse there is sometimes an overabundance of corrupted and even of noncorrupted blood. The symptoms of this disease are that the horse enjoys being rubbed vigorously, its droppings smell more than usual, its urine becomes thick and reddish, and its eyes are sometimes reddish and weepy.[228]

Pustules sometimes arise throughout its body which, due to the horse's coat, can be felt more readily than seen.

56 The horse sometimes loses its desire to eat due to a fullness of its vessels. At this time, therefore, the horse has to be bled from the vein that descends in the middle of its neck. It should be bled with a lancet that is wide but not thick and in proportion with the horse's strength and age. For, if the horse is five or more years old, three or four *librae* should be removed.[229] If, however, it is weak and young, one and a half *librae* or at most two should be removed. If this draining is neglected, many discomforts result which are ascribed various names depending on their differing actions. The horses sometimes incur ulcer-like things which wander over the flesh and perforate the skin in various spots. They sometimes incur itching [*prurigo*], *impetigo*, *serpigo* [ringworm], or sometimes *scabies* [mange].[230] Then, with one horse infected, the others dwelling with it are infected in three ways.[231] One occurs because horses occasionally nip at each other with their teeth and when one is infected, the others become infected from its saliva and breath. The second method occurs when one rubs itself on a place where another has rubbed itself, for the infected humor is spread on the place and this infects the other horse when it touches the same spot. The third method occurs because the air is infected by breathing and this one infected horse easily infects all the others in a stable since horses are warm and moist and bodies of that sort infect easily. An indication of this is that boys, who are warm and moist, are forbidden from approaching people sick with infected blood lest they be infected.

57 If the blood from a horse's wound flows too fast, it is helpful to place over the spot a bit of felt which has been burned entirely or partially and moistened with nettle juice. Also the fungus which some people call wolf bladder [*vesica lupi*] may

---

228. One suspects kidney problems, leading to blood in the urine.

229. A *libra* is, technically, a pound. But capacity measures, both wet and dry, were frequently reckoned by this weight. Thus, DuC (5:97) lists a *libra* of wine and of beer. Scanlan (106) estimates that a *libra* is equal to a scant pint.

230. Disease terms, of course, are dangerous to translate exactly. *Impetigo* would normally indicate eruptions accompanied by scabs, *serpigo* is generally taken as ringworm, and *scabies* is like our generic term "mange."

231. "Infected": *corruptus*, the same word used just above for "corrupted blood." The description that follows, however, is so close to our modern sense of how disease spreads that the modern "infected" seems in order. Once more one is struck by the fact that A. has clearly gone to the *experti*—grooms, handlers, and trainers—who would be in a position to make the observations needed to draw these conclusions.

be applied.²³² The powder of this fungus may be mixed with the feces of a pig that has been fed on grass or on field herbs.²³³ When these are well ground, they should be spread, fairly warm, over the spot and kept bound on until the third day. Ground horse dung is also good for this, as is the dust of an old wimple or cloak, if ear dust is added, and is often cast into the wound until the blood stops.²³⁴ For all abscesses and wounds in horses, care must be taken lest moonlight fall over the horse by night. This is especially true for the injured spot, since moonlight very frequently causes death in an injured horse.

Lampers [*lampistus*] is a horse disease arising from a superabundance of blood.²³⁵ A turgid swelling occurs at the top part of the mouth near and in between the teeth to the extent that the depressions between the teeth protrude, causing the horse to let its food fall from its mouth. If these swellings are large, they should be slightly cauterized with a red-hot, thin-bottomed instrument bent in a curve to the shape of an S. If, however, they are small, the one that is furthest from the middle should be bled with a blade until the blood exits or else a cut should be made through the middle of the depression itself.

Foscellae are swellings that arise in a horse's mouth on the lips opposite the outermost teeth. These swellings grow black in the center. They arise from eating cold, harsh plants, which lie for a long time on the lips and jaws. They do not allow the horse's food to be positioned over the affected spot, just as happens in a horse with *lampistus*. The cure for this is as follows. A slender instrument should be made in the shape of a stylus in front and curved into a hook. It should be quite sharp. With this the skin of the *foscella* should be lanced in the middle and drained.²³⁶ When the spot is dry, the skin covering the *foscella* should be cut away in a circle with sharp scissors or with a scalpel. It then heals.

58

---

232. Brown (1995, 366) states that the nettle *Urtica dioica* is "an astringent, diuretic, tonic herb that controls bleeding," and he cites many uses against unwanted bleeding such as excessive menstruation, nosebleed, and hemorrhoids. *Vesica lupi*: In his notes to *De veg.* 6.343, Jessen identifies this as *Lycoperdon bovista*, and Brown (1995, 306) and Vickery (1995, 298) state that the spores of a related plant, *L. perlatum* (*L. gemmatum*), are a staple in folk medicine for stopping unwanted bleeding.

233. The "powder" or "dust" (*pulvis*) of the mushroom might either be the mushroom itself, dried and ground, or its spores. The latter is perhaps meant for two reasons. First, *pulvis* is not normally used for a ground-up medicament, and second, the next sentence begins with *haec*, which from context must be ntr. pl., "these things."

234. Once again A. uses *pulvis*. It is unlikely that this means ashes, as A. constantly uses *cinis* for that item. The second item, "ear dust," is *pulvis auricularis*. While admittedly an ingredient that smacks more of charms than of medicine, it is probably unwise to change the text to read *aurichalais* with Scanlan (108). Note that canine ear scrapings were part of a cure at 22.35(16).

235. In English, the disease is also called "lampas." It consists of granular adhesions to the palate behind the incisor teeth, usually the result of harsh fodder.

236. "Drained": this use of *traho* is confirmed in Vegetius, a late Roman veterinary author of a *Mulomedicina*. Another known expert was Jordanus Ruffus, Frederick II's marshal of horses, whose *De medicina equorum* represents a first attempt to introduce veterinary medicine to the nobility in written form (Gaulin, 1994; Prevot, 1991).

Barbules are dry swellings which sometimes occur on a horse's palate and which have the shape of the conical breasts of some small beast or other.[237] When they reach a size longer than that of a small seed of grain, they keep the horse from eating. The cure for this disease consists of taking a hook prepared in the way we have just described and draining the barbules, and likewise in cutting them away from the palate with forceps or scissors.

59 A disease sometimes arises on a horse's tongue which results from putrid food generating putrid and phlegmatic blood. For the food, digested by the stomach, always draws the better juices to itself by its heat and it turns them into blood in the veins. The horse is then warmed and nourished by its heat and humor. When the blood is bad it is sometimes drawn away from the heat of the horse's chest to its throat and tongue, drawn there by the motion of its tongue and the pressure of the bridle, thereby putrefying the tongue. The symptoms of this sick tongue are that the tongue is skinned by a sticky phlegm which also flows out of the horse's mouth and that the veins beneath the tongue turn black. It sometimes happens that the disease descends to the feet, whereupon the horse can barely stand.[238]

The cure for this consists of first scraping away the ulcerousness and stickiness from beneath the tongue. Afterward, mix two spoonfuls of soot and one of salt, grind well, mix with one bulb of garlic, and rub the spot with the mixture.[239] After this, cut the two veins which are beneath the tongue and, on the fourth or fifth day, bleed the horse from the neck vein in proportion to its strength. But some say that at the outset of this disease, about one-half ounce of blood should be removed. If, however, the disease has descended into the feet, the horse should be bled within the third or fourth day over the *sotular* of the foot inside and out, in whatever foot you like.

You should also know that a horse's head becomes emaciated and dried out if it is frequently rubbed and washed with cold water before it is seven years old. Its neck, however, thickens better and its hair grows better if it is wetted thoroughly and often with warm water near its shoulders and if the hair is scratched with the fingers. Do the same near the head, but with cold water rather than warm, since the head should be thin.

60 *Stina* is a disease which afflicts a horse's neck. There is also a certain inflexible instrument called a *stina* and the disease is called *stina* by metaphor, since the horse

---

237. *Barbulae* has come into our language as barbules, "little beards," describing the ragged appearance of the growths. A. ignores the etymology and uses the udder of an animal such as a cow for comparison.

238. This description is not inconsistent with hoof-and-mouth disease.

239. "One bulb": *caput,* lit., "head." A.'s meaning is not certain here. In *De veg.* 6.284, he talks of garlic cloves as *dentes,* "teeth," without using *caput*. It seems most likely, however, that he means an entire garlic bulb here, using *caput* in the sense of "principal part," rather than in the sense of "top" (i.e., the flower). Further, if a single clove is a "tooth," then "head" becomes an appropriate designation for the entire bulb. Garlic was commonly used in simples, and it does seem to have some bactericidal effectiveness (Brown, 1995, 234; Talalaj and Czechowicz, 1989, 146).

cannot bend its head to this side or that without pain and can take its food only at intervals, snatching, as it were, whatever it takes from the ground quickly.

The disease arises from weight being put on the shoulders, from the extension of the nerves, and from repletion.[240] The cure for this consists of raising the horse's mane forcefully in the hand and of piercing each side of the neck near its hump with a red-hot awl, so that the flesh near the neck is burned a bit without touching the nerves. Do this five times along the length of the neck and put a bandage made of hemp, linen, or hairs from a horse's tail through each hole, leaving it there for fifteen days with a thread arranged to draw off the fluid.[241] However, some, by way of a cure for this disease, make many cauterizations on the left side of the neck near the apex of the neck and do not apply the bandages. In whichever method is used, the first or the second, the neck and shoulders of the horse should be washed and fomented with lukewarm water for all fifteen days, all day long.

Turtae are a kind of subcutaneous abscess arising on the surface of the flesh and have the shape of the bread called a *turta*.[242] This disease is especially caused by a subcutaneous superfluity of blood and putrid humor. They sometimes arise when the flesh is injured by some blow or other. The cure for it consists of cutting the skin over the middle of the *turta*. Afterward, when the swelling has gone down and the humor that was within it has dried up, it is massaged and thoroughly rubbed and in this way all the humor is pressed out. Then, let a drain be inserted, cutting as much of the skin as necessary until a crack descends to the healthy skin. Whatever putrescence or putrid flesh is found should be removed and the wound filled with tow.

This is done all day long until the wound is perfectly healed. If, however, it is noticed that a collection of such putrescence is occurring again, the skin should be split again and the pus squeezed out with the fingers. The horse should be bled from each side of the neck on the first, seventh, ninth, and fifteenth days. This should be done with careful consideration, considering the proportion of the humor from which the abscess arises and the other accidental traits from which this disease sometimes arises.

"Fig" [*ficus*] is a bad gelling of humor which grows outside the skin, without hair, with a reddish, bluish, or dusky color, and with the shape of a ripe fig. From this fact it takes its name.[243] It arises from too much subcutaneous blood.

The cure for this disease consists of this. If a wide lump of a *ficus* appears on the neck or on any other limb, then a cut should be made wider than the skin which forms the shape of the *ficus* and it should be opened up well so that a rounded hole is created in the middle. Between the *ficus* and the good skin place a paste made from

---

240. "Nerves": as always, this includes tendons and sinews.

241. That is, a drain.

242. *Turta: torta*, etymologically indicating the bread was twisted or braided. We know the form in English as "tart." Scanlan (110) suggests the disease is "a carbuncle or multilocular subcutaneous abscess, often caused in stabled horses by the highly contagious Staphylococcus aureus."

243. Probably the growth called the angleberry, or sarcoid, which is caused by the Papova virus (Scanlan, 111).

very glutinous, fine flour. If paste is not at hand, use potter's clay lest the healthy skin be harmed. After this make a small cake out of green horehound that has been well pounded and heated over a hot rock or wide iron utensil.[244] Apply this cake when it is quite warm to the *ficus* and hold it on as long as it is warm. When it cools off apply another warm one until the *ficus* begins to turn white. Finally, make some flat cakes [*tortelli*] out of cress—either the green or the water variety—and the bark of a walnut tree.[245] These heated cakes are to be pressed on to the *ficus* each in turn until the lump of the *ficus* is made to look like the healthy skin and until the putrefied humor is fully allowed to drain. Then the *ficus* is filled with quicklime dust or dust of the chalk used to make parchment. It should be cauterized with an instrument equal in width to the *ficus*, hollowing it out down to the good, live skin, while being careful that none of the nerves which happen to be there, no muscle, member, or joint, is touched with the instrument. Then, fresh cow dung or the dung of a cock or pigeon, well ground and mixed with soap, should be bound on for two days. The spot should then be smeared once a day with some warm ointment like *pentamiron* or something else like it until it is healed.[246] It should then be washed with cold water and fomented.

63 If, however, the swelling of the *ficus* is broad at its top both in length and breadth, but near the healthy skin is narrow, take a silk thread and a bristle or hair from the tail of a virgin foal which has not copulated yet. Having cooked these things ahead of time and braided them, make a cord with which the skin surrounding the *ficus* can be bound very tightly. If it comes loose, tie it again until what is held tightly by the thread falls off on its own.

If, however, *ficus* should reappear, a circle of paste or of the stickiest potter's clay, just as we mentioned above, should be applied between the healthy and injured spot. The *ficus* should be split and then warm honey, for this is a good cleansing agent, should be poured in while being very careful lest the warm honey drip over the healthy skin.[247] After a brief delay, clean away both the honey and the circle and heal the injured spot, as has been said above.

---

244. "Cake": *tortula*, a diminutive of *turta* discussed just above. From the description and the mode of cooking it, it must have resembled its later descendant, the tortilla. Whether the *tortula* is the same as the *tortellus* (see below) is unknown. On horehound (*marubium*) see *De veg.* 6.389. Brown (1995, 308) claims it is an antiseptic and that it is used "externally for minor injuries and skin eruptions." Talalaj and Czechowicz (1989, 315) add that it contains up to 7 percent tannins and steroids.

245. The "green cress" may be the *nasturcium* of *De veg.* 6.393, identified *ad loc.* by Jessen as *Lepidium sativum*, or garden cress. It is good, A. says, at dying up putrescences. "Watercress," *nasturtium aquaticum*, is identified as *Nasturtium officinale* by Jessen (1867, 729), and when ground and applied as a poultice, it reddens the skin as does a mustard plaster (Mitchell and Rook, 1979, 234). In this case, when applied with the warm "cake," it must have increased blood supply to the area. On the walnut (*nux*) see *De veg.* 6.147f.

246. From the Gr. *pentamyron*, "five myrrh." A recipe is given at 22.85.

247. Majno (1975, 115–20) attests to the ancients' use of honey to cover dressings as early as 1650 B.C. and concludes that it was "practically harmless to the tissues, aseptic, antiseptic, and antibiotic."

If many such growths arise on the body of the horse, it should be bled in proportion to its strength.

Moreover, if by chance the spot is full of nerves [sinews], caution should be exercised not to bathe the nerve in cold water, because the nerve itself is naturally cold and is turned torpid by the water and putrefies. If it is necessary that the nerve itself be cut, it is better to cut it through rather than to puncture it or crush it with a stone. For the very great pain of a puncture impedes the healing process more than does cutting through it unless it is a case of the large nerves which cannot be cut.

The cure for a nerve cut in this fashion should be accomplished with warm things and the spot should be fomented with warm penetrating agents such as oil, lard, and honey, all well cooked. A poultice should be made from the powder of the laurel berry and cumin. It should be mixed with honey and applied and the opening should be kept open until all the pus comes out.

If, however, the nerve is damaged by the blow of a stone, or by the horse's falling, or should it be impaired and torpid for some other reason, it should first be rubbed forcefully with warm water and warm wood ashes, fomented, and have pressure put on it. It should then be smeared and forcefully rubbed with whatever warm ointment can be found.

But if the flesh is wounded and the nerve has been struck and rendered putrescent, it is useful to apply a poultice over the injury made of bean or barley flour cooked with honey and wine until thick.

A poultice made of honey and the roots of dwarf elder [*ebulum*], marsh mallow [*altea*], bryony [*brionia*], and lily, and bound over the injured nerve also does great good for this.[248] Attention also must be paid if the nerve is cut lengthwise or obliquely, for it is not easily joined and will, perhaps, be impossible to join. When, however, it happens that such an incision has occurred, take the long vermin from the ground called earthworms [*lumbrici terrae*], grind them well, mix with honey, warm them a bit at the fire, and apply them with no other intervening medication. This is of use.

Cutting the veins is occasionally performed on a horse and the veins are cut transversely like a saw cuts wood. The incision should be done as follows to keep the blood or rheum from flowing to the members which are weak, such as the eyes, feet, and other members. The skin should first be fomented at the spot of the incision with warm water and, having been first shaved of hair, should be rubbed for a long time with the hands so that it opens up a bit. The skin should then be raised above the other skin and should be split along the length of the vein that must be split. The vein should then be separated from the flesh and split itself. If it is thick and filled, as

---

248. On the herbs and plants see, in order, *De veg.* 6.267, 285, and 245. Dwarf elder will reappear frequently in the recipes that follow. Brown (1995, 347) reports that flowers of the common elder may have anti-inflammatory properties. On mallow, see comments at 22.32. There are many species of bryony, but Talalaj and Czechowicz (1989) list several useful properties of black bryony, *Tamus communis* (66), and white bryony, *Bryonia dioica* (312). Brown (1995, 304) claims that *Lilium candidum* (madonna lily) "soothes and heals damaged or irritated tissues."

much blood as is necessary should be taken from the split. It is then raised with a stick made of soft wood to a height of two fingers and is then bound with a soft thread on each end of the part of the vein that must be cut. When the cut is made, the ends of the veins on each end should be slightly cauterized. Both the threads and the ends of the veins should hang outside the wound so that the vein between the ligatures can putrefy and both the threads and the pieces of the vein that has been cut can be easily removed. If, however, the blood gathers in a member, especially the foot, the vein should be tied off at its lower section before it is removed and not at the section that leads to the heart. It ought then to be opened and the blood will come forth.

66   If the nerves or muscles of the shoulder are injured by the prick of a spur or for any other reason, or if they just swell, the hair should first be shaved off around it. The injured place should then be frequently smeared with warm oil or liquid lard.[249] Then grind wormwood leaves or the tender small branches of that plant or of dwarf elder and mix with butter.[250] Bind on the poultice thus made and keep the hole of the puncture open for a while and it will be cured. If, however, a swelling of this sort persists for a long while without the benefit of medication, open the skin cautiously at its two edges where the swelling stops. Insert a drain, being careful of the nerves and veins, and move this until the collected pus is removed.

If, however, the horse is swollen near the neck due to a puncture of the shoulder or flank, it ought to have a drain put in its chest [*inzonetur*]. Then, be careful not to put the drain too near the shoulder.

Swelling frequently occurs when cold water enters through punctures in the skin of a horse made warm by movement and which has rested for two or three days thereafter without working and sweating. The treatment for this consists first of washing the swollen spot, and then thoroughly fumigating it with dwarf elder, ligustrum [*livisticus*], or wormwood so that the injured spots sweat freely.[251] Then wheat bran should be mixed with the lees of wine or of good beer or with dwarf elder juice or the leaves of the black elder [*sambucus*] until the mixture thickens.[252] Bind this onto the injured spot. Then, if they are needed, two or three drains should be cautiously applied between the horse's shoulders and thighs in order not to injure the nerves.[253]

---

249. *Sagimen lardi*: *sagimen* seems to be reserved for liquid animal fats and is used at 3.142 to describe the oil from a sperm whale's head. On the problem with A.'s vocabulary in this area, see Glossary, s.v. "Fat." To help the reader the Latin is often left in the text.

250. Wormwood, *artemisia absinthium*, would be more recognizable by A.'s *absinthium*, absinthe, the source of the poisonous but delightful liqueur so loved by Toulouse-Lautrec. See the interesting description offered by Mann (1992, 105–9). It does have a history of use for external bruises and bites (Brown, 1995, 243).

251. Two of these ingredients have already been discussed. Scanlan (115) identifies *Livisticum* as privet hedge, *Ligustrum vulgare*, what most people call simply hedge or ligustrum. This has been known to induce vomiting in animals. Jessen (1867, 726) identifies it as *Ligusticum levisticum* L.

252. *Sambucus*: Jessen (1867, 741) identifies this as *Sambucus nigra*, a black elder. Cf. *De veg.* 6.220 and see 22.71, where the *alnus* is also a black elder.

253. "Thighs": perhaps "hips."

For, unless this is done cautiously, there is injury to either the nerve leading from the penis to the chest along the middle of the stomach or to the great vein on its flank.

If the horse is injured between its cinches because it has gone in harness too long, or if the great vein on its flank is punctured and the blood cannot drip out due to the narrowness of the punctures and it swells as a result, the following ought to be done. Five days after the pus is ripe extract it through a hole on the skin in such a way that the pus is forced to leave by the pressure of the fingers.

There is a horse disease which some call *radunculus.* It is a wide swelling, red in color, occupying both skin and flesh to the very bottom. It is accompanied by a great, continuous fever and a pulsation of the skin at that spot. This occurs in cases of wounds, punctured nerves and muscles, and in large injuries to the back and side of the animal. It also sometimes occurs without any outside injury, just as it sometimes occurs in a human at a spot where the humors flow together, seeing that the humors flowing together are in a state of motion and remain in a tumult. Such a member should be touched only with cold things so that that member or neighboring ones will not be stricken with greater problems and so that the humors will not flow inwardly and cause the animal's death. These things will occur if warm things, which are known to attract humors, are applied. One ought, therefore, to repel such humors that flow together through the use of repellent and evaporative poultices, either to restrict them or dry them out. This disease arises from a superabundance of bloody humors flowing to an injured spot which, due to weakness, cannot repel them from itself, and can change them into neither evaporation nor pus. *Radunculus* differs according to the variety in the quality and nature of the humor which is dominant in the body.

In the course of treatment, however, one should know above all else the origin for all its causes so that the medications can be applied in accordance with what the disease [*causa*] demands.[254] For each disease one should therefore know what is proper to apply in the beginning stage, the stage of its growth, the stage when it holds steady, and the stage of its abatement. This is especially true for swellings, because swellings occur in different members for different causes. They therefore bring on serious dangers and very often cause bodily weakness unless assisted by the proper medicines. If, however, it is proper to remove some blood because there is too much, do the bleeding in a part opposite from the disorder. That is to say, if the lower righthand side is troubled, take the blood from the upper left side. But if the upper part is troubled, let it be taken from the lower left.[255] This advice is to be followed if the flow of the disease is at

---

254. An interesting problem. The Latin reads *oportet cognoscere originem omnium causarum,* which we have dutifully translated. Yet surely this is not what it meant originally, and for A. the use of *causa* here to mean "disease" is rare. The use was common in Vegetius (L&S, s.v. *causa,* F), who, directly or indirectly, is a source for much of what A. knew about horses. In its original sense, then, the passage undoubtedly meant that in order to cure something one needed to know the origin of all diseases. But to the scholastic, *causa* had an entirely different meaning. The unanswerable question, then, is what A. thought he was saying here. To help the reader follow the puzzle, *causa* is rendered simply as "cause" in this passage.

255. For a good discussion of the theory behind these practices, see Gil-Sortres (1994).

the beginning or at the growth stage. If, however, it is in decline or the stage when it holds steady, make the withdrawal from the member nearest to that part. For example, if the swelling descends to the femur or to the knee, do the bleeding in the foot. But if the swelling is of long standing and the heat in it has passed, the force of the pain has diminished, and the color has changed, and yet still a thick humor clings to the spot due to its own thickness as if it were some gluey mass which cannot be got rid of any other way, then the spot should be scarified fairly deeply when the horse is warm and the thick humor and blood should be removed by the application of a cupping glass. The scarified spots themselves should then be treated just like the wounds.

A medication by way of a general poultice is also made which is very effective for cutting away all swellings, scarifications, *radicunculi,* pains, and burns made by an iron and since turned hard. It is also good for chilling and softening warm swellings. Take any amount of wormwood with the yolk of an egg and the same amount of pure liquid fat [*sagimen*]. Mix them and add a small amount of the finest barley or oat flour. Apply this to the spot in pain. Apply it warm in winter, cold in summer.

69 There is another poultice which cures *radunculus* and swellings and which heals wounds. This is formed by mixing two measures of celeriac juice and two of wine with one of old fat [*anxugia*] cleansed of its salt.[256] Mix them until the fat liquefies and strain it through vessels (for then the salt will descend to the bottom). When it is cooled, mix two measures of raw honey with it and to all this add the best sort of wheat flour until it thickens. Apply this warm to the *radunculus* and it will get better. Another poultice is used for dispersing a swelling with its skin intact. The fine roots are cut off a leek, cut fine, and rubbed in liquid fat [*sagimen*], and this is bound on while still warm.

There is still another poultice used to repel the swelling and to compress it when the skin of the *radunculus* is intact. Take three measures of ram dung that has never been wetted with water, two measures of the juice of the blessed herb, that is, *gariofilata,* and the same number of raw egg yolks, one measure of pure sheep fat [*sepum*], one of the best rye or wheat flour, and mix all these together.[257] Cook for a while and bind the whole thing on warm.

70 Strangles [*stranguilina*] is a disease that takes its name from the fact that it constricts all the passages of the horse's throat which lead the breath from the depth of the chest to the nostrils with the heaviness of a cough. It is caused by putrid or inappropriate feed made too thick from water. The phlegm collects in the chest and body of

---

256. "Celeriac": *apium*. Jessen (1867, 704) identifies this as celeriac, *Apium graveolens,* a very redolent plant, and points to A.'s equine cures in his explanation. Curiously, the plant was found in King Tutankhamen's tomb (Brown, 1995, 240).

257. "Blessed herb": *herba benedicta,* the English avens plant, *Geum urbanum.* Jessen (1867, 706; and on *De veg* 6.470) equates this plant with *gariofilata.* Cf. 22.81, where A. gives yet another synonym, *cicuta.* In modern parlance most species of *Cicuta* are highly poisonous, so the synonymy is suspect. Brown (1995, 288) and Talalaj and Czechowicz (1989, 333) point to the use of *Geum urbanum* to control bleeding and reduce inflammation, and as a wound dressing.

the horse and much of it stays there when the horse is at rest.²⁵⁸ It is also occasionally caused from dry food eaten with dust or from drink that is too cold. It is also caused in cold weather when the horse has taken too much cold water and has stood too long in a cold place without any warming and without a covering. This is especially true if before this it was tired and empty of food. The blood should in no way be drawn until the phlegm is ripe, the rheum from the nose and mouth stops without a cough, and the breath exits as it does from a healthy animal.²⁵⁹ For if perhaps the blood should flow, it is with difficulty that the humor is purified of the coldness that arises as a result of the lack of blood unless perchance the blood was exceptionally abundant in it beforehand. This disease quickly brings the horse to ruin if it has labored under it recently, that is, within the last eight days. It should also be noted that a horse suffering from this disease frequently is free of it within twelve days or else the disease passes over into glanders and the horse is then in danger.²⁶⁰ These, then, are the causes from which the strangles arise.

It also arises on occasion from too much association with another horse suffering from the strangles. This is due to its breath, as has already been said above.

The cure for this disease is as follows. Take the bark of a black alder tree [*alnus*], which grows along the banks of streams.²⁶¹ Clean it well of its exterior superfluities and place it in a new pot. Pour clear water over the top and boil it until it is almost gone. Having poured it over again and still a third time, boil it the same way until it is almost consumed. Then put in an amount of liquid lard [*sagimen lardi*] equal to the remaining water.²⁶² Strain the whole thing and purify it. Place it in a dish or some other vessel, and then with some instrument or other put it in the nostrils of the horse, binding the head of the horse firmly in an upright position with a halter or with reins until the liquid flows from its nostrils into its head. Then it should be fed such warm herbs as are good at warming and thinning out humors. Bran is very good for this. If, however, it is winter, let it eat the herb called groundsel and a thin porridge made of wheat bran and warm water. If, however, the strangles is caused by a cold in the

71

---

258. Perhaps "causes great lethargy in it."

259. "Blood should in no way be drawn": the verb is *attraho*, which should indicate drawing blood toward a place. Moreover, the verbs *extraho* and *subtraho* are regularly used for blood extraction, as they just were in 22.68. Yet what follows clearly seems to indicate blood loss.

260. "Passes into glanders": *in morvellam transibit.* Scanlan (119) offers "it may enter a phase of delayed after-effects," but he does so without comment. *Morvella* is not found in the standard lexicons, but it is almost surely allied with the Fr. *morve,* "snot" or "glanders," and the phrase *cheval morveaux,* indicating a horse suffering from glanders. Glanders, described below at 22.77, is marked by a steady mucus discharge from the nostrils.

261. Jessen (1867, 702), in identifying the *alnus* as *Alnus glutinosa,* the black elder or common alder, refers to this passage, though he mistakenly says it concerns the dog. Brown (1995, 235) notes that the herb is "an astringent tonic herb that encourages healing of damaged tissues," and he specifies that the bark is used "externally for throat, mouth, dental infections, and scabies." Talalaj and Czechowicz (1989, 332) specify the bark as a cure for pharyngitis.

262. A troubled passage, not helped by the ungrammatical *quae* in St., missing in the Borgnet text. Omitting the *quae* allows for the translation proffered here.

head or if a dry cough shakes the chest, the horse should be given the herb which is called periwinkle with water to swallow for three days.[263] Moreover, warm aromatics which also moisten the internal parts should be put near it to arouse its heat. Also, the leaves of the dwarf elder and ground groundsel should be placed on it and they will do some good. Also, warm tiles and other warm things should be put on over it along with the aforementioned herb and moist linen bandages which are wrapped around the tiles. These are then placed around the horse and the herb is placed in a vessel. From this the nostrils of the horse are fumigated and its head is covered so that the vapor can go no place other than to the nostrils. It is fumigated this way for a long time and if the vapor [*fumus*] should happen to run out too quickly, put warm water in its nose.[264] Then, before it takes any other food, mallow root which has been ground with unsalted butter or which has been mixed with oil [*sagimen*] should be put in the horse's throat for it to swallow. Give the horse a porridge as well made of mallow and groundsel leaves. In this instance it is also useful to put well-cleaned linen or hempen drains on it, passing them through the horse's neck on either side of its esophagus [*canna gulae*] so that each drain is three fingers' breadth away from the other.[265] Leave these drains in until they fall off on their own or until the horse is cured of its strangles.

72 *Cancer* is a disease which eats widely at the skin along with the flesh.[266] It has a dusky color since it is generated from black and thick blood. If this disease is eating at the lip of the horse, dry hemp seed very well and sprinkle a very fine powder made from it over the lip twice a day until it is healed.[267] Meanwhile, however, keep the horse away from any wetting with water and, if it is necessary, take blood from the left part of its neck.

Frenes is a disease so called because in this disease there is a very great flow of the humors which destroys the horse's kidneys [*renes*] and renders them immobile.[268] The beast itself, however, falls to earth as if it has epilepsy and the humors thus rush to the heart. It therefore sometimes dies within two hours. This affliction occurs more often in warm than in cold weather on account of the warming of the humors.

---

263. "Periwinkle": *semperviva*, "ever-alive," probably so named because it is an evergreen. Cf. *De veg.* 6.447, where Jessen identifies it as *Vinca minor*. This is the plant from which vincristine, a chemotherapeutic agent for certain cancers, was first isolated. Both it and *Vinca major* have been used for nosebleeds, sore throats, and mouth ulcers (Brown, 1995, 369).

264. Perhaps "let it be put with hot water in its nose."

265. Hemp was widely used for rope, of course, but *Cannabis sativa* has also been reported to have analgesic, anti-inflammatory, and sedative powers (Brown, 1995, 253; Talalaj and Czechowicz, 1989, 167).

266. Not necessarily cancer in the modern sense, but rather a skin condition that spread in a fashion reminiscent of a crab [*cancer*].

267. "Twice a day": *bis in die*; note that physicians to this day use the abbreviation *b.i.d.* to indicate "twice a day."

268. The etymology is clever, but probably incorrect. The word is more likely related to the Gr. *phrēn* ("midriff," and thus the "mind" in early Greek thought), itself related to *phrenēsis*, or frenzy.

The treatment for this disease is as follows. The thick vein which lies between both thighs and the vein which is beneath the tail should be incised in a direction away from the rump for a length of four fingers so that the blood can be withdrawn from the rump.[269] This should be done quickly since it is dangerous to put off such an evacuation in such an affliction. The blood, moreover, should be allowed to flow almost to the point of harm since immoderate repletion demands immoderate evacuation. If, however, the kidneys are weak and after a few days cannot grow stronger, perform two cauterizations along the middle of the kidneys of the horse. To keep the burned places from showing hair loss, constantly apply cooked chervil [*cerefolium*] ground with fat [*adeps*].[270]

Cornu is a disease so called since the skin on the flesh of the back or previously injured flesh alone seems as hard as horn [*cornu*].[271] This usually happens when the horse is injured and swells from an excessive load and then has also borne a heavy load on itself before a swelling has subsided. It also occasionally occurs when its back has been injured, has swollen, and has lost its skin, and the horse is then ridden before it is entirely healed; from this great exertion its back gives off a warm, moist sweat. When this happens, the horse should be allowed to stand at peace for three or four hours without removing its saddle until its back has cooled. Then, with its saddle still in place, it should be ridden again. For *cornu* sometimes occurs when the hairs on the back are too long, for then they collect the sweat and are made shaggy. Then, if by chance some hard object like a stone, a piece of wood, or a folded cloth lies between the saddle and its back, then the load rocks back and forth and the horse is injured as a result.

If, however, a *radunculus* befalls the *cornu* and occupies the horse's sides and back before the *cornu* has been completely or partially removed, the animal will die or barely survive and will do so with great suffering.[272]

The treatment for this disease consists of first shaving the hair off all around and then, if the *cornu* lacks an opening, the *radunculus* should be scarified with many scars over the whole *cornu* and all the way to its bottom so that the injured blood can come forth. If, however, the swelling has been untreated for a long time, the spot should then be fomented for two or three hours with warm water before the scarification in order to thin out the blood. The water should be that in which celeriac, hellebore, or *myr*, which by another name is called hen's bite [*morsus gallinae*], has been cooked.[273]

---

269. Vague Latin yields the following alternatives: "thighs" is *coxae* and thus might be hips or even legs, but certainly the rear ones are meant; "in a direction away from the rump . . ." might indicate a place four fingers' distance from the rump, but in that case the use of *per* is suspect.

270. Chervil, *Anthriscus cerefolium*, has been used as an anti-inflammatory (Brown, 1995, 239).

271. Probably a very hard callous.

272. Perhaps "occupies the *cornu*'s sides and back."

273. Normally, hellebore would refer to *Helleborus* spp. (Ranunculaceae). These are very poisonous. *Morsus gallinae*: see *De veg.* 6.333; Jessen (1867, 728–29) tentatively identifies this as *Stellaria media*, common chickweed, used externally for skin complaints (Talalaj and Czechowicz, 1989, 94; Vickery, 1995, 65). Mabberley (1987, 270–72) identifies henbit [*sic*] as *Lamium amplexicaule*, a nettle that is somewhat toxic when ingested.

Afterward, a poultice made of dwarf elder, celeriac, and the leaves of the black alder [*sambucus*], well ground up and warmed with a bit of liquid fat [*sagimen*] or some wine, should be bound on it. Or else apply a warm poultice made with hen's bite juice or that of groundsel or the finest rye or wheat flour, mixed with four or five eggs. After this, pierce the skin of the *cornu* in different spots with a hot awl. Then cover it with a coarse linen or hemp cloth so that the cloth extends beyond the *cornu* to a width of four fingers in all directions. Afterward, heat a piece of bacon lard stuck on a hazel or willow branch over the fire so that the melted grease, running down the stick over the cloth that is placed over the *cornu,* drips through the pierced spots into the *cornu* and thus penetrates through the cloth into the inner parts of the *cornu.* Then, having put another, clean cloth over the first one, ride the horse until it breaks a good sweat. Later, should it be necessary, anoint the horse with a feather on the spot of the *cornu,* using the same fat, collected in another container. Do this until the *cornu* can be lifted off, and afterward put a poultice over it made of snails crushed up with their shells. Do this until the *cornu* is healed. If it should be necessary to ride the horse, remove the poultice for a while. After the *cornu* has been removed, fill the sore once a day with lint that is not very irritating but which is very clean. The lint of linen or hemp is very effective in healing sores or wounds and in preventing dead flesh from developing. Be careful, however, not to let the lint get wet.[274]

75   Dead flesh appears on a horse's wound when the treatment for the wound is drawn out or when the wound is not treated with appropriate medicines.[275] The symptom of this affliction is that this flesh sometimes moves beyond the skin, is not like the other flesh, and is not readily sensitive to touch. If it moves beyond the skin and is hard, and if veins and nerves are not in the way, either cut it with a very sharp razor down to the good flesh or else put Greek nettles [*urtica Graeca*] into it, for this eats away at dead flesh. Then apply either fresh cow dung or lint moistened with egg yolks and bind them on for three days. Before this, burn the wound a little with a heated instrument and afterward apply saliva.

Dead flesh is also removed without an instrument in the following way. Take three parts of quicklime, two of seashells [*concharum ostreorum*], one part salt, and one of deer horn. Grind them very fine. Mix them together with strong lye or human urine (the urine of a virgin boy is best for this). When mixed, shape into the form of a loaf of bread, cook it in the oven for a while, and then reduce it to a powder. Put this powder once a day over the dead flesh. If living flesh should appear in one part of the wound faster than in another, fill the part with the living flesh with lint that has been moistened with saliva and apply the powder to the other part until living flesh appears everywhere. Then apply lint bandages that have been slightly moistened with butter

---

274. Although lint seems rather a messy bandage to moderns, Majno (1975, 116) states that in ancient Egypt the lint was applied "alone and dry, without bandages, perhaps to promote drainage by capillary action."

275. This granulated tissue is often known by the name "proud flesh."

or liquid fat [*sagimen*] until the living flesh is equal to the flesh or the skin. Then heal it as has been discussed above concerning the healing of wounds.

There are those who maintain the following. If a person takes the bones from the legs or flank of a horse, the horn of a stag or ram, and old cuttings from shoe leather and grinds each up, making a powder from each, and if an equal amount of each powder is then taken, mixed thoroughly together, and saved for use, it is very effective at drying up wounds and eating away dead flesh. Some even add oregano and the bark of an oak to these ingredients.[276]

If the animal has to bear a pack in the interim, the lint and the poultice should be removed for a while and, when the work is over, the wound should be washed with warm wine, urine, seawater, salted water, or water in which oregano, celeriac, horehound [*marrubium*], or dwarf elder has been cooked. The place should be washed with this water warmed and then one of the things mentioned above should be bound on it again.

There is another poultice used to do away with bad flesh. If the white of eggs and old soap are liquefied, placed into powdered quicklime, mixed well, and then applied moderately thickened and quickly, the bad flesh will get better.

Farcy [*farcina*] is a disease that takes its name from stuffing [*farciendo*], taking the name from the overly moist flesh and immoderate repletion.[277] Some call it *vermis* [worm] since the superfluous humor in the flesh and skin causes holes that resemble wormholes. It is caused by blood which too often produces rheum outside the veins. It is sometimes caused by a great blow or impact and the resultant black-and-blue mark. This is especially so if they are not treated within two months, and if they are located in hollow areas and are between the shoulders and on the sides. It is sometimes caused by the bite of another horse that is suffering from farcy.

The treatment for this must be provided diligently and as quickly as possible since, if it occurs on muscles, nerves, bone joints, or hollow spots, it is cured with difficulty. This is the treatment. Determine whether the infirmity is occurring in the front part of the body and arises from an abundance of blood. If so, bleed it from the vein on the other side of its neck. Then moisten three handfuls of avens [*gariofilata*], three of plantain [*plantago*], three of agrimony, and one of radish root with water from a well or a ditch.[278] Give this to the horse to swallow or make a poultice from avens and radish root taken in equal quantities and ground together with a bit of old soap or

---

276. "Oregano": also known as wild marjoram. Cf. *De veg.* 6.398. Some oaks contain up to 20 percent tannic acid (Brown, 1995, 338) and could be effective in drying wounds if applied topically. Vickery (1995, 264) reports that "oak bark was commonly used as a cure for sore shoulders in horses. The bark was boiled and the sores washed with the water."

277. Also called glanders. As A. indicates below when he mentions its spread by bite, it is infectious and is caused by *Actinobacillus mallei*. See notes at 22.70.

278. Plantain is a common laxative and would thus address the issue of repletion (Mabberley, 1987, 460; Brown, 1995, 331). Agrimony has anti-inflammatory effects and radish contains the antibiotic raphinin, an effective agent against many fungal and bacterial infections (Brown, 1995, 331, 231).

honey. Bind this over the holes, having first shaved off the hair. Alternatively, make a powder from the powder of burned *atramentum,* quicklime, soap, and a bit of honey, and place this within the holes.[279] Do this twice a day, in the morning and evening, until the infirmity dries up. If, however, the holes are too narrow, open them up a bit with a razor. Meanwhile, have the horse eat barley straw, wheat, or hay. Keep it away from oats and from drinking water.

If, however, the spot is not in the hollows of the bones or muscles but rather is in a fleshy spot, do as follows. Rather than putting a poultice on the outside, it is better to cut into it with a razor and to lay bare all the hidden fleshiness down to the very bottom of the farcy. Afterward, burn the spot with a hot instrument and then apply a well-made poultice consisting of barley or wheat flour, ground together with egg yolks and agrimony or leeks. Some people fill all the holes with bran mash and then burn them to their very bottom with a hot awl.

78 If the horse is given barley grains or those of some other grain, or some kind of bean is served it for its fodder, and if it chews these grains either too little or not thoroughly enough, it overfills its stomach this way. Then the natural heat of the stomach, because of its overfilled condition, is not able to digest these things and the horse cannot hold its head up well, holding it too far away from itself. It is also thirsty, with an excessive desire to drink. For this reason, many who have a lot of experience with horses first give their horse, following a hard bit of work, chaff or hay to eat before they drink. They then throw the grains, scattered in front of them two or three times. The horses cannot take up many of these at one time before their hunger abates somewhat and they are accustomed to chew thoroughly. If the grain they dole out is too hard, they moisten it well three or four times. Now if, as mentioned before, the horse is overfilled, the treatment is to keep it away from water until it voids and makes urine. If it is not kept from water in the manner stated here, it will suffocate from too great a swelling of the grains and from the moisture of the water or else it will suffer from diarrhea and will have distress when it defecates. Now if it has not produced any dung or even urine for one or two days, take two parts of dwarf or common elder [*ebulus/sambucus*], one part of spearwort root [*flammula*], and one of mallow, and cook well in water.[280] When strained, give one cupful to the horse to drink. Then work the horse a bit to heat it up and then cover it. But if, within six hours, it has eliminated nothing, give it another cup of the aforementioned concoction. Now there are those who, on that same day, insert their hand into the horse's backside and draw out the grains, thus opening the passage, but the first method is more reasonable and healthy.

79 Worms sometimes grow very numerous in the stomachs of horses. The symptom of this infirmity is that the horse frequently rolls on its sides and tries to scratch its

---

279. *Atramentum* can mean several things, such as carbon black and ink. A. may mean, though, the mineral he discusses at *De min.* 5.1.3 (Wyckoff, 1967, 242–44).

280. *Flammula: Ranunculus flammula,* a tentative identification made by Jessen (1867, 716) in reference to this passage. Buttercups often contain a bitter and irritating compound, ranunculin, that is converted to protoanemonin, an active compound that induces diarrhea if eaten.

stomach with its hind legs. Its hair also stands erect as if bristling and the horse grows more slender than usual. Unless it is given aid quickly, before the worms perforate the intestines, it rarely or never recovers. The worms arise from bad food along with a lack of drinking water.

This is the treatment for this disease. Take all the intestines of a young chicken, place them whole and warm in the horse's throat, holding the horse's head up until it swallows them. Do this for three days, always in the morning. Meanwhile, do not allow the horse to eat anything or to drink more than a little bit or nothing at all until the ninth hour.[281] Likewise, if the horse does not wish to drink, take one handful of black alder [*sambucus*] and one handful of the white birch twigs [*mirica*].[282] Cook them strongly in water, strain through cloth, and place it beneath the body of the horse if it is unwilling to drink.

Some who are skilled in the art of horse care mix the tips of juniper [*savina*], southernwood [*abrotanum*], and broom plant [*genesta*], broken up fine, in the horses' food for them to eat and they give them salted water to drink.[283] In addition, three or four handfuls of rye are allowed to lie in water for an hour and afterward these same handfuls are placed in a spot on the bare earth where the wind cannot enter until they are seen to sprout and to germinate. Then, they are cut very finely and one handful is thrown to the horse to eat on an empty stomach, one handful each day for three successive days.

If the horse is unable to urinate and is less lively than usual, and if a swelling appears on its stomach, this most certainly indicates a threat to its life.[284] This infirmity arises sometimes when the horse has travelled a long time after it has had a desire to urinate and has not been permitted to do so. It sometimes arises from sudden cooling following excessive heat and sometimes from some other reason. It is treated in this fashion. Take one handful of sweet flag [*achorus*] root, one of dwarf elder, one of agrimony, and one of either celeriac or chervil [*cerefolium*], and cook them well in spring water.[285] Afterward, toss two or three cupfuls of that water into the horse's throat for it to swallow. Then lead it over the field at a moderate pace until it begins to sweat and then rub it down briskly and forcefully with your palm beneath its belly

80

---

281. "Ninth hour": that is, until about midafternoon.

282. In CL the *myrica* was a tamarisk tree, but for A. it is the white birch. See *De veg.* 6.107.

283. "Juniper": *Iuniperus sabina* (Jessen, 1867, 740), generally seen as too toxic for internal use, although its poison, podophyllotoxin, was used as an insecticide. "Southernwood": *Artemisia abrotanum*; cf. *De veg.* 6.280 and Jessen (1867, 700). This too has been used since antiquity to repel insects and to kill intestinal worms, such as threadworms in children. "Broom plant": Jessen (1867, 720) links *genesta* to the modern species name *Genista*, referring to several Old World bushes related to the broom plant. *Genista tinctoria*, dyer's greenwood, is a purgative emetic. The combination of all these plants might have been beneficial, but care would have been needed to avoid an overdose. See Brown (1995), 243, 287, 299; and Talalaj and Czechowicz (1989), 132, 272, 283.

284. Scanlan (127) suggests strangury, marked by an inability to urinate, often with an accompanying bladder infection.

285. *Achorus*: see Jessen (1867), 700–701.

and around its loins. Then lead it to a place where horses usually urinate. Do this frequently until it urinates and then graze it near spring streams or in meadows. Pay attention, however, to the fact that the more empty the horse is the more it exerts itself in its urination. Be careful, therefore, not to ride it quickly after such a strain, for some nerve or other might happen to leap out of position or be weakened unless gradual movement is employed to help the horse's limbs return to their accustomed position following such a strain.

81   Itch [*prurigo*] is so called from its itching [*pruriendo*] or burning [*ardendo*], and in this disease the horse wants to scratch itself with its teeth and takes pleasure in standing stock-still. It arises from putrid blood with burned, bloody phlegm mixed in, especially when it is autumn and the horse is forced to work too hard with too little food and drink and is rested too quickly thereafter. It also is caused when the horse grows fat quickly after a period of hunger and is not bled and when, after it has worked and raised a sweat, it is not covered at night, and is not cleaned and purged after its work is done. It sometimes also arises as the result of being suddenly cooled when, after it was worked and raised a sweat, it is allowed to stand at rest uncovered in a cold place or in the cold air.

Prurigo first begins near the flesh on the neck as ulcers and then spreads over the entire skin and body of the horse, causing its hair to fall out. Unless it is quickly treated, it turns into mange [*scabies*].

The treatment for a disease of this kind is as follows. If its cause was an overabundance of blood, the horse must be bled and the spots suffering from *prurigo* must be rubbed with warm liquid fat [*sagimen*] and on the third day the horse must be washed well with a lye wash. Make the lye wash from the ashes or cinders from a burned handful of barley, mixed with water. The washing can also be done with strong beer or water mixed with horehound [*marubium*] or *mirica* or the blessed herb [*herba benedicta*], which some call *cicuta* [water hemlock], or ivy seeds, or dwarf elder twigs along with their tips, for in them a natural power flourishes.[286] Cook all these for a long time and then wash the spots having *prurigo* well. Then rub them well with the cleansing device which they call a *strigil* [currycomb] and after, when the spots have dried, anoint them on the next day with an ointment made in this way.[287] Take the roots of red *campestris* [madder roots] and those of water hemlock and cook them with the liquid just mentioned until they grow soft.[288] After this, discard whichever of the roots is hard and mix the soft ones well with old lard and using this, anoint the

---

286. On *herba benedicta* and *cicuta,* see 22.69, with notes. Ivy, *Hedera helix,* has been used externally for skin conditions, but it may also create just such sores through contact (Talalaj and Czechowicz, 1989, 109).

287. *Strigil*: a Greek implement, shaped like a flat, curved, dull scraper, which was used in the gymnasium to scrape the body clean of sand and sweat using olive oil as a lubricant. Although the name has endured, the function surely changed. Latham cites a use of the term to indicate a horse truss, a saddle, and a currycomb. The latter seems in order here.

288. "Red *campestris*": *De veg.* 6.429 and 2.179, and Scanlan (129), for possible identifications.

spots on the horse with *prurigo* in the sun or at the fire. Likewise, to treat this same *prurigo,* take watercress and red horehound [*marubium*] (for this possesses very great efficacy) and mix them, ground quite well, with soot. Using this, rub the horse well on the injured spot.

Scabies is a sort of infirmity on a horse's skin and is called *scabies* because it produces things that are like fish scales or flakes of iron.[289] It arises from too much blood that is putrid and from the same sort of accidental causes as *prurigo.* It also is caused on occasion from exposure to another horse with *scabies* when they scratch each other with their teeth. It sometimes is contracted when they eat together and one horse eats in the same spot where the horse with *scabies* had first taken its food. It also happens when by chance they rub in turn on the same place or when the horse with *scabies* and the healthy one are cleaned with the same cloth or currycomb one after the other.

The treatment for this disease is as follows. If the *scabies* is severe, the horse should first be bled. The spots with *scabies* should then be cleansed with a currycomb until they bleed a little and they should then be washed frequently with a strong lye wash. The lye wash is made in this manner. Take three parts of ashes from an ash tree, two of cinders from bean straw, and one part quicklime, and mix well.[290] Place them in a jar that has a hole in the bottom, crush well, and shake. Strain through ashes what comes out through the small hole and collect it in a clean jar. If you want to test the goodness of what you have made, gently put in a hen's egg tied to a string. If the egg floats, the mixture is good, but if it sinks, it is not potent and is bad.[291] This is the lye wash which some call *capitellum.*[292] Take great care that so strong a lye wash does not get onto the healthy spots because it will quickly remove the hair there. After the injured spot has dried from such a bathing, anoint the spot in the sun or at a fire, using an ointment which contains equal amounts of powdered sulfur, alum, and black hellebore, using a pound and a half of each.[293] Also use the powder of the herb called horse foot [*pes equi*], or horse hoof [*ungula caballina*], and mercury, using three ounces each.[294] Add three pounds of old lard and combine it as follows. Grind the mercury with a bit of the lard until it loses its color. Then add the powders of the aforementioned items to

---

289. *Scabies* is normally translated as mange. But, as in the case of *lepra* in humans, the term surely covered a wide variety of ailments.

290. On the ash tree, *fraxinus,* see *De veg.* 6.108.

291. Something has been left out of the recipe. Since liquid is implied in the final product (e.g., it is strained through ashes, the egg floats in it), at some point water must be added to the jar with the hole in the bottom. Scanlan (130) adds this step to his translation, but it is not in the Latin.

292. *Capitellum*: lit., "little head," indicating the top of a column, a flower, or the like. It may indicate the best, as in the original sense of "acme" or even the British use of "capital"!

293. Cf. *De min.* 4.1.1, on sulfur (Wyckoff, 1967, 204–6), and *De min.* 5.1.4, on alum (Wyckoff, 1967, 244–45).

294. Jessen (1867, 748) identifies *ungula caballina* as *Asarum europaeum,* asarabacca. Brown (1995, 244) knows only of its use as a warming agent, and Talalaj and Czechowicz (1989, 54) claim it is toxic, even in low doses. *Pes equi* is surely not a separate plant but a synonym for *ungula caballina.*

the rest of the lard and grind together well. When it is all combined in this fashion, put it aside for use.

83   An oil is also made for healing *scabies* in this manner. Take the inner bark which is between the wood and the exterior bark of the tree which some call *bul* and which in Latin is called *mirica* [white birch] by some but *fibex* by others, with the bark proper being called *liber*.[295] It is from this inner bark that candles are sometimes made for illumination.[296] The purified bark is cut fine and is placed in a new pot that is well covered and has three or four holes in the bottom. This is put over a slow fire. Dig into the hearth another new pot, well glazed, to hold whatever liquid drips into it without burning it. Let its mouth be directly below the holes of the other pot. Then grind well two parts of the stickiest potter's clay with a quarter part of horse manure, using this potter's clay to smear very thoroughly where one pot is joined to the other. This is done to prevent the fire from entering them and to keep any steam or vapors from escaping. Now if some vapor does escape, smear it again with the potter's clay. After this, place a fire of coals or dry wood around the upper jar and then, after the fire is out, raise the lower one out of the ground. Using what is in this bottom pot, anoint the places on the horse with *scabies* with a cloth onto which this oil has been placed.

Still another oil is made from ground stag's horn and from a strained mixture made of ash wood sap, the dried pith of the black elder [*sambucus*], the sap of the wild apple tree, and the bark of the blackthorn [*nigra spina*] that is used to make ink.[297] This oil is extracted using two pots in the manner just described. It is very effective against all horse *scabies* and all diseases of the skin. When the horse is anointed, it should be done three times a day or more if it is necessary, and from the first day of anointing up until the sixth or ninth day, keep guard lest the ointment wash off or be smeared from the rubbing of its body.

84   There is another horse disease which is called *superos* [spavin] and which occurs when a sort of amalgamation of material amasses in dry limbs where, through the horse's natural heat and the heat of its motion, it becomes as hard as bone.[298] *Superos* arises especially in dry limbs that have been exercised in strong motion and it arises in the vicinity of joints since too much melancholic nourishment is attracted to the bones from the heat of motion. Now, the joints expel this because it is excessive and the efficacy of the heat dries and hardens it near the joints. From this the nerves grow rigid and the movement of the joint is impaired since, whereas it once was limber, it is now rendered tight due to the *superos,* and the horse begins to limp. When, however,

---

295. *Bul* seems to be a contracted form of *betula*, the birch. On *mirica* and *fibex* as the *Betula alba*, see Jessen (1867, 716). Some scholars believe that tree bark was used by the early Romans as a writing surface, giving rise to their word for book, *liber*.

296. Several plants actually produce enough waxy substance to be used in candles. The most famous are the Chinese tallow tree and the American wax myrtle or bayberry.

297. "Blackthorn": Jessen (1867, 744) identifies it from this passage as the modern *Prunus spinosa*.

298. Spavin is an equine disease that results in lumps on the hock, which interfere with gait. The Latin name means, lit., "over-bone," and often these lumps are accretions of bone tissue.

it is in its first stages of development, it is noticed most quickly when the horse's legs are being bathed, for then, when the hairs are pressed down and lie there from the wetness of the water, the amalgamation of the humors in a swelling near the limber joints is apparent. There is also another symptom that precedes that one, namely, that the leg grows perceptibly warmer near the joint than in other areas of its body. This is a signal that the *superos,* by its drawing heat, is beginning to erupt. A horse that is thus disposed to *superos,* or which has it, often stumbles and trips on rough spots because it has difficulty bending.

The disease is treated in this fashion. The hairs standing upright are first shaved off. The area is then anointed many times with the unguent called *pentamyron* and is rubbed well.[299] And a warm tray made of stag's horn and boxwood is often put over it so that the ointment might permeate well to the *superos* due to its heat.[300] The ointment called *pentamyron* is made as follows. Take two measures of celeriac of old harvest.[301] Take also two parts of "egg fat" [*sagiminis ovorum*], which is made in the following manner. Take yolks that have been cooked solid within the eggs, grind them, and cook them over a slow fire in an iron pan, constantly moving them around with a spoon lest they stick to the pan, until the dregs are well separated from the oil [*sagimen*] and stick together. Then take this oil along with two parts of honey drop; one part of virgin, pure wax; one of pine pitch drop; and five parts of laurel oil.[302] Laurel oil is made as follows. Grind the berries fine and cook for a long time in water; then strain through a strong, thick linen cloth and press out forcefully using two sticks. When the water is cooled, use a feather to collect the fatty substance floating on the surface and put it aside for use. Place all of these five things just mentioned together over a fire until they liquefy and, when liquefied, strain through a cloth.[303] Then anoint the horse's rigid nerves with this ointment, especially over the *superos*. When the *superos* grows directly on the joints themselves, it is not advisable to apply fire or a very corrosive poultice. Rather, anoint it as we have explained. For we have seen those who, in their inexperience, injure the nerves by putting iron over the joints.[304]

85

---

299. On *pentamyron* see also 22.62.

300. A *tabula* made of boxwood and horn sounds rather like a writing tablet.

301. "Celeriac of old harvest": *veteris annonae apii*. This odd phrase is to be preferred, perhaps, to Scanlan's unusual taking of *annona* as seed. The term should rather refer somehow to grain, and it is used to mean fodder at 22.91f. Likewise, his translation of *apium* here as parsley but above (e.g., 22.74) as celeriac is probably not wise. A. undoubtedly had the same herb in mind, be it parsley or celeriac.

302. The strange terminology seems to be explained as follows. A. could not simply say "two drops of honey or pitch" since the recipe is a proportional one. He means to say that for every one part of egg fat, put in two drops of honey or pitch. Thus, if you used six eggs, you would use twelve drops of honey and pitch. Of course, as is commonly the case, we are not told some important information, such as how many eggs to use, unless "two parts of egg fat" implies the oil from two yolks. Brown (1993, 329) reports that pine oil is widely used in massage oils.

303. A. has actually listed six ingredients: parsley, egg fat, honey, wax, pine pitch, and laurel oil. Perhaps the honey and wax are a single item, viz., honeycomb.

304. A. most likely is referring to a process of cauterization.

86 Moreover, if the *superos* is on the joint, it can be removed well in the following way. Having shaved the hair off the skin over the *superos,* pierce the skin in several spots with a thin awl in such a way that the perforation reaches the core of the *superos.* Then split a stick in the middle and place a stout linen or hemp cloth in the split. Wrap ram dung and a bit of salt in a knot in this cloth, and warm it well in warm honey or butter and liquid fat [*sagimen*] taken in equal amounts. Using the split stick, press the knot in the cloth, still warm, onto the *superos* until the skin begins to grow white. Keep the spot of the *superos* from water for seven days, being sure that no water reaches it.

Further, to treat the same thing, grind together two parts of clean lime, one part of strong soap, and one of salt. Place this on a tile over a fire of coals until it is thoroughly burned up and put the ashes onto the skin, binding it warm to the *superos,* leaving it bound there for a day and a night. Scarify the skin of the *superos* in several places, having first shaved away the hair, and put a piece of leather made from an old *sotular* in the hole, doing so carefully so that the powder does not reach the healthy skin and injure it.[305] After the things mentioned above have been taken away, anoint it all over with honey or some other ointment.

87 Likewise, some shave the hair, pierce the *superos* with a heated awl, and then rub the entire *superos* forcefully and for a long time using a green, moderately thick, hazel twig with the bark stripped off.[306] Then, a warm poultice made from egg yolks cooked very hard is bound on each day for three days, and they apply the leather piece we just mentioned around the *superos.* Then, five or nine little disks are made of radish root and placed over a warm tile. Then, each is bound successively over the *superos* until each grows cold. After the *superos* recedes, soothing agents are used to heal the skin.

The method of determining whether lime is "quick" is as follows. Take some lime that sticks together well, in the size of an egg, and put it, covered, beneath some water. If within a short time it has disappeared and the water is warm from this and smokes a bit, it is called "quick." If it does not do this, it is not so called.[307]

88 *Attactus* is an infirmity named after "touching" [*a tangendo*], for when the rear foot touches the tendon [*nervus*] of the front foot which is on the inside of the leg, the lower leg near the tendon swells a bit and sometimes not.[308] From that point on the horse develops a hitch in its gait. It is treated as follows. First shave the hairs that are standing up, then scarify the area so that the coagulated blood can get out from there, being careful not to harm the tendon in the leg with the scarification. If the *attactus* is recent, then bind over the spot a live cock, warm, with its heart still beating, and with

---

305. The piece of old leather (*taco veteris sotularis*) is doubly interesting. First, it is lexically interesting, in that *taco* is not that common a term. Second, it seems to indicate that the *sotular,* at least here, is skin rather than horn (the frog of the hoof, perhaps?; cf. Glossary, s.v.).

306. Up to this point, A. has used the word *superos* as if it were in the third declension. He now begins to use the form *superossus* (or *-um*), as if it were in the second.

307. "Quick" lime in Latin is *calx viva,* "living lime." Cf. the English phrase "the quick and the dead."

308. Scanlan (134) identifies this disease as "splints," named after the tendons connecting the splint bone to the cannon bone in the lower part of the foreleg.

all its innards, but split down the middle of its back. But if the *attactus* has been present for a fairly long time without medication, apply fern roots that have been well ground with a bit of honey or warm butter.[309] Or take marsh mallow roots which have been cooked in water beforehand until they soften and then ground with a little ointment or some fat [*pinguedo*], and apply this like a poultice. Or take two spoons of liquid fat [*sagimen*], three of soot and honey, and one of salt, a small amount of vinegar or of the dregs of old beer, and one handful of hemp or linen lint. Grind all these together and bind them on like a poultice, warmed once a day until the pain abates. Then, with a thin, curved cautery iron, make some cauterizations, small and long so that the nerve is not harmed by them. Afterward, the injured skin should be treated by binding on lukewarm, thick lard. If, however, the skin is broken too much by a warm poultice, treat it with the ointment called *pentamyron,* whose recipe we just passed along.

Sometimes, by passing through very cold, icy water up over its knees, the horse takes a chill in the tendons of its knee and leg. If it is chilled immediately after it leaves the water, the tendons are rendered inflexible because of their contraction and rigidity. The horse loses its usual speed until the tendons are relaxed again by means of heat. This is why horses being watered during the winter should not be led into the water very much, but can be so led a great deal in summer.

There are some lesions which occur on the hind legs and feet and which some people call "mules" [*mulae*].[310] They are produced in cold weather, when the horse, making its way on a difficult, muddy road, works very hard and afterward stands the entire night with wet and muddy feet in a place where there is little or no straw, standing on the bare earth or on stone. For then, if the horse is young (under six years old), its humors descend to its rear legs and in the face of the subsequent cold they are congealed there. Sometimes they generate swellings from the foot up to the knee. Sometimes they even split open, as if the leg had been scarified. But at other times, when it is in its beginning stages, the affliction lacks swelling, but it can be recognized by bristling, for the hairs on the horse's foot and leg stand up like pig bristles. It mostly occurs in winter, but sometimes in the spring, especially at its start. It rarely happens in the summer, but sometimes even occurs in autumn, just around its end.

The treatment for this disease is to shave the hairs and to scarify the area inside the hoof in such a way that only the skin is cut in order not to harm the nerves or arteries. But before scarification the legs and feet should be fomented for a long time with warm water to thin the humors out. Afterward, take two measures of quicklime, two

---

309. Brown (1995, 333) relates that *Polypodium vulgare* (polypody) was prescribed by Dioscorides as a laxative and that it was used as a poultice for sprains and fractures.

310. *Mulae* is more technically "she-mules" and was so named because it is a foot disease. Cf. the English name for a style of slipper, "mules," which is itself derived from an old French word for slipper. The Fr. for "chilblains" is still mule. It would appear that the disease, though equine, has nothing to do with mules, for its ultimate etymology is linked with *mulleus,* a shoe worn only by high-ranking Roman magistrates. It was red and in this is linked more to the *mullus,* red mullet, than to any equine. Likewise, chilblains is also called pernio in English, from the Latin *perniō,* chilblain on the foot, a derivative of *pern(a),* haunch of the leg.

of salt, two of the finest rye flour, and three of soot, and grind them with vinegar or wine. Then tie on a warm poultice made of this over the area of the pain.

If, however, this infirmity has persisted for some period of time, the foot should be cut to the rear, above the joint, to enable the humor to come out, much as viscous and coagulated gum comes out of a tree. The skin opposite the knee should also be split, using a sharp piece of wood shaped like an awl so that the nerve which is found there and is shaped like a barley grain can be raised up and drawn outside a bit.[311] Once it is drawn out, put on a poultice made of wormwood or celeriac, or celeriac roots with old lard mixed in with well-ground and well-cleansed hemp or linen tow. Put this poultice on the spot and open all the swellings and the veins on both the inside and outside surfaces of the legs after the fashion of a bleeding.

A horse is sometimes injured by a stick, rock, or some other hard thing on the rear of its foot, below the hoof, that is, the *sotular*, without any resultant swelling. Then, it is split open by the hoofbeat and a foul humor drips out since all pain incites rheum. When the lower members are rheumy this way they need both cold and warm help. This infirmity is treated as follows. Take one spoonful of honey, three of soot, a few spiders' webs, and nettle shoots which are around its stem. Grind well and after a poultice has been made, bind it on warm until the flow of rheum is checked. The same ailment can be treated by grinding together *atramentum* powder, two egg yolks, two leeks with their leaves or with four parts of the finest niter, and four of *atramentum* well mixed with old lard. Liquefy them all together and bind them warm to the foot. Or take the warm excrement of a human or dog which has eaten dry food for one or more days prior (wheat or barley bread, or bones, for example) and bind this for three days and keep the injured spot from getting wet. Some, however, split the hoof and skin a bit in both directions below the division of the hoof to treat this infirmity, allowing the noxious humor to leave.

The illness called *infundatura* occurs in the horse as a result of eating a lot of fodder when it is led too quickly and then drinks immediately thereafter with excessive thirst or desire before its food has been well digested.[312] However, if it urinates and defecates, one can tell that the food has been digested. It even strikes if, after heavy exertion, the horse is fed pure, good fodder or some other food and swallows many ears of its fodder but barely breaks them open with its teeth, and swallows them along with their husks and straw with great appetite and on an empty stomach without having been watered. This is why, as we said a bit above, some people, after the horse has worked hard, scatter the food little by little in front of their horses three or four times, thus adding more food gradually. If a horse is hot after great exertion, they do not allow it to drink a lot right then and there. Instead, they cover the horse lest it become suddenly chilled, because horses often catch *infundatura* as a result of all these things. For food and drink that are taken too quickly after labors of this sort are mixed with humors,

---

311. Scanlan (136) suggests that the barley-shaped object is "a sesamoidean extension of the suspensory ligament."

312. Scanlan (137) identifies this as colic. Just below, in 22.92, A. gives the alternate name of *infusio*.

especially with the blood, because the heat of movement attracts it. This is especially so in the horse because it has wide veins and spongy flesh and the blood therefore descends to its feet and produces this illness. But if it flows between the skin and the flesh, it causes itching. Thus, one should be careful that horses not drink immediately after heavy exercise without first resting for the space of an hour.

The symptoms of *infusio* (that is, *infundatura*) are as follows. When the horse walks, it lurches as if it were walking over glowing coals. When it stands, its feet tremble, nor does it stand with its limbs straight out but rather somewhat contracted. It continually wants to lie down and it will not be able to raise its hindquarters off the ground because of their weight. It is almost as if it were being held back with a bridle, and it just about collapses on its rear knees.

The treatment is as follows. If the infusion has been produced on the same day, grind well the herb *ganda* [weld], which dyers use.[313] A full scoop of this should be given to the horse to swallow with a little water. The herb tormentil does the same thing for every sick animal if it is given to it ground up to swallow with a little water.[314] If the horse has not been cured by the second day, then take blood from the neck up to the point where the animal cannot stand because of its weakness. While the blood flows, permit the horse to stand in cold water over its knees for the space of an hour, and after, for two days morning and evening, let it stand in cold water up to its belly for the space of three hours. Throughout these days do not allow it to drink or to eat fodder—but it can eat straw or hay that has been well moistened—and let it stand in a cold place. On the fourth day mix wheat bran with warm water and give it to the horse to suck down. Once this has been tasted, gradually give it something to drink. If it is not cured by this within three or four days, draw blood from it, from both temples between the eye and the jaw. Bind its head with a halter while the blood flows. Then, over subsequent days, do the same things in the way just described.

For the same problem make a lye wash, not very strong, from the ashes of barley straw and beans, and mix this with wood ash or cinders until it is moderately thick. Then make slender bindings out of hay or straw and, having soaked them in the lye wash you just made, bind them tightly around the horse's legs continuously from hoof to knee so that one circle touches the next, with the result that the entire leg is wrapped in spiral fashion. Also, while the legs are being bound, put some of the aforementioned lye wash over the leg between the bindings. After the *infundatura* has descended, the hoof should be thinned to the front and the veins opened to let the blood out.[315]

---

313. *De veg.* 6.352. A bright yellow dye is produced from the herb called weld, or dyer's rocket, *Reseda luteola.*

314. Tormentil: *tormentilla*; bloodroot, *Potentilla tormentilla,* a shrub whose root is very astringent (Jessen, 1867, 747; Brown, 1995, 334).

315. "Descended": A. used this verb just above to describe the descent of the disease to the feet. Yet the positioning of the phrase seems to tempt us to translate, with Scanlan, "after it has abated." Yet this is not a common use for A., who, at 22.68, calls this phase of a disease its *declinatio.*

93   There is also an infirmity which some call "curve" [*curva*] since the horse bends its leg in a curved fashion, hindering its flexibility. What it is, however, is a swelling on the leg near the joint, either in front of the knee or in back above the knee, produced by concussion or impact with some hard object. This is how it is treated. The hairs are shaved and the entire swollen area is scarified with many deep blows. But do this cautiously so as not to hurt the nerve or the joint. Then rub it forcefully with a stick of green hazel wood. Then, the skin which is over the *curva* should be raised forcefully with an iron hook and a slender but strong thread should be inserted from the middle of the swelling all the way down to where it ends. Then, with a very sharp iron instrument make one round hole or more, through which the humor caused by the blow can exit. Then take strong soap with a bit of salt added, rub it forcefully, and it will be cured.

    Let these, then, be the things said by us about the treatment of horses.

94   There are things from a horse which are good for human medicine as well. For if the sweat of a horse is mixed with wine and is drunk by a pregnant woman it expels the fetus. If the hair from a horse's neck, which is called the horse's mane, is cut off, it removes the desire for intercourse from the horse.[316] Horse sweat also swells the face and brings on quinsy and foul-smelling perspiration. If a mare senses the smoke from a candle or an extinguished lamp, she will miscarry and the same holds true for certain other pregnant animals. If horse manure is used warm and with vinegar as a poultice, it checks the flow of blood whether it is burned or unburned.[317] If the flow is recent, even smelling the manure checks it. If daggers or swords are warmed and placed in horse sweat and they drink deeply of that humor, they become so poisonous that whatever a person wounds bleeds continually until it dies. If some woman cannot conceive and if mare's milk is given to her to drink unawares, and if a man couples with her directly after this, she frequently conceives. It is said that if the horse's hair is attached to the door of a house, it prevents mosquitos from entering the house through the door. The teeth from a male horse, placed either under or above the head of a snorer when he is asleep, keeps him from snoring. Teeth of a yearling [*pullus equinus anniculus*] hung on a child whose teeth are falling out will make their departure fast and pain-free. If a pregnant mare should cross the trail of a wolf, she gets angry. It is also said that if a horse follows the tracks of a wolf or a lion for a long time, its feet fall asleep and grow numb to such an extent that it cannot move.

95   (39) EQUICERVUS: The elk comes in two genuses. One is the one about which we had much to say in previous books and is the one which we call the *elent* in German.[318] Solinus, however, says the elk is an animal of the East and of Greece whose

---

316. Although *equus* is either generic or masc. in this context, at 6.118, A. cites Avic. as saying mares are especially lustful and that their lust can be curbed by cutting their manes. The belief was commonplace in the bestiary tradition (Pliny *HN* 8.66.164; George and Yapp, 1991, 113; McCulloch, 1960, 128).

317. A.'s grammar fails him here. His Latin may mean that the cure works whether or not the mixture is burned or whether or not cauterization is used.

318. Cf. 2.22, with notes. Cf. ThC 4.35 for similar language.

males, but not its females, have horns. It has a mane on its neck descending all the way past its shoulders and it has a beard under its chin. It has feet [*soleae*] like a horse and a body about the size of a deer.[319]

(40) EALE: The yale, according to Solinus, is a beast that is roughly like a horse. It has a tail like an elephant's tail, a black color, the jaws of a boar, and horns more than a cubit long.[320] They say that its horns are adapted to any sort of movement because they are not rigid but rather move from their roots the way a member moves on its joint.[321] Thus, when it fights it sometimes puts one horn forward and folds the other one back, so that if one is blunted or harmed, the other one can come up to the defense. It is an aquatic animal and enjoys river waters.

(41) ENYCHYROS: The *enychyros* is an animal of the East, roughly the size of a bull, and it is somewhat similar to a bull.[322] But they have long hair descending on both sides of their shoulders and these hairs are softer and shorter than horse hairs. The color of its body is between black and red, tending more toward the black. Its hair is like wool on its other limbs. The voice of this animal is like that of the bull and its horns curve strongly inward and are useful for fighting. However, the hairs of its forehead make its forehead shaggy, hanging down here and there over its eyes. It lacks upper teeth, as does the bull, and its legs are covered only sparsely with hair. On its hooves it has horny *sotulares*, just like a horse. Its tail is short compared to its body. It paws the earth like a horse. It has tough skin that tolerates blows very well, sweet-tasting and suitable flesh, and as a result hunters pursue it. While it is fleeing a hunter it sometimes stops and fights when it is tired out by fleeing. At such a time it ejects its dung over a space of four paces because when frightened it suffers diarrhea accompanied by great flatulence. In any case it naturally has a lot of dung. It is remarkable that when this animal nears the time for giving birth, many animals of the same species gather around the one in labor and, defecating prodigiously, they gather the individual droppings into the shape of a wall around the one in labor.

---

319. Solinus 19.19 is speaking of the *tragelaphos* (cf. Isid. *Orig.* 12.1.20 and Vinc. 18.55). Both A. and Solinus are fairly clear in stating that it is a type of deer (*cervus*). See George and Yapp (1991), 80–81.

320. Cf. Solinus 52.35, Pliny *HN* 8.30.73, ThC 4.36, Jacob. 88 (182), and Vinc. 19.37. Note that Pliny says it is about the size of a hippopotamus, but Solinus changes the "river horse" to just a horse and his diction is as poor as the MS readings. To equate the yale with a rhinoceros (Stadler, 1400; Scanlan, 140) or the Indian water buffalo (George and Yapp, 1991, 111–12) is misleading. While there is some chance it may have begun as one of these beasts, it surely did not remain one for long once it reached the bestiary tradition. Surely, by the time it became a common heraldic animal it had entered into the realm of mythical beasts (Druce, 1911; Hope, 1911). McCulloch (1960, 190–92) discusses an alternative name of *centicore* found in a French bestiary. Cf. PsJF 364, "De cale."

321. See the delightful illustrations in George and Yapp (1991), 112 and Payne (1990), 46–47.

322. Cf. ThC 4.37 (*henichires*) and Vinc. 19.53 (*enchires*). Stadler suggests this is a version of Ar. *HA* 630a19f. and its description of the *bonasos*, and he cites 2.31 and 8.204 as other appearances of the animal but without the name. The *bonasus* is also A.'s *bonachus*; cf. 22.24(12). Stadler identifies the animal as *Poephagus grunniens*, better known as *Bos grunniens*, the "grunting cow," or yak.

(42) EMPTRA: The *emptra* [marmot], as some people call it, is a small animal in Germany and is an animal whose male and female both store up food in the summer on which to live during the winter.³²³ They hide these stores of food in piles in the ground. The female is naturally prodigal with and greedy for food. The male, on the other hand, is stingy, and he therefore throws the female out of the chamber where the stored food is on account of her greed. He then blocks up the entrance and does not allow her to enter. She, however, prepares a hidden entrance on the other side, to the rear, and, devouring the food, consumes prodigious amounts of it. The food they gather together is hay. As a result of all this, when they emerge in the spring the female is fat and the male is wasted away with hunger.³²⁴ This is the animal they call the mountain mouse [*mus montanus*] and it is only found in the mountains. It is the fairly large mouse that is seen in our lands.

(43) ERICIUS: The *ericius*, the *erinacius*, and the *cyrogrillus* are the same animal [hedgehog].³²⁵ This animal is spiny all over because, due to the coldness of its complexion, it has many viscous superfluities which are turned into spines. It lives in holes, can predict the weather, and is afraid of wind. This is why it makes two or four openings to its chamber, and whenever it determines that the wind is going to blow it blocks the opening through which the wind comes and takes itself off to the opposite quarter. Among the quadrupeds it has as a unique trait that its testicles are internal, over the kidneys, like a bird. Its copulation is therefore swift and it alone is said to have two anuses for the casting off of its excrement. Its flesh is said to have drying and loosening qualities, to comfort the stomach, to loosen the stomach, and to encourage urination. It is of use for those who are disposed to the leprosy known as elephantia.³²⁶

Sometimes, when the hedgehog is disturbed before it urinates, it is so infected by the backflow of urine that even if it evades a hunter its spines putrefy. This is why it is pursued after it has urinated by those who have skill.³²⁷ The hedgehog is fat and bears

---

323. This mountain mouse is the Alpine marmot, *Marmota marmota*, a large rodent that does indeed live at fairly high altitudes in Europe (*AAZ*, 116–17). Cf. 1.41; 7.82; 8.8. Cf. ThC 4.38 (*hemtra*), citing the *Liber rerum*, and Vinc. 19.54. Diefenbach (1857, 201) gives the vernacular form of *empster* as a synonym of *mus montanus*. Cf., however, *hamester* as a synonym of *cricetus* at 22.47(30). Yet Keller has shown that *marmotte* in French actually means "mountain mouse" (Grzimek 11:206).

324. Marmot fat is still a staple of folk remedies (Grzimek 11:207). Grzimek also speaks about the marmot's escape tunnels, the fact that the marmot blocks the entrance to its hole prior to hibernation, and the normal emaciation of a marmot following hibernation. His account (11:210–11) validates much of what A. says here.

325. See 22.46(28) for the *cyrogrillus*. The link of names here is paralleled at ThC 4.39 and Barth. 18.50–51 (who does say that the *herinacius* is smaller than the *ericius*). The identity of the hedgehog in this passage is secure. For ancient sources and bibliography, see *AAZ*, 85–86. Cf. Pliny *HN* 8.56.133f., Vinc. 19.55 (*erinacius*) and 19.59 (*hericius*), and PsJF 373.

326. *Utilis est ad lepram elefanticam*: showing again that *lepra* covered many diseases. Cf. 1.608 and 18.98.

327. Clearer in ThC 4.39.27f., based on Pliny *HN* 8.56.133–34, who says that the hedgehog, knowing it is going to be caught, urinates on itself to ruin its spines. Thus, those hunting it for its hide wait until it has urinated before going after it.

a resemblance to a piglet when it is skinned. The ashes of a hedgehog which has been burned with its spines and mixed with pitch bring hair back over scars. Because of its spines the hedgehog copulates with the female while standing upright, as we have stated in earlier discussions about animals' natures.[328]

This animal is said to have many medicinal applications. Its flesh has great loosening powers. If its flesh is dried and is drunk with oxymel, it is good against fleshy dropsy, swelling, jaundice [*citrina*], paralysis, kidney pains, and flow of the humors to the viscera. If its liver is dried in the warm sun on top of a tile, it is just as good as its flesh for consumption and kidney pains. If a poultice is made of this, it is good against a tendon which has contracted, against pain arising in the belly from excessive gas, and against difficulty digesting. It is good against those things we have mentioned when it is eaten roasted and is eaten warm by the one afflicted. Take salted hedgehog, boil it in water until the fat [*pinguedo*] separates from it, and smear a stick with its grease [*adeps*]; place the stick in a home, room, or bed, and fleas will be attracted to it. If an eyewash is made of its bile and is put in the eyes, it is good against leprosy. If someone suffering difficulty in urination takes one ounce of its kidneys with the water from black chickpeas, it immediately loosens both dysentery and cough. Regular eating of the hedgehog generates urinary gout [*guttatio urinae*] and is very good against infantile flatulence and to cure those who wet the bed.[329]

It is also said that if one ounce of the right eye of a hedgehog is fried with alder oil and is placed in a red-copper vessel and is then made into an eyewash, nothing will keep one who wishes to see at night from seeing in the shadows as well as he does by day. But the left eye, fried and placed in a vessel, brings sleep to a person when it is placed in his ear with the tip of a stylus.

99

If the *hyricius* which is called the "mountain" or the one called "marine" is burned, it purges and thins fine material.[330] Some use it for ulcers that very bad flesh has invaded. If the skin of the wild hedgehog [*ericius silvestris*], burned and mixed with liquid pitch, is rubbed on someone with hair loss [*allopicia*], it is cured. Also, burned, ground hedgehog cures a fistulous ulcer if it is placed into it. If its blood is mixed with honey and a gargle is made from it with hot water, it removes thickness and hoarseness of voice. Moreover, if blood is taken from a decapitated hedgehog and is smeared, mixed with equal amounts of oil, on the body of a man who is unaware of it, he is then unable to be with all women for up to a month. And if the flesh is cut

---

328. Cf. 15.18.

329. "Urinary gout": cf. 22.102, where *mingere guttatim* means to urinate drop by drop. This meaning may pertain here as well, suggesting, with Scanlan (143), urinary flow.

330. The *hyricius marinus* is, of course, the sea urchin. One should remember that many, if not most, land creatures were thought to have a water-dwelling counterpart. The phrase *hyricius montanus* is unusual, however, and to make things more difficult, A. will soon mention the *ericius silvestris* and *domesticus*. These last two commonly mean "wild" and "tamed" for A. and might even imply that they were raised on farms. Surely the number of uses A. lists for the animals implies a fairly steady supply. For the plethora of names, some of which might be synonymous, see Barth. 18.60, who says that there are *hericii sylvatici, et alii terreni, et alii sunt aquatici*.

from a beheaded hedgehog and then hung on a beast or person suffering difficulty in urination, that beast or person will immediately urinate.

100 The mountain hedgehog [*hericius montanus*] is better than the domestic one, has needle-like spines, and is about the same when it comes to cures. It is also better for eating and greatly helps the stomach, is stronger when it comes to settling the belly and to inducing urination. Every *ericius,* if burned and placed in an area having superfluous flesh, removes the superfluity of flesh. It is also a medicament for hair loss [*allopicia*]. Take the ash of sea urchins [*ericii marini*] and some red *gallae* and some bitter almonds.[331] Then take an amount of mouse dung equal to half of the existing ingredients and grind the whole thing thoroughly with vinegar. Put it in the area of the hair loss and this cures it. It is equally good if someone takes the ash, or the ash of its skin, or the ash of its head, or that of its inner organs, with bear fat [*sepum*]. The humor will straightway begin to flow back after the spot is gently rubbed until it grows red. Sea urchin, mixed with medicaments for mange, strengthens them in its effects.

Pliny relates a common story that the hedgehog carries grapes or apples by rolling itself in them and then carrying them away stuck on its spines.[332] He says that when it is afraid it removes itself behind its armament by rolling up, but that when it is put in warm water it relaxes its limbs, delighted by the water, and reverts to its proper shape.

101 (44) ERMINIUM: The ermine, which some call the *ermelinum,* is a small animal shaped like the weasels.[333] During the winter it is white as snow, but in the summer it is tawny like the weasel. But it is always white on its belly, while the very end of its tail is dark black. It hunts mice and birds and eats meat. Those who take delight in their clothes are decorated with its pelt.

(45) FALENA: They say the *falena* [mongoose?] is an animal living in the deserts of Libya which is hostile to arrogant people who vaunt themselves in front of it. It tears at those it overcomes but spares those who exhibit humility in its presence, sometimes allowing them to get away.[334]

---

331. Scanlan (144) identifies these *rubeae gallae* as "gallnuts," the puffballs that can grow on the leaves and twigs of oak trees. Cf. *De veg.* 6.206–7, where A. describes these growths and the "worm" they produce (larval gall wasps). He goes on to relate how the appearance of the *gallae* could predict the winter and to list several other uses.

332. Pliny *HN* 8.56.133, but only listing apples. For a delightful illustration, see George and Yapp (1991, 62), showing apples only. Some authors, following the *Physiologus,* relate that the hedgehog brings food to its young in this fashion (Curley, 1979, 24–25).

333. Indeed, the scientific name of the ermine is *Mustela erminea,* the same genus *Mustela* as the weasel (*AAZ,* 67–68). On the name, see Sanders (415–16). Cf. ThC 4.40 and Vinc. 19.55.

334. Stadler identifies this as a mongoose, *Mungos ichneumon,* and Scanlan (145) follows him. Yet there is little here to support this identification. Cf. the clearer account in ThC 4.41. Also cf. Vinc. 19.56. The *falena* may, in fact, be the lion. The description of it devouring the haughty but sparing the meek is repeated of the lion below at 22.107(58) and was a commonplace in the tradition (George and Yapp, 1991, 48). The form *fal(a)ena* may well be corrupted, therefore, from *leaena* in Latin or *leaina* in Gr., perhaps through the Arabic.

(46) FURO: The French call the *furo*, or the *furunculus*, the *furettus* [ferret].³³⁵ It is a small animal, bigger than a weasel, having a color between white and the color of boxwood.³³⁶ It is wild and bold, driving rabbits out of their holes into nets. It is hostile toward all animals, either out of anger as some say or because of the blood it drinks, though it does not eat flesh. It is a prolific breeder, producing seven or sometimes nine young at once. It is said to copulate while prone. During the mating season the female, if she lacks a male, swells and dies.³³⁷ She is pregnant for forty days. A newborn stays blind for thirty days and when it has reached the age of seventy days it hunts.

This animal is, so they say, medicinally useful. If one ounce of a male ferret's penis is ground, mixed with a measure of *garum,* and drunk, it is very useful for those who suffer bladder problems and for those who urinate with difficulty or drop by drop.³³⁸ If this animal bites someone and the bitten spot is an off-black color and the pain grows in it, then the cure for this is to drink pure wine and to place a poultice made out of ground onions and leeks on the spot every hour. But it is even better if the poultice is made out of leeks and out of fig leaves and ground cumin.

If the ferret's bile is drunk, the person dies, unless medical help is sought for the drinker. But the brain of a ferret, fried and drunk with vinegar, is said to cure epilepsy. If an abscess which appears behind the ear is anointed with its blood, it does some good. If its marrow is burned and its ashes are taken with liquefied wax and lily oil, it is good for the same thing.

(47) FURIOZ: *Furioz* is the Arabic name for an animal with exceptionally intemperate appetite, so much so that the animal often incurs danger for the sake of food.³³⁹ It has a short life span because of the excessiveness of its corruption. It is, however, very heated in its copulation, so much so that it seems to attack the female. When it cannot have intercourse, it cries out.³⁴⁰ It performs copulation prone, on top of the female, rather like humans and the genus of monkeys.³⁴¹

(48) FELES: Pliny says this is a beast which dwells in caves, small in size but large in malice and tricks.³⁴² It tosses its excrement out of the cave and covers it with dirt so its dwelling place will not be discovered by any trace of it. It preys on such beasts

---

335. Cf. ThC 4.52 and Vinc. 19.57. All the words are ultimately built on the Latin stem for "thief" though an ancient name was *viverra* (*AAZ,* 195–96).

336. Boxwood is generally pale, white to yellowish in color.

337. This strange statement is true. Hence, all female ferrets sold as pets are spayed before sale.

338. *Garum* is a fermented fish sauce and was a staple of the Roman diet (*FAZ,* 156–58).

339. On the basis of Ar. *HA* 540a10, Stadler identifies this as the *ailouros,* probably the common cat, though A. probably had no firm identification in mind. Cf. Vinc. 19.57. ThC 4.43 offers the form *furionz* and adds many details.

340. Cf. the description of the *furiomus* at 5.11. These may be the same animal.

341. The same fact is attributed to the *furunculus* by Vinc. 19.57.

342. According to Stadler, the ultimate source is Pliny *HN* 10.94.202, which describes the *pardus,* or leopard. *Feles* is the pl. of *felis,* cat, and the text has the air of a marginal comment that crept into a text somewhere. Stadler identifies it as *Felis ocreata domestica,* a wild cat.

as it can with a low and cunning crawling motion of its body. When it sees they are suitably located it leaps on them, kills them, and eats them.[343]

(49) FINGA: *Fingae*, as Pliny says, are dark-colored animals of Ethiopia with twin breasts on their chest like a human. They are not so very fierce that they cannot be tamed nor so docile that they do not harm those harming them. They live peacefully with those that are peaceful with them.[344]

(50) GLIS: The dormouse is a well-known animal.[345] It is variegated in color, being gray on the back and white on the belly. It has shorter hair and more tender skin than the one that is truly called variegated.[346] It is an animal from the mouse genus, feeds on the juice of fruits, and wanders in the forests. It sleeps for the entire winter and stays awake during the summer. While it sleeps it grows fat. This is why during the autumn in the vicinity of Bohemia and Carinthia the country folk prepare little storehouses in the forest. These animals take up residence in these in very great numbers and are then collected in them for human consumption. Its fat [*arvina*] is good against paralysis, according to Pliny.[347]

(51) GALI: The *gali* [weasel] is an animal that eats mice and as a result it also fights with serpents who eat mice as well.[348] When it has overcome the serpents it eats them, and afterward it eats rue against the poison. This animal lives in burrows and it makes the doors to the north and south so that no matter from what direction the wind may come it is safe against the wind.

(52) GENETHA: The genet is a beast which is a bit smaller than a vixen.[349] It has a color midway between black and yellow, with black spots scattered throughout. It is

---

343. A. seems to describe a crouching animal that relies more on trickery than speed to catch its prey. In this latter respect, it does not seem to resemble a cat, whereas in the hiding of its excrement it does. What began in the MS as a cat has, through various translations and over time, lost some of its characteristic traits.

344. *Fingae* is clearly a version of the "sphinxes" (*sphingae*) of Pliny *HN* 8.30.72 and reappears below at 22.120(76). George and Yapp (1991, 91f.) and McDermott (1938) discuss fully the problems involved in identifying the many simians in the bestiary tradition and offer a tentative identification of the eastern rhesus macaque. Stadler suggests the Diana monkey and Scanlan (146) adds the chimpanzee to the mix. Cf. ThC 4.45 and Isid. *Orig.* 12.2.32.

345. Cf. ThC 4.46, Isid. *Orig.* 12.3.6, Vinc. 19.131, PsJF 373, and Barth. 55. The dormouse was an ancient Roman delicacy (*AAV*, 60–61).

346. That is, the *glis varius* of 7.82, 160.

347. Cf. Pliny *HN* 30.26.86.

348. *Gali* is a perfect transliteration of the "modern" pronunciation of the Gr. *galē*, weasel. Cf. Ar. *HA* 609b28f., ThC 4.47, and Vinc. 19.58. Stadler's note identified the animal as *Mustela putorius*, which Scanlan followed, identifying the animal as more of a polecat. Yet Stadler himself changed the identification in his corrigenda to *Mustela rivalis*. The description fits very well with the standard medieval bestiary information on the weasel, *mustela*. An excellent picture of a mustelid with a mouse in its mouth, chasing off two snakes, is found in George and Yapp (1991), 67.

349. Cf. ThC 4.48 and Vinc. 19.58. The identification is Stadler's. The genet, *Genetta genetta*, looks rather like a cat and is common to the Mediterranean basin (Keller, 1909, 157–58; *AAV*, 74–75).

mild mannered as long as it is not harmed with injuries. It seeks its food on the banks of waterways.

(53) GUESSELIS: *Guesseles,* or *roserulae* as they are commonly called, are a kind of mouse whose dung has the odor of musk.[350] Even the skin of the mouse has this odor. This mouse is tawny on its back but white on its belly. It lives in meadows and on riverbanks and sometimes in houses if verdant meadows lie nearby. This animal gathers its dung together in a heap as if it knew it was of some aid to it to do so.

(54) IBEX: The ibex is an animal of the goat genus, tawny in color, and very abundant in the German Alps.[351] In size it is larger than a large goat, with quite huge horns burdening its head. So large are these that when it falls from the cliffs it catches its entire body on them. It is a good climber of cliffs and when it cannot climb any farther to evade a hunter, it sometimes turns and tries to throw the hunter off. But a skilled hunter leaps on its back with his legs spread apart and grasps its horns with his hands, and thus sometimes gets away, let down off the cliff.[352] But many things have already been said about this animal in what has gone before.

(55) IBRIDA: The "hybrid" is a quadruped that is dual-genused [*bigenerum*], that is, born of two genuses, namely, of the wild boar and the domestic pig.[353] This is just as the *tyturus* is born from a female sheep and a he-goat. Conversely, the *musino* is born from a she-goat and a ram.[354] For animals which have the same gestation period, an appropriately sized uterus, and are not very far apart as to shape intermingle, as we have said in what has gone before.[355]

---

350. Scanlan (148) follows Stadler in identifying the animal as a musk shrew, *Crodidura russula.* Yet if one follows the common transliteration of "gu" into "w," the resultant "wuessel" is very close indeed to "weasel," and ThC 4.49 says it is a *mustela* (weasel) rather than a *mus,* mouse. Cf. Vinc. 19.103, who offers the form *rosurella.* Latham offers *roserella* as weasel fur, so used ca. 1200 A.D.

351. "German alps": i.e., *Allemania,* Germany south of the Danube and west of Bavaria, including what today is Switzerland. Some (e.g., White, 1954, 29) identify this animal as a chamois. But the Alpine ibex, *Capra ibex,* seems a sure identification (St., ad loc.; Scanlan, 148; George and Yapp, 1991, 81; *AAZ,* 94–95). A., however, muddies the waters. At 8.42 he speaks of the *caper agrestis* and says both that its proper name is *ibex* and that *caper agrestis* and *caper montanus* are equivalent terms. At 5.11 he speaks of the *caper montanus* and says its proper name is *ibex.* All is well until one reads 12.229, where the *caper montanus* is said to be called the *gemze* in German, and this is the chamois (see notes ad loc.). Cf. Vinc. 18.59 and PsJF 291 (s.v. *ciconia*), who confuses *ibex* with *ibis.*

352. Medieval artists liked to portray the ibex leaping headfirst down from a crag (White, 1954, 29; McCulloch, 1960, 132; Payne, 1990, 33). George and Yapp (1991, 81) give the impr3ession that the horns break the fall, but the Latin of A. and of ThC 4.50 can be better interpreted to read that the animals catch themselves on the horns. The entertaining detail of the bareback-riding hunters might be A.'s addition. One wonders if this is a tall tale related to the curious Dominican by hunters.

353. Isid. *Orig.* 12.1.60 is the source for this passage, but Isidore is speaking in general terms and says "hybrids come from boars and pigs." This subsequently became a proper name.

354. Both names are from Isid. *Orig.* 12.1.60. *Tityrus* is a Gr. proper name commonly given to shepherds and to characters in bucolic poetry, especially in Vergil's *Eclogues.* The identification of it as a hybrid animal is from Isid., but his source is unclear. *Musino* is a version of Isidore's *musmo,* otherwise *musimo,* itself of unsure identification.

355. Perhaps 18.48f.

(56) ISTRIX: The porcupine is an animal that is commonly called the "spiny pig" [*porcus spinosus*].[356] It belongs to the hedgehog genus and lives next to the sea but sometimes in the mountains. It lies dormant during the summer and comes out of its lairs during the winter in contrast to many animals.[357] It has quite a temper and has long spines on its back which it sometimes hurls at will against dogs and humans.

(57) IENA: The hyena is an animal about the size of a wolf which often inhabits the sepulchers of the dead.[358] It also gladly frequents horse stables.[359] It learns from frequent listening and calls people and dogs by name. It kills the ones it has fooled by calling them and devours them. Sometimes it also deceives people with a wretching sound of vomit. It is said that hunting dogs that have come in contact with its shadow lose their bark. It is also said that it can change its color at will. Some also say that every animal which crosses its path sticks to its tracks. Jewelers also relate that this beast bears a precious gem in its eyes, or to be more accurate, in its forehead.[360] The hairs on the neck of this animal are like those of a horse's mane. Some say that its vertebrae are so rigid that it cannot bend its neck without twisting its entire body. Jorach also says that it is sometimes male and sometimes female and that it collects poison on its tail. But this Jorach often lies.[361]

(58) LEO: There are three genuses of lions.[362] The short ones have very hairy necks and these are weak. The thin ones are, as it were, made up out of leopards and these ones are timid. The long ones are strong.

It is an animal that rejoices in great and liberal things. This is why it is called king of the wild creatures [*rex ferarum*]. Because it is prone to share its prey, it thus does not deign to provide for tomorrow and it scorns returning a second time to the leavings of its food, allowing it instead to be taken by any animal whatever and especially by

---

356. *Istrix* is close to the Gr. *hystrix*, remembering that the aspiration had dropped out. Cf. ThC 4.52, Vinc. 19.63, Pliny *HN* 8.53.125, and Solinus 30.28.

357. Pliny states clearly that the porcupine (*AAZ*, 153–54) hibernates in winter. ThC also has it backward.

358. The hyena is an animal of long standing from antiquity (*AAZ*, 92–93) to the bestiary tradition (McCulloch, 1960, 130–32; George and Yapp, 1991, 58–59; White, 1954, 30f.). The description (mane, stiff way of carrying the neck) is rather accurate. Cf. Ar. *HA* 594a32f., Pliny *HN* 8.44.105f., Solinus 27.23, ThC 4.53, Barth. 18.59, Arnoldus Saxo 2.4.22d, Jacob. 88 (182), PsJF 374–75, and Vinc. 19.61–62.

359. *Stabula equorum*: an interesting error found both in ThC and A. It is a misreading of Pliny and Solinus, both of whom say hyenas frequent *stabula pastorum*, sheep pens.

360. Cf. A.'s statements at *De min.* 2.2.8. Pliny *HN* 37.60.168 tells us that these stones, placed under a person's tongue, bestowed the gift of prophecy.

361. Ar. himself had problems with the story that the hyena was, in alternate years, male and female. Cf. Peter Damian *De bono religiosi status* 19 (*PL* 145:780A). The origin of this story lies in the fact that hyena fetuses are bathed in the womb with male hormones. As a result the female's genitals become masculinized. What seems to be a fully formed penis is really an enlarged clitoris, and the vagina forms a pseudoscrotum (McMillen, 1996).

362. Compare this and what follows to A.'s long treatment of the lion at 8.198f. The lion was the first animal in the *Physiologus* (Curley, 1954, 3f.) and was a favorite of the bestiary tradition (*AAZ*, 108–11; White, 1954, 7f.; McCulloch, 1960, 137–40; George and Yapp, 1991, 46–49). Cf. Barth. 18.63, ThC 4.54, Jacob. 88 (176–77), PsJF 375–78, and Vinc. 19.66–75.

humans. It does not attack humans unless it is very hungry and has not found anything else. It tears to pieces those which encounter it and provoke it but it sometimes spares those which prostrate themselves and beg its indulgence. Although it rarely is afraid, it runs in wondrous fashion from the scorpion and avoids it. It is said to be very frightened of a white cock. When it is tamed it is disciplined by beating the cub. When it is overfilled with food, especially if it has taken the food prior to fleeing at a run, it casts out the food by putting its claws in its mouth. Moreover, it only barely accepts a second meal if the first one has been digested.

Pliny and Solinus say that when it has something stuck in its foot or mouth that is impeding it, it begs a human it runs into to remove the impediment. And if this is a person seeking escape from the lion, it closes off the roads to keep the person from doing so. They also say that when the lioness desires intercourse but the male is not capable of copulation because it is very hot, the lioness commits adultery with a pard [*pardus*] and the lion detects this by the odor.[363] But she, before she returns to the lion, washes herself in running water and thus disposes of the odor of adultery. But I think this is false.

When it is beseeching, the lion holds its tail still, but when it is angry, it first strikes the earth with its tail and afterward strikes its back, whereupon it leaps. When it is angry, it is consumed by internal heat. But of its members, its heart is especially hot.

In proportion to its bodily size the lion is stronger, larger, and has harder bones than the other animals. This is why, when the bones are struck together, sparks fly out.[364] They have marrow only in their thighs and lower legs. It considers the *onager* an enemy and hunts it out of hate.

It is also said to suffer almost continuously with quartan fever. But this is certainly false because nature produces no animal unless it has the balanced complexion which belongs to its species and in which it is healthy. Now, it is occasionally sick, but when it is it hunts the monkey, eats it, and is cured.[365] Sometimes if the blood of a dog is drunk it too cures it. Its fat [*adeps*] is warmer than that of all other animals and if a person is smeared with it he puts to fight every animal and even serpents. Its neck bones (that is, its vertebrae) are connected and its flesh is everywhere as tough as sinew [*nervus*]. As a result it cannot look backward. Its inner organs are like those of a dog, as are its teeth, except that they are bigger. Externally it is like a cat, and when it walks over rocky ground it retracts its claws to spare them as if they were its weapons.

---

363. The union of this adultery is the aptly named leopard. The identity is, of course, mythical, though George and Yapp (1991, 54–55) suggest the Asian leopard. McCulloch (1960, 150–51) points out that a tradition begun in Pliny has the lion mating with a female pard and that sometimes a panther is the one to mate with the lioness. For the pard, see 22.109(59) and 131(89). Note too the *alfech* of 22.15(2) which is said to be the product of the union of a lion with a leopard. The confusion of names for such cats has a long history (*AAZ*, 107,147).

364. Cf. *QDA* 2.2.

365. An interesting corruption. Pliny *HN* 8.19.52 relates that the only malady a lion may suffer is going off his food. The cure is to tie some monkeys to the lion, which will enrage it by their pranks. Thus fired up, the lion eats the monkeys and all is well for the lion, although not for the monkeys.

The flesh of a lion is warmer than canine flesh and is dry as well. Taken as food it is good for paralytics, and because it is digested slowly it generates flatulence and cramps when eaten. Clothing that is wrapped in its hide will be free from moths. Also, if its skin is placed together with that of a wolf, it takes the hair off the wolf's skin and does the same for others. If a fumigation is made using its fat [*sepum*] near a water source or if any part of it comes into contact with the water, it keeps wolves from drinking it. If its fat is ground with leeks so that it overcomes the odor of the leek, and if a body is anointed with it, wolves never come near it. But if its fat is liquefied and the fence around some sheep is smeared with it, wolves and predators will never come near it. If the tooth of a lion which is called the canine tooth is hung on the neck of a child before its teeth fall out and at the start of its secondary teeth, it will render him free of toothache. Lion fat, mixed with other unguents, erases spots on a person. The lion's blood, rubbed over a skin lesion [*cancer*], cures it. If a little lion's bile is drunk it cures jaundice. If its liver is placed in wine and is drunk, it removes liver pain. If its brain is eaten, it causes insanity.[366] But if the brain is instilled with some pungent oil into the ear, it is good for deafness. Its testicle, given ground with roses and taken in drink, brings on sterility. It does the same if it is eaten roasted or raw. Its excrement, drunk in wine, causes one to abhor wine.[367]

(59) LEOPARDUS: Some call the leopard the same in species with the pard [*pardus*] even though it is both different and similar. Since a leopard is made up of a lioness and a pard, it is reddish with black spots scattered throughout.[368]

It is an animal of intense anger, and when it grows ill it seeks to eat the blood of a wild goat and it seeks human excrement for medicine. What is wondrous is that it takes delight in camphor and guards the camphor tree to keep any from being picked.[369] When it is tamed, if it does not catch its prey on its third or fourth leap it becomes angry, so much so that unless it is placated with blood it sometimes leaps on the hunter.

The flesh of this animal is warm and dry but the fat [*pinguedo*] is thick with pungency. It is useful for paralytics and, when roasted, for those suffering from heart beat and head turn.[370] If the afflicted one gets just a whiff of it he recovers and does well. Its fat [*sepum*], mixed with laurel oil, cures a person afflicted with mange [*scabies*]. Its blood is good for swelling of the veins if the affected spot is rubbed with it heated. A small amount of its bile, taken with water, halts generation and brings on sterility.

---

366. "Eaten": lit., "taken," a verb that also can imply drinking.

367. Scanlan (132) suggests that this may be an early cure for alcoholism.

368. Cf. Pliny *HN* 8.17.41–43, Isid. *Orig.* 12.2.11, Ambrose *Hex.* 6.4.26, ThC 4.55, Barth. 18.65, PsJF 378, and Vinc. 19.76. Stadler indicates that the medicinal uses are from Rhazes. On the pard, see 22.106(58), with notes. On general confusion identifying big cats see *AAZ*, 107; Nicholas, 1999.

369. On camphor, see *De veg.* 6.300.

370. These ailments are troublesome. One would think, for example, that "heart beat," *pulsus cordis*, was actually desirable and that *revolutio capitis* often necessary. Scanlan's suggestions (152) of cardiac palpitations and torticollis (a twisted neck) are as good as any.

Fresh bile is a deadly poison and lacks any drying action.[371] If the weight of a gold piece is taken of it, it kills on that very day. That is why someone who drinks an ounce of it encounters death on the following day when he is asleep. Its right testicle, eaten by a woman whose menstrual periods have ceased, restores them. If, however, she eats them regularly, her periods will become frequent. The leopard loves wine, so much so that when it is inebriated it is captured. Leopard brain, mixed with the juice of the *eruca,* strengthens intercourse if the male's penis is anointed with it.[372] If its marrow is drunk it relieves pains of the vulva.

(60) LEPUS: The hare is a well-known animal, thriving in its speediness.[373] It has rear legs that are longer than its front ones and it therefore climbs a mountain faster than it descends it. It has hairy feet and is a timid animal even though it has a large heart.[374] But it has cold blood and a cold heart. This is why it goes out to feed only at night. It never grows very fat, and when it is fed in an enclosure and does not move around, sometimes its right kidney is covered with fat [*sepum*], whereupon it dies. Among animals having teeth in both jaws it alone has rennet, and this is better the older the little hare becomes. In very cold lands the hares are white, as they are in the Alps, but in other lands some of them grow white in the winter but revert to their natural color in the summer.[375] They can hardly ever be ousted from places where they have customarily lived and this is due to their multiple impregnation. They say that there are no hares in Cytacha, just as no deer, boars, goats, or bears live in Africa.[376]

It is a simple animal, with its only defense found in flight and in sleeping with its eyes open. They say that the weasel plays in a cunning manner with this animal and that when the weasel has tired of the game, it seizes the hare by the throat and, holding on tightly, holds it in place. The hare runs but cannot free itself, and the weasel finally kills the tired animal and eats it.

Hare flesh generates thick blood and causes black bile to multiply. However, it does dry and thin and it therefore is good against pain in the viscera and it stops diarrhea. If an enema is made of its flesh fried with oil, it stops diarrhea and ulcers of the intestines. If its flesh is eaten roasted in an oven or on a pan, it acts the same way. If very bad, shadowy black spots are anointed with its blood, they are removed. If the head of a hare, burned and ground with vinegar, is anointed on one with hair loss

---

371. "Lacks any drying action": perhaps "does not dry out."

372. *Eruca*: Latham identifies this as charlock, skirret, or white pepper. Scanlan (153) uses nettles, but at 22.124 translates *eruca* as colewort. Jessen (1867, 715) identifies the plant as *Eruca sativa,* garden rocket. Cf. *De veg.* 6.329.

373. Cf. ThC 4.65, PsJF 382–83, and Barth. 18.66. On treatment in the bestiaries, cf. George and Yapp (1991), 63–64. For ancient sources, cf. *AAZ,* 81–85.

374. The hare has, in fact, a hairy body. The comment is spawned by a literal translation of its Gr. name, *dasypous,* "hairy foot." For the relationship between timidity and the large heart, see 1.582; 13.35; and *QDA* 13.2.

375. A. probably refers to the arctic hare and then to the more local snow hare, *Lepidus timidus,* also called the Alpine hare and blue hare (Grzimek, 12:437–38).

376. Cytacha is the Ithaca of Ar. *HA* 606a2f.

[*allopicia*], the condition is cured. If the head is roasted and the brain eaten, it is good against trembling that is caused by illnesses. If hare liver is dried and if an epileptic takes an ounce of it, it helps. Hare droppings, liquefied with vinegar and smeared on coarse ulcerations or on *impetigo* from which yellow water is coming, cures them. If hare droppings are placed on a woman who has never given birth, she will never do so as long as she has it on her. And if a little bit of this is placed in the vulva, it dries out the menses and it dries the womb excessively. If a person has a toothache and places one of a hare's teeth in the area that hurts, it takes away the pain. The ashes of a hare burned whole are good against kidney stones. If, after her period, a woman drinks hare rennet for three straight days, some people say that it prevents conception and it sometimes helps in this regard, according to Avicenna.[377] If, however, she puts this in her vulva after her period, it is always of help in conception. Hare bile, mixed with clear honey, is good against "white of the eye."[378] Hare lung, placed on the eyes, helps them, but ground and used as an ointment, it heals feet.

(61) LEUCROCOTHA: Some say that the *leucrocotha* is a beast made up out of many beasts, for it has a body like that of an ass, the hindquarters of a deer, the breast and legs of a lion, and the head of a camel. But its mouth opening reaches all the way to its ears. It has split hooves, teeth like a lion, and it imitates the human voice. It also surpasses all beasts in speed.[379]

(62) LEONCOPHONA: According to Solinus, this is a moderate-sized beast which, when captured, is burned to ashes. If its ashes are scattered on lions' tracks, the lions are killed when they touch just a little of it. Therefore, the lions hunt this beast with a natural hate and crush it when they catch it. However, lions become lifeless from biting it. The beast itself sprays its urine, which the lion shuns, against the lion.[380]

(63) LACTA: The *lacta* is an animal that lives in sepulchers and which delights in the cadavers of the dead.[381]

(64) LAMIA: The *lamia* is a large, very cruel animal which comes out of the woods at night, enters gardens, and breaks trees, scattering them about. For it has strong arms suited for every sort of act. When people come upon it, according to Aristotle, it

---

377. Cf. Avic. *Can.* 2.2.395 (not 397, as in St.). On A. and contraception, see Noonan (1986), 205–7, 211.

378. *Albugo oculorum*, lit., "egg white of the eyes," is a disease also mentioned at 23.103(40) and seems reminiscent of 19.15. It may be cataracts.

379. Cf. Pliny *HN* 8.30.72, Jacob. 88 (181) (*cencrocota*), and ThC 4.62. Most identify the animal as a sort of hyena and some details suit this identification.

380. Solinus 27.21, from Pliny *HA* 8.56.136. The name, more accurately *leontophonos* (cf. PsJF 379), is Gr. for "lion-killer." Whatever truth lies behind the story is unknown. Cf. also Isid. *Orig.* 12.2.34, Jacob. 88 (184), ThC 4.63, and Vinc. 19.77.

381. The description is very similar to the opening sentence of *iena* at 22.106(57), even down to the redundant "cadavers of the dead," which imitates that passage's "sepulchers of the dead." Cf. ThC 4.64, citing the *Glossa ordinaria* on Lev., and Arnoldus Saxo 2.4.22d on the hyena, citing Jorach.

fights with them and wounds them with its bite. One who has been wounded, however, is not healed from its bite until he hears the voice of the same animal roaring.³⁸²

This animal delights in living in deserted, ruined places. It has something of a woman's shape to its face and is quite devoted to its young when it nurses them. Some say, however, that there are some *lamiae* in Chaldea of the same size as a he-goat and which are domesticated and which are rich in milk.³⁸³

(65) LAUZANY: The *lauzany*, according to Solinus, is an animal so fierce that it fends off lions themselves. Moreover, it hunts all beasts, especially those which prey on others. It hates humans most of all but it spares its own genus.³⁸⁴

(66) LINX: The lynx is a well-known animal which has such sharp vision that according to poetic tales it can penetrate solid bodies.³⁸⁵ It has a serpent-like tongue which it extends to an unusual length, it turns its head in a circle, and it has protruding eyes and large claws. It is longer than a wolf but has shorter legs. It is a swift leaper and lives by hunting though it is harmless to humans. Its neck is variegated with almost every color on it, and in winter it is hairy while in summer it is almost

---

382. Cf. 5.15, with notes, where the *lamia* makes another appearance in a passage taken from Ar. *HA* 540b18f. Ar.'s *lamia* is a shark, and it was also the name of a fierce bogey used to frighten Greek children. Cf. *GF*, 144, and Keller (1909), 1:156. In medieval texts the *lamia* acquired a number of demonic traits from the conflation of several traditions, which included not only Gr. but also oriental and biblical sources. For one tradition, the *lamia* is a wild and dangerous quadruped. In another, based on the association of the *lamia* with the figure of Lilith in Isa. (34.14), the *lamia* acquires the characteristics of an *incubus* and steals children in the night. In a third, the *lamia* is a domestic animal. For a discussion of the *lamia* in medieval texts, cf. Lecouteux (1981), Resnick and Kitchell (2006).

383. ThC 4.56 is more detailed, and parts of his report are worth quoting here: "We do not know if these are the *lamie* about which Jeremiah says in the *Lamentations* [Lam. 4.3]: 'The *lamie* have bared their breasts and have nursed their young.' . . . This animal is called the *lidit* [Lilith] in Hebrew and the Jews think it was one of the furies, which were called the Parcae, since they spare [*parcant*] no one. I have heard from someone that *lamie* are beasts in the Orient in the vicinity of those areas which contain the Tower of Babel in the field of Sennaar [LXX. *Sennaar*, Heb. *Shinar*; cf. Gen. 11.2]. And these beasts are larger than goats and are replete with milk. They are domesticated by humans, are led to pasture, and are useful because of their abundant milk." The association of the *lamia* with the figure of Lilith stems from Symmachus's Gr. translation of Isa. 34.14. The claim that the Jews view Lilith/*lamia* as one of the Furies can be traced back at least to Jerome's *Commentaria in Isaiam* (*PL* 24:373Df.). For most medieval Christian theologians, the *lamia* also became a figure of the Jews themselves: the *lamia*'s human face but bestial body came to express the Jews' human appearance but bestial nature. Cf. Vinc. 19.65, Rabanus Maurus *Commentarii in Jeremiam* 4 (*PL* 111:1249C) and Paschasius Radbertus *In Lamentationes Jeremiae* (*PL* 120:1205f.). The origin, however, of the tradition that the *lamia* is also a domestic beast prized for its milk is still unclear.

384. ThC 4.57 gives the form *lanzani* and cites Solinus and Jacob. 88 (176). No appropriate Solinus passage can be found with this information and no identification of the animal has been made. Yet it is tempting to associate it with the *alfech* of 22.15(2), which, A. tells us, is also called the *lunza* in the vernaculars of his day. The *alfech* is a hybrid that is extremely fierce, preys on wolves, and is specifically said to attack humans. The name is sufficiently close to warrant consideration. While we believe the *lunza* is the cheetah, a slightly more ferocious version, with a changed name, it could easily have become the *lauzany*/*lanzani*. Cf. PsJF 375, "lauzam."

385. Isid. *Orig.* 12.2.20, ThC 4.58, Jacob. 88 (179–80), Vinc. 19.79–80, Barth. 18.67, PsJF 379, and Peter Damian *De bono religiosi status* 21 (*PL* 145:781C).

bare. The *ligurius* stone is said to arise from its urine but it is said to begrudge this to humans so that it therefore hides its urine under sand.³⁸⁶

(67) LINTISIUS: The *lintisius* is an animal with a dog for a father and a she-wolf for a mother and which imitates the coloration and habits of each. It is fierce and very swift.³⁸⁷

(68) LUPUS: The wolf is a well-known animal, fierce and crafty.³⁸⁸ They say about it that if it sees a person before the person sees it, it takes the person's voice away. But if the human sees it first, the human takes away the wolf's boldness, and if the person has lost his voice, he gets it back by loosening his cloak.³⁸⁹ If a wolf follows someone and that person stops and makes some sign between himself and the wolf, the wolf stops, fearing a snare. As it goes through leaves it licks its paws, wetting them, so that its approach will not be heard. It is sometimes seen that a wolf takes willow leaves and twigs in its mouth and hides itself that way, with only its mouth sticking out, so that it can snatch up goats which come to graze on the willow leaves. It naturally hates sheep and it kills not just enough for itself but as many as it can. When it hears people following a sheep it has carried off, it carries it along uninjured lest, bitten and injured, it might impede its flight.

Wolves only copulate twelve times throughout the entire year and their penis is bony. At such a time they are even fiercer. When they howl, one sings out first and the others join in. When they are very hungry they sometimes fill up on the earth which is called *glis* [fuller's earth]. When filled to satiety with this they try to bring down a horse, ox, or some other strong animal and when they have done so, they vomit out the earth and return to their prey. If sometimes they have some prey and another victim presents itself, they attack the second one, as if forgetful of the one in front of them. They hide whatever is left over from their meal by burying it. Still, they divide it equally for those who took part in the hunt and if there is some left over they call the others up with a howl. Although it is said that they do not bark, this is false. The stone called the *syrothes* is said to arise in the wolf's bladder.³⁹⁰

---

386. The name of the stone, in Gr., simply means "lynx's urine." George and Yapp (1991, 49) have an illustration from a bestiary showing the stone being created (cf. White, 1954, 22). The stone has a long history (Wyckoff, 1967, 101–2; on *De min.* 2.2.10; McCulloch, 1960, 141; *AAZ*, 114–15.).

387. ThC 4.61 gives the form *lincisius* and cites the *Experimentator*. PsJF 382 offers *linciscus*. Isid. *Orig.* 12.2.28 (*lycisci*) and Vinc. 19.98 (*lyciscus*) bring us closer to the Gr. root *lyk-*, indicating "wolf." Isidore cites Pliny as a source, and at *HN* 8.61.148 we are told that the Gauls deliberately bred wolves and dogs together. Pliny gives no name to the offspring, however. It should be noted also that *Lyciscus* was a name given to dogs in antiquity (Vergil *Eclogues* 3.18) and at least once to an evil-tempered harridan (Juvenal 6.123).

388. The wolf, which played a lively part in everyday life in the Middle Ages, is treated often in the bestiaries (White, 1954, 56–61; George and Yapp, 1991, 50–51; McCulloch, 1960, 188–89). Cf. ThC 4.60, Vinc. 19.82–88, PsJF 379–82, and Barth. 18.69. For A.'s moralizing use of the wolf in his own theological works, see Anzulewicz (2013).

389. The diction and content of this passage identify it as from Ambrose *Hex.* 6.4.26, but Pliny *HN* 8.34.83 has contributed also. Cf. *AAZ*, 199–201.

390. Pliny *HN* 11.83.208 gives the name as *syrites*.

The wolf gulps down meat rather than eating it and this is why it does not grow fat or need much to drink, because drink does not assist undigested food. Wolf dung is dry and is ejected with flatulence, frequently over the white spine bush [*alba spina*], and its urine is odiferous.[391] Wolves sometimes eat humans, but rarely. But after they have tasted human flesh, then they seek out humans because of the sweetness of their flesh. Wolves take their pups with them when they flee, as do dogs. They have weak viscera and when these become sick, they eat herbs, especially *dracontea*, to sharpen their teeth.[392] Other, healthy animals do not eat herbs, with the exception of the human and the bear, if they are meat eaters that is. When wolves get old they find it difficult to hunt and then they approach villages and they incur danger as if disdaining life, for they are long-lived and in old age lose their teeth.[393] They can sustain many blows on the outside of their bodies and when wounded they do not mingle with the pack because they would be killed by those who fear discovery because of the blood of the wounded ones. So the wounded ones, going their solitary way, lick up their own blood lest they be caught. I myself have experienced this. The bite of a rabid wolf is very dangerous, just as is the bite of a rabid dog. It is a bold animal, crafty, and able to be tamed. If it is, it plays like a puppy but it does not lay aside its hatred toward hunters, lambs, and the small animals that wolves eat.

They say that Ethiopia has wolves which have multicolored manes.[394] Among wolves the short wolves are braver than the others. They kill dogs which they find out alone and then bury them.

The brain of a wolf and of a dog grow larger and smaller with the phases of the moon, although this happens in all animals. The heart of a wolf, dried and preserved, is said to become very aromatic. Wolf's flesh is said to be colder and fouler-smelling than that of a dog. The hatred that exists between the wolf and sheep is rooted in all its members, so that musical strings made of sheep gut and mixed with those made of wolf gut will not produce a sound. Drums made of wolf hide deafen those hearing them. If the penis of a wolf has the name of a man or a woman tied to it, he or she will not be able to have intercourse until the knot is undone.[395] A garment made of the wool of an animal the wolf has eaten will always have lice. If the tail of a wolf is hung in the stalls of oxen, it prevents the oxen from eating. If one spoonful of the liver of a wolf, dried and ground, is drunk, it is good against liver pain for a person no matter what their complexional makeup is.

---

391. *Alba spina*: perhaps hawthorne. Cf. *De veg.* 6.93, 210, 223.

392. Cuckoopint, *Arum maculatum*. Cf. *De veg.* 6.290. For the claim that the wolf eats *dracontea* to sharpen its teeth, see *QDA* 8.13.

393. The implication here is that they wish to die. ThC 4.60.64f. makes it clear that the wolves know there is easy prey near civilization.

394. Almost surely a hyena, notable for its mane and decidedly canine appearance. Cf. Solinus 30.27 and Pliny *HN* 8.52.123.

395. "Of a man": reading Borgnet's *viri* for Stadler's *vivi*, "of a living person."

117 The philosopher Gyrgyr says that the livers of all animals are good against liver pain.[396] This is why a vulture which is suffering liver pain hunts for large birds and eats their livers. Galen says that if wolf lung that is cooked, dried, ground, and mixed with capers and the milk of a beast of burden is drunk, it is good for those suffering weakness of breath.[397] If the head of a wolf is hung in a pigeon house, no cat, ferret, or any animal harmful to pigeons will approach. If someone should carry the heel of a she-wolf on his lance, and other lancers should come against him, they will not harm him as long as the heel remains on the lance. Wolf bile, mixed with a grain of musk and held to the nostrils at the beginning and in the middle of the month, is good for someone who has epilepsy. When the bile is mixed with rose oil and is smeared on someone's eyebrows, he will be beloved by the women when they go walking with him. If someone mixes the right testicle of a wolf with nut oil and gives it to a woman to put in her vulva with a bit of wool, it will take away from her the desire for intercourse, even if she is a whore. Wolf's blood, mixed with oil and instilled in the ear, is of use in deafness. If someone carries with him the teeth, skin, and eyes of a wolf, he will be victorious at court if he has a lawyer and he will be rich among all nations. If wolf's penis is roasted in the oven and sliced, and part of it is chewed, it immediately arouses desire for intercourse. If its tooth is hung on someone, he will lose all fear. A large drum made out of wolf's skin and beaten causes other drums to beat as well.[398]

If the flesh of a wolf is ground and is cooked with a bit of pepper and with honey whose froth has been removed, it is good for those suffering from colic. If the stomach of those suffering this is bound in wolf skin, it is effective, and if the sufferer regularly sits on this skin, it is good for him. Wolf excrement that is cooked with thin white wine and is drunk is very effective for those with colic and it is said that if a wolf's skin is suspended on the hip and is bound there with a thread made from fleece a wolf has bitten, it is good for the same thing. If the right eye of a wolf is suspended on a boy, it removes fear from him, as do its teeth and skin. If its tail is buried in a town, it prevents wolves and weasels from entering it as long as it remains there.[399]

118 (69) LUTER: The otter is a familiar animal, longer than a cat but with a broader body, a long tail, and short legs.[400] It is dusky in color but its pelt has a shininess to it, which is why the trim on clothes is made from it. It lives in burrows above water and lives by hunting fish. Although it is a breathing animal, it nevertheless stays under water a long time and it therefore sometimes dives and enters fish traps for its prey.

It so fills its burrows with a multitude of fish that it taints the surrounding air. It is sometimes made tame, whereupon it drives fish into nets. It is a most playful animal when it is tamed. It has sharp teeth and is very prone to bite.

---

396. Scanlan (158), citing Thorndike (1929, 2:718), suggests Girgith of Babylon, also called Germath, an author of astrological works. Anzulewicz (2013, 22, n.30) suggests instead an Islamic physician from al-Andalus, Sulaymān Ibn Ḥassān Ibn Jujul, known by the Latin name Gilgil.
397. "Capers": Borgnet reads *pipere*, pepper.
398. Although, from the statement A. made at 22.116, no one will hear it.
399. "Town": *villa*, always a difficult word; perhaps "estate."
400. ThC 4.66, s.v. *De lutere qui et lotter dicitur*. Cf. Vinc. 19.89.

Its flesh is cold and foul-smelling. Its skin is said to have an effect against paralysis. It eats food other than fish, but it loves fish best of all its food.

(70) MULUS: The mule is a familiar animal, fit for work, taking after the ass more than the horse.[401] The hinny [*burdo*] is the opposite, since the male's seed is the active and formative force.[402] Thus, the mule has the voice of an ass but the hinny of a horse. Nevertheless, each is generated from semen that is quite different in complexion and this is the reason for the sterility in them, as has been discussed at length in preceding books.[403] Mules do reproduce, however, on occasion in warm lands in which the external heat tempers the internal coldness of the ass.

If the hoof of a mule is used to fumigate a house, mice flee it. If a person takes mule marrow in an amount equal to the weight of three *aurei,* he will be struck dumb. If two mule testicles are bound together in mule skin and are hung on a woman, she will not conceive as long as she shall have them on her person.

The other things concerning this animal have been said in preceding books.

(71) MONOCEROS: They give the name *monoceros* [unicorn, rhinoceros] to an animal that is composed from many animals. It has a terrifying bellow, the body of a horse, the feet of an elephant, a pig's tail, the head of a deer, and in the middle of its forehead it bears a horn which is beautiful for its wondrous splendor. The horn is four feet long and is so sharp that it easily pierces everything it strikes with one blow. The animal is almost never able to be tamed and hardly ever comes into the power of men while still alive, for seeing itself beaten, it kills itself in a rage.[404]

(72) MOLOSUS: The *molosus* is a huge beast found in many places. It has a large mouth opening and huge, strong teeth which stick out. For this reason it fights long and hard. Although it is horrifying and hostile toward people, it nevertheless fears the blows of children and flees them.[405]

(73) MARICON MORION: The *maricon morion* is, as we have said in previous books, a beast seen but rarely in the East.[406] Its size is that of a lion and its color is reddish. It has three rows of teeth in its mouth. While it has the feet of a lion, its

---

401. Cf. Isid. *Orig.* 12.1.60, ThC 4.68, PsJF 383, and Barth. 18.70; and see George and Yapp (1991), 114–15. For ancient sources, see *AAZ,* 126–30.

402. The *burdo* is the result of the union of a stallion and a she-ass. Cf. 1.104, 16.137. It also gave its name to an organ pipe (19.43).

403. Cf. 16.130f.

404. It is always difficult to know when a medieval author is speaking about the unicorn of legend or the rhinoceros of fact. This passage seems more like the former than the latter (Shepard, 1956; George and Yapp, 1991, 86–89; McCulloch, 1960, 179–83). For *monoceros,* cf. Pliny *HN* 8.31.76, Solinus 52.39, *AAZ,* 161–63, ThC 4.69, Vinc. 19.91, PsJF 384, 392–93, and Jacob. 88 (179). The actual animal is the Indian rhinoceros, or *Indicus onager,* of 22.126(84).

405. Cf. ThC 4.70, citing Ald. *Aen.* 1.12 and Vinc. 19.91. While this creature surely originated in the famed Molossian hounds of ancient Greece (Hull, 1964, 29–30; *AAZ,* 52), it is a mistake to insert the word "dog" into the translation (Scanlan, 161). Neither A. nor ThC gives any indication that they thought the *molosus* was other than a wild beast.

406. Cf. 2.49, with notes. The animal is the ancient manticore, the next animal in A.'s list. Stadler points out that this description is ultimately derived from Avic. *DA* 1.3v.

face, eyes, and ears are those of a human. Its tail is that of a wild scorpion and its call imitates trumpets and a human's voice. It has the speed of a deer and it eats people, having deceived them first.[407]

(74) MANTICORA: The manticore is an animal composed of many animals. It has the face of a human, gray eyes, the body of a lion blood-red in color, the tail of a scorpion that is spiked with a strong stinger, and a voice that is so sibilant that it imitates the sound of pipes and trumpets. It eats human flesh quite greedily, and, according to Pliny and Solinus, it has three rows of teeth in its mouth. It is quite like the preceding animal.[408]

(75) MUSQUELIBET: The *musquelibet* [musk deer] is an animal of the East about the size of a small roebuck [*capreola*]. Platearius says that an abscess grows in its groin out of the humors that have collected there. When it is full-grown, it strikes its groin on a tree, and the gore which runs down and grows hard there is called *muscus* [musk]. While its flesh and dung are called musk, this gore is nonetheless the noblest. The musk itself, if it has lost its strength, recovers it from the stench of excrement and in latrines. It is good for fainting, weakness of the heart, and pains in the brain, liver, and stomach.[409]

(76) MAMONETUS: The *mamonetus* is an animal that is smaller than a monkey. It is dusky on the back and white on the belly, with a long and shaggy tail. It has a neck that is as large as its head and for that reason it is tied up at the hips rather than on its neck. It has a nose that is distinct from the mouth, as does a human, and that is not continuous with it, as is a monkey's. It is called *spinga* among the Italians.[410] It fights with implacable enmity with monkeys and although it is unequal in strength, it nevertheless wins out over them due to its boldness. It is native to the East, but lives well in our climes and is often seen. It has a round head and a face that is closer to the human's than to a monkey's.[411]

(77) MIGALE: The *migale* is the name of a small, poisonous animal which has some energy in its youth but which grows constantly and increasingly torpid as it gets

---

407. In this it resembles the hyena, who also is said to imitate human voices to lure people to their deaths. Cf. the *centrocata* of PsJF 364.

408. Pliny *HN* 8.30.75, citing Ctesias (see notes to 2.49); cf. Solinus 52.37, ThC 4.72, PsJF 385, and White (1954), 51–52. See also George and Yapp (1991), 51–53, with ill. 26; and Benton (1992), 21–23.

409. Cf. ThC 4.73 and Vinc. 19.93. The identity as *Moschus moschiferus* is fairly certain. The "abscess" is in fact a musk pouch that emits a strong scent during the mating season (Grzimek, 13:157).

410. *Spinga*: One first thinks of the "sphinx" of 22.103(49). Cf., however, 22.137(99).

411. Cf. ThC 4.74 and Vinc. 19.90. The description is quite detailed, with the flavor more of observation than hearsay. Scanlan's suggestion (162) of a Diana monkey has much to commend it, but cf. George and Yapp (1991), 91. The emphasis placed on the nose tempts one to give an identification of the Proboscis monkey (*Nasalis larvatus*), but that monkey has only recently been kept successfully in captivity (Grzimek, 10:459–61). Note that this passage implies they were kept as pets and that perhaps A. saw them during his stay in Italy. On monkeys as pets, see Walker-Meikle (2012), especially fig. 4, depicting a chained monkey.

older.⁴¹² It has a cruel spirit but hides this and lures animals on, killing them with its poison if it can. It does this to horses, mules, and especially to pregnant mares.

(78) MUSIO: The *musio* [cat] is a familiar animal which some call the *murilegus* [mouse-catcher] and others *cattus* from its catching [*capiendo*] or its cleverness [*astutia*].⁴¹³ It catches mice which it espies with its eyes that glow like coals in the night and which can make them out in their shadowy burrows. In its mating season it seeks out solitude and therefore at that time turns into a wild creature as if it were exhibiting shame. It takes delight in cleanliness and for this reason imitates the washing of a face by licking its front paws and then, by licking it, smoothes out all of its fur. These animals often fight for possession of the boundaries of their hunting area. They also kill serpents and toads, but do not eat them, being harmed by the poison unless they drink some water quickly. This animal loves to be lightly stroked by human hands and is playful, especially when it is young. When it sees its own image in a mirror it plays with that and if, perchance, it should see itself from above in the water of a well, it wants to play, falls in, and drowns since it is harmed by being made very wet and dies unless it is dried out quickly. It especially likes warm places and can be kept home more easily if its ears are clipped since it cannot tolerate the night dew dripping into its ears. It is both wild and domesticated. The wild ones are all gray in color, but the domesticated ones have various colors.⁴¹⁴ They have whiskers around their mouths and if these are cut off they lose their boldness.⁴¹⁵

Mention has already been made of their medicinal purposes under the letter C.

(79) MUSTELA: The weasel is a familiar animal, sort of a long mouse [*mus longa*], as it were, although in shape it is closer to the marten [*martarus*] than the mouse.⁴¹⁶

---

412. ThC 4.75, citing the *Glossa* and a commentary on Lev. by Brother Hugo of the Dominicans. The relevant passages of the *Glossa ordinaria* discuss the *mygale* in reference to Lev. 11.30 and describe it either as a mouse or a lizard (*PL* 114:816B–C). Cf. Jerome *Liber Levitici* (*PL* 28:312C), Barth. 18.73, and Vinc. 19.132. In origin this is the shrew, from the Gr. *mygalē*, but it is doubtful if A., ThC, or Hugo knew exactly what it was. For example, a shrew's energy is very high in level as an adult. Cf. A.'s description of a shrew at 7.122 and his use of *sorex* just below at 22.123(80). The *migale* has been identified as a weasel as well (George and Yapp, 1991, 66f.), and at 3.96 A. says the *migale* is a male *mustela*.

413. Something has gone awry with the etymologies, based loosely on Isid. *Orig.* 12.2.38, who starts with an allusion to *captura* (catching) and then says it might be because they see so well in the dark, citing the rare verb *cattat*. ThC 4.76 says the allusion to cleverness is through the Gr., and McCulloch (1960, 102) points out that this really comes from Isid., who relates the cat's ability to see in the dark to the Gr. verb for "burn," *kaiesthai*. Cf. Barth. 18.74 (*murilegus*) and PsJF 384.

414. George and Yapp (1991), 115: "The [bestiary] texts say nothing about the cat's relationship with man or about different coloured cats." This one quote illustrates well the difference between the bestiary tradition and that of A. and ThC. See also Rassart-Eeckhout (1997) and Walker-Meikle (2012, 10–13).

415. "Whiskers": *granones*, a Latin form of the MHG *gran* (Lexer, 1:1068; Sanders, 416). ThC 4.76 is forced to call these "long hairs around its mouth." ThC also adds an early notice of purring: "They delight in being stroked by the hand of a person and they express their joy with their own form of singing."

416. Cf. Isid. *Orig.* 12.3.3, ThC 4.77, PsJF 384–85, and Vinc. 19.133–35. On the marten see 7.49. "Weasel" should be taken as sort of a generic term since most people would be hard-pressed to differentiate casually between weasels, ferrets, polecats, stoats, and the like (George and Yapp, 1991, 66–68; McCulloch, 1960, 186–88; *AAZ*, 193–96). Note that A. did differentiate the marten carefully at 22.125(81).

It has rather short front teeth which a mouse does not have. Likewise, it is hostile to mice, which would not be the case were it of the mouse genus.

There are two genuses of it, the larger and the smaller. The larger one is white on the belly and tawny on the back. It lives in burrows in the earth and in cracks in rocks. When fighting with serpents it fortifies itself with wild rue [*ruta agrestis*].[417] It moves its young from place to place to keep them from being found. It is hostile to chickens, both sucking their eggs and killing them. It is tamed very easily. Weasels do not pass over into the island of Proselena and if they are brought there they die. Likewise, if brought to Boeotia they flee.[418] Solinus says that it kills the basilisk, then dies when it does.[419]

A weasel placed on a scorpion bite helps greatly. If a weasel is salted it is good against epilepsy. Its ashes open pores and are therefore good against gout if rubbed on with vinegar. If its flesh is dried and drunk with rue, it is good against the bite of all animals. If the skins of the male are treated and then written upon and hung on demoniacs and those under a spell, it does some good. If its heel is taken from it while it still lives and is placed on a woman, she will not get pregnant as long as it is there.

(80) MUS: The mouse is a well-known animal and has very many genuses.[420] The wild variety lives on the ground in fields and has two colors, red and black. The other one, the domestic, lives in houses and granaries.[421] It is black. There is a very small, red genus, with a short tail and a high-pitched voice. It is properly called the *sorex* [shrew] and is poisonous. It is therefore not caught by cats. There is a large genus which we call *ratus*.[422] It dwells in trees and has a dusky color with black spots on its face. There is also the *corilinum*, which eats filberts.[423] This one is reddish with a bushy tail. And there are many other genuses.[424]

---

417. Cf. *QDA* 8.10.

418. The island is Pordoselene, between Lesbos and Asia Minor, and the source is Pliny *HN* 8.83.226. The weasels are only said not to cross the road. The next phrase concerning Boeotia actually refers to moles in Pliny. Scanlan (163) follows Stadler in identifying the larger weasel as the stoat, *Mustela erminea*, and the smaller as the European weasel, *Mustela nivalis*.

419. Solinus 27.53. On the basilisk, see 25.9.

420. Cf. ThC 4.78, Barth. 18.71, PsJF 385, and Vinc. 19.126–30 (*mus*) and 19.136 (*rattus*). George and Yapp (1991, 65–66) discuss the many types of mice and rats described in the bestiaries. Cf. McCulloch (1960), 143. For antiquity, see *AAZ*, 123–26, 159–60.

421. Here the sense of "wild" and "domestic" rings true for A., as the former, *agrestis*, incorporates the stem for "field," while the latter, *domesticus*, contains the stem for "house."

422. For mention of a trained rat, see 21.22, with notes.

423. *Corilinum* means "hazel" and thus refers to filberts. The animal in question is a dormouse, the *Muscardinus avellanarius*. *Avellanarius* contains the word that A. uses for filberts (*avellana*).

424. Throughout the *DA*, A. calls various animals a "kind of mouse." To list a few: *mus marinus*, "sea mouse," 24.44(81); *mus montanus*, "mountain mouse," the marmot, 1.41, 7.82, 8.8, 22.96(42); the *cycel*, or ground squirrel, 8.104; *mus caecus*, "blind mouse," the mole, 7.123, 8.13, 22.143(105); *mus parvulus rubeus*, "small red mouse," the shrew, 7.122; *mus terrenus*, "earth mouse," the mole, 22.143(105); *mus varius*, "variegated mouse," the loir, 7.160. Even the preposterous *aforon* of 8.106 is said to be a mouse.

As King Alexander says, there are also mice in the East equal in size to foxes which prey on humans and other animals. In Arabia a large mouse is found that has front feet a palm's breadth wide and rear feet about a digit wide.[425]

This animal uses its tail to embrace during intercourse and, alone of the animals that have saw-like teeth, it ruminates. During intercourse its urine is said to be poisoned, and if the urine touches a human, it eats its way to the bones. Every genus of mouse is born from the earth, even though it also is produced from intercourse within its own genus.[426] For this reason, when it rains in Egyptian Thebes, very many of them are produced. Some in this genus are found that are white, and white stones are found in their excrement. These are very prolific breeders.

Every genus of mouse feeds on seeds and bread, eating hard food more readily than soft. When a mouse finds a lot of cheeses it tries them all and then eats from the best one. Its liver grows when the moon is full and becomes smaller during each interlunar period.

Mouse flesh is warm, a little bit fatty, and possesses the power of expelling melancholy.[427] If a mouse is eaten roasted, it dries up the saliva flowing from the mouth of children.[428] If a mouse is split and placed over a scorpion bite, it quiets the pain, and when it is placed over warts, it eliminates them. If a mouse head is hung in a cloth and is carried by one with a headache, it quiets it. If the head is hung on a person with epilepsy, he is cured. If its dung is placed with some salt and a bit of oil over the nostrils, it loosens the stomach. Galen says that mouse droppings are good against hair loss [*allopicia*] if the droppings are ground and then prepared with the juices of *eruca*, nasturtium, onions, and garlic, and if a poultice is made of all this and is applied to the spot.[429] If a powder made of house mice [*domorum murum*] is mixed in a jar with laurel oil and a poultice is made from this and put on the *allopicia* and if the spot is afterward rubbed with garlic every day, hair will arise as a result of doing this.

If a mouse is fed yellow arsenic worked in as a paste into some flour, it will die. If litharge is prepared with hellebore and flour and if the mice eat some of it, they will die.[430]

---

425. The source for these wondrous creatures is the spurious letter of Alexander the Great to Aristotle (Kübler, 1888, 101). It is interesting that ThC 4.78.33f., who is supposed to have been the source for much of A.'s animal lore, has this story but cites Jacob. 88 (184) and that neither of these two mentions Alexander at all.

426. Isid. *Orig.* 12.3.1 links, if fantastically, the etymology of *mus* with *humus*.

427. Curiously, we are in a position to test A.'s judgment of the taste of mouse flesh. The famed naturalist Farley Mowat once restricted his diet to mice to see if his observation that wolves subsisted entirely on the rodents during bad hunting periods was valid. At first he gutted the little creatures before eating them but found he craved fats. When he ate them whole, with their abdominal fat, he was satisfied. He reports: "The taste of the mice . . . was pleasing if rather bland. As the experiment progressed . . . I was forced to seek variety in my methods of preparation." He even ends his chapter with a recipe for *Souris à la Crème* (Mowat, 1963, 76–78). Apparently Mowat's devotion to science was no less than A.'s.

428. Perhaps when teething?

429. On *eruca*, see 22.109(59).

430. On yellow arsenic (orpiment) and litharge (lead oxide), see *De min.* 5.5 and 4.5, respectively.

(81) MARTARUS: The marten is an animal with the shape of a weasel and the size of a cat, but it is longer and has shorter legs.[431] It is tawny on back but white on the belly and throat like the weasel. It has shorter claws than the cat and has two genuses, one called the beech marten and the other the fir marten. The one that builds its nests in the fir tree is prettier by far.[432] Both genuses intermix in any combination, but the beech tree variety seeks marriages with the one called the fir tree variety, as if because of its great nobility. The pelt of either one is precious and remains beautiful for a long time. All genuses of this animal as well as the weasel have in common that when angry they become very foul-smelling and that they are animals that are almost never at rest, even when they are tamed.

(82) NEOMON: *Neomon* is the Greek name for the beast which is called *suillus* ["piggish"] in Latin since it has bristles instead of hair.[433] It distinguishes between food that is good for it and poisonous food by the food's odor. This beast hunts serpents and when it fights with an asp it holds its tail erect and the asp sees this as a threat. It thus transfers the asp, deceived in this way, into its power.

(83) ONAGER: The onager [*onager*] is a wild ass [*asinus ferus*] and, according to some, on the fifteenth day of March it brays twelve times at night and twelve times in the day to indicate the equinox.[434] There are wild ones in Africa and the males preside singly over single herds of females. The females are very lusty and for this reason become tedious to the males. It is said that the male is jealous and therefore nips off the testicles from its son when it is born, and for that reason the sons are laid up in hiding places by the mothers.[435] When a hunter pursues an onager with dogs, it emits odiferous excrement which is delightful to the dogs and thus detains them around it while it flees to safer places. It, more than all other animals, flees human company. It

---

431. ThC lacks an entry for the marten and indeed it is little treated by the medieval authors. Cf. 22.133(96).

432. These species would apparently be the beech or stone marten (*Martes foina*) and the pine marten (*Martes martes*).

433. It is clear by comparing Isid. *Orig.* 12.2.37 that this is a mangled form of the *ichneumon,* which is both a mongoose and an enemy of the crocodile, as a variant of the *ydrus* (McCulloch, 1960, 129; George and Yapp, 1991, 100–101). The supposed other name of *suillus* is also taken from Isid., who himself is quoting Dracontius, who says that pigs kill snakes. Isidore adds that they raise their tails before doing so. Cf. ThC 4.79, who quotes Isid. correctly, and Rabanus Maurus *De universo* 8.1 (*PL* 111:225D), who reports on both the *ichneumon* and the *suillus*. Cf. PsJF 394.

434. *Ferus* is, lit., "fierce" and is unusual for A., who prefers *silvestris* in this context and as found in ThC 4.80. A. is, in fact, accurately quoting Isid. *Orig.* 12.1.39, who derives the etymology through the Gr. *onos*, "ass," and *agrios*, "wild," which is *ferus* in Latin. Cf. Barth. 18.76, who has misread and gives his readers *asinus verus*, "the true ass." McCulloch (1960, 144–45) reports that in one bestiary not only is the ass said to bray twelve times to mark the equinox but the monkey is reported to urinate seven times. Cf. Peter Damian *De bono religiosi status* 23 (*PL* 145:783B), PsJF 389, and Vinc. 19.94–96. See also George and Yapp (1991), 84; White (1954), 82–84; and Curley (1979), 15.

435. This story traces back to Pliny *HN* 8.46.108 and Solinus 27.27. Curley (1979, 15) renders the foul deed charmingly: "the father will break their necessaries so that they produce no seed." A MS at Trinity College shows the deed in progress with the colt whining piteously (George and Yapp, 1991, 85).

endures thirst very poorly and seeks very clear water to drink. When aroused by lust, they stand on cliffs, drawing in the wind to moderate the heat of their lust.

(84) ONAGER INDICUS: The Indian onager [Indian rhinoceros] is different from the one just mentioned.[436] It has great size and strength and bears a huge horn in the middle of its forehead. As if by way of displaying its courage, this animal sometimes breaks rocks off cliffs, for no other reason than as a demonstration of its strength. It has very sharp, solid hooves.

(85) ONOCENTAURUS: The ass-centaur, so they say, is a composite animal, for it has the head of an ass and the body of a human. Some are said to be found with the body of a horse and the upper parts of a human.[437] It is thick with bristles and has hands adapted for any activity, and they sometimes begin to talk although they can never form the human voice perfectly. They throw rocks and branches at those that pursue them.[438]

(86) ORIX: The oryx is an animal like a roebuck, about the size of a he-goat. It has a beard under its chin, lives in the desert, and is easily tricked into a snare. It is numerous in the deserts of Africa and has as a particular trait the fact that its hair is turned toward its head. It is born in that part of Africa which lacks water.[439]

It is also said that if water from its bladder is carried off, no matter how small the amount, it is a remedy against a long thirst. This is why robbers in Gaetulia [Morocco], a land that lacks water, always carry some with them.[440] This animal rejoices in wondrous fashion at the rising of the Dog Star [Sirius] because it grows strong then.

(87) ORAFLUS: The *oraflus* [giraffe] is an animal that surpasses all other animals in the appearance of its coloration.[441] In its forequarters it is so tall with its head extended that it attains a height of twenty cubits. In its hindquarters, however, it is

127

---

436. Cf. the *monoceros* of 22.119(71). Here, however, we can be rather sure of an identification of *Rhinoceros unicornis*. Ar. *HA* 499b18–19 used the name *onos Indikos,* "Indian ass." Cf. ThC 4.81 and Vinc. 19.95. The Indian and Javan rhinoceros have a single horn, while those in Africa and Sumatra have two.

437. The ass-centaur is linked with the siren in the *Physiologus* (Curley, 1979, 23–24) and in bestiaries (George and Yapp, 1991, 78–79). Cf. Isid. *Orig.* 11.3.39, Jacob. 88 (182), Vinc. 19.97, ThC 4.82, PsJF 385–86, and Barth. 18.77 (who, citing a gloss on Isa. 34.14 that can be traced back at least to Haymo of Halberstadt's *Commentarium in Isaiam* [*PL* 116:893], breaks the word strangely and makes it half bull, *est animal monstrosum ex tauro et asina procreatum*). The animal surely worked its way into the tradition from the Vulgate's *onocentaurus* (Isa. 34.14).

438. This is commonly shown in Gr. and Roman representations. Most centaurs were wild creatures, as in the famed battle between the Lapiths and Centaurs. A few, such as Chiron, were gentle and learned. On centaurs in general see King (1995), 141–43.

439. Barth. 18.78 speaks learnedly about the translation of Isa. 51.20, where the sons of Israel are said to be ensnared like the Hebrew *tho,* some sort of hoofed creature, translated into the Vulgate as *oryx.* Barth. is wrong, however, in saying the animal is unclean (cf. Deut. 14.5). Cf. ThC 4.83 and Vinc. 19.98, who have many more details than does A. Since there is no mention here about the startling horns of today's oryx, *Oryx beisa,* A. was probably unaware of its appearance.

440. Pliny *HN* 10.94.201.

441. Cf. ThC 4.84. Cf. 22.16(5), where the giraffe is the *anabula,* and 37(19), where it is the *camelopardulus.*

lower, about the size of a deer. It has the feet and the tail of a deer but the head and extended neck of a horse, although a little smaller. It has many colors, although white and red are the most common. When it sees people looking at it and admiring it, it turns itself this way and that, offering its best view. This animal has been seen in our times and is called *seraf* in Arabic.[442]

(88) OVIS: The sheep is a familiar animal.[443] It is gentle and harmless, and it is useful for its fleece, skin, flesh, milk, viscera, and excretions [*egestio*], and for the dung [*fimus*] in which it takes its rest.[444] It has compassion for its own species, so that when a healthy sheep sees a sick one laboring in the heat, the healthy one places itself in front of the sun and makes a shadow for the sick one. In all the multitude of the flock, a lamb recognizes its mother's bleating (and thus the lamb [*agnus*] takes its name from recognizing [*agnoscendo*]).[445] The mother recognizes the lamb from the odor of its hindquarters. When a lamb nurses it moves its tail vigorously. Sheep that shake the ice from their tails are the stronger ones and they last longer and do better. Those that do not shake it off die quickly. Those sheep that are in warm, dry regions have very wide tails and hard wool. Those that are in cold, humid, and briny regions have soft wool and narrow tails.

The bellwether [*vervex*], the leader of the flock, strikes most forcefully with its horns and has a very hard forehead. Thus some say, falsely, that the bellwethers do this because they are goaded on by a worm that lives in their head. Their horns are frequently twisted and they sometimes have more than two. A ram is said to sleep for half the year on one side and for half on his other side. It flees a wolf even if it has never seen one before.[446] The ram is said to have a natural aversion to small lambs and to take pleasure in those that are more advanced.[447] Age is also said to give him comfort. If its horn is bored through near the ear, it loses its ferocity, and if its right testicle is tied or cut off, it is said to produce female offspring. Thunder causes miscarriages in solitary sheep but not in those that live in a flock. When the north wind is blowing, they conceive males, but females are conceived when the south wind blows. This is why the lambs born in winter are better. They conceive twice in warm, humid lands

---

442. ThC 4.84.10–11 specifically says that Frederick II had been given a giraffe by the sultan of the Babylonians. Cf. Vinc. 19.97. On "seraf," also mentioned under *anabula*, cf. Schühlein (1661). Cf. Laufer (1928).

443. This passage follows closely Pliny *HN* 8.72.187f. Cf. Isid. *Orig.* 12.1.9, ThC 4.85, Barth. 18.79, PsJF 386–89, and Vinc. 18.69–77, 93. See George and Yapp (1991), 117–19; and White (1954), 72–74.

444. Normally, *egestio* is bowel movements and *fimus* simply manure. Here, the former is probably more general and the latter specific. Note that sheep's urine is used below in the simples. *Fimus* can also mean simply "muck" in CL. It should be pointed out that ThC has *digestiones* for A.'s *egestiones*.

445. From Isid. *Orig.* 12.1.12.

446. Whether a lamb will flee a wolf if it has never had experience of one before is taken up in some detail at *QDA* 8.3.

447. At Pliny *HN* 8.72.188, the source for this sentiment, it is clearer that the ram waits to mate with older ewes because Pliny uses the fem. *agnas* over the perplexingly masc. *agnellos* of A. ThC 4.85.54–55 gets it right.

such as Mesopotamia. Sheep that have short legs and plucked bellies are the more noble kind.⁴⁴⁸

Soft, curly fleece is better and all the descendants of a ram follow his type of fleece. Those that drink a great deal, and that do so in turbid water, grow fat more often. Too much fat, however, is an impediment to their impregnation. The milk of black sheep is the best, but in goats it is just the opposite. Of all the animals, the sheep gives milk the longest in proportion to its size since it gives milk for eight months. The color of the offspring follows the color of the vein that is beneath the tongue.⁴⁴⁹ Those that are fed in dry places are healthier, whereas those in marshy areas are rendered sick.⁴⁵⁰ They eat better when they are enjoying the sounds of music. A sheep lives for ten to twelve years and gives birth for eight. In the spring those that have grazed on "honey-dew" [*ros melleus*] die.⁴⁵¹ In the fall they die from burst insides if they stuff themselves with ears of wheat and drink immediately thereafter. They pluck out herbs by the roots and destroy trees, and their bite is deadly to many plants. A sheep is ejected from its home with difficulty and if it is ejected, returns.⁴⁵² If young sheep are first to mate, they are predicting the evil of a plague that year. If, however, the old ones seek intercourse after their time, they are predicting the same thing.

Some have said that a sheep is fattened not on food but on water alone, and for this reason it should be fed on salt for five days in a row and in the autumn should be fed on salted gourds since these things bring on thirst. They also say that if it eats salt after a birth, its udders swell, and that if food is denied it for three days and it is afterward sated, the sheep is fattened the more for it.

Those that are suffering from watery eyes and who are anointed with sheep's brain are helped. Sheep dung, if made into a poultice with vinegar, removes warts and marks left by the pox [*variolae*]. Mixed with liquid wax and rose oil, it cures a burn from a fire. The warm skin of a newly flayed sheep, laid on a spot where blood has collected at the site of a large blow, dissolves the blood. It also gives great comfort to those who have been whipped and may cure them in a night and a day. It is also helpful for the abscess that arises at the base of the ear [boils], provided the entire poultice is made from the dung of a sheep mixed with chicken or goose fat. The urine of a red or black sheep is of great assistance in fleshy dropsy [*ydropsis carnosa*] and is good for dropsy

---

448. "Plucked": an interesting passage. Stadler reads *vellitae*, "plucked," from the verb *vello*, while Borgnet has *vellicosae*, "fleecy," which captures the sense. ThC 4.85.67 reads *ventris vestitus*, "clothed of belly," a variant reading found in the source, Pliny *HN* 8.75.198. Pliny in turn may have pinched the phrase from Columella, a first-century writer on agriculture, who said sheep are *ex omnibus animalibus vestissimum*, "the most clothed."

449. Cf. 3.98 and 6.109.

450. Cf. 3.141.

451. Scanlan (169) points out that "honey-dew" is the excretion of aphids. On a similar "dew," see *De veg.* 2.121, 132, and 6.113.

452. In its original form this statement makes clear that sheep will not leave and perversely return to a burning structure (ThC 4.85.92f.).

when mixed with honey.⁴⁵³ If two ounces of sheep dung are drunk with a cooked mixture of woodbine, it is good against jaundice.⁴⁵⁴ If a person's spleen is rubbed externally with vinegar and burned sheep's dung, its size is reduced. Light warts and superfluous flesh are removed by the application of a poultice made of sheep's dung with honey. It also removes *formica* and, when mixed with wax and vegetables, it cures the burn from a fire.⁴⁵⁵ If the marrow of a lamb is liquefied at a fire along with nut oil and if white sugar is added, and if it is then distilled over Arabian jasmine [*sambuchum*?] and is drunk, it dissolves bladder stones and helps one urinating blood and those with pains in the penis, bladder, and kidneys.⁴⁵⁶ If an area of the body that *cancer* has befallen is anointed with its bile, it gives clear relief.

If you want the figs on a tree to ripen, bury the horns of a ram at its roots and they will do so quickly. If you want to do away with a disease that is in your sheep and is killing them, take the stomach of a ram and cook it with wine and water. Then mix this in water, give it to the sheep, and it will remove the disease.

(89) PARDUS: Because of their similarity to the panther's varied coloration, pards are said to be a sort of panther.⁴⁵⁷ Thus, some say that they are sometimes produced from dogs and panthers. The *pardus*, however, is common in Africa where, due to a lack of water, many animals gather at the streams. There lionesses commit adultery with pards and produce lions, but base ones. The pard is a bloodthirsty animal, hunting by means of leaps rather than by running. It sometimes hides among leaves and thickets, and, trusting in its swiftness, hunts birds.⁴⁵⁸

(90) PANTHERA: The panther is an animal that is entirely adorned with various colors.⁴⁵⁹ Its spottedness is in the form of eye-shaped circles on a tawny background, edged sometimes in white, sometimes in blue.

They say that this animal is easily tamed and that it gives birth infrequently and with difficulty due to the length and sharpness of its claws, which frequently injure

---

453. Scanlan (170) identifies "fleshy dropsy" as "brawny edema."

454. Jessen (1867, 708) cites this passage in identifying *caprifolium* as woodbine, *Lonicera periclymenum*.

455. *Formica*, lit., "ant," describes a maddening itch (see 1.607, with notes). "Vegetables" is an attempt to make sense out of A.'s *olo* (the reading is in both St. and Borgnet).

456. On *sambuchum* see 22.41(24) and *De veg.* 6.115.

457. The identity of the pard is a muddled issue (cf. *AAZ*, 107–8, 147). Is it the male animal that mates with a lioness to produce a leopard (cf. 22.109[59]), another name for a leopard, or, as Pliny *HN* 8.23.62–63 would have it, a male panther? To choose would be to impose a precision upon the authors that they themselves did not possess. See McCulloch (1960), 150–51; and George and Yapp (1991), 54–55, with ill. 28. For parallel texts, see Isid. *Orig.* 12.2.10, ThC 4.86, Barth. 18.81, Vinc. 19.101, PsJF 390–91, and Rabanus Maurus *De universo* 8.1 (*PL* 111:220A).

458. ThC 4.86.10–11, following Pliny *HN* 10.94.202, and describing rather well the habits of leopards. The comment about birds, however, was originally made about ordinary cats.

459. Cf. Pliny *HN* 8.23.62, Solinus 17.8, Isid. *Orig.* 12.2.9, *Physiologus* (Curley, 1979, 42–45), ThC 4.87, Barth. 18.80, Vinc. 19.99–100, PsJF 389–90, and Jacob. 88 (177). The panther had a long and popular history in the bestiary tradition (White, 1954, 14–17; McCulloch, 1960, 148–50; and George and Yapp, 1991, 53–54). While certain identification is impossible, the main choice seems to be the African leopard.

the mother's womb. Due to the heat of its hunger, this animal, as do the other sharp-clawed animals, often overeats. At this time it takes itself to its cave and sleeps for a long time. When it awakens, a fine-smelling vapor emits from it and, according to Pliny, other animals follow this odor in herds.[460] But, as we have shown in our investigation *On Sense and the Sensed,* this is false, for no other animals apart from the human either take joy in or are saddened by smell.[461]

It is a vocal animal, giving frequent calls in its desire for intercourse. At this some animals of its own or of a related genus flock to it. This animal is said to be hostile to the *draco* and the *draco* is said to flee from it to the remote parts of its burrow.[462] Some also say that it has the shape of the moon on its right shoulder and that this parallels the growth of the moon, waxing and waning along with it.

(91) PIRADER: The *pirader* [reindeer] is an animal the size of an ox, with branched horns like a deer, the footprint of a boar or pig, the color of a bear, and thick, shaggy hair.[463] They say that when frightened this animal changes its color to match that of bodies placed around it. Thus, if it is lying among white rocks it turns white, and if in a thicket it turns green. Hermes says it has this in common with the basilisk and that using imitation of this sort it often eludes a hunter.[464]

(92) PEGASUS: Pegasus, whose shape the astronomers say is in the sky, is a composite animal that has its origin in Ethiopia. It is very large and horrible, having the front part of a horse. It also has wings like an eagle but much larger. It has a horned head that is so monstrous that many animals are terrified just by looking at it. It does not fly by raising itself aloft on its wings but rather it strikes the air with them to provide speed to its running. It is hostile to all animals, but especially to humans.[465]

(93) PILOSUS: The *pilosus* ["hairy man"] is a composite animal, human above and a goat below. But it has horns on its forehead and belongs to the monkey genus. It is quite monstrous and sometimes walks erect and is tamed. They claim that it lives in the deserted regions of Ethiopia and that sometimes it is captured and led to

---

460. Pliny *HN* 8.23.62.

461. *De sensu et sens.* 2.12. But see 22.126(83), where the onager's excrement is said to be "delightful to dogs."

462. *Draco,* or dragon, is treated at length by A. at 25.25(27).

463. A.'s *pirader* is better known in the texts as the *parandrus, tarandrus,* and at 22.135(98) as the *rangyfer.* The identification is not absolute (for example, some bestiary texts place it in Ethiopia), but it is a fairly safe one. Cf. Pliny *HN* 8.51.123–24 (Scythia), Solinus 30.25, ThC 4.88 (*pirander*), Jacob. 88 (182), PsJF 390 (*parander*), and Vinc. 19.101, 109. George and Yapp (1991, 85–86) offer a good summary for identification and an illustration; cf. McCulloch (1960), 150.

464. Reindeer (called caribou in North America) are a mix of brown and white, becoming predominantly white in winter. More curiously, their eyes become blue in winter as well, apparently the only mammals to do so (Stokkan, 2013).

465. Pegasus, of course, is completely mythical. Born from the body of Medusa when Perseus beheaded her, Pegasus was later captured by Bellerephon and he used it to slay the Chimera. Pegasus was not horned, but the rest of the description is accurate enough. Cf. Pliny *HN* 10.70.136, who denies it; Solinus 30.29, who says it is a bird having nothing equine about it save the ears; ThC 4.89 (*semivolucre*); and Vinc. 19.102.

Alexandria. A dead one was brought preserved in salt to Constantinople. Many such composite animals are found, as we have said in preceding books.[466]

(94) PAPIO: The *papio* is an animal that is quite numerous near the city of Caesarea.[467] It is a little bigger than a fox but has the habits of a wolf. When they are in a group these animals howl, with one leading the way and the others responding in a group after him. They are shaggy like foxes and when one of this species has died, the others howl around as if mourning the dead. Their voices are so loud that they are reckoned to be nearby even when they are far off.

This animal is said sometimes to enter the sepulchers of men when hungry and to feed on their bodies. It is, as it were, composed of a wolf and a fox.

(95) PATHYO: The *pathyo* is an animal which naturally has a wondrous decoration.[468] It has a strong purple color, so radiant that it seems to sparkle with it. When it is dead it retains the redness but its radiance is subdued. It is the size of a dog and very readily tamed, and it takes great pleasure in dainty foods. Certain of the ancients thought this animal had something of the divine in it. The bones of this animal are very strong and hard and its sinews are so strong that they can be dislodged only with great force.

(96) PUTORIUS: The *putorius* [mustelid] is an animal defined not by species but by genus.[469] It has this name because it smells quite bad, especially when it is angry.[470]

---

466. This entry is a combination of several things. The classical figure of Pan was human above and goat-like below. Satyrs (Scanlan, 172) generally had no horns and were often, though not exclusively, shown with equine nether parts. It is also described in terms that would fit several kinds of baboon, some of which were tamed for domestic chores by the Egyptians. Likewise, the *pilosus* belonged firmly in the tradition of the "wild man" (Friedman, 1981). ThC 4.90 also calls them *homines sylvestres* and cites as a source the *Glossa* on Isa. (Vulgate Isa. 13.21 and 34.14 refer to the *pilosi*) and Jerome's life of St. Paul the hermit, and gives their alternate names as satyrs, fauns, and incubi. He then goes on to describe, in fairly good baboon terms, a specimen that was brought in his day to the king of France, and he says it liked wine and cooked food, had good table manners, and generally behaved itself. Cf. Isid. *Orig.* 8.11.103, Vinc. 19.102, and Barth. 18.82. Cf. also the *pes pilosus* of 13.87.

467. The most famous ancient city bearing the name Caesarea was located about midway between present-day Tel Aviv and Haifa on Israel's Mediterranean coast. Herod enlarged a city existing on this site and renamed it Caesarea in about 13 B.C. The city would later serve as the capital of both Roman and Byzantine Palestine. Another Caesarea, however, was the capital of Roman Cappadocia. Before Cappadocia became a Roman province, the city was known as Mezigah or Megizah. ThC 4.91 identifies the latter Caesarea by placing it in Cappadocia but adds little to this account. Stadler and Scanlan after him (172) identify this animal as the golden jackal, *Canis aureus*. It certainly resembles a jackal in several of its traits. Cf. Jacob. 88 (176), who calls these animals *canes silvestres*.

468. ThC 4.92 and Vinc. 19.102. Stadler suggests, and Scanlan follows, an identification with the mandrill, a baboon whose coloration surely makes it a prime candidate, with bright blue to violet cheek prominences, pink to crimson sitting pads, and vividly colored patches of bare skin on its face and rear. Another candidate, based on its ability to be tamed, might be the sacred baboon shown in Egyptian art, *Papio hamadryas*. It also has large red sitting pads.

469. Apparently a generic description of all mustelids (animals resembling weasels), such as skunks, weasels, and martens. Cf. 22.104(51) and ThC 4.93 and Vinc. 19.102.

470. A. passes on an etymology connected with Isid. *Orig* 10.148: *putorem enim foetorem dicit*, "stench is called fetor."

It is what we call a *martarus* [marten] in all of its forms. And the weasel [*mustela*] is of this same genus, as is the ermine [*herminium*].[471] All the animals of this sort are hostile to mice and chickens and when they catch a chicken, they first take off its head and brain so it cannot cry out.

(97) PIROLUS: The *pirolus* [squirrel] is an animal also called by another name, the *spiryolus*. It is a bit larger than a weasel but is not longer. As some say, it only differs from the *varius* with respect to its locale.[472] For the one in Germany is red when old but black in the first year of its life, whereas the one in Poland has red mixed with gray and the one in parts of Russia is totally gray. Thus, no significant difference is found between the *varius* and the *pirolus* in shape, size, habits, and food.

These animals are like others of the mouse genus, having two very long lower teeth, being very restless, building their nests in trees, having long, shaggy tails, and leaping from tree to nearby tree, moving their tails as if using a rudder. When they move they drag their tail along behind, but when they sit they hold it erect behind their backs. When they take food, they take it in their front paws, using them like hands, as do others of the mouse genus, and then they place it in their mouths. Their food is nuts and fruit and their flesh is sweet and good.

(98) RANGYFER: The *rangyfer* [reindeer] is an animal produced in the northern regions near the North Pole [*polum arctchycum*], as well as in parts of Norway and Sweden and in regions that have a greater latitude.[473] It is called *rangifer* as if for *ramifer* [branch-bearing]. It has approximately the shape of a deer but with a larger body. It has very great strength and is quite swift in flight. It has three rows of horns on its head and each has two horns itself, with the result that its head seems to be surrounded with a ticket. Two of these are larger than the rest and are located where a stag has its horns. In a full-grown beast, these sometimes attain a size of five cubits and have twenty-five branches. The two in the middle of its head, however, are broad like the horns of a gazelle [*dama*], and they are surrounded with many short branches. It has two others on its forehead turned toward the front, but these seem more like bone. It uses all of these to fight wild beasts which confront it.

(99) SIMIA: The monkey is a familiar animal which, as we have said before, has many species.[474] The one that is found most often bears a similarity to the human in

---

471. On the marten, see 22.125(81). On the ermine, see 22.101(44) and Sanders (415–16).

472. The many name forms seem to correspond with the many colorations, all quite natural, of the squirrel. On the *varius*, see 22.149(111). For the name form *spiriolus*, see 3.96 and 12.197. Cf. ThC 4.94 (*pirolus*), Neckam 2.124, PsJF 393 (*scurulus*), and Vinc. 19.102. George and Yapp (1991, 60–61) add the forms *cyrogrillus* (but cf. 22.46[27], with notes) and *sciurellus*, and they have a fine illustration of what they say is the European red squirrel. For a tame squirrel, see Walker-Meikle (2012, fig. 8).

473. See the notes at 22.132(91), as well as ThC 4.95 and Vinc. 19.103.

474. This section owes much to Pliny *HN* 8.80.215f and as such is prone to the general confusion that existed in ancient literature concerning monkeys (*AAZ*, 118–22). The monkey and its behaviors fascinated a multitude of authors. It was popular in the *Physiologus* (Curley, 1979, 39) and in the bestiary tradition (White, 1954, 34–35; George and Yapp, 1991, 91–92; McCulloch, 1960, 86–88). Cf. ThC 4.96, Vinc. 19.106–8, Jacob. 92 (219), and Barth. 18.94. A modern must be careful not to exclude apes when reading about these "monkeys," as many of them were specifically said to lack tails.

the shape of its head and ears and in its hands, feet, and reproductive parts. It also has breasts on its chest as does a woman, but it is hairy. It has less hair on its face and has broad buttocks, with a place on them prepared, as it were, for sitting. It is imitative of humans and mimics human actions. For this reason it often falls into the hands of hunters. For when a hunter is able to be seen by monkeys dwelling in trees or cliffs, he sets out many shoes and puts on some of them, binding them on with force within sight of the monkeys. Or he might pretend to smear his eyes with birdlime that he has set out. Then, when the hunter lies hidden, the monkeys are deceived and are captured while putting on the shoes or smearing their eyes with the birdlime.[475]

These beasts frequently bear two offspring, but the one that is loved is hugged and carried in its arms. The one that is loved less sits on the back of the mother, and when the mother is sometimes chased by a hunter, she is forced by necessity to cast aside the son (that is, the young one) that she is carrying in her arms so that she can flee. The one clinging to her back, however, cannot be shaken loose and if she should happen to escape with that one she begins to love the one she formerly loved less.[476]

The monkey is a tricky animal with bad habits. However much it might be tamed, it is always rabid and it imitates the bad rather than the good human traits.[477] It is playful with the small offspring of humans and dogs, but it sometimes strangles unguarded boys, sometimes hurling them off a height. It collects vermin on heads and clothing and eats them. It remembers a wrong a long time and, as we said in previous books, although it resembles a human externally, it has no resemblance with him internally, and even less than nearly all the other beasts.

When it has been tamed and is living in a house and it gives birth, it shows its offspring to the people as if they were beautiful and pleasing. It has breasts on its chest as does a woman since nature gave it hands with which it can lift its young to its chest as a woman can. Other animals, however, have their breasts below so that they can be reached by their young.

Just as we have said quite often in preceding books, there are very many genuses of monkeys. One seems to be composed of a monkey and a wildcat [*agrestis cattus*] because it has the face of a monkey, varicolored with two black spots on its jaws, but it has a very long gray tail that is black at the end. Some call this animal a *spinga* while others call them *cythosicae symiae*.[478]

---

475. These stories have a long pedigree. Cf. Pliny *HN* 8.80.215 and Solinus 27.56. Ael. *DA* 17.25 says that the hunters put on regular shoes but leave out shoes weighted with lead.

476. Rowland (1973, 8–14) deals at some length with the symbolic usage of this belief and has an excellent illustration.

477. *Rabidus* normally means infected with rabies. Here it must mean savage and may stem from Solinus 27.58.

478. Pliny *HN* 8.80.216 and Solinus 27.58f. begin to talk of a variety of apes: *cynocephali* ("dog-headed apes," baboons), *satyri*, and *callitriches* ("pretty-haired," from the Gr.). The sphinx, callitrix, and satyr are discussed by George and Yapp (1991), 91–92. See also 22.120(76) and McDermott (1938).

There are others, however, that have very gracious and kindly faces. They are different from other monkeys, having beards and very long tails. These are said to be Ethiopian and if they are removed from Ethiopia they live but a short time. There are Indian ones that have completely white bodies. They are bearded and have wide tails. These are hunted with arrows by the Indians. When they are tamed, they are good at any trick, almost as if they had been created for tricks alone.[479]

The members of the monkey are said to be effective in many medications. But since this has not been personally experienced by me and since it is more fitting for another investigation, it seems better to pass it over here.

(100) TIGRIS: The tiger is produced in the regions inhabited by the Hircani.[480] It is amazingly swift and fierce. It is about the size of a greyhound and grows even larger. It is multicolored, a black color with tawny stripes intervening in waves, as it were.[481] It has sharp claws and teeth and a multipart, split foot that is separated into many divisions. It produces many young and if a hunter should sometimes capture them, he can scarcely escape, even with the protection of a ship. If the ship stops far off and the mother tiger has followed it, the hunter throws off one of the many young when the mother is near. While she is busy with carrying this one back to the cave, the hunter goes on farther. If she returns a second time and again to follow the hunter, he again returns one of the many young to the mother until finally, getting completely away, he keeps some of the young. Other hunters have glass spheres with them which they throw to the mother. When the mother looks into the sphere, the reflections of her young appear, since she is looking into a sort of mirror. Thus, by throwing sphere after sphere, they delude the mother who thinks, due to the motion of the sphere, that her young one is moving. But when she attempts to nurse her young one and breaks the sphere with her feet, she understands that she has been deceived, and the hunter, after the mother has been fooled several times in this way, gets away to civilization or to his ships and she loses her young.[482]

(101) TAURUS: The bull is a familiar animal. When it is not castrated it is very fierce, fighting with its horns, always grazing alone, apart from the herd, but not too

---

479. "Trick" here is *ludus*, and one is tempted to translate as "play," in keeping with Pliny *HN* 8.80.215, but cf. 22.36(17).

480. Hyrcania was a land bordering the Caspian Sea in antiquity. The Hyrcanian tiger was frequently mentioned and, because of its location, was probably the Siberian tiger (Keller, 1909, 1.62). Cf. Pliny *HN* 8.25.66, Solinus 17.4, ThC 4.97, Barth. 18.102, Jacob. 88 (180), PsJF 395–96, and Vinc. 19.112.

481. This careful description might bespeak firsthand knowledge. Note the spotted tiger on a MS from the Bodleian (George and Yapp, 1991, 56) and one with bar-like markings (Rowland, 1973, 150). See also McCulloch (1960), 176–77. Solinus 17.4 describes Hyrcanian tigers as "known by their spots."

482. This delightful story is found in Pliny *HN* 8.25.66, who just has the hunter tossing back cubs, and Ambrose *Hex.* 6.4.21, who specifies glass spheres, as does Peter Damian in *De bono religiosi status* 14 (*PL* 145:775), with an appropriate religious interpretation. Cf. Jacob. 88 (180) and Jorach as cited by Arnoldus Saxo 2.3.22a–b. The entire legend has been studied in depth by MacGregor (1989) with interesting illustrations. Both illustrations mentioned in the previous note are also of this scene. Glass spheres would have the effect of shrinking the image. Other versions seem to specify the use of mirrors (George and Yapp, 1991, 55–56). Jacob. has the odd *clipeos vitreos*, "glass shields."

far from the cows.⁴⁸³ When it is castrated, it is made more fit for bearing the yoke and it exhibits a special affection for its comrade in bearing the yoke, even to the extent that it grazes alongside it with great pleasure, calls for it with a lot of mooing and, if it should happen to lose track of it, seeks it by moving about.⁴⁸⁴

Throughout its genus, this animal plucks grass with its teeth, but it leaves the roots alone and can do no harm with its teeth. It makes the cow pregnant with a single thrust because she cannot endure the hardness of his penis for long.

If cow's milk is first mixed with water it is said to turn into a stone through coagulation.⁴⁸⁵ It is also said that bulls grow fat if they are washed often with warm water.

The signs of a noble bull are a fierce forehead, a threatening face, strong planting of the front feet, pawing at the earth, an erect tail with a little bit of a curve, and the fact that it always presents its horns for a blow. The bull's head is its strongest member and it therefore bears its horns on its head and these horns are bound to the yoke. It also has a neck made of strong, cartilaginous flesh, and some therefore tie the yoke to its neck.⁴⁸⁶ This animal has harder sinew cords than any other of the animals in its entire genus. But the bull [*taurus*] has harder ones than the castrated ox [*bos castratus*].⁴⁸⁷

This animal has dry, melancholic flesh that is difficult to digest, but it provides good and very excellent nourishment and, when kept from work, the animal grows fat. If, however, it is at leisure and amid pleasure before it goes to work, it dies more quickly. It takes pleasure in both standing in and drinking clear water, but it grazes readily in marshy areas.

The horns of a bull are softer than those of a cow. If its hoof was not completely formed in its mother's womb, the young one dies, and thus, those born before ten months die. If in a particular summer the cows have numerous births, it is a sign that the following winter will be a very rainy one.

---

483. Much of what follows is from Pliny *HN* 8.70.176f. Cf. ThC 4.98, Vinc. 18.87–91, PsJF 395 and 400–401 (*vacca*), and Barth. 18.98, 107. The bull, cow, and ox were common animals in the bestiary tradition (Rowland, 1973, 43–48; George and Yapp, 1991, 104–5; McCulloch, 1960, 98–99).

484. "Seeks it by moving about": the exact sense of *quaerat motu corporis* is unsure. The model for this statement is Isid. *Orig.* 12.1.30, who makes no mention of the body, simply stating that the animal moos. It might merely mean that the animal is restless and it is so translated here.

485. This is badly mangled from Ar. *HA* 575b9f., who states that a cow's first milk, when set, is hard as a rock unless it is mixed with water.

486. The advantage of the horns is that it relieves pressure on the neck and thus the windpipe. Up to eight oxen would be hitched to the heavy medieval plows. This section is derived from Pliny *HN* 8.70.179, which specifically states that Alpine cows, though small, are yoked at the head and not the neck.

487. "Sinew cords": *nervos cordis* is a most vexing phrase. Scanlan (177) takes *cordis* as the gen. sing. of *cor* and translates it as "has harder cardiac cords," drawing an analogy with the heart bone. This is unlikely. *Cordis* may also be the dat. or abl. pl. of *corda,* sinew; and after comparison with the parallel texts of ThC and Barth., it is clear that sinews and cords are being discussed. Still, the phrasing leaves much to be desired.

Cows are afflicted with a gout [*podagra*] that is not easily curable and they also suffer from a disease of the lungs whose symptom is drooping of the ears.⁴⁸⁸ According to some, drinking bull's blood is deadly.⁴⁸⁹ The bull alone of all the animals is made eager for work by use of the goad. Bull blood does not coagulate quickly, but cow's milk, due to its thickness, does so easily.

We have said in preceding books that the cows of some areas are so large that a person milking them is required to stand.⁴⁹⁰ These cows give a great deal of milk. In very warm regions their monthly cycle appears every five months. At other times it is removed with the urine. The indication of this is that the urine of all female animals is thinner when they are pregnant than at any other time.⁴⁹¹

When there are two flocks within a herd they fight and the victor mounts the cows.⁴⁹² And when one is tired out from too much intercourse, the one that was conquered in the first place returns and fights. When he wins, he mounts the cows. If the skin of a bull or cow is split open and is blown into, and if the animal is fed thereafter, it grows fat more quickly.

Warm bull's blood alleviates broken bones. Its bile, mixed with honey, extracts an embedded weapon or thorn and is good as an eyewash against "hoof eye" [*ungula oculorum*].⁴⁹³

If the words of Solinus the philosopher can be believed, there are bulls in India that have a dusky color, a powerful head, thick bristling hair, and a mouth opening that extends from ear to ear. They are so swift that they almost seem to fly and their hair is turned backward. They have horns that they can move as they wish, bending them either forward or backward for battle and they extend or retract them. They eject their dung behind themselves very violently and with its heat and stickiness they hold back hunters and dogs. They have backs and skins so hard that they pay weapons

---

488. Scanlan (178) suggests blackleg disease or foot rot for the first and bovine tuberculosis for the second.

489. The origin of this legend may lie in a plant toxin whose ancient name was "blood of the bull" (Kitchell and Parker, 1993).

490. At 3.176, A. had said this of the cows of "Ambardoz," and then he made mention of the large cows he knew in Frisia, Holland, and Seeland.

491. These last few statements are poorly written and vaguely stated. From ThC 4.98.65f. it would seem that A. wishes to say that the menses can appear every four or five months (A.'s *per quinque et quinque* is odd) and that normally the menses are excreted along with the urine.

492. An interesting corruption. ThC 4.98.69–70 preserves what *should* have been said here: *cum duo tauri fuerint in grege, pugnant inter se*, "when two bulls are in a herd, they fight one another." A. has *cum etian duo greges sunt in armento pugnant*. Since *grex* and *armentum* are synonyms, one is clearly a gloss for the other, yet both crept into A.'s text, perhaps through his source.

493. Scanlan (178) suggests that this is crow's feet, basing the interpretation on a translation of *ungula* as claw. Latham (s.v. *ungula*) suggests a tumor of the eye. If *ungula* is to be taken rather as "hoof," it might refer to some horny growth on the eye, perhaps even cataracts.

no mind. They have such rage that when captured they sometimes kill themselves in their fury.⁴⁹⁴

Some say that in the East there are oxen which have only one horn and others that are armed with three on their foreheads. These do not have a divided hoof but rather a horny and solid *sotular* like a horse.

(102) TRANEZ: The *tranez* is a small animal with a reddish color. It is about the size of a rabbit and is remarkably contentious and spirited, a sign of which is that it has received from nature a bony helmet for the protection of its head and brain.⁴⁹⁵

(103) TRAGELAFUS: The *tragelafus,* as befits its name, is an animal composed from a stag and a goat.⁴⁹⁶ It has branching horns like a deer and a bearded chin like a goat. On its chest it is shaggy and it is a hardy animal, ready to fight all that oppose it. It has its origin in the region that is called Fasida.

(104) TROGODYTAE: *Trogodytae* are a kind of animal that have very long horns reaching down from the head, in front of the jaws, all the way to the ground. It has, however, a very long neck and therefore, with its neck bent and twisted down, it places the length of its head and of its horns onto the ground. Its entire forehead thus lies on the ground and it therefore takes its food in a method contrary to that of all other animals. For they are not able to eat in any other way due to the length of their horns. They thus eat only the long herbs, which they seem to cut off rather than to bite out by the roots, as do other animals.⁴⁹⁷

(105) TALPA: The mole is a small animal of the mouse genus which is also called both the ground and the blind mouse [*mus terrenus et caecus*].⁴⁹⁸ It has quite short legs and sharp claws, with five digits in front and four in the back. It is black in color, with

---

494. Solinus 52.36. The ultimate source of this creature is to be found in Pliny *HN* 8.30.74–31.77, where the *eale,* wild bulls, and an Indian *bos* are described. A.'s creature has traits of all three. Cf. Isid. *Orig.* 12.1.29, Jacob. 88 (183–84), PsJF 395, and ThC 4.99, who gives the animal its own entry. Might this be the rhinoceros?

495. Despite this rather clear description and despite the fact that ThC 4.100 cites Pliny as a source, this animal cannot be identified.

496. The etymology is from Isid. *Orig.* 12.1.20. Pliny *HN* 8.50.120 (cf. Solinus 19.19) says it lives only by the Phasis River (A.'s Fasida), which runs into the Black Sea, but in Greece it was thought of mostly as another composite, fabulous animal (cf. LSJ, s.v.) such as the hippocamp or hippogriffon. Stadler suggests the elk as an identification, while George and Yapp (1991, 81) prefer the goat called the tur, which, however, does not have deer-like horns. For other medieval citations, see ThC 4.101, Barth. 18.99, PsJF 394, and Vinc. 19.113. Today the genus *Tragelaphus* contains medium to large antelopes with spiral horns.

497. Cf. ThC 4.102 and Vinc. 19.113. These accounts are based on Pliny *HN* 11.70.125, where, in a discussion of horns, Pliny mentions the cattle that belong to the fabled troglodytes (cave dwellers). Their horns point to the ground and they must therefore eat with their necks at a slant (Aiken, 1947, 218–19).

498. Cf. Isid. *Orig.* 12.3.5, ThC 4.103, Vinc. 19.137–39, PsJF 396, and Barth. 18.100. The mole was frequently in the bestiaries and is often well illustrated (George and Yapp, 1991, 68–69; White, 1954, 95–96). McCulloch (1960, 143) notes that Jerome's commentary on Isa. 2.20 was a source for much of the common text (*PL* 24:55). The mole's many odd traits led to its symbolic use (Rowland, 1973, 126). The animal held a fascination for A. Note A.'s experimentation on a mole at 1.140 and at *QDA* 1.4, where he refutes the idea that the mole lived only on pure elements.

soft hair that is thick and short. It has places for eyes but has no eyes and its skin thus has no hair in these places.[499]

It feeds on insects, but I have personally experienced that it willingly eats toads and frogs. I have found that a mole held onto a large toad from beneath the ground and that the toad, as it fled, pulled the entire body of the mole out of the ground and cried forcefully on account of the mole's bite. I have also personally experienced that both frogs and toads eat a dead mole. The mole eats the vermin called earthworms. If it is starving it eats the roots of plants, especially those of grains.[500]

This animal is generated from earth that has been rained upon and rendered putrid.

They say that if a mole is burned to ashes and the ashes are smeared with the white of an egg over a leprous spot, it does a great deal of good. The blood of a killed mole, if smeared over a spot that has lost its hair, causes the hair to return.[501]

This animal is not able to live for a long time outside of the ground and has very good hearing. It thus can hear worms tunneling along in the ground from far off. It does this not from the goodness of its hearing but rather from the continuity of the earth that is moved. For if a very long hole is made beneath the earth and if someone were to speak into it, he would be heard far off, just as one who speaks through shafts or through hollowed-out logs is heard however far away he is.[502] This is the reason that the mole makes many hollow tunnels around itself in every direction.

(106) UNICORNIS: The unicorn is an animal that has but moderate size for its strength.[503] It has a boxwood-like color and the hoof part of its foot [*ungula pedis*] is

---

499. The back feet have five toes, but one is quite stunted. The animal can see with its rudimentary eyes.

500. The diet of moles is more varied than one might expect (Gorman and Stone, 1990, 40–42). They generally feed on slugs, insects, and especially worms that have fallen into their tunnels. Small amounts of vegetable matter are found in some moles' stomachs, including truffle. In captivity they eat baby mice, and gamekeepers report they snatch pheasant eggs. As recently as 1968 a report was published that bears repeating here, as it sounds as if A. himself wrote it: "One evening in early autumn . . . I was walking in a grass field. . . . Suddenly a large frog emerged from the wood, going at a good pace in long hops and closely pursued by a mole which was hunting it like a spaniel. . . . In four or five yards it caught the frog, turned it over on its back and . . . started eating its stomach while the poor victim kicked and struggled in vain. This all happened right in front of me and only a few feet away" (quoted by Mellanby, 1973, 63).

501. Cf. Pliny *HN* 30.7.19–21, where the mole is said to be a favorite of magicians and healers since nature has given it powers to make up for its blindness and dreary life. A.'s recipes, however, are not from this source. ThC 4.103.9f. has the same recipes.

502. "Shaft": *hasta*, lit., "spear," and also used for such objects as the staff of a banner or a spit, but here clearly some long, hollow object.

503. As is common, what follows is an amalgamation of factual data on the Indian rhinoceros and the mythical unicorn (Shepard, 1956). Cf. Isid. *Orig.* 12.2.12–14, *Rhinoceron a Graecis vocatus . . . idem et monoceron, id est unicornus*. The reference to boxwood is persistent and is at least as old as Pliny *HN* 8.29.71. This animal, under many names, had a long tradition from Ctesias, throughout antiquity, to *Physiologus* (Curley, 1979, 51) and the bestiary (George and Yapp, 1991, 86–89; McCulloch, 1960, 179–83; White, 1954, 78), and it acquired many symbolic uses (Rowland, 1973, 152–57). Most recently Cherry (1995b) has traced the history of unicorn lore in a well-illustrated study.

split into two parts. It lives in mountains or in deserts and has a very long horn on its forehead which it sharpens on rocks and with which it pierces even the elephant. Neither does it fear the hunter. Pompey the Great exhibited this animal in the games at Rome.[504]

They say, however, that this animal respects virgin girls so much that when it sees them it grows tame and is sometimes captured and bound in a trance near them. It is also captured when it is a young animal and is tamed in this way.

(107) URSUS: The bear is a familiar animal that is very moist and ugly.[505] It has a thick, shaggy coat and human-like extremities, and it sometimes walks upright like a human, but not for long. It has, however, strength in its arms and loins and uses its arms for many activities, using them to throw branches it happens to find at dogs, to tear up olive groves and beehives, and also to climb. It is weak in the head, however, especially in the *sinciput*.

It is both aquatic and land-going. The aquatic one is white and hunts beneath the water like the otter and beaver.[506] The land one, however, eats meat, plants, honey, fruit from trees, and has very foul breath. As we have determined in previous books, it is one of the animals that lies dormant in winter at the time of phlegm production.[507] It is therefore then that the she-bear gives birth because of the comfort of her heat. The limbs on the newborn seem incomplete due to the overabundance of humor. For this reason the she-bear licks the young one for a long time and cares for it. The bear has very little blood except around its heart. When its flesh is cooked it seems, contrary to the nature of all other flesh, to increase in size. This is because the great moisture found in it is set free through the heat of cooking and causes it to swell.

When fighting, it leaps on the animals, and if they have horns it grabs their horns in its hands and lays them low. It does not readily fight with humans, however, unless it has received a wound from one. It frequently licks its claws against bodily pains, almost in a rage. When wounded and pierced, it tries to heal its wound with herbs having a dry complexion. Against diseases it eats ants and the herb that is called *fleonus* in Greek but which in Latin is called "bear herb" [*herba ursi*].[508]

Some say that the bear always is growing and that as a result bears are sometimes found with a length of fifteen cubits. In our lands they come in three colors: white, black, and brown. They are captured in many ways, but it is not the place here to speak of these methods. When captured, they are sometimes tamed and then become

---

504. The games were held in 55 B.C.

505. "Ugly": *informe,* perhaps also implying a lack of definite form, as when it is born. The bear is common in the bestiary tradition (White, 1954, 45–47; George and Yapp, 1991, 56–58; McCulloch, 1960, 94–95) and also appears with some regularity in the fable tradition (Ziolkowski, 1993). Cf. Pliny *HN* 8.54.126–31, Solinus 26.7, ThC 4.105, Neckam 2.130–31, Barth. 18.110–11, PsJF 396–99,and Vinc. 19.116–20.

506. A. here provides one of the earliest descriptions of the polar bear. Cf. 1.99 and 7.146.

507. Cf. 7.89, 154f.

508. Cf. Ambrose *Hex.* 6.4.18, which cites the Gr. *phlomos,* the mullein.

quite playful. But they are easily aroused and when they are, they go on a rampage and kill people. They sometimes turn wheels by walking on them and thus draw up water from wells or else draw stones onto lofty walls by means of a block and tackle. I have frequently experienced this myself. This animal copulates like the human and the monkey, not like other quadrupeds.

(108) VESONTES: The *vesontes* [wisent] is an animal like a cow, with a bristling neck and mane like a horse. It is so stubborn and truculent that it can scarcely or never be tamed when captured.[509]

(109) URNI: The *urni* [aurochs] are the cows which we call *wisent* in German.[510] They have two huge horns which can hold a great deal of liquid and thus many people pour their drink into them and store it in them. When aroused, this animal can toss a person along with his horse with these horns.[511] Many things have been said concerning these animals by us in our preceding books.[512]

(110) VULPES: The fox is so called as if it were swift of foot [*volipes*].[513] It is about the size of a moderate-sized guard dog and has the reddish hair of the forequarters of a dog. It is so warm an animal that it warms the dead.[514] It has a shaggy, large tail. It is full of tricks and when a dog is following it, it leads its tail through the mouth of the dog and leads it to and fro and thus sometimes eludes it. When sick it eats the resin from a pine tree and is cured. Its life is prolonged this way as well.[515]

Jorach says that when the fox has fleas, it takes as much hair or soft hay straw in its mouth as it can and submerges its whole body in the water a little at a time, beginning with its tail, so that the fleas, escaping the water, climb onto its head. It then submerges its head little by little so that the fleas climb onto the straw which it has in its mouth. It then spits out the straw and gets out of the water.

---

509. Cf. 8.205f. Also cf. Pliny *HN* 8.14.38 (*bisontes*), Solinus 20.4 (*visontes*), ThC 4.106, and Vinc. 18.21. Cf. *AAZ*, 140–45.

510. Obviously, the terms "wisent" and "auroch" have become intertwined and, in fact, they remain confused today. The scientific world currently restricts wisent to the *Bison bonasus* of Europe. This animal closely resembles, and is a relative of, the American bison, itself incorrectly called a buffalo. It can be seen today on game preserves (Grzimek, 13.393–98, and color pl. 343). Confusingly, the term "aurochs" is sometimes used incorrectly for the European bison, but scientifically aurochs is restricted to *Bos primigenius*, a wild ox, not a bison at all. The MG name preserves its true nature: *urochs*, "ancestral ox." It was fairly plentiful in central and western Europe in A.'s day, but by 1400 it had nearly died out, lasting longer farther east. Yet in 1627 the last aurochs, a female, died (Grzimek, 13.368f., with color pl. 163–64). On the name forms, cf. Sanders (424) and the analysis of Lengerken (1955, 156–72) and Vuure (2005).

511. An aurochs drinking horn is depicted in Lengerken (1955, 166), as is a woodcut of the attack of an aurochs dating to 1495 (161). Cf. PsJF 401.

512. E.g., 3.120; 6.102; 13.32; 22.24(12), 49(36); and cf. 22.149(111).

513. Cf. Isid. *Orig.* 12.2.29, *Physiologus* (Curley, 1979, 27–28), ThC 4.108, Vinc. 19.121–23, Barth. 18.112, PsJF 399–400, and Neckam 2.125–27. See also White (1954), 53–54; George and Yapp (1991), 70–71; and McCulloch (1960), 119–20. Reynard the fox also had a strong hold on the medieval fable tradition (Ziolkowski, 1993).

514. This odd statement is apparently A.'s addition and is not readily paralleled. One wonders if he has not misunderstood *callidum*, "clever," for *calidum*, "hot."

515. Cf. Ambrose *Hex.* 6.4.19.

This beast takes shelter in burrows which, so they say, it trickily seizes from the animal called the badger. When the badger has constructed a burrow, the fox enters it and defiles it with the smell of dung. The badger, shrinking from the defilement, will not go in his burrow any more. The fox, however, lives in it and a fox is often caught by being dug out of it.[516]

This animal smells bad both in mouth and anus. For it eats flesh and preys on mice just as the cat does. It sometimes catches rabbits and hares and also preys on chickens.

They say that this animal dies if it eats almonds. It sometimes suffers from a warming of its head, especially in summer.[517] It is cured by having its blood flow to the outside, causing sores on its skin, whereupon its hair falls out. Since it is crafty, it sometimes pretends that it wishes to play with a hare and then seizes and eats it. For this reason many hares flee the fox as an enemy. It imitates the barking of a dog and growls when hungry. It sometimes pretends to be dead, lying prone and drawing in its breath imperceptibly while it lolls out its tongue. It then seizes and eats the birds that alight on it.[518]

The male fox has a bony penis and it copulates lying on its side, embracing the female who is also lying on her side.

If it is cooked with its skin and a paralytic is bathed in its warm broth, it provides comfort. If its flesh is burned to a powder and is given in wine to asthmatics, it is said to do some good. The oily part of its fat [*sagimen adeps*] is very good for earaches.

(111) VARIUS: The *varius* [a squirrel] is, as we have already stated, a small member of the *pyrolus* genus.[519] It is small like the *pirolus* and is white on the belly and gray or ash-gray on the back. Its coloration is delightful and its pelt is exceptionally useful in the ornamentation of clothes. It has the nest, habits, and food of a *pirolus*.

(112) ZUBRONES: The *zubrones* [aurochs], so they say, are of the cow genus.[520] They sometimes have a length of fifteen cubits. It has very large horns three cubits long, is off-black in color, and is so swift that once it has ejected its dung behind it, the animal spins around and can catch it on its horns before it can fall to the ground.

The animal lives in the northern forests and has such strength that it can toss a person and his horse with its horns. And when it tosses him it throws him aloft and catches him again until it kills him from the concussion.

---

516. ThC 4.108.11f. describes the use of foxhounds trained to dig out the foxes in their lairs.

517. Apparently a corruption. ThC 4.108.31f., citing the *Viaticum* of Constantinus Africanus, speaks of the *calefactio . . . epatis,* whereas in A. this is *calefactio . . . capitis*.

518. This was one of the most popular and persistent stories about the fox and is frequently illustrated (White, 1954, 53; George and Yapp, 1991, 13, 19). Most sources say the fox draws in its breath forcefully, holding it to give the impression of a bloated corpse.

519. Cf. 22.134(97). Stadler identifies this as *Sciurus vulgaris varius,* whose coloration is much like this. Cf. ThC 4.109 and Vinc. 19.114.

520. See the discussion at 22.146(109). The name is of Polish origin (modern Polish *zubrzyca*). Cf. Sanders (424) and Scanlan (184), who point out correctly that A. travelled in Polish territory (see our introduction for the map "Travels of Albertus Magnus as Preacher of the Crusade"). Cf. ThC 4.110, who tells us some people incorrectly called these aurochs "tigers," and Vinc. 19.125.

This animal ejects its dung with such a violent flatulence that it renders useless a dog or hunter touched by it. It can only be captured using the trick of a trap or by a hunter walking around the periphery of a large tree while the animal follows it and he stabs it in the side with a hunting spear.[521]

(113) ZILIO: The *zilio* [hyena], as we said in preceding books, is an animal the size of a wolf which is most cruel in the death of humans and other beasts.[522] This animal imitates the human voice and calls out like humans in the thickest parts of the forest. It then eats the people that seek it out. Using a similar trick it imitates the barking of dogs and eats them.

It is a composite animal, made up of a hyena and the monkey that is called the *maritonmorion*. It enters sepulchers and eats the bodies of the dead.[523]

Let these things, then, suffice concerning the nature of quadrupeds since, if there are others, they can easily be discovered in the things discussed above.

---

521. "Trap": *fovea,* tending to indicate a pit. An engraving of 1596 shows an aurochs hunt that could serve as an illustration of A.'s description of stabbing from behind trees (Lengerken, 1955, 168).

522. Cf. 7.46, where the name is *zalio.*

523. Cf. 22.119(73) and 2.49f.

HERE BEGINS THE TWENTY-THIRD BOOK ON ANIMALS

1. In this book the nature of birds will be treated specifically, and since every scientific [*physica*] investigation moves from the general to the particular, we will first speak in general about the nature of birds. Afterward, moving according to the order of the Latin alphabet, the birds will be set forth by name in accordance with their species and types. Though it is granted that this procedure is not entirely philosophical insofar as in it the same thing is repeated many times because one and the same thing may pertain to many birds, it nevertheless is an effective procedure for easy teaching and many of the philosophers have held to this procedure.

Since, however, we have treated the generation, food, habits, members, and eggs of birds in a general way in our previous books, we have to speak here of only those things in which birds correspond to other animals and those in which they differ from them. Similarly, we will speak of those matters in which they correspond to one another and in which they differ.

We state, therefore, that birds, along with all other animals, have as a distinctive trait that they possess a sensible soul, which is the very thing through which, as we have said elsewhere, an animal is constituted. And since this sensible soul is perfect in them (insofar as it is in them in accordance with not part of but with the totality of their potencies), so it is that two potencies are generally present in all birds. These potencies are sense and motion, which are the ones the ancients used to say differentiated an animate object from an inanimate one. And since whichever animals have those senses which come through an external medium also have the senses that come through an internal medium, thus it is that every bird has all the senses, although this is not a reversible statement.[1] For it would not have the power of flight except for reaching a distant thing and it would not move to the distant object unless it perceived something through an external medium at a distance.

2. Since birds, therefore, throughout their entire genus have all the senses and motion too, it is clear through those things which have been said in the *De anima* and in the *De motibus animalium*, as well as in the *De principiis motuum animalium*, that birds are perfect animals.[2] They also have the organs of these powers of the soul in their body, since they have a head through which sensation and motion flow from

---

1. The "five senses" of Ar. are characterized as requiring an external medium (e.g., air), while the internal senses (the common sense, imagination, etc.) require none. See Steneck (1980) and Dewan (1980) for discussions of A.'s views on both types of senses and references.

2. Cf. *De anima* 3.4.7–8; *De mot. an.* 1.2; and *De principiis motus processivi* 2.1.19–30. According to Geyer's prolegomena to *De principiis motus processivi* (xxiii, n. 2), *De mot. an.* is not part of the Aristotelian paraphrases. Rather, it was written by A. "out of his own ingenuity" before having access to the Aristotelian text of *De motu animalium* (698–704a3), which is paraphrased as *De principiis motus processivi* by A. This latter work was originally intended to be Book 22 of *DA*.

the heart. They also have a heart to which the sensation and motion return just as if to the principle of sensation and motion. They also have spirit which is the bearer of either power [*virtus*]. It is clear therefore that birds attain the perfect sense of "animal."

This animal, that is, the bird, differs from "animal" in general in this. An animal is defined as such due to its participation in the sense of touch. The bird, however, throughout the entire range of its kind, participates in all the senses and in progressive motion, as has been shown before. For the terms "bird" [*avis*] and "flyer" [*volatile*] are interchangeable and the term *volatile* means "having some motion or other of flying [*volandi*]." Now, the wing is an organ of this sort of motion, and nature does not give an organ except for some use, for otherwise nature would abound in superfluities—something that cannot happen. Thus, animals that are unmoving with respect to place do not have organs for progressive motion. It is clear, therefore, that the bird differs from animals in general by shining abundantly in the three senses which arise through an external medium as well as by sharing in the organs of progressive motion.[3]

A bird is like a walking quadruped in that a quadruped shares in those things which belong to motion and sense, but it differs in that each quadruped does not have every sense. The mole, for example, does not have the sense of sight, whereas every flying creature participates in sight and every other sense. It differs, moreover, from every other quadruped in the number and types of its motion. The flying creature participates in a greater number of motions, namely, those of flying and walking, whereas the quadruped has only walking. And it even differs in the manner of its walking motion, because walkers, whether running or merely going ahead a little, make their way with two motions. But the flying creature in walking and running uses one motion, as does the human, but only when hopping does it use two, again just like the human.

In general, every bird has a shape different from every other genus of animal. For every bird generally has a round breast, a round head, and a columnar neck, and the lower members of the bird have a pyramidal shape extending from its breast. And while in every other animal the lower members or those more to the rear are larger than those to the front, in the bird, throughout its genus, its smallest members are those to the rear, ending, as it were, in a point, and the anterior members are larger. This is a sign of the lightness of the flesh of birds and of their spirits.

Throughout their genus the flesh of birds is much more purged, for in other animals purgation occurs through hairs and a large body, but this purgation is accomplished for the modest body of the bird by an ordinary purgation, which occurs for both the larger ones and the flyers by means of the plumage. And thus the flesh of a bird is generally lighter and more digestible than the flesh of other animals.

Many differences exist among the birds themselves. But that which most generally differentiates the flying class is that a bird has feathered wings and another flying creature has membranous wings. Moreover, the bird having feathered wings has only two

---

3. Note that here A. seems to put together the three senses of touch, smell, and taste as one, probably on the basis of flesh as the medium, thus making three external senses instead of the more usual five.

of them. One that has membranous wings sometimes has four. And whatever has two wings moves along by using each wing like oars. An oar, however, moves from above to below and then pushes forward. Such is the motion of the two wings of a bird. Sometimes, however, when certain large birds are flying just a little, they do not seem to strike their wings from top to bottom, yet they do this imperceptibly since they do fly a little. But when they fly quickly, they repeat this sort of motion often.

5 Birds have other differences from each other. For all birds have nostrils on their beak near their head and far from the opening of the beak. But nonflying animals have their nostrils almost continuous with their mouths. Whereas birds are alike in this, they differ in that, although they do not submerge their beak up to the nostrils when drinking, nevertheless some drink while in continuous contact with the water whereas others drink a little and elevate their mouth, and repeat this action several times. Examples include the genuses of chickens and aquatic birds.

They are alike in that no bird urinates, even though every bird is seen to drink. But a bird that eats meat rarely drinks and then but little, on account of the moistness of its nourishment. A bird that eats seeds, however, drinks more due to the dryness of its nourishment. Still, no bird drinks more than is sufficient for the mixing and passing of its nourishment. Therefore, none of them urinate.[4]

It is a common feature to every bird that it lays eggs and to none that it gives birth to live young. This is due to the narrowness of its body and especially down below where it is narrower than is adequate for parturition, as is clear from what has been said before. Therefore, the bird was able to go without breasts either above or lower down, since these would have impeded its double mode of motion.

The bird [*avis*] is so called since it keeps to no path [*via*] in its flight, being, as it were, pathless.[5] Thus it is that augurs look to the birds' flights since they are not toward a single direction like the movement of the walking animals.

6 Birds generally call more than other animals. This is due to the lightness of their spirits. Therefore, very small birds call all the more, so much so that many of them are musical due to the lightness of their spirits which are easily stretched forth toward every desire. They call most when they desire copulation, and thus it is that the males among them call more than the females, since the females are heavier and more frigid than the males.

There are other differences among birds, for one might be aquatic and another not. But there is no aquatic bird which remains continuously on the water or which produces young on the water. The aquatic bird is so called because it seeks its nourishment on the water. Although it is a light bird, as we have said, it is necessary for it to breathe because of the warmth around its heart. Without it, it would not have mobility in two motions. For we know that heat is the cause of motion and cold the cause of immobility. Every bird, therefore, breathes, although it is granted that one may hold its breath longer than another.

---

4. Cf. Neckam 1.80, but contrast Fred. *DAV* 1.43.
5. Isid. *Orig.* 12.7.3.

These are many differences indeed among the birds in all the genuses that have been enumerated. But, since many things have already been said on these matters in previous books, we should now turn our discussion to speaking about the specific properties of birds.

(1) AQUILA: The eagle [*aquila*] is so called from its sharpness [*acumen*].⁶ For it possesses three sharpnesses: namely, of vision, of wrath, and of its hunting tools, that is, its talons and beak. Every eagle flourishes due to its sharpness of vision, but especially that one which is called the "noble eagle" as well as *herodius* in Latin, named, as it were, the hero of the birds.⁷

It is a very large bird and is entirely black, although in its old age it becomes an ashy color on its back and over its wings. It has bright yellow feet with long, powerful talons. It has a large, ash-colored beak that tends toward a black color. The feathers of its large, wide wings are straight, spread out a bit at the end of the wing, and curve back on the top. This bird is a high flyer and has a vision so sharp that it can gaze into the orb of the sun. For this reason it is said to hold up its young and to eject those who are unable to look into the orb of the sun without tears.⁸ They say that another bird gathers up the ones which have been ejected. This bird is called the *fehit* in Greek and the *fulica* [coot] in Latin.⁹ That this is false is shown in that the bird called the *fehit* is a type of dove and does not nourish an eagle.¹⁰ The *fulica*, however, is a type of aquatic bird which is called the black diver [*mergus niger*] and is smaller than a duck. This one also in no way nourishes an eagle since neither of these birds has common food with any genus of eagle.

---

6. Isid. *Orig.* 12.7.10.

7. Ar.'s *chrysaetos*, the golden eagle. George and Yapp (1991, 141–42) identify the *herodius* as a gyrfalcon. St. identifies it as a golden eagle, *Aquila chrysaetos*, which appears in beautiful form in the Lindesfarne Gospels, fol. 209v (Yapp, 1981, 84–85). See also *GB*, 2–16; Pollard (1977, 76–79); *BAZ*, 2–4. Classical and medieval sources frequently discuss the eagles, but identification of exact species is often difficult. The relatively inaccessible location of their eyries, plus the fact that eagles were rarely used for falconry in Europe (Fred. *DAV* 2.2) because they were too heavy to be carried on the fist and they frightened other birds of prey, both contributed to the confusions. Identifications were further complicated by the variations in appearance within a species, considerable size differences between the sexes, and the fact that the young frequently bear little resemblance to the parents until after the first molt. Add to this unreliable reports by untrained observers and the folk and mythic traditions surrounding the eagle and the picture is much more confused than is true with falcons. Major classical sources are Ar. *HA* 618b18–19a14 and Pliny *HN* 10.3.6–10.6.18. Other sources include Ambrose *Hex.* 5.18.60, Neckam 1.23, Barth. 12.1, ThC 5.2–3, Vinc. 16.32–37, PsJF 269–71, and *Physiol. Latinus*, s.v. For numerous illustrations of various bits of eagle lore, cf. García Arranz (1996, 144–220).

8. A story appearing, among other sources, at Ar. *HA* 620a1ff. and Pliny *HN* 10.3.10. For many other classical references, see *GB*, 8–9. It is also found in most of the medieval sources, including A.'s *QDA* 8.17.

9. A.'s *fehit* appears as *phēnē* (lammergeier, or bearded vulture) in Ar. *HA* 619b24–35 and ThC 5.2, and as the synonymous *ossifragus* in Pliny *HN* 10.4.13. The *fulica* is the coot, *Fulica atra* L. (Gr. *phalaris*; *GB*, 298). Cf. 23.111(45).

10. The source of A.'s confusion can only be conjectured, but several Gr. names for doves and pigeons may be involved, including *peleia, phaps, phassa,* and *phatta*. Cf. 8.223, with note.

8 Different people give different reasons for the expulsion of the young from the nest. Some say that the eagle throws them out in this way because it recognizes that its race is being debased in such offspring.[11] These people affirm, moreover, that the female eagle sometimes mates with a male of another genus rather than with a male *herodius*.[12] But since we judge this to be improbable, others say that an eagle of another genus, having broken the eggs of the *herodius,* puts its eggs in their place. Therefore, since sometimes the young of the eagles are alien to it, they lack their approval. Whereas, however, the eagle is a bird of greatest ferocity and has a kind of foreknowledge by which it can tell when an enemy approaches its nest even when it is absent, it seems likely that no genus of bird possesses such boldness that it would dare to approach the eagle's nest. This is especially so since placing eggs in a foreign nest is a trait of only the vilest and most ignoble birds which are not capable of incubating their own eggs, and such birds never come near the nest of the *herodius.* Others say that the *herodius* itself places its own eggs under an eagle of another genus, intermingling them with the eggs of the eagle under which it places the eggs. And when the young emerge, the *herodius* returns out of a natural tenderness and sorts through its own and the other's young by inspecting them.[13] It then nourishes its own and ejects the others and the ejected ones are then nourished by that eagle under whom the *herodius* placed its eggs in the first place. Now, I would judge this the more probable explanation, if it had been based on personal experience, for hatching eggs involves work and emaciation as a result of fasting. This is something the *herodius* does not endure easily on account of its mobility and the abundance of food which it is accustomed to have.[14] Nevertheless, it is possible to say that the reason for the ejection is not the adulterous nature of the young or even the mixing in of eggs of another nature. Rather, just as it happens in other animals that the potency of nature is weakened, so it happens in the young of the *herodius.* And when the *herodius* senses this in its young's eyes, it casts it off as useless.

9 Certain of the philosophers say that the *herodius* produces two or three young and that two always come from one egg and a single young comes from the other, claiming that the *herodius* lays only two eggs.[15] I think this is false. Although this eagle has a large body, it has but little power in its semen and it therefore produces only one or

---

11. Cf. Ambrose *Hex.* 5.18.60 and *Jacob.* 92 (p. 219). The ejection from the nest most frequently occurs in eagles when, due to differences in hatching time, there is a marked difference in the size of the chicks. This leads to the phenomenon of "Cain and Abel," wherein the larger chick continually attacks the smaller and, in most cases, kills it. Even when the mother is present, she does not intervene.

12. An erroneous idea possibly arising from the considerable size difference between the sexes and differences within the species in coloration.

13. The eagle was a traditional symbol of tender care of the young. Cf. Deut. 32.11.

14. For this explanation, see 6.46.

15. Cf. ThC 5.2, who attributes the idea of two chicks from one egg to Ar. (*HA* 563a17–21), apparently misinterpreting Ar.'s statement from Musaeus that "it lays three, hatches two, and cares for one," which is followed by the statement that on occasion a brood of three has been observed.

two eggs, or conceivably three if it is young.¹⁶ Out of each egg there is produced but one young unless it occurs by error, as happens in other birds. But when more than two are created, it frequently ejects them due to the difficulty in nourishing them. This has been shown to be true many times in many birds. However, as I recall that I have said in my previous books, I sought experiential proof of this fact among many bird catchers and, through the course of many years, have never found a *herodius* to have more than one young. This is because it needs a great deal of food, which has to be sought far and wide by the older *herodii*. If, however, two are ever found, it will be in the northern regions near the forest and the sea where an abundance of prey is found, both of fish and birds and small wild creatures from the sea and forest.

This eagle has glowing eyes that are yellow beyond all measure, to the extent that the white of the eye presents the appearance almost of a topaz and the pupil of the eye the appearance of a very clear, black sapphire of compressed blackness.¹⁷ The area of its eyebrows is a bony ridge projecting a bit over its eyes so that with the eyebrow's shade they seem to concentrate their vision. Thus it is that when one is domesticated, it sees its prey, a rabbit, lying in a thicket before it is able to be spotted by a human or by the dogs.

I have not proven through experience what Jorach and Aldhelm have to say about this eagle.¹⁸ For they say that when this eagle grows old, at a time when its young are grown and know how and are able to hunt, the eagle seeks out the bubbling action of a pool that is clear and wide flowing. It flies directly over it, straight up into the third interval of the sky, which we called the *aestus* [fiery] in our book *Meteora*.¹⁹ When it has caught on fire there and is on the point of burning up, it suddenly lowers and pulls back its wings and rushes into the coolness of the pool so that by means of the external, restricting cold its internal heat may be multiplied in its marrow. It then rises from the pool and flies back to the nest, which it has in the neighborhood. There, wrapped in the wings of its young, it breaks into a sweat and thus burns off the appearance of old age. With its old feathers cast aside, and adorned again in new ones, it is sustained on the prey brought by its young until it is renewed again.

I do not know what to say to this except that the wonders of nature are many. But it does not agree with what I have observed in two *herodii* which are in our land. But these were domesticated and, as such, were changed with respect to the manner of other birds of prey.

Certain ones say that this eagle, like other birds, has a sharply hooked and curved beak and that it applies this, as well as its talons, to a rock to sharpen them when these

---

16. For A.'s own experience of this phenomenon, see 6.50.
17. "Topaz": see *De min.* 2.2.3 on the *chysolitus* (Wyckoff, 1967, 82–83).
18. Ald. *Aen.* 5.2. What follows is a common story in medieval sources, including *Physiologus* (Curley, 1979, 12–13; Carmody, 1939). It is possibly based on a combination of the tradition from Psalm 103.5 (102.5 Vulgate) and the molting of eagles. Most sources include the notion of bathing three times in the pond, perhaps analogous to the tradition of Christ arising on the third day.
19. Perhaps *Meteora* 1.1.8. On the *aestus,* see *DCM* 2.3.2.

tools of hunting become dull.[20] This has indeed been proven to be true. At any other time, however, birds of prey do not sit for long easily on rocks, lest the tips of their talons break. In our lands, however, this genus of eagle is found sitting and making its nest on sharp rocks, but always with an intervening layer of grass, dust, mud, or the pelts of animals it preys upon. For this reason the more skilled among the falconers do not allow their birds of prey to stand for long on a hard, dry branch without bark [*nudum*], but rather they lay down a cloth or hide for the talons of their birds. The *herodius*, moreover, has in common with other birds of prey the fact that it almost always looks after its feet when it is left to itself, unless it is seeking food or is holding itself onto something or onto its prey. It frequently draws the talons of its feet through its beak to sharpen them for grasping its prey.

11 Certain others say that an eagle that is pregnant, as it were, places a stone called an *echytes* or *gagates* among its eggs, and it is used to control the heat. We have spoken about the stone in our book on stones.[21]

Others say that the eagle places two stones called *indes* in its nest, without the presence of which the eggs cannot come to life. Whether this is true is unknown. But it is certain that certain birds place stones among their eggs. We have watched for an entire year domestic cranes place a stone among their eggs, but they chose these stones indiscriminately from stones found by chance.[22] It is, however, not known whether if they had had freedom of flight they would have sought out a proper, more useful stone. Since by chance they were not able to find a stone more useful for them, they accepted whatever they found lest nothing at all be placed among their eggs. The habits of birds differ in such matters.

Moreover, the *herodius* is the hero of birds, as its name implies. It very often shares its prey with other birds, but should there not be enough, it will deplume and devour the bird nearest it.

This bird is not easily provoked by birds striving against it, but once it is provoked, it feigns patience until its tormentor comes closer, in all confidence, and then it tears its captive to pieces.[23] This bird flies very high in order to spot its food and because of other birds challenging it, since sometimes certain trivial birds, having flown above it, will pluck out its feathers. If, however, any large birds such as cranes or storks should fly over it, they would kill it with the sharp point of their beaks. It therefore flies above them all. Although the *herodius* is called the king of the birds, it is not so

---

20. Cf. ThC 5.2, which says that they break their beaks. Actually, both this and A.'s version are incorrect, as the eagles (and other raptors) have to keep their beaks worn down. Falconers must periodically trim the beaks of their birds. A.'s version is closer to the actual case, but it is the same sort of misconception as in the notion of cats "sharpening their claws" when they are really blunting the nail tips.

21. *De min.* 2.2.5, 7 (Wyckoff, 1967, 87–88, 93). Cf. Pliny *HN* 10.4.12 and 36.39.149. Also cf. Jacob. 91 (p. 197), Barth. 16.39, and ThC 14.28 and 5.2 (where he also cites the *Experimentator* as a source).

22. Wyckoff (1967, 87, n. 1) attributes this story to the accidental inclusion of stones in the crane's nest, built on rocky ground.

23. For an example of the eagle's cunning, see 8.13.

called because it imitates "true rule" [*verum regimen*], but rather it is named from the violence of its tyranny. It is master over all insofar as it dominates and devours all. This is why all birds fight against it.

This bird is greatly irascible and proud, as are all other birds of prey. It therefore flies alone except when involved in the generation or education of its young. Then it flies above its young, teaching them to fly and hunt. However, when the young have become adults, it expels the very young it has nurtured from the area of its own habitation. Every bird of prey has this in common, and especially the *herodius* and the falcon. For this reason, falcons and *herodii* that have been ejected by their parents and are caught in the nets of bird catchers in lands where their nests are never found are called *peregrini* [wanderers].

A domesticated *herodius* cannot be carried about sitting on a hand. Rather, it sits, once the hand is offered, supported on the arm of the falconer which is entirely covered from the shoulder to beyond the hand with a deerskin. And when it is provoked into flight, it ought to be let go since it would otherwise hurt the falconer's arm. Therefore, it ought to be carried with its eyes covered or, when it wants to fly, let go. When it is called back, it returns to the falconer's arm and then rests easier, having either captured its prey or having been frustrated.[24]

You should know, however, that Aristotle and certain other philosophers place the eagle and the vulture in the same genus of bird, naming the *aquila* from its sharpness [*acumen*] and the vulture from its flying [*volandum*] and its capturing [*capiendum*].[25] On account of this, some say that the vulture is the most noble genus of bird. This is not our usage, however, for in our lands the vulture is a very big, lazy, ignoble bird about which we will speak in the following passages.

There is a great diversity of color, size, and habits among the eagles in our lands. After the *herodius,* the most noble eagle is the one which catches geese, swans, and other large birds of this sort.[26] It also catches hares and rabbits, especially when it has young. This eagle is smaller in size than the *herodius* and is multicolored, with white and ash-colored feathers mixed together. In its tail, however, the next-to-the-last feathers seem white and there is white in the feathers at its anus. Its tail is, moreover, extremely short. Throughout the entire range of its kind, the eagle genus has a short tail and wide and large wings, as opposed to the long, pointed ones that falcons possess throughout their genus. The genuses of eagles are generally large, with a slightly oblong beak and large, yellow feet.

In our land six genuses of eagles are found, of which the *herodius* is the first. Second in nobility after it is the one that catches geese and swans. We have already spoken about both of these. Third is one which, in our lands, takes pleasure in sitting

---

24. Having just used *falconarius,* A. then uses *auceps,* a word he usually reserves for a fowler, but which changes sense in a context such as this.

25. Isid. *Orig.* 12.7.10.

26. See Ar. *HA* 618b18f. and Pliny *HN* 10.3.7 for *pygarus,* the white-tailed eagle or erne, *Haliaetus albicilla.*

on the trunks of trees and is therefore commonly called in our land the trunk eagle [*truncarum aquila*].²⁷ It captures small animals if they come its way, as well as ducks and on occasion geese and their like. It is smaller in size than the two just spoken of and has an ash-gray color.

14   The fourth is one that catches fish. It is multicolored, being white on the belly, black on the back, and with black spots on its crop. It has one foot like that of a goose for swimming and the other with sharp talons for seizing.²⁸ It roosts in trees over rivers and pools hunting fish.²⁹ The fifth is very small. It is called by some "bonebreaker" [*fragens os*] since, when it has eaten the flesh off the bones, it carries them on high and allows them to fall over a rock.³⁰ It then sucks the marrow from the broken bones. This type as well is small and multicolored.

The sixth type of eagle is rare. Nevertheless, it is found in the Alps and on the banks of the Rhine River, as we have frequently experienced ourselves. It is totally white, the color of snow, and in size it is almost the equal of the *herodius*, but is neither as noble nor as swift. It lives off the hunting of hares and rabbits and other small animals of this sort, to the extent that it sometimes seizes small dogs and even young foxes or piglets. It is reported sometimes to seize fish which swim on the surface of the water.³¹

15   Every genus of eagle has in common that it readily gathers pieces of fox pelt, whether it tears them off a captured fox or finds them by chance, and that it lays its eggs in these pelts or in some other soft fur.

Pliny says that the northern eagle wraps its eggs in a fox pelt and hangs them in the sun on the limb of a tree until, brought to maturity by the heat of the sun, the young come forth.³² He also says that the eagle does not incubate them but rather the young emerge from the heat of the pelt and the sun, and then the eagle comes to them for the first time. I have experienced this to be quite false, for in Latvia, where there are quite large and fierce northern eagles, we discovered almost nothing of such behavior.³³ Rather, the eagles hatch their eggs and feed their young with fish, birds,

---

27. St. identifies this as *Aquila pomerina*, the lesser spotted eagle, native to Germany. The Latin name is intriguing, as it has the flavor of other entries which appear to be a Latin translation from the MHG. While Suolahti lists no *baumadler*, he does list a *baumfalck*, identifying it as *Falco subbuteo*, and lists an equation with the *herodius* in an older MS (1909, 344).

28. St. identifies this as the osprey, *Pandion haliaetus*. The goose-like foot is fanciful, but the osprey's feet do have sharp spicules on their underside for seizing slippery prey. Cf. 2.70.

29. See Yapp (1981, 115) for a plate showing an eagle fishing.

30. Apparently a confusion of two birds. Only the huge lammergeier (*phēnē*) exhibits the bone-breaking behavior described. Perhaps a young lammergeier is the source of the confusion. On this "bonebreaker," see 8.105.

31. This is apparently the northern goshawk, *Accipiter gentilis*, which exhibits a "white phase" and feeds as described by A. It is one of the larger hawks.

32. Pliny *HN* 10.50.97, actually speaking of a hare's pelt. Cf. 8.104.

33. "Latvia": actually *Livonia*, which OL describes as south of Estonia and east of Latvia. A. travelled to this general area as a preacher of the Crusade. For his description of these fierce eagles, see 7.30 and 23.109.

and beasts. Therefore, the inhabitants of the land take the young out of the nest, place them beneath the trees and block up their anuses, whether by binding them or stitching them together. In this way they ruin the birds' appetite. The older eagles, however, bring in no fewer prey in the form of birds and fish and the inhabitants take these and feed their families with it. In the interim, however, they open the young birds' anuses lest they die and in this way they take much of the eagles' prey for a long time. One man who is worthy of belief told me that before the young of one single eagle's nest were grown, he took more than thirty ducks, over one hundred geese, and about forty hares, as well as many large fish, whose number he did not recall.

Let these be the things said, therefore, about the genuses of eagles.

(2) ACCIPITER: The discussion of the *accipiter* must be placed after that of the genus of eagles since the birds have almost the same habits and rapacity.[34] The *accipiter* is also called the *astur* due to the natural astuteness [*astus*] it possesses, for it almost always lies in wait and flies near the earth, contrary to the usual custom of falcons. When it captures a bird, it almost turns itself around and takes the bird upward.[35]

The *accipiter* is entirely multicolored, but in its first year it has yellow and black spots. Later on, however, it is mottled with white and black spots and these are whiter and blacker in accordance with how many times it has molted. Its feet are yellow with large talons, although not as large as the eagle's. Its head is rounder than the eagle's and its beak is curved and is proportionally smaller than the eagle's but longer than the falcon's. On its back it has a few white spots but more black ones. Its wings are, proportionally, more pointed than those in the eagle genuses but less so than those of the falcons. It is, however, a very irascible bird and as a result usually flies alone, except at the time of mating. It lays three, four, or at most five eggs and, although it is larger, it has almost the same shape as the sparrow hawk [*nisus*], which is called the *spervarius*.[36] It is smaller than the trunk eagle but larger than the eagle that catches fish.[37]

When this bird dwells in the wild, its prey is frequently domestic fowl, especially the hen and the duck, and it devours these birds on the spot. It captures crows and birds of that sort and sometimes hares. When it catches a hare, it fixes its left foot in the ground and holds its prey with its right. Then, as quickly as it can, it plucks out its eyes and in this way kills it. The domesticated *astur*, however, captures larger birds as well, taking on boldness from the person urging it on and from the dog trained to hold onto the prey with it. Then it seizes the crane, goose, heron, and other birds of

---

34. The northern goshawk, *Astur gentilis* (Gr. *hierax*; see *GB*, 114–18, Pollard, 1977, 80–81, *BAZ*, 66–68), the most favored hawk used in medieval falconry. The goshawk is a predator of more varied tastes than most falcons, making it a good "table food" provider for the falconer. For sources, see Ar. *HA* 620a17-b5 and Pliny *HN* 10.9.21–10.10.24. Cf. Fred. *DAV* 2.5 (W&F, 112), Barth. 12.2, Neckam 1.24–23, Jacob. 92 (p. 221), ThC 5.10, Vinc. 6.18–21, PsJF 271, and Isid. *Orig.* 12.7.55–56. Also see Oggins (1980, 446–49) for parallel descriptions from modern authorities, testifying to the general accuracy of A.'s descriptions.

35. "Takes it": *accipit*, a pun on *accipiter*.

36. *Accipiter nisus*, the European sparrow hawk. Cf. 23.129(83).

37. Cf. 23.13–14, with notes.

this sort. With no difficulty it captures the larger duck, the diving duck [*mergus*], and the coot, capturing those birds in great numbers and without any great effort.

When this bird is sick, its feathers and wings fairly bristle and its wings droop as well. It also calls a good deal due to its feeling of languor. It is a sign of its being indisposed if it vomits out its food undigested and if it does this rather often. For this then signifies that it has a defect in its stomach and crop. However, it also sometimes suffers from fullness as does a human. At such a time it has dull vision, a heavy and slow flight, and does not desire food, wishing rather to sleep a good deal and to rest. It allows prey presented to it to get away and sits on the earth. When recalled with food to its master's hand, it does not return quickly and when called does not pay heed to the one calling. It sometimes suffers from excessive thinness. Then, bereft of its natural moisture, its wings develop spots which they call the "signs of hunger" [*signa famae*]. At this time, the wings are easily broken, it is not able to prolong its flight, and it lays aside its boldness, following only small prey while calling a good deal and constantly desiring to return to its master. It sometimes also suffers from constipation and blockage. Then it stays sluggish and desires neither food nor prey. It sometimes suffers from lice due to corruption of its humor. It sometimes suffers from fevers. This is recognized from the bristling, sadness, and trembling of the bird. We will assign a medicine for all these and for other diseases in the chapter on falcons in which we will set forth the medicines for all birds of prey at the same time.[38]

18   You should know that the nature of this bird is especially strengthened in the northern regions and that these birds are stronger and larger there. Since they are quite noble, they capture prey not for food but rather out of a sense of glory and because they delight in tyranny. Now, if they desire anything from the prey, they are partial to the heart and thus they tear any bird they capture down the side, extract the heart, and devour it.[39] Sometimes they desire the brain, and, having extracted it from the head, devour it, rejecting all the rest. The noble northern ones eat crabs with great pleasure, although they do not hunt them. Therefore, these, when domesticated, capture large birds better for their masters than all the *asturs* and receive a snack of crabs from their masters as a reward.

Let these things which have been said about the *astures* or *accipiteres*, then, be sufficient. Some people also call these birds *accipenseres*.[40] Pliny says that an effective medicine against pains of all the limbs is *accipiter* boiled in oil. For the flesh of the *accipiter* is very sweet and light due to the wholesomeness of its diet.[41]

---

38. Cf. 23.79–107.

39. This is in contrast to Ar. *HA* 615a4–5 and Pliny *HN* 10.10.24, which state that they do not eat the hearts of birds (Thompson, 1910, ad loc.).

40. *Accipenser* also appears at 6.47, where it is tied to an "Arabic" name. Yet the *accipenser* is a sturgeon, appearing in A. as *accipender* at 24.10(10). The origin, then, of this "hawk" is unknown.

41. Cf. 23.51, for A.'s classification of the order of nobility in falcons, apparently based partly on the Symmachus letter (Rigault, ed., 1612), perhaps via ThC 5.51. See also Oggins (1980), 450–51. For Pliny's remarks, see *HN* 29.38.125.

(3) ARPYA: The harpy, according to a certain man of no great authority and whose statements are not proven by experience, is a bird of prey having hooked talons and a human face.[42] It is said to live far off in a desert, near the Ionian Sea, in a land called Stropedes, and is said to be raging often from hunger and to be rapacious.[43] This bird sometimes kills a person it meets in the desert and, when it sits near the sea and its face shines in the water, it will catch sight of its own face, which is similar to a human's. Seeing it has killed one that is like itself, it is sad and grieves for the rest of its life because it has killed a human. But these things are not borne out by experience and seem to be fabulous, especially the things a certain Adelinus relates, as well as Solinus and Jorach, who are speaking about the members of animals.[44]

(4) AGOTHYLEZ: The *agothylez* is so named in Greek. In Latin we call it the *caprimulgus* [goat-milker].[45] It is a large bird, with a wide beak, which is common in eastern areas. It seeks out goats full of milk, places itself beneath them, and sucks out their milk. As a result of this there arises both a drying up of the milk in the teats and a dulling or even a blinding of the goat's sight.

(5) ARDEA: The *ardea* [heron], which some call the *ardeola* and others the *tantalus,* is a bird which takes its name, according to some, from the fact that it flies high and thus has a lofty [*ardua*] flight.[46] For they say that this bird flies high above the clouds when it senses a storm is coming, and that in this way, by its lofty flight, predicts the storm. We also know that when it flees the *accipiter,* this bird will fly high if it can. Others, however, say that the *ardea* is so called from "burning" [*ardere*], because its excrement burns [*adurit*] whatever it touches. This it has in common with other aquatic birds which live by hunting fish.

This bird, moreover, has an ash-gray color, is smaller in size than a crane, has a long neck, and, like other aquatic birds, has intestines that are not greatly convoluted. It is for this reason that it emits undigested food just like the diving duck [*mergus*].

---

42. The harpy (Gr. *harpē,* "the snatcher") is, of course, the mythical half-bird, half-woman creature who "snatched" the souls of the young who died prematurely. Thompson (*GB,* 55f.) has collected the ancient references; see Pollard (1977, 190–91) for a survey of the harpy in Gr. literature and art. For medieval implications, see Rowland (1978), 74–77. Cf. Vinc. 16.94 and ThC 5.4.

43. The Stropedes are actually islands inhabited by large colonies of gulls and shearwaters, leading to the identification of the birds of Diomedes with the shearwater. Cf. 23.40(33) and 23.42(37). Pollard (1977, 101, 164) also discusses the relationship with the harpy myths. The description provided here appears to be a conflation of the birds of Diomedes with a vulture or buzzard.

44. Cf. ThC 5.4.1. The Jorach reference does not appear, as is frequently the case, to come from Arnoldus Saxo. The Solinus reference seems spurious. King (1995, 148–52) has recently studied the harpy legend.

45. Gr. *Aigothelas* (*GB,* 24–25; Pollard, 1977, 50–51). *Caprimulgus europaeus,* the goatsucker. Cf. Ar. *HA* 618b2–9, Pliny *HN* 10.56.115, ThC 5.5, and Vinc. 16.24. Note that here ThC cites as his source "Michael [Scotus] who translated Ar.'s *De animalibus.*" For this bird see also 8.25.

46. Cf. Isid. *Orig.* 12.7.21 for the etymology. Cf. Jacob. 92 (p. 220), Ambrose *Hex.* 5.13.43, ThC 5.6, Vinc. 16.38, PsJF 272, and Neckam 1.63. For modern interpretations see *GB* 102–3, Rowland (1978), 79–81, and Yapp (1981), 15–16. In mythology Tantalus was punished in the netherworld by having to stand in water yet be unable to drink. The mythical allusion is most apt for a wading bird.

Therefore, this bird is always hungry, even for rotten and unhealthy meat. I consider as false that which some say, namely, that the heron mates with pain so great that it sheds tears of blood from its eyes and that the female conceives and bears the eggs with no less pain.[47] I consider this false, since I have often seen with my own eyes herons mating and laying eggs and have been able to prove none of these things. Like any other long-legged bird, the heron mates over the female's back with his legs bent so that the male's feet are at the female's head and his knees are over her back, toward the anus. Then, holding himself on by the motion of his wings, he touches the place of conception on the female and pours his semen into her.

This bird builds its nest in a flock of its own kind [*genus*], but does not fly in flocks since the *accipiter* and other birds of prey lie in wait for their young, and these cannot be defended unless many herons are always to be found near the nests. With their excrement, however, they dry out all trees in which they nest. If an *accipiter* ever tries to capture a heron, it turns its anus toward the *accipiter* and shoots forth putrescent excrement, which, should it touch the *accipiter*, rots his feathers.[48]

21 Three genuses of herons are found in our land. One is ash-gray in color, with a sharp beak and a long neck.[49] The second is totally white and in shape is entirely like the first, but is better feathered than it is.[50] The third has a longer neck and its beak is round in front so that it is like a circle on top of a circle. For this reason it is called a *cocliarium* [spoonbill].[51] It is totally white. Pliny adds a fourth, the *monoculus*, which he says has but one eye on one side only.[52] It is thus rather easily captured by a hunter who puts a snare on its blind side. But it seems that what he says is false and contrary to nature. For, just as two wings and two feet grow from the sides, so do two eyes. Reason does not allow that one eye is formed from one side and another is not formed from the other side. To be sure, this Pliny says many things that are entirely false and in such matters his words should not be given consideration.

22 (6) ANSER: The *anser* [goose], or *auca*, is a familiar bird.[53] It is aquatic, as its webbed feet show. Among the aquatic birds, however, it is larger than the swan and larger than the one we have called the *volmare* in preceding books.[54] This bird is found

---

47. Cf. Pliny *HN* 10.79.164 and Ar. *HA* 609b24.

48. So widespread was this story that Otto (1962, no. 17884) cites a two-line proverb: "The ibis does not forsake his beak nor the heron his anus. Neither does a person abandon his vice while he is practicing it."

49. Traditionally identified from Ar. *HA* 609b22 as the common heron, *Ardea cinerea* L. Cf. Gr. *herodios* (*GB*, 102–4; Pollard, 1977, 68–69).

50. Usually identified as *A. alba*, an egret. Stadler's identification is of the great blue heron, *A. herodias*, which is white when young.

51. *Platalea leucorodia* L., the spoonbill. Gr. *leukerodios* (*GB*, 193).

52. Pliny *HN* 11.52.140.

53. Cf. ThC 5.8, Rabanus Maurus *De universo* 8.6 (*PL* 111:241A, 248B), Vinc. 16.29–31, PsJF 273, and Neckam 1.71. See also Yapp (1981, 23 and 107, Pl. 14), Rowland (1978, 67–70) and Arnott (2007, 30–31). *Auca* is elsewhere found as *anca*, just as *anser* is found as *aucer*.

54. *Volmare*: the pelican, *Pelicanus onocrotalus* L. *Volmare* is a German word (Stadler, 1655).

in our land both in the forest and domesticated. The forest-dweller is found in five species in our land. There is a large goose of an ash-gray color which is called a *gragans* [gray goose] among the Germans.[55] Another is of the same color and shape but is smaller and flies higher and farther, due to the lightness of its body.[56] The third is totally white except for the ends of its wings where four or five feathers are very black.[57] This genus is small and flies long, far, and high. The fourth genus is smaller than this one. It has a goose's beak but has a head the color of a peacock, except that it lacks a crest of feathers. On its back it is ash-gray turning into black. On its crop it is black and is ash-gray on its belly. The common people [*vulgus*] say this one is born from a tree.[58] Pliny adds a fifth genus of goose, calling it a *comage*. He says its fat is quite a help to the preparation of the famous medication called *comagum*.[59] It is made from the fat of this goose mixed with cinnamon and hidden away in a bronze vessel beneath the snow until it is congealed from the cold. It is thus macerated and ripened. Others add another genus which, so they say, is the largest bird after the ostrich. They say it inhabits the Alps and the northern wastelands, where human habitation is rare.[60] But we have never seen it, unless it is the one which we call the *volmare,* for this is larger than a swan.

More than all these genuses of geese, it is the swan which possesses most of the characteristics of the genus, both in the shape of its beak and in the type of feet and style of its life. It differs in size, but inflates itself up like a goose in time of battle.

There is, moreover, a domestic goose which is found in almost every color found in the wild goose. A white one is found and an ash-gray one, as well as one spotted with these two colors. But a black one, a green one, or one of another color not mentioned is not found.

Every goose has in common that it is full of chatter, scarcely ever keeping still. They hear the slightest sound and sleep but little and, for this reason, while the guardians of the Capitol slept and their enemies were about to capture it, it was a goose that aroused them by its clamor. In memory of this a golden goose was ritually revered by the Romans for a long time.[61]

All wild geese keep to a lettered order in flight as do cranes.[62] When they fly they commit themselves to the blast of the wind in which they can most easily fly. There-

---

55. *Anser anser* L.

56. *A. fabalis* Lath.

57. *Chen hyperboreus* Pall.

58. The barnacle goose, *Branta laucopsis,* or the brant goose, *Branta bernicla* L. Cf. 23.31(19), *barliates.*

59. Pliny *HN* 29.13.55. The name *comage* is a corruption of the name of a district in Syria, Commagene, where this medication is prepared. Commagene is also the name of an herb used in the preparation. The same confusion is found in ThC 5.8, but Vinc. 16.31 presents it correctly.

60. Stadler conjectures that this is the Canadian goose, *B. canadensis.*

61. Pliny *HN* 10.26.51, concerning the awakening of the ex-consul Manlius in 390 B.C. by the sacred geese just in time to save the capitol from a surprise attack by the Gauls. Cf. Ambrose *Hex.* 5.13.44 and Isid. *Orig.* 12.7.52.

62. Cf. 1.58, with notes.

fore, many people predict winds, coldness, and rainstorms by their flight. They fly on high and know that the higher wind is always stronger than the lower one. They almost never allow humans to approach them and they move from place to place as do cranes, but they do not hide. When they are tired from flying, they feel pain. They cry out when flying and all geese fly with their necks extended. In this they differ from the heron, which flies with its neck drawn up.

24 All birds called geese feed in swamps to the north. They begin to mate after the winter solstice and to lay eggs in the first part of the spring. There are sixteen eggs at most. They say that if eggs are taken away before hatching, the goose will lay even more until the loss is corrected.

The genuses mentioned will interbreed. In our lands the goose that is like a peacock mated with and produced young from a domestic goose. All the young followed the father's coloring, but were bigger in size.

The life span of a goose is very long. We have seen a goose raised as a domestic animal that was more than sixty years old.[63] Also, a wild goose which was caught in our lands was not able to be softened even after three full days of continuous boiling. It was so tough that it could not be cut with a knife and no other beast even wanted to taste it. The flesh of the goose is usually cold and dry, hard and melancholic, and indigestible.

25 (7) ANAS: The duck is a familiar bird in our lands. It is wide in the back, smaller than a goose but larger in size than a coot [*pullus aquaticus*].[64] In our lands it is both wild and domesticated.

There are many types of wild duck in our lands, but they are all alike in the shape of their bill and feet. The bill is wide and serrated and it is not very long but opens quite wide for sifting mud. The feet are reddish and webbed, adapted for swimming. The wild duck in our land is of one of two colors. One true type is gray on its back and belly, and on its neck the male has radiant feathers like a peacock's with a coloring composed of green and azure, with a white band around its neck.[65] On its wings, near the belly, it has a shining green spot. The female, however, tends more toward a black, ashy color. Her voice is lower [*grossior*] and, indeed, in every genus of duck the male's voice is higher pitched.[66] This genus, then, of the aforementioned color, itself has two genuses, namely, a larger and smaller one, and each is well known in our lands.[67] A

---

63. Geese are among the longer-lived birds, but this seems to be an exaggeration received from "a man worthy of belief."

64. Identification is difficult here, as in the occurrence (by name only) at 2.75. A. is apparently thinking of some small waterfowl, perhaps the teal (Gr. *boskas*) of Ar. *HA* 593b17, where it is mentioned as smaller than the duck. Cf. ThC 5.9, who quotes the nonextant *Liber rerum* "avis per aliquantulum gallo maior," which is repeated in Vinc. 16.27. General discussions of ducks are also found in Isid. *Orig.* 12.7.51, PsJF 273, and Neckam 1.67. On the ancients and the duck, Gr. *nētta* see *GB*, 205–6, Pollard, 1977, 65–66, *BAZ*, 146–48.

65. The mallard (Scanlan, 204).

66. "Lower": Just as the thicker strings on a stringed instrument produce a lower pitch, so too the female's "thicker" [*grossior*] voice is lower.

67. *Anas crecca*, a teal (Scanlan, 205).

third type, however, is multicolored, composed of white spots and the aforementioned coloration. It is a sort of diver [*mergus*]. It is both larger and smaller than a duck, but has the same shape. This diver differs from the duck, not in shape as has been said, but in color and lifestyle. As a diver it catches fish beneath the water whereas a duck takes its food from plants and vermin hidden in the plants. It also takes small fish if it seizes them in the plants.

The domestic duck is found in both white and ash-gray as well as in a blend of these two colors. It is, however, a cold and melancholic bird, but with flesh less tough than that of the goose.

(8) ACHANTIS: The *achantis* [linnet?] is a small bird, like a sparrow.[68] It eats grasses and their seeds and builds its nest in thornbushes. When it sees the grass is being eaten by horses and asses and that the thornbushes, in which they nest, are being jostled, it becomes hostile to the horses and asses. It sits on their backs, biting them as hard as it can and it mocks them by imitating the whinny of the horses. This is a bird with a sort of derisive anger.[69]

(9) ASSALON: The *assalon* [merlin] is a small bird, the size of a sparrow.[70] It is hostile to ravens and breaks their eggs. It is also hostile to foxes and plucks out their hair. A raven, seeing this, helps the fox against their common enemy.

This bird eats the flowers of thornbushes and its seeds. It therefore hates an ass eating either this thornbush or the wild thistle and it sits on the back of the ass, pricking him as hard as it can with its beak.

(10) ALAUDA: The *alauda* [lark] is a bird taking its name from praise [*a laude*] since its music celebrates quiet, warm weather.

There are two kinds, namely, one with a plain head and one with a crest.[71] This last is also called the *galerica* [helmeted] or *cristata* [crested].[72] Its color is ash-gray and it is a bit larger than a sparrow. On the rear toe of its foot it has a claw of immense length. It lives in fields and not in the woods, eating seeds and vermin. The male is the musical one and has many melodies. It is the first among the birds to announce summer and at dawn it makes known the day with the praise of its song. It abhors rain showers and storms, and it so fears the hawk that it flees to the bosom or hands of people or sits on the ground and allows itself to be captured. It sings in its ascent as it flies in a circle,

---

68. Gr. *akanthis* (*GB*, 31–32; Pollard, 1977, 52–53), the linnet, *Acanthis cannabina*, or the goldfinch (Scanlan, 205).

69. This story is based on Ar. *HA* 610a4–7. Pliny *HN* 10.95.204–5 has the *aegithus* at war with the ass and later mentions the *acanthis* as living in thornbushes. Aiken (1947, 218–19) discusses this conflation, also found in ThC 5.12 and Vinc. 16.22.

70. Gr. *aisalōn* (*GB*, 30; Pollard, 1977, 81), traditionally identified as the merlin, *Falco aesalon* L. Ar. *HA* 609b30–35 gives the story of the merlin as the opponent of the fox and raven. Pliny *HN* 10.95.204–5 repeats the idea of enmity but has the *aesalon* breaking the raven's eggs and the *aesalon's* young as the prey of foxes. The story becomes conflated with the *achantis*, which appears next in Pliny *HN* (see note at 23.26(8) and Aiken, 1947, 218–19).

71. Probably the common skylark, *Alauda arvensis* L., and the crested lark, *A. cristata* L. Gr. *korydalos* (*GB*, 164–68; *BAZ*, 116–18).

72. Pliny *HN* 11.42.121.

and when it descends it at first does so a little bit at a time until finally, pulling in its wings, it falls suddenly, like a rock. In its fall it gives forth a song.[73]

When it is domesticated it sings in its captivity and moves its wings, demanding with a certain gesture to go forth into the free air. But if it is held captive for a long time, as it often is, it goes blind in one eye. And I have often experienced that in the ninth year it goes blind in the other eye. This bird is useful in medicine.[74]

(11) ALCIONES: The *alciones* [halcyon birds, kingfishers] are small birds, reaching a size a little bit larger than a sparrow.[75] They are dark blue in color with purple wings, although they have some white feathers mixed in. They have a long neck and live on fish. Very rarely do they appear on the sea at any time except near the solstices and they produce young in wintertime. At that time they build round nests on the sand by the sea in the shape of a sphere. It is so strong it can scarcely be broken even with many blows.

Some say that if the sea is stormy and if the halcyon's eggs are lying exposed, tranquility immediately results, and they assert that for seven days this bird hatches its eggs and for seven more raises the young until they are grown. They say that the favor of having these fourteen serene days in wintertime, at the time of their parental concern, is a gift from Jupiter. Sailors call these the halcyon days.[76]

(12) AERIFYLON: The *aerifylon,* they say, is a bird of prey and among them is the most noble. Its behavior is like that of falcons.[77] It is called the *aerifilon* because it seeks out the "air" and the lofty regions of the air in its flight and its hunting.[78] It is even said to go beyond the clouds and to throw down birds it has happened to come upon and kill. Thus, when these falling birds are sometimes found on the ground, it is not known whence they fell or how they were killed. This is because the *aerifilon* flies so high that it cannot be seen.

This bird has reddish feathers, a long tail, very large talons and legs, and is a bit larger than an eagle. It seizes roebucks upon which it first sets its sight and then falls upon them. Having pierced the head, it dashes out the brain and in this way kills it. It builds its nest on the highest mountains. There a young one is sometimes caught and

---

73. ThC 5.14 cites the *Experimentator.* The same details are found in Neckam 1.68. Cf. Vinc. 16.24. and PsJF 271–72.

74. Pliny *HN* 30.20.62.

75. Gr. *alkyōn, kerylos* (*GB,* 46–51; Pollard, 1977, 96–98; *BAZ,* 11–12). The bird and its nest are described in Ar. *HA* 616a14–34, identified as the kingfisher *Alcedo atthis.* The same details are repeated in Pliny *HN* 10.47.89–10.48.91, together with the mention of the halcyon days, also found in Ar. *HA* 342b4–16. For many other references to the classical tradition of the halcyon, see Gresseth (1964).

76. Jupiter had changed the grief-stricken couple Halcyone and Ceyx into kingfishers. Cf. Ovid *Meta.* 11.410f. The normal number of halcyon days is eleven, but cf. Ar. *HA* 542b2f. Likewise, the nest is often said to float. Cf. Ael. *DA* 9.17.

77. Perhaps *Falco cherrug,* the saker falcon. See ThC 5.16 (*aeriophilon,* commonly called *aelion*) and Vinc. 16.23. A. discusses the saker in more detail in 23.51–52. Clearly, however, the *alietus* of Barth. 12.3 is not the same as ThC's *aelion,* since Barth., after citing a gloss on Deut. 14, concludes that it is the same as the *nisus* (sparrow hawk).

78. This apparently Gr. word means "air lover," yet a Gr. source for it is not known. Neither does the adj. *aerophilos* seem to exist.

trained to the point that it stands on its perches or homes without any binding. This bird hunts with its mate like the falcon and, after it has been trained, it does not go away from its master but captures and returns all its prey to him. For this reason it is said to be the most noble.

(13) AVES PARADISI: The Egyptians call certain birds "birds of paradise."[79] They are the size of geese and are called "paradise" due to their decoration, for no color seems to be missing on them. If snared, these birds do not cease moaning until they either die or are given their freedom. Once freed, however, they sing so sweetly that they bring pleasure to everyone hearing them. The dwelling place of these birds is above the Nile, which is reported to flow from paradise.

In the same place there are some other, dusky birds the size of jackdaws which are also called "paradise" because it is not known where they are born or where they come from. These birds change places from time to time, as do many genuses of birds about which we have made sufficient mention in previous books.

(14) BUBO: The *bubo* is a bird from the owl genus [*noctua*] and is so called in imitation of its voice.[80] Of all the birds it has the largest eye openings. This bird lives by preying on those that walk about at night, like mice or on occasion hares and other small animals of this sort. By day it seeks out shadowy caves, hollows of trees, mountain caverns, or the shady places of buildings not frequented by people. If it ever does appear by day it has its feathers plucked out by birds that fly in daylight. It is therefore set out by fowlers near their nets so that, by its presence, other birds might be caught.[81]

Concerning this bird, Pliny says, although it is improbable, that it leaves the egg backside first.[82] If this is true, the reason for it can only be the heaviness and largeness of its head. On account of this, the head itself tips downward and remains immobile for a long time. Its backside, however, is small and abbreviated as if the bird itself has been truncated.

This bird has curved talons and a sharp, curved beak just like the talons and beaks of almost all birds of prey. It is multicolored in its feathers and is largest of those that fly at night.

(15) BUTEUS: The *buteus* [buzzard] is a blackish bird of prey which we call the *brobuxen* in German.[83] It is slow of flight, has hooked talons and a curved beak, and is about the size of a kite. It hunts frogs, mice, and small birds that are slow or sick and unable to fly, and it sometimes seizes the young of birds. It is quite sweet and tasty to eat if it is prepared roasted.

---

79. Of doubtful identification, although the janissaries wore plumes which may have come from what we call today a bird of paradise (*GB,* 257, "Ryndake"). Cf. ThC 5.17 and Vinc. 16.39.

80. Gr. *Byas* (*GB,* 66–67; Pollard, 1977, 81–82). *Bubo maxlmus* (*Bubo bubo*), the great horned or eagle owl. Cf. Isid. *Orig.* 12.7.39, ThC 5.18, Vinc. 16.42, Barth. 12.5, PsJF 287, and Jacob. 92 (p. 222). Cf. *BAZ,* 23. For illustrations of owls, cf. García Arranz (1996, 503–410.

81. Ar. *HA* 609a12–16.

82. Pliny *HN* 10.18.38.

83. *Buteo buteo* L. is the common buzzard (Gr. *triorchis*; *GB,* 286–87). Cf. ThC 5.19, Vinc. 16.43, and Pliny *HN* 10.9.21.

(16) BUTORIUS: The bittern is a bird similar to the heron in shape and size, but different in color since the bittern seems to be the color of the earth.[84] It has a long neck which it draws in by curving it and which it draws out by extending it, like the *ardea*. It lives on fish and therefore has been allotted long legs so it can hunt near riverbanks and shorelines. It stands so immobile when it hunts that you would think it was dead or an inanimate thing, and when it feels it has been snared, it stands immobile and it wounds the incautious and improvident hunter with its beak (which is quite strong, like the heron's) as he is taking him up. This bird has a wondrous odor when it is being roasted and its blood is of great use for those with the gout.

This bird, in the spring when it is its mating season, makes a horrible sound, like that of a horn, and it cannot do this unless it sinks its beak in muddy water, so that its voice bursts forth like thunder.

(17) BISTARDA: The *bistarda* [great bustard] leaps twice [*bis*] or three times [*ter*] before it can lift itself off the earth. It therefore takes its name from this fact.[85] In size and shape it is like an eagle, with a curved beak and hooked talons. But on its wings and tail it is white, while on the rest of its body it is multicolored like the eagle. It eats flesh but does not prey on flying creatures. Rather, it eats carcasses it has found or it kills some innocent animal like a lamb or little hare.[86] It does not accomplish these things alone, but a good many of these birds come together and dare to attack. When it is hungry, it even eats grasses and especially delights in chickpeas, peas, and other legumes. This is a rare thing for birds who eat flesh. Due to its slowness, it does not place its nest on high but rather lays its eggs on the earth at the time when the crops are ripe.

(18) BONASA: A false story is told concerning the *bonasa,* a bird common in parts of Germany, bigger than a partridge, and which we call a *haselhun* [hazel hen].[87] It has the color of a partridge, black on the outside but possessing white, very tasty, and tender flesh inside. The story says that during its mating season it does not copulate like other birds, but rather the male runs about until it begins to foam at the mouth. The female, taking this foam in her mouth, conceives and then lays eggs and cares for the young like other birds. We have already shown that this is false in previ-

---

84. The bittern, *Botaurus stellaris* L. (Gr. *asterias*; *GB,* 57). Cf. ThC 5.20 and Vinc. 16.43.

85. This etymology seems to be ThC's (5.21). The *avis tarda* of Pliny *HN* 10.29.57 is carried over to Isid. *Orig.* 12.7.13, who equates it to the Gr. *gradipus* (also found as *bradypus*), a rendering of *bradypetes* (slow foot), a mistranslation of *avis tarda* (*GB,* 65). The bustard actually is nearly as fast as the ostrich, so *GB* suggests the origin of *avis tarda* as some misreading of a foreign word. It is notable, however, that the walk of this rather large bird is stately and its flight powerful but slow, either of which could have been the source of the misunderstanding. Cf. Jacob. 91 (p. 222), *gradipes,* and Vinc. 16.41, *bistarda.* Cf. also 23.144(113).

86. The bustard is not considered a bird of prey, although its great size and flight may give the impression that it is one.

87. *Tetrastes bonasia* L., the hazel hen or hazel grouse (Gr. *tetaros*; *GB,* 281–82). Cf. ThC 5.22 and Vinc. 16.41.

ous sections of this book. For what is taken into the mouth goes into the stomach, is changed and gives nourishment, and is eliminated through the bowel.[88]

(19) BARLIATES: Certain ones lie when they say that the *barliates* [barnacle geese], which the people call *boumgans,* that is, "tree geese," are so called because they are said to be born in trees, hanging from the trunk and branches and nourished on the sap that is in the bark.[89] They also say that these animals are sometimes generated from rotten logs in the sea and especially from the putrescence of fir trees, maintaining that no one has ever seen these birds copulate or lay eggs. Now this is entirely absurd for, just as I have said in preceding books, I and many of my friends have seen them both copulate and lay eggs as well as nourish their young.[90]

This bird has a head like a peacock's but has black feet like a swan, with toes that are joined by a membrane for swimming. They have an ash-gray black color on their backs and are whitish on their bellies. They are a little smaller than geese.

(20) CALADRIUS: The *caladrius* or *caladrion* is, as certain ones say, a totally white bird inhabiting a region of Persia.[91] Nonetheless, it is found there but very rarely because it has many trying to capture it. Because of its powers of augury, it is a bird many kings demand and even Alexander, the king of kings, is said to have acquired one. When presented to a sick person, it indicates all the conditions of his disease and it is said to cure quite a few. If it is held up to a sick person and if it turns its face and eyes on him, it indicates he will be cured.[92] This is because it indicates that the sub-

---

88. Lit., through the privy or drain (*secessum*).

89. *Branta bernicla* L., the brant goose, or *B. leucopsis,* the barnacle goose. Cf. Neckam 1.48, Vinc. 16.40, ThC 5.23, PsJF 287–88 (*berneka*) and Gerald of Wales 1.11 (O'Meara, 1982, 41–42) or 1.15 (Dimock, 1867, 47–48). The legend has been thoroughly treated by Allen (1928), 10–108. Allen discusses the various origins of the myth (105f.) and concludes first that the supposed feathers seen on barnacles clinging to driftwood were in fact the *cirri,* plumose appendages on the barnacle. He then cites other scholars who note that the story seems to have originated in such northern countries as England, Ireland, and Scotland, and that the brant goose does not winter in these lands. Therefore, locals never got the chance to see them mate and lay eggs. Cf. Lugt (2000). See also Yapp (1981), 32, fig. 18; George and Yapp (1991), 133–34; and White (1954), 267–68, who also offers a delightful illustration of the geese being born. On *boumgans* (MG *baumgans*), see Suolahti (1909), 417–19.

90. Cf. 5.6. It is a curious sidelight that several groups, mostly in the northlands, argued that since the barnacle goose was born of sea creatures it was not really a bird and that, partaking in the nature of fish, it could be eaten with impunity during the Lenten fasts. The belief grew to the extent that Thomas of Cantimpré asserted, wrongly, that Pope Innocent III issued a bull at the Third or Fourth Lateran Council officially condemning the belief. A. must have been aware of the bull. Note also that medieval rabbis, as early as 1140, had been forced to rule whether the animal was fish or fowl (Allen, 1928, 16–17; Lugt, 2000, 381–91). All of this recalls the similar argument made for beaver tail, claiming that as it partook of the nature of fish, it was fair fodder for Lent. Cf. 22.40(22).

91. Gr. *charadrios* (*GB,* 311–14; *BAZ,* 27–28), uncertain but often identified with the stone curlew, *Charadrius oedicnemus* L. Cited in the *Physiologus* (Curley, 1979, 7–9; Carmody, 1939, 15–17) as mentioned in Deut. 14.18 and cf. PsJF 294. *GB* loc. cit. says that this is a confusion of names. See Druce (1912) for a complete discussion of the tradition of the *caladrius*.

92. Jacob. 90 (p. 192) attributes this story to Alexander the Great and Saint Brendan. The identical quotations appear in ThC 5.24. It may also be related to a confusion with the *icterus* of Pliny *HN* 30.28.94.

stance [*materia*] of the infirmity is evaporating and is being consumed by it. It turns itself so forcefully toward this vapor that it is held by it and is even infected by it. But afterward, flying into the air at the proper time, it consumes the substance of that infirmity. If, however, it is held up to a sick person and turns its face and eyes away from him, it signifies that he will die. This is because the substance is held tightly in his body and does not attract the bird with its odor. The augury of birds, however, is not the business of the present investigation. The interior parts of this bird's thigh have power against dullness of the eyes.[93]

(21) CYNAMULGUS: The *cynamulgus* is a bird which lives in Ethyopia, in both the first and second *climata*.[94] It weaves its nest out of the finest cinnamon on the outermost small branches of the tallest trees. The region's inhabitants, since they cannot climb to it due to the height of the tree and the fragility of the branches, knock the nests down with arrows weighted with lead and then collect the cinnamon. This little bird is not disemboweled but is eaten with its innards, due to the aromatic nature of the things it feeds on.

(22) CIGNUS: The swan is a familiar bird which is called *cignus* in Latin from its singing [*a canendo*]. In Greek it is called after their word *holon* [entire] because it is entirely white.[95] Although it has an ash-gray color in its first year, it nevertheless becomes very white after a year. Its flesh, however, and especially its feet, are black. Its flesh is tough, as is that of all large aquatic birds.

This bird is of the goose genus. It has a serrated bill like a goose, toothed like a sickle. With its bill it strains the mud to find food and it divides what is found with its teeth. When it goes from place to place to seek food, it swims with one foot and, raising the other, places it above, near its tail, so that it can direct itself toward the wind.[96] If, however, it plans to swim far away, it swims in a flock and flies. Then the ones to the rear place their heads on the backs of the leaders toward the front. This bird dwells more in still waters than in rivers and it does not easily tolerate geese or birds of any other genus which feed on the same food as it does. It feeds on grasses, vermin, eggs of fish and similar creatures, and seeds from crops. When tamed, the front joint of its

---

93. "Interior . . . thigh": *interiora femoris* is subject to many interpretations. Is it the inner meat (Scanlan) or the marrow of the groin (Clark, 1992, 229, n. 4)?

94. Ar. *HA* 616a6–12 is in all likelihood the original source of all but the final sentence here (Gr. *kinnamomon orneon*; *GB*, 142–43; Pollard, 1977, 102; *BAZ*, 97–98). The same general story is in Hdt. 3.111, except that pieces of meat are left which break the nest. Pliny *HN* 10.50.97 uses Ar.'s version, but cites Hdt. at 12.42.85, combining both versions and dragging the phoenix into the picture. Solinus 33.15, ThC 5.24, Isid. *Orig.* 12.7.23, Vinc. 16.51, and Pliny *HN* loc. cit. all have the location as Arabia. The original Ar. gives no specific locus. For references to confusion of the *cynamulgus* with the phoenix, see *GB*, 143. Note the name *fothehokoz* in PsJF 296.

95. Gr. *kyknos* (*GB*, 179–86; Pollard, 1977, 64; *BAZ*, 122–24). Cf. ThC 5.26, Ambrose *Hex.* 5.22.75, Jacob. 92 (p. 222), Vinc. 16.49–50, Neckam 1.49, PsJF 294, 303 (*kikiz*), and Barth. 12.11. The etymology and most of the details at 23.33 about singing (including mention of the Hyperboreans) are found in Isid. *Orig.* 12.7.18–19. The "singer" is probably the whooper *Cygnus musicus*, which makes sounds both with its wings and vocally (*GB*, 183). For illustrations, cf. García Arranz (1996, 271–305).

96. The image is of the swan using its foot, propped up on its back, to move the tail, like a sail into the wind. This odd placement of the foot is a fact.

wing is cut off. It is a poor walker, an excellent swimmer, and a mediocre flyer. It alone among the large aquatic birds is musical. For this reason a story says that in the Hyperborean areas it is claimed that swans sing along with songs of singers and fiddlers. But what we have experienced is that they sing only in time of pain and sorrow and they thus should be said to be lamenting rather than singing. This is what the poet testifies to when he says: "when the fates do call, the white swan, laid out in the moist grass,/ At the shoals of the Menander sings."[97]

At mating time, the males lean their necks on the females by way of a caress. Then the male ascends the female and projects his semen into her. It is said that the female receives the semen with pain and that therefore, after copulation, she flees from the male. This is false because he put in her only that humor which is received with pleasure. She flees because at this time the desire for copulation ceases. After copulation, moreover, both the male and female dip themselves in the water, as do aquatic birds. This is because the vapor of concupiscence is running throughout their flesh and they are seeking to purge it. A proof of this is that all birds have bristling feathers after copulation and they shake themselves with their feathers raised and spread. Other, hairy animals do the same. This bird lays eggs in its third year. It makes its nest near water and greatly cares for its young. It fights fiercely for them and, while fighting, it puffs itself up like a goose.

(23) CARISTAE: *Caristae* are birds which, as Solinus and Jorach say, fly unharmed through flames, burning neither their feathers nor body. But these philosophers tell many lies and I think this is one of their lies.[98]

(24) CICONIA: The *ciconia* [stork] is a familiar bird and is so called because it makes a noise like its name, by clashing together its beak.[99] This bird is multicolored, composed of black and white. It has black wings, but its tail and other parts are white. There is, however, a stork genus that is entirely black on the back and almost white on the belly, but this one does not build nests near areas of human habitation but rather builds only in marshes of remote areas.

This bird is more aquatic than land dwelling. It eats fish, worms, frogs, and some nonpoisonous snakes. It also eats mice with great pleasure, as well as flesh and eggs fried in oil, but it does not eat the poisonous toad. When it has caught an animal, it first most often draws it through its beak and crushes the bones which are in it. It then swallows the whole thing in one gulp and holds it in its crop until it is softened, whereupon it is sent to the stomach. When this bird feeds it young, it regurgitates the softened food from there and serves it to its delicate young. When this bird is unable to feed them all, it sometimes throws out one of its young. The common people claim that the storks are offering it to the master of the house on whose roofs they reside, as a sort of rent.

---

97. Ovid *Heroides* 7.1, also quoted by Neckam. The "swan song" at the approach of death is found in Ar. *HA* 615b2–5. Many other references to it are in *GB*, 182–83. Pliny *HN* 10.32.63 denies the swan song. See also Arnott (1977) and McCulloch (1959).

98. Cf. Solinus 11.15, ThC 5.27, and Vinc. 16.46.

99. Isid. *Orig.* 12.7.16–17. Cf. Ald. *Aen.* 4.2, Ambrose *Hex.* 5.16.53–55, Jacob. 92 (p. 220), Neckam 1.64–66, Barth. 12.8, and Solinus 40.25–27.

This bird has a very strong beak and fights very fiercely for its nest. When it wishes to fight, it places one wing down in front of its leg as a sort of shield, and it sometimes fights until many have died. It even pursues eagles and other birds of prey if it is the season for the young. If it alone is not sufficient it calls on others so that sometimes many of them gather as if in a battle line, ready to fight. It makes a sound by rapidly striking together its beak. It does so when it greets its mate or when it rejoices over a victory or over the prosperity and feeding of its young. Then it sounds for a very long time and it frequently twists its head front and rear over its back. It also sounds when in fear, but not for long, and then it calls for help. At night, if it senses something hostile or unusual, it puffs itself up.

It is said that this bird nourishes its parents for just as long as it was nourished by them. It was, therefore, called the "pious bird" by the ancients.[100]

Those who speak in a popular manner [*vulgariter*] say that these birds go away in winter to areas of the East and that there is a plain in Asia in which they wait for each other and they lacerate the one who comes in last and holds back the line.[101] They likewise say that when they are going away, they are led by crows and are defended by them to the extent that the crows return wounded and covered with blood.[102] That this is a sure falsehood is shown by three reasons. First, the *climata* from east to west are of the same makeup with respect to heat and cold. Therefore, it would be of no help if they were to betake themselves to the East.[103] Second, it has been proven that they go totally away from our [land], remaining in no habitable section of it. Third, as has been determined in previous books, since the bird is cold and uses cold, sticky food, it is necessary for it to lie quiet during the period when phlegm is produced. As for the fact that where they lie hidden has not been discovered, that is because they lie hidden in the deserted areas of swamps and caverns, just as do other creatures which lie hidden.

It is likewise said that the male of this bird can sense adultery by means of an odor unless its mate has first bathed in fountain waters. This has been shown to be false and fabulous.

(25) CHORETES: *Choretes* are birds that fight with ravens without any treaty.[104] At night, however, they stop fighting but they fight at every other time. They also steal young from each other.

(26) CALANDRIS: The *calandris* [lark] is a bird similar to the *alauda* [lark] in color and similar in all respects in shape, but it is larger in size. It changes the mod-

---

100. Cf. Ar. *HA* 615b23–28 and Ambrose *Hex.* 5.16.55. See *GB*, 223, for many other citations of this tradition.

101. Pliny *HN* 10.31.62.

102. Cf. Ambrose *Hex.* 5.16.53 and Jacob. 92 (p. 220).

103. In fact, storks fly south and east seeking warmer climates.

104. Probably the golden oriole (*chloreus*) of Pliny *HN* 10.95.203, which is said to fight with ravens while they are searching for each other's eggs at night. Gr. *chloreus* (*GB*, 332–33; Pollard, 1977, 51). Cf. ThC 5.29. The *icterus* of Pliny *HN* 30.28.94 is usually considered to be the golden oriole also, and Pliny gives it a further name ad loc., *galgulus*.

ulation of its voice like the *alauda,* but its voice is hoarser [*grossior*] and higher. It is much more sonorous as if it needed nothing but to sing. It is, therefore, shut up in cages for its song.[105]

(27) CORVUS: The raven is a familiar bird.[106] It is the most perfect and the largest of those which are of the *corvini* genus. It is totally black, has a strong beak and body, and is extremely noisy. It lays its eggs before springtime and hatches them out before the thunder sets in. It occasionally throws out one of its young due to weariness from feeding them.

When this bird is tamed it sometimes speaks and imitates the songs of domestic birds. It steals a lot, but, if compelled with threats, it sometimes makes restitution. It is, so the stories say, a friend of the fox but an enemy of the ass. Thus it is that it sometimes blinds an ass, rendering it useless, and bites its back.[107] The *achyton,* however, is a bird friendly to the ass but unfriendly to the fox.[108] Therefore, these two birds are unfriendly toward each other.

I have seen on occasion a hunting raven which seized a partridge and other wild ravens. But it seized the ravens only by relying on the help of a human near itself.

(28) CORNIX: The crow, or *cornicula,* is a bird of the raven genus.[109] In its genus, the male feeds the female who is sitting on the eggs, and this it has in common with ravens, for this is a particular trait more of the genus than of the species. The parents also follow the young for a long time even after it is already flying and they feed it out of a love of piety.[110]

This bird is immensely hostile to birds of prey and is, therefore, sometimes torn to pieces by them. It is fit for auguries and incantations but this ought not to be dealt with at the present.

(29) CORNICA: The *cornica,* according to Pliny, is a very large bird in regions of the East.[111] It has a lung almost the size of a calf's lung, and it is soft and full of blood. It therefore drinks a great deal, more than other birds. It has few and small feathers and its wings are small as well.

---

105. At 21.29 the form is *galandra,* either *Melanocorypha calandra* L., a well-known mimic, or the crested lark, *Galerida cristata,* Gr. *korydalos* (*GB,* 164–68; Pollard, 1977, 49). Cf. ThC 5.30, PsJF 290–91, and Vinc. 16.44. However, *alauda* usually refers to the crested lark, so the first choice seems best.

106. *Corvus corax* L., Gr. *korax* (*GB,* 159–64; Pollard, 1977, 26–27; *BAZ,* 109–13). Cf. ThC 5.31, Vinc. 16.61–63, Jacob. 92 (p. 220), Ambrose *Hex.* 5.22.74, Neckam 1.61, Barth. 12.10, and Isid. *Orig.* 12.7.43. For illustrations, cf. García Arranz (1996, 307–32).

107. Ar. *HA* 609b5f.

108. Ar. *HA* 609b30–33. Cf. 23.26(9), *assalon,* with notes.

109. *Corvus corone* L. and the hooded crow, *C. cornix* L., Gr. *korōne* (*GB,* 168–72; Pollard, 1977, 25–26; *BAZ,* 113–16). Cf. Isid. *Orig.* 12.7.44, Ambrose *Hex.* 5.22.74, Neckam 1.62, Barth. 12.9, ThC 5.32, PsJF 291, and Vinc. 16.60.

110. Cf. Ar. *HA* 563b11–13, Pliny *HN* 10.13.30, and Ambrose *Hex.* 5.18.58.

111. An unidentified bird that does not seem to appear in modern versions of Pliny. The same information and source are given in ThC 5.33.

(30) CUGULUS: The cuckoo is, as we have said in preceding books, a composite bird as well as having two types.[112] One is a combination of the pigeon and *nisus*, that is, a *sparverius* [sparrow hawk].[113] Another bird of this genus is a combination of a pigeon and an *astur* [goshawk].[114] It has the beak and feet of a pigeon but feather coloration and wings like a *nisus* or an *astur*. It has habits that are composite as well. From the pigeon it has the fact that it does not prey upon other birds, whereas from the *nisus* and *astur* comes the fact that it lies in wait for the nests of other, weak birds. Therefore, other birds fight with the cuckoos when they lay their eggs. Now, a bird in whose nest a cuckoo's egg is put will nourish it and will take delight in its beauty to the extent that it itself is afflicted with hunger from feeding it.[115] Others sometimes condemn their own true offspring and the cuckoo sometimes, as Pliny claims, even kills the one nourishing it.

It is certain that this bird lies dormant at wintertime in the hollows of trees and rocks. But it is surely false that in winter it is nourished from food it gathers for itself in the summer. Rather, in winter it loses and changes its feathers.[116]

They say, moreover, that when this bird is heard for the first time by someone in the spring, if the one hearing it digs up the earth directly beneath the entire sole of his right foot and sprinkles it in his beds and in other places, this earth prevents fleas.

(31) COREDULUS: The *coredulus* is a bird so called because it lives by hunting and eats the hearts [*corda*] of those it hunts. It eats very little else of the prey it has caught.[117]

(32) COLUMBA: The pigeon [*columba*] is so called because it cultivates its loins [*lumbos colit*] through much procreation.[118] We have spoken a good deal about their varieties in previous books. We have mentioned that it has no bile in its liver, that it is innocent and harms nothing with its bill or claws, that it feeds on seeds or on things made from seeds and on nothing else, that it flies in flocks, that it distinguishes each genus of hawk [*accipiter*] by whether it seizes its prey in a tree, on the ground, or in the air, and it wisely flees to a place of security.

---

112. *Cuculus canorus* L., Gr. *kokkyx* (*GB*, 151–53; Pollard, 1977, 43–45; *BAZ*, 102–3). Cf. Neckam 1.72, ThC 5.34, PsJF 293, and Vinc. 16.52.

113. Ar. *HA* 563b14ff. and Pliny *HN* 10.11.25 discuss the metamorphosis of the cuckoo from the hawk. *GB*, 152–53, points out the superficial resemblances between the merlin and the cuckoo but concludes that the tradition has deeper roots, being perhaps related to the confusion in call between the hoopoe and cuckoo. This first variety is probably the *C. canorus* L.

114. Perhaps the great spotted cuckoo, *Coccystes glandarius*.

115. The host bird does feed the cuckoo long after the host is dwarfed by it. Cf. Ar. *HA* 618a8–30.

116. Ar. *HA* 563b18 comments on the cuckoo's disappearance in winter. The migration of *Cuculus canorus* takes it as far as South Africa.

117. An unknown bird mentioned by Isid. *Orig.* 12.7.34, perhaps in contrast to the *accipitres* at Pliny *HN* 10.9.24 which do not eat hearts. Also cf. ThC 5.35 and Vinc. 16.60.

118. *Columbus* sp., Gr. *peristera* (*GB*, 238–47; Pollard, 1977, 89–91; *BAZ*, 183–86). Cf. the bird described in Jacob. 92 (p. 221), Neckam 1.56, Barth. 12.6, ThC 5.36, Vinc. 16.53–59, *Physiologus* (Curley, 1979, 64–66; Carmody, 1939, 53–55), PsJF 288–90, and Isid. *Orig.* 12.7.61.

Some say there is a tree in parts of Asia called the *yperydyxyon,* which means "near the right side."[119] Pigeons delight in its very sweet fruit. There is, however, a snake there hostile to pigeons, but it cannot endure the odor or even the shade of the tree, and when it lies in wait for the pigeons, they seek refuge and find food in the tree.

Pigeons raise and rear their young in any month, and on those young not leaving the nest the parents inflict injury by beating them with their wings. Sometimes the male copulates over them and ejects them. They have warm flesh, and blood recently drawn from beneath their right wing is very sharp and capable of dissolving. It is therefore very effective for moist and inflamed eyes. Pigeon chicks are healthier and better in the spring and autumn since they are then fed on seeds that have either been sown or which have ripened. The pigeon has a groan instead of a song, and it is said that sometimes sitting alone, gazing at the golden splendor on its neck feathers, it is greatly delighted and celebrates by taking flight. For this reason Jorach says that the pigeon has something of an intellect.

(33) CHARCHOTES: The *charchotes* is a black bird of the genus of divers [*mergi*] which, when submerged, holds its breath for the amount of time it takes a person to walk a mile measured on the earth.[120] It is a voracious bird and very harmful to fish.

(34) CHORTURNIX: The [quail] is a bird we commonly call the *quiscula.*[121] The Greeks called it the *orthyges* because it is said to have first been seen on the island Orthygia. It is also called the *orthygometra* [straight measure] because it walks in straight lines.[122]

This animal is the smallest of the genus of chicken [*pullus*]. Immediately upon leaving the egg, it eats and is a good runner. Many believe about this bird that when it goes away it goes across the sea.[123] But this is shown to be a lie since in winter the bird is not found across the sea. It therefore lies dormant, as do other birds which digest their viscous superfluities.

In this bird the female is rare and it has the following particular trait. There are many males seeking a single female and therefore it is captured by the call of the

---

119. The *Peridexion,* based on *Physiologus* (Curley, 1979, 28–29; Carmody, 1939, 55–57), which says it is called *circa dextram* in Latin (Gr. *peridexion,* "toward the right"). Cf. A. in *De veg.* 6.198, which makes no mention of directions and where A. points out that he is very unsure of the entire story of the *Peridixion* tree.

120. Gr. *kataraktēs* (*GB,* 131–32), usually referring to the shearwaters *Procellaria diomedia,* or birds of Diomedes. See 23.19(3) and 23.42(37), and Pliny *HN* 10.61.126. Cf. Ar. *HA* 615a29. The origin of "black" in A.'s version (cf. ThC 5.37, Vinc. 16.68) is unknown, but the term does not come from classical descriptions, which usually stress the white color of the birds. The *mergi* of Ambrose *Hex.* 5.13.43 are given no color and seem to represent gulls.

121. *Coturnix coturnix* L., Gr. *ortyx* (*GB,* 215–19; Pollard, 1977, 61–62; *BAZ,* 161–63). Cf. ThC 5.38, PsJF 292–93), and Vinc. 16.64.

122. Isid. *Orig.* 12.7.64–65, Solinus 11.20, Neckam 1.70, and Barth. 12.7. The true etymology is unknown.

123. Cf. Ar. *HA* 597a23, Pliny *HN* 10.33.65–67, and Jacob. 92 (p. 222).

female. It is said to delight in seed from poisonous herbs as its food and for this reason a good number of people avoid eating quail.[124]

(35) CARDUELIS: The *carduelis* [finch] is a small bird which sits on the thistle [*carduns*]. In our land it is called *distelvinche,* but elsewhere it is called *stygeliz* in imitation of its call.[125] We have proven false by testing [*experimentum*] the story that this bird eats thorns and spines from thornbushes. It feeds on the seed of thistles, burs [*lappa*], teasel, and the like.[126] It also eats the seed of poppy, rue, and hemp, and whatever seeds it is eating, it strips away the shell with its beak and then feeds on the pure inner part. It also eats nuts in the same way, stripping away the shell from the meats.

There are many genuses of this bird, but three are most common in our land. One particular genus is ash-gray on its back, yellow along its sides, and in front of its beak on the face portion of its head it is red like cinnabar.[127] This one is the noblest. Another one is small and yellow and is commonly called the *cysych*.[128] The third is entirely flaming red on its chest and is commonly called *vinche*.[129] Some, however, add a fourth genus which sits on the flax plant and is therefore called the flax bird [*avis lini*].[130] It is smaller than the third genus and is sort of ash-gray on its back, as is the first, and on its chest it tends toward an ashy yellowness.

All these are musical, but especially the first and, after it, the second. The third is less musical than it and the fourth least of all. They all have in common that they fly in flocks. When they sit on the thistles, they are so stupid that one by one they are dragged by the neck with a snare tied to the edge of the bush, through a hole in some wall or other behind which the fowler is hidden, and the others do not flee. Likewise, when they are captured with a cleft stick and one is stuck, many gather to help and are captured. When placed in a cage, in order to get a drink they use their bills and feet to draw up a horn suspended beneath them. After they have drunk in this unnatural manner, they allow the horn to drop.

(36) CROCHYLOS: The *crochylos* [wren] is the smallest bird of all, which we call the *regulus* [little king].[131] Although small in body, it has great courage and tries to fight against the eagle.[132] It flies alone and feeds on vermin and spiders. It gives birth a good deal in the winter, content in one hole in a cave, or else it allows many comrades

---

124. Pliny *HN* 10.33.69. Barth. 12.7 adds that hellebore, the dried roots of which were used in many medicinal preparations, was a favorite of quail.

125. The goldfinch, called the *Distelfink* or the *Stieglitz* in MG.

126. "Teasel": *virga pastoris,* "shepherd's staff," *Dipsacus fullonum et silvestris.* Cf. *De veg.* 6.466.

127. The goldfinch, *Fringilla carduelis* L.

128. The siskin, *C. spinus* L., Gr. *akanthis?*

129. The chaffinch, *F. coelebs,* Gr. *spiza.*

130. Scanlan (219) identifies this as the lesser redpoll.

131. *Troglodytes troglodytes* L, Gr. *trochilos* (*GB,* 287–88), not to be confused with *Trochilos,* the crocodile bird, as was done by Pliny *HN* 8.37.90, Solinus 32.25 (*strofilos*), and Neckam 1.57 (*strofilos*); cf. ThC 5.40, PsJF 313–14 (*strophilos*), and Vinc. 16.65. Cf. 7.34.

132. Ar. *HA* 609b12f., 615a18f., and Pliny *HN* 10.95.203.

of its own genus to live in its nest at night so that the meager heat of its small body might grow strong and be enhanced by the society of many.

This bird, if defeathered and stuck on a small spit, rotates itself on a fire, and this has been proven by us through testing. This bird also is musical and sings especially at times of great, dry cold in winter.

(37) DYOMEDICA: The birds of Diomedes are the ones which King Diomedes called to augury.[133] They are white with fiery eyes, are the size of swans, and have sharp beaks with which they hollow out the softer parts of cliffs in which they build their nests. They build roofs above this out of tiles and dust which resemble a large cup. They position their exit to the East when they go out to feed and they return to the entrance to the nest which is directed to the West, coming back to their homes by a different route.

If they ever emit querulous calls, people say they are indicating auguries for themselves, for the king of the country, or for the destruction of either of them. These birds, as some say, greet Greeks approaching it but attack all other nationalities coming toward their nests.

(38) DARYATHA: The *daryatha,* as Aristotle says, lacks feet.[134] When it falls to earth, it crawls along on its chest with the elbows on its wings almost like a bat except that a bat has weak feet attached to its tail. This one appears only after a rain shower in the beginning of summer and it produces its young at that time. When the young grow strong it dies, leaving its young as successors to its miserable life.

(39) EGITTHUS: *Egitthus* is the name of a small bird which makes its nest in the dense parts of thornbushes.[135] It is an enemy of the ass because the ass, in order to rub itself, strikes against the bushes, shaking and knocking down its nest. Now the bird, hearing this ass braying, immediately prepares itself for revenge. It sits on the sores on the back of an ass and under its tail and pricks the ulcers, hollowing them out, and does not stop no matter how far the ass may run away from it.

---

133. Gr. *katarraktēs?* (*GB,* 131–32; Pollard, 1977, 101; *BAZ,* 37–38, 85–86). Isid. *Orig.* 12.7.28–29 largely follows Pliny *HN* 10.61.126–27 and Solinus 2.45.50, but adds that the Gr. name is *herodius*—obviously an eagle in A. (cf. 23.7(1)), which is repeated by Jacob. 90 (p. 190), ThC 5.41, PsJF 295, and Vinc. 16.68. Only the final sentence comes from Ps. Ar. 836a8ff.

134. Probably the alpine swift, *Cypselus melba,* Gr. *drepanis* (*GB,* 91). Ar. *HA* 487b27–32 says that it is weak-footed and strong-winged. Pliny *HN* 11.107.257 also says "short feet." ThC gives the same misquotation as A.

135. Probably the long-tailed titmouse, *Acredula caudata,* Gr. *aigithalos* (*GB,* 22–23) or *aigithos* (*GB,* 23–24; Pollard, 1977, 37–38), which nests in bushes. Based on Pliny *HN* 10.95.204–5, from Ar. *HA* 609a31–35. Cf. ThC 5.43 and Vinc. 16.69 (who adds the acanthis to the titmouse's enemies). Cf. 23.26(8) and (9).

CHAPTER I

On the Shape of Falcons

(40) While many want to know of the nature of falcons, we desire to describe it more precisely and will first describe the nature of their genus.[136] Then we will describe the diversity of their species, and third, we will set forth a discussion of the doctrines of falconry. Fourth and last, we will append a discussion of the infirmities of birds of prey along with their appropriate medicines.

Beginning, then, we say that the genus of falcons has four particular characteristics by which it is separated from other birds of prey. First is the shape of its body, second is its color, third its particular behavior, and fourth the sound of its call.

The shape within the genus is common to all. It has a rather heavy head, shortish neck and similarly short beak, a rather large breast with a sharp pectoral bone, longish wings, a shortish tail, and legs that are shorter and stronger than other birds of prey in respect to its body.[137]

Moreover, in determining the disposition of these members, we define those things which are common to the falcon genus with regard to the other genuses of birds of prey. Although we said its head ought to be heavy, enormity of head ought not to be praised, since this befits the shape of the night owls [*noctuarum avium*], which all have enormous heads and are timid. This is because the enormity of their heads points to a superfluity of material rather than to greatness of courage [*virtus*]. The same sort of understanding must be reached concerning a round head. A round head is not always a good thing for each animal, since on a concave sphere the spirits flow about too much and are badly mixed. But while in the *astur* [hawk] the elongated anterior of the head is narrowed a bit and ends on the beak, much as the eagle's head is narrowed, nevertheless the falcon's head is not elongated or narrowed, so that the beak seems, as it were, placed right onto a sphere. Its forehead is widened beyond roundness and the upper surface of its cranium lacks true roundness, while its jaw parts are well rounded

---

136. This extensive portion on falcons, representing about one half of the book on birds, has been of great interest over the centuries. It was included in the Vatican codex as well as the first printed edition of the famous *De arte venandi* of the Emperor Frederick II (Velser, 1596), reprinted (Schneider, 1788–89), and translated into German in 1736 by Johann Pacius. In more recent times it was translated into Fr. and appears with the text of Dancus Rex (1883) as well as, of course, in the various editions of A.'s *DA*. As A. himself indicates, the motivation for this extensive section was the interest of his audience. For recent work in this area, see Abeele (1990a; 1990b; 1991; 1993; 1994). For illustrations, cf. García Arranz (1996, 471–501).

137. Cf. the additions of Manfred, using Fred.'s notes, at *DAV* 1.18–29 (W&F, 119–28; unless otherwise noted, we shall use this translation). This paragraph and the preceding one are cited by Vinc. (16.71) from A. and hence may represent A.'s own conclusions. Fred. makes no such generalizations, but it is possible that he may have or intended to have included such in the *Liber animalium* cited by him at *DAV* 1.24, which Schneider (1788–89) suggests as a possible nonextant work of the emperor.

and short. Such an arrangement is the sign of a humid, choleric, well-moving, and very bold nature. It is indeed a particular characteristic of the falcon among the birds that it moves quickly against its prey and is not divided in purpose and dares more than it can do.

I speak similarly about its shortness of neck. In the genus, the falcon's neck is shorter than that of the eagle, *astur,* and *nisus*. But immoderate shortness is to be censured, since it testifies to either a phlegmatic frigidity or a melancholic dryness with a lessening of power [*virtus*] and it is a characteristic of the night owl.[138] As we said in our *Physonomya,* it cannot be that some animal participates in the shape of another without having some of its habits in accordance with their nature.[139]

Similarly, although we may say that in accordance with their nature the legs of falcons ought to be short, all those who know the nature of falcons praise long hips in them because, since the leg is moved from the hip, if the hip were short its disposition would be frigid and bad. Let, then, the hip be long and well feathered and the leg short and the foot well spread with strong toes, especially at the nodes of the joints, and let the talons be strong and more than a little curved toward the inside of the foot.

In entirely the same way, then, let us determine the shortness of the tail in the genus by a comparison with the *astures* and *nisi*. Too great a shortness tends toward a similarity with the *zueta* and *bubo* [eagle owl], which fly at night.[140] Let then the length be such that the wings, when folded, touch each other over the tail or barely touch at the very end of the longer anterior feathers. From this point of their meeting, the tail should not drop much nor should it hang down like the tail of an *astur* or *nisus,* since the length of the tail is always a sign of the humor of the descending *nucha* [spinal cord] and signifies timidity.[141]

CHAPTER II

On the Particular Color of Falcons

The color of falcons is common to the genus. On its face it has black spots on the jaws and white ones around the sockets of its eyes and on both sides of its beak.[142] It has black eyelids and a blackish ash-gray color on its cranium, back, the upper part of its neck, and the outside of its wings and tail.[143] In other places it is fitted with a variety

---

138. A mistake also made by Fred. at *DAV* 1.26 (W&F, 63), apparently in both cases coming from the failure to recognize that the owl's neck appears so short only because of the profusion of feathers in that area.
139. Cf. 20.42–62.
140. The sparrow hawk; cf. 8.15, *sivetta* (MG *Sperber*).
141. *Nucha*: cf. 1.257–59, 261, 267, 272; 3.161; and 12.90, 113, 128, 134.
142. The falcon's "moustache."
143. "Eyelids": superciliary shield. See W&F, 60.

of colors, descending in stripes, as it were, with the stripes broken off here and there. This variety always has black for one color, but in its first year the second color is a red or pale red, but each time it molts it gets whiter and whiter.

The color of its eyes is a yellow so violent that it approaches redness and the *acies,* or pupil, is black.[144] The best color of the feet is yellow, greatly tending toward white. The less white the feet are, the more ignoble is the falcon.[145] If they tend toward the color of the hyacinth or to sapphire, that is an ignoble color. It does not indicate nobility because that color is made by terrestrial smoke being scattered over the skin of its feet and indicates slowness and a melancholic timidity. Such falcons, therefore, when aloft stand suspended, and before they fire the blow, their prey escapes them. If ever such falcons are found to be good ones, this happens from much labor and exercise. So it is that sometimes a person is lazy by nature but is rendered good through a course of exercise.

The species of falcons differ among these stated colors of their feet and wings, as will be clear in what follows. But throughout its genus, the color of falcons, both in whole and in their parts, is as has been said. The color that is on their heads is somewhat the same as that of certain night-flying birds of prey and to this extent those birds approach the excellence of the falcons. But since shape tells more of the complexion than color, they are plainly removed from the nobility of the falcons.

Let, then, these things which have been said about color be sufficient.

CHAPTER III

On the Characteristic Behavior of the Falcon According to Its Genus

48 The characteristic behavior [*actus*] of the falcon among birds of prey is to be borne into its prey on a rush.[146] Therefore, the falconer should beware of showing the prey to the falcon immediately. Rather, he should let the falcon go only once its prey has been moved some distance away so that the falcon will not attack it too hastily. When it desires to take prey, it is the falcon's way to ascend with a rapid flight and then, with its talons folded up to its chest, to descend on the bird in a rush, with so strong an effort that it arouses in its descent the sound of a raging wind. In such an attack it descends not diametrically or perpendicularly but rather at an angle. By striking it with this sort of descent, it inflicts a long wound with its talons to the extent that the bird sometimes falls split from head to tail and sometimes it is found with its entire head taken off.

---

144. Cf. Fred. *DAV* 1.24 (W&F, 59–60).

145. Generally, the feet become whiter with age. For the foot color of a young falcon, particularly a peregrine, see 23.69.

146. The diving velocity of the peregrine approaches 200 miles per hour. This sentence and the following one are copied by Vinc. 16.71, with citation.

The motion of a falcon is that of ascent and that of descent and between each of the two motions there are necessarily two periods of rest. It is the trait of a good falcon to interpose almost no period of rest between its ascent and descent.[147] But after it has descended, it sometimes hedges off and impedes the bird by flying below it until it is struck by an ally who has ascended. Thus, the best fowling occurs when two or more allied falcons help each other.[148]

It happens rather often that a higher falcon follows above until it sees that the bird is in a position appropriate for striking. But care must be taken lest he grow accustomed to sort of standing still with his wings uplifted, for this is a sign of timidity and a sign that he dares attack nothing save the reptiles of the earth.

Because it is the way of a falcon to strike with a blow from the chest, nature gave it, on the front part of the bone in its chest, a strong, triangular appendage whose prominent part is a right angle in the upper part of the chest.[149] It is a very tough and strong angle and the bird folds the talons of its feet over it and, with the rearmost talon, rends what it strikes. The one suddenly descending headlong is the best, but the one standing still on high with uplifted wings degenerates to the nature of the birds which are called *lanarii* by the philosophers but are commonly called *sweimere* in German.[150]

The falcon hunts well alone and its behavior is as has been described. But it hunts better with allies or an ally because in ascending and descending it is necessary that delay occurs, during which time the prey gets away if an ally does not hinder it.[151] This is why although the falcon is a wrathful bird like all the other birds of prey, and although it takes pleasure in being alone, it nevertheless takes pleasure in allies during a hunt due to their help, and it shares its prey with an ally without a fight, something neither the *astur* nor *nisus* does.

This, then, is the characteristic behavior of a falcon and is common to each falcon throughout the genus. Because, as we have said, it rashly tries everything it comes across, the falcon ought to have a hood covering its eyes when it is carried on the hand.[152] There should also be a time that is not for hunting, lest it try to fly too much. Since it has good wings and it desires to fly frequently, it therefore has to be held back by the falconer. Another reason for this is that when it has its eyes closed and then after the hood is removed something appears, it attacks all the quicker and more boldly, as if out of astonishment. In this way it is also rendered more submissive to humans and forgets any other ties it may have had.

---

147. See Fred. *DAV* 6.6–14 (W&F, 369–82) concerning "waiting on," or *circumvolare*, a highly desirable characteristic in a well-trained bird.

148. Cf. Fred. *DAV* 3.16–18 (W&F, 246–50).

149. Apparently, A. is describing the "wishbone," usually the *furcula* (forked bone). The "bone in its chest" is the sternum. Cf. Fred. *DAV* 1.32 (W&F, 71–72).

150. Cf. 23.71, *sweimere* (MG *Schweber*, "one who hovers").

151. Cf. ThC 5.51.

152. "Hood": A. uses the word *mitro* instead of the *capellum* of Fred. *DAV* 2.77–80 (W&F, 205–19), which discusses the use of the hood, apparently introduced by Fred. from the Arabs.

CHAPTER IV

On the Call by Which Falcons Call and Are Called

50 The sound of the falcon's voice is generally lower and more far reaching than that of the *astur* or the *nisus* and proceeds from the high pitched to the lower. When the falcon is called back by the falconer, it is not by a whettle but rather with a loud cry like that used to call back dogs.[153] Nor is it merely called back. A certain object made of four or more wings tied together to resemble a bird and with a piece of fresh meat bound on top is led around on a string.[154] Falcons, however, sometimes call too much and are then either angry or too worn out from hunger caused by taking away their food. At this time if they are angry they should be hooded with a head covering or, if they are worn out with hunger, they should be fed more. This is because a falcon that calls too much greatly harms the fowling since birds flee its cry even before the falcon is an appropriate distance for capturing them.

Likewise, when the falcons are called back by the falconer, they sometimes do not return, especially the ones known as mountain falcons [*montararii*]. This comes about for two reasons, of which one is the indignation of its wrath when its prey escapes. The other is its being too full, so that it scorns food. At this time, if it has been raised by a good falconer from its youth, this should not be of great concern since with its indignation set aside or when it gets hungry later on, it will return home of its own accord, for falcons and *nisi* have this amount of faithfulness that they return home like pigeons when they have been well raised and trained. I myself have seen falcons who used to come and go without bindings from the house and who used to come over the tables while we were eating, extending themselves in the rays of the sun in front of us as if they were flattering us. When it was time to hunt, standing on the roofs and at the windows, they went out into the air, flying over people and dogs toward the field, and when the falconer wished, they used to return at his recall.

If, however, they have not been well tamed, then the reason they do not return is a fear of humans. You should be careful of letting these birds go except when they are hungry, because then they are accustomed to return out of a desire for food.

Let, therefore, these things which have been said by us about the call with which they call and by which they are summoned be sufficient.

---

153. Fred. *DAV* 3.13 (W&F, 243) points out that the call was not used by British falconers, who could only give him the reason that it was the custom, which W&F (ad loc.) wryly point out is quite a sufficient British reason.

154. Fred. *DAV* 3.1 (W&F, 226) says that for the lure two wings are used when they are crane wings, four if heron wings (*DAV* 5.1; W&F, 317). On the lure, see also *DAV* 3.3–13 and 6.3–5 (W&F, 228–43 and 365–68).

CHAPTER V

*What the Seventeen Genuses of Falcons Are,
and About the First, Which Is Called "Sacred"*

Of the ones that concern us, there are ten genuses of noble falcons, three of ignoble falcons, and three mixed with noble and ignoble. There is also one mixed which, since it is not entirely from ignoble parentage, is found to be very useful for hunting.

The first genus of noble falcons is the noblest of all and is the one which some call "sacred."[155] Symmachus, however, calls it "British" [*Britannicus*] and others call it the *aelius*, as if from *aerius* ["of the air"], falcon.[156] Others also call it the *aeryphylus*, as if it were the falcon loving the air.[157] We have spoken about this one above saying that it loves to fly on high and that it does not care for small creatures.[158] It has coarse, knotty legs, talons more cruel than the eagle's, a terrible appearance, brightly flaming eyes that turn from yellow into red, a large head, and a very strong beak. It also has yokes or large foldings of its wings which are always shown when flying and this one alone among the falcons has a somewhat long tail. It is almost the size of a large eagle and no eagle or any other of the birds of prey flies beneath it due to fear. As soon as the other birds see the sacred falcon, they flee, crying out, to the dense parts of the trees or to the ground and they allow themselves to be taken by hand rather than return to the free air.

51

---

155. The saker, *F. cherrug*, considered the best of the desert falcons by falconers (W&F, 516) and a native of southern and central Asia. About the size of the peregrine, it often has inferior plumage, leading Fred. *DAV* 2.22 (W&F, 121) to comment that it should be judged by other criteria. Its smaller feet (in comparison with the peregrines') often led it to be valued below the peregrine. See W&F (526–27), who reproduce the ranking from the *Boke of St. Albans* cited in the classic work of Swann and Wetmore (1924–38), where the saker is ranked below both the gyrfalcon and several peregrines as the appropriate bird for a common knight.

The identification, with modern names, of medieval falcon and hawk species is made somewhat uncertain by the same factors discussed in the notes at 23.7(1). A. generally uses the masc. or ntr. pronoun, although usually the female of the species was much preferred over the male, since she is larger and a fiercer hunter. In English treatises, the name falcon was generally used only for the females, while tercel (*tertiarius*, *terliolus*) was used for the male, apparently because he is one-third smaller (W&F, 627). In the hawks (*accipitur* and *nisus*), A. specifically notes the difference in names of the male and female but does not differentiate among the falcons. A variety of authorities have attempted over the years to identify definitively A.'s falcons, including Killerman (1910), Stadler, W&F, Lindner (1962), and Oggins (1980). Many identifications are obviously, then, somewhat conjectural.

156. The description is taken largely from the so-called Symmachus letter (ed. Rigault, 1612), a major portion of which is contained in ThC 5.50. The saker and the merlin are the only falcons whose descriptions are the same in A. and ThC. Lindner (1962, 1:21) attributes the Symmachus letter's authorship to ThC, but there are differences between ThC and the two forms of the letter edited by Rigault (1612). See Oggins (1980, 443–44, with notes) for discussion of ThC and his relation to the Symmachus letter.

157. *Aeryphylus*: "loving the air," a translation of the Gr. name.

158. Cf. *aerifylon*, 23.27(12).

They fly two at a time and are rendered tame because they sit on one perch and follow a person as if they knew only how to be with a human. There is no large bird which it does not immediately drive out. Nor is it enough for them to drive one out but as many as they meet they drive out. They even seize roebucks and tear their eyes and brains with their talons.

52 They desire to be fed very delicately, always on fresh and very healthy flesh, but especially on hearts and brains. These should be so fresh that they are still warm with the heat of life itself, having been brought from an animal just then expiring. They eat almost as much as large eagles.

The genus of falcon is regal, flying the furthest and following prey for a span of two or three or even four or six hours. It hunts better with allies but still hunts very well alone. It delights in people and hunting dogs and hunts the more willingly with them present, as if it gloried in its strength in their presence.

It has those things which we have said are common to every genus of falcon: the spots on the face, the shape, the characteristic behavior, and the call. Its call, however, is frightening, and it rarely calls. When it is recalled, the one calling it back should shout in a resounding voice because it flies high and far. Also, the lure [*reclamatorium*] should be quite large in order to be seen from afar. If, however, it does not return quickly, there will be no danger, for it is accustomed to return home on its own.

This, then, is the first and noblest genus of falcons.

CHAPTER VI

On the Gyrfalcon

53 The second genus of noble falcons is the gyrfalcon, a genus only one step removed from the genus of sakers.[159] The gyrfalcon in its shape, color, behavior, and call has the perfect nature of a falcon, but in size it is larger than an *astur* and smaller than an eagle. It is called a gyrfalcon because of its spiraling [*gyrando*], since while spiraling far, it attacks its prey fiercely.[160] It does not care for small game, but it attacks large birds such as cranes, swans, and their like.

This is a very beautiful falcon, with a tail that is not long in proportion to its body. It has very handsomely shaped and strong yokes on its wings, and has smooth,

---

159. *Falco risticolus* (*F. gyrfalco* L.). We must concur with Oggins (1980), 453, who concludes that this chapter does not appear to be based on other written authorities. Cf. ThC 5.44, who bases his discussion on the *Liber rerum* and says that the gyrfalcon is called the *herodius*, which for A. is clearly an eagle. A large proportion of Fred. *DAV* books 3 and 4 is devoted to training and hawking with the gyrfalcon, generally the highest ranked and most prized of the falcons, being reserved for emperors, kings, and princes (W&F, 510).

160. An etymology concurred with by some modern experts. See Newton (1896), s.v.

knot-free legs, and strong enough talons, especially the hindmost ones. It hunts alone but does better with another. In comparison to the other falcons, it stands more erect and with its feathers well arranged. In hunting, it follows the prey a long way and the hunter therefore needs a fast horse on which to follow it and swift hounds trained for this who will help him when it throws them prey.

This falcon must especially be trained not to seek its prey on water since it frequently is at a distance from the falconer and can be harmed in the water. It therefore should not be sent off after prey along the length of a body of water but rather should be held until the birds appear away from the bank and over the fields. The gyrfalcon should then be sent off from the water's side toward them because then, on account of the motion of the gyrfalcon, the birds do not dare to return to the water. If, however, the gyrfalcon is let go from the field side, then the birds flee to the water and either evade the gyrfalcon or, if struck, fall into the water. Then the falcon, following its prey, is harmed or drowned, and even if it escapes, it is rendered timid on account of this injury arising from its hunting.

Other falcons and *astures* do not fly well beneath this gyrfalcon. Even the eagle does not easily meet up with it.

The gyrfalcon likes to be fed with fresh delicacies and healthy flesh still warm with the warmth of life. By delicacies I mean those things that are near the heart, because these are the more easily digested. The proof of this is that when it is wild and takes prey, it is found to eat nothing of its prey except the heart and afterward eats those things near the heart toward the right wing. Rarely is this toward the left wing, and then only in birds with a warm complexion such as the pigeon, wood pigeons [*palumbes*], and birds of this sort. That it likes to feed on fresh meat is a proof that when wild, no genus of falcon, *astur*, or *nisus* returns to the remains of its prey. It ignores what is left over from one feeding and when it has to feed again, it attacks new prey. The genuses of eagles do not do this. Also, falcons do not sit on a cadaver as do the genuses of eagles and kites. Therefore, flesh still warm with the warmth of life pleases the gyrfalcon. This is proven in that a wild gyrfalcon begins to eat its prey before it kills it. It delights also in healthy flesh and I therefore do not approve of those who take a wing [*coxa*] of a live hen and later, on the second day, take off another part to be served to the falcon, because certainly the body of the hen is more healthy when it remains whole and is infused with feverish heat, since fever is aroused in every body in the generation of health. Therefore, food of this sort is not good for the gyrfalcon, which is a delicate bird. The wise falconer ought to strive to follow nature in the art of feeding as much as he can and nature shows this perfectly when the gyrfalcon is left to itself as a wild bird. In this way the gyrfalcon makes progress in its own natural vigor. If, however, it happens otherwise, the bird fails, becomes infirm, and dies.

These, then, are the two genuses of falcons which have been spoken of. They are the first and the noblest among those which have come to us.

CHAPTER VII

On the Genus of Falcon Called the "Mountain"

55 The mountain genus of falcon is third in the nature of nobility. It is short and very compact in the body and has an especially short and compact tail. Its chest is very round and large and its strong legs are short with respect to the size of its body. Its feet are knotty with strong talons, and it is in the habit of frequently caring for its feet. On its back and the outside of its wings it has an ash-gray color and that color, as the bird goes through its changes of feathers [moltings] over the years, is rendered clearer and grows paler with a small, dusky variation interposed.

This genus of falcon is fierce, has bad habits, and is irascible with an inconsistent anger. It is therefore rare that a falconer is found who knows all its customs. It is thus the teaching of Ptolemy, the King of Egypt, that it rarely be held on the hand except at dawn and at a time for hawking [*aucupatio*] or hunting [*venatio*]. At other times it should be kept in a very dark room and a bright, nonsmoky fire should be kindled in the room two or three times a night. Do not let it be held on the hand except at the aforementioned times unless it has to be fed, for from this it becomes softened to the falconer's hand as if to a good thing, it gets trained, and it lays aside its wrath. When it is angry, however, nothing should be refused it by the falconer, for by this means its wrath is broken.

This genus of falcon has the thickness of a hawk or *astur*, although it is shorter than one. It has very pale feet and legs that are almost scaly with overlapping scales. Its shape when it stands is like a pyramid from shoulders to tail, if you imagine a pyramid slightly compressed at the back.

56 This genus of falcon becomes especially indignant over the escape of its prey, so much so that it sometimes attacks and slashes the face or head of the falconer calling it back. It also slashes the face of the horse on which the falconer sits and it sometimes attacks a dog. Sometimes one of the falcons attacks another. In such an instance, the falconer ought to be very patient and not turn the bird away. He should patiently ignore the recall until the anger has been laid aside and the falcon's spirit has soothed. Even if the mountain falcon has been recalled and does not return to the lure, it should not be a matter of great concern, except that it might be captured by someone else, for once its wrath has been put aside, it will return to the falconer's home of its own accord. This genus of falcon is not to be rejected because of behavior like this, because it is wonderfully bold in attacking large birds, so much so that it sometimes attacks and kills an eagle.

Ptolemy teaches that a wise falconer must be on his guard, for if the falcon is frequently sent off beyond its strength against strong birds when its anger is heated up, it unheedingly hurls itself to its death. Certain of our friends have seen just such an occurrence in the Alps. A mountain falcon was coming off a cliff, attacking the bird called the *perdix* [partridge], when an eagle, in front of which the aforementioned *perdix* flew, seized the prey. At this, the mountain falcon, ill-enduring the insult and goaded by the aforementioned eagle, tried to seize its stolen prey. But when after trying for some time it was unable to accomplish its intention, it ascended very high and, coming down in a rush, gave a blow to the head of the eagle, destroying both it and itself. This sort of rage must always be guarded against in the mountain falcon.

This falcon is found more frequently than the two aforementioned genuses of falcon. It rejoices marvelously in its fierceness and falcons of this genus are, therefore, frequently found for whom wounding and striking down one bird is not sufficient. Rather, they rejoice in the fact that they knock down many and delight in this fierceness so much that they forget food and are possessed by the cruelty of killing birds.

Let these things said about the mountain falcon be sufficient.

CHAPTER VIII

On the Falcons Called Peregrines

The genus of falcon called peregrine follows in fourth place, in respect to nobility, to the aforementioned genuses. It is called "peregrine" for two reasons, of which one is true since it always wanders [*peregrinatur*] from land to land, flying through almost all lands. The second reason is more in accord with the opinion of the falconers, namely, that the location of its nest is unknown and is not found by falconers, since it is captured in flight far from the place of its birth. I have heard the reason for both of these explanations from a very experienced falconer who lived in a wilderness area for many years near the highest peaks of the Alps. He said that these falcons which we call peregrines constructed their nests on the highest, sheerest cliffs of the mountains and that no approach lay open to the nest except if a person is let down on a rope from above, down from the peaks of these mountains. He said this letting down often had to be done with a rope 100 feet long, sometimes 150 feet, and sometimes even 200 or 300, and that frequently it was impossible, either because of the distance or because of the roughness of the cliffs.[161] He said that on account of this difficulty there arose the opinion that these falcons' birthplace is unknown. He also said that he had very often seen old ones bring food to their young in the cavities and clefts of such mountains. He added that it is the custom of the falcons that after the young are fully grown, the parents expel them from the place of their habitation because so few birds are found there. Therefore, the young immediately go forth into the plains

---

161. Cf. 6.54, where a similar procedure is followed, in that a person is lowered in a basket on a rope in order to find an eagle's nest.

where there are many birds, and they fly throughout the lands having no set dwelling place.

58 I have seen these falcons caught by two methods and the aforementioned hermit told me a third. They are caught by one common method in almost all lands. A wide net is spread out so that, with a cord, it is easily turned over that to which the fowler directs it. In front of that is stretched a cord to which is bound the red lanner, which people commonly call the *sweimer*, and to the cord hanging onto it is bound, with an extended cord, a bird or some wool or feathery thing resembling a bird so that when it is dragged and shaken by the fowler with the extended cord, the *sweimerius* seems to be following the bird to prey upon it. This cord is often struck in this way and a falcon, which by chance is passing through, sees it. He descends in a rush, intending to seize the prey from the lanner and, thus deceived, falls into the net.

I have seen another method that is much better. A construction is made of two pieces of wood in the shape of a cross. Over their ends two other pieces are curved into semicircles. Thus, the ends of the semicircles are attached to the ends of the wood of the cross, and amidst the quarter arcs of the circle other pieces are curved to the place where the semicircles intersect and are attached below onto pieces of wood that come out from the angles where the cross intersects, until only three or four digits separate each piece of wood. Each of these pieces of wood is filled with snares from top to bottom. This container is made to a height of seven or eight feet and to a width of five or six feet. In it is placed a kind of cage, six feet high and four feet wide, so that only one foot separates the sides of the containers from each other when one is inside the other. In the interior one there are pieces of wood placed in the manner of planks from bottom to top. Six or seven birds are put inside the inner cage and they constantly move up and down on the planks.

59 This whole thing is fashioned immovably over the wall or gate of a castle or in an open field. Then, a falcon that is passing by sees the birds and, wanting to capture them, is itself captured in the snares. I have seen the best falcons taken in this way.[162] The hermit did not say anything else should be done, except that a bird can be put in front of the net without the lanner. The falcon, desirous of the bird, hurls itself into the net. He said that with this method he had caught falcons every year.

---

162. We are grateful for the information provided by Steven Bodio and David Zincavage, who read this passage and identified A.'s trap as an upright version of a trap most commonly referred to as a bal-chatri or shikra trap. The trap originated in India and knowledge of its use may well have travelled from there to the West through Arabic sources. The bal-chatri trap consists of a cage which resembles an inverted basket made of pliable sticks and covered with a mesh of some sort. A bait bird is inserted into the "basket" and a large number of individual loop snares are placed over the surface of the basket. The trap is rated as one of the most efficient and thoroughly humane devices a falconer can use. Bub (1991, 195–96) has excellent illustrations of its shape in both old and modern forms, while Beebe and Webster (1964, 178–79) describe its use today in America. A.'s version differs only in that it is placed in the air on a frame, ostensibly to be more attractive to those raptors which strike on the wing. For other traps, see W&F, 433–49.

This falcon, which is commonly contained in almost every land, is smaller in size than the *montanarius,* has a short tail, long wings, a heavy head, a long thigh, and a short leg. If it has knotty legs, it is better. It ought to have legs and feet that are getting white, and it has good behavior when it is well fed. Its prey is, as a rule, the duck, and if rendered very bold by a wise falconer, it will seize an *ardea* [heron] and sometimes a *grus* [crane], but this is the extent of its boldness.

Let these things be enough said about this genus.[163]

CHAPTER IX

On the Gibbous Genus of Falcon

There is, moreover, a very noble falcon, fifth in nobility in the genus of falcons, which the aforementioned hermit called the gibbous [*gybosus,* hump-backed]. He showed me the three best of this genus that he had with him and claimed to have sold many others.[164]

This genus of falcon, then, has a very small sized body but possesses wondrous courage and boldness, and employs a vigorous flight when it attacks prey. Its size is slightly larger than that of the *nisus* which the common folk call the *sparverius* [sparrow hawk]. On its face it has spots like the peregrine and other genuses of falcon.

It is called gibbous since, with its short neck, its head is barely visible in front of the yoke of its wings when it folds them up alongside its body. For the size of its body, it has a large head, a very short, rounded beak, very long and high-rising wings, a short tail, strong thighs, and legs that are somewhat long compared to the proportionate relationship of its other members. The legs are also somewhat scaly, almost resembling the scales that appear on the sides of the bellies of serpents and lizards. A gibbous's feet are knotty at the point of the toe joints, especially toward the inner side of the feet. Gibbous falcons have flaming, glowing eyes and in overall color they are like the other falcons called peregrines. On top of their skull their head is fairly flat and it does not project out at the rear of their heads. Rather, the head is just about continuous with the neck. The gibbous is readily tamed and has a good disposition. It makes its nest in inaccessible cliffs as does the peregrine genus and it is captured when flying from its nest as is the peregrine. It has such boldness and vigor that it takes down wild geese, herons, and cranes. This genus is exceptionally swift and rises

---

163. M1 includes the following statement: "I add only what was told me by the aforementioned hermit, that this genus of falcon was not called peregrine but rather rock falcon, giving as reason for the name that it was accustomed to sit on rocks and take a rest."

164. Another species of uncertain identification. Oggins (1980, 455–56) lists the several options: (1) the hobby, *Falco subbuteo,* a medieval favorite; (2) the "light form" of Eleonora's falcon, *F. eleonarae*; and (3) a confusion between Eleonora's falcon and *F. peregrinus brookei,* a smaller, dark-colored race of peregrines. There is something to be said for all the various identifications but none can stand unchallenged.

so high that it eludes human eyesight. This falcon is not content to bring down one bird but rather wounds many, and during a hawking session it seeks to have several allies due to its own smallness and the largeness of the birds it hunts.

61 The aforementioned falconer told me something quite memorable about this genus of falcon. For he related that he sold three of the aforementioned gibbous falcons to a certain nobleman. While he was testing the falcons, taking the hermit with him, they happened upon some white wild geese. When the falcons were released, the geese flew very high, but the three falcons which had been released outstripped the geese in their ascent until they were lost to the sight of all who were there. After a while the aforementioned nobleman was lamenting his falcons as lost when gradually the geese, weakened by their wounds, began to fall all around them. They found more than twenty geese had fallen and at length the falcons too were recalled and returned to the lure. Moreover, all the geese were mortally wounded as if they had been split open in various parts of their bodies with a knife. The reason for this is that this falcon, much like the other genuses of falcons, does not strike immediately upon its descent but rather begins to ascend again, coming out of its plunge, and at that moment, with its hind talon positioned in front of its chest, it strikes, producing a long, lethal wound. Frequently it strikes so forcefully that it tears off its own talon and wounds itself seriously in the chest, or even dies.

62 This genus of falcon is the one which is always spreading out its wings as if to take flight. Its spiritedness is greater than its strength. It prefers to be fed with very fresh flesh, still steaming with the heat of life, and this indeed does it a lot of good. If, however, it is sometimes fed with other flesh, let the falconer take care that the flesh is light, like the flesh of hens.[165] It should also be fresh, or at least not smelling bad, having been kept fresh in cold water.[166] For birds of prey have quite weak stomachs with a thin membrane [*pellis*] and their stomachs are easily turned by bad food. This is why, for the slightest reason, they reject and vomit forth undigested food. This happens especially when they are fed heavy and melancholic flesh or flesh which is prone to go bad.

This genus of falcon likes to be carried on the hand for a long time early and late because, once it has grown accustomed to the human hand, it most readily sits on the hand and readily returns to it.

Let, then, the things we have said about the gibbous genus of falcons be sufficient.

---

165. "Hens": *altile*, an unusual usage for A. Scanlan's "fattened hens" is also possible since *LLNMA* cites a source where the term indicates fattened animals.

166. An interesting insight into premodern ways of preserving perishables.

CHAPTER X

On the Black Falcons

The black falcon [*falco niger*] possesses the sixth grade of nobility among the falcon genuses.[167] It is a bit smaller than the peregrine falcon, but in shape it is in all respects similar, even though it differs in color since on its back and on the outside of its wings and tail it has all over a certain dusky blackness. On its chest, belly, and sides it has a dusky variegation. On its face it has falcon spots [*guttae falconariae*] which are very black, with an intense black color surrounded by a certain swarthy, dusky pallidness.[168] However, it has the legs, talons, and beak of a peregrine. This falcon comes quite close to the shape of the bird called the *butherius,* which we mentioned previously.[169]

Emperor Frederick, following the statements of William, the falconer of King Roger, said that it was first seen in the mountains called the Gelboe, in the fourth *clima*.[170] The young birds, expelled by their parents, had come from there to the mountains of Salamis in Asia. The descendants of these first ones, expelled in turn, came to the mountains of Sicily and thus spread throughout Italy.[171] They recently appeared in the Alps and in the Pyrenees, and these genuses have even spread into Germany although they are rare there since few are found.

These falcons have the rearing and boldness of peregrine falcons.[172] But they seem to have a choleric complexion and whatever of the earthy and choleric that has been burned is converted into feathers and this is why they are black. As their age progresses, however, they grow whiter from their annual moltings. It is probable as well that in warm *climata* they are blacker than the ones born in our lands, for in warm

---

167. Stadler (1465, 1620) tentatively suggests an identification as the English hobby, *Falco subbuteo.* Oggins (1980, 457) opts for a smaller form of peregrine, *F. peregrinus minor.* Remember that *niger* can mean simply "dark" rather than necessarily "black."

168. The grammar of this sentence is fraught with problems. The general sense seems clear, however.

169. A. probably refers to the *buteus,* a buzzard, discussed at 23.29(15), but he seems to have misremembered the name, corrupting it with the name of the bird that follows the *buteus,* the *butorius,* a bittern. Cf. also 23.71.

170. Roger II of Sicily ruled from 1130 to 1154. His falconer, Guillelmus (William), wrote a new treatise on hawking (Tilander, 1963, 6–9, 118f.). Gelboe may be identical with the Gelboa/Gilboa of *OL* (2.136) which is identified as Jebel Fukua, a mountain on the border between Israel and Jordan. If this is so, then, "Gelboa" is merely a transliteration of the Arabic *jebel,* "mountain." Tilander's text of *Guillelmus* (1963, 158) reads "Gabuel" and lists three other MS variants for the name other than "Gelboe." Gelboe is used at 23.138(101) to mean a nesting ground for parrots.

171. Salamis was the ancient capital of Cyprus. *Guillelmus* (Tilander, 1963, 158) gives the route as Gabuel-Sclavia-Poland.

172. "Rearing": *nutritura,* a rare word for A. It might indicate patterns of nurturing the young or may be allied to the verb *nutrire,* which, in falconry terms, can be translated "to feed up" (W&F, 623). From its use in the title of Chapter XV (23.71), it carries a sense of raising and training the falcons.

places the egg-semen [*semen ovale*] is decocted with a strong heat and as a result that which is produced from it is blacker.[173] However, in cold regions it has more watery egg moisture [*aqueum humidum ovale*], which, on account of its clarity, is a cause of whiteness and, when scattered throughout the same earthiness, is a cause of variegation.[174] When, however, we consider two things which are external to animate bodies, namely shape and color, shape indicates a greater conformity or differentiation with respect to the species than does color. For we have seen jackdaws which, due to the coldness of their dwelling place, were born white, and the same is true for ravens. Yet the shape of the ravens gave a true indication concerning whether they belonged to the species of jackdaws or ravens. According to this argument it seems that this genus of falcon is very similar to the genus of peregrines, even if it does differ in color, especially since we have said that the parents in this genus force their offspring to travel [*peregrinari*] from the place they were born.

Let, then, these be the things said about the black genus of falcons.

CHAPTER XI

On the Nature of the White Falcon

65 The white falcon claims for itself the seventh genus of falcon. It comes from the North and the Atlantic Ocean [*Mare Occeanum*], from the regions of Norway, Sweden, and Estonia, and the neighboring forests and mountains.[175]

This falcon is an off-white, but in a variegated way, much in the way that the one we have spoken about is black. The special cause of its whiteness is the cold and humor of the place in which it is found. On its back and wings it is off-white. In other places, however, it has spots (that is, drops) that are very white, with other, pale drops interspersed.[176] In size it is larger than the peregrine falcon, being very similar to the white lanner that flies in the field hunting mice.[177] It is so similar that some falconers say that this falcon was first born with a peregrine falcon as its father and a white lanner as its mother, but its boldness shows that this is not true. For it is very bold and good, not lacking anything of the falcon's nature and having nothing of the behavior of a lanner. When it ascends to hunt, it does not hover suspended like a lanner, but

---

173. This passage makes it fairly clear that the "black" falcon is really to be thought of as a "dark" falcon.

174. Reading, with Borgnet, *terrestri* for Stadler's grammatical, but almost untranslatable, *terrestre*.

175. Almost surely the "Greenland" gyrfalcon, *Falco candicans*, the imperial falcon of medieval times, highly prized and relatively rare. W&F, 509–10, Stadler (1620), and Oggins (1980, 458) all concur in this identification. For *Mare Occeanum* as Atlantic Ocean, and not Arctic Ocean as Scanlan (240) renders it, see OL (2:473). A.'s Toledan contemporary Ibn Sa'id (d. 1286) also discusses the white falcon (cf. Eisenstein, 1993) and supports its location around the Atlantic Ocean.

176. A. prefers the term *gutta*, drop, for the more common *macula*, spot.

177. The white lanner is discussed below at 23.71f.

it strikes immediately like a falcon. Further, the shape of its feet, talons, beak, and entire body indicates its falcon nature even though it has heavier, knottier legs than the black falcon. But this befalls it from its moister complexion, which fills it up more and thickens its legs more than the legs of the other one, which is dry and choleric in complexion.

The mere fact that this falcon seems to have less boldness than the black one and that it is less swift due to the coldness and humor of its complexion comes nowhere near to establishing a sufficient proof of mixed parentage. For the power [*virtus*] of every animate being is increased by the size of the subject, when its natural size does not exceed the boundaries of its natural size. This falcon therefore has more power than the black one and it takes this boldness from confidence in its inborn power. So if it is not as fast in flying as the black one, it nevertheless lasts longer in following birds, and this is held in the highest regard for a falcon. Dusky, warm bodies are porous and open and thus the spirits, which are the carriers of power, evaporate easily from them. Therefore, they tire and grow weak even though they are agile by nature. On the other hand, white, cold bodies have dense flesh and, because they are very moist, they have many spirits. These spirits, due to the compactness of their flesh, do not dissipate by evaporating quickly and they therefore endure a long time during exertion. Moreover, the complexional moisture pours inwardly into the muscles and nerves, and the heat which is created by motion does not allow the nerves and muscles to harden.[178] They therefore remain movable and useful for flight for a long time. In respect to these things, then, the white falcons compensate for the fact that they seem to be ranked behind the black ones.

Let these things, then, which have been said about them be sufficient since in their training and hunting little or nothing seems to separate these from the others.[179]

CHAPTER XII

On the Nature of Red Falcons

The eighth grade in the genus of falcons is held by the falcon which is called red [*rubeus*] by the ancients, not because it is totally red but because the drops, which are white in others, are red in this genus.[180] There are black dots interspersed as in

---

178. As is common, "nerves" refers to sinews and tendons.
179. "Training": *nutritura*. For this sense, cf. 23.64, with note.
180. Another species of uncertain identification. Stadler (1466, 1620) suggests the common kestrel, *Falco tinnunculus*. But the small size of the kestrel (two-thirds that of a peregrine) seems to negate the suggestion in light of A.'s statements here. The kestrel is not a useful falcon for sport, taking only sparrow-sized birds, mice, and so on, and was little used, hence unlikely to be even considered seriously by a medieval falconer. Oggins (1980, 458–59) offers three of the smaller peregrines, *F. peregrinus brookei*, the Barbary falcon (*F. peregrinus pelegrinoides*), and the red-naped shaheen (*F. peregrinus babylonicus*). None is a perfect choice.

the others. This falcon does not appear red on its back or on the exterior of its wings; only when it spreads its wings does a dusky red appear in them. Certain writers on the nature of falcons claim mendaciously that this genus of falcon is a bastard form, saying that it was created from a red lanner and a falcon. This is entirely absurd since it has no similarity with the lanner at all apart from color. Rather, the cause of the redness is a weak heat spread over the surface of its body which heats up the smoky moisture that is expelled to help in the generation of feathers.[181] And this complexion produces midpoints lying between white and black.[182] Just as nature, which is the cause of order, does in others, so in this genus it does not go from extreme to extreme except by going through midpoints. Therefore, in the genus of falcons, she touches upon this midpoint, for other middle colors are not suited to the genus of falcons. Green, hyacinth-blue, yellow, and colors of this sort deviate from a bold and rapacious nature. Green is an indication of an extreme, watery frigidity (as when it appears in urine) or it indicates a consuming, devitalizing heat. Hyacinth-blue, however, indicates an airy, evanescent complexion, whereas yellow indicates a destructive bile. While two of these colors are found in birds and the third in the peacock, they nevertheless are unsuited to boldness and to quick flight.

68 This falcon is not large, being a bit smaller than a peregrine, but it has strong talons, feet, and beak. It is very agile in flight but is not sufficiently perseverant. It is tamed often and is rendered better after two or three moltings. It does not have as long a life as the others and it therefore ought to be fed with very fresh food, still steaming with the warmth of life. But this should not be done too much or too many times a day, only in the morning and evening. One ought to be strongly on guard to keep it from things that alter its complexion since it is easily altered. Even though this falcon at first attacks easily, one should not force it beyond measure to hunt birds since a red complexion is easily harmed and overcome by exertion. For this reason, advancing age and a mild tempering of its inflamed moisture very often aid this complexion, especially through the molting of its feathers. For a red feather of this sort is fragile and soft and does not long sustain the force of flying without breaking. The wise falconer ought to pay attention to all these things.

Let these things, then, be sufficient warning about the red falcon.

---

181. Cf. 19.38–42 on color in animals.

182. For the claim that red is an intermediate between the two contraries, white and black, see Boethius's *In Ciceronis Topica* 4 (*PL* 64:1119).

CHAPTER XIII

On the Falcon That Has Azure Feet

The ninth genus is already departing from the quality of the noble falcons and is the genus composed of those with hyacinth-blue or azure feet.[183] In size and shape it is like or the same as the peregrine falcon, but its back and the exterior of its wings are not as black and this genus is whiter on its chest than the peregrine falcon. Its wings also are not as long as the peregrine's, but its tail is a bit longer and its voice is sharper since it has more of the feminine or phlegmatic complexion. Its boldness in attacking birds is also much less, for the genus with azure feet rarely attacks birds larger than the magpie or little crow [*cornicula*], whereas peregrines and other larger falcons attack whatever birds they please.[184]

For this reason it happens that when the azure-footed one rises on high to give a blow, it begins to hover and to hang suspended on its wings on account of its timidity and it does not rush headlong against the bird. This it gets from its nature. With training, however, and with the encouragement of the person helping it, it often acquires greater boldness but not that of a true falcon. There is a similarity between the courage of soldiers and birds of prey. Among soldiers there is certainly no lack of those who are weak by nature and less bold due to their complexion. These, however, very often perform noble triumphs due to the science of warfare, from being accustomed to frequent victory, from experience in attacking the enemy as well as in waiting for him, striking, laying ambushes for him and turning him aside, and from the confidence of their friends urging them on. It is the same for this genus, for while it may be timid by nature, it is rendered bolder and better than it is by nature through skill (which it receives from a wise falconer), through being accustomed to knocking down and holding birds, and through confidence of the falconer who is positioned next to it. In just this way, due to boldness of this sort, the *sparverius* (that is, *nisus*) [sparrow hawk] sometimes attacks birds much stronger than itself, such as the smaller and occasionally the larger duck. Nor is it a marvel that the lanner genuses—whose natural timidity and slowness are such that they only attack mice or perhaps the young of birds that are not yet flying but are running about on the ground or are lying in the nest—can, as a result of this method, be made into birds which attack and capture other birds.

The means by which a timid bird can be rendered bold through training will be clear later on.

69

---

183. Cf. 8.108, where the names *pes plavus* and *iaccinctinus* are used, and ThC 5.50.13f.: "The fourth genus has sky-blue [*ceruleum*] feet and legs and from this fact it takes its name." Oggins (1980, 459–60) surveys the evidence for the generally agreed identification of this bird as an immature peregrine.

184. Or "peregrines and the other falcons attack whatever larger birds they please."

CHAPTER XIV

On the Small Falcon Called the Mirle

70 The tenth and last genus of falcon is the smallest in size and is called the *mirle,* popularly called the *smirlin* [merlin].[185] Although this genus is inferior to the ones introduced before in size, it is second to none of them in boldness in proportion to its strength. This is especially true when its spiritedness is prepared by means of skill, familiarity, and hope of help from a skilled falconer near it. William the Falconer says that with such birds he has sometimes captured a crane. Birds that are suited to its strength are larks and, at most, the partridge and pigeon, unless, as we have said, its strength has been assisted by art.

This genus of falcon is spotted on its face like all the others, has very long wings in respect to its body, and a commensurate tail. Its legs and feet are smooth and yellow. It is smaller than the sparrow hawk [*nisus*] and about the same size as the sparrow hawk [*nisus*] which people call the *muscetus*.[186] When it is wild, it captures goldfinches [*carduellus*] since it is very swift and also crafty, and it both captures while on the ascent and descends while striking, just like other genuses of falcons.

Since it is known to almost everyone, let these things which have been said suffice for its identification.

CHAPTER XV

On the Three Genuses of Lanners and Their Rearing

71 There are three genuses of ignoble falcons which ancient fowlers—as the friends of Ptolemy, Aquila, Symmachus, and Theodothyon relate—say are called lanners [*lanarii*] rather than falcons. Imitating this work, certain of the Germans call them *lanere*

---

185. *Falco columbarius,* often called the pigeon hawk (Oggins, 1980, 460). The name merlin was usually reserved for the female, the male being called a jack merlin. On the MHG *smirlin,* see Sanders (418) and Suolahti (1909), 338–39.

186. Cf. 23.109, where we learn that the *musceta* is a male *nisus*. The normal English term is "musket" and the Latin *muschetus* is also found (W&F, 623).

in their own idiom.¹⁸⁷ Certain others are accustomed to call them *sweimere*.¹⁸⁸ They are a kind of *butherius* which hunts for mice in the fields and they have different colors.¹⁸⁹ There is a white one and a black one the size of the falcons and a smaller red one which acts like the *mirle*.¹⁹⁰

When they are young, these birds, like all boys [*pueri*], are timid and have almost no boldness. This is due to their humor and their dull heat. When, however, they have molted two or three times and their natural cowardice has taken the medicine of training, they are turned into birds that capture pigeons and ducks.

This is the training which must especially be held to. In the first year, when they are tamed, they must be fed only with live birds, and when they have defeathered them a bit, the birds should be set free from the lanners' feet, and they must be allowed to flee, not in flight at first, but on the run. When the lanners have learned to attack most of the time, the birds should be allowed to flee with a mixed motion of flying and running. When again the lanners have learned this through frequent repetition, the bird should be allowed to flee in slow flight, the flight, that is, of a bird whose wing feathers have been shortened. It should finally be allowed to flee in full flight, and throughout all these stages the lanner should be urged on and incited most strongly with a loud voice, and it should be helped to hold the bird by the falconer, because it takes its boldness from these things. Then, in the second year, the small birds should be substituted with rather larger ones and in the third with larger ones still. By means of this method every bird derives both experience and boldness for catching whatever birds the falconer wishes. Now, while it is granted that all this is not needed by the falcons of the first nobility (these being those of the first eight

---

187. See Suolahti (1909), 330, 356. The sing. is *laner* (Lexer, 1820). The text evidently refers to the apocryphal letter of Symmachus, *Epistola Aquilae Symmachi et Theodotionis ad Ptolemaeum regem Aegypti de re accipitraria*. The Latin original is lost, and the text is now known only in a Catalan version printed by Rigault (1612, 183–200). The title itself contains a confusion, however, and is sometimes represented as the letter of Aquila Symmachus and Theodotion, treating two individuals—Aquila and Symmachus—as if they were one. In fact all three—Aquila, Symmachus, and Theodotion—had translated the Hebrew Bible into Gr. early in this era (that is, centuries after the Septuagint). Early Christian biblical exegetes—e.g., Irenaeus, Origen, Jerome, and Eusebius—were clear that these were three distinct individuals (for a discussion of their knowledge and evaluation of the work of these early translators, see Bludau, 1923, 10–12, and Barthélemy, 1971). Medieval authors too, however, were aware that Aquila, Symmachus, and Theodotion were three individual translators living early in this era (Rabanus Maurus *De clericorum institutione* 2.44, *PL* 107:363–66), even when they were unclear whether the three translators were Jews or Judaizing heretics, e.g., Ebionites (cf. Hugh of St. Victor, *De scripturis et scriptoribus sacris* 9, *PL* 169:17B). The association of the three with Ptolemy, who, according to legend, had commissioned the Septuagint translation of the Hebrew text centuries earlier, is historically inaccurate.

188. On this "hoverer," cf. 23.49.

189. On the *butherius*, cf. 23.63.

190. ThC 5.50 gives two genuses of lanners, without color description, based on the Symmachus letter. It is not possible to make an accurate identification here, except that apparently several color phases of the lanner, *Falco biarmicus*, seem to be involved. Lindner (1963, 1:51) and Oggins (1980, 460) conjecture that the red may be the kestrel, *F. tinnunculus*.

genuses), it nonetheless certainly contributes to increasing the boldness and experience of them all.

Let, therefore, these be the things said about the training of falcons.

CHAPTER XVI

On the Four Genuses of Interbred Falcons and on the Manner of This Mixing of Falcons

72 Since any one of these genuses can be interbred with any other, many genuses of falcons are created. But there are four which have come down to us. The peregrine falcon often interbreeds with the one with hyacinth-blue feet and when this takes place with a peregrine father and a mother with hyacinth-blue feet, it detracts very little from its nobility because the sperm of the male is the operative agent, making and forming by means of the spirit which is the bearer of the male power into the subordinate *gutta* of the female.[191] The young which is produced reflects the father, although a bit of the azure color is scattered over its feet. When the mixture is the other way around, with an ignoble father and a noble mother, an offspring is produced which greatly takes after the ignoble father and has little of the nobility of the mother.

Now, just as we have said that interbreeding occurs between these two falcons, so it is the peregrines which frequently fly alone to all places that sometimes interbreed with the black lanners, and sometimes with the white and red ones. Their seeds become mixed and they move, change, and complete each other. This is due to the nearness of their sperm, their quite similar complexions, and the fact that they employ the same period of time for impregnating the eggs and for hatching them out. The offspring that is created survives and reproduces with one like itself, just as happens in the case of many birds and other animals, as is clear from what has been written in the previous books of this study.

This interbreeding occurs when falcons of species which are different but which have complexions close to each other come together at mating time and do not find a mate of their own species with whom they can mate. Although we have said that four genuses of such interbreedings of falcons have come down to us, reason demands that there are many and that more genuses of falcons can be formed on a daily basis. We think this is why such diverse genuses of falcons are found in diverse regions. For while *climata* can diversify the behaviors and colors of animals, it is the interbreeding of which we spoke that especially causes the diversity among species so similar. This is just as we have seen happen in the genuses of geese, dogs, and horses in our time.

73 Nor is it probable that this only happens when genuses of falcons interbreed; it is also probable that when falcons interbreed with goshawks, sparrow hawks, and the

---

191. On the *gutta* of the female, see 16.84–86.

genuses of eagles, various combinations arise that lead to the generation of composite birds of prey. But in the four genuses mentioned, we have said that this combination with the peregrine falcon especially occurs since the peregrines are expelled from their homeland by their parents and are frequently separated from each other on account of their prey and their temper. When they cannot find a helpmate of their own species, they then turn to as close a one from another species as they can find at mating time. When they turn to the azure-footed one, the offspring created is very like a peregrine. When, however, they turn to a black lanner, the result is an ignoble black falcon or one like the black one. When it turns to a white lanner, the one produced turns out like a white falcon. When it interbreeds with a red lanner, the offspring turns out similar to a red falcon in color and shape.

These mixed falcons are more easily helped by training than those which are entirely ignoble, especially if the father is noble, for they take much of their nobility from the other parent. This is as we have already said in the preceding books of this study.

If, however, it is turned around and the mother is noble and the father ignoble, what is created will be less noble. Through good training, however, it will improve, especially after one or two years and with the method of stimulation mentioned a bit before.

Let these things, then, be what should be said about the genuses and natures of falcons, for by means of these, other things can be known.

CHAPTER XVII

On the Regimen Needed to Produce the Domestication,

Boldness, and Health of Falcons

Having finished with these matters, let us consider the regimen and medication of falcons.

The regimen of falcons is divided into three others—those of taming, health, and illness. Now, that of taming has two goals, one of which is training it to grow used to the hand of a person. The other is that it be made bold and swift in the capture of birds.

The first regimen is carried out by never feeding the falcon anywhere but on the hand. From this, as Symmachus says, it will grow accustomed to the hand and cherish it as a result of the kindness there that brings around its spirit. When it first has to be tamed, let it wear its hood [*pileum*] from before daybreak, and let it be held on the hand until the third hour, at which time it should be given a hen's leg.[192] When it has eaten it, place it on the grass and put water before it so that it can bathe if it wishes.

---

192. See Fred. *DAV* 2.49–61 (W&F, 157–84). It is particularly interesting that Fred.'s method of taming a bird involves "seeling" (sewing the eyelids shut), in addition to what Fred. calls the newer and more efficient method of hooding (p. 158).

Afterward, allow it to stand in the sun until it has thoroughly cleansed itself and then let it be sent to a dark place until evening time.[193] Then let it be held on the hand until the first sleep, when it is to be put back in a dark place.[194] Light a clear-burning fire or a lamp in front of it throughout the entire night until dawn. Afterward, hood it and stand with it a while in front of the fire.[195]

You should know that those which attained full growth in the nest are better and have stronger feathers.[196] If, however, they are sometimes taken from the nest before they are fully grown, let the bird be placed in a nest resembling as closely as possible that from which it was taken. Let it be given chicken flesh frequently because it is temperate. Let it also be given fresh bear meat occasionally because this produces and strengthens wings. Unless it is handled in this fashion, it grows weak in the wings and legs to the extent that a wing or leg gets broken. It is a great help if it is not touched with a hand before it is fully grown. Afterward, however, when it is fully grown, let it grow accustomed to the hand and to the hood [*capellum*], as has been stated, by doing it every day. Be careful from the beginning lest it ever encounter any harshness on the hand; always let it find it a kindness as well as a source of caresses.

75 It is good for its boldness for it to catch, hold down, and kill live birds, doing so to the cry by which it is usually urged on by the falconer. As we have said above, let the falcon catch these birds again, as they frequently, with the falconer's cleverness, get out from beneath the falcon's talons. Then let the falcon be left to its own devices as it overcomes the birds, but be careful lest the birds harm it with their claws or beak, for if a young falcon feels pain from the birds, it will be made cowardly. If, however, it overcomes and kills them frequently and without harm to itself, it will acquire boldness and cruelty toward birds. Let these things take place always with the encouragement of the falconer and in the presence of the dogs. The birds should be changed continuously, and stronger ones should be provided as time goes by.[197]

When it has been readied for bird hunting, let the matured falcon be let go after the birds at dawn, when the sun has just risen. If it is found to be bold and well disposed to hunt, let it be cared for and kept in the same care and circumstances in which it is. Let it feed on the birds which it took at that time until it is full. Let this occur at every hunt for three or four days.

---

193. "Cleansed": *purgaverit,* perhaps indicating defecation. Scanlan's "dry and preen its feathers" (248) and Kotsiopoulos's "to dry" (1969, 65) are surely incorrect.

194. "First sleep": perhaps referring to the monastic habit of rising in the middle of the night to chant matins and lauds and then returning to sleep (Glossary, s.v. "Hour"). The term is also found in Dancus Rex 16.7.

195. "Hood": *pillea,* imperative. Since the technical jargon of falconry is of interest, we will try to provide it when appropriate.

196. As W&F comment (xl), falcons hatched by hens are of little value and the young should be left in the nest as long as possible, conclusions that are generally accepted in modern falconry.

197. See Fred. *DAV* 2.2, which condemns the use of live birds and emphasizes use of the "lure."

If, however, it happens to be found to be slow and not willing in the first hunt, take it back to the hand and stop the hunt. On that day, feed it only with half a cock's leg and put it in the dark. On the next day, place half a cock's leg in cold water and keep it there until the third hour.[198] Also put in three of the purgatives which the Germans commonly call *guel* and which are sometimes made out of feathers but are better made of cotton.[199] At the third hour give the falcon the half leg along with the three purgatives just mentioned. When it has eaten them, put it in the dark until evening and then feed it again with purgatives as described. In the morning, go out to the hunt and if it captures boldly and greedily, keep it in just this very state of bodily starvation. If, however, it does not even then show itself as willing for the capture, take it out of the hunt again. On that day, give it to eat only three of the purgatives we have mentioned, taken from cold water, and nothing else. And if, on the next morning, it is still unwilling to capture, feed it with the leg of a small chicken that has been placed in strong vinegar with three purgatives made of cotton. Fed in this way, place it in the dark until evening. Afterward, hold it on the hand until the first sleep. Then heat some water and bathe it in warm water, placing it afterward, if it is not stormy, under the open sky in clear weather until morning. Then warm it at the fire on your hand. Go forth to the hunt then, for if it will not capture precipitously then, it is certainly sick and languid. This method, which we have taught here, is called "starving the falcons" [*maceratio falconum*]. Note that people make their purgatives differently from the way we spoke of. They take meat and, having placed it in strong vinegar, work it into a powder made of ground pepper, mastic, and aloe, and give it to the falcon. Such purgatives, however, should never be given to any bird of prey unless its innards are full of phlegmatic, viscous humor.

The regimen of health is this: that the wisest falconer will feed the falcon with the same food, at the same hours, and in the same amounts which the bird was accustomed to eat when wild.[200] Let this be done especially with light flesh still steaming with the warmth of life and let him keep the falcon between leanness and obesity, since too great a leanness weakens its strength, detracts from its boldness, leads to cowardice, and produces a noisy falcon. Thus, when it is let go from the hand, it sits on the earth near the falconer and cries. Obesity, on the other hand, causes slowness and disdain for the hunt. Let therefore the middle ground be kept so that the falcon does not fail to exult in its powers and so that the deprivation caused by its emptiness

---

198. This is the *carnes madefactas* of Fred. *DAV* 2.48 (W&F, 157), called "washed meat," a process used by falconers to extract the juices. The meat is then used to reduce the weight of the falcon and as a purgative.

199. *Guel* was a "casting," a "laxative dose of fur, feathers, wool, etc., given the hawk with her food to evacuate her bowels. This dose is afterward vomited up as an oblong or ovoid pellet enclosing any indigestible food lying in the gut" (W&F, 616). Cf. Fred. *DAV* 2.48. Cf. MG *Gewolle*, "made from wool." The Latin is faulty here, but from a statement just below we can be fairly sure that the *guel* was also soaked along with the chicken leg.

200. See Fred. *DAV* 2.33–35 (W&F, 129–36) on the feeding regimen.

urges it on against prey, not as the result of a defect but of the desire roused by natural hunger. This best occurs when it is not fed a second time unless the preceding food has been digested and excreted. In these matters let the skilled person use his judgment, since some falcons capture better if they surpass the mean and tend toward obesity rather than when they are between the mean and leanness. Others, to the contrary, capture better if they are thin. None, however, captures when worn out by hunger, just as none captures well if sated with too great an obesity.

Moreover, falcons have diverse complexions and very many diverse species. Those which are black have the sign of melancholia. These should be given bloody food which is warm and moist, such as the flesh of chickens, pigeons, kids, and the like.[201] If they are undergoing medication, use something like pepper, aloe, *paulinum,* or something of this sort.[202]

White ones are phlegmatic, frigid, moist, and have bad chyme. One should give them warm and dry things in their food as well as in their medication. Examples are the flesh of he-goats, little crows, kites, the jackdaw, sparrow, and the like, as well as pepper, cinnamon, *galanga,* and the like.[203] The one which has red feathers is full of inflamed blood and to these cold and moist things should be given, since cold things destroy dry things. Examples are chicken flesh, aquatic birds, and, on occasion, crayfish [*gambari*]. Also included are purging cassia and pith of the tamarind, and all these things should be given in vinegar.[204]

There are also certain falcons which have both noble and ignoble members of the same species. For these, adhere to the following regimen. The noble falcon in any species of falcon is the one with a moderately thick head, flat on top but round elsewhere. It has a curved, fairly thick beak that is moderately long, full shoulders, long wing feathers and legs, and wide feet that are spread and thin, and it is one which is accustomed to care for its feet frequently. The ignoble one in the same genus is the one lacking one or more of those traits. The ignoble one, however, is sometimes as good or better at the hunt than the noble and the falconer ought to consider this.[205]

The falconer ought to take care in guarding their feet and talons, lest he ever allow a falcon to stand or sit on anything except living rock or a wall. Let him be careful lest it stand on lime or on limed cement. For this reason I do not approve that certain

---

201. The Latin here is marked by several bits of almost impossible grammar. Yet the sense is clear from comparison with Borgnet.

202. Latham lists *paulinum* as a laxative drug, citing a source contemporary with A. The translation offered may be incorrect. The Latin reads *piper aloes paulino.* The first and third words are abl., as they should be after the verb *utor. Aloes* should properly be gen., and the sense would be "*paulinum* [made?] of aloe," as in *aloes lignum* (*De veg* 6.11–15). Yet Borgnet reads *aloe,* a proper abl. Note that *LLNMA* also lists *aloes* as a prepared ointment.

203. *Galanga,* an aromatic shrub, is identified by Jessen (1867, 395) as *Alpinia galanga* at A.'s description of it at *De veg.* 6.114.

204. The *gambarus* may be a lobster but is most likely a crayfish or, less likely, a crab. Cf. the form *gamarus* at 4.18; 15.35; 21.37, 46. *Gambarus* may be found at 24.22(31), 38(67), 39(71), 41(74).

205. See Fred. *DAV* 2.22 regarding the saker, where he makes the same judgment.

people keep their falcons on perches and others in wickerwork cages. For art should imitate nature and wild falcons are always found to sit on rocks or on the ground.

Let him be careful of feather breakage by strengthening them with warm water every third day lest they dry out too much. On these same days feed them a bit of aloe since it comforts and purges the stomach and intestines and strengthens the feathers. If too much humor is the reason the feathers do not have their requisite strength, let the flesh which they are fed be placed first for two hours in a juice made from ground radish [*rafina*] and earthworms. These both dry out and harden the feathers. But let him be especially careful lest the feathers of the wings or tail are harmed by any violent injury.

These things, then, and things of this sort are to be considered in the regimen of falcons and other birds of prey as well.

CHAPTER XVIII

On the Different Cures for Illnesses in Falcons According to William the Falconer

Let us now speak of the medicines for diseases of birds of prey and, in order to do this in a more orderly fashion, let us put first the medicines for falcons, then those for goshawks and sparrow hawks, and third let us treat the medicines that are proper to each in common.[206]

The first illness of falcons is a complaint of the head which in humans is called *soda*.[207] The sign of this is that the falcon closes its eyes and moves its head in different directions. At this time, take lard and mix it with ground pepper, and give it to the falcon to eat. On alternate days, let a bit of aloe with chicken meat be given because this complaint of the head arises from a stomach vapor and is therefore cured by having food such as this purge the stomach.

When, however, it opens its beak as if it suffers from stretching and shaking and it strikes its foot with its beak or vice versa, it is a sign that it has a bad humor in its head.[208] It should therefore be cauterized with a silver or golden needle at its nostrils so that the humor can come out. When the humor has come out, let the cauterized spot be smeared with olive oil or with butter if oil is not at hand.

When the bird sneezes and sprays water from its nostrils, it is a sign of an overly moist brain. The cure for this is to crush three stavesacre seeds with the same number

---

206. The source for these cures actually seems to be Dancus Rex (Tilander, 1963, 63–93). Oggins (1980, 443–44) discusses A.'s misattribution of sources and the parallels between this chapter and Dancus Rex are discussed by Lindner (1962, 1:25).

207. Latham lists the earliest use of *soda* as 1235 A.D. and traces its origin to the Arabic. See Scanlan (253).

208. "Stretching and shaking": cf. 6.22 and 19.22 for these *extensiones et alices*. Dancus Rex 2 (Tilander, 1963, 64) says that it opens its beak and moves its side and flanks.

of pepper seeds in a stone or copper mortar.²⁰⁹ The powder made from this should be softened with strong vinegar and put into the nostrils and onto the palate of the bird with cotton. Afterward, feed it with chicken meat.

When the falcon's neck is swollen, it is a sign that a warm gout is flowing into the members of its neck.²¹⁰ The neck must be plucked and it must be bled from the auricular vein so that the gout may be drawn out, causing the growth of replacement feathers.²¹¹ Let a frog be given to the falcon to eat. If it eats it, it is certainly well.

80 When it has a swollen gullet (that is, the artery which is called the *canna*) and when it puffs up as if it is going to suffocate, it is most certainly suffering from rheum.²¹² This is the cure. Take one ounce each of peacock blood, nutmeg, black myrobalan, almond, *gariophilis,* cinnamon, and ginger, and make nine pills.²¹³ Let one be given daily at the third hour and afterward at the ninth hour. Feed the falcon the flesh of a mouse.

A signal of a malady in the kidneys as well as of the gout that pours into the kidneys is seen when a falcon is not able to leap and hurl itself far off the hand with its wings outspread and to return as birds are accustomed to do. This gout is called "deadly" [*mortalis*] by some.²¹⁴ Take the seeds which grow on the white thornbush called *hagedorn* in German and which are red.²¹⁵ Pound them and mix them with the hairs of a rabbit and mix the whole thing with cooked flesh. Feed the falcon with such foods for up to nine days. If it retains food such as this, it will get well.

---

209. A.'s *stafisagria* is *Delphinium staphisagria,* a larkspur whose seeds contain such a violent emetic and cathartic that it is used to poison fish in Asia. See *De veg.* 6.436.

210. "Gout": *gutta,* lit., a "drop," indicating that the bad humors drip into the affected area.

211. "Auricular vein": the *vena auricularis* is something of a mystery. Scanlan (254) identifies it as "a superficial vein located beneath the bird's ear," but he cites no source. The term is not in Fonahn. Etymology could support this idea, as in the *digitus auricularis,* or little finger, so named because it was used to clean out the ear. Yet one suspects a corruption since Dancus Rex 4 has *vena originalis,* itself a troubled term (discussed by Tilander, 1963, 266) and thus readily corrupted. Note moreover that the phrase with the vein's name is found in the margin, an addition by A.'s M1.

212. The windpipe is often compared to a reed (*canna*) and was thought of as an artery because it carried *spiritus.* The word translated here as "gullet" is *guttum* and is almost surely corrupt. Dancus Rex 5 uses *gorgia,* crop, but some MSS entitle that section "On the gout (*guta*) which arises in the *gorga*" and the two words may have become conflated in later versions. Scanlan (254) identifies this *rheumatismum* as the disease of falcons known as frounce or roup.

213. "Nutmeg": *muscata,* the *Myristica moschata*; see *De veg.* 6.146. "Black myrobalan": Jessen identifies *mirabolanus* as *Terminalia chebula,* an eastern tree, known for a variety of fruits; see *De veg.* 4.162f., 6.141f. Scanlan (255) takes this further and identifies the compound phrase *mirabolanis chebolis* used here as identifying a specific type of myrobalan, which is the fruit borne by this tree. *Gariophilus*: identified by Jessen (1867, 720) as *Caryophyllus aromaticus,* a fragrant, clove-like plant. Scanlan's identification (255) is that for its classical manifestation and is taken from L&S (802).

214. Scanlan (250) identifies the disease as botulism. Tilander (1963, 262) merely says it is a disease of the bird's head and kidneys.

215. MG *Weissdorn, Crataegus oxyacantha,* known in English as hawthorn. See Lexer (1:1143).

There is another gout which is called *silera* by some.[216] When it begins to run through a falcon's body, it has the nature of a poison, and the ends of the falcon's beak and talons begin to grow white. This is cured by capturing the black snake called the *tyrus* and by cutting off a palm's width from its head and tail.[217] These parts are thrown away but what is left is to be roasted in a new pot and the fat which is liquefied is to be collected and given warm with peacock flesh to the falcons for eight successive days. Afterward, take a piglet and, removing its hair with hot water, give the tender part of its chest along with a small mouse to the falcon to eat. If it digests this food well, it is without a doubt healthy.

When a falcon raises its foot and constantly strikes its leg with its beak, you should know it has salty gout [*gutta salsa*]. At such a time bleed the vein which is between the leg and thigh and it will be healed.

When a falcon is beset by lice, take some quicksilver and mix it with human saliva and with ash until the quicksilver becomes still.[218] After this, add some old lard onto the whole thing and mix it all together. Then smear the head of the falcon and put on some ribbons, tying them to its neck and limbs. At this the lice will die. Another way: Take pepper and sesame seeds, grind them together and cook them in a new pot, add water, and then wash the bird and it will be made well. Or another way: Cook larkspur [*stafysagria*] in water and allow the bird to bathe in it. Afterward, spread a white linen cloth beneath it on top of the grass or a rock, for the bird will shake out all the lice onto the cloth. Every bird of prey does the same.

The sign that a falcon has a fever is if his feet are warm beyond all measure. Then, take aloe and chicken fat, place it in strong vinegar and give it to the bird intermittently to eat. Another way is to give it the shelled creature [*testudo*] called the *limax*. If it keeps this food down, it will be healed.[219]

When a falcon captures food and tears it with its beak but right away or shortly thereafter regurgitates it unchanged by digestion, then it certainly has that hard viscosity in its crop, stomach, and intestines called "stone" [*petra*] by some. Then, make a powder of *gariofilum* and sprinkle it on the flesh of a sparrow.[220] Give the flesh to the

---

216. DuC, citing this passage, lists *silera* as a form of flux. Dancus Rex 7 offers the name *philera* in all its medieval spellings. Tilander (1963, 250–51) discusses all the forms in some detail and describes it as a disease of the liver and gall bladder, an interpretation borne out by A.'s comments at 23.105. Below, at 23.94, citing William, the form is given as *fellera*. Therefore, despite the multiplicity of forms, the etymology is probably linked to the stem for "bile" (Scanlan, 273). None of the forms appear in Latham², s.v. *gutta*, where many varieties of this disease are listed. W&F (618) list a disease which may have affected the name of this one, viz., *filaria*, intestinal parasites (i.e., "thready ones").

217. On the deadly *tyrus*, see 25.42(59).

218. The Latin is ever so more descriptive: "until the living silver dies."

219. St. conjectures a helix (edible land snail). On *testudo* and *limax*, see 4.38.

220. Scanlan (256–57) suggests this is the "pip" and represents a respiratory disease. W&F (624) suggest an "internal avian disease accompanied by indurations of the intestines." Tilander (1963, 268) gives *lapis* as a synonym and describes it as swelling of the bird's crop. Part of the confusion is because A. here combines two entries in Dancus Rex 11–12, viz., *petra in magone* (stone in the crop) and *petra in fundamento* (stone in the fundament).

bird every other day and on the other days give it a young pigeon or a purgative.[221] If it will keep to this diet, it will be healed. When it does not defecate for a long time it is a sign of the same illness. Give it then the heart of a pig with pigs' bristles cut up fine into it for up to three days and the stone will be freed.

If a falcon is afflicted with worms in its belly and something of the same appears in its excrement, let filings of iron, especially of purified steel, be sprinkled over pork flesh and given the falcon for up to three days and it will be healed.[222]

82

If a *tinea* is consuming its feathers, take red wax, *muscatum*, the yellow myrobalan, rock salt, gum Arabic, and grains of wheat.[223] Place all this in strong vinegar and leave it there for nine days in a vase or basin. After this, put it up in a bottle and from this wash the falcon (or any other bird of prey) daily until you feel it appears well and cleansed. Afterward, wash the bird with rose water and it will be healed when it is placed out in the sun after the bath. Place equal amounts of the aforementioned ingredients in the vinegar. But use far more of the wax than the others for this is what holds all the others together.

Some say there is another way.[224] If the *tinea* is drawn out of the skin with a needle and then washed with aloe, and then after this *tinea* is removed the falcon's wing is cleaned in a bath with rose water, it is very effective.[225] Take care, however, that it does not touch itself with its beak as long as the aloe is on it, because it is very harmful.

The falcon sometimes suffers an infusion [*infusio*] like a horse.[226] An indication of this is when it does not accept its food and has enlarged eyes as if they were raised

---

221. "Young pigeon or a purgative": *pipio sive purgatorium*. Pipio as "squab" is common; see Tilander (1963), 269 (*pipio*) and 272 (*pupio*). Yet why does the treatment call here for either something mild or a harsh purgative? The opposites are discomforting. Scanlan's (257) "pip-pellet" is thus tempting, and one way to translate the Latin is "a *pipio* (that is, a purgative)." Yet Scanlan's appeal to German equivalents to explain this unusual sense of *pipio* may fail in light of the parallel text of Dancus Rex 11, which says plainly, *da ei pupionem*, "give it a young pigeon."

222. A. discusses steel at *De min.* 4.8 (Wyckoff, 1967, 235–36).

223. On *muscata*, see 23.80, where the black myrobalan is also discussed. At *De veg.* 6.141, A. tells us that there are many types of the fruits of this tree, saying the yellow variety is immature. Scanlan (257) identifies it as the "cherry plum," *Prunus cerasifera*. On rock salt, see *De min.* 5.2 (Wyckoff, 240).

But what, exactly, are these all intended to cure? Tinea is a difficult word to pin down (Beavis, 1988, 136–37). It most commonly means "moth," and especially "clothes moth," as at 26.41 (43). Yet the picture of falcons being eaten away by moths does not bear serious credence. Scanlan (257–58) therefore opts for mite or tick, but this is unlikely. The solution lies in the second cure A. lists below, where the *tineae* are picked off the bird with a needle. This is most ineffective for the very small mite but very effective for worms, and is so prescribed at 22.33. Rereading 26.34(43) makes it clear that A. believes the larval stage alone is born from the fabric and only later turns into a moth. What we have here, then, is a worm-like pest representing the larval stage of some other creature. Cf. 23.87, 93, 98, 105.

224. It is interesting that the first cure is from Dancus Rex 14 (Tilander, 1963, 78–79). The "needle cure" is to be compared, however, to *Guillelmus* 14 (Tilander, 1963, 146–49).

225. This passage is quite difficult and almost seems corrupt. Kotsiopoulos (1963, 69) omits it.

226. Cf. 22.91, where the form *infundatura* is used, for a description of this disease as brought on by improper diet following strenuous exertion. Dancus Rex 15 calls it *infundeso* and Tilander (1963, 256) lists several other variants.

up by windiness.²²⁷ Make some lye from vine branches and strain it three times.²²⁸ Then let the falcon's gullet be filled with it. Let it be until, by excreting it, it shows it has digested what it has eaten. Later, give it a lizard to eat. Or, another way: Take some warm wine with ground pepper and pour it into the gullet of the bird. Let the falconer hold the falcon until it excretes the food with which it was force fed and it will be relieved.

A falcon's feet that are swollen, but not as a result of a violent injury, signify gout. Against this take equal, one ounce portions of butter and olive oil and one ounce of aloe. Having mixed together all these, smear its feet three days and place the bird in the sun. Give it the flesh of a cat to eat and it will be better. Or, another way: Light some lint made from cotton paper and with this cauterize the sole of the falcon's foot.²²⁹ Place it on living rocks smeared with old lard and it will be relieved. Meanwhile, give it some mice to eat.

If a falcon combs itself and scratches itself with its feet and if it pulls feathers from its tail and casts them away, it is suffering from itch [*pruritus*]. To counteract this take goose and sheep excrement and aloe in equal portions.²³⁰ Put in strong vinegar and place in a brass vessel beneath clear skies and a warm sun for three days. If the heat of the sun is not to be had, place it before a mild fire. Bathe the entire falcon with this, feed it pigeon's flesh with honey and pepper, and place it in a dark place. Let this be done in this way for nine days and when a good feather appears, coming in on the tail, wash the bird with rose water and it will be cured. If, however, sharp and keen gout is showing in the bird, take the excrement of a goose or a pigeon and the bark of an elm root. Boil the root bark until the water turns red and then put in the excrement to soften. Then wash the falcon for three days and it will be given relief.

If a falcon should become wounded, take the white of an egg and olive oil. Mix these together and place them over the wound. Be careful that the injured place is not touched by water. When you wish to change what has been put over the wound, wash the place with warm wine. You should treat it this way until the wounded place acquires a crust or scale which closes the wound. If the falcon touches the wound

---

227. The vocabulary used here, inherited from Dancus Rex, may indicate enlarged, heavy, or coarse eyes. The mention of "windiness" seems to indicate a puffed up condition.

228. One presumes a lye wash is meant. Cf. the use of vine branches at 23.90.

229. "Cotton paper": *carta bombacina*. Scanlan (259) points out that the literal translation "silk paper" is unlikely and opts for "flax paper." Though paper made from linen rags was used in the Islamic world (*DMA* 12.698), Irigoin (1987, 390) states the truth of the matter succinctly: "To find a name for the new writing material Europeans often looked to its geographic origin. In the Byzantine Empire a paper book was called *bagdatikos* or *bambukinos,* signifying paper from Baghdad on Manbij (ancient Bambyke, on the Syrian border). This latter name was often confused with adjectives describing textile materials: *bombakinos* (cottony) or *bombukinos* (silken, 'bombazine')." Thus, since *Guillelmus* 16 (Tilander, 1963, 148) is probably the source of this passage, we are left to wonder whether he meant "cotton" paper or, less likely, Baghdad paper. On the confusion in the adjectives, see Latham², s.v. *bombycinus* and *bombyx,* where the equation with cotton is clear. On *coquo* as "cauterize," see 23.84.

230. Reading *accipiatur* for St.'s *accipiater*.

itself, place a bit of aloe there. If, perchance, it is wounded under the wing or on the chest, ribs, or thigh, put on a thick covering of tow that has been well scraped with a knife and do this until the bad flesh is eaten away.[231] After this, take incense and wax in equal proportions as well as tallow and resin. Soften all this in a pot near the fire and set aside. When it is needed, liquefy it at the fire and smear the bird with it. Let the smearing be done with a feather until the wound is closed with a scar.[232] If bad flesh appears, then apply Greek nettle [*urtica Graeco*] or the green from copper until it is eaten away.[233] Afterward, smear the place with white ointment and it will be healed.[234]

84   If a bird of prey has to be cauterized, it is the wisdom of the ancient Greeks that the first cauterization should be done beneath the tear duct of the eye, for this is a benefit to its vision.[235] The second one takes place on the top of the eye and is of benefit to its head. The third takes place over the joint of its wing and is effective against gout. The fourth takes place on the soles of the feet and is likewise effective against gout of the legs.[236] All these cauterizations are more useful if they take place in the month of March.

If putrid matter flows out of the falcon's nostrils and the falcon cannot eat, and the matter itself gives off a foul smell, there is an ulcerous spot [*fistula*] in that place and it should be treated as follows.[237] Pluck the back of the bird's head a bit and smear it with lard or, if you have none, with butter.[238] Cut the vein which goes from the nostrils to the eyes and cauterize the cut vein on the side away from the ulcer with a heated iron needle. The spot is to be thoroughly smeared each day with butter and the falcon put in a warm place for nine days and it will be healed.

---

231. "Tow": *stuppa crossa*, the exact phrase found in Dancus Rex 23. Tow is the fiber of flax, hemp, or jute, but *crossa* is more difficult. Scanlan (260) wishes to emend to *crassa* (coarse) or *grossa* (thick). Tilander (1963, 246) lists certain MSS of Dancus Rex which read *grossa*. What we have, then, is not a misprint but merely a medieval variant spelling.

232. "Scar": lit., "wart," *verucca*, probably emphasizing the raised nature of the new flesh.

233. Greek nettle is identified by Jessen as *Urtica urens* at *De veg.* 6.461. *Viride aeris* is the patina which forms on copper, sometimes called copper rust. For A.'s opinions on its formation and properties, see Wyckoff (1967), 119, 226.

234. Scanlan (260): "In the Middle Ages, white ointment was made from lard or lanolin with a small amount of beeswax. Later petrolatum derived from petroleum supplanted the animal fats and became the main ingredient in the white ointment of the Pharmacopeias." Tilander (1946/47, 179–82) quotes two recipes from the fourteenth and fifteenth centuries which are much more complex and he comments on the ointment's many uses, including that of female makeup, and on its origins.

235. "Cauterize" is the translation used here for *coquo*, lit., "to cook."

236. Here A. uses *guttam crurium*, indicating the lower leg (shin).

237. Scanlan (261) identifies the disease as *Trichomonas gallinae*.

238. A.'s text is most odd at this point, for *Accipiatur posterior pilus capitis* should mean "let the rearmost cap of its head be taken . . ." Scanlan's translation (260) cleverly takes this as a reference to the down feathers. Dancus Rex 27.3 says *Depila eum aliquantulum retro caput*, which forms the basis for the translation offered here.

If a falcon loses a claw from its roots, the claw does not grow back. The toe has to be bound to a white mouse that has been split open and it has to be smeared with marrow from the foot and toe of a pig until it is healed. Likewise, when a falcon is bathed, be careful that it is not put on rotten wood so that it is not poisoned. If, however, the falcon is poisoned, take some *tyriaca* and three pepper seeds and give them to the falcon with ground stone.[239] Watch over the falcon for nine days and then put *tyriaca* and pepper seeds in a clay vessel. Sprinkle the powder over some heat and feed this to the falcon.

If some wild creature bites a falcon, the place where the bite is should be deplumed. If the wound is small, it should be made larger with a razor. Then smear it with warm butter. Then make a mixture of frankincense, resin, wax, and tallow, and smear the bird with it until it is healed.

In this section, then, which we have set forth on healing, we have especially followed the expertise of William, the falconer of King Roger, adding a few things of our own.

CHAPTER XIX

On the Cures of the Illnesses of Falcons Which Are Different from What Has Gone Before, According to the Falconer of the Emperor Frederick

Some, however, understanding differently the teachings of the falconers of the Emperor Frederick concerning the cures of falcons, have thus made the following determinations.[240]

If the falcon's head has to be purged, take some very pure pitch the size of a bean and warm it in your fingers at the fire. Afterward, rub the falcon's palate with it until it sticks there. Next, take four seeds of stavesacre and the same number of white pepper and make a very fine powder. Place this over the pitch adhering to the palate of the falcon. Place what remains of the powder in the falcon's nostrils. When the sun has reached a good heat, set the bird out in the sun until all the badness in its head and the phlegm flowing from its head are purged in a frothy flow. For two hours a day, feed it with bland, sweet flesh.[241]

---

239. On the famous poison antidote called *theriaca*, see 1.102, 2.119. Yet another valid meaning is "treacle" (Latham, s.v. *theriaca*), which might better suit a simple of this sort. Cf. *De veg.* 6.455.

240. What follows is largely based on *Gerardus Falconarius* (Tilander, 1963, 198–229). On the identity of this falconer see Oggins (1980), 442–44.

241. "Sweet flesh": *suavi carne,* a corruption of *Gerardus* 3.6, *suina carne,* "swine flesh." See Tilander's apparatus (1963, 203).

To tighten up the palate if perchance some humor from the head is flowing into it, take equal amounts of old butter and celandine [*celidonia*] and give the powder made from them on warm flesh to the falcon.[242]

Against a spot on a falcon's eyes, take equal amounts of pepper and aloe and place them, ground, over the spot. If it is the right season for you to get the acacias which are the fruit of the wild thornbushes, drip three drops of the juice of the acacias onto the spot, for it will be very effective.[243]

If the lungs or trachea [*canna pulmonis*] of a falcon is injured, take one ounce each of sparrow and mouse excrement, five grains of white pepper, two ounces of rock salt, one ounce of raw wool, and grind it all. Mix it with six drops each of honey and pure oil and with nine drops of the milk from a mother nursing a male child she herself bore, add enough butter to suffice, and mix it all together. With the milk form three pills about the size of a filbert nut. Put this into the falcon's gullet and hold the bird on your hand for two hours so it will vomit out the entire potion. When it has vomited it all out, put it, after a bit, near water. If it drinks from the water, then a bit later feed it the lung and heart of a suckling lamb which is not yet eating grass and let the flesh be as warm as it can be. Later, feed it fully with sweet meat. In the late part of the day, give it sufficient portions of sparrow and chicken and it will be healed.

If it should be that it has *bulsus* (that is, it is infirm in the lungs), take one ounce of crushed orpiment and nine pepper seeds, crush them together, and serve them to the falcon with warm meat.[244] Another way is to take three pieces of lard, such as a falcon can swallow, and dip them in honey. Next, sprinkle iron filings over them and place them in the falcon's gullet. Do this for three days and serve it nothing else at all. On the fourth day, take a small pig and get it very drunk with strong, clear wine. Then, once it is drunk, warm its chest at a fire. When it is well warmed, strike its chest so that the blood ascends to the chest and is mixed with the wine. Then kill the pig and immediately dip the chest that is still warm into goat's milk. Feed the falcon for three days with such food and it will be healed.

If *anguillae* (that is, long worms) are eating at the falcon, take the well-washed *pudillum* (that is, intestine) of a tender chicken and make three knots in it, each about the distance apart as the midpart of a thumb.[245] Tie off each end firmly with thin

---

242. "Celandine": *Chelidonium majus*; cf. *De veg.* 1.176. A perennial herb having yellow flowers, also called swallowwort. Scanlan (262) discusses its medicinal effects.

243. "Wild thornbushes": *Acacia quae sunt pruna spinarum silvestrium.* Jessen (1867, 700) identifies *acacia* as *Prunus spinosus*.

244. Tilander (1963, 238) describes *bulsus* as difficulty in breathing caused by a lesion on the lung. Orpiment is discussed by A. in *De min.* 2.2.6 (Wyckoff, 1967, 92), where he tells us its other name was *falcones*. Its use here might, then, be sympathetic magic through the name.

245. *Anguillae* is, lit., "eels," an apparent absurdity which prompted the gloss. *Lumbrici,* for A., is normally earthworms. One would have expected *vermes*. Scanlan's (264) suggested identification of these pests as nematodes is insightful, for one particular nematode is called an eelworm. Note that *Gerardus* 8 (Tilander, 1963, 210) is entitled "If 'eels' are eating at the falcon internally." *Pudillus* is a variant of the more common (but non-Albertian) *bodellus/budellus,* which gave rise to the English "bowel." On nematodes in falcons, see Davis et al. (1971), 221.

thread and fill with very clear oil.²⁴⁶ Put this in the falcon's gullet in the same way as any other potion is usually placed. If, however, it is still bothered by worms the next day, take one part each of ivory shavings and of the dung of the Indian sparrow [*passer indicus*] (that is, the solitary sparrow [*solitarius*]).²⁴⁷ If this bird is not available, take the dung of another, common sparrow. Take one ounce of each, pulverize these well, and serve them with warm flesh to the falcon. On the third day, if the worms still persevere, remove the raw skin of the fish called the *tinea* [tench] (which the Germans call the *sligin*) and reduce it to ashes in a clay pot over charcoal, without flame or smoke.²⁴⁸ Take this, with ivory shavings and sparrow excrement, all in equal amounts, and grind them up. Serve these things to the falcon, mixed in with warm meat. If it is necessary, on the fourth day add iron filings and ground *git* to these things and serve the whole, pulverized, with warm meat.²⁴⁹

To counteract all diseases of the kidneys, give it, in the heart of a chicken, a powder from the *candria,* or if this is not available, powdered watercress [*nasturtium aquaticum*].²⁵⁰ Another way is to take equal amounts of the tree called *quercus* [oak] and of *bolus*.²⁵¹ Serve these ground up with the warm flesh of a chicken made drunk with wine and it will be cured of every ailment of the liver.²⁵²

87

Against lack of food and dryness, bring an egg to boil, without smoke, in goat's milk in a very clean pan until it is hard, and give this to the falcon to eat. If it digests this it will be relieved.²⁵³ This is good against any infirmity.

If *tineae* [external worms] are devouring the wings of the falcon, put some pure balsam in the openings out of which the feather fell. Know, then, that the *tinea* will be removed and a new feather will come in. Furthermore, take one ounce of the juice of pulverized oriental saffron [*crocus orientalis*], three spoonfuls of fresh goose excrement that has been strained through a cloth, and the same amount of very strong vinegar.

---

246. Note that the directions are impossible to follow in this order.

247. On the *passer solitarius,* see 23.137(99) and 14.2. St.'s identification is the *Monticola solitarius,* a blue rock thrush.

248. *Tinca tinca* (*Tinca vulgaris* Cuv.), the tench, a common freshwater fish that is known for its ability to survive out of the water for an extended time. Cf. MG *Schlei* and Sanders (438).

249. Jessen (1867, 720) identifies *git* as *Agrostemma githago* and as a synonym for the *nigella* of *De veg.* 2.135 and 6.396. This plant is the corn cockle. In older texts, such as Pliny, *git* was a form of coriander.

250. According to Tilander, *candria* (239, s.v. *camandrea*) is the plant "wall germander," *Teucrium chamaedrys.*

251. Scanlan emends the text from *bolus* to *ebolus,* dwarf elder, a common ingredient in A.'s simples. But at *De min.* 2.2.16 (Wyckoff, 1967, 114), A. tells the reader that *ramai* is another name for what is usually called *bolus armenicus,* a reddish clay, supposedly from Armenia and used for a variety of intestinal disorders.

252. The mention of liver here, whereas the entry began with kidney ailments, shows that *Gerardus* 9 and 10 have become conflated.

253. "Digests" (*digerit*) is an interesting variant. *Gerardus* 11.3 (Tilander, 1963, 214) states that the bird will be healed if *smaltaverit,* which means "defecates." At 23.81, the same verb came to A. properly as *egerit*; here the spelling became changed to *digerit.*

Put this with its flower into a brass pot until they thicken.²⁵⁴ Then thoroughly wash three times the places out of which the feathers fell, using some pure vinegar, and smear the places with this. There is also another way. Burn some leeches on tiles and make a powder of them. Put the feathers of peacocks over the smoke of a fire and make a powder from the soot that sticks to them, producing a similar weight of soot as of the foregoing one. Then make it a liquid with strong vinegar, but not too much. Then wash with strong vinegar the places out of which the feathers fell. Next, dip bits of lard into the mixture you made of the powders until the powder sticks to it. Twice a week, smear this around the places where the feathers fell out until the feathers return. Likewise, serve as food to the falcon the long hairs of a horse which have been ground very finely over meat. Likewise, soften with radish [*rafanus*] juice and strong vinegar a powder made of *piretrum*.²⁵⁵ Then smear over the worms and the bird will be healed. Still another way: Burn a toad, make a powder of it, and serve it to the falcon on its food. Serving the falcon iron filings with its meat for food is effective against the same problem.

If a falcon has growths on its foot, dry out the middle bark of a juniper, make a very fine powder of it, and give it to the bird with its food every other day for nine days and it will be cured.²⁵⁶

If it should suffer swelling of the feet, grind aloe together with the white of one egg. Then find a whetstone on which iron has frequently been sharpened and to which some of the iron still clings. Rub the aloe and egg white powder over this stone long enough that each sharpening which was done on the stone will transfer into the powder. Then place this on the swollen feet until it forms a crust and sticks to the feet. On the next day smear it thoroughly with soap and on the third day do as you see fit.

If bad flesh arises somewhere on the falcon's body, take equal amounts of lime and aloe, make a powder, put it on, and the bird will be healed.

To cure the feet of a falcon take equal amounts of milfoil [*millifolium*, yarrow], saxifrage [*saxifragium*], vervain [*verbena*], and plantain [*plantago*].²⁵⁷ Give this to the bird, ground to a powder and with warm meat, and it will certainly be cured.

If the falcon has an inordinate appetite, take equal amounts of mouse blood, honey, and parsley seed, and mix them together. Give them to the falcon and it will be cured.

---

254. For the saffron, *Crocus sativus*, see *De veg.* 6.297. The "flower" is a problem. Scanlan (265) takes this as the film which forms on vinegar. It could also be the flower of the saffron plant. Cf. *Gerardus* 12.5 (Tilander, 1963, 216).

255. Jessen (1867, 550), commenting on *De veg.* 6.411 and the plant *piretum*, equates it to the *piretrum* cited in this very passage, identifying both as *Achillea ptarmica*, sneezewort, which takes its name from its snuff-like qualities. On *rafanus*, see 23.94.

256. These growths may be the cystic tumors found on the feet of domesticated birds (W&F, 428). *Gerardus* 13 (Tilander, 1963, 218) calls them *porri*, "scallions," probably in reference to their shape. Scanlan (266) suggests a secondary infection in a cut on the foot.

257. In order, the plants are generally identified as *Achillea millefolium; Pimpinella saxafraga* (*De veg.* 6.452; Kotsiopoulos, 1969, 80); *Verbena officinalis* (but cf. Scanlan, 266). *Plantago* is a general term, for which "plantain" will suffice, but cf. Jessen (1867), 736.

Know likewise that when you take a falcon and wish to go out to the hunt, you ought to say "In the name of the Lord, may the flying creatures be beneath your feet," and when you take the falcon in the morning say "Whom the unjust man has bound, the Lord has set free by his arrival."[258] When the falcon is bewitched make a powder of a little frog and give it to the bird with warm meat.[259]

So that it will not desert its man, take parsley, black mint [*menta nigra*], and rock parsley [*petrosilinum*], and grind them together.[260] Serve it to the bird with warm meat.

So that the bird will not be injured by an eagle say "The lion of the tribe of Juda has conquered the root of David, alleluia!"[261] These last things, however, are not as reasonable as the first.

These, then, are the things which have been said about the cures of falcons following the experience of prudent people. The wise falconer, however, may add to or subtract from these in time as a result of experience as he sees expedient, given the complexion of his bird. For experience is the best teacher in such matters.

CHAPTER XX

On the Regimen for Goshawks That Are Ill,

According to the Expert Findings of the Emperor Frederick

In this chapter we will set forth the medication proper to goshawks [*astures*], which we called *accipitres* above.[262] We call every bird of prey used in hawking a falcon, since Theodotion, writing to Ptolemy the King of Egypt, placed *astures* (that is, *accipitres*) among the genuses of falcons. Whatever, then, has been omitted in the preceding books, we will set forth here so that the tract might be complete.[263]

---

258. The first charm is based on Ps. 8.7–9, though the text is not that of the Vulgate. The second is not from the Vulgate and is found in *Gerardus* 21 (Tilander, 1963, 228).

259. Scanlan (267) has an interesting note on *ranunculus,* which at first glance would be translated "little frog." He suggests that it may also mean a medicinal plant (*LSJ,* s.v.; Jessen, 1867, 739). *Gerardus* 22 reads *rumicem,* from *rumex,* sorrel. Note also that at 23.79 A. recommends a frog (*rana*) be given to a falcon after a cure has been applied. Moreover, A. uses *ranunculus* at 1.28 in a context which requires that "frog" be understood.

260. Jessen identifies black mint as *Mentha pulegium* at *De veg.* 6.422. On rock parsley, see 22.37.

261. Based on Rev. 5.5.

262. As A. indicates, *astur* and *accipiter* are to be taken as synonyms, though he prefers to use the latter term. For the purpose of this section, the former will be translated "goshawk" and the latter "hawk."

263. Theodotion is a reference to the so-called Symmachus letter. See notes at 23.55 and 71.

Following, therefore, the expert findings of Emperor Frederick, let us speak of what occurs if the goshawk is mangy [*bisticosus*] in its wings.[264] This is a defect producing various signs and is called *hungermal* in German.[265] This disease certainly arises from corrupt bowels and is especially a corruption at the base of the feathers. Make a mixture, therefore, of salt and human excrement, which has the force of *tyriaca*.[266] Dip the hawk's feathers in this, getting it into the base of the feathers where they are attached to the flesh. Do this and it will be healed. Then let it drink water, and dip the food it is fed in the juice of "Jupiter's beard" [*barba Jovis*].[267] Likewise, take mallow and savory [*saturegia*] and cook them a long time with pork fat.[268] Put this into the mouth of the hawk until you use up three spoonfuls of it. Afterward, give it the whole gall bladder of a pig or chicken, along with the warm lung of a pig. Until it is healed, wash it with water in the morning while it is hungry but toward evening feed it with butter.

If, however, the hawk is touched by the cold and is harmed in its chest due to that cold, take larkspur seeds and grind them in a mortar. Add peas and as much honey as is needed, then rub the palate of the hawk and put the hawk in the sun.

Likewise, take the seed of the herb called *radix* [radish] with equal weights of wild rue [*ruta*] and pepper, grind it, and mix it with honey.[269] Then make pills the size of a peppercorn and give it to the bird for three days. Do this as often as the cold touches it or when it produces excrement that is too loose due to the cold.

In the same manner, mix together the juice of horehound [*marubium*], ground pepper, a bit of honey, and sifted parsley seed so that there are two parts of juice to one part of honey.[270] When the bird is hungry and fervently desires its food, give this to it.

In addition, to cure the chest of a hawk, make a powder of mint, mix it with honey, and give it to the hawk in its food. Likewise, grind equal weights of mustard root and chervil and give it with milk and hyssop oil to the hawk in its food because it is very useful. Also, *nasturtium* mixed with honey and given in pork flesh is greatly effective.[271]

---

264. Although A. attributes this chapter to Fred., it is not a part of Fred.'s only extant work. Several times Fred. refers to books which have not survived, either through loss or because Frederick never had a chance to write them (W&F, lxxxvif.). Oggins (1980, 444) believes A. was merely passing on what he had learned from Fred.'s falconers. The sense of the Latin *bisticosus* is fairly clear from context, but the form is unparalleled. Latham² lists *bistucus*, citing Adelard of Bath's *De cura accipitrum*, from the century before A. Adelard apparently thought the disease was one of the lower intestinal tract.

265. Scanlan (268) makes a case for identifying this disease as today's *Hungerraude*, mange brought on by malnutrition.

266. In this context the sense of "treacle" may be best understood. See 23.84, with notes.

267. St. identifies this as a houseleek, *Sempervivum tectorum*, following Jessen at *De veg.* 6.288.

268. On *saturegia* see *De veg.* 6.449, where Jessen identifies it as *Satureia hortensis*. This is summer savory, used in cooking.

269. On the radish, *Raphanus sativus*, see *De veg.* 6.423.

270. On horehound, *Marubium vulgare*, see *De veg.* 6.389.

271. *Nasturtium* is identified by Jessen as *Lepidium sativum*, "pepperwort," at *De veg.* 6.393. If A. means "aquatic" *nasturtium*, then it is watercress.

If, however, the hawk is asthmatic, give it a ground up, fired tile, along with warm flesh and the blood of a he-goat for three days. Take the juice of the wormwood [*absinthium*] and place it between the skin and flesh of a hen's thigh, mix with the milk of an ass, and give this to the bird to eat.

When a hawk holds its food for three days in its gullet without digesting it, make lye from the ash of vine twigs and give it, well strained, to the bird for two days with warm meat. On three other days give it goat's flesh with butter and powdered mastic.

When it does not want to take the food brought to it but rather pushes away the proffered flesh from itself with its beak, put another flesh in its place by giving it the flesh of a crane. Put also one larkspur seed under its tongue and it will immediately vomit up the flesh it has taken in.

If you wish to loosen and purge it entirely, take the herb called *radix* [radish] in which there is not yet any green vein growing and cut it into three parts the size of the pinky finger and trim each to the size of a grain of barley.[272] Give this to the bird to eat, wrapped in butter, and then place the bird in the sun and it will be purged.

In order that the bird always be healthy and that its viscera never be bound up, cook the branches and boughs of mallow in water until the water is used up. After they have dried, grind them forcefully and put this in a vessel full of butter. Let them cook in this, and then strain this as wax is strained. Feed the hawk from the fat left over from the straining, one piece at a time. If, however, it refuses the fat, give it to the bird with the flesh of a cat.

There is still another fact proven by experience for the same condition. Take dwarf elder, rue, mallow, thyme, and rosemary [*ros marinus*] in a quantity greater than the others, or use savin if rosemary is not available.[273] Take also the fat of a pig that has never eaten acorns, grind the whole together, and bring it, mixed, to a boil in wine. Press it out afterward like wax and give it to the hawk in its food late at night and it will not get sick.

If the hawk has had many molts, set it in the mews at the beginning of January. But if it is a chick of the same year, send it to the mews at the beginning of July and feed it live birds if they can be found. Likewise, have it eat cloves and fennel seed with its meat.[274] Seek a molting house [*domus mutae*] fit for it and large enough. When it is finished with its molting, take it out of there. If it does not quickly cast off its feathers in the mews, cook along with some wheat that multicolored serpent which, among the others, has the least venom and which is called the *unc* in German.[275] Feed and water a hen with this wheat and the serpent broth. A hawk that has been restored with the flesh of this hen not only casts off its feathers but will expel a disease if it has one.

---

272. On the *digitus auricularis*, see 1.412.

273. On *ros marinus*, see Jessen (1867), 740. On *savina*, a medicinal juniper, *Iuniperus sabina*, see *De veg.* 6.121–22.

274. "Cloves": *gariofilos*. Jessen identifies this as *Caryophyllus aromaticus* L., at *De veg.* 6.115–17.

275. Sanders (445) points out that the modern *Unke* is a toad.

Well adorned with solid, new feathers, it will live a long time and will be continually healthy and good tempered.

Also, crush to a powder tiny river fish and sprinkle the powder over the meat that is to be given the hawk, and give it to the hawk to eat with mouse flesh. Then without doubt it will quickly molt. Pork loins dipped in lamb's blood, cut up, and given it will more quickly accomplish what has been said. A powder made of a burned green lizard does the same. Likewise, dig up from beneath the earth the seeds of the elderberry tree [*sambucus*] in September. Afterward, serve hens barley that is moistened with these and serve their flesh to the hawk. If you serve other flesh, dip it in the juice of the aforementioned seeds. This is very effective.

In the same manner, serve the hawk cut-up leeches, either by themselves or with flesh. If the bird refuses them, make a powder by burning them and serve it to the bird with its meat. Also, mice given to it alive or broken into tiny pieces and placed in its gullet are beneficial.

92 If, however, it breaks its feathers, cut another feather like it and insert it in the broken feather. If, however, it breaks a feather in the hollow part of the shaft of the feather, sew another hawk feather to it or, if you cannot get one, use some horn.[276] Do this with an iron or copper needle having four angles in the middle so that they are made sharp on both sides. You would, however, learn these things better by seeing them and through practice than through this book's instruction.

If you want your hawk to grow thinner, give it garlic ground with pennyroyal [*polegium*] or have it eat a lean meat portion of salted bacon which has stood in water all night.[277] Afterward, let it drink four times. If, on the other hand, you wish to fatten it, let it be idle for many days and give it pork loin and the hen flesh, both fatty, and let it always be fed by the same person. Let the person who carries it take it also on a horse travelling at a walk and let it frequently be given the brain of a gelded sheep or of a ram.

If it is harmed by the sun, put rose water in its nostrils, let it eat honey with goat's flesh, and then spit out some wine, exhaling it as it were, into its face.

If it is harmed by a storm when it goes after its prey, throw warm water on its shoulder blades, provided that you first spread open its feathers. You should pour so much water over it that it will drip, coming over its kidneys to its feet, for this will do it great good.

If it is ill-disposed in its viscera for hunting, let it eat owls [*noctuae*] that are still warm and bats for three days. If it wants to take it, let it eat three lumps of pork dipped in vinegar, for this removes its aversion for its food and works very effectively against illness in the head and chest.

If it hurts a toe joint, split a mouse and put the hawk's foot in its warm viscera, binding it tightly with a small bandage. If it is not healed by this, break the right

---

276. This process is called "imping" (Latin *imponere*, "to place in"), a standard technique of falconry (see W&F, 426).

277. On pennyroyal, *Mentha pulegium*, see *De veg.* 6.422.

hoof of a pig and smear the hawk's foot for three days with the marrow and it will be healed.

These, then, are the things written by us about hawks, following the expert findings of Emperor Frederick.

CHAPTER XXI

On the Regimen for Goshawks, According to the Expert Findings of William

Seeking again the cure for goshawks, we will set forth the teachings of William, who is most experienced with birds of prey.[278]

William says that we will cure a hawk that has asthma as follows. Take equal weights of cloves, cinnamon, ginger, cumin, pepper, aloe, salt, gum tragacanth [*dragantum*], and frankincense, and grind forcefully. Mix all together and place on a tile. Warm these at a fire and put this powder in the hawk's nostrils by blowing with a reed. Mix the rest of the powder with very clear lard or butter, an amount the size of a filbert, and when the lard has been well ground, place the mixture on the hawk's palate. Afterward, make the bird stand in the sun until it vomits up the portion. On the next day give it the same lard the size of a walnut to eat.[279] On the third day give it a red dove chick and on the fourth day bathe it and it will be cured.[280]

Against *tinea* [external worms] in the hawk, take yarrow and grind it. Put it in vinegar and mix it with goose excrement. Let these things stand mixed this way for three days, then place the whole thing on linen cloth. Press out the juice and smear it on the places having the worms, especially on the wings and tail. Then grind iron rust into powder and sprinkle it on the hawk's wings and tail three times a day for three days.

When the hawk throws up its food undigested, take cinnamon, cloves, cumin, and laurel leaves, taking all in equal amounts, and grind them. Put in a new pot with white wine and boil it a good deal so that only a little of the wine remains. Be careful, though, that while boiling it does not rise up and overflow the pot. Place what remains in a strong white cloth and twist out the juice. Pour into the gullet of the hawk as much as suffices of what you have pressed out and on that day you are to give it nothing to eat. On the next day pound and crush fennel, press out the juice, and dip the food you are going to give the bird in the juice.

---

278. William the Falconer (see 23.63, 79). William's text has been edited by Tilander (1963, 134–74), although William does not give many of the treatments listed by A. Note that in the first sentence, A. uses *accipiter*, but he uses *astur* throughout this section. For simplicity, "hawk" will be used for each, with the understanding that the more specific "goshawk" may have been intended.

279. "Walnut": Although *nux* can be generic for "nut," see Jessen (1867, 730) for references throughout *De veg.* to *nux* as specific for *Juglans regia*, the walnut.

280. Scanlan (273) suggests that a young wood pigeon or ringdove has a red breast and might be meant here.

Against the disease of hawks called *fellera*, which is the same as a repletion of corrupt humor, take an amount the size of a chickpea from the gall bladder of a she-bear, put it in the heart of a hen, and give it to the bird to eat for up to nine days before you put it in the mews.[281] Then take equal amounts of savin, rosemary, savory, betony [*betonica*], mint, sage [*salvia*], and a slightly larger amount of radish [*raphanus*], and grind them all together.[282] Add a bit of honey and give it to the bird to eat on an empty stomach. Afterward, put the bird in the mews.

Against the defect of stone in the hawk, take equal amounts each of cinnamon, aloe, clove, sugar, saxifrage, a worker bee [*gerulam*], and a cicada.[283] Grind them all and moisten with rose syrup. When you feed the hawk, give it two pieces of this the size of two beans with its meat.[284]

If the hawk throws up a *glanum* (that is, raw, undigested flesh), put one spoonful of lye made of wine shoots in its gullet.[285] If it is in distress, take violet syrup and mix it with cold water and put three spoonfuls of this in its gullet. After it has vomited and has returned to its old self, wash it in water, doing so in quite clear weather. It is very effective if a bit of ground rock salt is administered with the lye since it causes the bird to vomit in a cleansing manner.

If the hawk has gout [*gutta*], take as much Alexandrian laurel as the size of half a bean or of a filbert and give it to the bird to eat. On the third day, give it *tiriaca*.[286]

If it suffers from *aculei* (that is, sharp punctures), take pig bristles, cut them up fine, and serve them to the hawk sprinkled over its food.[287] Do this for up to nine days and afterward grind chervil, pour its juice over meat, and give the meat to the hawk to eat.

If it has froth at the eye from black bile [*melancholia*] that has gathered there, take white gentian [*siler montanum*] and water hemlock seed and place them over some coals.[288] With this, smoke some meat and feed the bird over such a fire as well, so that

---

281. See 23.80, with notes, where the same disease is called *gutta silera*. Here the disease is etymologically linked to *fel*, gall bladder or bile.

282. Betony: *Betonica officionalis* (*De veg.* 6.289). Sage: *Salvia officinalis* (*De veg.* 6.450). Radish: *Raphanus sativus* (*De veg.* 6.423). Cf. the *radix* of 23.89–90.

283. We have seen all these herbs before. *Gerula* is used by Pliny *HN* 11.10.24 to refer to a working bee, i.e., a "carrier" [*gero*] of honey, but A. uses the term nowhere else.

284. At this point the following was deleted from the MS: "If the hawk spits up raw food, take one and one-half seeds of wild or bitter grapes or take the juice of the grape. Grind it and moisten with warm water. Holding the hawk gently, open its mouth and put in some of this confection."

285. *Glanum* is undoubtedly related to the Latin *glans*, acorn, and refers to the shape of the cast-up object. Note that W&F (619) list the following among their specialized falconry terms: "Gleam: *n*. Material thrown up by the hawk after casting gorge."

286. On *tiriaca*, see 23.84.

287. Elsewhere, *aculei* refers to, for example, a bee's stinger.

288. At *De veg.* 6.448, Jessen identifies *siler montanum* as *Laserpitium siler*, known in English as white gentian or withy. Its names in other languages preserve the Latin stems (Garth van Wijk, 1911, 1:731). For the hemlock, see *De veg.* 6.317, identified by Jessen as *Cicuta virosa*, water hemlock, a potentially lethal plant.

the smoke gets into its crop and eyes. On the next day give it to eat a piece of aloe the size of half a bean. Give it also a green cricket [*cicada*] or a locust. If you cannot find a green one, grind up some that you had dried and saved, sprinkle this over some meat, and give this to the bird.

If the hawk has piles, take the small feathers of its wings and, having dried them, grind them up and sprinkle this over meat. Feed him nine times from this food.

If the hawk has inordinate thirst, take liquorice, rhubarb, betony, and violet syrup, put them in water, and leave them there for one night and morning.[289] Give the hawk this to drink every day for up to eight days, whenever it wishes to do so, and feed the hawk on frogs.[290]

If the hawk is thought to be bewitched, take sweet gale mushroom [*fungum mirti*], frankincense, asphalt, and date palm [*palma benedicta*], and place all these on a tile and use them to fumigate the hawk.[291]

If the hawk has a head ailment, take radish [*rafanus*], savin, rosemary, black elder, savory, mint, rue, sage, and betony, and mix them all together. Grind them thoroughly, add honey, and give three pills made from this to the bird to eat in the morning with a little meat. In the evening give him some the size of a filbert.

If you wish to fatten a hawk, feed it often with the flesh of a goose or dove. If it has lice, grind roman mint [*menta romana*], moisten with strong wine, and add larkspur.[292] If the weather is clear and warm, bathe the bird with this. If, however, it is turbid and cold, take the fat of a hen and put this in it. Then put the mixture over smoke for one night and on the next day smear the hawk with it on the wings and on the back above the tail.

If the hawk has growths, take leeches and put them on the growths. On the next day take the sap of the tree called the *celsa,* or the tasteless fig [*ficus fatuosa*], and smear the sores.[293] Then take the roots of the herb called wolf's foot [*brancha lupi*].[294] Grind and mix this with the sap of the *celsa* and put it on the growths. Let it stand on them

---

289. Cf. *De veg.* 6.126 (liquorice); 6.211 (rhubarb); 6.289 (*Betonica officionalis* L.).

290. The Latin may be read to say that the liquid is left to set for one night and then given every morning for eight days.

291. At *De veg.* 6.138, Jessen identifies *mirtus* as *Myrica gale,* sweet gale. Scanlan (276) notes correctly that the form *fungum mirti* is difficult and he suggests emending it to *fundum,* "cup." Yet all the recipes for falcon cures call for small measurements. It may be best to leave the text as is (Borgnet reads *fungum*) and to understand it as a reference to a mushroom that grows on or near this plant. "Blessed palm" is the date palm, *Phoenix dactylifera,* which got its name after being strewn before Christ on his entry into Jerusalem (John 12.13). Cf. *De veg.* 6.169–80.

292. *Menta romana*: Scanlan's reference to *De veg.* 6.146 is incorrect. At *De veg.* 6.386, A. discusses *menta* in general. Although Jessen, ad loc., mentions this passage, he gives no identification.

293. The grammar makes it clear that A. thinks the two names indicate the same tree. *Celsa,* besides meaning lofty, often indicates the mulberry (Latham²). *LLNMA* lists *celsum* as "fruit of the *sycamorus.*" Jessen (1867, 716) lists the names as synonyms and identifies the plant as *Ficus carica* (var. *caprifica,* "goat fig"); cf. *De veg.* 6.103–4. This is the wild ancestor of the cultivated fig trees and has inedible fruit.

294. Jessen (1867, 706) tentatively identifies this plant as *Leonuris cardiaca,* the bitter mint, "motherwort." The form is paralleled in the plant called *branca ursina* (Latham²).

for three days and three nights. Afterward, take the root of the plant called pig's tail [*cauda porci*] and boil it.[295] Give three bean-sized pieces to the bird to swallow early and late for nine days.

If a hawk has mange, take some old lard, sulfur, and quicksilver, and grind them together with some cloves and cinnamon. With this smear the mangy spot, either while at the fire or in the bath.[296]

Against pain in the eyes, take equal parts each of ginger, aloe, and frankincense. Grind them together well. Add the mixture to white wine, put it in a basin, and let it sit there for one night, after which put it on the eyes. Likewise, take aloe and white lead [*cerusa*] in equal amounts.[297] Scrape a little bit from the midpart of some old lard and mix all these things together. Put it on the bird's eyes late, when it is time to go to sleep.

98 If a hawk [*accipiter*] has a broken leg, chew some frankincense, some clay, cuckoopint [*serpentina*], and comfrey [*consolida*], and grind vigorously.[298] Mix with egg white and spread the whole thing out over a linen cloth. Straighten the broken leg and wrap it in the aforementioned cloth. Take a feather from a vulture's wing, split it, and make a splint [*incanna*] for the broken leg with it.[299] Let it stay this way for five days and nights.

Again, against stones and corruption of the kidneys in the hawk, take one part clary [*centrum galli*] and two parts pimpernel [*verbena*].[300] Grind them and press out the juice. Give a third of a spoonful of this to the hawk on an empty stomach. Leave it alone from morning until midday, and should it seem to be in distress give it three spoonfuls of diluted violet syrup or of honey with rose. On the fourth day, take knotgrass [*centinodia*] and *centinervia* (that is, *quinquenervia*), grind them, and press out

---

295. Cf. *De veg.* 3.9 and 6.331, where the name of the plant is *cauda porcina*. Cf. Jessen, ad loc., for the difficulty of identification.

296. The word *scabies* implies encrusted sores.

297. On white lead compounds see *De min.* 3.1.4 (Wyckoff, 1967, 162) and 4.3 (p. 212).

298. "Clay": *bolus*; cf. 23.87. Serpentina: *Arum maculatum*, if *serpentina* is *serpentaria*, as Jessen conjectures at *De veg.* 6.290. In this passage, A. gives *basilicus* and *dracontea* as other names for it (cf. 22.115). A modern name, if the conjecture is correct, is "cuckoopint." "Comfrey": *Symphytum officinale* (*De veg.* 6.332), long a favorite in natural cures.

Note that Kotsiopoulos (1969, 74) omits paragraphs 98 and 99 and that Scanlan mistranslates *mastica* as a form of *mastix*, "gum mastic." As unpalatable as it sounds to our ears, the word must rather be the imperative of the verb "to chew."

299. The verb *incanna* is of interest as it literally means to put in a reed or, in this case, the shaft of a feather. Scanlan (277) points out that the egg white would act as a stiffening agent for the "cast."

300. "Clary": *Salvia sclarea*, suggested by Jessen (1867), 709. On *verbena, Anagallis arvensus*, see *De veg.* 6.471.

one third of a spoonful of juice.³⁰¹ Put this in the crop of the hawk on an empty stomach and it will be healed.

If it is gouty [*podagricus*], grind spurge [*titimallus*] with honey and vinegar and a bit of lime, and bind it over the gout.³⁰² After it moves, spread on aloe with wine and it will be cured.³⁰³

Again, against both *fellera* and *tinea* [external worms] in a hawk, take snail shells, green twigs from bramble bushes, saxifrage, sage, olive leaves, the froth that a young colt spews forth from its nostrils when it is born, and the gall bladder of an eel.³⁰⁴ Mix all these together, put in a new pot, and burn it over a fire until it becomes a powder. In the morning, give some of this, about the size of half a filbert, to the hawk with a small bit of meat to eat, on an empty stomach. Take rhubarb and put it in water for one day. Give the hawk this to drink and do this three times a day for three days.

If the hawk has gout [*gutta*], take a well-fattened goose and remove its fat. Take also bear and fox fat. Skin a cat and, having removed the viscera and bones from the body, divide its flesh with a knife. Take a bit of wax, some labdanum, and some aloe wood [*xyloaloe*], and make a powder.³⁰⁵ Take the juice of the greater and lesser *policaria* [*policaria maior et minor*] and cut a white onion.³⁰⁶ Mix all these things together and put them in the stomach of a goose. Sew up the opening tightly and let sit for one day. Then roast the goose quite thoroughly and collect the rendered fat that falls from it in some earthen vessel. With this ointment, smear the gouty spot. This is good for all gouty animals.

99

---

301. Scanlan (278) offers some identifications, but these must be reconsidered. *Centinodia*: see *De veg.* 6.322, where Jessen tentatively suggests *Polygonum aviculare,* knot-grass. *LLNMA* and Latham², s.vv. *centinodia* and *corrigiola,* support this identification. The name means "hundred knots." The next two plants present a problem. Jessen (1867, 709) lists *centinervia* ("hundred sinews") and says "v. *Quinquenervia*," apparently equating the two (and, indeed, one way to read the Latin supports this). Frustratingly, there is no entry in his index for *Quinquenervia*. Latham², s.v., *centinervia,* identifies it as ribwort plantain, *Plantago minor,* citing medieval authors. His earlier work (Latham, s.v. *quinquenervia*) identifies *quinquenervia* as plantain and cites a source dating to 1200. The translation offered here is based on these identifications.

302. On spurge, *Euphorbia lathyris,* cf. *De veg.* 3.20, 6.476, identified by Jessen (1867), 747. Lime, of course, is not the citrus fruit, but the mineral.

303. "It moves": lit., "moves itself," referring, no doubt, to the bird's freedom to move about without pain.

304. On *fellera*, see 23.80, 94, 105. On *tinea,* see 23.82, 87, 93. "Colt": *poledrus,* an unusual word for A.

305. *Laudanum: Cistus cretici,* a gum, also called labdanum, made from the resinous juice of rockroses of the *Cistus* genus. Cf. *De veg.* 6.113, where A. says that another name for it is *gutta. Xyloaloe*: Jessen identifies this as *Aloexyli cigallochi* (*De veg.* 6.257–59).

306. Although Jessen (1867, 760) glosses the name from this passage, neither he nor Stadler offers an identification of these plants. Scanlan (279) sees a connection with the Gr. *polygala* ("much-milk") and identifies them as milkworts. Latham (s.v. *pulicaria*) identifies it as fleabane.

Likewise, to clarify the eyes, take sweet herb, aloe, and white lead in equal amounts.[307] Crush forcefully and put it on the fire with wax, oil, and lard in some clay pot until it liquefies.[308] Then mix with those things which have been mentioned. Next, mix them together, forcefully making them one.[309] Place some of the unguent made this way on the hawk's eyes morning and evening and they will be clarified.

Let these, then, be the things said about the cures for hawks in accordance with the expert findings of William.

CHAPTER XXII

On the Regimen for Domestication of Hawks [accipiter] *and on the Regimen for Hawking*

100 When you wish to hawk with a goshawk or sparrow hawk, be careful that the hawk [*accipiter*] is first well accustomed to the hand by taming it in this fashion.[310]

At first, it must be tethered both day and night. Holding it with a very long leash [*zona*], see that it flies to the hand frequently, and as often as it does fly to it, feed it so that it grows accustomed to the hand as a benevolent thing.

When you wish it to hawk, on the first day take a pigeon whose wings have been plucked, and see that it escapes from under the hawk's feet often. See also that the hawk recaptures it. Change the bird for one that flies better and better and exchange the weak and small bird for stronger and larger ones, just as we have already instructed concerning hunting with falcons. For, as far as this is concerned, there is but one method in the regimen for all birds of prey.

As often as it captures the bird, allow it to drink some of its blood to the encouragement and whistle of the falconer and in the presence of the dogs. In this way it takes on courage.

When you want to hawk by letting it go after wild birds, first feed it with the flesh of a tender calf or the tongue of a pig that has stood a little while in vinegar or urine. The next day, right at the crack of dawn, go forth to the hunt. Be careful, however, if you can, that you only cast it off face to face with the bird. See that the hawk sees the bird and is not standing far from it when it is cast off.

---

307. "Sweet herb": *dulcis herba*. This has not been identified. It may not be a proper name, but simply indicate "a sweet herb."

308. "Pot": *testa*, possibly "tile."

309. The text is in some confusion here. Stadler reads *commisce simul fortiter minando in unum*, which literally means "mix together, forcefully driving into one." Borgnet omits the troublesome *minando*, which means either to drive, as cattle, or to threaten.

310. Cf. 23.75–76 on training and taming falcons and Fred. *DAV* 2.9–61 (W&F, 157–84) on the early training.

Some say, however, that if you allow the meat about which I spoke to lie in urine, and if you feed the bird with this in the morning, and if you give it a little bit of the same meat at night, and if you feed it the following morning with the tongue of a piglet, and if, toward evening, you go to the riverbank area where there are birds, the hawk will have the courage to capture large birds.

When you see that the hawk sees its prey but does not seek after it, you know that the hawk is too fat and that it has feathers grown thick from fatness, since it has been fed too delicately. You should, then, take away part of its accustomed diet and give it a light and digestible diet, in smaller amounts than previously. For a desire to capture large birds will rise up in it out of a natural hunger. Keep it continuously on the diet which you have proven to be good for it. If it is too fat, thin it down with garlic ground with pennyroyal, but be careful not to make it weak or cowardly. From August to November, a hawk ought to be kept moderately fat and moderately thin, neither tending to fatness nor thinness. But from November on, it should be kept fat and strong.

During the day the hawk should be held on the hand for a long time. At the third hour let it be given one thigh from a chicken and then be allowed to bathe in water for an hour. Afterward, let it be in the sun until it arranges its feathers by oiling itself.[311] Then put it in a dark place until evening and put a woolen cloth over the perch on which it stands so that it does not injure its talons. From the hour of vespers until the first sleep, let it be held on the hand. Then place it on the perch with the cloth placed underneath as before. Burn a lamp in front of it throughout the entire night, and around daybreak sprinkle it with wine. After this, place it by a clear fire, and when day has broken, go forth to the hunt. If you see that it seeks after the birds, cast it off. If it does not do so, repeat what I said above. If it should capture something, feed it as much as it wishes to eat of the prey.

The exhalation, the bite, and the wounds from the talons of the hawk, falcon, and all birds of prey have to be strongly guarded against. This is especially true when they have been bathed and have arranged their feathers with their beaks, since a particular fatty substance adheres to the beak. They get this from their tails and it is poisonous.[312] These birds have poisonous breath, feathers, and feet. If, then, it strikes with its talons or beak, it can be dangerous, and there are already those who have died of the wound from this kind of blow. Note that if a hawk is more clamorous than it should be, it is to be fed with a bat full of ground pepper.

If it peeps as if from the illness of phlegm [*pituita*] and cannot cry in a clear voice, pierce its nostrils with a bronze pin.[313]

---

311. "Oils itself": *perungendo*, showing a clear sense of what the hawk actually does.

312. This refers to the secretion of the uropygial gland, also (falsely) believed to be poisonous by Fred. *DAV* 1.31 (W&F, 71), who calls the gland the *perunctum* and says that it collects a virulent moisture from the rest of the body which makes their beak and claws poisonous, spelling an even faster doom for their prey.

313. This is probably the disorder often called pip, named from *pipita*, "phlegm," and discussed at 23.103. When suffering from this disorder, a bird will make coughing sounds. For a discussion of the disease in aquatic birds in general, cf. 21.27, with note.

These, then, are the things concerning the cures of falcons and goshawks, which we have gathered from those who have pursued the study of such matters in our time. Nor can anyone reject as superfluous the fact that this serves the pleasure of many who are accustomed to take their sport with the birds of the sky. So, then, that our study might be more perfect, we will add here the studies undertaken by the ancients to those things that have already been said.

CHAPTER XXIII

On the Regimen for Illnesses of All Birds of Prey, According to Aquila, Symachus, and Theodotion

103 There are in existence letters of Aquila, Symachus, and Theodotion, men of very ancient times, written to Ptolemy, King of Egypt.[314] In them is contained the following teaching about the habits and cures of birds of prey in general.

If there is illness in the head, and it may also be in the eyes, smear it very often with olive oil, especially if the pain is on the exterior parts of the eye.

If the white grows inside the bird's eye, put in a powder of fennel seed with the milk of a woman who has had a male child.[315]

If its eyes become murky through old age, cauterize it with a suitable silver or gold needle over the nostrils where the *sinciput* is joined to the middle of the beak, in between the eyes.[316]

If it has closed-up nostrils, blow into them a powder of pepper and larkspur using the hollow part of a small feather.

If it has rheum in its head, put some rue next to its nostrils and dip the meat which it will eat in the juice of rue. Likewise, put some garlic crushed with wine in its nostrils and let it sit the whole day in a dark place, and for one whole day allow it to go hungry.

If it has the phlegm [*pituita*] which some call *pipia* [pip], open its mouth, take its tongue, and rub it with a powder made of larkspur and spiced honey.[317] If the things that have been mentioned do no good, give it some butter to eat. The powder of dried cabbage is also effective against the same thing.

---

314. For this letter, cf. 23.71, with note. The version which follows is extremely close to that in ThC 5.50.48f.

315. Although *albugo* is the common name for the white of the eye, 22.111(60) seems to indicate it may also be a disease, and a white spot over the pupil, or a cataract, comes to mind. Yet the Latin is specific in indicating *in interiore oculi*.

316. At 1.111, A. defines *sinciput* as the rearmost part of the head, comparing it to the stern or poop of a ship, getting his anatomy backward.

317. On pip, cf. 23.102. "Spiced honey": *mel conditum*, perhaps honey that has been preserved.

If it is clamorous beyond all measure, take a bat, put ground pepper inside of it, and give it to the bird to eat. If a bat is not available, another bird prepared in the same way with pepper will do no harm against the same thing. Now, excessive clamorousness is a sign of illness, of weakness due to hunger, or that it has eggs that have been generated in it.

If the bird has a loss of appetite, give it live shrews or give it a young puppy born the day before, before it has sight.

If it frequently vomits up its food unchanged, take some scammony [*scamoneum*] weighing a quarter of an obol and the same amount of cumin.[318] Grind these and sprinkle fatty pork with their powder and give it to the bird in its food.

If, however, it is unable to eat the aforementioned flesh, take the white of an egg and place in it the aforementioned powder and place it in the bird's mouth.

Likewise, if it vomits up its food, take raw eggs and break them into goat's milk, cook the whole thing, and give it to the bird three times to eat and it will be healed.

If it begins to molt, exempt it from all work and feed it generously, for as often as it is hungry, just so often will it have signs of fractures or fractures themselves in its feathers. Green clods of earth, spread out beneath its feet, are good for the birds at this time.[319] The moderate heat of the sun is good for it, but too great a heat harms it.

If it has fevers, give it the juice of the mugwort [*artemesia*] three or four times with the flesh of a hen.[320] Likewise, for the same thing, bind its right leg tightly. Then, in the middle of the leg, a vein will appear. Bleed it lightly, for on the leg of such birds there are four veins. One is toward the front, one is lower down, and a third is more exterior. The fourth is toward the rear, above the largest talon.[321] The signs of a fever are if the wings droop and if the bird holds its head down low and if it trembles as if in the clutches of some sort of chill. Other signs are if it ignores its food without any other reason or if it greedily takes it in but has trouble swallowing and getting it down.

If, however, the bird is very thirsty, take a powder made of cabbage, lovage, and the stems of anise and fennel, and cook it with wine. Mix in one spoonful of honey and give a strained mixture of this to it to drink or else, if it does not wish to drink it, put it in its mouth. Or else, give it meat smeared with honey for one day and on the second day, meat smeared with cold oil treated with roses.

If it is having trouble with its gall bladder (the disease which we called *fellera* above), sprinkle over its food a powder made of the flowers or the buds of the willow.[322]

---

318. At *De veg.* 6.437, A. seems to indicate that by *scamonea* he means the powder made from a plant of the *Convolvulus* species. He claims it is potent for thirty years. The obol is a Gr. weight which ranged from 0.72 to 1.05 grams in antiquity.
319. "Are good": this translation follows Borgnet's text, reading *valent* for Stadler's *valenti*.
320. Jessen identifies *artemesia* as *Artemesia vulgaris* at *De veg.* 6.286.
321. This is the posterior-lateral claw, the "killing claw," analogous to the thumb.
322. Cf. 23.80, 94, 98.

If its wings droop except as a result of fever, take the blood and fat of a goose. With the blood, rub its wings in the sun but feed it with the fat. Better still, take laurel oil and, with the bird's wings lifted, smear the hollows beneath its wings with this.[323] Also smear its wings with the gall bladder of a pig and dip its food in the juice of pimpernel or sage.

If it has gouty wings, cook ground ivy [*hedera terrestilis*] in water and around the bird, next to its sides, bind the cooked and well-crushed leaves and dip its food in the water.[324]

If it has gouty feet or if *tineae* [external worms] are eating away its feathers, let it eat goat meat dipped in vinegar and rub its wings often with warm vinegar and with laurel oil.

If you wish to extract a broken feather without pain, take the blood of the small animal called the *gruile* or the blood of a rat and smear it on the place where the feather is and it will fall out.[325] Afterward, using honey that has been cooked down to a great thickness, make a shaft the size of the opening the feather stood in, put this in the opening, and a new feather will arise. Likewise, smear its feathers with the warm juice of the poppy herb and dip its food in the same juice.

If it has a broken bone in its leg or in some other place, bind warm aloe over it, and leave it alone for one day and night. Likewise, bind over it the excrement of a cock, cooked in vinegar.

If the bird is ill at ease on its perch or on the hand, cook some *miria* in water.[326] With the water, sprinkle its body and dip its food in it up to nine times.

If it is troubled by the disease called *rampa* [cramp], dip its food in the juice of the mugwort.[327] Likewise, rub its feet with the warm blood of a lamb or with tepid wine in which nettles have been cooked. Dip its food in the wine.

If it cannot defecate, give it the gall bladder of a cock or cooked white snails to eat.

---

323. One property of laurel oil is heating that to which it is applied. Cf. Pliny *HN* 23.43.86.

324. Jessen (1867, 714) identifies this ivy as *Hedera helix*, English ivy. The "gouty wings" may refer to the "blain," characterized by watery vesicles formed at the joint of the bird's wing.

325. *Gruile* is unidentified. Körting (1923, 495) notes the Spanish *grulla* and Portuguese *grulha*, descended from the Latin *gruicula*, a diminutive of *grus*, crane. But this hardly qualifies as a "small animal." Unfortunately, the reading is secure, being found in St., Borgnet, and ThC 5.50.94.

326. *Miria* cannot be identified, but ThC 5.50.100 reads *myrra*, which may indicate myrrh, a bitter gum resin from trees of the *Commiphora* genus and the likely candidate here, or maple wood (Latham, s.v. *murra*). In the latter sense it resembles the *bruscum* of Pliny *HN* 16.27.68, a passage similar in content to much of A.'s discussion of *murra* in *De veg.* 1.120–21 and 6.184. None of this, of course, tells us whether A. understood what *miria* was. This form is not included in the extensive list offered by Blatt (1957f.), s.v. *myrrha*.

327. Scanlan (285) asks if *rampa* might not be related to "cramp." His instinct is correct. Körting (1923, 805, s.v. "rapon") specifically links *rampa* with the Germanic stem *ramp*, itself seen in MG *Krampf*, and Lexer (2.340) shows the MHG forms *ramph* and *ramphe*.

If, however, it is too loose, give it a little bit of the juice of henbane [*iusquiamus*] to drink and dip its food in it.[328]

If it has lice, take wormwood sap or water in which wormwood has been cooked and, while the bird is sitting in the sun, pour this all over its feathers and body.

If it has stone in its belly, give it the fatty portion of lard and butter to eat.[329] For the same thing, wrap the herb aloe with powdered parsley in the hearts of small birds and give this to it to eat. In this way and in ways similar to this, you can accomplish something effective in every medication of birds.

If you want to have a fat bird, give it meat from a bull or pig.[330] But if you wish to have it thin, give it young hens soaked in water. If you want it in-between, feed it old hens.

If you want it to be ready for hawking, be sure that you produce a good crop in it and close it in a shadowy place with a small lantern brought in there, and then hawk on alternate days.[331]

If you want to catch hares or rabbits, see that you teach this to the bird in its youth. Also, bind the rings to its legs near its feet, leaving a palm's breadth distance from leg to leg.[332] In this way it will capture them without injury. A wild hawk [*accipiter*] becomes tame more quickly if it is allowed to get quite hungry and is a young bird, and if the place in which it is kept is neither too warm nor too cold. In the place they stand, always keep mint and sage. If it is a hawk, let it at first stand on willow leaves and later always on willow or fir wood. If they often drink bird's blood, they take in courage, boldness, and a desire for the capture. Let not a cold bath be denied them when they are fed. Neither should their wings be touched a great deal lest they droop.

107

---

328. *Iusquiamus* is *Hyoscyamus niger*, which yields a poison that is, as its English name implies, especially deadly to fowls and has properties essentially like belladonna. At *DCPE* 2.2.1, A. indicates that when its root is burned, its smoke will cause birds to collapse as if dead. Cf. *De veg.* 6.362–63, where A. shows that he is aware of the plant's poisonous nature. The dried leaves contain hyoscyamine and scopolamine and are used as an antispasmodic and a sedative.

329. "Stone" is the disease *lapis* described at 23.94.

330. "Bull": *masculus bos,* lit., "male cow," and thus, perhaps, a steer or bullock.

331. "Crop": *vesica gutturis,* lit., "throat bladder." Scanlan (286) tries to justify this as indicating the hawk's hood, but this is not in the standard hawking vocabulary. Moreover, at 13.80, A. makes it clear that the words form a synonym of *struma*, crop.

332. What are these "rings"? Their name, *gyri,* implies something shaped like a ring or that turns around in circles. To claim they are jesses (Scanlan, 286) is unlikely since jesses are made of leather, are not ring-shaped, and have an established name in Latin, *jacti.* Identification is to be found in the swivel, or *tornettum,* "turner," of Fred. *DAV* 2.40 (W&F, 140). The Latin says that the *tornettum* is a thing made of two rings which rotate (*gyrare*) around each other and then it goes on to describe them in some detail. Today these are called the varvels, two flat rings which took the place of a swivel in the old days of hawking. The rings often had the coat of arms engraved on them and were fastened to the ends of the jesses, whereupon the leash was passed through them (Glasier, 1978, 302).

CHAPTER XXIV

On the Two Other Genuses of Falcons, on Their Habitations, and on How They Differ from Hawks

108 These things, then, have been said by us on the nature of and regimen for falcons and hawks, following the statements of the ancients. But in addition to all the other genuses of falcons which have been introduced, there are still two other genuses of falcons found in our lands. One of these is the rock falcon [*falco lapidaris*]; the other is called the tree falcon [*falco arborealis*].

The first is midway in size and vigor between the peregrine and the *gybosus*. It is found on the cliffs of the Alps and has the same diet and lifestyle as the peregrine.[333]

The second is midway in size and vigor between the *gybosus* and the merlin, having the same lifestyle as the merlin.[334] These, therefore, should not be treated separately.

In addition to those genuses already mentioned, it may be that many genuses of falcons are found in widely scattered places. But adequate conjectural knowledge on the nature and diet of all of them can be deduced from the things that have been said here.

109 In addition, that which has been said in previous books has to be added here as well. Every animal is abundant in a place where there is an abundance of food peculiar to and proper for it. Therefore, since the particular food of this sort of bird comes from the hunting of birds, they are abundant in places where the birds they hunt are abundant. These birds are aquatic ones, for they are slower in flight and fleshier for eating. Therefore, the genuses of hawks, falcons, and eagles are abundant toward the north in places like Britain, Sweden, Latvia, and the neighboring regions of the Slavs, Prussians, and Ruthenians.[335] These regions are cold, and in cold lands bodies are large with much blood and spirit. From these arise boldness and fierceness, as has been determined in the study *On the Nature of Dwelling Places* [*De natura locorum habitabilium*].[336] Thus it is, then, that birds of prey in the regions mentioned are large, possessing great boldness and ferocity. In other places, however, they take on proportional vigor, size, and boldness.

---

333. Oggins (1980, 461) conjectures that this may be *Falco peregrinus brookei*, whose size and habits roughly correspond to the description. Elsewhere (8.108), A. mentions that this falcon is so called because it rears its young on rocky cliffs. On the German name, cf. Sanders (418) and Suolahti (1909), 344–45.

334. Perhaps, as Oggins suggests (1980, 461), this is the hobby, *Falco subbuteo*, which became popular during medieval times, ranking second among the smaller hawks only to the merlin (W&F, 517–18).

335. "Latvia" is given for *Livonia*, although the ancient and modern borders do not match exactly. The *Pruteni* lived on the southeast coast of the Baltic Sea. The *Ruteni* inhabited Galicia and the Ukraine.

336. *De nat. loc.* 2.3. Also cf. *DA* 7.30.

Nor should it be passed over in silence that Aquila, Theodotion, and their comrade Symachus call all genuses of hawks [*accipiter*] falcons. They also determine that there are four genuses of these. They place the goshawk [*astur*], which is first in size, in the first genus and the smaller goshawk [*astur minor*], which we call the tercel [*tercelinus*], in the second. They place the sparrow hawk [*nisus*] in the third and the musket [*muscetus*] in the fourth. That these men can in no way be agreed with is proven in that the tercel is found in the nest of the hawk [*accipiter*] and the musket in the nest of the sparrow hawk.[337] Therefore, the *accipiter* and the tercel differ not in species but rather in sex, since the *accipiter* is female and the tercel is male. The sparrow hawk and musket have the same type of differentiation, for the sparrow hawk is feminine and the musket masculine.

Let, then, these be the things said about birds of prey.

(41) FATATOR: The *fatator* [ringdove] is said to be a bird of the East which, being desirous of young, copulates twice a year. It does so first in winter, in January after the winter solstice, but the eggs frequently perish due to the winter freeze. It copulates a second time in summer, after springtime and the summer equinox, near the time of the summer solstice. These eggs do well and bear young.[338]

(42) FENIX: The phoenix is said to be a bird of Arabia, in regions of the East. So write those who examine theological and mystical matters rather than matters of natural science. They say that this bird, having no mixing of male and female, is alone in its species and that it comes in cycles of, and lives alone for, three hundred forty years.[339]

It is, so they say, the size of an eagle, with the crowned head of a peacock. It also has wattled jaws and around its neck is a purple color, but with a golden glow to it.[340] It has a long tail of purple hue but with some rose-colored feathers mixed in, just as some circles shaped like eyes are mixed into the peacock's tail. This variation is a wondrous beauty.

When it senses it is weighed down with age, it builds a nest in a tall, hidden tree situated over a clear pool. It builds it out of frankincense, myrrh, cinnamon, and other

---

337. The term "tercel" is used for the male of a hawk, especially of a gyrfalcon and a peregrine. The musket is a male sparrow hawk.

338. Of doubtful identification, Stadler conjectures that the *fatator* is a corruption of *phatta,* dove or pigeon. Cf. Ar. *HA* 562b6–8, which speaks of pigeons and doves laying eggs in spring, never more than twice a season, with the statement that the hens sometimes destroy the first egg laid. Cf. ThC 5.47 and Vinc. 16.73. Cf. *GB,* 300–302.

339. The story of the phoenix is of great antiquity. Cf. the evidence collected by Thompson (*GB,* 306–9), Broek (1972), Pollard (1977, 99–101), *BAZ,* 191–93, and Mermier (1989). Other major medieval versions include *Physiologus* (Curley, 1979, 13–14; Carmody, 1939, 20–21), Isid. *Orig.* 12.7.22, Ambrose *Hex.* 5.23.79, Jacob. 90 (p. 190), ThC 5.45, Vinc. 16.74, Barth. 12.14, Neckam 1.34–35, Peter Damian *De bono religiosi status* 11 (*PL* 145:773B–D), PsJF 306–08, and Rabanus Maurus *De universo* 8.6 (*PL* 111:246A). Cf. also Hugh of St. Victor 1.49 (*PL* 177:48–50) and Hugo of Fouilloy (Clark, 1992, 230–35). While it is generally agreed that Hugo of Fouilloy wrote the aviary portion of *De bestiis et aliis rebus* (Clark, 1992, 7–8, with notes), reference will occasionally be made to Hugh of St. Victor's work due to some differences in its text and its availability in *PL.* For illustrations, cf. García Arranz (1996, 333–61).

340. "Wattled jaws": *fauces cristatos*; cf. 23.115(51).

precious fragrances. It rushes into the nest and puts itself in the blazing rays of the sun which the resplendence of its wings intensifies until a fire is started. It thus burns and reduces itself to ashes along with the nest. On the next day, so they say, a worm is born in the ashes which, having assumed wings on the third day, is changed within a few days into a bird with its former shapes, whereupon it flies off.

They also say that the following once happened in Heliopolis, a city of Egypt. They say this bird, carrying these fragrances over a pile of logs used for sacrifices, burned itself up. Within view of the priest and by the previously mentioned method, it was, through two acts of generation, formed into a worm and a bird and flew away. As Plato says, "those things which are reported as written in the books of the sacred temples should not be scorned by us."[341]

(43) FETYX: The *fetyx* [dove?] is a bird that produces young twice a year and creates a good many young. Since it is small and has a brief life span, it therefore puts a great deal into its semen.[342]

(44) FICEDULA: The *ficedula* [woodcock] is so called because it eats figs [*ficus edat*]. It also eats, however, grapes and other sweet, tree fruits.

The *ficedula* is the name usually used by us for the bird we call *sneppa* in German.[343] It has a very long beak, and during the fall it flies among the trees at dusk and dawn. For this reason it is captured at that time in nets spread on high, for it is accustomed to use the same routes entering and leaving the forests.

(45) FULICA: The *fulica* [coot] is a black aquatic bird of the genus of divers [*mergi*]. It is smaller than a duck and lives in the sea and around still waters. It does not wander through widely scattered places, but rather it stays in the place of its birth. The *fulica* feeds on carrion and seems to take delight in a storm since during one it swims to the deep seawater and plays. It builds its nest in rocks around which water flows. There, it stores up a hoard of food with which it is so generous that it shares it with alien birds.[344]

---

341. The Plato reference is untraceable. The tone of the quote smacks of Hdt. 2.73, which relates the story from the priests of Heliopolis but specifically states that it is incredible.

342. The behavior noted here puts one in mind of the *fatator*, 23.110(41); and ThC 5.48 makes note of this. Thus, and in light of the notes at 1.44, it seems certain that this is another variant of *phatta* and is a dove or pigeon. It is equally certain that A. did not know this.

343. The etymology is from Isid. *Orig.* 12.7.73 and is accurate. Cf. Neckam 1.53, ThC 5.49, PsJF 295, and Vinc. 16.75. Cf. also the *nepa* at 23.130(85). Sanders (433) clearly identifies this bird as the woodcock, *Scolopax rusticola*, for which the MG is *Schnepfe*. Cf. Suolahti (1909), 273f., *GB*, 56–57, 261–62, *BAZ*, 217–18.

344. The *fulica* is explicitly said to be the Gr. *phēnē* by Ambrose *Hex.* 5.18.61, but his version is quite different. Isid. *Orig.* 12.7.53 mentions only that the bird has a taste for cadavers. It appears that the classical *phēnē* (*GB*, 303; *BAZ*, 188), usually identified with the huge lammergeier and definitely some sort of vulture, has been conflated with the coot (Gr. *phalaris*; *GB*, 298; *BAZ*, 182–83). Thus the little coot, *Fulica atra*, a shallow-water bird whose diet consists of water plants, seeds, and worms, with an occasional excursion to land after small reptiles and mice, becomes a deep-water-diving carrion feeder. *Physiologus Latinus* (Carmody, 1939, 39) and Hugh of St. Victor 1.58 specifically say that the *fulica* does not eat carrion. For other versions, cf. ThC 5.46, Vinc. 16.76, Peter Damian *De bono religiosi status* 16 (*PL* 145:776D), PsJF 295, and Rabanus Maurus *De universo* 8.6 (*PL* 111:248C).

(46) GRIFES: That griffins are birds is more a statement made by the histories than something asserted by the expert findings of philosophers or by proofs based on natural philosophy [*rationes physicae*].³⁴⁵ For they say these birds have, in the front, the shape of an eagle for their head, beak, wings, and front feet, although griffins are much larger. In their hind parts, tail, and rear legs they say they imitate a lion. The eagle talons it has are very long, but the lion ones are short but large, and drinking cups are made from them. Thus it is that both long and short griffin claws are said to be found.³⁴⁶ They live in the Hyperborean Mountains and are especially hostile to horses and humans.³⁴⁷ So strong are they that they carry off both horse and rider. They say that there are gold and gems, especially emeralds, in these mountains. They also say that griffins place agates in their nest on account of their special benefit.³⁴⁸

(47) GRACOCENDERON: The *gracocenderon* is said to be a black bird which, of all the flying creatures, employs copulation the least. For with one copulation in the summer, it fills the female with an abundance of young. It therefore does not copulate at any other time during the year.³⁴⁹

(48) GOSTURDUS: *Gosturdi* [thrushes] are birds resembling both thrushes [*turdi*] and earth.³⁵⁰ They fly in flocks with a sort of wave-like action, now high and now low. They are said to lay their eggs in the earth and only rarely to keep them warm. For this reason the common folk lyingly assert that the eggs are kept warm by toads. What is true, however, is that the modest heat of the sun is sufficient to keep them warm. When the chicks come forth, they are collected and solicitously provided with food by the mother, who previously, however, seemed to be less solicitous of the eggs.

---

345. Mention of the griffin is found as far back as the Bronze Age. Hdt. 3.16 reports that griffins fight with one-eyed men over gold; Pliny *HN* 7.2.10 adds the wings, Solinus 15.22–23 elaborates on their ferocity. Isid. *Orig.* 12.2.17 includes residence in the Hyperborean Mountains, the lion and eagle combination, and their attack on man and horse. Cf. also Neckam 1.31, Barth. 12.19, Jacob. 87 (p. 174), ThC 5.52, and Vinc. 16.90. Two interesting studies are those of Peacock (1884) and Mayor (1994). The latter makes a plausible argument for the remains of dinosaurs such as *Protoceratops* giving rise to the legend. If one compares the shape of the fossil's head with the examples of medieval griffins given by Benton (1992, 130–31), the argument seems persuasive. Most recent is the study, well illustrated, of Armour (1995).

346. The two sizes of claws are not found in ThC 5.52 or other sources. It may be based upon having seen such drinking cups, probably made from some sort of exotic horns, much as narwhal horns were sold as unicorn cups, good against poisoning.

347. The Hyperborean Mountains were located, mythically, in the far north.

348. *De min.* 2.2.1 (Wyckoff, 1967, 72) lists several of the benefits one can derive from agates, including dreams while one sleeps and bodily strength.

349. The name of this bird is apparently corrupted from the word *karakoeidōn* ("raven-like," gen. pl.) of Ar. *HA* 488b6. The birds so described are briefly mentioned as rarely copulating. Cf. ThC 5.53.

350. The thrush is the Gr. *kichlē* (*GB*, 148–50; Pollard, 1979, 34–35). Cf. also ThC 5.54 and Vinc. 16.87. It is clearer in ThC that the resemblance to earth lies in color.

(49) GRUS: The *grus* [crane] is a familiar bird, dwelling in every land.[351] Much has been said about it in preceding books because it alone, due to its earthiness, grows darker in old age.[352] For this reason its flesh is tough and when it is going to be eaten, it should lie in its feathers after it is killed for one day in summer and for two days in winter. In this way the flesh is rendered more tender. This bird flies in a lettered formation [*ordine literato*] and it flies high so that it can catch sight from far off of where it has to go.[353] Cranes therefore gain altitude by flying in a spiral for a long time while they call. Then they fly in a circle until it is clear where they want to go. They fly for long distances when they commit themselves to the blasts of the winds, but they rarely struggle against the force of the wind unless they are fleeing something.[354] They establish leaders and the leader calls a good deal so that they will all come together. When some are tired they all come down to the earth so that they can wait for each other. When they mark out a camp, they establish watchmen and every tenth one stands watch duty. While they are on watch, they lift up stones with their feet so that a watchman will be awakened by the fall of the stone should he fall asleep. Moreover, at the fall of a stone they all call out as if chiding the watchmen for sleepiness. The others, who sleep, hide their head on their back under their wings for quiet and, by alternating feet, they hold themselves up on one foot. When they descend on occasion to feed, the leader constantly looks about with his head upraised while the other ones feed. If he sees anything hostile, he defends all by clammering.

While flying from place to place, some take along stones with their feet and others swallow sand. Some also vomit up stones that have the splendor of brass [*auricalcum*].[355] Not all but only certain ones load themselves down with all these things, doing so lest they be lighter and swifter than the others and get separated from them by flying faster, in this way abandoning the formation.[356] When they are tired, though, they sometimes cast off the stones. These have been caught by sailors into whose ships they have fallen.

---

351. The crane was discussed throughout antiquity (*GB*, 68–75; Pollard, 1979, 83–84; *BAZ*, 52–54), especially the legendary wars of the pygmies and cranes (Muellner, 1990). Cf. Ar. *HA* 597a6f., Pliny *HN* 10.30.58–60, Solinus 10.11, Ambrose *Hex.* 5.15.50, Isid. *Orig.* 12.7.14–15, Jacob. 92 (p. 215), Neckam 1.46–47, Barth. 12.15, ThC 5.55, Vinc. 16.91–93, PsJF 299–301, and Hugo of Fouilloy 44 (Clark, 1992, 202–5). For its importance, and for illustrations, cf. Rowland (1978), 31–35; George and Yapp (1991), 123–25; Yapp (1981), 13f.; and McCulloch (1960), 105–6. For illustrations, cf. García Arranz (1996, 437–69).

352. Cf. 20.6.

353. "Lettered formation": *ordine literato*. For the meaning and earlier use of this phrase, cf. note to 1.58.

354. ThC 5.55.17–18 says in contrast (as does Ar. *HA* 597a31f.) that they always fly against the wind.

355. Classically, *auricalcum* refers to either yellow copper ore or brass made from it (L&S, s.v.). At *De min.* 4.6, Wyckoff (1967, 224) translates it as "brass." ThC 5.55.18f. mentions this ballast, saying that cranes have a stone in their stomach which some claim will turn into gold in a fire. ThC defends the belief, saying these stones are *auricalcum*, brought to Europe from far-off places, and that *auricalcum* is a stone from which gold is smelted.

356. At 1.39, A. praises the "government" of the cranes.

These birds care for and aid each other a great deal, which is why when one leader has become hoarse from his cries, another is chosen. When, moreover, they go from place to place by flying throughout the lands, they make their way to the equator and beyond, whereupon they gather in those areas from which the Nile flows.[357] There they fight with pygmies one cubit high which resemble humans in shape and language.[358] These pygmies also have tiny horses, as has been said in previous books.[359]

This bird is playful and derisive so that even when wild it plays and dances to the voice of a person imitating its calls.[360] It sometimes places a stone among its eggs. Many other things also pertaining to this bird can be compiled from things which were introduced earlier.

(50) GLUTIS: The *glutis* is a slow bird and takes its name from the length of its tongue [*lingua*].[361] When the stork, swallow, and other birds are migrating, it flies forth and tries to be, so they say, the first to go away. Afterward, however, having experienced the labor of flight, it is slowed and rendered one of the last.

There is another bird which they call the *cycraunium* and which is so concerned about migrating that it arouses the others even during the night watch and urges them on to complete the journey they have begun.[362]

(51) GALLUS: The cock is a familiar bird, crested on its head as well as on its jaws.[363] It has long claws on its legs and a tail shaped into a semicircle with curved feathers. It has other feathers bent like a semicircle on its back and neck. It has tender flesh, but it is tougher than the hen.

This bird is very combative when it comes to getting hens. Therefore, when a number of them come together, they fight. The one that is victorious copulates with the hens and raises up its head and tail by way of glorying in its victory, while the other one wastes away due to his servitude. They sometimes fight so fiercely that death ends the fight.

The bird copulates a great deal and therefore fills many hens. To fertilize one egg it copulates many times with the same hen and, if there are many cocks, they kill the hen with too much copulation.

---

357. ThC 5.55.36 tells us this is the winter migration. Cf. Ar. *HA* 597a4f.

358. For the language of pygmies, see Resnick and Kitchell (1996).

359. Cf. 7.62.

360. The dance of the cranes was widely commented on in classical literature (*GB*, 75).

361. The "tongue bird," Gr. *glōttis*, is variously identified with the wryneck, flamingo, landrail, and others (*GB*, 80–81). Cf. Ar. *HA* 597b17f., Pliny *HN* 10.33.67, ThC 5.56, and Vinc. 16.88. Note that the pun in Gr. is lost in the Latin.

362. Pliny *HN* 10.33.68 speaks of the *cychramus*, following Ar. *HA* 597b17, *kychramos*. It may be an ortolan bunting (*GB*, 187–88).

363. There naturally are many references to such a common bird as the cock. For antiquity, cf. *GB*, 33–34 and BAZ, 9–11. For our period, cf. Ambrose *Hex.* 5.24.88, Ald. *Aen.* 26, Isid. *Orig.* 12.7.50–52, Jacob. 91 (p. 197), Hugh of St. Victor 1.36, Neckam 1.75, Barth. 12.16, ThC 5.57, Vinc. 16.77–80, PsJF 296–98, and Hugo of Fouilloy 41 (Clark, 1992, 180–87). Rowland (1978), 20–28; McCulloch (1960), 104; Yapp (1981), 19–20; and George and Yapp (1991), 155–56, all discuss the bird. For illustrations, cf. García Arranz (1996, 363–403).

This bird easily senses the changes in the weather caused by the sun's movements.³⁶⁴ It therefore marks off the hours with its song, and when it sings at night it holds itself erect, beats its wings, and shakes itself so that it can sing all the more wake-fully.³⁶⁵ When the hens die, the cock sometimes wastes away. When the cock sleeps, it places itself high up and whichever of the hens is the most wanton sits closest to it all night.

Decrepit, old cocks have more tender flesh, and if there is any viscosity in it, cooking the flesh consumes it. Therefore, broth made from decrepit cocks is very good for asthmatics and for those suffering a heart defect. The lion is said to fear a white cock greatly, either because of the opposition of their species or because it is said to be similar to a basilisk.

Some say that a decrepit cock generates an egg on its own and places this in excrement. They say that the egg lacks a shell but has a strong enough skin to stand up to the strongest blows and that this egg is fertilized by the heat of the dung to form a basilisk.³⁶⁶ This is a serpent like a cock in all respects except for having a long serpent's tail.³⁶⁷ Now I do not think this is true, but it was said by Hermes and has been accepted by many on the authority of the one saying it.

(52) GALLINA: The hen is a well-known bird, a slow flyer with moderate flesh. It is inordinately attentive to the young of its species, to the extent that it becomes ill with a sharp voice resulting from its attentions to its chicks.³⁶⁸ It does not care whose eggs it keeps warm and is even full of concern for the young of a different species. It lays many eggs and it both approaches and leaves its nest with a cry. If it is agitated and suffers harassment, it nevertheless finishes its song after it has found security.

It is said that it cannot be bitten by a snake on the day it has laid an egg and that its flesh is medicine for those who have been bitten.³⁶⁹ The egg is completed on the eleventh day and is not fertilized by a single admixture of sperm but rather by continuous copulation.

---

364. "Weather": *aura,* and thus, perhaps, "breezes."

365. The attempt to determine some stimulus other than light for the cock's crowing has a modern ring to it. Compare the approach of Ambrose *Hex.* 5.24.88.

366. Cf. 25.18–19(13) for this serpent, both in its "real" and mythical forms. The mythical basilisk was thought to be born from a cock's egg (cf. Neckam 1.75). See White (1954, 168f.) for a discussion of the history of this idea.

367. An excellent illustration from the *Hortus Sanitatis* and showing just these traits, with the addition of eyes all over the body of the beast, is reproduced by Best and Brightman (1973), 84.

368. The hen was not as common a subject in this literature as the cock. See Rowland (1978), 77–79; Yapp (1981), 19–20; George and Yapp (1991), 155; ThC 5.58; Vinc. 16.82–86; and Barth. 12.18.

369. Pliny *HN* 29.17.61 says that chickens that have eaten a bug (*cimex*) are immune to the bite of an asp and their flesh is beneficial to those bitten. The verb *ederint* has been read as the verb "to bring forth" (*edo, edere, edidi, editum*) instead of "to eat" (*edo, esse, edi, esum*). The result is that both ThC 5.58.52–54 and A. have turned the bug into an egg to accommodate the verb (Aiken, 1947, 219).

They sometimes lay wind eggs. Cocks are generated from round eggs and hens from elongated ones, although Aristotle seems to oppose this idea.[370]

Eggs ten days old are successfully incubated, as are those a few days old, up to four days old. Older or newer eggs are proven to do less well.[371] In summertime, in warm locations, they hatch out on the nineteenth day, and in the wintertime, on the twenty-ninth day. Eggs that are good for incubation have bloody veins on the fourth day, whereas those that appear clear at the pointed end when held up to the sun's rays at that time are not fertile. Full, fertile eggs sink in water, but others float. It is very harmful if they are moved around by hand since the veins and humors are ruined by tipping them upside down. An indication of this is that when a hen lays her eggs in a hidden nest, all the eggs are fertile. When they are touched by human hands, though, most of them are ruined. The chick's head is turned toward the point of the egg and its entire body toward the rest of the egg. The chick is born on its feet like the other chicks of birds.

When the chicks hatch out, the hen gathers them under her wings and even attacks the kite or any other assailants on their behalf. The hen seeks its food by scratching around with its claws and it then calls its chicks to what has been found.

Some hens lay eggs in which there are twins, but one of them crowds the other and sometimes, with the webs broken, one chick with two bodies is produced.[372] Many other things about eggs and hens have been said in preceding books as well.

This should be added, however. Pliny says that if gold, cut very fine, is placed in a hen's members, the members consume the gold and thus seem to be a gold poison.[373]

It should also be noted that the yolk of an egg laid during a full moon washes stains out of cloth. If it has been laid at another time, it does not clean away the spots. Some say the reason for this is that there is a fattened drop in the middle of the yolk which is in existence from the first generation. This then takes on a heat from the great light of the moon which moves the moisture that penetrates and breaks up the spots, something it can do at no other time.

(53) GALLUS GALLINACIUS: The capon is a castrated, effeminate cock, which therefore neither reproduces nor crows.[374] Now, there are certain genuses of harsh

---

370. Ar. *HA* 559a28f. has the females coming from the longer, more pointed eggs. Michael Scot's translation reverses the situation and A. was thus convinced that Ar. had reversed the truth. The matter was not helped by ThC 5.58.13–15, which cites Pliny *HN* 10.74.145, which also reversed the situation. This passage lets us see once more two of A.'s strongest traits. First, he was ready to put his own observations ahead of the received tradition. Second, his "seems" shows, as always, his respect for the First Philosopher.

371. The point is that once they are laid, they must be brooded over within this time frame. Cf. Pliny *HN* 10.75.151–52, which says only that they must be brooded over within ten days of being laid.

372. "Webs" is to be taken in the sense of "membrane." Cf. Glossary, s.v. "Web." ThC 5.58.46f. describes the two-bodied chicken in greater detail, telling us it was born in France.

373. Pliny *HN* 29.25.80 indicates that gold cooked with chicken will disappear into the chicken, showing that chicken flesh is a "poison for gold."

374. Cf. Barth. 12.17, ThC 5.59, Vinc. 16.79, and Isid. *Orig.* 12.7.50. Neckam 1.75 is at pains to tell us that this is not the true name for the capon but rather for the rooster, basing his argument on the Justinian Code.

nettles which are deadly to a hen's chicks and which she tries to pull out by the roots with her beak. She sometimes labors so hard pulling that she bursts internally. Nevertheless, a capon with a defeathered chest and a belly burned by the nettles cares for the chicks afterward, taking pleasure in their soft touch on his itching skin. Since it has been afforded pleasure this way, it is attracted and it therefore always cares for and looks after the chicks, feeding them and leading them around. I have seen this as a fact and have marvelled at it.

It is said of the capon that after six years it bears a stone called the *electorium* in its liver.[375] From then on it does not drink and, therefore, a man wearing this stone on his person is said not to thirst.

The capon has good flesh, more solid than a hen's. The ancients called the *gallinacius* a *papo,* but we moderns [*moderni*] call it a *capo.*

(54) GALLUS SILVESTRIS: The wild cock is called a *fasianus* [pheasant].[376] It is most beautiful, adorned with both ruby and fiery-red feathers, with some green mixed in on the head. But it neither has a crest on its head nor a spur [*spicula*] on its legs even though it is quite bold. It has an azure color on its neck and body as well as a bit of yellow, earthy color.[377]

It is a silly bird and easily deceived. If a white, rectangular cloth is stretched between four rods or poles so that it does not bend and if it bears the picture of a ruby-red-colored pheasant, the pheasant is so amazed at the picture that it does not notice the hunter behind it and is thrust into a net prepared well in advance.

A rectangular net is also set up in the way mentioned and is held up with a pole in such a way that it can fall at the slightest movement. To the pole holding up the net is attached a rope stretched beneath snow or leaves, running to the fowler's hiding place. Beneath the net a bait of oats is placed, this being the pheasant's food. The net is then dropped on the pheasants congregated there.[378] A pheasant is also caught in snares stretched over the paths on which it passes from the forest to the water.

This bird has a less decorated hen. It has very tender flesh, white and moderate.

(55) GARRULUS: The *garrulus* [jay] is named for its garrulousness and the Germans call it a *hester.*[379] It is so painted with colors that no color seems to be missing, and toward the lower part of its wings it has the most beautiful azure.

---

375. This story, found also in ThC 5.59, is attributed by A. to *Jacobus et Liber lapidarius.* The name derives from the Gr. *alektryōn,* "rooster." Cf. Pliny *HN* 37.54.144, Jacob. 91 (p. 197), *De min.* 2.2.1 (Wyckoff, 1967, 73), and Arnoldus Saxo 3.35b. The "stone" could be some sort of fibrous growth or something in the crop.

376. *Phasianus colchius;* cf. *GB,* 298–300; *BAZ,* 186–87; Pliny *HN* 10.67.132; Isid. *Orig.* 12.7.49; Neckam 1.42; ThC 5.60; Vinc. 16.72; PsJF 309; Rowland (1978), 133–34; and Yapp (1981), 30–31. On the older Germanic names for the bird, readily Latinized as *fasianus,* cf. Suolahti (1909), 226–28.

377. Stadler's *lazurini* is a clear error for *azurini.*

378. Yapp (1981, 28–29, fig. 15) perfectly illustrates this technique, with partridges as the prey.

379. This is *Garrulus glandarius,* the European jay, far more colorful than the American bluejay (Coombs, 1978, 178 and Pl. 8). Cf. *GB,* 146–48, and Pollard (1977), 55. Isid. *Orig* 10.11.4, ThC 5.61, Hugo of Fouilloy 50 (Clark, 1992, 220–25), and Vinc. 16.89. Of moderns, see Rowland (1978), 87–88, and Yapp (1981), passim. On the German name, MG *(Eichel)haher,* see Sanders (431–32).

This bird cries at all sorts of voices and imitates the voices of all.³⁸⁰ For this reason it is called the *marcolfus* by some.³⁸¹ When kept in cages it not infrequently copies the speaking voices of humans.³⁸² It sometimes rages so much in its anger that it destroys itself by hanging itself on forked limbs.

(56) GRACULUS: The rook is a black bird of the raven genus with a large beak that is white where it joins the head. It does not eat carrion. It builds its nest sociably in high trees so that there are many nests in one tree.³⁸³

It is a clamorous bird, especially at mating and hatching times. Its chicks are fit to eat when skinned of feathers and skin.

(57) IBIS: The *ibis,* whose name in the genitive case is *ibidis,* is multicolored when found along the Nile in Egypt, but is black in Ethiopia.³⁸⁴ It is a large bird, mimicking in many ways the nature of the stork. But it is not a stork because although it has a long beak, the beak is curved.

This bird fights with a particular serpent which is also called *ybis* but whose name is declined *ybis, ybis, ybi.*³⁸⁵ It fights with it because it has power over every venomous creature and brings serpent eggs to its chicks as a greatly desired food. Occasionally, when these *ybes,* serpents of Arabia, fly from Ethiopia, the ibises make their way

---

380. Cf. Ar. *HA* 615b19–20.

381. Suolahti (1909, 202–3) cites this passage and discusses its application.

382. Cf. *GB,* 147, for many classical references to this talent.

383. Cf. Isid. *Orig.* 12.7.45, ThC 5.62, Vinc. 16.89, Hugh of St. Victor 1.45, PsJF 301, and Hugo of Fouilloy (Clark, 1992, 220–25). Although not definite, the identification of the *graculus* as a rook, *Corvus frugilegus,* is fairly secure. Scanlan (297) points out correctly that juveniles often have a bare patch at the point where their lower mandible joins the head (cf. Goodwin, 1986, 78). The *graculus* does not eat carrion, and rookeries are commonly found, heavily populated, in tall trees (80). More tellingly, at 6.50, A. gives us the Germanic equivalent of *graculus* as *ruck*.

Some would argue that the bird here is a jackdaw, *Corvus monedula*; cf. *GB,* 155–58, and Pollard (1977), 27–28. Yapp (1981, 57) argues that *graculus* indicates a jay rather than a jackdaw, and Clark agrees (1992, 221, n.1). A. is quite specific in denoting the genus of this bird as "of the raven genus [*corvini generis*]," but even in modern terms ravens and jackdaws all belong to the family *Corvidae.* Neither eats carrion. Yet the jay is clearly identified in the previous entry through its Germanic name and it has a smallish bill (Goodwin, 1986, 195, fig. 6). The light area where the bill of the *graculus* meets the head is not readily discernible in either bird, but there is some lightening on the jackdaw (Goodwin, 1986, 73). Diefenbach (1857, 267), s.v. *graculus,* shows that the confusion is not merely modern, for he lists as synonyms words that indicate the rook and the jay. Since identification of the jay, for A., is clear, then rook seems the best choice here.

384. A. makes it clear that he, unlike ThC 5.63, both knows the correct grammatical forms and can distinguish the bird from the snake listed below. ThC lists *ibis, ibicus.* The stem *ibic-* refers, in fact, to a goat.

The most commonly known ibis is the sacred ibis, *Threskiornis aethiopicus,* which is white but has a black head, neck, and legs. This may fit A.'s *varia.* Less likely is that the variegated ibis may be the hermit ibis, the 'glossy ibis' of Hdt. 2.75f., which does feed on serpents (*GB,* 108–9; *BAZ,* 73–75). Scanlan (297) identifies this as A.'s black, or "dark," ibis. Other sources on the ibis include Pliny *HN* 8.97, Solinus 32.32, Isid. *Orig.* 12.7.33, Neckam 1.55, Jacob. 90 (p. 190), ThC 5.63, PsJF 302, and Vinc. 16.96.

385. A. dutifully gives the nom., gen., and dat. sing. cases to show the stem is different from that for the name of the bird.

together with them in an amassed battle line and devour them.³⁸⁶ As a result the bird's eggs and flesh are made poisonous.

Even though it is an aquatic bird, it does not enter the water but instead stays near the water and gathers small fish, cast-off carrion, and other animals it finds, especially snakes.

When this bird is constipated, it takes the food out of its anus with its beak, giving itself an enema by injecting salt sea water into its posterior, in this way relieving itself. This is how, according to Galen, from seeing things of this sort among ibises and cranes, the use of the enema syringe was discovered.³⁸⁷

(58) IBOZ: The *iboz* is a bird, so they say, of the eastern regions which is strong and an object of hatred for horses.³⁸⁸ It has a whinny like a horse but its voice is horrible and terrible to hear. The bird, by calling, drives the horses away from their pastures because it eats grass like a horse. Because of jealousy and avarice, the horse expels the same bird from the pastures when it settles on the meadows.

(59) INCENDULA: The *incendula* is a strong bird of the East, a member of the raven genus, which fights with the owl [*bubo*]. Because it sees more clearly by day than the owl, it overcomes the *bubo* by day and breaks and eats its eggs. At night, however, when the *bubo* has the sight advantage, it attacks the *incendula*, breaking its eggs and scattering its nest, repaying it for its malice.³⁸⁹

(60) IRUNDO: The swallow is a familiar bird. It is small, black on its back and wings, white on its belly, and red beneath its throat. It takes careful note of the seasons of its arrival.³⁹⁰ They say that this bird and the mouse cannot be trained, but I have often seen them tamed and flying to the hand as do other birds.³⁹¹ They construct their nests skillfully and lay eggs twice a year.

---

386. This story comes from Hdt. 2.75–76, where winged serpents fly from Arabia into Egypt. Thus, A.'s *Arabiae* is difficult here. Scanlan (298) takes it to indicate "to Arabia," an unlikely use of the dat. case. Taking it as a gen., as translated here, makes for better grammar but inferior word order.

387. Cf. Pliny *HN* 8.41.97. This odd habit led the bird to be a paradigm for filthiness in antiquity, leading Ovid to entitle one of his polemical poems *Ibis*. Cf. Galen *Introductio seu Medicus* 1 (Kühn, 14.675). The Latin *clyster* can indicate both the enema and the syringe used to administer it; cf. 11.19, 22.45.

388. Clearly based on the *anthus* of Ar. *HA* 609b15f. Thompson (1910, note ad loc.) discusses the origin of the legend, repeated in Pliny *HN* 10.57.116. This bird also appears in ThC 5.64, PsJF 301, and Vinc. 16.154. Cf. *BAZ*, 15.

389. This bird represents a conflation of Pliny *HN* 10.17.36, which mentions the unidentified "firebird," *incendaria avis,* later the *bubo,* and later still an unnamed bird in Nigidius that breaks eagle's eggs. Cf. ThC 5.65, PsJF (*nicedula*) and Vinc. 16.110.

390. The Latin is crabbed, but the true sense (viz., that swallows carefully mark the seasons for their migratory patterns) may be gleaned from the similar phrasing of ThC 5.66.24–25, which cites Isid. *Orig.* 12.7.70. Scanlan's (299) translation is not supported by the evidence. Cf. PsJF 302–03 and *BAZ*, 28–30. For illustrations, cf. García Arranz (1996, 405–35).

391. The mouse seems strangely out of context here, but the passage is from Pliny *HN* 10.62.128. Moreover, cf. the candle-holding mouse of 8.229 and 21.22. Birds were common medieval pets (Walker-Meikle, 2012).

Four genuses of them are found: the ones in homes, the ones outside in walls, the ones on the ground above rivers, and the ones called "marine."[392] They strengthen their nests with straw and mud so that they are firm. All the ones that build nests in walls close the nests, except for an opening through which they enter. Sparrows sometimes take the nests away, but the swallows, coming in an army, bring mud in their mouths and quickly close them up in the nest, suffocating them.[393]

It is said of this bird that if the eyes of their chicks are gouged out, they grow back and the chicks regain their sight.[394] Also, if a spot arises in their chick's eyes, the mother smears them with the juice of the *celidonia*.[395]

Blood taken from under its right wing is a cure for eyes, but its excrement, if warm, blinds the eyes.[396]

The swallow eats flesh, and especially preys on flies and bees. It gathers its food on the wing as it encounters it in the sky. It is an excellent flyer but a poor walker.[397] It has a long, forked tail. It is garrulous and sings to announce the day.

Its chicks in August sometimes bear a stone in their liver or stomach which people call the *celidonius*. I have found the stones to be of different colors.[398] It is said that the chicks bearing a stone can be recognized in that they sit with their faces turned toward each other, whereas the others sit with their posteriors turned toward each other. The stone-bearing chicks cast their droppings off the edge of the nest.

Many other things have been said about swallows in preceding books on animals as well.

(61) IPSIDA: The *ipsida* [kingfisher] is a beautiful bird called the *isvogel* in German.[399] On its back it has a color midway between green and blue, which, when the sun's rays hit it, seems to be sapphire-colored. On its chest it has the color of glowing coals. It flies near water and preys on fish and grubs.

Concerning this bird, they say that when it molts its feathers each year, its skin is taken off and fixed on a wall. But I have proven this not to be true for several of them.[400]

---

392. In order: the chimney swallow, *Hirundo rustica*; the house martin, *Delicion urbica*; the sandmartin, *Riparia riparia*; and either the cliff swallow, *Hirundo rupestris,* or the common tern, *Sterna hirundo.*

393. Cf. 8.19, where three genuses of swallows and their enemies are mentioned.

394. Cf. Ar. *HA* 563a15 and above at 2.101.

395. Cf. Pliny *HN* 8.41.98. *Chelidonia*, celandine or "swallowwort," *Cheldonium maius*, takes its name from the Gr. for swallow, *chēlidōn*. Cf. *De veg.* 1.176.

396. Stadler alertly notes the allusion to Tob. 2.11 (Vulgate).

397. Cf. Ar. *HA* 487b24f.

398. On "swallowstone," cf. *De min.* 2.2.3 (Wyckoff, 1967, 79–80), where A. mentions two colors, one black and one reddish brown, and describes how he had seen some stones extracted from the birds in the month of August. Cf. also Pliny *HN* 11.79.203, 30.27.91–92 (as a remedy), and 37.56.155 (two types).

399. For the bird and the German name, cf. 7.170, with notes. The kingfisher is *Alcedo ispida.* Cf. ThC 5.68 (*isida*), and Vinc. 16.154. *Ispida* is a simple transposed form of *(h)ispida*, "bristling."

400. This statement is in contrast to ThC 5.68.8, which says "whether true or false, I know not."

Those who follow auguries say that if this bird is kept in a treasure house, it increases the treasure and abolishes poverty.

(62) KYRII: The *kyrii* is a bird that lives on prey. It reproduces well and it happily raises not only its own chicks but also the chicks of the eagle which have been cast off due to difficulty in feeding them.[401]

(63) KARKOLOZ: The *karkoloz* is a lazy bird which neither incubates its own eggs nor raises its chicks.[402] Rather, it puts its egg stealthily into the nest of a dove, having first broken the dove's eggs. The chicks are then incubated and raised by the doves.

(64) KOMER: The *komer* is a bird which produces and raises young five or six times a year.[403]

(65) KYTHES: *Kythes* are birds that call in different voices, so much so that they change voices almost every day.[404] They make nests in trees out of wool and hair and there they raise their chicks. When the acorns are ripe, they amass a hoard of them.[405] When the chicks are grown, they put their parents back in the nest they have just left so that they will work no more and they feed them out of a natural piety.[406]

(66) LARUS: The "sea gull" is a bird of the kite genus which both swims in the water and flies in the air in search of prey.[407]

---

401. Surely based on the *phēnē* of Ar. *HA* 619b23–26, usually identified as the lammergeier (*GB,* 303). Cf. ThC 5.70, which has the same passage but calls the bird the *kym* and cites Ar. as the source. That the *kym* is the lammergeier is shown from 6.46.

402. This is clearly a cuckoo. St. plausibly suggests Ps. Ar. 830b11 as a source, but any passage with the form *kokkyges* ("cuckoos," nom. pl.) would suffice.

403. The form *komor* is found in ThC 5.72, which cites Ar. and calls this an "Arabian bird." St. asks if this can all be traced to Ar. *HA* 568a17, where the fish called the *kyprinos* is said to bear five or six times a year.

404. Cf. ThC 5.73, Vinc. 16.100, PsJF (*kike*). The source is Ar. *HA* 615b19–23, which discusses the jay, Gr. *kitta*. Curiously, the *-es* ending reflects the modern Gr. pl. noun ending of *-a*.

405. In fact, the jay in question is named *Garrulus glanclarius,* "garrulous acorn-gatherer." The scientific name thus perfectly mirrors A.'s two main traits for the bird. On its diet of acorns and its propensity to hoard them, cf. Goodwin (1986), 197–99.

406. This section on "natural piety" comes from the passage in Ar. immediately following the one on the jay's voice, Ar. *HA* 615b23f., which discusses the piety of storks and bee-eaters toward their parents.

407. ThC 5.74 tells us that the source for this entry is the "Gloss on that place where unclean birds are prohibited in the Law." At Lev. 11.16, the *larus* is forbidden. For the gloss, cf. *Glossa ordinaria PL* 114:814D, which epitomizes Rabanus Maurus *Expositiones in Leviticum* 3.1 (*PL* 108:357A). Also cf. Vinc. 16.101, PsJF 304, and Hildegard of Bingen's *Physica* 6.26 (*PL* 197:1299). What, then, is a *larus?* It transliterates the Gr. *laros,* a rather broad term which included, in antiquity, at least several kinds of sea gulls (*GB,* 192; *BAZ,* 131–33). Suolahti (1909, 352–53) discusses the names which are similar to Hildegard's *musar* with mixed results. It may be, he concludes, a type of buzzard, or, by being related to the MG *Möwe,* be related to the sea gull. A.'s description seems to have the same ambivalence about it.

(67) LUCIDIAE: *Lucidiae* are birds with feathers that glow in the dark.[408] Therefore, having thrown their feathers ahead of them, they point out their paths and this is why they have taken this name.

(68) LUCINIA: The *lucinia* [nightingale] is a bird which, when sitting on its eggs, alleviates the boredom of the long night with sweet songs. It also announces the day in song. It also brings to life its eggs, when they are born, with a song. This is believed to be the nightingale [*phylomena*].[409]

(69) LINACHOS: The *linachos,* as Aristotle says, is a bird that strikes its young before their feathers are fully in and forces them to look into the orb of the sun. If any of their eyes tear up, it ejects them and feeds the others.[410]

It is a bird of prey, living at the seashore. It hunts birds swimming in the sea there and they, seeing the *lynachus,* submerge themselves under the water out of fear. But the *linachos,* lying in wait over the surface of the water, forces them to be killed. Then, because they are borne up by their feathers, they float. It then catches the birds and takes them off.

(70) LAGEPUS: The *lagepus* [ptarmigan] is a bird which, in accordance with its name, has hare's feet. It also has hair instead of feathers and flies poorly. It lives, therefore, in hollows beneath the ground. When it sometimes leaps out for prey, upon capture of the prey it immediately returns to the cave and eats it. This bird is not domesticated and, if captured, dies. It rots very quickly because of its poor complexion.[411]

(71) MILVUS: The kite is a familiar bird the size of a hawk [*accipiter*], with reddish feathers and a curved beak and talons.[412] It is bereft of strength and courage except

---

408. Pliny *HN* 10.67.132 talks of nameless birds in the Hercynican forest (Black Forest) of Germany whose feathers had these properties. ThC 5.75 cites as sources the *Liber rerum* and Solinus 20.3, making it clear that the feathers serve as lanterns for those people who possess them. Cf. Vinc. 16.101. *Lucidiae* may be a proper name or merely mean "shining."

409. This is the nightingale [*luscinia*] of Ambrose *Hex.* 5.24.85. Cf. Isid. *Orig.* 12.7.37, Rabanus Maurus *De universo* 8.6 (*PL* 111:247A), Hugh of St. Victor 3.33, ThC 5.76, PsJF 304, and Vinc. 16.102. Cf. also 23.137(100) for the *phylomena*. On the name, cf. the Italian *lusignuolo* and Old French *lousignol* (Körting, 1923, 612, s.v. *lusciniola*).

410. The *aliaetos* of Ar. *HA* 620a1–12, a sort of sea eagle. The story is also told by A. at 8.106 with variant names. Cf. ThC 5.77 and Vinc. 16.103.

411. This is the ptarmigan, sometimes called a willow grouse, *Lagopus lagopus*. Pliny *HN* 10.68.133, comparing the foot of the *lagopus* to that of a hare (the bird's name means "hare-foot" in Gr.), says its covering serves the same thermal purpose. However, his *non extra terram eam vesci facile* ("outside that region it is not easily fed") becomes misinterpreted in ThC 5.78 and here to mean that the bird eats underground (i.e., "not outside the earth") and an appropriate cavern is even added (cf. Aiken, 1947, 29–30). A. adds that it flies poorly.

412. *Milvus milvus,* the red kite. Much of this section comes from Pliny *HN* 10.12.28. For ancient lore on the kite, cf. *GB,* 119–21, Pollard (1977, 39–40), *BAZ,* 76–77. Cf. Barth. 12.26; ThC 5.79; Vinc. 16.108; Rabanus Maurus *De universo* 8.6 (*PL* 111:252D); Hugh of St. Victor 1.40; and Hugo of Fouilloy (Clark, 1992, 206–7). For its medieval lore, cf. Rowland (1978), 93–96; McCulloch (1960), 135–36; Yapp (1981), 34; and George and Yapp (1991), 146.

that it preys upon domestic chickens. It is a gouty [*guttosa*] bird and suffers gout [*podagram*].⁴¹³ It is, therefore, timid and for this reason sometimes lies quiet. Near meat markets it preys on flesh and keeps a watch out for carcasses. It lies quiet most often near the solstices, for then it is more gouty [*podagricus*] and gouty [*guttosus*].

(72) MAGNALES: *Magnales* are large birds of the East with black feet and beak.⁴¹⁴ They do not harm humans, but they prey upon fish in rivers, pools, and other waters, and eat them.

(73) MELANCORIFUS: The *melancorifus*, so called because it is deadly [*mortificat*], is a bird, as Pliny says, of small size but great fertility. For it sits on more than twenty eggs and raises all the young. It feeds them so diligently that they are quite fat in the nest. When they fly from the nest they follow their mother in hoards. Nor does she desert them until they are ready to care for themselves.⁴¹⁵

(74) MORFEX: The *morfex* [cormorant] is an aquatic bird which is called the *scolucherem*.⁴¹⁶ It is black, has a serrated beak and strong claws, and dives beneath the water and catches large fish, especially eels. It builds its nests in groups in trees near water and feeds its chicks (which are voracious) with fish. It is said of these chicks that, when it is time for them to fly from the nest, if they sense they are too weighed down by recently eaten food, they vomit it so as to be light for departing and to be able to fly. The ones that do not vomit sometimes die. The excrement of this bird dries out the trees on whose branches it falls. When this bird is drenched, it spreads its wings to the sun to dry while sitting on posts and trees. It just barely raises itself up when it begins to fly, dragging its tail in the water for a long time. For this reason it is called by some the "wet ass."⁴¹⁷

(75) MEMNONIDES: Some people name the *memnonides* after some birds which the Egyptians name after a place. They fly in flocks from Egypt to Ilium, to the tomb of Memnon, a Pythagorean philosopher. They always do this in the fifth year

---

413. A. is using *podogra* to indicate gouty feet and *gutta* as a more general term for any inflammation of a joint, what we might call a rheumatism, or some other condition caused by the dripping of bad humors.

414. Jacob. 90 (p. 191) describes the bird without giving its name. Cf. also ThC 5.80, who gives Ar. as his source. The name, of course, is simply composed of *magnus* + *ales* = "big bird."

415. A.'s name is a version of the Gr. *melankoryphos*. Thompson (*GB*, 195–96) suggests several possible identifications, including a titmouse, the blackcap warbler, or the blackheaded bunting. The passage is an elaboration of Pliny *HN* 10.79.165 or Ar. *HA* 616b4–9. It is worth noting that the etymology was added in the margin by Stadler's M1.

416. *Phalacrocorax carbo*, a cormorant. The German name evokes memories of the *schalvehorn* of 5.28, q.v. for references. The MG is *Sharbe*, and it is again interesting to note that Stadler's M3 wrote *scherb* in the margin. Cf. Suolahti (1909), 393–97.

417. The vulgarity is A.'s; it is not found in ThC. The Latin, *humidus culus*, is a bit more vulgar than rendered here. On the name, cf. Sanders (435) and Suolahti (1909), 396.

and, when they have flown around for two days, they enter into a fight on the third day, cutting each other with their beaks and talons. They then go back to Egypt.[418]

(76) MEAUCAE: *Meaucae* are birds named in imitation of their voice.[419] They are bigger than ducks and are slender, with a short neck and feet, an ash-gray color, gray eyes, and a beak that is partly yellow and partly red. It is always calling "meauca" and is very eager for carcasses, especially those of humans. It therefore rejoices when there is a storm. It also preys on small animals, however. Among sea folk many other birds are also called *meaucae* after these birds.

(77) MERILLIONES: *Merilliones* [merlin falcons] are small birds but are still birds of prey and have something of the falcon's nature.[420] They fly as allies against prey and sometimes are taught to knock down a swan, doing so in the hope of a human's assistance. One attacks the swan's head, two sit on its wings, and one attacks its chest. By overwhelming it they cast it down to be captured by the falconer. When wild, however, they pursue small birds.

In color and shape, they are almost like blackbirds [*merulae*] except that they have curved feet, talons, and beaks to seize with.

(78) MUSCICAPAE: *Muscicapae* are so named because they catch flies [*muscas capiunt*]. They feed on these and grubs alone.[421]

It is a larger bird than a turtledove or a pigeon, with a lanner's color on its wings. It is slow in flight and has the feet and beak of a swallow. It opens its mouth very wide because with this gape it surrounds the many flies which enter its mouth for its humor.

(79) MEROPS: The bee-eater is a bird of the woodpecker [*picus*] genus whose chattering the augurs especially observe. It is somewhat green and therefore is called the green woodpecker [*picus viridis*]. It has on its back, though, some dark blue and on its chest it is off-red and it is whitish on its belly.[422]

---

418. The fabled birds of Memnon. There are several versions of the myth, but all turn around the slaying of Memnon, the son of Dawn and Tithonus and king of Ethiopia, in the Trojan wars. Many Gr. vase paintings depict his mother bearing his body away after he had been killed by Achilles. The bird is traditionally identified with the ruff, *Machetes pugnax*, which engages in mock battles with other males of its species in the general area of the Hellespont (Pollard, 1977, 101; *GB*, 200–201). The story appears in Pliny *HN* 10.37.74, Solinus 40.19, Isid. *Orig.* 12.7.30, ThC 5.83, and Vinc. 16.104.

419. The bird is clearly a sea gull but the exact type is unclear. Scanlan (303–4) suggests Andouin's gull, *Larus andouinii*, basing his identification on the color and on the bicolored bill. The name is onomatopoetic and is reminiscent of the MG *Möwe* (Suolahti, 1909, 397f.; Körting, 1923, 641, s.v. *mauwa*). Cf. ThC 5.84 and Vinc. 16.104.

420. Cf. ThC 5.75 and Vinc. 16.106. The name is one of many for the merlin: *mereella, merulus, emerlio, smerilio,* and others.

421. A modern genus is *Muscicapa*; perhaps the common spotted flycatcher, *Muscicapa griseola*, is meant (*GB*, 189). Cf. ThC 5.86 and Vinc. 16.109.

422. The Gr. *merops* is usually taken as *Merops apiaster* (*GB*, 201–3; Pollard, 1977, 46–47), but the description of this bird in Ar. *HA* 615b25f. does not really seem to coincide with the present passage, which perhaps refers to the green bee-eater, *Picus viridis* (Scanlan, 304). Cf. Pliny *HN* 10.51.99, ThC 5.87, and Vinc. 16.106.

With a natural cleverness it digs a hollow in the earth six feet deep. There it builds its nest and raises its young until they are fully grown.[423]

(80) MERULA: The *merula* [blackbird], they say, is so named because in antiquity it was called the *modula*, giving forth songs and music [*modulos*], as it were.[424] Thus it is that some say that a blackbird was taught artificially to sing so well the nine types of notes out of which every song is composed that no person could perfectly imitate it. As if glorying in the fact, that bird often formed the notes in order in the presence of people. Some, however, say that the *merula* is so named because it flies *meir*, that is, alone.

It sings in spring but in winter it stammers.[425] It is a black bird and a bit dusky on the chest. Enclosed in a cage it sings longer due to fatness and then, against its nature, eats meat, and due to this it sings all the more readily.[426] Contrary to the habit of birds, it molts not its feathers but rather its beak.[427] Its beak and feet are a yellowish-red that shines greatly. Although it is black everywhere else, in parts of Achaia it is white.[428]

There is, however, a genus of blackbird called the *caprimulgus* [goat milker], since it enters stables and sucks at the she-goats. Due to this the she-goats get sterile udders and blind eyes. This is as we have said in earlier parts of this book.[429]

The Romans also call the solitary sparrow [*passer solitarius*] the *merulus stercosus* [dung-laden blackbird] since it lives in old ruined latrines. But we will follow up on this below.[430]

(81) MONEDULA: The *monedula* [jackdaw] is named for money taking [*monetam tollens*] or delighting in money [*monetam diligens*].[431] It is a black bird and on the top of its head there is a bit of black tending to a shiny ash-gray color. It is a pleasing bird and whatever silver or gold money it finds, it takes away and hides.[432] It copies human voices when trained to do so as a young bird. It learns and copies these voices more readily in the morning, as is true for every other bird. It has black feet and

---

423. Ar. and Pliny both have the young feeding the parents, the reversal occurring both in this text and ThC. Cf. Aiken (1947, 220) for a brief discussion of this change.

424. *Turdus merula*, the European blackbird, Gr. *kossyphos* (*GB*, 174–76; Pollard, 1977, 35–36). Cf. ThC 5.88, Vinc. 16.107, Hugh of St. Victor 1.43, and Hugo of Fouilloy (Clark, 1992, 214–17). The etymology is loosely based on that of Isid. *Orig.* 12.7.69. The bird had many symbolic uses and was often depicted in illustrations. Cf. Rowland (1978), 10–14; Yapp (1981), 65; and McCulloch (1960), 96–97.

425. Pliny *HN* 10.42.80 and Ar. *HA* 632b16–17 say the bird is silent in the winter.

426. Rowland (1978, 11) has a nice illustration of a caged blackbird.

427. Pliny *HN* 10.42.80 says that the beak of young blackbirds is turned (*transfiguratur*) into an ivory color. ThC or his source renders this as *rostrum . . . mutat in candorem*, "changes into a whitishness." Apparently, *in candorem* disappears along the way and the birds are left shedding their beaks.

428. For a discussion of the famous and perhaps mythical white blackbirds, see Pollard (1977), 35–36, and *GB*, 174–75.

429. Cf. 23.19(4).

430. Note the change to the masc. for the bird's name. On this special variety, cf. 23.137(99).

431. On the jackdaw, *Corvus monedula*, in antiquity, see *GB*, 155–58, Pollard (1977, 27–28, *BAZ*, 104–05, and the comments on the confusion with the *graculus* at 23.120(56). Cf. Isid. *Orig.* 12.7.35; Jacob. 92 (p. 220); ThC 5.89; Vinc. 16.109; Yapp (1981), 57; and Rowland (1978), 86–87.

432. Pliny *HN* 10.41.77.

delights in having its head scratched. For this reason if its flesh is eaten, it induces itching of the head.

(82) MERGUS: *Mergus* [diver] is not so much a species of bird as a genus containing many species.⁴³³ The birds, however, that are generally called *mergi* are multicolored like the woodpecker and have the appearance, as far as size, beak, and feet are concerned, of ducks. They take in air by breathing and can therefore stay under water no longer than they can hold their breath. The chicks of this bird are so strong that as soon as they leave the egg they can feed themselves without their mother.

It is said that at the time of a storm these birds take themselves away from the sea to safe places on shore and by so doing give advance warning of storms.⁴³⁴

(83) NISUS: The *nisus* [sparrow hawk] is a bird also called the *sparvarius*.⁴³⁵ It is much smaller than a hawk [*accipiter*] but similar to it in color. It is named for the struggle [*nisus*], that is, the attempt it makes for prey, for it tries to catch birds such as the pigeon, duck, and little crow [*cornicula*], which are stronger than it is. It hunts alone, spurning allies, and I have found by experience that if there are two of them, one attacks the other and the bird gets away.

It is said that in winter it holds a live bird beneath its feet for its warmth and that in the morning, mindful of the favor, it allows it to go away alive, but I have not experienced this. This bird's diet has been established in things said before.

(84) NOCTUA: The *noctua* [owl] is not, as some falsely claim, the *nocticorax*.⁴³⁶ For the *noctua* is a bird with multicolored feathers on its legs, a huge head, and a curved beak, and it is larger than a sparrow hawk [*nisus*]. Due to the gray color of its eyes, it cannot bear the sun but rather flies at night capturing mice and *bruci*.⁴³⁷ If it does appear during the day, its feathers are plucked out by other birds. But it defends itself on its back with its beak and talons.⁴³⁸ The hawk [*accipiter*] helps it if it is nearby on account of the similarity of their genus. The color mix of the *noctua* tends toward white and it has a harsh voice.⁴³⁹

---

433. Cf. ThC 5.90, Vinc. 16.105, PsJF 304, and Barth. 12.25.

434. ThC cites Ambrose *Hex.* 5.13.43, but the same content is found in Isid. *Orig.* 12.7.54.

435. The sparrow hawk, *Accipiter nisus,* was discussed at 23.109. Cf. ThC 5.91, Vinc. 16.112, Neckam 1.29.32, PsJF 304–054, and Fred. *DAV* 2.29 (W&F, 127).

436. No certainty seems possible given the many names for "owl" in the medieval panoply and the writers themselves seem confused. Complicating matters is the confusion of two names. Isid. *Orig.* 12.7.40–41 uses *noctua* more generally and says specifically that *nicticorax ipsa est noctua,* a statement echoed by ThC 5.92.2 (*nocticorax* is A.'s spelling). Rabanus Maurus *De universo* 8.6 (*PL* 111:247A) clearly states that some call the *noctua* a "sea crow" (*corvus marinus*) but that it is not the same as a *bubo,* and later in the same passage he tells us that the *noctua* was rendered as *glaucus* in the Septuagint. Hugh of St. Victor 1.43 equates *noctua* and *nycticorax.* Cf. Vinc. 16.111. George and Yapp (1991, 148–50) discuss at length the *nicticorax/noctua* problem. On the *nyktikorax,* cf. *GB,* 207–9, BAZ, 152–53, and the additional work of Oliphant (1914), 61–62.

437. The *brucus* is a type of locust. Cf. 26.11(8).

438. Cf. Pliny *HN* 10.19.39.

439. The terms for coloration here are unusual for A. and difficult to pin down. It may mean: "The variegated color pattern of the *noctua* tends toward whiteness."

The *nicticorax,* however, is a night raven [*corvus noctis*]. It has a variety of black colors and its call is "cho." When it calls, it turns in a circle by turning itself around in the air. It is smaller than the *noctua,* but it has the same habits and hunting patterns and avoids the light as the *noctua.*[440]

(85) NEPA: The *nepa* [snipe] has a long beak and has the coloration of a partridge on its back and that of a sparrow hawk [*nisus*] on its belly.[441] It sticks its bill into mud and searches for its food of vermin. If, on occasion, it becomes stuck, with its bill placed in too deeply, it digs at the mud with its feet and frees itself. It rests by day, flying about at dusk and dawn, at which times, therefore, it is caught in nets placed aloft. It has flesh that is sweet to eat. Some also call it the *fiscedula.*[442]

(86) ONOCROTALUS: The *onocrotalus* [pelican] is a bird of prey with a long beak which it puts in water or in mud on land and, using it like a horn, emits a horrible call.[443] There are two genuses of them, aquatic and forest dwelling [*silvestris*], and they each, alone of the birds, lack a spleen.[444] It has large sacks on its throat to which food is first sent to be softened and is later sent to the stomach to be digested. It has no other receptacle for food and is therefore said by some to chew its cud.[445]

(87) OTHUS: The *othus* [an owl] is a bird smaller than the *bubo* but larger than the *noctua.*[446] It has, however, the appearance of a *bubo* and is therefore called a *bubo*

---

440. Thompson (*GB*, 207–9) outlines the history of the confusion of these two birds and identifies the *nycticorax* as a horned or long-eared owl, *Strix bubo,* but admits of a wide number of options. Scanlan (307) identifies it as a nightjar, *Caprimulgus europaeus.*

441. *Scolopax rusticola,* the European woodcock or snipe (*GB,* 261–62).

442. Cf. *ficedula,* 23.111(44).

443. Commonly identified as *Pelecanus onocrotalus,* this is the roseate or white pelican. Yet Thompson (*GB,* 212) cites Aldrovandi who, in language reminiscent of this passage, identifies it as the bittern, centering on its booming call. The pelican itself is mute. Cf. the myriad identifications under the name in Diefenbach (1857, 396). Suolahti (1909, 300–301), speaking of the *Purpurhuhn,* "purple-bird," found in the oldest German biblical glosses, points to Lev. 11.18, where the *onocratalus* is mentioned just before the *porphyrio* (cf. Deut. 14.17–18). Of interest in this regard is Barth. 12.29 and ThC 5.99 which equate the *pelicanus* with the *porphyrio.* Cf. A.'s *porfirion* (flamingo?) at 23.132(91). For the *onocratalus,* see Isid. *Orig.* 12.7.32, Jacob. 90 (p. 191), Barth. 12.28, ThC 5.94, Vinc. 16.113, Rabanus Maurus *De universo* 8.6 (*PL* 111:246D), PsJF 305, and Hugh of St. Victor 1.27. For illustrations, cf. García Arranz (1996, 627–56).

444. *Silvestris* is often used by A. as an antonym of *domesticus,* but here "forest dwelling" seems preferable. Scanlan's "land bird" (308) is tempting in light of the statements A. makes at 23.132(90) but unsubstantiated. The "land" pelican might be an invention of the Middle Ages, conjured up to justify the verse of Ps. 101.7 (Vulgate): "I am made like the pelican of the desert." Rowland (1978, 132) points out that medieval clergy would recite this psalm at least once a week. Ar. *HA* 506a13f. discusses real and apparent lack of spleens in animals and lists the *aigokephelos,* "goat-head," as the only bird really lacking a spleen. This name itself is often taken as a corrupt form of another name for an "eared" owl (*GB,* 25).

445. Garbled from Pliny *HN* 10.66.131, which says that *onocratali* hold their food in their mouths and then pass it to the stomach like a ruminant does. Comments such as these, coupled with the "forest" version of the bird and compared to the clearer description of the *pellicanus* at 23.132(90), lead one to question whether A. knew the true nature of this bird.

446. This bird is clearly an owl, and certainly one of the "horned" variety as depicted in Yapp (1981), 38, figs. 22 and 23. Scanlan (308) suggests *Asio otus.* Thompson (*GB,* 339–40) withholds identification of the Gr. version, the *ōtos* ("eared"), and has ancient references. Cf. ThC 5.95 and Vinc. 16.117.

by some. It flies by night, preys on mice, and appears to have ears because of some feathers that are raised up like ears. Its cry is "hu, hu," as if a man gripped by cold were calling. These and other night birds are said by augurs to predict death when they persistently increase the number of their cries.

(88) OSINA: The *osina*, which some of our people call a *volmarus* [pelican], is a white bird, the same size or larger than a swan.[447] It has a very long, strong beak, and a large sack hangs down from the lower part of its bill in front of its chest. In this it collects its food, fish, swallowing them a bit at a time, until it digests them. It lives in extensive waters well stocked with fish because it quickly depletes other waters of fish even if there is a large supply of them.

(89) ORYOLI: Orioles are, as Pliny says, called this by the common folk from the sound of their voices.[448] They have a golden color except that they are speckled on their wings with yellowish markings. They are members of the woodpecker genuses and are called *widewali* by the Germans.[449]

These birds build nests of wool and down so skillfully that they seem to copy the skill involved in making a cap. They hang the nests by cords on the thin, farthest branches of trees and as a result the nests seem to be hanging in midair. The nest seems to have, in all respects, the appearance of a ram's testicles. The bird makes a tunnel in this for its entryway.[450]

(90) PELLICANUS: The pelican is named for its white skin [*pelle cana*] since it has white feathers.[451] It is said to live in Egypt near the Nile, and there are said to be two genuses of pelicans: an aquatic one that lives on fish and a land-dweller that lives on vermin and snakes.[452] It is said to delight in crocodile's milk, which a crocodile spreads over the mud in swamps. This is why the pelican follows the crocodile.[453]

132

---

447. Unlike the *onocrotalus* of 23.131(86), we can here be fairly sure that A. knows he is speaking of a pelican (*GB*, 212), although both Sanders (424) and Suolahti (1909, 406) hold out the possibility that some sort of gull is meant. Cf. ThC 5.96, who calls it the *osma*.

448. *Oriolus oriolus,* the golden oriole, the name deriving from the Latin *aureolus,* "golden." Cf. Vinc. 16.115 and ThC 5.97.

449. Suolahti (1909, 169–71) discusses this name in depth.

450. Cf. the description of the nest given by Pliny *HN* 10.50.96–97, itself a somewhat confused passage. The two descriptive passages, comparing the nest to a hat and to a ram's testicles, seem to be A.'s additions to the tradition.

451. Cf. *BAZ,* 172–73. On the etymology, cf. Isid. *Orig.* 12.7.26. For other "pelicans" in A., cf. 23.131(86) and (88). The pelican has a long history in animal lore. Cf. *Physiologus* (Curley, 1979, 9–10; Carmody, 1939, 17–18), White (1954), 132–33, ThC 5.98, Vinc. 16.127, Jacob. 90 (p. 191), Neckam 1.73–74, Barth. 12.29, Peter Damian *De bono religiosi status* 15 (*PL* 145:776A), Rabanus Maurus *De universo* 8.6 (*PL* 111:251A), PsJF 312, Hugh of St. Victor 1.33 and 2.27, and Hugo of Fouilloy 38 (Clark, 1992, 168–71). For the rich symbolism surrounding the pelican, cf. notes on 23. 131(86). Cf. also *GB,* 231–33; Rowland (1978), 130–33; McCulloch (1960), 155–57; and George and Yapp (1991), 137–38. Yapp (1981, 160, Pl. 41) offers an excellent illustration drawn from life.

452. On this "land" pelican, cf. notes to 23.131(86).

453. At 1.227, A. tells us he has seen and examined two crocodiles. He surely had discovered that they were not mammals. Perhaps "milk" here is to be understood in its other sense of milt, the milky substance that male fish spread during fertilization.

They say that this bird kills those chicks which are hostile toward it and that after they are mourned a while, it recalls them to life with blood produced by biting its own breast. In the same way it brings to life those that have died from the bite of the snake that preys on its young. It is so weakened from the wound and from the blood that it stays in the nest and the chicks are forced to go forth for food for themselves and their mother. If there are some that do not wish to feed their mother, either out of laziness or impiety toward her, she ejects them when she has regained her strength and allows the pious ones which cared for her to follow her.[454] These matters, however, are more read about in stories than proven by experience through natural science.

(91) PORFIRION: The *porfirion* [osprey? flamingo?] is a bird, as some say, of the outer regions which has one foot like a goose for swimming and the other with separated toes, like a land bird.[455] This bird alone among the others has the habit that it drinks water by drawing it up in its foot, and that it puts food in its mouth with its foot.[456] It has to drink, moreover, at every mouthful of food since, due to weakness in its appetite, the food does not go down any other way. The better specimens of these birds are those possessing a large beak and long legs. But these things are more to be read in stories than proven by experience.

(92) PAVO: The peacock is a familiar bird with a small serpent-like head, crowned with long feathers.[457] It has a sapphire-colored, long neck, a red head, and the chest also has a sapphire color and is shiny. It has reddish wings and an ash-gray back tend-

---

454. This tradition is found in most of the medieval works (cf. the sources listed in *GB,* 233). The scene was frequently depicted: cf. George and Yapp (1991), 137, fig. 91; Clark (1992), fig. 10b; and Benton (1992), 22. The scene still adorns the state seal of Louisiana. Despite A.'s scepticism, the story is actually based on pelican behavior, misobserved and misunderstood. Pelican chicks fight viciously with each other, often drawing blood which can be observed on the brooding parent. Moreover, they have a habit of fainting near feeding time only to wake up after a period of seeming lifelessness. For bibliography and further discussion cf. Kitchell (2015, 133–35).

455. Originally this bird was the Gr. *porphyrion, Porphyrio veterum,* the purple waterhen (*GB,* 252–53), but by the time it reaches A. it seems to have changed character. ThC 5.99 seems to know of two birds by this name. He cites John the Philosopher for the unusual disposition of its feet and goes on to describe it as a bird of prey, which, taken with A.'s description of the feet at 2.70, and that of Gerald of Wales 12 (O'Meara, 1981, 42–43), makes one think of an osprey. But an osprey certainly does not fulfill the rest of A.'s description of the *porfirio.* We have mentioned above, in the notes to 23.131(86), a possible connection to the pelican, but ThC goes on to refute the belief that the *porphirio* is a *pelicanus.* Thompson (*GB,* 253) suggests that A. had the flamingo in mind, and much of the description seems to fit this bird. In the end, perhaps, we should not put more stock in the bird than A. does, who consigns it more to the realm of fable than fact.

456. This is based on Pliny *HN* 10.58.129.

457. The peacock, *Pavo cristatus,* has a long history in animal literature. For ancient references, cf. *GB,* 277–81, Keller (1909), 2:148–54 and *BAZ,* 236–38. For medieval sources, cf. White (1954), 149; ThC 5.100; Vinc. 16.122–25; Neckam 1.39–40; Barth. 12.31; Hugh of St. Victor 1.55; PsJF 308–09, and Hugo of Fouilloy 59 (Clark, 1992, 244–51). For its symbolic uses and for illustrations, see Rowland (1978), 127–30; Yapp (1981), passim; and McCulloch (1960), 153–54. For illustrations, cf. García Arranz (1996, 601–26).

ing toward red.⁴⁵⁸ The male has a long tail with plume feathers, and on the end of the feathers it has circles that have the splendid green of *crisolitus* and which are adorned with the colors of gold and sapphire.⁴⁵⁹ In cold, moist places there are also found white peacocks which present the colors mentioned but as if against a white cloth.⁴⁶⁰

The peahen [*gallina pavonis*] does not have any of the peacock's decoration except for a crown on her head and a bit of the sapphire color on her neck. As for the rest, she has a dusky redness and an earthy color.

The peacock's gait is simple and soft, its voice loud and terrifying. When heard, it puts to flight the serpents the peacock eats when it finds them.⁴⁶¹ This bird delights in its beauty. Especially in the spring, in its desire, it spreads and raises its tail. When the female comes near, in its desire it makes a sound with its tail feathers as if they were rustled by the wind.

It lays eggs in its third year, which is when it takes on its decoration. It lays three times a year and unless the eggs are laid on soft straw they are easily broken. The peahen [*pava*] hides her eggs from the cock since he, from too great a desire to sit on them, tramples and breaks them.⁴⁶²

When it is watched and is in the sun it shows its decoration, but it hides it as much as possible at other times. At night it is said to cry out because, not seeing its decorations, it thinks it has lost its beauty. It has white, tough flesh that lasts a long time without spoiling.⁴⁶³ If the chicks of a peacock get totally wet in the cold, they die.

(93) PERDIX: The partridge is a familiar bird whose color is reddish interspersed with long, black markings.⁴⁶⁴ It is called a *perdix* from the sound of its voice and

---

458. A.'s Latin is frustrating at this point, with too many ablatives in a row. The biggest problem is that peafowl simply do not have red heads and yet A. says he knows the bird well. ThC 5.100.4–5 says that the neck is long and sapphire-like (bright blue) in color; this coloration is correct. ThC then goes on to discuss the reddish body feathers, which do in fact lie below the wings (cf. Grzimek, 8.29, with color plate). One can only guess that in a moment of carelessness the "red" of the wings became misplaced in A.'s text. Note, however, that a poem in Ziolkowski (1993, 108, 291) describes the peacock as having a "breast like an emerald. In regal style the peacock shines crimson."

459. Wyckoff (1967, 82), commenting on *De min.* 2.2.3, states, "Albert's *chrysolitus* appears to be chrysolite, a pale green variety of olivine."

460. Ziolkowski (1993, 109, 244–45) presents a poem composed by Conrad of St. Avold which tells of the death of a white (not albino) peacock caused by an owl.

461. Grzimek (8:32): "The peacock enjoys a special reputation in India as an exterminator of cobras. In fact it enjoys eating young cobras. As a result, this poisonous snake soon disappears from peacock-inhabited territories. . . . Its predilection for snakes and its warning calls have made it a highly valued and popular bird in its home country, probably long before it was introduced as an ornamental bird in parks and gardens in other parts of the world."

462. Pliny *HN* 10.79.161 remarks, "the male peacock breaks the eggs, out of a desire for the females sitting on them." A. or his source has mistranslated.

463. The peacock was frequently cooked in ancient Roman times and Augustine tells us that a slaughtered peacock he had stored lasted more than a year without decay. Cf. *De civitate dei* 21.4.1 (*PL* 41:712).

464. For older citations, cf. *GB*, 234–38, *BAZ*, 309–10, Pollard (1977, 60–61) and Arnott, (2007, 174–76). Ambrose *Hex.* 6.3.13, Jacob. 92 (p. 221), Isid. *Orig.* 12.7.63, *Physiologus* 32 (Curley, 1979, 46–47), Neckam 1.43–45, Barth. 12.30, ThC 5.101, Vinc. 16.128–30, and Hugh of St. Victor 1.50. Alternate names include *cubaia* and *cubeg* (*DA* 5.18, with notes) and *cubez* (PsJF 294, 310).

from the perdition it suffers.⁴⁶⁵ Being a deceitful bird it steals other birds' eggs and sits on them. Its hope is frustrated, however, because when the chicks hear their genuine mother calling, they leave their nurse with her empty labor and go back to their natural mother.⁴⁶⁶

This bird makes its nest in the thick parts of thornbushes, and when someone nears the nest or the chicks, the mother presents itself as if too weak to fly and run so that the other follows her and is drawn rather far from the nest and chicks. Then the mother flies away, leaving the other one deceived. Many birds, both large and small, do this. Moreover, the chicks pull up little clumps of earth with their feet and hide beneath them.

Certain ones say falsely that the partridge, hen, peacock, and goose conceive orally.⁴⁶⁷ This is because feeling the sex drive, they turn to each other and, just as the doves do, they excite the males to copulation by rubbing them. They copulate and take in semen below, just as other birds.

It has as a characteristic that it smells bad at the time of intercourse because of the heat of desire which loosens the anus in them. It is through this that they take the semen into the pathway to their wombs, and its opening is directed toward the anus in all creatures that do not urinate. In those that do urinate, it is directed toward the mouth of the bladder. During the time they do not copulate, partridges turn their tails toward one another. They produce a lot of dung and, therefore, Pliny says, they do not grow fat.⁴⁶⁸ Its flesh is very temperate and tasty.

It is said to have a dry brain and that on account of this it is forgetful, even forgetting its own nest. But this is false and not said in accordance with natural science [*physice*] because memory is much strengthened by dryness. Moreover, a partridge which returns to its nest shows that it remembers it.

(94) PLATEA: The *platea* is, as Pliny says, a bird which flies against those birds which are in the sea and live on fish.⁴⁶⁹ It bites and clings to their heads, forcing them to give up the prey they caught by vomiting it. Sometimes it fills up on shellfish and, when it feels weighed down by the shells, it regurgitates them, throws away each shell, and eats the creature [*ostreum*] that lies within.⁴⁷⁰

---

465. Cf. Isid. *Orig.* 12.7.63. Neckam 1.44 makes the play on words even better, saying *perdix perdit ova*, "the partridge steals eggs."

466. This story probably arises from Ambrose *Hex.* 6.3.13 interpreting Jer. 17.13, "The partridge cried out, gathering the birds she did not hatch."

467. Perhaps a misconstruing of Ar. *HA* 541a26–30, which says that the female gets impregnated by simply smelling the male and also that partridges keep the mouth open and tongue out while copulating. Cf. Pliny *HN* 10.51.102.

468. Cf. Pliny *HN* 11.85.212, which, however, does not mention any connection with proliferation of dung. Scanlan's (312) suggestion that corruption of the Pliny text led to A.'s reading is plausible.

469. *Platea* means "wide" in Gr. The present entry stems from Pliny *HN* 10.56.115, and the bird is assumed by Cuvier (B&R, ad loc.) to be the spoonbill, *Platalea leucorodia* (GB, 193). This is, however, a wading bird, which generally sifts its food from the mud of marshes and swamps, not the open sea. Rackham (1940) translates Pliny's *platea* as "shoveller duck." Cf. ThC 5.103 and Vinc. 16.133.

470. The shellfish have opened in the presence of the bird's internal heat, much as they do when they are cooked. The bird has found an easier alternative to flying high and dropping them on rocks.

(95) PLUVIALES: Golden plovers are said to be birds that, even though they are quite fat, live on air alone.⁴⁷¹ But it has already been proven by us that air is not an element that can be incorporated into the body as nourishment. Moreover, those who say that nothing is ever found in the stomachs of these birds have put their trust in an inadequate proof. For the reason is that it has no other intestine except for the *ieiunum,* in which nothing is ever found. This is true for many animals and we have set forth the reason for it in the preceding parts of this book.⁴⁷²

It is said that this bird is caught by shooting leaded arrows beyond them in the air. Then, they grow fearful and fly lower, falling into the nets stretched out near the ground.

(96) PICA: The magpie is a familiar bird.⁴⁷³ It is multicolored with white and black, but on its tail the black color shines with a green tint mixed with sapphire.

This bird makes a nest and covers it with thorns on the outside. It builds two openings on opposite sides of the nest for entering and exiting. It frequently has two nests to fool those watching it and make them wonder in which nest it has its eggs.

It is said that this bird sometimes moves its eggs from nest to nest in the toes of its feet.⁴⁷⁴

It is, as we have said in preceding books, a playful bird whose chicks, when skinned and eaten, are said to sharpen eyesight.

Much has been related in preceding books about this bird.

(97) PICUS: The woodpecker is not a species but a genus of bird.⁴⁷⁵ One is marine dwelling and another forest dwelling.⁴⁷⁶ They both hollow out the bark of trees thinking that food is inside it. The whole woodpecker genus has the trait of opening a blocked hole with a particular herb. Not one of the authors whose writings have come down to us has said that he knew what this herb was.⁴⁷⁷

---

471. St. (1639) identifies the *pluvialis* as *Charadrius pluvialis* L., the golden plover, an identification supported by Suolahti (1909), 271–72. Cf. ThC 5.102 and Vinc. 16.133.

472. On this intestine, whose name means "empty," cf. 1.558.

473. *Pico pica* L. Cf. ThC 5.104, Vinc. 16.131, Neckam 1.69, Rabanus Maurus *De universo* 8.6 (*PL* 111:247C), PsJF 311–12, and Hugh of St. Victor 3.32. The magpie's ability to talk and its thievery made it the subject of frequent discussion: Rowland (1978), 102–5; McCulloch (1960), 141–42; Yapp (1981), passim (cf. esp. 107, Pl. 14); and George and Yapp (1991), 172–73, with ill. 125.

474. The Latin, and thus the English, is awkward here, reflecting a shortening of Pliny *HN* 10.50.98, which says that the bird cannot wrap its claws around the eggs. The bird therefore gets a stick, glues one egg to each end of it, and flies off with two eggs in balance to the new nest.

475. Cf. Isid. *Orig.* 12.7.47, ThC 5.105, Hugh of St. Victor 3.32, and Rabanus Maurus *De universo* 8.6 (*PL* 111:247D). See Rowland (1978), 180–84; McCulloch (1960), 190; and Yapp (1981), 54–55. Yapp also reproduces an excellent illustration of a green woodpecker listening to an angel's apocalyptic call (107, Pl. 14).

476. At first glance, the *picus marinus* would seem to be a misreading or miscopying. ThC 5.104 speaks of the *picus martius,* the "bird of Mars" mentioned by Pliny *HN* 10.40.20. Yet at 6.5 (q.v. for notes), A. elaborates on the *picus marinus,* equating it with the otherwise unknown *acoz.*

477. Pliny *HN* 10.20.40 reports the vulgar belief that after shepherds drive wedges into the nesting holes of woodpeckers, the birds use an herb to cause the wedges to slip out of the trees. Later, at 25.5.14, Pliny attributes the wedge story to Democritus, who told it to Theophrastus. Rowland (1978, 182) has an illustration of an improbably shaped woodpecker bearing an herb toward the wedge in its nest. ThC has the herb making arrows drop out, reminding one of dittany (Pliny *HN* 8.41.97).

This bird does not sit on rocks in order not to dull the points of its talons.[478] While there are many genuses of this bird, the most beautiful is the one which is reddish on top of its head, yellow on the chest, green on the neck, blue on the wings, and shiny on the tail.[479] Some, however, are black and quite large.[480] This bird sometimes talks quite perfectly, which is why someone made this verse about a woodpecker:[481]

> A small, talking bird, I salute you divine one, by voice
> If you saw me not, you would deny I am a bird

(98) PASSER: The sparrow is a small, familiar bird with an ash-gray color.[482] It is a warm bird and is therefore gluttonous, and it eats barley more readily than any other food, quickly separating the hull from the seed. If overcome with gluttony, it sometimes eats the chaff with the seed and chokes. It has great powers of copulation, to the extent that in one hour it copulates perhaps twenty times. Its flesh incites a person to lust and induces constipation because its flesh is warm and dry and, as a result, the sparrow fattens but little.[483] It is an active bird due to its consumption of superfluous moisture.

Male sparrows are distinguished by their black and white spots. They live but a short while due to their copulation and activity, which dry up their vital moisture [*humidum vitale*].[484] The females are an ash-gray color all over and live longer. They say this bird suffers from epilepsy because it eats henbane seeds.[485]

There are two genuses of them. One is more gray on its head and both builds its nest and lives on roofs. The other is less red on top and builds its nest in tree hollows.[486]

(99) PASSER SOLITARIUS: The "solitary sparrow" [blue rock thrush] is a black bird smaller than a blackbird.[487] It is a musical bird and is called solitary because it never gathers together with any of its own genus, except at mating time. It dwells

---

478. Cf. Pliny *HN* 10.21.42.

479. This would be *Picus viridis,* the green woodpecker (*GB,* 136–37; Pollard 1977, 47–48).

480. Probably *Dryocopus martius,* the great black woodpecker (*GB,* 92–93).

481. The "someone" is Martial 14.76, and the source is Isid. *Orig.* 12.7.46. Unfortunately, the initial word in the poem is *pica* (magpie), not *parva* (small). It thus was easily misattributed to the *picus.* Woodpeckers, of course, do not mimic speech. Several other words are incorrect in A.'s version.

482. On the sparrow in antiquity, see *GB,* 268–70 and *BAZ,* 225–28. Cf. Neckam 1.60, Barth. 12.32–33, ThC 5.106, Vinc. 16.120, Hugh of St. Victor 1.20, 26–32, PsJF 310–11, and Hugo of Fouilloy 23, 30–37 (Clark, 1992, 146–49, 156–69). On its symbolism and illustrations, see Rowland (1978), 157–60; George and Yapp (1991), 178–79; and Yapp (1981), 107, Pl. 14.

483. Cf. Pliny *HN* 30.49.141.

484. See Glossary for a discussion of the radical, or vital, moisture (s.v. "Radical moisture").

485. Henbane, *Hyoscyamus niger* (*De veg.* 6.362–63), earned its name from the fact that its flowers contain a substance poisonous to fowl. Barth., citing Constantine, seems to say that female sparrows eat henbane seeds without harm and then goes on to say that they sometimes get leprosy and suffer epileptic fits.

486. Respectively, the house sparrow, *Passer domesticus,* and the tree sparrow, *Passer montanus.*

487. *Monticola solitarius* (*GB,* 190–91; Pollard, 1977, 52). Cf. the description at Ar. *HA* 617a11f.

in ruined walls and allies itself to other sparrows. It flies with them to eat, despising entirely those of its own genus.

(100) PHYLOMENA: The nightingale is a small, familiar bird, named from *phylos* and *menos* because it loves sweet songs or it is named from *philos* and *mene* because each contends with the other in song. It would sooner give up its life than lay down its song defeated.[488] It is small of body, but its breathing has such a great vital force that it wondrously gives forth its melodious and complex song.[489] Now it is drawn out long in a continuous breath and then it is varied as if with the breathing of an inflected voice. Then it is punctuated with an abrupt sound and is finally linked to a convoluted breathing. The sound is full, low, high, complex and drawn out, exalted and depressed, imitating almost every musical instrument.

I have experienced myself that it flies toward singers if they sing well and that while they are singing the bird listens quietly. Afterward, as if trying to be victorious, it responds and sings back to them. They also provoke one another to sing in this way.

Pliny falsely states that this bird loses its voice and changes its color after copulation.[490] For we ourselves have often seen the bird singing while it sits on its eggs.[491]

(101) PSYTACUS: The parrot is a totally green bird with a bit of a golden ring on its neck.[492] These birds are found in India and Arabia and in desert areas of warmer climates where it rains but little. Its tongue is long and broad and it therefore forms articulate sounds very well when it is taught to do so from its youth.[493] It has a curved, strong beak, so strong that it latches onto a rock with it and saves itself with its beak as if using some sort of support.[494] It has a very strong head and, when it is disciplined, it is struck on the head with a little iron bar.

---

488. The Gr. stem *phil-* denotes love. The *menos* stem A. adduces in the first example is in reality *melos*, "song," and is reflected in the true Gr. name of the bird, *philomēlē*. The second example can be understood to read that the bird "loves strife," from the common *men-* stem, indicating anger or wrath. Neither etymology is in Isid. *Orig.* For this bird in antiquity, see *GB*, 16–22 and *BAZ*, 1–2. Cf. ThC 5.108, Vinc. 16.102, PsJF 310, and Neckam 1.51.

489. Compare what follows to Pliny *HN* 10.43.81–85.

490. This would seem to be based on a misinterpretation of Pliny *HN* 10.43.85: *mox aestu aucto, in totum alia vox fit, nec modulata aut varia* ("later when the heat has increased, their song becomes quite different, without modulation or variation"). *Aestus* has apparently been interpreted as the heat of sexual excitement and Pliny's previous mention of a totally white nightingale has become muddled as well. The same basic error is found in ThC.

491. St. contains the unattested form *adhaeree*. Borgnet omits this and substitutes *adhuc* (*sederet*). It is perhaps an error for *adhaerere*, suggesting that the bird sings when it sits to cling to its eggs.

492. St. identifies this as *Palaeornis torquata*, the ring-necked parakeet, yet the name is normally used more generically for "parrot." See Pliny *HN* 10.58.117f. and Solinus 52.43–45 as the source of much that follows. Cf. Neckam 1.36–38, ThC 5.109, Vinc. 16.135, and Rabanus Maurus *De universo* 8.6 (*PL* 111:246C). The *afastaga* of PsJF 272 has some of the traits of a parrot.

493. Cf. Isid. *Orig.* 12.7.24 and Jacob. 90 (p. 190). For A.'s treatment of animal language, see Resnick and Kitchell (1996).

494. This is badly garbled from Pliny, who merely says that whereas its feet are weak, its beak is so strong that it uses it to break its falls.

It uses its foot as a hand to feed itself. When it drinks, it stretches out toward the water while hanging from its feet with its tail up and head down. It does so because it cares greatly for its tail and often arranges it with its beak. It cannot endure rainwater but drinks and tolerates other water. It is therefore said to build its nest in the Gelboe mountains where it rarely rains.[495]

This bird loves to talk with boys and every bird learns to speak more easily from them. Its voice is natural and quite loud. It speaks so perfectly that if you did not see the bird you would think it is a person speaking.

(102) STRUTIO: The ostrich is a bird of the deserts of Libya, but it has still been seen rather often in our parts.[496] A young one is ash-gray and totally well feathered, although its plume feathers are not strong. In its second year and thereafter, its body is laid bare as it completely loses its feathers on its thighs, neck, and head. But it is protected from the cold by a tough skin, and the very black feathers of its back turn into a sort of woolly substance. It has very large thighs, fleshy legs, a white skin, and only two toes on its foot like a camel. It is therefore called the *cameleon* by some Greeks and the *asida* by others.[497]

It is tall and can measure up to five or six feet from its feet to its back. It has a very long neck, a goose-like head, and a beak that is quite small in respect to its body.

It is said that this bird eats and digests iron. But I have not experienced this to be so, since I have often spread out iron for several ostriches and they have not wanted to eat it. They did greedily eat rocks and large, dry bones that were broken into smaller pieces.[498]

This bird is said to be stupid and unable to fly, but it does run quickly with its wings outspread a bit. It has some projections [*spicula*, spurs] on the inner elbows of

---

495. See 23.63, where an argument is offered for equating Gelboe with Jebel Fukua, on the border between Israel and Jordan.

496. Although the ostrich, *Struthio camelus,* has been known from earliest times (Laufer, 1926), the Greeks never quite found a name for it, as *strouthos* really means "sparrow," although the ostrich was often called such things as the "big sparrow" and the "Libyan sparrow" (*GB*, 270–73; *BAZ*, 228–31). Cf. Jacob. 92 (p. 220), Neckam 1.30, Barth. 12.33, ThC 5.110, PsJF 313–14, and Vinc. 16.140. For illustrations, cf. García Arranz (1996, 221–244).

497. Yapp (1981, 55, fig. 45) offers an excellent illustration of the ostrich as a mix between camel and bird. A.'s two names are equally interesting. The first is a variant of the Gr. *struthocamēlos* ("camel sparrow"), which also came into Latin (Pliny *HN* 10.1.1). The second is more problematical and surely is not a corruption of *strouthos,* as Scanlan (316) suggests. Keller (1909, 2:174–75) discusses its names at some length but offers nothing to help with *asida.* ThC 5.110.1–2 says Isidore gives *assida* as the bird's Gr. name (cf PsJF 274), but an examination of Isid. *Orig.* 12.7.20 reveals no likely source for this name, even by corruption, and no Gr. word seems a likely candidate. Yet the word is fairly common (cf. McCulloch, 1960, 146). Diefenbach (1857, 55) cites some glosses which give the form *asida* as a synonym for "ostrich." Latham[2] (s.v. *assida*) lists a twelfth-century bestiary use of the word, unnecessarily obelizing the form, and DuC (s.vv. *asida* and *assida*) lists further examples, one of which makes it clear that the form is neither Latin nor Gr. The male ostrich has a rich, black hue against which its white wings are displayed—might *asida* be related to the Arabic word *aswad* (*swada,* fem.) for "black"?

498. The ability of the ostrich to eat iron was widely believed from antiquity on (Keller, 1909, 2:175; *GB*, 272). Note that Emperor Frederick II had ostriches in his travelling menagerie (Kantorowicz, 1931, 311).

its wings with which it strikes those it leaps on. This bird lays its eggs in the month of July, hiding them in the sand.⁴⁹⁹ The eggs hatch out from the heat of the sun, as do the eggs of many other animals. It does not return to them since it cannot keep them warm with its naked body. Sometimes, though, it guards them and watches over the place where they are lying. Therefore, the false rumor has arisen that they warm the eggs with their eyesight.

These are the things I have observed about the ostrich, a creature which seems to me to be not so much a bird as midway between a bird and a walker [*gressibilis*].⁵⁰⁰

(103) STRIX: The *strix* [owl] is a nocturnal bird about which Lucan makes mention, saying "both the warm *bubo* and the nocturnal *strix* complain."⁵⁰¹ This bird is commonly called the *amma* from its loving [*amando*], since it loves its chicks and is said to be alone among the birds in producing milky humor for them.⁵⁰²

(104) STURNUS: The starling is a familiar bird, black with a bit of pale grayness interspersed.⁵⁰³ It has a wide tongue and speaks quite perfectly. It flies in flocks, pressed quite tightly together, with each bird trying to get to the center of the formation out of fear of the hawk.⁵⁰⁴ When a hawk approaches from above or from the side, they flap at it with their wings. If it is flying below, they attack it with their excrement. They settle on beaches and swamps and are constantly with herds of cows due to the food they gather up from the cow dung. They have dry, tasty flesh.

(105) TURTUR: The turtledove is a familiar bird that cherishes modesty.⁵⁰⁵ It has an ash-gray color but has a reddish color sprinkled on its back on the surface of its feathers. It is a good flyer and is good on its feet as well.

---

499. A. does not make Ar.'s error of saying that ostriches lay more eggs than any other bird (*HA* 616b3f.), an idea arising from the fact that several ostriches lay in the same nest.

500. Cf. Ar. *Part. An.* 697b14f., where much the same conclusion is reached.

501. As is often the case with A.'s names for owls, exact identification is elusive. St. (1646) suggests *Strix flammea* L., modern *Tyto alba*, a barn owl. Thompson (*GB*, 268) merely claims it as an "owl." In fact, the *strix* has a long tradition which often assigns it tendencies more bat-like (as in the mention of milk just below) than avian. The matter has been studied intensively by Oliphant (1913; 1914). Pliny *HN* 11.95.232 rejects the idea that *strix* gives milk and confesses that he does not know what sort of bird it is. B&R (ad loc.) conjecture that the animal may be the vampire bat based on certain reports that it sucked blood.

The quote is from Lucan *Pharsalia* 6.689 via Isid. *Orig.* 12.7.42 and is from the wonderful scene where the witch is conjuring up her brew. Into the brew go several sounds, and the calls of these two birds are among them. The "warm owl" sounds odd because A.'s version has *tepidus* instead of Lucan's *trepidus*, "restless, anxious." On this bird in the medieval tradition, see ThC 5.111 and Vinc. 16.137.

502. The form *amma* is first found in Isid. *Orig.* 12.7.42 (Oliphant, 1914, 50) and is of unsure origin. It is frequently found elsewhere (*LLNMA*, s.v. *ama*; Oliphant, 1914, 50f.). The usual medieval sense of *ama* is that of a wine jar. Scanlan's (317) equation to (h)*ama* is unlikely, although, if referring to a bat, could refer to the gripping "hooks" on their wings. The answer, whatever it may be, lies with Isid. or his sources.

503. *Sturnus vulgaris*, the common starling. See *GB*, 334–35, and Pollard (1977, 38–39) for ancient lore. Cf. ThC 5.112, Vinc. 16.140, and Hugh of St. Victor 4.9.

504. Cf. Pliny *HN* 10.35.73.

505. *Turtur turtur* L., *Streptopelia turtur*. For its long tradition, cf. *GB*, 290–92, and Pollard (1977), 57–58. The reference to its *pudicitia* derives from Isid. *Orig.* 12.7.60 and refers not to its chastity (Scanlan, 318) but to its monogamy. At *QDA* 1.14, however, A. does refer to its chastity. For illustrations of doves, cf. García Arranz (1996, 543–600).

It is said that after it takes its first mate, this bird does not know a second one.[506] It produces and raises young twice in the spring and even lays a third time if the first or second eggs are taken away. In this it is like the doves. During the harshness of winter it lies dormant and they say that it does so in hollow fissures with its feathers gone.[507] They say the same thing about other birds, but this has not been proven. They say that a proof of this is that birds which lie dormant are not found to molt, but this is not sufficient proof, for birds molt in two ways. Some, like hawks, cast off their feathers all together at the same time, whereas others, like the magpie and raven, do it gradually.

It has dry, warm flesh and is found to be the most harmless and forbearing of all the birds. It is named for the sound of its voice, that is, *turtur,* and it moans as it calls, as do the other genuses of dove.[508] It announces springtime by its arrival.[509]

(106) TROGOPALES: The *trogopales,* as Pliny says, is a bird of Ethiopia. It has a head like a phoenix except that it has the horns of a ram and it overcomes all hostile birds with the fierceness of these horns. The rest of its members are rust colored.[510]

(107) TURDUS: The thrush is an ash-colored bird with a fairly small body which builds its nests out of mud in trees.[511] Within ten days after it has conceived eggs, it lays them and keeps them warm.[512] It sits on clumps of earth of the fields of those who are keeping the sabbath.[513] Its flesh has a good taste.

---

506. See Ar. *HA* 613a14f., which then appears in virtually all accounts; cf. Ambrose *Hex.* 5.19.62–63; White (1954), 145–46; Neckam 1.59; Barth. 12.34; ThC 5.113; Vinc. 16.143–44; *Physiologus* (Curley, 1979, 56–57; Carmody, 1939, 49–50); Rabanus Maurus *De universo* 8.6 (*PL* 111:248D); Hugh of St. Victor 1.20–25; and Hugo of Fouilloy 23–29 (Clark, 1992, 146–57). The bird thus becomes a symbol of a proper Christian marriage (Rowland, 1978, 46–48; McCulloch, 1960, 178–79) but was generally poorly drawn (Yapp, 1981, 46).

507. This belief represents a collation of data from Ar. *HA* 593a17–18 and 600a20–25, Pliny *HN* 10.35.72, and Isid. *Orig.* 12.7.60.

508. Cf. Isid. *Orig.* 12.7.60.

509. Cf. the famous quote from the Song of Solomon (Cant. 2.11–12), "Winter is ended . . . and the voice of the turtledove is heard in the land."

510. The source for this marvelous bird is Pliny *HN* 10.70.136, which does not claim the *tragopan* is from Ethiopia. That claim is made a few lines previous of griffins. The text is further corrupted in that Pliny, describing the bird, says it has a "purplish" head (*capite phoeniceo*): this comes to A. as meaning the head of the phoenix. Note that some of these details are not present in Solinus 30.29.

Tragopan means, lit., "goat-pan," a reference to the horns. Pliny judges the bird to be mythical. Rackham (1940) translates as "bearded vulture," while *B&R* (2:530), following Cuvier, opt for the horned pheasant. Thompson (*GB,* 285) suggests one of the hornbills.

511. Cf. *GB,* 148–49. The present passage is based on Pliny *HN* 10.74.147. Cf. Ar. *HA* 559a5f. The thrush had little place in the bestiaries, but it did have some symbolic overtones nonetheless. Cf. ThC 5.115, Vinc. 16.142, Rowland (1978), 174–77, and Yapp (1981), 64–65.

512. ThC 5.115.3f. specifies that the eggs are incubated internally for ten days before they are laid.

513. Again, ThC 5.115.4f. is of use in interpreting this phrase. Though the grammar is odd at this point, the passage states: "These birds signify those who are ready for every good deed. Once an idea for doing good has been formed in their heart they immediately set their hands to bringing it about and they bear good fruits." Curiously, Rowland (1978, 176) cites an instance of the bird being a symbol of sloth. Since *sabbatizo* can merely mean "to rest," one should perhaps not rule out this interpretation.

(108) TURDELA: The *turdela* is a songbird [*avis musica*] which exhibits the same size and about the same color of the *turdus*.[514] It is commonly called *droschele,* sings in the spring, and has a varied voice and good flesh, fit to eat.[515] It has an ash-gray color, turning at its breast to a mottled yellow. It is very easily tamed.

(109) VESPERTILIO: The bat is so called from using its wings in the evening [*vespere alis utens*].[516] It flies in the evening and is a sort of flying mouse.[517] It has a mouse's head but the shape of the head is that of a dog's head and the head is sometimes found with four ears.[518] It has saw-like teeth, unlike those of a mouse where there are two pairs of very long front teeth but rather like those of a dog, which has long canine teeth. In a tiny voice it mimics the growling and barking of a dog more than the squeaks of a mouse.[519] It has a hairy body with tawny hair and membranous wings, on the elbows of which it has one finger with a sharp claw. It holds itself with this as it clings to walls. It has a very wide, membranous tail, and on the tail are two foot-shaped objects with five toes and sharp claws which it uses to support itself below when it clings to walls. It does not sit or stand like other animals but rather, when it is not flying, it is either hanging in shady clefts in walls or lying in caverns.[520]

Its food consists of flies and gnats, which it seeks by flying at night.[521] It also eats flesh, however, and causes damage to bacon that is hung up.[522] It grows larger in warm climates than in cold and thus it is said in Alexander's letter that they are as big as the

---

514. St. (1649) identifies this as *Turdus musicus* L., the song thrush (*GB,* 287). Scanlan (319) opts for the redwing, but cf. next note. Isid. *Orig.* 12.7.71 talks of a *turdela,* but his details do not correspond with A.'s account (cf. Vinc. 16.142). ThC does not list the bird.

515. Cf. MG *Drossel,* "thrush." See Sanders (426) and Suolahti (1909, 66–68), who also support the identification of this bird as the song thrush.

516. Both Isid. *Orig.* 12.7.36 and Ambrose *Hex.* 5.24.87 mention only the evening and not the wings. See ThC 5.116 for an etymology resembling A.'s. Cf. Hugh of St. Victor 3.34, Rabanus Maurus *De universo* 8.6 (*PL* 111:254D), Vinc. 16.146, PsJF 317, and Barth. 12.38. The bat is not frequently portrayed in the bestiary tradition (McCulloch, 1960, 94) and had a generally poor reputation (Rowland, 1978, 6–9).

517. Cf. MG *Fledermaus.*

518. This rather awkward sentence apparently is meant to indicate that the mouse-shaped head is equipped with a long, canine snout filled with a dog's teeth; this meaning is amplified in the next sentence. The phrase "four ears" undoubtedly refers to the fact that some species have a very long, prominent tragus. The tragus is the small cartilaginous knob at the bottom of the human ear, and in these bats it is elongated, usually standing upright in front of the pinna (shell-like, wider part) of the ear. See the excellent illustration in Hill and Smith (1984), 11.

519. A.'s choice of noise terms is unusual. The word translated "squeak" is *sibilus,* more properly associated with snakes and usually rendered "hiss." The word for "growl" is *grunitus,* "grunt," and is usually applied to swine. The latter could be a confusion in both ThC and A. with *gannitus,* "growling." Cf. Isid. *Orig.* 12.7.36 and Barth. 12.38: *animal murium simile, non tam voce resonans quam stridore.*

520. "Lying" is *iacet* and implies a prone or supine position. ThC 5.116.3–4 says it lies (*iacet*) in caves as if dead. All of this implies uncharacteristic behavior. It may instead be an error for *latet,* "lies hidden," "keeps out of sight." Cf. Barth. 12.38, *in rimulis parietum se abscondunt.*

521. Cf. Pliny *HN* 10.81.168.

522. Barth. 12.38 reports, from a gloss on Isa. 2.20, that they lick dust and suck the oil of lamps.

doves in India and that they fall into the faces of people to wound them and sometimes even carry off their limbs.[523]

Pliny says that the blood of a bat, mixed with wild thistle, is a special remedy against snake bites.[524] Avicenna says that ointment made from the fat of a bat retards the growth of a girl's breasts if it is used before the appearance of the breasts and if at their appearance it is used continuously until the age when the breasts are grown.[525]

(110) VANELLI: *Vanelli* [lapwings] are the birds called *simphalides* by some.[526] They have a crest on their head like a peacock and have a shiny green neck, with the rest of the body being variously colored.

As soon as this bird has seen a human, however far the person may be from its nest, it immediately comes out of the nest and cries at him, giving away the nest with its cry. It is about the size of a dove.

(111) ULULA: The *ulula* [an owl] is named after the wailing [*ululatus*] and lamenting which it copies with its voice.[527] According to augurs, when it calls it is announcing sad tidings, but when it is quiet or is either sitting or flying high to the right and is quiet, it is announcing prosperity. It is a bird that avoids the light.

(112) UPUPA: The hoopoe is a familiar bird that sleeps in winter like the bat.[528] It has a crest of feathers on its head and is beautifully multicolored throughout the rest

---

523. "Alexander's letter": apparently the letter known as the *Secreta secretorum,* a pseudo-Aristotelian Arabic work translated into Latin in the twelfth century. The larger part of the work is a handbook or letter from Ar. to Alexander on statecraft (*De regimine dominorum*; cf. 7.134, with note), but accretions over time transformed the work into an encyclopedia treating medicine, astrology, physiognomy, and the occult sciences. For a discussion of this work, see Eamon (1985), 27–29.

524. Pliny *HN* 29.6.83.

525. Avic. *Can.* 2.2.739.

526. St. identifies this bird as *Vanellus cristatus* L. (*Vanellus vanellus*), variously called the peewit, green plover, or lapwing, a crested bird. He apparently does this based on the description of the call, comparing this passage to those which describe the bird he calls the *gywit* (1.37; 8.89; cf. Sanders, 433). Latham (s.v.) supports the identification; cf. Suolahti (1909), 264–67. On its depiction in MSS, see Yapp (1981), 52, fig. 43 and Pl. 10. Cf. ThC 5.117 and Vinc. 16.145.

Although St. prints s<t>imphalides, it is clear from ThC, who offers both *sinphalides* and *simphalides,* that the name was corrupt. It is thus unclear whether A. connected these birds with the mythical birds routed by Hercules (*GB,* 273–74; Pollard, 1977, 98–99). ThC further refers to Pliny when mentioning the *simphalides,* and it is interesting to note that at *HN* 11.44.121, Pliny specifically says the Stymphalian bird had a crest.

Finally, the description of the *vanellus*'s actions is certainly based on firsthand observation. In fact, the intrusive, garrulous nature of this bird gave rise to the MG verb *kiebitzen* and thus to the Yiddish word transferred to common English usage, "to kibitz." Both derive from *Kiebitz,* the MG name of this bird.

527. Cf. Isid. *Orig.* 12.7.38, Hugh of St. Victor 3.27, ThC 5.118, PsJF 317, and Vinc. 16.147. Stadler tentatively suggests that this refers to *Athene noctua,* a small screech owl in Greece, but his text markings indicate that A. inserted some comments into the text. He may, therefore, have had a local bird in mind. On varieties of the name, which ultimately is connected to "owl," see Suolahti (1909), 314.

528. *Upupa epops* was well known from antiquity onward (*GB,* 95–100; Pollard, 1977, 45–46; *BAZ,* 45–46). See Isid. *Orig.* 12.7.66, *Physiologus* (Curley, 1979, 14–15; Carmody, 1939, 21–22), ThC 5.119, Jacob. 92 (p. 221), Vinc. 16.148, Hugo of Fouilloy 57 (Clark, 1992, 238–41), Rabanus Maurus *De universo* 8.6 (*PL* 111:252B), PsJF 316, and Barth. 12.37. In the Middle Ages it was sometimes confused with the lapwing due to its crest (Yapp, 1981, 52; Best and Brightman, 1973, 56).

of its body.⁵²⁹ When lying dormant in winter it is mute, but in spring it is clamorous, with only one call.

It builds its nest out of human dung and for this reason its young stink.⁵³⁰ After it has raised its chicks, it loses its feathers, by way of changing them, in the same nest. During this period it is nourished by its children. It is said that it goes blind in its old age but that its chicks smear its eyes with an herb known to it and it recovers its sight.⁵³¹

If a person's temples are smeared with the blood of a hoopoe when it is time to sleep, they cause him to see terrible dreams. Wizards also seek out the hoopoe and its members, especially the brain, tongue, and heart. But we are not dealing with this here, for it is proper to look into this in another investigation.

(113) VULTUR: The vulture is a familiar bird. It is very large but heavy, and therefore it just barely gets off the ground after three or more leaps. Therefore it is often captured before it gets aloft.⁵³² I myself captured one by attacking it, but it had eaten a good deal of a carcass. This same heaviness is the reason that, unless it is afraid, it more readily sits on the ground than in a tree. This is why it was called the *gradipes* ["goes on foot"] by the ancients.⁵³³

This bird hunts from midday until nighttime and it rests from morning until midday. When it raises itself up off the ground, it flies well and high and some thus say the *vultur* is so called from its flying [*volando*], while others say it is from its greedy will [*voluntas*].⁵³⁴ On high it both sees and, with its remarkable sense of smell, senses the stench of flesh and carcasses over vast expanses of territory.⁵³⁵

Pliny and others state that no one has seen a vulture's nest and they are therefore thought to come from the other side of the world into our land.⁵³⁶ This is false, because

144

---

529. It is interesting that here, as at 7.34 and 8.89, A. uses *galea*, lit., "helmet," for the more common word for "crest," *crista*. Cf. Isid. *Orig.* 12.7.66, *cristis extantibus galeata*.

530. Cf. Ar. *HA* 616a35f. *GB,* 97, cites many instances of this odd statement (e.g., Isid. *Orig.* 12.7.66) and traces its origin to the hoopoe's habit of searching in dung for insect food, the fact that the female never leaves the nest during incubation of the eggs, and a "peculiar secretion of the bird's." Suolahti (1909, 14) lists two telling Swiss nicknames of the bird—*Stinkhahn* and *Schissdreckvogel*. Hugh of St. Victor 1.51 even offers an interesting Gr. etymology on the bird based on this trait. His source is not, apparently, Isid. Jacob. 92 (p. 221) describes the bird as *spurcissima . . . semper in sepulchris vel stercoribus*.

531. The hoopoe was thus a symbol of piety as far back as the Egyptians (Ael. *DA* 10.16; cf. McCulloch, 1960, 126–27, and Rowland 1978, 81–83). In many versions all the activities described here occur in the bird's dotage. More commonly the young are said to lick the adults' eyes into sight and they even are said to pluck out the old feathers, helping in the molting process.

532. No specific species can be identified here, as this probably is a generic coverage. Cf. Isid. *Orig.* 12.7.12, *Physiologus* (Curley, 1979, 47–49), Ambrose *Hex.* 5.20.64, ThC 5.120, Jacob. 92 (p. 222), Neckam 1.41, Barth. 12.35, PsJF 315–16, and Vinc. 16.149–53. For the wealth of stories concerning vultures from antiquity on, see *GB,* 82–87.

533. Cf. 23.30(17), with notes.

534. Cf. Isid. *Orig.* 12.7.12.

535. Cf. Fred. *DAV* 1.10 (W&F, 22), which reports that it is through vision, not smell, that vultures locate their carrion.

536. Cf. Pliny *HN* 10.7.19.

vultures build nests every year in the mountains which are between Trier and the Civitas Wangionum [City of the Wangiones] which is now called Wormacia [Worms].[537] So much do they do this that the land reeks from the carcasses carried there.

It is also said that vultures do not employ copulation. This is entirely false since they are very often seen to copulate there.

When these birds follow armies and people, they forecast future slaughter and pestilence.[538]

According to Pliny, a person bearing a vulture's heart with him, hung at his side, is safe from the bites of beasts and snakes, and the odor of their wings burned in a fire drives off snakes.[539]

(114) ZELEUCIDES: Pliny says that the *zeleucides* are the birds for whose assistance the inhabitants of the land and mountains of Dabyn prayed to God when their crops were being laid waste by locusts. These birds came then and killed the locusts, but it was not learned whence they came or where they went.[540]

Let these things, therefore, be those said about the nature of birds.

---

537. On the names, cf. *OL*, s.vv. This would locate the vultures just to the east of the Moselle River.

538. Cf. 8.98, 112, and Pliny *HN* 10.7.19.

539. Pliny *HN* 29.24.77.

540. The passage in question is Pliny *HN* 10.39.75, where the name is given as *seleucides,* apparently a reference to one of the many to follow Alexander the Great in ruling much of Babylonia, Media, and Susiana. Thompson (*GB*, 258–59, with ill.) traces the whole history of the bird and cites Cuvier, who identified it as *Pastor roseus,* the rose-colored pastor, a bird rather like a starling and famed for its locust-devouring capabilities. The mountain in Pliny is Mount Cadmus, a mountain of Caria. Cf. Ambrose *Hex.* 5.23.83, ThC 5.121, and Vinc. 16.154.

HERE BEGINS THE TWENTY-FOURTH BOOK ON ANIMALS

Which Is Especially About Aquatic Animals

CHAPTER I

On the Nature of Aquatic Animals in General and All Other Things, Set Forth in the Order of the Latin Alphabet

In the preceding book we have already discussed the flying creatures which, in the genus of animals, seem to be more perfect than the swimming creatures. It is logical, then, to continue on in this book to the swimming creatures. While it is granted that we have already spoken at length concerning their nature both in general and in particular, certain things must nevertheless be said here as well, first in general and then in particular. Although, as we have said in previous books, this might not be philosophical, and even though the same fact is often repeated, in such matters it nonetheless instructs the uneducated all the more. The aquatic animals are then set forth here, each with its own name, and the expositions of names set forth in previous books are considered as each of the animals is described with its characteristics.

We say in general, then, concerning all aquatic animals, that they are of a moist nature and that both their natural habitat and diet show this. For each thing is maintained in a place like-natured to itself and is nourished by food of the same complexion as the principles out of which it is constituted through generation.

It is agreed, then, that such creatures are, in most cases, cold. An indication of this is that they either lack blood or have but little of it. However, a few of them do have a neck, penis, womb, or breasts. Whichever of them also give birth from a womb to live young resembling themselves have cone-shaped breasts near their joints and also have wombs and penises.[1] But these appendages do not hang down on the outside. Rather, the creatures have internal penises and emit them at the moment of coitus, as do the dolphin and whale.[2] The breasts are cones of a sort, to which certain milk ducts are led from the body, having a point of interchange with the inferior epigastric vein [*rivertis vena*].[3]

---

1. "Near their joints": *iuxta iuncturas*; based on a confusion from the Gr. Cf. notes to 2.76.
2. Cf. Ar. *HA* 509b9–10 and Pliny *HN* 11.110.263.
3. On the *rivertis vena*, cf. Fonahn, 2813, and the use above at 1.440, 9.4, and 18.86.

Not one of the aquatic animals is ever seen to copulate in an unnatural manner and is never moved to copulate with any other save an animal of its own type.[4] When they lay their eggs, they rub their bellies together or the female leads the way, laying her eggs, and the male follows, spreading his milt over them. For this reason, most of the eggs perish, either because they do not exit at the contact of the rubbing or because they descend in such a way that the milt of the male does not touch them. This is why the number of fish does not exceed all bounds, even though the fishes' eggs would seem to be exceedingly numerous.

Of the egg layers, the large fish lay their eggs on rocks or in the mud, while the small ones lay them on the roots of aquatic plants. All marine fish have in common, however, that they prefer to lay their eggs at the entry to fresh water, due to the food flowing out there and the freshness of the water.[5] Many genuses of aquatic creatures guard their eggs to keep them from being devoured by others. Now, it is granted that all aquatic animals prey upon each other in turn, and especially the marine genuses. An exception is the *fastaleon* [gray mullet], which Aristotle likewise excepts since it does not eat flesh.[6] Fishermen have told me that the same thing is found in rivers with respect to the fish called the *barbellus* [barbel].[7] In any event, they all spare their own young until they are fully grown out of a natural desire in the species for the preservation of nature.

Moreover, it is a trait common to almost all fish to sleep but little and with their eyes open (for they have no eyelids, having hard eyes instead) and that in their sleep they move nothing but their tails slightly.[8] This occurs due to the coldness of the place where their digestion occurs which thus evaporates but little to the head.[9]

Moreover, although they live in deep water both near riverbanks and near the seashores, the flesh of those which are near the shores is firmer and drier and thus more healthful.[10] The flesh of those dwelling at the bottom is moister and softer, except for those very large fish with an earthy nature, like those called marine monsters [*beluae marinae*] by some. It is common for all fish to wander about before ovulation, seeking a place fit for it, and at the time of ovulation to go forth in pairs, male and female, to complete the eggs.

Wide fish, because they are moist and phlegmatic, are fattened more when the south wind is blowing, whereas long fish are fattened more during north winds.[11] The

---

4. "Unnatural manner": *unnaturaliter* does not indicate "perversely" but rather that the fish mate only with those of their own genus, not producing any crossbreeds. Cf. Barth. 13.26, citing Ar.

5. Cf. Ar. *HA* 567b15f., 598b5; Pliny *HN* 9.19.49.

6. Ar. *HA* 591a8f.

7. A type of catfish of the Moselle, *Barbus vulgaris,* mentioned frequently by A. (e.g., at 1.228 and 2.78). On the present subject, see 7.22.

8. Cf. Ar. *HA* 537a1f.

9. The importance of the rise of animal spirits to the brain is discussed at 12.132. The role of heat in life processes is generally discussed at 20.10–20. Cf. also 12.122–24. Summaries of the role of the heat of the heart relative to sensation and to the brain are found in Steneck (1980, 284–86, and app. B).

10. Cf. Ar. *HA* 598a3–8.

11. Cf. 7.100.

cause of this is surely that the south wind, by a similarity of complexion, increases the moistness in the wide fish. It thins out this moistness with its heat and draws it into their members. With its warmth, it loosens the long, naturally warm fish which the north wind constricts. Then, vaporizing the moisture circulating in them, it fills them up and fattens them. Moisture of this sort is the reason that in many of them the females are bigger than the males. Also, as Pliny says, that certain of them have such light heat that, if it were not for the weight of their enormous heads, they could not sink under the water.[12]

The hardness of the eyes of fish is the reason for their poor vision.[13] On account of this, they discern none save the extreme colors and of the middle colors only the clear ones, such as flaming red and bright yellow.[14] This is the reason that the more skilled fishermen do not dye their hooks with these colors when they fish. Fish are caught better at night and especially at dawn for the same reason, since they see less then and are moved to eat.[15] This is also why they are caught better in muddy water than in clear, for they see the net less well. Since, however, they are perfect animals, they abhor and flee old traps which have the odor of fish imbued in them and they likewise flee old nets and shun them more than new ones. However, if a device is imbued with the odor of a food of which the fish is fond, the fish will be caught more with that one than with any other.[16] This is the reason that nearly all types of fish die from oil and sulfur, for such an odor and taste are poisonous to them.[17]

Moreover, since each complexion rejoices in one like itself (unless, as Galen says, it has suffered *discrasia*), fish are harmed by dryness and birds rejoice in it, whereas fish do well in showers but birds are distressed by them.[18] When the cold, however, exceeds the limits of their temperament [*crasis*], a stone is occasionally generated in the innermost part of their brains and they die all the quicker due to the coldness of winter.[19]

Moreover, the generation of fish occurs in many forms, with some reproducing by means of copulation, while some are generated spontaneously, some from foam, others from putrescent slime, and others each in their own way, as has been set forth in our foregoing books. Hence, there happen to be many types of fish and shellfish [*ostrea*] in the depths of the sea unknown to us.

---

12. Pliny *HN* 11.46.129.

13. Cf. Ar. *Part. An.* 657a33f. and Pliny *HN* 11.55.152.

14. At *QDA* 4, annex. 14, A. identifies white and black as the extremes, while red and all other colors constitute the *medii*, or middle colors. As such, the middle colors are produced from these extremes, that is, white and black. Cf. *Phys.* 1.3.1.

15. Or "move out to eat"; cf. Ar. *HA* 602b5–10. *Moveo* in A. can be used in the passive to emulate the Gr. middle voice.

16. Cf. Ar. *HA* 534a12-b10.

17. Ar. *De sensu et sens.* 444b34f.

18. On *discrasia*, cf. 7.93, with note.

19. Cf. Ar. *HA* 601b28f. and Pliny *HN* 9.24.57. B&R (2.392) cite Cuvier as identifying this stone as an otolith, which is particularly large in some fish, notably the sciaena, whose otoliths were worn as amulets to prevent and cure colic.

Further, a fish with skin or scales can have its age told by their hardness or softness.[20] All fish of this sort have gills through which they spew forth the water that has been taken in. They take in this water for cooling purposes much as breathing animals take in air. It is an amazing fact that a sea fish is nourished on the fresh water which it absorbs out of the salt water but that when it is cooked in salt water it is not found to be salted, since its flesh was born to attract fresh water. A freshwater fish, however, absorbs salt water since it cannot tell the difference between salt water and fresh. When it is cooked in salt water, it is found to be salted. This is the reason, then, that a saltwater fish seeks fresh water and a freshwater fish absorbs salt water.

6 Their interior members which generate and receive humor (such as the liver in some of them) are as fat as they are due to the freshness which they draw in. Because of this they are used as kindling for fires.[21] An indication of this is that if lint or a piece of wood is rubbed with the liver of a sea fish, it burns as if it had been rubbed with oil. This is why fish come to the surface during a light shower, because they are seeking the freshness of the rainwater.[22] This is especially true when *manna* falls with the dew and showers, for they are greatly fattened from this.[23] They avoid storms, however, because of the violent tossing about they endure. They desire quiet and an indication of this is that fishermen create ditches [*fovea*] near shores and banks. The water let into these ditches stands quietly, and the fishermen fix stakes into the ditches so that it will be quieter still. Sometimes they fix them to the water itself so that fish, desiring quiet beneath them, are captured.[24]

The claim of some persons that all fish are suitable for eating is false and is shown to be so in the case of the *stupefactor* fish [electric ray], which is of so cold and poi-

---

20. Cf. Pliny *HN* 9.33.69.

21. A reference to the oiliness of certain fish livers. Most adults of a certain age remember too well the oil extracted from cods' livers. The "freshness" is *dulcedo*, which refers to the "fresh" (lit., "sweet") part of "fresh water," *aqua dulcis*.

22. Ar. *HA* 601b9–27.

23. The *Sinonoma Bartholomei* and the *Alphita* tell us that *manna* is a sweet fluid which is made by the dew falling on particular plants at certain times. Cf. Mowat (1882), 29; (1887), 110. A. does not seem to treat it in *De veg*. The ultimate reference, of course, is to the miraculous feeding of the Israelites in the desert as described in Exod. 16.14f.

24. A *fovea* is, technically, a ditch, and as such it was often used to mean "trap" on land. A. elsewhere uses *fovea* in connection with fish: 24.26(39), a hole to hide in; 24.44(81), a hole dug to lay eggs in; 24.49(100), a clear synonym for "cave." The source of this description is ThC 7.1, which makes it clear that a ditch is cut near the river and water is let into it. When A. (and ThC) tells us that the fishermen put stakes in the ditches, it might mean they built a sort of fence or breakwater around the rim or in front of the ditch. This would create the still water the fish sought and, at rest, they could easily be caught by conventional means. Stakes in the ditch itself would break up and quiet any wave action. His second statement, *aliquando infigunt (ligna) ipsi aquae,* is less clear and unfortunately has no corresponding version in ThC. Might he mean that a wooden shelter was floated, under which the fish then sought protection and were subsequently caught? Or was it something akin to a sea hedge (Lipeksaar, *DMA* 5.68)? A third alternative is that this served as some sort of fishpond. Barth. 14.54 tells us that many *fovea* were filled with water, now flowing and now still, and that within these ditches fish and certain reptiles (!) were nourished. Many monasteries cultivated fishponds which had to be stocked regularly and such devices would be of great use (Sahrhage and Lundbeck, 1992, 61–64). In addition to ThC, cf. Vinc. 17.26. Both add that this method was most effective in winter, and one should note that fish would be in great demand during Advent.

sonous a nature that it immediately takes away sensation and life from a person if it is touched for too long.²⁵

These things, then, have been said in general by us about fish. But, in closing, it ought to be noted that we are not distinguishing here the genuses of marine fish from the river ones, nor do we distinguish the marine ones among themselves.²⁶ For those matters, the things said in the preceding books are sufficient. But we will proceed with the discussion of all aquatic creatures, following the order of the alphabet as we did for the land and flying creatures.²⁷

(1) ASLET: The *aslet* [chimera] is a fish which I have seen. It has a body and skin rather like that of the fish which is called *ray* in French and *rocho* in German.²⁸ Its body, however, tapers down into a single long tail with fins at the end which is forked like the tails of other fish. Over the back of its tail is a large fin which, since its motion is crosswise, from side to side, is adapted not for swimming but for steering. Fins, however, which are adapted for propulsion in swimming are moved from front to back to propel the fish forward. This is just like the motion of the oars on a ship.

Moreover, this fish has two rows of double gill openings on each side. On both sides it has two sets of three gills, each above the other, ascending from the belly to the back of the head. It has two huge fins placed on the anterior width of its body (on the shoulders, as it were), and these are formed and articulated out of many jointed connections, in every respect like those of a bird's wing. One is on one side, one on the other, and they move from the back to the belly like birds' wings.²⁹ In exactly the same way it has two other fins placed on its sides below the place where its body begins to narrow down to the base of its tail, and these have the same free motion. On its belly it has two cartilaginous feet which do not have joints, one on one side and one on the other, and on the soles of these feet are some confused divisions into toes.³⁰

7

---

25. Cf. 8.114 and 24.58(127).

26. ThC devotes his chapter 6 to *Monstra marina*.

27. Many specific references will be supplied for specific fish, but the interested reader is also referred to an overview of the place of fish in the bestiary tradition in George and Yapp (1991, 202–11), who also offer useful charts of names and contemporary illustrations.

28. Stadler identifies this as the *Chimera monstrosa* L., a fish common to the Atlantic and found off the shore of Germany. This ray is commonly called the "rabbit fish" and has a poisonous barb dangerous to fishermen (Grzimek, 4:126–29). Scanlan (329) identifies the fish as "the giant ray, also called manta ray," but the manta prefers "the warm-temperate zones of all oceans and the Mediterranean" (McCormick, Allen, and Young, 1963, 276) and it is less likely that A. saw it in person. On the name forms, see 1.91 and Sanders (440).

29. Cf. McCormick, Allen, and Young (1963, 252–53) for an illustration of the swimming motion of a typical skate followed by an image from 1560 showing the skate with bird-like wings.

30. Throughout this passage one can vividly imagine A. examining the body of the ray at dockside. The "feet" are undoubtedly claspers, present only on the male and allowing him to attach himself to the female to ensure internal fertilization. They have many shapes but developed out of the pelvic fins. One name for them is *ptergygopodium*, "fin foot," and it is worth noting that *Chimera monstrosa* has deeply forked claspers (Harder, 1975, 1:221–24). Dodd (1983, 61) offers a close-up picture of the three-pronged clasper of a mature *Raja clavata* which indicates how easily the prongs could be taken for toes.

(2) ALLECH: The *allech* [herring] is extremely abundant in the ocean that touches parts of France, Great Britain, Germany, and Dacia.[31]

It is a fish about one palm long which, as long as it swims in a school, cannot be caught due to its numbers. But it is caught when, after the autumnal equinox, the schools split up. Even then, when the fish have been enclosed in many huge seines tied together, the two lines sometimes must be cut because the nets cannot be pulled in.[32]

This is a scaly and tasty fish, having no intestines except the *ieiunum*. Thus nothing is found in its stomach, and it is for that reason that some say, falsely, that the *allech* lives solely off the simple element of water. We have pointed out in other places that this is false.[33]

(3) ANGUILLA: The *anguilla* fish [eel] is named for the snake [*anguis*] because it has its shape.[34] It is a long fish, even held to attain the length of thirty feet in the Ganges River. It is not scaly but has a viscous and thick skin, which, when stripped off, renders the fish all the better. This is especially true when the fish is roasted, because then its excessive moisture is dried out.[35]

It is said that there is no male and female in this fish, although I have heard two times before from people worthy of belief that two eels were caught in parts of Germany and each of them had great quantities of thread-like creatures in its womb. When the mothers were cut open, a large number of the creatures came forth.[36]

The eel is found neither in the Danube nor in the waters flowing into it and it is said that if eels are placed into it, they die. Many eels are found in all other waters of Germany, however.

The eel fears thunder and at such a time will swim from the bottom of the water to the surface. Thus, if at this time a net is dragged through the ponds in which there are eels, almost all will be captured, so many in fact that the water will be emptied of them. Some say the eel is born from mud, some say from earthworms, while others say from the superfluities of fish.[37]

---

31. "Dacia" is the name given to the Scandinavian province of the Dominican Order; "Teutonia" is the Dominican province of central Germany (over which A. was made prior provincial in 1254); and "Anglia" is the province encompassing all of Great Britain. A. more usually employs "Brittania" and "Germania" rather than the Dominican terms. On the *allech,* cf. ThC 7.5 and Vinc. 17.30. Neckam 1.77 tells an interesting tale of a white bird (a gull?) which serves as a guide for herring schools. Cf. Saint-Denis (1947, 45) for the variety of name forms. For a quick overview of medieval fishing see Hoffmann (2005).

32. The herring trade was important in northern Europe at this time (Sahrhage and Lundbeck, 1992, 64f.).

33. See *QDA* 1.4.

34. The etymology is from Isid. *Orig.* 12.6.41. Cf. ThC 7.2, Vinc. 17.31, Rabanus Maurus *De universo* 8.5 (*PL* 111:238C), and Hugh of St. Victor 3.55 (*PL* 177:108B).

35. As is common, the culinary comments appear to be A.'s additions to the text.

36. "Great quantities of thread-like creatures": the text is somewhat ungrammatical at this point. The translation aims at what A. probably intended based on 6.87–88.

37. Eel reproduction was frequently discussed in antiquity (*GF,* 59–60). These ideas persisted until the late nineteenth century when the life cycle of the eel finally became known. For many centuries the larval form had been called leptocephalus and was not recognized as a larval eel.

Some also say that in the Ganges River, where there are huge eels, there are certain worms that have bifurcated front legs like crabs and attain a length of six cubits. These seize elephants and drown them.[38] In the winter, the eel lies dormant in the mud. It lives a long time out of water, especially when the north wind is blowing.

It frequently comes out of the water to go to a field where peas or chickpeas have been planted, but it is unable to creep through ashes or dry gravel.[39] Due to the symmetry and homogeneity of its body, it generates easily and lives for a long time when cut in half. We have given the reason for this elsewhere.[40] Although it can reproduce in mud and do well in it, it nevertheless shuns muddy water and is happiest in clear water.

(4) ALFORAZ: The *alforaz* is a fish which is produced from the putrescence of mud. It is generated in the mud without water, after the fashion of a worm, and then, when the water rises, this worm grows into the shape of a fish.[41] Sailors say of it that if it has rotted away, even as far as its eyes and head, it is reborn again when the water comes in. They also say that after this it lives a long time, whereas before a regeneration of this type its life span is brief.[42]

(5) ASTARAZ: The *astaraz* is produced in spring and autumn in the foam caused by rain showers. It swarms in this foam like the worms which arise from the dung of animals. It is also said to arise from mud.[43] This fish cannot endure the bright light of the sun and flees to the shade of trees which extend over the water. They do like warmth, however, and seek this by swimming on the bottom near the riverbank. During a dry period they are plentiful, but when there are many warm rainstorms, they are also plentiful.[44]

---

38. These eel/worms are prevalent in legend as the result of a long history of corrupted texts which can be traced from one author to the next (Kitchell, 1993, 352–56). Cf. 24.26(38), ThC 7.3, Vinc. 17.31, and Jacob. 90 (p. 192).

39. Cf. Pliny *HN* 9.37.73.

40. On the eel's ability to remain alive when cut, see 24.41(74). The implication is also that it regrows lost parts, and this is stated explicitly at *QDA* 14.6.

41. "After the fashion of a worm": *ad modum vermiculi*. Sense seems to require the translation "generated into the shape of a worm," but this use of *ad modum* would be unusual for A. ThC 7.3.3–4 reads *in modum*. It was commonly believed that worms were created parthenogenically from the earth. Cf. 24.12(19), where a turtle's tail is said to be *ad modum serpentis*.

42. Stadler conjectures that *alforoz* is a corruption of a passage by Ar. discussing spontaneous generation among fish wherein the Gr. *aphros* (sea foam) gives rise to the generic name for such fish, *aphyē*, lit., "unborn" (*GF,* 21–23). Cf. Ar. *HA* 569a28-b4, ThC 7.3, and Vinc. 17.29.

43. As with the *alforaz*, the *astaraz* seems to be the product of a misunderstanding of Ar.'s discussion of spontaneous generation (cf. Ar. *HA* 569b15–20).

44. A.'s Latin is curious. We have added the "also" to provide some clarity. It is hard to justify the two seemingly opposed criteria for abundance. Cf. ThC 7.4, who only mentions abundance when the year has been wet and warm, as does Ar. *HA* 569b21. The origin of the name *astaraz* may perhaps be the constellation Arcturus, as found in Ar. *HA* 569a28-b4, quoted above (Gr. *astēr* = star).

(6) ALBYROZ: The *albyroz* has such a strong skin that soldiers fit its skin to their heads underneath their helmets in order to receive without injury any sort of blows from swords.[45]

(7) ARIES MARINUS: Pliny says that the *aries marinus* ["sea ram"; killer whale?] is a fish which lies in ambush behind ships standing still on the sea. When it lifts its head out of the water, it looks for those entering the water to swim and then takes them under.[46]

(8) AUREUM VELLUS: The "golden fleece" [pinna] is a sea animal like a sponge, and it is in fact of the genus of sponges but more rare.[47] It is also softer and wool-like, and its fibers take on the splendor of gold when it is spread and stretched out, just as a fleece does. When it is in the sea it moves with a contraction and expansion movement just like a sponge, but it is rarely found. It was, however, found in Phrygia and it served as the cause of the Trojan War between the Greeks and the Phrygians, just as Dares Phrygius has said in his *History of Troy*.[48]

(9) ABARENON: The *abarenon* [sand smelt] is a fish which is fertile with many eggs, but it does not lay them unless it rubs its belly in contact with the rough sand, and then it raises and completes its young in this sand.[49]

(10) ACCIPENDER: The *accipender* [sturgeon] is, as Pliny says, the only one of all the fish which has scales pointed forward toward its mouth, covering its entire

---

45. The name's origin is unclear, but A.'s description (though unbeknownst to him) resembles that of the sponge of Achilles in Ar. *HA* 548b2f. Of course, a sponge makes a better shock absorber than a fish skin and this sponge was considered to be of high quality (*GF*, 23–24; Saint-Denis, 1947, 1). Cf. ThC (*albirez*) and Vinc. 17.29 (*albirem*).

46. Identification of the "sea ram" is unsure but probable. *Aries marinus* translates well the *thalassios krios* of the Gr. tradition (GF, 132–33). This passage is based on Pliny *HN* 9.67.145, and the aggressive behavior could describe *Orca gladiator*, the killer whale whose white spot, behind the eye, might have given the impression of horns. Ael. *DA* 15.2 compares the white spot of the *krios* to a diadem; cf. Pliny *HN* 9.4.10, "rams with a white streak resembling horns." George and Yapp (1991, 96, ill. 61) show an intriguing creature that is surely a cross between a walrus and an orca.

Rackham (1940, 3:260–61) and Saint-Denis (1947, 9–10) suggest a dolphin, but the aggressive behavior attributed to the sea ram in most texts seems better to fit *Orca*. Scanlan (332) also suggests *Orcinus orca*, the grampus. Note that the white spot sometimes curves, the better to resemble a horn (Ford et al., 1994, 20, fig. 3). Cf. ThC 7.7 and Vinc. 17.32.

47. *Pinna nobilis* is a bivalve mollusk famous for its byssus, a tuft of long, silky fibers by which it anchors itself. These fibers have been spun into a luxurious cloth for over two thousand years and some of the best are golden yellow in color (*GF*, 200–202). Cf. Pliny *HN* 9.66.142, Ambrose *Hex.* 5.11.33, and ThC 7.8.

48. "Dares" is a name found but once in the *Iliad* (5.79), where it belongs to a Trojan priest of Hephaestus. In late antiquity (fifth century A.D.?), a Latin work purporting to be a translation of Dares' account of the Trojan War appeared. A rather boring work, it was, however, most popular during the Middle Ages, substituting for the then unavailable masterpiece of Homer. In this account, the Trojan War arose due to a longstanding insult to the Trojans by the Greek crew of the Argo under Jason who were seeking the Golden Fleece.

49. Comparison with Ar. *HA* 571a6f. reveals that this fish originally was the Gr. *atherinē*, *Atherina hepsetus*, a tiny, smelt-like fish (*GF*, 3–4). Cf. ThC 7.9 and Vinc. 17.29 (Gr. *atherinē*). The corrupted name seems to contain a reference to its relation to the sand and *ab arena* could mean "from the sand."

body. It is a sweet fish, suitable for eating, and was in great demand among the ancients.⁵⁰

(11) AMIUS: The *amius* [bonito] is, so they say, a rock-dweller which bears a stone inside itself. It is a most beautiful fish, having purple-colored stripes on its sides, and the entire rest of its body is stippled and marked with diverse and pleasing colors.⁵¹

(12) HAMGER: The *hamger* [garfish] is, so they say, almost identical with the *amius*.⁵² It is long and rounded, and it is called the *gervisch* in German.⁵³ It is white, with good flesh, is shorter than an eel, and has a slender, long, red, bird beak in place of a mouth. Inside, it has a green spine, and the flesh which touches this spine also is green.

(13) AFFORUS: The *afforus* fish, owing to its slight size, cannot be caught with a hook.⁵⁴

(14) AUSTRALIS: The "southern fish" [*australis piscis*] is a fish that takes up a wave of the waters in its mouth. It is born when the Pleiades are setting, since this is the time of rainstorms.⁵⁵

---

50. Cf. Pliny *HN* 9.27.60; *FAZ,* 312–14. The sturgeon, of course, does not have scales facing forward, but it has longitudinal laminae in place of scales which do not overlap as scales do. Thompson (*GF,* 8) suggests that this idea may come from a corruption of Pliny's text where *squamis ad os versis* (scales turned toward the mouth) may represent a misunderstanding of *squamae in os conversae* (scales converted into bone). Cf. ThC 7.10, Vinc. 17.29, and *ezox* and *huso* at 24.32(49–49a). Cuvier (*B&R* 2:398–99) identifies Pliny's *accipenser* as perhaps the relatively small and relatively rare (to the Romans) *Acipenser ruthanus* rather than the larger *Acipenser sturio,* and he is followed by Scanlan (332). Cf. Saint-Denis (1947), 1–3. Finally, cf. A.'s avian *accipenser* at 6.47 and 23.18.

51. Both the name and the identification come from the Gr. *amia, Pelamys sarda* (*GF,* 13–4; Saint-Denis, 1947, 45). The term A. uses, *saxatilis,* is generally translated as "rock-dweller" (cf. Isid. *Orig.* 12.6.33). He may, however, be following ThC 7.11, which clearly states that these fish are *saxatiles* due not to the rocks they dwell near but to the rock they bear in their heads. The "stones" they bear refer, of course, to their otoliths (cf. 24.4), which are more prominent in some varieties of fish than in others. Cf. Vinc. 17.57.

52. This is the garfish, *Belone acus* (*GF,* 31–2; Gr. *Belone*), whose bones are notably bright green. A. fails to note (unlike ThC 7.12, *haniger*) that the similarity to the *amius* is in color, not in form. Sanders (443) discusses the name forms and links the first to a stem meaning "pointed" and the second to a stem meaning "small."

53. Cf. Sanders (443) on the name.

54. This appears to be another corruption of *aphros* from Ar. *HA* 569a25f. Cf. 24.9(4), 9(5), and 10(9). Cf. ThC 7.13, Isid. *Orig.* 12.6.40, and Barth. 13.26.

55. This rather vague passage is a version of Isid. *Orig.* 12.6.32, where (although this passage is also confused) the name *australis* is apparently linked etymologically (albeit poorly) with *os, oris* (mouth), and with *oritur* (lit., "it arises, is born"). ThC 7.14 follows Isid. closely, keeping the etymological thrust of the passage. The Pleiades set in October-November, heralding the beginning of bad weather. Cf. Barth. 13.26. The phrase "takes up a wave of the waters" (*aquarum undam ore suscipit*) lends itself to many equally vague interpretations in A., ThC, and Isid. alike. The sense is probably "causes waves to arise" or "takes water into its mouth," and the description seems more fitting of the wind named Australis than of a fish. Nonetheless, Stadler (1607) identifies the animal as *Pelecus cultratus* L., a mollusk.

(15) ARANEA: The *aranea* ["spider"; weaver fish] is a sea fish with spines on the flaps of its gills with which it strikes those approaching it.[56]

(16) ABYDES: The *abydes* is a sea animal whose first abode and food are found in the sea.[57] Later, its shape is changed and it turns into a land creature. It then comes out of the water and seeks its food on land. When its shape is changed, its name is changed, and it is called an *astoyz*.[58]

(17) HAHANE: The *hahane* is a sea animal more gluttonous than all other sea creatures and it is said not to have a separate and distinct stomach. It is a predator and all its food is converted into fat. Thus its stomach swells out beyond all reason. When this animal fears danger, it folds its skin and fat over its head, hiding its head like a hedgehog. If the thing causing the fear should remain for a long time, it is sustained by eating part of its own flesh until the danger goes away.[59]

(18) BELUAE: According to Pliny, *beluae* are certain sea creatures of the eastern sea so large and savage that they arouse storms from the bottom of the deep and constitute dangers for ships, and they even led the ships of Alexander to danger.[60]

---

56. This fish is clearly the *drakōn* of the Greeks, the *draco marinus* of Pliny *HN* 9.43.82, generally identified as the weaver fish, *Trachinus dracol araneus* (Saint-Denis, 1947, 9). Indeed, the name "weaver" (also found as "weever") is a reference to the spider, perhaps alluding to the fish's poisonous spines (*GF,* 56–57). It is interesting to note that, lacking scientific terminology, A. literally says the spines are on the "earflaps" of its gills. Cf. 4.20 and 24.26(39), as well as ThC 7.15 and Vinc. 17.32.

57. Stadler suggests another corruption of *aphros* and is followed by Scanlan (333). Yet Ar. *HA* 487b3f. is almost surely the source, for it speaks of creatures which first live in water and then change shape to become land-dwellers. The Ar. passage itself is textually corrupt (Peck, 1965, ad loc.) and offers several forms (e.g., *aspides*) which could have given rise to *abydes*. Cf. also 1.30, where the Ar. text also made its way into A. Essentially the same passage is found in ThC 6.2 and the poor creature thus joins the ranks of marine monsters.

58. The original Ar., uncorrected by Peck, says the *oistros,* gadfly, is produced.

59. Stadler suggests Ar. *HA* 571a28f. as a possible source for this sea creature, but the passage is far from the mark. The true origin of *hahane* is Ar. *HA* 591b1f., where the characteristics of several fish are described. One of these fish is *hē channē,* "the sea perch" (Saint-Denis, 1947, 21–22). In transmission the other names have dropped out and all their characteristics have been given to this one fish. For a fuller treatment of the process of corruption, see Kitchell (1993), 350–52. The animal has several names: ThC 6.3 (*ahune*), Vinc. 17.101 (*ahume*), and Barth. 13.26 (*habatue*). It is gratifying that A. came closest to the original Gr. The bellies cast out of the mouths of the fishes are, in fact, their swim bladders, which often inflate when they are brought up from the depths.

60. This section is based, confusedly, on Pliny *HN* 9.2.5. Pliny says that solstice storms stir monsters (*beluae*) from the depths and he compares the *beluae* to schools of tunny fish that were so great that Alexander's ships had to assume a battle line to get through them. *Belua* is a generic term, meaning "monster," and not the name of a specific animal. Cf. ThC 6.4.

(19) BARCHERA: The *barchera* [sea turtle] is a genus of hard-shelled animal under which the sea tortoise [*tortuca maris*], along with other species, is included.[61] This tortoise's mouth is so hard that it breaks rocks. The tortoise hunts small fish and is sometimes caught when it is fooled by some of these bound to a line. On occasion it comes out and grazes on grass. It grows so large that its shield has been found measuring eight or nine feet.[62] It has a horny substance on its head, as does the land tortoise, and this surrounds it like a helmet.[63] If when it comes into the light of day its shield-like back becomes dried out, it is not able to bend as it wishes afterward until the shield grows soft again over a long period of time.[64]

The fishermen of Germany and Flanders call this animal "the soldier" [*miles*] because it has a shield and helmet.[65] Its shield is formed as if from five plates.[66] It has four multitoed feet and a tail like that of a serpent.

(20) BOTHAE: *Bothae* [flounders] are of the family of flatfish [*pecten*].[67] In one part, namely on their back, they are black, stippled with red spots. On the stomach, however, they are white and have little fins all around the edge. They become fattened when the south winds blow. When the fish senses a fisherman, it descends to the bottom, clings to the dirt there, and stirs up the water lest it be seen.

(21) BORBOTHAE: The fish called *borbothae* [burbots] are river- and lake-dwellers.[68] They are somewhat similar to the eel but are shorter with large bellies. They seek the depths to such an extent that, in the lake of Germany which flows into Lake

---

61. This is based on Pliny *HN* 9.12.35–39, which discusses tortoises and turtles alike, using *testudo* for each. A. thus sounds entirely modern here, calling the *barchera* a *genus* and the tortoise a *species*. But he loses the advantage when he begins to use "tortoise" in his description. The origin of the name *barchera* is unknown. It is not, as Stadler suggests and Scanlan (334) accepts, a plausible corruption of *hai thalattiai chelōnai* (the sea turtles) from Ar. *HA* 589a26f. It must, rather, be related to the same (Arabic?) *barcor-* stem found repeatedly where the Gr. had the word for murex shell (e.g., 4.38, 68; 5.60; 7.85; 13.111). Might the stem, therefore, indicate "shell"? Cf. 24.58(126) and 60(138), and ThC 6.5 (*barchora*), Jacob. 90 (p. 193), and Vinc. 17.102.

62. "Shield": A. uses this odd term to begin a string of military analogies.

63. "Horny substance": *cornu* could also be translated "horn," but no appendage exists on these creatures which could be so misconstrued by a casual observer.

64. "Able to bend as it wishes": *ad nutum flectere* is, in the first place, unusual diction for A. Second, it is far from Pliny, who says they cannot dive and are unwillingly borne away on the water.

65. Elsewhere, A. mentions two sea creatures called the "soldier." The first group seems to concern a sea turtle: 7.77; 24.58(126) and 60(138). The second group discusses a crab of India: 4.20, 24.22(31). The name is mentioned but not discussed by Sanders (413).

66. "Plates": A. uses the word *asser*, which actually means "beam, plank, or board."

67. Stadler identifies this fish as *Rhombus maximus*, the European flounder or turbot (*GF*, 223), whose red spots help to serve as camouflage. Cf. ThC 7.16 and Vinc. 17.35. The term *pecten* is usually a scallop in CL (*GF*, 133). However, at 24.47(94), A. tells us that he uses the term *pecten* (lit., "comb") to indicate a round, flat fish, probably of the halibut or turbot types.

68. The burbot, *Lota vulgaris*, is a common cold-water fish of Europe, reaching a length of about one meter or more. Its liver stores fat and is still considered a delicacy (*GF*, 168). Cf. ThC 7.17 and Vinc. 17.35, and also the *solaris* at 24.52(110).

Constance, these fish have been caught on hooks to depths of up to three hundred feet below the surface.[69]

This fish is called the *alrutten* or *alquappen* by some in German, while others call it *lumpen*.[70] It has a sweet flesh, a viscous and not very thick skin, and a large, round, very sweet liver.

It is said that when this fish has passed beyond its twelfth year, it attains its maximum size, at which time it is called a *solaris*.[71]

(22) BABILONICI: According to Theophrastus, Babylonian fish [mudskippers] are found around Babylon, living in places where fresh water collects from the rivers flowing into hollow areas in the sea.[72] The fish venture forth from these hollow areas to eat. They have heads like those of the sea frogs and the rest of their body is like a *rubia*.[73] Their gills are like those of other fish and they move on little fins and with a frequent motion of their tail.

(23) CETUS: The whale is the biggest fish which is seen, the female of which is called the *balaena*.[74]

It is a fish of many genuses. Some are hairy and these are the biggest. Others have a smooth skin and are the smallest ones. Two types are seen and captured in our sea. One has the opening of its mouth lined with very large, long teeth. So large are the teeth that very often they are found to a length of two cubits, sometimes three and on occasion four, but more often one. The two canines are especially longer than the others and are hollow inside, as is a horn, after the fashion of an elephant's teeth and

---

69. Actually, no lake flows into Lake Constance (the Bodensee), so A. may be referring to one of the "fingers" of the lake, the *Untersee* or the *Obersee*. The word translated here, and at 23.57, as "feet" is actually *passus*, lit., "pace." Yet to equate it with a pace, five Roman feet, would give, by Scanlan's (335) reckoning, a fishing line some 1,452 feet long, about 600 feet longer than the lake is deep. While Scanlan finds this "acceptable," one must remember 23.57, where someone is dangled over a cliff on a 300 *passus* rope, a passage in which Scanlan translates *passus* as "yards." Either option is too long to be considered seriously. It is better to take *passus* in its secondary sense as a footstep and thus a "foot." Cf. Diefenbach (1857, 415), who links *passus* as a gloss with an early German passage which read *funff schuch weit lang oder breit*.

70. On the Germanic names, cf. 2.104, 7.85, and Sanders (437–38). On burbot as a food, *FAZ*, 64–65.

71. The name *solaris* implies "sunfish," and at 24.52(110) we are told it likes to sun itself. One immediately thinks of the marine sunfish, as it is a huge fish, but it does not inhabit rivers.

72. Stadler identifies this as the species *Periophthalmus*, the mudskipper, the astounding fish which "walks" on its fins across mud flats. The original of this citation is Ps. Ar. *De mirabilibus auscultationibus* 72 (*On Marvelous Things Heard*; Dowdall, 1913, 835b7–14).

73. On the sea frog, cf. 24.50(101). *Rubia* is more difficult to explain. *De veg.* 6.429 discusses *rubia* as a medicinal plant (*Rubia tinctorum*). However, the mental image offered by a creature that is part sea frog and part plant, while entertaining, is startling. Cf. ThC 7.18, where the rest of the body is likened to the body of *cabiones*, paralleling nicely the *kobiòs* (goby/gudgeon) of Ar. (*GF*, 137–39).

74. Elsewhere, A. uses *balaena* merely as a generic term for whale (Saint-Denis, 1947, 13–14, 20). Cf. Isid. *Orig.* 12.6.6, followed by several medieval authors, where the male whale is said to be called the *musculus*. On the whale in antiquity, *AAZ*, 197–99).

the boar's teeth called tusks. They seem to be used for fighting. This genus of whale has a mouth suited for chewing.⁷⁵

In our time another genus has been seen on occasion. It had a mouth suited for sucking, being toothless like a lamprey [*murena*]. This genus was a little bit smaller than the other and had a much better flesh.⁷⁶

Neither, however, has gills, for each breathes through a fistular canal, like a dolphin.

Each of these two types which has a smooth skin also has a dense, black hide. Above its eyes, which are very large (so large that the socket of one eye easily holds fifteen men and sometimes twenty), there are horn-like appendages resembling eyebrows, more or less eight feet in length, depending on whether the fish is larger or smaller. These horn-like appendages are shaped like the large scythe used to cut grain. There are two hundred fifty over one eye and the same number over the other. On the wider side they are rooted in the skin and on the narrower side they are separated. They are not set up so that they stand straight up, erect from the body, but lie arranged from the base of the eye back toward the fish's temples so that they give the appearance of one wide mouth like a large fan. This fish uses this thing for an eye covering in the time of a great storm.⁷⁷

It has a huge mouth and when it breathes it spews forth from its mouth a great deal of water. This sometimes swamps and sinks small boats. It has large fins which are shaped like those of the dolphin and it has a bifurcated tail more than twenty-four feet in width when the fish has grown to its normal size.⁷⁸ It has long curved ribs at the thickest part of the fish and these are about the size of the rafters in large houses.⁷⁹ I am calling rafters [*tigna*] those things onto which beams are nailed, to which the tiles of a house are nailed.

15

---

75. Scanlan (336) suggests that this toothed whale is the walrus. Indeed, the length of the canine teeth suggests this (a cubit is something akin to 2.5 feet). But A. specifically says the whole mouth is rimmed with long chewing teeth and at 24.19 he seems to describe a walrus as a separate sort of creature. The present animal must be a toothed whale. A. may have seen or heard about walrus tusks, which were said to have come from whales, leading to the enormous length given to their teeth in this passage. The sperm whale has impressive rows of conical teeth set in its lower jaw, up to a maximum length of 8 inches and 2 to 3 pounds each in weight. These teeth are not hollow, however, as the pulp area closes up as the animal ages. Thus, the very long teeth which A. saw may be the "tusk" (in reality an adapted tooth) from a narwhal. These have the requisite length and have a hollow pulp area throughout their length. Narwhal tusks were commonly sold during the Middle Ages as unicorn horns, useful in their ability to detect poisons.

76. While this is clearly a baleen whale, the exact species cannot be known. It appears A. observed this animal himself at dockside.

77. A.'s description perhaps refers to the right whale, whose baleens number 250 plates to a side (Ellis, 1980, 72). The word translated "eyebrow" (*cilium*) may, in the loose terminology of the anatomy of the eye, refer more to eyelashes or to the eyelid (cf. notes at 1.139, 161), but "eyebrow" may fit the intent best. A standard picture of a right whale (e.g., Ellis, 1980, 79) shows the line of its mouth beginning high above its eye and arching down. If the baleen is visible, it can indeed resemble an eyebrow, arching above the eye. The longest of the "teeth" of the baleen may reach 7 feet and they are composed of basically the same substance as nails, claws, and the like. Cf. Barth. 13.26.

78. Cf. Pliny *HN* 9.4.11 for the source of the measurement.

79. Cf. Pliny *HN* 9.2.7.

16   The ancients write that this fish occupies four *jugera* of land at the widest part of its belly. We have never been able to verify this from any of the fishermen who have seen many of them on many occasions.[80] But what we have verified is that at its largest, and when divided into flesh and bones, it comprises 300 cartloads. Such large ones are rarely captured, but frequently there are ones brought among us which are between two hundred and one hundred fifty cartloads, a little more or a little less.

This fish has testicles and a penis inside its body, like the dolphin, and it extends them at the moment of copulation. The female has a womb positioned just about like that of a woman. When copulation occurs, the female places herself beneath the male, as does a woman to a man and the she-dolphin to the male dolphin. Copulation is rapid, as is the case for all animals with internal testicles.[81] Whatever of the sperm is cast free (for there is a great deal of it and it is not all retained inside the womb of the *balaena*) is sought after by physicians. It is called *ambra* and is highly valued against gout and paralysis.[82]

The whale has its penis and testicles within so that they do not hinder its swimming and so that their power is not destroyed by the constant coldness of the water. Some say that after a single instance of copulation, the whale is rendered impotent for copulation with the *balaena* again and that he then plumbs the depths of the sea and grows, becoming so fat he is equal in size to the islands of the sea. But I do not think this is true, and those with experience do not tell such tales.[83] When, however, whales fight for the *balenae* and for their young, the conquered one flees to the depth of the sea. He remains there for some time, out of fear, and since he stays immobile, he grows quite fat. This fish has lard on its back like the pig and has a great deal of fat, especially in its head around the marrow of the brain [*cerebri medulla*].

---

80. A *juger* or *jugerum* was, in theory, how much land a yoked team could plow in a day and is commonly translated as "acre." The Roman version is generally given as 120 by 240 Roman feet, approximately half a modern acre. Scanlan (337) incorrectly lists this as 240 by 240 feet. A.'s conception of the exact area involved is perhaps unknowable. An English acre, for example, varied from 10,240 square yards to 2,308.73 square yards (see *DMA* 1.45). Whatever measurement a surveyor might have used, the size would be greater than that of any known whale. A.'s skepticism is noteworthy, as is his reliance on the firsthand observations of his *experti*.

81. "During mating the pair swim with flippers touching, belly to belly; as a rule the male on his back beneath the female, but sometimes the pair lie sideways" (Lockley, 1979, 71).

82. A. is referring to ambergris, a waxy substance from the intestine of the sperm whale, probably caused by the irritations arising from swallowed squid beaks. Ambergris is often found floating in the Indian Ocean and is used in perfume manufacturing. The formation of *ambra* from excess whale sperm is found in both Jacob. 87 (p. 173) and Barth. 13.26. True amber was usually called *electrum* (Gr. *elektron*), but A. calls it *lambra* at *De min.* 2.2.17 (Wyckoff, 1967, 121). For the history of beliefs concerning ambergris, see Dannenfeldt (1982).

83. A. here indirectly refutes such common tall tales as those of whales scooping sand onto their backs so that they could imitate islands and lure sailors to their deaths; cf. Isid. *Orig.* 12.6.8, ThC 6.6, and Barth. 13.26. B&R (2.359) comment on the propensity of the ancients and the early church fathers to attribute fantastic size to large fishes. Cf. the vivid illustration in George and Yapp (1991, 95, ill. 60).

Many have been captured in my time. One, in fact, was taken in Frisia, near the place called Stauria.⁸⁴ Its head, when pierced through the eye with a lance, yielded eleven jars of oil, each one scarcely able to be carried by one man.⁸⁵ I myself have seen both the jars and the oil. The oil is very clear and pure after it has been purified. Another was captured, beyond Utrecht, toward Holland, whose head yielded forty jars of oil. The lard of this fish is what is called *graspois*.⁸⁶

The *balaena* conceives one offspring and nurses it.⁸⁷ It follows its mother for a long time, even up to the age of three or four years.

When this fish is hemmed in within some narrow deep place and is surrounded by ships, it dives to the bottom and then suddenly emerges and sinks the ships. It is taken most frequently when it follows the herring [*allec*] with excessive greediness, beaches itself on the shore, and then cannot return to the water. One beached itself in just this way a little while ago on the shore of Frisia. When the inhabitants found this out, they were afraid that the tide might come back and they would then lose the fish. They therefore bound it with all the ropes they were able to get from all over their island, fixing stakes deeply into the earth, and tying the ends of the ropes to rocks, nearby houses, and other buildings. Upon the return of the sea, however, the fish, assisted by the water, broke all these things and returned to the ocean with the ropes, leaving the inhabitants mourning the loss of their ropes. But, on the third day, since it had not eaten, it followed another school of herring and, along with its ropes, again lay in the same place as before. The ropes were recovered by the inhabitants then, and

---

84. Frisia consisted of the North Sea coast roughly between the mouths of the Rhine and Elbe (cf. modern Friesland). Stauria is modern Staveren, about 100 km north of the city of Utrecht in the Netherlands, situated directly on the Zuider Zee.

85. Cf. 3.142, with notes, for further details on this event. It is possible that two separate events are being described. The whale is probably a sperm whale.

86. "Beyond Utrecht, toward Holland": translation of this small phrase, *ultra Traiectum versus Hollandiam,* is full of problems. Scanlan (338) is correct in stating that *Ultra Traiectum* was a name for Utrecht. Indeed, the modern name stems from that form. But the evidence does not seem to support his translation, "in Utrecht towards Holland." First of all, *ultra* here clearly seems to be a preposition governing the acc. *Traiectum,* which was frequently used alone for Utrecht (*OL* 3:505). That Stadler so took it is clear because he does not capitalize it. Second, if we capitalize *Ultra* and make it part of the name, then we need some sort of preposition governing the entire name of *Traiectum*. Third, Utrecht is many kilometers inland and no whale is likely to have been caught there. It seems best to keep *ultra* as a preposition and translate the entire phrase as rendered here. If *Hollandia* is taken as equivalent to the later earldom (*GHW* 117a C/D 1), then we are once more in modern Friesland, directly on the North Sea. A whale entering the area between the breakwater islands opposite Friesland and the shore would have been easier game than those encountered in the open sea. It is on these islands that the story of the beached whale A. recounts just below seems to have occurred. Note that A. visited this area from 1254 until 1255 when, as prior provincial, he was visiting all the priories in his province. He thus would have been to Utrecht and Friesland (e.g., Leeuwarden), and we can imagine him taking notes while engaged in conversation with local fishermen.

Graspois was more commonly written as *craspois* in older French and is, lit., "fat fish." As such it referred to the whale itself, but it often was used as "fish fat," i.e., "oil" (Tobler and Lommatzsch, 1925, 2:1015–16; Godefroy, 1881, 2:357).

87. Cf. Ar. *HA* 566b7.

the fish was killed and divided up. When the neck was split from its head, broken by the weight of the head, the break gave forth a sound as if a house were falling in ruins.

18 Whales are usually captured by the fishermen who hunt them in our seas in two ways.[88] In one method, upon which many fishermen agree, they sail in small boats holding three men each to the place where they presume the whales to be. Two of them do the sailing and the third stands in the boat prepared to strike. He holds an instrument whose shaft is made of ash wood for lightness. At the end of the shaft near the hand of the striker is a hole, and through this a strong, very long cord is passed. This is coiled in the boat in such a way that it leaves easily and without tangling, following the instrument to which its end is attached. The point, at the lower end of the instrument, has a triangle like a barbed arrow, and the triangle has a very sharp, polished tip so that it will penetrate more easily. The two edges come to a point, meeting as does the sharpest razor, and the surface of the triangle is smooth and very well polished.[89] In the middle of the side opposite the bottom sharp angle there arises, perpendicularly, a piece of iron joined to the side, a cubit or a bit more in height. Here there is a socket to which the shaft we spoke of is fixed.

The third man, standing in the boat, holds the instrument extended in his hand, and when the whale is found, the fishermen call up as many comrades as possible, all similarly prepared. They then strike the whale as it swims on the surface of the water, for it has driven fish up to this point from the bottom. They then strike it deeply with all their might. When their missiles have been thrown and are stuck in the wound, the fishermen immediately withdraw. If the whale heads for the open sea as soon as it receives the blows, they cut the cords and bemoan their labors and expenses as lost. If, however, it immediately dives to the bottom, then it is severely injured from its wounds. It rubs itself on the bottom on account of the saltiness in its wounds, and in so doing it drives the point deeper and deeper into its body. The fish's blood immediately indicates its labor as the blood bubbles toward the surface from the bottom. The whale, gradually weakened from loss of blood and seeking the firmness of the bottom, approaches the shore until it begins to be visible out of the water and, then, surrounded by all the inhabitants with a multitude of boats and lances, it is killed. The other method is the same except that the lance is fixed in the whale not by the stroke of a man but by the blow of a crossbow [*balista*].[90] A cord is sent through this one as well, just as we said concerning the other.

---

88. On medieval whaling, see Szabo (2008).

89. "Edges": lit., "lines," *lineae*. A. seems to mean the two outer edges of the "triangle," opposite the base, to which the shaft is attached. Scanlan (339) translates *novacula* as dagger, but "razor" is more accurate. The thin head is, lit., "sharp as a razor" (note that in chapter 4 of *Moby-Dick*, Queequeg shaves with his harpoon).

90. The ancient *ballista* was a siege weapon. In medieval Latin the term came mostly to mean a crossbow. There is nothing here to indicate if this crossbow was held to the shoulder or fixed to the bow of the boat.

Those, however, which are the hairy whales, and others as well, have the longest tusks.[91] They use these to hang from the rocks of cliffs when they sleep. At such a time a fisherman, coming close, separates as much of its skin as he can from the blubber near its tail. He passes a strong rope through the part he has loosened and he then ties the ropes to rings fixed into a mountain or to very strong stakes or trees. Then, by throwing rocks at the head of the fish from a large sling, he wakes it up. Aroused, the fish, as it tries to escape, draws off its skin from its tail down along its back and head and leaves it behind. Later, not far from that place, it is captured in a weakened state, either swimming bloodless in the water or lying half-alive on the shore.[92]

Strips of this animal's skin are very strong and are used for lifting huge weights on pulleys. They are always displayed for sale in the marketplace at Cologne.

These, then, are the facts which we have experienced concerning the nature of whales. We are passing over the things which the ancients wrote, since they do not find agreement among the experts.

(24) CLANCIUS: The *clancius,* which is also called the *glanis* [sheatfish], is a clever fish which steals bait by turning its jaws sidewise, without swallowing the hook.[93]

(25) CONGRUS: *Congrui* are common fish which some call sea eels (*anguillae marinae*) and which are abundant in the sea of Normandy.[94]

It is rather a long fish and is round, having a very white, delicious flesh, but one which is indigestible. It is thus said to cause leprosy.[95]

(26) CARPEREN: *Carperen* are common river and lake fish [carp].[96] They are fat and sweet but are not healthful. They have a soft skin and are proportionately thicker than they are long. They are reddish and very scaly, with strong scales. The roasted tongue of this fish is very delicious.[97]

This fish is extremely clever at avoiding nets. Sometimes it leaps across them, other times it digs down into the bottom, and at other times, holding to the grass by its mouth, it holds itself back so as not to be dragged along by the net. At other times,

---

91. Here A. is definitely talking about walruses, *Odobenus rosmarus,* whose strong hides are still sought by hunters.

92. The thought of the walrus standing still for this treatment leads one to wonder whether A. had been told a "tall tale" by his sources.

93. See the notes at 8.119. On the origin of A.'s *clancius* by way of textual confusion from Pliny *HN* 9.67.145, see Aiken (1947), 211. Cf. ThC 7.20 (*clautius*) and Vinc. 17.40.

94. This is the conger eel, *Conger vulgaris,* and the name is ultimately from the Gr. *gongros* (*GF,* 49–50). Cf. *gonger* at 24.34(55) as well as ThC 7.21, Vinc. 17.46, and Barth. 13.26.

95. In Galenic theory, the cause of leprosy is an excess of black bile (cf. 1.608). In *De melancholia* (Kühn, 19:699–720), Galen advises those suffering from excess black bile to avoid beef, tuna, rough bread, aged cheese, cabbage, dried lentils, and foods of acrid or sharp quality. The eel's indigestibility, perhaps associated with the process of pickling it, seems to be the source of A.'s advice.

96. Stadler identifies this as *Cyprinus carpio* (cf. *GF,* 135–36; *FAZ,* 74–75). Cf. ThC 7.23 and Vinc. 17.40. On the name, cf. Sanders (439) and the MG *Karpfen.*

97. The fleshy palate of the carp was often, as Ar. *HA* 533a28–30 warns, mistaken for a tongue. It is still considered a great delicacy and is still called "carp's tongue." Aldrovandi reckoned it an aphrodisiac (*GF,* 136).

coming down from the surface, it fixes its head in the bottom so strongly that it is passed over by the net, which touches nothing except its tail.[98]

This fish reproduces in one set of waters, but grows and lives in another. It does best on a clayey bottom that has first been sown with wheat, has then had clay sprinkled over it, and has then been covered with water.[99]

It is said of this fish that when the female feels that she is pregnant with eggs and that the time of laying them is at hand, she excites the male with a motion of her mouth to spread his milt and then she lays her eggs. The milt of this fish is very thick. The story, however, which certain people relate, that the female takes up the milt in her mouth in order to conceive eggs in the future, is entirely false and has been disproved by us elsewhere.[100]

(27) CAPITATUS: The *capitatus* [chub] is a common fish and is very numerous in the rivers of Germany and France.[101] It has a head almost equal in its size to the rest of its body. Its body is black or dusky and it lies hidden beneath rocks. It is rarely half a foot long and is good to eat, but with tough flesh. It is shaped like a club and is found in greater quantities in the Danube and the waters flowing into the Danube than in any other waters.

(28) CORVI MARIS: *Corvi maris* [sea crows] are fish so called because they sometimes make a noise like that of crows, not with voices, but with their gills and chest.[102] It is, moreover, through sounds of this sort that they give themselves away. Certain fish in the Aceolus River make a noise in this way.

(29) COCLEAE: Snails [*cocleae*], both aquatic and terrestrial, are called *limaces* and *testudines* from the spiraling of the shell which they inhabit.[103] The aquatic ones are found to be very large and their shells [*coclea*] are dotted with certain spines of

---

98. The carp's diet consists of vegetation as well as crustaceans and insects. Its habit of burrowing in the mud makes it hard to net.

99. Carp have been raised in fish preserves for many centuries. On carp as food, cf. FAZ, 74–75.

100. Cf. 5.18.

101. Stadler identifies this fish as *Cottus gobio* L. (bullhead) or the *Leuciscus cephalus* L. (common chub). Scanlan (342) follows this identification. Thompson, however, sees the fish as *Cyprinus cephalus* L., a chub known for its large head (*GF*, 37–38).

102. Based on content and the mention of the Greek river Acheloos, the ultimate source for this is Ar. *HA* 535b16–18, where several calling fish are named. None is specifically called the raven, however. Cf. Isid. *Orig.* 12.6.13, ThC 7.26, and Vinc. 17.46.

103. *Coclea* most commonly means simply "shell" or "shellfish." Thus Scanlan (342) translates as "gastropod." Yet according to Stadler's notation, A. has inserted into his text the phrase "are called *limaces* and *testudines*" and thus seems to wish to define *coclea* fairly narrowly here. What, then, are these two creatures? Comparison with passages such as 4.38–39, 42–43, 89, and 17.76 seems to show that the *testudo* and the *limax* were different from most *concylia* (A.'s more common generic word for shellfish) in that they had a head, a single shell, and a foot that formed the closure to its shell which it used to move from place to place. It seems that despite his confusing terminology (cf. Glossary, "Shellfish"), A. here wants us to think that *coclea* are creatures such as snails and limpets. Note also that Stadler's text is inaccurate, reading *limaces ei testitudines* for what obviously is *limaces et testitudines*. Cf. Borgnet's text, ThC 7.27 (clearly describing snails), and Vinc. 17.45.

shell-like material.¹⁰⁴ The animals themselves lack any defined shape to their head and are therefore said to have no eyes, having instead certain soft, horn-like appendages with which, so they say, they try out beforehand the path which they wish to traverse, moving themselves along by contracting and relaxing their bodies. On these horns there are two black dots which perform the function and action of eyes. These animals sometimes leave their shells [*coclea*], and these and all other genuses of shelled creatures [*testae*] and crabs grow larger under a waxing moon and smaller when it is on the wane.¹⁰⁵ This is because the moon moves their moisture.

(30) CONCHA: *Concha* [shellfish], from which *conchilia* are named, are hard-shelled creatures having shells formed like ships, possessing a rounded stern behind, a pointed prow, a keel at the belly, and a raft-like shape on its distended and elongated sides.¹⁰⁶ There are two shells enclosing each other, with a black or dusky color to the outside and the splendor of a pearl to the inside. The fish is found at the place where the two shells meet. A tiny animal called the *nauplius* enters the shell.¹⁰⁷ It has the shape of a cuttlefish but is quite small indeed and only enters for the sake of play. The animal inhabiting the shell, being conscious of its companionship, opens its shell to it. Pearls are very often found in these *concha* in contact with the shells.

(31) CANCRI: Crabs are aquatic animals contained in the genus of soft-shelled creatures. Now philosophy might call only those large, rounded sea creatures "crabs" [*cancri*] and call the others with tails *gambari* [crayfish] and the large sea variety *locustae* [lobsters], but, speaking here in the common way rather than philosophically, we are using the general name for crabs which the common people usually employ, using it for all animals with a soft shell and many feet which move backward when swimming and forward when walking.¹⁰⁸

---

104. The double use of *coclea*, referring to both animals and housings, only worsens the confusion.

105. Cf. Barth. 13.26. ThC 7.23 tells us that all fishes' brains grow or shrink as the moon waxes or wanes.

106. It is not easy to penetrate the confusion surrounding the terminology used for shellfish (q.v. in Glossary) because terms for shelled creatures are used both specifically and generically (Saint-Denis, 1947, 26). Moreover, *concha* is clearly given as a ntr. pl. here, whereas it is usually a fem. sing. (the pl. form *conchae* is found in ThC 7.22 and Vinc. 17.44). Further, in this passage from A., the form *concha* is used consistently for the animal and the form *conca* for its shell, a consistency to be mistrusted because of its rarity in medieval orthography. The Latin is given to aid the reader.

The origin of this particular *concha* appears to be Pliny *HN* 9.49.94, where he describes an unnamed animal and its passenger the *nauplius*, saying it is a *navigeram* (a sea-goer) and that *concham esse carinatam* (its shell is hulled). Thus, a generic term has given rise to the name of a specific animal.

107. On the *nautilus* and *nauplius*, cf. *GF*, 172–75.

108. Thus the crayfish is often *cancer fluvialis*, "river crab," as at 4.22f. Its other, more common name is *gamarus* (4.18, 15.35, 21.37). The fuller name for the lobster is *locusta maris* (e.g., 1.68, 93, 4.18). For a comprehensive treatment of the sources and textual corruptions behind A.'s treatment of the *locusta maris* (and the *karabo*), see Kruk (1985) and cf. *FAZ*, 105–6. See George and Yapp (1991, 205–7) for a rare illustration of a crab.

The moisture of these animals increases when the moon is waxing and, although they are dark when alive, they turn red when cooked. During a full moon, two stonelike semicircles grow on them internally beneath their eyes which, if ground up and given in a drink, are said to strengthen the heart.[109]

There are, moreover, many genuses of crabs. One, called the soldier [*miles*] due to its speed, is found in India and is said to have no superfluity in its body.[110] The cause of this is not that it does not eat, for this is impossible, but rather that it feeds the way plants do, solely by absorbing moisture. Solinus says that in the western sea there are crabs so large that they drown people and these crabs have a hard back like that of a crocodile.[111]

Crabs lie dormant in winter for five months. Coming out in spring, however, they shed their shells as a snake does its skin. Certain people say, however, that they frequent sunny shores in the winter but go back in groups to their preferred waters in the summer. During the winter they are harmed by the cold, but during autumn and spring they grow fat, especially during a full moon.[112]

Below the place where the tail is connected to the body, the males have four long projecting rods which the females lack. The tail of the male is rounder, fuller, and thicker. That of the female is thinner, emptier, and broader, almost compressed. The female's eggs are strongly compressed in her body at first. They then exit through an opening, adhering to some short rods beneath her tail until they are grown. These eggs are of medicinal use against the venoms of serpents.[113] When the male and female copulate, the male first ascends the back of the female and the female turns herself over onto her back. Then the male copulates with her by rubbing himself on her as the eggs exit.[114]

A crab kept in milk lives for many days without water.

---

109. ThC 7.19 and Vinc. 17.37 stipulate that these "lapides" appear when the crabs are old and specify that they help heal pricking pains, "punctiones," of the heart.

110. *Miles* is also a nickname for a turtle at 7.77 and 24.12(19) and for the swordfish at 24.35(60). Cf. Sanders (413) on the name. The present passage originally derives from Ar. *HA* 525b7f., probably through Pliny *HN* 9.51.97. Both locate the crabs in Phoenicia; the former calls them *hippoi* (horses) and the latter *hippoe*. See notes to 4.20 for the confusion of the Gr. *hippoi* (horses) and *hippeis* (cavalrymen). Apparently, then, "cavalrymen" has come into ThC and A. as *miles*, perhaps with the medieval sense of mounted soldier, or knight. Phoenicia has become "Iudea," which either A. or his source miscopied as "India." Cuvier (*B&R* 2:425) conjectures that these are the long-legged crabs commonly called spiders, the *Macropodia* and the *Leptopodia* of Linnaeus. Thompson (1910, ad 525b7) identifies the "horsemen" as *Cancer cursor* Belon. (*Ocypoda cancer* L.).

111. No appropriate passage in Solinus can be found, but mention of the huge size and hard backs is also in Jacob. 90 (p. 192), saying they exist in the "western sea."

112. Cf. Ar. *HA* 601a16, Pliny *HN* 9.50.95, and Barth. 13.26.

113. Crabs in general, but particularly freshwater crabs, were believed to be poison antidotes. Cf. Pliny *HN* 9.51.99, 32.19.53.

114. Cf. Ar. *HA* 541b19–29.

The anatomy of the crab has already been determined in preceding books of this study.[115]

Ambrose relates that the crab likes to enter the shells of oysters [*ostreum*] to eat them.[116] On the other hand, fearing to be caught within the shell [*conca*] and to be crushed, the crab does not dare to enter. It therefore keeps watch until the oyster opens its shell to take pleasure in the rays of the sun. Then, with its front claw, which is split, the crab puts a pebble within the shell, not allowing it to close.[117] It then feeds upon the oyster [*ostreum*].

(32) COCODRILLUS: The crocodile is one of the water creatures and has the shape of a lizard in all particulars except that it does not have quite so round a tail and has ridges on its tail.[118] When it is fully grown it reaches a length of up to twenty cubits. I myself have seen two of approximately sixteen feet and another one of eighteen. Its skin is wrinkled and is so strong that it is as if it were protected by a shield. The crocodile is in the water by night but during the day comes out onto land to feed. It is slow due to the shortness of its feet. Its mouth opening extends as far as its ears (if it had ears).[119] Its teeth are very strong and it lacks a tongue. Although there are many kinds of crocodiles, they all move their lower jaw except the *tenchea*, which thus has the strongest bite.[120]

It lies so still in the sun that it seems to be dead. When it opens its mouth in a great yawn, small birds fly in and clean its teeth. Sometimes, shutting its mouth, it swallows some of them.[121] It lies in wait for all animals, but especially for the *bubali* [buffalo], and, for its part, the *bubalus* tramples the crocodile as it lies in the sand.[122]

---

115. Cf. 4.28–29, 34; 14.11–13.

116. This passage adds still more uncertainty to A.'s vocabulary concerning shellfish. The Latin reads *Cancer libenter intrat conchilia ostreorum et comedit conchilia* and thus *conchylium* is both the shell and the creature within the shell. *Ostreum* is translated "oysters" here, although any shellfish could be meant.

117. Oppian *Halieutica* 2.167–79; Isid. *Orig.* 12.6.51; Ambrose *Hex.* 5.8.22; Barth. 13.26; Jacob. 90 (p. 193). The Loeb editor of Oppian notes ad loc. that crabs sometimes scrape sand over oysters to make them open their shells.

118. "Ridges": *pinnae*, lit., "fins," and thus linking it more closely to the fish. The crocodile fascinated both the ancient (*AAZ* 37–42) and the medieval mind; see Solinus 32.22, Isid. *Orig.* 12.6.19–20, *Physiologus* (Curley, 1979, 53–54; Carmody, 1939, 35), Jacob. 88 (pp. 182–83), Barth. 18.32, ThC 6.7, Vinc. 17.106, and Hugh of St. Victor 2.8, 3.55 (*PL* 177: 60C, 105B). It was a popular creature in the bestiaries (White, 1954, 49–51), was laden with symbolism (McCulloch, 1960, 106–8; Rowland, 1973, 55–58), and was frequently, if poorly, drawn (George and Yapp, 1991, 97–99, Pl.3; Benton, 1992, 82–83). Gradually, experience began to construct better descriptions of the crocodile; see Malkiel (2016).

119. A. has added the disclaimer, based no doubt on his personal observations.

120. Cf. Ar. *HA* 516a26–29, where an immobile lower jaw is specified, and Pliny *HN* 8.37.89, where a mobile upper jaw is specified. *Tenchea* is a corrupted word from the Arabic for "crocodile"; cf. 1.227.

121. See Aiken (1947, 211–12) for a discussion of the nature of the errors that occur in this section. This bird is generally recognized as the Gr. *trochilos*, the Egyptian plover, *Pluvianus aegyptius* (*GB*, 288–89).

122. In this context one would imagine that the *bubalus* is a water buffalo. See 3.120 for notes on this name. Scanlan (345) identifies this as a hartebeest.

The crocodile kills a human when it can but, as certain people claim, it afterward cries for the same person.

Prostitutes make an ointment from its excrement with which they smooth out wrinkles on their faces, but when the face is washed it becomes even more wrinkled than before.[123]

These monsters live in the Nile River, in other rivers, and in the rivers of India.[124] When they are filled with fish, they lay their heads on the banks as if asleep. The bird which cleans its teeth and which it sometimes swallows is called the *crochilos* in Greek and the *regulus* in Latin.[125]

There are certain people, according to Pliny, who live on an island of the Nile. They are called the Tyn(ti)ri and have small bodies.[126] These people know the trick of swimming above the back of the monsters and leaping onto their backs. They lead them captive, bridled as it were with sticks in their mouths, to the shore, and there force them on occasion to regurgitate the prey which they have most recently taken in.

(33) CAHAB: The *cahab* is said to be a marine animal having small feet with respect to the rest of its body.[127] It has one long foot which it uses in place of a hand and with which it brings food to its mouth and digs up plants. Its feet are made of a cartilage and are formed in the shape of a calf's foot. This animal breathes, and when it exhales, it returns to the air, spewing forth water as does the dolphin and the whale.[128]

(34) CRICOS: The *cricos* [hermit crab] is, they say, an animal which has a large, long left foot but an extremely small right foot.[129] It has a shell-like, soft, black hide, with scattered red spots. During calm weather, it walks about unattached. At such a time, it leans its whole body on its left foot, dragging the other one behind itself. When the weather is cloudy, however, it clings without moving to the rocks. It is almost impossible to drag this animal off whatever animal it has affixed itself to.

---

123. Cf. Pliny *HN* 28.28.109.

124. Isid. *Orig.* 12.6.19 gives the Nile. India may have entered the text in the same way that it did at 24.22(31). ThC 6.7 and Jacob. 88 (p. 183) read *Casaree Palestine*, i.e., "Iudea." Has it also been miscopied here as "India"?

125. The *regulus* is *rex avium* ("king of birds"). Cf. Pliny *HN* 8.37.90, Barth. 18.32, and ThC 6.7, the latter likening the "luring" of the bird to usurers luring the poor.

126. A. apparently wrote the name simply as "Tynri." Stadler expands the name in his edition, based on Pliny *HN* 8.38.92. Rackham (1940, 3:66) gives the modern name of the island as Denderah. ThC 6.7.48 has the correct name.

127. Stadler suggests this animal is a seal and Scanlan (346) claims the name is a corruption of *phōkē*. The latter is quite unlikely. With the reference to feet, one thinks first of the *cahab*, or ankle bone (cf. 1.297, with notes), and thus of an Arabic origin for the name. Cf. ThC 6.9 and Vinc. 17.103.

128. Stadler also suggests that the present passage is reminiscent of Ar. *HA* 498a32f. and that ThC 6.9 does cite Ar., but the resemblances are slight. Yet seals are not herbivores. Is it possible that this animal is the dugong? The dugong inhabited the Red Sea and Indian Ocean and thus an Arabic source may lie behind it.

129. Cf. Ar. *HA* 529b20f., whose *karkinion* has been shortened to *cricos*. On possible species, cf. *GF,* 104–5. Cf. ThC 6.10 and Vinc. 17.105.

(35) CELETHI: The *celethi* [selachians, cartilaginous fishes], as we have said in foregoing books, are a genus of animal which contains many species beneath it. It first produces an egg and then gives birth to a live young [*animal*], which is formed in the likeness of the parent. Because of the weight of its head, it sleeps so deeply that it can be caught by hand when it is asleep.[130]

(36) CHYLON: They say that the *chylon* [gray mullet] is a sea animal which does not eat but which feeds in the manner of one which absorbs viscous moisture.[131] Thus it is that nothing is found in its stomach.

(37) CANES MARINI: The sea dogs [*canes marini*, seals] are hairy like dogs. They have shining eyes and short feet in the front. In the rear, however, their skin is pinched together and thin, formed more similar to a rudder-like tail than to feet. Their walk is very weak. They are greatly given to biting and they follow large fish, biting and killing them. They hunt in packs, with large numbers of their own kind, herding many fish into one place.[132]

(38) CAERULEUM: The *caeruleum* is blue both in name and fact. It is an aquatic animal which the Ganges River supports. It has two arms, each quite fierce and each two cubits long. It seizes large beasts with these at the mouth of a river and drags them to the bottom.[133] This animal is not found in our waters.[134]

(39) DRACO MARIS: The sea dragon [sea snake?] is, as Pliny says, a beast shaped like a serpent.[135] It has, however, two fins for use in swimming. With a single thrust of the fins it swims through vast expanses of ocean due to their great strength. It is, moreover, a venomous creature and whatever fish or other animals it bites die. Sometimes,

---

130. Ar. *HA* 537a30. Cf. ThC 6.11 and Vinc. 17.104. The name is a corruption of the Gr. *selachē*. On *animal* as "live young," see 1.77.

131. Ar. *HA* 591a22f. states that the *kestreus*, or gray mullet, feeds not on other fish but on seaweed and sand. Cf. ThC 6.12 and Vinc. 15.105. The name *chylon*, however, is probably from *chelōn*, given at Ar. *HA* 543b14f. as a type of *kestreus*. Cf. *GF,* 108–10, 287–88.

132. The "sea dog" is surely not a porpoise (Scanlan, 347). It is a direct translation into Latin of the German *Seehund* and is a seal. Cf. 1.33, 99, and 5.15, with notes. Note that the *canes marini* of Pliny (*HN* 9.55.110, 70.151–53, etc.) are usually taken to be dogfish or sharks (Saint-Denis, 1947, 17–18). Cf. Vinc. 17.103 and ThC 6.13, both of which are much more fabulous in content.

133. "Large beasts": Pliny *HN* 9.17.46 and Solinus 52.41 both read "elephants," as does Jacob. 90 (p. 192).

134. Scanlan (347) points out that there is a pale cetacean in the Ganges called the "susu," but the source of the *caeruleum* is to be found in the world of textual corruption rather than in the waters of the Ganges. See 24.8(3) and the bibliography cited there for the source of this fictional beast and compare the notes offered by Saint-Denis (1947, 15–16). Cf. also 24.49(99), ThC 6.14, and Vinc. 17.104. The disclaimer of local habitation is A.'s addition to the text.

135. This creature is traditionally identified as the weaver fish or sea spider, on which cf. *Aranea*, 24.11(15), and *GF,* 192–93. Cf. Saint-Denis (1947, 33). But Ar. *HA* 505b5f. and 621a3–5 seem to be behind this passage and certainly imply a snake-like beast. We may have, then, a simple sea snake, some species of which are quite venomous. Scanlan (347) opts for the moray eel. Vinc. 17.114 quotes Isid., saying it has spines looking toward its tail to inject poison, hence its name. The weaver fish has such spines on its gill operculi. ThC 6.15 says it has *pinnas* ("feathers," "fins") instead of *alae* (wings), apparently differentiating it from the land-bound *draco*.

when it is dragged to shore in fishermen's nets, it suddenly makes a pit in the sand in order to be concealed from sight.[136] The ashes from its bones heal toothaches.[137]

(40) DELPHINUS: There are many genuses of dolphin in the sea.[138] However, the one which is most often seen in the sea is an animal with dark skin, a short head, and solid teeth in its mouth like the molars of a pig. It loves humans and any member of its own species so much that when the young are swimming in a school, two large dolphins are placed as guards.[139] Also, if one of its own species has died, the others guard it until the sea itself casts it ashore, lest it be devoured by other fish.[140]

In the large Mediterranean Sea, which touches on Italy on two sides, a school of dolphins gathers together and accompanies fishermen as they are going to sea to fish. At sea, arranged in the shape of a crown, they bunch up the fish, drive them along, and force them into the nets. Thus, the dolphin is not eaten by the Italians and the fishermen do not hunt them but rather give the dolphins part of the captured fish. The Italians call these dolphins *tumberelli* since, tumbling about before the ships, they spout forth water.[141]

This animal, by leaving the bottom for the surface at the time of a storm, gives notice that a storm is on the way.[142]

It is an animal, so they say, of great age. They even say that, once, a certain dolphin with a cut-off tail was ascertained to have lived one hundred years or more.[143] The dolphin copulates after a space of ten months. Its copulation is swift, since it has its testicles and penis inside itself. The female lies on her back beneath the sea and receives the semen into her womb, which is like that of the whale and of a woman.

This animal takes joy in music. Thus, it was dolphins who picked up a certain citharist named Arion, who had played his cithara and was thrown into the sea. They held him up, pushed him along, and brought him to shore. At another time, a certain

---

136. Cf. Pliny *HN* 9.43.82.

137. Cf. Pliny *HN* 32.26.79.

138. This entire section on dolphins is based on Pliny *HN* 9.7.20–11.34, probably through Solinus 12.3–13 (Saint-Denis, 1947, 31–32; Stebbins, 1929; *AAZ*, 53–57). Cf. ThC 7.29, Neckam 2.28–29, Barth. 13.26, Hugh of St. Victor 3.55 (*PL* 177:105C), and Rabanus Maurus *De universo* 8.5 (*PL* 111:237D-238A). Cf. White (1954, 200–201) and George and Yapp (1991, 96), with a totally inaccurate MS illustration.

139. Cf. 8.211f.

140. Cf. Ar. *HA* 631a16–20.

141. A.'s words in Latin make clear the derivation: *tumberellos . . . tumbantes*.

142. ThC 7.29 attaches this information to a separate type of dolphin, the only one he classes with his fish. The bulk of his information on dolphins is in 6.16. Rabanus Maurus *De universo* 8.5 (*PL* 111:238A) calls these dolphins *Simones* (cf. Barth. 13.26); whether he knew it or not, he was reflecting the Latin pet name for dolphins (Pliny *HN* 9.7.23–8.25; *GF*, 53). *Simo* means "snub-nosed," and one must recall that the Mediterranean dolphin was not bottle-nosed.

143. Pliny *HN* 9.7.32, Solinus 12.4, and Neckam 2.28 all say thirty years, but ThC 6.16, citing the *Experimentator*, says up to one hundred forty years. Apparently, notches were cut into the tail as identification, but use of *amputata* here shows the story had changed.

king of Caria had caught a certain large dolphin. A great number of dolphins followed the captured one to shore, conducting themselves in the manner of mourners. When the king of Caria saw this, he ordered that the captured dolphin be set free. The dolphins, who had congregated, received him in a sort of dance and led him away from the shore. They also say that if a person who has eaten the flesh of a dolphin and still has it in him falls into the water, he is devoured by dolphins if they are present. If, however, he has not eaten the flesh of a dolphin, he is returned to the shore by the dolphins even if he is dead and has been killed.

They say that when Augustus held the principate, a boy had fed a dolphin to such an extent that it would take the food from his hand. It eventually bore the lad through the sea on its back, taking him where he wished to go and then returning him to shore. Finally, when the boy died and did not appear on the seashore as was his custom, they say the dolphin died from grief. Something similar to this is said to have happened in the sea off the city of Yponensis in Africa, and near this same city, not far off, another boy named Henanus is said to have done much the same thing with a dolphin, as is written in the histories of the Persians.[144]

In our sea, however, lying near Germany, the dolphins are captured and eaten. This is why they flee humans.

There is, moreover, another genus of dolphin which, so people say, has its mouth in the middle of its body and is so fast that none of the marine creatures would escape it if it were not for the fact that it catches its prey turned over on its back so that its mouth faces upward.[145] The fish can escape then. This genus hates its young so much that the male would devour them if the female did not hide them.[146] The mother therefore hides the young and leads them with her until they are fully grown. After they are grown, she begins to hate them so much that with the male she would devour them if they could not defend themselves with their own strength.

There is yet another genus of dolphin, people say, in the Nile River. It is crested, with a row of spines arranged on its back like the teeth of a saw.[147] This genus places itself beneath crocodiles as they swim by and kills them by cutting open their soft bellies with the spines on its back.

(41) DIES: The "day" [mayfly] is a fish which is born, grows to full size, and dies all in the same day. It has two feet and feathery wings. Some say that it lives longer

---

144. A. is compressing two stories into one; cf. Pliny *HN* 9.8.26f. and Solinus 12.9f. The first story was told of the coastal city of Hippo Diarrhytus and the second of Iasus in Babylonia. The boy's original name was Hermias.

145. Cf. Ar. *HA* 591b25f. and *Part. An.* 696b24, which claim that all selachians, dolphins, and cetaceans feed in this position. The statement has generated much commentary.

146. Aiken (1947, 212) explains the origin of this peculiar statement.

147. Cf. Pliny *HN* 8.38.91, where the fin is shaped like a knife. Both Neckam 2.29 and Barth. 13.26 mention the saw, the latter citing Isid. *Orig.* 12.6.11.

than one day after its birth but that it does not live more than one day after it is fully grown.¹⁴⁸

(42) DENTRIX: The *dentrix*, or *peagrus* [bream], is a fish that takes its name from the fact that it has many large teeth with which it feeds on shellfish.¹⁴⁹ This fish is called the *dentatus* by people today. It has teeth positioned in the front of its mouth, both above and below, of equal length, almost in the shape of the teeth called incisors in a human. It has a lower jaw formed from two bones, as is that of a human. It is a scaly fish, no bigger than a hand can hold. It is shaped like the fish called the *monachus* [monk] except that it does not have quite so thick a head and is a little wider in the body. It is abundant in the Apulian Sea, in Sicily, and near Rome.¹⁵⁰

(43) EQUUS MARIS: The "sea horse" is a sea animal, the front part of which is in the shape of a horse and whose hind part tapers off into a fish. It is hostile to many marine creatures and its food consists of fish. It fears humans greatly, however, and can do nothing out of the water, as it immediately dies when taken out of the water.¹⁵¹

(44) EQUUS NILI: The "Nile horse" [hippopotamus] is an aquatic beast and is like the crocodile in both genus and nature. It has legs and teeth like a crocodile but they are much larger.¹⁵² It is a very cruel animal and very eager to kill people and to capsize ships. With one foot fixed on the land, it pushes the other at a ship and, with the force of it, either turns the ship over or pierces it. It cannot be easily captured, except with nets made of chains, and once it is captured it can be killed only with iron hammers.¹⁵³ It has a strong thick skin, almost a cubit thick.¹⁵⁴

---

148. Cf. 1.98, based on Ar. *HA* 552b21–3 and A.'s *DG* 2.3.5. *Dies* is apparently a simple translation of the Gr. *ephēmeron*. On the problems of identification, cf. *GI*, 157–58, and Beavis (1988), 88–89. It is included as a fish probably because its egg masses were observed to float on streams. ThC 7.28 and Vinc. 17.47 add the detail that it is bloodless. Jorach, as cited by Barth. 13.26, gives the *ephemeron* only a three-hour life span.

149. Cf. 24.49(100). This is surely a bream and the name *peagrus* is corrupted from the Gr. name for the bream, *phagros* (*GF,* 273–75). Cf. Isid. *Orig.* 12.6.22–23, ThC 7.30, and Vinc. 17.47. Barth. 13.26 calls it *phagion*. Cf. Saint-Denis (1947, 32, 80–81) on the *dentix* and the *pager* (*phagrus*); *FAZ,* 60–62.

150. The term *monachus* is all too prevalent in the medieval piscatory vocabulary. Cf. notes to 1.228. Perhaps in this case the chub is the best identification possible.

151. This sea horse is undoubtedly a fictitious animal, in no way resembling the modern usage of the term. At first one might think this is a description of the ancient hippocamp (Saint-Denis, 1947, 36). ThC 6.18 cites Ar. as his source, and Stadler's suggestion of Ar. *HA* 589a27 refers to a passage that mentions a number of animals which cannot live outside the water. The animals listed are the "sea turtle, the crocodile, the hippopotamus, the seal and some smaller creatures." If this is the source, then the "river horse" (hippopotamus) has somehow become a "sea horse." Some later commentator could have appended the description of a hippocamp. Isid. *Orig.* 12.6.9 offers almost identical wording; cf. Vinc. 17.115.

152. The Latin is ambiguous and "larger" may refer to either the animals or the appendages, as ThC 6.19 would seem to indicate. Cf. Pliny *HN* 8.38.92–94 and Vinc. 7.115. Sanders (412) claims plausibly that *equus Nili* is a translation of the German *Nilpferd*. Yet the description leaves one to wonder whether A. divined that this creature was a hippopotamus.

153. Cf. Jacob. 88 (pp. 184–85).

154. A.'s Latin is again sloppy, but it is preferable to take *fortissima* as referring to *cutis* (skin) than to the animal (which is ntr.), as does Scanlan (351).

(45) EQUUS FLUMINIS: The "river horse" [hippopotamus] is said to be an aquatic animal in parts of the East, at home both on land and water. It has hair like a horse but cloven feet with hooves like those of a cow. It has an upturned face, the tail of a pig, and it whinnies like a horse. Its skin is very thick. Its size, however, is like that of an ass.[155]

(46) EXPOSITA: According to Pliny, the *exposita* is a beast that was once found in that part of the sea which touches upon Joppa in Judea.[156] It has many long teeth and a layer of fat five cubits thick. It is of the family of whales [*cetus*].[157]

(47) HELCUS: The *helcus* is the *vitulus marinus* ["sea calf"; seal], about which we have spoken a great deal in preceding books.[158] It has a hairy pelt marked with white and black spots. This animal gives birth on land and feeds its whelps by nursing them at its breasts. It does not lead them to the sea for twelve days. The seal is difficult to kill unless the blow is directed to the vicinity of the brain. In its sleep it snores so deeply that its snoring resembles mooing. Of all other animals, it is afflicted with the heaviest sleep. For this reason, the flippers on its right side, used for swimming, are said to cause sleep if they are laid against a person's temple.[159] The hairs on a hide taken from one of these animals will, no matter where the hide is, indicate by the composition and erectness of its hairs the ebb and flow of the tide. The hides of other hairy sea creatures do the same thing.

(48) ESCYNUS: The *escynus* [sea urchin] is a member of the genus of crabs and is about half a foot in length. Around their crown, they are white and have a kind of spine in place of feet.[160] Their mouths are in the middle of their bodies and they are of a glass-

---

155. Cuvier (*B&R* 2:290) commented on how inaccurate the descriptions of the hippopotamus were even after it had been exhibited in Rome. Cf. *AAZ*, 87–88, Ar. *HA* 502a9–15, Pliny *HN* 8.39.95–6, ThC 6.20, Vinc. 17.115, Barth. 13.26, and Rabanus Maurus *De universo* 8.5 (*PL* 111:237B) (*equus fluvialis*). For other hippopotamuses in A., cf. 24.30(44), *equus nili*, and 24.35(61), *ipodromus*.

156. Joppa is the modern-day port city Jaffa. Jaffa was conquered by Latin knights during the first crusade in 1099 and became the capital of the Latin crusader kingdom of Jerusalem. Destroyed by Saladin's brother in 1196, Emperor Frederick II and King Louis IX of France rebuilt and fortified the town in the sixth (1228) and seventh (1248) crusades, respectively. By 1260, however, Jaffa had passed into Mongol hands.

157. Cf. Pliny *HN* 9.4.11, which relates how one Marcus Scaurus brought to Rome the skeleton of the sea beast, to which Andromeda was exposed (*exposita*). The source of the name is thus clear (Aiken, 1947, 210). Cf. ThC 6.21.

158. Cf. 2.87, 90; 5.13; 6.68–69; 12.177; 13.97; 14.87; etc. This account is based largely on Pliny *HN* 9.15.41–42 (cf. *AAZ*, 166–67). Cf. ThC 6.22, Vinc. 17.116, and Rabanus Maurus *De universo* 8.5 (*PL* 111:237B). Scanlan (351) makes a strong case for deriving *helcus* from the Gr. verb *helkō*, to drag, and compares the Germanic *selhos*, which gave us "seal." A. is always fascinated by the seal's grace in water but clumsy gait on land.

159. Again, the paucity of technical vocabulary available to medieval natural philosophers reveals itself. A. uses *pennae* to denote the flipper—a word also used for a fish's fins and the erect scales on a crocodile's tail.

160. This first paragraph represents a confusion of Pliny *HN* 9.51.100 and Ar. *HA* 530a30–31a7 (Aiken, 1947, 212; *FAZ*, 196–97)). The "crown" for example, is *corona*, corrupted from a place name, Torone.

like color, being approximately the shape of a scorpion. In place of teeth, they have heavy spines in their mouth. They produce five eggs which are very bitter. It is therefore very poisonous and is not able to be eaten without entirely destroying the eater.

This fish announces storms in advance. Feeling that the substance [*materia*] of the winds is being elevated from the bottom, it seizes a stone and stabilizes itself with it as if with an anchor. When sailors see the *escynus* dragging this rock, they secure their ships with anchors.[161]

32 This fish, the *escynus,* delays ships, as we have stated in preceding books. Clinging to the bottom side of a ship, even a ship of two hundred feet or more in length which is fully armed, it holds it back. No matter what the force of the winds is, the *escynus* prevents the ship from being moved, whether by force or by skill.[162]

(49) EZOX: The *ezox* is a fish which some call a *lahse* [salmon]. Certain people, however, also use the name *esox* for that large fish of the Danube and of certain waters flowing into the Danube which the Hungarians and Germans [*Alemanni*] call a *huso*.[163] The one which is called the *lahse* has the shape and color of a *salmo* except that it has a lower jaw bent back to its upper jaw, much as the upper beak of an eagle is bent down to its lower beak. The *lahse*'s lower jaw, however, is not longer than its upper, and the upper jaw is equipped with a hole into which it receives the lower jaw. It does not have flesh quite as red or as tasty as the *salmo*.[164]

(49) HUSO: The *huso* is a fish [sturgeon] which is not scaly and has the shape of a *sturio*.[165] Its skin is white and soft, and it is entirely lacking scales and bones. When

---

161. Cf. Pliny *HN* 9.51.100, Ambrose *Hex.* 5.9.24, Neckam 2.34 (*echinis*), ThC 7.31, and Vinc. 17.58.

162. Cf. 2.83. This passage represents a conflation of the sea urchin and the *remora,* and was discussed by Aiken (1947, 212–13). This story is found in Barth. 13.26, ThC 7.31, Vinc. 17.58, Peter Damian *De bono religiosi status* 13 (*PL* 145:774D), and Hugh of St. Victor 3.55 (*PL* 177:108B), often with elaborate etymologies.

163. The *ezox* is definitely the Atlantic salmon, *Salmo salar,* and takes its name from the Gr. *isox* (*GF,* 95; Saint-Denis, 1947, 37; *FAZ,* 290–91). Scanlan (352) prefers to see it as a pike. Cf. Isid. *Orig.* 20.2.30, Neckam 2.42, ThC 7.32, and Vinc. 17.87 and 17.53. On *lahse,* cf. Sanders (443) and the MG *Lachs,* whence "lox." *Huso* normally indicates a sturgeon. ThC says the Swabians call the *esox* a *husen.*

164. *Huso* is dealt with immediately below. *Salmo* is given its own entry at 24.51(104). What, then, is the difference between a *salmo* and an *esox?* This question could form the basis of a study in itself, but we propose here that the terms may be various phases of the same fish. In fact, most nationalities have a host of names for the salmon to reflect various stages in its development and changes in domain (Mills, 1989, 10–14). The salmon is born in the rivers; migrates to the seas, where it is a firm, fit fish; and returns to the rivers to mate. After salmon return to the river their appearance remains the same for a while, but it then changes dramatically (Mills, 1989, 92f.; Netboy, 1974, 26f., 1980, 36f.). They cease to eat and they lose scales and body tone, rendering them less palatable. Their color changes and the males' lower jaw actually adds bone mass to outstrip the upper jaw. Thus, even though A. states at 24.51(104) that the *salmo* is both a river and an ocean fish we could tentatively posit that *salmo* refers to the salmon either living in the ocean or recently returned to the rivers and that *ezox/esox* is the spawning, altered, river version.

165. Stadler identifies the *huso* as *Acipenser huso,* common to the Danube. This fish was quite a delicacy throughout antiquity (*FAZ* 312–14). Scanlan (353) identifies it as the beluga, *Huso huso,* famous for its caviar. Sanders (414) links *huso/hese* to MG *Hausen,* "sturgeon." The *sturio* is treated below at 24.51(105). Cf. ThC 7.32 and Vinc. 17.87.

it is fully grown, it is found having a length up to twenty-four feet and it is smaller in proportion to what it lacks in age. It has no internal bones except in its head. Instead of a bony spinal cord, it has cartilage which has a large, empty opening as if it had been hollowed out with a borer from head to tail. There are no bones at all in its body and its fins are joined together with cartilage. The taste of the flesh from its back is comparable to veal and that from its belly to pork. It has lard [*adeps*] intermixed with fat [*pinguedo*], as does a pig. The *sturio* agitates this fish by rubbing itself against it, and thus it is that they are often caught together. It is not found in any waters other than the ones which were named.[166]

(50) ERICIUS: The *ericius* is a sea fish [sea urchin] which has its head and mouth below and the point of exit for its superfluity above.[167] This is the opposite of the arrangement found in other animals. It uses its spines for feet and is a cause of horror to other fish. Pliny says it has flesh as red as cinnabar.[168]

(51) EXOCHINUS: The *exochinus* is a fish in Arcadia, according to Pliny. But it is more properly an aquatic animal, since it comes out of the water to dry land for sleep and is unable to live without sleep.[169]

(52) ERACLEYDES: The *eracleydes*, according to Theophrastus, are fish living in the sea near Heraclea. This fish seeks fresh water so much that digging into the earth, it passes from one body of water to another. It sometimes lies hidden in caverns where it has found underground rivers flowing.[170]

(53) FOCA: The *foca* [seal] is the "sea ox" (*bos marinus*) and is the strongest animal in its genus, fighting all animals of its species. It does so even to the extent that it kills its own mate and then, mating itself to still another, kills her. It thus successively mates with many until it dies on its own or is killed by one of the females. It even kills its young unless they defend themselves. It lives on prey and therefore fights its fellow predators.[171]

---

166. E.g., the Danube and the others named at 24.31(49).

167. Stadler identifies this as *Echinus esculentus*. Cf. 13.113–14 and 24.31(28). For parallel texts, cf. Isid. *Orig.* 12.6.57 (*iricium*), Vinc. 17.58, and ThC 7.33.

168. Pliny *HN* 9.51.100.

169. This is the Gr. *exokoitos*, "out-sleeper" (*GF*, 63–64; Saint-Denis, 1947, 37–38), coming to A. from Pliny *HN* 9.34.70, which says that this blenny (or goby), found in Arcadia, came out of the water to sleep. Cuvier (*B&R* 2:406) makes the identification, saying that these are fish that can stay out of water in seaweedy rocks for many hours. Cf. ThC 7.34 and Vinc. 17.53.

170. The source of this entry is Ps. Ar. *De mirabilibus auscultationibus* (*On Marvelous Things Heard*) 73 (Dowdall, 1913, 835b15f.) and follows directly the entry on the Babylonian fish which appears at 24.13(22), also attributed to "Theophrastus." Cf. also Pliny *HN* 9.83.176 and ThC 7.36. Cuvier (*B&R* 2:471) suggests this is a loach, *Cobitis fossilis*, which often hides in mud and can survive for an extended time after water is gone. It is often found in drained marshes and dried-up riverbeds. Heraclea was in Pontus, on the southern shore of the Black Sea.

171. *Foca* is clearly from the Gr. *phōke*, seal, and is a corrupted version of Ar. *HA* 608b22f., which says that seals fight, male against male, female against female, when food is scarce (Saint-Denis, 1947, 15). Scanlan (354) sees hints of the mating battles of male seals. A. usually calls the seal a *koky* and regularly glosses this with *vitulus marinus*, "sea calf." If the calf is the *vitulus*, then perhaps the *bos*, "ox," is the large male. For the "sea ox," cf. 1.615; 5.13; 6.68; and 7.77. Cf. also ThC 6.23 and Vinc. 17.116.

(54) FASTALEON: The *fastaleon* [mullet] is a sea animal that does not fight any animal, since it does not eat flesh, but instead it feeds only on plants.[172]

(55) GONGER: According to Pliny, the *gonger* [conger eel] is a large, robust fish which carries on hostilities and wars with the moray eel [*muraena*], the octopus, and other fish. It is so strong that it can pierce the tough octopus with its teeth.[173]

(56) GOBIO: The *gobio* [goby] is a tiny fish, almost round in shape.[174] It has white scales and is dotted with scattered black spots. This fish is healthy while on a sandy bottom.[175] Some people say that in summer it is weakened by small worms which it bears in its stomach.

(57) GRANUS: According to certain people, the *granus* [stargazer] is an ocean fish which, contrary to the usual arrangement in fish, has one eye on the top of its head which it always keeps open against ambush.[176]

(58) GALALEA: The *galalea* [dog fish], so they say, is a sea animal which, contrary to the custom of all other animals, drags its young out forcefully when it feels that they are alive in its womb. If they are sufficiently developed, it nurses and nourishes them. However, if they are not developed, it places them back in the womb and cares for them until they are fully developed.[177]

(59) GARCANEZ: They say that the *garcanez* [sheatfish] is a river animal whose female, running about like a vagabond, does not care for but rather neglects her young. The male, however, stands by those the female has neglected and builds a circle of wood around them lest they lie exposed to fish that are their enemies. It then sometimes vocalizes loudly out of the water so that animals hostile to it might be scared off. If, during a period of such care, the male should fall into a net, it bites through the net

---

172. It is clear that *fastaleon* is the gray mullet, the Gr. *kestreus* (*GF,* 108–10), but the name form is not readily seen as the result of corruption of the Gr. name. In light of other, similar forms, one might suspect a common link, perhaps to the Arabic, though Schühlein has no entry for it. For a sample of forms, cf. *fastoroz* (6.91), *fastoreon* (6.85, 90), *fastoreoz* (8.32), *fastoroz* (4.82, 100), *fastaroz* (7.76, 98; 8.113, 120), *fastaleon* (6.81), and *fastaniz* (5.36). Cf. Ar. *HA* 591a17f., ThC 6.24, Vinc. 17.54, and Barth. 13.26.

173. Pliny *HN* 9.88.185. This is *Conger vulgaris,* and it takes its Latin name from the Gr. *gongros* (*GF,* 49–50; Saint-Denis, 1947, 27; *FAZ,* 125–26yjn n). Cf. Ar. *HA* 610b15, ThC 7.38, Vinc. 17.46, and Barth 13.26. *Mur(a)ena* can indeed mean a lamprey, as Scanlan translates (354), but because the conger is a saltwater-dweller it is best to see here its other attested meaning of moray eel (*GF,* 162–65).

174. From the Gr. *kōbios* (*GF,* 137–39; Saint-Denis, 1947, 43–44). There are numerous species called "goby," but if A. had Ausonius's *Mosella* 130f. in mind, he might have equated this fish with the gudgeon. Cf. ThC 7.39 and Vinc. 17.56.

175. This could mean that it itself is healthy or that it is a healthful food.

176. If *granus* is a scribal error for *uranus,* then this fish is *Uranoscopus scaber,* the white rascasse, whose eyes (two) are indeed on top of its head. *Uranus* is but one of its many names; cf. *GF,* 98–99. Cf. Pliny *HN* 32.24.69, Isid. *Orig.* 12.6.35, ThC 7.40, Vinc. 17.56, and Neckam 2.24.

177. This is probably the dogfish, Gr. *galeos* (*GF,* 39–42; Saint-Denis, 1947, 39–40), rather than the whale (Gr. *phalaina*), as conjectured by Stadler. In addition to the closer name, the description of the unusual birth, taken from Ar. *HA* 565b24f., reflects the selachian's being internally oviparous but externally viviparous. Cf. ThC 6.25 and Vinc. 17.116. It is also found in Ambrose *Hex.* 5.3.7, with the change that the young of all live-bearing sea animals are taken back into the womb in time of danger. Note that the use of *lactat* indicates a mammal and thus A. too was probably confused on the matter.

and escapes due to the zeal it has for defending its young. It does all this not so much out of strength as from boldness.[178]

(60) GLADIUS: The *gladius* [swordfish] is the fish which our countrymen call the *miles* [soldier].[179] This fish has the skin of a dolphin and the shape of a sturgeon, except that the skin is smooth instead of rough. Also, the body does not gradually taper down to a narrow point where the body ends at the tail. Rather, a wide and bifurcated tail is affixed directly to the body, as if the body were cut short.[180] It is called the *gladius* because its nose is long, more than a cubit and a half in length, and it has both the point and shape of a sword. The nose is made of a black substance which is harder than horn but softer than bone. Beneath its nose, however, it has a mouth adapted not for sucking like a sturgeon but for chewing like the salmon. Beneath, its chin is triangular and toothed.

But it is with its sword that it kills fish and, so they say, punctures ships.[181] I have seen one of these fish dead and whole and have examined it with my own hands. It is a very fat fish, having fat (*adeps*) on its back like a pig.

(61) IPODROMUS: The *ipodromus* [hippopotamus] is a water monster, as Pliny says, which Scaurus brought to the Roman games along with five crocodiles.[182] This creature lives in the Nile River near the equator and is especially found in India. It is found on the land as well and is equally at home on both land and water.[183] It is sometimes bigger than an elephant. It has a turned-back snout, cloven hooves like a cow, a twisted tail, curved and scabrous teeth like the teeth of wild boars, but the back and whinny of a horse.[184] At night it grazes on crops, moving toward them backward so as to hide its footsteps and so that hunters will not be lying in wait when it returns.

---

178. This passage is apparently based upon Ar. *HA* 568a22-b17 and/or Pliny *HN* 9.75.165, describing *Siluris glanis,* the sheatfish, equivalent to the American catfish (*FAZ,* 299–300). See Ambrose *Hex.* 5.5.14, ThC 6.26, and Vinc. 17.55. Cf. 8.119, where *glanios* may contain a clue to *garcanez*.

179. *Miles* also is a nickname for a turtle at 24.12(19) and 60(138), and for a crab at 24.22(31). Cf. Sanders (413) on the German name and Saint-Denis (1947, 41–42) on the Latin.

180. The swordfish, *Xiphias gladius*. Cf. *GF,* 178–80, *FAZ,* 316–17, Pliny *HN* 32.6.15, Isid. *Orig.* 12.6.15, ThC 6.27, and Vinc. 17.55. Cf. also 24.60(139).

181. This behavior is reported by Pliny *HN* 32.6.15 and Ael. *DA* 14.23. An Associated Press report (November 11, 1979) gives curious support: "Fiji—A three-foot-long swordfish attracted by a light in a fishing boat leaped from the ocean and speared the fisherman in the chest, narrowly missing his heart." Scanlan (355) quotes an instance of a ship pierced twenty-two inches deep by the fish.

182. What follows is based on Pliny *HN* 8.39.95, 40.96. Other coverages of the hippopotamus are found in Ambrose *Hex.* 5.1.4, ThC 6.28 (*ipothamus*), Vinc. 17.136–37, Barth. 13.26, Hugh of St. Victor 3.55 (*PL* 177:105B), Rabanus Maurus *De universo* 8.5 (*PL* 111:237B), and Jacob. 88 (p. 183). Cf. also 24.30(44, 45). The games in question were given by Marcus Scaurus as aedile in 58 B.C.

183. Thus, some medieval authors, such as Hugh of St. Victor cited above, classify the beast among the *amfivia*.

184. A. reads *dentes prurigeneos,* as does ThC 6.28. It is a misreading of the Latin of Solinus 32.30, *aprugineis dentibus* (boar-like teeth), a shortening of Pliny's (*HN* 8.39.95) *dentibus aprorum aduncis*. Note that Isid. and Neckam get it right.

This beast goes about in a large herd and very eagerly seeks recently cut reeds or the spines of thornbushes. It rolls around in things such as these until a particular vein in its foot is wounded. It seeks to be made thinner from the bleeding of this vein. It takes a wound in its foot more readily than in other parts of its body and it diligently cares for the wound once it has been received.

This beast has an impenetrable back unless it is wet. Spears are turned out of the skin of this beast, as Pliny says.[185]

(62) IRUNDO MARIS: The *irundo maris* [sea swallow] is a fish that is very similar to the swallow that is a bird. It is, however, a fish that swims in the water and yet sometimes lifts itself up on its wings into the pleasant realms of the air. But it does not fly far without falling back into the water.[186]

(63) KALAOZ: The *kalaoz* [mullet] is an ocean fish for which rainwater is lethal. This is in contrast to the custom of all other fish, which are fattened during rainstorms. When a great deal of rain is descending, this one is blinded and then dies, since it is not able to find food.[187]

(64) KYLOZ: The *kyloz* [sea anemone] is a member of the shellfish genus [*ostreorum genus*] which adheres to rocks.[188] Its shell is like the shard of a vase but is very rough. The composition of its body is just like that of flesh. It is aware of all that approach it and its upper portion is unattached. Here it has two feet with which, as if with hands, it catches fish swimming above it over which it can prevail. If something should swim above it over which it cannot prevail, it is afraid and immediately draws itself back to the rock. It adheres to rocks with its rear feet and its mouth is in the middle of its body. There are, moreover, two types of this genus of shellfish. One is quite small and is good for the health, especially when its moisture is dried out. For this reason, people hunt it in wintertime and eat it salted during the summer as a guard against the heat.[189] The other is large and spotted with white. This one carries disease and is the undoing of anyone who eats it.

(65) KOKY: The *koky* [seal] is the sea calf [*vitulus marinus*]. We have already spoken about it above, saying that in Latin it is called the *helcus*.[190] Enough has been said

---

185. Pliny states that shields and helmets made of its hide are impenetrable unless wet, but he does not mention the spears. In fact the origin of this most unusual comment goes all the way back to Hdt. 2.71. The verb A. uses implies that the hide is so thick that the spear shaft is turned out of it on a lathe.

186. Either the flying gurnard, *Dactylopterus volitans,* or the true flying fish, *Exocoetus volitans,* but certainly directly translated from the Gr. *chelidon,* "swallow" (*GF,* 285–86; Saint-Denis, 1947, 49–50). Cf. Ar. *HA* 535b28, Pliny *HN* 9.43.82, ThC 7.41, and Vinc. 17.57.

187. Ar. *HA* 601b30f. mentions three fish in this connection: the *kephalos* (gray mullet), the *kestreus* (mullet), and the unknown fish called the *marinos* (*GF,* 108, 110, 159). Cf. ThC 7.42 and Vinc. 17.60.

188. This passage is based on Ar. *HA* 531a32-b18 and Pliny *HN* 9.68.146–47. The name appears to be a corruption of *akalēphē,* the sea anemone or sea nettle (*GF,* 5). Cf. ThC 7.43 (*kylok*) and Vinc. 17.60 (*cilos*).

189. Davidson (1976, 218, 299) discusses the edible sea anemone, which he identifies as *Anemonia sulcata,* saying they "can be fried or used in an omelet or for beignets."

190. Cf. 24.31(47). This is A.'s most common name for the seal, a corruption of the Gr. *phōkē.* Cf. Ambrose *Hex.* 5.2.6 and 5.3.7, Barth. 13.26, ThC 6.29, and Vinc. 17.118. Hugh of St. Victor 3.55 (*PL* 177:105B) uses the more accurate *phoca.*

about the anatomy of this animal in the preceding books and ought not to be repeated here.

(66) KYLION: The *kylion* [rascasse] is an animal which, in contrast to the arrangement found in other animals, has its liver placed on the left and its spleen on the right.[191]

(67) KARABO: The *karabo* is the same as the sea locust [*locusta maris,* lobster] and, except for the fact that it is very large, has the shape of a crayfish [*gambarus fluvialis*].[192] This animal gathers in groups and, having arranged itself in battle lines, fights with other lines of those of its own kind over food, the young, or females. It eats mud and small plants and, therefore, when it is split open, a great deal of mud is found in its body.

(68) LULIGO: The *luligo* [calamary, squid] is a sea animal which has scales and inhabits the bottom of the sea like a fish. Along with its scales it has fins on which it raises itself into the air like a bird. But it does not fly for a long time because it cannot endure the winds, and when it feels them it falls back into the water.[193]

(69) LUDOLACRA: They say the *ludolacra* [sea bass] is a marine animal having four fins, two on its face and two on its back. With these it is borne with great speed anywhere it wishes to go.[194]

(70) LOLLIGENES: As Pliny relates, *lolligenes* [calamary, squids] are fish living in the ocean near Mauritania, not far from the river Lixus.[195] They have a length of five cubits, and while swimming they often lift themselves out of the water like flying arrows.[196] They sometimes do this in such numbers that they sink ships in their path. This fish has two feet in front with which it brings food to its mouth, much as the lobster [*locusta*] does. Another genus of this fish is very rough and combative. It gathers in assembled battle lines, as does the *karabo* [lobster]. It has its head between its feet and its belly.

---

191. This fish is almost surely the same as the *granus* above at 24.35(57), viz., the stargazer, *Uranoscopus scaber*. The name is shortened from its most common Gr. name, *kalliōnymos* (*GF,* 98–99). This fish was famous for the amount of bile it produced and Ael. *DA* 13.4 has many quotes about it. Cf. Ar. *HA* 506b10, ThC 6.30, and Vinc. 17.117.

192. Cf. ThC 6.31, Vinc. 17.117, and Barth. 13.26. Cf. also 24.39(71). On the identity of the *locusta maris,* see Kruk (1985).

193. Although the passage is troubled, the animal being described is probably the Gr. *teuthos* or *teuthis* (*GF,* 260–61), and thus a calamary or squid, normally called the *loligo* (Saint-Denis, 1947, 56–59). The calamary's two lateral fins fit the description well, but there is no mention of its arms. It was known in antiquity for its ability to "fly"; see Pliny *HN* 9.45.84. The scales, which led Scanlan (357) to posit this as the flying fish, may have arisen from Ar. *HA* 524b7f., but this is not readily demonstrable. Cf. 24.38(70), ThC 6.32, Vinc. 17.119, and Ald. *Aen.* 1.18.

194. Based on Ar. *HA* 489b23f., this must be the Gr. *labrax, Perca labrax* (*GF,* 140–42). Ar. locates the first set of fins on the back, not the face. Cf. ThC 6.33 and Vinc. 17.119.

195. Cf. Pliny *HN* 32.6.15. Mauritania is roughly equivalent to today's Morocco and the river is generally identified as the Wady al-Khos. On the calamary, cf. 24.38(68). Cf. Isid. *Orig.* 12.6.47, ThC 7.44, and Vinc. 17.119.

196. Cf. Pliny *HN* 9.45.84. Indeed, the name of one prominent species is *Loligo sagittatus.*

(71) LOCUSTA MARIS: The sea locust [lobster] is called *karabo* in Greek.[197] As with the other crayfish [*gambari*], it lies dormant for five months in winter and, in spring and autumn, grows fat. They fight with horns which they bear erect when they feel safe but tilted back toward their sides when they are afraid.[198] They gather in battle lines, as we have said. They fear the octopus so greatly that, when placed next to one, they die.[199]

(72) LEPUS MARINUS: The sea hare is variously shaped. One, with a hairy coat and harder hair is found in the Indian Ocean, as Pliny says, and is so venomous that at the mere touch it induces vomiting and looseness of the stomach.[200] In our sea it is not so venomous.[201] An enemy of this fish and others like it is the extremely venomous fish called the *pastinaca* ["turnip"; stingray], for this fish pierces the others as if with an evil, poisonous weapon. Other fish so flee this one that, since there is no known remedy against its bite, they attach themselves to the roots of trees.[202]

There is, however, another *lepus* which has a head similar to that of a [land] hare but in the rest of its body is a fish. It is a good fish, with red skin and firm and indigestible flesh, which is said to cause leprosy.[203] We usually call this fish the *gernellus*.[204] It has four fins behind its head, two of which move along the length of the *lepus* and are long, like the ears of a hare. The motion of the other two is from its back to its belly, along the fish's height. It uses these to raise itself in front because of the weight of its head compared to the rest of its body.

(73) LUCIUS: The *lucius* [pike] is a common freshwater fish with a long snout, a large mouth opening, and jaws filled throughout with teeth.[205] It feeds on fish and it

---

197. Cf. 24.38(67), ThC 7.45, Vinc. 17.63, and Barth. 13.26. The Gr. word is *karabos*, which Thompson (*GF*, 102–3) identifies as "the spiny lobster, crawfish, or langouste." Cf. Saint-Denis (1947), 56; *FAZ*, 198; and Kruk (1985).

198. Cf. Pliny *HN* 9.51.98–99.

199. Ar. *HA* 590b14f., Pliny *HN* 9.88.185, and Barth. 13.26.

200. Antiquity admitted of two "sea hares" (*GF*, 142–44; Saint-Denis, 1947, 54–55). The poisonous sort listed here is identified by Thompson (*GF*, 143–44) as the globefish. The source is Pliny *HN* 9.72.155, but cf. Ael. *DA* 16.19. "Looseness of the stomach": *dissolutio*, perhaps "destruction" or "dissolving," as in the stomach lining.

201. A.'s local "sea hare" might be the sea slug (*GF*, 142–43).

202. The odd sense here results from an improper understanding and/or copying of Pliny *HN* 9.72.155, who says that the sting of *pastinaca arma ut telum perforat vi ferri et veneni malo* ("pierces armor like a weapon with the force of iron and the malevolence of poison"). This becomes *ut ferro malo veneni* in A. and *ut ferro veneni malo* in ThC 7.46. Pliny goes on to say that the stinger of the *pastinaca*, if driven into the roots of a tree, kills it (Aiken, 1947, 213).

203. It is unclear what traits a "bad" fish might have. There is almost surely a pun involved here on *lepus, leporis*, "hare," and *lepra*, "leprosy." Cf. 24.19(25), where the same claim is made of the conger eel's flesh.

204. This is probably one of the gurnards, as the Latin form suggests. There are several species (*GF*, 119–20), but all are noticeably red.

205. Stadler identifies this as the *Esox lucius* (*GF*, 151–52). Cf. Neckam 2.32–33, Vinc. 17.64, and ThC 7.48.

does not even spare those of its own genus. It has a stomach so closely attached to its gullet that it sometimes ejects it in its eagerness to swallow a fish.

It digests fish it has caught headfirst. On occasion it swallows a fish only slightly smaller than itself. It then gradually digests it headfirst, with a little bit of it hanging outside its mouth, drawing the parts of the fish in lengthwise until it has digested the entire thing.

When the female of this fish lays her eggs, she frequently ascends to the source of the waters because of the freshness of the waters there. This freshness is good for her imperfect eggs, which have to grow once they are cast out into the water.

If, however, it should happen upon a fish with rough scales and sharp spines, such as a perch, and devours it, it takes it by the head. But if it should take it by the tail, it cannot then swallow it due to the scales and spines, which, arranged in the opposite direction, prevent this. I myself have, moreover, seen and carefully observed that when it catches a fish, it first carries it transfixed with its teeth in its mouth sideways for a long time, and then, when the fish has died, swallows it.

(74) MURENA: *Murenae* are common fish, resembling serpents in the front portion of their bodies, but from the middle of their bodies to the end they are like eels [*anguillae*].[206] From that spot to the end of their tail, their entire body has little fins placed along the sides. It has a mouth adapted for sucking fluids and not for chewing. It is therefore false that they have teeth outside of their mouths as do the lobster [*karabo*] and the crayfish [*gambarus*].[207] It is also false that only the female *murena* exists and that she conceives from a male serpent, and that the serpent calls her forth from the water with a hiss, spits out his venom, and couples with her.[208] Now, the entire natural humor of a serpent is venom, and since it cannot spit up its natural humor, it cannot spit up its venom. Nor can the *murena* live out of water, whereas the serpent does live in water. Therefore, what is told is just a tale.

The *murena* is a fish found in the waters of France and Germany. But three genuses of *murenae* are found in the eastern part of Germany. One is found in the Danube and is very small, about as big around as a reed and not exceeding the length of a palm.[209] Another is larger than this one and is found in the northerly waters. It has a maximum length of a foot and a half and has nine spots on its body on both sides near the head.

---

206. *Murena* often refers to the moray eel, *Muraena Helena*, but it is also used to describe lampreys, whose mouth indeed is "adapted for sucking." The lamprey is still often vulgarly referred to as the "lamprey eel." Cf. *GF,* 162–65.

207. Cf. Pliny *HN* 9.39.76.

208. A story found throughout classical works (Pliny *HN* 9.39.76, 32.5.14; Ael. *DA* 1.50, 9.66; Oppian *Halieutica* 1.554–73) and medieval works (Neckam 2.41, Isid. *Orig.* 12.6.43, Jacob. 90 (p. 193), Barth. 13.26, Vinc. 17.71–72, and ThC 7.49). ThC cites Ambrose *Hex.* 5.7.18 as the source of the description of the spitting out of the venom.

209. Scanlan (360) identifies this as the brook lamprey, *Lampetra planeri,* a small freshwater species.

For this reason it is called "nine eyes" [*novem oculi*] by the local people.²¹⁰ The third kind is large, about the thickness of a person's arm and a cubit or more in length.²¹¹ It does not have "eyes."

This fish is tasty but not healthful. When it is eaten it must be pickled in hot spices and strong wine. It has no spines but instead has cartilage in place of a backbone.²¹² It has a very uniform body and thus lives for a long time when divided into parts.

(75) MUGILUS: The mullet is a comical fish because when it has hidden its head, it thinks its whole body is hidden, a trait it has in common with the ass.²¹³ It is a swift river fish. In the summer, due to an abundance of food, it lives in harmony with the pike, but in the winter, due to a lack of this same food, it fights with it.

(76) MARGARITAE: *Margaritae* ["pearls"] are members of the oyster genus.²¹⁴ They have a hard shell and dwell in shells which have the color of pearls. These oysters come toward the shore and collect dew from the sky.²¹⁵ If it is morning dew and is clear, and if the body of the oyster is well cleaned out and is in good condition, the oyster makes an excellent, well-rounded pearl from it, filled with a splendid whiteness like that of the clear moon. If, however, it is evening dew and the weather is cloudy, and if the body of the oyster is flawed and unpurged, it conceives and forms a murky pearl.

None has been found to date larger than one-half ounce.²¹⁶ Pearls are called *uniones* because a maximum of two are found at the same time, although generally only one is found.²¹⁷

If because of lightning, hail, or some other cause an oyster becomes afraid while it contains a conceived pearl, the pearl is pressed out of roundness, as it were, and loses the splendor of its color. A pearl is at first soft in the water and later hardens into stone. Oysters move about in groups for the purpose of drinking in the dew.²¹⁸ Pearls placed in vinegar dissolve and grow soft.²¹⁹

---

210. This is the migratory river lamprey, *Lampetra fluviatilis*. Note that Pliny *HN* 9.39.76 describes the "seven stars" on the right jaw in the shape of the constellation *Ursa Major* (*GF*, 164). These "eyes" are actually a description of the lamprey's gill openings. With Teutonic thoroughness, when the "stars" become "eyes," the two actual eyes (or the eye and nostril of each side) are added to the seven of Pliny and the name of the fish becomes *neunaugen* from A.'s day to the present (Sanders, 413). Cf. 14.79.

211. *Petromyzon marinus*, the sea lamprey, spawns in river waters.

212. As is, of course, true of all lampreys, hagfish, etc. "Spines" may refer either to small fish bones or to backbones.

213. The mullet is frequently discussed in medieval works (cf. Isid. *Orig.* 12.6.26, Barth. 13.26, Neckam 2.31, ThC 7.50, Vinc. 17.67, and Rabanus Maurus *De universo* 8.5 [*PL* 111:238B]). On the head hiding, cf. Ar. *HA* 591b4 and Pliny *HN* 9.26.59.

214. *Margarita* is normally used to denote the pearl, but here denotes an animal producing them. Cf. Isid. *Orig.* 12.6.49. Likewise, *ostrea*, "oysters," can also mean merely "shellfish," as it does elsewhere.

215. Cf. Pliny *HN* 9.54.107–8, Jacob. 90 (p. 193), and ThC 7.51. *FAZ*, 245–47.

216. The measurement is from Pliny *HN* 9.57.116.

217. On the *uniones*, see 1.143.

218. Oysters, of course, have no power of locomotion. Cf. Pliny *HN* 9.55.111.

219. See Pliny *HN* 9.58.120–22 and ThC 7.51 for the story of how Cleopatra won a wager with Anthony using this idea.

Pearls are found in three ways in our lands. They are sometimes found where the shells meet, sometimes in the oysters themselves, and sometimes among the rocks beneath which the oysters lie hidden. The ones from the East are the better ones. Crushed, they are beneficial for stomach disorders and, when carried, they increase chastity. When they are consumed, they give comfort to the heart.[220]

(77) MEGARIS: The *megaris* is a sea fish. It is not highly thought of in the area where it is caught, even though it has a better taste when fresh. But when it is salted and transported farther away, it is desired the more for its rarity.[221]

(78) MULLUS: The *mullus* [gray mullet] is a fish which the French call the *mullettus* and the Germans the *harder*.[222] It is a fish with a broad top to its head, it is about a palm and a half long, and it has very tasty flesh. This is why it is called *mullus*, since it is tender and sweet, soft [*mollis*], as it were. When eaten it diminishes the sexual drive and dulls the eyes, and one who eats it often reeks of fish. If this fish is killed in wine and if this wine is drunk, it takes away the desire for wine. An eel killed in wine does this also.[223]

(79) MULTIPES: The *multipes* [octopus] is a fish which is called *polipes* in Greek since it has many legs, namely, eight.[224] It is just as if eight serpents were joined into a single head. It is of the *malakye* genus. We have talked about it in foregoing books, saying that it builds a nest of wood and lays eggs in it stuck together in groups about the size of a nut. Afterward, it cares for them, whereupon they divide and become many little *polipi*. The female of this fish has a smaller head than the male, and when it cares for the eggs it becomes weak from not eating. The lobster [*karabo*], which is the sea locust, fears the octopus since it surpasses the lobster in cunning and vigor.[225]

---

220. Also reported by Barth. 16.62. Scanlan (361) points out that calcium carbonate is the chief component of pearls and thus forms an effective, if expensive, antacid.

221. Also described in ThC 7.52 and Vinc. 17.66. Neither Stadler nor Scanlan (362) identify this fish, and Lexer (1:2069) lists a *megar* as an unidentified sea fish. But Neckam, in *De laudibus divinae sapientiae* 3.454, says that boiled *megaris* pleases more than roasted flesh, and Wright (1863, 405, 508) glosses this as mackerel (cf. Blatt, 1959–69, 5:323–24). DuC lists a twelfth-century gloss on *megarus* as *macherel* and s.v. *megarus* Latham lists the forms *meragis* and (spurious) *mega*, linking them all to *makerellus*. At 7.76, A. himself offers *macarellis* as one of the local fish. We may be confident, then, that this fish is a mackerel and that A. is probably using a Latinized version of a local name for it.

222. Stadler identifies this mullet specifically as *Mugil chelo*, a thick-lipped gray mullet. For the extent of the mullet's popularity in antiquity, cf. *GF*, 161–62, and Saint-Denis (1947), 68–69. Cf. Isid. *Orig.* 12.6.25, ThC 7.56–58, Vinc. 17.69, and Neckam 2.31. On the German name, see Sanders (441), who claims that *harderen* is pl. at 7.76 and 8.32, but acc. here.

223. Cf. Pliny *HN* 32.49.138,

224. The octopus, *Octopus vulgaris,* came into the tradition with many names, but *multipes* is a literal translation of the Gr. *polypous*. On its traits, cf. *GF*, 204–8 and *FAZ*, 236–37. Cf. 24.49(100), as well as ThC 7.53 and Vinc. 17.68.

225. Perhaps "overcomes it with cleverness and energy." Cf. Pliny *HN* 9.88.185 and Barth. 13.26.

(80) MURICES: The *murices* [purple shellfish] are certain shellfish which lie dormant during the rise of the Dog Star and go forth at other particular times.[226] These shellfish bear a precious liquid used for dyeing in a white vein which they possess. They do not have the liquid when they are dead, as Pliny says, since they spew forth this juice along with their life.

Mucianus the philosopher is witness to the fact that when a ship laden with boys who were going to be castrated was on its way to King Periander, murexes adhered to the ship, stopped it, and caused it to stand still.[227] This is exactly, as we have said, how the *eschynus* [remora] stops a ship.[228]

(81) MUS MARINUS: The sea mouse [turtle] comes out of the water and, digging a hole, lays its eggs in the ground.[229] It covers them over, and when the young have come out of the eggs after the thirtieth day, it returns, digs them out, and leads them to the water.[230] The young are blind at first, acquiring the power of sight later.

(82) MULUS: The *mulus* [red mullet] is a fish which is produced only in that part of the northern ocean which borders on the West. There are many types.[231] One feeds on seaweed, shellfish, and mud, and this one is not noble.[232] The other, however, feeds on things which are found on the shores in clear weather. It is multicolored, very delicious, and noble.[233]

(83) MILAGO: The *milago* [flying fish] is a fish which, whenever a storm ceases, flies over the waves of the sea as a sign of joy.[234]

---

226. There are several species of murex, the most famous in antiquity being the one from which "royal purple" dye was made (*GF*, 209–18; Saint-Denis, 1947, 71–72). Cf. Pliny *HN* 9.60.125, ThC 7.54, Vinc. 17.73, Barth. 13.26, Hugh of St. Victor 3.55 (*PL* 177:110C), and Rabanus Maurus *De universo* 8.5 (*PL* 111:239A).

227. Gaius Licinius Mucianus was consul of Rome and a legate to Syria in 69 A.D. Pliny, from whom this reference comes (*HN* 9.41.80), used his book of *mirabilia*. Periander was tyrant of Corinth, ca. 625–585 B.C. He also figured in the dolphin and Arion story alluded to in 24.28(40). Pliny relates that for stopping the castration, the murex was worshipped in a temple to Venus on Cnidus.

228. Cf. 24.31(48), with notes.

229. Cf. the Gr. sea mouse, *mys ho thalassios,* which indicated individual animals: the mussel and a creature that may be a turtle (*GF*, 166–68). The "sea mouse" described here is conjectured by Cuvier (*B&R* 2:406) to be a small turtle (Saint-Denis, 1947, 72–75), perhaps *Testudo coriacea*. Cf. Ar. *HA* 558a8–10, Pliny *HN* 9.35.71 and 76.166, ThC 7.55, and Vinc. 17.74.

230. Ar. *HA* 558a8–10 and Pliny *HN* 9.76.166.

231. The identification of *mulus* as mullet is secure (*GF*, 264–68; *FAZ*, 280). Mullets are often called goatfishes because of their barbels, which resemble beards, a detail added by ThC 7.56. Cf. Vinc. 17.69.

232. Cf. Ar. *HA* 591a12 and Pliny *HN* 9.30.64.

233. *Mullus surmuletus* has 3 to 5 yellow bands on its red body and is somewhat larger than *Mullus barbatus*.

234. Either the true flying fish, *Exocoetus volitans,* or the flying gurnard, *Dactylopterus volitans* (*GF*, 285–87; Saint-Denis, 1947, 64–65). The two were, and still are, frequently confused. Cf. Isid. *Orig.* 12.6.36 (*millago*), ThC 7.58 (*mulago*), and Vinc. 17.66 (*milvago*).

(84) MONOCEROS: The *monoceros* [narwhal] is a sea fish bearing one horn on its forehead with which it is able to pierce fish and some ships. But it is a slow animal and, as a result, those it attacks are able to escape.[235]

(85) MONACHUS MARIS: Certain people say the sea monk [monkfish] is a fish occasionally seen in the British sea. It is a fish with white skin on top of its head, around which is a dark circle, like the head of a monk who has been recently tonsured. It has, however, the mouth and jaws of a fish. This animal entices those travelling on the sea until it lures them in. It then sinks them to the bottom and takes its fill of their flesh.[236]

(86) NEREIDES: Nereids are, as Pliny says, marine creatures that present a somewhat human shape. When someone is about to die, they mourn and wail to the extent that it is heard by the inhabitants.[237]

(87) NAUTILUS: The nautilus, as Pliny says, is a very remarkable sea monster, for it comes to the surface of the water and there lightens itself for easier sailing by discharging all its water through a tube.[238] Afterward, it bends back its two front arms and then extends between them a membrane of great thinness.[239] With its other arms, it rows and steers itself with its tail in the middle, and thus it sails in the free breezes.

(88) NASUS: The nose is a fish found in the Danube and in the waters flowing into the Danube. It is like the monk [*monachus*], but it is thinner and has a very thick nose.[240]

---

235. The narwhal is *Monodon monoceros*. The Gr. *monoceros* means "one horn" and frequently meant "unicorn" as well in other contexts. Indeed, the narwhal horn (in actuality a modified tooth) passed for a unicorn horn for many centuries, often being fashioned into anti-poison cups (Einhorn, 1976, 100–101, 244–47; Beer, 1977, 113–20; Pluskowski, 2004). Cf. ThC 6.35 and Vinc. 17.120.

236. Scanlan (363) suggests the monk seal, but this does not fit the description of the animal's body and habits. Stadler (1631) tentatively suggests *Monachus albiventer*. Sanders (cf. notes to 7.98) identifies the fish as a sort of carp, *Leuciscus cephalus*, which, curiously, Stadler reserves solely for the mention of *monachus* at 24.45(88) (*nasus*). Today's monkfish, described by A. at 24.52(111), is one of the squaloid sharks, genus *Squatina*, which bear close resemblance to rays (Davidson, 1976, 32). No one answer seems perfect and the chance that several reports are conflated into one is strong. Cf. Neckam 2.25, Vinc. 16.120, and ThC 6.34 (who moves the fish to monster status with a report of man-like upper parts).

237. Cf. Pliny *HN* 9.4.9, who offers several reports, one from Lisbon and one from the governor of Gaul to Augustus. In Greek mythology, nereids are young sea goddesses. However, Pliny describes them as bristling with hair all over, indicating that the reports he mentioned were probably based upon some sea mammal. Cf. ThC 6.36 and Vinc. 17.121.

238. On *Argonauta argo* and the many beliefs surrounding it, cf. *GF,* 172–75. Compare the present description to Pliny *HN* 9.47.88 and 49.94. Cf. also 24.21(30). Cf. Saint-Denis (1947), 75.

239. The texts of both Stadler and Borgnet are misleading here, printing *interea extendit*, "meanwhile it extends." A comparison with the original in Pliny and the parallel text of ThC 6.37, *inter ilia . . . extendit*, makes it clear that *inter ea* here should be printed as two words, with *ea* referring to *brachia* (arms). The description goes on to make it clear that the arms serve as masts for the "sail" formed by the membrane. Cf. 8.129 for the description of the same animal under multiple other names.

240. Stadler (1633) identifies this as *Chrondrostoma nasus*, common in northern rivers (Grzimek, 4:321). As mentioned in notes to 24.45(85), Stadler chooses here to identify this *monachus* as separate from the one he mentions there. Yet it may be best to take them as the same fish, for at 7.98 the *naso* is described by A. with the *monachus* as local fish.

(89) ORCHA: The *orcha* [killer whale], as Pliny says, is a huge sea animal which presents no definite shape save that of a huge mass of flesh.[241] This animal fights with pregnant *balaenas* [she-whales] or with seals or their calves, tearing them with bites of its very sharp teeth. The victims, not being able to resist, flee to the sea. The *orchae*, however, place themselves in the way and attack them or else they force them to hurt themselves on the rocks or force them into narrow shallows, where they become stuck and die.

(90) OSTREAE: *Ostreae* [oysters], as Pliny says, are a species of shellfish into whose shells crabs put stones and then devour their flesh, as we have said above.[242]

(91) PURPURAE: Purples, as Pliny says, are marine shellfish which live for more than seven years and lie dormant for thirty days at the rising of the Dog Star.[243] In the springtime, however, they gather together and secrete, almost in a froth, their precious liquid from dew they have drunk in. These shellfish reach full size in less than a year, and if they should grow to full size before a year, the humor which they produce is useless. The purple-gatherers therefore crush their shells by way of keeping them from growing rapidly. The liquid which they produce is the one with which the purple robes of kings are dyed and has the color of a darkening rose. A *purpura* returned to the sea half alive gets well and lives. It delights in foul odors and is attracted by them.

(92) PYNA: The *pyna* [pinna] is also a shellfish and Pliny says that they go about in pairs, male and female.[244] But the truth is that in such animals there is no male or female and they do not copulate. This shellfish opens itself up to the moon, and incautious fish which enter it for a meal are trapped inside and are consumed as a meal for the *pina*.[245]

(93) PUNGICIUS: The *pungicius* [three-spine stickleback] is the smallest of all the fish and has two little spines, one on each side at the base of its ventral fins.[246] The male of this fish is red beneath his throat, but the female lacks the red.[247] It does not have

---

241. This is *Orca gladiator,* the killer whale, the Gr. *oryx* (*GF,* 186–87; *AAZ,* 199; Saint-Denis, 1947, 77). Cf. the "sea ram" at 24.9(7). Cf. Pliny *HN* 9.5.12–13, ThC 6.39, and Vinc. 17.121.

242. Cf. 24.22(31), ThC 7.59, Jacob. 90 (p. 193), and Vinc. 17.76. It is well to remember that *ostrea* may also indicate "shellfish" in a general sense.

243. The word *purpura* translates well the Gr. *porphyrea.* To try to differentiate accurately between this animal and the *murex* of 24.44(80) is probably fruitless (*GF,* 209–18; Saint-Denis, 1947, 71–72, 92). This entire passage, based upon Pliny *HN* 9.60.124–62.135, is highly confused, both here and in ThC 7.60. Cf. Aiken (1947, 214–15) for a discussion of the source of the confusion. Cf. Vinc. 17.82–83 and Hugh of St. Victor 3.55 (*PL* 177:110C). On murex fishing and dyeextraction, Marzano (2013, 143–60).

244. This, like the *aureum vellus* of 24.10(8), is a member of the *Pinna* species of shellfish (*GF,* 200–202; Saint-Denis, 1947, 87). The reference to Pliny is to *HN* 9.66.142, but he states that the *pinna* is never without its companion crab. Cf. ThC 7.61 and Vinc. 17.79.

245. Also garbled from Pliny *HN* 9.66.142. Cf. *perna* 24.48(97) and Pliny *HN* 32.54.154.

246. The name *pungicius* is a variant of the more frequent *pungitivus* ("sticker") found in ThC 7.62, Vinc. 17.81, and Borgnet. The fish itself is generally agreed to be *Gasterosteus aculeatus,* a small but highly territorial fish. The identification of the three-spine stickleback is based upon his red belly. Some individuals have two or four spines and in most cases the third spine is quite small.

247. This coloration appears when the male is ready to fight during breeding season.

scales. Fishermen say about this fish that it generates on its own and also generates all other fish.²⁴⁸ They say this on the basis of experiential data [*per experimentum*], for these fish are immediately found in new fishponds as soon as the new year arrives. Then, after these fish, all other types of fish are found there in successive years, but without these fish having been placed in the fishpond.²⁴⁹

(94) PECTEN: The *pecten* is a round sea fish. It prefers moderate temperatures and does not endure extremes of either heat or cold. As Pliny says, its claws are like bone and shine at night.²⁵⁰ The bones of the fish which is in our waters and is called the *muruca* do the same thing.²⁵¹

(95) PORCUS MARINUS: According to Pliny, the sea pig is an edible fish which greatly resembles a pig.²⁵² Its head is like that of a pig and it has an unattached tongue, as does a pig. Almost its entire flesh passes over into fat and bulk. In its voice, however, it differs from a pig. On its back are certain spines which have a most powerful poison, but its gall bladder is an antidote against a puncture from these spines.²⁵³ These animals seek their food with great effort, for they root about on the bottom of the sea just as pigs do on land. They are thus not the ones which we usually call pigs and about which we spoke above.²⁵⁴

(96) PAVO MARIS: The sea peacock [parrot wrasse] is a fish decorated with the best colors on its neck and back, as is the real peacock.²⁵⁵

(97) PERNA: The *perna* ["ham"; pinna] is a sea animal of the genus of shellfish [*genus concarum*]. It is bright yellow in color, grows to a great size, and is known by its

---

248. "On its own" is more clearly given by ThC 7.62 as "without sperm." The female deposits the eggs in a nest built by the male; he darts through with great rapidity, depositing his milt, then drives her away.

249. That is, without stocking the *lacuna,* which may mean simply, "pond."

250. The only surety about the *pecten* in this passage is that A. never saw one. Part of his description is derived from Pliny *HN* 9.51.101, where the luminescence is an attribute of the creature called the *unguis,* "claw," a Latinized form of Gr. *onyx* (*GF,* 183). Thus, the *pecten*'s claws here do not, in fact, exist. Likewise, *pecten* regularly means scallop for A. (e.g., 4.47, 58, 89, 92; cf. Saint-Denis, 1947, 83). However, it is just as clear that A. seems to consider *pecten* a flatfish, for at 24.50(102) he calls *rumbus,* the turbot, a *pecten,* as does Neckam 2.40. Cf. also 7.76. Aiken (1947, 214) discusses the errors in this passage. Cf. ThC 7.63 and Vinc. 17.78.

251. The *muruca* was clearly a local fish (1.90; 2.80; 3.84; 7.76). Stadler (1632) tentatively suggests an identification of *Gadus morhua,* the Atlantic cod, and Latham, s.v. *morus,* lists endless variant names for the cod, one of which is *muruca.* However, this fish is not normally luminescent.

252. Cf. Pliny *HN* 32.9.19 for a sea pig which grunts when caught and is the largest of all. This is sometimes taken as *Orthagoriscus mola* L. (*Mola mola*), an ocean sunfish, but this fish is inedible. The sea pig is also mentioned by Isid. *Orig.* 12.6.12, Ambrose *Hex.* 5.2.6, ThC 7.64, Vinc. 17.80, and Barth. 13.26. Saint-Denis (1947, 90, 111) discusses both the *porcus* and the *sus* of the sea.

253. This would seem to derive from Pliny *HN* 32.19.56. The scorpionfish and the dory have been suggested as candidates for that fish. A. has clearly conflated the two, however.

254. A. may be referring to 1.103 or 5.66, but remember that in MG a porpoise is called a *Meerschwein* and that Lexer (1:2118) glosses the MHG *merswin* as a dolphin or a seal.

255. The parrot wrasse, *Scarus cretensis,* was as highly colored as it was highly prized by the ancients (*GF,* 238–41). Cf. ThC 7.65, Isid. *Orig.* 12.6.5, and Vinc. 17.78.

yellow and red fleece within its shell which is very valuable and from which precious ornaments for clothes and veils are made.²⁵⁶

(98) PISTRIS: The *pistris* is a sea animal which is, as Pliny says, very large and which is sometimes found in the ocean that touches on France. This animal sometimes raises itself up above the water like a very high column, appearing above the sails of ships sailing past. It then splashes them with a huge wave, causing great fear for those sailing aboard.²⁵⁷

(99) PLATANISTAE: According to Pliny, *plantanistae* are sea monsters which are found in the Ganges River in India and which also go out to sea.²⁵⁸ They have the snout and tail of a dolphin. They are sixteen cubits long.²⁵⁹ The monsters which are called *staciae* are their companions and have bifurcated claws, in which there is such force that they attack elephants entering the water and rip off the elephants' trunks with their claws.²⁶⁰

(100) POLIPUS: The *polipus* [octopus] is a marine animal which is called the *multipes* in Latin and about which much has already been said.²⁶¹ But Pliny adds that it is sometimes possessed of such strength that it snatches a person off a ship. While it cares for its eggs it makes a sort of chamber for itself. It fights with shellfish, and having laid hold of a rock, it uses it to make a hiding place [*umbra*] for itself and casts the rock into the unsuspecting shellfish. It thus prevents its closing and devours the creature inside.²⁶² I have heard from those experienced in such matters that the *dentrix* [bream],

---

256. Pliny *HN* 32.54.154 tells us the *perna* is a *pinna* and says it received its name from the fact that it stands in the sand, resembling a *suillum crus*, "pig's leg" (Saint-Denis, 1947, 85–86). *Pinna nobilis* (*GF*, 200–202) was the famous producer of the "golden fleece"; see 24.10(8). Cf. also 24.47(92). Cf. ThC 6.40 and Vinc. 17.122.

257. A conflation of two animals from Pliny *HN* 9.3.8, who first mentions the *pistris* (Saint-Denis, 1947, 91; *GF*, 219), probably the sawfish *Pristis antiquorum*. Then, in the same passage, Pliny mentions the *balena* (she-whale) as inhabiting the Indian Ocean and goes on to chronicle the *physeter* (Gr. *physētēr*; *GF*, 280–81), probably the sperm whale, as an animal of the "Gallic ocean" which bears itself up like a column and lets forth with a spray of water. Thus, while A.'s description is largely of the latter animal, the name seems mostly to be of the former. Cf. ThC 6.41 and Vinc. 17.122.

258. Cuvier (*B&R* 2:384) identified these as the dolphin of the Ganges and most have agreed (Saint-Denis, 1947, 88–89). This passage is based upon Pliny *HN* 9.17.46. Cf. ThC 6.42 and Vinc. 17.122.

259. A period is inserted at this point in the text to give the same division as found in ThC 6.42. Without the period, the passage reads "they have a dolphin's snout and a tail sixteen cubits long."

260. In a twist of fate that should give pause to all scholars who count the number of times they are cited, the name *staciae* actually is all that is left of one Statius Sebosus, an authority on fish who was cited by Pliny *HN* in his passage on the *platanistae* (Kitchell, 1993, 352–55; Aiken, 1947, 209). Cf. also 24.26(38).

261. Cf. *multipes* 24.43(79), Pliny *HN* 9.48.90f., Isid. *Orig.* 12.6.44, Ambrose *Hex.* 5.8.21, and Barth. 13.26. *Multipes* is a direct translation of the Gr. *polypous*, both meaning "many-footed." Cf. Saint-Denis (1947), 89.

262. Pliny *HN* 9.48.90, ThC 6.43, and Vinc. 17.124 all have the *polipus* using a rock to prevent the shellfish from closing his shell, as the crabs do at 24.22(31). Also, at this point in the text, Stadler notes that two other MSS of this work include an interpolation from 8.125 concerning the octopus's ability to change color.

also called the *dentatus*, fights with the *polipus*.²⁶³ Since it cannot drag the *polipus* out of its lair, the *dentatus* flutters about before the mouth of the cave as if it were dead. When the *polipus* sees this, it extends one of its arms to draw it in and devour it. But the *dentatus* suddenly bites off this arm. The *polipus* itself, when it is in difficulty, eats its own arms and they grow back.²⁶⁴ The *polipus* greatly avoids foul water.

(101) RANA MARINA: The sea frog [fishing frog] is one of the water animals. It has horns under its eyes and spines on its eyebrows. It sinks itself into the mud and stirs it up, and then, with the aforementioned horns and spines, pierces unsuspecting fish and devours them.²⁶⁵

(102) RUMBUS: The *rumbus* [turbot] is a fish very common in the sea of Italy and Greece. It is a member of the *pecten* [flatfish] genus and is totally round, almost like a circle, and it is surrounded by fins.²⁶⁶ It is partly colored black with red stripes and spots, and its entire surface is covered with extremely sharp spines which are bent a bit forward. This fish is a slow swimmer due to the width of its body. It sinks down on the bottom and stirs up the water and, in this way, pierces unsuspecting fish which approach it.²⁶⁷ Thus, even the swift mullet [*mugilus*] is found in its stomach. This fish is delicious, with tender and very sweet flesh.

(103) RAYTHAE: The *raythae* [rays], which are called *rays* in French, are fish of the *pecten* genus, so round that they sometimes reach a width of one or two cubits and the same in length.²⁶⁸ They have a long tail and on it are pointed little fins for swimming. They have horrible eyes and a most foul mouth just about in the area of the belly.²⁶⁹ They have tough, indigestible flesh and they are not appreciated for food, except in places where the fish itself is but rarely found. There, due to its rarity, it is sought after.

(104) SALMO: The salmon is a fish which is found both in the sea and in rivers, but it is not found in stagnant waters.²⁷⁰ Pliny says it was the best of all fish in older times, especially in Aquitania.²⁷¹ Now, however, it is proven to be better in the Rhine River and especially at Cologne. It reaches the size of one and a half to two cubits

---

263. The bream is discussed at 24.29(42). Most other sources (*GF*, 207) talk of the conger eel in this connection.

264. Pliny *HN* 9.46.87 explicitly denies this story.

265. The fishing frog is securely identified as *Lophinus piscatorius*. It was often discussed in antiquity (*GF*, 28–29; Saint-Denis, 1947, 93–94) and does indeed lure fish to its mouth. Cf. 8.113–15, ThC 7.66, Vinc. 17.85, and Rabanus Maurus *De universo* 8.5 (*PL* 111:237B).

266. The ancient name recurs in the scientific name, *Rhombus maximus* (*GF*, 223; Saint-Denis, 1947, 95). Cf. Pliny *HN* 9.67.144, ThC 7.67, and Vinc. 17.86. On the sense of *pecten* in this context, see notes to 24.47(94).

267. A.'s use of the verb *confodit* seems to imply that he thinks the fish strikes with its spines.

268. This appears to be a general description rather than a description of an individual fish. On the *raia*, see Saint-Denis (1947), 93. Cf. ThC 7.68 and Vinc. 17.84.

269. Given Albertian usage, the phrase *in loco ventris* would normally mean "which it uses as its belly."

270. Cf. discussion of the Atlantic salmon, *Salmo salar*, at 24.32(49, 49a) and Saint-Denis (1947), 269. Cf. Isid. *Orig.* 20.2.2, ThC 7.69, Vinc. 17.87, and Neckam 2.42. See also *FAZ*, 290 and Andrews (1955).

271. Pliny *HN* 9.17.44.

and the thickness of a palm or more. It is a slow fish but strong. When it comes to an obstacle in the form of a net or barrier, it puts its tail to its mouth and, bending itself in a circle, leaps over it. It has red, fat, sweet flesh, which is very filling and heavy. Its heart, if removed from the body, moves for the longest time of all other animal hearts.

(105) STURIO: The *sturio* [sturgeon] is a common fish which the ancients called the *stora*.[272] It has a long nose and it is a large fish, reaching a length of up to nine feet when it is fully grown. It is round, after the fashion of a nail, and has three rows of prickly little teeth on its skin all along the length of its body.[273] It has a mouth more adapted for sucking than for chewing and therefore nothing of a large nature is ever found in its stomach. Only a viscous humor is found there, which it takes in by sucking. It has white, tasty flesh and has no bones except in its head.[274] It has a yellowish fat and a large liver that is so sweet that if it were not tempered with its own bile, it would cause revulsion on account of its sweetness.

(106) SPONGIA: The sponge is one of the ocean creatures and has many types, as we have said in preceding books.[275] In their motion of contraction and dilation, they resemble animals. Certain of them, however, are immobile and, if removed from their rocks, regrow from their roots. Others move from place to place.[276] They live better in muddy waters and putrefy in clear waters. They eat mud, fish, and shellfish.

(107) SCOLOPENDRA: The *scolopendra*, as Pliny says, is a sea fish similar to the land animal which people call the *centipes*. When it has swallowed a hook, it vomits out all it has swallowed and then, with the hook cast out, takes it back in again.[277]

(108) STELLA: The star [starfish] is a fish which is shaped like a star, found in the eastern ocean.[278] Its flesh is inside, while outside, it is hard. It is said that it has a fiery heat inside of it, so that whatever it eats is found in its stomach cooked as hard as twice-cooked bread.[279]

(109) SUNUS: The *sunus* [sheatfish] is a fish which takes care of its young in marvelous fashion. The female completes her egg laying in three days, but the male guards the eggs for fifty days, attempting to overcome every animal which approaches it.[280]

---

272. This is *Acipenser sturio*, the Atlantic sturgeon (*GF,* 7–8). Cf. the discussion under *huso* at 24.32(49a). Cf. also ThC 7.70 and Vinc. 17.95. ThC tells us that *stora* is the name the "barbari" use.

273. Apparently the dorsal, lateral, and ventral rows of bony scutes.

274. Highly prized by the Romans at one time; see Pliny *HN* 9.27.60.

275. Cf. Pliny *HN* 9.69.148–50, Isid. *Orig.* 12.6.61, ThC 7.71, and Vinc. 17.91.

276. At *HN* 9.68.146, Pliny claims *urticae* can move, but he does not say the same about sponges.

277. Cf. Pliny *HN* 9.67.145. Cuvier (*B&R* 2:452–53) supports the identification of this as a large worm or annelid such as the nereids (Saint-Denis, 1947, 102), which have side bristles, strong jaws, and a trunk which can be disgorged. Scanlan (370) names the lugworm, *Arenicola marina*. Cf. 8.117, with notes; ThC 7.72; and Vinc. 17.88.

278. The starfish, Gr. *astēr* (*GF,* 19; Saint-Denis, 1947, 109–10), was easily studied in the shallows. Cf. ThC 7.73 and Vinc. 17.95. ThC claims it is in the western sea.

279. Cf. Pliny *HN* 9.86.183 and Ar. *HA* 548a7.

280. This large catfish, *Silurus glanis*, was known for its parental care (*GF,* 43–48). Cf. *claucius*, 24.19(24); *garcanez*, 24.35(59), 8.119; and ThC 7.74. Pliny *HN* 9.75.165 is the immediate source of this account and Pliny's *silurus* has somehow become a *sunus*.

(110) SOLARIS: The sunfish [burbot?] is a fish which likes to expose itself to the sun on riverbanks.[281] It has a large head, a wide mouth, black slippery skin like that of an eel, and an edible, tasty liver. It grows to a great length and immense size when it has lived for a long time because it grows slower than other fish.

(111) SCUATINA: The *scuatina* [monkfish] is a sea fish which the Germans call the sea puppy [*catulus maris*].[282] It has a length of five feet and a foot-long tail. Hidden in the mud, this fish kills other fish that are not on their guard. It has a skin so rough that, when dried, it is used to polish wood and ivory.[283] Its hair is short and black and is similar to the beards of fuller's grass, and is so tough that it can scarcely be cut with iron or steel.[284]

(112) SALPA: According to Pliny the *salpa* [saupe] is an obscene, vile, marine fish. It can in no way be cooked unless it is first beaten very forcefully and broken down with a club or stick.[285]

(113) SEPIA: The sepia [cuttlefish] is a well-known fish of the *malakye* genus.[286] The male of this fish is multicolored and has firmer flesh and is darker than the female. If the female is struck by a trident, the male helps her. But if the male is struck, the female flees. When this fish is afraid, it muddies the water by pouring forth a dark humor from itself. People say that this humor, if placed in a lamp and the lamp lit, causes those standing around to look like Ethiopians.

---

281. At 24.13(21) we learn that this is the name of the adult *borbotha,* or burbot. The burbot, Europe's only freshwater cod, is relatively slow growing and attains only about one meter in length, hardly the *longitudinem et vastitatem corporis magnam* of this passage. Cf. ThC 7.75 and Vinc. 17.93. Marine sunfishes (*Molidae* fam.) qualify as immense in size, but are not food fishes and do not inhabit rivers. The odd fact that this animal comes out onto the bank makes one suspect that it is in reality another animal entirely which folk wisdom said was an adult form of the burbot.

282. *Catulus maris* poses quite a problem. Stadler (1653) suggests "Seekatze?" and is followed by Scanlan (371). As Sanders (412) points out, this simply will not work since the form is unparalleled in German, and he adds that the analogous *Meerkatze* is a monkey. Yet note that Diefenbach (1857, 107) does list a Germanic gloss for *cattus* which says, too cryptically, "sea beast." The main difficulty lies in the fact that *catulus* does not mean "cat." It instead is regularly used for the young of a dog and, by analogy, of other mammals (Diefenbach, 1857, 107). The best English might be "whelp." Yet at 24.26(37) we already have a "sea dog," *canis marinus,* and the corresponding German *Seehund* is a perfect link to "seal." Whether, then, *catulus* is a slip for *cattus* or indicates a separate creature, the description and comparison of parallel passages makes it clear that this is *Squalus squatina,* the monkfish or angelfish (*GF,* 221–22). Cf. Pliny *HN* 9.67.144, ThC 7.76, and Vinc. 17.94.

283. Cf. Pliny *HN* 9.14.40.

284. The herb in question is fuller's teasel/thistle, *Dipsacus fullonum,* whose flower has curved, barbed bracts and is used to raise the nap in wool.

285. Pliny *HN* 9.32.68. The saupe (*GF,* 224–25) has no revolting habits. Pliny is merely saying that some fish are valued in one locale and scorned as disgusting (*obscenus*) elsewhere. *Vilis* probably entered the MS as a gloss to explain *obscenus,* for *vilis* can also mean "cheap" and Pliny is really talking about what drives prices for fish in various locations. Cf. ThC 7.77 and Vinc. 17.87.

286. *Sepia officinalis* was widely discussed in antiquity, of course (*GF,* 231–33; *FAZ,* 110). Much of what follows may be traced to Ar. *HA* 525a10–12, 550a10-b22, 608b18, and to Pliny *HN* 9.45.84. Cf. ThC 7.78 and Vinc. 17.89–90.

These fish swim in pairs, male and female, and they lay their eggs at any time. The female pours out the eggs and the male sprinkles his milt over them. The eggs are hard and are hatched out in forty days.

However, a great deal has already been said in the foregoing books concerning the nature of this fish and, therefore, let those things which have been said be enough.

(114) SPARUS: The *sparus* [sea bream] is a sea fish which takes its name from a throwable weapon of the country folk which is scattered when it is thrown. The fish is so called because it has the shape of this instrument.[287]

(115) SCORPIO MARIS: The sea scorpion [sculpin] is an animal resembling a scorpion in that it strikes the hand of one picking it up.[288] They say that if ten crabs are bound together and are thrown into a particular place in the sea, all the scorpions gather at that place.[289] It lays its eggs first in the spring and then again for a second time in the autumn.[290]

(116) SCARUS: The *scarus* [parrot wrasse] is a tasty fish which alone among the fishes chews its cud.[291] They say that this fish has only a few, closely arranged teeth and thus has need of rumination, although this is not the cause of it. They say that this fish was brought to Rome from far-off parts and that it multiplied in the Tiber.[292] This fish is held to be so clever that when it is caught in a trap or a net, it widens the holes by striking them with its tail and it struggles to get out backward. It is also said that if a *scarus* happens to be on the outside, it aids the trapped fish's efforts by tugging on its tail.

(117) SERRA: The *serra* ["saw"] is a sea beast with a huge body, very wide fins [*pinna*], and immense wing-fins [*ala*].[293] When this monster sees a ship under full sail on the ocean, it tries, with its wing-fins raised, to sail in competition with it, and when it has tried out this device for thirty or more stadia, it gives up, tired. Then, with its

---

287. The fish is a member of the family *Sparidae* (*GF,* 248–49; Saint-Denis, 1947, 107) and should not be confused with the gilthead. The etymology of the fish's name presented here is a garbled version of Isid. *Orig.* 12.6.31. Note that ThC correctly follows Isid. in stating that the name comes "from the act of scattering" (*a spargendo*), presumably the targets.

288. Cf. Isid. *Orig.* 12.6.17. The fish is *Scorpaena porcus* L. or *S. scrofa* (*GF,* 245–46), a fish whose threat was overlooked because of its delicious taste. It is an important ingredient in bouillabaisse and is often called the rascasse (Davidson, 1976, 145–47; Saint-Denis, 1947, 103–4). Cf. ThC 7.79 and Vinc. 17.88.

289. From Pliny *HN* 32.19.55, referring, however, to land scorpions.

290. Pliny *HN* 9.74.162, of the sea scorpions.

291. This is the well-known and highly esteemed *Scarus cretensis* (*GF,* 238–41; Saint-Denis, 1947, 100–102). Cf. Isid. *Orig.* 12.6.30, ThC 7.81, Vinc. 17.52, and Barth. 13.26. Cf. *megaris,* 24.43(77); *pavo,* 24.48(96).

292. Pliny *HN* 9.29.62–63.

293. A somewhat complicated entry. Pliny *HN* 9.2.4 discusses a *pristis* which is huge, as is this monster. One derivation of *pristis* is from a Gr. verb meaning "to saw" (*GF,* 219). Moreover, just a few lines previously, Pliny mentions but does not discuss the *serra,* probably the animal we know as the sawfish. Thus, exact identification is unsure (Saint-Denis, 1947, 91). Cf. also Pliny *HN* 32.53.144–45, Isid. *Orig.* 12.6.16, ThC 6.44, and Vinc. 17.127.

wing-fins lowered, it is said to sink to the bottom from its own weight, to the place where it first was.²⁹⁴

(118) SERRA MINOR: The lesser *serra* is a sea beast, as Pliny says, which has a hard head, crested like a toothed saw. Swimming underneath ships, it cuts them with this crest so that, when the water enters them and the men are drowned, it feeds upon their corpses.²⁹⁵

(119) SYRENAE: Poets' tales tell us that the sirens are sea monsters having their upper parts in the shape of a woman, with long, pendulous breasts with which they nurse their young. They have a horrible face, with long, tangled hair. Below, however, they have the feet of eagles and on their backs they have wings. To the rear is a scaly tail with which they steer themselves when they swim. When they appear they show their young and they emit a sort of sweet hiss with which they put those hearing it to sleep. Once they are asleep, the sirens tear them to pieces. But the wise ones pass by with their ears blocked and they throw them empty jars with which the sirens play until the ships pass by.²⁹⁶

(120) SCILLA: The poets likewise say that *Scilla* is a sea monster appearing most often in the sea which surrounds Italy.²⁹⁷ They say that it usually has the shape of a young woman, but it has a huge mouth opening and very sharp teeth. It is for this reason that it is called "*scilla*" that is, "head of a dog."²⁹⁸ It has a womb like that of a beast, the tail of a dolphin, and delights in flesh, being especially hostile to humans.

(121) SCINCI: The *scinci* [skinks] are aquatic animals that swim in the Nile.²⁹⁹ They are similar to crocodiles but are much shorter and more contracted. Cups treated with their flesh counteract the force of poison. They give advance warning of storms by entering their caves and opening that door of the cave which is opposite to the

---

294. Cf. *Physiologus* (Curley, 1979, 6–7; Carmody, 1939, 14–15).

295. Isid. *Orig.* 12.6.16 provides this passage, not Pliny. Cf. ThC 6.44–45 and Vinc. 17.127.

296. Extensive passages are found in *Physiologus* (Curley, 1979, 23; Carmody, 1939, 25–26); Isid. *Orig.* 11.3.30, 12.4.29; ThC 6.; Vinc. 17.129; Barth. 18.95; and Jacob. 90 (p. 191). Odysseus was the first to block his ears, in the twelfth book of the *Odyssey*, but he tossed no bottles. Remember that the siren of antiquity turned into the mermaid of later sagas and fables, a trend indicated in that whereas Pliny *HN* 10.70.136 discusses them as fabulous birds, A. includes them with his fish. In fact, A.'s description includes elements of both sorts.

297. "Surrounds": reading Stadler's *cingit*, but cf. Borgnet's *tangit*, "touches." Also faced by Odysseus, Scylla was the daughter of Phorcys, transformed by Circe into a sea monster with dogs around her middle. Across from her lay the whirlpool Charybdis. Scylla and Charybdis traditionally lay at the straits of Messina where the toe of Italy is opposite Sicily. Cf. Isid. *Orig.* 2.12.6, 11.3.32, 13.18.4, 14.6.32; ThC 6.47; Vinc. 17.127; and Neckam 1.68.

298. Gr. *skyllion*, "dog." This statement does not appear in any of the sources cited above. The punctuation used is that of the Borgnet ed.

299. This passage represents a conflation of two passages of Pliny: *HN* 8.38.91, on the *scincus*, and *HN* 8.58.138, on the weather-forecasting squirrel (*sciurus*). Stadler (1643) and Scanlan (373) identify the *scincus* as the skink, *Scincus officinalis*. The *scincus* and *stincus* of 24.55(122) were also frequently confused. Cf. Vinc. 17.128, who groups this passage and the next as his description of the *stincus*. ThC 6.48 seems to know the *scincus* only as described here, but he calls them *scinnoci*.

direction from which the wind is going to blow. Moreover, with their bushy tail, they block the opening which opens to the wind.[300]

(122) STINCUS: The *stincus* [starfish?] is a sea animal halfway between plant and animal. It is like a five-pointed star, with red skin, and it has slits in its middle with which it draws in nourishment.[301] It so excites lust that a person will eject blood in coitus and will not be satisfied even when lying together in the act of coitus.[302] Nor is there any cure for the lust except drinking lettuce juice.

(123) TESTUDINES: The Indian Ocean has turtles [sea tortoises] which have shells so wide that a few of them placed together as a covering provide adequate shelters for people. People occasionally sail in these shells between the islands as if they were boats.[303]

Pliny tells us that these turtles are captured in the following way: While they are enjoying the heat of the sun, they swim on the surface of the sea with their back entirely out of the water until they are so dried out they that they cannot dive. They swim this way unwillingly until they are captured by hand by those who see them. Some say that at night these turtles go forth to eat and, once filled, fall asleep swimming on top of the water. Then, when a number of people have gathered to capture them, three of them swim out to the turtle, and two take hold of it and turn the shell [*conca*] over so that the creature [*ostreum*] is lying on its back.[304] The third one throws a rope over its head or over the member it has in place of the neck, and the others, standing on shore, drag it to land.

Both the large and small turtle have no teeth, but they do have a beak with sharp edges arranged so that the upper jaw closes onto the lower one like the lid of a box. There is said to be such hardness in its jaw that it can even crush rocks.

---

300. Pliny *HN* 8.58.138 describes how the *sciurus* (squirrel) forecasts storms, opening a doorway on the side opposite the direction of the storm. The squirrel's bushy tail serves it as a covering. Essentially the same idea, but for the terrestrial *echinus,* or *ericium* (hedgehog), is found in Ambrose *Hex* 6.4.20.

301. Scanlan (373) follows Stadler (1646) in suggesting an identification of *Asterias rubeus,* a starfish known as the cross fish. The categorization of the creature as midway between animal and plant suits A.'s usage (21.38–39), but several of the details, such as its use as an aphrodisiac and as an antidote, are related by Pliny *HN* 8.38.91 of the Nilotic skink of the previous entry.

302. Isid. *Orig.* 17.9.43 says the satyrion root, which is shaped like the human male genitals, is an aphrodisiac and is commonly called *stincus/um*. This too may have added to the tradition.

303. This entire section is generally a paraphrase of Pliny *HN* 9.12.35–39. Scanlan (374) suggests an identification of the giant tortoise of the Seychelles, *Testudo gigantea* (*Geochelone gigantea*), but this is unlikely. This land-based relative of the Galapagos tortoise would not be found at sea as described below (Grzimek, 6:108; Ernst and Barbour, 1989, 250–51). The location near India and the capture at sea put one more in mind of the genus *Cheloniidae,* whose members are pelagic and reach impressive sizes (Grzimek, 6:109f.; Bustard, 1973, 28f., on the loggerhead), but the individual species probably cannot be named. It is also possible that Pliny was reporting several things at once—reports of turtle capture and accounts of tortoise shells brought to the mainland. Cf. Vinc. 17.131, ThC 6.49, Jacob. 90 (p. 192), and Barth. 18.105–6. Cf. also *barchera,* 24.12(19), and *tortuca maris,* 24.58(126).

304. A.'s language here is more indicative of a shellfish than a turtle.

Pliny says, moreover, that these animals copulate in the manner of cattle and that the female will not readily submit to copulation until the male puts some straws in her mouth when she has turned away.³⁰⁵ They lay their eggs when they have come out onto land. These eggs resemble goose eggs and the female produces a hundred or more. She digs a hole in the earth beyond the water and sometimes lies on the eggs with her chest during the nights. Some say that she incubates eggs with her sight, but this is false. They lead the young out and into the water in the space of a year.

(124) TYGNUS: The *tygnus* is a sea animal.³⁰⁶ This beast has a tail two cubits long. It produces its young in the sea and never on land, although it does come out onto the land for food. It enters the bank on its right hand and leaves it on its left hand. This happens because it sees better with its right eye than with its left.³⁰⁷ They come out of the water more readily when the north wind is blowing than at any other time.

Pliny says that they follow a ship out of curiosity at seeing sails. They are held in such a trance from this that they hardly flee when a trident is thrown at them. They lie dormant in winter. They get fat beyond all measure, to the extent that in the third or fourth year they generally die. According to Solinus, in Ethiopia there are beasts called *tigni* with a dusky color and two pendulous breasts with which they feed their young.³⁰⁸ He also says that the *tignus* is found in the Pontus, since it has fresher water than the other seas. If it enters rivers, it enters by the right bank and leaves by the left bank.

(125) TESTEUM: The *testeum* is a sea animal which, due to the hardness of its skin, is called *testeum* [shell-like].³⁰⁹ It develops such a thick and hard skin from the saltiness of the sea that its natural interior heat is unable to conduct respiration through its pores, and this is a cause of infirmity for the animal. It then seeks out fresh water, in which its skin is softened, and it is restored to its former good health. Having drunk this water, the animal again returns to the sea waters in which its skin becomes hardened. It changes itself this way frequently. Nevertheless, it cannot drink in the sea unless it is the fresh waters which are in the sea, a fact which we have proven in this investigation using a clay vessel placed in the water.³¹⁰

---

305. Confused from Pliny *HN* 9.12.37, itself an abstruse passage.

306. This "creature" represents a conflation of Pliny *HN* 9.17.44f., on the tuna fish (*thynnus*), and 8.30.72, on the Ethiopian sphinx (a kind of monkey). This, coupled with some misreadings or scribal errors, gave rise to a creature which A. surely did not think was a fish. Aiken (1947, 216) offers a detailed analysis of the passage. Cf. ThC 6.50–51 and Vinc. 17.132.

307. This trait of the tuna was often commented upon in antiquity (*GF,* 84).

308. Cf. Solinus 27.58, based on Pliny *HN* 8.30.72, where brief mention is given of *sphinges,* a type of monkey (McDermott, 1938, 104f.) which, in some MSS, was written as *spinges,* only to be corrupted into *tigni*.

309. Cf. the discussion at 4.6 on the terminology used here. It is difficult to think that A. felt "bivalve-mollusks" (Scanlan, 375; Stadler 1648) had this sort of mobility. Thorndike (1927, 2:541) interprets *testeum* as a turtle, though he misreports the experiment.

310. See 7.16 for the experiment.

(126) TORTUCA MARIS: The sea tortoise is the creature which is commonly called the soldier [*miles*] in Germany and is shaped like the land tortoise [*tortuca terrestris*], except that it grows very large. It is sometimes found to be eight cubits long and to have the shield of its back measure five cubits. It has long legs, and it has toes and claws stronger than a lion. It is not afraid to attack three people, but if placed on its back, it is rendered helpless because it cannot get up.[311]

(127) TORPEDO: The torpedo [electric ray] is the fish which we have called the *stupefactor* in preceding books.[312] Hidden in the mud, this fish seizes and devours fish that approach it.[313] It numbs [*stupefacit*] anyone touching it, no matter how fast he might withdraw his hand. It does so with such power that one of our comrades, just poking at it with the tip of his finger, touched it and just barely regained sensation in his arm within half a year by means of warm baths and unguents. Pliny and Isidore say that it numbs if only touched with a spear and that breezes blowing from this same fish numb those standing nearby.[314] It has, however, a very tender liver.[315]

(128) TREBIUS: The *trebius* is a black fish in summer but white in winter, as Pliny says. It grows larger in the ocean. When it is one foot long, it has fat to a depth of five fingers which, when salted, draws gold up from waters, for even though the gold has fallen into the deepest of wells, the fat makes it float up off the bottom. This fish constructs a nest out of seaweed and lays its eggs in this nest. The large ocean variety pierces ships with its extremely sharp beak.[316]

(129) TRUTHAE: *Truthae* [brown trout] are river fish living in torrential rivers that rush down mountains. They have scales and reddish flesh in summer, as does the salmon, but in winter they become white and less tasty. They have yellow, red, and black spots on their back.[317]

---

311. Cf. notes to 24.12(19), 56(123), and 60(138), as well as ThC 6.54, Vinc. 17.134, and Barth. 18.106. On the name, cf. Sanders (413). Several large sea turtles might lie behind the *miles*: the loggerhead, *Caretta caretta,* or leatherback, *Dermochelys coriacea*. Yet none of these sea turtles have the clawed feet found on so many land turtles (Ernst and Barbour, 1989, 119f.; Bustard, 1973, 17). Cf. *AAZ*, 186–88.

312. Cf. A. on the *barkys*, 8.114, which betrays the origin of this fish in the Gr. *narkē* (*GF,* 169–71; Saint-Denis, 1947, 115), *Torpedo marmorata*. Socrates was jokingly called a *narkē* since he numbed all those who came in contact with him (Plato *Meno* 80a). Cf. ThC 7.82 and Vinc. 17.96.

313. Pliny *HN* 9.67.143, Ar. *HA* 620b19.

314. Pliny *HN* 32.2.7, Isid. *Orig.* 12.6.45.

315. Pliny *HN* 9.67.143. Davidson (1976, 33) says electric rays "are not good to eat," yet Thompson (*GF,* 170–71) cites some evidence of its being eaten.

316. The *trebius* is not a fish at all but represents the confused name of one of Pliny's sources, Trebius Niger. Note how this passage begins: *Trebius est piscis niger*. In *HN* 9.41.80 a *murex* is discussed which has the fat and the gold-extraction power and is followed by the *maena* (Saint-Denis, 1947, 61–62) which changes its colors. In *HN* 9.42.81 the nesting behavior of the lamprey is described and it too is said to change color. Trebius is cited in each passage and, eventually, became the fish in A. One other passage, 32.6.15, may be involved for it states that "Trebius Niger tells us that the swordfish . . ." As *niger* means black, in this passage of A. we get the large, black ocean variety which pierces ships. Cf. the accounts of ThC 7.83 and Vinc. 17.97. Aiken (1947, 207–8) discusses the corruptions.

317. The fish is clearly the European brown trout, *Salmo trutta*. Cf. Isid. *Orig.* 12.6.6, Ambrose *Hex.* 5.3.7, ThC 7.84, and Vinc. 17.97. On the name, cf. 7.78, Körting (1923, 985, s.v. *tructa*).

(130) TIMALLUS: The *timallus* [grayling] is a fish that takes its name from a flower, for there is a flower called the *tymum* [thyme].³¹⁸ It is a sea fish with a pretty appearance, it is tasty, and, just as the flower is fragrant, it gives off pleasing odors from its body.³¹⁹

(131) VULPES MARINAE: Sea foxes are fish [fox shark] which, as Pliny says, occasionally swallow a hook. But they draw the hook down to a weaker part of the line, bite it off, and thus are not captured.³²⁰

(132) VIPERAE MARINAE: Sea vipers [lesser weavers] are small fish, one cubit long, having a little horn on their forehead whose blow is lethal.³²¹ Therefore, fishermen, when this fish is captured, cut off its head and bury it in the sand, reserving the rest of it for use as human food.³²²

(133) VENTH: The fish commonly called the *venth* [shad] is called *aristosius* in Latin on account of the innumerable tiny bones [*aristae*] which it has in its flesh.³²³ It is an inexpensive, small, food fish.

It is caught in this fashion: Nets are strung out either longitudinally or transversely on the water and, in front of the nets, floating on the water, is a bow-shaped apparatus. On its upper part is a bell, ringing, at the sound of which the fish come together in schools and fall into the net.

This fish has the color and shape of a *halsa* [shad], except that it has a very great number of hook-shaped bones on its spine as if its flesh were almost totally filled with needle-like spines.³²⁴

(134) VERGILIADES: According to Pliny, *vergiliades* [bream] are fish found in two lakes of Italy, Lake Como [*Lario*] and Lake Maggiori [*Iterbatiano*], which arise at the foot of the Alps.³²⁵ They appear only at the rising of the stars called *Virgiliae* [Ple-

---

318. The grayling, *Salmo thymallus* L. (*Thymallus thymallus*), takes its name from the *thymallos* ("thyme fish") of the Greeks (*GF*, 78–79; Saint-Denis, 1947, 113) and is common to the Rhone, Rhine, and Loire Rivers. Cf. Ambrose *Hex.* 5.2.5, Isid. *Orig.* 12.6.29, Neckam 2.46, ThC 7.85, and Vinc. 17.96.

319. This poetical detail seems to be just that, probably based on the attempt to account for the etymology of its name (*GF*, 79).

320. Generally agreed to be *Alopias vulpinus,* the Gr. *alopēx,* "fox" (*GF*, 12–13; Saint-Denis, 1947, 119–20). Cf. Pliny *HN* 9.67.145, from Ar. *HA* 621a11–15. Cf. ThC 7.86 and Vinc. 17.99.

321. Stadler (1651) and Scanlan (377) identify this as *Trachinus draco* (*GF*, 56–57), the lesser weaver, which has a sharp, highly poisonous spike on each operculum. Cf. ThC 7.88, Vinc. 17.98, and *aranea,* 24.11(15).

322. Cf. 2.82.

323. Sanders (441) studies the name forms (*vint* at 4.79, 7.76, 8.32) and identifies the fish as the twaite shad, *Clupea finta.* So, too, do Stadler and Scanlan (377). The shad's boniness was well known from antiquity (*GF*, 117; *FAZ*, 298) and the fish was undoubtedly important in the lucrative herring fisheries of the Middle Ages. Cf. ThC 7.89 and Vinc. 17.98.

324. Stadler identifies the *halsa* as the shad, *Clupea alosa.* The Latin vocabulary in this description is notable for its repetitive vagueness.

325. This passage is based on Pliny *HN* 9.33.69. Cuvier (*B&R* 2:405) thinks the fish may be a *Cyprinus*. Stadler vacillated in his identification (1549, 1664) but felt it was a chub (*Leuciscus*). Cf. ThC 7.90.

iades] and at all other times lie hidden. These fish have pretty scales that are as sharp as nails. They are small at their heads and then enlarge like the nails on sandals.[326]

(135) VACCA: The cow fish [horned ray?] is a sea animal which is large, strong, irritable, and dangerous. This animal is not oviparous but is viviparous. It produces a maximum of two young, but more often one. It takes this along wherever it goes since it cares greatly for it. The mother is pregnant for ten months. This animal has been proven, through amputation of its tail, to have lived for one hundred thirty years.[327]

(136) ZEDROSUS: The *zedrosus* is a sea beast and is a member of the whale genus in the Arabian Ocean. The doors and rafters of palaces are made from their bones, for their bones have been found to a length of forty cubits.[328]

(137) ZYDEACH: The *zydeach* [sea horse?] is said to be a sea animal which has a marvelous shape but is harmless.[329] It has the head of a horse but smaller, a body like that of a dragon in every respect, and a tail which is long but slender in proportion to its body. Its tail is twisted like that of a serpent. Its entire body is variously colored. In place of wings [*alarum*] it has fins [*pinnae*] like those of a fish, and it moves from place to place by swimming.[330]

(138) ZYTYRON: The *zytyron* [sea turtle] is, they say, a sea animal which the ancients called soldier [*miles*].[331] It is large and very strong. In its front part, this animal presents an appearance somewhat like that of a soldier in armor. Its head is covered with a helmet, which is formed from wrinkled, hard skin. Hanging down from its neck, as it were, is a long, wide, and tough shield [*scutum*], very firm and hollow on the inside.[332] Veins and nerves [sinews] go forth from its neck, as it were, and from its vertebrae, and it is by these that the shield is attached. The shield is triangular in shape. It has front legs rather like long, very strong, bifurcated arms, and with these it

---

326. It is clearer from Pliny than from A. that the fish are being compared to the nails used in hobnail sandals.

327. This entry seems to be a conflation of Ar. *HA* 566b1–27 and intermingles details of the fish called the *bous* and the dolphin. If, as seems natural, *vacca* translates *bous*, then the fish in question may be *Cephaloptera giorna*, the rare horned ray (*GF*, 34–35). Cf. Pliny *HN* 9.40.78, ThC 6.55, and Vinc. 17.135.

328. The name of this beast represents a corruption from Pliny *HN* 9.2.7 which turns the Gedrosi, who live by the river Arabis and make their homes from the bones of monsters, into beasts themselves (Aiken, 1947, 209–10). Cf. ThC 6.57 and Vinc. 17.138.

329. This passage offers a fairly accurate description of a sea horse, one of the members of the genus *Hippocampus* (*GF*, 93f.). The creature is mentioned by Pliny *HN* 9.1.3 as a remedy. Cf. ThC 6.58 and Vinc. 17.138.

330. A. often will use *ala* as a type of fin on a fish, but here it is almost as if he is clearly stating that the creature does not fly but swims.

331. Stadler identifies this as *Agonus cataphractus* L., while Scanlan (379) chooses *Caretta caretta*, the loggerhead turtle. Cf. ThC 6.59 and Vinc. 17.139. Cf. *barchera*, 24.12(19), and *tortuca*, 24.58(126).

332. The words chosen for the helmet imply a metal rather than a leather helmet. Mention of the "shield" hanging from the neck is apparently an attempt to discuss the plastron, the lower shell. *Scutum* is often used by A. for a turtle's shell, but here it has added meaning. If one imagines the turtle standing upright, then its lower plate could be said to be its shield and, indeed, in English "plastron" means the front part of a breastplate.

fights most bravely. When it is caught, it can scarcely be killed except with hammers. This animal is seen in the British Sea and is a member of the *tortuca* genus.[333]

(139) XYSYUS: The *xysyus* [swordfish?] is a sea animal. It is like no other and is of very great size, a member of the whale family. It has a monstrous head, an extremely deep mouth, horrible eyes, and in its entire body it is like no other animal.[334]

Let these things which have been said concerning aquatic creatures be adequate then. For these things, along with those which we have already said in our previous books, can suffice for understanding their natures. In addition to the other useful qualities which we have considered concerning the animals set forth here, much can be understood about the Greek and Arabic names set forth in our preceding books. For here their Latin names have been given, and their properties, customs, and natures have been set forth, and these make it sufficiently clear to what animals, named above in Greek or in Arabic, the description offered here belongs. It then can be understood to which animal the name belongs. Moreover, here also are found, in summary fashion and fitted to each animal, those facts which have been determined to belong to it and which, in the preceding books, were found in a scattered and general fashion.

---

333. On possible candidates for the exact species involved, cf. notes to 24.58(126).

334. The origin of this creature is surely to be sought in the Gr. *xiphias* (*GF,* 178–80), a fish we know as *Xiphias gladius*. Cf. 24.33(60) (*gladius*), ThC 6.60 (*xifus*), and Vinc. 17.138 (*zephius*).

HERE BEGINS THE TWENTY-FIFTH BOOK ON ANIMALS

Which Concerns the Nature of Serpents

CHAPTER I

On the Nature of Serpents in General

1. In this, the twenty-fifth book on animals, the nature of serpents must be dealt with.[1] In many ways they are like aquatic creatures in the composition of their body. They do not have bones but rather have "spines" like fish.[2] Serpents have scales on their belly which they use in place of claws and ones on their ribs which are in place of legs for walking or crawling.[3] For copulation, moreover, they do not have testicles, penises, or wombs, just as fish do not.[4] They have paths [*vias*] for the semen and eggs as fish do, although, in the manner of fish, they do not mix sexes.[5] For the serpents are joined to each other to unite the semen and egg paths and it is as if there were one body with two heads.[6] During this joining the female sends out her eggs, which cling together in a viscous humor, and the male pours his semen over them. This is the case for every serpent that is a true one. It is not the case for those that lay shelled eggs, such as the lizard [*lacerta*], since in ones such as these, the semen has to be inserted into the womb before the egg gets its shell. This is because semen falling over a shelled egg as it comes out would be of no use, since the semen would not reach its inside.

As does a fish, a serpent lays incomplete eggs which, like the eggs of fish, are completed at the touch of the male's semen. Therefore, serpents' wombs are long, as are those of fish, and it sends out the eggs continuously during copulation, as do fish.[7]

---

1. The material for this chapter is found largely in Ar. Generally the same content is found in Barth. 18.8, ThC 8.1, and Vinc. 20.3–6.

2. Ar. *HA* 516b19. Cf. 12.161.

3. Cf. Isid. *Orig.* 12.4.3.

4. Ar. *HA* 508a12, 509b4; *Part. An.* 697a10; *GA* 765a34. Cf. 18.13. Serpents do, however, have "hemi-penises," visible to the trained eye.

5. Ar. *GA* 716b16. The description which follows gives the impression that the fertilization is external, which it is not. Cf. also 2.98; 14.85; 15.11, 13.

6. Ar. *GA* 718a18–34; Pliny *HN* 10.82.169. Cf. 15.20, 22.

7. Ar. *HA* 511a18.

The heart of serpents is directly under their head, as is true for fish.⁸ Nor, as in fish, is it exactly pyramidal, but rather it is like a kidney.⁹ However, the serpent does not have a fixed tongue like a fish, but rather it has a very long, forked, black one which it sends forth beyond its mouth a great deal.¹⁰ Certain large and small serpents have livers, but the large ones have the gall bladder above the liver and the small ones have their gall bladders over their intestines.¹¹ This is because in fish and small serpents, the liver is not continuous but is divided according to the division of the intestine.¹² Therefore, the bile duct [*vas colericae*] also has to be present everywhere lest the bile that is generated ruin the creature's complexion.¹³

At first the eyes of a serpent are made of an incomplete moisture whose main part is not in the head and is not joined to the place the eyes are. In the part that is in the head is the eye's formative spirit [*spiritus*] and its formative power [*virtus*]. Therefore, if the eyes of a small, newborn serpent are pierced, they grow back, just as the eyes of a swallow grow back for the same reason.¹⁴ The tail of a serpent is generated mostly from food moisture [*humido cibali*], as is the tail of a lizard, and its body is also similar because it does not require a diversity of formative spirits and a variety of formative powers but rather each member is formed from another like it. Therefore, the serpent's tail, if cut, grows back again, as does the tail of a lizard.¹⁵

This is found to be so in all animals and members of similar complexions, unless the members cut off are the principal ones that differ from the others in shape and power, such as the head, heart, stomach, liver, and others of this sort.¹⁶

These, then, are the things we have said here in general about the composition of serpents' bodies, although these things also have been pointed out in preceding books.¹⁷

Moreover, the serpent, like the fish, has no neck.¹⁸ But it has as a unique characteristic the fact that it turns its head around while the rest of its body remains immobile.¹⁹

This characteristic had to have been given it so that it could look behind at the long body it drags along. For nature does what is best for every creature.

---

8. Ar. *Part. An.* 666b2f., *Resp.* 478b3, *HA* 507a3; Pliny *HN* 11.69.181.
9. Ar. 508a30. Cf. 2.99.
10. Ar. *HA* 508a23. Cf. 2.98; 12.205. At times the forked tongue appears as two tongues (cf. *QDA* 1.53).
11. Ar. *Part. An.* 676b17–23.
12. Ar. *Part. An.* 676b21–23.
13. Ar. *Part. An.* 677a26ff.
14. Ar. *HA* 508b4–6. Cf. 23.122(60).
15. Ar. *HA* 508b7.
16. Cf. *QDA* 14.7.
17. Cf. especially 2.82, 98–103, 120.
18. Ar. *Part. An.* 691b30.
19. Ar. *HA* 504a16, *Part. An.* 692a1. Cf. 21.44.

Every serpent eats poisonous flesh and, at other times, herbs. But the more poisonous the things it eats, the more poisonous is its bite.[20] It greatly desires wine and, when drowned in it, renders the wine totally poisonous if the serpent is poisonous.[21]

Although it drinks, it has no bladder.[22] Thus it drinks but little[23] and emits dry superfluities, few in number, since it eats little of substance. Rather, it sucks up as much moisture as it can and the spleen of a serpent is therefore small[24] since it takes in little of an earthy nature. Their natural humor is great since they suck up a lot of moisture.

Many serpents, but not all, lie dormant and sleep in the winter.[25] It is falsely said that every serpent is cold by the very nature of its complexion.[26] Hence, serpents with a warm complexion do not sleep. Those that do sleep in winter are so thinned by it that their skin grows loose. When they wake up they shed it, loosening it first at the face and then, forcing themselves into narrow holes, they shed it from their entire body.[27] Thus the serpent's youth returns, just as happens in the genuses of crayfish [*gambae*] and birds that molt.[28]

4

It ought not be passed over in silence that in cold places there are fewer serpents and other poisonous creatures than in other places. Thus it is that no serpents live on a particular island of the Hibernian isles.[29] This is so because such animals have little or no blood and in cold places their moisture congeals and thickens. Then they die and do not reproduce there since the place is not conducive to their reproduction. In warm places, however, their thick humor is thinned and, through digestion, is reduced to the vital spirit [*spiritum vitae*]. In those places, therefore, they grow. The proof of this is that in Nubia and India there are a large number of these sorts of animals, that they attain great sizes, and that their venoms are more deadly in such places than in cold ones.[30]

---

20. ThC reports that they eat meat and herbs *indifferenter* (indiscriminately).

21. Ar. *HA* 594a10; Pliny *HN* 10.93.198. Cf. 7.42. A. also discusses the appeal wine holds for serpents at *QDA* 7.18, and he describes an experiment he conducted himself in Cologne, where he fed a serpent wine and then watched it "stagger" across the room.

22. Cf. Ar. *Part. An.* 676a22f.

23. Ar. *HA* 594a7; Pliny *HN* 10.93.198.

24. Ar. *HA* 506a14f.

25. Ar. *HA* 599a31.

26. Cf. Isid. *Orig.* 12.4.39.

27. Ar. *HA* 600b23–34. Cf. 7.90.

28. This idea may have, in part, arisen from the fact that the cast-off skin in Gr. is *to geras*, which also means "old age." Cf. Curley (1979), 75. It also helps to explain the appeal of snake venom as an impediment to the natural aging process, as at *QDA* 7.31. *Gambae*: cf. 24.39(71).

29. I.e., Ireland. Cf. Solinus 22.3; Isid. *Orig.* 14.6.6.

30. Although, at 7.132 A. notes that the *tyrus* found in India, while it is the most poisonous creature, is also quite small. The passage here is perhaps based on Pliny *HN* 8.11.32–8.14.38, which discusses elephants in Ethiopia, larger elephants in India, where huge serpents are produced, and then elaborates on the theme. Nubia was just north of Ethiopia. Cf. also Pliny *HN* 8.14.36.

However, not all things that lack blood are cold as a result. For yellow things point to greater heat than do red ones and the moistures of many venomous creatures are this way, that is to say, are yellow.

The serpent, just as the name points out, creeps [*serpit*] on its ribs and is helped along by scales instead of claws. It has therefore been allotted movable ribs by nature and in many serpents they number thirty.[31]

Some of the serpents go about with their foreparts erect, to a height of up to a cubit and more or less in proportion to whether the serpent is larger or smaller. I have seen for a fact that many of them go about this way, especially the *tyrus*.[32] They also say that the serpent called the *phareas* goes forward erect on the strength of its ribs, even though they are more cartilaginous than bony in nature.[33]

I have seen them leap powerfully and long, and with the power of their scales they climb the tallest trees. While they are all long, they are nevertheless longer or shorter as befits their genuses and for two reasons. One reason is a natural condition which is allotted each animal in accordance with its species, since there is a limit to all things existing in nature as well as a rationale for size and natural growth. The second reason is the food and the place, warm or cold, which spreads out, a little more or less, the viscous moisture that is the material of their generation.

It is possible, therefore, to know the nature of serpents from these and similar things that have been determined in general along with those things considered above on the same matters.

CHAPTER II

On Understanding the Nature and Complexion of Serpents' Venom

While we have spoken about the nature of serpents in general, we ought to determine the nature of poison in its genus since the humor of a serpent is, for the most part, poisonous. Then we will know better the nature of poisons in their species.

Following the expert findings of philosophical tradition, we say that there are two species of poisons, on the basis of the two principles of their operation. One species of poison has its principle of operation in an active or passive quality inherent to it.[34] The

---

31. Ar. *HA* 508b3.
32. Cf. 7.43, 133; 23.42(59).
33. Cf. 25.38(49).
34. The remainder of this section and the following section are based on Avic. *Can.* 4.6.1.3. Snake venoms are generally either histolytic in action (vipers, adders, asps), destroying blood cells, tissues, and muscle with hemorrhages, etc.; or, they are neurotoxic (cobras, mambas), acting as a central nervous system depressant and inducing respiratory paralysis. Some venoms are not poisonous to humans but may produce local effects such as infections, gangrene, and so on.

other owes its principle of operation to a substantial form and the entire substance of its nature and species.

The first species of venom works in four ways. It can be corrosive, putrefying by means of a sharp heat in the moisture. Such is the poison of the sea hare.[35] It can be inflammatory, producing warmth by a heat that transforms the subtle moisture to its own nature. The poison *euforbium* is said to be of this kind.[36] It can be chilling, taking away feeling with a deadly cold, as opium is said to be. Finally, it can be one that blocks the paths of respiration in the body by means of the thickness of its cold substance which has an accidental sharpness. Burned lead is said to be of this kind.[37]

A poison working by means of its entire substance is one like *napellus* gum and leopard's bile.[38] These poisons are worse than those just mentioned.

Moreover, there are some poisons working their destructions only on one particular part, while others work on the entire body. One, for example, destroys the bladder, one the lungs, and another the whole body.

7. It is to be concluded from this that a poison that kills by changing the complexion, either by putrefaction or by operating on one particular member, need not kill immediately. Rather, it follows that the longer it is in the body, the worse its operation is and the more difficult a cure is. For there is no cure or escape from such a poison unless by an antidote opposed to that particular poison or by that which dissipates the poison, either by using it up, by sweating, or by vomiting.

It has to be considered carefully that a poison is sometimes aided in doing its harm by a heart with a warm complexion and it is sometimes hindered. If the poison is cold and has a thick substance and a dull quality, and if the heart is warm and the pathways of the arteries are wide open, then the warmth of the heart thins it out and the arteries draw it along all the faster with their heat and motion. It therefore comes more quickly in an unaltered and undigested state to the heart and kills. If, however, the heart is less warm and the arteries and paths are narrow, the poison is gradually dispersed and changed by the digestive heat and the pathways only draw it along already digested. It does not kill then since it has been altered before it reaches the heart. Likewise, if the poison is warm, the heart warm, and the pathways wide open, it is drawn along very rapidly before it can be changed, and it kills. If, however, the heart is cold and the pathways narrow, it sometimes changes and the animal is freed from death. But if

---

35. Cf. 24.39(72). On the sea hare, cf. also 14.79, with note.

36. Jessen (1867, 715) identifies *euforbium* as "*Euforbiae antiquorum, officinarum L.*" (cf. *De veg.* 3.65, 5.114), which produces an acrid gum resin used medically in former times as an emetic and cathartic. Scanlan (387) identifies the plant as *Euphorbia resinifera*.

37. See *De min.* 4.1.3 (Wyckoff, 211–12) on the process that was used to produce *cerusa* (white lead), and it comes to A. via Hermes's *Book of Alchemy* (cf. Wyckoff, note ad. loc.). Arsenic is a frequent byproduct of roasting lead ore, but A. says that it is (though sharp) hot and dry in its nature (*De min.* 5.5; Wyckoff, 245).

38. *Aconitum napellus* L. Cf. *De veg.* 6.391, "monkshood," whose dried tubers were once widely used as a respiratory and cardiac sedative. "Leopard's bile": *fel leopardi*. At 22.109(59), A. explains that the leopard's fresh bile is a deadly poison and even in a small amount causes sterility.

there is such cold that it cannot change, the poison lies still and putrefies the body's substance. This is why Galen says *napellus* is poison for a human but food for the *turdus* [thrush] and *passer* [sparrow]. For the *turdus* has a cold heart and narrow pathways and the poison is therefore changed into food for the *turdus* but not for a human.

There is also what Aristotle relates about a girl who was fed on poison.³⁹ At first she took only a little and then later a bit more and so on until she grew accustomed to poison like any other food. For the girl had a somewhat warm heart and narrow pathways and the poison was dispersed in her, being changed through the digestion of her food before it could reach her heart. She was, therefore, made poisonous herself through her diet, and her saliva and other humors killed everything coming near them. Moreover, those that copulated with her died. The doctor Rufus tells of this as well and it has happened not just once but often.

The symptoms of these poisons in those who have taken them in are as follows: If a biting pain, a gnawing sensation, pricklings, and a colicky feeling are felt, then you know it is a warm, sharp poison. If, however, a violent inflammation occurs, with a sudden onslaught of sweating, a reddening of the eyes, distress, and thirst, this signifies that it is a very warm poison. If deep sleep, stupor, and coldness occur, this signifies the poison is a cold one. If nothing happens other than a loss of strength, a cold sweat, and fainting, this signifies it is one of the poisons that operates from their entire substance and, as we just said, these are worse than the others. There is harm in all these if they remain in the body for long, for one must fear that they will come to the heart. This is clear from the studies of physicians. They make a person who has taken in poison vomit again and again and give him theriacs [*tyriacae*], *metridatum*, and other things of this sort which keep the poison from reaching the heart.⁴⁰

All these effects of poisons and many more exist in the actions of serpents and from these the nature of their poison is known.

Serpents are divided into three classes [*ordines*] by the ancient Greek wise men, following the operations of the poison as introduced here.⁴¹ They say that some have the most violent sharpness and their bite admits of no cure, killing in less than three hours. Others are of moderate sharpness and there are still others whose bite has the least strength since they do not have a poison that has to be treated, although they

---

39. A story variously attributed to Ar. and to Rufus of Ephesus. Cf. Thorndike (1923, 2:277) for a discussion and references. The story also appears above at 7.134. The claim that some people become immune to poison by regularly consuming it is found also in Ps. A.'s *DSM* (130) and serves to explain why menstruating women can contain the menses, thought itself to be poisonous, without themselves suffering death. Yet menstruating women and especially women who retain menses (postmenopausal women) were said to poison others with their undigested humors.

40. "Theriacs": a poison antidote, made from the snake *tyrus*. Cf. 22.45; 23.84, with note; 25.43; and Jacob. 89 (187). At *QDA* 4.9 (annex.), theriac is recommended as a cure for rabies, while at *QDA* 7.31 the proper, though dangerous, medicinal uses for poisons are discussed. *Metridatum: mithridaticum*, a plant ascribed to Mithridates (cf. Pliny *HN* 25.26.62) and supposed to be a poison antidote (L&S, s.v.). Mithridates, the king of Pontus who stabbed himself after losing to Pompey, fortified himself against poisoning by taking small doses over an extended time.

41. Cf. Avic. *Can.* 4.6.3.21.

may produce an ulcer and swelling with their bite. The bite of those with a moderate sharpness kills in one to seven hours.[42]

In the first class is a serpent called the *regulus* or *basiliscus* in Greek, for that represents *regulus* [kingly] in that language.[43] Also in the first class is the serpent that is called the *yrundo* because its color is like that of the *hirundo* [swallow].[44] The *regulus* kills when seen, even with the sound of its hiss. The *hirundo* has a length of about one cubit and kills in less than two hours. The serpent called the dry asp because of the excessive dryness of its skin is also of this class.[45] Its length ranges from three to five cubits and its color tends toward ashy-gray. Its eyes shine brightly like charcoal from the fire and it kills in two to three hours. The serpent called the *exspuens* [spitter] is from the same class.[46] It is so called because it kills with its sputum, which, with its teeth folded back, it spits out over those that come near it. It is the odor of the sputum that kills. The *exspuens*'s length is up to two cubits and its color, similarly, tends toward ashy-gray. It kills whomever it bites and spits on him before returning to its hole. The bite of these serpents is called "silent" [*surdus*] or "mute" [*mutus*] since it is not sensed or is scarcely sensed before it causes destruction. And no medicinal treatment is very effective against such as these unless the wounded member is immediately cut off or a cauterization [*combustio*] is effected deeper than the poison penetrated. Then, after these things have been done, one must pay diligent attention to vomiting, evacuations, and dispersions of the remnants of the poison.

Moreover, serpents of this first class are silent and are bow-shaped. They are called silent and mute for the reason just introduced. They are called bow-shaped because they bend themselves into the shape of a bow and strike anyone touching them from above, but they are unable to strike below themselves. Serpents such as these have many species which are very numerous in the lands of Egypt. Some of these have two horns and their color varies.[47] One is white and another is off-white, while one is red and another honey-colored, between red and yellow, and still another is ashy-gray in color. Others of them are like vipers and have teeth curved like hooks. The ones called *dracones* and which kill immediately belong to this genus.[48]

---

42. Although in this section the time it takes serpents of the second class to kill is spoken of in hours, later when discussing specific snakes—e.g., *arundutis,* 25.16(9)—it becomes extended to days, presumably reflecting a difference between the minimum time and the maximum time involved.

43. See 25.14(3), 18(13). Cf. Jacob. 89 (186). For an extended discussion of the basilisk's poison, see 7.132. This snake has been identified as the cobra (*Naja nigricollis,* or *Naja naja*). See Scanlan (389) and George and Yapp (1991), 199, 223.

44. Cf. 25.31(35).

45. Cf. 25.13(1).

46. This is a spitting cobra, whose poison is sent over a distance of several feet. Cf. 25.14(4). Scanlan (390) suggests the ringhals, *Hemachatus haemachatus* L. But see 24.41(74), where A. notes that serpents cannot spit out venom.

47. E.g., *Aspis cornutus,* the desert horned viper.

48. The *draco,* however, described at 25.25(27), is specifically not a poisonous genus. Cf. Isid. *Orig.* 12.4.5. Yet earlier at 1.65, A. describes the poison derived from the moisture of the *draco*; and at 7.131, he notes that under the genus *draco* there is a species of very poisonous *tyrus.* Perhaps the *dracones* here refer to that species.

The second class of serpents consists especially of the genus of vipers of which there are many quite different species. In this class are the vipers called *alesylaty* in Arabic and in the same class are the vipers called *quercinae*.⁴⁹ Also of the same class are the ones that cause thirst in whomever they bite, and many others about which we will speak below.

In the third class there are many genuses of serpents which we will follow up on below by name. These cause an ulcer with their bite and a putrefying inflammation unless it is treated with medicine. But it is not a poison that has to be treated, although diligent treatment of the ulcer should be employed.

Also there is great diversity in the sharpness and dullness of the poison in serpents. This follows their sex, male or female, even though they are in the same species. For males have fewer teeth, more poison, and sharper poison than the females. When it is said that the females are worse than the males, this is understood in respect to the size of their bite insofar as their bite has more teeth.

They also differ according to age, even when they are of one and the same species, for the old ones are worse than the younger ones due to their size. For big ones are worse than small ones when they are of the same species.

There is also a difference among them based on place because those living in waterless places and on mountains are worse than those inhabiting riverbanks, shore areas, and places with a good deal of water.

They differ again with respect to whether they are full or hungry, since the poison of full ones is less harmful than that of thirsting and fasting ones, especially if they have been without food for a long time and are famished.⁵⁰

They differ again with respect to the passions of their soul. For angry, provoked, and bold ones are worse than those of the same species that are differently dispostioned.

They are also differentiated according to types of weather. The venom of those in the same species is worse in summer than in winter and worse in warm, dry weather than at any other time.

Certain of the ancients thought that the venom of vipers is cold.⁵¹ This is an error, and they believed it from a false indicator, namely, that the extremities of one who has been poisoned grow cold. Now, this does not happen due to the coldness of the poison but rather because of the contrary effect which the poison has on the natural heat. It destroys it, and when it has been destroyed the extremities grow cold even though the heart is inflamed a bit from the inflammation of the poison.

---

49. Schühlein, 1661, notes that these two names apparently are related to oak trees and to the *ilicinus* of 25.31(36). He gives "root" and "tree snake" as appropriate translations.

50. The notion that hungry serpents have a more powerful venom may also explain why, as A. notes at *QDA* 7.17, apothecaries were known not to feed the serpents they had for sale for as long as two months. This information goes beyond Ar.'s comment (*HA* 594a21–24) that serpents can live a long time without food.

51. *Viperarum*, "of vipers," was added in the margin by A. Cf. 7.132.

There are those who say that the poison of the serpent called the *alesilaty* is especially cold, that it gathers together and congeals the blood of the heart and therefore causes a violent numbness. They also say that the serpent itself is very cold and that one indication of this is that it sleeps in winter like other animals with cold complexions. Winter, they say, adds heat to warm-complexioned animals since their stomachs are warmer in winter than in summer and warm-complexioned animals are, therefore, awake in winter. These people too have come to believe on insufficient proof and what they say, therefore, does not follow logically.[52] For the fact that the blood is congealed and sensation grows numb at the *alesylaty*'s bite can be because, as we said, the natural heat is destroyed. Then, sensation does not occur with unnatural heat, nor does expansion and thinning of the blood. Likewise, that it sleeps in winter can be due to its food or the disposition of its skin or location and not the coldness of its complexion. For the citrus wasp [*vespa citrina*], which has a very warm complexion, sleeps in winter and does not do so due to the coldness of its complexion.

13 We will, therefore, now speak following the order of the Latin alphabet and introducing a great number of serpents which are all reduced to the three differences introduced above. They have other modes of differentiation as well in their particular species and we will determine these under the serpents' names.

(1) ASPIS: The *aspis* is a serpent, so called because its name means poison in Greek and it spreads poisons with its bite.[53] It is also called the dry asp [*aspis sicca*] on account of the rough and tough dryness of its skin. Its length is between three and five cubits. Its color is ashy-gray and sometimes has a bluish tendency toward ashy-gray, as we said a little while ago. Its eyes are sparkling and it kills in two to three hours. It does not kill before the end of two hours unless the poison is aided by some accidental factor. If it bites someone, the person has a change of color, a numbing of the senses, and an increasingly frequent rattle in the throat.[54] His members grow cold, his eyelids close, and he sleeps deeply. These matters are certain and have been confirmed by experience. They say, though, that an asp's teeth extend out of its mouth like those of a boar.[55]

Solinus relates that the asp can only live with its mate and spouse since the asp left behind dies when its mate has died. Thus it is that an asp follows the slayer of its mate most persistently, does not allow him to rest even in crowds of people, and kills him if it can.[56]

---

52. "Follow logically": *non est necessarium,* lit., "is not necessary."

53. This is the Egyptian cobra, sacred to the Egyptians, and it was probably the asp of Cleopatra. Cf. Isid. *Orig.* 12.4.12, ThC 8.2, Vinc. 20.20, Neckam 2.114, Barth. 18.9, and Jacob. 89 (188). Cf. also Neckam 2.112, *Hypnale.* Cf. *AAZ,* 6–7.

54. These indicate the neurotoxic effects of the venom. This and the following sentence are from Avic. *Can.* 4.6.3.21, 25. The most complete, but often overlooked, Greek sources on venomous animals are Nicander's *Theriaka,* ca. 150 B.C.E. (Gow and Wells, 1953) and Philoumenos, ca. 180 C.E. (Wellmann, 1908).

55. In contrast to vipers, the fangs of the cobra do not fold back and down when the mouth is closed but rather remain in "striking" orientation.

56. Cf. Solinus 27.34, from Pliny *HN* 8.35.86.

They say that an asp in Egypt used to receive a fixed dole at the house of a certain man. After a while she bore young in the house of her benefactor. One of these killed the son of their benefactor and host. The mother, sensing this, killed her young asp offspring and never again dared to appear under his roof.[57]

They also say that an asp sometimes bears a precious stone on its head and that this stone wondrously guards it from a wizard.[58] For, feeling itself being bewitched, it presses one ear to the ground or to the stone and stops up the other ear with the end of its tail so as not to hear the spell of the one chanting it.[59]

(2) ANFYSIBENA: The *anfysibena* is the serpent the Arabs and Avicenna call *anysymen*.[60] Now just as Solinus lies about many things, so he lies when he says of this serpent that it has two heads.[61] For no animal has two heads naturally.[62] He was deceived because the serpent springs in two directions, namely to the front and to the back. This occurs because of the mobility its ribs have in either direction. Also, its two ends are equal in thickness and, as well, are equal in thickness to the middle of its body.[63]

It is also the small serpent which the Greeks call *amphim* on account of the weakness of its ends.[64] When it bites a person, great pain befalls him and the bite seizes the entire body in a short time by creeping through it. This one, like the preceding one, is a member of the first class.

(3) ARMENE: Absolutely no serpent, save only the *regulus* itself, is worse than the *armene*.[65] It does not differ from the *regulus* in its ability to cause injury for, like the *regulus*, it kills by its gaze and by the sound of its hiss. Whatever animal it bites is immediately destroyed and dies. Likewise, every animal that comes near it dies even if it is not bitten by it. In size it is larger than the *regulus* and its size is from one to one and one-half cubits. Its bite admits of no treatment whatever.

(4) ASYLUS: The *asylus* is a serpent of the first class and is the one we called the "spitter."[66] Its length is up to two cubits and its color is ash-gray, tending toward yel-

---

57. Phylarchus, cited by Pliny *HN* 10.96.208.

58. *De min.* 2.2.4 (Wyckoff, 86–87); Jacob. 89 (186). Jorach, cited by Arnoldus Saxo 3.10.29. Mention of the stone is from Pliny *HN* 37.57.158, *dracontias*.

59. This portion of the tale is from Isid. *Orig.* 12.4.12.

60. Avic. *Can.* 4.6.3.41 (*ankesimen*).

61. Solinus 27.29, after Pliny *HN* 8.35.85. Isid. *Orig.* 12.4.20; Neckam 2.118; Jacob. 89 (188); Barth. 18.8 (from Isid.); ThC 8.3; and Vinc. 20.19.

62. On the "monstrous" character of two-headed creatures, see 6.40, 12.227, 13.123, and 18.50.

63. Some claim that the origin of *amphisbaena* may have arisen from misobservation of the Indian sand boa, *Eryx johnii,* or to the legless lizards today classed as *Amphisbaenidae*. See Druce (1910); White (1954), 177–78; George and Yapp (1991), 199–200. A better option is the common blind snake (*Typhlops vermicularis*), readily observed in Europe (Kitchell, 2015, 135–36).

64. ThC 8.3 attributes the *amphin* to Solinus, but the word does not seem to appear in Solinus.

65. Avic. *Can.* 4.6.3.23 (*harmene*); Vinc. 20.19 (citing Avic.). This seems to be another part of the basilisk legend; see 25.9, 18(13).

66. Avic. *Can.* 4.6.3.26. Cf. 25.9, *exspuens*, with note. The symptoms described are those of a neurotoxic venom, as from cobras, producing the characteristic symptomology of respiratory paralysis. Cf. *AAZ*, 33–34.

low. It kills the one it bites before it can return to its hole. The one it bites remains without sensation, sleeping a very deep sleep. He also endures continual rattlings in his throat. His eyes close, his neck twists, and he suffers spasms and an irregular pulse. But at first he feels no pain except the bite and then he feels pain in his intestines. One bitten by this serpent constantly tries to put his hand to his throat to induce vomiting.

This serpent raises its head and spits its poison. The place of its bite is as small as a pinprick, has no abscess [*apostema*], and a bit of black blood runs out of it. The one bitten is, at the onset of the bite, beset with murkiness of the eyes and pain in his entrails and in the esophagus.[67] Later come the closing of his eyes and a deep sleep. One bitten does not live more than three hours after the bite.

15 (5) ANDRIUS: The *andrius* is a serpent of the first class which the Greeks call an *andris* when it is in the water and a *kedusudurus* when it dwells in the field or forest.[68] It is a smaller serpent than the silent asp [*aspis surda*], has a wide head, and is both a worse and more harmful serpent. Its bite begins with violent pain and the spot becomes very inflamed and then turns green and is eaten away. The bitten one suffers dizziness, a vomiting of fetid bile, irregular movements, and a loss of strength. He dies most often during the third hour, and if he makes it beyond the third hour it is because the serpent is the water one or because the complexion of the one bitten is very strong. If he escapes death, then he will be beset with illnesses from which he will never be freed.

(6) ASFODIUS: The *asfodius* is a serpent of the first class.[69] It is similar to the serpent called the *sabrin*, about which we will speak below.[70] The size of this serpent is from one to two cubits. It has a sand-colored body entirely speckled with black and white spots. It is a smaller serpent than the viper. It has a very small head and tail and, in most instances, is a sandy color. Black, red, and white members of this species are found. This serpent has white scales over its head that are cut and split, as it were. When it goes forward it makes a rustling noise that is like the rustle of dry leaves. This is due to the dryness of the skin on its belly. These serpents are slow movers and have uniform teeth.

This serpent is very bad and when it bites, all the visible pores which are on the bitten one's body as well as their pathways burst and they all drip blood as if from

---

67. "Esophagus": *os stomachi*. Cf. 1.242, 419, 549; 2.85, 91; 3.118, with notes. There are over ten species of spitting cobra.

68. Avic. *Can.* 4.6.3.29, *alesilati surda,* whose Arabic text Schühlein, 1661, refers to the original Gr. *hydros* (*Coluber natrix,* the ringed or grass snake) and *chersydros.* Cf. 25.23(21), *celydrus. Hydros* (L&S, s.v.) is identified as the grass or ringed snake, *Coluber natrix* (*Natrix natrix*), which has no venom and rarely strikes, employing instead the odorous secretions of its anal glands when threatened. *Chersydros* is an amphibious snake (L&S, s.v.).

69. Cf. Avic. *Can.* 4.6.3.36; Vinc. 20.18 (citing Avic.). The snake involved seems very likely to be the deadly *Echis carinatus,* the saw-scaled or carpet viper. A relatively small (ca. 0.5 meters) viper, but "it has a particularly devastating venom" (Brown, 1973, 10). The hemotoxic effects and hemorrhaging are particularly severe in this case, and it seems to fit well the Gr. name *aimorros* (transliterated from Arabic by Schühlein), a serpent that makes blood flow from everywhere.

70. Schühlein, 1661, sees this as the Gr. *kerastēs,* horned asp. Cf. 25.40(55).

ulcers once healed and then opened up again. This all occurs with continuous pain while vomiting and spitting blood. Sometimes the lower pathways and the "natural parts" break and blood drips from them as well as from the corners of the eyes.[71] The bitten one is thus overwhelmed by blood, by bloody spit, and by a flow of blood from his nose, as well as by stomach pain. The place where he was bitten ulcerates and grows black and from it flows a small amount of watery humor. The stomach is loosened and breathing arrested. He incurs difficulty in urinating, he loses his voice, and his members become lax and enfeebled. He falls into a state of oblivion and convulsions, his teeth fall out of his mouth, and he dies enduring pain such as this.

(7) ALTYNANYTY: The *altynanyty* are said to be serpents like the *cafezaty*—small, short, and tiny, but extremely shrewd, clever, and malicious.[72] They sometimes hide themselves above in the leaves of trees so that, as Avicenna says, they can hurl themselves down on those passing by beneath the trees and kill them. When they are thwarted in this clever trick, they come out of their holes and leap on passersby. In color they are on the reddish side. As a result of their bite a violent pain occurs which creeps throughout the entire body. This happens to kill the one they bite and they are thus of the first class.

(8) ARACSIS: The *aracsis* is an ill-natured serpent of the second class whose venom kills on the second day by eating at the liver and by rupturing the intestines.[73] The serpent is variously colored and not large in size.

(9) ARUNDUTIS: The *arundutis* and *cauharus* are serpents equal in size and harmfulness, but we will discuss the *cauharus* below.[74] Their length is up to one cubit. The *arunditis* [sic] is the color of sand and on its body there are traces of lines. One it bites is beset with violent pain at the place of the bite and a very large abscess appears from which slimy fluid flows. It sometimes kills on the third day, but there is sometimes a delay until after the seventh day. This serpent is therefore of the second class.

(10) AHEDYSYMON: The *ahedysymon* is a member of the dragon [*draco*] genus and belongs to the third class.[75] It has strong teeth and therefore crushes flesh by biting it, and by fixing its teeth in deeply, it gouges the flesh. Thus, one is more frightened by its wound, not so much because of the quality of the poison but because of the size and depth of the wound. Therefore, a very bad wound is made.

(11) ALHARTRAF: The *alhartraf* and *haudyon* [lamprey] are members of the dragon genus and are in the third class.[76] They are large, up to five cubits long. Their

---

71. "Natural parts": *partes naturales*. Perhaps the "privates," or genitals, as Scanlan (395) suggests.

72. Avic. *Can.* 4.6.3.40, *altararati*. Cf. Vinc. 20.18 (*alerati*, citing Avic.). This is the javelin snake, *iaculus*, of Pliny *HN* 8.35.85, Gr. *akontias*, mentioned also by Lucan *Phars.* 9.720, 822; and by Ael. *DA* 6.18. Cf. 25.30(32).

73. Avic. *Can.* 4.6.3.46, *arascis* (Arabic *al-raqša*), a white and black snake.

74. Avic. *Can.* 4.6.3.46, *amiudutus* (Arabic *ammudutis*; Gr. *ammodytēs*, sand burrower). The name "sand burrower" suggests that it is one of the small desert-dwellers, and Scanlan (396) suggests the desert viper, *Cerastes cerastes*.

75. Avic. *Can.* 4.6.3.54, *albedismon*, a nonvenomous but unidentifiable snake.

76. Badly distorted from Avic. *Can.* 4.6.3.57. Cf. Schühlein, 1662.

bite is followed by violent pain and excessive coldness. As a result, stupor and death can take a good number of days since the coldness of the poison kills more slowly. Sometimes the bite of this sort of dragon is cured. Let, then, information be given concerning others similar to these by means of the natures and properties of these two.

(12) HAREN: The philosopher Semeryon says that the *haren* is a serpent of the genus of sea dragons and Avicenna agrees with him.[77] It is like the serpent or dragon which is called *carnen* and which is a member of the sea dragons. It is very large, six cubits or more, black, a bit hairy near the head, and it makes a large, bad wound with its bite.[78] Its bite produces the same effects as that of the vipers.

(13) BASILYSCUS: The *basilyscus* is a serpent called *regulus* in Latin translation, for the word means the same as the Greek *basiliscus*.[79]

The reason for its name is twofold. First is that it seems to have a crowned head. This is because it has an appendage on top of its head spotted with white and hyacinth-blue, as if crowned with a kingly diadem with shining gems scattered throughout it.[80]

This serpent is two palms long and has a very pointed head.[81] Its eyes are red and its color is black turning to yellow. Its breath burns everything it approaches such that in the area of its hole nothing whatever lives [*oritur*]. This is because it dries out trees, herbs, and thickets, splits rocks, and infects the air so badly that when a bird flies toward the place in which the serpent stays, it immediately falls down dead. The same is true for other beasts and serpents, save only for the *armene*, which is joined to the *basiliscus* by a great affinity.[82]

The *basiliscus* also kills with its hiss, provided that it raised itself up, because the hiss is borne along with its spirit and the hiss kills so long as it is generated together with spirit. For that period during which the simple quality of sound continues without the spirit, I do not think that the *basiliscus* can kill.

It also kills with its gaze, for everything on which its gaze falls dies. Now Pliny and others say that the *basilyscus* does not kill a person by its gaze unless it sees the person first rather than being seen by him first and that the person's glance, should he see the *basiliscus* first, kills it.[83] But I do not think this is true, since it makes no sense, and nei-

---

77. See Avic. *Can.* 4.6.3.56. Perhaps an eel is meant (Scanlan, 396) or a sea snake. A. describes Semeryon (Symerion, or Semerion) as a Greek "wise man" of antiquity. Cf. 25.18(13), 21(16), 22(18, 20), 23(24), 25(27), 30(34), 31(35), 40(55), and 26.25(33). His identity, however, remains uncertain.

78. At 3.79, A. refers again to a *draco* with hair about its head, an apparent exception to his rule that only viviparous animals have hair.

79. Cf. 7.132. Both *basiliscus* (Gr.) and *regulus* (Latin) mean "kingly." Cf. Jacob. 89 (186), Isid. *Orig.* 12.4.6, and ThC 8.4. For a discussion of the confusion of names and attributes involving *basilicus, trochilis,* et al., see White (1980, 169, n.1). Cf. Neckam 2.120, Barth. 18.15, and Vinc. 20.22–24.

80. Pliny *HN* 8.33.78.

81. Cf. Avic. *Can.* 4.6.3.22; Pliny *HN* 8.33.78–79.

82. Cf. 23.14(3).

83. Pliny *HN* 29.20.66 remarks only that the gaze of the basilisk is fatal to a person if it looks at him. Cf. Isid. *Orig.* 12.3.6 and Jacob. 89 (186).

ther Avicenna nor Semeryon, who are natural scientists [*physici*] who speak through expert findings, relates this.

Neither is the reason that it kills with its gaze, as some say, the fact that rays come forth from its eyes and destroy whatever they fall upon. For it is not the opinion of the natural scientists that rays come forth from the eyes. Rather, the reason for the destruction is that the visual spirit [*spiritus visivus*] is dispersed very far due to the thinness of its substance.[84] It is this that destroys and kills everything.

All other serpents fear this serpent and flee it, except the one we have named, and even that one flees sometimes. For the body of whatever animal it has bitten liquefies and swells, a slimy fluid [*virus*] flows, and the animal dies immediately.[85] Moreover, if another person comes near the dead one, he dies too. And if it is touched with a long lance or branch, anyone touching that lance dies. This already happened to a certain soldier who touched it with his lance and his horse died likewise when, with the same lance, the soldier touched the horse's lip by chance.

This serpent is numerous in the country of Achohaz and in Nubia.[86]

It is said that the weasel kills it and that the inhabitants, overwhelmed by the large number of these serpents, send weasels into their holes.[87] The serpents flee the weasels and the weasels kill them. If this is true, it seems miraculous.

They also say that in a place where ashes of this serpent are sprinkled, spiders do not build nests and no other poisonous creatures appear.[88] Therefore, in ancient times, their ashes were sprinkled in temples.[89]

Hermes also says that silver that has been smeared with its ashes assumes the weight, solidity, and splendor of gold.[90]

Some say as well that there is a particular genus of *basiliscus* that flies, but I have not read about this in the books of the wise philosophers. Some also say that it is born from the egg of a cock, but this is most surely false and impossible.[91] When Hermes teaches that the *basiliscus* generates in a glass container, he does not mean the true *basiliscus* but rather a certain elixir used in alchemy by means of which metals are changed.

(14) BOA: The *boa* is a member of the dragon genus and is of the third class. When this *boa* is young it stays close to herds of *bubali* [gazelle] or cows which are very full of milk. Then, by sucking at one after the other for a long time, it grows to

---

84. On the visual spirit, see 1.359, 19.19. For A.'s criticism of the view that rays project from the eyes, see *De sensu et sens.* 1.5.

85. Avic. *Can.* 4.6.3.22; Pliny *HN* 8.33.78.

86. A corruption of Avic. loc. cit., *Arorch.* Schühlein, 1655, suggests it indicates the land of Turks.

87. Pliny *HN* 8.33.79, in somewhat altered form, apparently through Isid. *Orig.* 12.4.6. The story may originate from the mongoose and the cobra.

88. ThC 8.4.

89. Cf. Jacob. 89 (187).

90. "Hermes": that is, Hermes Trismegistus. For A.'s knowledge of Hermes, see Sturlese (1980).

91. Cf. 23.116(51).

an immense size, to the extent that afterward it lays waste to whole regions.[92] Once in Africa, as a Roman history relates, a *boa* 120 feet long appeared. Regulus, the Roman leader, fought it with ballistae and other such things with which they were accustomed to storm camps. Once it was killed, he took its skin and jaws to Rome for public display.[93] Its bite produces a badly cut wound, that is, one that is malicious and poisonous like that from other dragons.[94]

(15) BERUS: The *berus,* so they say, is a shrewd and worthless water serpent. It is of the first class, having a deadly poison. Some say that this serpent calls forth a *murena* and, having laid aside its poison, copulates with it.[95] This is a fable and is false, as we have said previously.[96]

(16) CORNUTA ASPIS: The horned asp, as Avicenna and Semerion say, is a serpent of the first class whose length is one to two cubits.[97] On its head there are two protuberances that resemble two horns. The color of its body is the color of ash and its belly has dry, hard scales which scratch over the earth and make a sound when it moves. Its teeth are equal in size and not curved backward. For the most part it lives in sandy places.

There is a genus of this serpent which is small [*parvus*] and therefore is called the *parvus,* since its horns are small and short. This arises from two reasons. Sometimes the horns of the large-horned asp fall off, and until they fully grow back, they are short. But this one is not called *parva.* But there is another reason, in accordance with nature, why the horns of one are always short, and that one is called the small-horned asp.[98] In this genus of serpent there are more species.

The symptoms of its bite are as follows. Pains like pinpricks or like nails driven in are felt at the spot of the bite. The bitten one is beset with the greatest heaviness in his body, his lips swell, and he endures dizziness, cloudiness of his eyes, and a loss of reason. Solinus also says that these serpents attack people with their horns, but this is not likely.

(17) CEREASTES: The *cereastes* is a serpent that lacks spines in its body, having cartilage in the place of spines.[99] It therefore has a more flexible body than other ser-

---

92. Solinus 2.33; *AAZ,* 156–58.

93. Pliny *HN* 8.14.37.

94. The same general details are found in several medieval works. Cf. Barth. 18.8, ThC 8.5, and Vinc. 20.25.

95. Ambrose *Hex.* 5.7.18–19 is probably the original source, but the immediate source under the name *berus* was the *Experimentator,* as cited by ThC 8.6. Cf. Vinc. 20.25.

96. Cf. 24.41(74).

97. Avic. *Can.* 4.6.3.27. The horned viper, *Cerastes cornutus.* Cf. Vinc. 20.27.

98. *Pseudocerastes fieldii,* the false horned viper, has a small pair of horns over the eyes. Scanlan (399) suggests the tentacled snake, *Herpeton tentaculatum.*

99. Isid. *Orig.* 12.4.18; ThC 8.9. "Spines": analogous to bones in serpents and aquatic animals. Cf. 3.2; 12.111–12, 139. Cf. *AAZ,* 26.

pents. It has eight horns on its head which are curved like rams' horns.[100] It is a small serpent the color of dust. For this reason it hides in the dust and with its poison kills sparrows who sit on its horns as if on twigs. It does the same to all other animals that tread upon it. Of the horned serpents, this is the one of the first class of harm.[101]

There are those who say that the *cereastes* horn sweats in the presence of poison and is therefore brought to the tables of noblemen and made into knife handles.[102] Fixed into the tables of noblemen, these betray the presence of poison. But this has not been sufficiently proven.

(18) CAFEZATUS: The *cafezatus,* as Avicenna and Semeryon say, is a serpent similar in many respects to a certain other one called the *altynanytus* and about which we have spoken previously.[103] These two genuses of serpents are small and short and hide themselves in trees in order to hurl themselves at passersby. They are warm and vile, and they tend toward a reddish color. Their bite creeps from the place of the wound throughout the whole body and kills. These are serpents of the first class.

(19) CAECULA: The *caecula* is, as Pliny says, a small serpent and it is said to be blind, but the force of its poison makes it a member of the first class.[104]

(20) CERYSTALYS: The *cerystalys,* as Semerion and Avicenna say, is a serpent of the viper genus and of the second class.[105] It is gray in color and is two cubits long. At the place of its bite there occur inflammations and blisters. A hardness appears and there occurs a flow of bloody, black fluid from the spot of the bite. The injured one experiences loss of reason, cloudiness of vision, and a fatal spasm.

(21) CELYDRUS: The *celydrus* is a serpent which they say is equally an aquatic and a land creature, as the name, a Greek compound, shows. For *chyron* is land and *ydor* is water.[106] It is a serpent whose back makes the land where it passes smoke due to the force of its poison. This is why Macer says, "Either the smoking backs breathe

---

100. Pliny *HN* 8.35.85 describes a cluster of four horns used by the *cereastes* to lure birds. Solinus 27.28 repeats the story, as does Jacob. 89 (189). Barth. 18.8 includes the new detail of the curved horns like those of a ram but mentions no source. ThC (citing Solinus) makes the number of horns eight. In fact, the horned serpents have only two horns (actually scales), although sometimes locals put a couple of hedgehog spines into the heads of the serpents and pass them off as four-horned (Boulenger, 1914, 179).

101. The venom of the snake is rarely fatal to man, although it is lethal to large birds and small mammals (Boulenger, loc. cit.). The effects were probably more severe in ancient times, however, with a higher probability of secondary infection and more severe shock effects.

102. ThC 8.9 (without source).

103. Avic. *Can.* 4.6.3.40. Cf. 25.16(7).

104. The "little blind-one," from Isid. *Orig.* 12.4.33; cf. Pliny *HN* 9.77.166. The snake in question is a member of the *Typhlops* genus, perhaps *T. vermicularis* or *T. braminus.* Each resembles a worm more than a snake and has eyes covered over by scales. The snakes are not venomous, but ThC 8.10 and Vinc. 20.26 also mention venom. See Grzimek, 6:359–60.

105. Avic. *Can.* 4.6.3.47; *triscalis.* Cf. *prester,* 25.34(43).

106. Cf. Isid. *Orig.* 12.4.24, *chelydros . . . et chersydros.* The same minor corruption is found in ThC 8.12, but not in Barth. 18.8 or Vinc. 20.26. The Gr. stem *chers* means "dry land," and *ydros* means "water."

forth venom, or the land smokes where the snake glides."[107] This serpent makes its way straight ahead in large measure since, if it twists itself very much while it goes, it makes a cracking noise.

(22) CENCRIS: The *cencris* is a serpent that always creeps in a straight line.[108] Hence Lucan says, "And the *cencris*, always to glide in a straight path."[109]

(23) CAUHARUS: On the basis of the testimony of Avicenna, the *cauharus* is a serpent similar to the *arunditis*.[110] Its length is a cubit and its color sandy. On its back it has traces of lines. On the wound caused by its bite there arise an extraordinary swelling and a most violent pain, and slimy fluid flows from it. It kills on the third to the seventh day and is therefore in the second class.

(24) CARNEN: The *carnen* [stingray], as Semeryon says, is in the dragon genus and is a very large serpent.[111] Its bite is like that of a viper.

(25) CENTUPEDA: The centipede is a serpent of the dragon genus, having many feet indeed.[112]

(26) DYPSA: The *dypsa* is the serpent whose name is translated into Latin as *situla* since the person it bites needs a *situla* [bucket].[113] For so greatly does he thirst that he dies from being thirsty and from drinking. For this reason it is called the "serpent causing thirst" by Avicenna and Semyrion the philosopher.

Concerning this serpent, Solinus says that it is not seen when it strikes, being so small that it is almost invisible.[114] I attribute this not to its size but to the fact that it lies hidden on the surface of the sand or in grass. Avicenna and Semyrion say also as follows: The length of the "serpent causing thirst" is one palm and there are many black markings on its body.[115] Its head is small and its neck thick and the creature begins to get continually narrower from its thick neck to its thin tail. Its shape is like that of a viper and the color of its rearmost parts as far as its tail tends toward black. When it goes forth it shakes its tail and holds it firm. Its origin [*generatio*] is in the land of Hasceni and Lokyati.[116] Some of the serpents of this genus live on the seashores.

When a person is bitten by it, his stomach burns and is so inflamed that it is not sated with water nor does he stop drinking water even though he evacuates nothing

---

107. Aemelius Macer was an Augustan poet whose lost poem *Theriaca* was surely the source for this line, quoted in Isidore (Morel, 1927, 108). Cf. Lucan *Phars.* 9.711.

108. Isid. *Orig.* 12.4.26. See Barth. 18.8, ThC 8.13, and Vinc. 20.26. Cf. 23.24(26), with note.

109. *Phars.* 9.712.

110. Avic. *Can.* 4.6.3.49. Cf. 25.16(9).

111. Cf. 25.17(12), with note.

112. "Serpent": *serpens,* used in the generic sense of "crawling thing." Cf. Isid. *Orig.* 12.4.33; ThC 8.11.

113. Cf. Isid. *Orig.* 12.4.13, Jacob. 89 (189), Neckam 2.116, Barth. 18.8 and 18.36, and ThC 8.15. On the classical *dipsas,* see 46–47.

114. Isid. *Orig.* 12.4.32, not Solinus.

115. Avic. *Can.* 4.6.3.38; Vinc. 20.34.

116. Arabic *Lûbija wa-al-šam* (Schühlein, 1659), i.e., Libya and Syria.

through urine or sweat, for his stomach is totally inflamed and the water runs into his veins.

(27) DRACO: *Draco,* according to Avicenna and Semeryon, is in the third class of serpents.[117] This is the class, as we have already said, that does harm when it bites, not from a poison that requires attention but from the wound. According to these philosophers, dragon [*draco*] is the name of a genus which contains many species within it.

These philosophers say that all dragons have very large bodies, so much so that the smallest in the genus is five cubits long and the largest is thirty or more, especially in parts of India.[118] They say a dragon has two large eyes and a protuberance under its lower jaw like a chin and that it has a great number of teeth.[119] The smaller ones are numerous in Nubia, but the larger ones are numerous in India.[120] For in Nubia and in Asia, ones four and five cubits long are found, but the Indian ones are the largest.

They have yellow and black faces and their mouths are exceedingly wide. Eye ridges cover their eyes and over their necks are scales. One was seen by Avicenna on whose neck, all along its breadth, hung long, coarse hairs descending like the mane of a horse. Dragons have three long, projecting teeth in the upper jaw and the same number in the lower. Their bite is accompanied by a small pain at the start, with the bite becoming inflamed afterward. The bite of the males is worse than that of the females. It has been learned that very large dragons are found not only in India but also in other areas.

This, then, is what of a more truthful nature is found about dragons in the expert findings of the philosophers. If, though, we were to follow the words of those who pass along the hearsay of the common people rather than point out the things discovered by natural science in their writings, then following Pliny, Solinus, and some others we would say that the dragon is the largest of all the animals of land or sea, that it has no poison, and that it has a crest on its head. We would say that its mouth is small for the size of its body and head, that it sticks out its tongue by constricting the channels of its esophagus [*canna*] and arteries, and that it gapes with its mouth but does no harm with its teeth.[121]

Its bite, however small it is, is very bad. They say this is not because of the dragon itself but because it eats deadly and poisonous things.[122] If it binds something in its

---

117. Avic. *Can.* 4.6.3.53; Vinc. 20.19 (citing Avic.).

118. The largest of the boidae, the reticulated python of India and the African rock python, may grow up to 35 feet in length. At 22.50(37), A. comments on certain *dracones* so large that they fight with elephants and drink their blood. Cf. *AAZ,* 156–57.

119. The teeth are typically larger than those of the vipers and recurved so that they constantly provide pressure, forcing the prey animal inside the snake. The jaw arrangement and teeth curvature are such that it is quite doubtful that the serpent could "change its mind" and eject a prey once it got started taking it in.

120. The sand boas (*Eryx* sp.) of arid northern Africa are usually 1 meter or less in size (Grzimek, 6.374).

121. Jacob. 89 (186), cited by ThC 8.16 along with Augustine. Barth. 18.37 cites Isid. *Orig.* 12.4.4 for the same information. Cf. Pliny *HN* 8.11.32ff., 29.20.67; Solinus 30.15. Canby (1995) has recently traced the history of dragon lore from antiquity to St. George.

122. The *Experimentator,* cited by ThC 8.16.

tail it kills it, and not even an elephant, given the size of its body, is safe when wrapped in its tail.¹²³ But I do not consider these things as sufficiently proven by experience.

Pliny says the dragon is nauseated in springtime and that it curbs this illness with the juice of the wild lettuce.¹²⁴ Its lair, so they say, is generally in rocky caves. It seeks a cold place on account of its bodily heat, from which it especially suffers when it flies. For a particular genus of dragon is said to have wings. It is likely that these are membraned and are not large ones, being rather modest. For it is not possible for such a great weight as the largest ones have to be borne and suspended in the air on wings.

They say also that these dragons are found in warm lands, especially in the ruins of the Tower of Babel and in its vicinity. They say that dragons' voices terrify people so much that they sometimes die. The sight of dragons is also terrifying. They say that they live a long time without food and that when they do eat they are not easily filled.¹²⁵

27 They say that during the stormy season, dragons burst forth from beneath the earth and fly into the air with their wings of skin spread wide. This would be more believable if the dragon had a large but short body. But since the body is long, it does not seem possible for it to be suspended by wings due to its length. Moreover, experienced men have handed down nothing about this while speaking in a philosophical manner. The genus of *stellio* [salamander] that is found winged does not have wings for flying but rather ones with which it can raise up the front part of its body so that it can move. This is just as was determined in preceding books of this study.¹²⁶ They say also that wherever a dragon is, it infects the air. They also say that it often lacks feet and creeps over the ground on its chest, and this is true. They add that a certain genus of dragon has feet but this is improbable since just a few feet would do no good over such a length. Some also say that a noble stone is cut from its brain but that the stone has no power unless it is drawn out while the dragon is still alive.¹²⁷ Some also say that dragon flesh is edible by Ethiopians because it cools them down. They also say that when a dragon is hot from flying, it cools itself down in elephant blood.¹²⁸

28 They say that dragons are afraid of thunder, and this animal is said to be struck by lightning rather often, just as, to the contrary, the eagle among the birds and the laurel among the plants are said never to be struck. Therefore, once the charms have been sung by the wizard, they say that a bag made of hide or a drum is struck in front of it, so that thinking it is thunderstruck, it is held in one place and allows the wizard access to it. They further say that once bound to the dragon, the wizard flies over vast

---

123. Jacob. 89 (186); Isid. *Orig.* 12.4.5; Pliny *HN* 8.11.32; Solinus 25.10. See the lively illustration in Benton (1992), 78.
124. Pliny *HN* 8.41.99.
125. ThC 8.16.
126. Cf. 2.68; 25.35(46).
127. ThC 8.16. Cf. Jacob. 89 (186); *De min.* 2.2.4 (Wyckoff, 86–87). The stone is *draconites,* originally from Pliny *HN* 37.37.115.
128. Cf. 22.50(37).

expanses of earth. Sometimes, however, the dragon is overcome by fatigue and falls into the sea, drowning itself and its rider. All these things, however, are said more based on rumor than on experience.

It is said that dragons are seen flying in the air which breathe forth shining fire, but this is impossible to my way of thinking unless there are those vapors which are called "dragons" and about which determination has been made in the book on weather.[129] These have been experienced as glowing, moving, and smoking in the air, and on occasion to fall in a ball into the waters, where they shriek like glowing iron. Sometimes when the vapor rises on a wind, they rise again from the water and burst forth and burn plants and everything they touch.[130] Because, then, of this ascent and descent, and the smoke that spreads like a mist at both ends in the shape of wings, the unskilled think this is a flying animal breathing fire.

(28) DRACO MARINUS: The sea dragon, as the philosophers Semyryon and Avicenna say, resembles the other dragon.[131] Its bite crushes and gouges the flesh but, since it is aquatic, is less poisonous than the bites of others.

(29) DRACONCOPEDES: The *draconcopedes* are what the Greeks call a large serpent of the third class and of the dragon genus which, they say, has the maidenly face of an unbearded man.[132] I have heard from people worthy of belief that just such a serpent was killed in our times in a forest of Germany and, until it rotted, was shown for a long time to all there wishing to see it. Its bite is like that of other dragons.

(30) EMOROYS: The *emoroys* is a serpent called by a Greek name formed from *emer*, which is blood, and *roys*, which is flow.[133] Its bite relaxes all the orifices of the veins and shakes out whatever blood is in the body of the one bitten. This is how it kills and it is of the first class. They say it is ash-gray in color and more than one cubit long.

(31) FALITUSUS: The *falitusus* is a multicolored serpent of the second class.[134] Its bite is like that of the vipers. A bite from it is followed by corruption and softening such as befalls a person with dropsy. There is a profound sleep, total oblivion, hunger, and sickness in the liver and colon.

---

129. *Meteora* 1.4.8.

130. A. appears to be describing a meteor.

131. Avic. *Can.* 4.6.3.54f. Of indeterminate species, but probably either a *muraena* (e.g., the moray eel) or a conger eel. It is certainly not a true "sea snake," as these are generally slim (unlike the image presented of *draco*) and highly venomous. Cf. 24.26(39).

132. ThC 8.17; Vinc. 20.33.

133. From Solinus 27.32 originally. Cf. Isid. *Orig.* 12.4.15, Jacob. 89 (189), Neckam 2.115, and Barth. 18.9. Cf. 25.15(6), *asfodius*, with note. The correct Gr. is *haima*, blood, and *rhois*, flow (compound form).

134. Avic. *Can.* 4.6.3.48; *famusus*. Schühlein, 1662, points out that the Arabic word could be read *fangarnios*, but that a change in a diacritical mark produces *qanhrinos*. Gr. *kegchrinēs*, a serpent with millet-like protrusions on its skin. The snake involved may be *Vipera aspis*, the European asp, with the "grains of millet" referring to its scales. Its color is often straw or brown and the action of its venom is similar to that described.

30   (32) JACULUS: Pliny and Jorach say the *jaculus* is a winged serpent named from "throwing" [*iaculando*].[135] It lies in trees and infects this fruit, and whatever eats the fruit dies and whatever meets face to face with the serpent is killed. Jorach says as well that there are two kinds of this serpent. One kills with a bite that lacks any painful feeling. The other is in that genus which wears down whoever touches it with continuous pain until he dies. These are both of the first class.[136]

(33) IPNAPIS: The *ipnapis* is a species of asp and of silent serpents.[137] Whomever it has bitten dies in his sleep. It is said that this is the one Cleopatra bound to her right side and which entered the tomb with her Anthony in order to end her life with him sweetly.

(34) HYDRA: The *hydra*, or *ydrus*, is a serpent that is the prettiest of all the serpents.[138] It is a member of the viper genus and is of the second class. It appears in the Nile River, and when the crocodile is asleep on the bank with its mouth open, the *hydra* wraps itself with slippery mud to better slide into it. It thus sinks into the crocodile's jaws and when the crocodile awakens, the *hydra* is swallowed. It then, creeping about, tears the entrails of the crocodile and exits.[139]

Semeryon speaks of its bite saying that its bite spreads from the place of the wound, widens, and the color of the bitten one becomes darker.[140] From the bite a great deal of very foul-smelling black fluid comes. A cure for it is prolonged and difficult. Pliny says that the ancients called the pain attendant on the bite *boa* since cow manure was very effective in its cure.[141]

31   (35) IRUNDO: The *irundo*, as Semeryon the Greek says, is one of the silent asps and is of the first class. Its color is like that of the *irundo* [swallow] and its length is around one cubit.[142] It kills within two hours and its bite is followed by coughing [*singultus*], a change of color, stupor, coldness in the limbs, deep sleep, closing of the eyelids accompanied by violent beating of the heart, and a great deal of pain.

(36) ILLICINUS: The *illicinus* is, as Semyryon says, a serpent that lives in oak trees [*in ilicibus*].[143] Whomever it bites suffers a loss of his skin. The bite likewise strips

---

135. Cf. Pliny *HN* 8.35.85; *AAZ*, 94. Jorach is cited by Arnoldus Saxo 2.10.29a. ThC 8.19 cites only Solinus 27.30. "Throwing": probably from Isid. *Orig.* 12.4.29, which quotes Lucan *Phars.* 9.720, *Iaculique volucres,* which can be interpreted either as winged or as flying (like an arrow, as ThC quotes from Psalms).

136. Cf. 25.16(7); Vinc. 20.37; Lucan *Phars.* 9.792, 822.

137. *Ipnapis*: more commonly *hypnale*; see Solinus 27.31, Isid. *Orig.* 12.4.14, Jacob. 89 (188), Neckam 2.112, Barth. 18.9, and Vinc. 20.37. Cf. 25.13(1), with note.

138. Pliny *HN* 29.22.72; *AAZ*, 91–92.

139. This story stems from confusion arising from a misreading or miscopying of Pliny *HN* 8.36.88 on the battle of the mongoose and asp, which concludes with a statement that the mongoose also vanquishes the crocodile. The same story is given in *Physiologus* (Curley, 1979, 53, 88) and the twelfth-century bestiary translated by White (1954, 179), as well as ThC 8.21.

140. Avic. *Can.* 4.6.3.31.

141. Pliny *HN* 28.75.244, cited by ThC, but the problem in question is the itch *scabies*, not a serpent bite. The *boa* is from *bos, bovis* (cow, ox).

142. Avic *Can.* 4.6.3.24.

143. Avic. *Can.* 4.6.3.42.

the skin off whoever touches the bitten one to care for him. Moreover, it has a quite vile odor, deadly to one in the presence of the dead person. There also occur as a result of the bite those things that result from the bite of vipers. This is one of the first class, therefore.

(37) LACERTAM: Some say that a *lacerta* [lizard] is a serpent due to the similarity of its body and tail to those of a serpent.[144] But it does not fit into any of the three classes mentioned above, since when a lizard or a *stellio* [salamander] bites, they leave behind the small teeth from their mouth stuck in the spot of the bite. These teeth are small, slender, and black like black hairs. Therefore, the spot does not stop hurting and itching until the teeth are removed with a saw or knife applied to them to cut them out.[145] Then the pain is quieted, just as Avicenna says. Oil and ash also draw out its teeth if put over them and if the spot afterward is sucked on and placed in warm water.

While, therefore, they may not be in one of the classes of serpents, their natures are nevertheless touched on here, due to their similarity of shape.

Jorach, in the book he compiled on certain animals, says that when the *lacerta* grows old and its eyes grow dim and it starts to go blind, it then faces the sun's rays.[146] It directs its eyes through some opening toward the sun at morning time until its vision returns and it sees perfectly as before. If this philosopher speaks truly in this matter, the clouding of the eyes is surely a result of a coldness restricting the eye fluid which the heat of the sun's light dissolves and thins out, restoring its vision.

Pliny says that *lacertae* go in pairs; if one is captured, the other attacks the captor.[147]

There is another little beast which is in the *lacerta* genus. Its shape is like that of a *lacerta,* but it has a black tail and feeds on forest spiders.

The *lacerta* lays eggs but does not sit on them. Some say that the mother devours her brood except for the most sluggish one and that this one later devours the parents, but this is entirely false.[148] The *lacerta* is said to have a hairy tongue, but it does not. Rather, its teeth are as small as hairs. Its tongue is split like a serpent's. In India there are *lacertae* twenty-four feet long with a shiny color.[149] One of my companions who is worthy of trust told me that in Provence [*Provincia*] and Spain [*Hispanica*] he has seen on occasion *lacertae* the thickness of a person's leg below the knee. They were not very long, were black, and they lived in holes beneath the ground. They leaped, by jumping up from the ground, onto animals and people as they passed by. Once, with one bite, they took off the entire jaw of a person.[150]

---

144. ThC 8.23, quoting Solinus (27.33?). On the many varieties of lizards in ancient sources, cf. *AAZ*, 111–14.

145. Avic. *Can.* 4.7.1.20, 4.6.2.6.

146. Arnoldus Saxo 2.9.28af.

147. Pliny *HN* 8.35.86, concerning the asp. Cf. Vinc. 20.55.

148. ThC 8.23.

149. Pliny *HN* 8.60.141.

150. Probably another tall tale, although it could refer to a monitor lizard (*Varanidae* sp.), not present north of Africa today but that could have ranged into Spain in medieval times. When pursued (as they frequently are for food), they can strike with great strength and rapidity. See Grzimek, 6:323f.

(38) MILIARES: The *miliares* is, as Semyrion says, a serpent which is the color of millet [*milium*], and it therefore took its name from it.[151] Whomever it bites suffers the ill effects of a viper's bite and we shall speak of these below when we speak about the bite of vipers. It is therefore a serpent of the second class.

(39) MARIS SERPENS: The sea serpent has many species.[152] It conforms to land forms because it has no feet, but like eels [*anguillae*] they creep along the bottom and swim when it is necessary, for they do have fins.

They are like land serpents in every way, except in the head, for their head is rough and hard. They are malicious but their poison is less harmful in water than if the same thing were on land.[153] These do not visit the bottom of the sea but rather swim near the shore and around ledges.

(40) NATRIX: The *natrix* is an aquatic serpent that infects the water in a well with its poison.[154] Hence Lucan says, "and the *natrix,* the violator of water."[155]

(41) NADEROS: They say the *naderos* is a serpent of the second class which is found in Germany.[156] It is two or more cubits long and is as thick as a person's arm below the elbow. Its belly tends toward a golden yellow and its back is somewhat green. Its poison spreads from the spot of the bite throughout the entire body unless the poison is cut out very quickly. If its tongue touches a rod or a sword, it is said to infect them, so much so that it changes the color of wood, as well as of an iron implement.

(42) OBTALIUS: The *obtalius,* as Jorach says, is in the asp genus and is of the first class.[157] Whatever animal it bites is weighed down with sleep and dies while sleeping.

(43) PRESTER: The *prester,* as Jorach says, is in the asp genus and of the first order of serpents.[158] It always goes about with its mouth open and smoking. Whatever the *prester* strikes is distended with corpulence and swelling as if it had dropsy.[159] After this swelling there follows putrescence and it thus dies. Hence Lucan says, "The voracious *prester,* stretching out its smoking mouth."[160]

---

151. Avic. *Can.* 4.6.3.44.

152. ThC 8.24. No snakes have fins, however, and Stadler conjectures that this is the needlefish, *Belone vulgaris.* Cf. 24.26(39), *draco maris.*

153. Perhaps based on Ambrose *Hex.* 5.2.6.

154. Cf. Isid. *Orig.* 12.4.25; ThC 8.25; Vinc. 20.36.

155. *Phars.* 9.720. The modern use of the term *natrix* is for a genus of *Columbrinae,* many members of which feed on frogs and toads and hence live around water, including *Natrix natrix,* the European grass snake.

156. ThC 8.26 and Vinc. 20.40. Stadler conjectures that this is *Vipera berus* L., the adder or northern viper. The size, color, and habitat seem to fit well.

157. Cited by Arnoldus Saxo 2.10.28d, *obtobans.* Cf. 25.30(33).

158. Based on Isid. *Orig.* 12.4.6; cf. *AAZ,* 155. See also Neckam 2.117, Barth. 18.9, ThC 8.27 (*pester*), and Vinc. 20.40. Cf. 25.24(26), with note. Jorach is cited by Arnoldus Saxo 2.10.28a.

159. Solinus 27.32; from Gr. *prēstēr,* lit., "bellows," hence the snake that causes inflation.

160. *Phars.* 9.722.

(44) PHAREAS: The *phareas* is a serpent that moves almost totally upright on its tail and on that part of its body near its tail.¹⁶¹ It therefore seems to make a furrow along the paths it travels. Hence Lucan says, "The *phareas,* content to furrow a path with its tail."¹⁶²

(45) RYMATRIX: The *rymatrix,* as Jorach says, is a serpent of the first class that rummages [*rymans*] through food and water, thus infecting them.¹⁶³ If anyone tastes the infected material, he immediately dies.

(46) SALAMANDRA: The salamander is a serpent; that is to say, it somewhat resembles a serpent. What the Greeks call a salamander, we call a *stellio*.¹⁶⁴ The opinions of the ancients are quite different on this animal. Some, namely Pliny and Solinus, say that the salamander is the same as the chameleon, that is to say, "earth lion."¹⁶⁵ It is a quadruped with the shape of a lizard. It lays eggs but has a composite face made up from that of a monkey and a pig. Pliny says that it has straight, long, hind legs that are attached to its belly, a broad, flexible tail that ends in a thin point, slightly hooked claws, a rough body, and skin like that of a crocodile.

They say that it has a kind of wool which does not burn in the fire since fire has no way into its pores.¹⁶⁶ However, I have found by testing it that a sample of this wool brought to us was not animal wool. Some say it is the wool of some kind of plant which I have not tested.¹⁶⁷ I have proven, however, that it is iron wool.¹⁶⁸ For where large amounts of iron are worked, the iron is sometimes split and a fiery vapor flies forth. When that is caught on cloth or by hand, or it sticks on its own to the roof of a building, it is like a dusky wool and is sometimes white. The wool itself and what is made from it does not burn in fire. This is what itinerant peddlers call "salamander wool."

---

161. Isid. *Orig.* 12.4.27. See also ThC 8.28; Vinc. 20.40. This is the asclepian snake (Gr. *pareias*), *Elaphe longissima* (*Coluber longissimus*), which hibernates an unusually long time, from early October until May (Boulenger, 1914, 143–44; Grzimek, 6:399–400).

162. *Phars.* 9.721.

163. Cited by Arnoldus Saxo 3.10.28d.

164. Jacob. 89 (187). See also ThC 8.30. For its use in art, cf. Garcia-Arranz (1990).

165. Pliny *HN* 8.51.120, who says a *chamaeleon* is like a lizard in shape and size, but with longer and straighter legs. Solinus 40.21 repeats Pliny. Jacob. 89 (187) gets the idea correct, as do Neckam 1.21 and Barth. 18.20. ThC is the only source who equates the two names, citing Pliny and Solinus. Cf. Vinc. 20.63.

Gr. *chamai* (adverb), "on the earth"; *leōn,* "lion." Cf. 2.64, where the *stellio maior* is identified as the chameleon.

166. Jacob. 89 (187).

167. Pliny *HN* 19.4.19, on asbestos. Barth. 18.20 attributes it to the salamander, citing Pliny, but apparently misreading or miscopying Pliny *HN* 29.23.76.

168. *De min.* 2.2.1 (Wyckoff, 69–70), where A. begins his discussion of *abeston* (Gr. *asbestos,* "unable to be burned"), saying that *abeston* is the color of iron, pointing out that it has the nature of salamander's down. In *Meteora* 4.3.17 he discusses salamander's down, which has the property of burning without being consumed because its moisture is completely sealed in it. The substance today is called slag wool and is often found at forges (*AAZ,* 164).

36   Many, following Jorach the philosopher, say that this animal lives in the fire.[169] This is false except insofar as Galen, in his book *On Complexions,* says that if it stays a brief time in a small fire, the fire makes no mark on it.[170] If, however, it stays a long time, it is burned up. Jorach says that if it is only a middling fire, it puts it out. This, however, is not because its life is in the fire but because it is a very cold animal, as Aristotle says, and has a very thick skin.[171] The fire is therefore not able to enter its pores. But if the animal remains in the fire a long time, the fire opens the pores bit by bit and burns up the animal. So great is the salamander's coldness that the fire is put out by its opposition if the fire is small and does not overcome the animal's qualities. Now, I have put this to the test in one similar to this one when I placed a spider with a thick skin and a cold humor on a piece of glowing iron and it lay there a long while before it twitched and felt the heat of the burning. I also held a small light up to another, large one, and it extinguished it as if it had been blown out.

The animal's gait is slow, due to its coldness, and its motion is like that of a turtle [*testudo*].[172] Its eyes are very deep, set as it were within deep hollow holes, and they never close. Pliny says that the eyes rotate entirely around.[173] It is a very large animal, but it is hollow. It has a loose skin and it is too thin due to its coldness and black bile [*melancholia*]. They say its liver is in the left side, differing from all other animals. It has no spleen since its black bile is distributed entirely through its body. It differs greatly from other animals with respect to its internal organs. Its gaping mouth is never closed and, moreover, it does not use it to obtain food or drink. Rather, it uses it to suck some humor or other, and nothing is therefore found in its organs. That it is said to live on dew or air is false. It has no blood save for a little around its heart and as a result is very timid.

Some say that it becomes every color put next to it, except white and red, but I do not think this is true. It is, however, true that it is one of the animals that is dormant in winter.

They also say that when this animal is in distress it recovers its health from laurel leaves. It scales trees and kills all the fruit, and all who eat the fruit die.[174] It lays eggs like other lizards and hens. This genus of animal is abundant in Asia.

37   (47) SALPIGA: The *salpiga* is said to be a serpent. It is not seen because of its smallness, but nevertheless its capacity for harm is very great.[175]

---

169. Jorach is cited by Arnoldus Saxo 2.10.28c. Cf. Isid. *Orig.* 12.4.36. Beginning with Ar. *HA* 552b16, which says that the salamander is not destroyed by fire but even puts it out, Pliny *HN* 10.86.188 elaborates, and by medieval times the salamander is said to live in the fire.

170. Cited by Arnoldus Saxo loc. cit.

171. Ar. *HA* 552b16. The phenomenon of escaping alive from and perhaps even quenching a small fire could have arisen from the milky liquid secreted by a frightened salamander.

172. "Turtle": *testudo,* although possibly a snail. Cf. 4.38, with note.

173. Pliny *HN* 8.51.122.

174. Cf. Isid. *Orig.* 12.4.36; Jacob. 89 (187).

175. Isid. *Orig.* 12.4.33, without a reason for it not being seen. ThC 8.31 says it is because it is small and that it is harmful. Cf. Vinc. 20.42.

(48) STELLIO: The *stellio* is a salamander. There is also an animal, a quadruped, called the *stellio* after its color, for its back is speckled with shining dots that look like stars [*stellae*]. It is so much an enemy of the scorpion that when it is seen by the scorpion, the scorpion is terrified and becomes immobilized.[176]

Pliny says that the *stellio* is a serpent whose poison is deadly and the cure for which is the ground flesh of a scorpion. He also says that wine in which a *stellio* has been immersed and killed removes "freckles" from the face. He says as well that the gall bladder of the *stellio,* ground up in water, causes weasels to gather by some hidden impulse.[177] The same man says that the *stellio* is very much the enemy of scorpions and that they prey upon each other out of a natural hatred. Because, however, the *stellio* is larger and stronger, the scorpion so fears it that, just seeing it, the scorpion is covered with a cold sweat.[178] Avicenna places the bile of the *stellio* in the same genus of harmfulness as the bite of the lizards.[179] For the *stellio* and the lizard are almost similar quadruped animals except that the *stellio* is slow and is wider on the back and tail than the lizard. It is also spotted, whereas the lizard is all of one color.

(49) SCAURA: The *scaura* is, they say, a serpent which, when it goes blind in its old age, enters a crack in a wall and, through an opening, directs its gaze at the orb of the sun until it regains its vision.[180] This is just as we have said above concerning the lizard. Either the reason for this is the same for each of them or, perhaps, the *scaura* is the genus of lizard that does this.

(50) SITULA: The *situla* is a serpent that has two species.[181] One is the *dipsas,* about which we have already spoken.[182] The other, however, is beautiful due to a variegation so wondrous that those gazing on it are held by wonder. It is so slow and small and has so fiery a poison that whomever it strikes burns away entirely. This serpent sheds its skin in winter.

(51) SIRENES: They say the *sirenes* are monstrous serpents with a sweet hiss, just like the sea sirens.[183] They are very swift and some of them fly. They have poison so effective that death follows the bite before pain appears.

(52) SERPS: The *serps,* as Jorach says, is a small serpent.[184] If a person steps on it he dies before he feels its poison. Concerning this serpent's poison they say that when

---

176. The description fits the fire salamander, which bears this sort of marks. Isid. *Orig.* 12.4.38. See also Barth. 18.92; ThC 8.32–33; Vinc. 20.65–66. Cf. 26.29(34).

177. Pliny *HN* 29.22.73.

178. Pliny *HN* 29.28.90.

179. Avic. *Can.* 4.6.6.20, 4.6.2.5–6.

180. Jacob. 89 (188); ThC 8.24; Vinc. 20.65. Cf. Isid. *Orig.* 12.4.37, *saura lacertus.* Gr. *sauros* means "lizard."

181. Solinus 27.30, *scytale.* See also Jacob. 89 (189); ThC 8.35; Vinc. 20.43. Cf. Aiken (1947), 221.

182. Isid. *Orig.* 12.4.13, *dipsas* or *situla.*

183. The *Experimentator,* cited by ThC 8.36.

184. Cited by Arnoldus Saxo 2.10.28d. See also Solinus 27.33; ThC 8.38; Isid. *Orig.* 12.4.17, 31; Jacob. 89 (188–89); Vinc. 20.42; Neckam 2.119; Barth. 18.9.

it strikes a person, it consumes his flesh and bones as if he were in the grips of a voracious flame.

There are, moreover, serpents in India with such large bodies, they say, that they devour stags. These swim in the very sea itself. There are others there, they say, that eat white pepper and bear precious stones in their heads. These fight among themselves every year and kill off each other to a great extent.[185]

They say that there are others in the east that have horns like those of rams and that they kill people by swinging at them with the horns.

(53) SPECTAFICUS: The *spectaficus* is, as Jorach says, a serpent at whose bite a person's flesh liquefies like oil and he dies.[186]

(54) STUPEFACIENS: The *stupefaciens* is an asp, as Jorach says.[187] Every animal, when it has seen it, dies from numbness even though this serpent is said to be slower and warmer than all other serpents. This is because it is said to molt its skin in the worst cold of winter.

Jorach, however, does not prove this very well when he says that it stupefies because of the effects of its coldness and gives the same reason for its slowness.

I do not think that it sheds its skin in winter due to its warmth but rather because this is the time for the generation of phlegm. Then the hard outer skin is loosened by the interior slipperiness of the phlegm, just as happens in all the serpents who lie dormant and sleep in winter.

(55) SABRYN: The *sabryn,* as the wise man Semeryon says, is a serpent with a sand-colored body, marked with blue and white spots.[188] Its length is the length of the horned serpent. This is similar in all respects to the one called the *afudius* and about which we spoke above.[189] The head and tail of this one, however, are small.

Five colors of serpents are found in this genus. Most are sand-colored but there are also various spotted ones, white ones, black ones, and red ones. All of these have white scales that are sectioned on the head, as it were, and the serpents make a loud noise due to the dryness of their belly skin. They are all slow movers, have teeth of equal length, and belong to the first class. The bite of each of them opens up all the natural pores such as the ears, eyes, nose, urinary and fecal tracts, and the mouth, and causes blood to flow from them all. They cause pains in the stomach and a slight, watery pus runs from all their bites. The stomach is loosened and the breathing is arrested. One incurs some difficulty in urinating, the voice becomes sharp, and the members grow

---

185. Jacob. 89 (189); Solinus 52.33.

186. Cited by Arnoldus Saxo 2.10.28d. The textual content is the same as Isid. *Orig.* 12.4.17 on *seps tabificus,* which appears as *spectavificus, spectaficus,* and *spectabificus* in various MSS used by Lindsay (1911). Not found in ThC or Vinc.

187. Cited by Arnoldus Saxo 2.10.29a, *comorreis* = *haemorrhois,* but the symptoms are more like the neurotoxic effects of a cobra bite. Cf. 25.13(1), 30(33).

188. Avic. *Can.* 4.6.3.36.

189. *Afudius: asfodius* at 25.15(6). Cf. the symptoms at 25.29(30). Perhaps the very deadly *Echis carinatus,* the saw-scaled or carpet viper. The "eyespots" on its back are large and quite distinct.

lax. There is an onset of forgetfulness, spasms occur, the teeth fall out, and suffering in this way he dies.

(56) SEYSETULUS: The *seysetulus* is a serpent almost like the ones that leap in two directions and which the Greeks call the *amphisibena*, as if it had two heads.[190] This, however, is not true, as we have pointed out above where we have already spoken of the reasons for this error. The bite of this serpent is like the bite of the others since this one is a particular type [*modus*] of the others, although it differs in that it seems so clearly to have two heads. Avicenna says these things about this serpent.

(57) SELFYR: The *selfyr*, as Avicenna says, is the name of a certain serpent of Egypt.[191] It has a wide head, a small neck, a blunt tail, and a rounded belly. On its head there are neither lines nor scales, but on its body there are lines of different colors. When these serpents go forward, they do not do so in a straight line. Rather, serpents of this species contract a great deal. A painful abscess follows their bite. At first the whole body is bruised. Then the whole, bruised body putrefies and the hair falls out. This serpent is, therefore, in the second class and is in the viper genus.

(58) TORTUCA: The tortoise [*tortuca*] is not a serpent but is placed here by us since it has a bit of a serpent's shape and because some people call it a "shielded serpent."[192]

It is, then, an animal that has a shield on its back and one on its stomach which are joined at four corners.[193] Between these the *tortuca*'s four legs come forth and they seem like the feet of the lizard, with five toes and claws. It has a serpent's neck and head and, if it is large, has a shield on its head as well. It has rough skin like a lizard and has a tail that comes out from beneath its two shields like a snake. It hisses in a higher tone than a serpent.[194] It lays eggs like those of a hen, and digging a hole it puts them in the earth until the young one comes out in the sun's heat.[195] There are two types of it in addition to the marine type, for there is a land *tortuca* and an aquatic *tortuca*. The land one has the more temperate flesh. Neither is poisonous but both, when eaten, make one fat. Some say, however, that if a person treads on the kidneys of a dead *tortuca*, he is infected with poison.[196] This animal, like the other egg layers,

---

190. Cf. Avic. *Can.* 4.6.3.45, *scisetati*. *Amphisibena*: cf. 25.14(2), with notes. Many snakes, when threatened, put their heads in the center of the coil and elevate the tail, to mimic the more vital (and protected) head.

191. Avic. *Can.* 4.6.3.51, *selsir*. Schühlein, 1662, notes the resemblance of the wound described at Ael. *DA* 16.40 in his descriptions of *seps*, but there is no other apparent correspondence.

192. Cf. ThC 8.40, *animal duorum pedum* [!] *caput habet ut buffo* ["an animal with two feet and the head of a toad"], making it sound as if he had no idea what the *tortuca* was. Cf. Barth. 18.106, who has the *tortuca fluvialis* ["river turtle"] as *mortifera . . . et venenosa* ["deadly and poisonous"]. At least he gives it four feet and a serpent-like head. "Shielded serpent": cf. German *Schildkrote*, "shield toad." Cf. *AAZ*, 186–88.

193. "Corners": *initiis*; s.v. Baxter and Johnson.

194. Cf. Ar. *HA* 536a8, "a low hiss."

195. Cf. Ar. *HA* 558a4–11, where the tortoise is said to brood over the eggs. A.'s version is correct, not Ar.'s.

196. ThC 8.40, citing Ambrose *Hex.* 5.10.31.

has a bladder.¹⁹⁷ Let these things, then, along with what has been said previously, be enough on the *tortuca.*

(59) TYRUS: They say the *tyrus* is found in the area of Jericho around the desert of the Jordan.¹⁹⁸ It is also in parts of Italy in the Apennine Mountains. Serapion, in his book *On Medicinal Simples,* says that it infects and destroys sound members but protects infected ones from poison.¹⁹⁹ These actions wherein like acts on like are also found, so the same Serapion says, in the *ceraste,* whose tongue, if taken out of the serpent and held close to poison, expels it so that the tongue grows wet as long as it is held on the poison. When it is in the serpent, however, it infects with poison whomever the serpent strikes.

The *tyrus,* which is common in our lands, is a serpent two or three cubits long and is black with two red lines on the sides of its back. It goes forward with a third of its body erect and is very swift. It leaps, up to ten feet, when it contracts its body.

It is a serpent that is hostile to birds, so much so that I have personally seen one scale a nut tree, chew, and devour the chicks of a magpie [*pica*], and fight with the older ones. When it held one of the old ones by the thigh, the serpent grew incautious and received so many blows in the head from the other that it fell dead from the nest and nut tree.²⁰⁰ It also eats the eggs of birds when it finds them. It is said to be hostile to animals as well, but I do not have the same sort of experience of that. This serpent in our lands, then, does not have a poisonous bite.

There is, however, a poisonous genus of *tirus* which is in the first class, even though with its flesh it cures poisoned members or ones infected with such as leprosy. For this reason *tiriaca* is made from its flesh and is effective against poison.²⁰¹ In just this way the gall bladder of a viper used in an eyewash does not act as poison even though, with its bite, the viper pours poison throughout the healthy members so that, from the spot of the bite, it infects and destroys the entire body.²⁰²

In Ethyopia, beneath the sign of Cancer, this serpent first strips off its skin from between its eyes, in the manner of the *draco,* so that it looks almost blind. It then strips it from its head and afterward, within the space of a day and a night, it lays it aside by passing its entire body through a narrow place and thus returns to its youth.²⁰³

The worst one of this genus of serpent is a particular, small genus, one cubit long, whose skin is prickly and hairy. Its bite is worse as well, putrefying the body, spreading from the place of the wound and thus killing.²⁰⁴

---

197. Ar. *HA* 541a10.
198. Jacob. 89 (187); ThC 8.43. See also Vinc. 20.45–47.
199. Arnoldus Saxo 2.10.28c.
200. Cf. 7.43.
201. Cf. 2.119, 7.133, 22.45, 23.84; ThC 8.43; Neckam 2.108; Barth. 18.8, 115; Jacob. 89 (187).
202. "Eyewash": *collirium.* Cf. 22.19, with note; 25.46.
203. Cf. 7.90. Cf. also Ar. *HA* 600b24f.; Pliny *HN* 8.41.99.
204. Perhaps *Echis carinatus,* the saw-scaled or carpet viper, which is about two feet long; cf. 25.15(6).

The *tyrus* of India, which is less than a cubit long, is more harmful than all the other *tiri*. It is of the viper genus since it first produces an egg internally and afterward bears a live young externally.

(60) TYLIACUS: The *tyliacus* is, as they say, a worm serpent belonging to the dragon genus.[205] It is called *tiliacus* after the *tilia* [linden tree] because it is said that it is born in the heart of the linden tree and that it feeds first on the heart [*medulla*] of the tree and then on the wood until the tree falls down.[206] It grows to a length and thickness of great size, and it preys on beasts and humans and devours them. It has the bite of a dragon. When it is captured it is necessary to use one of the sort of catapults that are used in storming castles.

(61) VIPERA: All genuses of vipers are in the second class of serpents.[207] The serpent, however, is narrower above the spot of the diaphragm and is broader below. It has a pointed head and a mouth full of teeth. The male, however, has only two teeth in the front of its jaws, that is, it has them in pairs, with two below and two on top corresponding to these. The female, however, has more, and in this, as Avicenna says, the bite of the male can be distinguished from that of the female.

Now, concerning this serpent [the viper], Jorach says that the female goes mad with lust and seizes the head of the male while the male is spitting.[208] Seizing it, then, in her mouth she severs it and conceives her brood from his spit. The young eat their way out of the mother's body and she, with her viscera torn, dies in birth. Jorach assigns as the cause for this that the exit point of her superfluity is only as big as a pinprick and she is therefore unable to conceive or give birth like other animals. But this is an impossible thing and as far as nature is concerned is absurd. It has also been disproved by us elsewhere.

Pliny, wishing to adapt the lie even more, says that she indeed conceives by decapitating the male, but she conceives many young and, due to the narrowness of her body, she can give birth to only one at a time. Since there are many of them—more than twenty in fact—boredom overcomes the last ones since they are not coming forth into the light and they then eat away at their mother's viscera. This is likewise a lie and impossible. For nature has never given an action for which she has not given the natural power and faculty, because otherwise she would be deficient with respect to what is necessary.

---

205. ThC 8.44, citing *Liber rerum*. Cf. the *Lintwurm* of 2.68.

206. Cf. *De veg*. 6.232.

207. Avic. *Can*. 4.6.3.32. The term viper historically covered a large number of snakes (*AAZ*, 190–91).

208. Cited by Arnoldus Saxo 2.9.28a, but cf. Jacob. 89 (188) and Isid. *Orig*. 12.4.11 (cited by ThC 8.45). The original story is from Pliny *HN* 10.82.169–70 and is also found in Neckam 2.105, *Physiologus* (Curley, 1979, 15–16), and White's twelfth-century bestiary (1954). Classically, it appears in Herod. 3.109 and Ael. *DA* 1.24. It may have originated from Ar. *HA* 558a30, "at times the young viper eats its way out from the inside of the egg" (Thompson trans.) or "eats through the membrane from inside and so gets out" (Peck trans.). The story as found in Herod. 3.109 merely has the little vipers "devouring her womb," with no mention of their eating their way out. See the illustration in George and Yapp (1991), 198.

All genuses, then, of serpents that first produce eggs and afterward bear live young do so because of the dual arrangements of their uterus. Vipers are so called as if they bear by some intrinsic force [*vis pariens*], since all others bear requiring some external assistance.[209] These are called vipers among the ancient Greeks, then.

46 Avicenna speaks of the things attendant upon the bite of this serpent. He says that from a bite caused by two or three teeth there issues forth at first blood and then a very hot, slimy fluid. Sometimes the blood is at first watery and then frothy, while afterward it is like copper rust and is then turned into the substance and color of the poison. The place is violently painful and the pain then spreads to the adjoining members. Afterward a warm, red abscess appears, having pustules and blisters as appear in a burn caused by fire.[210] It is then widened and the entire abscess becomes green near the bite and below the hole. The sick one then is beset by inflammation in his viscera, redness of the body, and rigidity. A cold sweat breaks out and his color is destroyed, tending toward green. There is a swaying of the head due to dizziness, rapid breathing, nausea, and coughing [*singultus*]. He sometimes vomits a choleric humor, encounters difficulty in urinating, and has a headache. There sometimes occurs a flow of blood from the nose, heaviness is felt in the back, a cold sweat breaks out, and there is a violent trembling and fainting. For the most part, he dies in three days, but sometimes lasts up to the seventh day.

Now, some say that the viper is like a human in the front part of his body and tapers off in the rear into a serpent.[211] This is entirely false unless it is to be understood as used fabulously by the poets by way of metaphoric adornment.

Certain other things are said about the viper's kidneys—that they kill when trod upon just as if the person were bitten.[212] It is also said that their skin when cooked with wine is good for the eyes and that their fat removes cloudiness from the eyes. These things are not denied, for the gall bladder is effective against "eye claw" [*oculi ungula*].[213] Some say that the male, out of reverence for the female, vomits up venom and calls her with a pleasant hiss and that she later cuts off his head.[214] This is an absurd and false theory.

---

209. Cf. Isid. *Orig.* 12.4.10. "External assistance": cf. 6.9, *extrinsecus est amminiculans*.

210. Avic. *Can.* 4.6.3.32. "Pustules": *buchor*. See Schühlein, 1657.

211. *Physiologus*, cited by ThC 8.45.

212. Another misreading or miscopying of Ambrose *Hex.* 5.10.31. Cf. *tortuca* at 25.41(58), with note. The error is found in ThC 8.45.

213. See Pliny *HN* 29.38.122. At 2.25, A. identifies the disease as a corneal ailment.

214. Ambrose *Hex.* 5.7.18.

HERE BEGINS THE TWENTY-SIXTH BOOK ON ANIMALS

Which Is About Vermin and Whose First Chapter Is on the Nature of Vermin

In this last book we will treat small animals that are *anaemys,* that is, do not possess blood.[1] We will do so first in general by repeating some things we have already said. Those animals that have no blood have some other humor which takes the place of blood for them. Such creatures are generally cold and consequently two accidental properties occur in these animals. One of these is that their bodies are everywhere cut into sections [*insecta*] and they are composed of rings that have their generation from a center, as it were. For those animals possessing blood are nourished by long canals that extend into the members in line-like fashion. But those animals which do not possess blood take the composition of their bodies from a humor drawn, in a sort of circular fashion, from the center of its body to its surface along the entire length of its body. Thus it is as if they were made up of hard, circular surfaces. They share this trait with the composition of serpents.

The other accidental property is that the spirits released from a temperate moisture of this sort are subtle and clear, without much heat mixing in and disturbing them. For this reason all such animals have refined operations and frequently dwell together peacefully and civilly. These accidental properties are found in them frequently.

Due to the great heat of the surrounding air and to the terrestrial element being forced to the outside, a shell-like hardness arises on these animals in rings on the outside of their bodies. As a result, they sometimes have to molt their skin. This is expressly evident in many of them whose skins are found cast off.

It follows from the viscosity of their internal humor and exterior "terrestrialness" that these animals either lie dormant in winter like the bee and wasp or they die, having been turned back to an egg-like nature. For if a continuous nutrimental humor were to occur in them, then their moisture would grow thickened in winter due to the weakness of their heat. Or else it would be multiplied at the time when phlegm is produced to such an extent that the natural heat would be suffocated. Or both of these may happen. Motion, then, during the winter brings no advantage to these animals.

These animals are also possessed of thin moisture. They cannot eat coarse foods, since with very few exceptions they all suck. They either suck natural moisures, as

---

1. The etymology is sound, the proper Gr. word being *anaimos.* The English "vermin" is used for the Latin *vermis* when it refers to a class of animals (cf. Glossary, s.v. "Vermin"). To translate as "insect" would be misleading. But note also that *vermis,* in its base meaning of "worm," often refers to the larval stage of a vermin's life.

bees do, or the humors of animals, as do flies.[2] A few, however, such as the caterpillar and locust, chew grasses, provided they are fresh. This is also why very few superfluities are excreted from them. And because this moisture is viscous and of a unifying nature like a chain, these animals live a long time without food. For their weak natural heat works a long time before it can dissolve and complete the moisture. Therefore, animals of this kind have a sort of storehouse of life in such a moisture and they can sustain life from it for a long time.

Moreover, since no part of a moisture of this sort is suitable for the nourishment of substance, some of these animals convert it for use as the home's covering and others make hunting implements out of it. Thus, silkworms make silk in which they wrap themselves and spiders make the webs with which they hunt.

Further, nearly all the animals of this sort have either no feet, many feet, or wings in addition to feet. Those without feet include some of those which are most lacking in heat and, removed from a principle of heat, can do almost nothing. These animals are made out of rings whose center encloses the moving heat which is directed along an axis throughout all the rings that make up the body of an animal of this sort. It is just as the moving force of a sphere is along its entire axis, moving from pole to pole through a central point. On this linear axis runs the power of that member which takes the place of a heart, doing so along with the spirit and the heat of the heart because it runs on an axis easier than in any other direction for two reasons. One of these reasons is that it is direct, for the movement of spirit and heat is more difficult if undertaken at an angle. The other reason is that this pathway is everywhere surrounded by circular members which protect and nourish it and these have the shape of a compressed sphere. For the spirit is nourished by its moisture and is protected by its rings, something that could not be if it were directed to a member of one nature and then were directed to legs, muscles, and the like. This, then, is why such animals have this sort of bodily disposition.

Those of them which have many legs have all their legs radiating out from below on their chest since their sluggish heat does not operate as a result of vicarious motion but rather in and of itself. When I say "vicarious," I mean that it operates the way the heart operates by means of the brain and the brain by means of the spinal cord [*nuca*].[3] Therefore, because the heat in these creatures is slight, their instruments of motion must be ordered with respect to the member which they have in place of a heart. An indication of this is that all animals of this sort grow rigid and become immobile at a small fright on account of the coldness of their hearts. Their multitude of feet is caused by their weakness in two respects, for their feet have little power because of their thinness and hardness. The fact that they have wings connected to their backs is because their food is scattered far and wide. They have only membranous wings because of

---

2. "Moistures": lit., "dews."

3. At 1.237, A. explains that the *nucha*, or spinal cord, is the "vicar" [*vicarius*] for the brain because it administers motion, on behalf of the brain, to the lower parts of the body. Cf. 1.356, 521; 12.128, with note.

the viscosity of the moisture out of which they are formed. This is also the reason that many of these animals have several phases of generation before they are perfected. For larvae [*vermes*] are made from eggs and the animals are then perfected from the larvae, as in the case of bees. Others repeat the process and are changed from larvae once more into an egg-like nature. This is what we see happen in the caterpillar, as we have shown by means of individual examples in preceding parts of this study.[4]

(1) APIS: The bee is a familiar insect whose generation occurs from eggs or an egg-like semen [*ovalis semen*] turning into larvae and then from larvae into bees.[5] Just as was determined previously, it is the opinion of Aristotle that the king of the bees is generally the mother of all the bees.[6] For evidence, he has the fact that semen is found only in its home.

Some of the bees are generated larger in size and are called males [*masculinae*].[7] They do not, so people say, have a sting, just as the king does not.[8] Pliny says that these bees are relegated to slave status to care for the honey that has been gathered together and to build the combs from wax.[9] Therefore, he says, when their duty is done, they are sometimes expelled by the others in the hive if there is a defect in the honey. If they do not wish to leave, they are killed.

The third bees are small, have a sting, and gather honey from natural dew and wax from flowers. The dew falls onto the flowers as if onto hollow vessels. The sun forces the thinnest part of the dew to evaporate, but the residue drips and falls to the bottom of the flower. It is gathered in a single place at the pointed end there and the bees thus stick their heads into the bottom of flowers.

The terrestrial element that is left behind when all the moisture has been evaporated is wax. The bee smears this on its thighs, collects the honey in its mouth, and carries it back to the swarm. Now, there is no reason wax should be the natural preserving medium for honey other than that wax is the terrestrial element in these flowers and in dew and that its viscous moisture was honey before it evaporated.

---

4. Cf. 5.31f. and 13.106f.

5. A.'s insistence that there are stages of the bee life cycle is notable. Since mating occurs in the air and egg laying is confined to a single individual deep in the nest, the specifics of bee generation were long guessed at but not known. For other medieval treatments of bees, see ThC 9.2, Vinc. 20.77–111, Neckam 2.163, Barth. 18.11, and Hugh of St. Victor 3.38 (*PL* 177:98). Thomas of Cantimpré (1627) was so taken with bees that he wrote the *Bonum universale de apibus* in which various traits of the bees are explained and then applied to life through the use of what Sarton (1927–48, 2:592) calls "absurd little stories for the edification of the clergy." Thorndike (1929, 2:381) calls the work "a tissue of monkish tales and gossip," but it still retains a certain charm. The bee, because of its industry and communal harmony, was commonly discussed and depicted in the bestiary tradition (White, 1954, 153–59; McCulloch, 1960, 95–96; George and Yapp, 1991, 217–19). For the ancient traditions, see the work of Fraser (1951).

6. Cf. Ar. *HA* 623b3f. and *GA* 759a8f.

7. Male honeybees are the drones and are larger than the female worker bees but smaller than the female queen.

8. Cf. Ar. *HA* 553b12 and Pliny *HN* 11.11.27. Both Ar. *HA* 553b6 and Ambrose *Hex.* 5.21.68 say that the king has a stinger but does not use it.

9. See Pliny *HN* 11.4.11–23.70 for much of what follows.

5   You should know that there are three genuses of bees which produce honey. One bristles almost as if with hair. It is bad, irritable, and is called the rustic one [*rusticum*]. Another is thin and it too is not the best. The third is thickset, well-rounded, columnar, and not long. This one is noble and produces a good deal of honey.[10]

Bees fly about in clear weather but in cloudy weather stay home to work and clean the hive. They gather up all the stinking matter in the hive and discard it during a leisure period when they are not scattered abroad about their business.

Bees stand guard and, when it is time for the swarm to set forth, one calls them forth as if with a trumpet by buzzing two or three times. When the king has emerged, he buzzes the trumpet again, and then, with the king flying off, they all fly off and stop where the king stops. The king they have is the one nature shows to be king for it is clearly larger and has a white spot on its head like a crown. Because it has short wings in respect to the rest of its body it does not fly well and is borne along by the others. When they build their homes the king's is the first built.

They all build hexagonal homes of uniform size, building their dwelling places first and then the storerooms for the honey. They build more homes than there are bees because of the young to come, thus attending to their dwelling places before they are born.

They sometimes carry water as well to moisten the honey and to use in smoothing down their homes. They carry it by getting their wings and mouth wet. Their strength lasts for seven years and sometimes more, even up to ten years. Some spiders, mice, and fumes are so harmful to them that they abandon their hives. They will sometimes also depart on account of a nearby bad smell.

6   Their particular food is honey, although they sometimes eat other sweet liquids. When smeared with oil they die, and they hate all fat since their wings and the rings of their bodies are destroyed by it. When it is time for sleep, one makes a noise and their buzzing then diminishes. Sometimes one makes a circuit and then, returning after this circuit, it buzzes like a trumpet. Then they all grow silent as a group.

It stings with its stinger and, as Avicenna says, it leaves the stinger in the wound.[11] It is the general opinion that the stinger goes ever deeper on its own until it is extracted. The unpleasantness of its sting is the same as that of a wasp. We will pursue this below when the wasp is discussed. A bee, however, that loses its stinger sometimes dies when its viscera are wounded or are cut away when the stinger breaks off. When, however, it loses only the stinger it escapes, although some say that even then it dies.

This, then, with the things that were said in previous places, can be sufficient for bees.[12]

7   (2) ARANEA: There are many genuses of spiders because the *rutela* is a species of spider, as we will show in our chapter on the *rutela*.[13] But we should now consider

---

10. There are so many species of bees of such widely divergent coloring that it does not pay to try to identify them precisely on the basis of the evidence offered here.
11. Avic. *Can.* 4.6.5.18.
12. Cf. 3.99, 4.71, 7.119, 8.141f., 17.51f.
13. Cf. 26.25(33).

that there is a differentiation of spiders in our lands based on hunting, work, size, and color.[14]

We call every vermin a spider which is round, has long legs that number eight, and is given to hunt animals.[15] The difference based on hunting is based on the fact that some hunt flies, some hunt creeping things [*reptilia*] like small lizards and caterpillars, some hunt fish, and some hunt insects by leaping at them.[16]

They are differentiated by work since some spiders weave a web and others do not. One that weaves attends to five points in its weaving. The first is the material to weave with. This arises from a superfluity of the food moisture [*humidum cibale*], and a spider therefore wastes away when it empties itself too greatly. Second, it pays attention to time, for all day long it reweaves a web broken in the morning; at dawn, however, it turns its attention to the movement of the animal it hunts.[17]

Third is the placement of the web. Some hang the web in the air where the animals they hunt have their pathways. Some hang it in the angle between two walls so that animals coming from either direction fall into the net-like web. This happens because small animals readily follow the surface of a wall to its corner. Fourth, it pays attention to the shape of the web and does so in two ways: namely, in the shape of the web as a whole and in producing lines with the threads between which it weaves the web. For some make a totally round web and suspend themselves in the middle of it, whereas others make a triangular one.[18] These make a web-like opening in one part of it in which they stay, waiting. Again, another type makes a reticulated web, stretching from thread to thread, as do almost all the larger spiders that make round nets. Some, however, weave a web-like fine cloth with the workmanship of a weaver. Examples include those which stretch their nets in the angles of walls.[19] Fifth is the manner of its work. Large ones attach the thread that is produced from the anus to their rear foot. Others, however, send the thread out from their mouths and weave with the front feet. These are the ones that make the dense web with the workmanship of a weaver.

---

14. Cf. ThC 9.3, Neckam 2.113, Barth. 18.10, Vinc. 20.112–17, Hugh of St. Victor 3.54 (*PL* 177:104A), and Rabanus Maurus *De universo* 8.4 (*PL* 111:235B–36D).

15. The important distinction based on the number of legs is not mentioned by Neckam or ThC. Vinc. gives the spider six legs (20.112), while Barth. (citing Avic.) gives it "six or eight, always an even number."

16. There are about one hundred species of fishing spiders that belong to the genus Dolomedes, sometimes called raft spiders on the mistaken belief that they build rafts from their silk. The nursery web spiders, the Pisauridae, can run over the surface of water and, like Dolomedes, can dive, remaining submerged to escape danger. These tend to eat flies. Leaping spiders are the Salticidae, with some being able to leap up to twenty times their body length (Bristowe, 1971, 152f., 186f.).

17. This text is almost completely ungrammatical and varies greatly from that in Borgnet, which, making more sense, stops at the semicolon. The whole should be compared to Ar. *HA* 623a21f., which states that a spider repairs its web at dawn or dusk, for this is when a creature is most likely to fall into the web.

18. Those with a round web are orb spiders, of the Argiopidae family.

19. Probably a sheet-web weaver, the Dipluridae.

8   Spiders are differentiated by size and by the shape of that size. Now, they are all made up of three parts, namely, a head, a chest, and the rearmost parts, which are the largest and which lie behind the point of intersection (that is, the waist [*succinctorium*]).[20] But some are large and sort of round, others are oblong like a compressed column, and some are small and thin.

They differ in color as well since some are ash-gray and some are totally green. Others are variously colored throughout their entire body, flecked with bright white and bright blue, both on its legs and on its body.[21] They also have several very white lines on their back, behind the waist.

There are three types of those that do not work in our lands.[22] One has long legs, is round, and runs in the grass. It sucks the moisture out of fruit and dead animals if it finds them.[23] Others sit in holes in the ground, leap onto small animals that pass by, and suck them.[24] These are black. Others run about on the water with uplifted legs and prey on mosquitos [*cinifes*] and tiny little fish and suck on them once they are caught.

Every spider, however, lays eggs and wraps its eggs in a web. Spiders sometimes carry the eggs with them at all times in a little pouch, as does the spider we said lies in wait in a hole in the ground.[25] When it carries its eggs it looks as if it were made up of two globes, one white and one black, for the eggs are very white. Other spiders, however, sometimes keep the eggs in their mouths, sometimes under their chests, and sometimes separate from them. I have personally seen all these ways.

The spider copulates at the end of spring. When it wants to copulate it drags in the male on a thread. The male is much smaller than the female, makes no web, and lives off the female's hunting. The spider lays its eggs in the fall and the small spiders come forth at the beginning of spring. They are so small that a large number of them hang down on one thread when the nest is moved.[26]

Let these things, then, along with what was said previously, suffice concerning spiders.

9   (3) ADLACTA: The *adlacta* is a vermin shaped like a locust, with long hind legs for jumping, but it is not a locust. It lays its eggs in the fall and the young come forth from them in the spring when the plants are fresh and tender.[27]

---

20. Note that "size" is *quantitas* and that the "parts" from which the animal is composed are also referred to as *quantitas*. On the *succinctorium,* variously "girdle," "cincture," or "waist," see notes at 8.143.

21. "Bright": *virulentus*; perhaps, as Scanlan translates (426), "sickly."

22. That is, they do not weave.

23. This sounds somewhat like the Phalangidae, the harvestman or daddy longlegs, which is a spider-like arachnid.

24. Perhaps the Salticidae, jumping spiders, or the Lycosidae, wolf spiders.

25. E.g., the wolf spiders, Lycosidae, which later carry the young spiderlings on their backs.

26. The picture of A. peering into nests, nudging them with his finger to count the number of spiderings dislodged, fits well our overall picture of him as one who first and foremost trusted his own observations.

27. Almost surely a grasshopper, as Stadler and Scanlan (427) attest. The name is strange and on the surface at least is close to the Latin stem *adlact-*, "to suckle." Cf. ThC 9.4 and Vinc. 20.120 (*bruchus*).

(4) BUFO: The toad is a four-footed vermin shaped like a frog, but it has an ash-gray color and a thick skin. Its skin is so thick that it wards off a forceful blow by its thickness and viscosity.[28] Its bite is as poisonous as a serpent of the second class and its dwelling place is beneath the earth. It lives on terrestrial humor and herbs and sometimes eats worms. It, in turn, is eaten by the mole.[29]

They say that the toad each day eats only as much of terrestrial moisture as it can take up in its hand at one time. This is not supported by experiential knowledge, but it is believed by the common people. A toad that is first wet down with water and then has salt put on it reacts by first puffing up, then croaking, and finally being consumed down to its bones.[30]

There is a particular genus of toad called "horned" [*cornutus*] after the sound of its voice.[31] It is a dusky gray color and is yellow on the belly. They sit in stagnant swamps and call one to the other. It is said that they do not call outside of France. But I have experienced that this is false because they call very shrilly throughout all of Germany.[32]

(5) BORAX: The *borax* is a species of toad which has a dusky color and is quite large, so much so that it sometimes attains a size of one cubit. It is found in warm lands and is accustomed to carry its young on its back from time to time.[33] This genus of toad usually bears a stone in its forehead for which it is killed. The stone is found in different colors. White is sometimes found and this is said to be the best. Sometimes an off-black is found and this is good when it has a yellow spot in the middle. A poisoned one is occasionally found and in my time an all-green one was found. Sometimes it is impressed with the picture of a toad.[34]

The *borax* fights with the spider just as the serpent does. For the spider comes from above, hanging on a thread, and pierces the brains of them both. The toad [*bufo*] gets angry, puffs up, and sometimes bursts due to the spider's poison. The *borax* rarely appears by day except in very isolated places and comes out of the earth only during rainy weather. But it sometimes comes out by night and it then travels freely on ways trodden by people. It loathes the flowers of the grapevine and rue and is put to flight by them, as are other poisonous creatures.

---

28. Cf. ThC 9.5, which says that a toad is "ranked among the other *vermes,* but it is a poisonous reptile." Cf. Neckam 2.121; Vinc. 20.57, 121; and Barth. 18.16.

29. A. expands on his observation at 22.143(105).

30. From the general tone of this passage with the stress on experimentation, as well as from Stadler's markings that indicate this is A.'s own addition, it seems likely that this "experiment" on the toad was A.'s own.

31. Taking the etymology from *cornu* as "horn" or "trumpet" rather than "horn" in its normal animal sense.

32. These "horn-toads" are not to be confused with the horned toad of the United States, which is a lizard of the Phrynosoma genus.

33. Stadler identifies this as *Alytes obstetricans,* a toad of southwest Europe most noted for the fact that the male carries the eggs on his back until they hatch. Cf. ThC 5.7, Vinc. 20.56, and Barth. 18.16.

34. The "toadstone" is called the *borax* at *De min.* 2.2.3 (Wyckoff, 1967, 75–76). Cf. Neckam 2.121 and ThC 9.7. The origin may lie in the *batrachites* of Pliny *HN* 37.55.149.

(6) BLACTAE: Cockroaches are insects which take their name from their color, since they dye black the hand of the person touching them.[35] This insect avoids the light given that it goes about only by night. This is the opposite of the fly, which seeks the light because it flies by day and rests in the shadows.

(7) BOMBEX: The silkworm is a vermin which weaves silk and about which we have said many things in preceding books.[36] There are, however, many genuses of silkworm. Pliny says there is one in Asia which makes a nest of mud which is so strong that it cannot be pierced with weapons. In this, he says, it gathers wax and weaves a web like a spider, sending forth threads that are used in girls' ornaments.[37]

(8) BLUCUS: The *blucus* is a black vermin, smaller than the *adlacta*. It is, in fact, the first offspring of the locust, and when it grows larger, it is called an *atthalabus*.[38] When it is full-grown it is called *locusta*, as if it were a long spear [*longa hasta*].[39] For it has two spear-like rear legs with which it raises itself up to jump and fly. The *brucus* [*sic*], however, since it is also immobile, eats away right down to the roots and leaves nothing green behind.

(9) CICENDULA: The *cicendula* [firefly] is one of the insects which shines by night. It has two external, hard wings, like a scarab beetle [*scarabeus*], but it is as small as a fly.[40] When it flies, it shines the most with its wings outspread. Likewise, when its spirit moves, it is made brighter, like a spark fanned by the wind. When it shines by day, its color is dimmed and becomes white. It is found in Italy more than in other areas.

(10) CENOMIA: The name *cenomia* is Latin derived from Greek, for it is the "dog-fly" [*musca canina*] and *cynos* is Greek for dog. It is a gray fly that bites dogs and other animals under their fur and draws up the blood. It especially sits on and does harm to dogs' ears.[41]

(11) CINIFES: *Cinifes* [mosquitos] are flying vermin with long legs.[42] They are tiny flies with beaks on their heads with which they pierce the skin of humans. They arise especially from a humor and are numerous near supplies of water. They follow the moist spirit that animals exhale, especially that of humans. This is the reason that, late

---

35. Cf. 4.56 for the dye. For the insect, cf. Isid. *Orig.* 12.8.7, ThC 9.8, and Vinc. 20.118.

36. Cf. 5.31; 8.138; 13.106; 15.44; 17.53; and 21.47. Other treatments include Ambrose *Hex.* 5.23.77, Isid. *Orig.* 12.5.8, Neckam 2.164, Barth. 18.17, ThC 9.9, and Vinc. 20.119.

37. Pliny *HN* 11.25.75. For the vexed question of just what species were meant by *bombyx* (Gr.) in antiquity, cf. Beavis (1988), 140–48.

38. The more common form of *blucus* is *bruc(h)us*, as at Lev. 11.22; it refers to an early stage in the locust's development. *Atthalabus* would seem to be the Gr. *attelabus*. On the difficulty in being precise with these terms, see *GI,* 135f., and Beavis (1988), 62–64. Cf. ThC 9.10 and Vinc. 20.120.

39. Cf. Isid. *Orig.* 12.8.9.

40. Cf. ThC 9.11, Vinc. 20.126, and Isid. *Orig.* 12.8.6. The insect in question would be a member of the species Lampridae. Being beetles, they have hard outer wings. The form of the name used here derives from the earlier *cicindela*, as in Pliny *HN* 18.66.250f. Cf. Beavis (1988), 175–77, and *GI,* 158–59.

41. For the etymology, cf. Isid. *Orig.* 12.8.12. Cf. ThC 9.12 and Vinc. 20.129. Stadler (1613) identifies this as *Stomoxys calcitrans,* a bloodsucking fly often called the stable fly.

42. Cf. notes at 1.96 for identification. Cf. Isid. *Orig.* 12.8.14, ThC 9.13, and Vinc. 20.159 (*scinifes*).

in the day, they fly in the air directly over the heads of those that are sweating. They are so troublesome, especially during the summer and late in the day around moist places, that sensitive people drape nets called canopies [*canopea*] around their beds.

(12) CULEX: The gnat takes its name from the very long, sharp stinger [*aculeum*] which it has in its mouth. This creature drills through the skin of animals and causes the blood to rise. It sucks acidic things more readily than sweet. It so loves the sun that it even burns itself up in its heat. This is why it bites the most when the sun is glowing hot at midday.[43]

(13) CANTARIDES: Blister beetles are green vermin with a shine on them the color of gold. They are born from the humor on leaves at the highest point of the branches of the ash and alder trees. They grow strong by eating away the leaves like caterpillars. They fly by day and by night they are grouped to form a single ball. These insects are gathered by physicians in August and are immersed in vinegar and saved for various medicinal purposes. When bound onto a person the insects cause pustules to form which in turn create useful cauterizations since the insects are strong drawing agents.[44]

(14) CRABRONES: *Crabrones* [hornets], as they are called by Pliny, are very large, long, yellow wasps with the cruelest of stings. So bad are these wasps that nine of them can kill a boy or a young horse.[45] These hornets make a useless honey in the hollows of trees and in the ground. They also make a bark-like wax that is terrestrial, dry, and not good for anything.[46] The shepherds say they have eyes behind their waist, and I have determined personally [*expertus sum*] that this is true. This has been determined in previous places and is the reason it is called the "blind bee" [*apis caeca*] by some.[47]

(15) CIMEX: The bedbug is a wide insect that gathers its strength in the cracks in walls near the beds of humans, whom it then bites. It is called a *cimex* after the herb called *cimex*, since it has its stench. It is commonly called the "wall louse" [*pediculus parietis*], however.[48]

---

43. *Culex* could, in antiquity, refer variously to mosquitos, gnats, and midges (Beavis, 1988, 229f.). Cf. Isid. *Orig.* 12.8.13, Barth. 12.12, Vinc. 20.127, and ThC 9.14.

44. This is *Lytta vesicatoria*, the blister beetle commonly called the Spanish fly. Scanlan (430–31) has an interesting note on its medicinal-aphrodisiacal properties. In antiquity there were many uses for the beetle's blistering properties, but its avowed efficacy as an aphrodisiac is quite a late development (*GI*, 92–93, ill. 20). Cf. Rabanus Maurus *De universo* 8.4 (*PL* 111:235C), ThC 9.15, and Vinc. 20.122–23.

45. On the hornet, *Vespa crabro*, cf. ThC 9.16, Vinc. 20.157, and Isid. *Orig.* 12.8.4 (*scabrones*). Pliny *HN* 11.24.71–75 relates that twenty-seven hornets (*ter novenae*, "three times nine") can kill a person. Cf. *GI*, 79, and Beavis (1988), 187f. Cf. 1.96, with notes.

46. "Bark-like wax": *ceras corticales*, a confusion of Pliny *HN* 11.24.71, *cerae autem e cortice*.

47. Cf. 4.65; 8.143.

48. *Cimex lectularius*, the bedbug. Cf. Peter Damian *De bono religiosi status* 25 (*PL* 145:785A), Rabanus Maurus *De universo* 8.4 (*PL* 111:236A), Hugh of St. Victor 3.54 (*PL* 177:105A), ThC 9.17, and Vinc. 20.126. For the etymology, cf. Isid. *Orig.* 12.5.17. The herb *cimex* does not appear in *De veg.* and Isid. only refers to a "certain herb." But Diefenbach (1857, 119) lists a *Cymex sambuci* as, apparently, a type of elder. Scanlan (431) suggests that the plant is a kind of bugbane, *Cimicifuga racemosa*. A.'s "wall louse" perfectly translates the MHG *Wantlus* (Lexer, 3:684), an instance of A. relying on the vernacular Sanders missed in his research. For A.'s entry on lice, see 26.22(31).

(16) CICADA: The *cicada* is an insect which we call a *grillius* [cricket], fitting the name to an imitation of its voice.⁴⁹ There are two types of them. One lives in warm places in the cracks in walls and sings at dusk and at night. The other is one that calls in the plants and in trees. Both are musical and if their head is cut off, both the head and the body live a long time. Since it is the spirit enclosed in its breast that forms its song, it thus sings more clearly at midday when the air is calmer. My companions and I have determined for ourselves that on occasion, if the head is cut off, it sings for a long time, making the sound in its chest as it did before.

Some call the *scabro*, which bears deer-like horns and flies in the forest, a *cicada*, but it is not well that they do so, since it is a *scarabeus* [beetle] and not a *cicada*.⁵⁰

(17) ERUCA: The caterpillar is a long insect, variously colored, having many short feet.⁵¹ This insect, like silkworms and spiders, at first spreads webs by wrapping them on the treetops. It deposits a great number of eggs there and from these the small caterpillars come forth at the beginning of spring. These enter houses in August, hang from the cracks in walls, draw a hard skin around themselves once they have shed the hairy skin they had, and live a life like that of a sleeping and senseless animal. Out of this almost egg-like nature, they take on another form, that of a winged creature. Then, in the fall, they mate during flight while clinging to each other for long periods of time. The female takes the male's tail into her bifurcated womb. The female then conceives so many eggs that it is as if she had been totally converted into them and she is therefore found dead, lying like an empty skin next to the eggs. She lays her eggs clinging to trees and plants because she first eats their leaves. Afterward, however, when the creature flies, it sucks on flowers using its long mouthparts, which it sometimes curves inward and at other times puts straight out.⁵²

(18) ENGULAS: The *engulas* [tick] is a vermin which is so named because it dips its entire head and throat [*gula*] into blood, which it always sucks. It draws out so much blood that it bursts. It clings to the lips of wolves and dogs and is called the "forest louse" [*pediculus silvae*] by some, but is commonly called a *theca*.⁵³

---

49. Cf. Isid. *Orig.* 12.3.8, Ambrose *Hex.* 5.22.76, ThC 9.18, Vinc. 20.124, Neckam 1.72 (quoting, with Isid. *Orig.* 12.8.10, the odd fact that cuckoo saliva gives rise to *cicadae*), and Hildegard of Bingen *Physica* 6.65 (*PL* 197:1310B). Cicada could also indicate a true cicada, however (*GI*, 113; Beavis, 1988, 91f.).

50. Here *scabro* does not indicate a hornet (as in Scanlan) but a beetle. Cf. notes at 1.96. Stadler (1585) identifies this as *Lucanus cervus*, the European stag beetle.

51. Stadler suggests that the larva of several butterflies, including the cabbage-eating *Pieris brassicae* larva, are the caterpillars in question. But A.'s usage throughout this work would tend to indicate he thought of *eruca* as a generic term for all caterpillars. Cf. Barth. 18.45, ThC 9.19, Vinc. 20.130, Hugh of St. Victor 3.54 (*PL* 177:105A), and Rabanus Maurus *De universo* 8.4 (*PL* 111:236B–D).

52. "Long mouthparts": lit., "trunk," *promuscida*, as A. uses regularly to refer to an elephant's trunk.

53. The behavior marks this insect clearly as *Ixodes ricinus*, a tick. The passage is based on Pliny *HN* 11.40.116, which does not give it a name. For the MHG *theca*, see Sanders (446). Cf. ThC 9.20 and Vinc. 20.130.

(19) FORMICA: The ant is a very small vermin which has as a trait particular to itself that it grows both in size and strength with increasing age.[54] It cares well for itself since it does not make food like the bee but rather collects and stores it. It collects and stores dried seeds, and if it seeks seeds too large for its strength, it cuts them up. They keep to their pathways in orderly fashion lest the ones going out should hinder the ones coming in by getting in their way. They dry out wet seeds so that they will not rot. They foretell the weather since, before a storm, they gather together in their homes. They are said to bear the dead out for burial. They so hate sulfur and wild marjoram that if these two are crushed and sprinkled over their homes they leave them.

With their bites they spread a poisonous humor that raises pustules. In old age some of them begin to fly. The ant sucks fruit and the bodies of animals it finds and takes nourishment from this. In reproducing, it first lays eggs, which break forth into white worms enveloped in pannicular-membranes. The ants are born from these once the worms are exposed to the sun on the surface. The ant is quiet in the winter, sustained by the food it provided for itself in the summer.

(20) FORMICALEON: The ant-lion is called "lion of the ants" [*leo formicarum*] and also by another name, *murmycaleon*.[55] Now, this animal does not start out as an ant, as some say. For I have seen myself and have pointed out to my companions that this animal has the approximate shape of an *engulas* [tick] and that it hides in the sand by digging a hemispherical hole in the sand at the bottom of which is the ant-lion's mouth. When ants cross it in pursuit of some prize, it seizes and devours them. We have watched this many times. It is said to steal food from the ants in winter since it gathered up nothing for itself in summer.

Nightingales also gather up ants and their eggs very greedily and they get stronger as a result of them when they are ill.[56]

(21) FORMICAE INDIAE: If we are to believe the things written in the letter of Alexander about the wonders of India, then there are ants in India as big as dogs or wolves, possessing four legs and hooked claws. They guard mountains of gold

---

54. The ant's frugal habits and ordered social life made it a very popular animal in the bestiary tradition. Cf. *Physiologus* (Curley, 1979, 20–22; Carmody, 1939, 22–25), White (1934), 96–99, Ambrose *Hex.* 6.4.16, Isid. *Orig.* 12.3.9, Jacob. 88 (184), Barth. 18.51, Rabanus Maurus *De universo* 8.2 (*PL* 111:228A), Hildegard of Bingen *Physica* 7.43 (*PL* 197:1336D), Hugh of St. Victor 2.29 (*PL* 177:75B), ThC 9.21, and Vinc. 20.131–34. Cf. George and Yapp (1991), 214–15, with MS ill. Much of what follows is a digest of Pliny *HN* 11.35.108–36.110.

55. The term "ant-lion" today refers to the larvae of insects in the family Myrmeleontidae. These larvae lie in wait in holes in the sand and devour passing ants with their formidable jaws. Cf. Isid. *Orig.* 12.3.10, Barth. 18.52, ThC 9.22, and Vinc. 20.135. The ant-lion of the *Physiologus* (Curley, 1979, 49) is a mythical, man-eating creature or, perhaps, a creature such as a honey badger (George and Yapp, 1991, 64, 214). For fine studies of this "ant-lion," see Druce (1923) and Gerhardt (1965), who notes approvingly that A. accurately describes its nature.

56. Cf. 23.138(100). The twelfth-century bestiary of White (1954, 47) follows Ambrose *Hex.* 6.4.26 in saying that bears eat ants to recover their health.

and tear to pieces people coming near there. But this has not been proven by actual experience.⁵⁷

(22) LIMAX: The *limax* [snail] is a slow vermin called the *testudo*. It is so called from the mud [*limus*] in which it is produced and nourished.⁵⁸ It lies dormant in winter and comes forth in spring. It has four horns, two short and two long, which it extends when it goes forward and which it retracts when it gathers itself in. The humor it has instead of blood effectively blocks the appearance of hair.

(23) LOCUSTA: The locust is so called from *longa hasta* [long spear] and not from *loco stans* [standing in place], as some say.⁵⁹ It is a vermin which comes forth and travels in a swarm from land to land for food. It has reached even the most distant regions like a cloud pushed along by the wind. It lays its eggs in the fall and these, in springtime, produce the *brucus* that devours everything. It has a head with the shape of a horse's head, and in front of its mouth there are two hard appendages which take the place of both lips and teeth. It is hard in front of its waist and soft behind it and has sort of thread-like horns [*cornua filaria*] instead of a tail. It has long legs in the back for jumping, as well as four others on each side for walking. It has four membranous, movable wings for flying, but a single intestine full of filth which is the juice of plants.⁶⁰

They sometimes move together in such numbers that they destroy all the fruits of the earth. It is held, therefore, as a law in many lands that at such a time the people are to go out into the fields and exterminate them. In the morning they sit unmoving, paralyzed by the night's chill. When the sun has arisen, however, they stretch their wings and all go forth in crowds with no king and no law to watch over their tyrannical government, for it is said that they sometimes devour one another.⁶¹

They say that the Parthians use the insect as food but perhaps it is not a locust of this genus.

(24) LANIFICUS: The *lanificus* [wool maker] is a vermin which is also called the *bombex* [silkworm] and which makes silk. They have the reproductive method of a caterpillar. At first, when it is a larva [*vermis*], it feeds on the leaves of the mulberry and makes fine silk. When, however, it takes lettuce leaves as its food it does not produce so fine a silk.⁶² This larva is white and large and elongated, with numerous feet. It has two

---

57. Scanlan (433) follows Stadler in referencing this animal to Pliny *HN* 11.36.111 and Solinus 30.23. For the letter of Alexander, see 22.123(80), with note. The story is actually as old as Herodotus 3.103–5, where an exciting tale is related of how the Indians actually steal the gold from these savage insects. How and Wells (1928, 1:289) gather up most of the threads of the story, and see also Kim (1997). Cf. ThC 9.23.

58. Isid. *Orig.* 12.5.7, ThC 9.24, Barth. 18.68, Vinc. 20.138. A. often uses *limax* and *testudo* interchangeably for "snail" even though the latter also means "tortoise."

59. Isid. *Orig.* 12.8.9 is the source of the first etymology, also found in Barth. 12.24 and Vinc. 20.139–43. Cf. Pliny *HN* 11.35.101–7.

60. Taking *qui* [masc.; "which"] with *sordes* ["filth"], despite the fact that its gender is fem.

61. The locusts are thus tacitly compared with the "higher" social insects such as ants and bees. Again A. has mangled the grammar, making a sing. participle agree with a pl. verb.

62. This reference to lettuce is A.'s comment and may reflect contemporary attempts to cultivate the silkworm.

soft white horns on its neck and it forms yellow and white silk around itself. All other colors are produced in the silk by craft. Wrapped up, then, within its sack, around which is the silk, it turns toward an egg-like nature and bursts forth into a winged flying creature, weak and much smaller than it had been as a larva. After a copulation that lasts three days, all the males die. The female, however, lays a great number of eggs, placing them on a white cloth. After this she dies herself. This cloth, however, is put aside for the whole winter in a place where the cold cannot freeze the eggs. Then, in the spring it is set out beneath a clear sun and in the sun's heat, at which time larvae are born from the eggs. They again make silk from their food, the leaves of the mulberry, as did their parents. They extract the humor out of which they make the silk from their mouths.[63]

(25) MULTIPES: The *multipes* is an insect called a *centipes* by some. It is not that it has this many feet, but it is so called from hyperbole since in reality it has only forty-four feet, twenty-two on each side.[64] When it is small, this animal belongs to the genus of animals that enter ears.[65] Sometimes it grows to a palm's length, especially in warm climates. It walks forward, but when it is turned around it also walks backward. It has but little poison or pain in its bite, but when it bites the lip it causes it to swell up greatly.[66] It bites people while they are sleeping. When cut into parts, it lives a long time.[67]

(26) MUSCA: The fly is a familiar vermin with two wings and eight feet which is born out of the putrescence of animals and of their dung.[68] It is therefore numerous on animals. It sucks the humor and has two genuses. One has a pointed beak which pierces a person's skin and draws out the blood. This one is most usually found in fields near water and in forests. The other, found in dwellings, is almost the same shape as it but does not have a pointed beak. It sucks humors that it finds, being unable to penetrate the skin to extract humor. So this one sucks fruits and food and it is said that it taints fresh meat by sucking at it so that not long thereafter the meat begins to swarm with larvae. Both of these sharpen their wings with their hind feet and wash their heads and beaks with the front ones.

Large flies differ in color. Some are totally black, very dry, with a rough body. These emit a black superfluity that they frequently deposit on white cloths and walls.

---

63. Cf. *Bombex* 26.11(7), ThC 9.26, and Vinc. 20.138. Silk may already have been produced for two hundred years in Europe by the time A. wrote and Frederick II was actively engaged in its production (Scott, 1993, 150–51; Kantorowicz, 1931, 282–83).

64. Based on the number of legs, this would be a member of the order Scolopendromorpha. Cf. Rabanus Maurus *De universo* 8.4 (*PL* 111:235C), Hugh of St. Victor 3.54 (*PL* 177:104A), ThC 9.27, and Vinc. 20.145.

65. Apparently confusing it with the earwig.

66. Avic. *Can.* 4.6.5.21.

67. This is apparently A.'s addition, perhaps based on his investigations.

68. Flies have six feet. Cf. Ar. *Part. An.* 683a27f., which describes the three pairs. Cf. Rabanus Maurus *De universo* 8.7 (*PL* 111:258A), Hildegard of Bingen *Physica* 6.64 (*PL* 197:1309D), ThC 9.28, and Vinc. 20.147–48.

This is easily cleaned off not during the year in which it was done but rather in the following year when it has dried out.

The other fly is also large but has white spots on its rear section. The superfluity of this one is white and it deposits it someplace with a black surface. I have seen one fly a long distance to a black cloth, throw off its white superfluity, and suddenly fly back to whence it came.

A fly eaten whole does no harm, but if broken up between the teeth and eaten thus it arouses great vomiting, especially in boys. When broken up and bound over a wound infected with poison, it draws the poison to itself.

Pliny also speaks of the fly called the *pyrella*.[69] It is large and abundant in Cyprus. It stays in ovens and lives on the fire and when it flies away from the spot of the fire it dies suddenly.

(27) OPIMACUS: They say the *opimacus* is a poisonous vermin which fights with the serpent and defeats it not with strength but by boldness of spirit and cleverness.[70] It accomplishes this by affixing itself directly beneath the serpent's head. It is small and cannot, therefore, be plucked off, and it thus kills the serpent.

(28) PAPILIONES: Butterflies are multicolored flying vermin. Some have purple wings, some white, and others hyacinth-blue, while still others possess a sort of reddish color. These are the ones that copulate in autumn after which the male dies. The female then lays the eggs and she herself dies in a similar fashion. After the winter, other flying ones are produced from these eggs. They have very long mouthparts [*promuscidae*], which are bent in toward themselves and which suck up the dew from flowers upon which they live.[71]

(29) PHALANGYAE: *Phalangyae* are members of the genus of small spiders and have a poisonous bite. They sometimes enter the noses of humans and cattle. When they enter the nose of a stag, Pliny says the stag is in trouble unless it quickly has crabs to eat.[72]

It lives in hollows and lays its eggs there. Symerion the philosopher says that their bite and that of certain others cause great windiness in the stomach, accompanied by the person's hair standing on end, chills in the extremities, and erection of the penis (or widening of the lips of the womb if it strikes a woman).[73]

(30) PULICES: Fleas are born from dust that has been warmed and moistened, especially if animal heat and the spirit exhaled from animals' bodies are mixed in.[74]

---

69. Pliny *HN* 11.42.119. Cf. ThC 9.29 and Vinc. 20.153.

70. The original form of the name, *ophiomachos,* indeed means "snake fighter" in Gr., and it would seem it was thought of in antiquity as a sort of locust or grasshopper (Beavis, 1988, 68–69). Cf. Lev. 11.22, ThC 9.30, and Vinc. 20.149.

71. Cf. ThC 9.31 and Vinc. 20.150.

72. The *phalangium* of Pliny *HN* 11.28.79 is a small, jumping spider, perhaps a wolf spider, though Beavis (1988, 44–56) offers several alternatives. Cf. also Pliny *HN* 8.41.97, ThC 9.32, and Vinc. 20.167.

73. Avic. *Can.* 4.6.5.11.

74. In the absence of sanitary practices and insecticides, the flea was an intimate part of everyday life in the Middle Ages. Cf. Isid. *Orig.* 12.5.15, ThC 9.33, Vinc. 20.152, and Barth. 18.87.

It is a small, round, black animal with a very sharp mouthpart [*promuscida*] with which it bores through the skins of animals and draws blood out through the hole. It draws other things to the surface as well so that the place it bit becomes red. It has "spears" to the rear for jumping and six other feet for walking.[75] It has a swift gait in proportion to its body and a pointed head. It draws in so much blood that its rear end continuously emits it, blackened and dried up. Its eggs, like those of lice, are nits, and it is sometimes found full of them. When they gather, a large one is always found with a small one and the small one is the male, the large one the female.

The ones that are born at the beginning of spring, in March or April, disappear in May, for May has no fleas or at least very few. Those that arise after May continue on into winter and are most vicious in August.

There is a particular genus of flea called the earth flea [*terrae pulex*]. This one eats plants when they first sprout from the seeds.[76]

Fleas are put to flight when, according to Avicenna, the house is sprinkled with a solution of *coloquintida*.[77] They then leap down and flee, and the reaction is the same to a boiled potion of blackberries. Some say that when goat's blood is poured in a hole in the room, all the fleas gather together there and die. They likewise gather at a log smeared with the fat of a hedgehog. They flee from the smell of cabbage and oleander leaves and this one is therefore called the flea-herb [*herba pulicum*].

(31) PEDICULUS: The louse is a vermin which is generated from the putrescence at the edge of a person's pores or which is amassed from it as it is warmed by the person's heat in the folds of his clothing or by that of an animal.[78] They are generated in a similar fashion in other animals, especially in birds of prey. The proof of this is that many ravenous boys and other ravenous people have lice when they eat fruit, especially figs. This is because the coarseness of their chyme generates many lice.

It is called *pediculus* from the many feet [*pedes*] it has, namely, six.[79] They take their coloring from the nature of the humor out of whose corruption they are generated. The lice in birds, called vulture lice [*vulturum pediculi*], are long, slender, dark, and have many feet. Those on people and sheep, however, are broad and thick.

---

75. That is, it has a leg configuration much like that of the grasshopper or locust. Cf. 26.17(23).

76. Perhaps *Haltica nemorum*, the flea beetle or fleahopper.

77. Cf. Avic. *Can.* 4.6.3.11. The plant is *Cucumis colocynthis* (*De veg.* 6.81f.), a bitter gourd from which a powerful cathartic was prepared. Pliny *HN* 20.8.14–17 seems to believe it is a universal panacea but does not mention fleas.

78. The text is in some confusion here and the sense depends on whether one adopts Stadler's *quae* or Borgnet's *qui*. Comparison with the text of ThC 9.34 is of little help because it varies significantly in content from A.'s text.

The intense interest shown in the origin of lice shows their prevalence given the level of medieval hygiene. Cf. Vinc. 20.151, Barth. 18.86, and Hugh of St. Victor 3.54 (*PL* 177:104D). One of the *Prose Salernitan Questions* (Lawn, 1979, 175) even asks why lice stream from a dead body (the lice leave as the body cools, seeking their next host).

79. Isid. *Orig.* 12.5.14.

There is a louse which Galen calls the vulture louse and which the Greeks call *memluket,* and this one is born in the groin of a human, in his hair, or in his armpits.[80] Its bite is only visible when it has grown greatly, and the bite is poisonous. For its bite sometimes causes blood to burst out of the anal vein, from the nose, from the stomach in vomiting, from the chest and lungs, and from the roots of the teeth. The swelling sometimes grows so large that it does not react to medicine. It is better to wash the bite with lettuce juice. Ground stavesacre [*stafisagria*] and mercury, mixed with oil, butter, or old lard, smeared on a swatch of cloth and carried among one's clothing kills lice efficiently, as many have found by experience.[81] If powdered mercury mixed with powdered lead is sprinkled over coals and then clothing is held over this or a person standing nearby takes the smoke into his clothing, it kills the lice.

(32) RANA: The frog is a four-footed vermin with the shape of a toad, but it has no poison. It has short front legs but long rear legs with long toes and a membrane between the toes for swimming. It has a tongue that sticks to its palate, and its call is therefore "coax," as it resonates from its throat to its mouth.[82] Because its vocal spirit does not emerge directly since the tongue is in the way, it creates two inflated vesicles at the side of the mouth. When it calls, it holds its lower lip on the surface of the water and its upper one out of the water. The only one who calls is the male calling for the female. It is said that it has its lips so tightly pressed together in August that they cannot be opened even with a tool. When it is injured out of the water, it produces a thin call like that of a mouse when it is frightened.

It copulates in spring and as a result of this copulation lays many eggs in the water in the spring of the next year. The female frog lies hidden in the midst of these. When the young come forth from the eggs, they have large heads and their bellies are right next to the heads. To the rear they have a tail with fins for swimming and this tail falls off after May, whereupon four feet are formed on the creature. During the winter it lies dormant out of water in warm crevices and sometimes in warm waters underground. In spring it goes out to the water. On occasion, when it has begun to be seized by the cold of autumn, it enters peoples' homes and sometimes creeps into a fold in one's garment over the groin or belly. They say that if the tongue of the aquatic frog

---

80. Stadler (ad loc.) points out that Avic. *Can.* 4.6.5.13 is the source for this Galenic reference and (1637) suggests *Phthirius pubis,* the crab louse, as an identification. Although we differentiate lice by their preferred spot of infestation (e.g., *Pediculus capitis, P. vestimenti*), those of earlier days tended more to identify them by the host. The source of *memluket* is not certain.

81. The seeds of *Delphinium staphisagria,* stavesacre, contain delphinine, still used sometimes to treat head lice. Cf. *De veg.* 6.436.

82. The use of the sound "coax" to describe a frog's croak is at least as old as Aristophanes's play *The Frogs.* Its signification was of interest to medieval grammarians and philosophers. Cf. Priscian *Institutionum Grammaticarum* 1.1 (1855–59, 1:6). For other descriptions of the frog, cf. Isid. *Orig.* 12.6.58–59, White (1954), 217, *Physiologus* (Curley, 1979, 60–61), ThC 9.35, Barth. 18.89, Vinc. 20.59–62, and Hugh of St. Victor 3.55 (*PL* 177:111A). The frog was little treated overall in the bestiaries (George and Yapp, 1991, 201; McCulloch, 1960, 120).

which calls is placed over the head of one asleep, that person will talk in his sleep and reveal secret things.[83]

There is also a frog which is called the *rubeta* or *rubetum* because it sits frequently in a *rubetum* [thornbush] or in patches of reeds.[84] This one's life is spent equally in the water and on land and many have spoken about this marvel. It has poison but only a small amount. When this genus is placed on a hook, it draws and attracts sea murexes [*purpura*] to itself. A little bone from its right side, when placed in a boiling pot of water, causes the water to stop boiling and it will not boil again unless the bone is taken out. Its ash, like that of a sea hare [*lepus marinus*], is a medication against their poison.[85]

There is also a small frog which lives in patches of reeds and which, if accidently eaten by a cow, swells the cow's stomach with such corpulence that it seems to burst.[86]

There is also a green frog which climbs trees and which predicts storms that are coming by its singing. It is silent at any other time. It is said that it makes a dog mute if it is placed in its mouth.[87]

(33) RUTELA: The *rutela,* as Galen, Avicenna, and Semeryon say, is an animal which is similar to the spider that hunts flies and it is therefore said to be in the genus of spiders. It is not a serpent, as some incorrectly consider it to be.[88]

According to Avicenna there are six types of this animal in its genus. The first has a rounded shape and the color of a grape, that is, tending toward black. It is called *albarbasyon* in Greek.[89]

---

83. In the *Book of Secrets* (Best and Brightman, 1973, 99) this is specified as a way to get a woman to tell the truth about her activities. The ultimate source is Pliny *HN* 32.18.49, which also specifies a woman.

84. Cf. Isid. *Orig.* 12.6.58 and Pliny *HN* 32.18.48–52 for what is basically a handbook of magical uses for frogs, much of which appears in what follows. *Rubeta* is often identified as a bramble-toad.

85. The Latin is vague here due, as often, to A.'s problem with pronouns. But it seems clear that A. intended to say that the ashes of the *rubeta* and of the sea hare, described at 24.39(72), both work against the poison of the *rubeta*. The whole thing is a confused version of Pliny *HN* 32.19.54, on which see Aiken (1947), 221–22.

86. This animal is given a separate entry by ThC 9.37. Cf. Pliny *HN* 32.24.75. Yet one is also put in mind of the Aesopian fable where the frog bursts itself trying to be as large as a cow, as in Horace *Satires* 2.3.314–20 (Ziolkowski, 1993, 27). Also, the cow's symptoms are remarkably like those attributed to the ancient *bouprestis,* generally thought to be a beetle (Gow and Scholfield, 1953, 195–96). Cf. 26.30(37).

87. ThC 9.38 lists this separately as the *corriens.* Cf. Isid. *Orig.* 12.6.59. Stadler (1641) suggests *Hyla arborea,* a common European tree frog that does suit many points of A.'s description (Grzimek, 5:435–36). The toxic reaction to the dog's tongue, however, also puts one in mind of certain toads.

88. The *rutela* is not related to the modern Rutelidae, a family of vegetable-eating beetles. The relevant passage is Avic. *Can.* 4.6.5.7–8, where the Arabic form *rutailā'* is given. On this form and those names that follow, see Schühlein (1660), who does his best to elucidate each. But also read what follows against Pliny *HN* 29.28.84f. and Nicander *Theriaka* 715f. Gow and Scholfield (1953, 184f.) offer excellent notes and identifications. It is quite likely that Pliny used Nicander and that Avic. borrowed from either or both. Cf. Vinc. 20.154. ThC 8.29 includes the *rutela* among his serpents.

89. Schühlein (1660) reads the Arabic *rauġion,* which is senseless. He suggests that it is an error for the Gr. *phalangion,* a venomous spider (Beavis, 1988, 44–56). Pliny *HN* 29.27.86 lists a grape-like spider called the *rhox,* perhaps the *Latrodectus mactans* (Beavis, 1988, 47, 55).

The second, called the *ancos*, is a bit wider in the body but still has a rounded shape. Around its neck there are some scales and over its mouth there are three protruding masses that are scattered and soft.⁹⁰

The third type is called the *murkyon*. It somewhat resembles the large bee and its color tends toward ashy-gray and its body is covered with small protruding red masses, especially on its back.⁹¹

The fourth type is called the *saguncloflon* because its whole body and head are hard. This one has wings like the large ants.⁹²

The fifth type is called the *suctyon* because it has a long, thin body on which there are spots, especially near the head and neck.⁹³

The sixth type is called the *furbul* since it has a long body of green color. It has a stinger under its neck, as it were.⁹⁴

These six types are accepted by the most ancient of the Greeks, like Semeryon, and certain others.

26   Galen, however, says there are twelve types of *rutela* and according to him the worst of these is the Egyptian.⁹⁵ Another is red, rather like the round spider. Another is white, with a rounded belly and a small mouth. Another is marked with stars [*stellata*] and has a sharp back with clear lines.⁹⁶ Another is yellow and hairy. Another is the "grape" [*uvea*], so named because it has the color of a grape.⁹⁷ There is another the middle of whose head and whose feet are short, with the feet curved back a bit. When it wants to sting, it first throws itself forward on its feet. When it wants to suck, it first vomits forth a small amount of fluid. It is somewhat thinner than the "grape." Another is called the "ant" [*formicalis*] since it resembles one. It has a red neck, black head, and

---

90. A.'s Latin is vague, but the sense is secured from Avic. Schühlein (1660) suggests this is the Gr. *lykos,* or wolf spider (Beavis, 1988, 51–52). "Scattered": perhaps "delicate," *rara*.

91. Schühlein (1660) reports that the Arabic reads *murmanqion*. This is probably related to Pliny *HN* 29.27.87, where the *myrmecion* is a spider with a red head like an ant and a bite as painful as a wasp's. Pliny's spider is marked with white spots, however. Beavis (1988, 200–201) discusses a Gr. insect called the *myrmēkion*, which is probably not a spider.

92. Schühlein (1660) corrects the Avic. text to yield *sqliru kefalos*, an Arabic version of the Gr. *sklērokephalon,* "hard-headed." Beavis (1988, 54) demonstrates that this is not a spider but a construct based on a corruption of a text that originally referred to moths. This accounts, then, for these "winged" spiders.

93. Schühlein (1660) emends the Arabic text of Avic. to yield *sfiqion,* a version of the Gr. *sphēkeion,* a wasp-like stinging insect (Beavis, 1988, 48). Pliny *HN* 29.27.86 describes a nameless *phalangium,* which differs from a hornet only in not having wings.

94. Schühlein (1660) once more corrects the text to yield a version of the Gr. spider called *kranokolaptēs,* a greenish, very deadly creature. Beavis (1988, 53–54) lists the ancient sources behind this creature and they are quite obscure, raising one's estimation of Avic. and his sources even higher. Beavis suggests that this creature as well was a moth before the tradition became corrupted.

95. The Galenic references are also from Avic. *Can.* 4.6.5.8.

96. Perhaps the *asterion* of Pliny *HN* 29.27.86, identified by Beavis (1988, 47) as *Latrodectus mactans*.

97. Pliny *HN* 29.28.86 describes the *rhox* as being like a grape. See notes on the *albarbasyon* at 26.25(33).

white back, and it is spotted with various colors.⁹⁸ Another is the red "wasp" [*vespalis*], which is like a wasp.⁹⁹ Another is the "grassy" [*herbalis*], which is round and small, with a small mouth, an off-white belly, and white feet with a great deal of hair.

The Egyptian, which was named first, is very nasty. It has a large belly and head and is like the animal that flies around candles. All these types of *rutelae* bite and, as Galen says, the bite of a *rutela* does not act like the bite of a scorpion, for the bite of the various types of *rutela* does not block up the veins and does not generally make defecation difficult.

Those who posit the first six types of *rutela* say that all six have in common that the place of their bite becomes abscessed. It is red at first and in many cases there occurs the coagulation of green blood in which there is an itching sensation.¹⁰⁰ Very great swelling and inflammation then occur in that place and coldness besets the sinewy [*nervosus*] members and the bones—for example, the knees, back, and shoulders. The body may become cold and tremble and the hair stands on end. The bitten one feels violent pain, lies awake, and cries out. His color turns yellow and it is seen that his eyes are moister than normal and they drip tears in frequent drops. A feeling like evacuation and emptiness is felt in the lower part of his belly and in the area of the groin. He naturally begins to expel a watery material above and below and sometimes something appears in the material that resembles a spider's web.¹⁰¹ Swelling occurs in his groin and testicles, and there is spasmodic contraction in his joints. Pain occurs at the cardiac orifice and there is nausea.¹⁰² The body emits a cold sweat and the head may hurt. His face also becomes yellow and there is heaviness in his body. Urine is passed with pain and there sometimes comes forth something that looks like a spider.¹⁰³

Galen says, speaking about each species, that the bite of the red one results in a small pain which is swiftly quelled. But the bite of the black one and of the multicolored one is followed by strong pain, with bristling of the hair, chills, trembling, heaviness in the thighs, itching, putrefaction, and softening of the belly, along with diarrhea.¹⁰⁴ The bite of the *stellata* is accompanied by strong pain, with itching, bristling of the hair, stupor, heaviness of the head, and softening of the body. From the "grape" comes violent pain at the point of the wound, chills in the entire body, bristling of the hair, trembling, spasms, a flow of cold sweat, loss of voice, stupor throughout the entire body, an abscess on the belly, tightness of the penis and groin, and involuntary

---

98. See notes on *murkyon* at 26.25(33).

99. See notes on the *suctyon* at 26.25(33).

100. "Green blood": the phrase is a bit disquieting. Perhaps it means pus-filled or bilious (*OLD*, 2072) blood. Scanlan's (442) "freshly drawn blood" is unlikely.

101. Cf. Pliny *HN* 29.27.86 for a description of the result of the bite of the *rhox*.

102. "Cardiac orifice": *os stomachi*, that is, where the esophagus joins the stomach. Cf. notes to 1.278.

103. Cf. Beavis (1988, 55–56) for a review of the ancient precedents for descriptions of spider bites. Again, Nicander seems to be the ultimate source.

104. "Bristling of the hair": *horripilatio*. Scanlan (442) suggests the idea of "goose flesh."

vomiting.¹⁰⁵ The black and smoke-colored one [*nigra et fumosa*] is bad-tempered and its bite is followed by stomach pain, frequent vomiting, headache, cough, and difficulty.¹⁰⁶ It kills swiftly. The bite of the yellow, hairy one results in very strong pain, a cold trembling, and swelling of the belly. This one very often causes death. Some say that the bite of the "grape" causes erection and tension of the penis accompanied by the discharge of sperm, spasms, and a cutting off of the voice. This, however, has not been adequately proven.

The bite of the "ant" is violent and results in a blistering of the body and a heaviness of the tongue. From the one like a wasp comes an abscess at the spot of the bite, as well as spasms, an overpowering deep sleep, and a weakness in both knees.

The bite of the small *herbalis* has the same effects as that of the "grape." But the Egyptian one is ill-tempered and causes a violent pain in the head and a very deep sleep.

(34) STELLAE FIGURA: The *stellae figura* ["star-shape"; firefly] is, as Pliny says, a vermin that shines at night like a star but only appears after great clouds when it predicts clear weather. They say that this vermin possesses such rigid coldness that it puts out a fire just like ice. If the flesh of a person comes in contact with the filth of this insect, all his hair falls out and whatever it touches turns green. They say of these creatures that they in no way give birth and they have no male or female among them. They are, therefore, born from putrefaction.¹⁰⁷

(35) SPOLIATOR COLUMBRI: The "snake despoiler" is a vermin that has a golden color with a shining green color interspersed.¹⁰⁸ It runs about in the dust on flat roads, sucking dry and devouring scarab beetles. They say that it is called the "snake despoiler" because, starting at the tail of a snake lying in the sun, it climbs the body. At first, it soothes the snake by rubbing its head, but afterward it bites it and will not stop doing so, even while the snake flees, until it penetrates the brain and takes its fill of the dead snake's flesh.

They say that when this creature is first born it is fed by its parents and afterward lies unmoving for a long time without food. It is finally roused to care for itself and looks after itself.

(36) SETA: The *seta* is a vermin about one cubit long that is so slender that it takes its name from a *seta*, that is, a hair from the mane or tail of a horse.¹⁰⁹ It is found to

---

105. "Tightness of the penis and groin": it is clear from statements immediately below that an erection is meant.

106. *Difficultas* seems to beg another word, and comparison with Avic. tells us it is difficulty in moving the bowels.

107. This charming insect is, in fact, a confusion from Pliny *HN* 10.86.188, where the salamander, spotted as if with stars, appears only during rainstorms, disappearing in clear weather. The errors are discussed by Aiken (1947), 222–23. The green coloration (*virorem*) comes from a misreading or miscopying of *vitiliginem* (a skin eruption). Cf. ThC 9.39 and Vinc. 20.168.

108. Stadler identifies this as *Calosoma sycophanta*, a predacious beetle in a family of the beetles that generally feeds on caterpillars. Cf. ThC 9.40 and Vinc. 20.168.

109. Stadler identifies this as *Gordius aquaticus*, an aquatic worm commonly called the "horsehair snake." Cf. ThC 9.41 and Vinc. 20.166.

be produced from standing water that is not very foul. It is so hard that it cannot be crushed underfoot and if it is boiled it does not become soft. When swallowed by a person, however, it robs him of life with agony and weakness. But if touched in any other way, it causes no harm. This vermin does not seem to have a head and swims in either direction. It is born by chance, from the hairs of horses, since horsehairs that have been put in standing water take on life and spirit and move, as we have proven by test [*experti sumus*] many times.

(37) STUPESTRIS: The *stupestris* is a vermin similar to a scarab which lies in the grass. When swallowed by cows it bursts their insides once it has reached their gall bladder. It is very numerous in Italy.[110]

(38) SANGUISUGA: The leech is a vermin of the swamp. It is familiar and is so called because it sucks the blood [*sanguinem sugat*] of animals.[111]

There are many types of this vermin, but in our lands there are entirely black and smooth ones, and ones that have red stripes on their backs and are a little bit wrinkled. These are the better ones. They do not have feet and no other member is to be seen on them. They are entirely columnar in shape.[112]

They are attached to bodies in order to suck out superfluous blood. Since, however, they also sometimes suck on poisonous creatures, it is to be feared that they might be infected with the poison. They should be cured first in a new pot by being sprinkled with a little salt so that they will vomit forth the poison and afterward they should be fed with a little bit of warm lamb's blood by having it sprinkled on them and two or three hours after this they are to be put on the body.

Some of them are invisible. They are thread-like and are drunk in with water. They cling to the throats of humans and animals and cause the blood to flow unceasingly.[113] If the spot to which one clings is covered over with salt, the leech immediately falls off.[114] If the place where it bit pours forth an excessive amount of blood, a leech is to be burned up in a pot and its ashes placed over the spot. This vermin makes a triangular wound. In its greed to suck sweet human blood, it spews forth what it has sucked up and straightaway sucks up fresher blood.

---

110. This is the *buprestis* of Pliny *HN* 30.10.30, which says it is rare in Italy. Cf. Isid. *Orig.* 12.8.5. This "cow-burster" is discussed at 26.24. Note too that a plant bore the same name but when used for a creature it indicated a type of beetle (Beavis, 1988, 173–75). Cf. ThC 9.52 and Vinc. 20.121.

111. Isid. *Orig.* 12.5.3.

112. These leeches belong to the class *Hirudinea*. The "better" one may be *Hirudo medicinalis,* the leech used by doctors. Cf. Rabanus Maurus *De universo* 8.4 (*PL* 111:235C), Hugh of St. Victor 3.54 (*PL* 177:104A), ThC 9.43, and Vinc. 20.155.

113. Scanlan (445) identifies these leeches as members of the genus Limnatus, stating that they can suck enough blood to cause anemia or even death.

114. In a famous scene from the *African Queen,* Humphrey Bogart ridded himself of leeches in exactly this manner.

31 (39) SCORPIO: The scorpion is a black, many-footed vermin which has large, bifurcated members in place of hands, like a crab.[115] It also has slender horns like a crab.[116] It alone, among the creased and sectioned animals, has a long tail with a nodule at the end, that is, at the last nodule of the tail. One of them has two stingers with which it strikes with a curved wound.[117] This is because it cannot strike unless it curves its tail back in the shape of an arc.

You should know that, as Avicenna says, the female is larger than the male, for the male is drawn out and thin whereas the female is fat and large.[118] The sting of the female is thin, but that of the male is thick.

It so happens that, as we have said, some of them have two stingers and these make two holes when they sting. The place of the sting grows cold and the rest of the body becomes warm and breaks out in a cold sweat.

There is, however, a particular winged genus of scorpion and it is large. But the wind often knocks it down as it is flying. This winged scorpion thus sometimes goes along with the wind from area to area and is carried by it from place to place.[119]

There is variation in the joints on the tail of a scorpion. There are some that have six joints on their tails and their malice is strong.[120] When the star called *Aquila* or *Vultur* in Latin but *Sahara* in Arabic rises with the sun, this scorpion's sting then kills.[121] Some scorpions have fewer joints in their tails.

32 The ancients say that there are nine colors of scorpion, namely, white, yellow, red, ash-colored, rust-colored, green, golden, those having black tails, and those whose tail tips are the color of wine. At the sting of these a needle prick is felt and harmful pain. There are also smoke-colored ones whose sting causes grinning [*risus*] and loss of reason.[122]

Many things occur at the sting of a scorpion. The place of the sting immediately becomes abscessed with a hard, red abscess and there is extensive pain. The one stung

---

115. Isid. *Orig.* 12.5.4, Rabanus Maurus *De universo* 8.2, 4, Hildegard of Bingen *Physica* 8.13 (*PL* 197:1344A–D), Vinc. 20.160–65, and Barth. 18.96.

116. Scorpions are arachnids.

117. Cf. Pliny *HN* 11.30.87.

118. Avic. *Can.* 4.6.4.2, which serves as the basis for much of the remainder of this section. The observation is accurate.

119. Cf. Pliny *HN* 11.30.88–89, probably a scorpionfly, one of the Panorpidae, harmless scavengers whose male has a tail that curves upward like that of a scorpion.

120. The "tail" of the scorpion is actually the last five elongated segments of the abdomen. The sixth "segment" may be the poison sac or, as Scanlan (446) suggests, "the last segment of the anterior mesosoma." The scorpion with fewer segments mentioned later in the paragraph probably is one with a smaller, less prominent (and less lethal) poison sac. The word for "segment" is *spondile*, normally used to refer to a vertebra.

121. Cf. Avic. *Can.* 4.6.5.2. The star is a bit of a puzzle. *Aquila* is a well-known constellation. *Vulturnus* was a southeast wind. Schühlein (1660) discusses the problems surrounding *Sahara*, suggesting it represents the star Sirius.

122. Scanlan (446) points out that scorpion venom can cause rictus and hallucinations.

is sometimes inflamed and sometimes chilled and he imagines he is being crushed in the pestles used to crush salt. Pains come suddenly and there are prickly feelings like those from a needle. There follow sweating, trembling of the lips, coldness, thick vomit which congeals immediately, bristling of the hair as well as its breaking, trembling and cold in the extremities, especially those which are near the wound. There follow softening of the entire body, protrusions on the groin, and rigidity in the penis. Swelling of the belly occurs and sometimes there is strong flatulence, especially if the sting is in the lower members. Abscesses occur on the armpits, there is a great deal of belching, and the victim's skin color changes.

Now if the scorpion has very strong malice, there will be very strong accidental traits as well. For the sting then is like a burned cauterization and the body undergoes total rigidity accompanied by coldness. A viscous fluid flows to the lips and congeals between them. A large amount of fluid runs from the eyes causing bleariness and becoming coagulated in their corners. Their forms are changed, an abscess appears on the penis, and the anus protrudes.[123] The tongue grows thick and the teeth grind, are closed tightly together, and do not open.

Galen says as well that the evil is great if the sting is over an artery. Fainting then occurs and if the poison finds the nerves, spasms occur. If it finds the veins, there is putrefaction.

The ones that are in Italy are small and do not cause great difficulty unless they make the wound in an artery.[124]

This animal lays eggs and they say that the father tries to devour the young but that those who hide themselves in the mother's haunches are safe, but this is false.[125] What I have observed of this animal is that when I immersed it in olive oil, it lived for twenty-one days walking on the bottom of the oil in a glass container. On the twenty-second day it died and bubbles were raised in every direction into the oil from the joints of its rings. The oil in which a scorpion has died and putrefied is, when mixed with vinegar, said to be a cure for its bite.[126]

(40) THAMUR: The *thamur* or *samyr* is, they say, the vermin by which glass and rocks are split and which ostriches use to split the glass in which their chicks have been

33

34

---

123. Stadler emends his text, changing the sing. verb *alteratur* to the pl. *alterantur* to agree with the pl. *formae*. The result is the vague sentence given above. The text of Borgnet also reads the sing. verb but has *forma*. This would translate: "The anus changes in form and protrudes." The original of Avic. *Can.* 4.6.5.3 is not helpful in choosing between the two, but Borgnet's reading produces the most coherent sense.

124. Cf. Pliny *HN* 11.30.89.

125. Cf. Pliny *HN* 11.30.91. Scorpions are solitary beasts, except at mating time. They are also highly cannibalistic and will devour their children, which are born live and ride on their mother for about a fortnight before scattering to their solitary lives. Note that part of this sentence, as printed in Stadler, is ungrammatical. The sense is clear, however.

126. The impetus behind this "experiment" might be Pliny *HN* 11.30.90, which states that oil is deadly to scorpions.

enclosed and thus extract them and the one Solomon used to split marble at will.[127] But this is a fable and I think it is one of the errors of the Jews.[128]

(41) TAPPULA: The *tappula* is a vermin which runs on four feet over the water. It is an aquatic spider.[129] It runs on land as well and, if pressed under water, does not get wet and does not drown. It seems to walk over water without getting its feet wet and it traverses expanses of water very quickly, although it sometimes grows tired, stops, and recovers its strength.

(42) TESTUDO: The snail [and slug?] is a vermin generated out of rotten vegetable matter and viscous dew. A white one is found, as are red, yellow, and black ones. The terrestrial viscosity of this last one hardens in a spiral into the shell in which it dwells.[130] It is said that when it is on occasion infected by the poison of a serpent, it cures itself with a familiar herb.[131] If salt is thrown over it, it almost totally liquefies and is turned into viscous water.

(43) TINEA: The *tinea* [clothes moth] is a vermin which gets its name from its tenaciousness [*tenendo*]. It arises from the vapors off of rotten wool and it eats the wool itself as well as hair.[132]

(44) TEREDO: The *teredo* [woodborer] is a vermin of wood which takes its name from the fact that it eats [*edat*] into the wood. Some say this insect is called a *termites* in Latin.[133]

---

127. According to 1 Kings 6.7, during the building of the Solomonic Temple in Jerusalem, no hammer, axe, or other iron implement was used to dress the stones for its construction (cf. Deut. 27.5). This raises an interesting question: How, then, were the stones prepared? According to rabbinic legend, Solomon had employed the *shamir*, a small worm with marvelous powers whose mere touch could split stones to any desired shape. See Ginzburg (1987), 1:33–34, and Patai (1988), 185–88. He was informed of the worm and its location by the chief of the demons, Ashmedai, who had been captured by Solomon's chief minister, Benaiah (cf. Babylonian Talmud Gittin 68a). According to ThC, this creature is called the *vermis Solomonis*. Because Mosaic law forbade the use of iron in building an altar to God, he adds that Solomon enclosed an ostrich's young in a glass enclosure. The mother ostrich then brought this worm whose blood would break the glass and free her young. Wise Solomon subsequently used it to split the stones. According to rabbinic accounts, it appears that this worm was entrusted to the hoopoe bird, which transported it. Cf. Babylonian Talmud Hullin 63a.

128. Cf. ThC 9.44 and Vinc. 20.170.

129. Stadler suggests that this creature is a member of one of two species: *Gerris* or their relatives the *Hydrometra*. Cf. ThC 9.45 and Vinc. 20.170. A. added the statement that it is a water spider, and Diefenbach (1857, 99) lists the form *capula* and gives the Germanic equivalent as *Wasserspinne*. A. may thus have thought the "four legs" were four pairs of legs.

130. Cf. Hildegard of Bingen *Physica* 8.18 (*PL* 197:1346A–B) for cures using snails, ThC 9.46, and Vinc. 20.172–73. It may be that given his specificity here for the black *testudo*, A. also uses the term to include slugs.

131. Perhaps a story that has crept in from *testudo* as tortoise. Cf. Ambrose *Hex.* 6.4.19, where a *testudo*/tortoise that has eaten snake entrails cures itself by eating marjoram. Cf. Pliny *HN* 8.41.98.

132. Isid. *Orig.* 12.5.11, Rabanus Maurus *De universo* 8.4 (*PL* 111:235D), Hugh of St. Victor 3.54 (*PL* 177:104C), ThC 9.47, and Vinc. 20.174. As with many vermin mentioned in this book, preventing them was of major importance in everyday life. Cf. 22.108 (protecting clothing) and 23.87 (*tineae* on birds).

133. Stadler conjectures that the larva of a woodborer is involved here. Cf. Isid. *Orig.* 12.5.10, Rabanus Maurus *De universo* 8.4 (*PL* 111:235D), Hugh of St. Victor 3.54 (*PL* 177:104C), ThC 9.48, and Vinc. 20.171.

It is born out of a corrupt humor in the wood. It eats its way out of the wood by making it sawdust and it exits through a hole. This humor occurs when wood is cut which has humor in it which has not matured.

It is said that termites are not born in wood in the East due to the dryness of the wood there. In our climates, however, that is, from the fifth to the seventh *clima*, *teredines* are found in all the trees. It is found less often, though, in the oak and the linden than in others.

(45) TATINUS: The *tatinus* is a vermin which is found in the fat of a pig when it goes bad. It is a hairy vermin.[134]

(46) URIA: The *uria* is a vermin found on pigs which gets its name from burning [*urendo*] since whenever they bite, blisters arise as if the place had been burned.[135]

(47) VERMIS: The *vermis*, as the name is commonly used, especially denotes the earthworm since it is said to be born solely from the exhalation of the earth.[136]

(48) VERMES CELIDONIAE: They say that there are Celidonian worms which live in the boiling hot waters in the Celidonian region and which, if transferred to cold waters, immediately die. Even though this is a well-known tale, it is more accepted by the common folk on word of mouth than it is something proven by definite experience.[137]

(49) VESPAE: There are many genuses of wasps, as we distinguished them in foregoing books.[138] They all collect useless honey and they generally build their nests in the ground or in mud walls. Some seek out dung for their food and others the juices of fish, whereas still others suck on meat.

The sting of a wasp, especially of the yellow one [hornet], produces heat more violently than that of a bee. Its sting or bite is followed by pain, redness, and abscess.[139]

---

134. Isid. *Orig.* 12.5.15 offers *tarmus*, the form offered by ThC 9.49. Stadler suggests that this interesting little creature is *Dermestes lardarius*, the pork or bacon beetle. Scanlan (448–49) points out that its larvae are hairy. Beavis (1988, 224) calls them simply "larvae in meat" and lists other references. Glosses cited by Diefenbach (1857, 574) seem to hint in a general sense to "maggot."

135. Stadler suggests this pest (also found as *usia*) is *Taenia solium*, the larva of the tapeworm. Cf. Isid. *Orig.* 12.5.16, Rabanus Maurus *De universo* 8.4 (*PL* 111:236A), Hugh of St. Victor 8.54 (*PL* 177:105A), and ThC 9.50. Diefenbach (1857, 630) gives the following synonyms for *usia: svinislus, sweynlaus, swinwrn, verkensluys*. If the "louse" stem is correct, some other, biting pest may be meant, and Beavis (1988, 119) thinks the *usia* may be *haematopinus suis*, pig lice. The blisters, then, are not the cysts of the tapeworm (Scanlan, 448) but actual sores arising from the bite.

136. Cf. ThC 9.52.

137. The name is reminiscent of the Chelidoniae insulae ("Swallow Islands") off the coast of Lycia (SW Turkey), but this tale is not linked to them. The clearest source seems to be Augustine's *De civitate dei* 21.2, but no name or location is given there. It may be that this is a confusion with or an extension from "swallowwort" (*Chelidonium maius*) of *De veg.* 1.176, which blooms in spring and dies when the cold comes and the swallows leave. Cf. Pliny *HN* 25.50.89–90. Cf. ThC 9.53.

138. Cf. Hildegard of Bingen *Physica* 8.69 (*PL* 197:1311A), ThC 9.51, and Vinc. 20.175–79. A.'s major entry on wasps is found at 8.184–97.

139. Cf. 4.65 and 26.13(14).

There is a particular genus of large wasp which has a black head. It has many stings and kills.[140] These are the most harmful and their sting sometimes leads to spasms and weakness of the knees. As a swelling arises from the bite of the small ones, these cause a blister to arise at the place of the sting, and they make the tongue heavy.

Let, therefore, these be the things that have been said by us about vermin.

36 The book of animals is now complete and in it the entire work of their natures has been completed. In it I have held as my governing rule that I have set forth the words of Peripatetics as well as I could. Nor can anyone detect that which I myself feel about natural science [*philosophia naturali*]. But rather, if he should have a doubt about something, let him compare the things said in our books with those sayings of the Peripatetics and then let him either criticize me or agree with me, saying that I was the interpreter and expositor of their learning. If, however, he should criticize me without having read or made comparisons, then it is clear that he criticizes out of hatred or ignorance, and I care but little for the criticism of such people as these.[141]

---

140. "Many stings": *multos aculeos,* seeming to imply not the ability to sting repeatedly but the actual possession of several stingers.

141. A. may be referring especially to the Franciscan Roger Bacon, who repeatedly chastised "the unnamed master," almost surely identified as A. (Hackett, 1980).

APPENDIX

Alternative Version of the Beginning of *De Animalibus*

From Cod. Vat. lat. 718 (ca. 1262–63), in Pelster (1935). This text is from a MS older than those MSS used by Stadler to reconstruct torn and missing folios from the beginning of A.'s autograph (Köln W258a); it illustrates development in A.'s work.

HERE BEGINS THE FIRST BOOK ON ANIMALS

Which Is on the Common Diversity of Animals

The First Tract, Concerning the Common Diversity of Animals in Their Members and Life

CHAPTER I

A Digression Declaring the Manner and the Order of the Instruction

The investigation on animals is the last one in natural philosophy because of the reason we gave in the *Physics,* namely, that an animal is that which, of the natural things, is more composite both in body and in soul.[1] For in the body it has the elements as well as their mixture, coagulation, complexion, composition out of heterogeneous materials, as well as a sharing with the soul of visible life, which is the perfect soul. And other natural things either do not have this composition or do not have it to this extent. Simple things do not have composition, and minerals and plants do not have it to the extent that animals do. And in the souls of animals there is also sensation and operation of visible life, things which contain in themselves (the way a tetragon contains a triangle in itself) vegetation and the operation of unseen life. On account, therefore, of both of these, then, the investigation of animals ought to be last in matters of natural science.

For all animals share in all these things which are known about animals, even though they may differ in the types of principles. Therefore, there should be a single investigation concerning all the diversity among animals, both according to their genuses and their species. For their principles are the same, whether they be of generation, of life, of nourishment, or of their lifestyles, even though they may differ in the manner in which they participate in these principles, as we have already said. This investigation ought, moreover, to be dealt with in two ways, for the things to be examined should first be set down and then the causes of the things to be examined should be investigated. For in every philosophy one should hold to this method. For in astronomy the cause of an eclipse of the sun or the reason for differences in its size are examined only if it first be known that there is an eclipse and that it has differences in size. It is the same for all other matters. Therefore, it is for this reason that we are dividing our investigation on animals in the following fashion. First we will speak of

---

1. Or "is more composite in natural things, both in the body and in the soul."

all the known differences of animals in their members, generation, food, customs, and other things. Afterward we will set out in order all the causes of these differences.

This will be a difficult matter for four reasons [*causa*]. The first is that some animals have an unknown life and they can be investigated only by studying those like them. The second is the smallness of certain animals, whose bodies cannot be split open through dissection. We should therefore argue about the members of this sort via the powers of their souls and sometimes by following a comparison of a likeness found in other animals. The third reason is the manifold diversity of animals which can scarcely be reduced to any common principles. The fourth is confusion among the books of the philosophers who passed on something about the investigation of animals before our time and who also named the animals using a variety of languages.

This is why we cannot know all the animals well. Still, as far as we are able, we will strive for orderliness and explication so that our discussion might be all the more clear.

We will follow Aristotle in accordance with our custom, introducing first the common differences in the animals, doing so first concerning their members, for in so doing we will give that which is stable in the bodies of animals. And since Aristotle's tract on animals is lengthy, our book is necessarily lengthy, for it will explain the book of Aristotle and it will interpose as well some things which he did not posit but which seem necessary for the investigation.

GLOSSARY

This glossary is provided as an aid to informed reading. Many of the terms included are ones that must, for various reasons, be treated with special care if the flavor and complexity of the original are to be retained.

Accipiter: Used both generically and specifically for hawks. When the term "hawk" appears, it is generally replacing this word. But note that *accipiter* is often given as a synonym for *astur*, which is usually translated as "goshawk." (Cf. 6.49, 7.30.) When necessary for clarity, the Latin is occasionally left intact.

Adeps: *See* Fat

Adiutorium: Lit., "helper." A. uses the term regularly to refer to the upper arm bone that is today called the humerus. The Latin is retained to keep the word distinct from A.'s *humerus*, which can, confusingly, indicate the clavicle, shoulder joint, or, on occasion, the humerus. See, e.g., 1.278.

Anulosi: *See* Ringed creatures

Astacus: From the Greek, meaning lobster or crayfish. In A., it sometimes means a crab (see, e.g., 7.92). On the confusion even in Greek times, see *GF*, 18–19.

Astur: *See* Accipiter

Breast: A.'s terminology is simple but often confusingly used. To convey the sense of the original, we use the following equations throughout: *mamilla* = breast; *uber* (*ubera*, pl.) = udder; *conus* = cone; *papula* = nipple. As examples, a cow has one *uber* and several cones, sometimes called nipples; a human female has two *mamillae*. To demonstrate the confusion caused by sloppy use of these terms, see 2.36–37.

Calcaneus: A difficult term, as are many of the words for parts of the foot. It originally meant "shoe," a sense sometimes found in A. (e.g., 2.24). In modern medical terminology, it is the heel bone and in A. it often has this meaning. It is also alternately equated with or differentiated from the *cahab*.

Call: *See* Voice

Calx: Another term for a hoof; *see* Solea

Celeti (var. *celety*, *kelete*): In general, this represents the Gr. *selachē* and thus is a generic term for selachians (cartilaginous fish, such as sharks, skates, and rays). At times, however, it is used in the singular and seems to represent an individual animal.

Chicken: *Gallina* very often indicates the domestic hen, as at 23.116(52) and when paired with *gallus*. But it also seems to be a generic term for A. (e.g., at 2.89), and thus often is translated as "chicken."

Chyle: *Chymus*, elsewhere *chylus*, is the semi-liquid contents of the digestive tract during the first stages of digestion. See 1.563; 3.100, 125; 20.16, 18 (*chymus nutrimentalis*); and 23.77(40). A direct borrowing from the Gr. See Hyrtl (1880), 114–16.

Chyme: *See* Chyle

Cincture: *Succinctorium*, a word commonly used by A. to denote the narrow points where the parts of an insect's body are joined. The Latin term's base meaning is that of girdling or belting, and it is translated as "cincture" because of that word's clerical associations (Latham; Nier.).

Claw: Frequently a translation of *unguis* (q.v.) but also of *ungula*. *See* Hoof

Clima (pl. -ata): These are the longitudinal bands into which ancient and medieval geographers divided the known world. As Tilmann (1971, 159–60) shows, A. divides the habitable regions of the earth into seven *climata* according to differences in latitude and longitude. See also Honigman (1929), Bunbury (1959, 2.4–11), and Dilke (1985, 177–78) on the Arabic influence.

Conceptum/us: *See* Embryo

Conc(h)a: From the Gr. *konchē*, which originally meant a shell as distinct from the animal within the shell and which also referred to various species (*GF*, 118). In *DA*, its use parallels that of *conchylium*. *See also* Shellfish

Conchylium (var. *conchilium*; pl. -ia): *See* Shellfish

Cow: *See* Ox

Crasis: *See* Discrasia

Creased: *See* Anulosi

Creation: *See* Embryo

Embryo: On medieval embryology, with its particular moral and theological concerns, see Demaitre and Travill (1980); Adelmann (1966); Needham (1959). A. uses a variety of terms to discuss the unborn. To facilitate study of these terms, *embri(y)o* is so rendered throughout the translation. "Fetus" is reserved for *fetus*, though this can also be used for the young creature itself. The less frequently used *creatura* is rendered "creation." *Conceptum/us*, a more general term indicating "the thing conceived" (see 2.118), is translated as "fetation" or is left in the Latin. In English, it is a broader term than "fetus" because it could mean a fertilized chicken egg but also be used as in 1.83, where we see the compound term *conceptum embrionis*. A.'s use of the ntr. form is also verified by 1.83; its more common, fourth declension form is used elsewhere. Throughout Book 15, *conceptus* seems on occasion to signify, simultaneously, the fetus and the act of conception. The less commonly used *partus*, when not referring to the actual act of parturition, is generally rendered as "newborn" or "young," as at 3.165f.; cf. 7.8.

Demaitre and Travill (1980, 410, 418) note that in *DA*, A. is frustrated in trying to reconcile discordant ancient authorities when treating matters of human conception, adding that the problem was "compounded by the difficulties inherent in each translation."

Experimentum: The *DA* is rich with words based on the *exper-* stem. The line between "to experience" and "to learn through experience" (that is, to "experiment") is not always clear. To enable readers to form their own opinions on the extent of A.'s "experimentation," we either use similarly based words in English or provide the Latin. The term "experiment" has generally been avoided as being too specific and carrying too many modern overtones.

Fat: A. most commonly uses *pinguedo*. Variant words (e.g., *adeps*, *sagimen*, *sebum*) are generally rendered in their Latin forms to facilitate study. Cf. 3.137.

Fetus: *See* Embryo

Fumes: A. uses *fumus* and *fumositas* to refer to a natural "smokiness" that exists in the body. In certain contexts, to modern ears "fumes" sounds better and implies a less dangerous quality than "smokiness" and is thus so rendered.

Genus: In A.'s use of *genus* and *species*, he has a rudimentary, if not a modern, sense of the distinction; see 1.20. His next lower subgroup is generally *modus*, which we render as "type." Note, however, that *genus* often simply means "type" or "kind."

Goshawk: *See* Accipiter; Astur

Gusanes: A term (sing. or pl.) most commonly used for a larval, worm-like stage from which a later, more perfected form of an animal comes. The form *gusanis* is also found (e.g., at 7.100). At 15.4 and 15.38, it is allied to a *vermis*. At 8.139, it is said to cling to and suck on larger animals. At 15.42, A. asserts it is Arabic. Körting (1923, 310, no. 2556) cites it as *cossanus*, giving Spanish and Portuguese parallels. Schühlein, 1658, takes this to indicate the Romance origins of the word. Monlau (1941) and Garcia de Diego (1955) derive *gusano* from the Latin *cossus*, a grub. See also the evidence gathered by Jessen in *De veg.* 721.

Gutta: Lit., "drop." But the term includes the idea of distillation, as when resin forms into a *gutta* on a pine tree. For A. it is often equated to the female fluid or sperm upon which the male sperm acts.

Halzum: An Arabic term denoting, generically, shellfish. See 4.37; Schühlein, 1658.

Hawk: *See* Accipiter

Hoof: *Ungula*, which is also sometimes translated as claw (see, e.g., 2.18); see 14.51f. *See also* Solea; Sotular

Hour: A. makes frequent mention of times of day, naming the hour and referring to periods of rising and going to bed. Medieval reckoning of time is rooted in the Roman system, which designated sunrise as the first hour of the day, or *hora prima*, and divided the rest of the daylight period into twelve equal hours. Midmorning was *hora tertia*, whereas *hora sexta*, or the sixth hour, fell about noontime. These temporal divisions came to be associated with the times of monastic prayer (the canonical hours): Matins, Lauds, Prime, Tierce, Sext, None, Vespers, and Compline. The length of an "hour" naturally varied by season. For further discussion, see Knowles (1933) and Leclercq (1975).

5.10ff.) *ascendere per saltum* (lit., "to mount with a leap") is used. For the sake of modern ears, "mount" is often used where "leap" appears in the Latin.

Lifestyle: *Regimen vitae*; equivalent to Ar.'s *bios* and, as A. points out at 1.25, encompasses the entire scope of an animal's activities. The etymology of *regimen* is related to *rego, -ere*, so it also implies order and rule. No facile English translation can encompass all its ramifications.

Lobster: Variously *karabo* (q.v.) or *astacus*.

Malachie/malakye: A. uses this consistently as a generic term for the cephalopod mollusks, such as the squid, cuttlefish, and octopus. A transliteration of the Gr. *malakia*, "soft-bodied ones." See *GF*, 155f.

Mandible: *Mandibula*; technically, the lower jaw, though often used generically for "jaw" and in such pleonasms as "the lower mandible." *See also* Maxilla

Maxilla, -ae (pl. Maxilla): Technically, the upper jaw. A. uses it also generically as "jaw," but in anatomical sections the Latin form and *mandibula* are used when it is useful to retain the effect of the original.

Melancholy: *Melancholia*; the Greek term means "black bile." Elsewhere called *colera nigra* or *nigra colera* (see, e.g., 1.611).

Member: A. freely uses two words for "part": *membrum* and *pars*. To reproduce the effect of the original (see 1.109, where the terms are used side by side), the former is regularly rendered "member" and the latter "part." Note, too, that *membrum* can be "limb." To generalize concerning A.'s usage, *membrum* is the preferred term for a body part, whereas *pars* tends to be used for something that is part of a more complex unit, as a finger is part of an arm. This is especially noticeable in the early sections of Book 12, but it is not, of course, a firm rule.

Meri/y: A.'s received Arabic term for the esophagus.

Mesenterics: *Meseriacae (venae)*; used for the system of veins and arteries supplying the lower intestinal tract.

Milt: *Lac*; lit., milk. Generally used by A. for the sperm of fish. Note that the English term for this substance, "milt," is itself related to "milk."

Mirach: According to Fonahn, 2077, this has several possible meanings: the abdominal wall; the abdomen; the muscles of the abdominal wall; the navel; and the peritoneum.

Monster: The term "monsters" may imply for medieval authors a deformed or an abnormal offspring, the product of demonic forces. A. acknowledged the power of astrological influences upon birth, but for him this did not imply that monsters were themselves evil—they might also be understood simply as "wonders" or "prodigies" of nature, as the etymology suggests ("something to be pointed out"). For example, see *De proprietate sermonum vel rerum* (Uhlfelder, 1954), a fourth- to fifth-century text that was frequently, though erroneously, ascribed to Isidore of Seville. There, a "prodigy" is confined to reference to celestial bodies; a "monster," to sublunar bodies (p. 76). A.'s view may be rather benign; as Demaitre and Travill (1980, 434) remark: "Throughout [A.'s] writings . . . runs the idea—perhaps not shared by much of twentieth-century society—that 'monsters' are perfect in being the effect of human generation and imperfect only in comparison with Nature's intention to make the best possible." For additional texts by A. on monsters, see *Phys.* 2.3.5.

Mount: *Ascendere*. *See* Leap

Nail: *See* Unguis

Nucha: Arabic for the "marrow" of the backbone, that is, the actual spinal cord itself. First defined at 1.257 but used regularly throughout.

Occiput: When the term is italicized in the text, A.'s Latin word is being reproduced. When in roman type, it is being used in its modern anatomical sense. The same distinction is made for sinciput.

Ostreum (pl. -trea, ntr.): A. also uses the masc. *ostreus, -ei* (4.36), and the fem. *ostrea, -eae* (13.121 and 24.46(90)). At 4.36 and 5.64, ostreum seems to mean the animal within the shell, but cf. 8.67. For the general confusion surrounding the term, see Shellfish.

Oviparous: *See* Viviparous

Ox: For the sake of uniformity, *bos* is regularly rendered "ox"; *vacca*, "cow"; and *taurus*, "bull." The terms are often somewhat confused; for example, A. will occasionally (e.g., 4.108; 6.101, 104) refer to the *femina vacca* ["female cow"], indicating that *vacca* can be both generic and specific. At 5.49, we are offered *vacula* as heifer and *bovulus parvus* as, apparently, a young steer.

Pannicular-membrane: A neologism to express the sense of *panniculus*, a common word A. uses for "membrane." In origin it means "bit of cloth," indicating the woven nature of the membranous tissue. In current medical terminology, it refers to a very specific type of membrane.

Part: *See* Member

Passion: *Passio*, which has several different meanings, all rooted in its origins in the verb *patior*, "to suffer or undergo." Thus, in addition to the modern usage (where passions are emotions), A. often uses the word as a correlative of *actio* to indicate a certain receptivity, passivity, or simply an action done to rather than by the subject. Less often it means an ailment, as at 7.111 (*passio yliaca*).

Pecten: Lit., a "fan" or "comb." Used in multiple ways in this work, most centering around the concept of the spread of the ribs of a fan. Thus, it refers to the central portion of the hand (1.121, 281f.) or foot (1.14, 298) and also to the groin area (1.445; 3.88). Elsewhere, it is the name of a flatfish (24.47(94)) and of the scallop (4.47, 58).

Pellicle: *Pellicula*; lit., "a little skin." This is one of A.'s many names for membranes. In modern usage, it is a thin, generally translucent membrane, but such precision should not be sought here.

Phantasy/phantasm: *Fantasia*; represents for A. one of five internal senses that collectively act upon the data provided by the five external senses (vision, smell, hearing, touch, and taste). The internal senses are the common sense, imagination, phantasy, estimation, and memory. Whereas the external senses are located in distinct and individual organs, the internal senses are to be found together in the brain. Each of the external senses receives sense data proper only to that sense (e.g., color to vision, odor to smell), while it falls to the common sense to *compose* discrete sense data to provide a composite "image" of the sensed object. Thus, the common sense joins the color (e.g., brown) and the smell (e.g., malodorous) of the object in order to provide the composite (brown *and* malodorous). The composite is imaged by imagination once the sense object is no longer present to the external senses and then is stored for future reference in memory. Estimation, or the estimative power, enables the animal to discern certain features in the object that are not directly given to the external senses but that accompany sense data, viz., *intentiones*. For example, a goat flees from a tiger, even if it has no prior experience of tigers, because it "estimates" that the tiger's intent is hostile. Phantasy is the power of composing or dividing such intentions or impressions. For a fuller discussion, see Steneck (1980).

Pigeon: Used regularly to translate *columba* in an attempt to keep *columba* and *palumba* (ringdove) distinct. Note, though, that the two are often confused, as at 5.41, where the *palumbes* is a member of the *columba* genus. On the difficulty in keeping the two straight, see *GB* 238f., 300f.

Pontic: "Pontic" implies a sharp bitter taste. The name is ultimately derived from the Greek and Latin for "rhubarb" and refers to the country Pontus on the south shore of the Black Sea.

Protuberance of the liver: *Gibbus/gilbus* is the same term used for the deformity of a hunchback's spine. A. defines it at 1.597 as the convex, rounded, upper surface of the liver. It and *sima* (q.v.) are used consistently.

Radical moisture: The concept of radical humor or moisture, as Hall (1971) and McVaugh (1974) point out, was transmitted to A. by Avic., who modified Galenic and older Greek notions. It represents the substantial moisture conveyed to a body with or by semen at the time of generation (3.102) and, as such, was understood to be essential to life—to the assimilation of food and to the generation of the bodily members (3.99). Over time, in the process of aging, the body's connatural heat, fueled by the radical moisture (*QDA* 13.13), diminishes the quantity of this moisture. This leads to the gradual desiccation of the tissues and internal organs. Disease or illness (e.g., fever) may hasten this process (*QDA* 7.29). The quantity of this radical moisture can be replenished by the conversion of food into this moisture during the digestive process, but it is never wholly replaced, resulting in the organism's decline and, ultimately, in its death.

Ratio: The meaning of this term is especially difficult to convey. In the broadest sense, it means "reason," both the human intellectual faculty and an objective divine order that exists in the world for the human mind to discover. It may also refer to the invisible or antecedent *causes* of events, as when we say that we are looking for the "reasons" for things. Platonic tradition also led medieval philosophers to understand *ratio* as the form that determines the nature of a thing and as the concept of a thing which is fashioned by the mind. The term frequently refers to the most basic sense of a thing. So many meanings can be provided for *ratio* that DeFerari (1960) required a six-page entry. The ambiguity of the term could not fail to create misunderstanding in medieval debates among pre-Scholastic thinkers (Cantin, 1972, 1977). Although Scholastic philosophers tried to give greater precision to basic terms like *ratio*, their efforts only increased the number of distinct meanings such words might carry.

Ringdove: Used regularly to translate *palumbus* and its cognate forms, *palumba* and *palumbes*. *See* Pigeon

Ringed creatures: Literal translation of *anulosi*, referring to the segmented bodies of insects. Akin, but not identical, to A.'s other term for insect bodies, *rugosus*, here rendered as "creased" but lit., "wrinkled."

Sagimen: *See* Fat

Sebum/sepum: *See* Fat

Secundina(e): A membrane; in the pl., "afterbirth."

Shellfish: Terms for shellfish were never clear-cut from antiquity onward. *Ostreon* (Lat. *ostreum*) in the original Greek meant both oyster and, generically, shellfish. *Konchylion* (Lat. *conchylium*) originally meant the shell itself. Yet both were used confusedly in antiquity. See *GF*, 118, 190–92, and Peck's (1965–70) note to *HA* 487a26. In A., the words are hopelessly muddled, often used variously in the same sentence (e.g., at 1.32, 4.1f., 17.69, 24.23(31)) or even as a compound term (e.g., at 4.3, 36–37, 43). In general, each term can be either generic or specific; context dictates the best sense. The Latin is given in the text as a guide. *See also* Conc(h)a; Conchylium; Halzum; Ostreum

Sima: When referring to the liver, *sima* is defined as the concave part of the liver where it rests over the roundness of the stomach (1.597). See also 3.35; 13.55.

Sinciput: *See* Occiput

Smoke, smokiness: *See* Fumes

Solea: A type of animal's hoof. At 14.51, A. defines it as "that horny part of the foot which they commonly call the *calx*." As with *calcaneus* and *sotular*, the word originally referred to footwear. A case can be made for *solea* tending to replace the *m nycha* of Ar., which indicated an animal with a single, solid hoof. See 18.70, 76f.

Sotular: For A., this is clearly a type of animal's hoof. The term is generally, but not exclusively, used for the foot of a horse (see 7.111) and is usually differentiated from a hoof (*ungula*). Frustratingly, though, we also find such things as *ungula equi* (2.32) and *sotulares vaccae* (2.56); see also 2.25, where an *ungula* is said to be like a *sotular*. The word is most commonly used in the Middle Ages to designate a human shoe and is found in a wide variety of spellings and forms (see *subtelaris*, lit., "under-ankle," at 7.111).

Sound: *See* Voice

Species: *Species* may mean "appearance" or the "likeness" of something. Often it carries both meanings at once insofar as a member of a species is normally like its fellow members in appearance. *See also* Genus

Sperm: This cognate of *sperma* is regularly used, even when it sounds odd to modern ears, in order to reflect A.'s actual word use.

Spiritual: Generally used to describe members, parts, or organs that have to do with the *spiritus*, the spirit or air circulating in the body. The spiritual members form what we would call the respiratory system. The word has no religious overtones in *DA*.

Talon: *See* Unguis

Tunic: *Tunica*; another type of membrane whose name invokes a metaphor to woven cloth.

Type: As a subset of species, *see* Genus.

Udder: *See* Breast

Unguis: No single English word translates *unguis* in its many meanings of "claw," "talon," and "nail" (toe- and finger-). *See also* Hoof

Vegetative soul: The vegetative soul (A. uses the terms *anima vegetabilis* and *anima vegetativa* seemingly interchangeably) is the lowest of the three soul-principles (the vegetative or nutritive soul; the sensible soul; and the rational soul) that A. understands to animate various life forms. As the lowest soul, the vegetative is also the most dependent on matter for its vital action, namely, the nourishment, growth, and propagation of an organism. Unlike the rational soul, whose function is intellection, the vegetative soul cannot exist apart from a body. It is also the most general or universal soul, common to plants, animals, and humans.

Veil: *Velamen*; yet another of A.'s many words for membrane.

Vermin: *Vermis*; often translated as "worm" or "insect." Although such a translation is sometimes correct when denoting a specific creature, it will not always do when it is used generically. As the title of Book 26 shows, *vermis* clearly includes such animals as spiders, toads, and frogs. In 26.1, A. seems to say *vermis* are "small, bloodless animals," and Scanlan follows this translation, but the inclusion of frogs and toads in this class somewhat belies this. The English "vermin" thus seems to best, albeit not perfectly, encompass the concept, and it is often so translated when the word is used generically. *See also* Larva

Virus: A generic term indicating a corrupt bodily fluid, generally the result of faulty digestion. In two common uses it designates snake venom and colostrum (the liquid that precedes lactation in mammals).

Viviparous: The term, as such, does not appear in *DA*. Its use in the translation is entirely a convenience because A.'s most common term, *animalia generantia sibi animal simile* ("animals generating a live young resembling themselves"), is too awkward to fit into certain sentences. There are places where "viviparous" will not do and the full phrase is thus written out (e.g., at 1.78, 176). On translating the second "*animal*" as "live young," see 1.77, with notes.

The emphasis is more on the appearance of the offspring as fully fledged, if small, versions of the adult than on the actual means of production. For homogeneity, the single verb, *ovo, ovare*, "to produce eggs," is generally translated as "is oviparous"; *vermes faciens*, is translated as "larviparous." See the Note on the Translation.

Voice: At 4.90, A. says that *vox* is sound produced with imagination. At *De anima* 2.3.22, he defines *vox* as not just any sound (*sonus*) but the sound of a creature that is animated by a sensible soul. This must be remembered when translating into English because a literal translation as "voice" or "vocalize" (*vocare*) may be misleading. Depending on the context, *vox* is occasionally rendered as "voice," "sound," or "call," but the literal translation is preferred. See 1.45f.

Web: *Tela*; a membrane, generally of fine texture.

Windiness: *Ventositas*, lit., a blowing of wind. It is used for situations ranging from flatulence to the propulsive power that propels ejaculation.

Windpipe: *Canna pulmonis*, "reed of the lungs," or simply *canna*. Synonymous with *trachea* or *trachea arteria*.

Zirbus: (var. *zyrbus, zirbum, zyrbum*). In technical usage, it is the omentum. It often seems, however, to refer to any sort of fat that protects an internal organ and, distressingly, sometimes seems to be a mere synonym for another type of fat (see, e.g., 1.572).

REFERENCES

Abeele, Baudouin van den. 1994. *La fauconnerie au Moyen Âge: connaissance affaitage et médecin des oiseaux de chasse d'apres les traites latines.* Paris: Klincksieck.
———. 1993. "Du faucon au passereau. La connaissance du comportement des oiseaux selon les traités de fauconnerie latins (Xe–XIVe s.)." In Bodson (1993), 215–28.
———. 1991. *Les traités de fauconnerie latins du Moyen Âge.* Louvain-la-Neuve: Diss. Université Catholique de Louvain.
———. 1990a. "Les traités de fauconnerie latins du XIIe s. Manuscrits et perspectives." *Scriptorium* 44: 276–86.
———. 1990b. *La fauconnerie dans les lettres françaises du XIIe au XlVe siècle.* Louvain: Leuven University Press.
Abramov, Dmitri. 2003. *'Liber de naturis rerum' von Pseudo-John Folsham—eine moralisierende lateinische Enzyklopädie aus dem 13. Jahrhundert.* Diss. Hamburg University.
Adair, Mark J. 1996. "Plato's View of the Wandering Womb." *Classical Journal* 91: 153–63.
Adelmann, Howard B. 1966. *Marcello Malpighi and the Evolution of Embryology.* Ithaca, N. Y.: Cornell University Press.
Aesop. (Perry, 1952).
Aiken, Pauline. 1947. "The Animal History of Albertus Magnus and Thomas of Cantimpré." *Speculum* 22: 205–25.
Albertus Magnus. *See under editors of individual works* (*see also* The Writings of Albertus Magnus *on p. xxxix*).
Aldhelm. *Aenigmata. PL* 89: 183–99.
Alford, D. V. 1975. *Bumblebees.* London: Davis-Poynter.
Allen, Edward Heron. 1928. *Barnacles in Nature and in Myth.* London: Humphrey Milford.
Ambrose of Milan. *Hexaemeron. Libri sex. PL* 14: 123A-274A.
André, J. 1949. *Études sur les termes de couleur dans la langue Latine.* Paris: Klincksieck.
Andrews, Alfred. 1955. "Greek and Latin Terms for Salmon and Trout." *TAPhA* 86: 308–20.
Anon. 1907. "International Fishery Investigations." *Nature* 75: 251–53.
Ansaldi, D. Mario. 1944. "Natura, origine et importanza del sangue in Alberto Magno." *Angelicum* 21: 306–25.
Anzulewicz, Henryk. 2013. "*Lupus rapax:* Allegorisches und Naturkundliches über den Wolf bei Albertus Magnus." *Reinardus. Yearbook of the International Reynard Society* 25: 11–27.

Aristotle. *De Generatione et Corruptione* (Forster and Furley, 1965).
———. *Generation of Animals* (Balme, 1972; Platt, 1910).
———. *Historia Animalium* (Peck, 1965–70; Balme, 1991; Thompson, 1966).
———. *Metaphysics* (Tredennick, 1961–62).
———. *On Marvellous Things Heard* (Hett, 1936).
———. *Parts of Animals* (Ogle, 1912; Peck, 1961).
———. *Physics* (Wicksteed and Cornford, 1934).
———. *Prior Analytics* (Cooke and Tredennick, 1967).
Armour, Peter. 1995. "Griffins." In Cherry (1995a), 72–103.
Arnaldi, F., et al. 1939. *Latinitatis Italicae Medii Aevi, inde ab A. CDLXXVI usque ad A. MXII Lexicon*. Brussels: Secrétariat administratif de l'U. A. I.
Arnott, W. Geoffrey. 2007. *Birds in the Ancient World from A to Z*. New York, Routledge.
———. 1977. "Swan Songs." *Greece and Rome*, ser. 2, 24: 149–53.
Ashley, Benedict M. 1980. "St. Albert and the Nature of Natural Science." In Weisheipl (1980), 73–102.
Auty, Robert, et al., eds. 1977–. *Lexikon des Mittelalters*. Munich: Artemis Verlag.
Avicenna, ca. 1500. *De animalibus per magistrum Michaelem Scotum de arabico in latinum translatus*. Venice: Joannes and Gregorius de Gregoriis, de Forlivio.
———. 1971. *Avicenna, Liber Canonis Medicine. Cum castigationibus Andree Bellunensis*. Brussels: Editions culture et civilization.
Bacon. 1859. *Opus Tertium*. In *Opera Quaedam Hactenus Inedita*, vol. 1, ed. J. S. Brewer. London: Longman, Green, Longman, and Roberts.
Badawi, Abdurrahman. 1968. *La transmission de la philosophie grecque au monde Arabe*. Paris: J. Vrin.
Baglioni, Silvestro. 1944. "Contributi di osservazioni e di esperimenti nella fonazione." *Angelicum* 21: 299–301.
Balme, D. M., ed. and trans. 1991. *Aristotle. History of Animals, Books VII–X*. Cambridge, Mass.: Harvard University Press.
———, ed. and trans. 1972. *Aristotle. De Partibus Animalium and De Generatione Animalium I with passages from II. 1–3*. Oxford: Clarendon Press.
Balss, Heinrich. 1928. *Albertus Magnus als Zoologe*. Münchner Beiträge zur Geschichte und Literatur der Naturwissenschaften und Medezin Hft. 11/12. Munich: Münchner Drucke.
Barstow, Anne Llewellyn. 1982. *Married Priests and the Reforming Papacy: The Eleventh Century Debates*. New York: Edwin Mellen.
Bartal, A. 1901. *Glossarium Mediae et Infimae Latinitatis Regni Hungariae*. Leipzig: Teubner.
Barthélemy, Dominique. 1971. "Eusèbe, la Septante et 'les autres.'" In *La bible et les pères*, 51–65. Paris: Presses universitaires de France.
Bartholomaeus Anglicus. (Pontanus, 1964).
Bartocetti, Vittorio. 1926. "Un interpretazione mistica della natura in S. Pier Damiano." *Scuola Cattolica* 7: 264–76, 337–52.
Beavis, Ian C. 1988. *Insects and Other Invertebrates in Classical Antiquity*. Exeter: University of Exeter.
Beebe, Frank Lyman, and Webster, Harold Melvin. 1964. *North American Falconry and Hunting Hawks*. Denver: North American Falconry and Hunting Hawks.
Beekenkamp, W. H., ed. 1941. *Berengar of Tours. De sacra coena adversus Lanfranc*. Kerkhistorische studien: behoorende bij het Nederlandscharachief voor Kerkgeschiedenis, deel 2. The Hague: Martinus Nijhoff.
Beer, Rüdiger Robert. 1977. *Unicorn: Myth and Reality*. New York: Mason/Charter.
Bell, Rudolph M. 1985. *Holy Anorexia*. Chicago: University of Chicago Press.
Bellamy, D. J., and Pfister, A. 1992. *World Medicine. Plants, Patients and People*. Oxford: Blackwell.

Bellunensis, Andree, ed. 1971. *Avicennae liber canonis medicinae cum castigationibus*. Brussels: Editions Culture et Civilisation.
Benjamin, Walter. 1992. "The Task of the Translator." In Schulte and Biguenet (1992), 71–82.
Benton, Janetta Rebold. 1992. *The Medieval Menagerie: Animals in the Art of the Middle Ages*. New York: Abbeville Press.
Benyus, Janine M. 1992. *Beastly Behavior*. Reading, Mass.: Addison-Wesley.
Bertram, Brian C. R. 1992. *The Ostrich Communal Nesting System*. Princeton: Princeton University Press.
Best, Michael J., and Brightman, Frank H., eds. 1973. *The Book of Secrets of Albertus Magnus*. Oxford: Clarendon Press.
Bknyi, Sandor. 1982. "Animals, Draft." In *DMA* 1: 293–98.
Blatt, Franz, ed. 1959–69. *Novum Glossarium Mediae Latinitatis ab Anno DCCC usque ad Annum MCC*. Hafnia: Ejnar Munksgaard.
Bloom, Allan, trans. 1968. *The Republic of Plato*. New York: Basic Books.
Bludau, August. 1925. *Die Schriftsfälschungen der Häretiker. Ein Beitrag zur Textkritik der Bibel*. Neutestamentliche Abhandlungen vol. 11.5. Münster: Aschendorff.
Bodson, Liliane, ed. 1997. *L'animal de compagnie: ses rôles et leurs motivations au regard de l'histoire*. Colloques d'histoire des connaissances zoologiques, 8. Liège: Université de Liège.
———. 1993. *L'histoire de la connaissance du comportement animal*. Colloques d'histoire des connaissances zoologiques, 4. Liège: Université de Liège.
Boese, H., ed. 1973. *Thomas Cantimpratensis: Liber de Natura Rerum*. Part 1. Berlin: De Gruyter.
Bono, James J. 1984. "Medical Spirits and the Medieval Language of Life." *Traditio* 40: 91–130.
Booth, Edward O. P. 1982. "Conciliazioni ontologiche delle tradizioni Platonica e Aristotelica in Sant'Alberto e San Tommaso." In *Sant' Alberto Magno l'uomo e il pensatore*. Studia Universitatis S. Thomae in Urbe, 15: 61–81. Milan: Massimo.
Borgnet, Auguste, ed. 1890–99. *B. Alberti Magni Ratisbonensis Episcopi Ordinis Praedicatorum Opera Omnia*. 38 vols. Paris: L. Vivès.
Bos, E. P., and Meijer, P. A., eds. 1992. *On Proclus and His Influence in Medieval Philosophy*. Leiden: Brill.
Bostock, John, and Riley, H. T., trans. and eds. 1893. *The Natural History of Pliny*. 6 vols. London: George Bell & Sons.
Boulenger, E. G. 1914. *Reptiles and Batrachians*. London: J. M. Dent and Sons.
Brett, Edward Tracy. 1984. *Humbert of Romans. His Life and Views of Thirteenth-Century Society*. Toronto: Pontifical Institute of Mediaeval Studies.
Breuning, Wilhelm. 1982. *Albertus Magnus. Sein Leben und seine Bedeutung. Albert—der Theologe*, ed. Manfred Entrich. Graz, Vienna, and Cologne: Styria.
Bristowe, W. S. 1971. The *World of Spiders*. London: Collins.
Brock, A. J., trans. 1979. *Galen. On the Natural Faculties*. Reprint of 1916 ed. Cambridge, Mass.: Harvard University Press.
Brody, Saul Nathaniel. 1974. *The Disease of the Soul: Leprosy in Medieval Literature*. Ithaca: Cornell University Press.
Broek, R. van den. 1972. *The Myth of the Phoenix according to Classical and Early Christian Traditions*. Leiden: E. J. Brill.
Brown, D. 1995. *Encyclopedia of Herbs and Their Uses*. London: Dorling Kindersley.
Brown, John H. 1973. *Toxicology and Pharmacology of Venoms from Poisonous Snakes*. Springfield, Ill.: Chas. Thomas.
Brunn, Émile zum. 1985. "La doctrine albertienne et eckhartienne de l'homme d'après quelques textes des 'Sermons allemands.'" *Freiburger Zeitschrift für Philosophie und Theologie* 32: 137–43.

Bub, Hans. 1991. *Bird Trapping and Bird Banding*, trans. F. Hamerstrom. Ithaca: Cornell University Press.
Bunbury, E. H. 1959. *A History of Ancient Geography among the Greeks and Romans, from the Earliest Ages till the Fall of the Roman Empire*. 2 vols., 2nd ed. New York: Dover.
Burger, Maria. 2015. *Alberti Magni opera omnia/Super I Librum Sententarum Distinctiones 1–3 Tomus XXIX,1: Super I Sententiarum*. Cologne: Aschendorf.
Burnett, Charles. 1992. "Scientific Speculations." In *A History of Twelfth-Century Western Philosophy*, ed. Peter Dronke, 151–76. Cambridge: Cambridge University Press.
Burren, A. M. J. van, ed. 1985. *Tussentijds. Bundel studien aangeboden aan W. P. Gerritsen ter gelegenheid van zijn vijftigste verjaardag*. Utrecht: HES Uitgevers.
Bustard, Robert. 1973. *Sea Turtles*. New York: Taplinger.
Cadden, Joan. 1993. *Meanings of Sex Differences in the Middle Ages: Medicine, Science, and Culture*. Cambridge: Cambridge University Press.
———. 1980. "Albertus Magnus' Universal Physiology: The Example of Nutrition." In Weisheipl, ed. (1980d), 321–39.
———. 1971. "The Medieval Philosophy and Biology of Growth: Albertus Magnus, Thomas Aquinas, Albert of Saxony and Marsilius of Inghen on Book 1, Chapter V, of Aristotle's *De Generatione et Corruptione*." Diss. Indiana University.
Campbell, Donald. 1926. *Arabian Medicine and Its Influence on the Middle Ages*. 2 vols. London: Kegan Paul, Trench, Trubner.
Campbell-Platt, Geoffrey. 1987. *Fermented Foods of the World. A Dictionary and Guide*. London: Butterworths.
Canavero, Alessandro Tarabochia. 1984. "A proposito del trattato *De bono naturae* nel *Tractus de natura boni* di Alberto Magno." *Rivista di filosofia neo-scolastica* 76: 353–73.
Canby, Sheila R. 1995. "Dragons." In Cherry (1995a), 14–43.
Cantin, André. 1977. "La 'raison' dans de la *De sacra coena* de Bérenger de Tour (av. 1070)." *Recherches augustiniennes* 12: 177–211.
———. 1972. "*Ratio et auctoritas* de Pierre Damien é Anselme." *Revue des études augustiniennes* 18: 152–79.
Carmody, Francis, ed. 1939. *Physiologus Latinus éditions préliminaires, versio B*. Paris: Librairie E. Droz.
Cary, George. 1956. *The Medieval Alexander*. Cambridge: Cambridge University Press.
Catania, Francis J. 1980. "'Knowable' and 'Namable' in Albert the Great's Commentary on the *Divine Names*." In Kovach and Shahan (1980), 97–128.
———. 1967. "Albert the Great." *Encyclopedia of Philosophy* 1: 64–66.
Centre universitaire d'études et de recherches médiévales (Aix-en-Provence, Bouches-du-Rhône). 1992. *Le cheval dans le monde médiéval*. Aix-en-Provence: Publications du CUERMA.
Chase, Frederic C., Jr., trans. 1970. *St. John of Damascus, Writings*. Washington, DC: Catholic University of America Press.
Chenu, M. D. 1968. *Nature, Man, and Society in the Twelfth Century: Essays on New Theological Perspectives in the Latin West*, ed. and trans. Jerome Taylor and Lester K. Little. Chicago: University of Chicago Press.
Cherry, John, ed. 1995a. *Mythical Beasts*. San Francisco: Pomegranate Books.
———. 1995b. "Unicorns." In Cherry (1995a), 44–71.
Christianson, Gerald, and Izbicki, Thomas M., eds. 1991. *Nicolas of Cusa in Search of God and Wisdom. Essays in Honor of Morimichi Watanabe by the American Cusanus Society*. Leiden: Brill.
Clark, Willene B. 1992. *The Medieval Book of Birds. Hugh of Fouilloy's Aviarium: Edition, Translation, and Commentary*. Medieval and Renaissance Texts and Studies.

Clark, Willene B., and McMunn, Meradith T. 1989. *Beasts and Birds of the Middle Ages: The Bestiary and Its Legacy*. Philadelphia: University of Pennsylvania Press.

Colgrave, Bertram, and Mynors, R. A. B, eds. 1969. *Bede's Ecclesiastical History of the English People*. Oxford: Oxford University Press.

Congar, Yves. 1980. "'In dulcedine societatis quaerere veritatem'—Notes sur le travail en équipe chez les Prêcheurs au XIIIe siècle." In Meyer and Zimmermann (1980), 47–57.

Cooke, Harold P., and Tredennick, Hugh, trans. and eds. 1967. *The Categories. On Interpretation. Prior Analytics,* by Aristotle. Cambridge, Mass.: Harvard University Press.

Coombs, Franklin. 1978. *The Crows*. London: Batsford.

Cornford, Francis M. 1975. *Plato's Cosmology: The* Timaeus *of Plato*. Indianapolis: Bobbs-Merrill.

Courtenay, William J. 1984. "Nature and the Natural in Twelfth Century Thought." In *Covenant and Causality in Medieval Thought*, 3: 1–26. London: Variorum Reprints.

Courtois, V., ed. 1956. *Avicenna Commemoration Volume*. Calcutta: Iran Society.

Cousteau, Jacques-Yves, and Diolé, Phillipe. 1973. *Octopus and Squid: The Soft Intelligence*. Garden City, N. J.: Doubleday.

Craemer-Ruegenberg, Ingrid. 1980a. "Die Seele als Form in einer Hierarchie von Formen." In Meyer and Zimmermann (1980), 59–88.

———. 1980b. "The Priority of Soul as Form and Its Proximity to the First Mover: Some Aspects of Albert's Psychology in the First Two Books of His Commentary on Aristotle's *De anima*." In Kovach and Shahan (1980), 49–62.

———. 1980c. *Albertus Magnus*. Munich: Beck.

———. 1980d. "Albert le Grand et ses démonstrations de l'immortalité de l'âme intellective." *Archives de philosophie* 43: 667–73.

Crombie, A. C. 1953. *Robert Grosseteste and the Origins of Experimental Science, 1100–1700*. Oxford: Clarendon Press.

Cunningham, Stanley B. 1967. "Albertus Magnus on Natural Law." *Journal of the History of Ideas* 28: 479–502.

Curley, Michael J., trans. 1979. *Physiologus*. Austin: University of Texas Press.

Dähnert, Ulrich. 1934. *Die Erkenntnislehre des Albertus Magnus gemessen an den Stufen der "Abstractio" mit einem ausführlichen systematischen sachverzeichnis und einer monographischen bibliographie Albertus Magnus*. Studien und Bibliographien zur Gegenwartsphilosophie, Hft 4. Leipzig: Hirzel.

Dalby, Andrew. 2013. *Food in the Ancient World, from A to Z*. New York: Routledge.

Dalby, David. 1965. *Lexicon of the Mediaeval German Hunt: A Lexicon of Middle High German Terms (1050–1500), associated with the Chase, Hunting with Bows, Falconry, Trapping and Fowling*. Berlin: DeGruyter.

Damon, S. Foster. 1930. "A Portrait of Albertus Magnus." *Speculum* 5: 102–3.

Dannenfeldt, Karl. 1982. "Ambergris: The Search for Its Origin." *Isis* 73: 382–97.

Davidson, Alan. 1976. *Mediterranean Seafood,* 2nd ed. Baton Rouge: Louisiana State University Press.

Davies, Malcolm, and Kathirithamby, Jeyaraney. 1986. *Greek Insects*. Oxford: Oxford University Press.

Davis, John W., et al., eds. 1971. *Infections and Parasitic Diseases of Wild Birds*. Ames: University of Iowa Press.

Davis, Michael T. 1982. "Mappa Mundi." In *DMA* 8: 120–22.

Dawson, Warren R. 1925. "The Bee-eater from the Earliest Times." *Isis* 7: 590–93.

De Asúa, Miguel. 2013. "War and Peace: Medicine and Natural Philosophy in Albert the Great." In Resnick (2013), 269–97.

———. 1991. "The Organization of Discourse on Animals in the 13th C.: Peter of Spain, Albert the Great, and the Commentaries on *De Animalibus.*" Diss. University of Notre Dame.
Deboutte, A. 1984. "Thomas van Cantimpré als auditor von Albertus Magnus." *Ons geestelijk erf* 58, 2: 192–209.
Deferrari, Roy J. 1960. *A Latin-English Dictionary of St. Thomas Aquinas.* Boston: Daughters of St. Paul.
Demaitre, Luke. 1996. "The Relevance of Futility: Jordanus de Turre (fl. 1313–1335) on the Treatment of Leprosy." *Bulletin of the History of Medicine* 70: 25–61.
———. 1985. "The Description and Diagnosis of Leprosy by Fourteenth-Century Physicians." *Bulletin of the History of Medicine* 59: 327–44.
Demaitre, Luke, and Travill, Anthony A. 1980. "Human Embryology and Development in the Works of Albertus Magnus." In Weisheipl (1980d), 405–40.
Dewan, Lawrence. 1980. "St. Albert, the Sensibles, and Spiritual Being." In Weisheipl (1980d), 291–320.
Dezani, Serafino. 1944. "S. Alberto Magno: L'osservazione e l'esperimento." *Angelicum* 21: 41–49.
D'Haenens, Albert. 1961. *l'Abbaye Saint-Martin de Tournai de 1290 á 1350.* Louvain: Bibliothèque, Université de Louvain.
Diefenbach, Lorenz. 1867. *Novum Glossarium Latino-Germanicum Mediae et Infimae Aetatis.* Beiträge zur wissenschaftlichen Kunde der neulateinischen und der germanischen Sprachen. Frankfurt am Main: J. D. Sauerlinder.
———. 1857. *Glossarium Latino-germanicum Mediae et Infimae Aetatis.* Frankfurt am Main: J. Baer.
Diels, Hermann. 1951. *Die Fragmente der Vorsokratiker,* ed. W. Kranz. 3 vols., 6th ed. Berlin: Weidmann.
Dilke, O. A. W. 1985. *Greek and Roman Maps.* Ithaca, N. Y.: Cornell University Press.
Dimock, James F., ed. 1867. *Giraldi Cambrensis: Topographia Hibernica et Expugnatio Hibernica.* Rerum Britannicarum Medii Aevi Scriptores. London: Longman.
Dirrigl, Michael. 1980. *Albertus Magnus: Bischof von Regensburg, Theologe, Philosoph, und Naturforscher (um 1193–1280).* Regensburg: Mittelbayerische Druckerei und Verlags-Gesellschaft.
Dodd, J. M. 1983. "Reproduction in Cartilaginous Fishes (Chondrichthyes)." In *Fish Physiology,* ed. W. S. Hoar et al., 9: 31–96. Orlando: Harcourt Brace Jovanovich.
Dorveaux, Paul, trans. 1913. *Le livre des simples médicines. Traduction française du Liber de simplici medicina dictus Circa instans de Platearius tirée d'un manuscrit du XIIIe siècle.* Paris: Société française d'histoire de la médecine.
Dowdall, L. D., trans. 1913. *Aristotle. De Mirabilibus Auscultationibus.* In *The Works of Aristotle,* vol. 6, ed. T. Loveday et al. Oxford: Clarendon Press.
Draelants, Isabelle, trans. 2007. *Le Liber de virtutibus herbarum, lapidum et animalium* (Liber aggregationis): *Un texte à succès attribué à Albert le Grand.* Florence: Sismel Edizioni del Galluzzo.
———. 2000. "Le dossier des livres 'sur les animaux et les plantes' de Iorach: traditions occidentale et orientale." In *Occident et Proche-Orient: contacts scientifiques au temps des Croisades: Actes du colloque de Louvain-la-Neuve, 24 et 25 mars 1997,* eds. Isabelle Draelants, Anne Tihon & Baudouin Van den Abeele, 191–276. Turnhout: Brepols.
Druce, George C. 1923. "An Account of the Myrmekoleon or Ant-lion." *Antiquaries Journal* 3: 347–64.
———. 1919. "The Elephant in Medieval Legend and Art." *Archaeological Journal* 76: 1–73.
———. 1912. "The Caladrius and Its Legend, Sculptured upon the Twelfth-Century Doorway of Alne Church, Yorkshire." *Archaeological Journal* 69: 381–416.

———. 1911. "Notes on the History of the Heraldic Jall or Yale." *Archaeological Journal* 68: 173–99.

———. 1910. "The Amphisboena and Its Connexions in Ecclesiastical Art and Architecture." *Archaeological Journal* 67: 285–317.

DuCange, Ch du Fresne sieur de. 1883–87. *Glossarium mediae et infimae latinitatis Conditum a Carolo du Fresne, domino Du Cange, auctum a monachis ordinis S. Benedicti, cum supplementis integris D. P. Carpenterii, Adelungii, aliorum, susque digessit G. A. L. Henschel; sequuntur Glossarium gallicum, Tabulae, Indices auctorum et rerum, Dissertationes,* new, enlarged edition, ed. Léopold Favre. 10 vols. Paris: L. Favre.

Eamon, William. 1985. "Books of Secrets in Medieval and Early Modern Science." *Sudhoffs Archiv* 69: 26–49.

Eckert, Willehad Paul. 1980. "Albert der Grosse als Naturwissenschaftler." *Angelicum* 57: 477–95.

Edgeworth, Robert J. 1987. "'Off-Color' Allusions in Roman Poetry." *Glotta* 65: 134–37.

Einhorn, Jürgen W. 1976. *Spiritalis Unicornis*. Münstersche Mittelalter-Schriften vol. 13. Munich: W. Fink.

Eisentstein, Herbert. 1993. "Zu drei nordeuropäischen Tieren aus Ibn Saìd's Geographie." *Acta Orientalia* 54: 53–61.

Ellis, Richard. 1980. *The Book of Whales*. New York: Alfred A. Knopf.

Ellis, Robinson. 1889. *A Commentary on Catullus*. Oxford: Clarendon Press.

Engel, Josef. 1970. *Grosser Historischer Weltatlas. II Teil. Mittelalter*. Munich: Bayerischer Schulbuch-Verlag.

Entrich, Manfred, ed. 1982. *Albertus Magnus: sein Leben und seine Bedeutung*. Graz: Styria.

Ernst, Carl H., and Barbour, Roger W. 1989. *Turtles of the World*. Washington, DC: Smithsonian Institution Press.

Farb, Peer. 1978. *Humankind*. Boston: Houghton Mifflin.

Fauser, Wilfried. 1982. *Die Werke des Albertus Magnus in ihrer handshriftlichen Überlieferung*. Münster: Aschendorff.

Favazza, Armando R. 1996. *Bodies under Siege: Self-Mutilation and Body Modification in Culture and Psychiatry,* 2nd ed. Baltimore: Johns Hopkins University Press.

Ferraro, Giuseppe. 1984. "Il tema pneumatologico nelle 'Enarrationes in Joannem' di Sant' Alberto Magno." *Angelicum* 61: 316–46.

Filthaut, Ephrem, ed. 1955. *De natura et origine animae, De principiis motus processivi, Quaestiones super de animalibus*. Vol. 12 of the *Opera Omnia* (Geyer, 1951–). Münster: Aschendorff.

———. 1952. "Um die *Quaestiones de animalibus* Alberts des Grossen." In Ostlender (1952a), 112–27.

Flower, Barbara, and Rosenbaum, Elisabeth. 1958. *The Roman Cookery Book: A Critical Translation of the Art of Cooking by Apicius*. New York: British Book Centre.

Foerster, Richard, ed. 1893. *Scriptores Physiognomonici Graeci et Latini*. 2 vols. Leipzig: Teubner.

Fonahn, A. 1922. *Arabic and Latin Anatomical Terminology: Chiefly from the Middle Ages*. Kristiana: Dybwad.

Forbes Thomas. 1972. "*Lapis bufonis:* The Growth and Decline of a Medical Superstition." *Yale Journal of Biology and Medicine* 45: 139–149.

Ford, John K. B., et al. 1994. *Killer Whales*. Vancouver: University of Vancouver Press.

Forster, E. S., trans. 1927. *The Works of Aristotle. Volume VII*. Oxford: Oxford University Press.

Forster, E. S., and Furley, D. J., trans. 1965. *On Sophistical Refutations. On Coming-to-be and Passing Away. On the Cosmos*. Cambridge, Mass.: Harvard University Press.

Fowler, W. Warde. 1918. "Aves Diomedae." *Classical Review* 32: 66–68.

Fraser, H. Malcolm. 1951. *Beekeeping in Antiquity*, 2nd ed. London: University of London Press.
Frawley, William, ed. 1984. *Translation: Literary, Linguistic, and Philosophical Perspectives*. Newark, N. J.: University of Delaware Press.
Freed, John B. 1977. *The Friars and German Society in the 13th Century*. Mediaeval Academy of America, Publication no. 86. Cambridge, Mass.: Mediaeval Academy of America.
Freeman, Kathleen. 1966. *The Pre-Socratic Philosophers*. Cambridge, Mass.: Harvard University Press.
———. 1962. *Ancilla to the Pre-Socratic Philosophers*. Cambridge, Mass.: Harvard University Press.
Friedman, John Block. 1981. *The Monstrous Races in Medieval Art and Thought*. Cambridge, Mass.: Harvard University Press.
———. 1974. "Thomas of Cantimpré. *De Naturis Rerum*. Prologue, Book III and Book XIX." In *Cahiers d'Études Médiévales, II. La science de la nature: théories et pratiques*, 107–54. Montreal: Bellarmin.
Fries, Albert. 1988. "Zur Problematik der Summa Theologica unter dem Namen des Albertus Magnus." *Franziskanische Studien* 70: 68–91.
———. 1987. "Zwei ungedruckte Quästionen des Albertus Magnus und die 'Summa Theologica.'" *Franziskanische Studien* 69: 159–72.
———. 1983. "Die Trinitätssequenz 'Profitentes' und Albertus Magnus." *Archiv für Liturgiewissenschaft* 25: 276–96.
Fuchs, J. 1970–. *Lexicon Latinitatis Nederlandicae Medii Aevi*. Leiden: Brill.
Führer, Markus L. 1991a. "The Contemplative Function of the Agent Intellect in the Psychology of Albert the Great." In Mojsisch and Pluta (1991), 1: 305–19.
———. 1991b. "The Theory of Intellect in Albert the Great and Its Influence on Nicholas of Cusa." In Christianson and Izbicki (1991), 45–56.
Galbraith, Georgina Rosalie. 1925. *The Constitution of the Dominican Order, 1216 to 1360*. Manchester: The University Press.
Galen. 1821. *De Methodo Medendi*. Vol. 10 of the *Opera Omnia* (Kühn, 1964–65).
García Arranz, José Julio. 1996. *Ornitología emblemática: las aves en la literatura simbólica ilustrada en Europa durante los siglos XVI y XVII*. Cáceres: Universidad de Extremadura.
———. 1990. "La Salamandra: distintas interpretaciones gráficas de un mito literario tradicional." *Norba-Arte* 10:53–68. Universidad de Extremadura: Servicio de Publicaciones. http://dialnet.unirioja.es/servlet/oaiart?codigo=107432
Garcia de Diego, Vicente. 1955. *Diccionario etimologico espanol e hispanico*. Madrid: Editorial S. A. E. T.A.
Gaulin, Jean-Louis. 1994. "Albert le Grand Agronome. Notes sur le *Liber VII de vegetabilibus*." In *Comprendre et maîtriser la nature au Moyen Âge. Mélange d'histoire des science offerts à Guy Beujouan*, 154–70. Geneva: Droz.
Geiger, L. B. 1962. "La Vie, Acte essentiel de l'ame, l'esse acte de l'essence d'apres Albert-le-Grand." In *Études d'histoire litteraire et docrinale*. Publications de l'institut d'études médiévales, 17, 49–116. Montreal: Institute d'études médiévales.
George, Wilma, and Yapp, Brunsdon. 1991. *The Naming of the Beasts: Natural History in the Medieval Bestiary*. London: Duckworth.
Gerald of Wales. (J. O'Meara, 1982).
Gerhardt, Mia I. 1965. "The Ant-Lion. Nature Study and the Interpretation of a Biblical Text, from the Physiologus to Albert the Great." *Vivarium* 3: 1–23.
Gerson, Lloyd P., ed. 1983. *Graceful Reason. Essays in Ancient and Medieval Philosophy Presented to Joseph Owens, CSSR*. Papers in Medieval Studies, 4. Toronto: Pontifical Institute of Mediaeval Studies.

Gerth Van Wijk, H. L. 1911. *A Dictionary of Plant Names*. 2 vols. The Hague: Martinus Nijhoff.

Geyer, Bernhard. 1973. "Albertus Magnus." *Encyclopedia Britannica* 1: 528–29.

———. ed. 1966. *Die Universitätspredigten des Albertus Magnus*. Munich: Verlag der Bayerische Akademie der Wissenschaften.

———. 1960–64. *Metaphysica*, 2 vols. Vol. 16, parts 1 & 2, of the *Opera Omnia* (Geyer, 1951).

———. 1958. "Die mathematische Schriften des Albertus Magnus." *Angelicum* 35: 159–75.

———, ed. 1951–. *Opera Omnia*. Monasterii Westfalorum: Aschendorff.

———. 1948. "Die handschriftliche Verbreitung der Werke Alberts des Großen als Maßstab seines Einflusses." In *Studia mediaevalia in Honor of Rev. Raymund Joseph Martin, O. P.*, 221–28. Bruges: De Tempel.

———. 1935. "Die ursprüngliche Form der Schrift Alberts des Grossen *De Animalibus* nach dem Kölner Autograph." In *Aus der Geisteswelt des Mittelalters*. Beiträge zur Geschichte der Philosophie und Theologie des Mittelalters, suppl. 3: 578–90.

Ghigi, A. 1944. "Correlazioni fra gli organi, le funzioni et l'ambiente nel trattato degli animali de Alberto Magno." *Angelicum* 21: 208–19.

Gieraths, Paul Gundolf. 1982. "Vita e Personalitá di Sant' Alberto Magno." In *Sant' Alberto l'uomo e il pensatore*, 211–44. Milan: Massimo.

Ginzburg, Louis. 1987. *The Legends of the Jews*, trans. Henrietta Szold. 7 vols. Philadelphia: Jewish Publication Society of America.

Glare, P. G. W., ed. *The Oxford Latin Dictionary*. Oxford: Clarendon Press.

Glasier, Phillip. 1979. *Falconry and Hawking*. Newton Centre, Mass.: C. T. Branford.

Godefroy, Frédéric. 1881. *Dictionnaire de l'ancienne langue française*. 9 vols. Paris: F. Vieweg.

Godley, A. D., trans. 1966–1971. *Herodotus*. 4 vols. Cambridge, Mass.: Harvard University Press.

Goergen, Donald. 1980. "Albert the Great and Thomas Aquinas on the Motive of the Incarnation." *Thomist* 4: 523–38.

Goldstein-Préaud, Tamara. 1981. "Albert le grand et les questions du XIIIe siècle sur le *De Animalibus* d'Aristote." *History and Philosophy of the Life Sciences* 3: 61 -71.

Goodwin, Derek. 1986. *Crows of the World*, 2nd ed. London: British Museum.

Gorman, Martyn L., and Stone, R. David. 1990. *The Natural History of Moles*. Ithaca: Comstock Publishing Associates.

Gotthelf, Allan, and Lennox, James G., eds. 1987. *Philosophical Issues in Aristotle's Biology*. Cambridge: Cambridge University Press.

Gow, A. S. F., and Scholfield, A. F., eds. 1953. *Nicander: The Poems and Poetical Fragments*. Cambridge: Cambridge University Press.

Grabmann, Martin. 1944. "Zur philosophischen und naturwissenschaftlichen Methode in den Aristoteleskommentaren Alberts des Grossen." *Angelicum* 21: 50–64.

Graesse, Johann Georg, ed. 1922. *Orbis Latinus*. 3 vols. Berlin: R. C. Schmidt.

Grant, Edward, ed. 1974. *A Source Book in Medieval Science*. Cambridge, Mass.: Harvard University Press.

Gray, Henry. 1977. *Anatomy, Descriptive and Surgical*, rev. American ed., from 15th English ed. New York: Gramercy.

Grayzel, Solomon. 1966. *The Church and the Jews in the XIIIth Century*, rev. ed. New York: Hermon Press.

Gregory, Justina. 2007. "Donkeys and the Equine Hierarchy in Archaic Greek Literature." *CJ* 102: 193–212.

Gresseth, Gerald K. 1964. "The Myth of Alcyone." *Transactions of the American Philological Association* 95: 88–98.

Griffith, Mark. 2006. "Horsepower and Donkey Work. Equids and the Ancient Greek Imagination." *Classical Philology* 101: 185–246, 307–58.
Grigson, G. 1955. *The Englishman's Flora*. London: Phoenix House.
Grimm, Jacob, and Grimm, Wilhelm. 1854f. *Deutsches Wörterbuch*. Leipzig: S. Hirzel.
Grönemann, Olaf. 1992. "Das Werk Alberts des Grossen und die Kölner Ausgabe der *Opera omnia*." *Recherche théologie ancienne et médiévale* 59: 125–54.
Gruner, O. Cameron. 1930. *A Treatise on the Canon of Medicine of Avicenna*. London: Luzac.
Grzimek, Bernhard, ed. 1977. *Grzimek's Encyclopedia of Ethology*. New York: Van Nostrand Reinhold.
―――. 1972–75. *Grzimek's Animal Life Encyclopedia*. 13 vols. New York: Van Nostrand Reinhold.
Guthrie, W. K. C. 1962. *A History of Greek Philosophy*. 6 vols. (1962–81). Cambridge: Cambridge University Press.
Haboucha, Reginetta. 1990. "Clerics, Their Wives, and Their Concubines in the 'Partidas' of Alfonso el Sabio." In Lemay (1990), 85–104.
Hackett, Jeremiah M. G. 1980. "The Attitude of Roger Bacon to the *Scientia* of Albertus Magnus." In Weisheipl (1980d), 53–72.
Hall, Thomas S. 1971. "Life, Death, and the Radical Moisture: A Study of Thematic Pattern in Medieval Medical Theory." *Clio Medica* 6: 2–23.
Halleaux, Robert. 1982. "Albert le Grand et l'alchemie." *Revue des sciences philosophiques et théologiques* 66: 57–80.
Hammond, N. G. L., and Scullard, H. H., eds. 1970. *The Oxford Classical Dictionary*, 2nd ed. Oxford: Clarendon Press.
Harder, Wilhelm. 1975. *Anatomy of Fishes. Part I: Text*. Stuttgart: E. Schweizerbartische Verlagsbuchhandlung.
Harting, J. E. 1964. *Bibliotheca Accipitraria: A Catalogue of Books Ancient and Modern Relating to Falconry, with Notes, Glossary, and Vocabulary*, reprint of 1891 ed. London: Holland Press.
Haskins, Charles Homer. 1960. *Studies in the History of Mediaeval Science*. New York: Frederick Ungar Publishing.
Hauréau, B. 1890–91. *Notices et extraits de quelques manuscrits latins de la Bibliothèque nationale*. 2 vols. Paris: Klincksieck.
Hegel, Georg Wilhelm Frederick. 1892–96. *Hegel's Lectures on the History of Philosophy*, trans. E. S. Haldane and Frances H. Simson. 3 vols. London: Routledge and Kegan Paul.
Henry, R. 1947. *Ctésias. La Perse, L'Inde: les sommaires de Photius*. Collection Lebègue, 7th ser., no. 84. Brussels: Lebègue.
Henry of Hereford. 1859. *Liber de Rebus Memorabilioribus sive Chronicon Henrici de Hervordia*, ed. A. Potthast. Göttingen: Dieterich.
Herrera, Maria Ester. 1994. "La Historia del 'diamante' desde Plinio a Bartolome el Ingles." In *Comprendre et maîtriser la nature au Moyen Âge. Mélange d'histoire des science offerts à Guy Beujouan*, 139–53. Geneva: Droz.
Hett, W. W., ed. and trans. 1936. "On Marvellous Things Heard," by Aristotle. In *Aristotle, Minor Works*, 237–325. Cambridge, Mass.: Harvard University Press.
Hill, Jane H. 1980. "Apes and Language." In *Speaking of Apes: A Critical Anthology of Two-Way Communication with Man*, ed. Thomas Sebeok and Jean Umiker-Sebeok, 331–51. New York: Plenum Press.
Hill, John E., and Smith, James D. 1984. *Bats. A Natural History*. London: British Museum.
Hinnebusch, William A. 1966–73. *The History of the Dominican Order*. 2 vols. New York: Alba House.
Hödl, Ludwig. 1986. "Die 'Entdivinisierung' des menschlichen Intellekts in der mittelalterlichen Philosophie und Theologie." In *Zusammenhänge, Einflüsse, Wirkungen: Kongres-*

sakten zum ersten Symposium des Mediävistenverbandes in Tübingen (1984), 57–70. Berlin: DeGruyter.

Hoffmann, Richard C. "Fishponds." In *DMA* 5: 73–74.

Holmér, Gustaf. 1966. "Traduction en ancien italien de quelques chapitres du *Liber de animalibus* d'Albert le Grand." *Studia Neophilologica* 38: 211–56.

Honigmann, Ernst. 1929. *Die Sieben Klimata und die Poleis Episemoi: Eine Untersuchung zur Geschichte der Geographie und Astrologie im Altertum und Mittelalter.* Heidelberg: Carl Winters Universitätsbuchhandlung.

Hope, T. E. 1971. *Lexical Borrowings in the Romance Languages: A Critical Study of Italianisms in French and Gallicisms in Italian from 1100 to 1900.* 2 vols. New York: New York University Press.

Hopkins, Jasper, ed. and trans. 1975. *Anselm of Canterbury*, vol. 4: *Hermeneutical and Textual Problems in the Complete Treatises of St. Anselm.* Toronto: Edwin Mellen Press.

Hossfeld, Paul, ed. 1987–93. *Physica*, 2 vols. Vol. 4, parts 1 & 2, of the *Opera Omnia* (Geyer, 1951–).

———. 1986a. "Albertus Magnus über die Ewigkeit aus philosophischer sicht." *Archivum Fratrum Praedicatorum* 56: 31–48.

———. 1986b. "Studien zur Physik des Albertus Magnus." In Zimmermann (1986), 1–42.

———. 1985. "Die Physik des Albertus Magnus, Quellen und Charakter." *Archivum Fratrum Praedicatorum* 55: 49–65.

———. 1983. *Albertus Magnus als Naturphilosoph und Naturwissenschaftler.* Bonn: Albertus-Magnus-Institut.

———, ed. 1980a. *De causis proprietatum elementorum.* Vol. 5, part 2, of the *Opera Omnia* (Geyer, 1951–), 48–106.

———. 1980b. "Die Lehre des Albertus Magnus von den Kometen." *Angelicum* 57: 533–41.

———. 1980c. "Die Arbeitsweise des Albertus Magnus in seinen naturphilosophischen Schriften." In Meyer and Zimmermann (1980), 195–204.

———. 1980d. "Senecas *Naturales quaestiones* als Quelle der *Meteora* des Albertus Magnus." *Archivum Fratrum Praedicatorum* 50: 63–84.

———, ed. 1980e. *De natura loci, de causis proprietatum elementorum, De generatione et corruptione.* Vol. 5, part 2, of the *Opera Omnia* (Geyer, 1951–).

———, ed. 1971. *De caelo et mundo.* Vol. 5, part 1, of the *Opera Omnia* (Geyer, 1951–).

Houwen, Luuk A. J. R., ed. 1997. *Animals and the Symbolic in Mediaeval Art and Literature.* Groningen: Egbert Forsten.

How, W. W., and Wells, J. 1928. *A Commentary on Herodotus.* 2 vols. Oxford: Clarendon Press.

Hudeczek, Methodius. 1944a. "De ratione quam habet S. Albertus cum problematibus scientiae hodiernae." *Angelicum* 21: 332–36.

———. 1944b. "De lumine et coloribus." *Angelicum* 21: 112–38.

Hull, Denison Binham. 1964. *Hounds and Hunting in Ancient Greece.* Chicago: University of Chicago Press.

Hünemörder, Christian. 1980. "Die Zoologie des Albertus Magnus." In Meyer and Zimmermann (1980), 235–48.

Huygens, R. B. C., ed. 1985. *Apologiae duae: Gozechini Epistula ad Walcherum. Burchardi, ut videtur, Abbatis Bellevallis, Apologia de barbis.* Corpus Christianorum Continuatio Mediaevalis, vol. 62. Turnholt: Brepols.

Hyland, Ann. 1990. *Equus. The Horse in the Roman World.* New Haven: Yale University Press.

Hyrtl, Joseph. 1880. *Onomatologia Anatomica Geschichte und Kritik der anatomischen Sprache der Gegenwart.* Vienna: Wilhelm Braumller.

———. 1879. *Das arabische und hebräische in der Anatomie.* Vienna: Wilhelm Braumller.

Imbach, Ruedi, and Flüeler, Christoph, eds. 1985. *Albert der Grosse und die deutsche Dominakerschule.* Special ed. of *Freiburger Zeitschrift für Philosophie und Theologie* 32.1/2.
Ineichen, Robert. 1993. "Zur Mathematik in den Werken von Albertus Magnus: Versuch einer Zusammenfassung." *Freibürger Zeitschrift für Philosophie und Theologie* 40: 55–87.
Irigoin, Jean. 1987. "Paper, Introduction of." In *DMA* 9: 388–90.
Irwin, Eleanor. 1974. *Colour Terms in Greek Poetry.* Toronto: Hakkert.
Jacquart, Danielle. 1990. "Medical Explanations of Sexual Behavior in the Middle Ages." In Lemay (1990), 1–21.
Jacquart, Danielle, and Thomasset, Claude. 1988. *Sexuality and Medicine in the Middle Ages,* trans. Matthew Adamson. Princeton, N. J.: Princeton University Press.
———. 1981. "Albert le Grand et les Problèmes de la Sexualité." *History and Philosophy of the Life Sciences* 3: 73–93.
Jacques de Vitry. *Historia Orientalis* (Moschus, 1971).
Jessen, Karl, ed. 1867. *Alberti Magni ex Ordine Praedicatorum De Vegetabilibus Libri VII,* 1982 reprint ed. Frankfurt am Main: Minerva.
Jones, W. H. S., ed. and trans. 1943. *Hippocrates,* Vol. IV. Cambridge, Mass.: Harvard University Press.
———, ed. and trans. 1923. *Hippocrates,* Vol. I. Cambridge, Mass.: Harvard University Press.
Jordan, Mark. 1992. "Albert the Great and the Hierarchy of Sciences." *Faith and Philosophy* 9: 483–99.
Kantorowicz, Ernst. 1931. *Frederick the Second, 1194–1250,* authorized English edition, trans. E. O. Lorimer. New York: Frederick Ungar.
Keller, Otto. 1909. *Die Antike Tierwelt.* 2 vols. Leipzig: W. Engelmann.
Ker, W. P., ed. 1961. *The Essays of John Dryden.* 2 vols. New York: Russell and Russell.
Keussen, Hermann. 1910. *Topographie der Stadt Köln im Mittelalter.* 2 vols. Dusseldorf: Droste.
Kibre, Pearl. 1980. "Albertus Magnus on Alchemy." In Weisheipl (1980d) 187–202.
Killermann, Sebastian. 1944. "Die somatische Anthropologie bei Albertus Magnus." *Angelicum* 21: 224–69.
———. 1910. *Die Vogelkunde des Albertus Magnus (1270–1280).* Regensburg: G. J. Manz.
Kim, Susan M. 1997. "Man-Eating Monsters and Ants as Big as Dogs." In Houwen (1997), 39–51.
King, Helen. 1995. "Half-Human Creatures." In Cherry (1995a), 138–67.
———. 1985. "From *Parthenos* to *Gyne:* The Dynamics of Category." Diss. University of London.
King, Judith E. 1983. *Seals of the World,* 2nd ed. Ithaca: Cornell University Press.
Kitchell, Kenneth F. 2015. "A Defense of the 'Monstrous' Animals of Pliny, Aelian, and Others." *Preternature* 4:125–51.
———. 2014. *Animals in the Ancient World from A to Z.* New York: Routledge.
———. 1993. "The View From Deucalion's Ark: New Windows on Antiquity." *Classical Journal* 88: 341–57.
———. 1989. "The Origin of Vergil's Myth of the Bugonia." In Sutton (1989), 193–206.
———. 1988. "Virgil's Ballasting Bees." *Vergilius* 34: 36–43.
Kitchell, Kenneth F., and Parker, Lin. 1993. "Death by Bull's Blood, a Natural Explanation." In *Alpha to Omega. Studies in Honor of George John Szemler on His Sixty-fifth Birthday,* ed. W. J. Cherf, 123–41. Chicago: Ares.
Kitchell, Kenneth F., and Resnick, Irven M., trans. 1999. *Albertus Magnus De Animalibus: A Medieval Summa Zoologica.* Baltimore: The Johns Hopkins University Press, 1999.
Kittler Ralf, et al. 2003. "Molecular Evolution of *Pediculus humanus* and the Origin of Clothing." *Current Biology,* 13: 1414–17.

Klemm, Fredrich. 1964. *A History of Western Technology,* trans. Dorothea Waley Singer. Cambridge, Mass.: MIT Press.

Kluge, Friederich. 1963. *Etymologisches Wörterbuch des Deutschen Sprache,* 19th ed., ed. Walther Mitzka. Berlin: De Gruyter.

Knowles, David. 1933. "The Monastic Horarium, 970–1120." *Downside Review* 51: 706–25.

Koch, Joseph. 1931. "Sind die Pygmäen Menschen?" *Archiv für Geschichte der Philosophie* 40: 195–209.

Köhler, Theodor Worlfram. 1991. "Anthropologische Erkennungsmerkmale menschlichen Seins. Die Frage der *Pygmei* in der Hochscholastik." In Zimmermann and Speer (1991), 2: 718–35.

Körting, Gustav. 1923. *Lateinisch-Romanisches Wörterbuch. Etymologisches Wörterbuch der romanischen Hauptsprachen,* 3rd ed. New York: Stechert.

Kotsiopoulos, George. 1969. *The Art and Sport of Falconry: Dancus Rex and Albertus, on Falcons.* Chicago: Argonaut.

Kovach, Francis J., and Shahan, Robert W., eds. 1980. *Albert the Great: Commemorative Essays.* Norman, Okla.: University of Oklahoma Press.

Krahe, Hans. 1964. *Unsere ältesten Flussnamen.* Wiesbaden: Otto Harrassowitz.

Kruk, Remke. 1985. "Aristoteles, Avicenna, Albertus en de *lucusta maris.*" In Burren (1985), 147–56, 346–48.

Kübler, B., ed. 1888. *Iuli Valeri Alexandri Polemi* Res gestae Alexandri Macedonis *translatae ex Aesopo graeco.* Leipzig: Teubner.

Kühn, C. G., ed. 1964–65. *Claudii Galeni Opera Omnia.* 20 vols., reprint of 1821–33 ed. Hildesheim: Brill.

Kuksewicz, Zdzislaw. 1989. "Die Einflüsse der Kölner Philosophie auf die Krakauer Universität im 15. Jahrhundert." In Zimmermann (1989), 287–98.

Kurdzialek, Marian, ed. 1963. *Davidis de Dinanto quaternulorum fragmenta: prolegomena.* Studia Medievistyczne, 3. Warsaw: n.p.

Latham, R. 1975. *Dictionary of Medieval Latin from British Sources.* London: Oxford University Press.

———. 1965. *Revised Medieval Latin Word-List from British and Irish Sources.* London: Oxford University Press.

Laufer, Berthold. 1928. *The Giraffe in History and Art. Anthropological Leaflet No. 27 of the Field Museum of Natural History.* Chicago: Field Museum of Natural History.

———. 1926. *Ostrich Egg-shell Cups of Mesopotamia and the Ostrich in Ancient and Modern Times.* Anthropology Leaflet 23. Chicago: Field Museum of Natural History.

Laurent, M. H., and Congar, M. J. 1931. "Bibliographie. Essai de Bibliographie Albertienne." *Revue Thomiste* 36: 422–68.

Lawn, Brian, ed. 1979. *The Prose Salernitan Questions, Edited from a Bodleian Manuscript (Auct. F. 3. 10).* London: Oxford University Press.

Layer, Adolf. 1981. "Albertus-Magnus-Bibliographie 1930–1980." *Jahrbuch des Vereins für Augsburger Bistumsgeschichte* 15: 71–112.

———. 1979a. "Biographisches Schrifttum zu Albert dem Grossen." *Jahrbuch des historischen Vereins Dillingen an der Donau* 81: 60–64.

———. 1979b. "Albert der Grosse und seine Schwäbische Heimat." *Jahrbuch des historischen Vereins Dillingen an der Donau* 81: 47–52.

———. 1979c. "Namen und Ehrennamen Alberts des Grossen." *Jahrbuch des historischen Vereins Dillingen an der Donau* 81:41 -43.

Leakey, Richard E. 1977. *Origins.* London: Macdonald and Jane's.

Le Bras, Gabriel. 1968. "Le mariage dans la théologie et le droit de l'eglise du XIe au XIIIe siècle." *Cahiers de civilisation médiévale* 11: 191–201.

Leclercq, Jean. 1975. "Experience and Interpretation of Time in the Early Middle Ages." *Studies in Medieval Culture* 5: 9–19.
Lecouteux, Claude. 1981. "Lamia-holzmuowa-holzfrowe-Lamich." *Euphorion* 3: 360–65.
Leibowitz, Joshua O. 1972. "Anatomy." *Encylopedia Judaica*, 2: 930–34.
Lemay, Helen Rodnite, trans. and ed. 1992. *Women's Secrets: A Translation of Pseudo-Albertus Magnus's* De Secretis Mulierum *with Commentaries*. SUNY Series in Medieval Studies, ed. E. Szarmach. Albany: SUNY Press.
———, ed. 1990. *Homo Carnalis: The Carnal Aspect of Medieval Human Life*. Center for Medieval and Early Renaissance Studies. Binghamton: SUNY.
Lenfant, Dominique, ed. 2004. *Ctésias de Cnide. La Perse. L'Inde. Autres Fragments*. Paris: Les Belles Lettres.
Lengerken, Hanns. 1955. *Ur, Hausrind, und Mensch*. Berlin: Deutschen Akademie der Landwirtschaftswissenschaften.
Leroy, Fernand. 1991. "Embryology and Placentation of Twins." In *Encyclopedia of Human Biology*, ed. Renato Dulbecco, 3: 305–15. San Diego: Harcourt Brace Jovanovich.
Lewis, C. S., and Short, C. T., eds. 1879. *A Latin Dictionary Founded on Andrews' Edition of Freund's Latin Dictionary*. Oxford: Clarendon Press.
Lexer, Matthias. 1869–78. *Mittelhochdeutsches Wörterbuch*. Leipzig. Hirzel.
Libera, Alain de. 1992a. "Albert le Grand et le platonisme de la doctrine des idées à la theorie des trois états de l'universal." In Bos and Meijer (1992), 89–119.
———. 1992b. "Psychologie philosophique et théologique de l'intellect: pour une histoire de la philosophie allemande en XIVe siècle." *Dialogue* 31: 377–92.
———. 1985. "Ulrich de Strasbourg, lecteur d'Albert le Grand." *Freiburger Zeitschrift für Philosophie und Theologie* 32: 105–36.
———. 1981. "Thèorie des universaux et réalisme logique chez Albert le Grand." *Revue des sciences philosophiques et théologiques* 65: 55–75.
———. 1980. "Logique et existence selon saint Albert le Grand." *Archives de philosophie* 43: 529–58.
Liddell, Henry George, Scott, Robert, and Jones, Henry Stuart, eds. 1968. *Greek-English Lexicon*, rev., 9th ed. Oxford: Oxford University Press.
Lindberg, David C., ed. 1983. *Studies in the History of Medieval Optics*. London: Variorum Reprints.
———, ed. 1976. *Theories of Vision from Al-Kindi to Kepler*. Chicago: University of Chicago Press.
Linden, Eugene. 1986. *Silent Partners: The Legacy of the Ape Language Experiments*. New York: Times Books.
Lindner, Kurt. 1963. *Von Falken, Hunden und Pferden: Deutsche Albertus-Magnus-Übersetzungen aus der ersten Hälfte des 15. Jahrhunderts*. Quellen und Studien zur Geschichte der Jagd, vol. 7, in 2 parts. Berlin: De Gruyter.
Lindsay, W. M., ed. 1911. *Isidori Hispalensis Episcopi Etymologiarum sive Originum Libri XX*. 2 vols. Oxford: Clarendon Press.
Littré, E., ed. and trans. 1853. *Oeuvres complètes d'Hippocrate*. 10 vols. Paris: Ballière.
Lloyd, G. E. R. 1987. "Empirical Research in Aristotle's Biology." In Gotthelf and Lennox (1987), 53–63.
Lluch-Baizauli, Miguel. 1989. "Sobre el commentario albertino a la Mystica Theologia de Dionisio." In Zimmermann (1989), 68–77.
Lockley, Ronald M. 1979. *Whales, Dolphins, and Porpoises*. New York: W. W. Norton.
Lugt, Maaike van der. 2000. "Animal Légendaire et Discours Savant Médiéval. La Barnacle dans tous ses etats." *Micrologus* 8: 351–94.

Luis of Valladolid. 1889. "Tabula Alberti Magni." In *Catalogus Codicum Hagiographicorum Bibl. Regia Bruxellensis, 1.2*. Brussels: Polleunis, Ceuterick et de Smet.

Mabberley, D. J. 1987. *The Plant Book. A Portable Dictionary of the Higher Plants*. Cambridge: Cambridge University Press.

MacGregor, Alexander P. 1989. "The Tigress and Her Cubs: Tracking Down a Roman Anecdote." In Sutton (1989), 213–27.

Madden, D. H. 1969. *A Chapter of Mediaeval History: The Fathers of the Literature of Field Sport and Horses,* reprint of 1924 ed. Port Washington, N. Y.: Kennikat.

Magie, David, trans. 1921. *The Scriptores Historiae Augustae,* 3 vols. Cambridge, Mass.: Harvard University Press.

Mahoney, Edward P. 1992. "Pico, Plato, and Albert the Great: The Testimony and Evaluation of Agostino Nifo." *Medieval Philosophy and Theology* 2: 165–92.

———. 1980. "Albert the Great and the *Studio Patavino* of the Late Fifteenth and Early Sixteenth Centuries." In Weisheipl (1980d), 537–63.

Majno, Guido. 1975. *The Healing Hand: Man and Wound in the Ancient World*. Cambridge, Mass.: Harvard University Press.

Malkiel, David. 2016. "The Rabbi and the Crocodile: Interrogating Nature in the Late Quattrocento." *Speculum* 91.1: 115–48.

Mandonnet, Pierre. 1931. "La date de naissance d'Albert le Grand." *Revue Thomiste* 36: 233–56.

Mann, J. 1992. *Murder, Magic, and Medicine*. Oxford: Oxford University Press.

Manzanedo, M. F. 1980. "Doctrina de San Alberto Magno sobre los seis últimos predicamentos." *Angelicum* 57: 433–76.

Marrone, Steven P. 1983. *William of Auvergne and Robert Grosseteste: New Ideas of Truth in the Thirteenth Century*. Princeton: Princeton University Press.

Martin, Ernest Whitney. 1914. *The Birds of the Latin Poets*. Stanford: Stanford University Press.

Martin-Darivault, H., trans. 1883. *Le livre du roi Dancus, texte français inédit du XIIIe siècle suivi d'un Traité de fauconnerie également inédit d'après Albert le Grand*. Paris: Librairie des Bibliophiles.

Marzano, Annalisa. 2013. *Harvesting the Sea: The Exploitation of Marine Resources in the Roman Mediterranean*. Oxford Studies on the Roman Economy. Oxford: Oxford University Press.

Mathes, D. Richard. 1982. "L'iniziatore di una scienza integrale." In *Sant' Alberto Magno l'uomo e il pensatore,* 43–58. Milan: Massimo.

Maxwell-Stuart, P. G. 1981. *Studies in Greek Colour Terminology. Volume I: ΓΛΑΥΚΟΣ*. Mnemosyne Supplement 65. Leiden: Brill.

May, Margaret Tallmadge, trans. 1968. *Galen on the Usefulness of the Parts of the Body*. 2 vols. Ithaca: Cornell University Press.

Mayor, Adrienne. 1994. "Guardians of the Gold." *Archaeology* 47: 52–59.

McCormick, Harold W., Allen, Tom, and Young, William. 1963. *Shadows in the Sea: The Sharks, Skates, and Rays*. Philadelphia: Chilton Books.

McCulloch, Florence. 1960. *Medieval Latin and French Bestiaries*. Chapel Hill, N. C.: University of North Carolina Press.

———. 1959. "The Dying Swan—a Misunderstanding." *Modern Language Notes* 74: 289–92.

McDermott, William Coffman. 1938. *The Ape in Antiquity*. Baltimore: The Johns Hopkins University Press.

McGonigle, Thomas D. 1980. "The Significance of Albert the Great's View of Sacrament within a Medieval Sacramental Theology." *Thomist* 4: 560–83.

McInerny, Ralph. 1980. "Albert on Universals." In Kovach and Shahan (1980), 3–18.

McMillen, Liz. 1996. "Gender-Bending Hyenas." *Chronicle of Higher Education,* 3 May, A12, 17.

McNair, Bruce. 1993. "Albert the Great in the Renaissance: Cristoforo Landino's Use of Albert on the Soul." *Modern Schoolman* 70: 115–29.
McVaugh, Michael. 1994. "Medical Knowledge at the Time of Frederick II." In *La scienza alla courte di Federico II. Micrologus: Natura, scienza e società medievali, II:* 3–18. Paris: Brepols.
———. 1974. "The 'Humidum radicale' in Thirteenth-Century Medicine." *Traditio* 30: 259–83.
Meersseman, Giles Gerard. 1931. *Introductio in opera omnia B. Alberti Magni, O. P.* Bruges: Charles Beyaert.
Mellanby, Kenneth. 1971. *The Mole.* New York: Taplinger.
Mercken, H. P. F. 1990. "Ethics as a Science in Albert the Great's First Commentary on the Nicomachean Ethics." In *Knowledge and Sciences in Medieval Philosophy, 3.* Proceedings of the 8th International Congress of Medieval Philosophy, Helsinki, 24–29 August 1987, 251–60. Helsinki: Finnish Society for Missiology and Ecumenics.
Meritt, Benjamin D. 1961. *The Athenian Year.* Berkeley: University of California Press.
———. 1928. *The Athenian Calendar in the Fifth Century.* Cambridge, Mass.: Harvard University Press.
Mermier, Guy R. 1989. "The Phoenix: Its Nature and Its Place in the Tradition of the *Physiologus.*" In Clark and McMunn (1989), 69–87.
Meyer, Gerbert, and Zimmermann, Albert, eds. 1980. *Albertus Magnus: Doctor Universalis 1280/1980.* Mainz: Matthias Grönewald.
Michener, Charles D. 1974. *The Social Behavior of the Bees. A Comparative Study.* Cambridge, Mass.: Belknap Press.
Migne, J.-P. 1844–55, 1862–64. *Patrologiae Cursus Completus. Series Latina.* 221 vols. Paris: Apud Garniere Fratres.
———. 1857–66. *Patrologiae Cursus Completus. Series Graeca,* 161 vols. Paris: Migne.
Miller, Konrad. 1895–98. *Mappaemundi: Die ältesten Weltkarten.* 6 vols. Stuttgart: J. Roth.
Mills, Derek. 1989. *Ecology and Management of Atlantic Salmon.* London: Chapman and Hall.
Mitchell, J., and Rook, A. 1979. *Botanical Dermatology: Plants and Plant Products Injurious to the Skin.* Vancouver: Greengrass.
Möhle, Hannes, trans. 2011. "Zur Methodologie der Naturwissenschaft. De animalibus 1.11 tr. 1." In *Albertus Magnus und sein System der Wissenschaften. Schlüsseltexte in Übersetzung Lateinisch-Deutsch,* eds. Hannes Möhle, et al., 229–77. Münster: Aschendorf Verlag.
Mojsisch, Burkhard. 1985. "Grundlinien der Philosophie Alberts des Grossen." *Freiburger Zeitschrift für Philosophie und Theologie* 32: 27–44.
Mojsisch, Burkhard, and Pluta, Olaf, eds. 1991. *Historia Philosophiae Medii Aevi.* 2 vols. Studien zur Geschichte der Philosophie des Mittelalters. Amsterdam: B. R. Grüner.
Molland, A. G. 1980. "Mathematics in the Thought of Albertus Magnus." In Weisheipl (1980d), 463–78.
Möller, Reinhold. 1963. *Hiltgart von Hürnheim. Mittelhochdeutsche Prosaübersetzung des "Secretum Secretorum."* Deutsche Texte des Mittelalters, Band 56. Berlin: Akademie-Verlag.
Mommsen, Theodore, ed. 1895. *C. Iulii Solini Collectanea Rerum Memorabilium.* Berlin: Weidmann.
Monlau, Pedro Felipe. 1941. *Diccionario etimologico de la lengua castellana.* Buenos Aires: Libreria El Ateneo.
Morel, Willy, ed. 1927. *Fragmenta Poetarum Latinorum Epicorum et Lyricorum praeter Ennium et Lucilium,* 2nd ed. Leipzig: Teubner.
Morse, Roger. 1963. "Swarm Orientation in Honeybees." *Science* 14: 357–58.
Morse, Roger, and Boch, R. 1971. "Pheromone Concert in Swarming Honey Bees (Hymenoptera: Apidae)." *Annals of the Entomological Society of America* 64: 1414–17.

Morse, Roger, and Hooper, Ted. 1985. *The Illustrated Encyclopedia of Beekeeping.* New York: E. P. Dutton.

Moschus, D. Franciscus, ed. 1971. *Iacobi de Vitriaco. Libri Duo, Quorum prior Orientalis, sive Hierosolymitanae, Alter Occidentalis Historiae Nomine Inscribitur,* reprint of 1597 ed. Westmead, England: Gregg International Publishers.

Moulinier, Laurence. 1989. "Un échantilon de la botanique d'Albert le Grande." *Médiévales* 16–17: 179–85.

Mowat, Farley. 1963. *Never Cry Wolf.* New York: Little Brown.

Mowat, J. L. G., ed. 1887. *Alphita. A Medico-Botanical Glossary from the Bodleian Manuscript, Selden B. 35.* Oxford: Clarendon Press.

———, ed. 1882. *Sinonoma Bartholomei. A Glossary from a Fourteenth-Century Manuscript in the Library of Pembroke College, Oxford.* Oxford: Clarendon Press.

Muellner, Lenny. 1990. "The Simile of the Cranes and Pygmies: A Study of Homeric Metaphor." *Harvard Studies in Classical Philology* 93: 59–101.

Munro, John H. 1988. "Silk." In *DMA* 11: 293–96.

Needham, Joseph. 1959. *A History of Embryology.* New York: Abelard-Schuman.

Negri, Giovanni. 1944. "La botanica di Alberto Magno." *Angelicum* 21: 192–207.

Netboy, Anthony. 1980. *Salmon.* Tulsa, Okla.: Winchester Press.

———. 1974. *The Salmon. Their Fight for Survival.* Boston: Houghton Mifflin.

Neugebauer, O. 1957. *The Exact Sciences in Antiquity,* 2nd ed. Providence, R. I.: Brown University Press.

Newton, Alfred. 1896. *A Dictionary of Birds.* 8 vols. London: A. and C. Black.

Nicholas, Nick. 1999. "A Conundrum of Cats: Pards and their Relatives in Byzantium." *Greek Roman and Byzantine Studies* 40: 253–98

Niermayer, J. 1954. *Mediae Latinitatis Lexicon Minus.* Leiden: Brill.

Noonan, John T. 1986. *Contraception: A History of Its Treatment by Catholic Theologians and Canonists.* Cambridge, Mass.: Belknap Press.

Noone, Timothy. 1992. "Albert the Great on the Subject of Metaphysics and Demonstrating the Existence of God." *Medieval Philosophy and Theology* 2: 31–52.

O'Boyle, Cornelius, ed. 1991. *Aggregationes de Crisi et Creticis Diebus: Medieval Prognosis and Astrology.* Cambridge: Wellcome Unit.

Oggins, Robin S. 1980. "Albertus Magnus on Falcons and Hawks." In Weisheipl (1980d), 441–62.

Ogle, William, trans. 1912. *De Partibus Animalium.* In *The Works of Aristotle,* ed. J. A. Smith and W. D. Ross. Oxford: Clarendon Press.

Oliphant, Samuel Grant. 1914. "The Story of the Strix: Isidorus and the Glossographers." *Transactions of the American Philological Association* 45: 49–63.

———. 1913. "The Story of the Strix: Ancient." *Transactions of the American Philological Association* 44: 133–49.

Olivieri, Dante. 1961. *Dizzionario Etimologico Italiano,* 2nd ed. Milan: Ceschina.

O'Meara, John J., trans. 1982. *The History and Topography of Ireland. Gerald of Wales.* New York: Penguin.

O'Meara, Thomas F. 1980. "Albert the Great and Martin Luther on Justification." *Thomist* 4: 539–59.

Oppenraaij, Aafke M. I. van, ed. 1992. *Aristotle,* De Animalibus. *Michael Scot's Arabic-Latin Translation. Part Three. Books XV–XIX: Generation of Animals.* Leiden: Brill.

Ostlender, Heinrich, ed. 1952a. *Studia Albertina. Festschrift für Bernhard Geyer zum 70. Geburtstag.* Beiträge zur Geschichte der Philosophie und Theologie des Mittelalters, Suppl. 4. Münster: Aschendorff.

———. 1952b. "Die Autographe Alberts des Grossen." In Ostlender (1952a), 3–21.

Otto, A. 1962. *Die Sprichtwörter und Sprichtwörtlichen Redensarten der Römer*, reprint of 1890 ed. Hildesheim: Olms.

Pacius, Johann Erhard, trans. 1756. *Friedrich des Zweyten Römischen Kaisers übrige Stücke der Bücher von der Kunst zu Baitzen, nebst den Zusätzen des Königs Manfredus aus der Handschrifft herausgegeben. Albertus Magnus von den Falcken und Habichten*. Onglsbach.

Page, T. E. 1965. *P. Vergilii Maronis* Bucolica *et* Georgica. London: Macmillan.

Pagnoni-Sturlese, Maria-Rita. 1980. "A propos du néoplatonisme d'Albert le Grand." *Archives de Philosophie* 43: 635–54.

Park, Katherine. 1980. "Albert's Influence on Late Medieval Psychology." In Weisheipl (1980d), 501–35.

Paravicini Bagliani, Agostino. 2013. "La Légende médiévale d'Albert le Grand (1270–1435). Premières Recherches." In *The Medieval Legends of Philosophers and Scholars*. Micrologus: Nature, Sciences, and Medieval Societies, 21. Florence: SISMEL: 295–368.

Patai, Raphael. 1988. *Gates to the Old City*. Northvale, N. J.: Jason Aronson.

Patterson, Francine. 1978. "Conversations with a Gorilla." *National Geographic* 154: 438–65.

Pauly, August Friedrich von, et al. eds. 1893–. *Paulys real-encyclopädie der classischen Altertumswissenschaft*. Stuttgart: J. B. Metzler.

Peacock, Edward. 1884. "The Griffin." *Antiquary* 10: 89–92.

Peck, A. L., trans. 1979. Aristotle. *Generation of Animals*. Cambridge, Mass.: Harvard University Press.

———, trans. 1965–70. *Aristotle. Historia Animalium, Vols. 1 & 2*. Cambridge, Mass.: Harvard University Press.

———, trans. 1961. *Aristotle. Parts of Animals*. Loeb ed., with *Movement of Animals and Progression of Animals*, trans. E. S. Forster, rev. ed. Cambridge, Mass.: Harvard University Press.

Pelster, Franz, S. J. 1935. "Die ersten beiden Kapitel der Erklärung Alberts des Grossen zu *De animalibus* in ihrer ursprünglichen Fassung. Nach Cod. Vat. lat. 718." *Scholastik* 10: 229–40.

———. 1922. "Alberts des Grossen neu aufgefundene Quaestionen zu der aristotelischen Schrift 'De animalibus.'" *Zeitschrift für katholische Theologie* 46: 332–34.

———. 1920. *Kritische Studien zum Leben und zu den Schriften Alberts des Grossen*. Freiburg-im-Breisgau: Herder.

Perry, Ben E., ed. and trans. 1965. *Babrius and Phaedrus*. Cambridge, Mass.: Harvard University Press.

———. 1964. *Secundus the Silent Philosopher*. Philological Monographs, 22. New York: American Philological Association.

———, ed. 1952. *Aesopica*. 2 vols. Urbana: University of Illinois Press.

Peter of Prussia, ed. 1621. *Albertus Magnus,* De adhaerendo Deo libellvs. Accedit eiusdem Alberti vita, *Deo adhaerentis exemplar*. Antwerp: Ex officina Plantiniana, apud Balthasarem Moretum, & viduam Ioannis Moreti, & Io. Meursium.

Pierantoni, Umberto. 1944. "Alcuni aspetti dell'opera biologica di Alberto Magno." *Angelicum* 21: 220–23.

Plath, Otto Emil. 1934. *Bumblebees and Their Ways*. New York: Macmillan.

Platt, Arthur, trans. 1910. *De Generatione Animalium*. In *The Works of Aristotle*, ed. J. A. Smith and W. D. Ross (1912). Oxford: Clarendon Press.

Pluskowski, Aleksander. 2004. "Narwhals or Unicorns? Exotic Animals as Material Culture in Medieval Europe." *European Journal of Archaeology* 7: 291–313.

Pluta, Olaf. 1985. "Albert von Köln und Peter von Ailly." *Freiburger Zeitschrift für Philosophie und Theologie* 32: 267–71.

Pollard, John R. T. 1977. *Birds in Greek Life and Myth*. London: Thames and Hudson.

Pomeroy, Sarah B. 1975. *Goddesses, Whores, Wives, and Slaves: Women in Classical Antiquity.* New York: Schocken Books.

Pontanus, D. G. B., ed. 1964. *De rerum proprietatibus,* by Bartholomaeus Anglicus. Frankfurt a.M.: Minerva, 1964. Reprint of the 1601 ed., published by W. Richter, Frankfurt, under the title *Bartholomaei Anglici De genuinis rerum coelestium, terrestrium et inferarum proprietatibus libri XVIII.*

Pouchet, F. A. 1853. *Histoire des sciences naturelles au moyen âge ou Albert le Grand et son époque considerérée comme point de départ de l'école experimentale.* Paris: Bailliers.

Preus, Anthony. 1977. "Galen's Criticism of Aristotle's Conception Theory." *Journal of the History of Biology* 10: 65–85.

Prevot, Brigitte, ed. 1991. *La science de cheval au moyen âge: Le traité d'hippiatrie de Jordanus Rufus.* Paris: Klincksieck.

Price, Betsey Barker. 1980. "The Physical Astronomy and Astrology of Albertus Magnus." In Weisheipl (1980d), 155–85.

Prinz, Otto, ed. 1967. *Mittellateinisches Wörterbuch bis zum ausgehenden 13. Jahrhundert.* Munich: Beck.

Priscian. 1855–59. *Institutionum Grammaticarum,* ed. Martin Hertzius. 2 vols. Leipzig: Teubner.

Purday, K. M. 1973. "Berengar and the Use of the Word *Substantia.*" *Downside Review* 91: 101–10.

Quinn, John M. 1980. "The Concept of Time in Albert the Great." In Kovach and Shahan (1980), 21–47.

Rabanus Maurus. *De Universo Libri XXII. PL* 111:9A-614B.

Rabassa, Gregory. 1984a. "Slouching Back toward Babel: Some Views on Translation in the Groves." In Frawley (1984), 30–34.

———. 1984b. "The Silk Purse Business: A Translator's Conflicting Responsibilities." In Frawley (1984), 35–40.

Rackham, H., ed. 1940. *Pliny, Natural History.* 10 vols. Cambridge, Mass.: Harvard University Press.

Ransome, Hilda M. 1937. *The Sacred Bee in Ancient Times and Folklore.* London: Allen and Unwin.

Rassart-Eeckhout, Emmanuelle. 1997. "Le chat, animal de compagnie à la fin du Moyen Âge? L'éclairage de la langue imagée." In Bodson (1997), 95–118.

Rath, Gernot. 1956. "Die Arabischen Nomina anatomica in der lateinischen Canonübersetzung." In Courtois (1956), 229–44.

Reeds, Karen. 1980. "Albert on the Natural Philosophy of Plant Life." In Weisheipl (1980d), 341–54.

Reichert, Benedictus M., ed. 1896. *Vitae Fratrum Ordinis Praedicatorum.* Monumenta Ordinis Fratrum Praedicatorum Historica, vol. 1. Louvain: E. Charpentier & J. Schoonjans.

Resnick, Irven M., ed. 2013. *A Companion to Albert the Great: Theology, Philosophy, and the Sciences.* Leiden: Brill.

———, trans. 2010. *Albert the Great. On the Causes of the Properties of the Elements* (Liber de causis proprietatum elementorum). Medieval Philosophical Texts in Translation, no. 46. Milwaukee: Marquette University Press.

———. 2002. "Talmud, *Talmudisti,* and Albert the Great." *Viator* 33: 69–86.

———, trans. 1994. *"On Original Sin" and "A Disputation with the Jew, Leo, Concerning the Advent of Christ, the Son of God": Two Theological Treatises of Odo of Tournai.* Philadelphia: University of Pennsylvania Press.

———. 1992. *Divine Power and Possibility in St. Peter Damian's* De divina omnipotentia. Leiden: E. J. Brill.

———. 1987. "Risus monasticus: Laughter and Medieval Monastic Culture." *Revue Bénédictine* 97: 90–100.
Resnick, I. M., and Kitchell, Kenneth F., trans. 2008. *Albert the Great's Questions Concerning Aristotle's 'On Animals.'* Fathers of the Church, Medieval Continuation 9. Washington DC: Catholic University of America Press.
———. 2007. "The Sweepings of Lamia: Transformations of the Myths of Lilith and Lamia." In *Religion, Gender, and Culture in the Pre-Modern World*, eds. Alexandra Cuffel and Brian Britt (New York: Palgrave Macmillan, 2007): 77–104.
———. 2004. *Albert the Great: A Selectively Annotated Bibliography (1900–2000)*. Medieval and Renaissance Texts and Studies, 269. Tempe, AZ: Arizona Center for Medieval and Renaissance Studies.
———. 1996. "Albert the Great on the 'Language' of Animals." *American Catholic Philosophical Quarterly* 70: 41–61.
Richards, Peter. 1977. *The Medieval Leper and His Northern Heirs*. Cambridge: D. S. Brewer.
Riddle, John M., and Mulholland, James A. 1980. "Albert on Stones and Minerals." In Weisheipl (1980d), 203–34.
Rigault, Nicholas, ed. 1612. *Rei accipitrariae scriptores nunc primum editi*. Paris: H. Drouart.
Roscher, Wilhelm Heinrich. 1883. "Die Vergiftung mit Stierblut im classischen Altertum." *Neue Jahrbücher für classische Philologie* 29: 158–62.
Rose, A. H., ed. 1982. *Fermented Foods*. Economic Microbiology, vol. 7. London: Academic Press.
Rowland, Beryl. 1978. *Birds with Human Souls*. Knoxville: University of Tennessee Press.
———. 1973. *Animals with Human Faces: A Guide to Animal Symbolism*. Knoxville: University of Tennessee Press.
Roy, Bruno. 1994. "Un Inédit d'Albert le Grand dans *l'Unguentarius* de Guillaume de Werda." In *Comprendre et mâitriser la nature au Moyen Âge. Mélange d'histoire des science offerts à Guy Beujouan*, 171–80. Geneva: Droz.
Royds, T. R. 1923. "Beekeeping in Classical Times." *Bee World* 5: 12.
Rudolph of Nijmegen. 1928. *Legenda Beati Alberti Magni*, ed. Heribert C. Scheeben. Cologne: Kölner Görres-Haus.
Ruello, Francis. 1980. "Le commentaire de *De divinis nominibus* par Albert le Grand." *Archives de philosophie* 43: 589–614.
Rutkin, H. Darrel. 2013. "Astrology and Magic." In Resnick (2013), 451–505.
Sahrhage, Dietrich, and Lundbeck, Johannes. 1992. *A History of Fishing*. Berlin: Springer Verlag.
Saint-Denis, Eugène de. 1947. *Le Vocabulaire des animaux marins en latin classique. Ètudes et commentaires, no. 2*. Paris: Klincksieck.
Salem, Sema'an I., and Kumar, Alok, trans. and eds. 1991. *Sa-'id al-Andalusi: Science in the Medieval World. "Book of the Categories of Nations."* Austin: University of Texas Press.
Sanders, Willy. 1978. "Albertus Magnus und das Rheinische." *Rheinische Vierteljahrsblätter*, 42, 402–54.
Sargeaunt, John. 1920. *The Trees, Shrubs, and Plants of Virgil*. Oxford: Blackwell.
Sarton, George. 1952–59. *A History of Science*. 2 vols. Cambridge, Mass.: Harvard University Press.
———. 1931. "Queries and Answers." *Isis* 15: 171–72.
———. 1927–48. *Introduction to the History of Science*. 2 vols., in 4 parts. Baltimore: Williams and Wilkins.
Savory, Theodore. 1959. *The Art of Translation*. London: Jonathan Cape.
Scanlan, James J., trans. 1987. *Man and the Beasts (De animalibus, Books 22–26): Albert the Great*. Binghamton, N. Y.: Medieval and Renaissance Texts and Studies.
Schäfke, Werner. 1980. *Albertus Magnus: Wissenschaftler, Politiker, Heiliger*. Cologne: Nachrichtenamt der Stadt Köln.

Scheeben, Heribert C. 1931a. *Albert der Grosse: zur Chronologie seines Lebens. Quellen und Forschungen zur Geschichte des Dominikanerordens in Deutschland.* Leipzig: Albertus-Magnus-Verlag.

———. 1931b. "Les écrits d'Albert le Grand, d'après les catalogues." *Revue Thomiste* 36: 260–92.

———, ed. 1928. *Rudolph of Nijmegen's* Legenda Beati Alberti Magni. Cologne: Kölner Görres-Haus.

Schipperges, Heinrich. 1980. "Das medizinische Denken bei Albertus Magnus." In Meyer and Zimmermann (1980), 279–94.

Schmid, Toni. 1952. "Albertus-Magnus-Fragmente in Schweden." In *Studia Albertina. Festschrift für Bernhard Geyer zum 70. Geburtstage.* Beiträge zur Geschichte der Philosophie und Theologie des Mittelalters, suppl. 4, 30–31. Münster: Aschendorff.

Schneider, Johann Gottlieb. 1788–89. *Reliqua librorum Friderici II imperatoris, De arte venandi cum avibus, cum Manfredi regis additionibus.* Leipzig: n.p.

Schneyer, J. B. 1964. "Predigten Alberts des Grossen in der Hs. Leipzig, University Bibl. 683." *Archivum Fratrum Praedicatorum* 34: 45–106.

Scholfield, A. F., trans. 1958–59. *Aelian. On Animals,* 3 vols. Cambridge, Mass.: Harvard University Press.

Schooyans, Michel. 1961. "Bibliographie philosophique de Saint Albert le Grand (1931–1960)." *Rivista de Universidade Católica de S o Paulo* 21, fasc. 37/38: 36–88.

Schulte, Rainer, and Biguenet, John, eds. 1992. *Theories of Translation: An Anthology of Essays from Dryden to Derrida.* Chicago: University of Chicago Press.

Schwaiger, Georg, and Mai, Paul. 1980. *Albertus Magnus. Bischof von Regensburg und Kirchenlehrer Gedenkschrift zum 700. Todestag.* Regensburg: Verlag der Vereins für Regensburger Bistumgeschichte.

Schwertner, Thomas M. 1932. *St. Albert the Great.* New York: Bruce.

Scott, Philippa. 1933. *The Book of Silk.* London: Thames and Hudson.

Scullard, H. H. 1974. *The Elephant in the Greek and Roman World.* Ithaca, N. Y.: Cornell University Press.

Seeley, T. D., et al. 1979. "The Natural History of the Flight of Honeybee Swarms." *Psyche* 86: 103–13.

Senner, Walter. 1980. "Zur Wissenschaftstheorie der Theologie im Sentenzkommentar Alberts des Grossen." In Meyer and Zimmermann (1980), 323–43.

Shah, M. H. 1966. *The General Principles of Avicenna's Canon of Medicine.* Karachi: Inter Services Press for the Naveed Clinic.

Sharpe, William D. 1964. "Isidore of Seville *On Man and Monsters.* A translation of *Isidori Etymologiarum sive Originum Liber XI: De Homine et Portentis."* Transactions of the American Philosophical Society 54: 38–75.

Sharples, R. W. 1984. "Some Medieval and Renaissance Citations of Theophrastus. Albertus Magnus and the *lumen animae."* Journal of the Warburg and Courtald Institutes 47: 186–90.

Shaw, James Rochester. 1975. "Scientific Empiricism in the Middle Ages: Albertus Magnus on Sexual Anatomy and Physiology." *Clio Medica* 10: 53–64.

Shepard, Odell. 1956. *The Lore of the Unicorn.* New York: Barnes & Noble.

Siegel, Rudolph E. 1968. *Galen's System of Physiology and Medicine.* New York: S. Karger.

Singer, Charles. 1957. *A Short History of Anatomy from the Greeks to Harvey.* New York: Dover.

Siraisi, Nancy G. 1980. "The Medical Learning of Albertus Magnus." In Weisheipl (1980d), 379–404.

Smith, William, ed. 1861. *Dictionary of Greek and Roman Biography and Mythology.* 3 vols. London: Walton and Maberly.

———, ed. 1854. *A Dictionary of Greek and Roman Geography.* 2 vols. London: Walton and Maberly.

Southern, R. W. 1986. *Robert Grosseteste. The Growth of an English Mind in Medieval Europe.* Oxford: Clarendon Press.
Spector, William S., ed. 1956. *Handbook of Biological Data.* Philadelphia: W. B. Sauders.
Stadler, Hermann, ed. 1916–20. *Albertus Magnus De Animalibus Libri XXVI.* Beiträge zur Geschichte der Philosophie des Mittelalters, 15 and 16. Münster: Aschendorfsche Verlagsbuchhandlung.
———. 1912a. *Vorbemerkungen zur neuen Ausgabe der Tiergeschichte des Albertus Magnus.* Sitzungsberichte der königlich Bayerischen Akademie der Wissenschaften: Philosophisch-Philologische und historische Klasse, I. Abhandlung. Munich: Verlag der Königlich Bayerischen Akademie der Wissenschaften.
———. 1912b. "Zur Charakteristic der gangbarsten Ausgaben der Tiergeschichte des Albertus Magnus." *Archiv für die Geschichte der Naturwissenschaften und der Technik* 3: 465–74.
———. 1909a. *Albertus Magnus von Cöln also Naturforscher und das Cölner Autogramm seiner Tiergeschichte.* Leipzig: F. C. W. Vogel.
———, ed. 1909b. *Alberti Magni Liber de principiis motus processivi.* Munich: F. Straub.
———. 1907. "Geschichtlich-zoologische Studien über des Albertus Magnus Schrift 'De animalibus.'" *Mitteilungen zur Geschichte der Naturwissenschaften und der Medizin* 6: 249–54.
Stahl, William H., trans. 1990. *Commentary on the Dream of Scipio by Macrobius.* New York: Columbia University Press.
Stahl, William H., and Burge, E. L., trans. 1977. *Martianus Capella and the Seven Liberal Arts. Vol. II: The Marriage of Philology and Mercury.* Records of Civilization, Sources and Studies, no. 84. New York: Columbia University Press.
Stamm, R. A. 1977. *The Nature and Function of Pair-Bonds.* In Grzimek (1977), 448–68.
Stange, Emil, ed. 1905–6. *Die Encyklopädie des Arnoldus Saxo, zum ersten Mal nach einem Erfurter Codex. II. De naturis animalium III. De gemmarum virtutibus IV. De virtute universali.* Königliches Gymnasium zu Erfurt, Beilage zum Jahresbericht 1905–6. Erfurt: Fr. Bartholomaeus.
———. 1885. *Arnoldus Saxo, der älteste Encyklopädist des dreizehnten Jahrhunderts.* Halle: T. G. Cramer.
Stannard, Jerry. 1980a. "Albertus Magnus and Medieval Herbalism." In Weisheipl (1980d), 355–77.
———. 1980b. "The Botany of St. Albert the Great." In Meyer and Zimmermann (1980), 345–73.
Stebbins, Eunice. 1929. *The Dolphin in the Literature and Art of Greece and Rome.* Menasha, Wisc.: George Banta Publishing.
Steele, Robert, ed. 1920. *Opera Hactenus Inedita Rogeri Baconi, Fasc. V. Secretum Secretorum cum Glossis et Notulis.* Oxford: Clarendon Press.
Steenberghen, Fernand van. 1980. *Thomas Aquinas and Radical Aristotelianism.* Washington, DC: Catholic University of America Press.
———. 1932. "Saint Albert le Grand Docteur de Église." *Collecteana Mechliniensia* 21: 518–34.
Stehkamper, Hugo, ed. 1980. *Albertus Magnus: Ausstellung zum 700. Todestag.* Historisches Archiv der Stadt Köln, 15. Cologne: Greven and Bechtold.
Steneck, Nicholas H. 1980. "Albertus on the Psychology of Sense Perception." In Weisheipl (1980d), 263–90.
Stokkan, Karl-Arne, et al. 2013. "Shifting Mirrors: Adaptive Changes in Retinal Reflections to Winter Darkness in Arctic Reindeer." *Proceedings. Biological Sciences / the Royal Society* 280.1773: 1–9. DOI: 10.1098/rspb.2013.2451
Strait, Paul W. 1982. "Cologne." In *DMA* 3: 480–82.

Strayer, Joseph R., ed. *Dictionary of the Middle Ages.* 13 vols. New York: Charles Scribner's Sons.

Sturlese, Loris. 1985. "Note su Bertoldo di Moosburg O. P., scienziato et filosofo." In Imbach and Flüeler (1985), 235–47.

———. 1980. "Saints et magiciens: Albert le Grand en face d'Hermès Trismégiste." *Archives de philosophie* 43: 615–34.

Suolahti, Hugo. 1909. *Die deutschen Vogelnamen.* Strasbourg: Karl von Truebner.

Sutton, Robert F., ed. 1989. *Daidalikon: Studies in Memory of Raymond V. Schoder, S. J.* Wauconda, Ill.: Bolchazy-Carducci Publishers.

Swann, H. Kirke, and Wetmore, Alex. 1924–38. *Accipitres (Diurnal Birds of Prey).* London: Wheldon and Wesley.

Sweeney, Leo. 1983. "Are Plotinus and Albertus Magnus Neo-Platonists?" In Gerson (1983), 177–202.

———. 1980a. "*Esse primum creatum* in Albert the Great's *Liber de causis et processu universitatis.*" *Thomist* 4: 599–646.

———. 1980b. "The Meaning of *Esse* in Albert the Great's Texts on Creation in *Summa de Creaturis* and *Scripta super Sententias.*" In Kovach and Shahan (1980), 65–95.

Synan, Edward A. 1980. "Albertus Magnus and the Sciences." In Weisheipl (1980d), 1–12.

Tabarroni, Andrea. 1988. "On Articulation and Animal Language in Ancient Linguistic Thought." *Versus Quaderni di Studi Semiotici* 47/48: 103–21.

Talalaj, S., and Czechowicz, A. S. 1989. *Herbal Remedies: Harmful and Beneficial Effects.* Melbourne: Hill of Content Publ.

Temkin, Owsei, trans. 1956. *Soranus' Gynecology.* Princeton: Princeton University Press.

Thesaurus Linguae Latinae. Editus auctoritate et consilio academiarum quinque Germanicarum Berolinensis, Gottingensis, Lipsiensis, Monacensis, Vindobonensis. 1900-. Leipzig: B. G. Teubneri.

Thijssen, J. M. 1987. "Twins as Monsters: Albertus Magnus's Theory of the Generation of Twins and Its Philosophical Context." *Bulletin of the History of Medicine* 61: 237–46.

Thomas Aquinas. 1963. *The Division and Method of the Sciences,* trans. Armand Maurer. Toronto: Pontifical Institute of Mediaeval Studies.

Thomas of Cantimpré. 1627. *Bonum Universale De Apibus.* Douai: Baltazaris Belleri.

Thompson, D'Arcy W. 1966. *A Glossary of Greek Birds,* reprint of 1936 ed. Hildesheim: Olms.

———. 1947. *A Glossary of Greek Fishes.* London: Oxford University Press.

———. 1918. "The Birds of Diomede." *Classical Review* 32: 92–96.

———, trans. and ed. 1910. *Historia Animalium.* Oxford: Clarendon Press.

Thorndike, Lynn. 1923–58. *A History of Magic and Experimental Science,* 8 vols. New York: Macmillan.

Tilander, Gunnar, ed. 1963. *Cynegetica Vol. 9. Dancus Rex. Guillelmus Falconarius. Gerardus Falconarius. Les plus anciens traits de fauconnerie de l'Occident.* Lund: Bloms.

———. 1946/47. "Maitre Aliboron." *Studia Neophilologica* 19: 169–83.

Tilmann, Jean Paul. 1971. *An Appraisal of the Geographical Works of Albertus Magnus.* Ann Arbor, Mich.: University of Michigan Press.

Tinivella, Felicissimo. 1944. "Il metodo scientifico in S. Alberto Magno e Ruggero Bacone." *Angelicum* 21: 65–83.

Tkacz, Michael W. 2013. "Albert the Great on Logic, Knowledge, and Science." In Resnick (2013), 507–540.

———. 1993. "The Use of the Aristotelian Methodology of Division and Demonstration in the 'De Animalibus' of Albert the Great." Diss. Catholic University of America.

Tobler, Adolf, and Lommatzsch, Erhard. 1925-. *Altfranzösisches Wörterbuch.* 10 vols. Wiesbaden: Franz Steiner.

Toynbee, J. M. C. 1973. *Animals in Roman Life and Art.* Ithaca, N. Y.: Cornell University Press.
Tredennick, Hugh, trans. 1961–62. *The Metaphysics,* by Aristotle, 2 vols. Cambridge, Mass.: Harvard University Press.
Tremblay, Bruno. 2006. "Modern Scholarship (1900–2000) on Albertus Magnus: A Complement." *Bochumer Philosophisches Jahrbuch für Antike und Mittelalter* 11: 159–194.
———. 1996. "A First Glance at Albert the Great's Teachings on Analogy of Words." *Medieval Philosophy and Theology* 5: 265–92.
Tugwell, Simon, ed. 1988. *Albert and Thomas: Selected Writings.* New York: Paulist Press.
Tummers, Paul M. J. E. 1984. "Albertus Magnus' View of the Angle with Special Emphasis on His Geometry and Metaphysics." *Vivarium* 22: 35–62.
Twain, Mark. 1887. *Following the Equator.* Hartford: American Publishing Co.
Uhlfelder, Myra L., ed. 1954. *De proprietate sermonum vel rerum: A study and critical edition of a set of verbal distinctions.* Rome: American Academy in Rome.
Vansteenkiste, C. 1988. "Il volume XVII del nuovo Alberto Magno." *Angelicum* 65: 271–86.
———. 1980. "I volumi XII e XIV del nuovo Alberto Magno." *Angelicum* 57: 542–56.
Vedamuthu, Ebenezer R. 1982. "Fermented Milks." In Rose (1982), 199–227.
Velser, Marcus, ed. 1596. *Reliqua librorum Friderici II imperatoris, De arte venandi cum avibus, cum Manfredi regis additionibus. Ex membranis vetustis nunc primum edita. Albertus Magnus De falconibus, asturibus et accipitribus.* Augustae Vindelicorum.
Vernier, J. M. 1992. "La definition de l'âme chez Avicenne et S. Albert le Grand. Étude comparative du *Liber de anima* d'Avicenne et du *De anima* de S. Albert le Grand." *Revue des sciences philosophiques et théologiques* 76: 255–78.
Vickery, R. 1995. *A Dictionary of Plant Lore.* Oxford: Oxford University Press.
Vincent of Beauvais. 1624. *Bibliotheca mundi. Vincentii Burgundi . . . Speculum quadruplex, naturale, doctrinale, morale, historiale. In quo totius naturae historia, omnium scientiarum encyclopaedia, moralis philosophiae thesaurus, temporum & actionum humanarum theatrum amplissimum exhibetur.* Douai: Baltazaris Belleri.
Von Loë, Paul. 1901. "De vita et scriptis B. Alberti Magni." *Analecta Bollandiana* 20: 273–316.
———. 1900. "De vita et scriptis B. Alberti Magni." *Analecta Bollandiana* 19: 301–71.
Vuure, T. van. 2005. *Retracing the Aurochs: History, Morphology and Ecology of an Extinct Wild Ox.* Sofia: Pensoft.
Wagner, Claus. 1985. "Albert's Naturphilosophie im Licht der neueren Forschung (1979–1983)." *Freiburger Zeitschrift für Philosophie und Theologie* 32: 65–104.
Walker-Meikle, Kathleen. 2012. *Medieval Pets.* Woodbridge: Boydell.
Wallace, William A. 1996a. "Albert the Great's Inventive Logic: His Exposition on the *Topics* of Aristotle." *American Catholic Philosophical Quarterly* 70: 11–39.
———. 1996b. *The Modeling of Nature: Philosophy of Science and Philosophy of Nature in Synthesis.* Washington, DC: Catholic University of America Press.
———. 1985. "Nature as Animating: The Soul in the Human Sciences." *Thomist* 49: 612–48.
———. 1984. "The Intelligibility of Nature: A Neo-Aristotelian View." *Review of Metaphysics* 38: 33–56.
———. 1980a. "The Scientific Methodology of St. Albert the Great." In Meyer and Zimmermann (1980), 385–407.
———. 1980b. "Albertus Magnus on Suppositional Necessity in the Natural Sciences." In Weisheipl (1980d), 103–28.
———. 1980c. "Galileo's Citations of Albert the Great." In Kovach and Shahan (1980), 261–83.
———. 1976. "Galileo and Reasoning *Ex Suppositione:* The Methodology of Two New Sciences." *Proceedings of the 1974 Biennial Meeting of the Philosophy of Science Association,* eds. R. S. Cohen et al., 79–104. Dordrecht/Boston: D. Reidel Publishing.

Walters, Mark Jerome. 1988. *Courtship in the Animal Kingdom.* New York: Anchor Books.
Walther, Hans, ed. 1963. *Proverbia Sententiaeque Latinitatis Medii Aevi. Lateinische Sprichtwörter und Sentenzen des Mittelalters und der Frühen Neuzeit.* Carmina Medii Aevi Posterioris Latina II/8. 9 vols. Göttingen: Vandenhoeck and Ruprecht.
Walz, Angelus, ed. 1951. *Beati Iordani de Saxonia Epistulae.* Monumenta Ordinis Fratrum Praedicatorum Historica, vol. 23. Rome: S. Sabina.
Walz, Angelus, and Scheeben, Herbert Christian. 1932. *Iconographia Albertina.* Freiburg: Herder.
Waszink, J. H., ed. 1975. *Timaeus a Calcidio translatus commentarioque instructus.* Corpus Platonicum medii aevi. Plato Latinus, vol. 4. London: Warburg Institute.
Wébér, Édouard. 1981. "Language et méthode négatifs chez Albert le Grand." *Revue des sciences philosophiques et théologiques* 65: 75–99.
Weisheipl, James A. 1982. "Albertus Magnus." In *DMA* 1: 126–30.
———. 1980a. "Albert the Great and Medieval Culture." *Thomist* 4: 481–501.
———. 1980b. "The Axiom 'Opus naturae est opus intelligentiae' and Its Origins." In Meyer and Zimmermann (1980), 441–63.
———. 1980c. "The Life and Works of St. Albert the Great." In Weisheipl (1980d), 13–51. Reprinted as "Albert der Grosse, Leben und Werke," trans. P. Gregor Kirstein, in Entrich (1982), 9–60.
———, ed. 1980d. *Albertus Magnus and the Sciences 1980.* Toronto: Pontifical Institute of Mediaeval Studies.
———. 1980e. *Thomas d'Aquino and Albert His Teacher.* The Etienne Gilson series, 2. Toronto: Pontifical Institute of Mediaeval Studies.
———. 1980f. "Albertus Magnus and Universal Hylomorphism: A Note on Thirteenth-Century Augustinianism." In Kovach and Shahan (1980), 239–60.
———. 1980g. "Albert's Works on Natural Sciences (*libri naturales*) in Probable Chronological Order." In Weisheipl (1980d), 565–77.
———. 1974a. "The Parisian Faculty of Arts in Mid-Thirteenth Century: 1240–1270." *American Benedictine Review* 25: 200–17.
———. 1974b. *Friar Thomas d'Aquino: His Life, Thought and Works.* Garden City, N. J.: Doubleday.
———. 1971. "The Structure of the Arts Faculty in the Medieval University." *British Journal of Educational Studies* 19: 263–71.
———. 1967. "Albert the Great." In *The New Catholic Encyclopedia*, 1: 254–58.
———. 1958. "Albertus Magnus and the Oxford Platonists." *Proceedings of the American Catholic Philosophical Association* 32: 124–39.
Weiss, Melchior. 1930. *Reliquiengeschichte Alberts des Grossen.* Munich: F. X. Seitz.
Wellmann, Max, ed. 1908. *Philumeni* De venenatis animalibus eorumque remediis; *ex codice vaticano primum edidit Maximilianus Wellmann.* Corpus Medicorum Graecorum X,1,1. Leipzig: Teubner.
Wetzelsberger, Jürgen and Henryk Anzulewicz. 2014. *Liber de principiis motus processivi: lateinisch, deutsch = Über die Prinzipien der fortschreitenden Bewegung.* Freiburg im Breisgau: Herder.
White, Terence Hanbury. 1954. *The Book of Beasts, Being a Translation in Full from a Latin Bestiary of the 12th Century.* New York: G. P. Putnam's Sons.
Wicksteed, Philip H., and Cornford, Francis M., trans. 1934. *Aristotle, the Physics,* 2 vols. Cambridge, Mass.: Harvard University Press.
Wiener, Leo. 1921. *Contributions Towards a History of Arabico-Gothic Culture*, vol. 4: *Physiologus Studies.* Philadelphia: Innes and Sons.
Wilder, Alfred. 1980. "St. Albert and St. Thomas on Aristotle's *De Interpretatione.*" *Angelicum* 57: 496–532.

Wiles, Maurice, and Santer, Mark, eds. 1975. *Documents in Early Christian Thought*. London: Cambridge University Press.
Williams, Steven. 1994. "The Early Circulation of the Pseudo-Aristotelian *Secret of Secrets* in the West: The Papal and Imperial Courts." In *Micrologus: Natura, scienze e società medievali, II: La scienze alla courte di Federico II*, 127–44. Paris: Brepols.
Wilmore, Sylvia Bruce. 1974. *Swans of the World*. New York: Taplinger.
Wilms, Hieronymus. 1933. *Albert the Great: Saint and Doctor of the Church*, trans. Adrian English and Philip Hereford. London: Burns, Oates.
Wippel, John F., and Wolter, Allan B. 1969. *Medieval Philosophy from St. Augustine to Nicholas of Cusa*. New York: Free Press.
Wisser, Richard. 1986. "Albertus Magnus. Ein Mensch auf dem Weg durch die Wirklichkeit." *Zeitschrift für Religions- und Geistesgeschichte* 38: 311–44.
Withington, E. T., trans. 1928. *Hippocrates III*. Cambridge, Mass.: Harvard University Press.
Wöllmer, Gilla. 2013. "Albert the Great and His Botany." In Resnick (2013), 221–67.
Wood, Casey, and Fyfe, F. Marjorie, trans. and eds. 1955. *The Art of Falconry, Being the* De Arte Venandi cum Avibus *of Frederick II of Hohenstaufen*, reprint of 1943 ed. Boston: Charles and Branford.
Woodburne, Russell T. 1983. *Essentials of Human Anatomy*, 7th ed. New York: Oxford University Press.
Wright, David F. 1980. "Albert the Great's Critique of Lother of Segni (Innocent III) in the *De Sacrificio Missae*." *Thomist* 4: 584–96.
Wright, M. R., ed. 1981. *Empedocles: The Extant Fragments*. New Haven: Yale University Press.
Wright, Thomas, ed. 1863. *Alexandri Neckam De Naturis Rerum Libri Duo with the Poem of the Same Author, De Laudibus Divinae Sapientiae*. Rerum Britannicarum Medii Aevi Scriptores, vol. 34. London: Longman, Green, Longman, Roberts, and Green.
Wyckoff, Dorothy, trans. and ed. 1967. *Book of Minerals*. Oxford: Clarendon Press.
Yapp, Brunsdon. 1981. *Birds in Medieval Manuscripts*. London: British Library.
Zambelli, Paolo. 1992. *The Spectrum Astronomiae and Its Enigma: Astrology, Theology, and Science at Albertus Magnus' Time*. Boston Studies in the Philosophy of Science, no. 135. Dordrecht/Boston: Kluwer Academic.
Zimmermann, Albert, ed. 1989. *Die Kölner Universität im Mittelalter: geistige Wurzeln und soziale Wirklichkeit*. Berlin: De Gruyter.
———, ed. 1986. *Aristotelisches Erbe im arabisch-lateinischen Mittelalter: Übersetzungen, Kommentare, Interpretationen*. Miscellanea Mediaevalia, vol. 18. Berlin: De Gruyter.
———. 1980a. "Albertus Magnus und der lateinische Averroismus." In Meyer and Zimmermann (1980), 465–93.
———. 1980b. "Albert le Grand et l'étude scientifique de la nature." *Archives de philosophie* 43: 695–711.
Zimmermann, Albert, and Speer, Andreas, eds. 1991. *Mensch und Natur im Mittelalter*. Miscellanea Mediaevalia, 21. 2 vols. Berlin: De Gruyter.
Ziolkowski, Jan M. 1993. *Talking Animals: Medieval Latin Beast Poetry, 750–1150*. Philadelphia: University of Pennsylvania Press.
Zirkle, Conway. 1946. "The Early History of the Idea of the Inheritance of Acquired Characters and of Pangenesis." *Transactions of the American Philosophical Society* 35, 2: 91–147.
———. 1936. "Animals Impregnated by the Wind." *Isis* 25: 95–130.

INDEX

A

Abrokaliz (Protagoras), 926
Adelinus, 1555
Aelian, 30, 36, 38
    works of
        Peri Zōōn Idiotētos, 38
Aemelius Macer, 37
    works of
        Ornithogonia, 37
        Theriaca, 37
Aeneas, 403, 1391–1392
Albert the Great
    works of
        Astronomica, 780, 1398
        De anima (On the Soul), 14, 41, 86, 94, 108, 110, 134, 466, 544, 720, 767, 868, 870, 898, 902, 937, 964, 1116, 1155, 1163–1165, 1167–1168, 1177, 1179–1180, 1185, 1188–1189, 1264, 1334, 1354, 1372, 1410, 1412, 1544
        De animalibus (On Animals), 14, 22, 29, 33–35, 38, 41–42, 45, 1087, 1131
        De caelo et mundo (On the Heaven and the Universe), 232, 454, 921, 1370, 1374, 1376
        De causis proprietatum elementorum (On the Causes of the Properties of the Elements), 511, 781
        De falconibus, asturibus et accipitribus, 33
        De generatione et corruptione (On Generation and Corruption), 966, 1123, 1299
        De intellectu et intelligibili (On the Intellect and the Intelligible), 1189, 1191, 1203, 1400
        De iuventute et senectute (On Youth and Old Age), 385, 407, 514, 912, 918, 921, 1293
        De memoria et reminscentia (On Memory and Reminiscence), 1424
        De mineralibus (On Minerals), 29, 32, 1360
        De morte et vita (On Death and Life), 69, 1325, 1403
        De motibus animalium (On the Movements of Animals), 94, 1400, 1413, 1436, 1544
        De natura boni (On the Nature of the Good), 8, 11
        De natura et origine animae (On the Nature and Origin of the Soul), 1177
        De natura loci (On the Nature of Geographical Places), 385, 505, 550, 612, 902, 912, 913, 915, 1294, 1375, 1622
        De nutrimento et nutribili (On Nourishment), 68, 372, 1167, 1349
        De sensu et sensato (On Sense and the Sensed), 124, 898, 958, 1192, 1342, 1337, 1379, 1394, 1420, 1531
        De somno et vigilia (On Sleep and Wakefulness), 938, 940, 1335–1336, 1412
        De spiritu et respiratione (On the Spirit and Respiration), 255, 971, 1365, 1379, 1401
        De unitate intellectus (On the Unity of the Intellect Against Averroes), 14
        De vegetabilibus (On Plants), 32, 45, 76, 823, 957, 1034, 1058, 1085, 1087, 1124,

Albert the Great (*cont.*)
    1135, 1145, 1200, 1205, 1209, 1244, 1273, 1362, 1408
  In IV Libros Meteorum/Meteora (On Meteorology), 377, 426, 650, 1181–1182, 1192, 1205, 1218–1219, 1288, 1326, 1349, 1360, 1366, 1382, 1390, 1549
  On the Common Activities of the Soul (Physica?), 62
  Peri Geneos/Pery Geneos (On Generation), 401, 1358, 1378
  Perspectiva (On Perspective), 1 20
  Physica (Physics), 22, 279, 375, 419, 530, 652, 812, 815, 983, 1057, 1178, 1180, 1195, 1217, 1375, 1377, 1385
  Physonomya, 1573
  Quaestiones super de animalibus (Questions on Animals), 35
  Summa de bono (On the Highest Good), 8
  Summa de creaturis, 8, 11
  Summa theologiae, 17, 22
  Super Ethica/Moralia (Questions on Ethics), 93
Albert von Peitengau, 15
Alchemists, 1219, 1229
Alchemy, 1182, 1721
Aldhelm, 1549
Alexander, Letter of, 1651, 1749
Alexander IV, Pope, 14, 15
Alexander Neckam, 40
Alexander of Aphrodisias, 41
Alexander the Great, 103, 211, 214, 225, 242, 647, 1471, 1525, 1563, 1664
Alexander the Myndian, 38
  works of
    Ornithiaka, 38
    Peri Zōōn, 38
    Thēriaka, 38
Alexander the Peripatetic, 1164–1168, 1173, 1175, 1178
Alfonso X, 39
Alkumon (Alcmaeon of Croton), 1246
Altymenon of Corinth, philosopher, 774
Ambrose of Milan, 38, 40, 614–615, 695, 1675
  works of
    Hexaemeron, 38
Ambrose of Sienna, 10
Anatomists, 345, 368
Anatomy, 46, 283, 342, 346, 360, 374, 376, 436, 440, 456, 458, 1004

aorta, 184, 191, 194, 219, 253, 265, 270–271, 280, 338, 351, 355, 937, 999, 1002, 1014, 1016, 1034, 1198
arteries, 134, 136, 141, 149, 189–195, 198, 201, 203, 205–206, 221–224, 227–228, 244, 358, 799, 805–806, 990–992, 1170, 1180, 1405
belly, 930–931, 949, 986–988, 993–994, 1008, 1020–1026, 1028–1031, 1033–1039, 1055–1056, 1064–1065, 1078–1079, 1092–1093, 1353, 1757–1758
bladder, 282–283, 1008, 1012–1014, 1035–1037, 1106–1109, 1313–1314, 1368
bone, 70–71, 88–92, 379–382, 875, 902–905, 927–928, 934–936, 945–948, 950–956, 1066–1070, 1223–1224, 1385, 1402
  complexion of, 397, 904, 916, 1218–1220
  generation of, 1223, 1360
  marrow, 381, 417, 927–928, 934–936
brain, 70–73, 87–92, 110–112, 116–117, 123, 126–129, 162, 178–184, 192–193, 242–251, 818, 850–851, 1126, 1202–1203, 1216, 1341, 1344, 1385, 1431, 1437
  Aristotle on, 158, 178, 936–944
  cells of, 347–348
  cognitive function, 72
  complexion of, 658, 916, 928, 936–937, 940, 1220–1221, 1234, 1346, 1348, 1403
  copulation, excessive, and, 1227–1228, 1235, 1348
  dissection of, 936
  dura mater in, 89, 123, 193, 201, 246–247, 250–251, 409, 538, 942, 947
  Galen on, 71, 158, 358, 368, 941
  location of, 959, 1055
  medulla of, 141, 413, 639, 897, 903, 917, 928, 944, 1668
  nerves divided in, 360–364, 471, 941, 944–945
  sensation in, 375, 938
  senses located in, 1056
  substance of, 941
  ventricles of, 179, 193, 244, 246–250, 940, 942–943
brain size, 269, 1221, 1350
breasts, 212–214, 290–291, 302–303, 425–429, 775, 778–779, 1063–1065, 1323–1324, 1534
cartilage, 91–92, 137, 206, 379–382, 904, 945–946, 954–955

diaphragm, 139, 146, 165, 186–187, 227–229, 257–259, 273–275, 341–343, 992–994, 1009–1010, 1016–1018, 1062, 1095–1099, 1199, 1365
ears, 109–115, 186, 898, 958–963, 1342–1344
esophagus, 132, 135–136, 138, 142, 148, 162, 171, 193, 200, 205, 207, 252, 257–258, 265, 271, 278, 329, 985–995, 1027–1028
eyelids, 97, 126, 159–160, 482, 540, 964–968, 1221–1222, 1342
eyes, 98, 898, 959, 961, 964–969, 1161, 1220–1222, 1290
  colors of, 101, 1332–1334, 1336–1340
  complexion of, 905, 1221, 1403
  composition of, 120–123
  copulation, excessive, and, 1227–1228, 1235
  humors of, 123–126, 917, 964, 1221, 1337–1341
  meninges of, 417, 959, 1339–1341
  physiognomy of, 98–108
fats, types of bodily, 53, 265, 410–413, 903–904, 927, 933–934, 1015, 1033–1034
feet, 868, 878, 946, 1055–1058, 1065–1072
gall bladder, 262, 273–278, 327–328, 359, 399, 823, 1010, 1029–1033, 1128, 1307, 1367
  dissection of, 1031
  Plato on, 1010
hand, 232–235, 868, 896, 946, 1055–1062, 1353, 1412
head, bones of, 113–119, 141, 378–379, 948
heart, 55, 70–73, 77, 90, 94, 138, 148, 158, 178, 184, 189–195, 223, 231, 248, 253–255, 266–273, 327, 850–851, 876, 899–904, 913, 915–918, 922–923, 928–932, 1126, 1161–1162, 1180, 1203–1204, 1307, 1323, 1365, 1368, 1378, 1386, 1404–1405, 1431
  Aristotle on, 178, 267, 1202
  complexion of, 916, 1313, 1327–1328
  Galen on, 71, 267, 371–375
  generation of, in fetus, 851, 997, 1204, 1209, 1213, 1216, 1278, 1291
  heat of, 268, 289, 366, 422, 651, 939
  location of, 268, 959, 988, 996–998
  origin of arteries and, 189
  origin of the veins and, 209, 351–352, 900, 996
  as receptacle for vital spirit, 928
  shape of, 270–271, 1370
  sizes of, 371, 999–1000
  spiritual power and, 242–244
  systolic and dyastolic action of, 189, 538, 932, 998, 1364, 1371, 1373, 1401
  ventricles of, 189, 198, 267, 270, 352, 370–371, 960, 998–999
heart-bone, 90, 157, 370, 998, 1356, 1472
intestines, 271, 277–278, 931, 1024–1029, 1368
  number of, 259–264, 1027–1029
jaw, 91, 117, 119, 307–310, 330–331, 969, 1070–1071, 1357, 1472, 1680, 1682
  composition of, 130–132
kidneys, 146–147, 280–282, 328, 350, 356, 405, 423, 835, 853, 923, 1012–1016
  digestion in, 923
  illnesses of, 1013
  location of, 534, 1007, 1014
legs, 154–157, 235, 240–242, 1239, 1436–1437
liver, 273–277, 800, 809, 818, 822, 850–851, 853, 939, 1009–1010, 1032–1033, 1126, 1161–1162, 1201–1202, 1216, 1365, 1389, 1431
  Galen on, 71, 1009–1010, 1202
  location of, 1007
  Plato on, 1010
lungs, 55, 72–73, 129, 136, 138, 148, 184, 190, 198, 205–207, 212, 242, 252–255, 266–272, 350, 824, 991–992, 1005–1006, 1161, 1201–1202, 1214, 1216, 1354, 1365, 1368
  complexion of, 903, 917, 1156
  as fan of the heart, 55, 73, 253, 255, 988, 992
  illnesses of, 1000
  spiritual power and, 242–243
muscles, 158–178
nerves, 158–189, 376–378, 944–947, 962–966, 1385, 1402, 1405
  complexion of, 904, 1220
  descending from the brain, 178–183, 1385
  generation of, 1222, 1360
  principle of, 374
nose, 126–130
pannicular-membranes, 138, 158, 178, 183, 184, 187, 188, 191, 193, 195, 198, 199, 201, 218, 221–222, 226, 228, 230, 246, 257, 266, 274, 279, 281, 374, 823, 848, 851–852, 993, 1360
penis, 73, 154, 172, 203, 216–225, 303, 1065–1066, 1088–1094, 1106, 1284
penis size, 844, 1094, 1106, 1207, 1235

Anatomy (*cont.*)
  sensory organs
    flesh, 898
    nerve, 898, 913
    olfactory canal, 127
    tongue, 131–136, 139, 898, 913
  skin, 284, 385–386, 1345
    complexion of, 904, 1220
    generation of, 1360
  skull, 946–947
    divisions of, 113–115, 247
    sutures of, 88–89, 379–380, 938, 948, 968, 1221
      Aristotle on, 89
  spleen, 278–280, 327, 350, 402, 922, 1006–1009, 1011, 1128, 1307
  stomach, 209–210, 330–331, 809, 822, 989, 1126, 1365, 1368, 1389
    composition of, 993–994
    digestion in, 259, 922–923
  tail, 1066–1067
  teeth, 307–312, 380, 955, 1341
    canines, 117, 119, 308, 579, 976, 1666
    functions of, 975–980
    generation of, 1223–1224, 1356–1357
      Democritus on, 1357
    incisors, 70, 118, 312, 1357
    juvenile, 1232, 1236
    molars, 70, 119, 131, 307, 309, 312, 976, 979, 1023
    number of, 119, 978
    types of, 978–980
    wisdom, 119, 978, 1357
  testicles, 73, 131, 172, 188, 194, 203–204, 216–222, 225, 228, 304, 815, 1088–1109, 1171, 1197, 1287, 1316, 1327
    location of, 1098
    piercing of, 386
  thorax, 148–154, 1062–1065
  throat, 205–208
  tusks, 307, 471, 976, 979, 984, 1453
  vagina, 203, 216, 220–221, 1107
  veins, 195–205, 899–900, 931, 1002–1004, 1218, 1224–1225, 1285, 1368, 1385, 1405
    complexion of, 904, 917, 1218
    hepatic, 348, 355
    mesenteric, 73, 193, 197, 228, 260–267, 273, 276–277, 280, 328, 923, 931, 1024
    origin of, 345–376, 817, 1002
      Aristotle on, 351–358, 365–371
    Averroes on, 365–371
    Avicenna on, 360–364, 367–371
    Galen on, 358–360
    physicians on, 360–363, 818
    pulmonary, 353
    spermatic, 131, 423, 1444
    vena cava, 190, 196–197, 202, 203, 219, 221, 255, 268–275, 278–281, 348, 352, 354, 356, 359, 405, 800, 822, 937, 1002, 1198
  vertebrae, 141–147, 347, 379, 382, 953
  vulva, 213, 216, 219–224, 304, 828, 836, 1065–1066
  webs, membraneous, 408–410, 1033
  windpipe, 127, 136–142, 148, 252–257, 265, 270–272, 326, 959, 986–992, 1354
    Aristotle on, 139
    composition of, 990
    location of, 989
  womb, 68, 154, 171, 188, 204, 213, 216–221, 283, 799–803, 1088–1090, 1207–1210, 1224–1228, 1292
    blockage in, 845, 1313, 1317
    chambers of, 218–220, 341, 1283, 1309
    complexion of, 1208, 1235, 1281–1283
    Galen on, 805–812, 1443
    health of, 827–834
    illnesses of, 841–844
    location of, 1095–1099
    mole of, 839–841, 846, 1321–1322
      Aristotle on, 847
    movement of, 1313, 1330
    test for openness of, 1227
    vital heat of, 1170
    wind in, 836
Anaxagoras, 71, 403, 1031, 1114, 1171, 1189, 1191, 1255, 1377, 1390, 1399
Animals (*See also* Birds; Marine animals; Quadrupeds)
  accidental traits of, 1332–1357
  aquatic, 55–56, 589–590, 652, 885, 935, 1655–1656
  complexion of, 926
  dolphin, 74, 83, 112, 303, 321, 337, 381, 425, 483, 551, 584, 590, 596–597, 621, 653, 759–761, 955, 1005, 1032, 1082, 1098, 1365, 1429, 1655, 1667, 1678–1679
  copulation of, 495, 1089, 1254
  epilepsy in, 472
  life span of, 557
  multiple births in, 1283

music and, 1678–1679
reproduction of, 1096, 1100, 1156
frog, 55, 57, 324, 479, 486, 632, 673, 676, 741, 1077, 1237, 1249, 1435, 1754–1755
hippopotamus, 297, 1680–1681, 1685–1686
horse (*equus aquaticus*), 590
seal, 57, 281, 293, 307, 425, 478, 551, 1005, 1031, 1069, 1082, 1365, 1429, 1677, 1681, 1683, 1686–1687
copulation of, 493
ears, absence of, 1344
skink, 1701–1702
whale, 74, 112, 303, 321, 337, 381, 425, 483, 551, 557, 584, 590, 620–621, 630, 653, 656, 955, 1005, 1082, 1365, 1429, 1655, 1666–1671, 1681, 1706
capture of, 1669–1671
copulation of, 1668
harpooning of, 1670–1671
killer, 1694
multiple births in, 1283
reproduction of, 1096, 1156
whale oil, 413, 1669
arachnids
scorpion, 467, 646, 1513, 1522, 1524–1525, 1733, 1757
generated from putrescence, 1104
stinger of, 1047
varieties of, 1760–1761
spider, 60, 293, 486, 502, 732, 737, 1161, 1436, 1740, 1745, 1762
bite of, 1757–1758
copulation of, 499
eggs of, 75, 1744
varieties of, 726–730, 1742–1744, 1755–1758
venoms of, 69, 727–729
webs of, 1743–1744
bloodless
anatomy of, 1035–1044, 1739–1740
divisions of, 433
complexion
environmental influences upon, 664–666
types of, 900–902
conception
impediments to, 837–838
superfetation in, 1314–1317, 1330
copulation
across species, 1304
act of, 62, 1309–1311
organs of, 1088–1103

pleasure at, 515, 571, 581
types of, 490–515
definition of, 1545
diets of, 654–656
domestic, 61, 883
boar, 1433
cat, 1469
copulation of, 572
elephant, 572
leopard, 686
lion, 754
eggs, wind, 837–838, 1212, 1401
friendship among, 678–680
generation of, 489, 1281–1330
genus of, 860–861, 1407–1408
habits of, 667–765
hair of, 384–385, 1344–1347, 1351–1353
hibernating, 622–626, 658–660
illnesses of, 628–641, 1461
larviparous, 74–75, 1103–1104, 1141, 1155–1158, 1251, 1277–1279, 1318
menstruation of, 574–575, 1288
migratory, 612–617, 657
molting, 626–627, 649, 661–664
multiple births in, 1283, 1286, 1306–1312
oviparous, 47, 74–76, 1085–1110, 1156, 1201, 1204, 1228, 1252
anatomy of, 1238
eggs of, 1246, 1260
generation of, 1089–1090, 1095–1110, 1237–1238
sensory organs of, 1070
puberty, effect on, 511, 1323, 1332–1333
reproductive members, 303–305, 336–343, 1090–1110, 1441
castration of, 339–340, 383, 1089, 1091, 1105, 1291
reptiles, 1433–1435
chameleon, 315–317, 1072, 1731
crocodile, 47, 80, 131, 292, 294, 314–316, 319, 331, 426, 480, 623, 671, 678, 686, 973, 1070–1071, 1434, 1675–1676, 1679, 1728, 1731
dragon, 47
lintwurm, 317
lizard, 47, 252, 294, 337, 409, 426, 486, 643, 714, 729, 960, 1006, 1029, 1089, 1156, 1433–1434, 1601, 1610, 1708, 1729
copulation of, 494
tongue of, 1069
newt, 1433

Animals (cont.)
- salamander, 60, 426, 1029, 1433–1434, 1726, 1729, 1731–1733
- toad, 729, 1435, 1523, 1539, 1606
  - varieties of, 1745
- tortoise, 626, 686, 951, 1006, 1012–1013, 1070, 1089, 1109, 1156, 1433, 1735–1736
  - testicles of, 1098
- turtles, 589, 1692, 1732
- respiration of, 650–652
- ringed, 48, 112, 435–436, 945, 1278, 1435–1438
  - anatomy of, 951–953, 1045–1048, 1207, 1740
  - centipede, 436, 1724
  - copulation of, 498–499, 1120, 1140–1141, 1259, 1271, 1438
  - cricket, 478
  - earthworm, 431, 673, 1272, 1277, 1438, 1485, 1539, 1763
  - egg laying of, 502–503
  - feet of, 1047–1048
  - generation of, 1156, 1157, 1260–1264
    - from putrescence, 503, 1085, 1103
  - genuses of, 465–468, 1102
  - illnesses of, 640–641
  - leech, 1475, 1606, 1759
  - respiration of, 1365
  - sensory powers of, 475–476
  - sexual differentiation in, 1102
  - silkworm, 502, 729, 1104, 1161, 1264, 1436–1437, 1740, 1746, 1750–1751
  - stinger of, 1046–1047
  - vocalizations of, 1366
- senses, external
  - hearing, 1411, 1415
  - smell, 1411
  - touch, 1412
  - vision, 1411
- sexual differentiation of, 484–486, 1085–1089, 1102–1103, 1144–1145, 1154–1155, 1211–1213, 1281–1294, 1327–1328
  - Democritus and Empedocles on, 1281–1287, 1301
- sleep of, 482–483, 1334–1336
  - dreams, 482, 1336
- strife among, 670–678
- viviparous, 74–77, 1197
  - embryo of, 76–77, 800–801, 1097, 1159–1161, 1170, 1205, 1209–1210, 1216, 1224–1225, 1234, 1281–1286, 1292–1295, 1315, 1320, 1323, 1327, 1335
  - formation of, 1161, 1367, 1399
  - gestation period of, 1324–1326
  - fetus of, miscarried, 1214
  - reproduction of, 1155–1156
  - vocalizations of, 62, 476–481, 511, 573, 973, 1323, 1353–1356, 1424

Anthony, Marc, 1728
Antiphon, the Sophist, 1183
Antyfon (Antiphon, the Sophist?), 1122
Antyseus, 308, 642
Apion, 38
Apollo, 50, 103, 113, 419, 674
Apothecary, 605, 727
Apulegius (Apuleius), 103, 674, 678
Apuleius
  works of
    On the God of Socrates, 1397
Aquila, 1590, 1618, 1623
Arabs, 159, 346, 348–349, 390, 444, 581, 614, 629, 642, 823, 1057, 1103, 1169, 1193, 1229, 1256, 1449, 1461, 1717
Architect, 1364
Arion (the Citharist), 1678
Aristotle, 10–11, 18, 26, 31–32, 37, 41, 48, 55, 62, 74, 84, 93, 98, 106, 116, 139, 189, 191, 230–231, 236, 257, 271, 285, 342, 351, 358–364, 372–376, 403, 412, 420–423, 438, 461, 473, 475, 486, 492, 496, 499–500, 528, 544, 551, 608, 638, 641–642, 671, 676–677, 685, 687, 695, 697, 701, 706, 720, 756, 759–760, 793, 798, 807, 811, 814, 820, 853, 876, 893, 915, 917, 930, 937, 949, 965, 998, 1016, 1025, 1031, 1038, 1054, 1079, 1112–1113, 1146, 1159, 1166, 1170, 1172–1175, 1186, 1188–1189, 1206, 1229, 1236, 1238–1239, 1241, 1305, 1307, 1314, 1321, 1329, 1400, 1516, 1571, 1629, 1635, 1656, 1713
  works of
    Analytica priora (Prior Analytics), 285
    Categoriae (Categories), 21
    De anima (On the Soul), 859, 871, 878, 1159
    De animalibus (On Animals), 19, 21, 39, 106, 139
    De caelo (On the Heavens), 21, 952
    De generatione animalium (On the Generation of Animals), 39–40

De generatione et corruptione (On Generation and Corruption), 21, 869
De interpretatione (On Interpretation), 21
De partibus animalium (On the Parts of Animals), 39–40
De somno et vigilia (On Sleep and Wakefulness), 361, 482
Historia animalium (History of Animals), 39—40
Metaphysica (Metaphysics), 20–21, 367, 866, 869
Meteora (On Meteorology), 21, 301, 908, 910, 933
Nicomachean Ethics, 20–21, 93
On Nourishment, 361
Peri geneos (On Generation), 869, 1178, 1326
Problemata (On Problems), 403, 1026, 1229, 1391
Physica (Physics), 19, 21, 420, 858, 861–865, 871–875, 882, 885, 1172
Topica (Topics), 878–879
Ps. Aristotle
  works of
    De plantis (On Plants), 19, 488
    De regimine dominorum (Guide for Lords), 647
    Liber de causis, 11
Armeria, King of Valentia, 1462
Arnoldus Saxo, 40
Arodotos, poet, 432
Arotimus, 544
Arsenic, red, 1229
Artist, 1402
  jeweler, 1512
  sculptor, 1176
Arts
  liberal, 772–773
  mechanical, 771–773, 890, 1420, 1425
  practical
    carpentry, 862, 1143, 1174, 1219
    glassworking, 1380, 1425
    gold refining, 1367
    handicraft, 872, 889
    instruments of, 1175, 1193, 1364, 1379, 1402, 1406
    sculpting, 1401
    shoemaking, 1060
    silk making, 1104
    weaving, 1091, 1347, 1406, 1743
    woodworking, 862–863, 1364, 1380
  principles of, 1417
Arybaros, philosopher, 1256

Asciepius, 50, 1445
Astrologers, 861
Astronomers, 1277, 1331, 1397, 1400, 1531
Astronomy, 1273, 1767
Astyages, 308
Athenaeus, 38
Atlas, 1362
Augurs, 526, 545, 580, 602, 671, 693, 702, 710, 717, 1546, 1637, 1641, 1652
Aulus Gellius, 38
Averroes, 19–20, 41, 351, 375, 420, 1189
Avicenna, 19–20, 41, 58, 94, 96, 119, 140, 192, 200, 211, 239, 294, 297, 299–300, 304, 308–309, 315–316, 346–351, 360, 364, 367, 370–371, 375, 391, 420, 422, 428, 431, 437–438, 440, 443, 461, 471–472, 475, 480, 486, 492–493, 496, 500–501, 523, 528, 532–533, 545–548, 551, 564, 566, 581, 594–595, 601, 613–614, 618, 636–637, 641–642, 644, 647–648, 669–672, 675–679, 685–689, 694–695, 700–701, 706–707, 720, 735, 743–744, 753–754, 756, 760–761, 792, 818, 979, 1072, 1104, 1167–1169, 1173, 1184, 1189, 1241, 1255–1257, 1263, 1438, 1441, 1443–1445, 1516, 1652, 1717, 1719–1721, 1725, 1727, 1729, 1733, 1735, 1737–1738, 1742, 1753, 1755, 1760
works of
  Canon of Medicine, 34, 40
  De animalibus (On Animals), 360

B

Babel, Tower of, 1726
Bacon, Roger, 1, 12, 19–20, 25, 27, 39
Balss, Heinrich, 42
Bardanus, King, 431
Bartholomew of Capua, 16–17
Bats (*See* Birds: bat)
Bede, the Venerable, 597
Beekeeping, 742, 1270
  tanging, 745
Beer, 1486, 1496, 1501
Berthold of Moosburg, 34
Birds, 599–604, 1072–1077, 1544–1654
  anatomy of, 68, 317–321, 334–336, 961, 1074–1077, 1208, 1245
  aquatic, 66, 206, 602–604, 697, 1022, 1073, 1426–1427
  egg laying of, 504

Birds (*cont.*)
    bat, 58, 112, 343, 1610, 1651–1652
    beak of, 977–978, 980, 1073, 1223
    bee-eater, 526, 602, 1637–1638
        nest of, 693
    bittern, 1562
    black diver, 1023, 1547
    blackbird, 507, 601, 1638
    bustard, great, 1562
    butcherbird, 601, 702
    buzzard, 1561
    capacity for instruction, 1425–1428
    capon, 1629–1630
    chaffinch, 509
    chicken, 63, 252, 320, 654, 688, 1305, 1495, 1524, 1533, 1542
        copulation of, 1238
            pleasure at, 535
        egg laying of, 525–526, 1240, 1628–1629
        eggs of, 528
    cinnamon bird
        nest of, 699–700
    cock, 1627–1628
    complexion of, 654, 1157, 1239
    coot, 321, 1023, 1547, 1624
    cormorant, 501, 673, 1636
    crane, 58–59, 66–67, 392, 485, 491, 615–616, 694–695, 715, 767, 887, 1626–1627
        colors of, 1351
        nest of, 692
    crop of, 930, 1022
    crow, 60, 547, 600, 672, 705, 707, 1318, 1352, 1567
        eggs of, 528
    cuckoo, 481, 547–548, 615, 672, 712, 765, 1568
        eggs of, 1239–1240
        nest of, 707–709
    diet of, 60
    diver, 1639
    domestic, 1427
    dove, 537, 712
    duck, 66, 206, 549, 603, 978, 1238, 1558–1559
        copulation of, 535
        nest of, 692
    eagle, 51–52, 58, 60, 62, 66, 81, 319, 383, 546, 599, 670–672, 675, 677, 678, 696, 710, 713–715, 1073, 1547–1553
        Aristotle on, 1551
        augury and, 717
        capture of, 547
        egg laying of, 545, 1548
        golden, 52, 544, 547, 549, 1247
        incubation of, 545
    egg laying, time of, 500–502, 507–509
    eggs, 75, 1304–1305
        colors of, 1242–1244
        completion of, 529, 1095, 1141, 1156, 1241, 1243
        complexion of, 1243
        corruption in, 531–532, 1237, 1246–1247
        double yolked, 542, 1305
        embryo, completion of, 536–543
        heart formed first in, 1248
        incubation of, 529, 1238, 1246, 1248
        miscarried, 1245
        sensible soul in, 1242
        shapes of, 1244
        time of hatching, 532–533
        white of, 532–534, 538–541, 1242–1248
        wind, 489, 530–531, 534, 537, 542, 550, 1141–1142, 1152, 1196, 1206, 1212, 1238, 1240–1243, 1257–1258
        yellow of, 1242–1250
    eyes of, 965–967
    falcon, 52, 335, 547, 599, 1337, 1572–1623
        breeding of, 1592–1593
        capture of, 547
        color of, 1573–1574
        genuses of, 1577–1607, 1622–1623
        medications of, 1595–1607
        nest of, 544
        training of, 1591–1597
        varieties of, 715
    feathers, 53, 966, 1070, 1073, 1345
        coloration of, 392–393, 1353
        generation of, 1030, 1238
    finch, 601, 703, 1570
    geese, 58, 64–65, 335, 338, 485, 549, 603–604, 616, 619, 1238, 1423, 1551–1553, 1557–1558, 1583–1584
        barnacle, 1563
        copulation of, 535
        eggs of, 527
        eggs, wind, 1241
        monstrous births among, 1305, 1313
        nest of, 692
    genus of, 52–53, 58–59, 886, 887
    goat milker, 677, 709, 1555, 1638
    golden plover, 1645
    goshawk, 716, 1607–1616, 1623

great diver, 697
griffin, 599, 1625
halcyon, 1560
harpy, 1555
hawk, 58, 66, 81, 327, 599, 674, 695, 707, 1553–1554, 1559, 1608–1616
  copulation of, 1226
  incubation of, 546
  training of, 1616–1618
  varieties of, 715–717
hazel grouse, 617
hazel hen, 1562
heron, 336, 501, 1555–1556
  varieties of, 676–677
hoopoe, 63, 321, 701, 1652–1653
ibis, 1631–1632
illnesses of, 628–629, 1618–1621
jackdaw, 547, 672, 704, 1318, 1426, 1638
jay, 1630–1631
kingfisher, 1560, 1633–1634
kite, 98, 327, 599, 601, 625, 673–674, 711, 1635–1636
  eggs of, 528
  incubation of, 546
lammergeier, 545, 714
land rail
  copulation of, 692
lapwing, 706, 1652
lark, 62, 509, 1427, 1559–1560, 1566–1567
linnet, 1559
magpie, 60, 62, 320, 336, 605, 672, 674, 699, 978, 1352, 1425, 1645, 1736
  eggs of, 528
merlin, 1559, 1590, 1637
migration of, 660–661, 691
molting of, 663–664
nightingale, 62, 480–481, 509, 548, 764, 767, 772, 1427, 1635, 1647
oriole, 1641
osprey, 318, 599
ostrich, 52, 59, 319, 533, 1083, 1648–1649
  eggs of, 527, 1649
owl, 51, 319, 327, 335, 600, 625, 672, 695, 698, 1561, 1573, 1640–1641, 1649, 1652
  varieties of, 713–714
parrot, 52, 320, 618–619, 761, 1425, 1647–1648
partridge, 63, 320, 334, 338, 765, 1199, 1226, 1643–1645
  copulation of, 496, 535–536, 549, 1238, 1255

egg laying of, 526
eggs of, 528, 1241
testicles of, 550
peacock, 64, 321, 1352, 1557, 1642–1643
  copulation of, 692
  egg laying of, 550
  kiss of, 537
pelican, 603, 1640–1642
pheasant, 320, 528, 617, 625, 1630
phoenix, 1103, 1623–1624
pigeon, 63, 327–328, 334–335, 480, 501, 508, 628–629, 1305, 1318, 1352, 1484, 1520, 1568–1569, 1579
  copulation of, 525, 542, 1238, 1255
  egg laying of, 542–544, 1240
  eggs of, 528
  life span of, 544, 690
  nest of, 689–690
  varieties of, 508, 690, 715–716
ptarmigan, 1635
quail, 328, 335–336, 479–480, 615, 1569–1570
  copulation of, 692
  egg laying of, 526
raven, 60, 63, 500, 547, 600, 643, 672–679, 705, 710, 717, 1426, 1567, 1631–1632
  copulation of, 1255
  incubation of, 546
  life span of, 705
  nest of, 710
ringdove, 59, 63, 65, 501, 508, 547, 602, 690, 765, 1318, 1623
  egg laying of, 526, 543, 1239
rook, 336, 547, 1631
rooster, 52, 480, 530–531, 534, 536–537, 761, 1199
sea gull, 603, 1074, 1634
sea woodpecker, nest of, 527
senses of, 1544–1545
sexual differentiation of, 485–486
snipe, 320, 1640
sparrow, 60, 328, 392, 492, 509, 555–556, 600–601, 674–675, 689, 765, 887, 1318, 1352, 1646, 1713
  eggs of, 528, 701
  life span of, 690
  nest of, 701
sparrow hawk, 334–335, 489, 547–548, 1238, 1553, 1639
starling, 58–59, 66, 320, 508, 619, 1425, 1649

Birds (cont.)
  stork, 52, 60, 81, 491, 546, 615, 672, 676, 686, 741, 1022, 1565–1566
    piety of, 615, 695, 1566
  swallow, 58, 60, 336, 392, 507, 600, 615, 625, 709, 741, 764–765, 978, 1082, 1318, 1462, 1632–1633
    nest of, 688–689
    varieties of, 675
  swan, 62, 485, 549, 603, 1073, 1427, 1564–1565
    copulation of, 1565
    incubation of, 546
    nest of, 527
  talons of, 1073
  testicles of, 1091–1094
  thrush, 1625, 1650, 1713
  titmouse, 600, 701, 1318
  turtledove, 63, 65, 338, 508, 600, 616, 673–674, 679, 690, 704, 765, 1649–1650
    chastity of, 690
    egg laying of, 526, 543, 1239
  vocalizations of, 480–481, 696, 697, 764–765, 1425–1428, 1546
  vulture, 60, 544–545, 599, 629, 671, 675–677, 696–698, 714, 1411, 1414, 1520, 1551, 1653–1654
    augury and, 717
    egg laying of, 545
    nest of, 544
    varieties of, 710–713
  wagtail, 548, 707
  warbler, 548, 672, 707
  water raven, 55
  woodcock, 615, 1624
  woodpecker, 319, 601–602, 618, 1645–1646
    nest of, 693–694, 700–701
  wren, 1570–1 571
  yellow hammer, 702
Blood
  choleric, 1126
  coagulation of, 378, 932–933
  complexion of, 905–911, 917, 937
  digestion of, 279, 1131
  formation of, 212–213
  generation of, 273, 275, 903, 917
  heart created from, 415, 1210
  male and female, differences in, 415–416
  menstrual, 76, 218, 220, 416, 418–419, 425, 427–428, 432, 513, 538, 575, 638–639, 775–777, 784–785, 797, 803, 827–829, 839, 842, 846–847, 853, 934, 956, 1065, 1109, 1111, 1130–1152, 1153, 1175, 1180, 1184–1185, 1194–1200, 1207–1208, 1233, 1238, 1283, 1316–1317, 1440
    absent in birds, 1240–1242
    absent in fish, 1240
    in animals, 1240
    Aristotle on, 805–806
    complexion of, 1281, 1288
    Galen on, 805–811
    impurity of, 800
    most abundant in women, 1199
    nourishes the embryo, 782, 815, 822, 1200
    Peripatetics on, 811–813
    retention of, 213, 1200, 1320
  natural, 924
  nature of, 413–416
  origin of, 345, 369, 374, 1034, 1210
  as seat of the soul, 94
  as ultimate food, 930–933, 1127, 1131, 1210
  unnatural, 395, 924
Bloodletting (bleeding), 349, 635, 637, 666, 848, 1469, 1487–1488
  as abortifacient, 1135
Bodies
  complexion of, 912–918, 1105, 1192, 1358–1369
  composition in, 894–896, 912–913
  elemental, 895
  fifth, 1370–1374
  formal principles of, 1393–1406
  inanimate, 1358
  macerated, 351
  material principles of, 1358–1381
  mineral, 1359
  principles of mixture in, 1381–1390
  proportions of, 1413
  spiritual, 1343, 1393
Boethius, 19
  works of
    De divisione, 21
    Topica, 21
Borgnet, Auguste, 18, 29, 34–35
Borgyan, 906
Brocotoz, the Philosopher (Herodotus), 1183

C

Caesar Augustus, 1679
Calias, 767

Candles, 31, 483, 768, 1123, 1498, 1504
Cato (Marcus Porcius Cato Uticensis), 795–796
Causes
  doctrine of four, 862, 1084
  efficient, 406, 862, 865–867, 869–870, 1132, 1215–1216, 1219, 1329, 1333, 1357, 1404
  final, 406, 862, 866–867, 874–876, 892–893, 1084, 1333, 1357
  first, 1393–1397, 1400–1401, 1404
  formal, 406, 866, 872, 876, 892–893, 1333, 1402
  material, 867–870, 872, 875, 920, 1084
  motive, 1084, 1087
  moving, 1154, 1333
  propter quid, 872–873, 1084, 1153, 1412
  secondary, 862
  types of, 1122–1123
  universal, 1396
Ceres, 1397
C(h)alcidius, 24, 1337
Chaldeans, 1397
Chaysomes, the sophist, 696
Chenu, M. D., 25
Clement IV, Pope, 15
Cleopatra, 1298, 1728
Clotho, 421
Comani (Cumani), 295, 427
Congar, Yves, 10
Conrad of Hochstaden, 13
Constantine the African, 41, 94, 116, 140, 401, 419, 814, 1440–1444, 1448, 1465
  works of
    De coitu (On Intercourse), 1440–1441, 1443, 1448
    Pantegni, 1442
    Viaticum, 1465
Constellations, 1165, 1168, 1177, 1331, 1384, 1397–1398
Costa ben Luca, 41, 1463
Creator, 135, 1440
Ctesias, 36–37
Cupid, 1183

D

Dares Phrygius
  works of
    History of Troy, 1662
Definition
  formal, 869, 896

genus and differentiae in, 877–888, 891–892
propter quid, 858
quid est, 858
Democritus, 729, 772, 869, 875–876, 995, 1209–1210, 1216–1217, 1225, 1228, 1230, 1235, 1281–1282, 1284, 1287, 1301, 1304, 1324, 1330, 1357, 1381–1383
Didymus of Alexandria, 37
Digestion, 922, 939, 1003–1004, 1021–1022, 1033–1034, 1126–1128, 1132–1133, 1175–1177, 1438
  completion of, 370
  epsesys, 1363
  fifth, 1177
  final, 243, 1022, 1167, 1173
  first, 369, 394, 949, 956, 1276
  fourth, 417–418, 423, 815, 1127, 1130, 1149, 1167, 1171, 1173, 1175, 1238, 1292
  heat as principle of, 1287
  impediments to, 1365
  imperfect, 404–405
  last, 657, 782, 1288–1289
  location of, 1128
  number of, 372
  optesys, 1363
  perfect, 405–408, 949–950
  process of, 930–931
  second, 265, 393, 406–407, 1148
  third, 73, 947,
  types of, 372–374, 922
Diogenes of Apollonia, 347–348
Diomedes, bird of, 501, 533, 660, 764, 1571
Dionysius of Viterbo, 10
Ps. Dionysius the Areopagite, 12, 1258
  works of
    The Divine Names, 11
Dissection, 30, 337, 351, 358, 368, 552, 555, 566, 598, 667, 749, 806, 936, 1031, 1081, 1108, 1225, 1283, 1286, 1309, 1335, 1437, 1768
Divination, 94, 1287
Dog Star, 517, 563, 624, 632, 633, 765, 1527, 1692, 1694
Dominicus Gundissalinus, 19

E

Earth, equator and poles of, 1330
Elements
  division of, 1382
  fifth, 1370

Elements (*cont.*)
  four, 894, 912, 1273, 1363, 1377, 1444
  mixture of, 1164
  qualities of, 895, 1165, 1174–1175, 1177
Empedocles, 41, 90, 414, 489, 866, 868, 872, 875, 906, 1114–1118, 1125, 1144, 1166, 1228–1230, 1235–1236, 1281–1284, 1286–1287, 1301, 1337, 1340, 1381, 1386, 1388–1389, 1395, 1442
Epicureans, 1397
Equinox, 612–614, 621, 624, 1232, 1279, 1526, 1623
  autumnal, 505, 1660
  vernal, 568, 1450–1451
Eradytis, the poet, 889
Esceny, King, 678
Esculapius, 1397
Ethiopians, 382, 385, 706, 1699, 1726
  hair of, 1346
  sperm of, 432, 1183
  white child from, 1113
Euclid
  works of
    Theorems, 1325

F

Favorinus, 105
First Master (Aristotle), 360–364
Fish
  anatomy of, 321–324, 328–329, 334, 1030, 1077–1082, 1238
  barbel, 132, 322, 473, 567, 595, 614, 623, 977, 1656
  blood in, 1155–1156
  bones of, 381
  bonito, 1663
  brain, anatomy of, 251
  bream, 1680, 1696, 1705
  burbot, 334, 624, 1665–1666, 1699
  carp, 132, 977, 1023, 1671–1672
  chub, 1672
  cod, 324, 386
  complexion of, 1078, 1240, 1430
  copulation of, 495–496, 555–556, 559–562, 1093, 1107, 1254
  egg laying, 505–506, 559–563, 1242, 1251–1254, 1258, 1656
  eggs, 1240, 1249–1250
    completion of, 551–552, 1095, 1156, 1237, 1257
    wind, 1254
  eyes of, 967, 1657
  flatfish, 79, 556, 619, 1665, 1697
  flounder, 1665
  flying, 1078, 1692
  garfish, 1663
  generation of, 489–507, 1208, 1657
    from putrescence, 563–566
    times of, 567–571
  genus of, 52, 887, 1023
  gills of, 322–323, 472–473, 590–591, 597–598, 1080–1081
  goby, 1684
  grayling, 1705
  gray mullet, 1656, 1677, 1691
  hake, 718
  herring, 59, 504, 567–570, 619, 680, 722, 1660, 1669
  illnesses of
    lice on, 632
    rheum, 631
    water boil, 629, 633
  mayfly, 1679–1680
  migratory, 620–622
  monkfish, 132, 630, 974, 977, 1693, 1699
  mudskipper, 1666
  mullet, 473, 567–568, 595, 619, 680, 718, 721, 1684, 1686, 1690, 1697
  nose, 1693
  parrot wrasse, 1695, 1700
  perch, 265, 1259, 1689
  pike, 54, 79, 265, 314, 322–324, 593, 596, 620, 656, 670, 973, 979, 1688–1690
  pinna, 1662
  rascasse, 1687
  red mullet, 1692
  reproduction of, 648–649
  rock-fish, 631
  salmon, 265, 269, 322–323, 334, 483, 1682, 1697–1698, 1704
  sand smelt, 1662
  saupe, 1699
  sea bass, 473, 1687
  sensory organs of, 471–475
  sexual differentiation of, 486, 1204, 1213, 1252–1254
  shad, 58, 472, 619, 680, 1705
  sheatfish, 329, 1023, 1671, 1684–1685, 1698
  sleep of, 482–483, 1656
  smelt, 680
  stargazer, 1684

sturgeon, 54, 265, 324, 345, 483, 656, 721, 955, 1023, 1081, 1662–1663, 1682–1683, 1698
   glue made from, 385–386
  teeth of, 977, 1023
  three-spine stickleback, 1694–1695
  trout, 620, 1704
  tunny, 621, 632
  turbot, 79, 1697
  vocalizations of, 720
  water-wolf (*lupus aquaticus*), 323, 504, 716
  weaver fish, 1664
  wombs, anatomy of, 551
Fishing methods, 451
  bait, 471, 474
  hooks, 719, 1657
  lures, 718
  nets, 55, 472, 590, 593, 625, 716, 1520, 1657, 1671, 1678, 1705
  night fishing, 483
  spearing, 483, 670
  tanging, 472, 474, 1428, 1705
  traps, 474, 590, 633–634, 1520, 1657, 1658
Food
  assimilation of, 371–374, 1020–1024, 1175–1177
  digestion of, 1025–1026, 1175, 1349–1350
  melancholic, 1011, 1224
  qualities of, 1222
  sanguinous, 1224
Foods
  broth, 213, 411
  cheese, 76, 299, 429
  cheese making, rennet for, 430
  coarse, 209, 371, 905, 1739
  manna, 1658
  meat, roasted, 213, 1451, 1515
  pumpkin, 776
Form, 27, 371–374, 420–423
  act and, 1178
  animal, 108
  Aristotle on, 1377
  celestial, 1375–1381
  elemental, 1372
  factive, 1154, 1195, 1274
  first, 1395
  harmonic, 1390
  imaginative, 941
  inchoate, 421, 1165, 1172
  intellectual, 1397
  light and, 1376–1384
  moving, 1420
  nature and, 1178
  operable, 1422
  operative, 1140, 1154, 1211
  Platonic, 421, 892, 1377, 1393
  principle of generation and, 1395
  sensible, 87, 88, 108, 826, 862, 941, 1177, 1221
  specific, 1188
  speculative, 1420
  substantial, 870, 880, 882–883, 886, 892, 1163, 1165, 1176, 1179, 1188–1189, 1203, 1386, 1409–1410, 1712
  universal, 371
  vegetable, 1177
Fowling methods
  bait, 675
  beaters, 716
  falcon used, 1576
  nets, 66, 618, 672, 696, 1551, 1561, 1582, 1630, 1645
  owl used, 672, 704
  snare, 1556, 1570, 1582
Frederick II, Emperor, 6, 1450, 1585, 1603, 1607–1608, 1611
Fronesys (Socrates's wife), 1299

G

Galen, 40–41, 71, 119, 158, 163, 170, 223, 267, 276, 351, 358–376, 396, 404, 419–420, 423, 530, 664, 804–813, 815, 819, 903, 917, 919–920, 924–925, 928, 941, 979, 1009–1010, 1108, 1150, 1202, 1204, 1206, 1309, 1404, 1441–1445, 1448, 1457, 1463–1465, 1520, 1525, 1632, 1657, 1713, 1732, 1754–1757, 1761
  works of
    De complexione (On Complexion), 1463, 1732
    On Accident and Disease, 1445
    On Sperm, 1441, 1443, 1447
    On the Care of the Members, 1441
Galileo, 31
Geometry, 285, 865
Gerard of Cremona, 19
Gerhardt, Mia, 42
Geyer, Bernhard, 2
Giles of Lessines, 13
God, 23–27, 1236, 1391, 1439, 1445, 1654
Godfrey of Bléneau, 10
Gods, Plato on, 1171
Grabmann, Martin, 11
Grosseteste, Robert, 19, 21, 25

Grattius
  works of
    Cynegetica, 37
Gregory the Great, Pope, 42
Gregory IX, Pope, 20
Guerric of St. Quentin, 10–11
Gyrgyr, philosopher, 1520

H

Hadrian, 104, 526
Halbamnan, poet, 109
Haly 'Abbās ('Alī ibn al-'Abbās al-Majūsī), 375
Heavens, 872, 1303, 1370, 1372–1376, 1380
  life of, 1373
  motion of, 1177, 1382–1383, 1400
  mover of, 1171
  powers of, 769, 1168, 1176–1177, 1187, 1303, 1382–1383, 1396, 1400
  revolutions of, 1325
Hector (of Troy), 403, 1392
Henricus Aristippus, 19
Henry of Hereford, 3, 6, 8, 12
  works of
    Liber de rebus memorabilibus, 3
Henry of Lauingen, 15
Heraclitus, 1358
Hermaphrodites, generation of, 1312–1313, 1443
  (*See also* Humans: hermaphrodite)
Hermes Trismegistus, 1445, 1531, 1628, 1721
Hermodicytes (Herodotus), 1254
Herodotus, 36–37, 39
Hesiod, 572, 1018
Hildegaard of Bingen, 40
Hippocrates, 93, 108, 113, 283, 423, 804, 845, 1227, 1442–1444
Homer, 353, 501, 579–580, 629, 643, 698, 711, 754, 1018, 1123, 1161–1162, 1287, 1350
  works of
    On Animals, 711
    Taking of Ilion, 629
Hossfeld, Paul, 29, 31
Hugh of St. Cher, 8, 13
Hugo of Lucca, 16
Humans
  aging, signs of, 284, 386–387, 1056, 1332, 1338–1340, 1349–1351
  baldness
    causes of, 388, 762, 1345, 1347–1350
    cure for, 1351
  beheading of, 921, 1018
  body
    growth of, 1300, 1309, 1319, 1336
    symmetry of, 1413
  bones, number of, 140, 145–148, 152, 154–157
  brain, size of, 938, 940
  complexion of, 86, 238–239, 914, 1057, 1293–1294, 1391–1392, 1403, 1413, 1441, 1444
  conception
    aids to, 1475, 1504. 1516
    Aristotle on, 812–820
    Galen on, 804–811
    impediments to, 830–837, 842–848, 1134, 1516, 1521
    solstice, influence upon, 525
  copulation of, 219, 499
    health or illness and, 1130
  deaf-mutes, 481
  differentiae of, 877, 891, 1407
  digestive powers of weakest, 229
  dreams of, 482–483, 803, 829, 835, 838–839
  embryo, 194, 214–220, 782, 785, 787–793, 798–802, 815–816, 830, 836, 842, 924, 934, 1225, 1322–1323, 1335, 1442
    Aristotle on formation of, 819
    development of, 820–824, 851–852, 997, 1320–1321
  female
    amenorrheal, 1227, 1320, 1515
    beards on, 131
    brain size, 939
    breasts, 1063
      location of, 1323
      puberty, effect on, 775–779
    childbirth, 994, 1134
      age for, 780
      labor, 854–856, 1464, 1469
      loss of appetite following, 1320
      midwives at, 790, 799, 802, 823, 1321
      miscarriage, 782, 786, 792, 794, 821, 841, 846–849, 995, 1233, 1313, 1320–1321
      pain of, 779–780, 788, 801, 840
    complexion of, 831, 835, 906, 914, 926, 1136–1138, 1319, 1338
    contraceptives, 805, 1519–1520
    dreams, sexual, 829, 835, 838–839
    flawed male and, 422, 1195
    lactation of, 428, 784–785, 803, 822, 840, 1322–1323
    lustfulness of, 581, 836, 1315–1316
    manual labor among, 1320
    masculine, 1227, 1234, 1328

masturbation, 776–778, 825, 829, 1108
menopause, 795
menstruation of, 574, 776–777, 781–782, 828, 840, 926, 1133, 1137–1138, 1198, 1313, 1316, 1320, 1323, 1469
    Aristotle on, 781
multiple births of a, 778, 791–794, 824–825, 836, 1138, 1311, 1320
nocturnal emissions of, 778, 1207
passions of, 669
penis of, 216
pregnancy, 219, 775
    abortion of, 848, 1135, 1477, 1504
    aids to, 784
    false, 791, 839, 846, 1321
    impediments to, 1227
    intercourse during, 792
    menstruation during, 1135
    morning sickness, 787–788, 816
    retention of menses during, 1320, 1324
    signs of, 783–788, 816, 849–850
    times of, 782
prostitutes, 777, 842, 1676
reproduction, age of, 514
sexual intercourse
    desire for, 776, 1441, 1520
    pain in, 844
    pleasure of, 535, 827, 850, 853, 1108, 1120–1121, 1134, 1137, 1146, 1151, 1207–1208
    positions in, 844–845
spermatic vessels in, 216, 806, 1107–1108
sterility in, 844, 1226–1228
superfetation in, 793–794, 1315, 1330
testicles of, 774, 806, 810, 816, 1107–1108, 1138
vulva, 194, 216, 219–221, 776, 783–784, 786–787, 844–846, 1314–1315
fetus, 220, 361, 427, 800, 801, 808, 1134, 1315, 1335
    complexion of, 1443
    first movement of, 822
    formation, time of, 785–787
    genital formation in, 786
    gestation period of, 787, 790, 1304, 1311, 1322
    injury to, 794
    position during childbirth, 801–802, 825–826
    removal of, 1469
    sexual differentiation of, 785–798, 853–854, 1443

four ages of, 119, 306, 918–922, 1056, 1129, 1328, 1348
hair, 1345–1349
    complexion of, 904
    curled with asses' dung, 1452
    eyebrows, 968–969
    facial, 131, 389–390, 1105, 1112
    generation of, 968
    genital, 294, 389, 424, 511, 774–775, 842, 1138, 1348
    graying, 386–388, 1341, 1345, 1349–1351
    loss due to pregnancy, 789
    physiognomy of, 95–97, 241, 269
hermaphrodite, 813, 1306, 1312–1313
hibernation of, 659–660
infants, 843
    brain in, 540, 1221
    crying in womb as evil omen, 1335
    dreams among, 803, 826
    dwarf, 830, 1233
    laughter of, 826, 1336
    nursing of, 213, 427–428, 430, 801–804, 1357, 1441, 1604
    premature, 1318–1319
    resemblance to parents, causes of, 1295–1303, 1328–1330, 1443
    sleep of, 1335
    teeth, first appearance of, 803
    umbilical cord on, 802
intellect of, 932, 1057–1058, 1399, 1413, 1445
laughter as proper predicate of, 1017–1018, 1447
male
    breasts, 779, 1063
    castration of, 389, 762, 844, 1105, 1287, 1348, 1444
    complexion of, 1319
    contraceptives, 844
    effeminate, 1227, 1328
    lactation of, 428
    masturbation of, 1207
    nocturnal emissions of, 511, 762, 778, 1136, 1207
    penis
        congenital defect in, 1314
        erection of, 223, 424, 1441
        puberty, effect on, 775–776
    reproduction, age of, 514, 780, 795
    seminal vessels, 776, 1170, 1197, 1292, 1300, 1302, 1327, 1441
    sexual desire in, 1242, 1441
    sexual impotence, 777

Humans (cont.)
   sexual intercourse
      aids to, 1454, 1515
      pleasure of, 1208
      sterility in, 1226–1228
   as microcosm of the world, 937, 1187, 1413
   monstrous births among, 1295, 1303–1307, 1313, 1444
      polydactylism, 1306
      Siamese twins, 1313
   as most perfect animal, 1410
   nerves, number of, 189
   puberty, changes occurring with, 511, 774–779, 1313, 1443
   senses, nobility of, 1411–1412
   sensory instruments, 471
   sexual intercourse, 62, 93, 100, 213, 218–223, 225, 340, 775
      aids to, 1475
      as cause of aging in, 780
      desire is greatest among, 783
      excessive desire for, 776–780, 1300
      hair loss caused by, 389–391
      immoderate, effects of, 777, 1444
      impediments to, 1519
      pleasure of, 423, 535, 794, 796, 798, 814, 825, 850, 853, 1091, 1108, 1111, 1120–1121, 1134–1137, 1144, 1146, 1148, 1151, 1207–1208, 1235
   shame as proper predicate, 1446
   skin color, 1351–1352
   speech, 135–136, 477, 481, 972–973, 976, 1343, 1354
      defects of, 131, 135, 481, 973, 1417
   substance of, 859, 1153
   tail, absence of, 1066
   upright posture of, 957, 1056, 1066–1067, 1369, 1447
   voice, 62, 127, 774–777, 1343, 1354
      instruments of, 134–138, 961, 972–973, 976, 979
      physiognomy of, 106, 140
   wild men, 308
Humbert of Romans, 14–15
Humboldt, Alexander von, 31
Humors, 913
   bile, 240–241, 277–278, 327, 925, 1368, 1391
      unnatural, 399–404, 406, 661, 923, 925
   black bile, 195, 275, 280, 345, 395, 401, 810, 1391, 1732
      types of, 401–404
   breast milk, 425–432
   chyle, 273–276, 279–280, 376. 394, 405–406, 412, 443
   complexion of, 45, 48, 1402
      discrasia in, 628, 1125, 1379, 1389, 1657
   doctrine of four, 394–395, 912, 924, 1363, 1391, 1442
   four causes of, 406
   generation of, 393–408
   nutrimental, 399, 567 1294, 1367, 1435 1739
   phlegm, 118, 236, 284, 345, 818, 842, 903–904, 917, 924–925, 968, 1126, 1128, 1182, 1391–1392, 1734
      types of, 396–399
   radical moisture, 393, 395, 540, 584, 662, 920, 1184–1185, 1224, 1312, 1387, 1435
   red bile, 279, 345, 399, 402–404, 406–407, 416
      types of, 399–400
   sperm, 776–787, 791–798, 804–821, 824–825, 873, 949, 956, 1065, 1191, 1405
      act of the soul and, 1180
      Albert the Great on, 1127–1130
      Anaxagoras on, 1114–1117, 1123–1126
      Avicenna on, 1167
      begetting twins, 1114
      complexion of, 423–424, 904, 1118–1119, 1181–1183, 1195, 1235, 1295, 1328
      corrupted, 1195
      descends from the brain, 1227
      digestion of, 1093
      emission of, 1090–1094, 1137
      Empedocles on, 1114–1118, 1125–1126
      female, 76, 419, 791, 829, 831, 834, 842, 853, 1121, 1130–1131, 1139, 1178, 1184, 1203, 1207–1208, 1223, 1286, 1440
         Aristotle on, 419–423, 1108, 1150
         Galen on, 419–422, 530, 805–808, 1108, 1150, 1204, 1206
         physicians on, 825, 1135, 1150
      first generation of, 779
      first operations of, 1163
      formative power in male, 815, 1132, 1140, 1154, 1178, 1195, 1211, 1296, 1310
      fourth digestion and, 815, 1127, 1131, 1149, 1171, 1175, 1194, 1292
      Galen on, 1441–1442, 1447–1448
      generated from nutrimental moisture, 221
      nutritive power in, 1201
      origin of, 1111–1112, 1112–1114, 1117–1121
      principle of generation and, 1111–1112, 1123–1124, 1152, 1153, 1158–1159, 1180

production of, 510
quality of, 780–781
retention of, 1129
sexual differentiation and, 421–423, 785, 812–814, 824, 1115, 1118–1119, 1146–1147, 1169, 1194–1196, 1282–1291
small animal and, 1115, 1147, 1159, 1178–1179
sterility of, 842–845, 1207, 1227, 1234
Empedocles on, 1228–1230, 1235–1236
Stoics on, 1149
testing of potency, 1227
spermatic fluid
generation of, 417–420
yellow bile, 277, 910
Hunting methods, 1453
deer bladder, 684
spear, 1543

I

Illnesses (*See also* Medicines)
abscesses, *passim. See especially* 132, 226, 264, 275, 278, 404, 634–635, 803, 922, 1452, 1445, 1464, 1472, 1483, 1735, 1738, 1757–1758, 1760–1761, 1763
healed by spittle, 1447
asthma, 256, 1542, 1628
black morphew, 797
boils, 1529
cancer, 278, 796, 848, 1195, 1490, 1514, 1530
catarrh, 206, 397, 903, 937–938, 990
colic, 263–264, 399, 897, 925, 1029, 1032, 1368, 1464, 1473, 1520, 1713
consumption, 237, 426, 856, 1507
convulsions, 803–804
deafness, 481, 1514, 1520
diarrhea, 278, 635–636, 1472, 1515
dog hunger, 1024
dropsy, 230, 840, 842, 846, 1012, 1357, 1454, 1457, 1507, 1529, 1727, 1730
drunkenness, 1203, 1222
dysentery, 839–840, 1321, 1507
earache, 1542
elephantia, 278, 1506
epilepsy, 778, 791, 1057, 1203, 1222, 1330, 1335, 1451, 1457, 1477, 1509, 1516, 1520, 1524–1525, 1646
eye, 98, 1516
"eye claw," 1738
"webeye," 1466
"hoof eye," 1537
fever, 909
gall bladder as cause of, 1031
hectic, 211, 239, 1369, 1451
quartan, 630, 664
tertian, 664, 785, 919
flatulence, 229, 241, 264, 804, 850, 1454, 1479, 1507, 1761
formica, 278
gout, 796, 1452, 1467, 1469, 1507, 1524, 1562, 1668
headache, 842, 849, 1477, 1525
hemorrhoids, 230, 416, 781, 856, 1133, 1136. 1474
herysypila, 278
hylyaca, 1368
insanity, 1392, 1514
insomnia, 1392
jaundice, 278, 1463, 1473, 1507, 1514, 1530
kidney stones, 397–398, 410, 1000, 1446, 1516
leprosy, 239, 278, 605, 796, 1195, 1330, 1445, 1454, 1461, 1477, 1506–1507, 1539, 1671, 1688, 1736
lienteric flux, 1136
lunar influence on, 826
lungs, 137, 207, 256
madness, 939, 1392
mange, 1480, 1508, 1514
melancholy, 94, 256, 278, 796, 1057, 1417
molinsis (corruption), 1328
nosebleeds, 416, 508, 1452
old age, 1350
paralysis, 442, 1362, 1468, 1507, 1510, 1521, 1668
plague, 632, 1104, 1446
pox, 1529
pytuita, 1426
quinsy, 1464, 1504
rabies, 755, 1458, 1461, 1519
scabies, 278, 1461
scrofula, 1451
skin, 1349
sperm, retention of, 1129–1130
toothache, 1514, 1516, 1678
ulcers, 844, 848, 1507, 1515–1516
varicose veins, 278
vomiting, 231, 987, 1148, 1688
warts, 833, 844, 1525, 1529–1530
womb, suffocation of, 782, 806, 1097
Indians, 572, 679, 901, 915
Insects
ant-lion, 1749

Insects (*cont.*)
    ants, 59, 60, 62, 65–66, 73, 75, 293, 693, 726–727, 880, 1036, 1415, 1463, 1749–1750
        bites of, 727, 1749
        generation of, 1101
    beast-flies, 82
    bedbug, 1047, 1102, 1747
    bees, 55, 59–60, 65–68, 70, 73, 75, 292, 436, 475, 483, 612, 623, 656, 726, 750, 752, 767, 905, 975, 1035, 1262–1272, 1415–1416, 1436, 1612, 1739, 1741–1742
        anatomy of, 468
        Aristotle on, 743–745, 1263, 1270, 1741
        Avicenna on, 743, 1263, 1742
        complexion of, 1391, 1435
        generation of, 489–490, 1086, 1263–1272, 1741
            from putrescence, 503, 1101
        hives of, 733–746, 1265, 1270
        honey of, 732, 742–746, 1270, 1741–1742
            medicinal use, 744
        illnesses of, 640–641, 743
        larvae of, 1103–1104, 1119
        prognostication with, 747
        sexual differentiation of, 1263–1268
        stinger of, 393, 736, 741–742, 1047, 1741–1742
        varieties of, 465, 730–747
        wings of, 1045–1046
    black flies, 502
    blister beetle, 1747
    butterfly, 1035, 1752
    caterpillar, 56, 62, 300, 466, 502–503, 1261, 1264, 1277, 1436–1437, 1740–1741, 1743, 1748
        metamorphosis of, 1436
    cicada, 60, 465, 1612–1613, 1748
    cockroach, 1746
    corn weevil, 81, 1261
    cricket, 468, 478, 1613, 1748
    dog-fly, 82, 454, 466, 612, 656, 975, 1746
        copulation of, 499
    firefly, 436, 880, 883, 1746, 1758
    flea, 1047, 1416, 1461, 1463, 1507, 1541, 1568, 1752–1753
        generation of, 1086
    fly, 292, 975, 1035, 1415, 1436–1437, 1740, 1751–1752
        generation of, 1086
            from putrescence, 1101

    generation of, 1260–1262
    gnat, 656, 1651, 1747
    hornet, 467, 1158, 1475, 1747, 1763
    larvae of, *passim. See especially* 73–75, 489–490, 561, 737, 1086, 1101, 1103, 1155–1158, 1251, 1260–1267, 1278, 1741
        generated from putrescence, 503, 640–641, 737, 1086, 1101, 1261, 1751, 1753
    lice, 73, 489, 1477, 1613, 1621, 1753–1754
        generation of, 1086
    locust, 59, 465, 478, 726, 741–742, 837, 1047–1048, 1104, 1436, 1613, 1654, 1740, 1750
        generation of, 1101–1102, 1746
    mosquito, 503, 656, 975, 978, 1047, 1477, 1504, 1744, 1746–1747
    moth, 1514, 1762
    scarab beetle, 81, 331, 467, 980, 1103, 1746, 1758
    stinger of, 393, 466–468, 656, 975, 1035, 1044, 1046–1047
    termites, 320, 601, 641, 1261, 1762–1763
    tick, 1748–1749
    wasps, *passim. See especially* 55, 59, 70, 77, 81, 436, 467, 664, 726, 731–732, 741, 746, 1046, 1261–1262, 1353, 1436, 1716, 1742, 1747, 1763
        Avicenna on, 753
        generation of, 503, 1086, 1101, 1262, 1271–1272, 1472
        hives of, 750–752
        stinger of, 749–750, 1047
        varieties of, 465, 736, 741, 747–753, 1271, 1353, 1716, 1757, 1763–1764
        wings of, 1046
Instruction
    animals capable of receiving, 667–669, 1408, 1415–1416
    humans alone naturally susceptible to, 1419
    three degrees of, 1419
Intellect, 61, 65, 1191, 1398
    absent in brute animals, 64, 766
    active, 1393
    agent, 1193
    celestial, 1400, 1404
    divine, 1191, 1193, 1396, 1418
    error in the, 1421
    first, 1187, 1190
    light of, 1189
    moving, 1187, 1413

operative, 1193, 1203, 1412
practical, 1058, 1193, 1203, 1393–1394, 1399
separate, 1206, 1396
simple, 1417
speculative, 1039, 1446
Isaac Israeli, 1168, 1371, 1433, 1472
Isidore of Seville, 8, 40, 1467, 1475, 1704
Italians, 1078, 1322, 1449, 1552, 1678

J

James of Venice, 19
Jammy, Peter, 18
Jean de la Roche, 12
Jerusalem, Temple of, 429
Jews, 594, 1762
John Damascene, 135, 212, 667
John of Fidanza (St. Bonaventure), 12
John of Spain, 19
John of Vercelli, 1 5
Jorach, 1450, 1457, 1512, 1541, 1549, 1555, 1565, 1569, 1728–1734, 1737
  works of
    On Animals, 1450
Jordan of Saxony, 4–5
Josephus, Flavius, 429
Jupiter (god), 737, 795, 889, 1018, 1087, 1397, 1560
Jupiter (planet), 510, 513, 514, 780, 939

K

Kilwardby, Robert, 12

L

Law, Roman, 795
Legenda Coloniensis, 3
Light, 121–122, 126, 1399
  fifth body and, 1370–1380
  motion of, 110
  power of, 1168
  rays, angles of, 1165
  spectrum of, 1394
  vehicle for the soul and, 1373
Logic, 1333
  science of, 860
  syllogism
    middle term of, 861
    operative, 1421
    practical, 1421
  syllogistic argument, 810, 874, 1112, 1164, 1334
    enthymeme, 1421

Loxus, 94, 96, 99, 108, 113, 130, 133, 214, 225, 238, 241, 256
Lucan, 1649, 1724, 1730–1731
Luis of Valladolid, 3
  works of
    Tabula Alberti Magni, 3
Lyons, Council of, 16

M

Macer, Aemelius, 1723
Machomet (Mohammad), 493
Magic, 12, 1143
  incantations, 638–639
  magicians and, 325, 639
  necromancers, 573–574
  spells, 1445
  wizards, 1653, 1717, 1726–1727
Maimonides, Moses, 19
Manuscripts
  Köln W258a, 34–35
  Naples, Bibl. Naz. I B 54, 12
Mappa Mundi, 78, 312
Marcia (wife of Marcus Porcius Cato Uticensis), 795
Marine animals, 591–598
  calamary, 57, 435, 496, 723, 1036, 1053–1054, 1259, 1687
  carnivorous, 596–597
  cephalopods, *passim. See especially* 57, 433–435, 439–440, 945
    copulation of, 1101
  chimera, 1659
  conger eel, 582, 584, 593, 595, 1684
  crab, *passim. See especially* 82–83, 132, 434–435, 473, 593, 627, 951, 1037, 1049–1050, 1431, 1554, 1661, 1673–1675
    anatomy of, 1055
    claws of, 331, 1071
    hermit, 1676
    as poison antidote, 684, 1471
  crayfish, 437, 593, 1050, 1100, 1431, 1435–1436, 1596, 1673, 1687–1689, 1710
  crustaceans, 436–441, 441–448, 951
    anatomy of, 1049–1051
    copulation of, 497–498, 1100
    generation of, 522–523
  cuttlefish, *passim. See especially* 57, 70, 435, 437, 723–724, 952, 974, 1036, 1429–1432, 1699–1700
    anatomy of, 1053
    copulation of, 496–497

Marine animals (*cont.*)
  egg laying of, 506, 522, 1259
  dragon (*draco aquaticus*), 440, 633
  eel, *passim. See especially* 68, 322, 383, 597–599, 656, 1079, 1081, 1213, 1438, 1660–1661, 1671, 1684, 1689
    eggs of, 558, 564
      generated from larvae, 1277
      generated from worms, 564–567
    moray, 79, 382, 570, 595, 619, 631
    sexual differentiation of, 485, 1253
  genuses, number of, 433–435
  lamiae, 494, 1516–1517
  lamprey, 721, 1079, 1667, 1719–1720
  lobster, 57, 70, 80, 82–83, 441, 593–594, 631, 722, 974, 1037–1038, 1049–1051. 1431, 1436, 1673, 1687–1688, 1691
    copulation of, 522, 1256
    egg laying of, 455, 497, 522–523
    eggs of, 523, 1259
    hermaphroditism of, 1256
  mollusks, 449–458, 1038
    anatomy of, 1048–1049
    generation of, 518–519, 1272–1280
    sexual differentiation of, 484
    shell, generation of, 463–465
  murex, 519, 593, 631, 721, 1692, 1694
    dye produced from, 516–517
    egg laying of, 507
    generation, 516, 1274
  mussels, 519
  narwhal, 1693
  nautilus, 1693
  nereids, 1693
  octopus, *passim. See especially* 57, 465, 474, 518, 593–594, 648, 722–724, 952, 1036, 1428–1433, 1691, 1696–1697
    anatomy of, 1051–1055
    complexion of, 937
    copulation of, 497
    egg laying of, 506, 523–524, 1259
  oysters, 54, 56, 1675, 1690–1691, 1694
  pinna, 1662, 1694–1696
  ray, 74, 1697, 1724
    electric, 554, 718, 1658–1659, 1704
    horned, 282, 1706
  razor clam, 449–451, 588
  remora, 1692
  scallop, 456, 476, 479, 519, 1048
  sculpin, 1700

sea anemone, 462, 592–593, 1686
sea bass, 1687
sea bream, 1700
sea calf, *passim. See especially* 281, 327–328, 493, 620, 962, 1031, 1082, 1365, 1429, 1681, 1686–1687
  parturition of, 557–558
sea cow, 493, 557, 620
sea crow, 1672
sea dog, 83, 323, 464, 556, 1677
  copulation of, 495
sea dragon, 1677–1678, 1720, 1727
sea eel, 619, 1671
sea fox, 553, 556, 1705
sea frog, 80, 322, 590, 717–718, 1080, 1250–1251, 1697
sea gem, 619
sea hare, 79, 619, 1079, 1688, 1712, 1755
sea hedgehog, 337, 434, 1507
sea horse, 1680, 1706
sea locust, 57, 70, 80, 425, 441, 443, 522, 593, 722, 1687–1688
sea mouse, 1692
sea nettle, 461
sea ox, 1683
sea pig, 478, 1695
sea ram, 1662
sea scorpion, 504, 619, 1700
sea snake, 719, 1677
sea soldiers, 620
sea sow, 556, 619
sea spider, 443, 457
sea sponge, 56, 434
sea swallow, 479, 1686
sea tortoise, 493, 1003, 1665, 1702–1704
sea turtle, 493, 593, 620, 1665, 1706–1707
sea urchin, 450, 455, 459–460, 476, 519, 974, 1039–1041, 1347, 1508, 1681–1683
  eggs of, 507, 1039–1041
sea viper, 1705
sea wolf, 83, 551, 719–720, 760, 1005
  copulation of, 495
shark, 74, 553–554, 556, 1705
shellfish, 1673
skate, 79
snail, 476, 1275, 1672–1673
sponge, 434, 470, 588, 1042, 1432–1433, 1439, 1662, 1698
  varieties of, 520–522
squid, 57, 1036, 1587

anatomy of, 1053
copulation of, 496
starfish, 434, 469, 519, 1131, 1432–1433, 1698, 1702
swordfish, 1685, 1707
Marriage, 1446
Mars (god), 1397
Mars (planet), 513, 939
Martynyon, physician, 926
Matthaeus Platearius, 1472
Medicine, definition of, 862
Medicines
enema, 864, 1472, 1515, 1632
human excrement, 685
poison antidote
endive leaf, 1416
fox penis, 688
simples, 1178, 1329
theriac, 685, 701, 1473, 1603, 1713
Meersseman, Giles Gerard, 18
Mellissus (Musaeus), 349, 545
Members, *passim. See especially*
animal, 1020
Aristotle on, 141, 205
complexion of, 929
composite, 45
corruption of, 1365
formation of, 1277–1278
generation of, 1215, 1218–1224
heart, 1201
divine member and, 1056
power of, 1160, 1400–1401
principle of digestion and, 370, 651
principle of form and, 372, 1405
principle of life and, 988, 1056, 1180, 1214
principle of movement and, 375, 820, 988, 1216–1217
principle of nourishment and, 820, 988, 1056
principle of progressive motion and, 1434
principle of sensation and, 375, 820, 899, 931, 948, 950, 988, 1056
principle of the nerves and, 375, 1356
principle of the veins and, 953, 1210, 1218, 1220
homogeneous, 408, 956, 1207, 1209, 1211, 1213, 1218, 1402
instrumental, 875, 890, 1149
motion and, 291–298
natural, 1020

nonuniform, 47, 49–51, 70, 241, 895–896, 913, 1112–1117, 1125, 1174, 1179, 1361
nutritive, 92, 148–149, 165, 257, 986, 989, 994, 1016, 1020, 1202–1203, 1387
official, 896, 912–913, 1113, 1163, 1174, 1289, 1294
organic, 897, 904, 913, 956, 1216, 1353
principal, 327, 368, 819, 845, 899–902, 915, 1116, 1161, 1290–1291, 1296, 1381
radical, 805, 808–810, 1248, 1277
spiritual, 165, 257, 938, 989, 1020, 1062, 1417
uniform, 46–50, 69–70, 77, 90–91, 236, 243, 344–345, 433, 867, 895–900, 904, 912, 950, 955, 1112–1117, 1125, 1163, 1174, 1213, 1361
sensation in, 898
vital, 1020
Memnon, the Pythagorean, 1636
Mercury (planet), 940
Michael Scot, 19, 39–40
Monsters, 113, 133, 308, 542, 983, 1295, 1309
generation of, 1303–1307, 1312–1313
Democritus on, 1304
marine, 1656, 1696, 1700–1701
Scilia, 1701
sirens, 1701
minotaur, 1304
sirens, 1733
Moon
eclipse of, 861
female body and the, 781, 828, 1198
generation and the, 1273–1274, 1279–1280
incubation of eggs and the, 532
phases of, 1293, 1325–1326
births and, 513
power of, 939
revolutions of, 1331
weather patterns and the, 1293
Moors, 432, 797, 902, 1146
Motion, rectilinear, 1413
Movers
first, 71, 896, 1296, 1393, 1395, 1400, 1413, 1436
moved, 72, 1173, 1393
prime, 71, 1173, 1187
second, 1413
Mucianus, philosopher, 1692
Music, 285, 867, 1529
bourdon, 1354
chords, 1389

Music (*cont.*)
  cymbals, 1389
  drums, 1519–1520
  lullabies, 427
  lute, 1425
  neumes, 1355
  pipes, 1059, 1215–1216, 1406, 1471
  Pythagoras on, 1379
  viol, 771, 891

N

Nature, *passim. See especially* 618, 655, 693, 697, 717, 817, 1513
  abhors the infinite, 1103, 1265
  cannot err across an entire species, 99
  creates monsters, 1306
  does nothing uselessly, 1221–1222
  gradations in, 587, 652
  intelligence of, 1176–1177, 1186
  universal, 826
  well ordered, 1364
  wise, 1061, 1251
  works for the best, 1090, 1289, 1434
Necessity
  absolute, 27, 863–867, 874, 1357
  consequentiae, 865
  consequentis et rei, 865
  four types of, 874–875, 891
  hypothetical, 28
  material, 873
  mathematical, 866
  suppositional, 27–28, 863–867, 874–876, 892–893, 895
Nectanabus (Alexander the Great's father), 1444
Negroes, 643
Nemesianus, 38
  works of
    Cynegetica, 37
    De aucupio (*On Fowling*), 38
Neo-Platonism, 11
Nicholas the Peripatetic, 1011
Nigidius Figulus Paulus, 37

O

Odericus Francigenus, 10
Odo Rigaud, 12
Olympias (Alexander the Great's mother), 1444
Oppian, 38
  works of
    Cynegetica, 38
    Halieutica (*On Fishing*), 38

Order of Preachers, 4–5, 12–16, 19
Origen, 1436
Ovid, 301

P

Palemon, 94, 96, 99, 102, 105, 108, 113, 129, 133, 140, 211, 238, 240, 256, 284, 1061
Pamphilius of Alexandria, 37
  works of
    Peri botanōn (*On Plants*), 37
    Physika, 37
Paradise, birds of, 1561
Parthians, 295, 299, 300, 1750
Pelagonius, 38
Pelster, Franz, 35
Periander, King, 1692
Peripatetics, 70–71, 189, 205, 351, 375, 650, 785, 804–805, 808, 1146, 1149, 1159, 1176, 1184, 1189, 1200, 1205, 1327, 1356, 1374, 1380, 1405, 1764
Peter Gallego, 39
Peter Lombard, 8
Peter of Prussia, 3, 12
  works of
    Vita B. Alberti Doctoris Magni, 3
Peter of Spain, 39
Philip the Chancellor
  works of
    Summa de bono, 8
Philosophers
  ancient, 1215, 1333
  natural, 227, 435, 862, 869–870, 872, 876, 889, 892, 972, 1213–1215, 1301, 1303, 1333
  Platonic, 62
Photius, 37
Physician(s), 226, 278, 346, 349, 355, 360–363, 420, 428, 636–637, 653, 678, 727, 743, 757, 816, 818–819, 821–825, 832–833, 840, 847, 850–851, 862, 864, 869, 919, 926, 938, 944, 997, 1106, 1135, 1140, 1149–1150, 1158–1159, 1170, 1178, 1180, 1203, 1282–1283, 1304, 1373, 1378, 1392, 1668, 1713, 1747
Physiognomy, *passim. See especially* 46, 93–141, 211–225, 235–242, 256, 284–285
  Aristotle on, 93–96, 133, 211, 214, 225, 240–242
  Avicenna on, 94–96, 140, 211, 239
  Loxus on, 94–99, 238–241, 256
  Palemon on, 93–99, 238–240
  Plato on, 95, 129, 133, 140, 210–211, 239–241

Physiologus, 30, 38, 39
Pierre d'Ailly, 32
Pius XI, Pope, 2
Pius XII, Pope, 2
Planets, 506, 511, 513, 781, 1331, 1396–1398, 1444, 1447
Plants and trees, 947, 956
    acalips seed, 1461
    agrimony, 1493–1495
    almond tree, 733, 745, 747
    apple tree, 1498
    ash tree, 1497
    avens, 1493
    barley, 368, 845, 1456, 1479, 1494.1646
    bear herb, 1540
    betony, 1612–1613
    bitter vetch, 431–432
    black alder, 1489, 1492, 1495
    black elder, 1486, 1498, 1613
    black henbane, 844, 1612, 1646
    black mint, 1607
    blackthorn, 1498
    broom, 1495
    bryony, 1485
    cabbage, 1261, 1618–1619, 1753
    cabbage, Roman, 1199, 1204
    camphor tree, 1514
    celeriac, 1491–1493, 1495, 1499, 1502
    chervil, 1491, 1495, 1608, 1612
    chickpeas, 516, 635, 1507, 1661
    coriander seed, 1464
    date palm, 1613
    dittany, 1471
    dracontea, 682, 684, 1519
    dwarf elder, *passim. See especially* 1485–1486, 1490, 1492–1496, 1609
    elderberry tree, 1610
    fennel, 1477, 1609, 1611, 1618–1619
    fig, 704, 1509
    fig tree, 1252, 1509, 1526, 1641, 613
    fir tree, 1526, 1563
    flowers
        columbine, 1041
        poppy, 733–734, 744–745, 747
    food, absorption by, 931
    fruits
        apple, 1124, 1508
        pear, 1124
        quince, 1124
    garden rocket, 1135, 1475
    garlic, 845, 1227, 1274, 1482, 1525, 1617–1618
    grapes, 368, 1306, 1508, 1624, 1745
    groundsel, 1462, 1489–1490, 1492
    hellebore, 1491
    hemlock, 1496, 1612
    horehound, 400, 1484, 1493, 1496–1497, 1608
    horse foot, 1497
    ivy, 684, 1496, 1620
    jasmine, 1469, 1530
    juniper, 1495, 1606
    larkspur, 1599, 1608–1609, 1613, 1618
    laurel, 1485, 1499, 1514, 1525, 1611–1612, 1620, 1726, 1732
    leeks, 1488, 1494, 1502, 1509, 1514
    lettuce, 666, 687, 845, 1702, 1726, 1750, 1754
    lily, 1485, 1509
    linden tree, 744, 1737
    mandrake, 1058
    marjoram, wild, 475
    marsh mallow, 1461, 1485, 1501
    mint, 1477, 1607–1608, 1612–1613, 1621
    mugwort, 1619
    mustard root, 1608
    myrtle, 733, 744, 747
    nettle, 428, 1480, 1502, 1620, 1629–1630
        Greek nettles, 1492, 1602
    nut tree, Roman, 1129
    oak tree, 1493, 1605, 1728, 1763
    oleaster, wild, 1252
    olive tree, 60, 736, 744, 1252, 1540
    onion, 1509, 1525, 1615
    oregano, 686–687, 696, 743–744, 1493
    parsley, 514, 926, 1466, 1606–1607
    pear tree, 747
    pennyroyal, 1466, 1610, 1617
    pimpernel, 1614, 1620
    pine tree, 1541
    plantain, 1493, 1606
    polegium, 685
    pomegranate, 733, 744
    radish, 1493, 1500, 1597, 1606, 1608–1609, 1612–1613
    reproduction of, 588, 1119–1120, 1124, 1185
    rhubarb, 1615
    rocket plant, 1135, 1475
    rosemary, 1609, 1612, 1613
    rue, 687, 1510, 1524, 1608–1609
    rye, 1479, 1488, 1492, 1495, 1502
    saffron, 1227, 1605
    sage, 1612, 1613, 1615
    savin, 1612
    savory, 1613

Plants and Trees (*cont.*)
  saxifrage, 1606, 1615
  seeds of, 1244, 1399
  senses absent from, 1335
  southernwood, 1495
  spearwort root, 1494
  spinach, 1041
  stavesacre, 1597, 1603, 1754
  sweet flag, 1495
  tansy, 1461
  taste and touch in, 1433
  thyme, 1609
  tormentil, 1503
  vervain, 1606
  walnut tree, 1484
  water hemlock, 1496, 1612
  watercress, 1497, 1605
  wheat, 132, 404, 687, 845, 1486, 1494
  wheat bran, 1486, 1489, 1503
  white birch, 1462, 1495, 1498
  willow, 562, 733, 1492, 1518, 1619, 1621, 1635
  woodbine, 1530
  wormwood, 784, 1486, 1488, 1502, 1609, 1621
  yellow lavender, 1461
Plato, 2, 11, 24, 26, 41, 64, 95, 125, 129, 133, 140, 179, 210–211, 223, 239–241, 372–373, 375, 419, 421–422, 491, 513, 587, 814, 818, 892, 1010, 1060, 1110–1112 1114, 1140, 1161–1162, 1170–1172, 1190–1193, 1211, 1268, 1292, 1337, 1393, 1404, 1445, 1624
  works of
    Timaeus, 24
Platonism, 11, 23, 25–26
Platonists, 133, 419, 881
Pleiades, 501, 505, 556, 567, 597, 621, 623, 1663
Pliny the Elder, 37–38, 41, 71, 1434, 1449, 1451, 1453, 1470, 1473, 1508, 1509–1510, 1513, 1522, 1531, 1552, 1554, 1556–1557, 1561, 1567–1568, 1629, 1636, 1641, 1644, 1647, 1650, 1652–1654, 1657, 1662, 1664, 1676–1677, 1681, 1683–1688, 1692–1699, 1701–1705, 1720, 1723, 1725–1726, 1728–1729, 1731–1733, 1737, 1741, 1746–1747, 1752, 1758
  works of
    Historia Naturalis, 38
Plutarch, 38
Poisons, types of, 1711–1713
Polyclitus, 1127, 1173–1173
Pompey the Great, 1340
Popchydyus, the philosopher, 1445
Poppaea (Nero's mistress), 1451
Porphyry, 19, 420, 1146
Pouchet, Felix A., 32
Prester John, 641
Priam of Troy, 403, 1392
Proportions
  arithmetic, 1164, 1378, 1389
  geometric, 1164, 1378, 1389
  harmonic, 1295, 1388
  numerical, 1164
Protagoras, 189, 926, 1144
Ptolemy, 94, 581, 1580, 1590, 1607, 1618
  works of
    Centilogyum (*Centiloquium*), 581
Pygmies, 61–62, 112, 290, 480, 613, 616, 892, 938, 972, 980, 1057, 1416–1421, 1627
  hands of, 1422
  as irrational animal, 1416
  speech of, 477, 1417
  upright posture of, 1417
Pythagoras, 140, 220, 236, 420–421, 668, 717, 772, 926, 973, 1161, 1291, 1379, 1388, 1390
Pythagoreans, 72, 1362

Q

Quadrupeds, 605–611
  anatomy of, 326–328, 1056–1072
  antelope, 1465
  apes, 613, 886, 892, 1369
  ass, 51, 60, 85, 243, 250, 286, 289, 297, 309, 414, 515, 610, 674–676, 678, 968, 1020–1021, 1031, 1064, 1087, 1153, 1188, 1199, 1228–1233, 1236, 1423, 1450–1452, 1521, 1571
    complexion of, 914, 1232, 1236, 1450
    copulation of, 513, 581
    domestication of, 678
    illnesses of, 639, 1450
    milk of, 429
    parturition of, 1096, 1451
    shoes made from skin of, 1451
    wild, 1452, 1526–1527
  ass-centaur, 1527
  auroch, 574, 1541–1543
  badger, 622, 1475, 1542
    dog, 1475
    hog, 1475

bear, 60, 294, 296, 331, 606, 608, 658–659, 669, 675, 1508, 1519, 1531, 1540–1541
  hibernation of, 623, 625–626, 1540
beaver, 590, 607, 653, 979, 1105, 1467–1469, 1540
  castration of, 1467
  musk, 1467–1469
bison, 295, 301, 1455, 1541
boar, 63, 95, 307, 337, 576, 755, 979, 1453–1454, 1505, 1511, 1515, 1531
  capture of, 1453
  copulation of, 512, 515, 1089
  domestic, 515
  rutting of, 571–573
bonacus, 756–757
buffalo, 63, 268, 299, 427, 574, 1455–1456, 1675
bull, *passim. See especially* 63, 77, 279, 297, 304–305, 307, 341, 414, 482, 512, 608, 610, 643, 675, 859, 933, 976, 982, 1455, 1505, 1535–1538
  blood of, 378, 414, 1537
  castration of, 339, 512, 762, 1091, 1106–1107
  complexion of, 1536
  copulation of, 580–581, 1089
  gestation period of, 1304
  rutting of, 571, 573–574, 580
  voice of, 486, 512, 1354–1356
camel, 294, 296–297, 303–305, 307, 426, 610–611, 758, 981, 1021, 1031–1032, 1064, 1066, 1308–1309, 1449, 1456–1457, 1465, 1516
  castration of, 762–763
  complexion of, 926
  copulation of, 467, 492–493, 515, 758
  illnesses of, 1456
  milk of, 426, 429
  rutting of, 572
cat, 64–65, 268, 304–305, 307, 311, 491, 655, 968, 1021, 1066, 1336, 1469, 1520, 1523, 1526, 1542, 1601, 1609, 1615
chamois, 981, 984
chimera, 1473
cow, *passim,* 51, 62–63, 243, 281–282, 296, 302–305, 327–328, 385, 513, 575–576, 580, 609–610, 675, 681–682, 756, 763, 800, 968, 976, 1020–1021, 1024, 1064, 1098, 1156, 1314–1315, 1336–1337, 1352, 1448, 1484, 1536–1537, 1541–1542, 1721, 1755, 1759

copulation of, 492, 572–573, 580–581
gestation of, 580
horns of, 486, 610, 970, 976–977
illnesses of, 632, 636, 1537
milk of, 427, 429–432, 580, 803
mountain, 294, 299
multiple births in, 1448
voice of, 486, 512, 1354–1355
wild, 51, 981, 998
deer, 60, 63, 243, 268, 294–295, 300–301, 305, 311, 327–328, 511, 682, 687, 713, 763, 800, 976, 982, 984. 998, 1064, 1465, 1470–1473, 1475, 1505, 1515, 1516, 1521–1522, 1528, 1533
  antlers, 296–297, 301, 682–684, 1471
  copulation of, 492
  hunting of, 42, 685–686
  life span of, 1471
  stag, 66, 85, 378, 383, 414, 605, 682, 684, 979–980, 1032, 1471, 1533, 1538, 1734
  castration of, 762
  rutting of, 679, 1470
dog, *passim. See especially* 51, 60–61, 64, 287, 294, 303–304, 307, 309, 311, 328, 341, 379, 461, 482, 511, 605–607, 643, 653, 655, 756, 791, 968, 1024, 1057, 1062, 1064, 1111, 1199, 1226, 1230, 1306, 1308, 1318, 1336, 1343–1344, 1352, 1411, 1423, 1435, 1449, 1457–1464, 1474, 1512–1513, 1518–1519, 1526, 1540, 1542, 1592
  bird, 716, 1460
  breeding, 1460
  complexion of, 1459–1461
  copulation of, 492–494, 499, 513, 515, 563, 572–579
  gestation of, 578, 1234, 1304
  greyhound, 578–579, 645, 668, 1226, 1458, 1460
  hunting with, 308, 606–607, 981, 1343–1344, 1449, 1476, 1512, 1537, 1543, 1549, 1553, 1578
  illnesses of, 636, 685, 687, 1458, 1461–1464
  life span of, 579
  mastiff, 645, 668, 1459–1460
  parturition of, 579, 1215, 1306, 1459
  training of, 668–669, 1415, 1459–1460
dormouse, 60, 622, 800, 1510
elephant, *passim. See especially* 51, 61, 64, 268, 287–293, 302, 304, 311, 331, 337, 380,

Quadrupeds (*cont.*)
    383, 432, 481, 610, 963, 981, 1031, 1064, 1066, 1073, 1098, 1308–1309, 1315, 1476–1477, 1505, 1521, 1540, 1661, 1696, 1726
  breasts of, 1064
  capture of, 679–680, 1449
  complexion of, 1477
  copulation of, 493, 515, 758
  gestation period of, 1325
  illnesses of, 636, 639–640
  life span of, 611, 758, 1330
  military uses of, 679
  rutting of, 572, 758
  snakes drinking blood of, 1476, 1726
  sperm of, 1183
  testicles of, 1098, 1476
  training of, 679–680, 758, 1415
  trunk of, 51, 129, 969–970, 975, 1046, 1414, 147
  tusks of, 309, 380, 679
 elk, 286–287, 300, 687, 763, 982, 1449, 1504–1505
 ermine, 1508, 1533
 ferret, 492, 1509, 1520
 fox, 60, 63, 65, 100, 294, 622, 643, 675, 677–679, 687–689, 1199, 1215, 1226, 1318, 1336, 1541–1542, 1559
  copulation of, 492
 gazelle, 1474, 1721
 genet, 1510
 giraffe, 981, 1057, 1465, 1527–1528
 gnu, 1474
 goat, *passim. See especially* 51, 61, 109, 140, 385, 479, 482, 514, 608, 611, 643, 682, 713, 763, 977, 1019, 1021, 1064, 1098, 1337, 1465–1467, 1511, 1515, 1531, 1538
  blood of, and breaking rocks, 514, 1465
  copulation of, 512, 1089, 1129
  hermaphroditism of, 1307
  life span of, 577
  milk of, 428–429, 1529
  monstrous births in, 1306–1307
  mountain, 294, 492, 685, 687, 980–981, 983, 1467
  rutting of, 571
  stupidity of, 681
 hamster, 268, 1474
 hare, *passim. See especially* 63, 268, 294, 378, 393, 414, 430, 491, 578, 643, 670, 675, 713, 800, 1026, 1066, 1255–1256, 1317, 1346, 1473, 1515–1516, 1542
  complexion of, 1515
  hermaphroditism of, 1255–1256
 hedgehog, 385, 659, 1344, 1506–1508
  copulation of, 492, 1092, 1507
  mountain, 1507–1508
  prognostication with, 688, 1506
  testicles of, 1092, 1098
 horned, 294–301, 976–977, 980–984, 1019
  milk of, 578, 1026
 horse, *passim. See especially* 50–53, 56, 60–61, 70, 74–77, 85, 107, 243, 250, 286, 289, 294, 304, 307, 309, 327, 341, 482, 511, 605, 608, 610, 613, 632, 639, 655, 676, 681–682, 741–742, 873, 946, 968, 981, 1020–1021, 1031, 1057, 1066, 1087, 1129, 1156, 1186, 1188, 1238, 1308, 1337, 1341, 1351, 1423, 1455, 1465, 1477–1505, 1523, 1559, 1592
  charger (*dextrarius*), 581, 754, 1478
  complexion of, 914
  copulation of, 513, 581, 759
  gaits of, 1479
  illnesses of, 632, 636–638, 1461, 1480–1504, 1600
  life span of, 512–513
  mare, 304, 682, 759, 1199, 1229–1230, 1523
  intercourse during pregnancy, 792–793
  lustfulness of, 572, 581, 1316
  menses of, 1231, 1236, 1316
  milk of, 426, 429
  palfrey, 1478
  parturition of, 575, 1096
  racehorse, 1478
  rutting of, 573–574
  semen of, 1232
  stallion, 1065, 1229–1233, 1236
  superfetation in, 1315
  war-horse, 1478
  workhorse, 1478
 hybrid, 1511
 hyena, 99, 606, 1474, 1512, 1543
 ibex, 492, 685, 983, 1511
 lamia, 1516–1517
 leopard, 297, 303, 307, 311, 644, 669, 678, 685–686, 1231, 1352, 1449, 1512, 1514–1515
 lion, *passim. See especially* 51, 60–66, 85, 129–130, 133, 240–241, 287, 293–294, 303–

305, 307, 311, 381, 383, 393. 417, 491,
605, 655, 685, 753–756, 791, 859, 945,
955, 968, 1021, 1057, 1062, 1066, 1187,
1215, 1230, 1318, 1336, 1352, 1449,
1463, 1504, 1512–1514, 1516 1530
   capture of, 685
   complexion of, 926, 1392, 1313, 1463
   copulation of, 493
   genuses of, 1512
   poisonous bite of, 755
lioness, 1064, 1231
   adultery of, 1513, 1530
   offspring of, 1240
lunza, 1231, 1449
lynx, 51, 679, 1062, 1411, 1464, 1517–1518
manticore, 308, 1521–1522
marmot, 60, 670, 1506
marten, 607, 1523, 1526, 1533
mole, 60, 98, 112, 471, 642, 672, 962,
1538–1539
   dissection of, 30
   generation of, 1539
   sight of, 1545
mongoose, 1508
monkey, *passim. See especially* 51, 99, 112, 250,
268, 290, 302, 304, 311–313, 480, 702,
958, 1063, 1066–1067, 1408, 1419–
1421, 1435, 1460, 1470, 1522, 1531,
1533–1535, 1541, 1543, 1731
   capture of, 1534
   copulation of, 1422
   genus of, 1419–1420
   hands of, 970, 1422
   imitative nature of, 1416, 1419
   training of, 1415
   upright posture of, 304 1057
mouse, 51, 60, 112, 293, 343, 391, 608,
642–644, 659, 673, 676, 688, 714, 763,
768, 1006, 1031, 1308, 1417, 1423,
1469, 1476, 1508, 1510–1511, 1521,
1523–1525, 1533, 1542, 1561, 1591,
1598, 1610
mountain, 622, 670, 1506
mule, *passim. See especially* 85, 250, 309, 327,
575, 610, 681, 1020–1021, 1031,
1226–1230, 1236, 1423, 1449, 1452,
1521, 1523
   complexion of, 1233
   Democritus on, 1228–1230, 1235
   generation of, 581, 1137, 1226, 1521

   gestation period of, 1233
   menses of, 575, 1233
   sterility of, 1226–1236
musk deer, 1522
mustelid, 1532–1533
onager, 643, 1452, 1513, 1526–1527
oryx, 1527
otter, 590, 607, 653, 1468, 1520–1521, 1540
oviparous, *passim. See especially* 292, 313–317,
342, 1069–1072
ox, 63, 294–295, 299, 384, 574, 605, 1449,
1455–1456, 1461, 1465, 1519, 1536,
1538
   complexion of, 926
panther, 955, 1352, 1530–1531
pard, 1464, 1513–1514, 1530
pegasus, 311, 1531
pig, *passim. See especially* 61, 84, 133, 265, 279,
294–295, 297, 304, 307, 309, 311,
327–328, 331, 337, 341, 417, 479,
608–609, 642, 655, 677, 762–763, 800,
933, 978, 981, 1020–1021, 1024, 1064,
1233, 1308, 1453, 1476, 1511, 1600,
1608, 1731, 1763
   castration of, 762
   complexion of, 926, 1318
   copulation of, 499, 512, 515, 1089
   illnesses of, 226, 634–635, 1454
   life span of, 576
   monstrous births in, 1233, 1306
   multiple births in, 1318
pilosus ("hairy man"), 1531–1532
polar bear, 83, 653, 1352
polecat, 688
porcupine, 84, 1512
rabbit, 675, 713, 1317, 1473–1474, 1509, 1542
rat, 676, 1423, 1620
reindeer, 1531, 1533
rhinoceros, 955, 981–982, 1521, 1527
roebuck, 294, 300, 309, 675, 684, 713, 763,
1467, 1527, 1560
   rutting of, 571–574
seraph, 1449
sheep, *passim. See especially* 51, 61, 140, 608,
611, 632, 643, 682, 746, 763, 1013,
1015–1016, 1019, 1021, 1293, 1415,
1511, 1518, 1528–1530
   coloration of, 577–578
   illnesses of, 279–280, 1015, 1529
   life span of, 514, 577, 1529

Quadrupeds (*cont.*)
   milk of, 432, 1529
   monstrous births in, 1306
   pastorage of, 413
   prognostication with, 1448
   ram, 479, 673, 688, 1019, 1528–1530
      copulation of, 512, 571, 1089
      rutting of, 571
      sexual differentiation of, 577–578, 1528
      stupidity of, 681
      wool of, 384–385, 611, 678, 1261, 1346–1347, 1528
   shrew, 1524
   squirrel, 391, 714, 962, 970, 1474, 1533, 1542
   tiger, 401, 645, 1226, 1535
   unicorn, 297, 955, 982–983, 1521, 1539–1540
   viviparous, 287–292
   weasel, 391, 1416, 1508–1510, 1515, 1520, 1523–1524, 1526, 1533, 1721, 1733
   wildcat, 1534
   wisent, 295, 301, 403, 574, 998, 1541
   wolf, *passim. See especially* 51, 61, 63, 294, 305, 308, 605–607, 609, 643, 653, 668–669, 679, 684–686, 688, 716, 755–758, 968, 1062, 1215, 1226, 1308, 1318, 1336, 1352, 1423, 1453, 1464, 1474, 1504, 1512, 1514, 1517–1520, 1528, 1532, 1543
      complexion of, 1519
      copulation of, 492, 1518
      illnesses of, 636
      rutting of, 572
   yale, 1505

R

Rabanus Maurus, 40
Rhazes (Abū Bakr Muhammad ibn Zakariyyā al-Razī), 191, 192, 202, 204
   works of
      On Animals, 1454, 1463
Roger II, King of Sicily, 1585, 1603
Rudolf of Hapsburg, 16
Rudolph of Nijmegen, 1, 3
   works of
      Legenda Alberti Magni, 3
Rufus of Ephesus, 1713

S

Saturn (planet), 510, 513–514, 780, 939–940, 1444
Scheeben, Heribert Chr., 19
Schwertner, Thomas M., 13

Sciences
   ethics, 65
   moral, 874
   natural, 85, 93, 125, 158, 362, 398, 401, 463, 778, 808, 837, 858, 860, 863, 869, 871, 873–874, 880, 909, 1439, 1764
   knowledge
      demonstrative, 857–859
      of accidents, 857–861
   principles of, 1417
Sclavi, 295–296
Scotus, Michael. *See* Michael Scot
Secundus, philosopher, 759
Semeryon (Semyrion/Symerion), philosopher, 1720–1725, 1728, 1730, 1734, 1752, 1755–1756
Senses, 110, 114, 135, 238, 932, 1155
   common, 947, 1056, 1423
   Empedocles on, 1340
   estimation, 586, 769, 1418, 1422–1431, 1434–1435
   estimative power, 586, 767, 932
   external, 958–964, 1406, 1434
   hearing, 49, 109–112, 471–476, 1342–1344, 1390, 1422–1424
   imagination, 667, 698, 766–771, 1419
   intentions drawn from, 1415–1425, 1435
   internal, 1406, 1434
   memory, 1414–1424, 1428, 1435
   number of, 471
   prudence, 766–770
   smell, 126–127, 243, 471–476, 1342–1344
   taste, 134–135, 139, 181, 182, 238, 243, 314, 452, 454, 475
   touch, 69, 134, 139, 158, 185, 238, 243, 305, 475, 1408, 1436
   variations among, 1145
   vision, 107, 110, 121–127, 180, 471, 476, 1332, 1337–1344, 1389, 1408
Sensory organs, 1188, 1222, 1335–1336, 1407
   location of, 960–964
   shapes of, 1407
Serapion
   works of
      On Medicinal Simples, 1465, 1736
Serpents, *passim. See especially* 47, 84, 381, 604–605, 644–648, 970, 1433–1434, 1466, 1708–1738
   absence of feet in, 1069

anatomy of, 324–326, 332–333, 1030, 1071–1072, 1079–1081, 1089–1090, 1305, 1709–1710
aquatic, 1079
asp, 1526, 1714, 1716–1718, 1722, 1728
  domestication of, 1434
basilisk, 646–647, 1524, 1531, 1628, 1714, 1720–1721
bones of, 381, 935
cerastes, 300, 1722–1723, 1736
copulation of, 1094, 1438, 1708
draco/dragon, 47, 317, 383, 623, 671, 687, 1531, 1714, 1719, 1725–1727
eggs of, 1156, 1708
genus of, 84, 1713–1715
hydra, 1728
monstrous births among, 1305
reproduction of, 84, 342–343, 1096
sea, 1730
tongue in, 960, 973
tyrus, 84, 342, 469, 604–605, 623, 626, 646, 671, 677, 679, 1030, 1156, 1473, 1599, 1711, 1736–1737
venoms, 69, 646–648, 665–666, 1674, 1689, 1710–1716
  theriac, 342, 685, 701, 1473, 1603, 1608, 1612, 1713, 1736
  types of, 1711–1716
viper, 69, 74, 84, 342, 686–687, 1201, 1715, 1737–1738
vision of, 333
winged, 81, 686, 1631
Shaw, James Rochester, 33
Siger of Brabant, 1
Simonides, 500
  works of
    On Birds, 500
Sixtus IV, Pope, 17
Socrates, 41, 50, 103, 767, 876, 886, 1170, 1297–1299
Socratics, 1172
Solinus, 38, 81, 686, 1449, 1455, 1470, 1474, 1504–1505, 1513, 1516–1517, 1522, 1524, 1537, 1555, 1565, 1674, 1703, 1716–1717, 1724–1725, 1731
Solomon, King, 1762
Soothsayers, 128, 429, 1335
Sophronicus (Sophroniskos), 50
Soul, *passim. See especially* 860, 929, 1441
  accidents of, 668, 1443, 1446
  act of, 1002, 1115, 1164, 1409

analogous to fire, 936–937
animal, 373
  absence of reason in, 772
appetitive, 373
concupiscent, 372, 1010
as formal principle of animals, 870
illnesses of, 106
image of the world and, 1400
intellectual, 871, 1057, 1189, 1206, 1410
internal principle of life and, 1161
irascible, 1010
leap of, 1373
lower, 1168
moving, 1403
nutritive, 920, 1185, 1190, 1202, 1211, 1258, 1775
parts in, 872
powers of, 350, 586–587, 766–773, 891, 897, 899, 913, 1167–1171, 1176, 1179, 1187, 1191–1192, 1310, 1364, 1379, 1390, 1396, 1398, 1401, 1404, 1406, 1409–1422, 1429
principle of sensation and, 1186
rational, 1186, 1188–1192, 1202–1203, 1401, 1409–1411
  first active forty days after conception, 826
second incorporation of, 1445
sensible, 49, 538, 770, 1103, 1111, 1186, 1188, 1190–1193, 1202, 1205–1206, 1212–1213, 1401, 1409–1412, 1426, 1431, 1434, 1544
sensitive, 1213
species of, 1407
substantial form of a body, 1163
tripartite, Plato on, 818, 1112
vegetative, 920, 1185, 1188, 1190–1193, 1205–1206, 1211–1212, 1258, 1409–1412
vital, 373
world, 871
Speussipus, 419
Spirit, *passim. See especially* 189–190, 193, 221, 241, 248, 267, 271, 814, 816–818, 929, 1193–1198, 1203–1207, 1358, 1363, 1368, 1404, 1423
animal, 49, 107, 127, 244, 248, 290, 820, 850, 1381
generation of, 1364, 1378
as instrument of the soul, 818, 1364
motive, 247
motor, 928
natural, 93, 818–819, 850, 1357, 1379, 1381

Spirit (*cont.*)
  origins of, 1401
  principle of the soul and, 1193
  sensitive, 928
  visual, 107, 124, 180, 942, 964, 1341, 1721
  vital, 254, 585, 818–820, 850, 928, 991, 1000, 1170, 1206, 1213, 1357, 1379, 1381, 1401
St. Andrew, church of, 17
St. Dominic, 4
St. James, priory of, 10
St. Katherine, cloister of, 16
Stadler, Hermann, 34
Stannard, Jerry, 32
Stars (*See also* Constellations)
  incubation of eggs and, 532
  motion of, 1398
  number of, 1397
  procreation and, 46, 1105
  revolutions of, 1330
  substance of, 1153
Steenberghen, Fernand Van, 18
Stoics, 41, 408, 417–419, 1149, 1170, 1396–1397
Substance, *passim. See especially* 859, 875
  act and potency in, 871
  animal, 1154
  contingent, 1153
  corporeal, 1380
  eternal, 1153
  generation for the sake of, 1333
  incorporeal, 1380
  rational, 1203
  sensible, 1203
Sun
  eclipse of, 861
  as heart of heaven, 72
  motion of, 1394
  power of, 72, 373, 939, 1192
  rays of, 1369
  revolutions of, 1326, 1331
Surgery, 1313
Symmachus, Letter of, 1618
Symmachus, 1577, 1590, 1593, 1623

T

Tartars, 427
Tethys (goddess), 1397
Teutonic knights, 6
Thales the Milesian, 346, 403
Theodorus, 41, 1169

Theodotion, 1590, 1607, 1618, 1623
Theophrastus, 41, 420, 812, 1146, 1167–1170, 1189, 1414, 1666, 1683
  works of
    On Animals, 1169
Thomas Aquinas, 2, 12–14, 16, 26–27
Thomas of Cantimpré, 40, 42
  works of
    De natura rerum (On the Nature of Things), 40
Thorndike, Lynn, 2
Tolomeo of Lucca, 16–17
Trees. *See* Plants and trees
Trojan war, 353, 1662
Tropic of Cancer, 517
Tugwell, Simon, 5–6, 8, 17

U

Ulrich of Strasbourg, 1, 13, 15
University of Padua, 4, 32
University of Paris, 10
Urban IV, Pope, 15

V

Vegetius, 38
Venus (goddess), 1111, 1183
Venus (planet), 940
Vermin, 1436, 1438
  fumigation of, 1452
  generation of, 1265
  varieties of, 1739–1764
Virgin Mary, 5
Virtues
  intellectual, 1419
  moral, 65, 1410, 1420
Vita Fratrum Ordinis Praedicatorum, 3
Vulcan, 90, 414

W

Wallace, William A., 2, 29, 31
Weather
  honey production and, 1270
  incubation of eggs and, 533, 1246–1247
  miscarriages and, 681, 1528
  rainstorms, generation of, 938
  thunderstorms, 58
Weisheipl, James A., 2, 6, 8, 15, 17, 21–22, 35, 42
William of Auvergne, 12
William of Auxerre, 8
William of Conches, 25
William of Moerbeke, 19, 40

William of St. Amour, 14
William of Tocco, 12
William the Falconer, 1585, 1590, 1597–1603, 1611–1616
Wind
    fish development and, 1656–1657
    movement of, 1326
    sexual differentiation and, 1293
Wine, 101–103, 414, 503, 514, 604, 610, 619, 635, 640, 646, 744, 795, 804, 987, 1118, 1247, 1452, 1454, 1464, 1466, 1472, 1485–86, 1488, 1492–93, 1502, 1504, 1509, 1514–1515, 1530, 1542, 1601, 1604–1605, 1609–1615, 1617–1620, 1690–1691, 1733, 1738, 1760
    corrupted by hot weather, 1247
    desired by serpents, 604, 1710; parrots, 618
    mixed with water, 1118, 1229
    types: formic, 727; red, 404, 611, 742, 1391; white, 804, 1520, 1611, 1614

X

Xenophon, 37

Z

Zeno, philosopher, 982
Zodiac, 1383